THE ROUTLEDGE COMPANION
TO PHILOSOPHY OF PHYSICS

The Routledge Companion to Philosophy of Physics is a comprehensive and authoritative guide to the state of the art in the philosophy of physics. It comprises 54 self-contained chapters written by leading philosophers of physics at both senior and junior levels, making it the most thorough and detailed volume of its type on the market – nearly every major perspective in the field is represented.

The Companion's 54 chapters are organized into 12 parts. The first seven parts cover all of the major physical theories investigated by philosophers of physics today, and the last five explore key themes that unite the study of these theories.

    I. Newtonian Mechanics
   II. Special Relativity
  III. General Relativity
  IV. Non-Relativistic Quantum Mechanics
   V. Quantum Field Theory
  VI. Quantum Gravity
 VII. Statistical Mechanics and Thermodynamics
VIII. Explanation
  IX. Intertheoretic Relations
   X. Symmetries
  XI. Metaphysics
 XII. Cosmology

The difficulty level of the chapters has been carefully pitched so as to offer both accessible summaries for those new to philosophy of physics and standard reference points for active researchers on the front lines. An introductory chapter by the editors maps out the field, and each part also begins with a short summary that places the individual chapters in context. The volume will be indispensable to any serious student or scholar of philosophy of physics.

Eleanor Knox is Reader in Philosophy of Physics at King's College London. She works in philosophy of physics, particularly the philosophy of spacetime physics, and is also interested in issues of reduction and emergence, and how these two come together in quantum gravity.

Alastair Wilson is Professor of Philosophy at the University of Birmingham and Senior Adjunct Research Fellow at Monash University. He works on philosophy of physics, philosophy of science, metaphysics, and epistemology, with special interests in the philosophy of quantum theory and the metaphysics of dependence. He is the author of *The Nature of Contingency: Quantum Physics as Modal Realism* (2020).

ROUTLEDGE PHILOSOPHY COMPANIONS

Routledge Philosophy Companions offer thorough, high quality surveys and assessments of the major topics and periods in philosophy. Covering key problems, themes and thinkers, all entries are specially commissioned for each volume and written by leading scholars in the field. Clear, accessible and carefully edited and organised, *Routledge Philosophy Companions* are indispensable for anyone coming to a major topic or period in philosophy, as well as for the more advanced reader.

Also available:

THE ROUTLEDGE COMPANION TO SHAKESPEARE AND PHILOSOPHY
Edited by Craig Bourne, Emily Caddick Bourne

THE ROUTLEDGE COMPANION TO THE FRANKFURT SCHOOL
Edited by Peter E. Gordon, Espen Hammer, Axel Honneth

THE ROUTLEDGE COMPANION TO FEMINIST PHILOSOPHY
Edited by Ann Garry, Serene J. Khader, and Alison Stone

THE ROUTLEDGE COMPANION TO PHILOSOPHY OF PSYCHOLOGY, SECOND EDITION
Edited by Sarah Robins, John Symons, and Paco Calvo

THE ROUTLEDGE COMPANION TO MEDIEVAL PHILOSOPHY
Edited by Richard Cross and JT Paasch

THE ROUTLEDGE COMPANION TO PHILOSOPHY OF PHYSICS
Edited by Eleanor Knox and Alastair Wilson

For more information about this series, please visit:
https://www.routledge.com/Routledge-Philosophy-Companions/book-series/PHILCOMP

THE ROUTLEDGE COMPANION TO PHILOSOPHY OF PHYSICS

Edited by
Eleanor Knox and Alastair Wilson

Routledge
Taylor & Francis Group

NEW YORK AND LONDON

First published 2022
by Routledge
605 Third Avenue, New York, NY 10158

and by Routledge
2 Park Square, Milton Park, Abingdon, Oxon, OX14 4RN

Routledge is an imprint of the Taylor & Francis Group, an informa business

© 2022 Taylor & Francis

The right of Eleanor Knox and Alastair Wilson to be identified as the authors of the editorial material, and of the authors for their individual chapters, has been asserted in accordance with sections 77 and 78 of the Copyright, Designs and Patents Act 1988.

Library of Congress Cataloging-in-Publication Data
A catalog record for this title has been requested

ISBN: 978-1-138-65307-8 (hbk)
ISBN: 978-0-367-76961-1 (pbk)
ISBN: 978-1-315-62381-8 (ebk)

Typeset in Bembo
by codeMantra

CONTENTS

List of Figures, Boxes, and Tables   *xi*

Notes on Contributors   *xiv*

    Introduction   1
    Eleanor Knox and Alastair Wilson

SECTION A: THEORIES

PART I

Newtonian Mechanics   **6**

  1 Newtonian Mechanics   8
    Ryan Samaroo

  2 Formulations of Classical Mechanics   21
    Jill North

  3 Classical Spacetime Structure   33
    James Owen Weatherall

  4 Relationism in Classical Dynamics   46
    Julian Barbour

PART II

Special Relativity   **59**

  5 Relativity and Space-Time Geometry   61
    Tim Maudlin

Contents

6 The Dynamical Approach to Spacetime Theories   70
   Harvey R. Brown and James Read

7 Relativity and the A-Theory   86
   Antony Eagle

8 Relativistic Constraints on Interpretations of Quantum Mechanics   99
   Wayne C. Myrvold

PART III
General Relativity   **123**

9 The Equivalence Principle(s)   125
   Dennis Lehmkuhl

10 The Hole Argument   145
   Oliver Pooley

11 Relativistic Spacetime Structure   160
   Samuel C. Fletcher

PART IV
Non-Relativistic Quantum Theory   **181**

12 Bell's Theorem, Quantum Probabilities, and Superdeterminism   184
   Eddy Keming Chen

13 Quantum Decoherence   200
   Elise M. Crull

14 The Everett Interpretation: Structure   213
   Simon W. Saunders

15 The Everett Interpretation: Probability   230
   Simon W. Saunders

16 Collapse Theories   247
   Peter J. Lewis

17 Bohmian Mechanics   257
   Roderich Tumulka

Contents

PART V
Quantum Field Theory   **273**

18 The Quantum Theory of Fields   275
David Wallace

19 Renormalization Group Methods   296
Porter Williams

20 Locality in (Axiomatic) Quantum Field Theory: A Minority Report   311
Laura Ruetsche

21 Particles in Quantum Field Theory   323
Doreen Fraser

PART VI
Quantum Gravity   **337**

22 The Development of Quantum Gravity: From Feelings to Phenomena   339
Dean Rickles

23 String Theory   351
Richard Dawid

24 Quantum Gravity from General Relativity   363
Christian Wüthrich

25 Spacetime "Emergence"   374
Nick Huggett

26 The Problem of Time   386
Karim P.Y. Thébault

PART VII
Statistical Mechanics and Thermodynamics   **401**

27 Equilibrium in Boltzmannian Statistical Mechanics   403
Roman Frigg and Charlotte Werndl

28 Equilibrium in Gibbsian Statistical Mechanics   414
Roman Frigg and Charlotte Werndl

29 Quantum Foundations of Statistical Mechanics and Thermodynamics   425
    Orly Shenker

30 Entropy Asymmetry   439
    Arianne Shahvisi

SECTION B: THEMES
PART VIII
Explanation   **452**

31 Causal Explanation in Physics   454
    Mathias Frisch

32 Non-Causal Explanations in Physics   466
    Juha Saatsi

33 Mechanistic Explanation in Physics   476
    Laura Felline

34 The Explanatory Value of Selecting the Appropriate Scale(s)   487
    Lina Jansson

PART IX
Intertheoretic Relations   **497**

35 Nagelian Reduction in Physics   499
    Foad Dizadji-Bahmani

36 Phase Transitions   512
    Sorin Bangu

37 Universality   524
    Robert W. Batterman

38 Chance and Determinism   536
    Nina Emery

PART X
Symmetries   **549**

39 Symmetry and Superfluous Structure: A Metaphysical Overview   551
    Shamik Dasgupta

Contents

40 Symmetry and Superfluous Structure: Lessons from History and
    Tempered Enthusiasm   563
    Jenann Ismael

41 Permutations   578
    Adam Caulton

42 Gauge Theories   595
    Nicholas J. Teh

43 Time Reversal   605
    Bryan W. Roberts

44 Symmetry Breaking   620
    Elena Castellani and Radin Dardashti

PART XI
Metaphysics   **633**

45 Laws   635
    Marc Lange

46 Chance   644
    Mauricio Suárez

47 Holism   655
    Richard Healey

48 Dimensions   666
    Susan G. Sterrett

49 Fundamentality   679
    Steven French

PART XII
Cosmology   **689**

50 Why Is There Something, Rather Than Nothing?   691
    Sean M. Carroll

51 Time in Cosmology   707
    Craig Callender and C. D. McCoy

Contents

52 The Fine-Tuning of the Universe for Life   719
    Luke A. Barnes

53 Dark Matter and Dark Energy   731
    Melissa Jacquart

54 Evidence in Astrophysics   744
    Sibylle Anderl

Index   *753*

FIGURES, BOXES, AND TABLES

FIGURES

2.1	Two-dimensional tangent bundle	24
2.2	Plane pendulum	26
3.1	(a) An arrow u relating two points p and q of space. (b) "Scaling" an arrow u at a point p by various amounts, including flipping its direction. (c) Adding arrows u and v: here $u+v$ is the arrow relating p and r, where r is the point determined by q and v, and q is the arrow determined by p and u	34
3.2	Galilean spacetime may be thought of as consisting of copies of three-dimensional space stacked on top of one another to form a four-dimensional structure	36
3.3	The motion of a body is described by its 4-velocity, which is an arrow u with temporal length one	37
5.1	A four-dimensional representation of Newtonian space-time	62
5.2	Light cone structure associated with points of space-time	64
5.3	Trajectories through p with equal interval lengths	67
5.4	The "twin paradox" in Lorentz coordinates	67
8.1	The hypersurfaces considered in the proof	104
8.2	Spacelike separated regions between two spacelike hypersurfaces	111
8.3	A region common to two spacelike hypersurfaces	112
16.1	Three quantum states	248
16.2	Collapse for a single particle	249
16.3	GRW collapse for two correlated particles	250
17.1	Several possible trajectories for a Bohmian particle in a double-slit setup, coming from the left	259
22.1	Penrose's impossible triangle as a model for the problem of quantum gravity: locally (e.g., focusing on general relativity or quantum mechanics in isolation), we can make sense of picture, but not so globally in which the situation becomes paradoxical	340
22.2	Quantum theory as containing general relativity	342
22.3	General relativity as containing quantum theory	343
22.4	Quantum theory and general relativity as equally fundamental	344
22.5	Quantum gravity as containing both quantum theory and general relativity	344
25.1	A pictorial representation of a GFT quantum, $\hat{\Phi}^{\dagger}(g_1, g_2, g_3, g_4)|0\rangle$	381

29.1	The interplay between dynamics and macrovariables	431
34.1	A bead sliding down a frictionless static wire shaped as a helix	493
34.2	A Newtonian decomposition into gravitational and normal forces	493
36.1	Phase diagram for water in terms of temperature and pressure	513
36.2	A singularity in the free energy	514
36.3	Singularities in Helmholtz and Gibbs free energies ((a) and (b)) correspond to discontinuities in entropy and volume, respectively ((c) and d))	515
37.1	Cartoon PVT diagram for water	525
37.2	Universality of critical phenomena	526
37.3	Droplets inside droplets inside droplets	528
37.4	Blocking and averaging to yield a new (coarse-grained) effective system	532
37.5	Fixed point and universality class	532
37.6	Fixed point, universality class, and λ-transformation	533
40.1	Wall tiling with repeated pattern	565
40.2	A reflected scene in water	565
40.3	Circular patterns of cactus leaves	566
40.4	Nested patterns of tiles and architectural details in a mosque	566
40.5	Newtonian spacetime	569
40.6	Galilean spacetime	569
40.7	Planes of absolute simultaneity in Galilean spacetime	569
42.1	Two patches H_1 and H_2 intersecting in a belt H_{12} around the equator of a spherical spacetime	600
42.2	Gauge orbits within a constraint surface in phase space	602
43.1	Some properties of the time reversal operator T	606
43.2	Time reversal viewed as reversal of temporal orientation: the black lobe represents the "future direction" in each case	610
43.3	The Albert-Callender "pancake" approach to time reversal: order reversal of inert spatial slices with no temporal properties	612
43.4	Brave or cowardly? This instantaneous property of the soldier appears to depend on the direction of time	614
43.5	Time reversal of a harmonic oscillator's phase space: the pancake account only includes the first (order-reversing) transformation, leading to momentum and velocity in opposite directions; the standard account avoids this by including the second (instantaneous state) transformation as well	615
44.1	Plot of potential in equation (44.11)	626
46.1	Norton's dome	648
51.1	A depiction of York time in a spacetime with singularities	712
51.2	Aeons of time in a conformal cyclic universe	715
51.3	The arrow of time aligning with increasing entropy in "two-sided" universes	716
52.1	The Ptolemaic model of the Solar System	720

BOXES

48.1	Definition of the International System of Units (from Bureau International des Poids et Mesures)	675

TABLES

12.1 Possible assignments of properties to the two photons   187

13.1 Degrees of freedom of system and environment   205

37.1 Scaling exponents for different transitions   529

CONTRIBUTORS

Sibylle Anderl is Guest Researcher at the Institut de Planétologie et d'Astrophysique de Grenoble and works as a science editor for the *Frankfurter Allgemeine Zeitung*. She is the author of *Das Universum und ich: Die Philosophie der Astrophysik* (Carl Hanser, 2017).

Sorin Bangu is Professor in the Department of Philosophy at the University of Bergen, Norway. He has published on the relation between mathematics and physics, and recently edited *Naturalizing Logico-Mathematical Knowledge* (Routledge, 2018).

Julian Barbour supported himself as an independent theoretical physicist for decades. He is a past Visiting Professor in Physics at the University of Oxford and the author of *The End of Time* (Orion, 1999) and *The Janus Point* (Bodley Head, 2020).

Luke A. Barnes is Lecturer in Physics at Western Sydney University, Australia, researching cosmology, galaxy formation, and the fine-tuning of the Universe for life. He is the co-author (with Prof. Geraint Lewis) of *A Fortunate Universe: Life in a Finely-Tuned Cosmos* (Cambridge University Press, 2016) and *The Cosmic Revolutionary's Handbook* (Cambridge University Press, 2020).

Robert W. Batterman is Distinguished Professor of Philosophy at the University of Pittsburgh. He is the author of *The Devil in the Details: Asymptotic Reasoning in Explanation, Reduction, and Emergence* (Oxford University Press, 2002) and *A Middle Way: A Non-Fundamental Approach to Many-Body Physics* (Oxford University Press, 2021), and the editor of *The Oxford Handbook of Philosophy of Physics* (Oxford University Press, 2013).

Harvey R. Brown is Emeritus Professor of Philosophy of Physics at Faculty of Philosophy, Oxford University, and Emeritus Fellow of Wolfson College, Oxford, UK. He is the author of *Physical Relativity: Space-Time Structure from a Dynamical Perspective* (Oxford University Press, 2007).

Craig Callender is Professor of Philosophy and Co-Director of the Institute for Practical Ethics at the University of California San Diego. He is the author of *What Makes Time Special?* (Oxford University Press, 2017).

Sean M. Carroll is Research Professor of Physics at Caltech and External Professor at the Santa Fe Institute. His most recent book is *Something Deeply Hidden: Quantum Worlds and the Emergence of Spacetime* (E.P. Dutton, 2019).

Elena Castellani is Associate Professor in Philosophy of Science at the University of Florence, Italy. She co-edited *Symmetries in Physics: Philosophical Reflections* (Cambridge University Press, 2003), *The Birth of String Theory* (Cambridge University Press, 2012), and a special issue of *Studies in History and Philosophy of Modern Physics* on Dualities in Physics (2017).

Adam Caulton is Associate Professor and Clarendon University Lecturer at the Faculty of Philosophy, University of Oxford, and Fellow and Tutor in Philosophy at Balliol College, Oxford.

Eddy Keming Chen is Assistant Professor of Philosophy at the University of California, San Diego, an associate editor of the journal *Foundations of Physics*, and a fellow of the John Bell Institute for the Foundations of Physics.

Elise M. Crull is Assistant Professor of Philosophy at The City College of New York. Her books include *The Einstein Paradox: The Debate on Nonlocality and Incompleteness in 1935* (Cambridge University Press, 2021; co-authored with Guido Bacciagaluppi) and *Grete Hermann: Between Physics and Philosophy* (Springer, 2017; co-edited with Guido Bacciagaluppi).

Radin Dardashti is Junior Professor in Philosophy of Physics at the University of Wuppertal. His research focuses on the various methods used in theory development and assessment in modern physics. He is co-editor of *Why Trust a Theory? Epistemology of Fundamental Physics* (Cambridge University Press, 2019).

Shamik Dasgupta is Associate Professor of Philosophy at the University of California, Berkeley. He works in metaphysics, philosophy of science, and value theory.

Richard Dawid is Professor of Philosophy of Science at Stockholm University, Sweden. His research focuses on the philosophy of physics and the general philosophy of science. He is the author of *String Theory and the Scientific Method* (Cambridge University Press, 2013) and the co-editor of *Why Trust a Theory? Epistemology of Fundamental Physics* (Cambridge University Press, 2019).

Foad Dizadji-Bahmani is Associate Professor of Philosophy at California State University, Los Angeles. He works on issues in general philosophy of science, with a focus on intertheoretic relations and the realism debate, and on philosophy of physics.

Antony Eagle is Senior Lecturer in Philosophy at the University of Adelaide. He works in metaphysics, philosophy of science, and adjacent fields.

Nina Emery is Associate Professor of Philosophy at Mount Holyoke College and Affiliated Graduate Faculty at the University of Massachusetts, Amherst. She writes on topics at the intersection of metaphysics and philosophy of physics.

Laura Felline is an independent scholar. She lives in Seneghe (Sardinia) and her work focuses on the role of different (causal and non-causal) varieties of explanations in physics.

Samuel C. Fletcher is Assistant Professor in the Department of Philosophy at the University of Minnesota, Twin Cities, a Resident Fellow of the Minnesota Center for Philosophy of Science, and an External Member of the Munich Center for Mathematical Philosophy.

Doreen Fraser is Associate Professor of Philosophy at the University of Waterloo. She has published on topics including quantum field theory, Newton, underdetermination, and scientific realism.

Steven French is Professor of Philosophy of Science at the University of Leeds. He has published numerous books and papers in the philosophy of science and philosophy of physics, including mostly recently *Applying Mathematics: Immersion, Inference, Interpretation*, with Otávio Bueno (Oxford University Press, 2018), and *There Are No Such Things As Theories* (Oxford University Press, 2020).

Roman Frigg is Professor of Philosophy in the Department of Philosophy, Logic and Scientific Method at the London School of Economics and Political Science. He is the winner of the Friedrich Wilhelm Bessel Research Award of the Alexander von Humboldt Foundation. He co-authored, together with James Nguyen, *Modelling Nature. An Opinionated Introduction to Scientific Representation* (2020).

Mathias Frisch is Professor at Leibniz University of Hannover. He is the author of two books: *Inconsistency, Asymmetry, and Non-Locality: A Philosophical Investigation of Classical Electrodynamics* (2005) and *Causal Reasoning in Physics* (2014).

Richard Healey is Professor of Philosophy at the University of Arizona. He is the author of *The Quantum Revolution in Philosophy* (Oxford University Press, 2017) as well as *Gauging What's Real* (Oxford University Press, 2007).

Nick Huggett is LAS Distinguished Professor of Philosophy at the University of Illinois at Chicago. His work focuses on philosophy of physics, especially regarding theories of quantum gravity. His recent work includes *Everywhere and Everywhen* (2009), *Beyond Spacetime* (2020), *Philosophy Beyond Spacetime* (2021), and *Out of Nowhere* (forthcoming).

Jenann Ismael is Professor of Philosophy at Columbia University. Her books include *Time: A Very Short Introduction* (Oxford University Press, forthcoming), *How Physics Makes Us Free* (Oxford University Press, 2016), and *Essays on Symmetry* (Routledge, 2001).

Melissa Jacquart is Assistant Professor at the University of Cincinnati and Associate Director of the UC Center for Public Engagement with Science in Cincinnati, Ohio, USA.

Lina Jansson is Associate Professor in the Department of Philosophy at the University of Nottingham. She works on issues related to explanation, laws of nature, and confirmation.

Marc Lange is Philosophy Department Chair and Theda Perdue Distinguished Professor of Philosophy at the University of North Carolina at Chapel Hill. His most recent books are *Laws and Lawmakers* (Oxford University Press, 2009) and *Because Without Cause: Non-Causal Explanations in Science and Mathematics* (Oxford University Press, 2017).

Dennis Lehmkuhl is Lichtenberg Professor for History and Philosophy of Physics at the University of Bonn. He is also one of the editors of the Collected Papers of Albert Einstein published by Princeton University Press, and a Visiting Associate of the Einstein Papers Project at the California Institute of Technology.

Peter J. Lewis is Professor of Philosophy at Dartmouth College. He is the author of *Quantum Ontology* (Oxford University Press, 2016).

Tim Maudlin is Professor of Philosophy at New York University and is the Founder and Director of the John Bell Institute for the Foundations of Physics. His recent books include *Philosophy* *of Physics: Space and Time* (Princeton University Press, 2012), *New Foundations for Physical Geometry* (Oxford University Press, 2014), and *Philosophy of Physics: Quantum Mechanics* (Princeton University Press, 2019).

C. D. McCoy is Assistant Professor of Philosophy at Underwood International College, Yonsei University. His research falls within the philosophy of physics and general philosophy of science, and is especially focused on philosophical issues in modern cosmology.

Wayne C. Myrvold is Professor in the Department of Philosophy at the University of Western Ontario. He is the author of *Beyond Chance and* (Oxford University Press, 2021).

Jill North is Professor in the Department of Philosophy at Rutgers University. She is the author of *Physics, Structure, and Reality,* (Oxford University Press, 2021).

Oliver Pooley is Fellow and Tutor in Philosophy at Oriel College, Oxford, and an Associate Professor in Philosophy in the Faculty of Philosophy at the University of Oxford.

James Read is Associate Professor in the Faculty of Philosophy, University of Oxford, and a Tutorial Fellow at Pembroke College, Oxford.

Dean Rickles is Professor of History and Philosophy of Modern Physics at the University of Sydney. His recent books include *Covered in Deep Mist: The Development of Quantum Gravity, 1916–1956* (Oxford University Press, 2020) and *What Is Philosophy of Science?* (Wiley, 2020).

Bryan W. Roberts is a philosopher of physics, Associate Professor of Philosophy, Logic and Scientific Method, and Director of the Centre for Philosophy of Natural and Social Science at the London School of Economics and Political Science.

Laura Ruetsche is Louis E. Loeb Collegiate Professor of Philosophy at the University of Michigan. Her *Interpreting Quantum Theories* (Oxford University Press, 2011) shared the 2012 Lakatos Award with David Wallace's *The Emergent Multiverse* (Oxford University Press, 2012).

Juha Saatsi is Associate Professor at the University of Leeds. He works on various topics in philosophy of science, and he has particular interests in the philosophy of explanation and the scientific realism debate.

Ryan Samaroo is Associate Faculty at the University of Oxford and is currently serving as an advisor on scientific methodology to the Government of Canada. He is interested in the foundations of physics and mathematics, with a focus on the structure of theories.

Simon W. Saunders is Professor of Philosophy of Physics at the University of Oxford, and a Fellow of Merton College, Oxford. He is the lead editor (with Jonathan Barrett, Adrian Kent, and David Wallace) of *Many Worlds? Everett, Realism, and Quantum Theory* (Oxford University Press, 2010).

Arianne Shahvisi is Senior Lecturer in Ethics at the Brighton and Sussex Medical School. She has broad philosophical interests, including feminist philosophy, bioethics, philosophy of science, and social epistemology.

Orly Shenker is Eleanor Roosevelt Chair in History and Philosophy of Science and Director of the Sidney M. Edelstein Centre for History and Philosophy of Science, Technology and Medicine at the Hebrew University of Jerusalem. With Meir Hemmo, she is the co-author of *The Road to Maxwell's Demon: Conceptual Foundations of Statistical Mechanics* (Cambridge University Press, 2012) and the co-editor of *Quantum, Probability, Logic: the work and influence of Itamar Pitowsky* (Springer, 2019).

Susan G. Sterrett is the Curtis D. Gridley Distinguished Professor of History and Philosophy of Science at Wichita State University. She has published widely on models, including *Wittgenstein Flies a Kite: A Story of Models of Wings and Models of the World* (Pi Press, 2005).

Mauricio Suárez is a philosopher of science and probability. He edited *Probabilities, Causes and Propensities in Physics* (Springer, 2011), and is the author of *Philosophy of Probability and Statistical Modelling* (Cambridge University Press, 2020). He is currently Full Professor at Complutense University of Madrid, and a research associate at the Centre for Philosophy of Natural and Social Science at the London School of Economics and Political Science.

Nicholas J. Teh is Associate Professor in the Department of Philosophy and the History and Philosophy of Science Program at the University of Notre Dame. His research interests include the philosophy of physics, issues surrounding scientific representation, modeling and idealization, and the philosophy of painting.

Karim P.Y. Thébault is Senior Lecturer in Philosophy of Science in the Department of Philosophy at the University of Bristol. His research is principally within the philosophy of physics, with a particular emphasis on time and symmetry in classical and quantum theories of gravity.

Roderich Tumulka is a faculty member in the mathematics department of the Eberhard Karls University of Tübingen. He works on foundational questions in quantum mechanics, quantum field theory, and quantum statistical mechanics.

David Wallace is A.W. Mellon Professor of Philosophy of Science at the University of Pittsburgh. He has written on a wide range of topics in philosophy of physics, including statistical mechanics, symmetry, spacetime, quantum mechanics, quantum field theory, and quantum gravity.

James Owen Weatherall is Professor of Logic and Philosophy of Science at the University of California, Irvine. He is the author of *The Misinformation Age: How False Beliefs Spread* (with Cailin O'Connor) and *Void: The Strange Physics of Nothing*, both published by Yale University Press.

Charlotte Werndl is Professor for Logic and Philosophy of Science in the Department of Philosophy at the University of Salzburg, Austria. Her book, *The Legacy of Tatjana Afanassjewa: Philosophical Insights from the Work of an Original Physicist and Mathematician* (Springer, 2020), was edited together with Jos Uffink, Giovanni Valente, and Lena Zuchowski.

Porter Williams is Assistant Professor of Philosophy at the University of Southern California. He works primarily in the philosophy of science and the history and philosophy of physics, particularly quantum theories.

Christian Wüthrich is Associate Professor of Philosophy at the University of Geneva. He has two forthcoming monographs, *Out of Nowhere: The Emergence of Spacetime in Quantum Theories of Gravity* with Nick Huggett and *Time and Again: On the Logical, Metaphysical, and Physical Possibility of Time Travel* with J.B. Manchak and Chris Smeenk.

INTRODUCTION

Eleanor Knox and Alastair Wilson

What is philosophy of physics? We can give plausible-looking but uninformative answers of the form "the intersection of physics and philosophy," but any serious investigation into the character of the field has to look at its historical context. Depending on what we identify as the same intellectual tradition, we obtain very different conceptions of philosophy of physics and of its beginnings. Attempts to understand the most general features of the physical world probably go back beyond recorded history, but by the first millennium BCE, the Babylonians and Greeks were engaged in systematic theorizing and hypothesis testing about the natural world and in putting forward global metaphysical explanations for natural phenomena. Should we count the atomism of Democritus as a contribution to philosophy of physics, or to pure metaphysics? We leave that to the reader to decide.

Physics itself did not emerge as a recognizable discipline, distinct from the broader category of natural science, until after the Renaissance in Europe. Philosophical reflection accompanied the new discipline from the beginning. In the writings of Galileo, Newton, and Leibniz, we can recognize central questions that still animate contemporary philosophers of physics. Works like the Leibniz-Clarke correspondence (1717) and du Châtelet's commentary on her translation into French of Newton's *Principia* (1756) served as interpretive guides to the new physics and formed the core of a nascent foundational literature. Does this period mark the start of philosophy of physics proper? An interesting question no doubt, but not one we will address here: our focus is on contemporary philosophy of physics. We aim to take a pluralist approach to what counts as philosophy of physics, one which links the nature of the field to the varied institutions within which it is taught and researched.

By 1900, physics and philosophy had been institutionalized in Western universities as distinct disciplines within faculties of science and humanities, respectively, but there were clear examples of cross-overs between these disciplines: Mach, Poincaré, Broad, Russell, Noether, and Eddington had very different academic careers but they worked on the same cluster of topics. For the first decades of the 20th century, physics was respected within philosophy, and vice versa. That relationship has become strained at times in the post-war world; generations of physicists were raised with a distrust of philosophy under the enduring influence of logical positivist and logical empiricist traditions. In recent decades, though, philosophy of physics has found institutional security as a subfield in its own right. Today there are numerous undergraduate and postgraduate degree programs which offer courses and qualifications in philosophy of physics, substantial research clusters at leading universities, well-established recurrent conferences, prizes, specialist journals, and the inevitable concomitant rankings. Our goal with the Companion has been to provide a representative cross section of this wide-ranging research activity.

Limiting our focus to the contemporary practice of philosophy of physics does not of course guarantee a unified conception of what philosophy of physics is and how it should be pursued. Philosophers are fond of prescriptive definitions and divisions, and physicists are not much better. So it will hardly surprise the reader to hear that, even in a field as small as the philosophy of physics, methodological disagreements are rife. Given its essential engagement with so mathematical a field as physics, some gatekeeping is inevitable: without a solid understanding of the physics, things can go badly off the rails. But we are of the view that imposing divisions has largely hindered rather than helped the field: at times mathematical technical ability is mistaken for philosophical (or indeed physical) insight, while at others towering metaphysical structures are built on flimsy physical foundations. Likewise, philosophy of physics is at its most important and influential when it looks to the future, but work on contemporary theories suffers if the lessons of history are ignored. This volume is constructed with the assumption that good philosophy of physics can be done in physics departments, philosophy departments, HPS departments, and independently. The very best work in contemporary philosophy of physics combines a deep understanding of the physics with philosophical, and sometimes historical, subtlety: we hope some of that work is showcased here.

While a form of pluralism is at the heart of this volume, some carving up of the field is inevitable. Our chosen primary division, between theories and themes, aims to divide without excluding, and thereby to capture important work even where it falls at the edges of philosophy of physics. Many, but not all, philosophers of physics consider themselves "philosophers of X", where X is a theory or family of theories: quantum mechanics, general relativity, or the like. The first half of this volume aims to capture a range of this kind of theory-focused work with respect to the main theories that form the core of modern physics: specifically, classical mechanics, relativistic mechanics, quantum theories in their various forms, statistical mechanics, and thermodynamics. But this focus on particular theories can obscure the wider philosophical questions that arise from physics; much exciting work, particularly in recent years, lies at the boundary between philosophy of physics and other areas of philosophy. The second half of the volume encompasses this kind of thematic work: philosophy that reflects on the discoveries and the practice of physics to reach conclusions about topics of wider philosophical interest such as explanation, causation, or reduction.

We hope that this two-fold structure helps to make the book suitable for a wide audience: for the physicist interested in the foundations of their field, for the philosophy of physics graduate student looking for a thesis topic, but also for the philosopher seeking to understand how physics bears on their areas of interest. We've aimed for a range of levels of accessibility within the sections, and at least some chapters in each section should be readable by an advanced philosophy undergraduate without physics training. That said, in areas such as quantum field theory, authors inevitably presuppose more specific knowledge of physics. In others, like spacetime physics, important results cannot be expressed without a certain level of technical sophistication. Typically, chapters increase in degree of technicality through each part, especially in the "Theories" section. But generally, we have not tried to impose too much uniformity on the style or structure of the chapters; part of the strength of philosophy of physics, as we have tried to highlight, is diversity of individual approach among its practitioners. We hope that this shows through in the varied character of the contributions.

It is traditional in an introduction like this to opine on the future of the field. Such crystal-ball-gazing has its drawbacks: 20 years ago, for example, one might not have pegged the study of Newtonian spacetime for a comeback. Nonetheless, recent years have seen substantial developments in our understanding of various Newtonian spacetime structures and the relations between them. But some trends in the discipline are easier to spot. Several spring from the renewed relationship between physicists and philosophers mentioned above: as physics pushes new boundaries, new philosophical problems come to light, and old ones spring back up from the dusty corners to which physics tried to relegate them. A host of new physics research programs, and the grants that accompany them, involve physicists asking for active input from philosophers. Some of these new research programs are featured in this volume. Developments in quantum gravity, or perhaps the lack of them, raise questions about progress in science, as well as revealing underlying disputes along old philosophical lines (it is hard to do quantum gravity without implicitly taking a stand on the measurement problem, for example; in cosmology, no external observer is available to collapse the wavefunction). But quantum gravity also reveals exciting new philosophical questions, for example: what might it mean for the spacetime of our experience not to be fundamental? How should we think about the "dualities" posited by string theory? In a theory without a primitive notion of time, do we need a primitive notion of causality to replace it? The philosophy of cosmology is likewise a relatively young field, and one that looks set to grow in the coming years. Here the pressing questions are often obviously connected to more general philosophical issues: how do we deal with "one-shot" theories for which experimentation is impossible? What role might anthropic reasoning play in such a theory? Can non-empirical values play a role in theory choice here that they might not play elsewhere?

It will not escape the reader's attention that many of the questions addressed in this Companion seem to sit in general philosophy of science and/or in metaphysics rather than in the philosophy of physics, narrowly conceived. That is certainly true, although it's also true that dealing adequately with these problems in the context above requires a good understanding of both physics and philosophy of physics. But the expertise that physicists sometimes lack and seek elsewhere is often distinctively philosophical. So here, perhaps, is the moral for a graduate student in search of a topic: philosophers of physics can sometimes underestimate the importance of their general philosophical training, but it is also the unique skill they bring to the table. Work at the outward-facing edges of philosophy of physics is as difficult and significant as technical and theory-centered work.

Inevitably in a project of this scale, we have not managed to achieve as uniform a coverage as we would have liked. In particular, we regret the lack of chapters on quantum information theory and black holes. We hope to fill some of these gaps in future editions, as well as keeping the treatment of cutting-edge topics up to date. We've aimed to take a snapshot of the philosophy of physics community which is reasonably diverse with respect to geography, disciplinary affiliation, gender, and other social variables; however, there is no getting away from the fact that philosophy of physics as it is currently practiced is lacking in demographic diversity. We hope and expect that this situation will improve in the coming years. Philosophy of physics, overall, is in as good a shape as it has ever been, and we hope that this Companion will provide a useful resource for existing practitioners and for newcomers alike.

Section A

Theories

PART I

Newtonian Mechanics

Introduction to Part I

In both physics and philosophy of physics, our study usually begins with Newtonian physics. Whether it's resolving the forces on objects on a slope, or contemplating the justification for Newton's postulation of absolute space, the theory provides a fertile ground for pedagogical exploration. And within physics, of course, knowledge of this theory is crucial; despite being superseded in various domains by relativistic and quantum physics, Newtonian physics remains accurate enough in many domains that it still provides us with many of our best physical models. But why should we study the *foundations and philosophy* of a superseded theory? What might the aims and outcomes of such an enterprise be?

One thing is clear: if we think of the philosophy of Newtonian physics as an exploration of the metaphysics of a non-actual possible world governed by Newtonian laws, the ink spilled on the topic is wasted. After all, Newtonian theories can't account for the existence of stable matter. So what justifies the enduring popularity of the subject? In part, the answer lies with the theory's familiarity: after all that exposure in school we take ourselves to have an intuitive understanding of Newtonian mechanics. This means that thinking about the foundations is more straightforward; even though in many ways the structures of Newtonian theories are more complex than those of relativistic ones, the concepts come more easily. Moreover, the connection between Newtonian theory and empirical results is largely unproblematic. This value goes well beyond pedagogy. In order to think about theories that test the very bounds of our physical and mathematical understanding, we need clear heuristics for theory interpretation: how should we think about differences that are, in principle, not empirically detectable? What role do concepts like "force" or "inertia" play in our theories? What kinds of consideration determine what we should say about the spacetime of a theory? The chapters in this section suggest answers to many of these questions, and hence hold morals for further theories discussed in subsequent chapters.

In Chapter 1, Ryan Samaroo explores the conceptual structure of Newton's theory. One might be forgiven for thinking that Newton's primary achievement lies in giving an empirically accurate model of terrestrial and astronomical motion, but Samaroo argues that the conceptual achievements of the theory go beyond this: Newton provides principles that have a special character and that articulate and apply in a special way concepts of quantity of matter, motion, space, and time. While these principles and the concepts they articulate require adjustment in the light of more recent theories, they represent nonetheless a paragon of the kind of theorizing required for modern physics as we understand it.

Samaroo works primarily in the familiar framework of forces and acceleration used by Newton himself, but part of the enduring importance of the theory stems not from Newton's own mathematical framework, but from the great re-renderings of the theory offered by Lagrange and Hamilton in the late 18th and early 19th centuries. The Lagrangian and Hamiltonian formulations of classical mechanics look, at least on the surface, quite different, both from each other and from Newton's original theory. The later formulations are particularly important because they provide a framework not just for Newtonian physics, but for the relativistic and quantum theories that were later developed. But although all three formulations are usually thought of as notational variants, the relation between them is far from obvious. Jill North's chapter explores these relations and asks widely applicable questions about the notions of theory, formulation, equivalence, and fundamentality.

The 19th century held further mathematical developments of relevance to Newtonian mechanics, although it took some years for this relevance to be widely recognized. Geometrical techniques more commonly used in relativity can be equally well applied to Newtonian theories, and can help to elucidate the spacetime structures of the theory. Jim Weatherall's chapter explores these structures; using geometrical ideas more usually encountered in relativity, he considers what we should say about Newtonian spacetime structure. This involves a modern look at the debate between Leibniz and Clarke over the existence of absolute substantival space. As Howard Stein pointed out in 1967, thinking about the structure of Newtonian spacetime allows us to understand that it's possible for both Leibniz and Clarke to be right: Newton overreached in postulating absolute space but was nonetheless correct that some absolute structure is required by his theory to support the existence of absolute accelerations.

While Newton may have been right about the need for absolute structures in his own theory, contemporary strategies do not entirely close the debate between *substantivalism* (which holds that space or spacetime is an entity in its own right) and *relationism* (which holds that space is reducible to the relations between bodies). What becomes clear is that for Leibniz (and later Mach) to be correct, a new, relationist theory is needed which can replicate the empirical results of Newtonian mechanics without commitment to spatiotemporal structure. Mach's legacy had to wait until the early 1980s for a spatially relationist theory to emerge through the work of Julian Barbour and Bruno Bertotti. In Barbour's chapter for this volume, he explores the possibility of going beyond their original theory, and developing a classical theory which gives a truly relational account of both space and time. This chapter is more technical than the others in this section, so the reader not familiar with the physics and mathematics is advised to start with the earlier entries.

1

NEWTONIAN MECHANICS

Ryan Samaroo

The theory that we now know as "Newtonian mechanics" is Newton's science of matter in motion, and its philosophical significance, in a sentence, is this: Newton gave us more than just an empirically successful theory of mechanics – he gave us an account of what *knowledge* of the physical world should look like, one that remains with us. But what is this account of physical knowledge? What is it that remains with us? Various answers to these questions have been given and they concern the *methodological character* of the laws of motion. What is methodologically rational about them? What is their distinctive feature? These are the questions on the table.

The structure of this chapter is as follows. I will begin by introducing the laws of motion, the relations among them, and the spatiotemporal framework that is implicit in them. Then I will turn to the question of their methodological character. This has been the locus of philosophical discussion from Newton's time to the present, and I will survey the views of some of the major contributors. A theme running through this section is that there is something in the spirit of Kant's analysis of Newtonian physics that is worth preserving, though distilling what that is is an open problem. I will conclude by showing that while Newtonian mechanics motivates a number of philosophical ideas about force, mass, motion, and causality – and through this, ideas about space and time – the laws are themselves the outcome of a philosophical or critical conceptual analysis. Therefore, taking some care to understand how the theory grew out of Newton's analysis of the conceptual frameworks of his predecessors and contemporaries is valuable for its insights into the nature of that activity.

A word about the scope of this chapter is in order. It is worth recalling that *Principia* contains two theories: the theory of mechanics and the theory of universal gravitation. The former is found in a few pages right at the start of *Principia* in "Axioms, or The Laws of Motion," immediately following Newton's articulation of a few basic notions in "Definitions." The latter is a derived theory within the mechanical theory, that is, once that theory has been extended, through a number of assumptions, to encompass planetary systems. My focus will be on the theory of mechanics.

What is this theory? It is a theory of causal interaction: it is about motion and the forces producing motion. Newton dealt with matter in resisting and non-resisting media. My focus will be the mechanics of point particles in non-resisting media – the most basic subject matter with which the theory deals. But it is worth mentioning, if only in passing, the formulations of Newtonian mechanics of Euler and Cauchy. (See Truesdell (1977) for details.) And there are still other formulations of Newtonian mechanics, notably those of Lagrange and Hamilton. These formulations are based on the principle of least action, and they incorporate insights into the conservation of momentum and energy. They reveal a deep layer of structure exhibited by physical systems of many kinds and make them amenable to a similar treatment; in this way, these formulations extend Newtonian mechanics and greatly increase its computational power. They also provide a point of contact between classical and quantum theory. I will not discuss any of these formulations – my focus will be on "old-fashioned" Newtonian mechanics. (See Curiel (2014) and North (2019) for a detailed account of the relation of the Lagrangian and Hamiltonian formulations to classical systems and each other.)

It should also be noted that there is a reconstruction of Newtonian gravitation, patterned on Einstein's theory of gravitation, in which the basic account of motion is inseparable from the gravitational field: Cartan's reconstruction. On that reconstruction, gravitation is not a force causing acceleration but a manifestation of the curvature of space-time; as in Einstein's theory, the trajectories of free particles are geodesics (or "straight lines") of the (curved) space-time. Cartan's proposal is in several respects a natural and instructive way of thinking about Newtonian theory. (See Malament (2012) for details.) But I will deal only with Newtonian mechanics, independent of the gravitation theory.

1.1 Background: The Theory and the Spatiotemporal Framework Implicit in It

Newton's theory of causal interaction has three axioms, the laws of motion:

> Every body perseveres in its state of being at rest or of moving uniformly straight forward, except insofar as it is compelled to change its state by forces impressed.

> A change in motion is proportional to the motive force impressed and takes place along the straight line in which that force is impressed.

> To any action there is always an opposite and equal reaction; in other words, the actions of two bodies upon each other are always equal and opposite in direction. (Newton, 1726 [1999], pp. 416–417)

The laws, taken together, define and interpret the concept of inertial motion. This concept is the backbone of the theory, and, by examining how it is articulated, we will come naturally to all core aspects of the theory. The first law defines an ideal force-free trajectory, one from which a particle can be deflected by the action of some force, an objective cause. The first law alone, however, does not provide an account of inertial motion since we do not yet have a definition of force – it is a precondition for such a definition. It is the second law that defines and interprets the concepts of force and mass, tying them to acceleration. The acceleration of mass is the measure of the action of some force. The second law expresses a criterion for distinguishing free particles from particles acted upon by a force. Now, one might be tempted to suggest that the second law alone is enough for giving an account of inertia – one might suggest that the first law is a limiting case of the second, that is, when there is no force impressed. But the first law associates or coordinates a free particle with a particular kind of trajectory, a straight line. Hence, the first and second laws are interdependent.

The first and second laws provide a complete account of inertial motion, provided that one is interested only in ideal point particles. For actual bodies – bodies that are themselves composed of particles – the third law is a necessary condition for inertial motion. The forces among the constituent particles must be equal and opposite, failing which the body by its internal forces will accelerate of its own accord. (See the Scholium to the Laws of Motion, where Newton gives a proof of this; see also Samaroo (2018).) This is what the third law establishes and it is the basis for formulating a principle of conservation of momentum: in an isolated system the total momentum is conserved. Hence, the third law is also a criterion for distinguishing free particles from those acted upon by a force. What should be clear from this brief account is that inertia depends on all three laws for its articulation. It should also be clear that the laws of motion are mutually complementary. Only taken together do they determine Newton's theory of causal interaction.

Now, we can associate with any particle in inertial motion a reference frame. A reference frame is a space, one in which we can describe the motions of bodies in the space among themselves, for example, using a coordinate system. But we can also perform mechanical experiments and calculate their outcomes using the laws of motion. In any such space, the outcomes will be the same. These "inertial frames" that are picked out by the laws of motion are the basis for empirical investigation in Newtonian mechanics.

With this brief account of Newtonian mechanics in hand, let us consider the spatiotemporal framework that is implicit in it. The laws of motion define inertial motion as that state in which a body unacted upon by forces, or on which the net force is zero, moves in uniform rectilinear motion. In other words, a body in inertial motion moves *equal distances in equal times*. In this way, the laws of motion define an ideal clock that marks the "equable flow" of time. Furthermore, it is implicit in the theory that all inertial observers, and all ideal clocks, will measure proportional time intervals and agree on which events are simultaneous.

The concept of space that is implicit in Newtonian mechanics is tied to the inertial frame concept. As we have seen, an inertial frame is one in uniform rectilinear motion – one furthermore in which the outcomes of mechanical experiments, calculated using the laws of motion, are the same. And Newton noted that the same outcome would be obtained in any frame in uniform rectilinear motion relative to *it* – this is the Galilean principle of relativity. By way of this invariance property, we obtain an equivalence class of inertial frames – the class of frames in which the outcomes of mechanical experiments are the same. The equivalence-class structure so determined *is* the structure of the space-time of Newtonian mechanics.

It is important to note that this structure admits of no distinction between rest and uniform motion – both are states of inertial motion. By contrast, Newton held that while inertial frames are empirically indistinguishable, they are not theoretically equivalent. They move with various velocities relative to what he called "absolute space," even if those velocities cannot be known. It was only in the 19th century, through the work of Neumann (1870), Thomson (1884), Lange (1885), and others, that absolute space was shown to be superfluous.

It is sometimes said that the concept of space-time has its origin in Einstein's special theory of relativity in 1905. But something that should be evident from the foregoing is that already in Newtonian mechanics there is a concept of space-time. The very concept of an inertial trajectory, which is the basis for Newton's theory of causal interaction, appeals not just to places and times but to *places connected at times*. And not only does an inertial trajectory connect places at times but it connects them in such a way that certain states of motion are well defined. (See Stein (1967), DiSalle (2006), and Malament (2012) for careful accounts of this; see also Earman (1989) and Weatherall (2016) for other accounts of the spatiotemporal framework that Newtonian mechanics is sometimes held to motivate.) To summarize, what we find in the laws of motion is not only an account of force, motion, and causality but also a spatiotemporal framework.

1.2 The Methodological Character of the Laws of Motion

The theory, now introduced in overview, is a paragon of empirical success. What is the rational justification for this success? This is the question with which methodological analysis is concerned. A methodological analysis asks two basic questions. What *kind* of principles are the laws of motion? What is their *role* in the conceptual framework of physics? For example, are the laws entirely determined by empirical evidence? Or do they reflect elements of choice, for example, considerations of simplicity? Or is their methodological character more complex? And, if so, what is their character?

One answer is that given by Hume (1777 [1975]), who took Newtonian science to be a revolutionary advance – he took it as the model for his "science of human nature." Hume regarded the laws of motion as empirical generalizations that are inductively derived from constant and regular experience, that is, from a set of empirical facts. Consider the following remark about the second law: "Geometry helps us apply this law ..., but the law itself is something we know purely from experience, and no amount of abstract reasoning could lead us one step towards the knowledge of it" (*Enquiry*, IV). It is useful to recall Hume's claim that all objects of human reason are either "relations of ideas," of which his examples are the propositions of arithmetic and geometry, or "matter of fact," namely contingent empirical propositions such as "the Sun will rise tomorrow." Evidently, Hume regards the laws of motion as matters of fact.

Hume's view of the laws of motion singles out an obvious feature, namely that experience has a role to play in their formulation. But there are a number of criticisms one might raise against his view. One might say that we do not know the laws "purely from experience." For example, there are no truly force-free bodies: inertial motion is an ideal state and we have no "impressions." Hence, one could hardly say that the first law derives from mere induction. One might also take issue with Hume's remark that geometry merely "helps us" apply the laws. The formulation of the laws *presupposes* a number of mathematical concepts, notably concepts belonging to Euclidean geometry and the calculus. In this way, one might argue that Hume's division of the objects of reason into relations of ideas and matters of fact is inapt for the analysis of the laws of motion.

Hume's failure to give a satisfactory characterization of the laws is naturally contrasted with Kant's. Kant's account is not without its own difficulties, but it captures a feature of the laws that has remained part of subsequent discussions. Much like Hume, Kant saw in Newton's theory not only a revolutionary scientific discovery but a revolutionary philosophical advance. He saw a basis for criticizing the reigning Leibnizian tradition in which concepts of force and motion, space and time, substance and causality are applied to the "intelligible" world of monads.

In the First Critique (1787 [1998]), Kant asked, how has science achieved universal assent, while philosophy is the subject of endless dispute? What distinguishes scientific reasoning from philosophical reasoning, so that the former leads to principles that are necessary and universal, whereas the latter remains arbitrary and particular? How can philosophy start on "the secure path of a science"? Kant argued that nothing less than a Copernican Revolution in philosophy is needed. No longer should philosophy be done after the fashion of Leibniz and Wolff, or following an earlier empiricism: philosophy's task is to reveal the structure of our faculty of understanding – the structure that the very possibility of knowledge implicitly presupposes. Kant's theory of the constitution of experience provides an account of the *concepts* of this faculty and, of particular interest to us, the *principles* both "constitutive" and "regulative" by which they are applied to possible experience. The principles are rules that the understanding imposes on the appearances, in order to submit them rules. Without such rules, experience would be impossible. We would have nothing but a chaos of sensory appearances. The principles are about the world, and therefore "synthetic," but known through transcendental deduction, and therefore "a priori." Kant regarded the account of the constitution of experience as the true subject matter of metaphysics.

In the *Metaphysical Foundations* (1786 [2011]), Kant held the laws of motion to be just such principles. They have a constitutive function: they determine the concepts of the objects of enquiry; they make it possible for objects of knowledge to be objects of knowledge. The laws of motion are constitutive not only of a particular conception of force, mass, inertia, and causal interaction but of a spatiotemporal framework relative to which true motion can be understood.

Breaking with the Leibnizian tradition, Kant argued that our metaphysical concepts of force and motion, causal interaction, and space and time have no content at all except through their "sensible" counterparts, that is, through their articulation in the laws of motion. Kant's account is a landmark in the theory of knowledge, but it is problematic in at least one respect: Kant took Newton's laws to be the only ones that constitute the concepts of force, mass, and motion, and, furthermore, space, time, and causality. But though physics did not end with Newton, the idea that the laws of motion and certain other physical principles have a constitutive function was developed in the work of Kant's successors, notably in the work of Poincaré.

The idea that the laws of motion are definitions has a special place in the analysis of their methodological character. What is meant by "definition" is important here. For example, we might take a Russellian notion of definition as a starting point. For Russell (1897), the constituents of a sentence expressing a proposition must have independently grasped meanings. On such an account, the terms appearing in the laws – "force," "mass," and "inertial motion" – would already have their meanings, independently of the theory determined by the laws.

But there is another way of thinking about the laws of motion: we might regard them as implicit definitions. We find this idea in the work of Poincaré and Duhem. Poincaré (1902 [1952]) argues against Russell's view that we can know the meanings of primitive terms directly, for example, by intuition or acquaintance. For Poincaré, the primitive terms are implicitly defined by the axioms in which they figure. The laws of motion, on his account, are definitions disguised as claims.

But there is a further aspect to his view: Poincaré pointed out that a geometrical framework must be presupposed for the construction of a mechanical theory and he claimed that we can choose any one of the geometries of constant curvature, namely Euclidean, Bolyai-Lobachevskian or Riemannian. For him, there is no fact of the matter about which of them is the actual space of experience, but, since the laws of mechanics will be simplest on a Euclidean background, he held that Euclidean geometry would always be preferred. He stressed that geometry, on its own, tells us nothing about the behavior of physical objects, only geometry together with physical laws. He held that a geometrical framework and parts of the laws can be chosen arbitrarily – all that is required is that the remaining part of the laws be chosen such that the resulting theory is empirically adequate. For these reasons, he claimed that the laws of motion are conventions.

The presupposition of a Euclidean background is evident in the first law of motion, where "straight" is understood in the Euclidean sense. The first law defines the trajectory of a force-free body as a straight line – it establishes a correspondence between a physical object and a geometric notion, as part of a particular way of constructing a mechanical theory. But, for Poincaré, it is important that assuming a Euclidean background does not preclude the possibility that the completed theory or another theory that is in some sense more fundamental may lead us to revise our presuppositions about geometry.

What about the second law of motion? Taken on its own, we cannot speak of the truth or falsity of the relation expressed in the law because there is no experiment that settles the question. To speak of the truth or falsity of the law would be to assume that there is something prior to Newtonian mechanics that provides an independent definition of force – at least a definition that forms part of an empirically adequate theory of mechanics. For example, we cannot say that the relation between force and acceleration expressed in the law is imprecise, since any imprecision that we might notice while measuring some particular force only suggests to us that we need to look for the forces contributed by some yet-unnoticed bodies. But to say that there is no experiment that settles the question of the truth or falsity of the relation is not to suggest that the force law is not empirically constrained. The law can only be evaluated as part of the entire system of mechanics that it helps define.

What about the third law of motion? Poincaré notes that there are no perfectly isolated systems, only nearly isolated systems. When we observe such systems, we see that the constituent parts interact with one another such that they satisfy the third law and the center of gravity of the system moves (nearly) uniformly in a straight line. Poincaré asks, could a more accurate experiment invalidate this? "What, in fact, would a more accurate experiment teach us? It would teach us that the law is only approximately true, and we know that already" (Poincaré, 1902 [1952], p. 105). The third law defines action and reaction to be equal and opposite – it expresses a criterion by which we can determine whether momentum is conserved in an isolated system.

Now, one might think that Poincaré's view that the laws of motion are conventions commits him to the view that they are arbitrary. But he is clear that the laws are not arbitrary:

> Are the laws of acceleration and of the composition of forces only arbitrary conventions? Conventions, yes; arbitrary, no – they would be so if we lost sight of the experiments which led the founders of the science to adopt them, and which, imperfect as they were, were sufficient to justify their adoption. (Poincaré, 1902 [1952], p. 110)

What we find in this passage is Poincaré's recognition that while the laws of motion reflect elements of choice, they are empirically constrained. The laws may of course be revealed to be bad definitions or to have only limited applicability, but they function nonetheless as implicit definitions of the basic concepts of mechanics.

Before pressing on, it is worth noting a (perhaps obvious) feature of laws of motion: they fail a condition of *observational non-creativity*, according to which, roughly speaking, a definition should have no observational consequences. This is a condition that definitions are required to satisfy if they are to be regarded as analytic. The laws of motion are definitions, but they are not analytic – they are empirically constrained.

The proposal that the laws of motion are correctly understood as definitions, and furthermore implicit definitions, takes us much of the way to later proposals. But the notion of implicit definition has its origin in 19th-century work in the foundations of geometry, where it is discussed without reference to physical theory. It is evident that the laws of motion have a feature that the axioms of geometry do not: not only do they implicitly define the concepts of mechanics but they interpret them. They coordinate theoretical concepts with empirically measurable correlates.

That certain principles have a defining and coordinating function was recognized by Reichenbach (1928 [1958]), even if he did not discuss the laws of motion explicitly. Reichenbach regarded relativity theory as well established, but not well understood. He sought to improve our understanding of it by revealing the physical presuppositions that underlie the application of relativistic geometry and chronometry. Specifically, he argued that their application depends on principles that he called "coordinative definitions." These definitions establish how the claims of a mathematical theory are transformed from mathematical truths into claims that can be revised on the basis of experience. To take a simple example, Euclidean geometry becomes a theory of applied or physical geometry by means of the principle of free mobility: practically rigid bodies undergo free motions without change of shape or dimension. This principle is a presupposition of our ability to perform the compass-and-straightedge constructions of Euclidean geometry, and in this way it controls the application of the theory.

Now, for Reichenbach, the main interest in identifying coordinative definitions resides in their capacity to isolate which among the assumptions that control the application of geometry and chronometry are *conventions* and which are *factual* claims. And it is central to his view that certain principles that control the application of geometry and chronometry are based on stipulations. For this reason, he held coordinative definitions to be arbitrary. We see this, for example, in his account of special relativity, in which he claims that the Einstein synchronization criterion rests on a stipulation about the to and fro velocities of light; hence, the synchronization of distant clocks is a matter of convention.

The interpretive function of coordinative definitions led Reichenbach to regard them as constitutive principles. Coordinative definitions serve to apply an uninterpreted conceptual framework – the pure concepts of the understanding – to the world of experience. But while Kant held Newton's laws to be the unique set of principles that constitute the conceptual framework of physics, Reichenbach regarded them not as absolute but relative. He recognized that experience might lead us to mutually inconsistent coordinations that are relativized to particular contexts of enquiry, but have nonetheless a constitutive function.

Reichenbach's notion of coordination is incorporated into the recent work of Michael Friedman (e.g., 2001, 2010). Friedman's account of the laws of motion is found in his analysis of Newton's and Einstein's gravitation theories. According to Friedman, a satisfactory methodological analysis of these theories requires us to distinguish between three levels of enquiry. The *first level* is comprised of principles that are epistemologically distinguished by the fact that they define a space of intellectual and empirical possibilities, and so determine a framework of investigation. They articulate theoretical concepts and their physical interpretations. The *second level* is comprised of empirical hypotheses that are formulable within the framework. The *third level* is comprised of distinctly philosophical principles that motivate discussions of the framework-defining principles and the transition from one theory to another.

Friedman calls the first-level principles "constitutive principles." He includes in this category both mathematical principles or presuppositions and coordinating principles. The mathematical principles define a space of mathematical possibilities; they allow certain kinds of physical theories to be constructed. Among other examples, we find the calculus, linear algebra, and Riemann's theory of manifolds. The coordinating principles, which Friedman understands in Reichenbachian terms, interpret theoretical concepts. They express mathematically formulated criteria for the application of concepts such as force, mass, motion, electric field, magnetic field, and others.

On Friedman's analysis, Newtonian gravitation has as its *constitutive component* Euclidean geometry, the calculus, and the laws of motion. This component defines the space of intellectual and empirical possibilities that allows us to conceive of gravitation as a force, and that makes it possible to formulate the law of universal gravitation – an *empirical hypothesis*. For this reason, Friedman regards the principles comprising this component as relativized but nonetheless constitutive principles. They are not a priori, as Kant held them to be, but they are prior to the development of hypotheses about particular systems.

Friedman's approach to the analysis of physical theories is intended as a corrective to Quine's (1951) account of scientific knowledge. Quine took aim at the logical empiricists' account of theories with its distinction between the analytic and synthetic components of a theoretical framework. Quine represented scientific knowledge as a web of belief in which no satisfactory analytic-synthetic distinction can be drawn. Strands of the web are not subject to confirmation or disconfirmation as individuals – the web is confirmed or disconfirmed as a whole. And he claimed that in the case of a derivation where the conclusion conflicts with experience, there is nothing to prevent us from holding on to the conclusion by revising the logical and mathematical principles that were assumed in the derivation. It is Friedman's principal goal to show that there are distinctions between the components of our frameworks of physical knowledge, and that these components are stratified. Friedman argues, furthermore, that certain components of our frameworks can hardly be said to be revisable, as Quine maintained. It makes little sense to speak of revising the constitutive component of a theory in the case of a conclusion that conflicts with experience since constitutive principles determine the framework of empirical investigation – the framework without which an empirical hypothesis could be neither formulated nor tested.

Friedman's proposal is significant for its restoration of the logical empiricists' idea that frameworks of physical knowledge are stratified. But I have argued (Samaroo, 2015) that Friedman's account of a constitutive principle is too broad: only coordinating principles should be regarded as constitutive. Friedman's inclusion of both mathematical and coordinating principles in the category of constitutive principles is intended to address Quine's contention that the mathematics involved in formulating a theory is just another element in the web of belief. Friedman argues that this view of the role of mathematics in physics fails to account for the way in which mathematics makes certain kinds of empirical theories intellectual possibilities; it fails to account for the way in which mathematics supplies some of the concepts required for formulating a theory and for deriving predictions. While I agree with Friedman about this, there are good reasons for taking constitutive principles to be only those principles that constitute or interpret theoretical concepts by expressing criteria for their application.

First, one might argue that including mathematical principles in a theory's constitutive component opens the notion of a constitutive principle to trivialization. One might argue that what is constitutive is relative to some particular formulation of a theory, and since what is constitutive in one is not constitutive in another, the very idea of a constitutive principle is undermined. For example, Newtonian mechanics admits of various formulations, some of which rest on radically different mathematical frameworks from others. Take, for example, the mathematical frameworks peculiar to analytic mechanics, which are very different from the one in which Newton worked. But, however the theory is formulated, Newtonian mechanics is the theory whose basic structure at least is constituted by the laws of motion.

Second, one might argue that including mathematical principles in a theory's constitutive component lends support to a main feature of Quine's account of theories. A Quinean might argue that if, e.g., the calculus and Euclidean geometry are constitutive components of Newtonian mechanics, then they are confirmed or infirmed along with the rest of the theory. Friedman argues against Quine that constitutive principles can hardly be said to be tested along with the empirical hypotheses whose formulation they permit – they are principles without which empirical hypotheses would make neither mathematical nor empirical sense, and without which no test would be possible. But the principles that establish Friedman's argument against Quine are not the mathematical principles, which, on their own, are subject to neither empirical confirmation nor disconfirmation, but the *coordinating principles* that interpret theoretical concepts and control and application of the mathematics. Hence, distinguishing the mathematical principles from the coordinating principles strengthens the case against Quine.

Most importantly, however, the inclusion of both mathematical principles and coordinating principles in a theory's constitutive component blurs the distinction between the theory's *factual* and *non-factual* components. By taking only coordinating principles to be constitutive, we can distinguish clearly between those components of our theories that are empirically constrained and those that are not; we can distinguish between those principles that define and articulate our epistemic relation with the world and those that are part of the formal background or language. The proposed limitation to the account of a constitutive principle is in no way intended to diminish the role of mathematical principles in the articulation and application of physical theories, but to clarify the fact that mathematical principles and coordinating principles have different criteria of truth. This proposal benefits the account of the stratification of our theoretical knowledge and allows a still stronger criticism of Quine's account to be given; it aims in this way to vindicate something close to the analytic-synthetic distinction that Quine rejected. The foregoing is only a brief overview; see Samaroo (2015) for a sustained critical analysis.

In light of these criticisms of Friedman's program, what remains of the notion of a constitutive principle? And what of the methodological character of the laws of motion? I have argued that the notion of a constitutive principle – a principle that constitutes or interprets a theoretical concept by expressing a criterion of its application – has something to offer the account of the laws of motion. On this account, the laws of motion express criteria for the application of the concepts of force, mass, and inertial motion – and those that depend on these.

It is worth noting that this account of the laws of motion – that is, of the laws as *empirical criteria* – is essentially Einstein's. In a short but suggestive article, Einstein (1919 [2002]) sketched a distinction between theories that provide a general framework for physics ("principle theories" or "framework theories") and specific theories constructed within such a framework ("constructive theories"). Although Einstein's focus was relativity theory, Newtonian mechanics and Einstein's special and general theories are all framework theories. That is, they provide frameworks of constraints in which physical quantities can be constructed and whose evolution can be determined. As Einstein put it, these theories are based on "empirically discovered . . . general characteristics of natural processes" and they express "mathematically formulated criteria" that physical processes satisfy (Einstein, 1919

[2002], p. 213). These criteria enable us to articulate theoretical concepts such as force, mass, inertia, acceleration, rotation, and simultaneity; furthermore, they motivate spatiotemporal frameworks. These theories must be presupposed for the construction of theories of special systems, for example, the theory of a point particle or that of a perfect fluid; in the Newtonian context, the theory of the gravitational field. So, while the frameworks articulated by Newtonian mechanics, special relativity, and general relativity are not a priori in any Kantian sense, they are prior in this particular sense. Nonetheless, the presupposition of these frameworks does not preclude the possibility that some new theory will motivate their replacement.

The laws of motion, in sum, are founded on experience and are in that sense synthetic, but they are not mere empirical generalizations, derived by induction. Nor are they synthetic a priori propositions, though they function as "constitutive a priori principles" in a particular sense of that term. They are certainly definitions, but they are not analytic in the sense of being true by mere stipulation or convention, because they are responsible to a body of observation and experiment, and to the pre-analytic concepts of which Newton gave an analysis. They are "analytic of" the concepts they determine. They implicitly define the basic concepts of mechanics that appear in them, but they do more than that: they interpret those concepts. They function as "coordinative definitions," but they are not arbitrary in the Reichenbachian sense of that term. Still, they have the constitutive function that Reichenbach sought to capture: they constitute or interpret theoretical concepts by expressing criteria for their application; furthermore, they control the application of a number of mathematical theories. More simply, perhaps, they are empirical criteria for the application of the basic concepts of mechanics. Newtonian mechanics, then, is a "framework theory" in Einstein's sense. It is a framework of investigation that is "prior" to the theories of special systems that we might pursue and evaluate, but it is evidently not a priori in the usual sense of the term.

1.3   The Laws of Motion as the Outcome of a Conceptual Analysis

I will close with a reflection on the nature of Newton's "activity" in constructing his theory of mechanics. I wish to make the following point: while the laws of motion motivated philosophical ideas about motion, force, and causality, they come from Newton's analysis of what is presupposed in the dynamical reasoning of his predecessors and contemporaries. More specifically, the laws are formulations of the principles that Newton thinks are explicitly, as in the case of the first law, or implicitly, as in the case of the second and third laws, presupposed in the reasoning of these figures – when they are reasoning properly, that is, solving problems successfully. In this way, the laws of motion are the outcome of a *philosophical* or *critical conceptual analysis* of the conceptual framework of the mechanical philosophers. I will give a brief account of a few of the main parts of that framework. This terrain has been covered by others and more carefully. I will introduce only as much as is needed to make my point.

The Early Modern current known as "the mechanical philosophy" represents the Universe as a mechanism, one subject to mechanical laws governing all matter and implying determinism. The central tenet of the mechanical philosophy is a principle of action or causality: all physical action is mechanical action, that is, action through pressure or impact. This is the basis for the reductionist view that all natural phenomena can be explained by mechanical processes. Some, though not all, of the mechanical philosophers held that all matter is composed of minute corpuscles and aggregates of them. Their configurations determine bodies' primary and secondary qualities. This, in overview, is the mechanical philosophy.

The mechanical philosophy was intended as a corrective to the hierarchical and teleological aspects of Aristotelian and scholastic-Aristotelian science that persisted. Its proponents took it to be a development of Galileo's program of mechanical explanation. For this reason, one might begin with Galileo – after all, it was his arguments for the heliocentric hypothesis, and his proposals along the way, that undermined much of Aristotelian physics. From Galileo, we might move from figures such as Mersenne and Gassendi, through Hobbes and Boyle, to Descartes, Huyghens, and Leibniz. But I will focus on Descartes, whose *Principles of Philosophy* (1644 [1983]) was the standard work in 17th-century natural philosophy and whose physical principles Newton sought to refute.

Let us consider a few aspects of the mechanical philosophers' views on inertial motion, force, and the conservation of momentum. I will organize my discussion around their embryonic versions of these concepts – the concepts that Newton would subject to a critical analysis.

The first set of ideas is bound up with inertial motion. Two ideas are commonly held: that all motion is relative; a certain conception of inertial motion. Consider first the relativity of motion, an idea with which Descartes, Huyghens, and Leibniz are generally associated, though they understood "relativity of motion" differently. Descartes' criterion of "true motion" is in many respects singular and worth distinguishing from the others: a unique standard of motion that is also relative – relative, that is, to immediately contiguous bodies, which, however, provide a univocal reference. This is unlike the Leibnizian view, or "standard" relativism, according to which *any* two descriptions that agree on the relative distances, and so on changes of instantaneous "situation," are equivalent. This might aptly be called a "general principle of relativity." Now, on the surface, Huyghens shared with Leibniz the view that all motion is relative. But where Leibniz defended a "general relativity," Huyghens recognized that determining accelerations and rotations implicitly depends on a *privileged* state of uniform rectilinear motion relative to which they can be referred. This was also recognized by Newton – see, e.g., his criticism of Cartesian motion in *De grav* (c1660 [2004]). Both Newton and Huyghens saw clearly that such a state of motion is necessary for a satisfactory expression of the principles of mechanics. It is the recognition of this privileged state of motion – a state of "true motion" over and above the merely relative motions – that led Huyghens to formulate the first law of motion, which Newton embraced. For Newton, then, the first law expresses what was explicit in Huyghens' work and implicitly presupposed in other 17th-century accounts. (See Stein (1977) for references to the original sources and for translations of previously unpublished fragments of Huyghens.)

Consider also the idea of inertial motion. This is commonly held to have come from Descartes, though it was again Huygens who first stated it properly. For all that, it is worth considering the view of Descartes, who helped lay the foundations of mechanics by taking motion and rest to be primitive states of bodies that do not require further explanation. We find this in his first two laws of nature:

> The first law of nature: that each thing, as far as is in its power, always remains in the same state; and that consequently, when it is once moved, it always continues to move . . .
> (*Principles*, II, §37)

> The second law of nature: that all movement is, of itself, along straight lines; and consequently, bodies which are moving in a circle always tend to move away from the centre of the circle which they are describing. (*Principles*, II, §39)

Many of the salient features of Newtonian inertia are there: a body at rest remains at rest; a body in motion perseveres in its state of motion; bodies move in straight lines. But the concept of motion that Descartes articulates in the first and second laws differs from Newtonian inertia in that motion and rest are different states. Descartes also fails to make clear the connection between motion and force: there is, for example, no recognition that in the cases of both rest and motion the net force on a body is zero. Nor is there any requirement that a body's state of motion be *uniform with respect to time*, that is, unaccelerated. The requirement of uniformity is essential: without it, there is no notion that a body moving inertially moves *equal distances in equal times*, and hence no notion of an ideal clock that marks time. On Descartes' account, there is no concept of inertial motion in any Newtonian sense, and hence no basis for articulating viable concepts of force and mass. Newton's formulation of the first law of motion reflects his understanding of the very features that the Cartesian account lacks; his formulation of the law is an explication of what a satisfactory account of force demands.

The second set of ideas I will consider is concerned with the concept of force or action. Here, too, Cartesian physics was a starting point for Newton's analysis. Descartes sought to free mechanics from the hierarchical and teleological aspects of scholastic science; he set out to rid physics of qualitative properties and to reduce everything to "certain dispositions of size, figure, and motion." For all that, Descartes discusses force in a way that is reminiscent of the scholastic-Aristotelian framework of impetus and resistance: he appeals to the power or tendency needed to maintain bodies in their state of rest or to keep them in rectilinear motion. There is a question whether a body's tendency to move or to remain at rest is an essential, God-given characteristic or whether it derives from extension and from the interactions of bodies among themselves.

Newton's critical analysis of Cartesian force follows naturally in several respects from his articulation of the concept of inertial motion. Newtonian inertia is a necessary presupposition for saying what an *objective cause* or *force* is: it is that which deflects a body from its uniform rectilinear trajectory. Through this account, we also gain the concept of inertial mass, which is understood as a body's resistance to changes to its state of motion. All three concepts – inertial motion, force, and inertial mass – are articulated at once. Only with Newton's laws, therefore, do we find an explication of force that is a full realization of the mechanical philosophers' ideal: a physical quantity known only through its *effects*. As is well known, however, Newton's account of force encompasses the action of fields of force on distant matter – something the mechanical philosophy cannot comprehend.

The third set of ideas is bound up with the conservation of a certain "quantity of motion" in collisions and interactions. Here, again, Cartesian physics is in the background. This is the subject of Descartes' third law:

> The third law: that a body, upon coming in contact with a stronger one, loses none of its motion; but that, upon coming in contact with a weaker one, it loses as much as it transfers to that weaker body. (*Principles*, II, §40)

Descartes' account of the quantity of motion that is transferred from one body to another in a collision is clarified in §43. He takes that quantity to be the product of the size and speed of the body, where "size" appears to be understood as volume or bulk; there is no mention of the vector quantity velocity, only the scalar quantity "speed."

Descartes' account falls short of Newton's in several respects – see, e.g., the seven rules of impact (§46–52) where this is manifest – but the principle at issue is an embryonic version of the principle of conservation of momentum. Descartes' account is significant for being one of the first attempts at formulating the principle. But, for Descartes, the conservation of momentum has as much a theological as an empirical foundation: when God created the Universe, He gave to all bodies a certain quantity of motion, a quantity that He preserves at every successive moment, even when it is transferred (§62).

For a strictly empirical account of the conservation of momentum, we need to look elsewhere. The germ is already there in the work of Galileo and his contemporaries. But the first systematic accounts are found in the work of Wallis (1668), Wren (1668), and Huyghens (1669) on the laws of impact: Wallis dealt with inelastic collisions; Wren and Huyghens dealt with elastic ones. Their work is mentioned by Newton in the Scholium to the Laws and the third law of motion is based on it. Given two bodies A and B, the third law of motion defines their interaction as $F_{A \, on \, B} = -F_{B \, on \, A}$, and, in an isolated system, the corresponding expression for the conservation of momentum is $\frac{dp_A}{dt} = -\frac{dp_B}{dt}$. In this way, Newton not only incorporates the contributions of his contemporaries but explicates them in his theory of mechanics.

What should be clear from the foregoing is that the laws of motion, far from being radical, are implicitly and explicitly presupposed in the work of the mechanical philosophers. This was Newton's reason for taking them to be axioms. Newton's activity, then, is an eminently philosophical one: it is a critical conceptual analysis of confused concepts; it is also an analysis of successful practice that aims to discover the principles on which that practice depends. This culminates in the explication of the basic concepts of mechanics and the articulation of criteria for their application. These criteria are the basis for an empirically adequate theory of mechanics. We find in Newton's construction of his theory an exemplar of an approach to conceptual analysis that has been at the heart of the foundations of physics, at least since Galileo, and at the heart of the analytic tradition, at least since Frege. Conceptual analysis, so understood, is the practice of identifying central features of a concept by revealing the assumptions on which use of the concept depends. (This way of expressing the basic idea of conceptual analysis is due to Demopoulos (2000, p. 220).) This practice proceeds by examining the use, misuse, and limitations of pre-existing concepts – in the case of interest to us, inertia, force, and mass – and revealing the assumptions on which their pre-analytic use depends. Therefore, while Newton's theory of mechanics motivates philosophical ideas about matter and motion, space-time and causality, and philosophical debates about them, its philosophical significance goes deeper: the theory is a reminder of what conceptual analysis in the foundations of physics might aspire to. The laws of motion are the result of that analysis.

References

Curiel, E. (2014). Classical mechanics is Lagrangian; it is not Hamiltonian. *British Journal for the Philosophy of Science*, 65: 269–321.

Demopoulos, W. (2000). On the origin and status of our conception of number. *Notre Dame Journal of Formal Logic*, 41: 210–226.

Descartes, R. (1644 [1983]). *The Principles of Philosophy*, translated by V.R. Miller and R.P. Miller. Dordrecht: Reidel.

DiSalle, R. (2006). *Understanding Space-time*. Cambridge: Cambridge University Press.

Earman, J. (1989). *World Enough and Spacetime: Absolute and Relational Theories of Motion*. Cambridge, MA: MIT Press.

Einstein, A. (1919 [2002]). What is the theory of relativity? In M. Janssen et al. (eds.), *The Collected Papers of Albert Einstein*, Vol. 7. Princeton, NJ: Princeton University Press, pp. 206–215.

Friedman, M. (2010). Synthetic history reconsidered. In M. Domski and M. Dickson (eds.), *Discourse on a New Method: Reinvigorating the Marriage of History and Philosophy of Science*. Chicago: Open Court, pp. 571–813.

Friedman, M. (2001). *Dynamics of Reason: The 1999 Kant Lectures of Stanford University*. Stanford: CSLI Publications.

Hume, D. (1777 [1975]). *Enquiries Concerning Human Understanding and Concerning the Principles of Morals*, reprinted from the 1777 edition with Introduction and Analytical Index by L.A. Selby-Bigge. Oxford: Clarendon Press.

Huyghens, C. (1669). A summary account of the laws of motion. *Philosophical Transactions of the Royal Society*, 4: 925–928.

Kant, I. (1787 [1998]). *Critique of Pure Reason*, translated and edited by P. Guyer and A. Wood. Cambridge: Cambridge University Press.

Kant, I. (1786 [2011]). *Metaphysical Foundations of Natural Science*, edited by M. Friedman. Cambridge: Cambridge University Press.

Lange, L. (1885). Ueber das Beharrungsgesetz. Berichte der Koniglichen Sachsischen Gesellschaft der Wissenschaften zu Leipzig, *Mathematisch-physische Classe*, 37: 333–351.

Malament, D. (2012). *Topics in the Foundations of General Relativity and Newtonian Gravitation Theory*. Chicago: University of Chicago Press.

Neumann, C. (1870). *Ueber die Principen der Galilei-Newton'schen Theorie*. Leipzig: B. G. Teubner.

Newton, I. (c1660 [2004]). De gravitatione et æquipondio fluidorum. In A. Janiak (ed.), *Philosophical Writings*. Cambridge: Cambridge University Press, pp. 12–40.

Newton, I. (1726 [1999]). *The Principia: Mathematical Principles of Natural Philosophy*, translated by I.B. Cohen and A. Whitman. Berkeley: University of California Press.

North, J. (2019). Formulations of classical mechanics. In E. Knox and A. Wilson (eds.), *The Routledge Companion to the Philosophy of Physics*. London: Routledge.

Poincaré, H. (1902 [1952]). *Science and Hypothesis*, translated by W. Greenstreet. New York: Dover.

Quine, W.V. (1951). Two dogmas of empiricism. *The Philosophical Review*, 60: 20–43.

Reichenbach, H. (1928 [1958]). *The Philosophy of Space and Time*, translated by M. Reichenbach and J. Freund. New York: Dover.

Russell, B. (1897). *An Essay on the Foundations of Geometry*. Cambridge, Cambridge University Press.

Samaroo, R. (2015). Friedman's thesis. *Studies in History and Philosophy of Modern Physics*, 52: 129–138.

Samaroo, R. (2018). There is no conspiracy of inertia. *British Journal for the Philosophy of Science*, 69: 957–982.

Stein, H. (1967). Newtonian space-time. *Texas Quarterly*, 10: 174–200.

Stein, H. (1977). Some philosophical prehistory of general relativity. In J. Earman, C. Glymour, and J. Stachel (eds.), *Foundations of Space-Time Theories. Minnesota Studies in the Philosophy of Science*, Vol. VIII. Minnesota: University of Minnesota Press, pp. 3–49.

Thomson, J. (1884). On the law of inertia; the principle of chronometry; and the principle of absolute clinural rest, and of absolute motion. *Proceedings of the Royal Society of Edinburgh*, 12: 568–578.

Truesdell, C. (1977). *A First Course in Rational Continuum Mechanics*. New York: Academic Press.

Wallis, J. (1668). A summary account given by Dr. John Wallis, of the general laws of motion, by way of Letter written to him to the Publisher, and communicated to the R. Society, Novemb. 26. 1668. *Philosophical Transactions of the Royal Society*, 3: 864–866.

Weatherall, J. (2016). Maxwell-Huygens, Newton-Cartan, and Saunders-Knox Spacetimes. *Philosophy of Science*, 83: 82–92.

Wren, C. (1668). Theory concerning the same subject; imparted to the R. Society Decemb. 17. Last, though entertain'd by the Author divers years ago, and verified by many Experiments, made by Himself and that other excellent Mathematician M. Rook before the said Society, as is attested by many Worthy Members of that Illustrious Body. *Philosophical Transactions of the Royal Society*, 3: 866–867.

Further Reading from the Editors

For an accessible introduction to these topics with extracts from classic readings, see N. Huggett, *Space from Zeno to Einstein* (MIT Press, 1999). Chapters 7 and 8 discuss Newton, but earlier chapters also give historical context. For a collection with a range of papers on Newton's philosophy, see Cohen, I.B. and George E. Smith (ed.), *The Cambridge Companion to Newton* (Cambridge: Cambridge University Press, 2002). Chapters 1, 2, 4, and 8 are particularly relevant. Two papers on Newton with an historical focus are Rynasiewicz, "By their Properties, Causes and Effects: Newton's Scholium on Time, Space, Place and Motion – I. The Text" (*Studies in History and Philosophy of Science* 26 1995) and his "By Their Properties, Causes and Effects: Newton's Scholium on Time, Space, Place and Motion—II. The Context" in the same volume.

2

FORMULATIONS OF CLASSICAL MECHANICS

Jill North

2.1   Introduction

Classical mechanics is the physical theory with which we are most familiar, the one we first encounter in school. Philosophers tend to regard classical mechanics as metaphysically unproblematic. At first glance, it does appear straightforward: the theory is fundamentally about particles, with intrinsic features like mass,[1] that move around in three-dimensional space in response to various forces, which arise via interactions between the particles. It seems as though if any physical theory is metaphysically perspicuous, classical mechanics is. But the theory is not as clear-cut as it initially seems. Our familiarity misleads us.

The reason is not just that classical mechanics ultimately runs into the kind of trouble that presaged quantum mechanics. Even taking it to be the true fundamental theory of a world,[2] classical mechanics does not offer as candid a picture of things as we tend to think. One reason for this is that there are different formulations, which are generally claimed to be equivalent by physics books, but which are at least not *obviously* equivalent – neither in terms of the mathematical structure they use, nor in terms of the physical world they describe.

What I want to do in this chapter is to outline the three leading formulations of classical mechanics, and to raise some questions about them, the chief one being: are these genuinely equivalent formulations, as usually thought? If so, in what sense are they equivalent? If not, in what way(s) do they differ? Another way to put the focal question of this chapter is by means of a title of Mark Wilson's (2013): "What Is 'Classical Mechanics', Anyway?" Indeed, since the terms "classical mechanics" and "Newtonian mechanics" "are used virtually synonymously" (Spivak, 2010, p. 7), one aim of this chapter is to suggest that it is not right to do so. There are different versions of classical mechanics, which might even amount to distinct theories. A related aim is to show that there are interesting philosophical questions that arise in the context of classical mechanics. Classical mechanics merits the attention of philosophers, who often disregard it as either too perspicuous or too outdated to warrant much discussion.[3]

Although this chapter is limited to classical mechanics, a host of general questions in the philosophy of physics and science are touched upon, such as the following. What is the right notion of theoretical equivalence: when are two scientific theories mere notational variants? How do we interpret a scientific theory: how do we figure out the nature of the world according to a theory? When faced with different theories or formulations, how do we choose which one to adopt? Indeed, must we choose?

2.2 Three Formulations

I will outline the three main formulations of classical mechanics – Newtonian, Lagrangian, and Hamiltonian mechanics – in relatively standard ways, before turning to some questions about them.[4] My focus will be on the dynamical laws and the quantities that appear in them. This is where much of the action lies in comparing and contrasting the different formulations.

2.2.1 Newtonian Mechanics

Newtonian mechanics might be the only formulation one comes across, the others typically not introduced until more advanced college courses. In the Newtonian mechanics of point particles – point-sized physical objects with intrinsic features like mass[5] – two sets of coordinates specify a system's fundamental state at a time: the positions and velocities (or momenta) of all the particles. Assuming the particles are free to move around in three-dimensional physical space, these coordinates will each have three components, one along each spatial dimension.

For a system consisting of n particles, the total state is specified by means of $6n$ coordinates: three coordinates for the position and three coordinates for the velocity of each particle in the system. It turns out to be extremely useful to represent all the possible states of a system in a mathematical space called the *statespace*, each point of which represents a different possible fundamental state of the system. Since we need $6n$ coordinates to specify the state of a system, the statespace will have $6n$ dimensions.

Different curves through the statespace represent different possible histories of the system, different sequences of fundamental states over time. (The curves are parameterized by time.[6]) These histories will be given by a theory's dynamical laws, in this case, Newton's second law:[7]

$$\Sigma \mathbf{F}_i = m_i \mathbf{a}_i = m_i \ddot{\mathbf{x}}_i. \tag{2.1}$$

Here, $\Sigma \mathbf{F}_i$ indicates the sum of the forces – which are vector quantities, written in bold – on a given particle labeled by i (i ranges from 1 to n, for n particles in the system); m_i is the particle's mass; $\mathbf{a}_i$, or $\ddot{\mathbf{x}}_i$, is the particle's acceleration, the second derivative of its position with respect to time, which is also a vector quantity. (A dot over a quantity indicates a derivative with respect to time of that quantity.) In other words, $\Sigma_{j \neq i} \mathbf{F}_{ij} = m_i \mathbf{a}_i$, where $\Sigma_{j \neq i} \mathbf{F}_{ij}$ is the sum of the forces on the given particle due to all the other particles (both in the system and external to it).

The above is a vector equation. There is one such equation for each particle in each component direction – three equations per particle in three-dimensional space. These equations can be grouped together into one master equation, which says how the point representing the state of the entire system moves through the statespace over time. Given the initial state of a system and the total forces acting on it, integrating (twice) yields a unique solution or history: the laws are deterministic.[8] A solution picks out a trajectory in the statespace, which represents the paths of all the particles through ordinary physical space.[9]

Equation 2.1 is the fundamental dynamical equation of the theory. Newton's second law, mathematically represented by this equation, predicts the motion of every particle, in any situation. What forces there are will depend on the types of particles involved, and to calculate the forces, we will need additional rules, like the law of gravitation. But this one dynamical law predicts any system's behavior, once given those forces.

Two other laws of Newtonian mechanics as standardly presented are important to the theory as a whole, but will play a less central role here. Newton's first law says that an object continues with uniform velocity unless acted on by a net external force. This law helps define what it is for an object to not accelerate, or to travel inertially (with the second law saying what happens when an object is subject to a net force that yields an acceleration). Newton's third law specifies the nature of forces. It is often stated in "action-reaction" form: for every action there is an equal and opposite reaction; when one object exerts a force on a second object, the second simultaneously exerts a force equal in magnitude and opposite in direction on the first. This law tells us that forces come in pairs, as the result of interactions between two objects. It "describes the forces to some extent" (Feynman et al., 2010, sec. 9.1), with the particular force laws further indicating that forces do not depend on anything other than the types of particles involved and their spatial separations, and that they are central forces, directed along the line between the particles. (Conservative forces, derivable from a potential.)[10]

2.2.2 *Lagrangian Mechanics*

In Lagrangian mechanics, two sets of what are called *generalized coordinates* characterize systems' fundamental states at a time: the generalized positions, q_i, and their first time derivatives, the generalized velocities, $\dot{q}_i$ (*i* from 1 to *n*, for *n* particles in the system). As in Newtonian mechanics, we need $6n$ coordinates to completely specify the state of a system of *n* particles: three generalized position coordinates and three generalized velocity coordinates per particle. But unlike in Newtonian mechanics, these do not have to be ordinary position and velocity coordinates. (They are called generalized positions and velocities by analogy to ordinary positions and velocities.) Generalized coordinates can be any set of independent parameters that completely specify a system's state.[11] Generalized positions can have units of energy, or length squared, or an angle, or can even be dimensionless. We can use any kind of coordinates that are suited to a system, the choice typically guided by the number of degrees of freedom of the system[12] and the topology of the spatial region in which the particles are free to move around. For a pendulum, for example, we might use the angle θ the suspending string makes with respect to the vertical as the generalized position, with $\dot{\theta}$ being the generalized velocity (as we will see in Section 2.3).

The Lagrangian statespace is a $6n$-dimensional space with the structure of a tangent bundle. This space comprises a $3n$-dimensional space in which we represent the generalized positions (called the *configuration space*), plus the $3n$-dimensional tangent space at each point (to represent the generalized velocities, which are tangent to the generalized positions). Each point in the statespace picks out a generalized position and generalized velocity for each particle in the system. Standard labels are Q for the configuration space (the "base space" of the tangent bundle), $T_q Q$ for the tangent spaces (the "fibers," one for each q in Q), and TQ for the entire statespace, sometimes referred to as the velocity phase space. Notice the configuration space is what represents the physical space the particles move around in. Given the freedom in generalized coordinates, this representation needn't occur in an obvious way, yet the structure of physical space will still be coded up in the structure of Q.

The dynamical laws, called the Euler-Lagrange, or simply Lagrange, equations, say how the point representing a system's state moves through the statespace over time, given a scalar function called the Lagrangian, L. At each point in the statespace, this function assigns a number, typically equal to the system's kinetic energy, T, minus its potential energy, V.[13] Although this gives the Lagrangian as defined on TQ, we can think of this function as coding up information about particles' ordinary spatial features, those that are relevant to their energies, so that it is ultimately about goings-on in three-dimensional space. The motion of an *n*-particle system in three-dimensional space is then given by $3n$ second-order equations, one equation for each particle in each direction – one for each degree of freedom (three per particle in three-dimensional space):

$$\frac{d}{dt}\left(\frac{\partial L}{\partial \dot{q}_i}\right) - \frac{\partial L}{\partial q_i} = 0. \tag{2.2}$$

Given L, these equations uniquely determine the motion for an initial state characterized by the generalized position and generalized velocity of each particle in the system. A solution, found by integrating, gives a function or trajectory on Q, which represents the motions of all the particles through physical space. (Solutions are curves through TQ, which are projected onto Q.)

To get a feel for the Lagrangian statespace, picture the statespace for a particle moving on a one-dimensional circle: Figure 2.1. (Keep in mind that this is "just about the only easily visualized nontrivial TQ" (José and Saletan, 1998, p. 94); with more degrees of freedom, things quickly become difficult to picture.) This is a two-dimensional space, each point being picked out by two coordinates $(q, \dot{q})$. The circle represents the different possible values of the generalized position coordinate, the lines represent the different possible values of the generalized velocity. Curves through this space represent different possible histories of the system, different sequences of generalized positions and velocities over time. The figure could represent the statespace of a point-mass pendulum, for instance, with the circle representing the values of θ and the lines the values of $\dot{\theta}$.

Briefly note three interesting, interrelated differences between the Lagrangian and Newtonian formulations.[14] First, in Lagrangian mechanics, a scalar energy function is what determines a system's motion, whereas in Newtonian mechanics, the motion is given by the forces, which are vector quantities. Second, Lagrangian mechanics takes a more "holistic" approach to describing systems' motions, in terms of the energy of the system as a whole. By contrast, the Newtonian formulation "is intrinsically a particle-by-particle description" (Sussman and Wisdom, 2014, p. 3), given in terms of the forces on each individual particle due to every other particle. Third, Lagrangian mechanics is a more coordinate-independent formulation of the dynamics, in that we can substitute any kind of coordinates for q and $\dot{q}$ in equation 2.2. The central equation of Newtonian mechanics, however, contains an implicit preference for Cartesian coordinates, those in which it has the form of equation 2.1. We can of course use other kinds of coordinates, but the form of the equation will differ (contrast equation 2.1 with the form in polar coordinates, e.g.: Taylor (2005, eq. 1.48)). This is not the case in Lagrangian mechanics: "Lagrange's equations, unlike Newton's, take the same form in any coordinate system" (Taylor, 2005, p. 237). (The form of an equation is the form as a function of its variables, a standard notion in physics.[15])

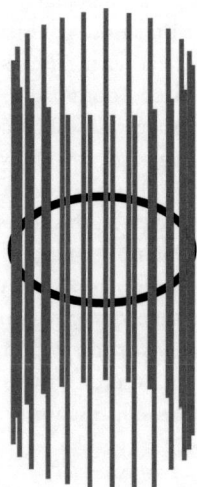

Figure 2.1 Two-dimensional tangent bundle

2.2.3 Hamiltonian Mechanics

Hamiltonian mechanics shares a special kinship with Lagrangian mechanics, more so than with Newtonian mechanics. Here, too, a scalar energy function determines the motion, and the central equations are formulated in terms of generalized coordinates. There are also some notable differences. Hamiltonian mechanics uses a different energy function and a different kind of generalized coordinate, with the result that the dynamical equations and statespace also differ.

The Hamiltonian coordinates are called *canonical coordinates*. These are the generalized positions, q_i, and the generalized momenta, p_i. (Again, i ranges from 1 to n for n particles, three of each coordinate per particle in three-dimensional space.) The Hamiltonian statespace is the cotangent bundle of configuration space, T^*Q: the configuration space, Q, together with the cotangent space, T^* (dual to the tangent space), at each point in Q (to represent the generalized momenta, which are covectors, or one-forms). This is a $6n$-dimensional space, each point of which picks out a generalized position and generalized momentum for each particle in the system. It is often called the momentum phase space, or simply the phase space.[16]

The scalar function that describes a system's motion is called the Hamiltonian, H, which is (typically[17]) equal to the total energy of the system – the sum of the potential and kinetic energies, instead of the difference between them, as in Lagrangian mechanics. The dynamical laws are a set of $2n$ first-order equations, two equations for each particle in each direction; two equations for each degree of freedom:

$$\dot{q}_i = \frac{\partial H}{\partial p_i}, \quad \dot{p}_i = -\frac{\partial H}{\partial q_i}. \tag{2.3}$$

These equations, called the Hamiltonian or canonical equations, uniquely determine a system's motion given an initial state specified by the canonical positions and momenta of all the particles in the system.

The Hamiltonian and Lagrangian formulations are both more coordinate-independent than the Newtonian formulation. Each of them is given in terms of generalized coordinates, with the result that the dynamical equations retain their form regardless of which coordinates we use. The reason is that the Lagrangian and Hamiltonian functions, which determine the motion, are scalar functions. In Newtonian mechanics, by contrast, vector quantities – forces – determine the motion. Although vectors are coordinate-independent objects, their components change with the coordinate system. (Vectors can be defined by means of how their components transform under coordinate changes.) And as Feynman puts it, "The general statement of Newton's Second Law for each particle. . . is true specifically for the *components* of force and momentum [or acceleration] in any given direction," since "any vector equation involves the statement that *each of the components is equal*" (Feynman et al., 2010, sec. 10.3, 11.6; original italics). Scalars are even more coordinate-independent than that, being completely unaffected by coordinate changes, not even "altering component-wise." (The form of a scalar function such as L or H may change with the coordinate system, but not the scalar value, nor the form of the equation in which L or H appear.)

2.3 Example: Plane Pendulum

Briefly work through a simple example to get a feel for the different flavor of each formulation. Consider a vertical plane pendulum, which moves through two spatial dimensions, as shown in Figure 2.2. (Assume the usual idealizations: frictionless, massless, rigid suspending string; point-mass bob; negligible air resistance; uniform gravitational field.) Use each formulation to find the equation of motion for the pendulum, the equation that describes the position of the bob as a function of time. We will see that each formulation yields the same equation of motion, but by means of different routes.

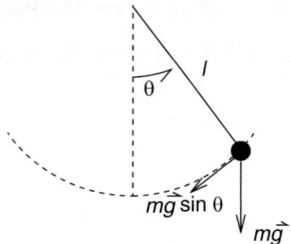

Figure 2.2 Plane pendulum.

To use Newton's law, equation 2.1, first choose a rectangular coordinate system. Let y be in the radial direction, with x in the direction tangential to the path of the bob. Resolve the forces on the bob into their components in this coordinate system. There are two forces on the bob: the tension directed along the string, and the downward–directed gravitational force. The component of the gravitational force in the direction of the acceleration along the path – the tangential force – is $mg \sin \theta$, where θ is the angle the string makes with respect to the vertical, as shown in the figure.

There are two component equations of Newton's law, one for each direction of our coordinate system: $F_x = ma_x$ and $F_y = ma_y$. Plugging in the relevant force components yields $F_x = -mg \sin \theta = ma_x$ (the negative sign because the gravitational force points downward) and $F_y = T - mg \cos \theta = ma_y$, with T being the tension in the string. Note that $a_y = 0$; as a result, we effectively ignore this second equation when solving for the equation of motion. (T has no component in the direction of nonzero acceleration: it is merely a "constraint force.")

The arclength, which measures the distance traveled by the bob along the curved path, is given by $s = l\theta$. The second derivative of this quantity, $\ddot{s} = l\ddot{\theta}$, is the acceleration along the path. Plug into the x-component equation of Newton's law, and we obtain the following equation of motion for the pendulum:

$$-g \sin \theta = l\ddot{\theta}. \tag{2.4}$$

We obtain the same equation of motion, in a different way, using Lagrangian mechanics. We could use rectangular coordinates as we did above; but things are simpler if we instead use generalized coordinate θ, with $\dot{\theta}$ being the generalized velocity. We can plug these coordinates directly into equation 2.2 to find the solution. We can effectively treat θ and $\dot{\theta}$ as ordinary position and velocity coordinates, respectively, and, perhaps surprisingly, this yields the right answer.

First calculate the Lagrangian, $L = T - V$. The kinetic energy $T = \frac{1}{2}mv^2 = \frac{1}{2}m(l\dot{\theta})^2$. (The arclength is $s = l\theta$, the velocity its first time derivative.) The potential energy $V = -mgl \cos \theta$, setting the zero at the height of the pivot point where $\theta = \frac{\pi}{2}$. (Gravitational potential energy $= mgy$, with y being the vertical distance from a chosen zero.) Thus, $L = \frac{1}{2}m(l\dot{\theta})^2 + mgl \cos \theta$. Calculate the following derivatives (in effect treating θ and $\dot{\theta}$ as independent variables, even though one is really defined as the time derivative of the other): $\frac{\partial L}{\partial q} = \frac{\partial L}{\partial \theta} = -mgl \sin \theta$ and $\frac{\partial L}{\partial \dot{q}} = \frac{\partial L}{\partial \dot{\theta}} = ml^2\dot{\theta}$, so that $\frac{d}{dt}\left(\frac{\partial L}{\partial \dot{\theta}}\right) = ml^2\ddot{\theta}$. Finally, plug into equation 2.2: $ml^2\ddot{\theta} - (-mgl \sin \theta) = 0$, i.e., $l\ddot{\theta} + g\sin \theta = 0$, which, rearranged, is equation 2.4.

In Hamiltonian mechanics, we first find the Hamiltonian, $H = T + V$. Given L above, we can see that $H = \frac{1}{2}m(l\dot{\theta})^2 - mgl \cos \theta$, but we need to rewrite this in terms of canonical coordinates. To find the generalized momentum, p_θ, which is "conjugate" to the position variable θ, use this equation: $p = \frac{\partial L}{\partial \dot{q}}$, often taken to be the definition of the generalized momentum.[18] Using the equation $p_\theta = \frac{\partial L}{\partial \dot{\theta}}$, we find that $p_\theta = ml^2\dot{\theta}$, so that $\dot{\theta} = \frac{p_\theta}{ml^2}$, which we can use to eliminate $\dot{\theta}$ from the expression for H. Thus, $H = \frac{1}{2}m\left(l\frac{p_\theta}{ml^2}\right)^2 - mgl \cos \theta = \frac{p_\theta^2}{2ml^2} - mgl \cos \theta$. Now we can find the equation of motion for the pendulum using the Hamiltonian equation $\dot{p} = -\frac{\partial H}{\partial q}$; that is, $\dot{p}_\theta = -\frac{\partial H}{\partial \theta} = -mgl\sin\theta$. Differentiate $p_\theta = ml^2\dot{\theta}$ to obtain $\dot{p}_\theta = ml^2\ddot{\theta}$, and plug into the equation for $\dot{p}_\theta$ to obtain $ml^2\ddot{\theta} = -mgl\sin\theta$; i.e., $l\ddot{\theta} = -g\sin\theta$, which again yields equation 2.4.

2.4   Equivalent Formulations?

We find the same equation of motion for the pendulum regardless of which formulation we use. This turns out to be true in general. It is often simpler to use Lagrangian or Hamiltonian mechanics rather than Newtonian mechanics, since we do not have to calculate the various component forces on each particle. Nonetheless, it is generally agreed that each formulation suffices for describing the motion of any classical mechanical system.[19] The difference seems to be merely a matter of calculational convenience.

Indeed, physics books typically state, and go on to prove, an equivalence among the three formulations, by showing that their dynamical equations are all inter-derivable.[20] A typical route is to begin with Newton's laws, derive the Lagrangian and Hamiltonian equations from them, and then show that the derivation can go the other way. Thus, José and Saletan, at the beginning of their chapter on Lagrangian mechanics, following the one on Newtonian mechanics, write, "In this chapter we show how the equations of motion can be rewritten....We should emphasize that the physical content of Lagrange's equations is the same as that of Newton's" (1998, p. 48). They then show that Hamilton's equations, in turn, can be derived from Lagrange's, and vice versa, concluding that these all "contain the same information" (1998, p. 207). Another book concludes, "From the point of view of the physicist this division [into the three formulations] is rather artificial.... The segregation is based entirely on the mathematical methods used" (Talman, 2000, p. 163). It certainly seems like these are simply "alternative statements of the laws" (Marion and Thornton, 1995, p. 213), with "nothing new...added to the physics involved" (Goldstein et al., 2004, p. 334) as we pass from one formulation to another. That is the standard view: the three formulations are completely equivalent, mere notational variants; they say all the same things, just in different ways.

I want to urge caution in adopting the standard view. The alleged equivalence is not as straightforward as the above statements would have us believe. The reason is that there *are some* differences among the formulations, and it is not obvious that they are as superficial as usually thought. Draw a rough distinction between two kinds of differences: mathematical and metaphysical. I won't go into these in detail, but will point to places where there is a case to be made that the differences go deeper than ordinarily claimed.

2.4.1   *Mathematical Differences*

It is important to keep in mind that two things can be similar or equivalent in some ways while differing in other ways. Two objects can share a shape yet have different colors or patterns. Two spaces can share a distance structure yet differ in whether they have a privileged location. In mathematics more generally, two mathematical objects are considered equivalent when there is the relevant structure-preserving mapping between them, in which case they are said to be equivalent with respect to that structure. Two such objects can still differ with respect to other kinds of structure.

All of which is to say that, even if the three formulations of classical mechanics are equivalent in all the ways that physics books suggest, the formulations could still be inequivalent in other ways. The question is whether they are equivalent, full stop. The answer depends on whether what differences there are *matter* in any way.

There is one patent mathematical difference among them: the formulations use different symbols, in equations that do not "look" the same. The standard view is that this difference does not matter. Consider the change from Cartesian to polar coordinates to describe a Euclidean plane, or from one set of Cartesian coordinates to another that is rotated or translated with respect to the first. Some things will be different when we switch to the other coordinate system – the points will get different numerical labels, for example – but we know that nothing has *really* changed. The plane remains the same; we have simply used a different, equally legitimate way of describing it. The standard view is that the differences among the three formulations of classical mechanics are just like the differences among the coordinate-based descriptions of the plane: just a change in the coordinates or variables being used to describe the very same physics.

However, there are some reasons to question this idea. Take Newtonian mechanics, on the one hand, and Lagrangian and Hamiltonian mechanics, on the other. The latter are comparatively coordinate-independent formulations of classical mechanics. This suggests that they more directly get at the nature of classical mechanical reality, apart from our descriptions of it – just as the metric tensor on the Euclidean plane, rather than any coordinate-dependent distance formula, more directly captures the intrinsic structure of the plane. (The familiar form of the distance formula stemming from the Pythagorean theorem, $d = \sqrt{\Delta x^2 + \Delta y^2}$, for instance, assumes Cartesian coordinates and won't work in other types of coordinates, even though the distance between any two points is the same regardless of the coordinate system.) This, in turn, suggests that we have reason to prefer these formulations. Physics prizes coordinate-independence, and with good reason.[21] Since there is freedom in which coordinate system to use, any choice we do make will be arbitrary – a conventional choice made from among equally good descriptions. (Recall the different coordinate systems for the plane.) We can be misled into thinking that coordinate-dependent features, which rest on an arbitrary choice in description, reflect genuine features of reality.[22] A formulation that is independent of coordinates is then preferable, other things being equal, when it comes to figuring out what physics says about the world. So even if the equations of the three formulations are inter-derivable in some sense, there is also a sense in which the formulations are not mathematically on a par, a sense in which they are not *completely* equivalent. Some of them may more directly represent physical reality than others.[23]

We can go further. For the way in which the formulations differ in their reliance on coordinates suggests particular physical differences among them. Newtonian mechanics contains an implicit preference for Cartesian coordinates, the kind of coordinates in which its core equation takes the standard form. A preference for Cartesian coordinates, in turn, is indicative of a Euclidean metric structure. This suggests that the spatial structure of a Newtonian world is Euclidean. (Newton himself, of course, assumed such a structure.) Lagrangian mechanics, which allows for a wider range of coordinates in describing classical systems, does not constrain the spatial structure in the same way. This suggests that the physical space of a Lagrangian world has a "looser" metric structure. (I explore this difference, which will be reflected in the theories' statespace structures, in North (2021, ch.4).) Hamiltonian mechanics allows for even greater freedom of coordinates than that. (In particular, it allows for coordinate changes that mix up the ps and qs, whereas in Lagrangian mechanics, since $\dot{q}$ is defined as the time derivative of q, there is no allowable transformation in which these coordinates "get intermingled" (Taylor, 2005, p. 538n10).[24]) As a result, the Hamiltonian formulation does not require a metric structure, but only a lesser type of structure akin to a volume measure. (I explore this difference in North (2009).)

I'd go so far as to suggest that there is a hierarchy, in order of increasing mathematical structure, from Hamiltonian to Lagrangian to Newtonian mechanics – a mathematical inequivalence among the three. (In the above-mentioned writings, I argue that less such structure is in general a reason to prefer a theory.) If we take a theory's mathematical structure seriously in telling us about the nature of the physical world, then this mathematical difference should reflect a similar hierarchy in the physical structure of the world(s) each theory describes – a physical inequivalence among them. In other words, these may not be wholly equivalent formulations, neither mathematically nor physically, contrary to the standard view.[25]

2.4.2 *Metaphysical Differences*

Since the dynamical equations and basic quantities of the three formulations are inter-derivable in ways that physics books claim, you might want to conclude that the different formulations are simply "mutually supporting, compatible perspectives on the phenomena of mechanical motions" (Wilson, 2007, p. 179). That, once again, is the standard view.[26] But there are other differences among the formulations, what I call here "metaphysical" ones, that could lead to a different conclusion. (Don't let the term mislead you: these differences arguably matter to *physics*.) Although no theory wears its metaphysics on its sleeves, on a natural way of interpreting the formulations, they differ from one another in potentially significant ways. All assume a fundamental ontology of point-mass particles with relative positions. Beyond that, each one offers a fairly different picture of the world, given the different quantities that appear in their respective dynamical equations. (What follows are some initial suggestions; the metaphysics of the three formulations has not been much explored in the literature.)

First compare Newtonian mechanics, on the one hand, with Lagrangian and Hamiltonian mechanics, on the other. Newtonian mechanics "describes the world in terms of forces and accelerations (as related by the second law)" (Taylor, 2005, p. 521), where "force is something primitive and irreducible" (Lanczos, 1970, p. 27). Lagrangian and Hamiltonian mechanics describe systems in terms of energy, with force being "a secondary quantity" derivable from the energy (Lanczos, 1970, p. 27). According to Newtonian mechanics, the world is fundamentally made up of particles that move around in response to the various forces between them. According to Lagrangian and Hamiltonian mechanics, particles move around and interact as a result of their energies. Although energy and force functions are inter-derivable in ways that physics books will show (albeit under certain contestable assumptions: note 10), these are nonetheless prima facie different pictures of the world, built up out of different fundamental quantities, with correspondingly different explanations of the phenomena. the Schrödinger and Heisenberg formulations of nonrelativistic quantum mechanics are generally considered inter-derivable, yet you might not want to regard them as wholly metaphysically equivalent even so; many philosophers take only the former to directly or perspicuously represent what is going on physically, for instance. (You might think that Lagrangian and Hamiltonian mechanics can be seen as fundamentally forced-based, given in terms of "generalized forces." However, generalized forces are so-called by analogy to ordinary forces. It isn't clear that they count as regular forces of the Newtonian kind.)

There are potential metaphysical differences between the two energy-based approaches as well. In Lagrangian mechanics, generalized velocities are defined as the first time derivatives of the generalized positions. This suggests that positions are the only truly fundamental dynamical features of the particles, the velocities being defined in terms of them. In Hamiltonian mechanics, however, the canonical positions and momenta are both independent variables, neither being defined in terms of the other: both seem to be fundamental. (This, in turn, may amount to an "impetus" view in the medieval tradition, with further metaphysical repercussions: Arntzenius (2000, sec. 4). This assumes that the second equation of Hamiltonian mechanics is not a definition of the generalized momentum, as often claimed, but a further fundamental dynamical law.) Another difference is that the Hamiltonian is typically equal to the total energy of a system, whereas the Lagrangian is the difference between the kinetic and potential energy. Perhaps this, too, amounts to a genuine difference.[27]

In fact, there is a range of potential views on what's fundamental to each of the formulations, and it is not clear which is correct. It is an open question whether, on any of them, ordinary three-dimensional space is fundamental, or whether what we usually think of as the merely abstract, high-dimensional statespace (or the configuration space) is. Relatedly, it is open whether particle features like positions and momenta are fundamentally defined on the low- or high-dimensional space. (Compare the debate in quantum mechanics over the fundamentality of the high-dimensional space of the wavefunction versus ordinary three-dimensional space.) Within energy-based approaches, it is open whether the energy function, L or H, is fundamental, or whether instead the potential and kinetic energies are; or indeed whether any energy quantity is fundamental, rather than the particle positions and velocities in terms of which the energy is standardly defined; or whether all of these might be fundamental. Analogous questions arise for Newtonian mechanics: are total forces or component forces fundamental?[28] For that matter, can Newtonian mechanics be seen as a fundamentally energy-based theory, given the inter-derivability of the different quantities?[29] Finally, are any of these genuinely distinct possibilities, or are they all equivalent – just different, equally legitimate ways of describing the same physical reality, analogous to the different coordinate-based descriptions of the plane? Although physics books generally assume the latter, certain metaphysical views will say that only one description gets at the real or fundamental properties (Lewis, 1983; Sider, 2011).

In all, it seems very much an open question whether the three formulations of classical mechanics are genuinely equivalent, mere notational variants of a single theory, as usually thought. There is a case to be made that the differences are significant enough to render them more like distinct theories, with different accounts of what the physical world is like. All of this warrants further investigation.[30]

Notes

1 Also charge, although there is a question whether electromagnetic features ought to be considered part of the domain of classical mechanics; see for instance note 10.
2 Of course, because of the previously mentioned troubles, it is not clear that classical mechanics can be a true fundamental theory of a world, but set that aside here.
3 A recent book-length exception: Sklar (2013).
4 There are other varieties I don't discuss, such as formulations in terms of Poisson brackets, Hamilton-Jacobi theory, or four-dimensional spacetime geometry.
5 This is the fundamental ontology assumed here. Wilson (2013) discusses the classical mechanics of rigid bodies and continua and complications involved in trying to encompass all of these within a single theory. See Hall (2007, sec. 5.2), Esfeld et al. (2018), Allori (forthcoming) on the non-standard idea that particles don't have fundamental intrinsic properties.
6 Alternatively, time can be included as an additional dimension of the statespace.
7 Another familiar version of the law, ordinarily seen as equivalent to the above, is given in terms of momentum: $\Sigma \mathbf{F} = \dot{\mathbf{p}}$. See Hicks and Schaffer (2017) on whether these are equivalent.
8 Whether the theory really is deterministic is an interesting question. Apparent counterexamples are in Earman (1986) and Norton (2008); further discussion is in Malament (2008) and Wilson (2009).
9 Standard statespace constructions effectively assume the existence of physical space. See Belot (1999, 2000) on reconstructions that aim to do away with this assumption.
10 There are questions surrounding the further restrictions that forces be central and conservative. It is usually thought that nonconservative forces, like frictional ones depending on velocity, arise from fundamental conservative ones. As Feynman notably put it, "there are no nonconservative forces!" (2010, sec. 14.4). Newton himself did not restrict forces in this way; Feynman suggests it is an additional empirical posit. [The updated online version of this book no longer contains this sentence.] The restrictions are assumed in standard proofs of energy conservation and other theorems. (This is one place the question of electromagnetic features (note 1) comes into play. Consider the magnetic force on a moving charge, which does not satisfy these restrictions.) Concerns over the above have led the odd physicist to doubt the equivalence of the different formulations of classical mechanics: Lanczos (1970, 77 n1); Gallavotti (1983, ch. 3). See also Hertz (1899) and Wilson (2009, 2013, forthcoming) on these and other reasons to doubt their equivalence.
11 There are some mild constraints on generalized coordinates (José and Saletan, 1998, sec. 2.1.2). Wilson (2009) points out that the idea of generalized coordinates, as well as the requirements on them, is not as straightforward as usually assumed.
12 The number of degrees of freedom is the number of independent parameters "necessary and sufficient for a unique characterization" of the system (Lanczos, 1970, p. 10).
13 Standard examples in which it does not have this form come from outside the point-particle mechanics assumed here. See José and Saletan, (1998, sec. 2.2.4) and Goldstein et al. (2004, sec. 7.9) for examples from electromagnetism and special relativity.

14 See Lanczos (1970) for discussion of these and other differences. See Butterfield (2004) for an extended discussion of Lagrangian mechanics in particular.

15 See Brading and Castellani (2007, p. 1343).

16 A Hamiltonian statespace can in fact have a more general structure than this: North (2009).

17 See Goldstein et al. (2004), Taylor (2005, sec. 7.8) for conditions under which this holds.

18 The above is an instance of a Legendre transformation, which can be used to change back and forth between Hamiltonian and Lagrangian coordinates, energy functions, and statespaces: see Lanczos (1970, ch. 6), Arnold (1989, sec. 3.14), José and Saletan (1998, ch. 5).

19 Or so I assume here, setting aside reasons for hesitation on this point (note 10).

20 Examples: Arnold (1989); Marion and Thornton (1995); Hand and Finch (1998); José and Saletan (1998); Talman (2000); Goldstein et al. (2004); Taylor (2019); Baez and Wise (2019); see also Feynman (1965, ch. 2).

21 Lanczos notes of the Lagrangian equations that they "stand out as the first example of that 'principle of invariance' [a kind of coordinate-independence] which was one of the leading ideas of 19th century mathematics, and which has become of dominant importance in contemporary physics" (1970, p. 117).

22 Einstein once said that the main reason it took him so long to develop general relativity is that "it is not so easy to free oneself from the idea that co-ordinates must have an immediate metrical meaning" (Schilpp, 1970, p. 67).

23 All that said, the role of coordinates in physics is more subtle and complicated than the above discussion might suggest: see North 2021.

24 There is a mathematical transformation between them (note 18), but even it "leads one to suspect that there actually is a nontrivial difference between L and $\dot{q}$ on the one hand and H and p on the other" (José and Saletan, 1998, p. 217).

25 Opposition to this conclusion, for different reasons, can be found in Swanson and Halvorson (2012); Curiel (2014); Barrett (2015). Barrett (2019) points out how our judgments about the relationship between the theories will depend on what we take to be their core structures, with different views on their structures leading to different such judgments.

26 Following Coffey (2014), the standard view may more accurately be put as that Newtonian mechanics accurately represents classical mechanical reality, with Lagrangian and Hamiltonian mechanics being mere reformulations of it.

27 Baez and Wise (2019, ch. 1) tries to distinguish these physically.

28 Cartwright (1983, ch. 3) argues against the reality of component forces.

29 Wilson (2007) defends the existence of Newtonian forces against various objections.

30 Some further investigation is in North 2021, especially Chapters 4 and 7.

References

Allori, V. (forthcoming). Fundamental objects without fundamental properties: A thin-object-oriented metaphysics grounded on structure. In D. Aerts, J. Arenhart, C. De Ronde, and G. Sergioli (eds.), *Probing the Meaning and Structure of Quantum Mechanics*. World Scientific.

Arnold, V.I. (1989). *Mathematical Methods of Classical Mechanics* (2nd edition). New York: Springer. Translated by K. Vogtmann and A. Weinstein.

Arntzenius, F. (2000). Are there really instantaneous velocities? *The Monist*, 83: 187–208.

Baez, J.C. and Derek K. Wise (2019). *Lectures on classical mechanics*. Available at: http://math.ucr.edu/home/baez/classical/texfiles/2005/book/classical.pdf.

Barrett, T.W. (2015). On the structure of classical mechanics. *British Journal for the Philosophy of Science*, 66: 801–828.

Barrett, T.W. (2019). Equivalent and inequivalent formulations of classical mechanics. *British Journal for the Philosophy of Science*, 70: 1167–1199.

Belot, G. (1999). Rehabilitating relationalism. *International Studies in the Philosophy of Science*, 13: 35–52.

Belot, G. (2000). Geometry and motion. *British Journal for the Philosophy of Science*, 51: 561–595.

Brading, K. and Castellani, E. (2007). Symmetries and invariances in classical physics. In J. Butterfield and J. Earman (eds.), *Handbook of the Philosophy of Science: Philosophy of Physics, Part B*. Amsterdam: Elsevier, pp. 1331–1367.

Butterfield, J. (2004). Between laws and models: Some philosophical morals of Lagrangian mechanics. Unpublished manuscript. Available at: http://philsci-archive.pitt.edu/1937/.

Cartwright, N. (1983). *How the Laws of Physics Lie*. Oxford: Oxford University Press.

Coffey, K. (2014). Theoretical equivalence as interpretative equivalence. *British Journal for the Philosophy of Science*, 65: 821–844.

Curiel, E. (2014). Classical mechanics is Lagrangian; it is not Hamiltonian. *British Journal for the Philosophy of Science*, 65: 269–321.

Earman, J. (1986). *A Primer on Determinism*. Dordrecht: D. Reidel.

Esfeld, M. and Deckert, D.-A. with Lazarovici, D., Oldofredi, A. and Vassallo, A. (2018). *A Minimalist Ontology of the Natural World*. New York and London: Routledge.

Feynman, R. (1965). *The Character of Physical Law*. Cambridge, MA: MIT Press.

Feynman, R.P., Leighton, R.B. and Sands, M. (2010). *The Feynman Lectures on Physics: New Millennium Edition*, volume 1. New York: Basic Books.

Gallavotti, G. (1983). *The Elements of Mechanics*. New York: Springer-Verlag.

Goldstein, H., Poole, C. and Safko, J. (2004). *Classical Mechanics* (3rd edition). Reading, MA: Pearson Education.

Hall, N. (2007). Humean reductionism about laws of nature. Unpublished manuscript. Available at: http://philpapers.org/archive/HALHRA.

Hand, L.N. and Finch, J.D. (1998). *Analytical Mechanics*. Cambridge: Cambridge University Press.

Hertz, H. (1899). *The Principles of Mechanics Presented in a New Form*. London: Macmillan and Co.

Hicks, M.T. and Schaffer, J. (2017). Derivative properties in fundamental laws. *British Journal for the Philosophy of Science*, 68: 411–450.

José, J.V. and Saletan, E.J. (1998). *Classical Dynamics: A Contemporary Approach*. Cambridge: Cambridge University Press.

Lanczos, C. (1970). *The Variational Principles of Mechanics* (4th edition). New York: Dover.

Lewis, D. (1983). New work for a theory of universals. *Australasian Journal of Philosophy*, 61: 343–377.

Malament, D.B. (2008). Norton's slippery slope. *Philosophy of Science (Proceedings)*, 75: 799–816.

Marion, J.B. and Thornton, S.T. (1995). *Classical Dynamics of Particles and Systems* (4th edition). Florida: Harcourt, Brace and Company.

North, J. (2009). The 'structure' of physics: A case study. *Journal of Philosophy*, 106: 57–88.

North, J. (2021). *Physics, Structure, and Reality*. Oxford: Oxford University Press.

Norton, J.D. (2008). The dome: An unexpectedly simple failure of determinism. *Philosophy of Science (Proceedings)*, 75: 786–798.

Schilpp, P.A., editor. (1970). *Albert Einstein: Philosopher-Scientist* (3rd edition). La Salle, IL: Open Court.

Sider, T. (2011). *Writing the Book of the World*. Oxford: Oxford University Press.

Sklar, L. (2013). *Philosophy and the Foundations of Dynamics*. Cambridge: Cambridge University Press.

Spivak, M. (2010). *Physics for Mathematicians: Mechanics I*. Publish or Perish, United States.

Sussman, G.J. and Wisdom, J. (2014). *Structure and Interpretation of Classical Mechanics* (2nd edition). Cambridge, MA: MIT Press.

Swanson, N and Halvorson, H. (2012). *On North's 'the structure of physics'*. Unpublished manuscript. Available at: http://philsci-archive.pitt.edu/9314/.

Talman, R. (2000). *Geometric Mechanics*. New York: John Wiley & Sons.

Taylor, J.R. (2005). *Classical Mechanics*. Sausalito, CA: University Science Books.

Wilson, J. (2007). Newtonian forces. *British Journal for the Philosophy of Science*, 58: 173–205.

Wilson, M. (2009). Determinism and the mystery of the missing physics. *British Journal for the Philosophy of Science*, 60: 173–193.

Wilson, M. (2013). What is 'classical mechanics' anyway? In R. Batterman (ed.), *The Oxford Handbook of Philosophy of Physics*. Oxford: Oxford University Press, pp. 43–106.

Wilson, M. (forthcoming). Newton in the pool hall: Subtleties of the third law. In C. Smeenk and E. Schliesser (eds.), *The Oxford Handbook of Newton*. Oxford: Oxford University Press.

Further Reading from the Editors

Lawrence Sklar's *Philosophy and the Foundations of Dynamics* (Cambridge: Cambridge University Press 2013) is an excellent book-length treatment of classical mechanics which covers some of the issues discussed here. For a subtle approach to classical physics, rooted in a deep understanding of particular examples, look at Mark Wilson's work, for example "What is 'Classical Mechanics' Anyway?" in Robert Batterman, ed., *The Oxford Handbook of Philosophy of Physics* (Oxford: Oxford University 2013). For two sides of the debate over whether Lagrangian or Hamiltonian mechanics is more fundamental, see Jill North's "The 'Structure' of Physics. A Case Study." (*Journal of Philosophy* 106, 57–88, 2009) and Erik Curiel's "Classical Mechanics Is Lagrangian; It Is Not Hamiltonian." (*British Journal for the Philosophy of Science* 65, 269–321, 2014).

3

CLASSICAL SPACETIME STRUCTURE

James Owen Weatherall

3.1 Introduction

One often associates *spacetime* – a four-dimensional geometrical structure representing both space and time – with relativity theory, developed by Einstein and others in the early part of the 20th century.[1] But soon after relativity theory appeared, several authors, such as Hermann Weyl (1952 [1918]), Élie Cartan (1923, 1924), and Kurt Friedrichs (1927), began to study how the spatiotemporal structure presupposed by classical, i.e., Newtonian, physics could be re-cast using the methods of four-dimensional geometry developed in the context of relativity.

These reformulations of classical physics were initially of interest for the insight they offered into the structure of relativity theory.[2] But this changed in 1967, when Howard Stein proposed that the notion of a "classical" spacetime could provide important insight into Newtonian physics itself – including on issues that engaged Newton and his contemporaries regarding the status of "absolute" space and motion. On Stein's reconstruction, Newton's oft-decried "absolutism" about space amounted to the claim that the laws governing the motion of bodies presupposed that there are (unobservable) facts about which bodies are at rest. This idea can be reconstructed as a claim that space and time together have a certain geometrical structure, now known as *Newtonian spacetime*. Leibniz's response, meanwhile, that there was no discernible difference between whether all of the bodies in the world were at rest or in uniform rectilinear motion can then be taken as an argument that Newton's laws assume more structure than can be supported on metaphysical grounds; Leibniz, it seems, believed space and time had the structure of *Leibnizian spacetime*, which, in a certain precise sense, is *less structure* than Newtonian spacetime.[3]

Perhaps the most striking and influential aspect of Stein's paper was his argument that in fact the spatiotemporal structure presupposed by Newton's laws of motion was somewhat less than Newton imagined – though somewhat more than Leibniz would have accepted. This intermediate structure has come to be known as *Galilean spacetime*.[4] Remarkably, Galilean spacetime provides the resources needed both to avoid Leibniz's famous shift argument,[5] which exposes the unobservability of absolute uniform rectilinear motion, and to accept Newton's famous bucket argument, which Newton and many others have taken to show that at least *some* absolute motions are empirically detectable.

The remainder of this chapter will proceed as follows.[6] I will begin by making some remarks about "space" in classical physics. I will then introduce the notion of Galilean spacetime and discuss several senses in which it is the spacetime structure presupposed by Newtonian physics. Next I will discuss Newtonian spacetime and Leibnizian spacetime. I will conclude by briefly discussing some other ideas concerning classical spacetime structure, including the recent proposal by Saunders (2013) that one should take a structure strictly intermediate between Galilean and Leibnizian spacetimes to be what is really presupposed by Newtonian physics.

3.2 Space

Before discussing classical space*time*, we first consider the geometry of space – something that all of the structures I will discuss below agree on. *Space*, in what follows, will be understood as a collection of (infinitesimally) small *places*, i.e., locations where small bodies may be situated. We will call these locations the *points* of space.[7] (Since most physical objects are not vanishingly small, their locations cannot be represented by single points of space; in general, bodies occupy regions of space, i.e., collections of contiguous points.) This collection of points is understood to be structured, in the sense that there are various further relations defined on them. It is these relations that one aims to characterize when one speaks of "spatial structure" – or, *mutatis mutandis*, "spatiotemporal structure."

For example, space is three-dimensional. We make this idea precise by introducing the notion of an *arrow* between pairs of points of space.[8] (See Figure 3.1a.) First, pick any two points of space, p and q. We suppose one can always draw a (unique) arrow whose tail begins at p and whose head lands at q, and likewise, one can always draw an arrow whose tail begins at q and whose head ends at p. Conversely, we suppose that any arrow whose tail begins at p ends somewhere in space, i.e., at some point. This means that we can think of any point and an arrow originating at that point as uniquely picking out another point of space; and we can think of any ordered pair of points as uniquely determining an arrow.

We will call arrows that originate at a point p *arrows at p*. We suppose that given an arrow at p, one can identify the "same" arrow – i.e., an arrow with the same length and orientation – at any other point. Thus, it makes sense to speak of an arrow without mentioning the point at which it is based.

Now pick any point of space, p. We define two basic ways of manipulating the arrows at p. For one, we assume we can always "scale" any arrow. (See Figure 3.1b.) That is, given any arrow u (say), one can uniquely identify an arrow that points in the same direction as u, but which is twice as long (say). We will call this arrow $2u$. Likewise, one can identify an arrow that points in precisely the opposite direction as u; we will call this $-u$, short for $(-1)u$. Given any arrow u at p, $0u$ is the (unique) arrow from p to itself.

The second operation one can perform on arrows at a point is to add them to one another. Consider two arrows u and v at p. We define the arrow $u + v$ to be the unique arrow from p to the

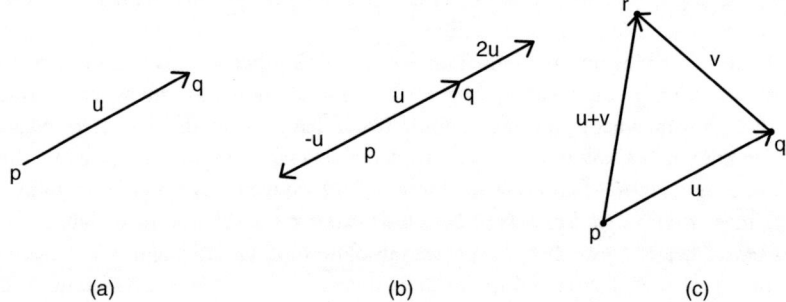

(a) (b) (c)

Figure 3.1 (a) An arrow u relating two points p and q of space. (b) "Scaling" an arrow u at a point p by various amounts, including flipping its direction. (c) Adding arrows u and v: here $u + v$ is the arrow relating p and r, where r is the point determined by q and v, and q is the point determined by p and u.

point r, where r is defined as follows: first, consider the point q determined by p and u; then take v, and let r be the point determined by q and v. (See Figure 3.1c.) In other words, we understand $u + v$ to be the arrow taking us from p to the point we would reach by first following u and then following v. We assume that for all u and v, $u + v = v + u$ – i.e., following v and then u leads to the same place as following u and then v.[9]

I can now say what it means for space to be three-dimensional. Space is three-dimensional just in case at any point p, there exist three arrows, x, y, and z, which are such that (1) none of these three can be constructed by any process of scaling or adding the other two; and (2) one can construct any arrow v at p just by scaling and/or adding together x, y, and z. The important point, here, is that for ordinary space, any collection of arrows with these two properties always has exactly the same number of elements: namely, three.

There is a bit more structure that we will take space to have. First, given any two points p and q, we assume we can say how far apart they are: that is, we have a notion of *spatial distance*.[10] In other words, we assign a *length* – a non-negative number, which is zero only for the arrow taking a point to itself – to the arrow v between p and q, which we will write $||v||$. Likewise, given two arrows u and v, we can assign an angle – a number between 0 and 2π – to them, where an angle of 0 (or 2π) means the arrows point in the same direction; an angle of π means they point in opposite directions; and an angle of $\pi/2$ or $3\pi/2$ means they are *orthogonal*, i.e., the angle between u and v is precisely the same as the angle between u and $-v$ (or, equivalently, between $-u$ and v). Finally, we assume that this notion of length satisfies the following two conditions: (a) for any real number a and any arrow u, $||au|| = |a| \cdot ||u||$, where $|a|$ is the absolute value of a; and (b) if u and v are orthogonal, then $||u + v||^2 = ||u||^2 + ||v||^2$.

3.3   Galilean Spacetime

We now turn to our first spacetime structure: *Galilean spacetime*. To characterize Galilean spacetime, we begin much as in our discussion of space: Galilean spacetime consists in a collection of points, now understood as locations not only in space, but also in time. Rather than "places," these locations are *events*, in the sense that they represent occurrences in a small region of space for an instant of time. They are not necessarily *significant* events: an event might be a speck of dust existing at a moment, or even the occurrence of nothing at all.

Just as ordinary extended objects are not located at single points of space, neither are objects that are extended in space – say, a rope – nor objects that persist through time – say, a particle or a person – represented by single points in spacetime. Instead, these are represented by various sorts of curves and surfaces, as described below.

This collection of events is once again structured. In particular, Galilean spacetime is a four-dimensional "space" of events, in much the same way that space is three-dimensional. That is, we suppose that any pair of spacetime points p and q are uniquely related by an arrow; and that given any point p, and any arrow v, there is a unique point q such that v is the arrow from p to q. Now, though, we suppose that at any point p of Galilean spacetime, one can find four arrows – t, x, y, and z – with the properties that (1) none of these four can be constructed by any process of scaling or adding the other three; and (2) one can construct any other arrow by scaling and/or adding together t, x, y, and z.[11]

Now consider any two points p and q in Galilean spacetime. There are several additional relations that hold between them. One such relation is *temporal distance*, t, which assigns a real number to any ordered pair of events – or, equivalently, a *temporal length* to any arrow relating events. This number represents the duration between those two events. Temporal distance has the following properties. First, if u and v are arrows, then $t(u + v) = t(u) + t(v)$, i.e., if p and q are related by u, and q and r are related by v, then the temporal distance between p and r is the sum of the temporal distance between

p and q and the temporal distance between q and r. Likewise, given any real number a and any vector u, the temporal distance satisfies $t(au) = at(u)$, where on the left-hand side of this equation we are applying the scaling operation, and on the right-hand side we are just multiplying real numbers.[12] Finally, we can always find a (non-unique) collection of three arrows, x, y, and z, with the properties that (1) none of these three can be constructed by any process of scaling or adding the other two; and (2) the temporal distance assigned to each of these arrows, and thus all other arrows that can be constructed by scaling and adding them, is zero.

If the temporal distance from p and q is positive, we say that q is in the *future* of p; if it is negative, we say it is in the *past* of p; if it is zero, then p and q are *simultaneous*. Now let p be a point, and let x, y, and z be three arrows satisfying (1) and (2) in the previous paragraph. Then any point q related to p by an arrow that can be constructed by scaling or adding x, y, and z will be simultaneous with p, and all events simultaneous with p can be found in this way. Thus, the events simultaneous with p form a three-dimensional "space." We take such collections to represent space at a time.

Finally, given space at any time, Galilean spacetime includes a notion of spatial distance between those events, and a notion of angle between arrows relating those events, that satisfy all of the conditions we placed on spatial distance above. Note, however, that we do not have any notion of spatial distance between points that are not simultaneous – a caveat that will turn out to be important in what follows.

Summing up, we can think of Galilean spacetime as an infinite collection of copies of three-dimensional space, stacked on top of one another to form a four-dimensional structure.[13] (See Figure 3.2a.) Each copy of space has all of the structure we described in the previous section, including the relations of spatial distance and angle. Moreover between any two slices, we have a notion of duration. It is in this sense that Galilean spacetime may be taken to represent space and time – i.e., to deserve the name "spacetime."

Having described this structure, we can now say the sense in which it is the "right" structure for Newtonian physics. The key is to understand what Galilean spacetime allows one to say about motion. Consider a particle – i.e., some vanishingly small body that we understand to "persist" in the sense of existing over time. We represent this particle by a collection of events – specifically, by a *curve* through spacetime with the property that it intersects each spatial slice no more than once.[14] Such

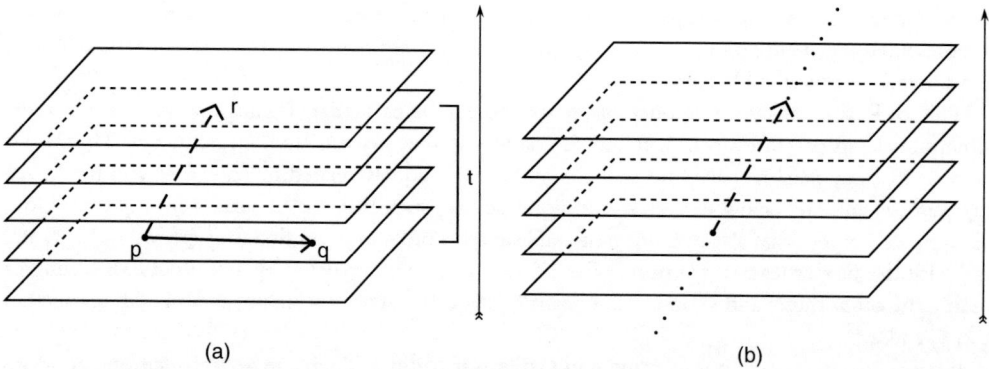

(a) (b)

Figure 3.2 Galilean spacetime may be thought of as consisting of copies of three-dimensional space stacked on top of one another to form a four-dimensional structure. (a) Here points p and q are simultaneous, and so the arrow between them has temporal length zero, but non-zero spatial length; the point r is not simultaneous with p or q and so it has a non-zero temporal distance from both, and its spatial distance to p and q is not defined. (b) Here we depict the straight line through p determined by a given arrow of unit temporal length.

a curve is called a "trajectory" or a "world-line." The idea is that the (single) point of intersection between the particle's world-line and any spatial slice represents the location of the particle at that time. The whole world-line, then, represents the history of the particle over time: it consists in the collection of places, at successive times, that are occupied by the particle. In other words, it represents the *motion* of the particle through space.

Extended bodies are represented, similarly, by "world-tubes" that are bounded in each slice of space; the intersection of the world-tube with each slice represents the configuration of the body in space at that instant. In what follows, I will focus on the particle case for simplicity, though much of what is said carries over because one can generally associate a "center of mass" curve with extended bodies, which characterizes their mass-averaged motion.

We now turn to Newton's laws of motion. We first remark that, in Galilean spacetime, we have the resources to characterize a special kind of trajectory: namely, a "straight line through spacetime." (See Figure 3.2b.) Given any point p and any arrow v with temporal length one, we define the *straight-line trajectory* through p with *4-velocity v* to consist of all the points one can reach from p by scaling v. (We say 4-velocity because the arrow v relates points in four-dimensional spacetime.) These trajectories describe motions through spacetime wherein a body moves in some fixed direction at a constant velocity. Such trajectories play an important role in Newton's theory, encapsulated in his first law of motion.

> **Newton's First Law**: In the absence of any external force, massive bodies will follow straight-line trajectories.

This law establishes a tight connection between two classes of curves: the *mathematically* privileged curves picked out by the structure of Galilean spacetime, and the *physically* privileged curves picked out by the "default," force-free motions of bodies. These default trajectories are known as *inertial* trajectories.[15]

Before proceeding, let us briefly comment on the role of 4-velocity here. We usually understand velocity as a quantity with some direction (and magnitude) in *space*; 4-velocity, meanwhile, has direction in *spacetime*. These are different. To recover velocity as we usually conceive of it — i.e., *3-velocity* — we need to introduce an *observer*, O, which is an idealized measuring apparatus, situated somewhere in space and with its own inertial state of motion (represented by some trajectory with 4-velocity, which we also denote O). 3-velocity at a time as determined by this observer is given by $u - O$,

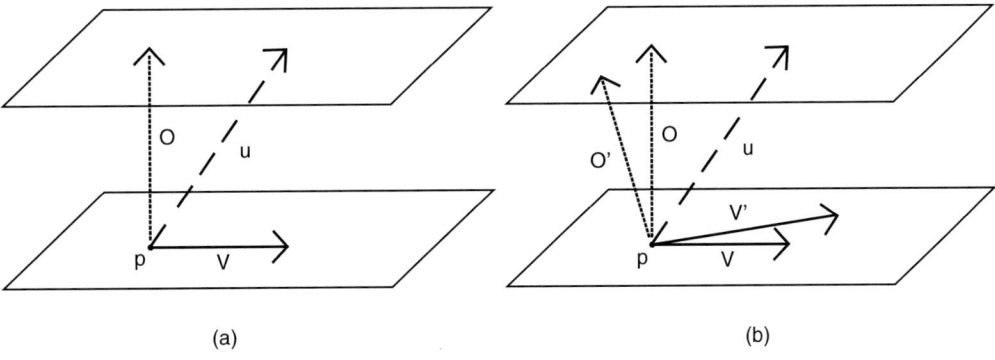

(a)   (b)

Figure 3.3 The motion of a body is described by its 4-velocity, which is an arrow u with temporal length one. (a) To recover ordinary "3-velocity" one introduces an observer O with its own 4-velocity; the (relative) 3-velocity of the body at p is given by $V = u - O$, which is an arrow with temporal length zero. (b) Two different observers, O and O', with different states of motion, will determine different 3-velocities, V and V'. It is in this sense that (3-)velocity is *relative* in Galilean spacetime.

where u is the 4-velocity of the body at that time. (See Figure 3.3a.) This will always be a vector with temporal length zero (since O and u both have temporal length one), and so 3-velocity relative to any observer has spatial length, representing the "speed" of motion.

The important point about 3-velocity is that it is an essentially *relative* notion: that is, its value at a time depends on the state of motion of the observer. Different observers, with different states of motion, would attribute different 3-velocities to a body. (See Figure 3.3b.) Indeed, in Galilean spacetime, the notion of an "absolute," i.e., non-relative, 3-velocity does not make sense. But this does not mean that all measures of motion are relative. In particular, 4-velocity may be characterized independently of any observer.

Thus far, we have focused on straight-line trajectories. But not all trajectories are straight: in general, the motion of a body will change over time. We capture this by associating an *instantaneous 4-velocity* with a body at each instant. Given two (necessarily non-simultaneous) points p and q on a body's trajectory, we can define the change in 4-velocity from p to q to be the difference $v(p) - v(q)$, where $v(p)$ is the 4-velocity at p and $v(q)$ is the 4-velocity at q; by considering the change in 4-velocity between pairs of points that are ever closer together, and scaling appropriately, we can define the instantaneous rate of change of the particle's 4-velocity, or its *acceleration*, at each point of its world-line. The acceleration will itself be an arrow at p with temporal length zero. Acceleration is important in Newton's second law of motion.

> **Newton's Second Law**: If a body of mass m is subject to an external force F at a point, then the acceleration a of the body at that point will satisfy $F = ma$.

Thus, we see that an impressed force, represented by an arrow F, causes a body to deviate from inertial motion, i.e., to accelerate in the direction of the force. The mass of the body is what determines the magnitude of the acceleration. It is important to emphasize that acceleration as just defined in Galilean spacetime is an absolute quantity – it is defined without reference to any observer – which is crucial to Newton's second law because force is meant to be an objective, absolute quantity (i.e., not observer dependent), and it would not make sense to have a law that equates a relative quantity with an absolute quantity.

For completeness, we include Newton's third law, although it is not essential for the discussion (and we suppress commentary for space reasons).

> **Newton's Third Law**: If body 1 exerts a force F on body 2, then body 2 exerts a force $-F$ on body 1.

Finally, although it is not strictly part of Newtonian mechanics, it is worth noting that Newton's laws provide a framework for detailed studies of particular forces. Newton, for instance, focused on gravitation.

> **Newton's Law of Universal Gravitation**: A body of mass m_1 located at point p will exert on a body of mass m_2 located at (simultaneous) point q a force given by

$$F = \frac{Gm_1 m_2}{||r||^3} r$$

> where r is the arrow from q to p, $||r||$ is the spatial length of that arrow, and G is Newton's constant.

It is worth emphasizing that this law invokes temporal length (to define simultaneity) and spatial distance (to define distance between the bodies). To further relate this force to the acceleration of a body, of course, one also requires the full four-dimensional arrow structure of spacetime, as discussed above in connection with Newton's second law. Thus, we see that all of the structure of Galilean spacetime enters into Newton's laws of motion and his law of universal gravitation.

3.4 Newtonian and Leibnizian Spacetimes

Although the language may be anachronistic, Newton, in the *Principia* and elsewhere,[16] took for granted the structure of Galilean spacetime. But as Stein (1967) convincingly argues, Newton also believed that spacetime had further structure: namely, what Newton called *absolute space*, which is the structure needed to say whether two places, at different times, are "the same." In other words, Newton believed there was a basic matter of fact about whether any given object remains in one place over time – i.e., at rest – or whether it moves.

As we have just seen, this is precisely what Galilean spacetime does *not* provide. There, one can say whether a body is *accelerating*, i.e., if its velocity is changing over time. But one cannot say that the body is at rest or moving (at a constant velocity).

To represent this further structure, we need a way of identifying points in space at different times: that is, a way of saying that "here" now is the same as "here" five minutes ago. We do so by choosing some arrow ξ, of unit temporal length, as special. This arrow represents Newton's absolute space as follows: two events p and q are at the *same place* (at different times) if and only if the arrow from p to q is some multiple (possibly zero) of ξ. Likewise, a particle is *at rest* (relative to absolute space) if its world-line is the straight line determined by ξ. More generally, a particle with constant velocity (in the sense of Galilean spacetime) has *absolute velocity* $||\xi - u||$, where u is the arrow of temporal length one determining that particle's world-line. If we take Galilean spacetime, and we add to it this further structure of a privileged arrow ξ, we arrive at *Newtonian spacetime*.[17]

There are several reasons why someone – even someone who accepted all of Newton's physics – might worry that the world does not exhibit the full structure of Newtonian spacetime. One particularly influential reason for skepticism, most famously associated with Leibniz's arguments in his correspondence with Clarke (Leibniz and Clarke, 2000, pp. 18f.), is that the structure of absolute space, as characterized by ξ, has no observable consequences. In particular, one could take any system of bodies and imagine setting the entire system in motion, at some constant speed (relative to ξ) in some fixed direction, and show that, by the lights of Newton's own theory, the relative motions of the bodies would be unchanged.

Newton fully understood this, of course: Corollary 5 to the Laws of Motion (Newton, 1999 [1687 / 1713 / 1726], p. 423) demonstrates precisely this fact. But Leibniz took it to have great significance. He concluded, on its basis, that space and time could not have the structure of absolute space.

But what structure *would* Leibniz attribute to space and time? It is not entirely clear that a satisfactory or complete answer is to be extracted from Leibniz's writings. But as Stein (1977) argues, and Earman (1977, 1979, 1986, 1989a,b) develops, one might begin with Leibniz's claim, in various places, that *all* motion is relative – i.e., there are facts only about how the relative configurations of bodies change over time.[18] In other words, it would not make sense to say anything more about the motion of a given body than to describe the rates at which its distance from other bodies is changing over time.

Both Newtonian spacetime and Galilean spacetime provide the resources to describe these changes. One can consider, for instance, at each time a collection of point-like particles in space, whose relative configuration is represented by the places at which they are located (and the arrows between those places); and one can describe, by word-lines through spacetime, the changes in relative position of these bodies over time. But one can also say more than this: for instance, in both of these structures, there is a fact about each particle's acceleration – not as a relative matter of the rate of change of the relative velocity of one particle with respect to another, but as an absolute matter. Moreover, this is precisely what someone who thinks all motion is relative would deny.

Thus, to characterize Leibniz's views, we need to *excise* from Galilean spacetime the structure that allowed us to distinguish a special class of motions – namely, the arrows relating non-simultaneous points. We are left with *Leibnizian spacetime*, which is a collection of events with a notion of temporal distance between pairs of events, such that the collection of all points simultaneous with any given point has the structure of three-dimensional space as described in Section 3.2, but where we have no arrows between non-simultaneous events.[19]

We saw above that Galilean spacetime – and thus, Newtonian spacetime – provided the resources needed for Newtonian physics. Can we say the same of Leibnizian spacetime? In short, no – for reasons Newton himself pointed out.[20] Indeed, Newton offered at least one case in which absolute motion *would* be empirically testable. Imagine a bucket partially filled with water. Newton argued that if the bucket is rotating (absolutely) then the surface of the water will appear curved; if the bucket is at rest, then the surface of the water will appear flat. Newton argued that this would be the case, according to his theory, even if there were nothing in the universe but the bucket and the water, and so the behavior of the water in the bucket could not have anything to do with merely *relative* motion.

This thought experiment has been very influential, and Newton himself believed it settled the issue in favor of absolute space – i.e., Newtonian spacetime. But it is not quite what it seems. The key to understanding the bucket experiment is to realize that any rotating object is accelerating. In particular, each little bit of the bucket is constantly changing velocity, because it is changing the direction in which it is moving. Thus, Newton's thought experiment is not a way of measuring absolute motion in the sense of measuring absolute velocity. Instead, it is an experiment to determine (one kind of) absolute *acceleration*. In other words, the bucket experiment is an argument that we need at least the structure of Galilean spacetime to make sense of Newtonian physics; meanwhile, Leibniz's shift argument against Newtonian spacetime, that there are no empirical tests of absolute *velocity* in Newtonian physics, still stands – as it must, in light of Corollary 5. Taken together, then, we are pushed to Galilean spacetime as a spacetime structure intermediate between Newtonian and Leibnizian spacetimes.

3.5 Maxwell-Huygens and Newton-Cartan Spacetimes

At the end of the previous section, I observed that in order to accommodate Newton's bucket thought experiment, one needs more structure than Leibnizian spacetime provides. We went from this remark to the conclusion that one needs the structure of Galilean spacetime (at least) to support Newtonian physics. But one might worry that this conclusion is too fast: Galilean spacetime makes all acceleration absolute, whereas the bucket thought experiment concerns only a very special kind of accelerated motion: namely, *rotation*. Perhaps rotation must be absolute to accommodate Newtonian physics, but does it follow that all acceleration is absolute?

This worry is buttressed by the fact that, immediately after Corollary 5 to the Laws of Motion, Newton proves another result: Corollary 6 to the Laws of Motion establishes that if a system of bodies is undergoing uniform linear acceleration – i.e., all of the bodies have, in addition to their other motions, some fixed acceleration in a given spatial direction – then their motions relative to one another would be indistinguishable from the case in which that acceleration was absent (Newton, 1999 [1687 / 1713 / 1726], p. 423ff). Corollary 6 makes precise a certain sense in which absolute linear acceleration has the same status as absolute velocity.

Very recently, a number of authors have taken up the question of whether Corollary 6 motivated adopting some alternative spacetime structure, intermediate between Leibnizian spacetime and Galilean spacetime, as the structure presupposed by Newtonian physics.[21] For instance, Simon Saunders (2013) argues that a spacetime structure that he calls *Newton-Huygens spacetime* is the proper setting for Newtonian physics.[22] Newton-Huygens spacetime is like Leibnizian spacetime – i.e., there are notions of spatial and temporal distance, but no arrows between non-simultaneous events – except one has, in addition, a standard for whether a system of bodies is rotating over time.[23] But one does not, in general, have a notion of absolute acceleration.

There are reasons to be cautious about accepting this move, however. For one, it is more radical than it may at first appear. In Newtonian physics, forces are not relative: either a body experiences a force or it does not.[24] But then, by dint of Newton's second law, acceleration must also not be relative, since it is incoherent to say that an absolute quantity is proportional to a relative quantity.[25] It follows that to accept Newton–Huygens spacetime, one needs to revise both the conceptual and mathematical foundations of Newtonian physics.[26] How significant a revision this amounts to, and whether it can succeed, is a topic of ongoing debate.

A second reason to be cautious is that another response to Corollary 6 is available. In particular, Eleanor Knox (2011) has argued that Corollary 6 supports a move to a different theory of gravitation, sometimes known as *geometrized Newtonian gravitation* or *Newton-Cartan theory*, developed by Cartan (1923, 1924) and Friedrichs (1927).[27] Geometrized Newtonian gravitation is a theory with the same empirical consequences as Newtonian gravitation, but set in a spacetime structure importantly different from any of those discussed thus far: it is a theory in which spacetime is *curved* by the distribution of matter in the universe, and where the motion of bodies in spacetime is influenced by that curvature.[28] The details of geometrized Newtonian gravitation are beyond the scope of this chapter, but one point is worth emphasizing. In geometrized Newtonian gravitation, while there are (in general) no arrows between non-simultaneous events, there is nonetheless a standard of absolute acceleration.

Acknowledgments

This chapter is partially based upon work supported by the National Science Foundation under Grant No. 1331126. I am grateful to David Malament for helpful comments on an earlier draft.

Notes

1 The idea of "spacetime" was actually introduced by Minkowski (2013 [1911]), several years after Einstein first introduced relativity theory.

2 And indeed, they remain of interest for this reason: see, for instance, Friedman (1983), Malament (1986a,b, 2012), Earman (1989b), Fletcher (2014), Barrett (2015), Weatherall (2011a,b, 2014, 2017c,d), and Weatherall (2018) for philosophical discussions of the relationship between relativity theory and Newtonian physics that make heavy use of this formalism.

3 For more on the notion of "structure" being used here, see Barrett (2015) and Weatherall (2017b). Newtonian spacetime was introduced in Stein's original 1967 paper; Leibnizian spacetime was introduced by Stein somewhat later (Stein, 1977), and then developed by Earman (1977, 1979, 1986, 1989a,b). Note, however, that there are subtle differences in Earman and Stein's understanding of what Leibnizian spacetime captures (Weatherall, 2020), and that Earman, in particular, believed that Leibnizian spacetime includes an implicit commitment to some form of "substantivalism" (see endnote 7) that Leibniz would have denied; he suggested that one should move to an algebraic framework to better reflect Leibniz's views. For replies to this proposal, see Rynasiewicz (1992) and Rosenstock et al. (2015).

4 What we now call Galilean spacetime was first discussed by Weyl (1952 [1918]). It is not clear whether any of Newton's contemporaries were in a position to recognize this intermediate structure, though Stein (1967, 1977) and others have suggested that Christian Huygens came closest. See Stan (2016) for a recent discussion of Huygens' views on rotation, which is the context in which he most closely approached the idea of Galilean spacetime.

5 What I call the shift argument concerns setting the whole world in motion at a constant velocity; it is sometimes known as the "kinematic" shift argument, to distinguish it from the "static" shift argument, wherein one considers shifting the entire universe by some fixed amount (Pooley, 2013); see also endnote 7.

6 My presentation will not be particularly historically sensitive; nor will it be technical. For more historical detail, but with the same basic perspective, see Weatherall (2016c) and (especially) references therein; for technical details, see Weatherall (2016b). For a treatment at a level comparable to the one attempted here, which also develops ideas from relativity theory, see Geroch (1981); a more philosophical perspective is offered by Maudlin (2012).

7 There is a subtle metaphysical issue here, concerning whether we understand these "points" or "places" to themselves be physical objects, existing (metaphysically) prior to and independently of bodies (*substantivalism*); or if instead they characterize something about relations between bodies (*relationism*). Nothing I say here should be understood to be taking one or the other of these positions for granted. Perhaps more importantly, the classical spacetime structures I describe here are supposed to provide insight on a different issue, concerning the character of *motion*, rather than, in the first instance anyway, the metaphysics of space or time. See Weatherall (2018, 2020) for more on this perspective; for further discussion and other perspectives, see Earman (1989b), Belot (2000), and Pooley (2013).

8 These arrows will make space a (three-dimensional) *affine space*. See Malament (2009) or Weatherall (2016b) for details.

9 The arrows at each point are to be understood as forming a vector space; and there is a canonical isomorphism between the vector spaces at each point. Vector spaces are required to satisfy some additional conditions that I do not mention, but these are all met by the mental model meant to be invoked by this description. See the texts already cited for details.

10 This notion of spatial distance consists in a Euclidean metric on the vector space of arrows between points.

11 In other words, Galilean spacetime consists in a four-dimensional affine space of events, with further structure to be described.

12 That is, temporal duration is a linear functional acting on the vector space associated with Galilean spacetime.

13 More precisely, Galilean spacetime is a four-dimensional affine space endowed with (1) a non-vanishing linear functional t on its associated vector space; and (2) on each affine subspace by the subspace of vectors to which t assigns 0, a Euclidean metric h.

14 We intend by "curve" a map from (some open subset of) the real numbers into Galilean spacetime that is continuous and at least twice differentiable relative to a topological and differential structure canonically determined by the affine space structure. For details, see Weatherall (2016b).

15 See Earman and Friedman (1973) for a discussion of the status of Newton's first law of motion; see DiSalle (2016) for a discussion of inertial frames more generally.

16 See, especially, the Scholium to the Definition, i.e., the Scholium on Space, Time, Place, and Motion (Newton, 1999 [1687 / 1713 / 1726], pp. 408–415).

17 In other words, Newtonian spacetime is Galilean spacetime as described in endnote 13, with a privileged vector of unit temporal length.

18 It is not universally accepted that Leibniz was committed to Leibnizian spacetime: see, for instance, Roberts (2003); see also endnote 3. For more background on Leibniz's views on physics (and philosophy of physics), see, for instance, Garber (1995), McDonough (2014), and references therein.

19 More precisely, Leibnizian spacetime is a three-dimensional affine bundle over a one-dimensional affine space, where each fiber of the bundle is endowed with a Euclidean metric h and there is a one form t on the base space representing temporal distance.

20 Again, see the discussion in the Scholium to the Definitions (Newton, 1999 [1687 / 1713 / 1726], pp. 408–415).

21 Questions about classical spacetime structure in light of Corollary 6 are hardly new: see, for instance, Stein (1977) and DiSalle (2008); see also Malament (1995) and Norton (1995) for an older discussion of the "relativity of acceleration" in Newtonian physics in a different context. But whether Corollary 6 provides an argument against Galilean spacetime has been of particular interest recently (Saunders, 2013; Knox, 2014; Weatherall, 2016a, 2017a; Wallace, 2016, 2017; Teh, 2018; Dewar, 2018).

22 Newton-Huygens spacetime was first introduced by Earman (1989b) under the moniker *Maxwellian spacetime* (Earman, 1989b); Weatherall (2016a) and others have called it *Maxwell-Huygens* spacetime.

23 In other words, Newton-Huygens spacetime is Leibnizian spacetime endowed with a *standard of rotation* (Weatherall, 2017a).

24 Curiously, Leibniz, too, seems to have taken force to be absolute in this sense, which may raise issues for his views on space and time as reconstructed here. See Roberts (2003), Garber (2012), and McDonough (2014) for more on Leibniz on force.

25 As Stein (1977, pp. 19–20) puts it in a very nice discussion of precisely these issues, absolute acceleration is a *vera causa* in Newtonian physics.

26 This is a project that Saunders (2013) and Dewar (2018) have undertaken.

27 For background on geometrized Newtonian gravitation, see Trautman (1965) and Malament (2012, Ch. 4). For more on the relationship between geometrized Newtonian gravitation and ordinary Newtonian gravitation, see, in addition to those resources, Glymour (1980), Knox (2014), and Weatherall (2016d); for discussions of the relationship between geometrized Newtonian gravitation and gravitation in Newton-Huygens spacetime, see Weatherall (2016a), Wallace (2016), and Dewar (2018).

28 In this, it is like general relativity. See Wald (1984) or Malament (2012) for textbook treatments of general relativity; see also the chapter of this volume on relativistic spacetime.

References

Barrett, T. (2015). Spacetime structure. *Studies in History and Philosophy of Modern Physics*, 51: 37–43.

Belot, G. (2000). Geometry and motion. *British Journal for the Philosophy of Science*, 51(4): 561–595.

Cartan, E. (1923). Sur les variétés à connexion affine, et la théorie de la relativité généralisée (première partie). *Annales scientifiques de l'École Normale Supérieure*, 40: 325–412.

Cartan, E. (1924). Sur les variétés à connexion affine, et la théorie de la relativité généralisée (première partie) (suite). *Annales scientifiques de l'École Normale Supérieure*, 41: 1–25.

Dewar, N. (2018). Maxwell gravitation. *Philosophy of Science*, 85(2): 249-270.

Dewar, N. and Weatherall, J.O. (2018). On gravitational Energy in Newtonian Theories. *Foundations of Physics*, 48: 558–578. Available at: arXiv:1707.00563 [physics.hist-ph].

DiSalle, R. (2008). *Understanding Space-Time*. New York: Cambridge University Press.

DiSalle, R. (2016). Space and time: Inertial frames. In E.N. Zalta (ed.), *The Stanford Encyclopedia of Philosophy*, Winter 2016 edition. Available at: https://plato.stanford.edu/archives/win2016/entries/spacetime-iframes/.

Earman, J. (1977). Leibnizian space-times and Leibnizian algebras. In R.E. Butts, J. Hintikka (eds.), *Historical and Philosophical Dimensions of Logic, Methodology and Philosophy of Science*. Dordrecht: Reidel, pp. 93–112.

Earman, J. (1979). Was Leibniz a relationist? *Midwest Studies in Philosophy*, 4(1): 263–276.

Earman, J. (1986). Why space is not a substance (at least not to first degree). *Pacific Philosophical Quarterly*, 67(4): 225–244.

Earman, J. (1989a). Leibniz and the absolute vs. relational dispute. In N. Rescher (ed.), *Leibnizian Inquiries. A Group of Essays*. Lanham, MD: University Press of America, pp. 9–22.

Earman, J. (1989b). *World Enough and Space-Time*. Cambridge, MA: The MIT Press.

Earman, J. and Friedman, M. (1973). The meaning and status of Newton's law of inertia and the nature of gravitational forces. *Philosophy of Science*, 40: 329.

Fletcher, S.C. (2014). *On the reduction of general relativity to Newtonian gravitation*. Unpublished manuscript.

Friedman, M. (1983). Foundations of Space-Time Theories: Relativistic Physics and Philosophy of Science. Princeton University Press, Princeton, NJ.

Friedrichs, K.O. (1927). Eine invariante Formulierung des Newtonschen Gravitationsgesetzes und der Grenzüberganges vom Einsteinschen zum Newtonschen Gesetz. *Mathematische Annalen*, 98: 566–575.

Garber, D. (1995). Leibniz: Physics and philosophy. In N. Jolley (ed.), *The Cambridge Companion to Leibniz*. Cambridge: Cambridge University Press, pp. 270–352.

Garber, D. (2012). Leibniz, Newton, and force. In A. Janiak and E. Schliesser (eds.), *Interpreting Newton*. Cambridge: Cambridge University Press, pp. 33–47.

Geroch, R. (1981). *General Relativity from A to B*. Chicago, IL: University of Chicago Press.

Glymour, C. (1980). *Theory and Evidence*. Princeton, NJ: Princeton University Press.

Knox, E. (2011). Newton-Cartan theory and teleparallel gravity: The force of a formulation. *Studies in History and Philosophy of Modern Physics*, 42(4): 264–275.

Knox, E. (2014). Newtonian spacetime structure in light of the equivalence principle. *The British Journal for the Philosophy of Science*, 65(4): 863–888.

Leibniz, G.W. and Clarke, S. (2000). *Correspondence*, trans. and ed. by Roger Ariew. Indianapolis, IN: Hackett Publishing Co.

Malament, D. (1986a). Gravity and spatial geometry. In R.B. Marcus, G. Dorn and P. Weingartner (eds.), *Logic, Methodology and Philosophy of Science*. Vol. VII. New York: Elsevier Science Publishers, pp. 405–411.

Malament, D. (1986b). Newtonian gravity, limits, and the geometry of space. In R. Colodny (ed.), *From Quarks to Quasars*. Pittsburgh: University of Pittsburgh Press, pp. 181–201.

Malament, D. (1995). Is Newtonian cosmology really inconsistent? *Philosophy of Science*, 62(4): 489–510.

Malament, D.B. (2009). *Notes on geometry and spacetime*. Unpublished lecture notes. Available at: http://www.socsci.uci.edu/ dmalamen/courses/geometryspacetimedocs/GST.pdf.

Malament, D.B. (2012). *Topics in the Foundations of General Relativity and Newtonian Gravitation Theory*. Chicago: University of Chicago Press.

Maudlin, T. (2012). *Philosophy of Physics: Space and Time*. Princeton, NJ: Princeton University Press.

McDonough, J.K. (2014). Leibniz's philosophy of physics. In E.N. Zalta (ed.), *The Stanford Encyclopedia of Philosophy* (Spring 2014 edition). Available at: https://plato.stanford.edu/archives/spr2014/entries/leibniz-physics/.

Minkowski, H. (2013) [1911]. Space and time. In F. Lewertoff and V. Petkov (eds.), *Space and Time: Minkowski's Papers on Relativity*. Montreal: Minkowski Institute Press, pp. 111–125.

Newton, I. (1999) [1687 / 1713 / 1726]. *The Principia: Mathematical Principles of Natural Philosophy*, edited and trans. by I. Bernard Cohen and Anne Whitman. Berkeley, CA: University of California Press.

Norton, J.D. (1995). The force of Newtonian cosmology: Acceleration is relative. *Philosophy of Science*, 62(4): 511–522.

Pooley, O. (2013). Substantivalist and relationalist approaches to spacetime. In R. Batterman (ed.), *The Oxford Handbook of Philosophy of Physics*. Oxford: Oxford University Press, pp. 522–586.

Roberts, J.T. (2003). Leibniz on force and absolute motion. *Philosophy of Science*, 70(3): 553–573.

Rosenstock, S., Barrett, T. and Weatherall, J.O. (2015). On einstein algebras and relativistic spacetimes. *Studies in History and Philosophy of Modern Physics*, 52B: 309–316.

Rynasiewicz, R. (1992). Rings, holes and substantivalism: On the program of Leibniz algebras. *Philosophy of Science*, 59(4): 572–589.

Saunders, S. (2013). Rethinking Newton's *Principia*. *Philosophy of Science*, 80(1): 22–48.

Stan, M. (2016). Huygens on inertial structure and relativity. *Philosophy of Science*, 83(2): 277–298.

Stein, H. (1967). Newtonian space-time. *The Texas Quarterly*, 10: 174–200.

Stein, H. (1977). Some philosophical prehistory of general relativity. In J. Earman, C. Glymour and J. Stachel (eds.), *Foundations of Space-Time Theories*. Minneapolis: University of Minnesota Press, pp. 3–49.

Teh, N. (2018). Recovering recovery: On the relationship between gauge symmetry and trautman recovery. *Philosophy of Science*, 85(2): 201–224.

Trautman, A. (1965). Foundations and current problem of general relativity. In S. Deser and K.W. Ford (eds.), *Lectures on General Relativity*. Englewood Cliffs, NJ: Prentice-Hall, pp. 1–248.

Wald, R. (1984). *General Relativity*. Chicago: University of Chicago Press.

Wallace, D. (2016). *Fundamental and emergent geometry in Newtonian physics*. Available at: http://philsci-archive.pitt.edu/12497/.

Wallace, D. (2017). More problems for Newtonian cosmology. *Studies in History and Philosophy of Modern Physics*, 57: 35–40.

Weatherall, J.O. (2011a). On (some) explanations in physics. *Philosophy of Science*, 78(3): 421–447.

Weatherall, J.O. (2011b). On the status of the geodesic principle in Newtonian and relativistic physics. *Studies in the History and Philosophy of Modern Physics*, 42(4): 276–281.

Weatherall, J.O. (2014). What is a singularity in geometrized Newtonian gravitation? *Philosophy of Science*, 81(5): 1077–1089.

Weatherall, J.O. (2016d). Are Newtonian gravitation and geometrized Newtonian gravitation theoretically equivalent? *Erkenntnis*, 81: 1073–1091. ErkenntnisPublished online. doi:10.1007/s10670-015-9783-5.

Weatherall, J.O. (2016a). Maxwell-Huygens, Newton-Cartan, and Saunders-Knox spacetimes. *Philosophy of Science*, 83(1): 82–92.

Weatherall, J.O. (2016b). *Space, time, and geometry from Newton to Einstein, feat.* Maxwell, lecture notes from the 2016 MCMP summer school in mathematical philosophy; available on request.

Weatherall, J.O. (2016c). *Void: The Strange Physics of Nothing*. New Haven, CT: Yale University Press.

Weatherall, J.O. (2017a). *A brief comment on Maxwell(/Newton)[-Huygens] spacetime*. Available at: arXiv:1707.02393 [physics.hist-ph].

Weatherall, J.O. (2017b). Categories and the foundations of classical field theories. In E. Landry (ed.), *Categories for the Working Philosopher*. Oxford: Oxford University Press. Available at: arXiv:1505.07084 [physics.hist-ph].

Weatherall, J.O. (2017c). Conservation, inertia, and spacetime geometry. *Studies in History and Philosophy of Science Part B: Studies in History and Philosophy of Modern Physics*, 67: 144–159.

Weatherall, J.O. (2017d). Inertial motion, explanation, and the foundations of classical space-time theories. In D. Lehmkuhl, G. Schiemann and E. Scholz (eds.), *Towards a Theory of Spacetime Theories*. Boston, MA: Birkhäuser, pp. 13–42. Available at: arXiv:1206.2980 [physics.hist-ph].

Weatherall, J.O. (2018). Regarding the 'Hole Argument'. *The British Journal for Philosophy of Science*, 69(2): 329–350 Available at: arXiv:1412.0303 [physics.hist-ph].

Weatherall, J.O. (2020). *Some philosophical prehistory of the Earman-Norton hole argument*. Studies in History and Philosophy of Modern Physics, 70: 79–87.

Weyl, H. (1952) [1918]. *Space Time Matter*. Mineola, NY: Dover Publications.

Further Reading from the Editors

The classic paper which introduced the idea of Newtonian spacetime is Howard Stein's "Newtonian space-time" (*The Texas Quarterly* 10, 174–200, 1967). John Earman's *World Enough and Space-Time* (The MIT Press, Cambridge, MA. 1989) surveys many of the arguments about classical spacetime. More recent interest in Newton-

Cartan spacetime can be traced back to (among others) David Malament's "Is Newtonian cosmology really inconsistent?" (*Philosophy of Science* 62 (4), 489–510, 1995). Saunders' "Rethinking Newton's Principia" (*Philosophy of Science* 80 (1), 22–48, 2013) and Knox's "Newtonian spacetime structure in light of the equivalence principle" (*The British Journal for the Philosophy of Science* 65 (4), 863–888, 2014) give different perspectives on how to think about Newtonian spacetime structure in the light of accelerative symmetries.

4

RELATIONISM IN CLASSICAL DYNAMICS

Julian Barbour

4.1   Relational Arenas

Mach (1960) said that the universe is given "only once, with its relative motions alone determinable." He also said, "It is utterly impossible to measure the changes of things by time. Quite the contrary, time is an abstraction at which we arrive by means of the changes of things." This chapter, based on work with collaborators, will propose a concrete relational theory that reflects these aphorisms.

For this we need a framework in which the universe is "given." Mach was reacting to Newton's concepts and beliefs, which included infinite extent of the material universe. That cannot be concretely "given," but N self-gravitating mass points in Euclidean space can. They form a dynamically closed system with counterpart in general relativity: a spatially closed universe.

Besides relative masses $\bar{m}_a = m_a/m_{\text{tot}}$, $m_{\text{tot}} = \sum_a m_a$, $a = 1, \ldots, N$, and ratios of the distances between them, nothing but the symmetries of Euclidean space is used. The extra structures that Newton introduced with absolute space and time play no role in the universe's dynamics but emerge for its subsystems, expressing their relation to the whole: the universe and its evolution. The relational principles developed for Euclidean space can be carried over with little change to Riemannian geometry.

Euclidean geometry is based on the similarity group Sim. Suppose the (Cartesian) coordinates $\mathbf{r}_a^1, \mathbf{r}_a^2$, $a = 1, \ldots, N$, specify two sets of particle positions. Hold the coordinates of set 1 fixed. If, using the translations, rotations and scaling (dilatation) transformations of Sim set 2 can be brought to "overlap" with set 1, $\mathbf{r}_a^1 = \mathbf{r}_a^2$ for all a, then the two figures are *similar*. If the condition can be achieved using translations and rotations alone, then the two figures are *congruent*. In absolute space, two congruent or similar figures not in overlap are different states. Leibniz Alexander (1956) argued that, since nothing intrinsically observable differs, absolute space is a fiction. We formalize this intuition by comparison of N-body spaces.

First, the standard $3N$-dimensional Newtonian configuration space Q, which is N copies of Euclidean space equipped (for convenience) with Cartesian coordinates. Identifying all the configurations $q \in$ Q that translations and rotations make "Leibniz-identical," we obtain the $(3N-6)$-dimensional *relative configuration space* R Barbour (1974). Taking the process further by identifying configurations related by dilatations, we come to the $(3N-7)$-dimensional space S of possible shapes $s \in$ S of the universe: *shape space* (introduced in Barbour (1999) and used in Barbour (2003)).

A law of relational dynamics in either **R** or **S** requires at least three particles: at least one dependent variable and an independent variable. For it Newton introduced absolute time t. But Mach's aphorisms require us to extract time from relative change. In a universe of a single particle, no relative change can occur. In the two–particle R, an assumed external scale allows an interparticle separation but nothing to serve as independent variable. In S, for which no external scale is defined, we do not even have a dependent variable with only two particles; they either do or don't coincide. But the separations of three particles fix two shape-defining internal angles of a triangle, the bare minimum for shape dynamics.

I do not see how, at our present level of understanding, the elimination of absolute elements could be taken beyond Sim, for which angles are defined, to the general linear group (for which they are not).

In S all quantities are *dimensionless*:mass ratios and ratios of distances. Because all measurements yield ratios,[1] one may call S *the empirical space*.

In Newtonian dynamics, a solution is fixed by specifying initial coordinates $\mathbf{r}_a$ and velocities $\dot{\mathbf{r}}_a$. This uses the extra Newtonian structure not present in S and absolute time. To eliminate all trace of Newton's absolute elements, we propose:

The Fundamental Postulate of Relational Dynamics. *An initial point and an undirected line through it in S should uniquely determine a solution.*[2]

Here "undirected line" eliminates the notion of velocity, which has magnitude and direction. The two directions from a point on a line have equal status and no magnitude.

4.2 Newtonian Dynamics in Shape Space

Suppose *the law of the universe* is what we have in the Newtonian N-body problem, $N \geq 3$. To what extent would it satisfy the fundamental postulate? Of course, even if it does not it must not come into conflict with observation. As for all scientific theories, empirical adequacy is also required in relational dynamics.

The N-body problem is usually defined by $3N$ mass-weighted Cartesian coordinates $\sqrt{m_a}x_i^a$, $i = x, y, z$. However, these hide heterogeneity that is at the heart of relationism: $3N - 7$ degrees of freedom (dofs) define the instantaneous *shape* of the system, three fix the center of mass, three fix the orientation, and one that turns out to be exceptionally important the overall scale defined by the root-mean-square length ℓ_{rms} using the dimensionless masses $\bar{m}_a = m_a/m_{\text{tot}}$:

$$\ell_{\text{rms}} = \sqrt{\sum_{a<b} \bar{m}_a \bar{m}_b\, r_{ab}^2} \equiv \frac{1}{m_{\text{tot}}} I_{\text{cm}} = \frac{1}{m_{\text{tot}}} \sum_a m_a\, \mathbf{r}_a^{\text{cm}} \cdot \mathbf{r}_a^{\text{cm}}, \tag{4.1}$$

where $r_{ab} = |\mathbf{r}_a - \mathbf{r}_b|$, and in I_{cm}, the center-of-mass moment of inertia, the $\mathbf{r}_a^{\text{cm}}$ are the positions relative to the system's center of mass. The dimensionless masses $\bar{m}_a$ are not relationally problematic since they are constants and, by making enough observations, internal observers could determine their constant values. In contrast, four non-shape dofs (all but the three translational dofs) together with Newton's time show up problematically. Let us consider what observers in S will "see."

If the Newtonian N-body problem does faithfully describe the universe, a curve representing its empirically observable history is obtained by taking a Newtonian solution and simply abstracting way all the data except the $3N - 7$ that define the successive shapes. Call them s_i. Such a mapping from Q to a curve c in S is many-to-one and eliminates much redundant information. The parametrization of c by time is completely washed out. The $3N$ Newtonian coordinates taken to represent an initial condition project to $3N - 7$ shape dofs s_i. The representation of initial velocities is even more restricted. Of the $3N$ that represent them in Q, $3N - 1$ define a direction in Q and one defines a magnitude. It is absolute time that permits its introduction – speed is distance divided by time. Speeds cannot be represented in S. There are just curves and undirected lines, represented by $3N - 8$ numbers, tangent to them.

Since the curve in S encodes all the objectively given data, i.e., data that can be determined by observers within the universe, no true information can be lost by the "projection" to S. We obtain a relational representation of Newtonian dynamics. Will it satisfy us? Since relational intuition rejects the absolute elements that Newton introduced, we need to see if and how any of them show up "objectionably" in the relational arena S. We have agreed that the history of the universe is a smooth curve in S. Our fundamental postulate characterizes the kind of law which should determine the curve. It is allowed to use only the irreducible minimum of geometrical structure, namely invariants of S: angles, ratios of interparticle separations, and, more generally, any dimensionless ratios that define the shape. In contrast, Newton introduced absolute elements: lengths, times, orientation and position of the N-body center of mass. Of these, only the last is relationally harmless: the Galilean relativity of Newtonian mechanics ensures that the absolute position and velocity of the center of mass cannot be determined by any observations within an N-body system.

The relationally "objectionable" elements in the Newtonian N-body problem are certain dimensionless ratios that can be formed from Newtonian initial data but cannot be represented by a point and direction in S. We note first that, by the velocity decomposition theorem (Saari, 2005), the center-of-mass kinetic energy T_{cm} can, at any given instant, be decomposed uniquely into three parts: $T_{cm} = T_s + T_r + T_d$, respectively the kinetic energy in change of shape, overall rotation, and overall change of size (dilatation). Only the first is relationally "acceptable." The rotational term is associated with angular momentum $\mathbf{L}$ and is unambiguously "objectionable." Through the magnitude of $\mathbf{L}$, it brings in not only the dimensionless ratio T_r/T_s, which cannot be encoded in a shape and direction in S, but also two more that define the direction of $\mathbf{L}$. All three of these relationally problematic quantities are eliminated by best matching (Section 4.4), which automatically ensures that the universe as a whole must have vanishing angular momentum. Newtonian theory based on absolute space cannot make this prediction. The best-matched universes with $\mathbf{L} = 0$ can perfectly well contain dynamically isolated subsystems with angular momentum provided the sum for all of them is zero.

The dilatational term T_d, from which the ratio T_d/T_s can be formed, would seem to be relationally problematic in the same way, but here, as I shall explain (Section 4.5), the problem is resolved in a way that is both subtle and remarkably interesting.

So far, I have identified four "objectionables." They are all associated with the way elements of Sim can, without changing anything intrinsic, change the $\mathbf{r}_a$ in absolute space (in modern terms, in an inertial frame of reference).

Absolute time gives rise to the fifth and final "objectionable." It is manifested in initial data. In the N-body problem, these take the form

$$\mathbf{r}_a, \ \dot{\mathbf{r}}_a, \ a = 1, \ldots, N. \tag{4.2}$$

With the initial data (4.2) Newton's laws generate a solution that projects to some curve in S. The new initial data set

$$\mathbf{r}_a, \ c\dot{\mathbf{r}}_a, \ a = 1, \ldots, N, \tag{4.3}$$

created by multiplying all the velocities in (4.2) by the common constant c will generate a new solution that will in general project to a different curve in S because the multiplication by c changes the value of the dimensionless ratio $\epsilon = T_{cm}/V$, where V is the potential energy. The value of ϵ will vary along the solution curve in S unless the total center-of-mass energy is zero: $E_{cm} = 0$. Then $\epsilon = -1$ along the entire curve in S. Thus, if our fundamental postulate is to hold, we must have $E_{cm} = 0$.

This is not the same kind of argument for concluding that $E_{cm} = 0$ as best matching (Section 4.4) is for the vanishing of the angular momentum, but both arise from elimination of the effects of Newton's absolutes. It may also be noted that, like vanishing of the magnitude of the angular momentum, $E_{cm} = 0$ is a scale-invariant (unit-independent) condition, whereas $E_{cm} = E_0$ is not.

In summary, there are five relationally problematic quantities in standard Newtonian mechanics: three associated with angular momentum, which defines a magnitude and an orientation in absolute space; one with dilatation (the ability of the scale variable associated with a Newtonian system to increase or decrease); and one with energy, the value of which can be changed due to the role of absolute time. If none of these play any role in determining the shape-space solutions of the law of the universe, we can say it is relational.

4.3   From Newtonian to Relational Dynamics

The N-body problem is usually derived from the principle of least action. One specifies initial and final points q_i, $q_f \in Q$ in the Newtonian configuration space and the time $t_f - t_i$ for the system to pass between them. Then along all possible smooth paths joining q_i and q_f one finds the one that extremalizes the action $\int dt (T - V)$, where T is the Newtonian kinetic energy

$$T = \sum_a \frac{m_a}{2} \frac{d\mathbf{r}_a}{dt} \cdot \frac{d\mathbf{r}_a}{dt}, \tag{4.4}$$

and V is the potential energy.

Now the time t is an external independent parameter, but any real clock is made of material parts of the universe and the movements of its hands are Mach's "changes of things" from which we abstract a measure of time (Barbour, 2009). Relational dynamics cannot employ an external time difference $t_f - t_i$. One does not attempt to specify any time difference required for this transition (temporal relationism) but only two distinct configurations. History does not "happen in time." The history of the universe is not a spot of light moving along a curve in a configuration space. It is the curve. Any notion of duration (and direction of time) that we can obtain must be deduced from the curve. Computationally, it is however convenient to parametrize the points on the curve in R by an arbitrary monotonic parameter λ.

Moreover, one specifies that the initial and final configurations are not in Q but in either R (r_i and r_f) or S (s_i and s_f). That implements spatial relationism by eliminating absolute orientation. In fact, empirical adequacy will lead us to make the passage to dynamics in S differently. Also without some intrinsic justification the direction of time implied by "initial" and "final" is nominal, dynamics being time-reversal symmetric. The principle of least action clearly needs to be reformulated in a "sharpened" relational form.

One obvious way to proceed is to define a metric on either R or S and require the history of the universe to be a geodesic with respect to it. Such a geodesic is fixed by specifying two distinct points or, equivalently, in the spirit of the fundamental postulate, a point and direction. In Barbour (1974), I proposed such a theory. In fact, I later learned that, in essence (though without the elimination of an external time), the theory had already been proposed and rediscovered many times in the 20th century.[3] The key idea, which Bertotti and I (still unaware of the earlier proposals) developed in Barbour and Bertotti (1977), is to replace the Newtonian kinetic energy (4.4), which is a sum of single-particle contributions, by a sum of two-particle contributions dependent only on relative distances (rd):

$$T_{\text{rd}} = \sum_{a<b} \frac{m_a m_b \, \dot{r}_{ab}^2}{r_{ab}}, \tag{4.5}$$

where the dot denotes differentiation with respect to time (or an arbitrary labeling parameter λ in the timeless implementation of Barbour and Bertotti (1977)). The denominator r_{ab} is introduced on plausible physical grounds in order to have the same distance dependence as in the Newton gravitational potential:

$$V_{\text{New}} = -\sum_{a<b} \frac{m_a m_b}{r_{ab}}. \tag{4.6}$$

The theory based on the relational kinetic energy (4.5) has several attractive and interesting features and, as I said, was discovered and rediscovered independently numerous times during the 20th century. However, as Reissner and Schrödinger already knew and Bertotti and I emphasized, the theory based on (4.5) predicts that experimentally observed inertial mass must be anisotropic. Modern experiments of Hughes-Drever type (see the article by Will in Barbour and Pfister (1995)) now rule this out to an extraordinarily high accuracy.

Although the theory based on (4.5) in the timeless form of Barbour and Bertotti (1977) eliminates the three "objectionables" associated with angular momentum and meets four of the five requirements of the fundamental postulate, it does so at the price of being, empirically, completely inadequate.

To avoid the mass-anisotropy problem, Bertotti and I introduced what we called the *intrinsic derivative* (Barbour and Bertotti, 1982) but I now call *best matching* (Barbour, 1994, 1999). The basic idea is outlined in the next section.

4.4 Best Matching

Because satisfactory treatment of dilatations is subtle, I will illustrate the key idea behind best matching using only the freedom to make transformations of the Euclidean group, i.e., translations and rotations, though it can easily be extended to dilatations to create a theory in S. The idea is very intuitive; it surprises me that it does not seem to have been considered hitherto. The key point is that the generators of translations and rotations are used to move, in an auxiliary representational space, one configuration relative to another. Let the coordinates $\mathbf{r}_a$ and $\mathbf{r}_a + d\mathbf{r}_a$ represent two slightly different configurations of N particles of masses m_a, $a = 1, \ldots, N$, in Euclidean space. This suggests that the particles have been displaced by $d\mathbf{r}_a$ in the transition from the first to the second configuration. However, let us translate and rotate (about its center of mass) the second configuration by infinitesimal amounts so that it becomes $\mathbf{r}_a + d\mathbf{r}_a + \delta\mathbf{a} + \mathbf{r}_a \times \delta\omega$. This leaves the interparticle separations r_{ab} in the second configuration unchanged, so there has been no intrinsic change in it. However, the Newtonian action associated with the two configurations has been changed from $\sum_a m_a\, d\mathbf{r}_a \cdot d\mathbf{r}_a$ to

$$\sum_a m_a\,(d\mathbf{r}_a + \delta\mathbf{a} + \mathbf{r}_a \times \delta\omega) \cdot (d\mathbf{r}_a + \delta\mathbf{a} + \mathbf{r}_a \times \delta\omega). \tag{4.7}$$

Thus, two configurations that are infinitesimally slightly different can give rise to arbitrarily different contributions to the action depending on the choices made for $\delta\mathbf{a}$ and $\mathbf{r}_a \times \delta\omega$. By Galilean relativity the displacement $\delta\mathbf{a}$ does not lead to any changes observable within the system, but the rotation $\mathbf{r}_a \times \delta\omega$ does. It is the explanation for the three "objectionables" associated with angular momentum. Is there any way to eliminate the ambiguity present in (4.7)?

Luckily, the answer is yes. This is because (4.7) is a positive-definite quantity and has a unique minimum. It is found by using the generators of translations and rotations until (4.7) is minimized. This happens when the two configurations are "best matched," or brought as close as is possible to congruence. Let us denote the changes from an initial configuration to the slightly different one in the best-matched position by $d\mathbf{r}_a^{\mathrm{bm}}$.

Let us now see what our relational action principle looks like when we take into account the relativity of position and Mach's intuitively persuasive aphorism about time. I have already noted it can be implemented by a timeless geodesic principle. One that allows us to take into account forces has the form (the subscript Jac will be explained shortly)

$$\delta A_{\mathrm{Jac}} = 0, \quad \delta A_{\mathrm{Jac}} = 2\int d\lambda\sqrt{WT_{\mathrm{Jac}}}, \quad T_{\mathrm{Jac}} = \sum_a \frac{m_a}{2}\frac{d\mathbf{r}_a^{\mathrm{bm}}}{d\lambda}\cdot\frac{d\mathbf{r}_a^{\mathrm{bm}}}{d\lambda}, \tag{4.8}$$

where W is a function that depends only on the r_{ab} and serves as a potential from which forces can be obtained; λ is any smooth monotonic curve parameter. A key property of (4.8) is its reparametrization invariance. The action is unchanged by any parameter transformation $\lambda \rightarrow \lambda'(\lambda)$ that maintains the monotonicity and smoothness condition and has $\lambda' = \lambda$ at the limits of the integration. There is no external time.

The equations of motion that follow from (4.8) are

$$\frac{d}{d\lambda}\left[\sqrt{\frac{W}{T_{\text{Jac}}}} m_a \frac{d\mathbf{r}_a^{\text{bm}}}{d\lambda}\right] = -\sqrt{\frac{T_{\text{Jac}}}{W}} \frac{\partial W}{\partial \mathbf{r}_a}. \tag{4.9}$$

It is immediately clear that this equation takes its simplest form if the arbitrary parameter λ is chosen, as it always can be, such that

$$W = T_{\text{Jac}}. \tag{4.10}$$

The equation of motion then takes the form of Newton's second law and the specially chosen λ, which we now denote by t, becomes indistinguishable from Newton's absolute time with W playing the role of the potential energy:

$$m_a \frac{d^2 \mathbf{r}_a}{dt^2} = -\frac{\partial W}{\partial \mathbf{r}_a}. \tag{4.11}$$

From (4.10) we obtain an explicit implementation of Mach's aphorism by writing the increment of the distinguished label t in the form

$$dt = \sqrt{\frac{\sum_a m_a\, d\mathbf{r}_a^{\text{bm}} \cdot d\mathbf{r}_a^{\text{bm}}}{2W}}. \tag{4.12}$$

Here, the increment of Newton's time turns out to be the mass-weighted sum of the true differences between two nearly identical configurations divided by the total potential. Time is indeed an abstraction obtained from "the changes of things." Moreover, potentials are only defined up to additive constants, so, introducing the Newton potential (4.6), we can write $W = E - V_{\text{New}}$, so that, suitably rearranged, (4.10) takes the form

$$T_{\text{Jac}} + V_{\text{New}} = E. \tag{4.13}$$

This looks like the statement of the energy theorem, but in relational dynamics it becomes *the definition of time*. Moreover, (4.8) takes the form of Jacobi's principle (Lanczos, 1949). For further discussion, see Barbour and Bertotti (1982), and Barbour (2009), including the role of the parameter defined by (4.12) as ephemeris time in astronomy. Note also that the ephemeris time (4.12) has length dimensions $[\ell]^{3/2}$, so it is only meaningfully defined if there is a notion of length. How this emerges will be explained at the end of Section 4.5.

Let us now consider the effect of best matching. The calculations confirm what intuition suggests: the centers of mass of the two configurations are brought to coincidence and the net rotation of the second configuration relative to the first about the common center of mass is zero. More precisely, since angular momentum is the mechanical measure of the rotation, it is forced to be zero. There is a further condition beyond best matching that must be satisfied if this condition is to be maintained throughout the evolution: the potential must be a function invariant under translations and rotations, i.e., a function on R. This is a useful criterion for the selection of possible relational potentials and is satisfied by the Newton potential (4.6). Indeed, it has long been recognized that the non-relational part of the N-body problem resides in its kinetic energy; the potential is relationally "kosher."

The formal equations of the best-matching procedure, expressed in terms of rates of change rather than finite differences, can be found in Barbour and Bertotti (1982); Barbour (1994) and show how the total angular momentum of the universe must be zero. Because our principles do not exclude the presence of the constant E, which in the Newtonian representation appears as the total energy of the universe, we cannot yet say that best matching forces it to be zero (though the fundamental postulate does).

What happens if best matching is extended to dilatations. This was done in Barbour (2003), to which I refer the reader for details and here merely summarize the results. Every book on dynamics discusses angular momentum, the expression for which in the center of mass of the N-body problem is $\mathbf{L} = \sum_a \mathbf{r}_a^{cm} \times \mathbf{p}_a^{cm}$, where $\mathbf{p}_a^{cm}$ is the center-of-mass momentum of particle a. This is a measure of the rotation in the system in absolute space. Rather surprisingly, its analogue $D = \sum_a \mathbf{r}_a^{cm} \cdot \mathbf{p}_a^{cm}$, which measures the amount of expansion in the system, is seldom considered and had no name before I called it the *dilatational momentum* in Barbour (2003).

When one extends best matching to dilatations as in Barbour (2003), one obtains the result one would expect, namely, D must vanish. However, for this two conditions must be satisfied. First, the total energy must be zero. At the first glance, this seems good: our principles are more restrictive, eliminating one more "objectionable" parameter. A point and undirected line in shape space do now determine a solution uniquely; all five objectionables are eliminated, and we have a "dynamics of pure shape."

But there is a catch. The second condition is that the Newton potential must be replaced if the condition $D = 0$ is to be maintained throughout the evolution. This has to do with what is called homogeneity of a function. A function $f(x_i)$ of arguments a_i is said to be homogeneous of degree k if $f(ax_i) = a^k f(x_i)$, where $a > 0$ is a constant. The Newton potential is homogeneous of degree -1. Now if $E = 0$ the remaining term $\sqrt{V_{New} T_{Jac}}$ in (4.8) is homogeneous of degree $1/2$. The integrand is not scale-invariant – it changes under dilatations. In Barbour (2003), this defect is rectified by replacing the Newton potential by $I_{cm}^{-1/2} V_{New}$, which, being homogeneous of degree -2, balances the $+2$ degree of the kinetic term and allows the condition $D = 0$ to be maintained. As shown in Barbour (2003), this theory reproduces Newtonian gravity extremely well in subsystems but introduces additional forces that maintain I_{cm} constant: the size of the system cannot change. The problem with this is that I_{cm} is the analogue of the volume of the universe in general relativity. Modern cosmology tells us the universe is expanding. Therefore, we cannot model the universe with a theory for which I_{cm} is constant.

4.5   Janus-Point Theories

The resolution of the problem just presented came from a most unexpected place with very interesting possible consequences that had never been envisaged by Bertotti and myself myself (1982) when we introduced best matching. The key insight relies on the first qualitative result obtained in dynamics. In a justly famous prize-winning essay on the three-body problem published in 1772, Lagrange proved, among other things, that if the center-of-mass energy of the system is non-negative, $E_{cm} \geq 0$, then

$$\ddot{I}_{cm} > 0. \tag{4.14}$$

Jacobi later extended this result to the N-body problem with $N > 3$. The two-line proof is given in Barbour, Koslowski and Mercati (2014). It uses Newton's second law and two properties of the Newton potential V_{New}: its homogeneity of degree -1 and the fact that $V_{New} < 0$. Note that our fundamental hypothesis requires $E_{cm} = 0$, so this is compatible with the result (4.14).

The all-important relation (4.14) shows that if $E_{cm} \geq 0$ then the function $I_{cm}(t)$ is concave upward and $I_{cm} \to \infty$ as $t \to \pm\infty$. This means that at least one of the particles must escape to infinity in either limit $t \to \pm\infty$, so that a system with $E_{cm} \geq 0$ is unstable. This raised the question of whether the solar system, which has negative energy, is stable. This was the prize problem posed by the king of Sweden and won by Poincaré in the paper (Poincaré, 1890) in which he discovered chaos.

What is relevant for us is not this but that the slope of the function $I_{cm}(t)$,

$$\dot{I}_{cm} = 2D = 2\sum_a \mathbf{r}_a \cdot \mathbf{p}_a, \tag{4.15}$$

is monotonic and varies from $-\infty$ to $+\infty$, passing through zero just once at what may be called *the Janus point* because it divides the solution into two qualitatively similar oppositely facing halves.

The significant consequence of (4.14) for relational dynamics is the possibility it raises of creating a framework in which our fundamental postulate of relational dynamics is implemented, namely a point and undirected line in S do uniquely determine a solution. Recall that standard Newtonian dynamics leads to the appearance of five dimensionless quantities in S that thwart this. Best matching cleanly eliminates the three associated with angular momentum. We can consistently eliminate the fourth associated with a non-vanishing energy because energy is conserved in Newtonian dynamics: if $E_{cm} = 0$ at some initial instant it will remain zero. The only "objectionable" quantity that remains is D (4.15), which, if we retain V_{New}, is always non-zero except at the unique Janus point. This means that at any general point on the solution curve in S of a Newtonian solution with vanishing energy and angular momentum a point and undirected line do not encode enough information to predict the further evolution. One needs just one more number, namely the one that encodes the ratio T_d/T_s of the amount of kinetic energy in change of size to the amount in change of shape.

However, the really significant thing is that at the unique Janus point, which exists on every solution curve, we do have $D = 0$. In turn this means that at this point the ratio T_d/T_s vanishes and *a point and undirected line do determine a solution*. Since every single solution has a Janus point, we can uniquely specify them all by giving the position of their Janus point in S and the undirected line of the solution curve at that point. This enables us to realize our fundamental hypothesis. For the expression of this in phase-space terms, see Barbour et al. (2015).

It is helpful to clarify the role that scale and time play in this implementation. They are both emergent. As is explained in Koslowski et al. (2016) for a general-relativistic model analogous to the 3-body problem, the equations that determine the evolution curves in S are autonomous: they can be solved without any reference to scale or time. Then, once a solution curve in S has been obtained, one can specify at any point on it a purely nominal non-vanishing value of I_{cm}. For given dimensionless mass ratios, this determines all the other lengths (the r_{ab}) at that instant. At it, one can ascribe a time t and, because the scale is given, the ephemeris rate at which it shall flow. Moreover, from the already known curve in S one can then determine I_{cm} and all other lengths together with the time everywhere else on the solution curve in the Newtonian configuration space Q. Absolute scale does not emerge from the shape-space evolution but relative scale does. But this is expressed through a dimensionless ratio and is compatible with the principles of relational dynamics provided the presence of a Janus point allows their implementation.

Mach said time must be abstracted from the changes of things. We see that relational dynamics achieves more and is specific about what the "things" are. They are the universe's successive shapes. From their evolution, it is a two-step passage to time (duration). It is first necessary to abstract relative scale, the more important quantity, and only then, with a notion of (relative) length defined, the ephemeris time (4.12).

Although, at the time of writing, the details are still being worked out by my collaborators and published material as yet do not go beyond (Koslowski et al., 2016), the non-Newtonian "pure-shape" theory with $D = 0$ throughout the evolution will play a significant role in the equations that determine the evolution curves in S. First, the "pure-shape" theory is based on a metric whose increment ds can be used to parametrize these curves. Second, the shape-space equations can be expressed in terms of the instantaneous shape and the direction chosen for evolution together with a variable that expresses the non-vanishing of the ratio T_d/T_s in the Newtonian representation. Moreover, the deviation between a Newtonian solution curve and the pure-shape geodesic that best approximates it at a given point p is least when p is the Janus point of the solution for that is where both theories have $D = 0$. Away from the Janus point, the deviation between the two theories gets ever greater.

4.6 Phenomena Explained by Relational Dynamics

We first consider what can be explained by the implementation of relational dynamics through best matching. We recall that it brings two configurations that have a slight intrinsic difference to a unique "overlap" position in which their incongruence is minimized. Now this can be done for any two successive configurations in the solution curve. If we start with any one of them, we can best match all the successive configurations along either direction of any given solution curve into the ongoing least-incongruence position. Taking, as usual, spatial dimensions to be represented by horizontal axes, Bertotti and I called this *horizontal stacking* (Barbour and Bertotti, 1982). Taking the vertical direction to represent the evolution, we then supposed a *vertical stacking* in accordance with which successive configurations are separated vertically by means of the distinguished ephemeris time (4.12). As shown, this creates a framework indistinguishable from the (center-of-mass) frame Newton introduced through his absolute space and time. In it, the particles evolve exactly in accordance with Newton's laws. The framework is uniquely determined up to trivial nominal options: the orientation of the axes, their handedness, and the origin and rate of time (definition of the second).

Whereas Newton's frame is completely recovered and its existence explained by the relational best matching, there are two important dynamical differences. First, the universe must have vanishing angular momentum, which does not follow from Newtonian dynamics. It must also have vanishing total energy. In the conceptual setting of Barbour and Bertotti (1982), Bertotti and I did not give an adequate justification for this. It needs the fundamental hypothesis. Hypotheses are standard in theoretical physics: one makes an assumption, argues from it logically to firm predictions, and then makes observations to see if they are confirmed.

Let us consider some of the predictions of best-matched relational dynamics that go beyond the already noted prediction of an emergent global inertial frame of reference, including units of length and time. Because, as shown in Barbour and Bertotti (1982), the equations of motion in that frame are identical to Newton's, it follows that any collection of n particles, $n \ll N$, sufficiently far removed from other particles to be isolated in the usual Newtonian sense will have its own center-of-mass inertial frame of reference. The difference now is that such subsystems can have any values of their standard Newtonian conserved quantities, including angular momentum.

This leads to a remarkable structural feature, the invariable plane. It passes through the center of mass of the subsystem with normal fixed in the global frame and pointing along the angular momentum of the subsystem. All angular momenta of any such systems that form will maintain constant angles with each other. The centers of mass of these subsystems will move inertially within the global frame. Within each subsystem, wherever it is and whatever its motion, the laws will always be Newton's. Galileo's principle of relativity will hold. All of the subsystems will generate their own ephemeris times (4.12), all of which will "march in step" (Barbour, 2009), i.e., the ratios of their rates will remain constant. The behavior of subsystems is especially striking when $n = 2$, when they form bound Kepler pairs. Each of these forms a compass of directions, a clock whose ticks are the period of the pair, and a rod, the length of which is the semimajor axes. For any two Kepler pairs the ratios of the rates and lengths stay constant as does the mutual inclination of their directional compasses. Finally, the total angular momentum and energy of the system both must be zero. This all appears as miraculous fine-tuning, but it is a direct consequence of the principle of least action as strengthened (significantly) by our fundamental postulate. All the architectonic features of Newtonian dynamics are explained by the "principle of relational least action" and the additional relational predictions are made.

In fact, it does even more, since it shows (Barbour et al., 2014, 2015) that the breakup into subsystems with their remarkably correlated properties must happen to a greater or lesser extent in all solutions of the relational N-body problem. The relational theory also bids fair (Barbour et al., 2014, 2015) to show that arrows of time are dynamical in origin, i.e., necessary, and nothing whatever to do with selection of special solutions by the imposition of some very particular solution in the past as proposed by Boltzmann (1897):

> The second law of thermodynamics can be explained mechanically ... if one assumes that the universe is a mechanical system and it, or at least a very extended part of it surrounding us, started to evolve from a very improbable state and still remains in such a state. (my translation)

In this paper, others in 1895 and 1896, and his final comments on the topic of the observed continual growth of entropy at the end of his (1898), Boltzmann made his famous suggestion that our sense of the direction of time coincides with the direction of entropy growth in the part of the universe accessible, in both space and time, to our observations. In all these comments, he assumed that the universe as a whole is in thermal equilibrium and departures from it arise through statistical fluctuations. For the last 120 years, some such entropic assumption, not necessarily based on thermal equilibrium, has been accepted as the only possible explanation for the entropic arrow of time.

That a quite different explanation is possible was first noted in Barbour et al. (2013, 2014, 2015). The key observation, as already anticipated at the end of Section 4.5, is that the vanishing of D (4.15) at the unique Janus point divides every relational N-body solution into exactly two halves, on either side of which I_{cm} grows from some nominal minimal value (which can, if it occurs at a central configuration (Barbour et al., 2013), be zero) to infinity. This already suggests two arrows of time pointing in opposite directions away from the Janus point in every solution.

However, I_{cm} is a single global quantity whose increase is not directly observable within the system (it is not a dimensionless ratio), and its increase says nothing about how the observable shape dofs evolve. In fact, they are subject to the characteristic effect of gravity, which has an inherent tendency to convert an initially homogeneous matter distribution into one that is increasingly inhomogeneous. This happens through the formation of clusters. To measure the degree of such clustering, we introduced in Barbour et al. (2013, 2014) the scale-invariant dimensionless quantity

$$C = -I_{cm}^{1/2} V_{New}, \qquad (4.16)$$

which we called the *complexity*. (The minus sign is introduced to obtain a positive quantity by canceling the negative sign of the Newton potential.) This function is a sensitive measure of clustering because when N is reasonably large, its first factor $I_{cm}^{1/2}$ changes little if some particles cluster together whereas the Newton potential changes markedly and even becomes infinite in magnitude if two particles coincide.[4]

Now, not surprisingly and reflecting the time-reversal symmetry of the dynamical equations, the shape degrees of freedom behave in qualitatively the same way on the two sides of the Janus point. In each solution, the system is always in its most random but overall uniform state near the Janus point, where C typically has its smallest values. In either direction away from the Janus point C grows with fluctuations that lie between monotonically rising bounds. This reflects the growth of structure through the formation of subsystems that become ever better isolated, so that the scenario described above is realized of dynamical necessity. Two things work together to bring this about: the existence and necessary growth of the scale variable ℓ_{rms} (or equivalently $I_{cm}^{1/2}$) and gravity, which gives rise to clustering.

Now, back in the 1890s, Boltzmann (by assuming thermal equilibrium) was implicitly considering a system in a confined space, in which the scale variable necessarily has a bounded range,

$0 \leq \ell_{\mathrm{rms}} \leq L$. Under these conditions, arrows of time could only arise through deep statistical fluctuations of the entropy, on either side of which oppositely pointing arrows of time would be defined by entropy growth. But one can also define arrows of time by growth of structure, measured by the secular growth of the complexity C (4.12), which happens either side of the Janus point in all solutions of the relational N-body problem. Two things completely change the situation with regard to the explanation of the arrow of time (for full details see Barbour et al. (2015)): removal of confinement and non-trivial inclusion of gravity (Boltzmann only included gravity as an external potential in his important generalization of Maxwell's energy distribution law to include other forms of energy besides kinetic energy.) In fact, as argued in Barbour et al. (2015), unbounded growth of the scale variable in any system with Janus-point solutions will ensure the existence of arrows of time. For example, bidirectional arrows also arise in the wave motions of dissipationless fluids. The role of gravity is to determine the precise way in which the arrows are manifested.

One may argue, as did an otherwise supportive referee of our Barbour et al. (2014), that the insights into the origin of the arrows of time do not rely on the arguments for shape dynamics. Indeed, it is true that the results of Barbour et al. (2014) are in the first place simply consequences of Newtonian theory that had hitherto escaped notice in discussions about the arrow of time. Two responses can be made to this. First, the relational arguments do much more than provide the conceptual mindset that recognized the significance of Lagrange's 1772 result. They explain the entire framework of Newtonian theory and pinpoint the need for Janus-point solutions to get round the problem of the final, fifth "objectionable" associated with dilations. Moreover, the evolution in shape space can be continued smoothly through singularities of the scale variable. The first demonstration of this can be found in Koslowski et al. (2016), which shows that Janus points also exist in general relativity. It seems to me that the fact which is most likely to establish the fruitfulness of the shape-dynamic approach is that the scale and temporal evolution "rides piggy back" on the autonomous shape evolution. That is shape dynamics. Besides Koslowski et al. (2016), the relatively straightforward extension of the ideas developed here for particle dynamics to dynamical geometry as in Einstein's general theory of relativity is described, at least in the first steps, in Barbour et al. (2013).

Notes

This article was written while I was working on *The Janus Point* (Basic Books (2020)). The book extends and to some extent modifies the content of this paper. In particular, I would now say that the fifth Newtonian "objectionable" is not really eliminated by specifying initial data at the Janus point since, if an expanding universe and all its increasing variety is to be modelled, a single dynamical variable must be added to the variables that specify a shape and the direction in which the shape is changing. In *The Janus Point* I call this variable the *creation measure*. In chapters 16 to 18 I also consider what I call 'total explosions'. Unlike Janus point solutions in which the Newtonian size of the system is finite at its minimum and there are two qualitatively similar halves of such solutions, the total-explosion solutions are like 'half a Janus-point solution' and begin with zero Newtonian size. They are exceptionally interesting.

1 A length is the number of times the unit fits along an interval divided by one (the unit's length).
2 Readers familiar with my paper Barbour (2010) will note I no longer call this the Poincaré principle or make the distinction between its strong and weak forms. This is because the insights I had about the existence and significance of Janus points (discussed in Section 4.5) made the distinction irrelevant. The analysis of Poincaré on which Barbour (2010) is based was crucial for the eventual formulation of this new fundamental postulate. The difference is the formulation in terms of the data which distinguish solutions one from another rather than determine them at some point in the evolution (see Section 4.5).
3 See the translations in Barbour and Pfister (1995) of the earliest proposals [by Hofmann, Reissner and Schrödinger (!) shortly before he discovered wave mechanics]; Assis, another rediscoverer, gives interesting historical background.
4 It is interesting to note that C (4.16), which is about the most obvious choice that one could make to measure clustering in a scale-invariant manner, generates, when regarded as a potential, forces that only change the shape of the system, not its size.

References

Alexander, H.G. editor, (1956). *The Leibniz–Clarke Correspondence*. Manchester University Press.

Barbour, J. (1974). Relative-distance Machian theories. *Nature*, 249: 238.

Barbour, J. (1994). The timelessness of quantum gravity: I. The evidence from the classical theory. *Classical and Quantum Gravity*, 11: 2853.

Barbour, J. (1999). *The End of Time*. New York: Oxford University Press.

Barbour, J. (2003). Scale-invariant gravity: Particle dynamics. *Class. Quant. Gravity*, 20: 1543.

Barbour, J. (2010). The definition of Mach's principle. *Foundations of Physics*, 40: 1263.

Barbour, J. (2009). *The nature of time*. Available at: arXiv:0903.3489 [gr-qc].

Barbour, J. and Bertotti, B. (1977). Gravity and inertia in a Machian framework. *Nuovo Cimento*, 38B: 1.

Barbour, J. and Bertotti, B. (1982). Mach's principle and the structure of dynamical theories. *Proceedings of Royal Society (London) A*, 382: 295.

Barbour, J. Koslowski, T. and Mercati, F. (2013). *A gravitational origin of the arrows of time*. Available at: arXiv:1310.5167.

Barbour, J. Koslowski, T. and Mercati, F. (2014). Identification of a gravitational arrow of time. *Physical Review Letters*, 113: 181101.

Barbour, J. Koslowski, T. and Mercati, F. (2015). *Entropy and the typicality of universes*. Available at: arXiv: 1507.06498.

Barbour, J. and Pfister, H. editors, (1995). *Mach's Principle. From Newton's Bucket to Quantum Gravity*. Boston, MA: Birkhäuser.

Boltzmann, L. (1898). *Vorlesungen über Gastheorie, Zweiter Teil*, J A Barth, Leipzig. English translation (by Stephen G Brush): *Lectures on Gas Theory*, University of California Press (1964) and Dover, New York (1995).

Boltzmann, L. (1897). Zu Hrn. Zermelos Abhandlung "Über die mechanische Eklärung irreversibler Vorgänge. *Annalen der Physik*, 296: 392.

Koslowski, T., Marcati, F. and Sloan, D. (2016). *Relationism evolves the universe through the big bang*. Available at: arXiv:1607.02460.

Lagrange, J.L. (1772). Essai sur le Probleme des Trois Corps. *Oeuvres de Lagrange*, 6: 229.

Lanczos, C. (1949). *The Variational Principles of Mechanics*. Toronto: Toronto University Press.

Mach, E. (1883). *Die Mechanik in Ihrer Entwicklung Historisch-Kritisch Dargestellt*. English translation of 1912 edition: *The Science of Mechanics* (1960), LaSalle: Open Court.

Poincaré, H. (1890). Sur le probleme des trois corps et les quations de la dynamique. *Acta Mathematica*, 13: A3.

Saari, D. (2005). *Collisions, Rings, and other Newtonian N-Body Problems*. Providence: American Mathematical Society.

Further Reading from the Editors

For classic historical texts on relationism, see H G Alexander, ed., *The Leibniz-Clarke Correspondence* (Manchester University Press, 1956) and E. Mach, *The Science of Mechanics* (Open Court, LasSalle, 1960). Barbour and Bertotti develop their relational mechanics in "Gravity and inertia in a Machian framework" (*Nuovo Cimento* 38B, 1, 1977) and "Mach's principle and the structure of dynamical theories" (*Proc. R. Soc. A* 382, 295 1982). Barbour's "Relational concepts of space and time." (*British Journal for the Philosophy of Science* 33, 251–274, 1982) is also a classic paper. For a comprehensive view on relationist theories in classical mechanics and beyond, see Oliver Pooley "Substantivalist and relationist approaches to spacetime," in Robert Batterman, ed. *The Oxford Handbook of Philosophy of Physics* (Oxford University Press, 2013).

PART II

Special Relativity

Introduction to Part II

Newtonian physics may appear interpretationally unproblematic (although further inspection reveals that this is an illusion). Quantum mechanics wears its philosophical problems on its sleeve. Special relativity sits somewhere in between. The theory itself is beautifully simple: it can be derived from two empirical principles, and its basic logic and mathematics are explicable to a motivated high-school student. And yet it appears to upend deeply held intuitions about the nature of space and time. Moreover, because of its empirical underpinning, the results of special relativity are some of the most secure in all of physics – it's almost universally assumed that future theories must reproduce the symmetries of special relativity, at least at the empirical level. These features mean that it provides fertile ground for philosophical investigation, concerning both the nature of an axiomatic "principle" theory and the ways in which theories constrain our metaphysics.

Einstein derives special relativity from two empirically confirmed principles – the light postulate and the relativity principle. From these, he derives the Lorentz transformations, coordinate transformations which imply that moving rods contract and moving clocks dilate. But this methodology brings with it its own questions. In a 1919 letter to the *Times*, Einstein made the following much-analyzed comment about relativity:

> There are several kinds of theory in physics. Most of them are constructive. These attempt to build a picture of complex phenomena out of some relatively simple proposition. The kinetic theory of gases, for instance, attempts to refer to molecular movement the mechanical, thermal, and diffusional properties of gases. When we say that we understand a group of natural phenomena, we mean that we have found a constructive theory which embraces them.
>
> But in addition to this most weighty group of theories, there is another group consisting of what I call theories of principle. These employ the analytic, not the synthetic method. Their starting point and foundation are not hypothetical constituents, but empirically observed general properties of phenomena, principles from which mathematical formula are deduced of such a kind that they apply to every case which presents itself. Thermodynamics, for instance, starting from the fact that perpetual motion never occurs in ordinary experience, attempts to deduce from this, by analytic processes, a theory which will apply in every case. The merit of constructive theories is their comprehensiveness, adaptability, and clarity; that of the theories of principle, their logical perfection, and the security of their foundation. The theory of relativity is a theory of principle.

As Einstein points out, special relativity belongs to a class of theories that are deduced from empirical principles, in this case, the light postulate and the relativity principle. As such, the results of the theory are as secure as the much-evidenced principles from which they derive. But principles alone tell us little about the ontology of the theory – what is it about the world that makes special relativity true?

In particular, why do the objects we see around us universally obey the edicts of special relativity? Why do moving rods contract and moving clocks dilate? The cogency of Einstein's definition and division of principle and constructive theories has been much questioned but, at least for scientific realists, the central interpretational question remains.

The first two chapters of this section give contrasting answers to the above question. Tim Maudlin presents the orthodox answer – one on which Minkowski's 1908 geometrization of special relativity reveals the theory's true interpretation. According to this view, special relativity posits a spacetime with a particular geometry; realism about this spacetime is then justified by the power of spacetime geometry to explain the kinds of regularities expressed in Einstein's empirical principles. By contrast, Brown and Read advance a more recent view of special relativity, one on which rods, clocks, and other bodies conform to the rules of special relativity not because they are embedded in Minkowski spacetime, but because they are governed by Lorentz covariant dynamics (that is, their laws retain the same form when transformed according to the Lorentz transformations). The proponent of the geometrical effect will argue that the Lorentz covariance of the laws itself depends on the Minkowski geometry of spacetime. The advocate of the dynamical approach questions exactly how the structure of spacetime is able to explain the form of the laws.

Regardless of the ontology one takes to underpin relativity, its most crucial conceptual feature is the relativity of simultaneity. In Newtonian physics, and in common sense, there is an absolute matter of fact about what is happening at the same time as what – no matter what state of motion you and I are in, we should agree about which events in Australia, or on Alpha Centauri, happen at the same time that I click my fingers. One can think of Newtonian spacetime as made of a stack of distinct three-dimensional spaces, each representing an instant of time. Special relativity tells us that no experiment can pick out this absolute standard of simultaneity. Instead, a standard of simultaneity is only picked out relative to a reference frame – a system in a state of motion. Relativistic spacetime cannot be sliced into preferred or absolute three-dimensional "spaces at a time." This has obvious consequences for any metaphysics of time that depends on such a foliation. As a result, many physicists and philosophers alike espouse a "block universe" metaphysics, on which one thinks of spacetime as a single four-dimensional block in which all times are equally real. But does special relativity force us to this conclusion and how might the defender of an alternative temporal metaphysics, for example one with a notion of temporal passage, defend their view? Antony Eagle's chapter explores some options, with some morals for the ways in which physics might or might not constrain our metaphysics.

Relativity constrains the dynamics of our physical theories. To take special relativity seriously is to accept that fundamental physics is Lorentz covariant (or, at the very least, that the physics that operates at the phenomenological level is Lorentz covariant). And yet ordinary quantum mechanics, despite all its success, is not Lorentz covariant and thus is incompatible with special relativity. Wayne Myrvold's chapter explores this clash, including arguments that any properly relativistic quantum theory must be a quantum field theory. This chapter also examines the ways in which Bell's theorem constrains the possibility for a relativistically acceptable interpretation of quantum mechanics. The reader unfamiliar with these issues may wish to read parts of Chapter 4, particularly Eddy Keming Chen's chapter, before coming back to the final chapter of this section.

5

RELATIVITY AND SPACE-TIME GEOMETRY

Tim Maudlin

Space and time are theoretical entities. We cannot directly observe them: we observe the behavior of material entities and postulate space-time structure as part of the physical explanation of that behavior. The space-time structure appears in the framing of the fundamental laws, and the nature of that structure is, in a broad sense of the term, geometrical.

As an example, Newton's first law of motion reads: "Every body perseveres in its state either of rest or of uniform motion in a straight line, except insofar as it is compelled to change its state by impressed forces." In order for this law to have any content, there must be a distinction between states of rest, states of uniform motion in a straight line, and other states of motion. Newton defended an account of space and time that would support the first law in this way.

Newton postulated the existence of absolute space and absolute time. The absolute space was Euclidean, which allowed him to use Euclidean constructions and proofs in the *Principia*. Absolute space persists through time: one and the same point of absolute space exists at all times. This feature allows for the definition of absolute rest. A body is at absolute rest just in case it remains at the same point of absolute space through some period of time.

The persistence of points of absolute space also allows one to define the spatial trajectory of a body. Since the body is at some point of absolute space at each moment of time, there will be a collection of these points that constitute the path through absolute space that the body takes. That trajectory could be a straight line in the space, accounting for half of the characterization "uniform motion in a straight line." What remains is to characterize "uniform," which requires bringing in the structure of absolute time.

The geometry of absolute time is simpler than that of absolute space because time is one-dimensional, but there is non-trivial geometry nonetheless. Absolute time is more than just a sequential ordering of moments, but it also has a metrical structure so one can compare quantities of time that elapse between events. A uniform motion covers equal distances (in absolute space) in equal times. Therefore without both a distance metric on space and a temporal metric on time, Newton's first law makes no sense.

Absolute space and its persistent points are not observable entities; hence, neither is absolute motion. Nonetheless, there are observable consequences of the motions of Newtonian objects. The relative motions of objects are observable, and these are the differences of their absolute motions. More importantly, the absolute acceleration of an object requires the existence of a force, which can sometimes be observable.

The iconic example is Newton's bucket experiment: a bucket of water is hung from the ceiling by a cord, tightly twisted, and then let go. The rotation of the bucket reveals itself via the surface of the water. The concavity of the rotating water produces forces toward the center of the bucket, which are required to keep the water rotating. Similarly, tension in the cord connecting two massive bodies can betray the rotation of the system in absolute space and time even though space and time remain invisible.

Newton's theory uses the persistence of points of absolute space through time to associate features of motions with geometrical properties of lines in absolute space. We have already seen this in the characterization of uniform straight-line motion required for the first law. Similarly, the Second Law adverts to the change of state of motion of a body, a quantity well defined in absolute space and time since the instantaneous states of motion are well defined.

Newton postulated this geometrical structure to space and time in order to formulate his laws of physics. It is a consequence of those laws that some of the geometrical features – in particular the precise trajectories through absolute space – are not observable. But that alone does not constitute an objection to Newton's theory. It may motivate the search for a replacement theory, but such a replacement has to account for the same observable phenomena.

Newton thought in terms of space and time separately, but not in terms of space-time. The unification of space and time into a single entity is the first step toward a replacement of Newton's account, and we will approach it in steps. The first step is the construction of a four-dimensional representation of Newton's theory. Since absolute space persists through time we imagine making a copy of absolute space for each moment of absolute time, and then stacking all of these copies above one another in a four-dimensional Euclidean space. The copies are stacked up so that each point of absolute space is represented by a vertical straight line. Figure 5.1 provides a picture with one of the spatial dimensions suppressed.

Each leaf in Figure 5.1 represents an instantaneous moment in the history of absolute space. Particle A, which remains at the same point in absolute space, is at absolute rest. Particle B moves along a straight line in absolute space at a constant rate; C and D are particles in accelerated motion, orbiting each other. Both A and B instantiate the first law while C and D must be under a force directed at the midpoint between them.

So far all we have done is create a pictorial representation of Newton's theory. But the picture immediately suggests a way to formulate an alternative theory with less geometrical structure. Newton's first law makes no real distinction between A and B – they both obey the law of inertia – and in the picture each is represented by a straight line. The reidentification of individual spatial points at different times is not playing any role in the first law reconceived in this way: all the law says is that the trajectory through space-time of an object subjected to no forces is a straight line. The distinction between vertical and inclined straight lines, which corresponds to the distinction between absolute rest and uniform motion, is no longer required.

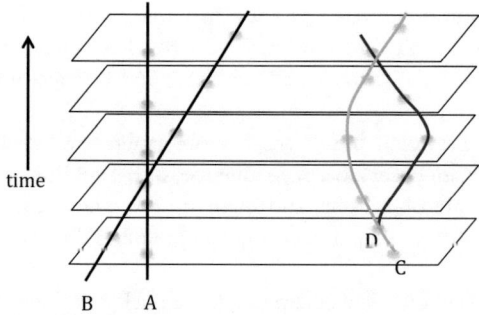

Figure 5.1 A four-dimensional representation of Newtonian space-time.

The essential conceptual change demanded for our new theory is the notion of space-time – a single entity made up of point events. Each event occupies a position in both space and time. Although it can be tempting to model point events as very fast physical happenings – a finger snap, for example – there is no assumption that any interesting physical occurrence happen at an event. Even a vacuum is full of events.

The entirety of events forms space-time, a four-dimensional manifold that contains all of the past and all of the future. Whereas Newton postulated a geometrical structure of space and a separate structure of time, our new theory attributes a geometrical structure to space-time. Part of that structure is Newtonian absolute time. Every event takes place at some absolute time, and each event falls into a class of events that all take place at the same time. Therefore, our new space-time divides into leaves of simultaneous events, just as in Figure 5.1. Each of these leaves, or simultaneity slices, has a spatial structure, which is again three-dimensional Euclidean space. Most importantly, there is a properly spatiotemporal structure that determines the collection of straight trajectories through space-time. The result is called *Galilean* or *Neo-Newtonian* space-time.

In Galilean space-time there is no notion of absolute rest or absolute motion. There is no fact about how fast anything is moving or in what direction. There are, of course, still the relative motions of bodies: they can get closer together or move farther apart as time goes on. But Newton's bucket experiment demonstrates that there is an objective, observable difference between a rotating and a non-rotating system, and that difference does not have anything to do with the relative motions of bodies. It is here that the space-time geometry comes in: the difference between accelerated and unaccelerated bodies is understood as the difference between bodies on curved space-time trajectories and bodies on straight trajectories. Particles C and D in Figure 5.1 must be subject to a force because their paths through space-time are curved.

We have seen how Newton appealed to the structure of space and time that he postulated when he formulated his first law. The first law can be reformulated for Galilean space-time as follows: "Every body perseveres in a straight trajectory through space-time, except insofar as it is compelled to change its state by impressed forces." The Second Law would then specify how an impressed force causes the trajectory to curve. The empirical content of Newton's Laws is recovered without the assumption of absolute space persisting through time. The key to constructing this alternative is regarding space and time as aspects of a unified space-time, and attributing exactly as much geometrical structure to this space-time as is needed to formulate the fundamental physical laws.

As we began by noting, space and time (or space-time) are theoretical rather than directly observable entities. The observable consequences of Newton's theory – relative motions of bodies and observable forces such as tensions – are identical to the observable consequences of Newton's Laws adapted to Galilean space-time. Given a choice between the two, one would have to appeal to some super-empirical criterion such as Occam's razor. Newton's theory postulates strictly more geometrical structure than the alternative. In a sense the Galilean space-time is built to mesh with Newton's Laws, although one has to reformulate the laws to see that.

The shift from absolute space and time to Galilean space-time presages the more radical shift from Galilean space-time to Minkowski space-time. The conceptual situation, however, is quite similar. Certain phenomena that cannot be accounted for in the Galilean structure naturally fit the Minkowski geometry.

The most revelatory phenomenon is the behavior of light in a vacuum. The trajectory of a light ray in a vacuum is independent of the state of motion of the emitter, which means that each point p in space-time has associated with it a locus of points that could be reached by light emitted at p, and a locus of points from which emitted light could reach p. The former set of points is called the *future light cone* of p and the latter the *past light cone*.

The question is what determines the past and future light cones. Since the light is traveling in a vacuum it does not seem possible to appeal to any matter to account for these special trajectories.

And if not matter, then what is left is the structure of space-time itself. Somehow space-time must have a geometry that determines a light cone structure. But there is no such structure in Galilean space-time.

Note that we are not starting with the assertion that the speed of light is constant. Although that proposition is often taken as an axiom of Relativity, it requires more complex considerations to explicate what is meant by the "speed" of anything. After all, Newton's account of space and time implied that objects have absolute speeds, but it was considered progress to change to Galilean space-time with no speeds. The situation in Relativity is subtle: massive objects have no objective speed in the theory just as in Galilean space-time but in a sense light has a fixed speed. We will come to speeds eventually, but for the moment we want to focus on this simpler behavior of light: all light emitted from a point in space-time (in a vacuum) travels on a fixed set of trajectories through space-time. In particular, two light rays emitted from the same point in the same direction will stay together: neither will outpace the other.

We can picture this physical fact in a diagram. We associate a light cone with each point in space-time (Figure 5.2). What we now are seeking is an account of the geometry of the space-time that yields this light cone structure as intrinsic.

Figure 5.2 requires some commentary. First, we are still interested in a four-dimensional space-time, but the picture is restricted to three dimensions. Just as Figure 5.1 eliminated one spatial dimension – turning three-dimensional simultaneity slices into flat planes – so this depicts a world with only two spatial dimensions. Now imagine setting off a flash bulb in a vacuum. Light spreads out in all directions forming an ever-expanding sphere. If the world were only two-dimensional, confined to plane, then the flash bulb would produce an expanding circle of light. It is this expanding circle that constitutes the future light cone. The past light cone shows all the ways that light could reach the point. (The light cone is just a collection of locations in space-time; there is no requirement that there actually be light on it.)

Note that no simultaneity slices are indicated in Figure 5.2. Nothing in the behavior of light requires the notion of simultaneous events. Therefore, Minkowski space-time has both more and less structure than Galilean space-time. The shift from the Galilean account to the relativistic account, unlike the shift from Newton's theory to the Galilean theory, is not a matter of weakening the geometrical structure. It is rather an entirely new structure.

Figure 5.2, being a picture drawn in Euclidean space, has some potentially misleading characteristics. For example, each of the cones in the diagram has a unique axis of symmetry, but in Minkowski space-time every straight line through event *p* that lies in the light cone is an axis of symmetry. The Euclidean angles in the diagram do not correspond to angles in Minkowski space-time. For example, each direction along the light cone is orthogonal to itself.

What other structure does Minkowski space-time have? We have seen how Galilean space-time affords the resources to state Newton's first law in terms of straight trajectories through space-time. Minkowski space-time does the same thing. Indeed, the structure of straight lines in Minkowski

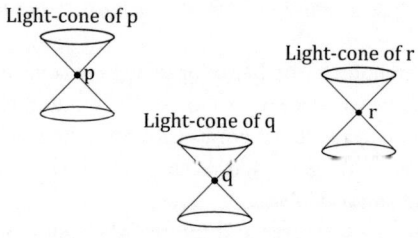

Figure 5.2 Light cone structure associated with points of space-time.

space-time can be read right off our diagrams: straight lines in the space-time are represented by straight lines in the diagram. The basic account of Newton's bucket experiment is identical: particles in the rotating bucket travel on curved trajectories through space-time and so require a force. The notion of acceleration gets translated into geometrical terms as the curvature of a trajectory.

What sort of geometry has both a light cone structure and a straight-line structure built into it? So far we have avoided reference to coordinates or coordinate systems in favor of a direct description of geometrical structure, but in order to go further we will have to introduce coordinates. A coordinate system allows us to label points with numbers and then calculate quantities from those numbers. The calculation is easier in some coordinate systems than others. Let's warm up with coordinates on Euclidean space.

Euclidean space has an intrinsic geometrical structure that can be studied without the use of coordinates, as Euclid did. But many proofs are simpler when expressed those terms. The most familiar system is Cartesian coordinates: two or three coordinate axes all at right angles to one another. In the three-dimensional case each point is labeled by a triple of real numbers (x,y,z). The thing that is so convenient about this system is how easy it is to calculate the distance between any pair of points. Let point p have coordinates (X_p,Y_p,Z_p) and q (X_q,Y_q,Z_q). Then the distance from p to q is

$$D_{pq} = \sqrt{\left(X_p - X_q\right)^2 + \left(Y_p - Y_q\right)^2 + \left(Z_p - Z_q\right)^2}.$$

It completely specifies the geometry of three-dimensional Euclidean space to say that it admits of a coordinate system in which the distances can be expressed in this way. Such a space is called *flat*. Curved spaces, such as the surface of a sphere, cannot be covered by Cartesian coordinates.

Discussions of Relativity can invite misunderstanding by conflating objective geometrical facts about space-time with coordinate-dependent facts. To get a sense of the difference consider the above equation. The distance D_{pq} is intrinsic to the space independently of any coordinates. But the values of $\left(X_p - X_q\right)$, $\left(Y_p - Y_q\right)$, and $(Z_p - Z_q)$ depend on how the coordinate system happened to have been set up. If the coordinate axes had been chosen with different orientations, these three numbers would have been different. But the new numbers, plugged into the equation, would yield the same distance.

Just as Euclidean space is a flat space, so Minkowski space-time is flat. Moreover, just as the structure of Euclidean space can be specified by the existence of special coordinates, so too can Minkowski space-time. We will make the analogy between the two cases as strong as possible.

Minkowski space-time is four-dimensional, so to make a comparison to Euclidean space we need to start with a four-dimensional Euclidean space. To cover this space we now need four coordinates, and so we will add t as the fourth: (t,x,y,z). There exist coordinates on our four-dimensional Euclidean space such that

$$D_{pq} = \sqrt{\left(T_p - T_q\right)^2 + \left(X_p - X_q\right)^2 + \left(Y_p - Y_q\right)^2 + \left(Z_p - Z_q\right)^2}.$$

The analogy to the three-dimensional case is obvious.

Minkowski space-time is also four-dimensional, and so it has to be covered by four coordinates. It has a fundamental objective (coordinate free) quantity that is the analog to the distance in Euclidean space. This quantity is called the *relativistic interval* between any pair of points. There are different quantities that are given that name, but the one that makes our analogy to Euclidean geometry clearest is given as follows:[1]

$$I_{pq} = \sqrt{\left(T_p - T_q\right)^2 - \left(X_p - X_q\right)^2 - \left(Y_p - Y_q\right)^2 - \left(Z_p - Z_q\right)^2}.$$

Minkowski space-time is completely specified by the condition that it admits of coordinates in which the equation for the interval takes this form.

The interval is somewhat analogous to a distance but there are important differences. Perhaps the most important is that because of the change of three plus signs to minus signs, pairs of distinct point events can have an interval of zero between them. For example, consider the event p with coordinates $(0,0,0,0)$ (the origin of the system) and q with coordinates $(1,1,0,0)$. $(T_p - T_q) = (X_p - X_q) = -1$ and $(Y_p - Y_q) = (Z_p - Z_q) = 0$, so $I_{pq} = 0$. Similarly the interval from p to $(1,0,1,0)$ is 0, and to $(1,0,0,1)$, and to $(2,2,0,0)$, and to $(3,2,2,1)$. It is not difficult to show that all of the points with $I_{pq} = 0$ from p, when plotted on Cartesian coordinates in Euclidean space, form a double cone originating at p. Suppressing the z dimension and plotting the points in Euclidean space yields exactly Figure 5.2: a double cone gets associated with each point in space-time.

We can now see how the intrinsic geometry of Minkowski space-time can help account for the behavior of light: it must follow from the laws of electrodynamics that the trajectory of light in a vacuum is a path with interval 0 between every pair of points. Those laws, of course, must be couched in terms of this same space-time structure. But without even writing down the laws we have seen that Minkowski space-time contains exactly the right sort of geometry to account for the phenomena. The trajectory of light in a vacuum is not determined by matter but by the intrinsic geometry of space-time itself.

The light cone at each point effects a division of the space-time into five distinct regions: the region inside the future light cone, on the future light cone, inside the past light cone, on the past light cone, and outside the light cone. While light in a vacuum propagates along the light cone, massive particles propagate into the interior of the future light cone. If we think of the future light cone of some event p as an expanding sphere, all massive particles that pass through p must stay inside the sphere. In that sense we can say that the "speed of light" forms an unachievable limit for massive particles even though we have not yet defined the speed of anything.

Speeds in Minkowski space-time are not intrinsic to the geometry. The speed of an object can be defined relative to a coordinate system laid down on the space-time, but different systems will typically attribute different speeds to the same objects. Things are slightly better if we restrict our attention to the special coordinate systems mentioned above, in which the interval is calculated by the analog to the Pythagorean law. Just as the special coordinates on Euclidean space are called "Cartesian coordinates," the special coordinates on Minkowski space-time are called "Lorentz coordinates," and the equations that take you from one Lorentz system to another constitute the Lorentz transformation.

Given a Lorentz coordinate system, the speed of an object relative to those coordinates can be defined. If the object goes through event p and then event q, its average speed can be defined in terms of the coordinates as distance traveled divided by time. Let p's coordinates be (T_p, X_p, Y_p, Z_p) and q's be (T_q, X_q, Y_q, Z_q). Then the average speed is given as follows:

$$\sqrt{(X_p - X_q)^2 + (Y_p - Y_q)^2 + (Z_p - Z_q)^2} / (T_p - T_q).$$

The condition that $I_{pq} = 0$ for the trajectory of a light ray in a vacuum implies that the speed of light in a vacuum is always 1 in a Lorentz coordinate system. This purely mathematical observation about coordinate speeds in a special class of coordinate systems does not by itself explain the outcome of concrete laboratory operations described as measuring the speed of light. But if physical analysis of the operations shows that they are recording the coordinate speed in some Lorentz frame then we know what the outcome has to be.

No massive particle has the same coordinate speed in all Lorentz systems. For them the situation is similar to Galilean space-time where there are no absolute velocities. Unfortunately, the "constancy of the speed of light" encourages locutions that suggest Newtonian absolute velocities, such as talk of particles "travelling near the speed of light." No trajectory inside the light cone is any "nearer" or "farther" from the speed of light than any other. Our Euclidean diagrams of the light cones can be misleading here as well: some of the interior points in Euclidean space are closer to the surface of the cone than others. In Minkowski space-time that is not true.

So far the only use we have made of the interval is to define the light cones, but non–zero values of the interval connect directly to observable behavior as well. The value of the interval between an event and any event inside its light cone is positive (given our conventions). Integrating the interval along the trajectory of a massive particle yields a length of the trajectory. That length is called the *proper time* along the trajectory, and holds the key to understanding how temporal structure appears in Minkowski space-time. An ideal clock measures the proper time along its trajectory rather than measuring any absolute time. Clocks are generally complex structures, whose behavior requires extensive analysis, but there are simple systems that function somewhat like clocks. In particular, unstable particles have a characteristic half-life: the time on average that it takes for half of the sample to decay. Such unstable particles can be used to measure the interval along their paths.

Muons, for example, typically decay into an electron, an electron neutrino, and a muon antineu-trino. On average, muons last 2.2 microseconds. In the Newtonian or Galilean regime, this would indicate how much absolute time passes, on average, from the creation of the particle to its decay. In Minkowski space-time, there is no absolute time but each muon has a trajectory with a length. In Figure 5.3, several straight trajectories through *p* with interval length 2.2 are indicated.

Although the trajectories appear in the diagram to have different lengths, they are all the same in the Minkowski geometry. In every set of Lorentz coordinates, though, the muons with greater coordinate speed last for a longer coordinate time. One practical consequence is that muons in a cyclotron last longer than muons at rest in the lab: the lengths of their trajectories are the same, but the corkscrew trajectory of the muon in the cyclotron goes higher up the space-time diagram.

This effect is the basis of the "twin paradox." We can illustrate the situation quantitatively using a set of Lorentz coordinates. Suppose two twins start out at the event with coordinates (0,0,0,0). One twin travels inertially (on a straight trajectory) to the event (10,0,0,0). The other twin travels inertially to (5,4,0,0), then changes direction, and travels inertially to (10,0,0,0) (see Figure 5.4).

A simple calculation shows that the interval length of the trajectory from (0,0,0,0) to (10,0,0,0) is ten years, and of each of the other two trajectories is three years. Therefore, the total elapsed time

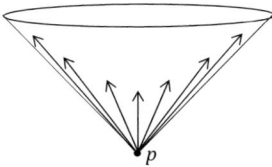

Figure 5.3 Trajectories through *p* with equal interval lengths.

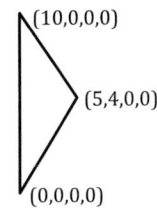

Figure 5.4 The "twin paradox" in Lorentz coordinates.

from the initial event to the final event is ten years along one path and six along the other: the twin who goes through (5,4,0,0) will be younger by four years when they meet again.

The physical analysis of the aging of a human body is a complicated business, and we only predict the effect by assuming that the body functions somewhat like an ideal clock. But the situation with our muons is simpler: the dynamical equations for the muon are couched in terms of the space-time geometry in such a way that the proper time along the muon path determines the probability of decay. This is hardly a surprise since the proper time is the most obvious geometrical feature that the path has.

What is it, then, to say that space-time geometry explains an empirical effect? The first step is to specify the geometry. In the case of Special Relativity, that means Minkowski space-time. Although it is often convenient to use coordinates for this purpose it is not strictly necessary, and in any case the geometry exists independently of any coordinates we may happen to use. The geometry could include simultaneity slices, light cones, straight trajectories through space-time, and metrical notions such as amount of elapsed time or the relativistic interval. The laws of the theory – the dynamics of matter – are specified in terms of this geometry. What makes a theory "relativistic," for example, is just that the dynamics require nothing but the interval for its formulation.

This same basic scheme applies in General Relativity. The geometry there depends on the matter and energy distribution via the Einstein field equation, and there are in general no coordinate systems in which the interval gets expressed in a simple uniform way in terms of the coordinates. But the resulting geometry still has a light cone structure, straight and curved trajectories, and a proper time length defined for trajectories of massive particles. Newton's bucket experiment is explained in the same way: the trajectories of particles in the rotating bucket are curved (accelerating), requiring a force, while the trajectories in a non-rotating bucket are straight. This notion of acceleration is defined by reference to the space-time structure itself, not by reference to motion with respect to other bodies that may exist.

The space-times of General Relativity are generically curved rather than flat. Gravitational phenomena are explained by the curvature, including extreme conditions such as black holes. Unlike the fixed and uniform structure of Minkowski space-time, General relativistic space-times have variable curvature and can differ geometrically in different locations. A gravitational wave, for example, is a propagating disturbance of the space-time structure. The recent detection of gravitational waves constitutes yet another verification of the predictions of General Relativity.

Space-time structure in General Relativity is connected to observable behavior in the same way as in Special Relativity. As a first pass one can say that objects subjected to no forces travel on straight trajectories; objects subjected to forces have trajectories bent in the direction of the force and proportional to it; light in a vacuum propagates along a light cone; clocks measure the proper time along their trajectories. Gravity itself is not a force but rather a manifestation of curvature in the geometry. Given the space-time structure one can use these principles to make empirical predictions such as the prediction for the twins. But these principles themselves can also be derived using dynamics that is framed in terms of the relativistic metric.

The detailed predictions of General Relativity are more accurate than those of Newton's gravitational theory. These include predictions for planetary orbits (especially Mercury, whose motion could not be accounted for using Newtonian resources), the "bending of light" passing near the Sun (due to the curvature of space-time), gravitational time dilation effects (an atomic clock raised up in a gravitational field shows more elapsed time than a clock that remains lower), and gravitational waves. It also accounts for some phenomena more simply than Newton does. Two bodies let to fall in a vacuum fall at the same rate no matter how massive they are. For Newton, this is because the force of gravity on an object is proportional to its mass: the heavier body is harder to accelerate, but gravity provides just enough greater force to it. In General Relativity the account is simpler: both bodies are traveling on parallel straight lines through space-time. Their masses don't come into the explanation.

Our account of Special and General Relativity has been in terms of the objective geometrical structure of space-time and how that structure figures into physical explanations via dynamical law. We have not mentioned anything being "relative," such as the "relativity of simultaneity." That is because simultaneity – two events "happening at the same time" – is not relative in these theories, it is simply non-existent. The simultaneity slices of Newtonian and Galilean space-time do not exist. Given a coordinate system one can look at events that are assigned the same t-coordinate, and call them "simultaneous in the coordinate system," but this notion has no intrinsic physical interest. Coordinate-dependent facts are not objective physical facts.

Nor does either Special or General Relativity endorse the claim that all motion is the relative motion of bodies. As Newton's bucket argument shows, the acceleration of a body can have observable effects even when there are no relative motions at all. Newton claimed that one could verify the rotation of a system of two masses connected by a cord by the tension in the cord even if the rotation occurred in a vacuum, with nothing to compare the masses to. This claim is validated by both theories of Relativity, where the acceleration is defined in terms of the curvature of the masses' trajectories in space-time. That curvature, in turn, is defined with respect to the space-time geometry.

Space-time is an objective physical structure in Relativity. In General Relativity the structure is influenced by the distribution of matter and energy in it, whereas in Special Relativity it is not. The structure is not directly observable, but like all such theoretical postulates it earns its keep by playing a role in the explanation of observable phenomena. Space-time provides the wherewithal to specify equations of motion for matter and fields. In the General Theory space-time itself is governed by such an equation. The relevant structure is described by a metric on the space-time that specifies the interval between points.

It is exactly because space-time is not directly observable that one should be cautious of assertions about its structure. But physics at present cannot do without a space-time structure, and the features appealed to in stating the relativistic laws governing matter are geometrical.

Note

1 The Interval is most commonly defined as the square of what is written above, and sometimes the square of the difference of the t-coordinate is subtracted from the sum of the squares of the other differences.

Further Reading from the Editors

Einstein's original presentation of special relativity, "On the electrodynamics of moving bodies" (*Annalen der Physik* 17, no. 891 (1905): 50), is a surprisingly accessible classic paper. For more on the kind of geometrical approach to relativity presented here, see Tim Maudlin's *Philosophy of Physics: Space and Time* (Princeton University Press, 2012). A contrasting view is presented in Brown, Harvey R. *Physical Relativity: Space-Time Structure from a Dynamical Perspective* (Oxford University Press, 2005). Wallace, David. "Who's afraid of coordinate systems? An essay on representation of spacetime structure" (Studies in history and philosophy of science part B: Studies in history and philosophy of modern physics, 2017) defends the use of coordinate-based approaches to physics. Friedman, Michael. *Foundations of Space-Time Theories: Relativistic Physics and Philosophy of Science* (Princeton university press, 1983) is a classic text for those who are comfortable with more technical detail.

6

THE DYNAMICAL APPROACH TO SPACETIME THEORIES

Harvey R. Brown and James Read

6.1 Introduction

In 1940, Einstein offered the following nutshell account of his special theory of relativity (SR):

> The content of the restricted relativity theory can ... be summarised in one sentence: all natural laws must be so conditioned that they are covariant with respect to Lorentz transformations. (Einstein, 1954, p. 329)

Einstein's innocuous-sounding statement actually represents on his part a significant departure from his 1905 "principle theory" approach to SR, based upon the relativity principle, the light postulate, and the isotropy of space. The shift in his thinking did not come about overnight. He had been disconcerted for many years by both the limitations of the thermodynamic template he had used in 1905 with its appeal to phenomenological principles, and the inordinate emphasis he had placed on the role of light – and therefore electromagnetism – in his theory. (Einstein also had a bad conscience about the "sin" of treating ideal rods and clocks as primitive entities in his 1905 formulation.[1])

The so-called "dynamical" approach to spacetime theories can be considered a development of Einstein's 1940 characterization of SR, and an attempt to extend it both to general relativity (GR) and to a broader context of spacetime theories. We defer the details until later in this chapter, but note in the meantime a feature of the 1940 statement common to the 1905 treatment: the absence of reference to the "geometry of spacetime," and emphasis on the (nongravitational) laws of nature.

The seeds of this dynamical approach were sown in the late 19th century by the great ether theorists George Frances FitzGerald, Hendrik Antoon Lorentz, Joseph Larmor, and Henri Poincaré. (It can never be overstated that the approach nonetheless does *not* postulate a preferred inertial frame!) The earliest post-1905 articulations can be found in the writings of such luminaries as Hermann Weyl, Arthur Eddington, and Wolfgang Pauli, and it is not unreasonable to put Einstein himself into the mix. Later allies include William F. G. Swann, Lajos Jánossy, John S. Bell, and Dennis Dieks.[2] Notable recent articulations within the physics literature are due to D. J. Miller and William N. Nelson.[3] In the last few decades, the approach has come to life within the philosophy of physics as a reaction to aspects of the influential "angle bracket school" of spacetime theories, first prominently exposed in the philosophical literature of the 1970s and especially the 1980s.[4] (Note, however, that debates over the primacy of coordinate-independent approaches in the foundations of spacetime theories should in general be divorced conceptually from debates regarding the dynamical approach to spacetime.[5]) The central role of geometry in the treatment of pre-general relativistic theories in this school has led some philosophers to the view that special relativistic effects such as length contraction and time dilation are ultimately explained by recourse to the geometric structure of Minkowski spacetime, and that all such explanations prior to the 1908 work of Minkowski are either misguided or incomplete. Such a view still has weighty defenders, as the chapter by Maudlin in this volume attests. Defenders of the dynamical approach take issue with this view.

A review of this size cannot do justice to recent criticisms of the dynamical approach, but at least its conceptual grounding can be laid out. To this end, we begin with Maxwell's 1865 theory of electrodynamics: the first special relativistic theory in the history of physics.

6.2 Maxwell's Electrodynamics

In saying that Maxwell's electrodynamics is (special) relativistic, we are referring to the standard field equations, possibly with source terms, but not in combination with the original (non-relativistic) Lorentz force law. (Strictly speaking, we are referring here to the formulation of the theory in terms of the 3-vector formalism later developed by Heaviside and Hertz. It is only with the 4-tensor formulation of Minkowski that the theory becomes manifestly Poincaré invariant.[6]) Despite this theory's unique role as the historical cradle of SR, the latter theory transcends it, as Einstein would insist in his mature years. SR amounts to a specific constraint on *all* the nongravitational interactions, not just the electromagnetic, and it transcends the exact form of the equations of these interactions when written in inertial coordinate systems. *What SR demands is simply that the equations governing all these interactions be Poincaré invariant.* It is satisfaction of this condition that makes Maxwell's field equations special relativistic.[7]

Looking forward to GR, Maxwell's name reappears in the context of the Einstein-Maxwell field equations. It is often thought that the fragment of these equations related to the electromagnetic field is "locally" equivalent to the original 19th-century theory. Although this is not entirely true for several reasons, this (more fundamental) general relativistic fragment is still locally Poincaré invariant.[8] We return in Section 6.4 to discuss further such local dynamical equations. In the meantime, we wish to counter what we see as two common confusions concerning the Maxwell theory as understood prior to Einstein's revolution, as a warm-up to defending the dynamical approach to spacetime theories.

(i) The first is that the theory is incomplete. The theory, the argument goes, does not specify explicitly the spacetime structure required for the very definition of the technical terms it employs. As John Earman affirmed in 1989, "...laws of motion cannot be written on thin air alone but require the support of various space-time structures" (Earman, 1989, p. 46). At first sight, this claim should strike one as puzzling. What physical effect do Maxwell's equations, in say 3-vector form, fail to predict, or leave ambiguous, that more explicit versions might not? None, if one bears in mind that in the 19th-century inertial frames were taken to be global. (More on this later.) Recall John S. Bell's perceptive remark, in his 1976 essay *How to Teach Special Relativity*, that "the laws of physics in any one reference frame account for all physical phenomena, including the observations of moving observers" (Bell, 2004, p. 77).

The incompleteness claim is tempting to those brought up in the aforementioned "angle bracket school," according to which a spacetime theory first specifies a triple structure $\langle M, A, \Phi \rangle$, where M is the spacetime manifold, A is a placeholder for geometric-object fields characterizing fixed spacetime structure (if any),[9] and Φ represents dynamical geometric-object fields. The theory then has the job of specifying which of the kinematically possible models (KPMs) built out of this structure are dynamically allowed (i.e., are dynamically possible models (DPMs)), the elements of A being unchanged in each kinematically (*a fortiori* dynamically) allowed model of the theory. Certainly, this geometrical approach has a number of merits, not the least of which is allowing for successive theories from Galileo to Newton to Maxwell to Einstein (SR and GR) to be formulated in a common formal language.[10] In the case of GR, there are of course no fixed fields A in the usual sense, and Φ stands for the metric field together with the energy-momentum tensor(s) associated with matter fields, if any.

This unified language typically involves dynamical equations that are generally covariant – i.e., which are form-invariant under arbitrary differentiable coordinate transformations. The traditional Maxwell equations – even in the form provided by Minkowski – are not form-invariant in this sense. Is this a defect? Here opinions are bound to differ. We see nothing intrinsically superior about general covariance; the key issue is which coordinate systems most simplify the form of the equations of the relevant dynamical theory. In the language of Friedman (1983, p. 60), in which coordinates is the "standard formulation" of the theory obtained? In the case of Maxwell theory, it is the inertial coordinates in which the Poincaré-Einstein synchrony convention is adopted, related by Poincaré transformations.

The argument for general covariance which rests on the claim that coordinate systems are mere labels of spacetime events, and that theories should be expressed in a label-general way, strikes us as naïve. How is it in SR, as understood by Einstein in 1905, that the Lorentz transformations – supposedly mere changes of labels – can encode the physical phenomena of length contraction and time dilation?

(ii) The second and related confusion about Maxwell's electrodynamics is that, prior to Einstein, the theory must be thought of as being defined in pre-relativistic spacetime. By this, we mean that at the very least, inertial coordinate transformations involving boosts are Galilean, not Lorentzian. Let's return again to Bell's 1976 essay, and its dynamical analysis of a hydrogen-like atom. Bell showed that if one chooses an inertial frame involving the Poincaré-Einstein convention, so that Maxwell's equations take their familiar form therein, the predictions concerning the shape and period of an inertially *moving* atom obtained by solving the equations in this frame will be consistent with those obtained by Lorentz transforming to the rest frame of the atom and using the standard form of Maxwell's equations in that frame. Admittedly, Bell's analysis used, over and above the Maxwell field equations, relativistic dynamics of the kind that features in the relativistic version of the Lorentz force law. But Galilean transformations, if they have anything like the operational significance we attribute to them today, are not easy to reconcile with Maxwell's field equations.

Here is another way to make this point. Consider the simple light clock or Langevin clock: two separated mirrors mounted in parallel on, and perpendicular to, a rigid straight rod. A pulse of light traveling parallel to the rod bounces back and forth between the mirrors. When at rest in the frame relative to which Maxwell's equations are valid, the period of the clock is approximately $2L/c$ where L is the distance between the mirrors, assumed to be large compared to the pulse width, and c is the light speed as measured in that frame. Maxwell's equations predict that c is constant, i.e., independent of the speed of the source and isotropic. It follows that if the Langevin clock is now boosted into a new state of inertial motion with speed v relative to the original frame, and in the direction of the principal axis of the rod, the period becomes (as calculated by the observer in the original "resting" frame) $2\gamma^2 L/c$, where γ is the familiar Lorentz factor $\left(1 - v^2/c^2\right)^{-1/2}$. It is assumed that the length of the rod is not affected by the boost, again as determined by the resting observer. Assuming more generally that the Galilean transformations hold, and have the standard operational meaning, it also follows that there can be no time dilation associated with moving clocks. When applied to the Langevin clock, this constraint implies that the proper period of the moving Langevin clock does not coincide with that in its resting state.[11] But note that, as Einstein himself stressed,[12] it is a fundamental condition for the very possibility of kinematics, such as exemplified by the Galilean transformations, that ideal clocks are boostable, meaning that as long as the forces producing the boosts are weak enough, the proper periods of such clocks in their equilibrium states are invariant under the boosts. Therefore, the Langevin clock is a counterexample to this principle, and this conclusion should also shake our faith in the notion that Maxwell's equations can be combined with Galilean kinematics.

Such considerations are, however, entirely absent in attempts within the angle bracket school to formulate a generally covariant version of "classical" Maxwellian electrodynamics within a spacetime structure which purports to be consistent with the Galilean law of the transformation of velocities – attempts to provide a rigorous systematization of presumably what Maxwell himself thought was going on.[13] Of course, it is acknowledged that such theories are hard to reconcile with the null results of ether wind experiments, especially that of Michelson and Morley in 1887. The issue here is not whether this version of electrodynamics is false but whether it is coherent. What is lost sight of within the geometric intricacies of such attempts is the fact that physical rulers and clocks are complicated dynamical objects subject to the very nongravitational interactions that are in play. In other words, what is overlooked is the fact that the distinction between kinematics and dynamics is not fundamental.[14]

Even the great Maxwell failed to see this. We expect undergraduates to imbibe in their first course on relativity theory a profound insight largely obscure to *all* the 19th-century giants, including Maxwell, Lorentz, Larmor, and Poincaré: the physical meaning of inertial coordinate transformations. It was Einstein in 1905 who was the first to understand the physics of such transformations,[15] and the fact they are neither *a priori* nor conventional. This achievement in our view outstripped his treatment of length contraction, time dilation, and relativity of simultaneity – effects that were, after all, already in the air at the turn of the century.[16] Nor was this achievement based on a temporary state of philosophical derangement on Einstein's part associated with crude operationalism. No wonder he was unimpressed two decades later when Bohr and Heisenberg picked up what they thought was the Einstein banner and applied crude operationalism in their disturbance theory of indeterminacy in quantum theory.

6.3   Explanation

One of the most significant claims that the dynamical approach calls into question has to do with the nature of the explanation of the so-called kinematical relativistic effects of length contraction and time dilation. No one, to our knowledge, questions that these effects are a consequence of the Poincaré invariance of the laws governing the nongravitational interactions that are responsible for the forces of cohesion within "rigid" rulers and "ideal" clocks. The bone of contention is whether the buck stops there.

6.3.1   *Symmetries and the Dynamical Approach*

Consider again a spacetime theory specifying a triple $\langle M, A, \Phi \rangle$, with fixed spacetime structure, and hence containing geometric-object fields associated with the entry A. Following Earman (1989, p. 45), a *spacetime symmetry* is defined as a mapping from M onto itself that preserves (leaves invariant) each element of A. In contrast, a *dynamical symmetry* is defined as a mapping from M onto itself such that if $\langle M, A, \Phi \rangle$ is a dynamically allowed model of the theory, then so is the model obtained by dragging along the elements of Φ induced by the mapping. (Earman defines both spacetime and dynamical symmetries in terms of diffeomorphisms; this restriction excludes discrete symmetries associated with time reversal and space reflection (parity). In our view, it is not clear on what grounds symmetries involving discrete rather than continuous (Lie) groups are so excluded. See Section 6.3.2.)

In the case of Minkowski spacetime, the spacetime symmetries are the isometries of the Minkowski metric; the dynamical symmetries are related to the Poincaré group of coordinate transformations which preserve the form of the equations of motion when written in inertial coordinates. Note that if the notion of spacetime symmetries is cashed out in terms of the transformation properties of fixed fields A, then the above analysis cannot be applied to e.g., GR, or the alternative formulation of SR presented at Pooley (2017, p. 120), for such theories feature no fixed fields. A change of nomenclature broadly suffices to solve this difficulty – see Pooley (2013, §3.1), according to which "spacetime symmetry groups" are "groups of transformations that *preserve spatiotemporal structure*," and a "dynamical symmetry group" "is a group that *preserves the form of the equations* that express the dynamical laws." (Note, however, that this particular choice of nomenclature presupposes that the geometrical objects associated with "spacetime" may be picked out from the triple structure associated with a given theory; such an assumption is questionable on the dynamical approach – see Brown (2005, ch. 9).)

In his famous discussion in 1989, Earman claimed that every spacetime symmetry is a dynamical symmetry, and *vice versa* (Earman, 1989, ch. 3, §4). He stressed that this double requirement is not a matter of definition, but represents "adequacy conditions" for the theory in question to be "well-tuned." Now it is well known that Newtonian mechanics set in "Newtonian" spacetime has dynamical symmetries that are not spacetime symmetries. Earman argued that Newtonian space-time should be replaced by "neo-Newtonian" (sometimes: 'Galilean') spacetime, essentially on the grounds of Occam's razor: the former has absolute structure which is in an important sense idle. As for whether a spacetime symmetry may not be a dynamical symmetry, Earman found it hard to envisage such a possibility.

There is a sense in which the advocate of the dynamical approach can agree with this latter verdict of Earman. However, there is also a sense in which, for the advocate of the dynamical approach, the former scenario – *viz.*, cases in which dynamical symmetries outstrip spacetime symmetries – is equally difficult to envisage. The reason for this is that, as discussed elegantly by Myrvold in a recent paper (Myrvold, 2019), the advocate of the dynamical approach regards the coincidence of spacetime and dynamical symmetries not merely as an "adequacy condition" on a given theory, but rather indeed as holding analytically, "in virtue of considerations of meaning" (Myrvold, 2019, p. 7). By proceeding in this manner, any mystery regarding how spacetime structure is supposed to "explain" the dynamics of matter is dissolved.[17]

Myrvold tells us that spacetime structure *just is* that fixed structure A the spacetime symmetries of which coincide with the dynamical symmetries of the theory in question. Though we broadly concur with this presentation, it is possible that Myrvold's account might give the impression that the advocate of the dynamical approach is interested principally in conceptual analysis – spacetime symmetries *just are* dynamical symmetries; spacetime structure (ultimately reducible to symmetries of dynamical laws) *just is* that structure which manifests those symmetries.[18] It is important to note, however, that advocates of the dynamical approach may resist being thus tarred with the analytical brush. Rather, the principal aim of this approach is to account for the *chronogeometricity* of metric structure – that is, to answer the question of why that structure is surveyed by rods and clocks built out of matter fields; this is achieved by stating that the metric field in such theories is *reducible* to dynamical symmetries. There is a sense in which the term "spacetime" is redundant to this goal; we anticipate that some advocates of the dynamical approach would be happy to excise it. (When viewed in this manner, the *spacetime functionalism* of Knox can be understood as augmenting the dynamical approach, essentially with the functionalist criterion that "spacetime" should be identified with whatever structure in a given theory has chronogeometric significance.[19])

6.3.2 *The Geometrical Approach*

Having sketched some key aspects of the dynamical view, let us turn now to a diametrically opposite approach. It is sometimes argued that the symmetries of Minkowski spacetime *explain* the fact that the dynamical laws are Poincaré invariant. Such a view is implicit in Friedman's 1983 book.[20] Though not himself not an advocate of the angle bracket school generally, Michel Janssen wrote similarly in 2002:

In Minkowski spacetime, the spatio-temporal coordinates of different observers are related by Lorentz transformations rather than Galilean transformations. Any laws for systems in Minkowski spacetime must accordingly be Lorentz invariant. (Janssen, 2002, p. 499)

More recently, Tim Maudlin has likewise stated:

The Minkowski geometry takes exactly the same form described in [any] Lorentz coordinate system (by the symmetry of Minkowski spacetime), and the laws of physics take exactly the same coordinate-based form when stated in a coordinate-based language in any Lorentz coordinate system (because the laws can only advert to the Minkowski geometry, and it has the same coordinate-based description). (Maudlin, 2012, pp. 117–118)

In a similar vein, Norton has argued that inertial motion is a result of the geodesic structure of Minkowski spacetime:

The loose talk surrounding relativity theory could lead one to think that empty space is nothing at all. That idea is not sustainable in special relativity. In the theory, empty space is rich in structure. Part of that structure is the existence of special motions, the inertial motions.[21]

Readers may recall Bell's 1976 discussion of a string connecting two rockets undergoing identical accelerations. Maudlin argues that the length contraction responsible for the eventual breaking of the string can be traced to three circumstances:

The geometrical symmetries in Minkowski spacetime, the necessity of specifying dynamical laws in terms of the Minkowski structure, and the physical constitution that makes the thread a rigid body (Maudlin, 2012, p. 119)

It is somewhat ironic then that Bell was able (as Einstein would have been in 1905, but with different emphasis) to explain the string breaking effect without ever referring to Minkowski spacetime; indeed one of the aims of Bell's increasingly well-known paper was to warn against "premature philosophising about space and time" (Bell, 2004, p. 80) when teaching special relativity. Was he misguided?

Fixed geometric structures in physics are not limited to spacetime. Consider, for example, the Lobachevskian geometry of the velocity space in special relativity that is associated with the non-commutativity of (non-aligned) boosts and hence the Thomas precession effect, or the curved metric of the projective Hilbert space in quantum mechanics associated with the phenomenon of geometric phase.[22] These different geometric structures (discovered after the physical effects to which they are associated) are rather elegant mathematical representations of aspects of the physics of boosts and of Schrödinger evolution, respectively, but arguably do not constitute an *explanation* of these effects. Why should it be any different in the case of absolute spacetime geometry?

Then there is the question of how the explanation suggested by Janssen, Maudlin, and others is supposed to work. Does Minkowski spacetime somehow "act" on the dynamical geometric-object fields? Friedman made it clear in 1983 that Minkowski spacetime is in his view a real physical entity.[23] Is it then capable of action? It is certainly not capable of *reaction*, because it is a fixed structure. If it acts in some causal way, it is a curious business: no terms relating to the action appear in the dynamical laws when expressed in their standard configuration. (Hence, the possibility of a relativistic theory – Maxwell's electrodynamics – appearing decades before spacetime was a glint in either Poincaré's or, more importantly, Minkowski's eyes.) The use of vague expressions like the necessity of laws to "advert" to the Minkowski geometry does little to dispel the mystery. As we shall see, Janssen himself does not embrace any such causal claim, speaking instead of a "common origin inference" to Minkowski spacetime structure, understood as universal Lorentz invariance construed as a *kinematical* constraint.[24]

Consider also the standard model of particle physics (following Brown and Sypel (1995, p. 256)). In this model, the strong and electromagnetic forces are known to conserve parity (i.e., the associated laws are invariant under the discrete symmetry of spatial reflections); not so for the weak force, which violates parity and time-reversal invariance. In order to "explain" weak force dynamical laws, advocates of the geometrical approach presumably must introduce extra spacetime structure, over and above the Minkowski metric, in order to ensure that spacetime and dynamical symmetries in the sense of Earman align – at least, if such discrete symmetries are also subject to Earman's "adequacy conditions"(see above). The extra structure involved in the case of the weak interactions would be a continuous choice throughout the manifold M determining which half of the null cone represents the future, and an antisymmetric tensor field which similarly provides a continuous choice of "right handed" versus "left handed" orthonormal triads of spacelike vectors at each point in M.[25] But, having introduced this structure, the question arises: why does the strong force, for instance, not also "advert" to this spacetime structure? Some further nuancing of the notion of geometric explanation is surely needed here.[26]

Note that it is not just the weak interactions that seem to call for this augmented Minkowski spacetime. In local relativistic Quantum Field Theory (QFT), it is a theorem that for systems of identical particles confined to two spatial dimensions, if either spatial reflection in a line or time reversal is a symmetry of the theory in all its superselection sectors, then the quantum statistics are associated with the ordinary permutation group, i.e., either Bose-Einstein or Femi-Dirac. See Frohlich (2009, p. 56). However, remarkably, the possibility of "fractional" or "braid" statistics ranging between Bose-Einstein and Fermi-Dirac is known to be self-consistent in the case of two-dimensional confinement, in a way that it is not when electrons explore three dimensions (Leinaas and Myrheim, 1977). This phenomenon is not a mere theoretical nicety – it is apparently displayed in two-dimensional electron gases in a transversal external magnetic field exhibiting the fractional quantum Hall effect (see e.g. Prange and Girvin (1990)). Space reflections and time reversal are not symmetries of such electron gases (Frohlich, 2009, p. 56) – and so, the same concerns as articulated in the previous paragraph would seem to apply.

Another awkwardness in the geometric approach is exposed in the case of the de Broglie-Bohm hidden-variable quantum theory. In its relativistic treatment of quantum fields, the equation governing the quantum degrees of freedom is Poincaré invariant; however, the guidance equation governing the sub-quantum degrees of freedom is Galilean invariant, and in general involves instantaneous action at a distance. From the geometric perspective it seems that the theory invokes two distinct spacetime structures – Minkowski and Galilean[27] – though the latter is "hidden."[28] It is unclear how the geometric explanation of dynamical symmetries is supposed to work in this case.[29] To reject the theory on the grounds that its spacetime structure is ambiguous, or degenerate, would surely be too high-handed!

It is well known that Newton, unlike the post-1921 Einstein, attributed no powers of action to absolute space, not even to constrain the inertial motion of force-free bodies. It is noteworthy that in attempting to account for the universality of Poincaré invariance in terms of Minkowski spacetime, Janssen accepted in 2009 that this geometric structure has no existence independent of the matter fields subject to the dynamical laws of interest, so its action is not causal (Janssen, 2009, p. 28). Janssen gives a number of overlapping accounts as to what it is, one being a kind of nomic necessity inspired by Marc Lange's analysis of "meta-laws" in physics (Lange, 2007). Rather, however, than pursue this line of thinking, we wish to make two points. The first is that there is nothing in Minkowski's own writings suggesting that he was providing an explanation for the universal nature of the Poincaré group; on the contrary, it was the latter on which he based his geometrical insights.[30] Second, the dynamical approach outlined in Section 6.3.1 constitutes a straightforward means of resolving these mysteries.

6.3.3 *The Dynamical Approach, Reprise*

It should be clear that the dynamical approach to spacetime theories is to be set in contrast with the above geometrical account. One further aspect of the dynamical approach is now worth emphasizing: in the context of SR, Poincaré invariance of the dynamical laws is, on this view, to be understood as a *brute fact*: no "explanation" of such a result via appeal to spacetime structure is mandated; nor is it clear how such an account could proceed.[31] In a theory in which all dynamical equations for non-gravitational fields manifest the same symmetries – again, such as SR – spacetime structure may be viewed as a *codification* of such symmetries.

Having thus presented the dynamical approach, it is worth dispelling one possible source of confusion. While the geometrical approach is often endorsed by members of the angle bracket school (such as Friedman (1983, ch. IV)), and advocates of the dynamical approach often do not view as illegitimate the issuing of coordinate-dependent presentations of physical theories (cf. Section 6.2), the debate between coordinate-dependent versus -independent presentations should be considered orthogonal to the debate between the geometrical and dynamical approaches. This can be illustrated through two points: on the one hand, authors such as Janssen can be viewed as embracing (a certain version of) the geometrical approach alongside coordinate-dependent presentations; on the other, we shall see in Section 6.5 that there may be some advantages to viewing the dynamical account through the lens of coordinate-independent approaches to spacetime theories.

A criticism due to Norton (2008) is that the dynamical, or "constructive," approach tacitly assumes an already-existing spacetime endowed with topological properties. Pooley (2013) subsequently pointed out that the goal of the dynamical project was never to excise all pre-geometric structure; Menon (2019), however, has recently gone further, arguing that this excision can be achieved. See also Stevens (2015, 2020) in this connection.

6.4 General Relativity

In the context of theories with non-dynamical metric fields such as SR, one crucial aspect of the debate between the dynamical and geometrical perspectives may be put straightforwardly: for the advocate of the former, the metric field (in the case of SR, the Minkowski metric field) need not be regarded as an ontologically distinct and primitive entity; rather, it is a codification of certain brute facts about the dynamical laws governing matter fields (namely, facts about their symmetries). By contrast, for certain advocates of the latter, the metric field in such theories is an ontologically distinct and primitive entity, its presence explaining certain facts about the dynamical laws governing matter fields (namely, the fact that they all manifest certain symmetries).

This situation changes in important ways on moving to theories with dynamical metric fields, such as GR. Since in such cases the metric field comes equipped with its own associated field equations (in the context of GR, the Einstein field equations), this field is not straightforwardly *ontologically reducible* to the matter fields, as per the dynamical approach in the context of spacetime theories with non-dynamical metric fields. (To claim that the metric field is reducible to the matter fields in GR is to endorse a certain form of *relationism* about the metric field; there are profound difficulties with implementing this program in GR. An obvious illustration of this difficulty can be found in the existence of vacuum solutions in the theory.) Indeed, advocates of both the dynamical and geometrical perspectives in the context of theories with dynamical metric fields are in agreement on this point.

That said, there remain crucial differences between the two camps, in particular regarding the *chronogeometric significance* of the metric field. On the dynamical perspective, the dynamics of the metric field tell us that it is "just another field": "Nothing in the form of the equations *per se* indicates that [the metric field] is the metric of space-time, rather than a $(0, 2)$ symmetric tensor which is assumed to be non-singular" (Brown, 2005, p. 160). How, then, does the metric field acquire its chronogeometric significance in GR? For the proponent of the dynamical approach, the metric field "earns its spurs by way of the strong equivalence principle" (Brown, 2005, p. 151). That is to say, locally in GR – assuming that one's experimental apparatuses are sufficiently insensitive that curvature effects may be ignored – not only does the metric field take the form of the Minkowski metric field (i.e., is diagonalizable, and invariant under Poincaré transformations), but dynamical equations for matter fields take a Poincaré invariant form.[32] This, some advocates of the dynamical approach maintain, is an important condition for rods and clocks built out of matter fields to survey locally the metric field. Moreover, if an equivalent form of the strong equivalence principle is available in spacetime theories with dynamical metric fields other than GR, then the same analysis may go through.

By contrast, advocates of the geometrical approach in the context of spacetime theories with dynamical metric fields may maintain that the metric field has a *primitive* connection to spacetime geometry – and that in the local regime in which curvature effects may be ignored, the dynamical laws governing matter fields are locally *constrained* to be invariant with respect to the local symmetries of this field, in the same manner as for the geometrical picture in the context of theories with non-dynamical metric fields, such as SR.

6.4.1 Problem Cases for the Geometrical Approach

In this section, we present what we take to be two problem cases for the geometrical approach in the context of spacetime theories with dynamical metric fields. To begin, consider the *Jacobson-Mattingly theory* (presented in e.g. Carroll and Lim (2004), and Jacobson and Mattingly (2001)[33]), in which the action for a coupled Einstein-Maxwell system is augmented with an additional term (via a Lagrange multiplier field λ), imposing (as a field equation, via variation with respect to λ) that the vector potential A^a be locally timelike:

$$S_{\text{JM}}\left[g_{ab}, A^a, \lambda\right] = \int d^4x \sqrt{-g}\left(R - \frac{1}{4}F^{ab}F_{ab} + \lambda\left(g_{ab}A^aA^b - 1\right)\right). \tag{6.1}$$

The first term is the Einstein-Hilbert action; F_{ab} is the Faraday tensor associated with A^a. The imposition of this Lagrange multiplier term means that, in the Jacobson-Mattingly theory, the dynamical behavior of nongravitational fields does *not* reflect the local (Poincaré) symmetries of the metric field. Given this, however, we appear to have in our possession a problem case for the geometrical approach.

Turn now to our second case: Bekenstein's bimetric TeVeS ("Tensor-Vector-Scalar") theory, presented in Bekenstein (2004). As discussed in Brown (2005, §9.5.2), in this theory the metric field which is surveyed by rods and clocks, the conformal structure of which is traced by light rays, and the geodesics of which correspond to the motion of free bodies, is not the "fundamental" metric field g_{ab}, but rather a less "fundamental" metric field $\tilde{g}_{ab}$, constructed from the other matter fields in the theory (Brown, 2005, p. 174). Indeed, the TeVeS theory presents another case in which the local symmetries of the dynamical laws do *not* mirror the local (Poincaré) symmetries of the "background" metric field – a necessary condition for the geometrical approach to go through.[34]

6.4.2 Two Miracles

Seemingly, the advocate of the dynamical perspective in the context of GR is committed to the existence of two unexplained input assumptions in the foundations of the theory – recently dubbed the two "miracles of GR" (Read et al., 2018, §5):[35]

1   All nongravitational interactions are locally governed by Poincaré invariant dynamical laws.[36]

2   The Poincaré symmetries of the laws governing nongravitational fields in the neighborhood of any point coincide – in the regime in which curvature can be ignored – with the symmetries of the dynamical metric field in that neighborhood.

(1) holds in SR: it tells us that the dynamical laws governing all matter fields are Poincaré invariant; the advocate of the dynamical approach takes this provisionally to be a brute fact;[37] the advocate of the geometrical alternative attempts to rationalize this by appeal to Minkowski spacetime. (1) still obtains in the neighborhood of any point in the spacetime manifold in GR.[38] However, in the case of GR there also exists an ontologically autonomous metric field, and this leads to (2): why is it – assuming that curvature terms can be ignored (which, to repeat, depends upon the sensitivity of one's experimental apparatuses relative to the strength of curvature effects in the neighborhood of those apparatuses) – that the symmetries of the dynamical laws governing nongravitational fields in a suitable neighborhood of any $p \in M$ coincide with those of the dynamical metric field in that neighborhood? Again, the advocate of the dynamical approach may postulate this as a brute fact. By contrast, the advocate of the geometrical approach may attempt to argue that the ontologically primitive metric field explains the form of the dynamical laws governing matter fields; however, as in SR, they face an outstanding burden to delineate how this is supposed to work. For further discussion regarding these "miracles of GR," see Read et al. (2018, §5).

6.5   Coordinate-Independent Approaches

Let us turn now in greater detail to a question raised in Section 6.3.3: to what extent is the dynamical perspective reconcilable with the coordinate-independent, "angle brackets" approach in the foundations of spacetime theories? Though this question has been addressed to some extent in Pooley (2013, §6.3.2) and Wallace (2017, §6), it is worth reflecting further on the matter. In this section, we first consider theories with non-dynamical metric fields (Section 6.5.1), before turning to theories with dynamical metric fields (Section 6.5.2).

6.5.1   *Non-Dynamical Metric Fields*

Consider the theory of a massive Klein-Gordon field φ, as specified in the coordinate-dependent approach via

$$\eta_{\mu\nu}\partial^\mu\partial^\nu\varphi + m\varphi = 0. \tag{6.2}$$

In (6.2), $\eta_{\mu\nu}$ denotes the components of a matrix $\mathrm{diag}\,(-1, 1, 1, 1)$. The choice of coordinate basis used in the above is (partly) arbitrary: this is captured in the fact that (6.2) is invariant under all Poincaré transformations, i.e., coordinate transformations of the form $x^\mu \rightarrow \Lambda^\mu{}_\nu x^\mu + a^\mu$, with $\Lambda^\mu{}_\nu \Lambda^\lambda{}_\sigma \eta_{\mu\lambda} = \eta_{\nu\sigma}$ and the a^μ arbitrary constant vectors. As elaborated in detail by Wallace (2017, §§2–3), on the so-called *Kleinian conception of geometry*, dynamical equations such as (6.2) may be understood as *defining* certain spacetime structure, via their transformation properties – so, for example, (6.2) defines a Minkowski spacetime structure via the fact that it is invariant under Poincaré transformations, in accordance with the tenets of the dynamical approach.[39]

Keeping the above in mind, suppose now that one seeks to reconcile the dynamical perspective in the context of such theories with the coordinate-independent approach. In the coordinate-independent picture, the theory in question is expressed through the introduction of the fixed geometrical object η_{ab} – a Minkowski metric field fixed identically in all KPMs of the theory; DPMs are then picked out by

$$\eta_{ab}\nabla^a\nabla^b\varphi + m\varphi = 0. \tag{6.3}$$

Since, however, this object is introduced in order to capture (in a coordinate-independent way) that the form of (6.2) is preserved only under Poincaré transformations, we may understand *this object* η_{ab} to codify the symmetry properties of (6.2) – this first equation being understood to be fundamental, on the dynamical approach.

In this way, we understand fixed metric fields such as η_{ab} to be ontologically supervenient upon the (dynamics of) matter fields (in this case φ); in doing so, we reconcile the coordinate-independent approach with the dynamical perspective. On this account, although the structure η_{ab} appears in the KPMs of the theory in question, it is *not* to be understood as ontologically autonomous. Of course, the picture of introducing KPMs and *then* DPMs remains highly unnatural on the dynamical view. Nevertheless, there is nothing erroneous in introducing these classes of models *post hoc*, while continuing to reconcile the primacy of coordinate-dependent dynamical laws for matter fields. (Indeed, there exists a precedent in the angle bracket school of including non-fundamental geometrical objects in the KPMs of a given theory, for convenience – consider e.g. the inclusion of a derivative operator ∇_a over and above a timelike vector field σ^a in the KPMs of Newtonian mechanics set in "Newtonian" spacetime.[40])

6.5.2 *Dynamical Metric Fields*

Consider a minimally coupled,[41] general relativistic theory of the massive scalar field, analogous (as far as possible) to the special relativistic theory presented above. In the coordinate-dependent approach, such a theory may be specified through writing down the following two dynamical equations[42]:

$$G_{\mu\nu} = 8\pi\, T_{\mu\nu}, \tag{6.4}$$

$$g_{\mu\nu}\nabla^\mu\nabla^\nu\varphi + m\varphi = 0. \tag{6.5}$$

Here, $G_{\mu\nu}$ is the Einstein tensor associated with the metric field $g_{\mu\nu}$; $T_{\mu\nu}$ is the stress-energy tensor associated with φ. Since in this case the metric field is dynamical, we have a theory for $g_{\mu\nu}$ and φ *together*.[43] The facts that (a) $g_{\mu\nu}$ is *locally* diagonalizable to read diag $(-1, 1, 1, 1)$; (b) locally and in a regime in which curvature effects may be ignored (this depends crucially upon the sensitivity of the experimental apparatuses available relative to the strength of curvature effects – see Read et al. (2018, §4)) $g_{\mu\nu}$ has higher derivatives vanishing; and (c) in this regime (6.5) takes a Poincaré-invariant form, contribute to $g_{\mu\nu}$ being surveyed locally by rods and clocks built out of the φ field – thereby affording it chronogeometric significance.

On the coordinate-independent approach, one characterizes this theory as follows. First, KPMs are picked out by triples $\langle M, g_{ab}, \varphi \rangle$; then, DPMs are those KPMs the geometrical objects of which obey the dynamical equations

$$G_{ab} = 8\pi\, T_{ab}, \tag{6.6}$$

$$g_{ab}\nabla^a\nabla^b\varphi + m\varphi = 0. \tag{6.7}$$

So be it – but ask now how one would express the dynamical perspective in coordinate-independent language in this case. Again, equations written in a coordinate basis – here (6.4) and (6.5) – are to be considered primitive. In this case, however, on moving to the coordinate-independent formalism, no new geometrical objects are introduced.[44] This makes clear an important point already noted in Section 6.4: in this case, metric field structure is not merely a codification of the symmetries of the given dynamical laws – rather, such structure *already exists* in (6.4) and (6.5).

Thus, consideration of the coordinate-independent formalism arguably illustrates a point in the context of spacetime theories such as GR to which the advocate of the dynamical approach assents. Perhaps more importantly, though, further merits of deploying this formalism emerge when one considers the *local* forms of (6.4) and (6.5). Considering (6.5) locally and in a regime in which curvature terms may be ignored, this equation takes a form invariant under Poincaré transformations.[45]

Thus, applying the Kleinian conception of geometry to this new, *local* equation, one again extracts (in exactly the same manner as in Section 6.5.1) Minkowski spacetime structure – and again, on the dynamical approach, this structure can be understood as a codification of the symmetry properties of that local dynamical law.

Locally, therefore, one might introduce a Minkowski metric field $\tilde{\eta}_{ab}$ to codify the symmetries of this dynamical law. But note that the dynamical g_{ab} field remains in this theory! Moreover, locally (and in a regime in which curvature effects can be dropped), g_{ab} also resembles the Minkowski metric field.[46] One test of whether g_{ab} has local *chronogeometric significance*, then, is whether the dynamical metric field g_{ab} coincides locallly – in the appropriate regime in which curvature effects may be ignored – with the metric field which codifies the symmetries of the dynamical laws, $\tilde{\eta}_{ab}$.

Putting things in this way – i.e., introducing a new geometrical object taken to codify the symmetries of the original (nongravitational) dynamical laws – is deeply ingrained in the coordinate-independent approach; nevertheless, it makes perspicuous the tests of chronogeometric significance envisaged by advocates of the dynamical approach in the cases in which the metric field itself is a dynamical entity. We conclude, therefore, that not only is the dynamical approach *compatible* with the coordinate-independent approach in this context; but moreover, there may be some *advantages* to its use, from the point of view of presenting the dynamical approach.

Let us attempt to demonstrate this stronger claim via two final points. First, consider again bimetric theories with *two* dynamical metric fields. Using the above formalism, the metric field in these theories which may lay claim to chronogeometric significance is precisely that which coincides locally with the metric field $\tilde{g}_{ab}$ which codifies the local symmetries of the dynamical laws. Second, thinking in this manner constitutes a good framework for consideration of scenarios in which there exists a misalignment between metric field structure and symmetries of the dynamical laws – for in such cases one may say that the dynamical metric field does not coincide locally with the metric field which codifies the symmetries of the nongravitational laws.

6.6   Outlook

We have sketched the progression of the dynamical approach – from its origins in the development of SR, to its opposition to the contemporary "geometrical" approach, to the manner in which it plays out in GR. Moreover, we have demonstrated that the perspective is not incompatible with certain coordinate-independent approaches in the foundations of spacetime theories. Examples of recent interest in the dynamical approach are found in Knox (2011, 2014, 2019), Menon (2019), Myrvold (2011), Pitts (2019), Stevens (2015), Wallace (2017), and Weatherall (2017); it is our hope that its framework continues to offer insight to those working in the foundations of spacetime theories.

Acknowledgments

We are indebted to Eleanor Knox and Alastair Wilson for their invitation to write this review; and to Eleanor Knox (again), Dennis Lehmkuhl, Tushar Menon, Simon Saunders, David Wallace, and especially Oliver Pooley for many hours of invaluable discussions on the dynamical approach. During the course of working on this chapter, J.R. was supported by an AHRC studentship, and by a senior scholarship at Hertford College, Oxford.

Notes

1 For discussion of these misgivings, see Brown (2005, §7.1) and particularly Giovanelli (2014).

2 For details concerning all these historical claims, see Brown (2005, chs. 4, 7).

3 See Miller (2010) and Nelson (2013, 2015). The notion that dynamical considerations are more fundamental than spacetime geometry is also found in Peres (1962) and the work of F. W. Hehl; see in particular Obukhov and Hehl (1999) and Hehl and Obukhov (2003).

4 See in particular Michael Friedman's classic (Friedman, 1983).

5 See Section 6.3.3. A recent brief defense of something like the dynamical view by Craig Callender, inspired by a 2011 paper due to Geroch (2011), is found in Callender (2017, section 7.4).

6 This is not to say that the Minkowski metric cannot be *derived* from Maxwell's equations; early recognition of this point in connection with the vacuum equations is found in Peres (1962). For further discussion, see Obukhov and Hehl (1999).

7 In Brown (2005, pp. 146–147), this condition is referred to as the *big principle*.

8 See Read et al. (2018).

9 In light of subtleties regarding the notion of an "absolute object," and the question of whether GR itself has such objects (see Pitts (2006)), in this chapter we follow the nomenclature of Pooley (2017) in referring to the objects picked out by A as "fixed fields."

10 This is achieved in unprecedented detail in Friedman (1983); see also Earman (1989) and Malament (2012).

11 See Brown (2005, p. 44).

12 See Einstein (1910) and Brown (2005, pp. 30, 81).

13 See Friedman (1983, ch. III, §5).

14 For a more detailed critique of "classical electrodynamics," see Brown (1993a).

15 Well, there was Keinstein in 1705, but that's another story. See Brown (2005, ch. 3).

16 See Brown (2005, ch. 4).

17 See also in this connection Acuña (2016).

18 Cf. Knox (2019, p. 2).

19 For more on Knox's approach, see Knox (2011, 2014, 2019), and Read and Menon (2019).

20 See, e.g., Friedman (1983, ch. VI, §4).

21 Norton (2021, ch. 2). For a critique of this kind of reasoning, see Brown (2005, §2.2.5). (This section attributes the same reasoning to a 1976 paper by Graham Nerlich, though this was based on a misinterpretation of Nerlich's point; see Nerlich (2010).)

22 For more details, see Brown (2005, pp. 134–136).

23 See Friedman (1983, ch. VI, §4).

24 See e.g. Janssen (2002, 2009).

25 See Wald (1984, p. 60) and Huggett (2000).

26 For an attempt to defend the geometrical approach on this point, see Read (2020).

27 See e.g. Malament (2012, ch. 4) for technical details.

28 For further details see Passon (2006) and Struyve (2011).

29 For related comments, see Myrvold (2019).

30 See Brown (2005, ch. 8).

31 Of course, it may be possible to explain this fact by appeal to a deeper theory than general relativity – see e.g. Read (2019).

32 These are significant subtleties in this vicinity – see Read et al. (2018, §§3–4) for recent discussion.

33 In fact, the version of the Jacobson-Mattingly theory discussed in this paper is a special case of that presented in Carroll and Lim (2004); Jacobson and Mattingly (2001).

34 For further detailed discussion of bimetric theories in defense of the dynamical approach, see Pitts (2019). Pitts also cites massive scalar gravity as a case in which "the chronogeometrically observable ... metric isn't clearly the One True Geometry" (Pitts, 2019, §5).

35 A similar "miracle" in the context of Newtonian theory would be the proportionality of gravitational and inertial masses – cf. Weatherall (2011).

36 That such universality cannot be taken for granted is highlighted in Peres (1962), section IV.

37 Recall endnote 31.

38 See Read et al. (2018, §3).

39 See also in this connection (Brown, 2005, p. 9).

40 Our thanks to Tushar Menon for this point.

41 For the definition of minimal coupling, as well as detailed philosophical discussion of the procedure, see Brown and Read (2016, §IV) and Read et al. (2018, §3).

42 We set $G_N = c = 1$.

43 Cf. Pooley (2017, p. 115).

44 The reason for this is part of the special character of GR: (6.4) and (6.5) were generally covariant (i.e., held in an arbitrary coordinate basis) to start off with.

45 This is demonstrated in e.g. Read et al. (2018, §A).

46 Cf. Read et al. (2018, §3).

References

Acuña, P. (2016). Minkowski spacetime and Lorentz invariance: The cart and the horse or two sides of a single coin? *Studies in History and Philosophy of Modern Physics*, 55: 1–12.

Balashov, Y. and Janssen, M. (2003). Presentism and relativity. *British Journal for the Philosophy of Science*, 54(2): 327–346.

Bekenstein, J.D. (2004). *An alternative to the dark matter paradigm: Relativistic MOND gravitation, invited talk at the 28th Johns Hopkins Workshop on Current Problems in Particle Theory*, June 2004, Johns Hopkins University, Baltimore, MD. Available at: arXiv:astro-ph/0412652.

Bekenstein, J.D. (2004). Relativistic gravitation theory for the MOND paradigm. *Physical Review D*, 70(8): 083509. Available at: arXiv:astro-ph/0403694.

Bell, J.S. (2004). How to teach special relativity. In *Speakable and Unspeakable in Quantum Mechanics* (2nd edition). Cambridge: Cambridge University Press, pp. 67–80.

Brown, H.R. (1993a). Correspondence, invariance and heuristics in the emergence of special relativity. In S. French and H. Kamminga (eds.), *Correspondence, Invariance and Heuristics: Essays in Honour of Heinz Post*, Dordrecht: Kluwer Academic Press, pp. 227–260. Reprinted in J. Butterfield, M. Hogarth and G. Belot (eds.) *Spacetime*, International Research Library of Philosophy, Dartmouth Publishing Company, Sept. 1996, chapter 7.

Brown, H.R. (1993b). Aspects of objectivity in quantum mechanics. In J. Butterfield and C. Pagonis (eds.), *From Physics to Philosophy*, Cambridge: Cambridge University Press, pp. 45–70. E-print: PITT-PHIL-SCI 223.

Brown, H.R. (2001). The origins of length contraction: The FitzGerald-Lorentz deformation hypothesis. *American Journal of Physics*, 69: 1044–1054. Available at: E-prints: arXive:gr-qc/0104032; PITT-PHIL-SCI 218.

Brown, H.R. (2005). *Physical Relativity: Spacetime Structure from a Dynamical Perspective*. Oxford: Oxford University Press. Corrections were made in the 2007 (paperback) edition.

Brown, H.R. and Lehmkuhl, D. (2016). Einstein, the reality of space, and the action-reaction principle. In P. Ghose (ed.), *Einstein, Tagore and the Nature of Reality*, London and New York: Routledge, pp. 9–36.

Brown, H.R. and Pooley, O. (2001). The origins of the spacetime metric: Bell's Lorentzian pedagogy and its significance in general relativity. In C. Callender and N. Huggett (eds.), *Physics Meets Philosophy at the Plank Scale*, Cambridge: Cambridge University Press, pp. 256–272. Available at: E-print: arXive:gr-qc/9908048.

Brown, H.R. and Pooley, O. (2006). Minkowski space-time: A glorious non-entity. In D. Dieks (ed.), *The Ontology of Spacetime*, Amsterdam: Elsevier, pp. 67–89. Available at: E-prints: arXiv: physics/0403088; PITT-PHIL-SCI 1661 (2004).

Brown, H.R. and Read, J. (2016). Clarifying possible misconceptions in the foundations of general relativity. *American Journal of Physics*, 84(5): 327–334.

Brown, H.R. and Sypel, R. (1995). On the meaning of the relativity principle and other symmetries. *International Studies in the Philosophy of Science*, 9(3): 235–253.

Callender, C. (2017). *What Makes Time Special?* Oxford: Oxford University Press.

Carroll, S. and Lim, E. (2004). Lorentz-violating vector fields slow the universe down. *Physical Review D*, 70: 123525.

Earman, J. (1989). *World Enough and Space-Time: Absolute Versus Relational Theories of Space and Time*. Cambridge, MA: MIT Press.

Einstein, A. (1910). Le principe de relativité et ses conséquences dans la physique modern. *Archives des Sciences Physiques et Naturelles*, 29: 5–28; republished in *The Collected Papers of Albert Einstein, Vol. 3, The Swiss Years: Writings 1909–1911*, M. J. Klein, A. J. Kox, J. Renn, and R. Schumann (eds.), Princeton, NJ: Princeton University Press, 1993.

Einstein, A. (1954). *Ideas and Opinions*. New York: Bonanza.

Friedman, M. (1983). *Foundations of Space-Time Theories*. Princeton, NJ: Princeton University Press.

Frohlich, J. (2009). Spin, or actually: Spin and quantum statistics. In B. Duplantier, J.-M. Raimond and V. Rivasseau (eds.), *The Spin*. Basel: Birkhäuser Verlag, pp. 1–60.

Geroch, R. (2011). Faster than light. In M. Plaue, A.D. Rendall and J.R. Pulham (eds.), *General Relativity: Proceedings of the Lorentzian Geometry Conference in Berlin*, volume 49 of AMS/IP Studies in Advanced Mathematics, American Mathematical Society, Providence RI; pp. 59–70.

Giovanelli, M. (2014). 'But one must not legalize the mentioned sin': Phenomenological vs. dynamical treatments of rods and clocks in Einstein's thought. *Studies in History and Philosophy of Modern Physics*, 48, Part A: 20–44.

Hehl, F.W. and Obukhov, Y.N. (2003). *Foundations of Classical Electrodynamics: Charge, Flux and Metric*. Boston, MA: Birkhäuser.

Huggett, N. (2000). Reflections on parity nonconservation. *Philosophy of Science*, 67(2): 219–241.

Jacobson, T. and Mattingly, D. (2001). Gravity with a dynamical preferred frame. *Physical Review D*, 64: 024028.

Janssen, M. (2002). *COI* stories: Explanation and evidence in the history of science. *Perspectives on Science*, 10(4): 457–522.

Janssen, M. (2009). Drawing the line between kinematics and dynamics in special relativity. *Studies in History and Philosophy of Modern Physics*, 40: 26–52.

Knox, E. (2011). Newton-Cartan theory and teleparallel gravity: The force of a formulation. *Studies in the History and Philosophy of Modern Physics*, 42: 264–275.

Knox, E. (2014). Newtonian spacetime structure in light of the equivalence principle. *British Journal for the Philosophy of Science*, 65(4): 863–880.

Knox, E. (2019). Physical relativity from a functionalist perspective. *Studies in History and Philosophy of Science Part B: Studies in History and Philosophy of Modern Physics*, 67: 118–124.

Kretschmann, E. (1917). Über den physikalischen Sinn der Relativitätspostulate. *Annalen der Physik*, 53: 575–614.

Lange, M. (2007). Laws and meta-laws of nature: Conservation laws and symmetries. *Studies in History and Philosophy of Modern Physics*, 38: 457–481.

Leinaas, J.M. and Myrheim, J. (1977). On the theory of identical particles. *Il Nuovo Cimento B*, 37(1): 1–23.

Malament, D. (2012). *Topics in the Foundations of General Relativity and Newtonian Gravitation Theory.* Chicago: University of Chicago Press.

Maudlin, T. (2012). *Philosophy of Physics: Space and Time.* Princeton, NJ: Princeton University Press.

Menon, T. (2019). Algebraic Fields and the Dynamical Approach to Physical Geometry. *Philosophy of Science*, 86(5): 1273–1283.

Miller, D.J. (2010). A constructive approach to the special theory of relativity. *American Journal of Physics*, 78: 633–638.

Myrvold, W.C. (2019). How could relativity be anything other than physical? *Studies in History and Philosophy of Modern Physics*, 67: 137–143.

Nelson, W.M. (2013). *Relativity Made Real: Under the Hood of Einstein's Theory* (2nd edition). Charleston, SC: CreateSpace Publishing.

Nelson, W.M. (2015). Special relativity from the dynamical viewpoint. *American Journal of Physics*, 83(7): 600–607.

Nerlich, G. (2010). Why spacetime is not a hidden cause. In V. Petkov (ed.), *Space, Time and Spacetime. Physical and Philosophical Implications of Minkowski's Unification of Space and Time.* Berlin, Heidelberg: Springer-Verlag, pp. 181–192.

Norton, J. (2008). Why constructive relativity fails. *The British Journal for the Philosophy of Science*, 59(4): 821–834.

Norton, J. (2021). ebook *Einstein for Everyone.* Available at: https://www.pitt.edu/ jdnorton/teaching/index.html.

Obukhov, Y.N. and Hehl, F.W. (1999). Spacetime metric from linear electrodynamics. *Physics Letters B*, 458: pp. 466–470.

Passon, O. (2006). *What you always wanted to know about Bohmian mechanics but were afraid to ask.* Available at: http://philsci-archive.pitt.edu/3026/1/bohm.pdf.

Peres, A. (1962). Electromagnetism, geometry and the equivalence principle. *Annals of Physics*, 19: 279–286.

Pitts, J.B. (2006). Absolute objects and counterexamples: Jones-Geroch dust, Torretti constant curvature, tetrad-spinor, and scalar density. *Studies in History and Philosophy of Modern Physics*, 37(2): 347–371.

Pitts, J.B. (2019). Space-time constructivism *vs.* modal provincialism: Or, how special relativistic theories needn't show Minkowski chronogeometry. *Studies in History and Philosophy of Science Part B: Studies in History and Philosophy of Modern Physics*, 67: 191–198.

Pooley, O. (2013). Substantivalist and relationist approaches to spacetime. In R. Batterman (ed.), *The Oxford Handbook of Philosophy of Physics.* Oxford: Oxford University Press, pp. 522–586.

Pooley, O. (2017). Background independence, diffeomorphism invariance, and the meaning of coordinates. In D. Lehmkuhl, G. Schiemann and E. Scholz (eds.), *Towards a Theory of Spacetime Theories.* New York, NY: Birkhäuser, pp. 105–143.

Prange, R.E. and Girvin, S.M., editors. (1990). *The Quantum Hall Effect.* Graduate Texts in Contemporary Physics. New York, NY: Springer-Verlag.

Read, J. (2020). Explanation, geometry, and conspiracy in relativity theory. In C. Beisbart, T. Sauer and C. Wüthrich (eds.), *Thinking About Space and Time: 100 Years of Applying and Interpreting General Relativity*, Einstein Studies series, vol. 15, Basel: Birkhäuser.

Read, J. (2019). On miracles and spacetime. *Studies in History and Philosophy of Modern Physics*, 65. 103–111.

Read, J., Brown, H.R. and Lehmkuhl, D. (2018). Two miracles of general relativity. *Studies in History and Philosophy of Modern Physics*, 64: 14–25.

Read, J. and Menon, T. (2019). The limitations of inertial frame spacetime functionalism. *Synthese.* (Forthcoming.)

Stevens, S. (2015). The dynamical approach as practical geometry. *Philosophy of Science*, 82(5): 1152–1162.

Stevens, S. (2020). Regularity relationalism and the constructivist project. *British Journal for the Philosophy of Science*, 71(1): 353–372.

Struyve, W. (2011). *Pilot-wave approaches to quantum field theory. Journal of Physics: Conference Series*, 306(1): 012047. Available at: https://arXiv:1101.5819v1 [quant-ph].

Sypel, R. and Brown, H.R. (1992). When is a physical theory relativistic? In D. Hull, M. Forbes and K. Okruhlik (eds.), *Proceedings of the 1992 Biennial Meeting of the Philosophy of Science Association*, volume 1, Philosophy of Science Association, East Lansing, Michigan, pp. 507–514.

Wald, R. (1984). *General Relativity*. Chicago: University of Chicago Press.

Wallace, D. (2017). Who's afraid of coordinate systems? An essay on representation of spacetime structure. *Studies in History and Philosophy of Science Part B: Studies in History and Philosophy of Modern Physics*, 67: 125–136.

Weatherall, J.O. (2011). On (some) explanations in physics. *Philosophy of Science*, 78(3): 421–447.

Weatherall, J.O. (2017). Conservation, inertia, and spacetime geometry. *Studies in History and Philosophy of Science Part B: Studies in History and Philosophy of Modern Physics*, 67: 144–159.

Further Reading from the editors

Einstein's original presentation of special relativity, "On the electrodynamics of moving bodies." (*Annalen der Physik* 17, no. 891 (1905): 50), is a surprisingly accessible classic paper. The dynamical approach was first articulated in the philosophical literature in Brown (1993a), Brown and Pooley (2001, 2006) and then developed in Brown's *Physical Relativity: Space-Time Structure from a Dynamical Perspective* (Oxford University Press, 2007). Tim Maudlin's *Philosophy of Physics: Space and Time* (Princeton University Press, 2012) presents a different, geometrical view. Friedman, Michael. *Foundations of Space-Time Theories: Relativistic Physics and Philosophy of Science* (Princeton University Press, 1983) is a classic text for those who are comfortable with more technical detail, and represents the kind of orthodoxy that Brown and Read are reacting to. Volume 67 (2019) of *Studies in History and Philosophy of Modern Physics* contains a special section on Brown's dynamical approach to relativity – the papers assembled there are an excellent guide to discussions around the view.

7

RELATIVITY AND THE A-THEORY

Antony Eagle

The special theory of relativity (STR) is widely supposed to be in tension with theories of time which give a special significance to the present moment. In this chapter, I will develop and explore the prospects for resolution of this tension.

Overview. In Section 7.1, I will explain A-theories of time and introduce the key pieces of ideology on which they rely. In Section 7.2, I will introduce just enough of STR to enable us to bring that theory into contact with the A-theory. In Section 7.3, I will develop the tension between STR and the A-theory. In Section 7.4, I will consider A-theoretic responses that preserve the adequacy and completeness of STR; in Section 7.5, I will consider A-theoretic responses which accept STR, so far as it goes, but take it to be incomplete with respect to tensed facts. I conclude in Section 7.6 with my evaluation of the overall prospects for the A-theory in light of STR.

7.1   The A-theory

A first argument. It is a mundane observation that not everything which happens had to happen. The actual concrete world might have been different in various ways, and if it had been, different things would have happened. But what is the case actually is what is the case *simpliciter* – it's just the way things **are** (rather than the way things are in, or according to, such-and-such a possibility). So the way things are is not the way they must necessarily be. In this sense, reality (the way things are) is contingent.

A second argument. It may seem equally mundane to observe that not everything which happens always happens. The actual concrete world was (and will be) different in various ways; and when it was (will be), different things were (will be) happening. But what is the case now is what is the case *simpliciter* – it's just the way things **are** (rather than the way things are at, or according to, such-and-such a time). So the way things are is not the way they are permanently. In this sense, reality (the way things are) is temporary.

It is important to be clear on how to understand the conclusion of this second argument. It claims that there is a basic distinction between the things which happen in reality – or **really happen** – and those which merely did or will happen. Moments of time differ from one another concerning what happens when they accurately represent reality, or **are present**. So what really happens genuinely changes over time as the character of present reality changes. Let us call the philosophers who endorse this conclusion, thus understood, **A-theorists** (following some rebarbative but entrenched terminology due to McTaggart, 1908).

There are many different views that all agree on the A-theory, so characterized. One is **presentism** – the view that concrete reality is temporally unextended, but that this momentary reality changes its character as different times are successively present (Sider, 2001; Markosian, 2004;

Bourne, 2006). (Presentists disagree over the reality of other times, but all deny the existence of other times that are distinct parts of concrete reality.) Everything that happens, happens in the one existing present moment. However, since the character of present reality **was** different, and **will be** different, those events are not permanent happenings. Another view is the **moving spotlight** – the view that while reality is temporally extended, there is an objectively privileged moment such that everything which is happening according to that moment is really happening *simpliciter* (Cameron, 2014; Deasy, 2014). There is real change in reality, but unlike on the presentist view, there is no change in what things exist – rather, there is change in what is true *simpliciter* of what exists.

*   *   *

The contrasting view, which we may call the **B-theory**, is that reality is not temporary. On this view, the correct way to understand phenomena of apparent temporariness, such as natural language tense, involves a temporally extended reality which exhibits internal variation between its distinct parts, rather than change in the character of a momentary present. Every event in this temporally extended reality is real and thus really happens: the way things are is the way that temporal reality is *in toto*. There are no events of which it is true that they exist, while false that they are really happening. A description of reality which depicts the global pattern of variation in happenings from time to time and place to place can be a complete and correct account of what is true *simpliciter*, even if it doesn't give any special role to what is true of the present moment.

B-theorists reject the second argument with which we began; yet almost all of them accept the first argument. At which point do those arguments diverge? A typical B-theorist might say: the claim from which the A-theory appears to follow – namely "the way things are is not the way they are permanently" – does not in fact entail the A-theory. For *the way things are* contains a present tensed *are*. The B-theorist regards tense as a surface phenomenon of natural language sentences, which is not reflected in the underlying propositions those sentences express. Those propositions, say the B-theorist, have their truth values permanently. They will typically say that tensed phrases are **context-sensitive**: what *the way things are* expresses, when uttered at *t*, is something like THE WAY THINGS ARE AT *t*. According to the B-theorist, not every time is such that the way things are then is the way things are at *t*. So the B-theorist says: in every utterance of the second argument, the key claim *the way things are is not the way they are permanently* expresses a truth. But the truth expressed is not one that entails the A-theory. The truth expressed is something like THE WAY THINGS ARE AT *t* IS NOT THE WAY THINGS ARE AT ALL TIMES. This is true because the B-theorist accepts variation from time to time in temporally extended reality. But to entail the A-theory, there would have to be a context-insensitive use of *the way things are* on which it always means the same thing; namely, it denotes the totality of temporal reality. According to the B-theory, the premises in the second argument don't involve any such use – and if we were to stipulate that *the way things are* is to be understood invariantly, the key premise would be false, since it would express something the B-theorist denies, namely, that the way things are *in toto* is not the way they are permanently.

There are many arguments, philosophical and linguistic, that aim to adjudicate this dispute between the A-theory and the B-theory. For example, one major discussion concerns precisely this issue of whether tensed language is best understood as involving context-sensitivity (Partee, 1973; Lewis, 1980; King, 2003; Brogaard, 2012). This discussion embraces another issue, involving the apparent commitment of the A-theory to the idea that because reality changes over time, what is true concerning reality also changes over time, so that propositions vary in their truth value from time to time. This view, known as **propositional temporalism**, has proponents who think only temporal propositions can be the objects of our temporary propositional attitude ascriptions (Prior, 1959), or only temporal propositions can explain the non-redundancy of temporal operators like *it was the case that* when applied to natural language sentences (Kaplan, 1989). It has detractors who think that temporal propositions actually don't successfully play the proposition role, e.g., in belief reports (Richard, 1981).[1]

In this chapter, I will focus on another line of argument against the A-theory. This argument is that there is a certain consequence of any A-theory that is, given other seemingly unobjectionable premises, incompatible with the special theory of relativity (STR). The consequence of particular interest to us is as follows:

(1)   There is an absolute fact of the matter about which events are simultaneous.

The notion of **really happening** marks, according to A-theorists, an absolute distinction between present events and others. It is obvious that presentists think all events which are really happening are simultaneous, since that view says there is only one moment of time. But non-presentist versions of the A-theory also make use of a single moment such that everything which happens according to it is really happening (and things which merely were or will happen according to it are not really happening).[2] So we may use this notion of really happening to define an absolute notion of simultaneity, specifically:

(2) p is simultaneous with q (they belong to the same moment of time) iff the way things are is such that p is really happening and q is also really happening.

Given this definition, and the A-theorists commitment to the absoluteness of "really happening," (1) follows. It is clear that (1) does not follow from the B-theory, because that theory maintains that many non-simultaneous pairs of events exist, and that many non-simultaneous events are jointly really happening, so they could never accept definition (2) — even if they were to adopt the ideology of "really happening."

7.2   Special Relativity from A to B

The mathematical and physical content of STR goes well beyond what I can present here.[3] Thankfully the purported conflict between STR and the A-theory stems from very general features of physical geometry which do not depend on addressing many important matters of detail.

STR is a theory of the geometrical structure of space and time together: Minkowski spacetime. The theory postulates an underlying manifold of spacetime points. The geometrical structure of this manifold can be given by specifying a distance (an **interval**) between any two points, just as we do for ordinary Euclidean spacetime. The details of this distance function are however strikingly different. In Euclidean spacetime, there is just one point that is zero spatiotemporal distance from a given point: itself. But in Minkowski spacetime, there is a non-trivial set of points that have a zero interval from any given point. Moreover, points can even have a negative interval from a given point. We can use the sign of the interval to tell us something about the global structure of spacetime. At any given point p, there will be a collection of points which have zero interval from p – these are **lightlike separated** from p. There will be a collection of points which have a positive interval from p, which are **timelike separated** from p. Moreover, there will be a collection of points which have a negative interval from p, which are **spacelike separated** from p. This threefold classification is systematic. The lightlike separated points from p form the surface of (the four-dimensional analogue of) a double cone, which is called the **lightcone**. The spacelike separated points are those which lie outside the lightcone, and the timelike separated points are those which lie strictly within the lightcone.

The terminology is suggestive. While only the spatiotemporal interval is absolutely given in the underlying geometry, and the interval between two points isn't a spatial or temporal distance, it can nevertheless be decomposed into spatial and temporal components. However, this decomposition is not unique: many different spatial and temporal distance components correspond to the same spatiotemporal interval. We can say this much, however:

- Whenever the interval between p and q is timelike, the temporal component of every decomposition of that interval is non-zero; though it has no determinate value, it is determinately non-zero.

- Whenever the interval between p and q is spacelike, the spatial component of every decomposition of that interval is non-zero; though it has no determinate value, it is determinately non-zero.

- Whenever the interval between p and q is lightlike, the spatial and temporal components of every decomposition of that interval are equal.

Accordingly, the timelike separated points from p are those that can be reached from p by simply waiting for some time to elapse, at least for suitable voyagers. More precisely, any two timelike separated points are such that there possibly exists an object moving in a way not subject to external force (freely falling, or moving **inertially**) whose trajectory passes through both points.

The points lightlike separated from p, falling on the lightcone, are those where the spatial distance matches the temporal distance on every decomposition, so that the interval is 0. Intuitively, points on the lightcone correspond to the trajectory of something moving at some absolute velocity: if we apply the same decomposition process to every point on some lightlike line originating at p, we will obtain a set of points of uniformly increasing and matched temporal and spatial distance from p. It is a basic truth of STR that it identifies light as the thing which moves with this absolute velocity. The lightcone structure around a point p is by itself sufficient to determine the permissible trajectories for a light ray passing through p. The possible trajectories for light rays do not depend on the state of motion of their source or any possible observer. This is at variance with our ordinary experience. If we imagine p to be the location of the emission of a physical object, the prior state of motion of the emitter will have a significant effect on the subsequent trajectory of the object. The same is not true of light, which can move only on the paths laid out for it by the underlying light cone geometry.

The physical significance of spacelike separation is dual to that of timelike separated points. These are the points of determinately non-zero spatial separation from one another. Though the spacetime interval doesn't uniquely determine which points are zero temporal distance from one another, there is scope to use the notion of spacelike separation to define regions that intuitively correspond to moments of time:

(3) A **moment of time** is any region comprising spacetime points T such that

a. for any $p, q \in T$, p and q are spacelike separated – T is **achronal**.
b. T is maximal: every possible timelike line intersects it once (and only once, given a).
c. *Optional*: T is **flat** – there is a timelike line l such that for any $p, q \in T$, p and q lie on a straight line at right angles to ("orthogonal to") l.

This definition – setting aside the optional condition (3c) for now – identifies moments of time with **Cauchy surfaces** (Earman, 1995, p. 44), and involves only absolutely given geometrical facts. This definition of moments of time is well-behaved, relativistically – indeed, it is well-behaved in general relativity as well as STR. So defined, moments of time contain no events that are determinately separated by a positive temporal interval. Moreover, since each possible observer, whether point sized or larger, intersects any moment of time T just once, and every point on their trajectory not in T determinately occurs at some non-zero temporal distance from the points in T, none of which determinately occur at some non-zero temporal distance from any other point in T. Therefore, the moment of time does serve to mark, for every possible object, a division of that object's career into earlier and later; an achronal region contains no events such that one is strictly already over before the other one occurs. Intuitively, then, these regions behave in many ways like classical moments of time.

The "optional" condition c requires that a moment of time should be a **flat** Cauchy surface, or a **hyperplane**. (A Cauchy surface is a region such that for any two points within it, there is a decomposition of their interval that makes the temporal separation between them zero. A hyperplane is such that there is a decomposition of the interval d such that for any two points within the region, the temporal separation between them according to d is zero. The difference is in quantifier scope.) If we insist on this flatness condition (Bacon, 2018), there is a proof that such regions are uniquely well-behaved as moments of time: *belonging to a hyperplane* is the only non-vacuous equivalence relation definable from the underlying geometry which does not privilege any one point within the moment of time and which is invariant under appropriate symmetry transformations (due to Malament, 1977). In general, Cauchy surfaces need not be invariant under such transformations, which is just to say not all of them are flat. Hyperplanes are often called "hyperplanes of simultaneity," because they are associated with Einstein's operational account of simultaneity (Einstein, 1905). This operational definition, like the Malament proof, presupposes that moments of time should be flat. We need not resolve this issue. Below, when I refer to definition (3), the reader with a preference should include or omit the optional condition as they see fit.

Definition (3) is adequate for defining a moment of time in classical spacetime too. But the definition permits moments of time to have features which classical times do not. For example, any moment of time in STR includes some points within it which have positive temporal separation under some decomposition of the interval separating them into spatial and temporal components. But this is not true for every decomposition, so it is false that any moment of time contains points which are determinately at some temporal distance.

More strikingly in STR, unlike classical spacetime, each point p lies within many moments of time. There are many Cauchy surfaces (maximal sets of mutually spacelike separated points) including p. We thus have a tension that arises when identifying the best referent in STR for the classical conception of a moment of time at which p. There are maximal achronal regions containing p, which can be picked out just using the underlying geometry of STR. But there are too many of them, according to the classical conception, which says that each spacetime point occurs in exactly one moment. We may accommodate the classical conception only if we select a moment of time for each point, despite there being no geometrical motivation within STR for the selection. Insofar as the A-theory is committed to the classical conception of moments of time, embodied in (1), this tension is at the heart of the difficulty STR poses for the A-theory.

7.3 A Puzzle for STR and the A-theory

Definition (3) of a moment of time uses only mathematical and geometrical notions that are well-defined in STR, since we can define a maximal achronal region in terms just of spacelike separation, timelike lines, and intersection. So, **being a moment of time** is relativistically invariant, as is **belonging to the same moment of time**: if p and q fall within the same maximal achronal region, that is because of the fundamental geometry rather than depending on some specific way of coordinatizing it or representing it. The relation among events of BEING SIMULTANEOUS just is the relation of those events **occurring at or belonging to the same moment of time**. It follows that if two events are simultaneous, that is a fundamental (coordinatization-invariant) geometrical fact in STR.

This may sound surprising to those raised on the idea that simultaneity is relative, that "it is of the essence of the theory of special relativity that absolute simultaneity as such does not exist" (Saunders, 2002, p. 280). The appearance of conflict is deceptive, but it is instructive to explore it. The standard view about simultaneity that Saunders is invoking is something like the following:

(4) Necessarily, **absolute simultaneity** is:

    a. a **simultaneity** relation, which never holds between events not belonging to the same moment of time, and always holds between events belonging to the same moment of time;

    b. an **equivalence** relation (reflexive, symmetric, transitive) between points of spacetime; and

    c. a **unique** and **absolute** relation.

Note that these conditions are jointly satisfiable in classical physics, because the relation HAVING ZERO TEMPORAL SEPARATION is well-defined and invariant in Galilean spacetime geometry. It can be used to specify moments of time, is an equivalence relation, and is unique and absolute. This definition of absolute simultaneity is satisfied in classical spacetime.

Recall the A-theoretic account of simultaneity in (2). If the ideology of "really happening" is acceptable, we can use it to define a simultaneity relation which meets the conditions under (4) for absolute simultaneity. The relation of "co-happening" is uniquely defined using only absolute notions and it is an equivalence relation. It is by construction a simultaneity relation, since of course the basic premise of the A-theory is that a single moment of time is distinguished in containing all and only the really happening events (or being the moment of time relative to which all and only that which is true *simpliciter* is true). So the most natural A-theoretic conception of simultaneity is a notion of absolute simultaneity.

Saunders' claim that absolute simultaneity does not exist rests on a further principle (2002, p. 283):

(5) If it exists, absolute simultaneity is relativistically definable.

Simultaneity is a basic temporal relation. Methodologically, such a relation ought to be identifiable within our best theory of temporal structure. As Hawley puts it, any legitimate simultaneity relation "would show up in our best scientific theories: ... this is just the sort of thing you'd expect science to tell you about" (Hawley, 2009, p. 511). If (5) is to be respected by the A-theory, we will need to find a relativistically definable relation which is coextensive with A-theoretic simultaneity.

The obvious candidates don't work. We have seen that BELONGING TO THE SAME MOMENT OF TIME is relativistically definable. So it meets condition (5). It is not an absolute simultaneity relation, however, because it is not transitive on our domain of moments of time and so it fails to be an equivalence relation. For there being a moment of time to which p and q belong, and there being a moment of time to which q and r belong, do not entail that it is the **same** moment of time being talked about, since q belongs to more than one moment of time. Indeed, we can readily pick p and r to be timelike separated and yet there be for each of them some moment of time that they share with q.[4]

However, the relation p AND q BELONG TO T for each specific T is an equivalence relation (which is yet more reason to identify each T with a moment of time, since each moment determines a relation which is formally a simultaneity relation). But of course the many relations p AND q ARE SIMULTANEOUS ACCORDING TO T are explicitly relative to selection of a moment of time, and the selection is not determined by geometry alone (no moment is privileged by the spacetime structure). If we allow them all to be absolute simultaneity relations, we fail to satisfy uniqueness. But there are no grounds internal to spacetime geometry to privilege one of these relations.

In view of the failure of these candidate relativistic definitions of absolute simultaneity, the A-theorist has a few options.

- **Conciliatory approaches** attempt to alter the definition of a moment of time (3) – which is not, strictly speaking, part of STR but rather a bridge principle connecting STR to ordinary ideology – so as to make BELONGING TO THE SAME MOMENT OF TIME an absolute simultaneity relation.
- **Antagonistic approaches** say that the failure to identify a relativistically invariant relation co-extensive with the **co-happening** relation is a problem for STR. These approaches further divide over what kind of problem this is.

  - **Supplementing** approaches say that STR is not inconsistent with absolute simultaneity, but it – perhaps in common with other physical theories of spacetime – is incomplete with respect to A-theoretic facts.
  - **Revisionary** approaches say that STR is inconsistent with absolute simultaneity, so it ought to be (and will be, they hope) replaced by some successor theory which does feature a basic absolute simultaneity relation.

I will say no more about revisionary approaches. These approaches agree with the B-theorist that STR and the A-theory are incompatible; they disagree only over whether some successor physical theory will be more hospitable. Proposals about what future physics will hold are speculative, but existing discussions do not uniformly encourage hope for the A-theory (Monton, 2006; Callender, 2007; Wüthrich, 2012; Skow, 2015, §9.5).

7.4 Conciliatory Approaches

Conciliatory approaches trace the problems for the A-theory to the fact that each point of spacetime is part of many moments of time, according to (3). This is what prevents the relation p AND q BELONG TO THE SAME MOMENT OF TIME from being an equivalence relation. If we can come up with another account of *moment of time* without this problem, perhaps we can reconcile STR and the A-theory. Malament's result discussed in Section 7.2 shows that any rival relativistic definition of a moment of time has to give up some plausible features we expect moments of time to have. It is unsurprising that plausible alternative candidates are nevertheless counterintuitive.

One motivation for the A-theory is that the objects of our present experience are apparently more real than objects of merely past or future experience. It is commonsense that my present experiencing falls within a moment of time consisting of just those events that are in some idealized sense perceptible now.

> When we gaze into the night sky, I suggest, what we observe is the actual state of the universe, not some causal remnant of its former state. We gaze at the star Sirius and observe its state; not some Sirius-trace which is the antecedent of its actual present state. The latter supposition would suggest that there is some actual contemporaneous state which we cannot know now, but will know later, and I think that special relativity shows us that this is mistaken. (Godfrey-Smith, 1979, p. 241)

Godfrey-Smith in effect defends this proposal:

(6) A **moment of time** is any region comprising points lying on the same backward light cone from some point of spacetime.

In the Malament proof, this definition is excluded by the condition that moments of time should yield a temporal line of symmetry in spacetime.[5] While this condition is eminently plausible, some suggest that technical considerations favor loosening the requirement that moments of time should be temporally symmetrical. In particular, it has been argued that only definition (6) allows observers moving relative to one another who meet at a point to agree (operationally) on which moments of time that point belongs to (Sarkar and Stachel, 1999).

But (6) is a poor definition. If interpreted at face value, each moment belongs to many moments of time, no more permitting an "absolute" definition of simultaneity than (3). Simultaneity remains non-transitive, since we can pick timelike separated p and r such that p lies on the past lightcone of q, and q lies on the past lightcone of p. The proposal (which boils down to the proposal that p and q belong to the same moment iff they are lightlike separated) is intrinsically worse than the previous one, since it entails that events which are determinately temporally distant belong to the same moment, and that events which are potentially at zero temporal separation cannot belong to the same moment. According to (6), moments of time are not achronal.

But its defenders will say that using past lightcones to define moments of time, to be used in the standard way to define simultaneity, doesn't capture their intent. Rather, they want to start with a spacetime point presumed to be present, and extend simultaneity ("joint presentness") out from it. Moments of time play no role. On this view, q is **present for** p just in case q belongs to the past lightcone of p. But then if q is present for p, p cannot be present for q, since p will lie on the forward lightcone of q. Whatever else it may be, this is not a simultaneity relation – it is not even symmetric.

These intuitions mustered by supporters of (6) actually emphasize the privileged role of single spacetime points: those at the apex of past light cones conceived of as the locus of present experiences of events on those lightcones. If we respect those intuitions about present experience, we ought to identify the present moment at p with the set of points (i) experienceable at p and (ii) not determinately temporally separate from p. It is trivial that this yields p as the unique point belonging to the same moment as p. That is:

(7)   A moment of time is any region consisting of a spacetime point (Sklar, 1987, p. 302; Savitt, 2000, pp. S567–S568).

Simultaneity which makes use of definition (7) meets our conditions (4) and (5) on relativistic simultaneity. But it is also a vacuous relation, which is ruled out in the Malament proof by the observation that at least two distinct points should belong to any moment of time. Nevertheless, it has its defenders:

> in the theory of relativity, the only reasonable notion of "present *to a space-time point*" is that of the mere identity-relation: present to a given point is that point alone – *literally* "here-now" (Stein, 1991, p. 159).

Compare: "the present instant, properly speaking, does not extend beyond the here" (Robb, 1921, p. 13).

It is strikingly counterintuitive to say that all that is really happening at any moment is a single point-sized event. Definition (7) has been said to be "obviously untenable" (Saunders, 2002, p. 286), and in conflict with the "the common sense picture that motivates the A-theory [that] the present is spatially extended and so public and shareable" (Gilmore et al., 2016, p. 109).

But perhaps more troubling for the A-theory is how to account for change in what is really happening over time. There is one moment of time (point-region of spacetime) which corresponds to how reality **is**. Part of how reality is consists of how it currently was and currently will be. The only viable semantics, given (7), for the past tense *was* is given as follows: *it was the case that* ϕ is true iff there is a moment of time q in or on the past lightcone of the present moment at which ϕ.[6] Suppose ϕ is true only at some moment spacelike separated from the present moment. This is true: *it will be that it was that* ϕ. But it is not true that ϕ; nor was it true that ϕ; nor will it be true that ϕ. This conflicts with platitudes about time, such as that if something hasn't yet been, but it will have been, then it will come to be (Bacon, 2018). Some dismiss such platitudes – "so what?" says Skow (2015, p. 166). But violate too many platitudes, and the "moments of time" we end up with aren't worth the name.

7.5 Supplementing Responses

The upshot of the previous section was that no rival account of moments of time was both viable and more accommodating to the A-theory than definition (3). Supplementing responses to the problem of absolute simultaneity adopt that definition, but appeal to some additional facts – not grounded in the geometry of STR – to pick out that moment of time such that the events occurring at it are just those that are **really happening**. These views thus deny assumption (5), and argue that STR lacks the resources to capture the A-theoretic facts about absolute simultaneity. It is in this sense that Saunders and many others hold the A-theory to be "inconsistent" with STR: that A-theory ideology in (2) requires a notion of simultaneity that rejects the adequacy of STR as encapsulated by (5).

A model of STR yields a tenseless mosaic or manifold of all that ever happens and the spatiotemporal relations between these happenings. What it is for a model to be true is for the overall structure of spacetime and its occupants to correspond to the model. Facts represented in this manifold collectively form a supervenience basis for modeling the paths and changes in material things, predictions of relativistic measurements, and the trajectories of light signals. But contrary to Hawley's remark quoted in Section 7.3, it would be quite inappropriate for temporary facts – such as those about what is really happening now – to appear in this representation of what is permanently true. Nor do "dynamic" facts about the coming into being over time of the tenseless manifold appear in the manifold.[7] But that doesn't mean that those temporary facts are not facts – only that it is not part of the remit or ambition of STR to represent them.

The most common kind of supplementing response is to simply add the A-theoretic ideology to STR:

> There is a region of the manifold in which events are really happening, it includes [my present location] and many other points, and it does not coincide with any region that is geometrically distinguished, according to SR. (Zimmerman, 2011, §3)

Models of this theory involve at least an STR manifold together with a specification of the moment of time N which is the uniquely accurate moment.[8] The purest version of such theories takes the A-theoretic ideology to be in perfectly good standing in its own right, and uses (2) to pick out the privileged moment, regardless of whether the privileged moment has any further special physical features or special material contents. Such a theory adopts existing explanations in STR wholesale, but adds new explanatory capacities, since now the whole panoply of tensed facts can be given grounds in this additional A-theoretic structure.[9]

Such supplementing views admit a body of facts which STR was already sufficient to explain, including all the physical facts about material objects, their nature and behavior at every point, and the causal relations between regions of spacetime. Such facts remain constant and are explained in the same way regardless of which moment is privileged. This gives rise to two related objections.

- An epistemic objection: the physics of measurement devices, including our own sensory apparatus, supervenes just on the manifold and its contents. So the A-theoretic structure would be undetectable (Savitt, 2000, p. S570): concerning the model which happens to privilege the actual present, "it is impossible to know that it, rather than some alternative..., is the true one" (Skow, 2015, p. 157). Indeed, if there is a privileged present, no one could know that there is, since pure STR models without a privileged present are explanatorily adequate to physical experience.

- A theory which posits a privileged moment in addition to pure spacetime geometry is "guilty of a commitment to surplus geometrical structure ... which is standardly taken to be a theoretical vice" (Gilmore et al., 2016, p. 109). The sole reason to postulate A-theoretic structure is to accommodate tensed facts – no physical grounds motivate this addition, and it is methodologically vicious because "scientific methodology is always against superfluous pomp" (Callender, 2007, p. 67).[10]

Against the methodological objection one might reply: we accept plenty of ordinary objects that are not explicitly mentioned in the physics – there is structure at the level of dogs and people and material artifacts that genuinely exists even though it is not unambiguously identifiable with the structure of (collections of) point particles (Zimmerman. 2008, p. 219). A commitment to rocks and clouds over and above pluralities of rock- and cloud-parts is anodyne, but nevertheless seems to fall foul of this supposed methodological principle.

However, we have a considerable amount of evidence for the existence of rocks and clouds. If the methodological principle is weakened to say that it is theoretically vicious to accept surplus structure without evidence, few will question it. The problem for the supplementing response is that the epistemic objection does seem to undermine our evidence for tensed facts. If our experience supervenes on the physical, and that experience is thus insensitive to whether STR or some supplemented model of STR is accurate – an insensitivity which both A-theorists and B-theorists agree exists – we might wonder: how can that experience be evidence for the additional A-theoretic structure? This is not simply an objection that the experience would be the same regardless of whether STR is supplemented or not (Price, 1996, pp. 14–15). The objection is that the **explanation** of the experience is the same, so that the postulated privileged moment is explanatorily idle with respect to precisely the experiential phenomenology that is supposed to motivate the postulation. As Savitt puts it, "If the present [moment] is indeed so elusive, I find it difficult to imagine what aid or comfort it could be to a metaphysician" (Savitt, 2000, p. S570).

Of course, the present isn't **entirely** elusive. If my present experience is of p, then I know that the present moment is an achronal region including p. Our knowledge of the privileged moment will not extend further than this however; our experience is casually connected only with things that were (within the past lightcone of p) and every moment of time which might be the privileged one agrees that those things were. And, of course, if the only point invoked in the account of my current experience is the point event of that experience p, then it must be noted that even the B-theory gives that point some special significance for my current thought and talk, albeit significance of a merely indexical sort.

The bolder response for the A-theorist is to reject the idea that the explanation of my present experience is a physical one. This would be to bolster the A-theory by appeal to a more thoroughgoing incompleteness of physics with respect to phenomenal experience. If STR is supplemented only with a privileged moment, the role of that additional structure is purely epiphenomal with respect to explanations of experience. If we don't like that, we can discard the excess structure, and with it the A-theory – or we can reject the adequacy of physical explanations of tensed experience in favor of a robustly non-physicalist A-theory on which the conscious experience as of a privileged present moment is both veridical and has no complete physical explanation. This may be a consistent proposal, but it certainly has the misfortune of yoking the A-theory to a speculative philosophy of mind whose substantive content appears to go well beyond the relatively banal grounds offered in favor of the A-theory in Section 7.1.

These orthodox supplementing A-theories are consistent with STR and the A-theory. If our prior confidence favors the A-theory, the availability of such packages may be enough to allow us to continue to maintain that belief in light of scientific evidence. Perhaps the tenability of the A-theory given STR is enough for some A-theorists (Zimmerman, 2011). But nothing in these packages looks like a positive reason to come to accept the A-theory, since the substantive explanatory work done by the supplemented theories is parasitic on that explanations provided by unsupplemented STR.

I wish to briefly mention an unorthodox supplementing theory (Bacon, 2018), on which A-theoretic moments of time are not to be identified with regions of spacetime at all, but are instead ways of mapping families of temporary properties to hyperplanes at which permanent correlates of those properties may be instantiated. These mappings are obviously not part of the manifold, and so this is a supplementing theory. The details are unfortunately too complicated to discuss here. The theory is arguably egalitarian about hyperplanes, since a time associates every hyperplane h with some tensed properties sufficient to correctly describe the entire manifold as if from the perspective of h. How then can we say that some events in the manifold are really happening while others are not? Bacon says that at each time, each hyperplane h is such that the events on it have a temporary property RH_h, but each timelike line is associated with a context that picks out one such property as the referent, in that context, of "really happening." Given my present context, just one hyperplane is really happening given the present time; but for every other hyperplane, there is a context (some of which are monstrous) where "really happening" picks out the property had just by events at that hyperplane. Whether this contextualist version of the A-theory provides enough metaphysical "oomph" to satisfy A-theoretic intuitions is debatable, but it does indicate the contortions that A-theories which try to respect egalitarianism about moments of time are forced into.

7.6   Whither the A-Theory?

There is no knock-down argument from STR against the A-theory, though we ought to lower our posterior confidence in that theory in light of relativistic physics. The A-theoretic picture of a wave of simultaneous happenings that sweeps through time, grounding tense in the four-dimensional manifold, is not straightforwardly compatible with STR, a tension dramatized by attempts to locate or impose a relation of absolute simultaneity on Minkowski spacetime. But A-theories more or less deserving of the name can be defended by weakening our aspirations for how absolute simultaneity is to be found in STR, by rejecting (5).

Over time, however – and especially given semantic and metaphysical difficulties for the A-theory and related doctrines such as propositional temporalism – it appears that attempts to marry the A-theory with STR are more trouble than they are worth. If our most widely adopted theories of truth conditions involve truth relative to worlds but not times, and more of us have prior confidence in "block universe" interpretations of the physics, there is decreasing pressure to find any fundamental theoretical role for moments of time – let alone some special metaphysical significance for a specific moment of time. These trends in philosophical views are no doubt accelerated by the widespread consideration of STR, and the clear difficulties in locating moments of time that behave as we wish in that theory. But "one can only extract so much metaphysics from a physical theory as one puts in" (Sklar, 1987, p. 291). We can put enough A-theoretic metaphysics into the pure theory of spacetime geometry to get out a consistent theory. But we don't need inconsistency to appear in order to realize we might be better off overall with the B-theory.[11]

Acknowledgments

Some of the material in this chapter has been presented at the Universities of Oxford, Adelaide, Sydney, and Melbourne, and the Munich Centre for Mathematical Philosophy; I'm grateful to audiences on those occasions for comments and questions. The chapter was substantively finished while I was visiting the MCMP funded by a fellowship from Alexander von Humboldt Foundation. While making final revisions I was funded in part by an Australian Research Council Discovery Project (DP200100190). I am grateful to all the named institutions for their support.

Notes

1 Zimmerman (2005) contests the claim that the distinction between A- and B-theories lines up with the debate over propositional temporalism.

2 Indeed, any consistent A-theory which involves the happening of non-simultaneous events – such as a "moving floodlight" – seems to be forced to adopt a broadly B-theoretic account of property ascriptions and change for those non-simultaneous happenings (involving covert temporal indices or the like).

3 My own presentation is very much influenced by the clear and accessible discussion in Geroch (1978). The chapters by Maudlin and Brown in this volume give more detail as well as some insight into the controversies over the interpretation of STR.

4 These observations about transitivity are reminiscent of Putnam's (1967) argument, though he considers the "real for" relation rather than simultaneity, which choice obscures rather more than it illuminates.

5 Each moment of time determines a past and future; if we systematically swap a given past and future, the only region that should remain invariant is the present moment of time.

6 Any rival semantics for *was* which classified some spacelike separated points as past could be used to extend the present from a single point to an extended region of mutually spacelike separated points, and would end up closer to definition (3).

7 Like an ordinary mosaic, the finished image doesn't betray the order of its construction.

8 Depending on details of how the tense operators are to be handled, we may also need to be inegalitarian about which moments of time wholly to the past (future) of N are involved in grounding truths about what was (will be). Zimmerman, for example, argues for a privileged foliation as well as a privileged moment.

9 Some argue that such additional structure can be motivated by consideration of physics going beyond STR: Dolby and Gull (2001) – but see Eagle 2005 – and Forrest (2008). We here stray into the territory of revisionary theories, however.

10 Sometimes this methodological objection is raised as the principle that the basic truths – including those about absolute simultaneity – ought to be invariant under spacetime symmetries (Earman 1989, pp. 45–47; Skow, 2015, p. 148). This boils down to a bare insistence that only conciliatory A-theories can be right, so is dialectically inappropriate.

11 Turner (2020), published as this chapter was being finalised, makes some related arguments.

Bibliography

Bacon, A. (2018). Tense and relativity. *Noûs*, 52(3): 667–696, doi:10.1111/nous.12187.

Bourne, C. (2006). *A Future for Presentism.* Oxford: Oxford University Press.

Brogaard, B. (2012). *Transient Truths.* New York: Oxford University Press, doi:10.1093/acprof:oso/9780199796908.001.0001.

Callender, C. (2007). Finding 'real' time in quantum mechanics. In W. Craig and Q. Smith (eds.), *Einstein, Relativity and Absolute Simultaneity.* London: Routledge, pp. 50–72.

Cameron, R. (2014). *The Moving Spotlight.* Oxford: Oxford University Press.

Deasy, D. (2014). The moving spotlight theory. *Philosophical Studies*, 172(8): 2073–2089, doi:10.1007/s11098-014-0398-5.

Dolby, C.E. and Gull, S.F. (2001). On radar time and the twin "paradox". *American Journal of Physics*, 69(12): 1257–1261, doi:10.1119/1.1407254.

Eagle, A. (2005). A note on Dolby and Gull on radar time and the twin "paradox". *American Journal of Physics*, 73(10): 976–979, doi:10.1119/1.1994855

Earman, J. (1989). *World Enough and Space-Time*, Cambridge, MA: MIT Press.

Earman, J. (1995). *Bangs, Crunches, Whimpers and Shrieks: Singularities and Acausalities in Relativistic Spacetimes.* Oxford: Oxford University Press.

Einstein, A. (1905). Zur Elektrodynamik Bewegter Körper. *Annalen Der Physik*, 17: 891–921.

Forrest, P. (2008). Relativity, the passage of time and the cosmic clock. In D. Dieks (ed.), *The Ontology of Spacetime II.* Amsterdam: Elsevier, pp. 245–253.

Geroch, R. (1978). *General Relativity from A to B.* Chicago: University of Chicago Press.

Gilmore, C., Costa, D. and Calosi, C. (2016). Relativity and three four-dimensionalisms. *Philosophy Compass*, 11(2): 102–120, doi:10.1111/phc3.12308.

Godfrey-Smith, W. (1979). Special relativity and the present. *Philosophical Studies*, 36(3): 233–244, doi:10.1007/BF00372628.

Hawley, K. (2009). Metaphysics and relativity. In R. Le Poidevin and R.P. Cameron (eds.), *The Routledge Companion to Metaphysics.* London: Routledge, pp. 507–516.

Kaplan, D. (1989). Demonstratives. In J. Almog, J. Perry and H. Wettstein (eds.), *Themes from Kaplan*. New York: Oxford University Press, pp. 481–563.

King, J.C. (2003). Tense, modality, and semantic values. *Philosophical Perspectives*, 17(1): 195–246, doi:10.1111/j.1520-8583.2003.00009.x.

Lewis, D. (1980). Index, context, and content. In S. Kanger and S. Ohman (eds.), *Philosophy and Grammar*. Dordrecht: D. Reidel, pp. 79–100.

Malament, D.B. (1977). Causal theories of time and the conventionality of simultaneity. *Nous*, 11: 293–300.

Markosian, N. (2004). A defense of presentism. In D.W. Zimmerman (ed.), *Oxford Studies in Metaphysics*, Vol. 1. Oxford: Oxford University Press, pp. 47–82.

McTaggart, J. (1908). I.–the unreality of time. *Mind*, 17(68): 457–474, doi:10.1093/mind/XVII.4.457.

Monton, B. (2006). Presentism and quantum gravity. In D. Dieks (ed.), *The Ontology of Spacetime II*. Amsterdam: Elsevier, pp. 263–280.

Partee, B. (1973). Some structural analogies between tenses and pronouns in English. *The Journal of Philosophy*, 70(18): 601–609, doi:10.2307/2025024.

Price, H. (1996). *Time's Arrow and Archimedes' Point*. Oxford: Oxford University Press.

Prior, A.N. (1959). Thank goodness that's over!. *Philosophy*, 34(128): 12–17, doi:10.1017/S0031819100029685.

Putnam, H. (1967). Time and physical geometry. *Journal of Philosophy*, 64(8): 240–247, doi:10.2307/2024493.

Richard, M. (1981). Temporalism and eternalism. *Philosophical Studies*, 39(1): 1–13, doi:10.1007/BF00354808.

Robb, A. (1921). *The Absolute Relations of Time and Space*. Cambridge: Cambridge University Press.

Sarkar, S. and Stachel, J. (1999). Did Malament prove the non-conventionality of simultaneity in the special theory of relativity? *Philosophy of Science*, 66(2): 208–220, doi:10.1086/392684.

Saunders, S. (2002). How relativity contradicts presentism. *Royal Institute of Philosophy Supplement*, 50: 277–292, doi:10.1017/S1358246100010602.

Savitt, S.F. (2000). There's no time like the present (in Minkowski spacetime). *Philosophy of Science*, 67(3): 563–574, doi:10.1086/392846.

Sider, T. (2001). *Four-Dimensionalism: An Ontology of Persistence and Time*. Oxford: Oxford University Press.

Sklar, L. (1987). Time, reality, and relativity. In *Philosophy and Spacetime Physics*. Berkeley: University of California Press, pp. 289–304.

Skow, B. (2015). *Objective Becoming*. Oxford: Oxford University Press.

Stein, H. (1991). On relativity theory and the openness of the future. *Philosophy of Science*, 58(2): 147–167, doi:10.1086/289609.

Turner, J. (2020). Why special relativity is a problem for the A-theory. *Philosophical Quarterly*, 70(279): 385–406, doi: 10.1093/pq/pqz051.

Wüthrich, C. (2012). The fate of presentism in modern physics. In R. Ciunti, K. Miller, and G. Torrengo (eds.), *New Papers on the Present*. Munich: Philosophia Verlag, pp. 91–132, http://arxiv.org/abs/1207.1490.

Zimmerman, D.W. (2005). The A-theory of time, the B-theory of time, and 'taking tense seriously'. *dialectica*, 59(4): 401–457 doi:10.1111/j.1746-8361.2005.01041.x.

Zimmerman, D.W. (2008). Defending an 'A-theory' of time. In T. Sider, J. Hawthorne and D.W. Zimmerman (eds.), *Contemporary Debates in Metaphysics*. Oxford: Blackwell, pp. 211–225.

Zimmerman, D.W. (2011). Presentism and the space-time manifold. In C. Callender (ed.), *The Oxford Handbook of Philosophy of Time*. Oxford: Oxford Universtiy Press, pp. 163–244.

Further Reading from the Editors

The tension between relativity and certain views of time was first articulated in H. Putnam, "Time and physical geometry" (*Journal of Philosophy*, 64(8): 240–247, 1967). A similar argument is put slightly differently in S. Saunders "How relativity contradicts presentism," (*Royal Institute of Philosophy Supplement*, 50: 277–292, 2002). M. Hinchliff, "A defense of presentism in a relativistic setting" (*Philosophy of Science* 67: S575–S586, 2002) is a response to these arguments. The original discussion of the A- and B-series can be found in J. McTaggart, "I.—the unreality of time" (*Mind*, 17(68): 457–474, 1908).

8

RELATIVISTIC CONSTRAINTS ON INTERPRETATIONS OF QUANTUM MECHANICS

Wayne C. Myrvold

8.1   Introduction

Though it is fairly noncontroversial that an empirically adequate quantum theory must be a quantum field theory, and must be able to treat of the relativistic regime of energies, much of the literature on interpretations of quantum theory has been focused on nonrelativistic quantum mechanics. There is some justification for this. Nonrelativistic quantum mechanics is conceptually and mathematically simpler, and does, after all, work very well in a low-energy regime. Any solution of the measurement problem, or any account of the ontology of quantum theory, must yield sensible results in this regime, and so nonrelativistic quantum mechanics can serve as a testing ground for the viability of interpretations.

It should not be assumed, however, that an approach to the interpretation of quantum mechanics will carry over straightforwardly to the context of a relativistic quantum field theory. There are two sorts of challenges to be met. One is adaptation of the approach to a theory with relativistic causal structure. Another challenge arises from the fact that relativistic quantum theories are *quantum field theories*, that is, quantum theories of systems with infinitely many degrees of freedom.

As we shall see below, each of these challenges can pose a serious difficulty for an approach to the interpretation of quantum mechanics. We will focus on four main avenues of approach: (i) additional beables theories, which include "hidden-variables" theories and modal interpretations, (ii) dynamical collapse theories, (iii) Everettian, or "many-worlds" interpretations, and (iv) non-realist interpretations, which deny that quantum states represent anything in physical reality independent of considerations of agents and their beliefs.

Summarizing briefly, the following conclusions will be drawn:

(i)   Additional beables theories face serious difficulties with relativistic causal structure. This can be dealt with by introducing a dynamically distinguished (albeit unobservable) foliation of spacetime, or else by making the additional beables relational, so that some quantities, such as pointer positions, that we think of as local beables are not local beables after all. Furthermore, modal interpretations that employ the spectral resolution of a system's quantum state to pick out the additional beables run into serious trouble when extended to quantum field theories, as they run the risk of picking out no non-trivial additional beables.

99

(ii)   Dynamical collapse theories can be adapted, in a fairly straightforward way, to relativistic causal structure. However, a collapse theory that respects the full spacetime symmetry of Minkowski spacetime and its causal structure can do so only at the price of introducing additional, non-standard quantum degrees of freedom that are unlike those that appear in familiar quantum field theories.

(iii)   Everettian interpretations fare better; they extend to the relativistic context with little or no adjustment.

(iv)   Some non-realist interpretations bypass difficulties with relativity by limiting the scope of application of quantum mechanics.

8.2   The Quantum Measurement Problem

If a quantum theory is capable, in principle, of affording a complete description of the physical goings-on in the world (or even just the laboratory), then it must be possible to model an experimental apparatus as a complicated array of quantum systems, and the interactions of the apparatus with the system of interest as a process conforming to the laws of quantum dynamical evolution. This, however, poses a problem, as typical interactions will result in a quantum state that is a superposition of terms corresponding to distinct experimental outcomes. The literature on what is (misleadingly) known as the interpretation of quantum mechanics deals with how to make sense of this.

Among approaches that attempt a realist construal of quantum theories, we can distinguish three main avenues of approach. One avenue accepts the usual, linear evolution of the quantum state, but does not saddle the quantum state with the burden of fully representing the physical world, but, rather, supplements it with additional structure. These are traditionally called "hidden-variables theories," but, as Bell (1981b, 1987a) has pointed out, this is misleading, since, according to a theory like this, it is via these additional variables that the world manifests itself. This is the reason we are calling theories of this sort *additional beables theories*. These include the de Broglie-Bohm pilot wave theory, and also modal interpretations.

Another avenue of attack modifies the dynamics of the quantum state so that, in appropriate circumstances, a process closely approximating the textbook collapse of the state vector occurs. This typically involves a stochastic modification of the dynamics, resulting in an indeterministic theory. The best-known theory of this type is the Ghirardi-Rimini-Weber (GRW) theory, on which collapses are instantaneous, discrete events; there is a continuous analogue, which is known as *continuous spontaneous localization* (CSL).

These two approaches modify the quantum formalism, either by adding additional structure or by modifying the dynamics. A third class of approaches retains the linear quantum dynamics, with its attendant superpositions of macroscopically distinct states, and attempts to construe a quantum state description as a complete description. These are *Everettian*, or *many-worlds* approaches to quantum theory.

An approach that does not fit neatly into these categories is the *relational interpretation* advocated by Carlo Rovelli. It is akin in some ways to Everett's original conception, which he called the *relative-state* interpretation. The relational interpretation differs from the relative state interpretation in not taking quantum states to be representational. For more on this, see Laudisa and Rovelli (2019), and references therein.

On any approach, empirical adequacy requires recovery of at least an approximation to the quantum probabilities for outcomes of experiments, insofar as these have been verified by experiment. This includes violations of Bell inequalities. It might seem that this, by itself, leads to an unavoidable conflict with relativistic causality. As we shall see, things are not so simple!

8.3 Relativistic Spacetimes

Compatibility with special relativity will be a concern in this chapter, and this amounts to the question of whether a theory can live happily in Minkowski spacetime, without adjoining any additional space-time structure, such as a privileged relation of distant simultaneity. But, of course, special relativity is false, and the spacetime in which we live and move and have our being is not Minkowski spacetime, but, as general relativity teaches us, a spacetime of variable and dynamic curvature. Minkowski space-time is of interest only as a local approximation, on scales at which curvature can be ignored. More-over, general relativity itself is, almost certainly, an approximation to a deeper theory that incorporates quantum gravity.

For this reason, we will be interested not only in Minkowski spacetime but in a wider class of relativistic spacetimes. The radical departure of special relativity from previous conceptions of spacetime lies in its rejection of a privileged global temporal order. In Galilean spacetime, for any event p, any other event either is in the past of p, the future of p, or is simultaneous with p, and the relation of simultaneity is, as one would expect, an equivalence relation. In Minkowski spacetime, events that are spacelike separated are neither to the past nor to the future of each other, and, although there will always be some inertial reference frame whose accompanying Lorentz coordinates assign the two events the same t-coordinate, this is not simultaneity in any physically significant sense.

In Minkowski spacetime, define a relation $\prec$ that holds between spacetime points p, q, when there is a future-directed curve from p to q that is everywhere timelike or lightlike. This relation is irreflexive (no point stands in that relation to itself), antisymmetric (if $p \prec q$ then it is not the case that $q \prec p$), and transitive.

Now define the relation $\sim$ of temporal unconnectedness that holds between two points p and q if neither $p \prec q$ nor $q \prec p$. Distinct points that are temporally unconnected are said to be *spacelike separated*. This relation is reflexive and symmetric, but not transitive. On the contrary, for any two points p, q such that $p \sim q$, one can always find a third, r, such that $p \sim r$ but $r \prec q$.

All of this remains true in the curved spacetimes of general relativity, provided that they are sufficiently well-behaved.[1] This is a striking difference between relativistic spacetimes and Galilean spacetime. In Galilean spacetime we have a relation $\prec$ of temporal precedence that is irreflexive and antisymmetric, but the corresponding relation of temporal unconnectedness is transitive, and hence spacetime can uniquely be partitioned into equivalence classes under this simultaneity relation.

In this chapter, we will be concerned with spacetimes equipped with a relation $\prec$ of temporal precedence, from which we define a relation $\sim$ of temporal unconnectedness. We will assume that $\prec$ is irreflexive and antisymmetric (no temporal loops). We will say that a spacetime's temporal structure is *causally relativistic* iff for any two points p, q such that $p \sim q$, there exists a point r, such that $p \sim r$ and $r \prec q$. Note that this is a condition on causal structure and not a condition of symmetry under translations or boosts (which, of course, are additional conditions that one might impose).

On the usual notion of causation, a cause must temporally precede its effects, and so, if an event p is a cause of an event q, we must have $p \prec q$. It is this that motivates the claim that special relativity forbids action at a distance; without it, there is no reason to think that causal relations between spacelike separated events are incompatible with relativistic spacetime structure. Given a causally relativistic spacetime, we will say that a theory is *causally local* if all causal relations that the theory affords respect the condition that the cause be in the past of the effect and, hence, that there are no cause-effect relations between temporally unconnected events. There seems to be no harm in assuming that any spacetime point to the past of a point q is potentially a locus of an event that is a cause of q. If we make this assumption, then the relations of temporal precedence and potential causal influence coincide, and we may, as is usual in discussions of spacetime structure, treat temporal structure and causal structure interchangeably.

The condition of causal locality – that is, the condition that there be no cause-effect relations between spacelike separated points – is independent of symmetries such as boost symmetries. One can invent theories that are causally local but not invariant under boosts, and one can invent theories that violate the condition of causal locality without picking out a preferred reference frame.

There is a tradition of invoking retrocausality – that is, backward-in-time causation – in explaining quantum phenomena. This approach has its roots in a suggestion by O. Costa de Beauregard (1976) and has been championed by a number of authors; see Friederich and Evans (2019) for an overview. The idea is to make causal connectability time-symmetric by permitting cause-effect relations in both temporal directions. On this approach, causal influences can propagate along timelike or lightlike lines in either temporal direction. Assuming that causal connectability is transitive, this turns the relation of causal connectability into the trivial relation that holds between any pair of spacetime points. Any approach of this sort is, therefore, straightforwardly causally nonlocal, in the sense we are using the phrase, as any two spacetime points, including those that outside of each other's light cones, are causally connected. As Kastner (2013, Ch. 6) has argued, an approach of this sort is capable of respecting the symmetries of Minkowski spacetime.

8.4   Quantum Theories in Relativistic Spacetimes

A quantum-mechanical theory is what results when one subjects a classical theory of a system with finitely many degrees of freedom (such as, for example, a finite number of particles) to quantization. A quantum field theory, however, results from quantization of a classical theory with infinitely many degrees of freedom. Introductory quantum field theory textbooks typically contain, in their early chapters, arguments that a relativistic quantum theory must be a quantum field theory. For a precise version of one such argument, see Malament (1996).

There are, therefore, two sorts of difficulties one might encounter when one seeks to extend to the relativistic domain an approach to the quantum measurement problem that has been formulated in terms of nonrelativistic quantum mechanics. One source of difficulties might lie in adapting the theory to relativistic causal structure. Another source of difficulties might stem from differences between quantum mechanics and quantum field theories.

Typically, relativistic quantum field theories are expressed in the Heisenberg picture, rather than the Schrödinger picture. The Schrödinger picture, or its analogue, can, however, be formulated in a relativistic spacetime. On this picture, we track the rate of change of an observable via a changing state vector. Take a foliation of spacetime into spacelike hypersurfaces (most conveniently, hyperplanes, but we can also consider nonflat surfaces), considered to be smooth, so that there is, at any point of each of these hypersurfaces, a vector normal to the surface. Integral curves of this vector field form a timelike congruence. Pick one hypersurface σ_0, and consider a field operator $\hat{\phi}(x)$. For each x_0 on σ_0, there will be a family of operators $\hat{\phi}(x)$ corresponding to points on the integral curve containing x_0, and, for any quantum state, one can consider how the expectation value of $\hat{\phi}(x)$ changes as one follows these timelike curves. This can be expressed, either in the Heisenberg picture, using the same state vector and different operators $\hat{\phi}(x)$ for different points on an integral curve, or in the Schrödinger picture, using the same operator $\hat{\phi}(x_0)$ for every spacetime point on the integral curve containing x_0, and different state vectors for evaluating expectation values of observables on different elements of our foliation.

There is also an analogue of the interaction picture, called the *Tomonaga-Schwinger picture*. On this picture, one writes the Lagrangian density as a sum of a free-field Lagrangian and a term incorporating interactions:

$$\mathcal{L}(x) = \mathcal{L}_0(x) + \mathcal{L}_I(x). \tag{8.1}$$

The operators employed are solutions of the free-field equations, and one associates, with each space-like Cauchy surface σ (whether flat or not), a state vector $|\psi(\sigma)\rangle$. Evolution from a surface σ to another, σ', differing by an infinitesimal deformation about a point x, satisfies the *Tomonaga-Schwinger equation* (Tomonaga, 1946; Schwinger, 1948):

$$i\hbar c \frac{\delta |\psi(\sigma)\rangle}{\delta \sigma(x)} = \mathcal{H}_I(x) |\psi(\sigma)\rangle . \tag{8.2}$$

Integration of this equation yields, for any Cauchy surfaces σ, σ', a unitary mapping from $|\psi(\sigma)\rangle$ to $|\psi(\sigma')\rangle$. This approach is not commonly discussed in recent textbooks but can be found in some older texts; see, e.g., Schweber (1961).

8.5 Additional Beables Theories

8.5.1 *De Broglie-Bohm Theory*

The best-known additional beables theory is the de Broglie-Bohm pilot wave theory. In this theory, ordinary physical objects consist of point particles, or corpuscles, whose motion is guided by the quantum wavefunction. The dynamical laws of theory are the Schrödinger equation for the evolution of the quantum wavefunction, and the guidance equation, which fixes the velocities of Bohmian corpuscles.

$$m_i \mathbf{v}_i(\mathbf{x}_1, \mathbf{x}_2, \ldots, \mathbf{x}_n, t) = \nabla_i S(\mathbf{x}_1, \mathbf{x}_2, \ldots, \mathbf{x}_n, t), \tag{8.3}$$

where $S/\hbar$ is the phase of the wavefunction:

$$S(\mathbf{x}_1, \mathbf{x}_2, \ldots, \mathbf{x}_n, t) = \hbar \, \mathrm{Im} \, \log \Psi(\mathbf{x}_1, \mathbf{x}_2, \ldots, \mathbf{x}_n, t). \tag{8.4}$$

It is assumed that, at some time t_0, the probability density for configurations of the corpuscles is given by the square of the wavefunction amplitude. If this holds for some time t_0, the dynamics, consisting of the Schrödinger equation and the guidance equation, ensures that it holds for all time. This property is known as *equivariance* of the Born-rule distribution.

8.5.1.1 *Causal Structure*

It can be seen from the guidance equation (8.3) that the velocity of any one of the corpuscles depends on an instantaneous configuration, and may, in principle, depend on the positions of arbitrarily many corpuscles, arbitrarily far away. Thus, the guidance equation requires a distinguished relation of distant simultaneity for its formulation.

As an illustration, consider a pair of spin-1/2 particles in the singlet state, spatially separated and sequentially subjected to Stern-Gerlach experiments for the same direction, say the z-direction, which we will regard as the vertical direction. The quantum state can be written as follows:

$$|\Psi(t)\rangle = \psi_{+-}(\mathbf{x}_1, \mathbf{x}_2, t) \, |z^+\rangle_1 \, |z^-\rangle_2 + \psi_{-+}(\mathbf{x}_1, \mathbf{x}_2, t) \, |z^-\rangle_1 \, |z^+\rangle_2 \tag{8.5}$$

Before the experiment, there is no correlation between position and spin-state, and the two wavepackets overlap completely.

$$\psi_{+-}(\mathbf{x}_1, \mathbf{x}_2, t_0) = \psi_{-+}(\mathbf{x}_1, \mathbf{x}_2, t_0). \tag{8.6}$$

The outcome of both experiments is determined by the position of the particle on which the experiment is conducted first. The Stern-Gerlach experiment on particle 1 separates the wavepackets ψ_{+-}

and ψ_{-+} along the z_1-direction. Since Bohmian trajectories cannot cross, there will be a plane or other two-dimensional surface in $\mathbf{x}_1$-space such that particle 1 will go in the $+$ direction if it is initially located above this plane, in the $-$ direction, if it is located below. After this experiment, there is no longer any overlap, in configuration space, between the two wavepackets, and the two particles are determinately located in only one of them. The second particle responds to its Stern-Gerlach apparatus accordingly, responding $+$ or $-$ depending on which wavepacket it is in.

This means that if the experiment on particle 1 is conducted first, the outcome of both experiments is determined by the initial position of particle 1. By parity of reasoning, if the experiment on particle 2 is conducted first, the outcome of both experiments is determined by the initial position of particle 2. These outcomes could differ, meaning that, for the same initial wavefunction and initial configuration of the particles, the outcomes of the experiments could differ depending on which experiment is conducted first. Thus, the theory requires there to be a matter of fact about which experiment is conducted first, even if the experiments are conducted at spacelike separation.

We might ask: is this an artifact of the way this theory is formulated, or is it inherent in the very project of formulating a theory of this type, on which particles have definite trajectories and probabilities over configurations are to be given by the Born rule, with dynamics for the particles that ensures that this probability distribution is equivariant?

As Berndl et al. (1996) have shown, the answer to this question is that a theory of this type must, indeed, invoke a privileged foliation, as it is impossible for the distribution postulate to be satisfied on all spacelike hyperplanes.

Suppose that we have a pair of spin-1/2 particles, initially both located in the same small region. The particles are then separated a large distance, and each is passed through a Stern-Gerlach apparatus oriented in the x-direction, which separates the wavepacket (in configuration space) along the x-axis of the particle it's interacting with. The beams are then carefully recombined, with care taken not to let the particles interact with anything that might induce decoherence, and then they are passed through Stern-Gerlach devices oriented in the z-direction. We suppose that all of this splitting, recombining, and resplitting of the two beams occurs at spacelike separation.

We consider four hypersurfaces (see Figure 8.1):

1   A hypersurface α, on which both particle 1 and particle 2 have their wavepackets split into $x+$ and $x-$ wavepackets.
2   A hypersurface β, on which both particle 1 and particle 2 have their wavepackets split into $z+$ and $z-$ wavepackets.
3   A hypersurface γ, on which the wavepackets are split along z_1 and x_2 axes, respectively.
4   A hypersurface δ, on which the wavepackets are split along x_1 and z_2 axes, respectively.

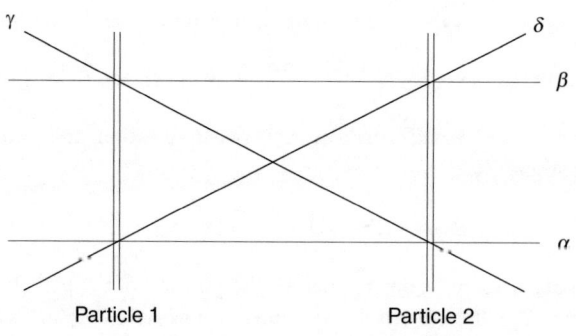

Figure 8.1 The hypersurfaces considered in the proof.

It is not, in fact, always possible to satisfy the distribution postulate along all four of these hypersurfaces. Suppose that our pair of spin-1/2 particles is prepared in the Hardy state:

$$|\psi\rangle_H = \frac{1}{\sqrt{12}} \left(|x^+\rangle_1 |x^+\rangle_2 - |x^+\rangle_1 |x^-\rangle_2 - |x^-\rangle_1 |x^+\rangle_2 - 3 |x^-\rangle_1 |x^-\rangle_2 \right) \qquad (8.7)$$

$$= \frac{1}{\sqrt{3}} \left(|z^+\rangle_1 |z^-\rangle_2 - |z^+\rangle_1 |z^+\rangle_2 + |z^-\rangle_1 |z^+\rangle_2 \right) \qquad (8.8)$$

$$= \frac{1}{\sqrt{3}} \left(|z^-\rangle_1 |z^+\rangle_2 - \sqrt{2} \; |z^+\rangle_1 |x^-\rangle_2 \right) \qquad (8.9)$$

$$= \frac{1}{\sqrt{3}} \left(|z^+\rangle_1 |z^-\rangle_2 - \sqrt{2} \; |x^-\rangle_1 |z^+\rangle_2 \right). \qquad (8.10)$$

Let us write the states on these four hypersurfaces, with the position degrees of freedom included, writing $|+\rangle$ for the particle being in the $+$ beam, $|-\rangle$ for its being in the $-$ beam.

$$|\psi(\alpha)\rangle = \frac{1}{\sqrt{12}} \left(|x^+\rangle_1 |+\rangle_1 |x^+\rangle_2 |+\rangle_2 - |x^+\rangle_1 |+\rangle_1 |x^-\rangle_2 |-\rangle_2 \right.$$
$$\left. - |x^-\rangle_1 |-\rangle_1 |x^+\rangle_2 |+\rangle_1 - 3 |x^-\rangle_1 |-\rangle_1 |x^-\rangle_2 |-\rangle_2 \right) \qquad (8.11)$$

$$|\psi(\beta)\rangle = \frac{1}{\sqrt{3}} \left(|z^+\rangle_1 |+\rangle_1 |z^-\rangle_2 |-\rangle_1 - |z^+\rangle_1 |+\rangle_1 |z^+\rangle_2 |+\rangle_2 + |z^-\rangle_1 |-\rangle_1 |z^+\rangle_2 |+\rangle_1 \right) \qquad (8.12)$$

$$|\psi(\gamma)\rangle = \frac{1}{\sqrt{3}} \left(|z^-\rangle_1 |-\rangle_1 |z^+\rangle_2 |+\rangle_2 - \sqrt{2} \; |z^+\rangle_1 |+\rangle_1 |x^-\rangle_2 |-\rangle_2 \right) \qquad (8.13)$$

$$|\psi(\delta)\rangle = \frac{1}{\sqrt{3}} \left(|z^+\rangle_1 |+\rangle_1 |z^-\rangle_2 |-\rangle_2 - \sqrt{2} \; |x^-\rangle_1 |-\rangle_1 |z^+\rangle_2 |+\rangle_2 \right). \qquad (8.14)$$

If the distribution postulate is satisfied on α, there is a probability 1/12 that both particles are in their respective x^+ beams. If the postulate is satisfied on γ, then in such cases particle 1 must be in its z^- beam on γ. If the postulate is satisfied on δ, then particle 2 must be in its z^- beam on δ. But this means that, on β, both particles will be in the z^- beams, which should have probability zero. This leads to the conclusion that: the distribution postulate cannot be satisfied along all foliations, and the dynamics for the particles must be choosy about which, if any, foliation, it is with respect to which equivariance is to be maintained.

The argument does not depend on preparing the Hardy state exactly, or that our theory yields quantum probabilities exactly. On a theory of this sort, there must be a joint probability distribution over the positions of the particles at each point of intersection of the hyperplanes considered that yields the correct correlations on each of the four hyperplanes. As demonstrated by Fine (1982a,b), violation of a Bell-type inequality for these variables is equivalent to impossibility of a joint distribution of that sort.

For example, if there is a joint distribution over these variables, we should have

$$Pr(x_1^+(\alpha), x_2^+(\alpha)) \leq Pr(x_1^+(\delta), z_2^+(\delta)) + Pr(z_1^+(\gamma), x_2^+(\gamma)) + Pr(z_1^-(\beta), z_2^-(\beta)), \qquad (8.15)$$

which is violated by the Hardy state.

A trajectory theory must choose a foliation along which the distribution postulate is to be satisfied. Moreover, it is fairly easy to see that, in order to maintain the distribution along this foliation, the dynamics must permit causal dependence of the behavior of particles on choice of experiment at spacelike separation.

Consider three scenarios.

A. For particle 1, operations proceed as above, recombining the beams and resplitting them into z^+ and z^- wavepackets. Nothing is done to particle 2 between α and β, and there is a split into x^+ and x^- wavepackets for particle 2 on both α and β.

B. Nothing is done to particle 1 between α and β, and it remains split into x^+ and x^- wavepackets on both α and β. Recombining and resplitting proceed as above, for Particle 2.

C. Recombining and resplitting proceed as above, for both particles.

Suppose that we have dynamics according to which if the probabilities for positions of particles match the Born-rule probabilities on α, they do so on β, for all three scenarios. Suppose, also, in scenarios A and B, the particle to which nothing is done does not hop from one of its wavepackets to another. That is, in scenario A, if particle 2 is in the x^+ beam on α, it is in the x^+ beam on β.

These conditions can be satisfied only if, for some initial configurations on α, transition probabilities – that is probabilities of a particle ending up in a certain position on β, given the configuration on α – for one particle depend on what is being done to the other particle.

To see this, suppose the contrary. Suppose that, for any configuration on α, the probability that particle 1 ends up in the z^+ beam on β is the same, whether scenario A or C is in effect, and that the probability that particle 2 ends up in the z^+ beam on β is the same, whether scenario B or C is in effect. Then, since, on Scenario A, particle 2 is in the same beam on α and β, if the Born rule is satisfied on β, it is also satisfied on γ. If transition probabilities for particle 1 are the same in scenarios A and C, it follows that, in scenario C, the Born rule is satisfied on γ.

Running the same argument with the roles of particles 1 and 2 reversed yields the conclusion that if in scenario B the Born rule is satisfied on β, and if transition probabilities for particle 2 are the same in scenarios B and C, it follows that in scenario C the Born rule is satisfied on δ. But, as we have seen, in scenario C, the Born rule cannot be satisfied on all four hypersurfaces.

8.5.1.2   *Bohmian Quantum Field Theories*

Even though there are good reasons to think that an additional variables theory must employ a preferred foliation, the question still arises whether there could be a theory of this sort that recovers the empirical content of relativistic quantum field theories. This would, of course, entail that the preferred foliation is empirically inaccessible. Two avenues of approach suggest themselves for extending Bohmian theories to the context of quantum field theories. One is to retain a basic particle ontology, and permit particle creation and annihilation. The other is to take classical field configurations as the basic ontology, and to provide a non–classical dynamics for the evolution of these field configurations. For an overview of these options, see Struyve (2011). Struyve concludes that the field approach works best for bosonic fields, but is problematic when one attempts to extend it to fermionic fields. For fermionic fields, the particle approach seems more appropriate.

The task of extending additional beables theories to the context of quantum field theories remains a work in progress. However, there do not seem to be in-principle obstacles to the creation of a theory of this sort that recovers the phenomena that are currently handled by quantum field theories.

8.5.2   *Modal Interpretations*

Modal interpretations supplement the quantum state with additional beables. A variety of interpretations along these lines have been proposed; see Lombardi and Dieks (2017) for an overview. Many of these share the feature that the nature of the additional beables depends on the quantum state. Given a quantum system α, which may be a subsystem of a larger system, one considers the reduced state of α, obtained by tracing out the degrees of freedom of the rest of the system. If the state of the larger system is a pure quantum state, the reduced state of α will be pure if α is not entangled with the rest of the system, mixed, if it is. One resolves the density operator representing the reduced state into its spectral components:

$$\rho_\alpha = \sum_i w_i\, P_i, \tag{8.16}$$

where $\{P_i\}$ are mutually orthogonally projections. In the non-degenerate case, in which all nonzero coefficients $\{w_i\}$ are distinct, this resolution is unique. The *basic modal rule* assigns definite values 1 or 0 to these projections; one of these projections represents a property definitely possessed by the system, the others, definitely absent. The probability that P_i is the possessed property is w_i.

8.5.2.1 *Difficulties Arising from Relativistic Causal Structure*

Can modal interpretations respect relativistic causal structure, or do they require a distinguished relation of distant simultaneity? Dickson and Clifton (1998) proved, for a broad class of modal interpretations, that it is impossible to furnish the theory with transition probabilities in such a way as to yield Born-rule probabilities for possessed values on arbitrary hyperplanes. Dickson and Clifton's proof relies on an assumption concerning the transition probabilities for possessed values, the assumption they call "stability," but, as Arnztenius (1998) pointed out, the stability requirement is dispensable and the core of the proof concerns the nonexistence of certain joint distributions yielding the appropriate Born probabilities as marginals. Taking these proofs as a starting point, Myrvold (2002, 2009) adapted the argument of Berndl et al. (1996), outlined above, to show that any modal interpretation, so long as the possessed values of a system are capable of being regarded as local beables, cannot have Born-rule probabilities for these possessed values satisfied on arbitrary spacelike hyperplanes. The argument at the end Section 8.5.1.1, above, can be straightforwardly adapted to show that these theories, also, must permit causal dependence of the behavior of the additional beables on choice of experiment at spacelike separation.

Undeterred, Berkovitz and Hemmo (2005a, 2005b, 2006) have proposed a relational modal interpretation. The interpretation assigns properties to systems that are not thought to be intrinsic to that system, but are possessed by it only relative to some other system. In a relativistic context, these relational values may depend on what hypersurface is used to define a global state of the universe, and it can happen, in the presence of entanglement at spacelike separation, that some property of a system, at a single position on its worldline, may have different values relative to different hypersurfaces. Thus, the possessed values are not to be thought of as local beables, and, as the authors argue in some detail, the no-go theorems don't apply.

8.5.2.2 *Difficulties Arising from Infinitely Many Degrees of Freedom*

In addition to the difficulties for modal interpretations arising from relativistic causal structure, there is another difficulty for attempts to extend modal interpretation to the context of relativistic quantum field theories, arising from infinitude of degrees of freedom.

There is, first, the issue of extending the basic modal rule to the context of a relativistic quantum field theory. We want to be able to apply the rule to local subsystems of the universe. If we consider a bounded region R of spacetime, and the algebra of operators $\mathcal{A}(R)$ formed from fields in R, there is a difficulty in applying the basic modal rule, since, as pointed out by Dieks (2000), these local algebras will not contain either a density operator representing the state, or projections onto finite-dimensional subspaces. Clifton (2000) provided a version of the modal rule that makes sense in the context of algebraic quantum field theory, and also showed that there exists a whole host of states for which the rule picks out no definite observables other than multiples of the identity (that is, observables that have the same value no matter what the state is). Earman and Ruetsche (2005) extend Clifton's result, and discuss the significance of these results for modal interpretations of quantum theory.

These trivialization results mean that a modal interpretation of this sort cannot respect relativistic causal structure and symmetries while accomplishing the goal of providing a solution to the measurement problem. The purpose of a modal interpretation is to have observables such as apparatus pointer observables take on definite values in physically realistic states. Obviously, this is not accomplished if no observables are definite other than multiples of the identity.

8.6   Collapse Theories

A dynamical collapse theory modifies the evolution of the quantum state. One approach that might be tried would be to replace the deterministic, unitary evolution of the quantum state by a deterministic and non-unitary evolution. There are arguments to the effect that, given plausible assumptions, this would lead to the possibility of superluminal signaling (Gisin, 1989; Simon et al., 2001).[2] For this reason, in this section we focus on stochastic collapse theories.

On a theory such as this, the dynamical laws do not specify a unique future state, given the state at a time t_0; rather, which state will be realized is, on such a theory, a matter of chance. The idea is to specify a probabilistic law that approximates the usual unitary, deterministic evolution within those domains where we have good evidence for it, but departs sufficiently from it in some situations – which should include experiments that lead to recorded outcomes – so as to closely approximate the textbook collapse postulate, according to which some definite outcome obtains at the end of an experiment, with probabilities as to which outcome is obtained given by the Born rule. One theory of this sort, with unitary evolution punctuated by discrete jumps, is the GRW theory, referred to by the authors as *Quantum Mechanics with Spontaneous Localization* (QMSL) (Ghirardi et al., 1986). Another is the *Continuous Spontaneous Localization* (CSL) theory (Pearle, 1989; Ghirardi et al., 1990).

Among the experiments to be accounted for are tests of Bell inequalities. Since our theory is required to produce correct probabilities for the outcomes of such experiments, we do not expect the probabilistic law to be one on which spacelike separated events are probabilistically independent. It might seem obvious, at first, that this dooms compatibility with relativity from the start.

Things are not so simple, however. Bell himself said that when he first encountered the GRW theory, "I thought I could blow it out of the water, by showing it was *grossly* in violation of Lorentz invariance" (Bell, 1989, 17). A careful examination led him to the opposite conclusion, namely, that, though couched in terms of nonrelativistic quantum mechanics, the theory exhibits "a residue, or at least an analogue, of Lorentz invariance" (Bell, 1987a, p. 10) in Bell (1987b, pp. 206–207). The idea is this. On the usual, Schrödinger, dynamics, for two (or more) systems, widely separated from each other and not interacting with each other, evolution of each system is independent, and we can employ separate time variables for each system. The Schrödinger dynamics do not require any notion of synchronization of the time variables. This is a feature that Bell calls *relative time translation invariance*. The question he addressed is: for systems like that, does the GRW theory retain relative time translation invariance, or do its probabilistic laws require there to be a matter of fact about the temporal order of events at a distance? Bell's verdict is that the probability of any sequence of GRW jumps is relative time translation invariant.

> I am particularly struck by the fact that the model is as Lorentz invariant as it could be in the nonrelativistic version. It takes away the ground of my fear that any exact formulation of quantum mechanics must conflict with fundamental Lorentz invariance. (Bell 1987a, p. 14, in Bell 1987b, p. 209)

In his Trieste lecture, delivered in November 1989, Bell discussed the prospects for a genuinely relativistic version of a dynamical collapse theory, and concluded that the difficulties encountered by Ghirardi, Grassi, and Pearle in producing a genuinely relativistic version of CSL, a theory that would be "Lorentz invariant, not just for all practical purposes but deeply, in the sense of Einstein, eliminating entirely any privileged reference system from the theory" (Bell, 2007, p. 2931), were "Second-Class Difficulties," technical difficulties, and not deep conceptual ones (see also Bell 1989, p. 25).

A relativistic collapse theory, if it is to yield even an approximation to the usual quantum probabilities, will include correlations between events at spacelike separation that violate Bell inequalities. The Bell inequalities can be derived from the condition that correlations be *locally explicable*. This has two parts: that the correlations be explicable in a certain sense, and the explanation be local. The explicability condition was taken by Bell to be the condition that correlations between events that are not in a direct cause-effect relation with each other be attributable to a common cause (see Bell 1981a, C255 in Bell 1987b, p. 152). This condition is violated in relativistic collapse theories, which involve probabilistic correlations between spacelike separated events that are *not* attributable in the usual way to events in their common past. Unlike cause-and-effect relations as usually conceived, the relation of probabilistic correlation between these distant events is symmetric, and does not require a temporal order between the events. For this reason, it is misleading to assimilate this relation to the causal relation and refer to it as nonlocal "influence" or "action at a distance." See Myrvold (2016) for further discussion.

As we shall see, there are difficulties associated with constructing a fully relativistic collapse theory. These difficulties arise, not because a collapse theory cannot be formulated in the absence of a distinguished relation of distant simultaneity, but rather from the problem of constructing a relativistic collapse theory with a stable vacuum state. In order to see how all this works, we start with a few words about collapse theories in general.

8.6.1 *Stochastic Quantum State Evolution*

It is generally accepted that any physically realizable change of the state of a quantum system must be a *completely positive* mapping of the system's state space into itself. Any such mapping can be represented by a set $\{K_i\}$ of operators, such that the density operator ρ representing the state undergoes the change

$$\rho \rightarrow \sum_i K_i \, \rho \, K_i^{\dagger}, \tag{8.17}$$

where

$$\sum_i K_i^{\dagger} K_i \leq \mathbb{1}. \tag{8.18}$$

If $\sum_i K_i^{\dagger} K_i = \mathbb{1}$, the operation is called a *nonselective operation*; if $\sum_i K_i^{\dagger} K_i < \mathbb{1}$, it is *selective*. Such a representation of a completely positive mapping of the state space into itself is called a *Kraus representation* of the mapping; and the operators $\{K_i\}$, *Kraus operators*.

Selective operations reduce the trace-norm of the density operator, but this is not an issue, as normalization is only a convention. With an unnormalized density operator, we compute the expectation value of an observable represented by an operator A via

$$\langle A \rangle_{\rho} = \mathrm{Tr}(\rho A)/\mathrm{Tr}(\rho). \tag{8.19}$$

In the simplest case, the set of Kraus operators is a singleton, and the evolution is deterministic; this includes the case of unitary evolution. However, one can also consider stochastic processes. Suppose that we have a set of operators $\{K_i\}$, such that the state vector undergoes a stochastic transition of the following form. For some i,

$$|\psi\rangle \Rightarrow |\psi'\rangle = K_i |\psi\rangle / \|K_i |\psi\rangle \|, \tag{8.20}$$

with the probability for which transition it undergoes given by

$$p_i = \| K_i \, |\psi\rangle \|^2. \tag{8.21}$$

Since these probabilities must sum to unity for every vector $|\psi\rangle$, we must have $\sum_i K_i^\dagger K_i = \mathbb{1}$. For each i, we have a selective operation. One of these yields the actual state. The transition to the mixture of these candidate states, which corresponds to the proposition that some one of these transitions has occurred, without specification of which, is given by weighted mean of these.

$$\rho \to \bar{\rho}' = \sum_i p_i \frac{K_i \rho K_i^\dagger}{\mathrm{Tr}(K_i \rho K_i^\dagger)}. \tag{8.22}$$

In (8.20), we indexed the possible state transitions via a discrete index i. We can also consider stochastic processes of a more general sort. Let $\langle \Gamma, \mathcal{F}, \mu \rangle$ be a probability space.[3] Suppose that we have a family of operators $\{K_\gamma\}$, for $\gamma \in \Gamma$, such that

$$\int_\Gamma K_\gamma^\dagger K_\gamma \, d\mu(\gamma) = \mathbb{1}. \tag{8.23}$$

These can serve as the operators that induce our state transitions.

This gives us a rather general schema for a stochastic process in a Hilbert space. For the moment, we will presume that we have a unique global time function; extension to a relativistic spacetime will be considered in the next subsection.

Any theory satisfying the following conditions will give us stochastic evolution of the state vector.

1   We assume a measure space $\langle \Gamma, \mathcal{F}, \mu \rangle$. The elements of Γ are to be thought of as possible complete histories, specifying events for all times. For each time t there is a σ-algebra $\mathcal{F}_t$, whose elements are to be thought of as sets of possible histories up to time t. For $t < s$, we will have $\mathcal{F}_t \subseteq \mathcal{F}_s \subseteq \mathcal{F}$.

2   Given a history up until time t, and a state vector $|\psi(t)\rangle$, the state at a later time $t+s$ is a random variable, defined as follows.

   (a)   We assume a family of operators $K_\gamma(s; t)$, which depend only on events up until time s (that is, if two histories γ, γ' agree up to time s, then $K_\gamma(s; t) = K_{\gamma'}(s; t)$), such that

$$\int_\Gamma K_\gamma^\dagger(s; t) K_\gamma(s; t) \, d\mu = \mathbb{1}.$$

   (b)   For some γ,

$$|\psi(t+s)\rangle = \frac{K_\gamma(s; t) \, |\psi(t)\rangle}{\| K_\gamma(s; t) \, |\psi(t)\rangle \|}.$$

   (c)   The probability distribution for the realized outcome is given by

$$\mathrm{Prob}(\gamma \in F) = \int_F \| K_\gamma(s; t) \, |\psi(t)\rangle \|^2 \, d\mu.$$

   for any F in $\mathcal{F}_s$.

On a theory of this sort, given a state $|\psi(t_0)\rangle$, for any later time $t_1 = t_0 + s$ there will be an actual state vector $|\psi(t_1)\rangle$, and its corresponding pure-state density operator

$$\rho(t_1) = |\psi(t_1)\rangle \langle \psi(t_1)|. \tag{8.24}$$

There will also be what may be called an *ensemble, mean* density operator, which is a weighted average of the various possibilities for $\rho(t_1)$, weighted by their respective probabilities.

$$\bar{\rho}(t_1; t_0) = \int_\Gamma K_\gamma(t_1; t_0) \, \rho(t_0) \, K_\gamma^\dagger(t_1; t_0) \, d\mu. \tag{8.25}$$

8.6.2   Collapse Theories in Minkowski Spacetime
8.6.2.1   Causality Conditions

It is useful to work within what may be called the *stochastic Tomonaga-Schwinger picture*. We start with field operators that are solutions of the standard Heisenberg-picture equations (these may be free or interacting fields). We associate with each spacelike Cauchy surface σ a state vector $|\psi(\sigma)\rangle$. Our theory should deliver a stochastic evolution of $|\psi(\sigma)\rangle$ to $|\psi(\sigma')\rangle$ whenever σ' is nowhere to the past of σ. This will go much as it did in Section 8.6.1.

1   We assume a measure space $\langle \Gamma, \mathcal{F}, \mu \rangle$. The elements of Γ are to be thought of as possible complete histories, specifying events for all times. For each spacelike Cauchy surface σ there is a σ-algebra $\mathcal{F}_\sigma$, whose elements are to be thought of as sets of histories to the past of σ. For two Cauchy surfaces σ, τ, with $\tau \preceq \sigma$, we will have $\mathcal{F}_\tau \subseteq \mathcal{F}_\sigma \subseteq \mathcal{F}$.

2   Given two Cauchy surfaces σ, τ, with $\tau \preceq \sigma$, let δ be the spacetime region between them. Given a state vector $|\psi(\tau)\rangle$, the state vector $|\psi(\sigma)\rangle$ is a random variable, which is defined as follows.

   (a)   We assume a family of operators $\{K_\gamma(\delta)\}$, which depend only on events in the past of σ (that is, if two histories γ, γ' agree on events in the past of σ then $K_\gamma(\delta) = K_{\gamma'}(\delta)$), such that

$$\int_\Gamma K_\gamma^\dagger(\delta) K_\gamma(\delta)\, d\mu = \mathbb{1}.$$

   We call these *evolution operators* for the region δ.

   (b)   For some γ,

$$|\psi(\sigma)\rangle = \frac{K_\gamma(\delta)\, |\psi(\tau)\rangle}{\| K_\gamma(\delta)\, |\psi(\tau)\rangle \|}.$$

   (c)   The probability distribution for the realized outcome is given by

$$\mathrm{Prob}(\gamma \in F) = \int_F \| K_\gamma(\delta)\, |\psi(\tau)\rangle \|^2\, d\mu.$$

   for any F in $\mathcal{F}_\sigma$.

Here, again, we can define a mean density operator for the state on σ, given a state vector on τ, with $\tau \subseteq \sigma$.

$$\bar{\rho}(\sigma; \tau) = \int_\Gamma K_\gamma(\delta)\rho(\tau)K_\gamma^\dagger(\delta)\, d\mu. \tag{8.26}$$

Consider two Cauchy surfaces, τ, σ, which coincide everywhere except on the boundaries of two bounded regions δ_1, δ_2, where σ is to the future of τ (see Figure 8.2). We must have a unique law of stochastic evolution from τ to σ, through $\delta_1 \cup \delta_2$, and this should coincide with the composition of the evolution through δ_1 and the evolution through δ_2, in either order. The necessary and sufficient condition for this is:

3   Evolution operators corresponding to spacelike separated regions commute.

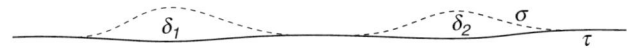

Figure 8.2 Spacelike separated regions between two spacelike hypersurfaces.

One difference between theories of this sort and a theory involving deterministic, unitary evolution is that for calculating probabilities of outcomes of experiments performed in some spacelike slice α that is common to two Cauchy surfaces τ, σ, it can make a difference whether the calculation is made using $|\psi(\tau)\rangle$ or $|\psi(\sigma)\rangle$. It is this feature that has given rise to concerns that a relativistic collapse theory must either be incoherent or exhibit radical contextualism about possessed properties. These concerns are misplaced; more on this in Section 8.6.2.2. However, it *would* be problematic if probabilities for events in α differed depending on whether one used $|\psi(\tau)\rangle$ or $\bar{\rho}(\sigma)$, which is the probabilistically weighted mixture of various possibilities for $|\psi(\sigma)\rangle$, given the state on τ. Moreover, if the evolution from τ to σ, through a region δ, spacelike separated from α, depended on a choice of experiments to be made, then, if different choices yielded different mixtures $\bar{\rho}(\sigma)$, the choice of experiment could be used for superluminal signaling. We, therefore, require $\bar{\rho}(\sigma)$ to yield the same probabilities for outcomes of experiments as $|\psi(\tau)\rangle$, for any allowable evolution through δ. The necessary and sufficient condition for this is (Arias et al., 2002):

4   Evolution operators pertaining to a region δ commute with all operators representing observables at spacelike separation from δ.

Conditions 3 and 4 ensure compatibility with relativistic causal structure.

8.6.2.2   Collapse Theory Ontology

If the region between two Cauchy surfaces τ, σ is a bounded spacetime region δ, then the transition from the state on τ to the state on σ can be attributed to events within δ. Quantum states, however, are in some sense "non local" – a better term is *nonseparable*. They cannot be regarded as supervening on local matters of fact. A quantum state encodes not merely local matters of fact but also correlations between distant regions.

This is reflected in the way that a collapse affects the global state. Let τ and σ be two spacelike hypersurfaces that coincide everywhere except on the boundaries of a bounded region δ, where σ lies to the future of τ. Let α be a bounded region that lies in both τ and σ, and hence is at spacelike separation from δ (see Figure 8.3). The two states can differ on probabilities assigned to observables pertaining to α. For example, consider the familiar EPR-Bohm scenario. Suppose the state on τ involves an entangled pair of spin-1/2 particles, one of which is located (to the extent that it can be located) within α, and suppose that in δ an experiment is performed on the other one, yielding, on σ, a near eigenstate of spin in some direction a that is correlated by the state on τ with spin of the other particle in direction b. Then the state on τ is a near eigenstate of b-spin for the second particle. Thus, the two states $|\psi(\tau)\rangle$ and $|\psi(\sigma)\rangle$ differ on probabilities assigned to experiments performed in the forward domain of dependence of α.

Relativistic causality precludes events occurring in a region δ from affecting *local* beables pertaining to regions at spacelike separation from δ. If one thought that the differing probabilities assigned to observables pertaining to α by the states $|\psi(\tau)\rangle$ and $|\psi(\sigma)\rangle$ were local beables, this would give rise to a serious worry about the conceptual coherence of a relativistic collapse theory (see, e.g., Maudlin 1994, p. 209 for worries along these lines).

Figure 8.3 A region common to two spacelike hypersurfaces.

But the reduced states of α obtained from the global states $|\psi(\tau)\rangle$ and $|\psi(\sigma)\rangle$ are not obviously local matters of fact about α, and, indeed, what these reduced states are depends on events at spacelike separation from α. What *can*, and *should* be regarded as the intrinsic state of the spacetime region α, that is, its state stripped of implicit references to goings-on at a distance from α, is the past light cone state. This is the limiting state obtained from a sequence of spacelike hypersurfaces that converge on the past light cone of α. This state will reflect collapse events to the past of α, but not collapse events at spacelike separation.

This has implications for what the local beables of the theory could be. Ghirardi and Grassi (1994, p. 419) proposed a criterion for property attribution in the context of relativistic collapse theories. If A is a local observable with support in α, a system possesses a property $A = a$ when the expectation value of P_a, the projection onto the subspace $A = a$, is extremely close to 1, evaluated with respect to the past light cone state of α.

Ghirardi, Grassi, and Benatti (1995) proposed, as an appropriate local beable for collapse theories, a smeared mass density, with the smearing taken over lengths scales large compared to the atomic scale but small on a macroscopic scale. For a relativistic theory, the corresponding quantity would be a smeared energy-momentum density, whose 00 component is mass-energy density, which, in a nonrelativistic context, is dominated by the rest mass. Combining the choice of mass density as local beable with the past light cone criterion for property attribution, the mass density that in a relativistic context is taken to be a local beable in a region α is the mass density defined by the state on the past light cone of α (Tumulka, 2007; Bedingham et al., 2014). See Myrvold (2016, 2018, 2019) for further discussion of ontological issues in collapse theories.

8.6.2.3 A Problem: Stability of the Vacuum

There is a difficulty with constructing a physically sensible collapse theory that respects the full symmetries of Minkowski spacetime. This has to do with the stability of the vacuum. A relativistic collapse theory must have a stable vacuum, which is not achievable via a straightforward extension of CSL to a relativistic quantum field theory.

If the theory were to produce excitations of the vacuum, it would have to furnish a probability distribution over possible excitations. The difficulty is the nonexistence of a bounded Lorentz-invariant measure over possible excitations.

This is readily seen in the simplest case, a free theory.[4] Let Δ be some subset of momentum space, and suppose that our collapse theory yields a certain probability $p(\Delta)$ per unit time of producing from the vacuum a particle with momentum in Δ. Lorentz invariance requires that there be the same probability for producing a particle with its momentum in any Lorentz boost of this volume.

This requirement cannot be satisfied. The problem is that there is no bounded invariant measure on the mass-shell consisting of all four-momenta of a given magnitude m. Any probability measure on the mass-shell, therefore, must break Lorentz symmetry. In the vacuum state, which is a Lorentz-invariant state, there is nothing that could be used to define such a measure. To generalize beyond free theories, the issue is the absence of a bounded Lorentz-invariant measure over the space of possible excitations of the vacuum.

The requirement of a strictly stable vacuum applies only to theories that respect the full set of spacetime symmetries of Minkowski spacetime. One could, for example, have a theory on a lattice, with relativistic causal structure. The rate of energy increase could, for such a theory, be kept finite. Whether it could be kept below the threshold of current observations would depend on the lattice spacing and other parameters of the theory; see Ghirardi, Grassi, and Pearle (1990, 1297) for discussion.

8.6.2.4 Impossibility of a Relativistic Collapse Theory Using Only Standard Degrees of Freedom

Though relativistic causality does not preclude the formulation of a relativistic collapse theory, there is a tension between the requirements of relativistic causality on a collapse theory and the need for a stable vacuum. It is possible to avoid an outright contradiction, but only at the cost of introducing nonstandard degrees of freedom, which do not obey anything like the usual laws of quantum evolution.

Here's the problem. Suppose we have a collapse theory with a stable vacuum state. Consider evolution from one surface τ to another, σ, through a bounded spacetime region δ. Stability of the vacuum means if the state on τ is the vacuum state, subsequent evolution is deterministic, in that there is only a single outcome that has nonzero probability, namely, maintenance of the vacuum. Deterministic evolution starting from vacuum means that, except perhaps for a set with zero probability of being realized, all collapse operators $\{K_\gamma(\delta)\}$ map the vacuum into multiples of the same vector. It can be shown (see Myrvold 2017 for details) that this will also hold for all standard states, where standard states are those that can be obtained by operating on the vacuum with operators of the sort that appear in standard quantum field theories. That is, for standard quantum field theories, the constraints imposed by relativistic causal structure, plus the constraint, imposed by Lorentz invariance, of a stable vacuum, are satisfied at the expense of rendering the theory deterministic, and thus incapable of suppressing the sorts of superpositions of macroscopically distinct states that collapse theories are meant to suppress.

This is not a no-go result for relativistic collapse theories *tout court*. Bedingham (2011a,b) and Pearle (2015) utilize a nonstandard field first introduced by Pearle (1993). This "index" or "pointer" field does not obey anything like an ordinary law of evolution; instead, its value at any point is independent of its value at any other point. The vacuum state $|\Omega\rangle$ of this theory is the ground state of the standard fields *and* this pointer field. Standard operators operating on the vacuum state leave the pointer field in its ground state. On Bedingham's theory the pointer field couples to the standard fields, so that if there are excitations of the standard fields, these produce excitations of the pointer fields. The theory tends to collapse toward eigenstates of smeared pointer-field operators, with accompanying collapse toward eigenstates of smeared matter fields.

One might also reject the requirement of having a theory that produces collapses in Minkowski spacetime. Okon and Sudarsky (2014, 2016, 2018) have proposed, in the context of semi-classical gravity, a modification of the CSL theory on which the collapse rate depends on local curvature. One might consider a modification of this proposal, according to which collapse rate goes to zero as curvature goes to zero (Bengochea et al., 2020). If there is a viable theory of this sort, it might satisfy the constraint of having a stable vacuum in Minkowski space, while producing collapse in the presence of matter.

8.6.3 Causal Quantum Theory

Standard quantum theory predicts correlations between outcomes of spacelike separated experiments that violate Bell inequalities. The crucial assumption of derivation of Bell inequalities is that the correlation be *locally explicable*, that is, they can be explained by reference to conditions in the past light cones of the experiments, such that a full specification of these conditions screens off the correlations. Experimental demonstrations of violations of Bell inequalities are usually considered to indicate correlations that are not locally explicable are a genuine physical phenomenon (see (see Myrvold, Genovese, and Shimony 2019 for an overview, and a discussion of the conditions required to reach this conclusion).

There is a loophole, however, identified by Kent (2005a), and called by him the *collapse locality loophole*. Tests of Bell inequalities that aim to establish nonlocality take care to ensure that the experiments involved take place at spacelike separation from each other, from choice of experimental setting to registration of result. A judgment of whether this condition is satisfied requires, therefore, a judgment of when an experiment is over. Typically, it is assumed that this occurs once the result has been permanently recorded in some macroscopically readable device. However, as Kent observes, on some proposals, the quantum state will remain in a superposition corresponding to distinct outcomes beyond this point. One such proposal is the suggestion that state reduction takes place only when the uncollapsed state involves a superposition of sufficiently distinct gravitational fields (Diósi, 1987; Penrose, 1989, 1996). Another is Wigner's suggestion that conscious awareness of the result is required to induce collapse (Wigner, 1961).

Kent introduces a class of theories, which he calls *causal quantum theories*, on which collapses are localized events, collapse probabilities are conditioned only on events in the collapse's past light cone, and, as a consequence, collapses at spacelike separation from each other are probabilistically independent, conditional on events in their past light cones. A theory of that sort differs in its predictions from standard quantum mechanics, but a decisive test requires experiments that are completed at spacelike separation from each other.

Kent concluded that, on a theory of the sort proposed by Diósi and Penrose, on which collapse occurs when only differing results correspond to sufficiently distinct gravitational fields, the condition that the experiments be concluded at spacelike separation had not been enforced in any of the experimental tests of Bell inequalities as of the time of writing (2005). The experiment of Salart et al. (2008) closed the loophole for the particular proposals of Penrose and Diósi, though, as Kent (2018) pointed out, altering the Penrose-Diósi threshold by a few orders of magnitude would render these proposals compatible with the results of this experiment.

No experiment to date has addressed the collapse locality loophole if the collapse condition is taken to be awareness of the result by a conscious observer. See Kent (2018) for proposals of ways in which causal quantum theory could be subjected to more stringent tests.

8.7   Everettian Interpretations

Everettian interpretations make use of minimal machinery: they leave the usual, unitary dynamics untouched, and add nothing to the quantum state. The complications arise in connection with painting a sensible picture of the world with this minimal palette, and with making sense of probabilities or a working substitute for them.

Because of this minimality, worries about compatibility with relativity are also minimal. If the theory to be interpreted is a relativistic quantum field theory, then there is no question about the dynamics being compatible with relativistic causal structure; it is. As there are no additional beables, there is no question about whether their dynamics is compatible with relativity. Some proponents of Everettian interpretations have taken this to be one of the chief attractions of interpretations of this sort (see, e.g., Vaidman 1995, 2016, 2018).

Given a bounded subset α that is part of many Cauchy surfaces σ, σ', *etc.*, we can consider the reduced state of α,[5] evaluated with respect to various Tomonaga-Schwinger state vectors $|\psi(\sigma)\rangle$, $|\psi(\sigma')\rangle$. If the evolution from one Cauchy surface to another is unitary, and if the generators of evolutions through regions spacelike separated from α commute with operators representing observables pertaining to α, then the reduced state of α will be the same whether calculated from $|\psi(\sigma)\rangle$ or $|\psi(\sigma')\rangle$ or any Cauchy surface containing α. Therefore, in that sense, on any quantum-state monist theory with deterministic, unitary dynamics, operations performed at a distance have no effect on states of affairs in α.

Is there nonlocality, in any sense, in an Everettian theory? It is not immediately clear how to pose the question. Unlike, say, the case of additional beables theories, we cannot ask (at least in any straightforward way) whether a choice of a parameter setting at one location affects the outcome of an experiment performed at another, because, on an Everettian theory, talk of *the outcome* of an experiment is unwarranted – experiments do not have unique outcomes. If the usual notions of causal dependence, such as act-outcome correlations, presuppose unique outcomes, then they cannot even be formulated in an Everettian context – and *ipso facto* they cannot be violated. Instead of saying that Everettian theories are causally local, perhaps it would be more apt to say: it is not the case that they are causally nonlocal.

Brown and Timpson (2016) have made a stronger claim about Everettian interpretations. Not only is there no dynamical nonlocality, they claim, there are no nonlocal correlations. The claim is that, in the Everettian context, talk of correlated outcomes of experiments makes sense only when the outcomes of the distant experiments are compared, at which time both experiments are in the past. A similar argument has been made of which quantum theories are but a special case, by Brassard and Raymond-Robichaud Robichaud (2013, 2019). Analysis of this claim is beyond the scope of this chapter; we note only that compatibility of Everettian interpretations with relativity does not ride on whether a claim of this sort is accepted or rejected.

8.8   Non-Realist and Pragmatist Interpretations

There is a long tradition of denying that the ontology of the physical world contains anything corresponding to a quantum state. In recent years, views of this sort have been championed by advocates of a position called, by its proponents, *Quantum Bayesianism*, or *QBism* (Caves, Fuchs and Schack 2002; Fuchs, Mermin, and Schack 2014; Fuchs and Schack 2015). This is one variant of a class of views commonly referred to as ψ-*epistemic* views.[6] A somewhat related, but different, view is the pragmatist view of quantum mechanics defended by Richard Healey (2012, 2017a, 2020). See, also, Healey (2017b) for an overview of related views.

For Healey, quantum chances for events are assigned much as they have been in the discussion, above, of collapse theories. The chance of an event, say, an outcome of an experiment, is indexed by the spacetime event from which the chance is to be ascribed. The chance $Ch_x(e_A)$ of event e_A is, in effect, calculated using what in a dynamical collapse theory would be the past light cone state, which incorporates experimental outcomes that have occurred in the past light cone of x. On Healey's view, "Any agent who accepts quantum theory and is (momentarily) located at space-time point x should match credence in e_A to $Ch_x(e_A)$," where $Ch_x(e_A)$ is the chance evaluated conditional on events in the past light cone of the point x (Healey 2016, p. 179). Healey regards these chances as objective, in the sense that it is an objective fact that these chances are optimal credences for an agent that has epistemic access to all and only the facts in the past light cone of x. But they are not to be thought of as localized physical magnitudes.

> As we saw, the chance of outcome e_A does not attach to it in virtue solely of its physical description: the *chances* of e_A attach also in virtue of its space-time relations to different space-time locations. Each such location offers the epistemic perspective of a situated agent, even in a world with no such agents. The existence of these chances is independent of the existence of cognizers. But it is only because we are not merely cognizers but physically situated agents that we have needed to develop a concept of chance tailored to our needs as informationally deprived agents. Quantum chance admirably meets those needs: an omniscient God could describe and understand the physical world without it. (Healey 2016, p. 183; see also Healey 2017c, pp. 175–176)

In Healey (2020), Healey argues that this view should be regarded as a form of scientific realism.

In their statement of QBism, Fuchs, Mermin, and Schack (2014) take a considerably more radical stance. First, they not only deny that quantum probabilities represent features of physical reality; for them, these probabilities are not objective in any sense – they are "personal judgments," not backed up by objective facts. Second, the scope of application of quantum mechanics is severely limited. In ordinary practice, physicists routinely use quantum mechanics to calculate probabilities about what will be observed by other people if they perform certain experiments, whether or not the agent carrying out the calculation will ever be aware of the results (which might, for example, not be realized within the agent's lifetime). In contrast, Fuchs et al. restrict an agent's assignment of probabilities to propositions concerning the agent's future experience.

The class of propositions to which the agent assigns probabilities all concern events along her own worldline, and, therefore, contain no spacelike separated events.

> when any agent uses quantum mechanics to calculate "[cor]relations between the manifold aspects of [her] experience," those experiences cannot be space-like separated. Quantum correlations, by their very nature, refer only to time-like separated events: the acquisition of experiences by any single agent. Quantum mechanics, in the QBist interpretation, cannot assign correlations, spooky or otherwise, to space-like separated events, since they cannot be experienced by any single agent. Quantum mechanics is thus explicitly local in the QBist interpretation.

And that's all there is to it. (Fuchs, Mermin, and Schack 2014, pp. 750-51)

There is no action at a distance, on this view, because there are no distances between events that the agent assigns probabilities to. The spacetime of events to which the theory is applied is, in effect, a one-dimensional timelike worldline (or else a narrow world-tube). There is no superluminal signaling, because there is no signaling of any kind (unless writing a note to your future self is said to be sending a signal to your future self).

Approaches that deny the reality of quantum states bypass some of the problems faced by realist approaches by limiting the scope of what is taken to be physical. In my opinion, however, such approaches must be regarded as promissory notes. If one denies that quantum states represent anything in physical reality, this raises the question of what *does* constitute physical reality. It does not seem, for example, that QBists wish to deny that distant places and other agents exist; they merely claim that such things are outside the scope of legitimate application of quantum mechanics. This raises the question of what physics *does* apply to them. If one says nothing at all, then ipso facto one says nothing that conflicts with relativity. The question still remains whether whatever it is that *does* exist can coexist peacefully with relativity.

8.9   Conclusion

There is a temptation to say that because of quantum nonlocality, as revealed by Bell's theorem, that quantum theory is flatly incompatible with relativity. As we have seen, things are not so simple. It is only additional beables theories, such as the de Broglie-Bohm theory and related theories, that must violate relativistic causal structure (at least, as long as the additional beables are local beables). Collapse theories can respect relativistic causal structure; unlike the additional beables theories, they need no distinguished relation of distant simultaneity. Full relativistic invariance, however, is possible only at the cost of introducing nonstandard degrees of freedom. For Everettian theories, relativity poses no special problem. For non-realist interpretations, it is difficult to render a verdict, as proponents of such views are usually less clear about what they do think *is* in the world than about what they think is *not*.

Acknowledgment

Support for this research was provided by Graham and Gale Wright, who sponsor the Graham and Gale Wright Distinguished Scholar Award at the University of Western Ontario.

Notes

1 There is no need here for a precise characterization of "well-behaved," but global hyperbolicity suffices for our purposes.
2 These assumptions may be questioned; see Kent (2005b).
3 That is, Γ is some set, $\mathcal{F}$ is a σ-algebra of subsets of Γ containing both the empty set $\emptyset$ and Γ itself, to be regarded as the measurable subsets of Γ, and μ is a non-negative countably additive set function on $\mathcal{F}$ with $\mu(\Gamma) = 1$.
4 This paragraph is based on the discussion in Pearle (2015, 2).
5 Those who want to insist that states are to be assigned only to regions with nonempty interior may take this to mean: the state of the causal domain of dependence of α.
6 See Harrigan and Spekkens (2010) for a discussion.

References

Arias, A., Gheondea, A. and Gudder, S. (2002). Fixed points of quantum operations. *Journal of Mathematical Physics*, 43(12): 5872–5881.
Arnztenius, F. (1998). Curiouser and curiouser: A personal evaluation of modal intepretations. In D. Dieks and P.E. Vermaas (eds.), *The Modal Interpretation of Quantum Mechanics*. Dordrecht: Kluwer Academic Publishers, pp. 337–377.
Bedingham, D. (2011a). Relativistic state reduction model. *Journal of Physics: Conference series*, 306: 012034.
Bedingham, D. (2011b). Relativistic state reduction dynamics. *Foundations of Physics*, 41: 686–704.
Bedingham, D., Dürr, D., Ghirardi, G., Goldstein, S., Tumulka, R. and Zanghì, N. (2014). Matter density and relativistic models of wave function collapse. *Journal of Statistical Physics*, 154: 623–631.
Bell, J.S. (1981a). Bertlmann's socks and the nature of reality. *Journal de Physique*, 42: 611–637. Reprinted in (Bell, 1987b, 139–158).
Bell, J.S. (1981b). Quantum mechanics for cosmologists. In C.J. Isham, R. Penrose and D.W. Sciama (eds.), *Quantum Gravity 2: A Second Oxford Symposium*. Oxford: Oxford University Press, pp. 611–637. Reprinted in Bell (1987b, 117–138).
Bell, J.S. (1987a). Are there quantum jumps? In C. Kilmister (ed.), *Schrödinger: Centenary Celebration of a Polymath*. Cambridge: Cambridge University Press, pp. 41–52. Reprinted in Bell (1987b, 201–212).
Bell, J.S. (1987b). *Speakable and Unspeakable in Quantum Mechanics*. Cambridge: Cambridge University Press.
Bell, J.S. (1989). Towards an exact quantum mechanics. In S. Deser and R.J. Finkelstein (eds.), *Themes in Contemporary Physics II: Essays in Honor of Julian Schwinger's 70th Birthday*. Singapore: World Scientific, pp. 1–26.
Bell, J.S. (2007). The Trieste lecture of John Stewart Bell. *Journal of Physics A: Mathematical and Theoretical*, 40: 2919–2933.
Bell, M. and Gao, S., editors. (2016). *Quantum Nonlocality and Reality: 50 Years of Bell's Theorem*. Cambridge: Cambridge University Press.
Bengochea, G.R., León, G., Pearle, P. and Sudarsky, D. (2020). Discussions about the landscape of possibilities for treatments of cosmic inflation involving continuous spontaneous localization models. *European Physical Journal C*, 80: 1021.
Berkovitz, J. and Hemmo, M. (2005a). Can modal intepretations of quantum mechanics be reconciled with relativity? *Philosophy of Science*, 72: 789–801.
Berkovitz, J. and Hemmo, M. (2005b). Modal interpretations of quantum mechanics and relativity: A reconsideration. *Foundations of Physics*, 35: 373–397.
Berkovitz, J. and Hemmo, M. (2006). A new modal interpretation in terms of relational properties. In W. Demopoulos and I. Pitowsky (eds.), *Physical Theory and its Interpretation: Essays in honor of Jeffrey Bub*. Dordrecht: Springer, pp. 1–28.
Berndl, K., Dürr, D., Goldstein, S. and Zanghì, N. (1996). Nonlocality, Lorentz invariance, and Bohmian quantum theory. *Physical Review A*, 53: 2062–2073.
Brassard, G. and Raymond-Robichaud, P. (2013). Can free will emerge from determinism in quantum theory? In A. Suarez and P. Adams (eds.), *Is Science Compatible with Free Will?* Dordrecht: Springer, pp. 41–61.
Brassard, G. and Raymond-Robichaud, P. (2019). Parallel lives: A local-realistic interpretation of "nonlocal" boxes. *Entropy*, 27: 87.
Brown, H.R. and Timpson, C.G. (2016). Bell on Bell's theorem: The changing face of nonlocality. In M. Bell and S. Gao (eds.), *Quantum Nonlocality and Reality: 50 Years of Bell's Theorem*. Cambridge: Cambridge University Press, pp. 91–123.

Caves, C.M., Fuchs, C.A. and Schack, R. (2002). Quantum probabilities as Bayesian probabilities. *Physical Review A*, 65: 022305.

Clifton, R. (2000). The modal interpretation of algebraic quantum field theory. *Physics Letters A*, 271: 167–171.

Costa de Beauregard, O. (1976). Time symmetry and interpretation of quantum mechanics. *Foundation of Physics*, 6: 539–550.

Dickson, M. and Clifton, R. (1998). Lorentz-invariance in modal interpretations. In D. Dieks and P.E. Vermaas (eds.), *The Modal Interpretation of Quantum Mechanics*. Dordrecht: Kluwer Academic Publishers, pp. 9–47.

Dieks, D. (2000). Consistent histories and relativistic invariance in the modal interpretation of quantum mechanics. *Physics Letters A*, 265: 317–325.

Dieks, D. and Vermaas, P.E., editors. (1998). *The Modal Interpretation of Quantum Mechanics*. Dordrecht: Kluwer Academic Publishers.

Diósi, L. (1987). A universal master equation for the gravitational violation of quantum mechanics. *Physics Letters A*, 120: 337–381.

Earman, J. and Ruetsche, L. (2005). Relativistic invariance and modal interpretations. *Philosophy of Science*, 72: 557–583.

Fine, A. (1982a). Hidden variables, joint probability, and the Bell inequalities. *Physical Review Letters*, 48: 291–295.

Fine, A. (1982b). Joint distributions, quantum correlations, and commuting observables. *Journal of Mathematical Physics*, 23: 1306–1310.

Friederich, S. and Evans, P.W. (2019). Retrocausality in quantum mechanics. In E.N. Zalta (ed.), *The Stanford Encyclopedia of Philosophy* (Summer 2019 ed.). Metaphysics Research Lab, Stanford University. Available at: https://plato.stanford.edu/entries/qm-retrocausality/

Fuchs, C.A., Mermin, N.D. and Schack, R. (2014). An introduction to QBism with an application to the locality of quantum mechanics. *American Journal of Physics*, 82: 749–754.

Fuchs, C.A. and Schack, R. (2015). QBism and the Greeks: Why a quantum state does not represent an element of physical reality. *Physica Scripta*, 90: 015104.

Ghirardi, G. and Grassi, R. (1994). Outcome predictions and property attribution: The EPR argument reconsidered. *Studies in History and Philosophy of Science*, 25: 397–423.

Ghirardi, G.C., Grassi, R. and Benatti, F. (1995). Describing the macroscopic world: Closing the circle within the dynamical reduction program. *Foundations of Physics*, 25: 5–38.

Ghirardi, G., Grassi, R. and Pearle, P. (1990). Relativistic dynamical reduction models: General framework and examples. *Foundations of Physics*, 20: 1271–1316.

Ghirardi, G.C., Pearle, P. and Rimini, A. (1990). Markov processes in Hilbert space and continuous spontaneous localization of systems of identical particles. *Physical Review A*, 42: 78–89.

Ghirardi, G.C., Rimini, A. and Weber, T. (1986). Unified dynamics for microscopic and macroscopic systems. *Physical Review D*, 34: 470–491.

Gisin, N. (1989). Stochastic quantum dynamics and relativity. *Helvetica Physica*, 62: 363–371.

Harrigan, N. and Spekkens, R.W. (2010). Einstein, incompleteness, and the epistemic view of quantum states. *Foundations of Physics*, 40: 125–157.

Healey, R. (2012). Quantum theory: A pragmatist approach. *The British Journal for the Philosophy of Science*, 63: 729–771.

Healey, R. (2016). Local causality, probability, and explanation. In M. Bell and S. Gao (eds.), *Quantum Nonlocality and Reality: 50 Years of Bell's Theorem*. Cambridge: Cambridge University Press, pp. 172–194.

Healey, R. (2017a). Quantum states as objective informational bridges. *Foundations of Physics*, 47: 161–173.

Healey, R. (2017b). Quantum-Bayesian and pragmatist views of quantum theory. In E.N. Zalta (ed.), *The Stanford Encyclopedia of Philosophy* (Spring 2017 ed.). Metaphysics Research Lab, Stanford University. Available at: https://plato.stanford.edu/entries/quantum-bayesian/

Healey, R. (2017c). *The Quantum Revolution in Philosophy*. Oxford: Oxford University Press.

Healey, R. (2020). Pragmatist quantum realism. In S. French and J. Saatsi (eds.), *Scientific Realism and the Quantum*. Oxford: Oxford University Press, pp. 123–146.

Kastner, R. (2013). *The Transactional Interpretation of Quantum Mechanics: The Reality of Possibiliy*. Cambridge: Cambridge University Press.

Kent, A. (2005a). Causal quantum theory and the collapse locality loophole. *Physical Review A*, 72: 012107.

Kent, A. (2005b). Nonlinearity without superluminality. *Physical Review A*, 72: 012108.

Kent, A. (2018). Testing causal quantum theory. *Proceedings of the Royal Society A*, 474: 20180501.

Laudisa, F. and Rovelli, C. (2019). Relational quantum mechanics. In E.N. Zalta (ed.), *The Stanford Encyclopedia of Philosophy* (Winter 2019 ed.). Metaphysics Research Lab, Stanford University. Available at: https://plato.stanford.edu/entries/qm-relational/

Lombardi, O. and Dieks, D. (2017). Modal interpretations of quantum mechanics. In E.N. Zalta (ed.), *The Stanford Encyclopedia of Philosophy* (Spring 2017 ed.). Metaphysics Research Lab, Stanford University. https://plato.stanford.edu/entries/qm-modal/

Malament, D.B. (1996). In defense of dogma: Why there cannot be a relativistic quantum mechanics of (localizable) particles. In R. Clifton (ed.), *Perspectives on Quantum Reality: Non-Relativistic, Relativistic, and Field-Theoretic.* Dordrecht: Kluwer Academic Publishers, pp. 1–10.

Maudlin, T. (1994). *Quantum Non-Locality and Relativity: Metaphysical Intimations of Modern Physics.* Cambridge: Cambridge University Press.

Myrvold, W., Genovese, M. and Shimony, A. (2019). Bell's theorem. In E.N. Zalta (ed.), *The Stanford Encyclopedia of Philosophy* (Spring 2019 ed.). Metaphysics Research Lab, Stanford University. https://plato.stanford.edu/entries/bell-theorem/

Myrvold, W.C. (2002). Modal intepretations and relativity. *Foundations of Physics,* 32: 1773–1784.

Myrvold, W.C. (2009). Chasing chimeras. *The British Journal for the Philosophy of Science,* 60: 635–646.

Myrvold, W.C. (2016). Lessons of Bell's theorem: Nonlocality, yes; action at a distance, not necessarily. In M. Bell and S. Gao (eds.), *Quantum Nonlocality and Reality: 50 Years of Bell's Theorem.* Cambridge: Cambridge University Press, pp. 238–260.

Myrvold, W.C. (2017). Relativistic Markovian dynamical collapse theories must employ nonstandard degrees of freedom. *Physical Review A,* 96: 062116.

Myrvold, W.C. (2018). Ontology for collapse theories. In S. Gao (ed.), *Collapse of the Wave Function: Models, Ontology, Origin, and Implications.* Cambridge: Cambridge University Press, pp. 97–123.

Myrvold, W.C. (2019). Ontology for relativistic collapse theories. In O. Lombardi, S. Fortin, C. López and F. Holik (eds.), *Quantum Worlds: Perspectives on the Ontology of Quantum Mechanics.* Cambridge: Cambridge University Press, pp. 9–31.

Okon, E. and Sudarsky, D. (2014). Benefits of objective collapse models for cosmology and quantum gravity. *Foundations of Physics,* 44: 114–143.

Okon, E. and Sudarsky, D. (2016). A (not so?) novel explanation for the very special initial state of the universe. *Classical and Quantum Gravity,* 33: 225015.

Okon, E. and Sudarsky, D. (2018). The weight of collapse: Dynamical reduction models in general relativistic contexts. In S. Gao (ed.), *Collapse of the Wave Function: Models, Ontology, Origin, and Implications.* Cambridge: Cambridge University Press, pp. 312–345.

Pearle, P. (1989). Combining stochastic dynamical state-vector reduction with spontaneous localization. *Physical Review A,* 39: 913–923.

Pearle, P. (1993). Ways to describe dynamical state-vector reduction. *Physical Review A,* 48: 913–923.

Pearle, P. (2015). Relativistic dynamical collapse model. *Physical Review D,* 91: 105012.

Penrose, R. (1989). *The Emperor's New Mind.* Oxford: Oxford University Press.

Penrose, R. (1996). On gravity's role in quantum state reduction. *General Relativity and Gravitation,* 28: 581600.

Salart, D., Baas, A., van Houwelingen, J.A., Gisin, N. and Zbinden, H. (2008). Spacelike separation in a Bell test assuming gravitationally induced collapses. *Physical Review Letters,* 100: 220404.

Schweber, S.S. (1961). *An Introduction to Relativistic Quantum Field Theory.* New York: Harper & Row.

Schwinger, J. (1948). Quantum electrodynamics. I. A covariant formulation. *Physical Review,* 74: 1439–1461.

Simon, C., Bužek, V. and Gisin, N. (2001). No-signalling condition and quantum dynamics. *Physical Review Letters,* 87: 170405.

Struyve, W. (2011). Pilot-wave approaches to quantum field theory. *Journal of Physics: Conference Series,* 306: 012047.

Tomonaga, S. (1946). On a relativistically invariant formulation of the quantum theory of wave fields. *Progress of Theoretical Physics,* 1: 27–41.

Tumulka, R. (2007). The unromantic pictures of quantum theory. *Journal of Physics A: Mathematical and Theoretical,* 40: 3245–3273.

Vaidman, L. (1995). On the paradoxical aspects of new quantum experiments. In D. Hull, M. Forbes and R.M. Burian (eds.), *PSA 1994: Proceedings of the Biennial Meeting of the Philosophy of Science Association, Volume One: Contributed Papers.* East Lansing, MI: Philosophy of Science Association, pp. 211–217.

Vaidman, L. (2016). The Bell inequality and the many-worlds interpretation. In M. Bell and S. Gao (eds.), *Quantum Nonlocality and Reality: 50 Years of Bell's Theorem.* Cambridge: Cambridge University Press, pp. 195–203.

Vaidman, L. (2018). Many-worlds interpretation of quantum mechanics. In E.N. Zalta (ed.), *The Stanford Encyclopedia of Philosophy* (Fall 2018 ed.). Metaphysics Research Lab, Stanford University. Available at: https://plato.stanford.edu/entries/qm-manyworlds/

Wigner, E. (1961). Remarks on the mind-body question. In I.J. Good (ed.), *The Scientist Speculates,* pp. 284–302. London: Heinemann.

Further Reading

Those interested in understanding the compatibility of Quantum Mechanics and relativity should start with the Stanford Encyclopedia of Philosophy entry, "Bell's Theorem," which contain an exposition of Bell's theorem, an overview of philosophical discussions, and pointers to the relevant literature. See also E. Keming Chen's "Bell's Theorem, Quantum Probabilities, and Superdeterminism" (this volume), and T. Maudlin's *Quantum Non-Locality and Relativity: Metaphysical Intimations of Modern Physics* (Cambridge University Press, 1994).

PART III

General Relativity

Introduction to Part III

In formulating special relativity, Einstein built on the achievements of Lorentz, Poincaré, and Fitzgerald. His search for a relativistic theory of gravitation was harder won. The result, general relativity (GR), is an extraordinary theory: one which takes a novel mathematical form and posits a spacetime geometry with its own dynamics. Well-used analogies with the geometry of rubber sheets and two-dimensional surfaces are helpful for getting a feel for the non-Euclidean geometry of GR, but ultimately fail to do justice to the strangeness of a curved and dynamical four-dimensional geometry, which for the most part defies intuitive or pictorial understanding. Engaging with GR requires a deepish dive into the mathematics and details of the theory, and this section will be more technical than earlier sections. We will see that the deepest philosophical puzzles of GR do not involve its surprising consequences for time and causality, but rather its founding principles, mathematical form, and lessons for future physics. This section focuses on those issues.

There is a Whiggish, textbook, version of the history of GR that goes something like this: in 1907, while still working in the patent office, Einstein had what he later called "the happiest thought of my life." He realized that the behavior of systems falling in a uniform gravitational field was identical to that of systems which did not experience gravity at all. This thought, the origin of the "equivalence principle", led him to realize that a relativistic gravitational theory could be found if we only reconceptualized inertial, or force-free, motion in such a way that we understood freely falling bodies to be moving inertially. This requires them to follow the geodesics (straightest possible lines) of a curved geometry. There is some truth here, and it's the view of this editor that something like the above does indeed give important insight into GR. But GR is a subject that requires and rewards historical subtlety. In his chapter on the equivalence principle, Dennis Lehmkuhl points out that Einstein's "happiest thought" was really the result of a desire to extend the relativity principle, and do away with a privileged state of inertial motion altogether. Although GR is commonly thought of as a theory in which spacetime comes to the foreground as a piece of dynamical structure, Einstein was in fact searching for a theory that would enact the vision of Ernst Mach, the arch-relationist.

GR is not a Machian theory. The geometry of spacetime is not determined by the distribution of matter, but instead has its own degrees of freedom. But questions about what plays the role of the relativity principle, and hence what the symmetry group of GR is, remain. Lehmkuhl discusses the many forms of the equivalence principle, some of which cease to make sense in the GR context (and hence only serve as a "midwife" at its birth) and others which reflect more general symmetries. These forms reflect two sides of GR. On the one hand, GR posits a richly structured geometry – one which generically has no global symmetries (although particular solutions may). On the other, it is generally covariant: that is, the equations can be written in arbitrary coordinate systems. This last feature is, at some level, an artifact of the use of differential geometry to formulate the theory – any theory is generally covariant once formulated in terms of differential geometry. But there is arguably a sense in which GR is "substantively" generally covariant, because it cannot be written in a non-generally covariant form.

This general covariance of GR can itself generate puzzles, the most famous of which is Einstein's "hole argument", which delayed the development of the theory by two years. In 1913, Einstein became puzzled by the fact that the general covariance of the theory allowed for arbitrary coordinate transformations (or their active counterpart: transformations of the manifold points) within a localized region (the "hole"). Concerned that this meant that no amount of knowledge of the system outside the hole could determine behavior within the hole, he originally abandoned the search for a generally covariant theory before realizing that the relevant transformations did not change the empirical content of the theory.

Resurrected in 1987 by John Earman and John Norton, the hole argument has led to a lively debate in the philosophy literature. Earman and Norton argued that the "manifold substantivalist," that is, someone who believes that the four-dimensional manifold represents real spacetime points, is committed to indeterminism. The debate, they claimed, had much in common with the original debate over substantivalism and relationism between Newton and Leibniz. And indeed it does. The literature that has arisen since has done much to clarify the options for thinking about spacetime symmetries and coordinate transformations. While a majority of authors disagree that substantivalism is ruled out by the hole argument, there is a great deal to be learned about the ways in which our mathematical models represent the world. Oliver Pooley's chapter guides us through aspects of this vast debate and makes some particularly helpful comments on the vexed issue of determinism.

As noted, any adequate treatment of the philosophy of GR will need to go deeper into mathematics and technical details than the previous sections. Sam Fletcher's chapter is no exception, and requires some familiarity with differential geometry. But it will reward the reader willing to do the work, for it gives a deep insight into the mathematical structures involved in GR. Much of the philosophical work required in interpreting GR involves thinking about how the various bits of mathematics represent spacetime. Because neither the mathematics nor the symmetries are particularly straightforward, there is much work to be done in understanding which pieces of structure are necessary in the formulation of a spacetime theory and which bits of structure determine which others. An understanding of these issues is crucial in considering alternative spacetime theories and many aspects of the quantum gravity theories discussed in Part VI.

9

THE EQUIVALENCE PRINCIPLE(S)

Dennis Lehmkuhl

9.1 Introduction: A Manifold of Principles

The equivalence principle of general relativity (GR), the Einstein equivalence principle, or the strong equivalence principle is a curious beast. Or rather, it is not really a beast at all, but a whole bunch of beasts, as there sometimes seem to be more versions of the equivalence principle than physicists and philosophers who have thought about it. For some it is one of if not *the* main pillar upon which GR rests, for others it is but a chimera, or a ladder that was helpful in finding GR but to be thrown away now that we have the theory in hand.

Despite the plethora of different versions of the equivalence principle, they can be divided into two classes; both classes originate with Einstein, though he himself only discusses the first class under this name. This first class postulates an equivalence between gravitational and inertial effects, or between the very essences (as Einstein put it in 1918) of gravity and inertia. This class typically allows for homogeneous gravitational fields, or certain gravitational effects, to be "transformed away" by changing the coordinate system. In contrast, the second class states that GR is "locally special relativistic," in the sense that locally there is no evidence of gravity; the different members of this class differ primarily in how they spell out what it means to be "locally special relativistic," and under which conditions the respective principle holds. I will call the first class of principles Einstein equivalence principles (EEPs), and the second strong equivalence principles (SEPs).[1]

Einstein himself introduced what others have come to call (a form of) the strong equivalence principle as a premise of the theory in his celebrated 1916 review paper in the following way:[2]

> [L]et us now introduce the following premise: For infinitely small four-dimensional regions the theory of relativity in the restricted sense [i.e., special relativity] holds, if the coordinates are suitably chosen.

Einstein goes on to elaborate that the suitably chosen coordinates are to be such that no gravitational fields appear; from elsewhere in the same paper (p. 802) we know that this means that the components of the affine connection are to vanish in those coordinates.

The Einstein equivalence principle appears much earlier than the strong equivalence principle in the 1916 review paper – in the section entitled "On the reasons to extend the relativity principle."[3] Rather than formulating it as a principle (as in previous and later papers), Einstein here introduces the EEP as a "well-known physical fact," and takes it as a reason to expect that the relativity principle applies not *only* to uniform motion: an observer has no way of distinguishing between being uniformly accelerated but not subject to gravity and being at rest but subject to a (homogeneous)

125

gravitational field. The similarity to Galileo motivating the relativity principle by appeal to the inability of an observer to distinguish between being at rest and moving with constant velocity is immediate. In Section 9.3, I shall return to how this principle was related to the other principles Einstein saw at the heart of GR.

Einstein was adamant in defending his version of the principle against much of the rest of the community, and against the claim that it was of *only* heuristic importance in the search for GR. As we shall see, for Einstein it was also intimately related to what he saw as the main result of GR: the unification of inertia and gravity in a sense to be specified.

In contrast, others argued that the Einstein equivalence principle is *false* according to GR, but that one version or the other of the strong equivalence principle (related to the local validity of special relativity) *does* hold in the theory.

In the following, I shall distinguish the two classes of the equivalence principle and the different versions within each class that have been proposed. I shall put particular emphasis on the role the respective principles play within GR according to their proponents. In both cases, the relationship to GR's Newtonian and special relativistic predecessors plays a major role. For many proponents of the EEP the link to Newtonian theory shows *how* GR unifies inertia and gravity and how it overcomes the distinction between real and fictitious forces; for many proponents of the SEP the local validity of special relativity imbues the $g_{\mu\nu}$ field with its chronometric significance (i.e., relates it to the measurements of rods and clocks).

But before I can elaborate on all that, I will speak about the beginning of it all, the precondition for this mess, if you will: the *weak* equivalence principle.

9.2   The Weak Equivalence Principle: How It All Began

The rest of this chapter will be about the Einstein equivalence principle and different versions of the strong equivalence principle. These two categories have often been seen as rivals: many authors seem to believe that you have to subscribe to one or the other. However, there is widespread consensus that the weak equivalence principle is logically weaker than both the EEP and the SEPs, and even a precondition to both of them.

The weak equivalence principle (WEP), too, comes in two major forms. One appeals to the Newtonian concepts of inertial and gravitational mass and asserts an equivalence between the two of them. The other form tries to get by without reference to these concepts, and instead asserts the "universality of free fall": all bodies fall at the same rate (of acceleration). The latter version is historically prior, partly because, when it was first formulated by Galileo, the concepts of inertial and gravitational mass (indeed, arguably the notion of "mass") did not yet exist. In this form the WEP is best seen as a generalized observation, akin to Galileo's principle of relativity, in that it generalizes and abstracts away from observations with a minimum of theoretical concepts.

One of the most careful ways of stating this version of the WEP is due to Clifford Will. He defines the weak equivalence principle thus:[4]

> [WEP1:] [I]f an uncharged test body is placed at an initial event in spacetime and given an initial velocity there, then its subsequent trajectory will be independent of its internal structure and composition.

Will goes on: "By 'uncharged test body' we mean an electrically neutral body that has negligible self-gravitational energy (as estimated by Newtonian theory) and that is small enough in size so that its coupling to inhomogeneities in external fields can be ignored."[5]

The second version of the WEP draws more strongly on Newtonian concepts. It says:[6]

> [WEP2:] For any body the gravitational mass of the body is equal to its inertial mass.

Note that both Ohanian and Will drop the restriction to test bodies and claim that WEP2 applies to *any* body.

What is the relationship between WEP1 and WEP2? Given careful definitions of the terms "inertial mass" and "gravitational mass," WEP2 implies WEP1.[7] However, especially given the restriction to test bodies present in WEP1 but not in WEP2, the reverse is not true.[8]

Loránd Eötvös made precise measurements based on the second way of thinking about the WEP, measuring the difference between the inertial and gravitational mass of a body using the torsion balance he himself had invented. The precision of these results had a direct influence on Einstein; he himself stated that Eötvös' results both inspired and were a precondition for the Einstein equivalence principle to hold.[9]

9.3 Gravity and Acceleration: Einstein Equivalence Principles

9.3.1 *Two Construction Sites Instead of One*

After Einstein completed the founding paper of what we today call the special theory of relativity in 1905, he faced a bit of a conundrum. On the one hand, it was immediately clear that the theory was in conflict with Newton's theory of gravity, which presupposed action at a distance and the absoluteness of simultaneity. Many of the early readers of Einstein [1905] were aware of this and immediately set out to develop a (special) relativistic theory of gravity.[10] But for Einstein, there was a second problem, a perspective that made him see the theory as but a stepping theory to a more general theory. Einstein was convinced that the correct theory would relativize not only motions at constant speed but *all* motions: it was supposed to be a matter of which frame of reference was chosen whether a body moved not at all, at constant speed, or in an accelerated manner. It was obvious that this second project demanded a major trick; after all we *feel* the effects of acceleration, we feel "inertial forces" when we are accelerated. However, it had long been known in classical mechanics that one could "create" inertial forces by transforming to particular coordinate systems, and that one could "transform them away" in the same manner. Indeed, the German word for "inertial forces" is "*Scheinkräfte*," which literally translates to "fictitious forces." Any conceptually minded student studying physics at a German speaking university, as Einstein had, would likely wonder about the exact status of these forces compared to allegedly "real" forces like gravity.

Thus, Einstein saw two construction sites, two projects spreading from his work of 1905: reconciling relativity theory with Newton's theory of gravity, and extending the relativity principle to arbitrary motion. This was the state of affairs that Einstein found himself in 1907 when, while performing his duties in what would be his final year as a patent clerk in Bern, Johannes Stark asked him to write a review of relativity theory. It was likely during the course of writing this review paper (or while he was gearing up to do so) that a thought occurred to Einstein that he would come to call "the happiest thought of [his] life" in 1922; see Vol. 7, Doc. 31, of the Collected Papers of Albert Einstein (CPAE from now on). Essentially, the thought was that the two construction sites are connected, that it might be possible to kill two birds with one stone. Einstein realized that it might be possible to reconcile Newton's theory of gravity with relativity theory *in such a way* that the relativity principle would be extended – not to arbitrary states of motion but to uniformly accelerated motion. Remember Galileo's thought experiment involving the ship: he argued that there is no way for an observer to tell the difference between being in the bowel of a ship at rest and being in the bowel while the ship is moving at a constant velocity. It would be possible to argue that it would be similarly impossible to decide whether the observer is in a uniformly accelerated elevator or in an elevator at rest; as long as he was subject to a homogeneous gravitational field in the latter case.

9.3.2 "Just" a Midwife?

It is clear that the Einstein equivalence principle was absolutely crucial for Einstein's path toward the general theory of relativity. Given this, and given that Einstein himself often emphasized the *heuristic* value of the equivalence principle, one might have some sympathy with Synge [1960] when he writes the following oft-quoted passage in the preface to his book on GR:

> I have never been able to understand this Principle [of Equivalence]. [...] Does it mean that the effects of a gravitational field are indistinguishable from the effects of an observer's acceleration? If so, it is false. In Einstein's theory, either there is a gravitational field or there is none, according as the Riemann tensor does not or does vanish. This is an absolute property; it has nothing to do with any observer's world-line. Space-time is either flat or curved, and in several places in the book I have been at considerable pains to separate truly gravitational effects due to curvature of space-time from those due to curvature of the observer's world-line ... . The Principle of Equivalence performed the essential office of midwife at the birth of general relativity ... . I suggest that the midwife be now buried with appropriate honours and the facts of absolute space-time faced.

First, let us note that Synge captures well Einstein's equivalence principle, in the second sentence quoted, if one adds the qualification of an indistinguishability between *homogeneous* gravitational fields and *uniform* accelerations. So what would Einstein have answered to this criticism? Fortunately we don't have to wonder, for both Friedrich Kottler (in 1916) and Max von Laue (in 1950) posed very similar questions to Einstein. We shall see in the following that Einstein essentially rejected both the identification of gravity with the curvature tensor and the distinction between real and fictitious gravitational fields that Synge presupposes in the above quote; and that in addition to the *heuristic* role the principle played in finding GR, Einstein saw it as having an *enduring* role in how it unified gravity and inertia. Thus, he saw the principle as bringing about a breakdown of the old distinction of an allegedly real force like gravity and an allegedly fictitious force like the centrifugal force. But before we discuss the enduring role Einstein saw for the equivalence principle, let us speak about its heuristic role in helping Einstein climb the mountain on which he found the gravitational field equations of late November 1915.

9.3.3 The Heuristic Role: Midway to a General Principle of Relativity

Synge has created one of the most memorable metaphors in the conceptual analysis of spacetime theories with his picture of Einstein's equivalence principle as a midwife: absolutely indispensable to bringing GR into the world, but not needed anymore once it has been brought to life. Indeed, Einstein himself stressed again and again the heuristic importance of the EEP in his search for what came to be GR. This role of the principle is intimately connected to Einstein thinking of it as a relativity principle. He clearly saw it as extending the special principle of relativity, which states that all inertial motions, including rest, are empirically indistinguishable and thus equivalent in an important sense. Likewise, Einstein's equivalence principle asserted that uniformly accelerated motion is empirically indistinguishable from rest, if the observer at rest is also subject to a homogeneous gravitational field. Thus, Einstein's equivalence principle as relativity principle was midway between the special principle of relativity that treats all inertial motions as on a par, and the general principle of relativity that does so with *all* motions, including non-uniformly accelerated motions.[11]

This role becomes immediately clear from almost any formulation of the equivalence principle in Einstein's texts; here is a particularly clear example:[12]

Starting from this limiting case of the special theory of relativity, one can ask oneself whether an observer, uniformly accelerated relative to K in the region considered, must understand his condition as accelerated, or whether there remains a point of view for him, in accord with the (approximately) known laws of nature, by which he can interpret his condition as "rest." Expressed more precisely: do the laws of nature, known to a certain approximation, allow us to consider a reference system K' as at rest, if it is accelerated uniformly with respect to K? Or somewhat more generally: Can the principle of relativity be extended also to reference systems which are (uniformly) accelerated to one another? The answer runs: As far as we really know the laws of nature, nothing stops us from considering the system K' as at rest. If we assume the presence of a gravitational field (homogeneous in the first approximation) relative to K'; for all bodies fall with the same acceleration independent of their physical nature in a homogeneous gravitational field as well as with respect to our system K'. The assumption that one may treat K' as at rest in all strictness without any laws of nature not being fulfilled with respect to K', I call the "principle of equivalence."

Note how within this short quote Einstein emphasizes thrice the approximate nature of our knowledge of the laws of nature, and how he generalizes the principle of equivalence twice within the paragraph, while naming a version of the weak equivalence principle as a precondition. Two years later, he would make yet another step of generalization. In Einstein [1918], Einstein started to distinguish between the equivalence principle and the general principle of relativity more strongly, while at the same time choosing different words to define the principle:

The equivalence principle: Gravity and inertia are the same in their very essence ("*wesensgleich*").

One might interpret this as Einstein moving from an *epistemological* form of the principle (it's about states that we can't distinguish) to an *ontological* one (it's about identifying gravity and inertia). However, in his 1921 Princeton lectures, the published version of which is the closest thing Einstein ever wrote to a textbook on GR, he related the wording of the 1916 version to that of the 1918 version:[13]

... there is nothing to prevent our conceiving this gravitational field as real. That is, we can consider as equally justified the point of view that K' is "at rest" and a gravitational field is present, and the point of view that only K is a "justified" system of co-ordinates and no gravitational field is present. The premise of the complete physical justification of this point of view we call the "principle of equivalence"; it is obviously suggested by the equality of inertial and gravitational mass and signifies an extension of the principle of relativity to co-ordinate systems which are in non-uniform motion relative to each other. Through this point of view one reaches a theory in which inertia and gravity are the same in their very essence (*wesensgleich*).

We see that in 1918 Einstein uses the term "*wesensgleich*" to describe the relationship between gravity and inertia established by the equivalence principle. It is notoriously difficult to translate. I chose above to translate *"wesensgleich"* as "the same in their very essence"; Norton translates it as "equality of essence," Janssen as being "of the exact same nature."[14]

However one translates the term, it is clear that it is at the core of what Janssen [2014], p. 177, calls "the mature version of the [Einstein] equivalence principle." To appreciate it fully we have to recall that Einstein had gone through an important development between 1916 and 1918, a development directly related to the question of what the term "gravity" or "gravitational field" that appears in every version of the equivalence principle means.

9.3.4 *Gravitational Fields as Fundamental*

Throughout his search for GR, Einstein saw himself in line with Mach's criticism of Newton's concept of absolute space. In GR, he wanted to eliminate from the category of fundamental constituents of the world what he saw as the successor of Newton's absolute space: the metric field $g_{\mu\nu}$. For a long time, Einstein thought that what makes a body move inertially according to GR should be entirely determined by the distribution of the other material bodies.[15] Like the equivalence principle, these thoughts were tightly connected to Einstein's aim of generalizing the relativity principle; after all, if the distribution of material bodies in the universe did determine which motions are inertial motions, then the distinction between inertial and non-inertial motions would become an entirely contingent one. It should be noted that though this line of thought resonates with the idea of relativizing all motion, it is not as straightforward to think of Mach's principle as a relativity principle as it is for the equivalence principle. In Einstein [1918], Einstein defined "Mach's principle" as the requirement that the metric $g_{\mu\nu}$ be uniquely determined by the mass-energy-momentum tensor $T_{\mu\nu}$. But if this were the case and $T_{\mu\nu}$ would thus, via the metric determining which curves are geodesics, also determine which motions are inertial motions, then there would still be a fact of the matter about which motions are inertial and which are not. It would be a contingent fact rather than an absolute fact as in Newtonian theory, but it would still be a fact, in contrast to what the general principle of relativity demands.[16] Likewise, Einstein expected that the inertial mass of each and every body should be determined by its relations to the other material bodies. He did not sharply distinguish between this demand and the demand that which motions are inertial should depend on the mass distribution. The connection might be that he saw inertial mass, like Newton, as the measure of resistance of a body against being diverted from inertial motion. This might have made it seem natural to expect that which makes inertial motion contingent to go hand in hand with making inertial mass contingent.[17]

Consequently, Einstein expected that there should not be non-trivial solutions to the Einstein field equations if the mass-energy-momentum tensor of all the matter in the universe is equal to zero.[18] Indeed, in Einstein [1917], Einstein suggested a modified form of the field equations (introducing the cosmological constant) in part to ensure that Minkowski spacetime (the only vacuum spacetime he knew about at the time) would not be among the solutions of the field equations. Alas, Willem de Sitter showed within months that even the modified field equations *did* have non-trivial vacuum solutions: GR allowed for non-trivial inertial structure, non-trivial spacetime structure, even in the absence of matter.

This brought about a severe change in Einstein's thinking, though one that is difficult to pin down. Einstein [1918] contains the first explicit definition of "Mach's Principle" as the demand that $T_{\mu\nu}$ was to uniquely determine $g_{\mu\nu}$, even though that principle had already come under pressure by de Sitter. However, we have to note that through much of the debate with de Sitter Einstein looked for a way to dismiss the solution as non-physical, which would have rescued Mach's principle to some extent.[19]

Indeed, this is how Einstein would come to operate during the 1920s: Mach's principle became a selection principle to choose between physical and non-physical solutions both of the original field equations and of the field equations with cosmological constant.[20] At the same time, he slowly came to acknowledge that his field equations as such demanded of him to see $g_{\mu\nu}$ as a fundamental element of reality, as something not reducible to something else, like the distribution of material bodies in the universe. Starting with Einstein [1919], Einstein began to embrace fields, and especially $g_{\mu\nu}$, as the fundamental stuff that constitutes material bodies, rather than as something that is reducible to the distribution of material bodies.

It has to be said that with regard to the electromagnetic field Einstein had already thought along these lines by 1909, as we see in a letter to Lorentz, in which he considers a non-linear generalization of the Maxwell equations to which he expected to find particle solutions, in particular solutions capable of representing electrons and photons.[21] So why did Einstein not think of $g_{\mu\nu}$ as equally fundamental, as on a par with the electromagnetic field, much earlier?

I believe that part of the answer is that Einstein saw $g_{\mu\nu}$ as the successor to Newton's absolute space, and thus, following Mach, as something that should not be fundamental in the next step of theoretical development. But another part of the answer may be that through relating gravity and acceleration, and thus gravitational forces and inertial forces, via the equivalence principle, Einstein had opened up the possibility of thinking of gravitational forces as just as "fictitious" as inertial forces.[22] Indeed, this was exactly Kottler's challenge, alluded to above. As we shall see in the following section, Einstein's answer to this challenge (and to Max von Laue's, which mirrors Synge's) will allow us to make the puzzle pieces fit together and to answer: (i) why Einstein did not start out thinking of $g_{\mu\nu}$ as on a par with the electromagnetic field, even though he used the analogy with the electromagnetic field equations all the time in his search for new gravitational field equations; (ii) why Einstein did not think that GR *reduced* gravitational forces to inertial forces; (iii) how he thought that GR *unified* gravity and inertia and how this point of view demanded an enduring role for the equivalence principle in GR.

9.3.5   The Enduring Role: Unifying Gravity and Inertia

The *heuristic* value of Einstein's equivalence principle was that it guided Einstein toward what he thought of as a general theory of the relativity of motion, embodied by the fact that the theory took its simplest form in a generally covariant representation.[23] Even Einstein might have admitted that once such generally covariant field equations were at hand, the equivalence principle had fulfilled *this* purpose. But the *enduring* value of the principle in Einstein's mind is connected to a *second role* of the principle, and it is connected to what Einstein regarded as the main achievement of the theory. The second role of the principle concerns the unification of inertia and gravity in GR, which Einstein saw as a direct analogue to the unification of electricity and magnetism in (special relativistic) electrodynamics.

But first let us come back to Kottler's challenge, hinted at toward the end of the previous section. Kottler claimed in 1916:[24]

> Since then, Einstein has abandoned the equivalence hypothesis. The reasons lie primarily in a particular perception of its results, which amounts to giving an independent existence to the forces of the gravitational field. *Here*, motion in a gravitational field will be seen as force-free. Thus, the law of inertia must be changed and gravitation be seen as a purely inertial phenomenon. This perception seems to me a strict consequence of the equivalence hypothesis; and thus can only be abandoned *together* with the latter. [...] The prime difference [of my approach as compared to Einstein's] is one of principle: the kinematical, rather than dynamical, conception of gravity.

So for Kottler the equivalence principle entailed that gravitational forces are *nothing but* inertial forces: the concept of inertia is fundamental, and gravity has been reduced to inertia. Unsurprisingly, Einstein resisted; for him the point of the equivalence principle was exactly the indistinguishability, and eventually the essential unity, of inertia and gravity. Michel Janssen has dubbed this Einstein's insight of the relativity of the gravitational field, in contrast to the relativity of motion:[25]

> Two observers in non-uniform motion with respect to one another can both claim to be at rest as long as they agree to disagree about whether or not there is a gravitational field. This requires a coordinate-dependent definition of the gravitational field, but, unlike modern relativists, Einstein opted for such a definition, representing the gravitational field by the so-called Christoffel symbols.

For Einstein, seeing the presence of gravitational fields as a coordinate-dependent state of affairs was not a price to be paid but a major achievement of the theory. It was tightly linked to how he interpreted the geodesic equation which, like Kottler, he saw as the successor to the Newtonian law of inertia, as a "generalized law of inertia." But for him, in contrast to Kottler, this generalization lay exactly in the fact that the clear separation of inertia and gravity was overcome:[26]

> Kottler complains that with regard to the equations of motion
>
> $$\frac{d^2 x_\nu}{ds^2} + \left\{ \begin{matrix} \alpha\beta \\ \nu \end{matrix} \right\} \frac{dx_\alpha}{ds} \frac{dx_\beta}{ds} = 0 \qquad (9.1)$$
>
> I interpreted the second term as the representative of the influence of the gravitational field on the point mass, whereas I interpret the first term as, so to speak, the representative of Galilean inertia. This, he claims, would introduce "real gravitational forces," which is supposed to contradict the spirit of the equivalence principle. [...] The labelling of the terms I introduced does not really matter though and was only meant to accommodate our physical habits of thinking. This is also true, in particular, for the concepts
>
> $$\Gamma^\nu{}_{\mu\sigma} = -\left\{ \begin{matrix} \alpha\beta \\ \nu \end{matrix} \right\}$$
>
> (components of the gravitational field) and t_σ^ν (energy components of the gravitational field). The introduction of these labels is in principle unnecessary, but for the time being they do not seem worthless to me, in order to ensure the continuity of thoughts... .

So Janssen is right: Einstein does operate with a coordinate-dependent concept of "gravitational field." But at the same time, what *makes* Christoffel symbols the representatives of the gravitational field is only the comparison to the predecessor of GR, Newton's theory of gravity; in GR taken by itself, there is only "the unity between inertia and gravity [as] expressed by the fact that the entire left side of [equation (9.1)] is tensorial (with respect to arbitrary coordinate transformations), whereas the two terms separately are not."[27]

9.3.6   Is There a Spacetime Free of Gravity?

Ehlers, Stachel, and Giulini have identified the affine connection with this unified inertio-gravitational field, partly following in the footsteps of Weyl, who saw the affine connection as a "guiding field" because of how it establishes the distinction between geodesic and non-geodesic motion, and because of how it "makes" particles move on geodesics.[28]

Giulini especially contrasts this view with that expressed by Synge, namely, that Minkowski spacetime corresponds to the gravity-free case, while a curved metric (i.e., a non-vanishing curvature tensor) signifies the real, invariant, presence of a gravitational field. Indeed, Synge thus implicitly upholds an objective distinction between inertia and gravity, which Einstein believed GR had overcome.

Luckily, Max von Laue challenged Einstein in a way very similar to Synge during a lengthy correspondence in the early 1950s. Von Laue was working on a revised edition of his classic relativity textbook from 1921, and his reengaging with the subject brought about a detailed correspondence between the two old friends. In a letter from 8 September 1950,[29] von Laue questioned Einstein's treatment of the rotating disk in the 1921 Princeton lectures. In particular, he challenged Einstein's use of co-rotating measuring rods in the treatment, and claimed that one should only use rods that move on geodesics. In Einstein's answer, currently dated as from 12 September 1950, he writes:[30]

> It is true that in that case the R_{iklm} vanish, so that one could say: "There is no gravitational field present." However, what characterizes the existence of a gravitational field

from the empirical standpoint is the non-vanishing of the Γ^l_{ik}, not the non-vanishing of the R_{iklm}. If one does not think intuitively *(anschaulich)* in such a way, one cannot grasp why something like a curvature should have anything to do with gravitation. In any case, no reasonable person would have hit upon such a thing. The key for the understanding of the equality of inertial and gravitational mass is missing.

Remember that the position considered in the first sentence is exactly that advocated by Synge. In what follows, we see Einstein rejecting it, emphasizing the heuristic role of the EEP, which takes the components of the affine connection Γ^l_{ik} as representing a coordinate-dependent gravitational field; but we also see him insisting that only in this way WEP2, the numerical equivalence of inertial and gravitational mass, can be understood. We will come back to the latter point. For now let us note that von Laue did not drop the issue. In a follow-up letter he attacks Einstein's interpretation of the geodesic equation as featuring components of the gravitational field. On 8 January 1951 he writes:[31]

> So in the general theory of relativity, *g*-brackets are supposed to represent something like the field strength of the gravitational field. Now let us look at the normal pseudo-Euclidean metric of the special theory of relativity, and transform to spatial cylinder coordinates, or indeed any other curved coordinates, without changing the temporal coordinate. Success: *g*-brackets show up in the geodesic equation. But surely it does not make physical sense to say that one has created a gravitational field by this purely mathematical operation.

Here von Laue makes the ingenious move of looking at the geodesic equation in special relativity, where in pseudo-Euclidean coordinates the geodesic equation looks just like Newton's law of inertia, rewritten in four-dimensional language. Von Laue points out that by transforming e.g. to cylinder coordinates, Christoffel symbols will enter the geodesic equation – which according to Einstein's position would correspond to the "creation" of a gravitational field in virtue of a coordinate transformation.

This sounds like the type of letter that Synge might have written had he received Einstein's previous letter. But note that both Synge and von Laue might not have objected to saying that by virtue of the coordinate transformation *inertial* forces had been "created." After all, creating inertial forces by coordinate transformations was an old hat in Newtonian physics; and they were not *real* forces, so an ability to create them by help of coordinate transformations was not seen as problematic. However, likewise in Newtonian physics, gravitational forces were supposed to be real forces that could not be brought about in this way. In his answer, Einstein compared this Newtonian conception of inertia and gravity with what he saw as that of GR. On 16 January 1951 he writes:[32]

> Now to the gravitational field. Here, one has to properly distinguish between different concepts. In Newtonian theory, everything that is built from the potential counts as the gravitational field. In particular, we would understand the first derivative of the potential as the field strength. In the relativistic theory of gravity, the gravitational field is everything built from the symmetric g_{ik}. Now, it is clear that we cannot do justice to the interpretation of the relativistic gravitational field by appeal to Newtonian theory. Of course, the interpretation of the field of a system that is accelerated and moving parallel to an inertial system (equivalence principle) was of the utmost heuristic importance, because such a field is equivalent to a Newtonian gravitational field with parallel force lines. In this case, the Newtonian field strength is equal to the spatial derivative of g_{44}. One can thus, if one wants to, speak of the first derivatives of the g_{ik}, i.e. of the affine connection Γ, as the gravitational field strength. Of course, these quantities are not tensorial. With this manner of speaking, it is indeed the case that introducing cylinder

coordinates in a Galilean space [i.e. Minkowski spacetime] would bring with it the appearance of field strengths. This is only a manner of speaking. But what is essential is that also in the case of a Galilean, i.e. Minkowskian, space there is, according to the general theory of relativity, a gravitational field, even if its field strengths vanish in the sense defined above. For in the theory of relativity the dimensionality of the field is the only thing that remains from the earlier, physically independent, (absolute) space. I think that most have not understood this main achievement of the theory.

Note the similarity to Einstein's answer to Kottler in Einstein [1916b], where Einstein spoke of the terms "gravitational field" and "Galilean inertia" as an unnecessary (but useful) attempt at accommodating our "habits of thinking." Similarly, in his letter to von Laue 45 years later Einstein sees these notions as just "a manner of speaking" when used in the context of GR.[33] But what is *essential* for Einstein is not to hold on to these Newtonian distinctions, but to accept that Minkowski spacetime is just as much a gravitational(-inertial) field as solutions to the Einstein equations for which curvature does not vanish. Thus, for Einstein there is no distinction anymore between "real gravitational forces" and "fictitious inertial forces." For Einstein, GR, if interpreted in light of the EEP, unified gravity and inertia, and thereby gave up the very distinction between gravity and inertia.

Now, does this mean that the unified field is real or fictitious? Pauli addressed this question when he wrote, right after pointing out that the geodesic equation in GR replaces Galileo's law of inertia:[34]

> In Einstein's theory, gravitation is just as much a *fictitious force* as the coriolis and centrifugal forces are in Newton's theory. (However, it is equally justified to say that in Einstein's theory neither of these two forces is a fictitious force.)

In GR, as understood by both Einstein and Pauli, the distinction between inertial structure and gravity, the distinction between gravitational and inertial forces, and, more generally, the distinction between real and fictitious forces are overcome through the equivalence principle. As a result, a spacetime without curvature has exactly the same status as a spacetime with curvature: it is a particular state of the unified gravitational-inertial field.

9.4   The Local Validity of Special Relativity: Strong Equivalence Principles

9.4.1   *Einstein's Version of the Strong Equivalence Principle*

The strong equivalence principle (SEP) has taken center stage in modern philosophy of physics primarily through the work of Harvey Brown. Brown's project started from the so-called dynamical approach to *special* relativity that he had pioneered together with Oliver Pooley.[35] In that context, Brown and Pooley argued that the non-dynamical Minkowski metric of special relativity can be reconceptualized as being merely a way of expressing features of the dynamical properties of matter field equations, especially their universal Lorentz covariance. It is then the universal Lorentz covariance of all these laws that brings about the chronometric significance of the metric, by allowing to interpret rods and clocks as "waywisers" of the Minkowski metric.

Brown [2007], chapter 9, then argues that *in GR* it is precisely through the strong equivalence principle, which guarantees the local validity of special relativity in GR, that the metric obtains its chronometric significance, its relationship to rods and clocks. In other words, since the SEP does not follow from the Einstein field equations, the equations governing the dynamics of $g_{\mu\nu}$, the geometric interpretation of $g_{\mu\nu}$ is made possible by a principle that goes beyond gravitational dynamics.[36]

I have already alluded to the fact that Einstein himself stated and discussed the role of the SEP in GR, though not under this name. One of the most telling discussions of the principle can be found in a little-known letter from Einstein to Paul Painlevé from 7 December 1921. There he writes:[37]

According to the special theory of relativity the coordinates x, y, z, t are directly measurable via clocks at rest with respect to the coordinate system. Thus, the invariant ds, which is defined via the equation $ds^2 = dt^2 - dx^2 - dy^2 - dz^2$, likewise corresponds to a measurement result.

The general theory of relativity rests entirely on the premise that each infinitesimal line element of the spacetime manifold physically behaves like the four-dimensional manifold of the special theory of relativity. Thus, there are infinitesimal coordinate systems (inertial systems) with the help of which the ds are to be defined exactly like in the special theory of relativity. The general theory of relativity stands or falls with this interpretation of ds. It depends on the latter just as much as Gauss' infinitesimal geometry of surfaces depends on the premise that an infinitesimal surface element behaves metrically like a flat surface element … .

It is easy to read this quote as Einstein seeing a (geo)metric interpretation of GR as absolutely essential to the theory. But it is important to note that in 1921 one of the things he was preoccupied with was to differentiate GR from the first attempt at a unified field theory, delivered by Hermann Weyl. Einstein believed that the main problem of Weyl's theory was that it gave up on the direct relationship between the line element and the readings of rods and clocks. However, Einstein came to concede Weyl's riposte that rods and clocks should come about as solutions of the fundamental field equations governing matter and gravity, rather than being treated as primitive entities.[38] Moreover, he came to actively oppose the idea that GR "geometrizes the gravitational field," or that it shows that the latter is reducible to spacetime geometry.[39]

Of course, Brown, whose position is partly foreshadowed by Einstein's letter to Painlevé, is far from believing that the local validity of SR in GR makes the chronometric interpretation *essential* to GR. Indeed, both Einstein and Brown can be interpreted as putting their finger on the fact that the SEP makes a chronometric interpretation of $g_{\mu\nu}$ *possible*, but that this interpretation is not brought about by the dynamical equations of GR, the Einstein field equations, but by an extra premise, the SEP.

9.4.2 *Different Versions of the Strong Equivalence Principle*

Though we have seen clearly that, under a different name, the SEP was present in Einstein's thought from early on, the honor of forging a clear relationship between the local validity of SR on the one hand and the Einstein equivalence principle on the other hand in the context of GR belongs to Wolfgang Pauli. In his seminal encyclopedia entry on relativity theory, Pauli writes:[40]

General Formulation of the Equivalence Principle. Link between Gravitation and Metric. Originally, the equivalence principle was posited only for *homogeneous* gravitational fields. In the general case we can formulate it as follows: *For every infinitely small region of the world (i.e. a region so small that the spatial and temporal variation of gravity can be neglected therein), there always exists a coordinate system $K_0(X_1, X_2, X_3, X_4)$ such that there is no influence of gravity either on the motion of mass points or on any other physical processes.* In short, in an infinitely small region of the world the gravitational field can be transformed away. […] It is obviously natural to assume that the special theory of relativity holds in K_0. […] In this sense, one can thus say that the invariance of the laws of nature with respect to Lorentz transformations continues to hold in the infinitely small.

Knox takes the italicized part of this quote as a starting point to distinguish between different ways of restating the principle.[41] Casting the net slightly more widely, one might distinguish between two broad classes: pointy SEPs and neighborly SEPs. The first class makes assertions about the local validity of SR at a point, the second about its validity in the neighborhood of a point.

Pauli's own version of the SEP is a member of the class of neighborly SEPs, distinguished from its pointy rivals by the claim that the coordinate system whose existence is postulated is taken to exist in some neighborhood of a point p of the spacetime manifold. Different versions of the neighborly SEP differ with respect to the demands placed on the coordinate system, and with respect to the demands placed on the neighborhood in question.

Regarding the coordinate system, many authors stipulate that in the coordinate system in question (I shall keep calling it K_0 for convenience) the "gravitational field" or "the influence of gravity" can be "transformed away" or "neglected." Other authors have complained about the talk of transforming gravity away, for, they say, if gravitational fields are real then one ought not be able to transform them away. Talk of being allowed to "neglect" the gravitational field in certain circumstances has caused less ire, presumably because it suggests that the principle is supposed to be merely an approximation; we shall come back to this. Other versions of the neighborly SEP, like Knox' own, try to avoid reference to "gravity" or "gravitational fields" altogether and instead demand that in K_0 the metric field should "take on Minkowskian form,"[42] or they demand that K_0 be a normal coordinate system, which is defined as a coordinate system in which the components of the affine connection vanish.[43]

Regarding the demands placed on the neighborhood in question, Pauli's version of a neighborly SEP states that the neighborhood for which K_0 can be defined is to be "infinitely small;" others speak of it being "arbitrarily small," which I take to be a cautious way of expressing the same idea. The general idea of neighborly versions of the SEP has been criticized by way of pointing out that tidal gravitational effects, or effects of curvature, cannot in general be transformed away or neglected in arbitrarily small neighborhoods. The two claims are related by the fact that in Newtonian theory tidal gravitational effects are described by equations involving second derivatives of the Newtonian gravitational potential, while in GR the same effects are described by equations involving second (covariant) derivatives of the metric, which form the curvature tensor. This observation has led to different reactions. As outlined in Section 9.3.2, it led Synge to claim that the SEP is false and should be abandoned altogether, and it led Ohanian [1977] and Ghins and Budden [2001] to versions of the pointy SEP.[44]

Another possible reaction is to claim that the neighborly SEP is only *approximately* true in GR: true if and only if tidal gravitational effects/spacetime curvature can be neglected *given* a particular region of spacetime, a particular physical system in that region, and an interest in particular properties or behavior of said systems. The most important thing to note is that the approximation is of a particular kind: it would be wrong to say that the SEP is approximately true for all regions and physical systems in question. It is taken to be approximately true for some of them, namely exactly those where curvature/tidal effects are negligible with respect to the properties of the physical system under investigation.

Even though both the pointy and the approximate neighborly SEP can be (and have been) motivated by criticizing quotations of Einstein's equivalence principle (as described in Section 9.3) and Pauli's version of the strong equivalence principle, it ought to be noted that both Einstein and Pauli were perfectly aware that there may be physical systems for which second derivatives of the metric, and thus the components of the curvature tensor, will not be negligible given a particular small region of spacetime and a particular question posed, given even an arbitrarily small but still extended region. An example of such a system, discussed in detail first by Eddington [1923], was recently taken up again by Read et al. [2018]: the wave equation for the electromagnetic field $F_{\mu\nu}$ that can be derived from the standard Maxwell equations contains second derivatives of $F_{\mu\nu}$, which in a general relativistic context brings about terms containing the curvature tensor. Thus, the wave equation becomes

$$F^c_{ab;c} = 2\left(F^e_{[b}R_{a]e} - R_{abcd}F^{cd} + J_{[a;b]} \right). \tag{9.2}$$

Going to a sufficiently small region (Knox), an infinitely small region (Pauli), or even a point (Ohanian) won't help with such a system if the components of the curvature tensor don't vanish at that point. For, as noted above, the components of a tensor cannot be transformed away, even at a point.[45] Pauli was most explicitly aware of this when he stated that in SR special relativity is locally valid *if* second derivatives of the metric (and hence curvature) can be neglected.[46]

So far so good: we have distinguished between quite a few versions of the SEP. But what is the role of the principle in GR? What property does a theory in which (a version of) the SEP holds gain by the principle holding, how does the principle constrain the theory? Equally present in Pauli (and Einstein) but more strongly emphasized in recent literature (and pioneered by Brown [2007] especially) is the idea that the SEP is about constraining the *non*-gravitational laws, the fundamental matter fields, and how they couple to the metric.

One example of focusing on this role of the SEP is Knox [2013], who proposes the following version of the (neighborly) SEP after dismissing other versions:[47]

> To any required degree of approximation, given a sufficiently small region of spacetime, it is possible to find a reference frame with respect to whose coordinates [K_0] the metric field takes Minkowskian form, and the connection and its derivatives do not appear in any of the fundamental field equations of matter.

Equally focused on the matter field equations is a recent version by Read et al. [2018]:[48]

> The dynamical equations for non-gravitational fields reduce to a Poincaré invariant form, with no terms featuring the Riemann tensor or its contractions, in a neighbourhood of any $p \in M$.

It is not entirely clear whether Knox [2013] believes that it is possible for *any* physical system to find a K_0 for which her SEP will be fulfilled (given a particular required degree of approximation and a sufficiently small region of spacetime); if so, then the electromagnetic wave equation (9.2) would be a counterexample.[49] However, Read et al. commit to the claim that the principle only holds approximately, where *how good* the approximation is depends on the strength of curvature, the strengths and dynamics of the matter fields under investigation, and the measurement apparatus used to investigate them.[50]

But is this all there is to say – the SEP constrains how the matter fields couple to the metric? What does the SEP teach us about the metric itself?

9.4.3   The Role of the Strong Equivalence Principle

Let us come back to Einstein's letter to Painlevé from 7 December 1921, discussed in Section 9.4.1. Its core is Einstein's statement that it is the locally Minkowskian form of the GR metric that allows us to discern local inertial coordinate systems, with the help of which the metric can be related to the measurements of clocks exactly like in SR.

Pauli and Brown likewise stated that it is via the local validity of special relativity within GR that the dynamical $g_{\mu\nu}$ of GR can be related to readings of rods and clocks, just like in SR.[51]

Equally important is the idea that through the local validity of SR we can "import" the SR notion of inertial frames into GR, which, according to Pauli, allows us to think of the motion of a mass point subject to gravitational fields as force free.[52] Brown [2007] and Janssen [2014] point out that in a sense this existence of local inertial frames, brought about by the local validity of SR, means that GR is *not*, truly, a general theory of the relativity of motion.

It is difficult to say whether Einstein and Pauli would have been convinced by this argument. After stating that the concept of inertial frames can be "imported" from the locally valid special theory of relativity, Pauli hastens to add that Galileo's law of inertia (and, one might deduce, the concept of inertial frames inherent in it) was to be replaced by the geodesic equation in GR. Einstein consistently spoke of the geodesic equation in GR as the "generalized law of inertia," even while stating that, by help of the EEP and the link to Newtonian physics described in Section 9.3.5, one can think of a particle moving on a geodesic as subject to gravitational forces. How exactly does all this fit together?

Note that in discussing the role of the SEP in GR we have now been led to return to the role of the EEP, albeit having gone to such lengths to distinguish them from each other. But distinguishing does not mean entirely separating. What exactly is the relationship between the EEP and (versions of) the SEP?

9.5   The Equivalence Principles as a Bridge between Theories

We have seen in Section 9.3.5 that the role of the Einstein equivalence principle, at least in Einstein's mind, was to unify the Newtonian concepts of gravity and inertia, concepts that Einstein himself thought were not necessary to GR in and of itself. Thus, one might look at the EEP as a bridge principle, a principle forming a bridge from GR to Newtonian theory, a bridge that allows us to see the shadows of Newtonian theory in GR.[53] But this bridge is not *just* about accommodating our "physical habits of thinking" in allowing us to keep operating with the terms "gravity" and "inertia"; it also implies that a curvature-free spacetime is just as "gravitational" as a strongly curved spacetime.[54]

While the Einstein equivalence principle can be seen as a bridge from GR to Newtonian theory, the strong equivalence principle can be seen as a bridge from GR to SR. But the nature of this bridge is different: while the EEP relates GR to a predecessor theory that is not really needed anymore once we have GR, the SEP has a double role. On the one hand, it likewise relates GR to another theory, SR; but contrary to Newton's theory of gravity that theory plays a role *within GR itself*. Therefore, the SEP is a bridge principle like the EEP, but here the bridge is to (part of) the internal structure of GR itself.

What about the practical role of the SEP and the EEP, respectively? We have seen that the role of the EEP is to unify the Newtonian concepts of gravity and inertia, and in turn to provide the possibility of interpreting the equations of motion, the geodesic equation, of GR in a certain way. The role of the SEP, as we have seen in Section 9.4.3, seems likewise interpretational: it allows us to carry over certain interpretational possibilities from SR. In particular, it allows us to transfer the interpretation of rods and clocks as waywisers of the metric tensor from the special case of the Minkowski metric to the case of a generically curved (but locally Minkowskian) metrics, and it allows us to interpret the frames of reference in which the metric is locally Minkowskian as local inertial frames in the sense of "inertial frame" we are wont to use from SR.[55] The local validity of SR allows a "trickling up" of interpretations from SR to GR. I said that this makes the role of the SEP seem interpretational, but we have to be careful not to see its role as "merely" interpretational. The SEP *explains* why rods and clocks can serve as waywisers of the metric field.

So both the EEP and the SEP can be thought of as bridge principles in the sense described above. But are the bridges themselves connected to one another? Pauli clearly thought of the SEP as a generalization of the EEP. And indeed, if we restrict ourselves to neighborly versions of the SEP (as described in Section 9.4.2), then the EEP lives quite close by. For given that neighborly versions of the SEP make statements about extended regions, the transformation between frames of references described in Einstein's original statement of the EEP is possible in that region. Thus, one might argue that the demand that the SEP approximately holds for a particular system in a particular neighborhood is a necessary condition for the EEP to be applicable to that system and that neighborhood.

The attentive reader may realize that in a way we have come full circle. After all, in Einstein's statement of the EEP (cited in Section 9.3.3), Einstein begins with "Starting from this limiting case of the special theory of relativity, one can ask oneself...," i.e., he motivates the EEP with the SEP.

Have we learned anything new? Well, we have seen different versions of the SEP, implications of both EEP and SEP, interpretations and cross-theoretic links made available because of them. True, Einstein may have packed all that into one paragraph, and it took us a whole book chapter to unpack it all and ponder its implications. But we're in good company: Einstein himself likewise wrote dozens if not hundreds of letters unpacking all this for his contemporaries.

Acknowledgments

I would like to thank especially Harvey Brown and John Norton for many illuminating discussions about different versions of the equivalence principle over the years; I could not have written this chapter without learning from them first. Furthermore, I would like to thank Patrick Dürr, Niels Martens, and James Read for very helpful comments on earlier versions of this chapter, and the Albert Einstein Archives for permission to quote from the correspondence between Einstein and Max von Laue.

Notes

1 Many modern sources call the second class (or a member thereof) "the Einstein equivalence principle," and many use the two labels as synonyms. However, Norton [1989] showed that Einstein not only carefully distinguished the two principles but also only called (a member of) the first class "equivalence principle" in the context of GR. Norton [2020] shows how Einstein transitioned from what we here call the EEP to the SEP for a while in the context of his 1912 scalar theory of gravity, only to return to the EEP in GR, now taking a version of the SEP as a starting point if not a precondition to motivate the EEP. This makes sense in that in the context of his 1912 scalar theory he came to think of the SEP as an emergency solution, only to be accepted in place of the EEP if there was no other choice, and clearly logically weaker than the EEP in the same sense that the WEP was logically weaker than the SEP. Both could thus serve as motivators and even preconditions for the EEP; but the latter was the only true equivalence principle (in the sense of a true love rather than a clearly true belief) in Einstein's heart.

2 Einstein [1916a] p. 777.

3 For a comprehensive discussion of the 1916 review paper see Sauer [2004]; for a beautiful facsimile reproduction and commentary see Gutfreund and Renn [2015].

4 Will [1993], p. 22.

5 Ohanian [1977], p. 904, calls something very close to WEP1 above "Galileo's principle." He follows up with the suggestion to define "test particle" by the limiting case of a particle of small size R ($R \to 0$) and small mass m ($m \to 0$), resulting in a vanishing gravitational self-field ($\frac{Gm}{Rc^2} \to 0$), alongside the vanishing of spin and higher multipole moments.

6 See again Ohanian [1977], p. 904 and Will [1993], p.22. Ohanian gives definitions of both "inertial mass" and "gravitational mass," quite in line with Mach [1872] in defining them both as ratios of relative accelerations.

7 Will [1993], p. 12 could be read as stating that WEP2 and WEP1 are equivalent, but I believe he would agree that given the more precise definition he gives on p. 22 entails that the arrow of implication goes only from WEP2 to WEP1. After all, WEP1 could be true if we did not even have the concepts of "inertial mass" and "gravitational mass."

8 It has been shown that some alternatives to GR, like Brans-Dicke theory, predict the inequality of inertial and gravitational mass for sufficiently large, extended objects. Thus, such a theory would violate WEP2 but not necessarily WEP1. Indeed, Brans-Dicke theory is an example of a theory in which WEP2 is violated while WEP1 holds. For details see Lehmkuhl [2017].

9 For particularly clear (and early) statement to this effect see the introduction of Einstein and Grossmann [1913]. For a careful reconstruction and analysis of Galileo's and Newton's conceptions of the weak equivalence principle, both linked to Newton's Corollary VI, which is derived from Newton's three laws of motion in his *Principia*, see Saunders [2013].

10 See Renn [2007b], Volume 3.

11 Norton has argued that it is possible to make precise Einstein's way of thinking of the equivalence principle as an extension of the relativity principle, and of this extension being "halfway" between the Lorentz covariance of special relativity on the one hand and the general covariance of GR on the other, even in the face of Kretschmann's objection that every theory can be brought into generally covariant form

(Kretschmann [1918]). See Norton [1993] and especially Norton [1992b], section 7.3. At the same time, it has been widely argued that GR did not indeed succeed in relativizing all types of motion, that it is indeed not a theory that implements the general principle of the relativity of motion. See Janssen [2014] for a comprehensive and accessible review.

12 Einstein [1916b], pp. 639–640; judged by Norton [1989], whose translation of the passage I adapt, as the clearest expression of the Einstein equivalence principle in Einstein's oeuvre. Note that what I call the ontological version of Einstein's equivalence principle below (and what Janssen [2014] calls the mature version) is much more concise, but also less self-contained.

13 My translation.

14 See Norton [1992a], note 42 and Janssen [2014], p. 219.

15 See Hoefer [1994, 1995], Barbour [2007], and Renn [2007a] for detailed descriptions of the role Machian thoughts (and different versions of "Mach's Principle") played for Einstein in the development of GR.

16 Note, however, that Einstein resisted seeing the metric field as *primarily* the successor of inertial structure as established by absolute space in Newtonian theory; more on this in the next section.

17 See Barbour [1990], Norton [1995], and Brown and Lehmkuhl [2017] for studies in how these ideas that Einstein always attributed to Mach differed from Mach's actual ideas; and how fruitful this flexibility in reading Mach became for Einstein in the development of GR.

18 This is strictly speaking an additional demand to expecting a unique determination of $g_{\mu\nu}$ by the energy-momentum tensor; for if there were *one* solution corresponding to $T_{\mu\nu} = 0$, the latter condition would have been fulfilled.

19 For more details on the debate between Einstein and De Sitter, see CPAE Vol. 8, the editorial note "The Einstein-De Sitter-Weyl-Klein-Debate," on pp. 351-357; and Smeenk [2015], section 4.

20 See section the Introduction to CPAE Vol. 13, p. xlv.

21 Einstein to Lorentz, 23 May 1909; Vol. 5, Doc. 163 CPAE.

22 Remember my digression on the German word for "inertial force" corresponding to "fictitious force" or "apparent force" in Section 9.3.1. It is also worth noting that the same argument allows to say that inertial forces have turned out to be just as real as gravitational forces as Pauli noted in 1921 (see the end of section 9.3.6).

23 See Einstein's answer to Kretschmann [1918] and Einstein [1918].

24 Kottler [1916], pp. 955–956, my translation.

25 Janssen [2012], p. 159. See also Janssen [2005] and Janssen [2014]. In these sources, Janssen also argues that Einstein did not indeed succeed in relativizing motion in the sense demanded by the general principle of relativity. Before the passage quoted in the main text he writes: "It is so obvious today that the general theory of relativity does not extend the relativity principle from uniform to arbitrary motion that it has become something of a puzzle how Einstein could ever have claimed it did ... . [...] [T]o a large extent, the solution of this puzzle is simply that what Einstein called the relativity of non-uniform motion is more appropriately called the relativity of the gravitational field."

26 Einstein [1916b], p. 641.

27 Einstein [1922], p. 51. For further analysis see Lehmkuhl [2014], section 4. Janssen appreciates the above; indeed, he calls the following statement the equivalence principle in its mature form: "There is only an inertio-gravitational field that breaks down differently into inertial and gravitational components depending on the state of motion of the person making the call." Janssen [2014], p. 178. (Note that Janssen [2014] was written in 2008, though only published in 2014; thus it was written well before Lehmkuhl [2014]).

28 Ehlers [1973], Stachel [1986, 1995, 2007], Giulini [2002], Weyl [1922], pp. 219–224.

29 Einstein Archive Call No. (in the following abbreviated to EA) 16-143.

30 EA 16-148; Translation from Norton [1989], p. 39.

31 EA 16-152, my translation. Note that when von Laue speaks of g-brackets he means the Christoffel symbols Γ^l_{ik}, i.e., the components of the affine connection.

32 EA 16-154, my translation. I quote in some detail because the letter is not yet available in the CPAE, and will not be for some decades.

33 See Norton [1989], section 8, for further analysis.

34 Pauli [1921], p. 709. I have translated "*Scheinkraft*" as fictitious force in this instance; the emphasis is Pauli's.

35 See Brown and Pooley [2001, 2004], For details on the dynamical approach see the chapter by Brown and Read in this book, Brown and Read [2018], and references cited therein.

36 Note that in contrast to the dynamical view in the context of special relativity, Brown does *not* claim that the metric field in GR can be reduced to the dynamics or the symmetries of the matter fields. Like Einstein

after 1917 (see Section 9.3.4), Brown accepts that $g_{\mu\nu}$ has irreducible dynamical degrees of freedom. What is at issue is how this newly dynamical field obtains its relation to the measurements of rods and clocks.

37 Vol. 12, Doc. 314 CPAE. Unfortunately, the letter was not selected for translation in the translation supplement to Vol. 12. Thus, the translation given here is my own; and I quote in more detail than I otherwise would.

38 See Giovanelli [2013, 2014].

39 See Lehmkuhl [2014] and Giovanelli [2016].

40 Pauli [1921], pp. 705–706 (Pauli's emphasis, my translation).

41 Knox [2013], p. 351. Knox motivates all of these types by the aim of restating the SEP in such a way that no reference to "gravity" or "gravitational field" is necessary. She distinguishes between what she calls "the geometrical SEP," attributed to Trautman [1966], which restricts the SEP to the claim that "all possible experiments determine the same affine connection"; the "pointy SEP" which I subsume under the first class of SEPs described in the main text; and her own "effective SEP," which I shall subsume under the second class about to be described.

42 Knox [2013], p. 352.

43 Arguably, most if not all of these statements can be translated into one another. If, following Synge, "gravitational field" is identified with "non-vanishing curvature," then the demand for "no influence of gravity" (Pauli) and that for the metric "taking on Minkowskian form" coincide. Likewise, if we restrict our attention to theories like GR where there is a unique affine connection and a unique metric which are compatible according to the condition $\nabla^{\mu} g_{\mu\nu} = 0$, then the demand for K_0 to be a normal coordinate system implies that the metric compatible with the connection looks like a Minkowskian metric in that coordinate system. The "looks like" is important here, for higher derivatives of the metric (second derivatives and above) don't necessarily vanish in normal coordinates (see Read et al. [2018] for details on this).

44 Norton [1989] has been widely lauded for clarifying what Einstein's own conception of the equivalence principle actually was. But the article did even more than that: it also constructed a major challenge to the pointy strong equivalence principle, a problem that Einstein had alluded to in correspondence with Schlick and that Norton formalized using modern differential geometry and concepts developed by Robert Geroch. Norton [1989], section 10, concludes: "The results of this section vindicate Einstein's objection to Schlick. If we understand the infinitesimal principle of equivalence [i.e., the SEP] to assert that special relativity holds at a point to second-order quantities only, then it follows that we cannot formulate special relativity's requirement that the world line of a free point-mass is a geodesic."

45 Note, however, that the curvature tensor needs more than a point in order to be defined. Given that the definition of the curvature tensor involves second covariant derivatives that only make sense in terms of parallel transport along a curve, one *does* need an extended (though arbitrarily small) region to define the curvature tensor. This is just a special case of the fact that properties of spacetime regions that are represented by vectors or tensors (rather than merely by scalars) cannot be defined at points alone; at least an implicit reference to points in the neighborhood is necessary. See Butterfield [2005] for details.

46 Pauli [1921], p. 707.

47 Knox [2013], p. 352.

48 Read et al. [2018], p. 12. The authors call this version of the SEP EP1', distinguished from EP2'. The latter does *not* demand the absence of curvature terms in the matter laws as formulated in the neighborhood in question.

49 One open avenue to rebut this would be to demand that the principle only holds for the *fundamental* equations governing the matter fields, in this case the Maxwell equations.

50 Note that they do not include any reference to a particular frame of reference in their formulation of the SEP, though the existence of a frame of reference in which the metric takes Minkowskian form follows if their version of the SEP is fulfilled.

51 See Pauli [1921], p. 708 and Brown [2007], chapter 9.

52 Pauli [1921], p. 708.

53 It is possibly needless to say this, but I don't mean "bridge principle" in the Nagelian sense of the term; the bridge formed here is between theories.

54 Recall Section 9.3.6, in which we saw Einstein making this point in correspondence with Max von Laue, and see also section 12 of Norton [1989].

55 For a pathbreaking paper specifying the circumstances under which matter fields in a curved spacetime "behave" locally as if they were in a Minkowski spacetime see Fletcher [2020].

References

Barbour, J. (1990). The part played by Mach's principle in the genesis of relativistic cosmology. In B.B. et al. (ed.), *Modern Cosmology in Retrospect*. Cambridge: Cambridge University Press, chapter 4, p. 47.

Barbour, J. (2007). Einstein and Mach's principle. In J. Renn and M. Schemmel (eds.), *The Genesis of General Relativity*, Vol. 3. Dordrecht: Springer, pp. 1492–1527.

Brown, H.R. (2007). *Physical Relativity. Space-time Structure from a Dynamical Perspective*, Oxford University Press.

Brown, H.R. and Lehmkuhl, D. (2017). Einstein, the reality of space, and the action-reaction principle. In P. Ghose (ed.), *Einstein, Tagore, and the Nature of Reality*. Routledge Studies in the Philosophy of Mathematics and Physics, London and New York: Routledge, pp. 9–36.

Brown, H.R. and Pooley, O. (2001). *Physics Meets Philosophy at the Planck Scale: Contemporary Theories in Quantum Gravity*, Cambridge: Cambridge University Press, chapter The origin of the spacetime metric: Bell's 'Lorentzian pedagogy' and its significance in general relativity, pp. 256–272.

Brown, H.R. and Pooley, O. (2004). Minkowski space-time: A glorious non-entity. In D. Dieks (ed.), *The Ontology of Spacetime*, Vol. I of *Philosophy and Foundations of Physics*. Amsterdam: Elsevier, pp. 67–92.

Brown, H.R. and Read, J. (2018). The dynamical approach to spacetime theories. In E. Knox and A. Wilson (eds.), *The Routledge Companion to Philosophy of Physics*. Routledge.

Butterfield, J. (2005). *Against pointillisme about geometry*. Available at: *arXiv preprint physics/0512063* .

Eddington, A.S. (1923). *The Mathematical Theory of Relativity*. Cambridge: Cambridge University Press.

Ehlers, J. (1973). *Survey of General Relativity Theory*, Vol. 38 of *Astrophysics and Space Science Library*. D. Reidel Publishing Company, pp. 1–123.

Einstein, A. (1905). Zur Elektrodynamik bewegter Körper. *Annalen der Physik*, 17: 891–921. Reprinted as Document 23 (page 275) of Volume 2 CPAE.

Einstein, A. (1916a). Die Grundlage der allgemeinen Relativitätstheorie. *Annalen der Physik*, 49(7): 769–822. Reprinted as Vol. 6, Doc. 30 CPAE.

Einstein, A. (1916b). Über Friedrich Kottlers Abhandlung "Über einsteins Äquivalenzhypothese und die Gravitation". *Annalen der Physik*, 356(22): 639–642. Reprinted as Vol. 6, Doc. 40 CPAE.

Einstein, A. (1917). Kosmologische Betrachtungen zur allgemeinen Relativitätstheorie. *Sitzungsberichte der Königlich Preussischen Akademie der Wissenschaften*. Reprinted as Document 43 (page 541) of Volume 6 CPAE.

Einstein, A. (1918). Prinzipielles zur allgemeinen Relativitätstheorie. *Annalen der Physik*, 55: 241–244. Reprinted as Vol. 7, Doc. 4 CPAE.

Einstein, A. (1919). Spielen Gravitationsfelder im Aufbau der materiellen Elementarteilchen eine Rolle? *Sitzungsberichte der Königlich Preussischen Akademie der Wissenschaften*, pp. 349–356. Reprinted as Vol. 7, Doc. 17 CPAE.

Einstein, A. (1922). *Vier Vorlesungen über Relativitätstheorie gehalten im Mai 1921 an der Universität Princeton*, F. Vieweg. Reprinted as Vol. 7, Doc. 71 CPAE; and in various editions as "The Meaning of Relativity" by Princeton University Press.

Einstein, A. and Grossmann, M. (1913). *Entwurf einer verallgemeinerten Relativitätstheorie und einer Theorie der Gravitation*. Leipzig: Teubner. Reprinted as Vol. 4, Doc. 13 CPAE.

Fletcher, S. (2020). Approximate local poincare spacetime symmetry in general relativity. In C. Beisbart, T. Sauer and C. Wüthrich (eds.), *Thinking about Space and Time: 100 Years of Applying and Interpreting General Relativity*. Einstein Studies. Dordrecht: Springer.

Ghins, M. and Budden, T. (2001). The principle of equivalence. *Studies in the History and Philosophy of Modern Physics*, 32(1): 33–51.

Giovanelli, M. (2013). Talking at cross-purposes: How Einstein and the logical empiricists never agreed on what they were disagreeing about. *Synthese*, 190(17): 3819–3863.

Giovanelli, M. (2014). 'But one must not legalize the mentioned sin': Phenomenological vs. dynamical treatments of rods and clocks in Einstein's thought. *Studies in History and Philosophy of Science Part B: Studies in History and Philosophy of Modern Physics*, 48: 20–44.

Giovanelli, M. (2016). '. . . But i still can't get rid of a sense of artificiality': The Reichenbach–Einstein debate on the geometrization of the electromagnetic field. *Studies in History and Philosophy of Science Part B: Studies in History and Philosophy of Modern Physics*, 54: 35–51.

Giulini, D. (2002). Das Problem der Trägheit. *Philosophia Naturalis*, 39: 343–374.

Gutfreund, H. and Renn, J. (2015). *The Road to Relativity: The History and Meaning of Einstein's "The Foundation of General Relativity", Featuring the Original Manuscript of Einstein's Masterpiece*. Princeton, NJ and Oxford: Princeton University Press.

Hoefer, C. (1994). Einstein's struggle for a Machian gravitation theory. *Studies in History and Philosophy of Science Part A*, 25(3): 287–335.

Hoefer, C. (1995). Einstein's formulations of Mach's principle. In J. Barbour and H. Pfister (eds.), *Mach's Principle: From Newton's Bucket to Quantum Gravity*, Vol. 6 of *Einstein Studies*. Birkhäuser, pp. 67–90.

Janssen, M. (2005). Of pots and holes: Einstein's bumpy road to general relativity. *Annalen der Physik*, 14(S1): 58–85.

Janssen, M. (2012). The twins and the bucket: How Einstein made gravity rather than motion relative in general relativity. *Studies in History and Philosophy of Science Part B: Studies in History and Philosophy of Modern Physics*, 43(3): 159–175.

Janssen, M. (2014). *'No Success like Failure...': Einstein's Quest for General Relativity, 1907–1920*. Cambridge University Press, chapter 6, pp. 167–228.

Knox, E. (2013). Effective spacetime Geometry. *Studies in History and Philosophy of Science Part B: Studies in History and Philosophy of Modern Physics*, 44(3): 346–356.

Kottler, F. (1916). Über Einsteins äquivalenzhypothese und die gravitation. *Annalen der Physik*, 45(16): 955–972.

Kretschmann, E. (1918). Über den physikalischen Sinn der Relativitätspostulate, A. Einsteins neue und seine ursprüngliche Relativitätstheorie. *Annalen der Physik*, 358(16): 575–614.

Lehmkuhl, D. (2014). Why Einstein did not believe that general relativity geometrizes gravity. *Studies in History and Philosophy of Science Part B: Studies in History and Philosophy of Modern Physics*, 46: 316–326.

Lehmkuhl, D. (2017). Introduction: Towards a theory of spacetime theories. In D.e.a. Lehmkuhl (ed.), *Towards a Theory of Spacetime Theories*. Basel: Birkhuser, Springer, pp. 1–11.

Mach, E. (1872). *Die Geschichte und die Wurzel des Satzes von der Erhaltung der Arbeit*. Calve'sche Buchhandlung, Prag. Translation as 'History and Root of the Principle of the Conservation of Energy', P. Jourdain, Chicago: Open Court (1911).

Norton, J. (1989). What was Einstein's principle of equivalence? In D. Howard and J. Stachel (eds.), *Einstein and the History of General Relativity*, Vol. 1 of *Einstein Studies*. Basel: Birkhäuser.

Norton, J. (1992a). Einstein, Nordström and the early demise of Lorentz-covariant, scalar theories of Gravitation. *Archive for History of Exact Sciences*, 45: 17–94. Reprinted in Renn, J.: The genesis of General Relativity, Vol. 3.

Norton, J. (1992b). The physical content of general covariance. *Studies in the History of General Relativity*, 3: 281–315.

Norton, J. (1993). General covariance and the foundations of general relativity: eight decades of dispute. *Reports on Progress in Physics*, 56(7): 791.

Norton, J. (1995). Mach's principle before Einstein. In J. Barbour and H. Pfister (eds.), *Mach's Principle: From Newton's Bucket to Quantum Gravity*, Vol. 6 of *Einstein Studies*. Boston: Birkhäuser.

Norton, J. (2020). Einstein's conflicting heuristics: The discovery of general relativity. In C. Beisbart, T. Sauer and C. Wüthrich (eds.), *Thinking about Space and Time: 100 Years of Applying and Interpreting General Relativity*. Einstein Studies. Dordrecht: Springer.

Ohanian, H.C. (1977). What is the principle of equivalence? *American Journal of Physics*, 45(10): 903–909.

Pauli, W. (1921). *Relativitätstheorie*, B.G. Teubner, Leipzig. Republished by Springer in 2000 with new annotation by Domenico Giulini.

Read, J., Brown, H.R. and Lehmkuhl, D. (2018). Two miracles of general relativity. *Studies in History and Philosophy of Science Part B: Studies in History and Philosophy of Modern Physics*, 64, pp. 14–25.

Renn, J. (2007a). The third way to general relativity: Einstein and Mach in context. In J. Renn and M. Schemmel (eds.), *The Genesis of General Relativity*, Vol. 3 of *Boston Studies in the Philosophy of Science*. Dordrecht: Springer.

Renn, J., editor. (2007b). *The Genesis of General Relativity. Vol. 1–4*. Dordrecht: Springer.

Sauer, T. (2004). *Albert Einstein's 1916 review article on general relativity*. Available at: *arXiv preprint physics/0405066*.

Saunders, S. (2013). Rethinking Newton's Principia. *Philosophy of Science*, 80(1): 22–48.

Smeenk, C. (2015). Einstein's role in the creation of relativistic cosmology. In M. Janssen and C. Lehner (eds.), *The Cambridge Companion to Einstein*. Cambridge University Press, chapter 7, pp. 228–269.

Stachel, J. (1986). What a physicist can learn from the discovery of general relativity. In R. Ruffini (ed.), *Proceedings of the Marcel Grossman Meeting on General Relativity*, Amsterdam, North-Holland, pp. 1857–1862.

Stachel, J. (1995). The history of relativity. In L. M. Brown, A. Pais and B. Pippard (eds.), *Twentieth century physics*. Institute of Physics Publishing [ua], pp. 249–356.

Stachel, J. (2007). The story of Newstein or: Is gravity just another pretty force? In J. Renn (ed.), *The Genesis of General Relativity*. Dordrecht: Springer, pp. 1962–2000.

Synge, J.L. (1960). *Relativity: The General Theory*. Amsterdam: Noth-Holland Pub. Comp.

Trautman, A. (1966). The general theory of Relativity. *Soviet Physics Uspekhi*, 9(3): 319–339.

Weyl, H. (1922). *RaumZeitMaterie* (5th edition). Dordrecht: Springer.

Will, C.M. (1993). *Theory and Experiment in Gravitational Physics*. Cambridge University Press.

Dennis Lehmkuhl

Further Reading from the Editors

Despite its central interpretational role, there is not as much philosophical literature on either of the equivalence principles as one might like. Ohanian, H. C. 'What is the principle of equivalence?', (*American Journal of Physics* 45(10), 903–909, 1977) is a classic paper in the physics literature. John Norton's "What was Einstein's principle of equivalence?" in D. Howard and J. Stachel, eds., Vol. 1 of *Einstein Studies* (Birkhauser, 1989) gives an excellent overview of the issues. Another take can be found in Ghin's and Budden's "The principle of equivalence" (*Studies in History and Philosophy of Modern Physics*, 32(1), 33–51, 2001). Chapter 9 of Harvey Brown's, *Physical Relativity: Space–Time Structure from a Dynamical Perspective* (Oxford University Press: Oxford, 2006) also includes a subtle treatment of equivalence principles.

10

THE HOLE ARGUMENT

Oliver Pooley

10.1 Introduction

Our best theory of space, time, and gravity is the general theory of relativity (GR). It accounts for gravitational phenomena in terms of the curvature of spacetime. In more mathematical presentations of the theory, solutions are standardly represented as n-tuples: $(M, g_{ab}, \phi_1, \phi_2, \ldots)$. The ϕs are objects that represent the assorted material content of spacetime (such as stars and electromagnetic fields). M and g_{ab} together represent spacetime itself. M is a differentiable manifold representing the four-dimensional continuum of spacetime points. g_{ab} is a Lorentzian metric tensor defined on M. It encodes some of spacetime's key spatiotemporal properties, such as the spacetime distances along paths in M. In particular, spacetime's curvature can be defined in terms of g_{ab}.

In its contemporary guise, the hole argument targets a natural interpretation of this mathematical machinery. According to the *spacetime substantivalist*, spacetime itself, represented by (M, g_{ab}), should be taken to be an element of reality in its own right, on (at least) equal ontological footing with its material content. John Earman and John Norton's version of the hole argument (Earman and Norton, 1987) aims to undermine this reading of the theory. In particular, it seeks to establish that, under a substantivalist interpretation, GR is radically – and problematically – *indeterministic*.

Earman and Norton's hole argument set the agenda for the wide-ranging debate that burgeoned in the late 1980s and early 1990s, and that has rumbled on in the decades since. The hole argument, however, did not originate with them. It was first advanced by Einstein (Einstein, 1914), as he struggled to find a theory of gravity compatible with the notions of space and time ushered in by his 1905 special theory of relativity.

By 1913 Einstein had already settled on a non-flat metric tensor, g_{ab}, as the mathematical object that would capture gravitational effects. The remaining task was to discover field equations describing how g_{ab} depends on material "sources," encoded by their stress energy tensor T_{ab}.

Einstein sought a theory that would generalize the *relativity principle*. The restricted relativity principle of Newtonian physics and special relativity only asserts the equivalence of all *inertial* frames: frames in which (*inter alia*) force-free bodies move uniformly in straight lines. The inertial frames are, however, physically distinguished in these theories from frames moving non-uniformly with respect to them. Einstein believed that fundamental physics should treat *all* frames of reference as on a par.

In 1913, Einstein came tantalizingly close to settling on the famous field equations that he would eventually publish toward the end of 1915. These equations are *generally covariant*: they hold in all coordinates systems within a family related by smooth but otherwise arbitrary coordinate transformations. Because such transformations include transformations between coordinates adapted to frames in arbitrary states of motion, Einstein initially believed that generally covariant equations embody a generalized relativity principle.

In 1913, however, he temporarily gave up the quest for general covariance. His original version of the hole argument convinced him that any generally covariant theory describing how g_{ab} relates to T_{ab} must be indeterministic. (For further details of Einstein's argument and its role in his search for his field equations, see Stachel (1989), Norton (1984), and Janssen (2014, §3).)

In this chapter, I focus on the hole argument as an argument against substantivalism. The next section reviews some technical notions standardly presupposed in presentations of the argument. Section 10.3 presents the argument itself. The remainder of the chapter reviews possible responses.

10.2 Diffeomorphisms

As characterized above, the general covariance of a theory is a matter of the invariance of its equations under smooth but otherwise arbitrary coordinate transformations. In order to make contact with contemporary discussion of the hole argument, we need an alternative formulation that dispenses with reference to coordinates.

Let $g(x)$ stand for some specific solution of a generally covariant theory T, expressed with respect to some specific coordinate system $\{x\}$. Let $g'(x')$ be a redescription of the same situation but given with respect to a new coordinate system $\{x'\}$. Since T is generally covariant, $g'(x')$ will also satisfy T's equations (provided that the coordinate transformation $x' = f(x)$ is smooth).

Understood in this way – as a map between descriptions of the very same physical situation given with respect to different coordinate systems – the transformation $g(x) \mapsto g'(x')$ is a *passive transformation*. The *active interpretation* of the transformation involves asking what the function of coordinates g' represents *when interpreted with respect to the original coordinate system*, $\{x\}$. (Note that this question presupposes something that one might dispute, namely, that it makes sense to talk of holding a specific coordinate system fixed, independently of any solution described with respect to it.)

When $g(x) \neq g'(x)$, $g(x)$ and $g'(x)$ will describe *different* physical situations (assuming no redundancy in the way the mathematical object g represents physical reality). Moreover, because of T's general covariance, g's status as a solution of T is independent of which coordinate system, $\{x\}$ or $\{x'\}$, it is referred to.

The distinction between active and passive transformations first arises in the context of coordinate transformations. The terminology is now used, however, in more general contexts. In particular, one may distinguish between what are labeled (somewhat misleadingly) active and passive *diffeomorphisms*.

Recall that solutions of GR are n-tuples $(M, g_{ab}, \ldots)$, where M is a differentiable manifold. To call M *differentiable* just means that it is equipped with structure that distinguishes, e.g., smooth from non-smooth curves. A *diffeomorphism* between differentiable manifolds M and N is a bijective map such that both the map and its inverse preserve such structure (e.g., under the map, the image of a smooth curve in M will be a smooth curve in N and vice versa).

Let d be a diffeomorphism from M to itself. There is no sense in which d *by itself* can be said to be active or passive: it simply associates to each point of M a (possibly distinct) point. To get further, we need to: (i) consider maps naturally associated with d that act on structures defined on M and (ii) distinguish between two types of structure.

The first type of structure includes coordinate systems: maps from M into $\mathbb{R}^4$ (assuming that M is four-dimensional) that preserve M's differentiable structure. (In fact, the differentiable structure of M is standardly defined extrinsically, via a set of preferred coordinate systems.) The second type includes objects defined on M that are intended to represent something physical: fields or other objects located in spacetime, or physically meaningful spatiotemporal properties and relations.

The definitions of the natural maps associated with d on such objects will differ in detail, depending on the type of object. The basic idea, however, is straightforward. If d maps $p \in M$ to q, we can define d's action on an object F via the requirement that the image object, d^*F, "takes the same value" at p as the original object takes at q. Therefore, for example, given a coordinate system $\phi : U \subseteq M \rightarrow \mathbb{R}^4$, we can define a new coordinate system $d^*\phi$ on the open set $d^{-1}(U)$ via the requirement that $d^*\phi(p) = \phi(d(p))$ for all $p \in d^{-1}(U)$.

A *passive diffeomorphism* corresponds to the case where one contemplates the action of the diffeomorphism on a coordinate system (or systems) while leaving the objects representing physically significant structure unchanged. It is the exact correlate of the passive coordinate transformation described above.

An *active diffeomorphism*, by contrast, leaves coordinate systems unchanged but acts on the objects representing physical structure. So, for example, if (M, g_{ab}) is a manifold equipped with a metric tensor field g_{ab}, (M, d^*g_{ab}) is the (in general) mathematically distinct object that results from applying to g_{ab} the active diffeomorphism d. (M, g_{ab}) and (M, d^*g_{ab}) are mathematically distinct because, in general, when $d(p) \neq p$, the value of d^*g_{ab} at p (I write: $d^*g_{ab}|_p$) does not equal the value of g_{ab} at p.

With this machinery in place, a revised, coordinate-free notion of general covariance, often labeled *(active) diffeomorphism invariance*, can be stated. Let the models of a theory T be n-tuples of the form $(M, O_1, O_2, \ldots)$. T is generally covariant if and only if the following condition is satisfied: if $(M, O_1, O_2, \ldots)$ is a structure of the relevant type and d is a diffeomorphism between M and N, then $(M, O_1, O_2, \ldots)$ is a solution of T if and only if $(N, d^*O_1, d^*O_2, \ldots)$ is also a solution of T. GR is generally covariant in this sense. (Whether this coordinate-free notion of general covariance is equivalent to the notion of general covariance given earlier in terms of coordinate transformations is a subtle business. For more discussion, see Pooley (2017), where the definition of diffeomorphism invariance is also refined to take account of the distinction between dynamical and non-dynamical fields.)

Finally, I define a *hole diffeomorphism*. Let M be a differentiable manifold and let H (the "hole") be a compact open subset of M. A diffeomorphism $d : M \rightarrow M$ is a hole diffeomorphism corresponding to H if and only if d is the identity transformation outside of H but comes smoothly to differ from the identity transformation within H. In other words, $d(p) = p$ for all $p \in M \setminus H$, but $d(p) \neq p$ for some $p \in H$.

10.3   The Argument

The core conclusion of the hole argument is that, under a substantivalist interpretation, any generally covariant theory such as GR is indeterministic. (To simplify exposition, I focus on GR as the paradigm generally covariant theory.) There is then the further question of whether this core conclusion counts against a substantivalist interpretation of GR: is the indeterminism in question problematic?

The argument for the core conclusion has three main premises: a claim about what substantivalism entails, a claim about what general covariance entails, and a claim about what it takes for a theory to be deterministic. The core conclusion can be resisted by calling into question each of these claims. They therefore merit careful individual articulation. Before that, however, a brief initial statement of the argument will help ensure that we do not lose sight of the wood for the trees.

Let $\mathcal{M} = (M, g_{ab}, T_{ab})$ be a solution of GR. Let $d^*\mathcal{M} = (M, d^*g_{ab}, d^*T_{ab})$, where d is a diffeomorphism from M onto itself. It seems that a substantivalist should take $\mathcal{M}$ and $d^*\mathcal{M}$ to represent distinct possibilities. This is because (i) the substantivalist regards the points of M as representing genuine entities, namely substantival spacetime points, and (ii) $\mathcal{M}$ and $d^*\mathcal{M}$ assign the very same points different properties. If p is such that $d(p) \neq p$, then (assuming g_{ab} possesses no symmetries) $\mathcal{M}$ and $d^*\mathcal{M}$ will ascribe different geometrical properties to p ($g_{ab}|_p \neq d^*g_{ab}|_p$) and may ascribe different matter content to (the neighborhood around) p (if $T_{ab}|_p \neq d^*T_{ab}|_p$).

We assumed that $\mathcal{M}$ is a solution of GR. It follows from GR's general covariance that $d^*\mathcal{M}$ is also a solution. This seems to mean that the (accepting the above reasoning, distinct) situations that $\mathcal{M}$ and $d^*\mathcal{M}$ represent are both physically possible.

Finally, suppose that d is a hole diffeomorphism. In particular, suppose that M is foliable by a family of acausal (with respect to g_{ab}) three-dimensional surfaces, let Σ be one such surface, and let H lie entirely to the future of Σ. (A foliation of an n-dimensional manifold is a family of disjoint $(n - m)$-dimensional submanifolds ($m < n$) whose union is M. In this case, we take $m = 1$. A surface is acausal if no two points in the surface lie to the absolute past or future of each other.) $\mathcal{M}$ and $d^*\mathcal{M}$ represent distinct situations (d is nontrivial within H) but, outside of H, they are identical: they assign exactly the same (spatiotemporal and material) properties to all the points of $M \setminus H$. In particular, the possible spacetimes that they represent are identical up to the instant corresponding to Σ but differ to its future. Since both spacetimes are physically possible according to GR, it seems that GR is indeterministic: fixing the laws and the entire spacetime up to Σ fails to fix what will happen (which points will have which properties) to the future of Σ.

That concludes our initial statement of the hole argument. Note that the indeterminism is (in a certain sense) radical. H can be freely specified. Similarly, d can be freely specified, so long as it preserves M's differentiable structure and so long as it smoothly reduces to the identity outside of H. This means that, for any compact region of M as small as one likes, completely specifying the properties of spacetime and matter outside of that region fails to fix the properties within the region (Earman and Norton, 1987, 524).

The rest of this section develops a slightly more careful version of the argument, designed to be immune to some of the less telling criticisms found in the literature.

I stated above that the core of the hole argument involved three main premises. In the presentation just given, these are: (1) that the substantivalist is committed to taking $\mathcal{M}$ and $d^*\mathcal{M}$ to represent distinct situations; (2) that GR's general covariance entails that the possibilities represented by $\mathcal{M}$ and $d^*\mathcal{M}$ are equally physically possible; and (3) that a theory is indeterministic if it both regards $\mathcal{M}$ and $d^*\mathcal{M}$ as representing distinct possibilities (in the way described) and regards the possibilities as equally possible. For each premise, we should identify the most defensible version that is strong enough to play the required role in the argument.

Premise (1) is closely related to what Earman and Norton dubbed the "acid test" of substantivalism. They wrote:

> If everything in the world were reflected East to West (or better, translated 3 feet East), retaining all the relations between bodies, would we have a different world? The substantivalist must answer yes since all the bodies of the world are now in different spatial locations, even though the relations between them are unchanged. (521)

They then went on to claim that the diffeomorphism "is the counterpart of Leibniz' replacement of all bodies in space in such a way that their relative relations are preserved" and concluded that substantivalists were necessarily committed to the denial of *Leibniz Equivalence*, which they defined as the thesis that *diffeomorphic models represent the same physical situation* (522). Here two models $(M, O_1, O_2, \ldots)$ and $(N, O'_1, O'_2, \ldots)$ are diffeomorphic just in case there is a diffeomorphism $d : M \to N$ such that, for each object O_i, $O'_i = d^*O_i$.

Although Earman and Norton conclude by making a claim about how substantivalists must interpret diffeomorphic models, the Leibniz-inspired scenario that they use to introduce their acid test *makes no mention of models*. Instead, their claim is directly about the physical situations that such models represent. They assert that, for any given situation, a substantivalist must recognize as genuinely distinct the situation where the entire material content of the universe is shifted three feet East relative to its location in the first situation. This very natural assumption went unquestioned by *both* the antisubstantivalist Leibniz and the substantivalist Clarke, in their famous *Correspondence*

(Clarke, 1717). If spatial locations are autonomous entities in their own right, doesn't one have to allow that two situations might be genuinely distinct in virtue of differing only in terms of which substantival places serve as the locations of various material bodies, even if everything else about the two situations is identical?

Premise (1) is, therefore, best thought of as the combination of two theses: one about the plurality of possibilities that it is alleged a substantivalist must acknowledge; and another about how particular mathematical objects represent those possibilities. Ultimately, it is only the first thesis that does essential work in the hole argument.

Let's label the two theses **Plurality** and **Models** and state them more carefully.

Plurality Suppose that P is a possible spacetime. The substantivalist is committed to a plurality of possibilities distinct from P that (i) involve the same pattern of spatiotemporal properties instantiated in P and contain the same material fields as P, but that (ii) differ from P (solely) over which spacetime points have which properties and serve as the locations of the common material content.

Now for **Models**. Suppose that $\mathcal{M} = (M, g_{ab}, T_{ab})$ can be taken to represent a possible spacetime P, and suppose that P' is a distinct but related possibility of the type contemplated in **Plurality**. Further, suppose that, while differing over how the common geometrical and material properties are distributed over their common set of spacetime points, P and P' do not differ over which collections of points count as smooth paths (i.e., they agree on differentiable structure). The second thesis in Premise (1) is that, for some suitable choice of diffeomorphism d, $d^*\mathcal{M} = (M, d^*g_{ab}, d^*T_{ab})$ must be interpreted as representing P'.

It is immediately clear, however, that this claim is needlessly strong. It is sufficient for the hole argument that $d^*\mathcal{M}$ *may* be so used. In other words, it is sufficient that there is a permissible *joint* interpretation of the models $\mathcal{M}$ and $d^*\mathcal{M}$ according to which $\mathcal{M}$ represents P and $d^*\mathcal{M}$ represents P'. The advocate of the hole argument can easily concede that $\mathcal{M}$ and $d^*\mathcal{M}$ are equally apt to represent either possibility, i.e., that they have the same "representational capacities" (Weatherall, 2018, 332). No more is required in order to articulate the argument than the following claim:

Models If $\mathcal{M} = (M, g_{ab}, T_{ab})$ can be chosen to represent a possible spacetime P then, *relative to that choice*, there is a permissible and natural interpretation of $d^*\mathcal{M}$ according to which it represents a distinct possibility P'.

Let us now turn to Premise (3), before revisiting Premise (2). In the spirit of the emendation to Premise (1), note that whether a theory is deterministic is, in the first instance, a matter of the range of situations that it judges to be possible and only secondarily a matter of how models might represent those possibilities (cf. Brighouse, 1994, 118).

Consider two possible spacetimes, P and P', differing in the way just contemplated. That is, suppose that P and P' involve the same global pattern of spatiotemporal properties and the same global pattern of material fields but that the two spacetimes differ, for some of their common spacetime points, over which of those points instantiate which of the properties common to both spacetimes. Further, suppose that P and P' are in every respect identical up to some global spacelike hypersurface and that their region of disagreement is confined to a "hole" to the future of that hypersurface.

If P and P' are both physical possibilities according to the theory, then the theory is, in one obvious and natural sense, indeterministic. A theory will fail to be deterministic if it is consistent with worlds that involve identical pasts but different futures. In the case at hand, fixing the entire past up to some instant (a region where P and P' match perfectly) fails to fix the future: according to the theory, spacetime's continuation might be that of P, or it might be that of P'.

Finally, consider Premise (2) again. GR's general covariance entails that $\mathcal{M}$ is a solution if and only if $d^*\mathcal{M}$ is a solution. What follows concerning the physical possibility (according to the interpreted theory) of the spacetimes that $\mathcal{M}$ and $d^*\mathcal{M}$ may be taken to represent?

Since we are not naively assuming that $\mathcal{M}$ and $d^*\mathcal{M}$ represent unique possibilities, we should not simply assert that both the spacetime represented by $\mathcal{M}$ and the spacetime represented by $d^*\mathcal{M}$ are equally possible according to GR. These definite descriptions do not pick out unique situations. Rather, the natural claim, in light of our reworked Premises (1) and (3), is the following:

Equipossible Suppose that $\mathcal{M}_1$ and $\mathcal{M}_2$ are both solutions to a theory T. If there is a permissible joint interpretation of $\mathcal{M}_1$ and $\mathcal{M}_2$ according to which $\mathcal{M}_1$ represents possibility P_1 and $\mathcal{M}_2$ represents possibility P_2 then if P_1 is physically possible according to T so is P_2.

10.4   Responses to the Argument

Our reworked premises entail the hole argument's core conclusion: according to the substantivalist, GR is indeterministic. Responses to the argument divide into those that accept this core conclusion and those that reject it. Responses rejecting the core conclusion can then be classified according to which key premise they reject.

For those who accept the core conclusion, the options are to *reject substantivalism* or to bite the bullet and *accept that GR is indeterministic*. In their paper, Earman and Norton favored the first position. Determinism, they concluded, "may fail, but if it fails it should fail for a reason of physics, not because of a commitment to substantival properties which can be eradicated without affecting the empirical consequences of the theory" (Earman and Norton, 1987, 525).

According to the traditional, bipartite classification, the alternative to substantivalism is *relationalism*. Relationalists deny the (autonomous) reality of spacetime points and analyze facts about spacetime itself as grounded in facts about spatiotemporal properties and relations instantiated by matter. Relationalism evades the hole argument by lacking the plurality of possibilities allegedly plaguing substantivalism: if spacetime points simply do not exist in their own right, there can be no differences between possibilities that concern only which points have which properties.

If relationalists deny that the manifolds in models of GR have a physical correlate, they owe us a positive alternative picture of what such models should be taken to represent. The simplest option is to view physical fields not as patterns of properties and relations instantiated by the points of substantival spacetime, but as extended objects in their own right, possessing infinitely many degrees of freedom. The role of the manifold is then to represent the continuity and differentiable structure of the fields themselves, and to encode which pointlike parts of one field are coincident with those of another. There are both philosophers and physicists who count as relationalists in this sense and who, to a greater or lesser degree, endorse the hole argument (see, e.g., Brown, 2005, 156; Rovelli, 2007, 1309–1310).

Some remain as skeptical of relationalism as of substantivalism and have sought a "third way" between the two. This was the program that Earman himself tentatively backed (1989, 208) but a genuinely novel reconception of the metaphysics of spacetime remains elusive. In particular, various attempts to articulate a "structural realist" approach to spacetime arguably collapse into variants of either relationalism or (more frequently) substantivalism (see Greaves, 2011).

The other option available to someone who accepts the core conclusion is to bite the bullet. Should substantivalists be embarrassed at being forced to view GR as indeterministic? In one sense the indeterminism is pernicious in that, for every possible spacetime, no matter how small a region one considers, the laws and the rest of spacetime fail to fix the state of that region. In another sense, however, the indeterminism is anodyne. Any two possibilities represented by models that differ by a hole diffeomorphism instantiate the very same global pattern of properties and relations. They are therefore *qualitatively* perfectly alike. Their differences involve only which particular individual space-time points instantiate which properties. In the terminology of modal metaphysics, the differences between the possibilities are purely *haecceitistic* (Kaplan, 1975; see also Caulton, this volume).

The substantivalist can urge us to recognize that determinism is not an "all-or-nothing affair" (Earman, 1986, 13). Given the past of a spacetime, GR might not fix which future individuals get to instantiate this or that qualitative feature, but it might nonetheless fix which qualitative features get to be instantiated. The substantivalist can claim, therefore, that, for all the hole argument has shown, GR is qualitatively or *physically* deterministic: given the past and the laws, all future physical facts might be fixed. The merely haecceitistic facts that fail to be pinned down do not, this substantivalist argues, count as the kind of features of the world that one should expect physics to have anything to say about (Brighouse, 1997). (Note that the hole argument's failing to show that GR is physically indeterministic does not entail that GR is in fact physically deterministic. See Earman (2007, §6) for a review of the wider question of whether GR is deterministic in senses other than that at stake in the hole argument.)

A further step would be to *reject* the notion of determinism presupposed in the hole argument. One would then block the core conclusion of the argument by denying Premise (3). Leeds (1995), for example, argues that whether a theory is deterministic is not a matter of which situations the interpreted theory classifies as possible. Instead, he claims, it is a matter of what sentences are provable within the language of the theory. In order for this strategy to work, it would need to be shown that the notion of determinism presupposed in the hole argument is not merely a possibly disfavored option among several but that it is somehow illegitimate. That seems like a tall order. Leeds himself concedes that his proposal can be read as offering just one more definition of determinism and, moreover, one that matches other definitions framed in model-theoretic or possibility-based terms (Leeds, 1995, 435). If the link between substantivalism and indeterminism is to be severed, Premises (1) and (2) are more promising targets.

Maudlin seeks to evade the core conclusion on the basis of a position he dubs *metrical essentialism* (Maudlin, 1989, 1990). Suppose model $\mathcal{M} = (M, g_{ab}, T_{ab})$ is apt to represent a possible spacetime and consider model $d^*\mathcal{M} = (M, d^*g_{ab}, d^*T_{ab})$ for some diffeomorphism d. Recall that Premise (1) of the hole argument was split above into two components: **Plurality** and **Models**. Maudlin accepts **Models** in at least the following sense: he accepts (in fact insists: see Maudlin, 1989, 84) that there is a permissible joint interpretation of $\mathcal{M}$ and $d^*\mathcal{M}$ according to which they represent (if one assumes substantivalism) different *ways for the world to be*. (I here borrow terminology from Salmon (1989).) But, according to Maudlin, these ways for the world to be are not both *ways that the world might have been*; they do not both correspond to genuinely *possible* worlds.

Let us stipulate that model $\mathcal{M}$ represents a possible world. The defining commitment of metrical essentialism is that spacetime points bear their geometrical properties and relations essentially. The value of the curvature scalar at the spacetime point represented – or "named" – by $p \in M$ is, therefore, one of that point's essential properties. Now suppose that $d(p) \neq p$. In that case, $d^*\mathcal{M}$ represents the very same point as having different geometrical properties, for the value of the curvature scalar at p in $d^*\mathcal{M}$ will (in the generic case) be different from its value in $\mathcal{M}$. It follows that, according to the metrical essentialist, $d^*\mathcal{M}$ represents a state of affairs that is not even metaphysically possible.

Note that the initial choice of $\mathcal{M}$ to represent the genuine possibility is arbitrary – consistently with their representational equivalence, one might equally well have chosen $d^*\mathcal{M}$. (This answers Norton's (1989, 63) charge that the metrical essentialist has to explain what distinguishes the "real" model from "imposters.") What the metrical essentialist insists on is that, *relative to that choice* and *relative to a natural and permissible joint interpretation of the models*, $d^*\mathcal{M}$ represents something metaphysically impossible.

Models had to be tweaked so as to be acceptable to metrical essentialists. Something similar is true of **Plurality**. In one sense, metrical essentialists block the hole argument by rejecting **Plurality**. Setting aside cases with nontrivial isometries that are not also symmetries of the matter distribution, the metrical essentialist recognizes at most one possible world corresponding to a given pattern of metrical and material properties and relations for any given collection of spacetime points.

To characterize the metrical essentialist as rejecting **Plurality** is, however, in some ways misleading. Maudlin endorses the intuition behind the "acid test"; he agrees with Earman and Norton that the substantivalist must view a Leibniz-inspired shift of all matter three feet East as generating a genuinely distinct possibility. His dispute with Earman and Norton is over their classification of diffeomorphisms as the natural generalizations of such shifts. Maudlin stresses that Leibniz shifts apply only to the matter of the universe; they leave the geometric properties of the individual spacetime points unaltered. He therefore sees the models $\mathcal{M} = (M, g_{ab}, T_{ab})$ and $\mathcal{M}' = (M, g_{ab}, d^*T_{ab})$ as representing the proper generalization of Leibniz shifts when the points of M are interpreted as naming the same spacetime points in each model (Maudlin, 1990, 552–553). Moreover, of course, if $\mathcal{M}$ is a solution of GR, then $\mathcal{M}'$ will, in general, not be (when $T_{ab} \neq \mathbf{0}$). Since at most one of the possibilities represented is physically possible (by the lights of GR), their distinctness does not threaten indeterminism.

Although $\mathcal{M}'$ does not represent a physically possible spacetime, Maudlin will judge that it does represent a *metaphysically* possible spacetime. The atypical case of spacetimes with symmetries is therefore revealing. In such cases, if d is an isometry, $\mathcal{M}'$ can represent a physically possible world genuinely distinct from that represented by $\mathcal{M}$ but one nevertheless qualitatively indiscernible from it. Maudlin thus accepts the meaningfulness of merely haecceitistic distinctions even if he denies that they (invariably) entail a plurality of genuine possibilities.

This suggests the following representation of the metrical essentialist's position. They accept:

Plurality* Suppose that P is a possible spacetime. The substantivalist is committed to a plurality of *ways for the world to be* distinct from P that (i) involve the same pattern of spatiotemporal properties instantiated in P and contain the same material fields as P, but that (ii) differ from P (solely) over which spacetime points have which properties and serve as the locations of the common material content.

But they reject:

Equipossible* Suppose that $\mathcal{M}_1$ and $\mathcal{M}_2$ are both solutions to a theory T and let P_1 and P_2 be ways for the world to be. If there is a permissible joint interpretation of $\mathcal{M}_1$ and $\mathcal{M}_2$ according to which $\mathcal{M}_1$ represents P_1 and $\mathcal{M}_2$ represents P_2 then if P_1 is physically possible according to T so is P_2.

With these tweaks, the metrical essentialist counts as someone who accepts Premise (1) but rejects Premise (2).

This regimentation highlights that metrical essentialists evade the hole argument's core conclusion only by rejecting what might seem like the obvious moral of the diffeomorphism invariance of the theory. Grant Maudlin that the points of the manifold M can be treated like proper names and that, so understood, models $\mathcal{M}$ and $d^*\mathcal{M}$ represent distinct ways for the world to be. What remains to be decided is whether these ways are both genuinely possible ways for the world to be. Having gone this far, however, it is a very natural further step to take the diffeomorphism invariance of GR as telling us precisely that both states *are* possible.

In postulating the real existence of spacetime points in the first place, the metrical essentialist is likely to be a scientific realist who is happy to take GR as a guide to ontology. Should not GR also be our guide as to which properties are essential to spacetime points? What the diffeomorphism invariance of GR appears to tell a haecceitist substantivalist is that the only properties essential to a spacetime point are those that it exemplifies as part of a set with the structure of a differentiable manifold (*cf.* Earman, 1989, 201).

A different criticism of metrical essentialism focuses on non-isomorphic models. One straightforward way of illustrating the dynamical nature of spacetime structure in GR is to assert, for example, that, had extra mass been present close to some point, then the curvature at that point would have been different (Earman, 1989, 201). How are metrical essentialists to evaluate such counterfactuals? They have to judge that the idea that the curvature could have been different from its actual value at *this very point* is as metaphysically absurd as, for example, the idea that Tim Maudlin might have been made of iron girders. Maudlin is forced to concede that "no model not isometric to the actual world can represent how *this* space-time might have been" (Maudlin, 1989, 89–90) but he insists that dynamically allowed models of GR that are not isometric to the actual world can represent genuine possibilities: "they are just different possible space-times, not different possible states of this space-time" (1989, 90). He goes on to allow that such possible spacetimes can be used to give a counterpart-theoretic explanation of the truth of Earman's counterfactual. As Brighouse observes (1994, 119–120), this is a rather unsatisfying blend of essentialism and counterpart theory.

In the face of such criticism, what positive reasons can metric essentialists offer for their position, aside from the *ad hoc* benefit that it avoids the indeterminism of the hole argument? Maudlin's answer appeals to Newton. In a somewhat obscure passage, which has inspired almost as many different interpretations as commentators, Newton wrote:

> The parts of duration and space are understood to be the same as they really are only because of their mutual order and position; nor do they have any principle of individuation apart from that order and position, which consequently cannot be altered. (Newton, 1684 [2004], 25)

According to Maudlin's gloss, Newton is saying that "the parts of space and time, being intrinsically identical to one another, [have] to be differentiated by their mutual relations of position. Parts of space bear their metrical relations essentially" (Maudlin, 1989, 86).

This is not especially compelling. Why should someone otherwise comfortable with haecceitistic distinctions think that intrinsically identical objects require (metaphysical?) "differentiation"? Rather than cleaving to haecceitism and avoiding indeterminism by way of an otherwise unmotivated essentialism, perhaps the substantivalist does better to embrace wholeheartedly the "structuralist" view that others (e.g., Stein, 2002, 272) read in Newton's cryptic remarks. If spacetime points are only "individuated" one from another by their spatiotemporal relations (i.e., by their positions in the overall network of spatiotemporal relations), a possible spacetime is exhaustively specified by a complete catalog of the qualitative facts concerning the full pattern of spatiotemporal relations that are instantiated by its points. According to this antihaecceitist (or *generalist*) point of view, there simply are no further "individualistic" facts concerning which objects possess which properties. At a fundamental level, reality, at least concerning spacetime points, is purely qualitative.

Despite the important differences between them, Butterfield (1989a), Maidens (1992), Stachel (1993, 2002, 2006), Brighouse (1994), Rynasiewicz (1994), Hoefer (1996), Saunders (2003), Pooley (2006) and Esfeld and Lam (2008) all endorse some kind of antihaecceitism, at least concerning spacetime points, whether on general philosophical grounds (as in Hoefer's case), or as a perceived lesson of the diffeomorphism invariance of the physics (as in Stachel's case). Whether acknowledged or not, these authors, in their commitment to spacetime points as entities not reducible to matter and its properties, count as substantivalists, albeit of a "sophisticated" variety (Belot and Earman, 2001, 228).

Sophisticated substantivalists reject the core conclusion of the hole argument by rejecting Premise (1). In particular, they reject **Plurality**. The distinct possibilities countenanced by **Plurality** are precisely possibilities that differ merely haecceitistically. That antihaecceitism and **Plurality** are incompatible is therefore immediate. One strand of criticism questions whether it is coherent to combine acceptance of spacetime points as entities in their own right with a denial that there are the substantive facts (about which such entities possess which properties) that would generate haecceitistic distinctions. Some argue that a fleshed-out metaphysical story explaining how this combination is possible is still to be given (Dasgupta, 2011, 130–135).

In addition to **Plurality**, sophisticated substantivalists reject **Models**, but this is a simple consequence of their rejection of **Plurality**, which **Models** presupposes. This might lead one to wonder whether there is a satisfactory response to the hole argument that rejects **Models** while disavowing metaphysics and remaining neutral with respect to **Plurality**. One can interpret Weatherall (2018) and Fletcher (2020) as defending positions of this kind.

According to **Models**, $\mathcal{M}$ and $d^*\mathcal{M}$ can be used to jointly represent physically distinct situations. The essence of both Weatherall's and Fletcher's views is that this use is not consistent with treating them as Lorentzian manifolds. Weatherall's starting point is that the physical interpretation of a theory's formalism should be consistent with our best understanding of the mathematics of that formalism. In particular, the models employed in a physical theory should count as (physically) equivalent just when they are equivalent according to the mathematics used in formulating those models (Weatherall, 2018, 331). Since isometry provides the standard of "sameness" in the mathematics of Lorentzian manifolds, it is a condition on any acceptable interpretation that it regard isometric manifolds, such as $\mathcal{M} = (M, g_{ab})$ and $d^*\mathcal{M} = (M, d^*g_{ab})$, as physically equivalent.

In order to bite against **Models**, this stricture needs to be understood not merely as insisting that any two isometric models are equally apt to represent any given possibility (something our formulation of the hole argument was careful to allow), but as ruling out a *joint* interpretation of them on which they are physically inequivalent in the sense that (so interpreted) they represent distinct physical possibilities.

Fletcher is explicit that such a use of the models is in conflict with treating them as members of the mathematical category of Lorentzian manifolds. Any aspect of a state of affairs that is represented by one such model *so conceived* must, he argues, be similarly represented by each isomorphic model. This is because isomorphic models are equivalent "as objects in that category": their being isomorphic just is a matter of there being a bijective map of a specific sort that preserves all of the structures constitutive of objects of that type. The consequence Fletcher draws is that any putative representational differences between such isomorphic models are "not reflected at all in the models themselves as members of [the] category they are taken to be [members of] – there is no mathematical correlate of those differences definable in the category" (Fletcher, 2020, 239–240).

For the sake of argument, let us concede to Weatherall and Fletcher that, on a natural understanding of GR's mathematical machinery, it cannot be used to represent haecceitistic differences between possible spacetimes. What follows for the hole argument? It becomes evident that all the heavy lifting in Premise (1) is performed by its metaphysical component, namely, **Plurality**. That thesis was not the outcome of a naive way of thinking about the mathematics of GR. It arose from an interrogation of the substantivalist's metaphysics. Weatherall's and Fletcher's reflections, therefore, leave it untouched.

In effect, both authors argue that if there are pluralities of merely haecceitistically distinct possibilities, the mathematical formalism of GR, correctly interpreted, is necessarily indifferent to differences between them. But this just means that GR does not distinguish between any two elements of such a plurality; both will count as physically possible according to GR or neither will. And that, of course, is just to admit that, according to any metaphysical view committed to such pluralities, GR is indeterministic. The indeterminism cannot be avoided by remaining loftily above the metaphysical affray.

10.5 Defining Determinism

Contrast the following two definitions of determinism for a theory T:

Det1 T is deterministic just in case, for any worlds W and W' that are possible according to T, if the past of W up to some timeslice in W is intrinsically identical to the past of W' up to some timeslice in W', then W and W' are intrinsically identical.

Det2 T is deterministic just in case, for any worlds W and W' that are possible according to T, if the past of W up to some timeslice in W is qualitatively (intrinsically) identical to the past of W' up to some timeslice in W', then W and W' are qualitatively identical.

(On these definitions, determinism is a matter of whether the entire history of a world up to some time fixes its future, given the laws. Alternative notions of determinism are easily obtained by considering whether a world at a time (or some other part of a world) fixes the remainder, given the laws. There are also reasons to focus on the conditions for a world, rather than a theory, to be deterministic (Brighouse, 1997, 468). For reasons of space, I ignore these complications.)

According to sophisticated substantivalists, there are no primitive trans-world facts about which objects in one world are identical to which objects in another. On their view, two possible worlds (or two proper parts of distinct worlds) that are not (intrinsically) identical differ qualitatively. Sophisticated substantivalists therefore interpret **Det1** and **Det2** as strictly equivalent.

According to straightforward (haecceitist) substantivalists, in contrast, worlds W and W' can differ not just by failing to be perfectly qualitatively alike but by failing to have the very same individuals playing identical qualitative roles. For them, therefore, **Det1** describes a criterion for determinism that is strictly stronger than that described in **Det2**. Absent sufficiently strong essentialist constraints on what is possible for spacetime points, straightforward substantivalists will judge GR to be indeterministic according to **Det1**.

Det2 corresponds closely to the definition of determinism offered by David Lewis (1983, 359–360). A related model-theoretic definition was defended by Butterfield (1989b), who argued that it captures the notion of determinism implicit in physicists' discussions of GR's determinism. A significant strand of the hole argument literature has targeted definitions akin to **Det2**, arguing that they misclassify as deterministic theories that are clearly indeterministic. Such criticism is an obvious problem for sophisticated substantivalists, for whom **Det2** is equivalent to **Det1**, but it is also a problem for straightforward substantivalists who, as noted above, might wish to distinguish a notion of physical determinism from determinism *tout court*. **Det2** might have seemed to adequately capture the former notion.

Problem cases for **Det2** were raised by Wilson (1993, 216) and Rynasiewicz (1994, 418), and have been discussed in detail by Belot (1995), Brighouse (1997), and Melia (1999). Here are two simple illustrative examples. In the first (adapted from Melia 1999, 660–661) our theory, T_1, governs the behavior of two types of particle: A particles and B particles. Consider a world that contains one A particle equidistant from two B particles, all at rest with respect to one another. Suppose that T_1 determines that, at some fixed and predictable time, the A particle will move at a fixed velocity towards one of the B particles. Intuitively, T_1 is indeterministic because, despite fixing the qualitative evolution of the situation just described, it fails to fix which B particle the A particle will move towards.

Imagine that we can label the particles in our toy world: a_1, b_1, and b_2. There appear to be two possible futures: one where a_1 moves toward b_1 and a second where a_1 moves toward b_2. Haecceitists will judge that T_1 is indeterministic according to **Det1**, for they recognize the merely haecceitistic distinctions between its being b_1 or b_2 towards which the A particle moves. According to **Det2**, however, the world is compatible with T_1's being deterministic: if there are two possible futures, they are qualitatively identical.

Oliver Pooley

In our second example (*cf.* Belot 1995, 191–192, and Melia 1999, 646–647), theory T_2 governs the decay of A particles into B particles. We suppose that everything qualitative about such decays (their spacetime locations, the momenta of the decay products, etc.) is fixed by the qualitative history of the world prior to the decay. Now consider a world governed by T_2 involving the simultaneous decay of two A particles each into a B particle. The haecceitist will judge that T_2 is indeterministic because, despite fixing the qualitative behavior of all decays, it fails to fix the identities of the decay products. Again, imagine that we can label the four particles and suppose that, in the world we are considering, a_1 decays into b_1 and a_2 decays into b_2. A world where a_1 decays into b_2 and a_2 decays into b_1, but where everything else is otherwise held fixed, might seem to be an alternative possibility compatible with T_2. While T_2 is therefore deterministic according to **Det2** (the qualitative nature of all decays is fully determined by the qualitative nature of the pre-decay state), a haecceitist will judge that the theory fails to be deterministic according to **Det1**.

Some philosophers (e.g., Belot, 1995) accept that both T_1 and T_2 manifest genuine indeterminism. They have reason to reject **Det2** and can rest content with a haecceitist understanding of **Det1**. They are likely to accept the hole argument's conclusion that substantivalist GR is indeterministic.

An antihaecceitist, however, will view the alleged indeterminism of T_2 as suspect: the purportedly distinct possibilities involved in the example are merely haecceitistically distinct. For many, however, the intuition that theory T_1 is indeterministic is harder to dispel. Is there a principled way for antihaecceitists to acknowledge that T_1 is indeterministic but to deny that T_2 is indeterministic?

Despite doubts recently expressed by Brighouse (2020, §4), it would seem that this can be done. Consider again the three-particle world governed by T_1. According to the haecceitist, there are in fact two such possible worlds: one in which a_1 moves toward b_1 and another in which a_1 moves toward b_2. According to the antihaecceitist, there is only one such world: it contains an A particle that moves toward one but not the other of two previously qualitatively identical B particles. But the antihaecceitist can (and, indeed, must) recognize two possible futures for two qualitatively identical *proper parts* of this world. Prior to the A particle's starting to move, there were two qualitatively identical but distinct (and overlapping) pairs composed of an A particle and a B particle. For convenience we can imagine labeling them "(a_1, b_1)" and "(a_1, b_2)" but, note, their distinctness involves no haecceitistic presuppositions. It is secured by the distinctness of b_1 and b_2, two particles coexisting in the same world and situated some distance apart from one another.

T_1's indeterminism can, therefore, be understood in terms of its failure to fix the (qualitative) future of every part of each world that it governs (Melia, 1999, 652). Take our two pairs of an A particle and a B particle. An exhaustive qualitative specification of such a pair up to the time at which the A particle moves will involve the complete specification of the qualitative history of the whole world up to that time together with a qualitative characterization of the pair's situation in this history. Up until the time at which the A particle moves, both pairs will satisfy exactly the same qualitative description. Such a specification, therefore, fails to determine whether one of the particles in the pair will move toward the other.

Note that indeterminism is still conceived, as it must be for the antihaecceitist, as a matter of the qualitative past failing to fix the qualitative future. In order to recognize T_1's indeterminism, one just needs to attend to proper parts of a world, in addition to the world as a whole. It turns out that it is relatively straightforward to provide alternative definitions of determinism that regiment the intuitions just described (see Belot 1995, 191, Definition 2; Melia 1999, §4.1). This need not mean that **Det2** should simply be jettisoned. Following Dewar (2016), one might go on to distinguish "determinism *de dicto*" (captured by **Det2**) from "determinism *de re*":

> ... with a little hindsight, it is *utterly unsurprising* that there should turn out to be two concepts of determinism. Determinism is a matter of whether there is one possibility or

more consistent with things being a certain way at a certain time; we have two species of possibility, *de dicto* and *de re*; so as a consequence, there are two species of determinism. (Dewar, 2016, 53–54)[1]

Acknowledgments

I am grateful to numerous colleagues for discussions of the hole argument over the years but, for especially pertinent recent discussion, I am grateful to Eleanor Knox, Sam Fletcher, and Jim Weatherall, and, for very helpful comments on an earlier draft, to James Read.

Note

1 Note that determinism *de re* in Dewar's sense is distinct from determinism *de re* as defined by Hawthorne (2006, 239). Hawthorne's notion of determinism *de re* explicitly requires that a world's past (together with its laws of nature) fixes all haecceitistic facts, in addition to all qualitative facts.

References

Belot, G. (1995). New work for counterpart theorists: Determinism. *The British Journal for the Philosophy of Science*, 46: 185–195.

Belot, G. and Earman, J. (2001). Pre-Socratic quantum gravity. In C. Callender and N. Huggett (eds.), *Physics Meets Philosophy at the Planck Scale: Contempory Theories in Quantum Gravity*. Cambridge: Cambridge University Press, chapter 10, pp. 213–255.

Brighouse, C. (1994). Spacetime and holes. In D. Hull, M. Forbes and R. M. Burian (eds.), *Proceedings of the 1994 Biennial Meeting of the Philosophy of Science Association*, Vol. 1, *Philosophy of Science Association*. East Lansing, University of Chicago Press: Michigan, pp. 117–125.

Brighouse, C. (1997). Determinism and Modality. *The British Journal for the Philosophy of Science*, 48(4): 465–481.

Brighouse, C. (2020). Confessions of a (cheap) sophisticated substantivalist. *Foundations of Physics*, 50: 348–359.

Brown, H. R. (2005). *Physical Relativity: Space-time Structure from a Dynamical Perspective*. Oxford: Oxford University Press.

Butterfield, J. N. (1989a). Albert Einstein meets David Lewis. In A. Fine and J. Leplin (eds.), *Proceedings of the 1988 Biennial Meeting of the Philosophy of Science Association*. East Lansing, Michigan: Philosophy of Science Association, pp. 65–81.

Butterfield, J. N. (1989b). The hole Truth. *The British Journal for the Philosophy of Science*, 40(1): 1–28.

Clarke, S. (1717). *A Collection of Papers, Which Passed between the Late Learned Mr Leibnitz, and Dr Clarke, In the Years 1715 and 1716*, London.

Dasgupta, S. (2011). The bare necessities. *Philosophical Perspectives*, 25: 115–160.

Dewar, N. (2016). *Symmetries in physics, metaphysics, and logic*. PhD thesis, University of Oxford.

Earman, J. (1986). *A Primer on Determinism*. Dordrecht: D. Riedel Publishing Company.

Earman, J. (1989). *World Enough and Space-Time: Absolute versus Relational Theories of Space and Time*. Cambridge, MA: MIT Press.

Earman, J. (2007). Aspects of determinism in modern physics. In J. N. Butterfield and J. Earman (eds.), *Philosophy of Physics*, Vol. 2 of *Handbook of the Philosophy of Science*. Amsterdam: Elsevier, pp. 1369–1434.

Earman, J. and Norton, J. D. (1987). What price spacetime substantivalism? The hole story. *The British Journal for the Philosophy of Science*, 38: 515–525.

Einstein, A. (1914). Comments on "outline of a generalized theory of relativity and of a theory of gravitation". *Zeitschrift für Mathematik und Physik*, 62: 260–261. Reprinted in Klein et al. (1995), pp. 580–582.

Esfeld, M. and Lam, V. (2008). Moderate structural realism about space-time. *Synthese*, 160(1): 27–46.

Fletcher, S. C. (2020). On representational capacities, with an application to general relativity. *Foundations of Physics*, 50: 228–249.

Greaves, H. (2011). In search of (spacetime) structuralism. *Philosophical Perspectives*, 25: 189–204.

Hawthorne, J. (2006). *Metaphysical Essays*. Oxford: Oxford University Press.

Hoefer, C. (1996). The metaphysics of space-time substantivalism. *The Journal of Philosophy*, 93(1): 5–27.

Janssen, M. (2014). 'No success like failure ...': Einstein's quest for general relativity, 1907–1920. In M. Janssen and C. Lehner (eds.), *The Cambridge Companion to Einstein*. Cambridge: Cambridge University Press, pp. 167–227.

Kaplan, D. (1975). How to Russell a Frege–Church. *The Journal of Philosophy*, 72: 716–729.

Klein, M. J., Kox, A., Renn, J. and Schulmann, R., editors. (1995). *The Swiss Years: Writings 1912–1914*, Vol. 4 of *The Collected Papers of Albert Einstein*. Princeton, NJ: Princeton University Press.

Leeds, S. (1995). Holes and determinism: Another look. *Philosophy of Science*, 62: 425–437.

Lewis, D. K. (1983). New work for a theory of universals. *Australasian Journal of Philosophy*, 61: 343–377.

Maidens, A. (1992). Review, Earman, John S. [1989]: *World Enough and Space-Time: Absolute versus Relational Theories of Space and Time*. *The British Journal for the Philosophy of Science*, 43(1): 129.

Maudlin, T. (1989). The essence of space-time. In A. Fine and J. Leplin (eds.), *Proceedings of the 1988 Biennial Meeting of the Philosophy of Science Association*. East Lansing, Michigan: Philosophy of Science Association, pp. 82–91.

Maudlin, T. (1990). Substances and space-time: What Aristotle would have said to Einstein. *Studies in History and Philosophy of Science*, 21(4): 531–561.

Melia, J. (1999). Holes, haecceitism and two conceptions of determinism. *The British Journal for the Philosophy of Science*, 50(4): 639–664.

Newton, I. (1684 [2004]). De Gravitatione. In A. Janiak (ed.), *Philosophical Writings*. Cambridge: Cambridge University Press, pp. 12–39.

Norton, J. D. (1984). How Einstein found his field equations: 1912–1915. *Historical Studies in the Physical Sciences*, 14(2): 253–316.

Norton, J. D. (1989). The hole argument. In A. Fine and J. Leplin (eds.), *Proceedings of the 1988 Biennial Meeting of the Philosophy of Science Association*. East Lansing, Michigan: Philosophy of Science Association, pp. 56–64.

Pooley, O. (2006). Points, particles, and structural realism. In D. Rickles, S. French, and J. Saatsi (eds.), *The Structural Foundations of Quantum Gravity*. Oxford: Oxford University Press, pp. 83–120.

Pooley, O. (2017). Background independence, diffeomorphism invariance, and the meaning of coordinates. In D. Lehmkuhl, G. Schiemann and E. Scholz (eds.), *Towards a Theory of Spacetime Theories*, Vol. 13 of *Einstein Studies*. Basel: Birkhäuser, pp. 105–143.

Rovelli, C. (2007). Quantum gravity. In J. N. Butterfield and J. Earman (eds.), *Philosophy of Physics*, Vol. 2 of *Handbook of the Philosophy of Science*. Amsterdam: Elsevier, pp. 1287–1329.

Rynasiewicz, R. (1994). The lessons of the hole argument. *The British Journal for the Philosophy of Science*, 45(2): 407–436.

Salmon, N. (1989). The logic of what might have been. *The Philosophical Review*, 98(1): 3–34.

Saunders, S. W. (2003). Indiscernibles, general covariance, and other symmetries: The case for non-reductive relationalism. In A. Ashtekar, R. S. Cohen, D. Howard, J. Renn, S. Sarkar and A. Shimony (eds.), *Revisiting the Foundations of Relativistic Physics: Festschrift in Honor of John Stachel*, Vol. 234 of *Boston Studies in the Philosophy of Science*. Dordrecht: Kluwer, pp. 151–173.

Stachel, J. (1989). Einstein's search for general covariance, 1912–1915. In D. Howard and J. Stachel (eds.), *Einstein and the History of General Relativity*. Boston, MA: Birkhäuser, pp. 63–100.

Stachel, J. (1993). The meaning of general covariance. In J. Earman, A. I. Janis, G. J. Massey and N. Rescher (eds.), *Philosophical Problems of the Internal and External Worlds: Essays on the Philosophy of Adolf Grünbaum*, Vol. 1 of *Pittsburgh–Konstanz Series in the Philosophy and History of Science*. Pittsburgh: University of Pittsburgh Press, pp. 129–160.

Stachel, J. (2002). "The relations between things" versus "the things between relations": The deeper meaning of the hole argument. In D. B. Malament (ed.), *Reading Natural Philosophy. Essays in the History and Philosophy of Science and Mathematics*. Chicago: Open Court, pp. 231–266.

Stachel, J. (2006). Structure, individuality and quantum gravity. In D. Rickles, S. French, and J. Saatsi (eds.), *The Structural Foundations of Quantum Gravity*. Oxford: Oxford University Press, pp. 53–82.

Stein, H. (2002). Newton's metaphysics. In I. B. Cohen and G. E. Smith (eds.), *The Cambridge Companion to Newton*. Cambridge: Cambridge University Press, pp. 256–307.

Weatherall, J. O. (2018). Regarding the "hole argument". *The British Journal for the Philosophy of Science*, 69: 329–350.

Wilson, M. (1993). There's a hole and a bucket, dear Leibniz. *Midwest Studies in Philosophy*, 18: 202–241.

Further Reading from the Editors

The paper that started the modern debate is Earman, J. and Norton, J. D. 'What price spacetime substantivalism? the hole story' (*The British Journal for the Philosophy of Science* 38, 515–525, 1987). For some subtle thinking about determinism, look at Butterfield, J. N. 'The hole truth' (*The British Journal for the Philosophy of Science* 40(1), 1–28, 1989). Hoefer, C. 'The metaphysics of space-time substantivalism' (*The Journal of Philosophy* 93(1),

5–27, 1996) is a nice presentation of the sophisticated substantivalist view. Oliver Pooley gives more details of his version of this view in his 'Points, particles, and structural realism', in D. Rickles, French, and J. Saatsi, eds., *The Structural Foundations of Quantum Gravity* (Oxford University Press, Oxford, pp. 83–120, 2006). In recent years, the debate has been revived by Jim Weatherall, with an emphasis on mathematical representation: 'Regarding the "hole argument"' (*The British Journal for the Philosophy of Science* 69, 329–350, 2018). A recent special issue of *Foundations of Physics* (Volume 50, 2020) is dedicated to the reignited debate.

11

RELATIVISTIC SPACETIME STRUCTURE

Samuel C. Fletcher

11.1   Introduction and Scope

In the broadest sense, spacetime structure consists in the totality of relations between events and processes described in a spacetime theory, including distance, duration, motion, and (more generally) change. A spacetime theory can attribute more or less such structure, and some parts of that structure may determine other parts. The nature of these structures and their relations of determination bear on the interpretation of the theory – what the world would be like if the theory were true (North, 2009). For example, the structures of spacetime might be taken as its ontological or conceptual posits, and the determination relations might indicate which of these structures is more fundamental (North, 2018). Different perspectives on these questions might also reveal structural similarities with other spacetime theories, providing the resources to articulate how the picture of the world that that theory provides is different (if at all) from what came before, and might be different from what is yet to come.[1]

My present twofold goal is to survey (in Sections 11.2 and 11.3) in particular the spacetime structures of the general theory of relativity – our best current theory of space, time, and gravitation – and (in Sections 11.4 and 11.5) the determination relations between them, including both what I take the "standard" account of these to be (in Sections 11.2 and 11.4) and alternative perspectives worth investigating (in Sections 11.3 and 11.5).[2] Along the way, many conceptual and technical questions arise that are worth further research. Some of these questions have been discussed extensively in the physics and philosophy literatures, some less so.[3] My hope is that they will stimulate readers' interests to pursue them.

11.2   What There Is: The Standard Account

The standard account of relativistic spacetime structure divides that structure into two classes: the manifold structure of events and the geometrical structure built upon it. This division does not entail that the two types of structures must be interpreted as ontologically independent (Pooley, 2006). Nevertheless, it will aid exposition to introduce these two types of structures separately; one only has to keep in mind that the interpretation of some of these structures as they are introduced – in particular, those of the manifold in Section 11.2.1 – may be incomplete until others are introduced in particular, those of the geometry in Section 11.2.2.

11.2.1   Standard Manifold Structure

The *events* of a relativistic spacetime consist in an uncountable set M; they represent all the potential or actual point-sized happenings – that is, without spatial or temporal extension – everywhere and everywhen in that spacetime. Consequently, collections of events represent processes, histories of particular objects, and so on. But, not *all* collections of events do so: first, one may intend to use the same relativistic spacetime model to describe different states of affairs, in which happenings occur at different points relative to one another; second, one typically restricts attention to continuously connected events to describe processes and histories.

To distinguish between continuous and discontinuous collections of events, one must use the *topological structure* of the spacetime's events. In particular, the events M are equipped with a topology $\mathcal{T}$ that makes them into a *topological four-manifold* (Lee, 2011). Most simply, a topological four-manifold is a topological space that is locally homeomorphic to $\mathbb{R}^4$ with its usual topology: for every point $p \in M$, there is a homeomorphism $\varphi : U \to V$, where $p \in U \in \mathcal{T}$ and V is an open subset of $\mathbb{R}^4$. Thus, a neighborhood of every event – some collection of events spatiotemporally proximal to it – can be described with four coordinates, anticipating their interpretation as three spatial coordinates and one temporal coordinate.[4]

The events' topology is required to satisfy several further conditions (Geroch and Horowitz, 1979, p. 218). First, it is *Hausdorff*, meaning that if two points of M are distinct, then there are neighborhoods of each that are disjoint. This is really an extension of the idea that, locally, events should have the continuous structure of a region of $\mathbb{R}^4$. Second, it is *path-connected*, meaning that for every $p, q \in M$, there is a continuous function $f : [0, 1] \to M$ such that $f(0) = p$ and $f(1) = q$.[5] If there were events not so connectible, they would not bear any spatiotemporal relations to each other at all, and could scarcely be said to be part of the same universe.[6] Third, it is *second-countable*: there is a countable base for $\mathcal{T}$, meaning that there is a countable subset $\tau \subset \mathcal{T}$ such that every open set in $\mathcal{T}$ can be expressed as the union of elements from τ. This condition prevents the collection of events from being too "large" or "spread out," and is equivalent (when conjoined with the Hausdorff condition) to the manifold being embeddable in a finite-dimensional Euclidean space.[7]

The topological structure T of a spacetime provides the sense in which certain sets S of events are continuously connected – those for which there is a path between every pair of elements in S. But it does not provide the structure needed to describe the directions of paths (i.e., tangent vectors), their rates of change, and so on. For these concepts, one needs to be able to do calculus on the manifold, which is provided by a *differentiable structure*. It will be convenient to define this structure in terms of (four-)charts, versions of which we have already encountered in expressing the sense in which a topological four-manifold is locally $\mathbb{R}^4$.

A *(four-)chart on M* is an injective map $\varphi : U \to \mathbb{R}^4$, where $U \subseteq M$ and $\varphi[U]$ is open in $\mathbb{R}^4$. An *atlas $\mathcal{A}$ for M* is a collection of charts whose domains cover M. Finally, a *C^k differentiable structure on M* (for $k \geq 1$) is an atlas whose charts are mutually *C^k compatible*: for every $\phi, \phi' \in \mathcal{A}$ with respective domains $U, U' \subseteq M$, if $U \cap U' \neq \emptyset$, then $\phi \circ \phi'^{-1} : \phi'[U \cap U'] \to \phi[U \cap U']$ is a C^k function – i.e., k times continuously differentiable – from an open set of $\mathbb{R}^4$ to an open set of $\mathbb{R}^4$. (Such compositions are often called *transition maps* or *changes of coordinates*.) A differential structure that is as large as it can be is called *maximal*: it contains *all* the charts that are mutually compatible.

The differential structure of spacetime events is standardly required to be smooth (C^∞), maximal, and compatible with its topological structure $\mathcal{T}$ in the following sense: the domain of every chart in $\mathcal{A}$ is an open set in $\mathcal{T}$. This compatibility condition ensures that the continuous and differentiable structures mesh with one another, in that the transition maps of the latter are also continuous with respect to the former. The maximality condition is also extremely important, for it ensures that there are no distinguished coordinate systems: each one that is compatible is on an equal footing to describe locally a collection of events. This is, in one sense, what characterizes manifolds as *geometrical* objects. Finally, the smoothness (C^∞) condition provides the structure to describe arbitrary orders of derivatives of quantities or fields, which, consonant with the maximality of the differentiable structure, does not depend on particular coordinate systems.

In sum, the standard manifold structure of a spacetime provides the means to describe how collections of events form continuous or even smooth collections by constraining them to be locally like $\mathbb{R}^4$. But this structure does not yet provide for concepts of temporal duration, spatial extension, and motion, whether straight or curvilinear. For these one must add geometrical structure.

11.2.2   Standard Geometrical Structure

The *covariant derivative operator* (or *affine connection*) ∇ of a relativistic spacetime provides a standard of constancy for general tensor fields on that spacetime. In particular, it allows one to describe a sense in which certain curves are straight, while others are curvilinear.

A *derivative operator* at a point $p \in M$ is a map from (q, r)-tensor fields $\Phi^{a_1 \cdots a_q}_{b_1 \cdots b_r}$ at p to $(q, r+1)$-tensor fields $\nabla_c \Phi^{a_1 \cdots a_q}_{b_1 \cdots b_r}$ at p satisfying the following conditions (Hawking and Ellis, 1973, §2.5):

1   It commutes with addition; e.g., $\nabla_c \left(\Phi^{a_1 \cdots a_q}_{b_1 \cdots b_r} + \Phi'^{a_1 \cdots a_q}_{b_1 \cdots b_r} \right) = \nabla_c \Phi^{a_1 \cdots a_q}_{b_1 \cdots b_r} + \nabla_c \Phi'^{a_1 \cdots a_q}_{b_1 \cdots b_r}$.

2   It satisfies the Leibniz rule for tensor multiplication; e.g.,
$$\nabla_e \left(\Phi^{a_1 \cdots a_q}_{b_1 \cdots b_r} \Psi^{c_1 \cdots c_s}_{d_1 \cdots d_t} \right) = \left(\nabla_e \Phi^{a_1 \cdots a_q}_{b_1 \cdots b_r} \right) \Psi^{c_1 \cdots c_s}_{d_1 \cdots d_t} + \Phi^{a_1 \cdots a_q}_{b_1 \cdots b_r} \left(\nabla_e \Psi^{c_1 \cdots c_s}_{d_1 \cdots d_t} \right).$$

3   It commutes with contraction; e.g., contracting the indices $a_i \to b_j$ of $\Phi^{a_1 \cdots a_q}_{b_1 \cdots b_r}$ (for any $i \in \{1, \ldots, q\}$ and $j \in \{1, \ldots, r\}$) and then applying ∇ is the same as contracting the indices $a_i \to b_j$ of $\nabla_c \Phi^{a_1 \cdots a_q}_{b_1 \cdots b_r}$.

4   For any scalar field $f : M \to \mathbb{R}$, $\nabla_a f = d_a f$, where d is the exterior derivative operator.

If ξ^a is the tangent vector field to a curve $\gamma : I \to M$ (where I is any interval of $\mathbb{R}$), then at points on the image of γ, $\xi^c \nabla_c \Phi^{a_1 \cdots a_q}_{b_1 \cdots b_r}$ is a directional derivative of the field $\Phi^{a_1 \cdots a_q}_{b_1 \cdots b_r}$ along γ. In particular, $\xi^b \nabla_a \xi^a$ measures (according to ∇) the extent to which ξ^b is constant along γ, is self-parallel or locally straight. It is so constant when $\xi^b \nabla_a \xi^a = \mathbf{0}$ at each point of the image, in which case γ is said to be a *geodesic*. Now, a relativistic spacetime assumes a single, globally defined (i.e., at every $p \in M$) derivative operator. The images of geodesics are then interpreted as *the* (locally) straight lines of events in a spacetime.

It is important to note, however, that the property of being a geodesic depends on parameterization: if γ is a geodesic, then for any diffeomorphism $\alpha : I' \to I$, $\alpha \circ \gamma$ is a geodesic if and only if α is linear. Conversely, if the tangent vector ξ^a satisfies the weaker self-parallel condition $\xi^b \nabla_a \xi^a = f \xi^a$ for some scalar function f on the image of γ, then there is always a diffeomorphism $\alpha : I' \to I$ such that $\alpha \circ \gamma$ is a geodesic. Such curves are called *pregeodesics*; they have the same images as the geodesics, but are less restrictive regarding parameterization. This interpretative importance motivates the definition of a *projective structure* on spacetime, which is the equivalence class of all affine connections that determine the same pregeodesics – they agree on which collections of points constitute locally straight lines.

Algebraically, the covariant derivative may not commute with itself. Geometrically, in determining which lines are locally straight, the affine connection can allow for initially parallel geodesics to converge or diverge, and to twist. These two features are encoded in the Riemann curvature and torsion tensors, R^a_{bcd} and T^a_{bc} respectively, which are uniquely defined by their action on vector fields X^a, Y^b, and Z^c:

$$R^a_{bcd}X^bY^cZ^d = X^b\nabla_b(Y^c\nabla_cZ^a) - Y^c\nabla_c(X^b\nabla_bZ^a) - [X, Y]^b\nabla_bZ^a,$$

$$T^a_{bc}X^bY^c = X^b\nabla_bY^a - Y^b\nabla_bX^a - [X, Y]^a.$$

Here, $[X, Y]^b$ is the commutator (or Lie bracket) between vector fields X^a and Y^a (Malament, 2012, pp. 44–45). In general relativity, one requires the torsion of the connection to vanish everywhere, but the curvature is general is not fixed. This entails that the derivative operator commutes on scalar fields f, i.e., $\nabla_a\nabla_bf = \nabla_b\nabla_af$ (Hawking and Ellis, 1973, p. 34).

The affine connection provides a standard of change for tensor fields defined within copies of the tangent and cotangent spaces at each event. But it does not provide the structure needed to distinguish time and space – i.e., describe durations and distances – or angles of incidence beyond parallelism. In general relativity, the structure that supports these concepts is, essentially, the *spacetime metric*, which is a smooth, symmetric field g_{ab} that is invertible – i.e., there exists an inverse metric g^{bc} such that $g_{ab}g^{bc}$ is the (1, 1)-identity tensor – and has Lorentz signature (1,3). Thus, one of the functions of the metric is an *inner product structure* on the tangent space at a point that assigns between any two non-zero vectors X^a, Y^a there a "cosine angle" $g_{ab}X^aY^b/|(g_{ab}X^aX^b)(g_{ab}Y^aY^b)|^{1/2}$, hence a *norm structure* on the space: $\|X^a\|_g = |g_{ab}X^aX^b|^{1/2}$. Based on this inner product, the elements X^a of the tangent space are divided into three classes:

- If $g_{ab}X^aX^b > 0$, X^a is *timelike*: it represents an instantaneous direction in the manifold of events corresponding to differences in time, hence an instantaneous trajectory for a positive-mass particle.

- If $g_{ab}X^aX^b = 0$, X^a is *null* (or *lightlike*): it represents an instantaneous direction in the manifold of events corresponding to an instantaneous trajectory for a zero-mass particle, such as a light ray.

- if $g_{ab}X^aX^b < 0$, X^a is *spacelike*: it represents an instantaneous direction in the manifold of events corresponding to a spatial direction.

In the tangent space, the null vectors form a double cone structure – the light cones – with the apex of the cones at the origin. The vectors inside the cones are precisely the timelike ones, while those outside are spacelike. Causal vectors are said to be those that are either timelike or null. Curves inherit these classifications, so that those with always timelike tangent vectors represent the possible *worldlines*, or locus of present events, of a positive-mass particle, and similarly null curves represent the possible worldlines of zero-mass particles. Spacelike curves represent, in some sense, spatial paths.

The significance of this classification is more apparent when one considers how the space-time metric provides an integration measure for assigning *magnitudes* to continuous (and piece-wise differentiable) timelike and spacelike curves. (Null curves are always assigned a magnitude of zero.) If $\gamma : [s_0, s_1] \to M$ is a timelike or spacelike curve with tangent vector field ξ^a, then $\|\gamma\| = \int_{s_0}^{s_1} \|\xi^a\|_g \, ds$. If γ has a unit tangent vector field – i.e., $\|\xi^a\|_g = 1$ everywhere it is defined then its magnitude is just $s_1 - s_0$. For spacelike curves, this is the spatial *length* of the curve. The magnitude of a timelike curve is interpreted as the *duration* of the curve, i.e., the time elapsed that a particle with the image of γ as its worldline would experience. In general, the magnitude of a curve does not just depend on its endpoints (if it has any); hence, it cannot in general be interpreted as a distance function between those points; this is the source of the well-known "twin paradox."[8]

In addition to providing for the structure for durations and lengths, the spacetime metric also does so for hypervolumes – in general, temporally extended volumes – through the structure of a volume element. A *canonical volume element* on M is a four-form ϵ_{abcd} which is normalized:

$$g^{a_1 a_2} g^{b_1 b_2} g^{c_1 c_2} g^{d_1 d_2} \epsilon_{a_1 b_1 c_1 d_1} \epsilon_{a_2 b_2 c_2 d_2} = -4!.$$

One can then use the volume form to define a positive-definite measure on M (Hawking and Ellis, 1973, §2.8). Note that four-forms always exist locally, but not always globally. (This depends only on the structure of the manifold, but not on the geometry.) So, when they do, the manifold is said to be orientable and a particular four-form is an *orientation (structure)*. However, without (the inverse of) a spacetime metric, there is no way to define the normalization condition to define a unique volume form.

Thus, the spacetime metric provides the structure for defining both "angles" and "volumes." The aspect of that structure pertaining only to the former is called the *conformal structure*. This is given simply by a continuous assignment of light cones to the tangent space of the manifold, or in other words, an equivalence class of metrics related to one another by an arbitrary positive smooth scalar field Ω. Given any g_{ab}, Ωg_{ab} is also a metric of Lorentz signature that assigns exactly the same vectors to be timelike, null, or spacelike at every point; hence, it also agrees on which curves are timelike, null, and spacelike (Malament, 2012, p. 125).

The spacetime metric (and its inverse) provides the means to determine a different standard of constancy for tensor fields along curves. For simplicity, consider the case of vector fields X^a, Y^a defined along the image of $\gamma : I \to M$. One can say that the norm of X^a, say, is constant along γ when $\|X^a\|_g$ is so constant, and that the "angle" between X^a and Y^a is constant along γ when $g_{ab} X^a Y^b$ is so constant. If constancy with respect to an affine connection ∇ implies constancy in this sense, ∇ is said to be *compatible* with the metric g_{ab}. In general, an arbitrarily chosen connection will not be compatible with a given metric, but in general relativity it is *required*. It turns out that the compatibility of ∇ with g_{ab} is equivalent to the condition that $\nabla_a g_{bc} = \mathbf{0}$, i.e., the inner product is preserved by parallel transport (Hawking and Ellis, 1973, p. 40).

Two other pieces of spacetime structure are typically assumed, although they are not definable for all relativistic spacetimes (Hawking and Ellis, 1973, §6.1). The first piece is a *spatial orientation*, which is a continuous assignment of three ordered linearly independent spacelike vectors to all events in spacetime. The order of the elements determines either a right- or left-hand rule in the usual way; spacetimes that are spatially orientable thus provide the structure to distinguish left and right globally.

The second piece is a *temporal orientation*. This is a continuous selection of one of the two null cones at every point, the interior elements of which are designated the *future-directed* timelike vectors, the others the *past-directed* ones, and similarly for the null vectors on the cones themselves. These features are inherited by timelike and null curves. A temporal orientation thus allows one to distinguish past from future. It can be determined by any continuous nonvanishing timelike vector field t^a on M: in the interior of whichever null cone it falls at a point is the future direction. Consequently, another such field t'^a yields an *equivalent* temporal orientation just when $g_{ab} t^a t'^b > 0$ everywhere; it yields the *opposite* orientation just when $g_{ab} t^a t'^b < 0$ everywhere instead.

When a spacetime is equipped with a temporal orientation, one can describe its *causal structure*. This structure consists in two binary relations on the events of M. Supposing that $p, q \in M$, then q is said to be in the *chronological* (respectively *causal*) future of p, written $p \ll q$ (respectively $p \leq q$), just when there is a continuous, future-directed timelike (respectively causal) curve $\gamma : [a, b] \to M$ such that $\gamma(a) = p$ and $\gamma(b) = q$. Similar definitions hold for the *chronological (causal) past*. The whole causal past of an event p is often interpreted intuitively as the total set of events that could "influence" p, and its whole causal future as those that it could "influence"; it is a surprisingly subtle matter to make this idea precise, however (Earman, 2014). Even when a spacetime does not have a temporal orientation, one can still define the two-place relation of chronological (causal) connectibility by dropping the directedness of the above definitions.

11.3   What There Is: Alternatives

The border between alternative *formulations* of general relativity and alternative relativistic *theories* of spacetime is vague. Even if one declared the above structures as a definite description of general relativity, it would still be undeniable that many proposed alternatives bear a strong family resemblance to it. Consequently, I have had to draw the line somewhere. My heuristic in doing so had been twofold. First, I consider alternatives that aim to accomplish at least as much as general relativity but with less – or at least different – structure. These are often motivated by conceptual or philosophical considerations that suggest that general relativity posits *too much* spacetime structure, or structure of the *wrong kind*. Second, I consider proposals that *augment* the standard spacetime structure of general relativity by *weakening* some standard constraint on that structure. These are typically motivated by the desire to represent a *wider* range of states of affairs using the theory.

Now, the division between and interpretation of these two heuristics is also vague; for example, within the scope of the latter I have not considered additions of extra "gravitational" fields as one finds, for example, in Brans-Dicke theory (Brans, 2014). But my hope is that the selection of alternatives presented – for manifold structure in Section 11.3.1 and for geometrical structure in Section 11.3.2 – will stimulate more work on the status of alternative structure in general. Regardless of whether any such alternatives should be adopted, such work would strengthen the conceptual and philosophical foundations of general relativity.

11.3.1   *Alternative Manifold Structure*

The first structure one might question is the collection of events M itself. There is a long history of skeptical arguments toward there being points of space (Forrest, 1996; Coppola and Gerla, 2013), some of which extend to points – events – of spacetime (Russell, 1927, Chs. XXVII–XXX). Many of these are of a metaphysical character; one with a more epistemological focus rests on the claim that one never observes directly the contents of events themselves, but rather spatiotemporally extended processes. Points are then idealized limits of such processes (Geroch, 1978, Ch. 1). However it might be motivated, there have been several approaches to point-free spaces, such as Whitehead's point-free geometry (Whitehead, 1919, 1920; Russell, 1927; Simons, 1987; Gerla and Miranda, 2008), mereotopology (De Laguna, 1922; Whitehead, 1929; Roeper, 1997; Casati and Varzi, 1999; Cohn and Varzi, 2003), and the theory of locales (Johnstone, 1982, 1983), which seek to describe regions by generalizing the algebraic structure of various topological concepts. As far as I am aware, these approaches have not been applied successfully to general relativity, for it is yet unclear how to recover the differentiable structure needed to describe to formulate the necessary geometrical structure in the first place (Arntzenius, 2012, Ch. 4). But those sympathetic to this idea might look to alternative methods for describing differentiable structure, such as pseudogroups (Kobayashi and Nomizu, 1963), ringed spaces (Hartshorne, 1977), or sheaves (Tennison, 1975), which depend on being given a topological space, then generalize these methods for point-free spaces – perhaps through the methods of topos theory (Mac Lane and Moerdijk, 1994).

Another alternative approach to events and manifold structure is that of smooth algebras (Nestruev, 2003). Roughly speaking, a smooth algebra is a purely algebraic formulation of the algebra of smooth real functions on a manifold.[9] Just as fixing a manifold determines its associated algebra of smooth real scalar fields, it turns out that fixing a smooth algebra determines a smooth manifold – with all the trappings of topological and differentiable structure – whose points are reconstructed as homomorphisms of the algebra into $\mathbb{R}$. Indeed, these two mathematical theories are equivalent, in the sense of being categorically dual to one another (Rosenstock et al., 2015).

The motivation for this structural alternative is usually not a philosophical preference for extended regions over points – indeed, there is a sense in which regions are still composed of points in the smooth algebra approach, even though points are themselves not primitive objects there. Rather, by eschewing direct postulation of manifold spacetime itself, it may seem to implement a sort of relationism or empiricism regarding spacetime structure: if the smooth algebra is interpreted as (in some idealized sense) the algebra of world observables, then spacetime events can be seen as merely coincident values of these observables (Earman, 1986, 1989). (I continue the discussion of this alternative in Section 11.3.2, since one can go further, adding the analogue of geometrical structure to smooth algebras.)

Another sort of concern comes from the foundations of mathematics, especially over the choice of set-theoretic axioms and the proper set-theoretic commitments of a scientific theory, and from empiricist considerations regarding the observational accessibility of real-valued quantities (i.e., we can only ever measure quantities with finite precision). There has been some work to see how far one can go letting the local structure of M be only that of the rationals ($\mathbb{Q}^4$) or real closed fields (Székely and Madarász, 2013; Székely, 2015), although these must be traded for some sacrifices in representational capacity (as one might expect from forgoing the richness of the real numbers). It remains to be seen whether these sacrifices are from the meat or fat of general relativity.

A different concern with the full structure of the real numbers and sets is that the introduction of numerical coordinates for the spacetime events unduly constrains the kind of structure they can have. Why should spacetime be modeled as a set-theoretic object? Synthetic differential geometry attempts to rectify this by providing an axiomatic rather than constructive (i.e., real-analytic) approach to geometry using the theory of topoi within category theory (Lavendhomme, 1996; Kock, 2006, 2010). In doing so, it provides for a rigorous foundation for infinitesimals within geometry, capturing much of their heuristic use in physics and the engineering sciences, at the expense of a weakening of classical logic to intuitionistic logic. Despite the geometrical tools of the theory being around since the 1980s, and despite its originator Lawvere's original intent to develop the theory as a foundation for continuum mechanics (1980), applications to general relativity have been only preliminary – see Heller and Król (2017) and references therein. Much more work needs to be done to understand what consequences, both conceptual and technical, such a formulation has for spacetime structure.

Even if one retains the standard topological and differentiable structures of the manifold in terms of sets and real numbers, one might still weaken some of the conditions those structures have been required to satisfy. I consider two of these regarding topological structure,[10] and two regarding differentiable structure. The first of these concerns the locally Euclidean nature of the manifold, that every point in M has a neighborhood U such that there is homeomorphism $\phi : U \to \mathbb{R}^4$. One can weaken this to add two further types of chart to the differentiable structure: those with boundaries, and with corners.[11] A chart with boundary (corners, respectively) is a homeomorphism $\phi_1 : U \to [0, \infty) \times \mathbb{R}^3$ ($\phi_2 : U \to [0, \infty)^k \times \mathbb{R}^{4-k}$ for any $k \leq 4$, respectively) with $U \subseteq M$. Boundary (corner) points in the manifold are those covered only by charts with boundary (corners).

On the one hand, boundary and corner points are unusual because they are events that are locally unlike typical events: they are the "last" events in one or more spatial or temporal "directions." Any curve passing into such a point in a direction "outward" from the manifold must end there. On the other hand, the singularity theorems of general relativity (for manifolds without boundary) guarantee that under quite generic conditions, a relativistic spacetime will contain incomplete inextendible timelike geodesics, curves that (also) cannot be extended but (from their parameterization) end prematurely.[12] Although this is a different sense in which the spacetime manifold "ends prematurely," it perhaps undercuts some of the motivation for excluding boundary or corner points in the first place. Furthermore, even setting those issues aside, manifolds with corners can still be used to model collections of events that are strict subcollections of the events of a relativistic universe.

The second topological condition that one could relax is the Hausdorff condition. There are two sorts of contexts where this has been motivated. The first comes from attempts to extend a given spacetime to make it as "large" as it can be – i.e., with as many events as possible. Recall (Hawking and Ellis, 1973, p. 58) that a spacetime (M, g_{ab}) is *extendible* when there is another spacetime

(M', g'_{ab}) (its *extension*) and a proper isometric embedding $\psi : M \rightarrow M'$; it is *inextendible* otherwise. It turns out that there exist extendible spacetimes with two classes of null geodesics. In one extension thereof, those in first class are extended and in fact made complete, while the second ones are not and remain incomplete, and vice versa for the second extension. It is not possible for the original spacetime to be extended in both ways at once, *unless* one relaxes the Hausdorff condition (Hawking and Ellis, 1973, §5.8).

The second context for allowing for non-Hausdorff manifolds comes from a program to represent in the model itself a way in which the future is open, i.e., there are many distinct continuations of local states of affairs. The most natural implementation of this proposal in general relativity would be to drop the Hausdorff condition, so that at certain collections of events – e.g., along the future light cones of certain events – the spacetime manifold itself splits into the different alternatives: "such a model is more in accordance with one's intuitive feelings of a determinate past and an indeterminate future than is our normal picture of a Hausdorff space-time" (Penrose, 1979, p. 594). Although this idea has not found much application in the mainstream literature on general relativity, there has been increasingly sophisticated philosophical work to develop this idea for special (Placek and Belnap, 2012; Müller, 2013) and general relativity (Placek, 2014). There are still a number of challenges (Earman, 2008), but the future is open regarding how this theory of branching spacetimes can be fleshed out.

Turning now to the topological and differentiable structure, there have been some proposals to weaken their compatibility. One way to do so is to keep the differentiable structure but change the topological structure. Hawking et al. (1976) proposed the path topology, the largest topology for which the timelike curves continuous according to the manifold topology are made continuous.[13] In general, this topology is larger than the manifold topology, but it was constructed explicitly to be more closely connected to and motivated by the geometrical – in particular, causal – structure of spacetime.

Another topology on spacetime events prompted by a similar but still distinct motivation is the Alexandrov (or interval) topology. For any $E \subseteq M$, let $I^+(E) = \{q \in M : p \ll q \text{ for some } p \in E\}$ be the (totality of the) chronological future of E, and similarly for the chronological past $I^-(E)$. Then the Alexandrov topology on M is the smallest topology in which each $I^+(E)$ and $I^-(E)$ is open for every $E \subseteq M$. In general, the Alexandrov topology is coarser than the manifold topology, but coincides with it just when the spacetime is *strongly causal*, which is the condition that every neighborhood (in the manifold topology) of $p \in M$ has a sub-neighborhood U such that no continuous causal curve in M intersects U more than once (Hawking and Ellis, 1973, p. 192).

Although both of these proposals provide an alternative topological structure for spacetime determined by some of its causal structure, they can only be cogently interpreted as *additional* topological structures (giving spacetime the structure of a so-called bitopological space) rather than as replacements.[14] This is simply because both of them *depend* on the manifold topology for their definition.

There are other ways to change the differentiable structure while retaining the standard topological structure. Incredibly, for topological manifolds of dimension at least four, it is sometimes possible to find a differentiable structure that is *not* diffeomorphic to the standard one that is so compatible. Indeed, the exotic version of $\mathbb{R}^4$ is homemorphic to the topological product $\mathbb{R} \times \mathbb{R}^3$ but *not* to the corresponding smooth product! These so-called *exotic smooth structures* are as mathematically surprising as they are difficult to interpret physically for a spacetime model. Nevertheless, some tentative conclusions have been drawn (Asselmeyer-Maluga and Brans, 2007). In certain cases the exotic structure can be localized to a region of the manifold, acting as a "source" for gravitation in the sense of admitting only of a connection that is non-flat in that region (Brans, 1994; Sładkowski, 2001; Asselmeyer-Maluga and Brans, 2015). Moreover, because the manifold of a globally hyperbolic spacetime can always be smoothly decomposed into $\mathbb{R} \times \Sigma$ for some smooth three-manifold Σ, it has recently been argued that spacetimes with exotic smooth structures for $\mathbb{R}^4$ provide a counterexample to the strong cosmic censorship conjecture (Etesi, 2015), roughly that generic spacetimes satisfying some relevant energy condition are globally hyperbolic (but cf. Earman (1995, pp. 45–46)). Much more work needs to be done to understand if there is an acceptable interpretation of these exotica and their bearing on other aspects of spacetime structure, such as the presence of singularities (Earman, 1995, p. 50).

11.3.2   *Alternative Geometrical Structure*

One way to introduce additional geometrical structure is by relaxing some of the constraints on the affine connection. For example, if the requirement that it be torsion-free is dropped, as Cartan (1922, 1923a,b, 1924) investigated in a series of papers,[15] the result, now known as Einstein-Cartan theory, allows for distinct dynamics for particles with intrinsic angular momentum (Hehl et al., 1976; De Sabbata and Sivaram, 1994; Trautman, 2006).[16] The torsion tensor is then related to a spin tensor representing the intrinsic angular momentum density of matter fields (and their interactions) in spacetime.[17] One can further drop the requirement that the connection be compatible with the metric (Hehl et al., 1995); the degree to which this occurs is represented by the nonmetricity tensor $\nabla_a g_{bc}$. The dynamical equations for this "metric-affine" theory are then specified by some Lagrangian function of the scalar curvature (as is the case in standard general relativity) and these new geometrical structures. Both the Einstein-Cartan and metric-affine theories have been suggested as better accounts of the "microstructure" of the dynamics of spacetime with matter (Stachel, 1999), in particular the symmetries found in theories of particle physics, but there is still some controversy regarding whether they are ultimately equivalent to theories with more complicated matter fields but with only the usual spacetime structure.[18]

Relaxations of the smoothness constraint on the metric tensor, although they don't introduce new geometrical structures per se, are worth mentioning. First, the metric need only be twice differentiable to define the Einstein tensor (for which see Section 11.4). Alternately, if it is only continuous but with locally square integrable weak first derivatives, then the equation is defined in a distributional sense (Hawking and Ellis, 1973, pp. 57–58).[19] When complemented with distributional matter sources of some (but not all) types, this allows one to model shock waves, thin shells of matter, cosmic strings, etc. (Geroch and Traschen, 1987), or to provide spacetime extensions unavailable when smoothness is demanded (Galloway and Ling, 2017); when the first derivatives of the metric are not Lipschitz continuous in particular, they allow for indeterministic geodesic trajectories (Fletcher, 2017).

There have been some proposals to try to *reduce* the geometrical structure that standard general relativity posits without losing any substantive possibilities or properties that the theory can represent. One that picks up its point of development from the smooth algebras discussed in Section 11.3.1 is that of Einstein algebras (Geroch, 1972): they augment smooth algebras with further structure that amounts to the standard geometrical structure of the spacetime metric. Earman (1986, 1989) had suggested this approach to provide a spacetime structure that blocks a formulation of the hole argument (Norton, 2015). It turns out that Einstein algebras are in a precise sense equivalent with Lorentzian geometries (Rosenstock et al., 2015), so it is less clear both that it avoids an analogous formulation of the hole argument and that the alternative structure it posits is really just different in appearance, not in substance (Rynasiewicz, 1992).[20]

An idea different from Earman's but toward the same goal is to take all the models of the standard spacetime structure and then form equivalence classes under the orbits of the isometry group (Iftime and Stachel, 2006). In practice, though, it seems one must almost always work with a representative from the class, and having many isometric models with the same representational capacities can be useful. Given that there are resolutions of the hole argument that do not require this awkward move (Weatherall, 2018; Pooley, 2013, §7), it is not clear how well motivated it is in the end.

Another type of quotient one might apply to reduce the geometrical structure is that of the homothety group, which, for a given spacetime metric g_{ab}, is just the one-parameter group of transformations given by $g_{ab} \mapsto \lambda g_{ab}$ for $\lambda > 0$. Although this is not motivated by the hole argument, it does capture the former proposal's underlying idea that perhaps there is surplus representational structure in general relativity, in this case with its representation of the units of spatiotemporal quantities. If one recalls from Section 11.2.2 that the spacetime metric determines both durations and distances, then homothetic transformations might plausibly be interpreted as scale transformations for these quantities. Indeed, homothetically related spacetimes are often glossed as differing only by a choice of units. If this is correct, it would be an attractive elimination of structure for comparativists about quantities (Dasgupta, 2013; Eddon, 2013), but more work needs to be done to assess its cogency.[21]

11.4   What Determines What: The Standard Account

The standard presentation of a relativistic spacetime achieves a remarkable reduction of structure: very much is determined by relativity little. In particular, one can do with just the following: $(M, \mathcal{A}, g_{ab})$. Here, M is the collection of events and $\mathcal{A}$ is an atlas for M with some particular properties, which determines all other manifold structures. For geometrical structure, the metric g_{ab} determines all else. If the manifold is orientable, then one needs to add only a temporal orientation t^a to the geometrical structure.

11.4.1   *Standard Determination of Manifold Structure*

Both the differentiable *and* topological structures of the manifold can be determined by an appropriate atlas. In particular, if the atlas satisfies certain analogues of the conditions on the topological structure discussed in Section 11.2.1, then the topological structure determined will also have those features (Malament, 2012, p. 4).

So, take $\mathcal{A}$ to be a maximal, smooth (C^∞ differentiable) structure on M satisfying the following conditions:

1   It is *Hausdorff*, i.e., for any distinct $p, p' \in M$, there are charts $\phi, \phi' \in \mathcal{A}$ with domains $U, U' \subseteq M$ such that $p \in U$, $p' \in U'$, and $U \cap U' = \emptyset$.
2   It is *path-connected*, in the sense that for every $p, q \in M$, there is a function $f : [0, 1] \to M$ such that $f(0) = p$, $f(1) = q$, and whenever $f(x) \in U$ for some chart $\phi : U \to \mathbb{R}^4$ in $\mathcal{A}$, then $(\phi \circ f)_{|f^{-1}[f[0,1] \cap U]}$ is a smooth function.
3   It contains a *countable differentiable substructure*, i.e., there is a countable differentiable structure $\mathcal{A}' \subset \mathcal{A}$.[22]

This is the atlas used in the standard presentation of a relativistic spacetime. Instead of requiring compatibility between this structure and a separately defined topology on M, the topological structure on M is rather determined as that *coherent* with the domains of the charts of $\mathcal{A}$, i.e., the smallest topology on M for which all the charts are continuous. The resulting topological structure then makes M a Hausdorff, connected, second-countable topological four-manifold (O'Neill, 1983, p. 23).

Two further features regarding determination of manifold structure are of note. First, *any* C^k differentiable structure $\mathcal{A}$ on a set M for $k \geq 1$ determines a unique, maximal C^∞ differentiable structure on M, in the sense that there is a unique such structure whose charts are compatible with those of $\mathcal{A}$ (Whitney, 1936). This is pragmatically important, for it permits one great economy in expressing a spacetime model: one need only describe some C^k differentiable structure or other, which can often be done with only a few charts. But this determination is not typically incorporated as a further reduction of manifold structure, for on its face it would tend to suggest misleadingly that certain charts are somehow representationally privileged based on the manifold structure alone.[23] Second, there the global existence of a derivative operator *implies* (and is implied by) the countable differentiable substructure condition (Geroch, 1971). In fact, this condition is implied by the existence of a spacetime metric. Therefore, one could leave that condition off of the list above as a redundancy.

11.4.2   Standard Determination of Geometrical Structure

On the standard account, the spacetime metric g_{ab} determines all the other geometrical structure described in Section 11.2.2 (except for temporal and spatial orientations, if they exist). I have already described how it determines the inner product (null cone) and norm structures on the tangent space at each point – hence, the conformal structure and the relation of causal connectibility as well. Once a smooth Lorentz-signature metric is specified, it also determines a unique globally defined torsion-free derivative operator, the *Levi-Civita* derivative operator, which can be explicitly constructed from the metric (and an arbitrarily chosen derivative operator). A unique derivative operator in turn determines a unique curvature tensor and a unique projective structure – the class of all pregeodesics on the spacetime. In fact, it determines the geodesics as well: these are just the pregeodesics whose tangent vectors satisfy the geodesic equation with respect to the Levi-Civita derivative operator.

The only geometrical structures not quite determined by the spacetime metric concern orientability of various sorts. It turns out, though, that any two of the following three conditions imply the third (Hawking and Ellis, 1973, §6.1):

1   Spacetime is orientable: there exists a global four-form on M.
2   Spacetime is temporally orientable: there exists a continuous timelike vector field on M.
3   Spacetime is spatially orientable: there exists a continuous spatial frame field on M.

So, when the spacetime is orientable, the metric determines the volume element (hence the volumetric structure of spacetime) according to its normalization condition. In this case, the addition of a temporal (resp. spatial) orientation as geometrical structure on M determines a spatial (resp. temporal) orientation on M. It is most typical to assume a temporal orientation, so that causal structure is then determined (as the orientation allows one to split the causal connectibility relation into two, based on past and future).

Readers with enough familiarity with the general theory of relativity may notice that I have not yet mentioned anything about matter fields on spacetime, or about Einstein's field equation, which connects their energetic properties with spacetime curvature. This is because, on the standard account, matter fields and their energetic properties are not a part of spacetime structure, even though there is a sense in which they are all intimately connected. To show this requires some preliminary definitions.

The Ricci tensor R_{ab} associated with an affine connection is defined as the contraction of the associated curvature tensor: $R_{ab} = R^c{}_{abc}$.[24] The scalar curvature field R, meanwhile, is defined from the Ricci tensor in the presence of a metric: $R = g^{ba}R_{ab}$. These two fields are combined into the Einstein tensor $G_{ab} = R_{ab} - \frac{1}{2}Rg_{ab}$. Einstein's field equation then states (in units where Newton's constant and the speed of light have been set to one) that $G_{ab} = 8\pi T_{ab}$, where T_{ab} is the energy-momentum tensor associated with all matter fields and their interactions.[25] It encodes the energy densities and three-momentum densities that would be measured relative to all observers at each event of spacetime (Malament, 2012, Ch. 2.5).

We see from Einstein's field equation that the energy-momentum tensor is in fact determined by the Einstein tensor, which is in turn determined by the spacetime metric, for the Ricci tensor is just that associated with the Levi-Civita connection. However, on the standard account this does *not* imply that the metric determines matter. For the matter fields on spacetime *also* determine the energy-momentum tensor, but not vice versa: distinct matter fields on a spacetime can generate the same energy-momentum tensor. From one perspective, it is just this overdetermination which makes general relativity a theory from which it is difficult to extract exact solutions, i.e., jointly consistent descriptions of manifold, metric, and matter according to the aforementioned structural constraints.

11.5   What Determines What: Alternatives

Before turning to various alternatives concerning what determines what, it will be helpful to review the converses of the standard determination relations outlined in Section 11.4. As the existence of exotic smooth structure (discussed in Section 11.3.1) attests, in general the topological structure does not determine the differentiable structure.[26] An affine connection ∇, even if it is torsion-free, does not determine a Lorentz metric in at least two ways. First, there may not be *any* metric g_{ab} such that $\nabla_c g_{ab} = \mathbf{0}$.[27] Second, even if there is such a metric, its signature is not in general determined; even if that is provided the metric is not in general determined, even up to homothety (Hall, 2004, Theorem 9.2). Similarly, the curvature structure does not determine the affine connection either.[28]

What about the conformal and projective structures? Naturally, the conformal structure only determines the metric up to a conformal factor – that is, the class $\Omega^2 g_{ab}$ for positive smooth scalar fields Ω.[29] Interestingly, though, the conformal structure does determine which null curves are pregeodesics – and vice versa – for it is only this class of pregeodesics which is invariant under arbitrary conformal transformations (Malament, 2012, p. 125). (This will play an important role in the discussion below of the approach by Ehlers et al. (1972).) Similarly, the projective structure does not determine the affine connection.[30] But, a famous theorem of Weyl (1921) shows that two spacetimes with the same conformal and projective structures are in fact homothetically related.

It is tempting to conclude that Weyl's theorem shows that conformal and projective structures determine metric structure up to a homothety, but this is not quite correct: one must *assume* in the first place that the conformal and projective structures already are determined by a metric. However, in a remarkable paper, Ehlers et al. (1972) investigated the circumstances under which one *could* make such a determination. In fact, they go much further than this, providing an informal axiomatic structure characterizing the aspects of this determination.

Here is a sketch of their framework: their first class of axioms establishes the manifold structure by supposing that each particle worldline has the structure of a one-dimensional manifold and that all events can be covered by radar coordinate systems. The second concerns the propagation of light, which establishes the conformal structure by assuming that the directions of light propagation bound those of particles into two connected components – what end up being the null cones. The third concerns the worldlines of free particles, which are assumed to have coordinate representations in terms of straight lines, thereby determining a projective structure up to a choice of torsion, assumed to vanish, according to which the worldlines of the free particles are geodesics. The fourth class of axioms postulates the compatibility of the conformal and projective structures, entailing that the worldlines of light rays are also geodesics of the projective structure. This compatibility then determines a Lorentz metric up to a conformal transformation. The resulting space, called a Weyl geometry,[31] only yields a relativistic spacetime up to homothety when one further condition is imposed: that there is no second clock effect. This is the statement, in accordance with experimental evidence, that the rates of ideal clocks (i.e., the scale factor of their affine parameter) are not path-dependent: two ideal clocks at an initial event whose rates are in sync will still be in sync at any common other event. Because of the multifaceted richness of their approach, many alternatives have been proposed that improve upon various aspects of their construction, such as the representation of projective structure, its compatibility with the conformal structure, and even the use of particles as primitive entities.[32]

These works all have a certain operational subtext that prioritizes material constituents over spatiotemporal structure. The most famous in this empiricist tradition emanates from Mach, whose appellated principle states that, in some sense, the matter content of the universe determines its spacetime structure, in particular its inertial structure (which in general relativity is implemented by the projective structure or the affine connection or some related structure).[33] It is still not yet clear how some determination of this sort could be made: specifying the energy-momentum T_{ab} in Einstein's field equation does not even determine the full Riemann curvature tensor, much less the affine connection or metric; moreover, for most matter fields, their associated energy-momentum tensors cannot even be stated without invoking the metric (Malament, 2012, pp. 159–160). Indeed, although Einstein was motivated by some version of Mach's principle in his development of the general theory of relativity, in the end he agreed it was not so implemented (Torretti, 1996, Ch. 6.2). Nevertheless, there is some indication that a version of the principle might hold in a restricted class of spacetime, in particular those with certain types of boundary conditions.[34]

A different approach in a similar spirit aims to show that the metric can be determined not from the distribution of energy-momentum of matter, but from the detailed dynamical equations of the matter fields themselves. Hehl and collaborators (Itin and Hehl, 2004; Hehl and Obukhov, 2005, 2006) have proposed a determination of the spacetime metric through an appropriate "premetric" version of electromagnetism. Naturally, this theory requires further assumptions and constraints which well deserve further attention. Perhaps as well the approach could be generalized to include any matter theory formalized in terms of quasilinear hyperbolic systems of equations, which surprisingly seems to include almost all classical fields of interest (Geroch, 2011).

A third approach is the "dynamical relativity" program of Brown (2006) influential among philosophers of physics, which seeks to show how the metric field encodes durations and lengths not by fiat, but through its coupling as a physical field to more ponderable matter fields. Although there continues to be controversy about how to interpret the program, one version, due to Knox (2013, 2018), takes it to claim that symmetry properties of the equations of motion for matter fields determine spacetime structure, which should be understood functionally as whatever provides inertial structure. Inertial structure, in turn, is understood through inertial coordinate charts generated from frame fields on timelike geodesics. This is supposed to succeed as a determination of spacetime structure because Ehlers et al. (1972) proved that "the full set of timelike geodesics (the inertial trajectories) . . . is sufficient to fix both conformal and projective structure. This in turn fixes metric structure up to a global scale factor" (Knox, 2013, p. 349), i.e., a homothety. However, even the cursory review of Ehlers et al. (1972) above shows that no such determination has been proven unless a number of substantive non-inertial assumptions are made. Projective structure itself (i.e., the inertial trajectories) does not determine conformal structure, and even given a Lorentzian metric and its Levi-Civita connection on a four-manifold, that metric is not the unique Lorentz metric compatible with that connection, even up to homothety. There is a subtlety in this last negative conclusion, however: the failure of uniqueness up to homothety depends on the spacetime having nontrivial symmetries.[35] Symmetric spacetimes are not a new problem for the dynamical approach (Norton, 2008; Pooley, 2013, pp. 573–574), but these considerations show that Knox's particular brand of functionalism about spacetime structure does not avoid it.

In addition to these approaches that seek to determine other structure from that of matter, there is another venerable tradition in which causal structure takes the role of matter. Robb (1914, 1921, 1936) showed how the metric of Minkowski spacetime can be determined up to homothety from a single relation of causal precedence on an uncountable domain, and some results are known regarding how this could be weakened to a countable domain (Sen, 2010, Part I). Zeeman (1964) revealed accordingly that the causal isomorphisms of Minkowski spacetime (M, η_{ab}) – i.e., the bijections such that for all $p, q \in M$, $p \ll q$ if and only if $\phi(p) \ll \phi(q)$ – are exactly the homotheties composed with the Poincaré transformations, i.e., the maximal homothety group for any spacetime (Hall, 2004, p. 293).

It is important to note that these results do not extend verbatim to general relativity, where one cannot hope that conformal structure alone determines the metric. But with some additional constraints, a version of Zeeman's theorem does. In order to describe it, one more definition is needed. A spacetime (M, g_{ab}) with a temporal orientation is *future-* (respectively, *past-)distinguishing* when, for all $p, q \in M$, $I^+(p) = I^+(q)$ (respectively, $I^-(p) = I^-(q)$) implies that $p = q$; it is *distinguishing* when it is both future- and past-distinguishing. Malament (1977) then showed that given any two distinguishing, temporally oriented spacetimes (M, g_{ab}) and (M', g'_{ab}), if $\phi : M \rightarrow M'$ is a causal isomorphism, then ϕ is in fact a conformal isometry. This result might be glossed informally as "causality implies the conformal group" or "causal structure determines conformal structure," but like with Weyl's theorem, one must take some care: the determination only holds for causal structures themselves determined from distinguishing spacetimes.

Acknowledgment

Juliusz Doboszewski, Laurenz Hudetz, Eleanor Knox, J. B. Manchak, James Read, and Gergely Székely have my gratitude for comments on earlier drafts.

Notes

1 For example, it might do so by showing that one spacetime theory posits less structure than another (Barrett, 2015). If there is a relevant sense in which the theories are nevertheless equivalent, then this structural surplus might be deemed superfluous (Weatherall, 2017).

2 Due to lack of space, I assume some familiarity with topological manifolds (Lee, 2011), differential geometry (Kobayashi and Nomizu, 1963, 1969), and the abstract index notation used in mathematical relativity (Wald, 1984; Malament, 2012), but have also endeavored to provide references where the elements I discuss may be studied in more detail, in particular in Hawking and Ellis (1973).

3 The approaches most similar to mine are perhaps Ehlers (1973) and Torretti (1996, Ch. 6.1); for a more technical but elegant presentation, see Trautman (1976).

4 I have not yet introduced the geometrical structure needed to support these interpretations; nevertheless, one might demand four-dimensionality from empirical observation or perhaps even from a priori considerations on the form of intuition (Carnap, 1922).

5 One usually finds the requirement stated that as a topological space M must be *connected* rather than path-connected – i.e., the only clopen sets are M and $\emptyset$ – but these two conditions are equivalent for manifolds and being path-connected has a more obvious interpretation.

6 Cf. the arguments of Lewis (1986) on how to separate worlds in the context of defending his modal realism.

7 For Hausdorff connected manifolds, being second-countable is also equivalent to being paracompact, being Lindelöf, and the global existence of a covariant derivative operator – see Section 11.4.

8 One can in certain circumstances define a Lorentzian "distance" function between (timelike- or spacelike-related) events as the infimum of the magnitudes of the (timelike or spacelike) curves between them, but this function does not satisfy the usual requirements of a distance function, nor can it always be globally defined on M. Nevertheless, one might still say that there is a *local Lorentzian distance structure* on the spacetime (Beem et al., 1996, Def. 4.25).

9 See also Penrose (1968, pp. 141–142) for an approach intermediate between the algebraic approach and the standard one with charts.

10 One comment on an additional topological assumption, that of second countability: if this is dropped, then there will not exist a globally defined derivative operator. However, the manifold will admit of locally defined derivative operators: could this be enough to formulate a viable analogue of general relativity?

11 Caution: there are many inequivalent definitions of these within the literature. Here I follow Joyce (2012), adapting some inessential details to match the rest of the exposition.

12 For more discussion of the singularity theorems, see Hawking and Ellis (1973, Chs. 8–10), Wald (1984, Ch. 9.5), and Earman (1995, Ch. 2.8)

13 Strictly speaking, they restricted their definition to strongly causal spacetimes, defined presently. For a review of related proposals, see Saraykar and Janardhan (2016).

14 See Heathcote (1988) for a discussion of the interpretive relevance of the path topology in particular.

15 For English translations of Cartan (1922) and Cartan (1923a,b, 1924), respectively, see Cartan (1980) and Cartan (1986).

16 Einstein-Cartan theory is thus in a sense also a generalization of teleparallel gravity (Aldrovandi and Pereira, 2012), in which one allows for torsion but assumes that the curvature vanishes, although this does not mean that these objects play the same role in that theory (Knox, 2011) or in the Einstein-Cartan theory

(Knox, 2013). More could be done to present these theories within a comprehensible unified conceptual framework.

17 A different sort of additional structure used to model particles with intrinsic angular momentum is spinor structure, which is definable on an orientable manifold just in case it is parallelizable, i.e., it has a continuous section of its (tangent) frame bundle (Geroch, 1968). Although spinor structure provides an elegant treatment of various topics, it is representable from tensor fields (Penrose and Rindler, 1984).

18 In the case of Einstein-Cartan theory, the equivalence may depend on the type of spin coupling (Trautman, 2006). For further extensive replies to objections, see Hehl and Obukhov (2007).

19 An intermediate position that avoids the use of distributions holds that the metric is everywhere C^1 but piecewise C^3 (Schild, 1967, §7).

20 Other presumably equivalent reformulations of standard general relativity (perhaps modulo some technical conditions) include those in terms of gauge theory (Blagojević and Hehl, 2013) – that is, using the mathematics of principal bundles – and geometric algebra (Lasenby et al., 1998). Within the former, the Cartan formalism emphasizes the use of frame fields and the exterior calculus.

21 When, say, matter theories formulated in a general relativistic spacetime introduce different spatiotemporal scales, it is no longer clear whether homotheties could still be interpreted as scale transformations. For such matter-augmented models, then, quotienting by the homothety group may identify models that are not representationally equivalent.

22 This is the analogue of the Lindelöf condition – see endnote 7.

23 The issue is that it is not transparent what the category of (not necessarily maximal) differentiable structures is supposed to be, i.e., what their relevant notion of isomorphism is. One might think that it is the same as maximal differentiable structures – namely, diffeomorphisms – so that two differentiable structure with a common maximal extension would be isomorphic, but then it becomes unclear just how they *determine* further structure.

24 Caution: not all references use the same index placement convention for the definition of the Ricci tensor, and similarly for the sign convention for the metric, the Einstein tensor (for which see below), etc. (Misner et al., 1973).

25 I have omitted the term Λg_{ab}, where Λ is the cosmological constant. Including it might change whether the energy-momentum tensor is determined by the metric, depending on its interpretation (Earman, 2003).

26 No general statement is yet known since the analysis of exotica in four dimensions has proceeded on a case-by-case basis, with major work so far focusing on non-compact four-manifolds.

27 It is necessary and sufficient that the connection's holonomy group be a subgroup of the orthogonal group corresponding to the metric signature (Schmidt, 1973) – see also Atkins (2008) for a different characterization. There are pointwise conditions any such connection must satisfy if it is to be a metric connection locally, but a connection being locally metric everywhere does not make it globally so.

28 Here the question is complicated somewhat by how one decides to treat torsion, and whether one wishes to determine the connection only or a metric connection in particular (which, in the mathematics literature, is known as the problem of prescribed curvature). See Hall (2004, Ch. 9.3) for a treatment of the problem assuming that the curvature structure is associated already with the Levi-Civita connection of some Lorentz metric.

29 Such conformally equivalent metrics also have the same Weyl conformal tensor, but the converse does not hold unless they are conformally flat, i.e., unless the Weyl tensor vanishes (Malament, 2012, p. 85).

30 Projectively equivalent connections have the same Weyl projective tensor, but the converse does not hold (Hall, 2013).

31 The reference is to Weyl's early theory, of which this geometry is a model. For historical analysis, see Ryckman (2004).

32 For commentary and references, see Ehlers (1973), Sklar (1977), Trautman (2012), and Pfister and King (2015).

33 There are in fact many different versions of this informal idea. For a variety of perspectives, see Barbour and Pfister (1995).

34 The shape dynamics program of Barbour and collaborators is also a Machian program which aims to describe relativistic spacetimes as a collection of scale-invariant Riemannian three-geometries; thus, it only seeks to recover globally hyperbolic spacetimes with certain boundary conditions. See Pooley (2013, §6.2) for an accessible review and references.

35 When the manifold is simply connected, the set of metrics on that manifold for which the desired functional determination is valid is in fact generic in the C^∞ open (or Whitney) topology on those metrics (Hall, 2004, pp. 258–259). But it is not yet clear whether this is the relevant topology with which to evaluate this question (Fletcher, 2016): does this verdict change if one uses a less problematic topology capturing global similarity (Fletcher, 2018)?

References

Aldrovandi, R. and Pereira, J.G. (2012). *Teleparallel Gravity: An Introduction.* Dordrecht: Springer.

Arntzenius, F. (2012). *Space, Time and Stuff.* Oxford: Oxford University Press.

Asselmeyer-Maluga, T. and Brans, C.H. (2007). *Exotic Smoothness and Physics: Differential Topology and Spacetime Models.* Singapore: World Scientific.

Asselmeyer-Maluga, T. and Brans, C.H. (2015). How to include fermions into general relativity by exotic smoothness. *General Relativity and Gravitation,* 47: 30.

Atkins, R. (2008). When is a connection a metric connection? *New Zealand Journal of Mathematics,* 38: 225–238.

Barbour, J.B. and Pfister, H., editors. (1995). *Mach's Principle: From Newton's Bucket to Quantum Gravity,* Vol. 6 of *Einstein Studies.* Boston, MA: Birkhäuser.

Barrett, T.W. (2015). Spacetime structure. *Studies in History and Philosophy of Modern Physics,* 51: 37–43.

Beem, J.K., Ehrlich, P. and Easley, K. (1996). *Global Lorentzian Geometry* (2nd edition). Boca Raton, FL: CRC Press.

Blagojević, M. and Hehl, F.W. (2013). *Gauge Theories of Gravitation: A Reader with Commentaries.* Singapore: World Scientific.

Brans, C.H. (1994). Localized exotic smoothness. *Classical and Quantum Gravity,* 11: 1785–1792.

Brans, C.H. (2014). Jordan-Brans-Dicke theory. *Scholarpedia,* 9(4): 31358. revision #151619.

Brown, H.R. (2006). *Physical Relativity: Space-Time Structure from a Dynamical Perspective.* Oxford: Oxford University Press.

Carnap, R. (1922). *Der Raum: Ein Beitrag zur Wissenschaftslehre.* Berlin: Reuther & Reichard. (Kant-Studien Ergänzungshefte, no.56).

Cartan, É. (1922). Sur une généralisation de la notion de courbure de Riemann et les espaces à torsion. *Comptes rendus de l'Académie des sciences,* 174: 593–595.

Cartan, É. (1923a). Sur les variétés à connexion affine et la théorie de la relativité généralisée, Part I. *Annales Scientifiques de l'École Normale Supérieure,* 40: 325–412.

Cartan, É. (1923b). Sur les variétés à connexion affine et la théorie de la relativité généralisée, Part I. *Annales Scientifiques de l'École Normale Supérieure,* 41: 1–25.

Cartan, É. (1924). Sur les variétés à connexion affine et la théorie de la relativité généralisée, Part II. *Annales Scientifiques de l'École Normale Supérieure,* 42: 17–88.

Cartan, É. (1980). On a generalization of the notion of Riemann curvature and spaces with torsion. In P.G. Bergmann and V. De Sabbata (eds.), *Cosmology and Gravitation: Spin, Torsion, Rotation, and Supergravity.* New York: Plenum, pp. 489–492. Trans. G. D. Kerlick.

Cartan, É. (1986). *On Manifolds with an Affine Connection and the Theory of General Relativity.* Naples: Bibliopolis. Trans. A. Magnon and A. Ashtekar.

Casati, R. and Varzi, A. (1999). *Parts and Places: The Structures of Spatial Representation.* Cambridge, MA: MIT Press.

Cohn, A.G. and Varzi, A.C. (2003). Mereotopological connection. *Journal of Philosophical Logic,* 32: 357–390.

Coppola, C. and Gerla, G. (2013). Special issue on point-free geometry and topology. *Logic and Logical Philosophy,* 22(2): 139–143.

Dasgupta, S. (2013). Absolutism vs comparativism about quantity. In K. Bennett and D.W. Zimmerman (eds.), *Oxford Studies in Metaphysics,* Vol. 8. Oxford: Oxford University Press, pp. 105–148.

De Laguna, T. (1922). Point, line and surface as sets of solids. *The Journal of Philosophy,* 19: 449–461.

De Sabbata, V. and Sivaram, C. (1994). *Spin and Torsion in Gravitation.* Singapore: World Scientific.

Earman, J. (1986). Why space is not a substance (at least not to first degree). *Pacific Philosophical Quarterly,* 67(4): 225–244.

Earman, J. (1989). *World Enough and Space-Time: Absolute versus Relational Theories of Space and Time.* Cambridge, MA: MIT Press.

Earman, J. (1995). *Bangs, Crunches, Wimpers, and Shrieks: Singularities and Acausalities in Relativistic Spacetimes.* Oxford: Oxford University Press.

Earman, J. (2003). The cosmological constant, the fate of the universe, unimodular gravity, and all that. *Studies in History and Philosophy of Modern Physics,* 34(4): 559–577.

Earman, J. (2008). Pruning some branches from "branching spacetimes". In D. Dieks (ed.), *The Ontology of Spacetime II.* Amsterdam: Elsevier, pp. 187–205.

Earman, J. (2014). No superluminal propagation for classical relativistic and relativistic quantum fields. *Studies in History and Philosophy of Modern Physics,* 48: 102–108.

Eddon, M. (2013). Quantitative properties. *Philosophy Compass,* 8(7): 633–645.

Ehlers, J. (1973). The nature and structure of spacetime. In J. Mehra (ed.), *The Physicist's Conception of Nature*. Dordrecht: Kluwer, pp. 71–91.

Ehlers, J., Pirani, F.A.E. and Schild, A. (1972). The geometry of free fall and light propagation. In L. O'Raifeartaigh (ed.), *General Relativity, Papers in Honour of J. L. Synge*. New York: Clarendon Press, pp. 63–84. Republished as Ehlers et al. (2012).

Ehlers, J., Pirani, F.A.E. and Schild, A. (2012). Republication of: The geometry of free fall and light propagation. *General Relativity and Gravitation*, 44(6): 1587–1609.

Etesi, G. (2015). Exotica or the failure of the strong cosmic censorship in four dimensions. *International Journal of Geometric Methods in Modern Physics*, 12: 1550121–1–1550121–14.

Fletcher, S.C. (2016). Similarity, topology, and physical significance in relativity theory. *British Journal for the Philosophy of Science*, 67(2): 365–389.

Fletcher, S.C. (2017). Indeterminism, gravitation, and spacetime theory. In G. Hofer-Szabó and L. Wroński (eds.), *Making it Formally Explicit: Probability, Causality and Indeterminism*. Cham: Springer, pp. 179–191.

Fletcher, S.C. (2018). Global spacetime similarity. *Journal of Mathematical Physics*, 59(11): 112501.

Forrest, P. (1996). From ontology to topology in the theory of regions. *The Monist*, 79: 34–50.

Galloway, G.J. and Ling, E. (2017). Some remarks on the C^0-(in)extendibility of spacetimes. *Annales Henri Poincaré*, 18(10): 3427–3447.

Gerla, G. and Miranda, A. (2008). Mathematical features of Whitehead's point-free geometry. In M. Weber and J. Desmond, William (eds.), *Handbook of Whiteheadian Process Thought*, Vol. 2. Frankfurt: Ontos Verlag, pp. 121–132.

Geroch, R. (1971). Spacetime structure from a global viewpoint. In B.K. Sachs (ed.), *General Relativity and Cosmology*. New York: Academic Press, pp. 71–103.

Geroch, R. (1978). *General Relativity from A to B*. Chicago: University of Chicago Press.

Geroch, R. (2011). Faster than light? In M. Plaue, A.D. Rendall and M. Scherfner (eds.), *Advances in Lorentzian Geometry: Proceedings of the Lorentzian Geometry Conference in Berlin*. Providence: American Mathematical Society, pp. 59–70.

Geroch, R. and Horowitz, G.T. (1979). Global structure of spacetimes. In S.W. Hawking and W. Israel (eds.), *General Relativity: An Einstein Centenary Survey*. Cambridge: Cambridge University Press, pp. 212–293.

Geroch, R. and Traschen, J. (1987). Strings and other distributional sources in general relativity. *Physical Review D*, 36: 1017–1031.

Geroch, R.P. (1968). Spinor structure of space-times in general relativity, I. *Journal of Mathematical Physics*, 9: 1739–1744.

Geroch, R.P. (1972). Einstein algebras. *Communications in Mathematical Physics*, 26: 271–275.

Hall, G. (2013). On the converse of Weyl's conformal and projective theorems. *Publications de l'Institut Mathématique*, 94(108): 55–65.

Hall, G.S. (2004). *Symmetries and Curvature Structure in General Relativity*. Singapore: World Scientific.

Hartshorne, R. (1977). *Algebraic Geometry*. New York: Springer.

Hawking, S.W. and Ellis, G.F.R. (1973). *The Large Scale Structure of Space-Time*. Cambridge: Cambridge University Press.

Hawking, S.W., King, A.R. and McCarthy, P.J. (1976). A new topology for curved space-time which incorporates the causal, differential, and conformal structures. *Journal of Mathematical Physics*, 17(2): 174–181.

Heathcote, A. (1988). Zeeman-Göbel topologies. *British Journal for the Philosophy of Science*, 39(2): 247–261.

Hehl, F. and Obukhov, Y. (2006). Spacetime metric from local and linear electrodynamics: A new axiomatic scheme. In J. Ehlers and C. Lämmerzahl (eds.), *Special Relativity: Will It Survive the Next 101 Years?* Berlin: Springer, pp. 163–187.

Hehl, F.W., McCrea, J.D., Mielke, E.W. and Ne'eman, Y. (1995). Metric-affine gauge theory of gravity: field equations, Noether identities, world spinors, and breaking of dilation invariance. *Physics Reports*, 258: 1–171.

Hehl, F.W. and Obukhov, Y.N. (2005). To consider the electromagnetic field as fundamental, and the metric only as a subsidiary field. *Foundations of Physics*, 35(12): 2007–2025.

Hehl, F.W. and Obukhov, Y.N. (2007). Élie Cartan's torsion in geometry and in field theory, an essay. *Annales de la Fondation Louis de Broglie*, 32(2–3): 157–194.

Hehl, F.W., von der Heyde, P., Kerlick, G.D. and Nester, J.M. (1976). General relativity with spin and torsion: Foundations and prospects. *Reviews of Modern Physics*, 48: 393–416.

Heller, M. and Król, J. (2017). Infinitesimal structure of singularities. *Universe*, 3: 16.

Iftime, M. and Stachel, J. (2006). The hole argument for covariant theories. *General Relativity and Gravitation*, 38: 1241–1252.

Itin, Y. and Hehl, F.W. (2004). Is the Lorentz signature of the metric of spacetime electromagnetic in origin? *Annals of Physics*, 312(1): 60–83.

Johnstone, P. (1982). *Stone Spaces*. Cambridge: Cambridge University Press.

Johnstone, P. (1983). The point of pointless topology. *Bulletin of the American Mathematical Society (New Series)*, 8(1): 41–53.

Joyce, D. (2012). On manifolds with corners. In S. Janeczko, J. Li and D.H. Phong (eds.), *Advances in Geometric Analysis*. Boston, MA: International Press, pp. 225–258.

Knox, E. (2011). Newton-Cartan theory and teleparallel gravity: The force of a formulation. *Studies in History and Philosophy of Modern Physics*, 42(4): 264 – 275.

Knox, E. (2013). Effective spacetime geometry. *Studies in History and Philosophy of Modern Physics*, 44: 346–356.

Knox, E. (2018). Physical relativity from a functionalist perspective. *Studies in History and Philosophy of Modern Physics* 67:118–124.

Kobayashi, S. and Nomizu, K. (1963). *Foundations of Differential Geometry*, Vol. 1. New York: Interscience.

Kobayashi, S. and Nomizu, K. (1969). *Foundations of Differential Geometry*, Vol. 2. New York: Interscience.

Kock, A. (2006). *Synthetic Differential Geometry* (2nd edition). Cambridge: Cambridge University Press.

Kock, A. (2010). *Synthetic Geometry of Manifolds*. Cambridge: Cambridge University Press.

Lasenby, A., Doran, C. and Gull, S. (1998). Gravity, gauge theories and geometric algebra. *Philosophical Transactions of the Royal Society A*, 356: 487–582.

Lavendhomme, R. (1996). *Basic Concepts of Synthetic Differential Geometry*. Dordrecht: Kluwer.

Lawvere, F.W. (1980). Toward the description in a smooth topos of the dynamically possible motions and deformations of a continuous body. *Cahiers de Topologie et Géométrie Différentielle Catégoriques*, 21(4): 377–392.

Lee, J.M. (2011). *Introduction to Topological Manifolds* (2nd edition). New York: Springer.

Lewis, D.K. (1986). *On the Plurality of Worlds*. Oxford: Blackwell.

Mac Lane, S. and Moerdijk, I. (1994). *Sheaves in Geometry and Logic: A First Introduction to Topos Theory*. Berlin: Springer.

Malament, D.B. (1977). The class of continuous timelike curves determines the topology of spacetime. *Journal of Mathematical Physics*, 18(7): 1399–1404.

Malament, D.B. (2012). *Topics in the Foundations of General Relativity and Newtonian Gravitation Theory*. Chicago: University of Chicago Press.

Misner, C.W., Thorne, K.S. and Wheeler, J.A. (1973). *Gravitation*. San Francisco: W.H. Freeman.

Müller, T. (2013). A generalized manifold topology for branching space-times. *Philosophy of Science*, 80(5): 1089–1100.

Nestruev, J. (2003). *Smooth Manifolds and Observables*. Berlin: Springer.

North, J. (2009). The "structure" of physics: A case study. *The Journal of Philosophy*, 106(2): 57–88.

North, J. (2018). A new approach to the relational-substantival debate. In K. Bennett and D.W. Zimmerman (eds.), *Oxford Studies in Metaphysics*, Vol. 11. Oxford: Oxford University Press, pp. 3–43.

Norton, J.D. (2008). Why constructive relativity fails. *British Journal for the Philosophy of Science*, 59: 821–834.

Norton, J.D. (2015). The hole argument. In E.N. Zalta (ed.), *The Stanford Encyclopedia of Philosophy* (Fall 2015 edition). Metaphysics Research Lab, Stanford University. https://plato.stanford.edu/entries/spacetime-holearg/

O'Neill, B. (1983). *Semi-Riemannian Geometry, with Applications to Relativity*. San Diego: Academic Press.

Penrose, R. (1968). Structure of space-time. In C.M. DeWitt and J.A. Wheeler (eds.), *Batelle Rencontres: 1967 Lectures in Mathematics and Physics*. New York: W.A. Benjamin, pp. 121–235.

Penrose, R. (1979). Singularities and time-asymmetry. In S.W. Hawking and W. Israel (eds.), *General Relativity: An Einstein Centenary Survey*. Cambridge: Cambridge University Press, pp. 581–638.

Penrose, R. and Rindler, W. (1984). *Spinors and Space-Time*, Vol. 1. Cambridge: Cambridge University Press.

Pfister, H. and King, M. (2015). *Inertia and Gravitation: The Fundamental Nature and Structure of Space-Time*. Cham: Springer.

Placek, T. (2014). Branching for general relativists. In T. Müller (ed.), *Nuel Belnap on Indeterminism and Free Action*. Cham: Springer, pp. 191–222.

Placek, T. and Belnap, N. (2012). Indeterminism is a modal notion: Branching spacetimes and Earman's Pruning. *Synthese*, 187(2): 441–469.

Pooley, O. (2006). Points, particles and structural realism. In D. Rickles, S. French and J. Saatsi (eds.), *The Structural Foundations of Quantum Gravity*. Oxford: Oxford University Press, pp. 83–120.

Pooley, O. (2013). Substantivalist and relationalist approaches to spacetime. In R. Batterman (ed.), *The Oxford Handbook of Philosophy of Physics*. Oxford: Oxford University Press, pp. 522–586.

Robb, A.A. (1914). *A Theory of Time and Space*. Cambridge: Cambridge University Press.

Robb, A.A. (1921). *The Absolte Relations of Time and Space*. Cambridge: Cambridge University Press.

Robb, A.A. (1936). *Geometry of Time and Space*. Cambridge: Cambridge University Press.

Roeper, P. (1997). Region-based topology. *Journal of Philosophical Logic*, 26: 251–309.

Rosenstock, S., Barrett, T.W. and Weatherall, J.O. (2015). On Einstein algebras and relativistic spacetimes. *Studies in History and Philosophy of Modern Physics*, 52: 309–316.

Russell, B. (1927). *The Analysis of Matter*. London: Kegan Paul.

Ryckman, T. (2004). *The Reign of Relativity: Philosophy in Physics, 1915–1925*. Oxford: Oxford University Press.

Rynasiewicz, R. (1992). Rings, holes and substantivalism: On the program of Leibniz algebras. *Philosophy of Science*, 59(4): 572–589.

Saraykar, R. and Janardhan, S. (2016). Zeeman-like topologies in special and general theory of relativity. *Journal of Modern Physics*, 7: 627–641.

Schild, A. (1967). Lectures on general relativity theory. In J. Ehlers (ed.), *Relativity Theory and Astrophysics*, Vol. 1. Providence: American Mathematical Society, pp. 1–104.

Schmidt, B.G. (1973). Conditions on a connection to be a metric connection. *Communications in Mathematical Physics*, 29(1): 55–59.

Sen, R.N. (2010). *Causality, Measurement Theory and the Differentiable Structure of Space-Time*. Cambridge: Cambridge University Press.

Simons, P. (1987). *Parts: A Study in Ontology*. Oxford: Oxford University Press.

Sklar, L. (1977). Facts, conventions, and assumptions in the theory of space-time. In J. Earman, C. Glymour and J. Stachel (eds.), *Foundations of Space-Time Theories*, Vol. VIII of *Minnesota Studies in the Philosophy of Science*. Minneapolis: University of Minnesota Press, pp. 206–274.

Sładkowski, J. (2001). Gravity on exotic $\mathbb{R}^4$ with few symmetries. *International Journal of Modern Physics D*, 10: 311–313.

Stachel, J. (1999). On the interpretation of the Einstein-Cartan formalism. In A. Harvey (ed.), *On Einstein's Path*. New York: Springer, pp. 475–485.

Székely, G. (2015). What properties of numbers are needed to model accelerated observers in relativity? In J.-Y. Beziau, D. Krause and J.B. Arenhart (eds.), *Conceptual Clarifications: Tributes to Patrick Suppes (1922–2014)*. London: College Publications, pp. 161–174.

Székely, G. and Madarász, J.X. (2013). Special relativity over the field of rational numbers. *International Journal of Theoretical Physics*, 52(5): 1706–1718.

Tennison, B.R. (1975). *Sheaf Theory*. Cambridge: Cambridge University Press.

Torretti, R. (1996). *Relativity and Geometry*, corrected edn. New York: Dover.

Trautman, A. (1976). A classification of space-time structures. *Reports on Mathematical Physics*, 10(3): 297–310.

Trautman, A. (2006). Einstein-Cartan theory. In J.-P. Françoise, G.L. Naber and S.T. Tsou (eds.), *Encyclopedia of Mathematical Physics*, Vol. 2. Oxford: Elsevier, pp. 189–195.

Trautman, A. (2012). Editorial note to: J. Ehlers, F.A.E. Pirani and A. Schild, The geometry of free fall and light propagation. *General Relativity and Gravitation*, 44(6): 1581–1586.

Wald, R.M. (1984). *General Relativity*. Chicago: University of Chicago Press.

Weatherall, J.O. (2017). Categories and the foundations of classical field theories. In E. Landry (ed.), *Categories for the Working Philosopher*. Oxford: Oxford University Press, pp. 329–348.

Weatherall, J.O. (2018). Regarding the 'hole argument'. *The British Journal for the Philosophy of Science* 69(2): 329–350.

Weyl, H. (1921). Zur Infinitesimalgeometrie: Einordnung der projektiven und der konformen Auffassung. *Nachrichten der Königlichen Gesellschaft der Wissenschaften zu Göttingen*, 1921: 99–112.

Whitehead, A.N. (1919). *An Enquiry Concerning the Principles of Natural Knowledge*. Cambridge: Cambridge University Press.

Whitehead, A.N. (1920). *The Concept of Nature*. Cambridge: Cambridge University Press.

Whitehead, A.N. (1929). *Process and Reality. An Essay in Cosmology*. New York: Free Press.

Whitney, H. (1936). Differentiable manifolds. *Annals of Mathematics*, 37(3): 645–680.

Zeeman, E.C. (1964). Causality implies the Lorentz group. *Journal of Mathematical Physics*, 5(4): 490–493.

Further Reading

Other reviews of some of the issues discussed here include Penrose, R. (1968), Structure of space-time, in C. M. DeWitt and J. A. Wheeler, eds., 'Batelle Rencontres: 1967 Lectures in Mathematics and Physics', W. A. Benjamin, New York, pp. 121–235; Ehlers, J (1973), The nature and structure of spacetime, in J. Mehra, ed., 'The Physicist's Conception of Nature', Kluwer, Dordrecht, pp. 71–91; Trautman, A. (1976), 'A classification of space-time structures', Reports on Mathematical Physics 10(3), 297–310; and chapter 6.1 of Torretti, R. (1996), Relativity and Geometry, corrected edition, Dover, New York. In addition, Geroch, R. and Horowitz, G. T. (1979), Global structure of spacetimes, in S. W. Hawking and W. Israel, eds., 'General Relativity: An Einstein

Centenary Survey', Cambridge University Press, Cambridge, pp. 212–293, is an especially accessible discussion of some aspects of global spacetime structure. Excellent general references for the mathematics of general relativity and its structure include Malament, D. B. (2012), Topics in the Foundations of General Relativity and Newtonian Gravitation Theory, University of Chicago Press, Chicago, and Hawking, S. W. and Ellis, G. F. R. (1973), The large scale structure of space-time, Cambridge University Press, Cambridge.

PART IV

Non-Relativistic Quantum Theory

Introduction to Part IV

Quantum theory is a framework for physical theorizing, a system for building theories rather than a complete dynamical theory in its own right, but it has been so successful that the vast majority of specific theories being actively explored in contemporary theoretical physics are quantum theories. Quantum theory was initially developed over the first four decades of the 20th century in the context of non-relativistic particle mechanics, and was extended to the context of relativistic field theories beginning in the late 1920s. Part V of this volume addresses the latter developments under the heading of quantum field theory. The notorious foundational puzzles posed by quantum theory can however already be understood in the non-relativistic setting.

The term "quantum theory" itself is contested. Proponents of Everettian quantum theory, for example, present themselves as taking quantum theory literally as a direct description of physical reality. Others maintain that quantum theory is a bare formalism that requires further interpretation, setting up the familiar dialectic where rival interpretations proliferate and present competing claims to be simultaneously empirically adequate and more intuitive (or more minimal, or otherwise philosophically more virtuous) than their rivals. This parade of alternative interpretations tends to obscure the clear sense in which Bohmian mechanics is a distinct physical theory from the dynamical collapse theory of Ghirardi, Rimini, and Weber, and in which both of these are distinct from Everettian quantum mechanics. These theories say straightforwardly different things about what the physical world is like. What the different approaches to quantum theory have in common is what Wallace (2012) refers to as the *quantum algorithm:* a recipe for assigning a quantum state to a system and using it to extract predictions about the behavior of the system. Where the approaches differ is in their explanation of how the quantum algorithm works and why it yields such empirically successful results.

Philosophy of quantum theory has a tumultuous history. Its early pioneers, in particular Bohr and Heisenberg, were unsettled by the unfamiliarity of the new theory and cast around for new philosophies with which to understand it. In the heady philosophical times of 1920s and 1930s Europe, attitudes to foundational questions were influenced both by the philosophy of logical positivism and by the intellectual temperaments of strong characters within the discipline. An approach known as the Copenhagen interpretation, championed by Bohr and Heisenberg but drawing also on Born's ideas, came to predominate; this approach still has a deep influence in the way physicists think about quantum theory. The Copenhagen interpretation denies any reality to the quantum state, and regards it instead as an encoding of what predictions are possible about measurement outcomes, where the latter are described in terms taken entirely from classical physics. Quantum theory, according to proponents of the Copenhagen approach, forces upon us a new way of understanding science that moves beyond the straightforward realism suggested by pre-quantum physics. The role of physics in directly representing the world is at least partly replaced with the role of determining what can and cannot be known (or even said).

James T. Cushing's *Quantum Mechanics: Historical Contingency and the Copenhagen Hegemony* (1994) and Mara Beller's *Quantum Dialogue: The Making of a Revolution* (1999) taken together provide a textured history of 20th-century quantum theory, each emphasizing the historical contingency of how foundations of quantum theory achieved a kind of consensus around the Copenhagen approach. A major theme is that these contingent factors may explain the sidelining of the realist approach suggested initially by de Broglie in 1927 and rediscovered by Bohm in 1952. This approach, known variously as the pilot-wave theory, de Broglie-Bohm theory, and Bohmian mechanics, can recapture all of the predictions of non-relativistic *n*-particle quantum theory. This apparent empirical equivalence did not play favorably for the theory within the foundational community; Heisenberg captured a widespread sentiment when he called the Bohmian potential an "ideological superstructure" (1958). The approach has since been rehabilitated and now attracts widespread support within philosophy of physics, although it is still largely ignored within theoretical physics. It is the subject of Roderich Tumulka's chapter in this part.

A key development that eventually led to a resurgence of philosophical work on quantum theory was the development of a variety of no-go theorems, including especially Bell's theorem in 1964, which can be formulated in an interpretation-neutral way. These theorems rule out certain classes of *hidden variable* or *psi-epistemic* theories – in particular, those theories where quantum probabilities are always a function of our uncertainty of some underlying *causally local* sub-quantum physical process. (Bohmian theories are hidden variable theories, but they incorporate a straightforward causal non-locality and hence are not ruled out.) Bell showed that quantum theory predicts correlations between spacelike-separated measurement events which cannot be explained in terms of pre-measurement individual states of spacelike-separated entangled systems. These striking correlations, later empirically confirmed in a series of groundbreaking contributions to "experimental metaphysics", place imposing constraints on any candidate realist approach that aspires to recover the predictions of quantum theory. In his contribution, Eddy Keming Chen presents a simple introduction to the Bell argument and the various proposals that have been put forward for avoiding the charge of non-locality.

For several decades in the 20th century, resistance to the Copenhagen approach organized itself around the *quantum measurement problem*. The quantum algorithm involves making a distinction between system and measuring device, and the Copenhagen approach elevates this distinction into one between the macroscopic world (classical, knowable) and the microscopic world (quantum, only partially knowable) – yet there is no objectively correct place to draw the line, no sharp line where quantum gives way to classical. The different interpretations are then represented as ways of "solving the measurement problem". This way of thinking about the task of philosophy of quantum theory has, however, been rendered largely obsolete by advances in our understanding of the process of *decoherence*, which provides us with an understanding of the measurement process from within the framework of quantum theory itself. Decoherence theory by itself doesn't answer any questions about the underlying metaphysics, and hence it requires supplementation to yield a complete explanation of the success of the quantum algorithm. However, it has yielded a wide range of models of interactions between individual quantum systems and various environmental degrees of freedom, and these models display the striking feature that the quantum state of the combined system rapidly becomes sharply peaked on histories corresponding to approximately classical behavior. This environment-induced suppression of quantum coherence is by now so well-understood that any plausible approach to quantum theory needs to incorporate it instead of offering a separate *sui generis* account of the quantum-classical divide. The chapter by Elise Crull surveys some different models of decoherence, and provides an introduction to the fertile new philosophical territory that progress in decoherence theory has opened up within philosophy of quantum theory.

Decoherence undermines much of the explanatory role that a substantive metaphysics of the quantum-classical divide would play, and it shows us that the process of measurement can itself be modeled quantum-mechanically; this tends to support approaches to quantum theory that leave it mostly unchanged and that broaden its scope to a wider range of systems. The approach originally suggested by Everett in 1957 takes this broadening of scope to the limit, hypothesizing that the evolution of the quantum state is entirely unitary and that quantum theory applies universally. At times derided as absurd, Everettian approaches to quantum theory have thrived and diversified in recent decades and they now attract widespread support within the foundational community (especially within the domain of quantum cosmology where Everettian quantum theory is typically presupposed). The two chapters by Simon Saunders present contemporary Everettian quantum theory – dividing the material into an account of the ontology of the Everettian multiverse and an account of the way in which quantum probabilistic phenomena are recovered (or emulated) within the Everettian worldview.

While Everettian quantum theory involves adding no ideological superstructure to quantum theory, it involves a great proliferation of quantum-mechanical histories – every physically possible macroscopic outcome of a quantum process is realized – and ongoing work in the foundations of quantum theory seeks well-defined and relativistically invariant alternative theories with a less inclusive ontology. The two most prominent such programs in the contemporary debate are dynamical collapse theories and Bohmian mechanics.

Dynamical collapse theories, as surveyed in the chapter by Peter Lewis, are a varied bunch; some, though not all, involve an ontology of local beables. What unites this family of theories is that they promote the effective collapse of the state, a central part of the quantum algorithm, to a real objectively probabilistic physical process of collapse that is distinct from the unitary evolution of the state. Dynamical collapse theories in general make predictions that deviate from those of quantum theory, but these differences are typically limited to domains that are inaccessible to current experimental techniques. The central difficulty faced by dynamical collapse theories is the threat of inconsistency with special relativity; the most up-to-date dynamical collapse models, however, aspire to Lorentz invariance.

Bohmian mechanics, in the contemporary version that is presented in Roderich Tumulka's chapter, is a generalization of the de Broglie-Bohm theory that offers a clear one-world ontology (typically made up of particles but sometimes of field values). Most versions of Bohmian mechanics are deterministic, although some incorporate stochastic elements. In them the quantum state is reimagined as a "quantum potential", as a physical field in a high-dimensional space, or alternatively as a highly complex law of physics; in either case, the quantum state is responsible for explaining the behavior of the entities which are taken to compose physical reality. The overall result is a theory which, although it involves non-local activity mediated by the action of the quantum potential, has a straightforward fundamental ontology of *local beables* – physical quantities, localized to regions of physical space, which unambiguously possess definite values.

12

BELL'S THEOREM, QUANTUM PROBABILITIES, AND SUPERDETERMINISM

Eddy Keming Chen

12.1 Introduction

As early as the beginning of quantum mechanics, there have been numerous attempts to prove impossibility results or "no-go" theorems about quantum mechanics. They aim to show that certain plausible assumptions about the world are impossible to maintain given the predictions of quantum mechanics, which can and have been empirically confirmed. Some of them are more significant than others. Arguably, the most significant is J.S. Bell's (1964) celebrated theorem of non-locality: given plausible assumptions, Bell shows that, in our world, events that are arbitrarily far apart can instantaneously influence each other.

Bell's theorem is most significant because its conclusion is so striking and its assumptions are so innocuous that it requires us to radically change how we think about the world (and not just about quantum theory).

Before Bell's theorem, the picture we have about the world is like this: physical things interact only locally in space. For example, a bomb dropped on the surface of Mars will produce immediate physical effects (chemical reactions, turbulences, and radiations) in the immediate surroundings; the event will have (much milder) physical effects on Earth only at a later time, via certain intermediate transmission between Mars and the Earth. More generally, we expect the world to work in a local way such that events arbitrarily far apart in space cannot instantaneously influence one another. This picture is baked into classical theories of physics such as Maxwellian electrodynamics and (apparently) in relativistic spacetime theories.

After Bell's theorem, that picture is untenable. Bell proves that Nature is non-local if certain predictions of quantum mechanics are correct. Many experimental tests (starting with Aspect et al. (1982a) and Aspect et al. (1982b)) have been performed. They confirm over and over again the predictions of quantum mechanics. Hence, we should have extremely high confidence in the conclusion that Nature is non-local: events that are arbitrarily far apart in space can instantaneously influence each other. (In the relativistic setting, it amounts to the conclusion that events that are space-like separated can influence each other.)

However, not everyone is convinced. In fact, there are still disagreements about what Bell proved and how general the result is. Some disagreements can be traced to misunderstandings about the assumptions in the proof. Others may be due to more general issues about scientific explanations and the standards of theory choice.

There are many good articles and books about Bell's theorem. (For example, see Maudlin (2011, 2014), Goldstein et al. (2011), and Myrvold and Shimony (2019).) In this short chapter, I would like to focus on two strategies that attempt to avoid the conclusion of non-locality. They are about (1) quantum probabilities and (2) super-determinism, both having to do, in some ways, with the philosophy of probability. First, I argue that solving the problem by changing the axioms of classical probability theory is a non-starter, as Bell's theorem only uses frequencies and proportions that obey the rules of arithmetic. Moreover, this point is independent of any interpretation of probability (such as frequentism). Second, I argue that a super-deterministic theory may end up requiring an extremely complex initial condition, one that deserves a much lower prior probability than its non-local competitors. Since both issues can be appreciated without much technical background and have implications for other subfields of philosophy, I will try to present them in a non-technical way that is accessible to non-specialists. The lessons we learn from them also apply to the more recently proven theorem (2012) of Pusey, Barrett, and Rudolph about the reality of the quantum state, which is in the same spirit as Bell's theorem. (Their theorem says that, under plausible assumptions, quantum states represent states of reality rather than merely certain knowledge about reality.)

12.2   Bell's Theorem

There are many versions of Bell's theorem and Bell inequalities. For illustration, in this section, we discuss a version of them by adapting a simple example involving perfect correlations discussed in Maudlin (2011, §1). (Another simple example, involving perfect anti-correlations, can be found in Albert (1992, §3).)

Under certain physical conditions, the calcium atom can emit a pair of photons that travel in opposite directions: left and right. We have laboratories that can realize such conditions. In this situation, we can set up polarizers on the left and on the right, as well as devices on both sides that detect photons that happen to pass through the polarizers. If a photon is absorbed by a polarizer, then the photon detector placed behind the polarizer will detect nothing. (Here, we assume that the photon detectors are 100% reliable. The idealization can be relaxed, and analyses have been carried out to show that the differences do not change the conclusion we want to draw.) Further, we can arrange the polarizers to be pointing in any direction on a particular plane. Each direction is representable by a number between 0 and 180, corresponding to the clockwise angle of the polarizer away from the vertical direction. Since either polarizer receives exactly one incoming photon, we say that the pair of photons agree if they either both passed or both got absorbed by the polarizers (so the photon detectors on both sides clicked or neither did); they disagree if one passed but the other got absorbed (so exactly one photon detector clicked).

When we carry out the experiments, say, by using 100,000 pairs of photons, quantum mechanics predict that we would observe the following:

- Prediction 1: If the left polarizer and the right polarizer point in the same direction, 100% of the pairs agree.
- Prediction 2: If the left polarizer and the right polarizer differ in direction by 30 degrees, 25% of the pairs disagree.
- Prediction 3: If the left polarizer and the right polarizer differ in direction by 60 degrees, 75% of the pairs disagree.

(The situation is a bit simplified. In actual experiments, the empirical frequencies will be approximately 25% and approximately 75% respectively and will increasingly approach them as we carry out more trials.) In the end, these statistics will be shown to clash with a plausible hypothesis of locality:

Locality Events arbitrarily far away cannot instantaneously influence each other.

Bell shows that the conjunction of Locality and the predictions of quantum mechanics leads to a contradiction. There are two parts in Bell's argument. The first part is based on the argument of Einstein et al. (1935), which is also known as the EPR argument.

In the context of our example, the EPR argument can be summarized as follows. First, the photon traveling to the left and the photon traveling to the right can be separated arbitrarily far away. Second, we can always place a polarizer in the path of the photon on the left and another in the path of the photon on the right. Third, according to Prediction 1, if the two polarizers point in the same direction, the pair of photons always agree, however far away they are from each other. Moreover, if we first measure the photon on the left and find that it passed the polarizer on the left, then we do not even need to measure the photon on the right if the polarizer on the right points in the same direction; we know the result—it will pass the polarizer on the right. Assume Locality: what happens to the photon on the (distant) left cannot instantaneously influence the photon on the (distant) right. So there is already a fact of the matter, before measurement, about the result on the right. Hence, Locality implies that there are facts of the matter about the polarization direction of the photon on the left and the photon on the right. In other words, their values of polarizations are predetermined.

Here is another way to see this. Given Prediction 1, since there is no way to "know" the directions of the two polarizers, the photons must already agree, even inside the calcium atom, how they would react to the polarizers come what may. That is, they must already agree whether to both pass or both get absorbed for polarizers pointing to any particular angle. For example, they must "agree" how to react when facing polarizers pointing at 0 degree, when facing polarizers pointing at 30 degrees, when facing polarizers pointing at 60 degrees, and so on. Otherwise, they would not be able to satisfy Prediction 1. However, such predetermined facts are not included in the quantum mechanical description using a wavefunction. So somehow these facts will be encoded in further parameters going beyond quantum theory. Indeed, the EPR argument aims to show that Locality implies that quantum mechanics is an incomplete description of Nature. (A famous example of a theory that adds additional parameters is the de Broglie-Bohm theory, but it is manifestly non-local in the particle dynamics. Therefore, it is not an example of the kind of local completion of quantum mechanics that EPR look for. Nevertheless, the non-local character of the de Broglie-Bohm theory was one of the motivations for Bell to investigate the generality of non-locality. See Bell (1964, §1) and Bell (1971).)

In short, what was shown by EPR and used in Part I of Bell's argument is the following:

Part I   Locality and Quantum Predictions $\implies$ Predetermined Values

In Part II, Bell shows the following:

Part II   Predetermined Values and Quantum Predictions $\implies$ Contradiction

We will see that predetermined values and quantum predictions lead to a contradiction with the laws of arithmetic (regarding addition, multiplication, and fraction). Recall that there are facts of the matter about the polarization properties of the pair of photons. But there are still two possibilities for each angle. For example, for polarizers pointing at 30 degrees, there can be two alternatives: both pass and both get absorbed. To simplify the example, we assume that the directions of the polarizers have only three choices (say, limited by the turning knobs on the devices): 0 degree, 30 degrees, and 60 degrees. Then for each choice of the angle of polarizer, there can be two possibilities for the pair: both pass (P) or both get absorbed (A). For example, they may both instantiate P_{30}, which means they will both pass if the polarizer is pointing at an angle of 30 degrees; they may both instantiate A_{60}, which means they will both get absorbed if the polarizer is pointing at an angle of 60 degrees. Since $2^3 = 8$, there are exactly eight choices for the assignments of properties in the two photons.

Eight Possible Assignments of Properties

	Left Photon	Right Photon	Feature	Percentage (%)
(1)	P_0, P_{30}, P_{60}	P_0, P_{30}, P_{60}	X	α
(2)	A_0, A_{30}, A_{60}	A_0, A_{30}, A_{60}		
(3)	A_0, P_{30}, P_{60}	A_0, P_{30}, P_{60}	Y	β
(4)	P_0, A_{30}, A_{60}	P_0, A_{30}, A_{60}		
(5)	P_0, A_{30}, P_{60}	P_0, A_{30}, P_{60}	Z	γ
(6)	A_0, P_{30}, A_{60}	A_0, P_{30}, A_{60}		
(7)	P_0, P_{30}, A_{60}	P_0, P_{30}, A_{60}	W	δ
(8)	A_0, A_{30}, P_{60}	A_0, A_{30}, P_{60}		

To satisfy Prediction 1, different pairs of photons can choose exactly one of these eight assignments. If a pair does not choose among these eight, then it can violate experimental results.

The eight assignments can be put in four groups as indicated in the table. Let us label the four groups with features X, Y, Z, and W, which we mention again in Section 12.3. Now suppose we have a large number of pairs of photons emitted from a collection of calcium atoms. (The larger the number, the closer empirical frequencies will approach the predicted percentages.) Assuming Locality, each pair must adopt one of the eight assignments listed above. Let α be the percentage of pairs that realizes either (1) or (2), β be the percentage of pairs that realizes either (3) or (4), γ be the percentage of pairs that realizes either (5) or (6), and δ be the percentage of pairs that realizes either (7) or (8). By the laws of arithmetic,

$$\alpha + \beta + \gamma + \delta = 100. \tag{12.1}$$

Moreover, each percentage number must be non-negative. In particular,

$$\gamma \geq 0. \tag{12.2}$$

Therefore,

$$\gamma + \delta + \beta + \gamma \geq \beta + \delta. \tag{12.3}$$

Unfortunately, this is inconsistent with the conjunction of Predictions 2 and 3. According to Prediction 2, if the angles of the polarizers on the two sides differ by 30 degrees, then we find photon disagreement 25% of the time. We run the large number of pairs of photons with the left polarizer pointing to 0 and the right pointing to 30. By inspection of the table, we know that pairs realizing assignments (1) and (2) will agree. Therefore, we know that α percent of the pairs agree. Moreover, we know that pairs realizing assignments (7) and (8) will also agree. That is another δ percent of pairs that agree. The only pairs that disagree will be those realizing assignments (3), (4), (5), and (6). That is $\beta + \gamma$ percent pairs that disagree. Hence,

$$\beta + \gamma = 25. \tag{12.4}$$

Similar considerations apply when we set the left polarizer at 30 degrees and the right at 60 degrees. Then,

$$\gamma + \delta = 25. \tag{12.5}$$

According to Prediction 3, if the the angles of the left and the right polarizers differ by 60 degrees, in our example that is when one is pointing at 0 and the other 60, then pairs of photons disagree 75% of the time. All disagreements come from photon pairs that realize assignments (3), (4), (7), and (8). Hence,

$$\beta + \delta = 75. \tag{12.6}$$

From the above three equations, since 50 is smaller than 75, we can conclude that

$$\gamma + \delta + \beta + \gamma < \beta + \delta. \tag{12.7}$$

But equation (7) is inconsistent with equation (3). We have arrived at a contradiction. Hence, the second part of Bell's argument is established. Together, Part I and Part II imply:

Locality and Quantum Predictions $\Longrightarrow$ Contradiction

Since quantum predictions have been confirmed to an extremely high degree, we should have very high confidence that Locality is refuted and that Nature is non-local. (Here, we take quantum predictions to be statistical—regarding empirical frequencies—rather than probabilistic.) Of course, we have made some implicit assumptions in the derivation:

(A)   The rules of inferences obey classical logic.
(B)   The laws of arithmetic are true.
(C)   Frequencies and proportions obey the laws of arithmetic.
(D)   There are no conspiracies in nature.

Strictly speaking, it is only by assuming (A)–(D) can we derive the contradiction from Locality and Quantum Predictions. We will return to these implicit assumptions in the next two sections. (Another assumption is the idea that each experimental outcome is unique and definite, which is denied in the Many-Worlds interpretation. See Section 12.7 for further readings. One can label this as the fifth assumption. However, this assumption is arguably already contained in our description of Quantum Predictions about empirical frequencies. If experimental outcomes are not definite, empirical frequencies wouldn't even make sense unless we state them in a different way, such as by pairing certain outcomes into a single branch and using "branch-weighted" frequencies.)

In this section, we have presented one version of Bell inequalities (in equation (3)) and explained how it is violated by the predictions of quantum mechanics (in equation (7)). (Bell's own version (1964) uses perfect anti-correlation and is stated in terms of expectation values. Clauser et al. (1969) provide a generalization of Bell's result that allows imperfect correlations.)

12.3   Quantum Probabilities to the Rescue?

Perhaps due to the significance of Bell's theorem, there have been many attempts that try to avoid the conclusion of non-locality by identifying some other "weak link" in the argument. (For some examples, see Further Readings in Section 12.7.) That is surprising, since the other assumptions are quite innocuous and *a priori*, as illustrated by the previous example.

One purported "weak link" is associated with the "implicit assumptions" about classical probability theory. One might suspect that the derivations of Bell's theorem require substantive assumptions about the nature of probability. Probability is notoriously difficult to understand. Hence, there may be room to revise our classical theory of probability given empirical data. The suggestion is that, instead of rejecting Locality, we can modify (or generalize) the classical axioms and algebraic structure of Kolmogorov probability theory to avoid the contradiction. (For example, see Fine (1982a), Fine (1982b), and Pitowsky (1989).)

However, the previous example serves as a counterexample. In the argument of Section 12.2, assumptions of classical probability theory do not even occur. Nor do they implicitly play any essential role. All we ever needed were proportions and how they arithmetically interact with each other (addition, multiplication, subtraction, and division). For example, Predictions 1, 2, and 3 are formulated in terms of percentages of pairs of photons. The four groups of possible assignment of properties receive percentages α, β, γ, and δ. We call them "percentages," which may remind readers of probabilities. But in our argument they merely represent proportions. To say that α percent of the pairs realize property assignments (1) or (2) is to say that the number of pairs having those properties is exactly α per 100 pairs. If we have 100,000 pairs in total in the collection, then that amounts to $1,000 \times \alpha$ pairs.

Since the percentages α, β, γ, and δ represent proportions, it is in their nature that they obey the laws of arithmetic, and their bearers (property assignments (1)–(8)) obey the rules of Boolean algebra. (Tumulka (2016) makes a similar point.) The fact that we are assuming, in the conditional proof, they have hidden properties does not matter at all. As such, proportions obey the axioms governing how we should count a finite number of things, which obey the Kolmogorovian axioms, which may also govern probabilities (according to some interpretations of probability). Nevertheless, that does not make proportions subject to various interpretational issues as probability does. Many other concepts also satisfy Kolmogorovian axioms, including mass, length, and volume of finite physical objects. Neither are they subject to the interpretational controversies surrounding the concept of probabilities. Probability faces a wide range of interpretational puzzles, and it is controversial what its axioms ought to be. Still, there are no similar difficulties with concepts of mass, length, volume, frequencies, or proportions.

Why is it in the nature of frequencies and proportions to obey the laws of arithmetic or counting finite number of things? This may seem like a question in the philosophy of mathematics. Fortunately, we do not need to settle those controversies to answer that question for our purposes here. The discussion about non-classical probability spaces and Bell's theorem is sometimes highly technical, and different proposals have been suggested to understand violations of the rules of Boolean algebra and Kolmogorov axioms. For our purposes, we can distill the central intuitions using the concrete example of Section 12.2. Suppose we have a large collection of photon pairs adequately prepared. Consider four features that each photon pair can have—X, Y, Z, and W—that are mutually exclusive and jointly exhaustive, and consider the following propositions:

(i)   The percentage of photon pairs having exactly one of the four features is 100%.

(ii)   The percentage of photon pairs having feature Z is non-negative.

(iii)   The percentage of photon pairs having either Y or W is the sum of the percentage of photon pairs having Y and the percentage of photon pairs having W.

(iv)   The sum of the percentage of photon pairs having the property (Y or Z) and the percentage of photon pairs having the property (Z or W) is well defined—a non-negative number.

Can these propositions be false? In particular, can they fail in the following ways?

(i′)   The percentage of photon pairs having exactly one of the four features is 115%.

(ii′)   The percentage of photon pairs having feature Z is −5%.

(iii′) The percentage of photon pairs having either Y or W is less than the sum of the percentage of photon pairs having Y and the percentage of photon pairs having W.

(iv′) The sum of percentage of photon pairs having the property (Y or Z) and the percentage of photon pairs having the property (Z or W) does not exist.

It is *a priori* that propositions (i)–(iv) cannot be false while propositions (i′)–(iv′) cannot be true. Propositions such as (i)–(iv) are sufficient to prove the violation of a Bell inequality (equation (3)) in Section 12.2. They are not dependent on any substantive theory or axioms about probabilities, because they are about proportions and not about probabilities. We do not need to appeal to assumptions about the nature of probabilities to prove that Nature is non-local.

A potential misunderstanding is that, to say the thing we just said, we must be endorsing a particular interpretation of probability—frequentism, according to which probabilities boil down to long-run frequencies. But that is a mistake. We can make judgments about those eight propositions without endorsing any particular interpretation of probability. To evaluate them, we do not have to settle the debate among subjectivism, frequentism, and the propensity interpretations. For example, one can be a subjective Bayesian about probabilities and still accept that frequencies, percentages, and proportions obey propositions (i)–(iv). One can even adopt the view that the actual axioms governing real probabilities are non-Kolmogorovian and involving non-Boolean algebra without denying that frequencies and proportions obey the rules of arithmetic. (Moreover, the actual evidence we use to support quantum theory consists in empirical frequencies, which obviously obey the classical probability axioms.)

However, not everyone would agree with our assessment. Fine (1982a,1982b) and Pitowsky (1989) seem to suggest it remains possible to save locality by revising classical probability theory. (See Malament (2006) for a clear introduction to this project. Feintzeig (2015) demonstrates further mathematical constraints.) The project has led to important and beautiful mathematical results that can shed light on the mathematical structures of impossibility theorems. Nevertheless, if the above analysis is correct, then the project of avoiding non-locality by revising probability axioms is a non-starter; it cannot get off the ground, no matter how ingenious or elegant the models of non-classical probability spaces are. No matter what changes we make to classical probability theory, they do not affect the conclusion of non-locality. The argument for non-locality does not rely on classical probability theory. We only need to use rules for counting relative frequencies and proportions.

Quantum probability (as an alternative to classical probability) is related to quantum logic (as an alternative to classical logic). Some people who want to keep classical logic may nonetheless be open to revise the axioms of probability to make room for locality. But as we just discussed, it is the axioms governing frequencies and proportions that need to be revised if one goes that route. Since they obey the axioms of arithmetics, and since the latter are closely related to logic, it is hard to see how to pursue this route without also revising logic in some way. (See Wilce (2017) for a survey of quantum logic and quantum probability theory.)

Therefore, we cannot save Locality by changing the axioms governing classical probability theory. Which probability theory is correct is an important question in the philosophy of probability but it is irrelevant to the question whether Nature is non-local.

12.4   Escape with Super-Determinism?

Another purported "weak link" in Bell's argument is associated with the assumption of statistical independence. The strategy is to allow systematic violations of statistical independence in favor of "super-deterministic" theories. (This is sometimes labeled as "conspiratorial theories.") In this section, we will try to understand what the strategy is and what difficulties it faces.

In Section 12.2, we assumed that the direction of the polarizer can be set independently of the collection of incoming photon pairs. We can, for example, use a mechanical device that randomly selects (say, based on certain digits of π) among the three choices—pointing at 0 degree, 30 degrees, and 60 degrees. That assumption—statistical independence—seems fundamental to scientific experimentation. Another way to see it is in terms of random sampling. Given any collection of photon pairs adequately prepared, and after the experimental setup is completed, we can perform random sampling on the collection and obtain a sub-collection that reflects the same statistical profile as the overall collection and any other sub-collection so randomly chosen. That is, if the sub-collection is such that 25% of them would disagree when pairs of photons pass through polarization filters that differ by 30 degrees, then the whole collection (and other randomly chosen sub-collection) would also have that property. In other words, the choice of the sub-collections can be made statistically independent of the experimental setup. Statistical independence enables us to apply the conjunction of Predictions 1, 2, and 3 to the collection as a whole (and to each sub-collection) and to deduce equations (4), (5), and (6), from which we derive a contradiction with inequality (3).

Without assuming statistical independence, the inference is not valid. We can construct an example in which the quantum predictions are all satisfied during experiments but there is no contradiction. Suppose we have 100,000 photon pairs to start with. Each photon pair realizes one of the eight assignments listed in the table. Suppose further that $\alpha = \beta = \gamma = \delta = 25$. We have three experimental setups:

(A)   Left polarizer at 0 degree, right polarizer at 30 degrees.
(B)   Left polarizer at 30 degrees, right polarizer at 60 degrees.
(C)   Left polarizer at 0 degree, right polarizer at 60 degrees.

From the collection of 100,000 photon pairs, we choose three sub-collections—(a), (b), and (c)—each with exactly 100 photon pairs. It turns out that when we send (a) through (A), 25% of them disagree; when we send (b) through (B), 25% of them disagree; when we send (c) through C, 75% of them disagree. (As before, this is an idealization. The fractions get closer to these numbers when we run the trials with more pairs.) This can be realized in the following way. In (a), 25 pairs are of type (3) and the rest are of type (1); in (b), 25 pairs are of type (5), and the rest are of type (1); in (c), 75 pairs are of type (7) and the rest are of type (1). That is, sub-collection (a) has exactly the kind of statistical profile required to be in agreement with quantum predictions for experiment (A); sub-collection (b) for (B); and the sub-collection (c) for (C). Hence, each sub-collection has the "right" statistical profile matching the experimental setup it goes through, but none of them has the statistical profile required by the conjunction of the three predictions. Moreover, none of the sub-collections is statistically similar with any other sub-collection. Still, the outcomes of experiments are consistent with quantum predictions. The problem is that the sampling is not random. Somehow, the choice of which photon pairs to send to which experimental setup is correlated with the choice of the experimental setup itself. In this case, equations (4)–(6) do not hold for the entire collection or any particular sub-collection, and $\gamma + \delta + \beta + \gamma$ is larger than or equal to $\beta + \delta$ without contradicting quantum predictions. In this case, $100 \geq 50$; no contradictions exist between outcomes of actual experiments and the assumption of Locality.

Such a violation of statistical independence would seem to require some extraordinary conspiracies in Nature. Not only does this have to be true for these particular setups, which is incredible already, we need there to be similar conspiracies for every such experimental setup, done by anyone, anywhere, and anytime. No matter where, when, and who to carry out the experiment, the strategy requires that no matter what random sampling method we use, the photon pairs with the "right" statistical profile should always find themselves at the "right" experimental setup. The randomization can be done by a deterministic device that decides, based on the digits of π, which photon pair goes into which sub-collection. The randomization can also use other mundane methods, such as the rolling of dice, flipping of coin, and the English letters in Act V of *Hamlet*. No matter what randomization method is used in experiment, the superdeterministic theory will require violations of statistical independence in such a way that the sub-collections will be statistically dissimilar to each other, rendering equations (4)–(6) false of each sub-collection and the entire collection. Nature conspires to hide its locality from us.

Such extraordinary features may be difficult to achieve in any realistic physical theories. Are there any physical theories that can do this? I am not aware of any worked out theory at the moment. However, some initial steps have been taken to investigate possible dynamics and toy models of superdeterminism. 'T Hooft (2014) provides an illustration. Hossenfelder and Palmer (2020) provide an up-to-date overview and some philosophical discussions. (Friederich and Evans (2019) review some "retrocausal" models that use backward-in-time causal influences.)

Superdeterminism faces many objections. An important criticism focuses on the fact that endorsing violations of statistical independence would be bad for science in one way or another. After all, the assumption of statistical independence is integral to ordinary statistical inferences. Shimony et al. (1976) argue that rejecting statistical independence would undermine the scientific enterprise of discovery by experimentation:

> In any scientific experiment in which two or more variables are supposed to be randomly selected, one can always conjecture that some factor in the overlap of the backwards light cones has controlled the presumably random choices. But, we maintain, skepticism of this sort will essentially dismiss all results of scientific experimentation. Unless we proceed under the assumption that hidden conspiracies of this sort do not occur, we have abandoned in advance the whole enterprise of discovering the laws of nature by experimentation.

Similarly, Maudlin (2019) suggests that rejecting it would make it impossible to do science:

> If we fail to make this sort of statistical independence assumption, empirical science can no longer be done at all. For example, the observed strong robust correlation between mice being exposed to cigarette smoke and developing cancer in controlled experiments means nothing if the mice who are already predisposed to get cancer somehow always end up in the experimental rather than control group. But we would regard that hypothesis as crazy.

These objections based on scientific methodology seem quite compelling to many people.

Recently, Hossenfelder and Palmer (2020) argue that there are multiple mistakes in this type of criticism. One of the mistakes is

> the idea that we can infer from the observation that Statistical Independence is useful to understand the properties of classical systems, that it must also hold for quantum systems. This inference is clearly unjustified; the whole reason we are having this discussion is that classical physics is *not* sufficient to describe the systems we are considering. (emphasis original)

We may have justification for applying statistical independence to classical systems such as experimental setups involving mice and cigarette smoke. But it does not logically entail that we have justification for applying it to quantum systems of photons and electrons. (What kind of justification do they mean here? I think they mean both epistemic and pragmatic justifications but the text is ambiguous.)

Their response does not seem to address the worry about scientific methodology. Statistical independence is not the kind of principles we try to empirically justify. Rather, it is part of the inductive principles that we presuppose in order to do science. That is, statistical independence is a precondition for empirical investigation by experimentation. It is not clear what would be an experiment that confirms or disconfirms it, and we may need to assume statistical independence to draw conclusions from the very experiment itself. It may be impossible to empirically justify statistical independence, but that does not suggest there is a problem for applying it in the first place. This follows from a more general observation that even if we cannot empirically justify induction, we are justified in using induction to learn about the world (See Henderson (2020) on Hume's problem of induction.) Hence, their response does not seem to answer the objections of Shimony et al. (1976) and Maudlin (2019).

Nevertheless, their response raises an interesting possibility. It is certainly logically consistent for a defender of superdeterminism to maintain that while small microscopic systems (such as electrons and photons) violate statistical independence, large macroscopic systems (such as mice) do not violate it for all practical purposes. That is, we may have reasons to think that the violations of statistical independence may be suppressed when we reach the macroscopic level. Hence, it is logically consistent for one to claim that statistical independence is false about microscopic systems but for all practical purposes true of macroscopic systems. In short, in ordinary situations when we experiment with mice, we can still use statistical independence; but we should not assume statistical independence when experimenting with electrons and photons (and other microscopic systems).

That is of course logically consistent. But we may ask what reasons do we have for thinking that it is true in the superdeterministic theory? One might appeal to decoherence as the mechanism for suppressing certain quantum effects from manifesting in the macroscopic domain (for more on decoherence, see Crull's chapter in this volume). But decoherence does not fit naturally in a superdeterministic theory. For one thing, decoherence is primarily about the behaviors of quantum states (represented by wavefunctions). However, typically a superdeterministic theory (such as the type favored by Hossenfelder and Palmer (2020)) does not regard the quantum states to be objective and does not postulate quantum states in the fundamental ontology. Moreover, it is unclear how decoherence can suppress violations of statistical independence. Decoherence explains the dynamical features that certain "branches" of the wavefunction do not interfere much with each other. Although the possibility is interesting, there is much work to be done to demonstrate its plausibility in a superdeterministic framework.

I would like to raise a different worry about superdeterminism. We may worry that superdeterminism of this sort is unlikely to result in a simple fundamental theory. (Here, by "unlikely" I mean unlikely in the epistemic sense: unlikely given what we know so far and absent any explicit empirically adequate models that show otherwise.) The constraints on empirical frequencies are so severe that it is difficult to see how it can be written down in any simple formula. (See Kronz (1990) for a related argument. See Lewis (2006) for a discussion of Kronz's argument as well as a new "measurement problem" for superdeterminism.) In order for the local theory to be compatible with the predictions of quantum mechanics, it would have to radically constrain the state space of the local theory so that only a very small class of histories will be allowed. (Such a constraint can be a joint effect of some lawlike initial conditions and the dynamical laws.) Not all arrangements of the local parameters will be permitted—otherwise one cannot guarantee perfect agreement with quantum predictions. What kind of constraints? They will have to encode as much information as the setup and non-local correlations. For example, they would need to entail that an experiment conducted today using randomization method based on the digits of π will somehow still result in statistically dissimilar sub-collections in such a way that produce the desired outcomes of experiments conducted at arbitrarily far away locations. Similarly, it will be the case for randomization based on the letters of Act V of *Hamlet*, the Chinese characters in the *Analects*, or the hexagrams of *I Ching*. No matter what randomization method we choose, the superdeterministic mechanism must ensure that the chosen sub-collection is somehow just the right one for a particular experimental setup. Since the randomization methods seem to have nothing in common, it is difficult to see how the constraints on initial conditions and dynamics can be simple at all. These give us reasons to think that they will be quite complicated.

A defender of superdeterminism may reply that there is a simple formula: just write down the usual Born rule of quantum mechanics and demand that the superdeterministic theory more or less respects that. It is not clear how to state the Born rule as a simple law in terms of objects accepted on superdeterminism. As mentioned earlier, typically a superdeterministic theory (such as the type favored by Hossenfelder and Palmer (2020)) does not postulate quantum states (represented by wavefunctions) in the fundamental ontology. After all, a non-separable quantum state may lead to non-local dynamics. However, the Born rule is stated in terms of the quantum state. Respecting the Born-rule statistics (or something close to it) is certainly a nice goal when trying to construct a local superdeterministic theory with a well-defined ontology and dynamics. The goal is simple (respect the Born rule where it is valid), but it does not follow that the underlying theory will be simple.

Because of the lack of simplicity, the constraints we need to impose in a superdeterministic theory will not look lawlike. Hence, such a theory can be quite complex and difficult to compete with other candidate theories that are far simpler. For example, 'T Hooft (2014)'s Cellular Automaton Interpretation requires the selection of an initial state of the universe, which may be extremely detailed and not at all simple. Here I take simplicity as a hallmark of fundamental laws of nature. A superdeterministic theory will likely postulate an extremely complicated initial condition (or complicated dynamical laws) that looks nothing like a fundamental law.

Hence, this problem of superdeterminism boils down to a violation of a familiar constraint on fundamental laws of nature. A fundamental law should not be too complex. When we evaluate competing theories we are judging them (in part) by the relative complexities of the fundamental laws. Among competing observationally equivalent theories, the more complex a theory is the lower prior probability we should assign to it. This corresponds to an objective Bayesian way of thinking about probabilities. However, complexity and simplicity come in degrees. Now, simplicity and complexity are notoriously vague. But they are indispensable theoretical tools when we confront observationally equivalent theories. For our purpose here, one can plug in any reasonable notion of simplicity and complexity for evaluating scientific theories.

In fact, some good physical theories do constrain initial states in order to explain certain widespread regularities. For example, in a universe with wide-spread temporal asymmetries, we postulate a low-entropy initial condition. That is now called the Past Hypothesis (Albert (2000)). We ought to subject the Past Hypothesis to the constraint of simplicity because it is a candidate fundamental law. It is a candidate fundamental law because it underlies many nomological generalizations such as the Second Law of Thermodynamics and does not seem to be further explained by the dynamics. (This will no longer be true if Carroll and Chen (2004)'s model can successfully explain time's arrow.) Fortunately, we have reasons to think that the Past Hypothesis is not extremely complex. Indeed, it can be specified in terms of simple macroscopic variables (the values of the pressure, density, volume, and energy of the early universe). In certain frameworks, it can even be specified in simple microscopic variables, such as Penrose (1979)'s Weyl Curvature Hypothesis or Ashtekar and Gupt (2016)'s initial condition for Loop Quantum Cosmology. In the density-matrix-realist framework, the Past Hypothesis can be replaced by the Initial Projection Hypothesis (Chen (2018)) that pins down a unique quantum microstate of the universe. It is interesting that a simple postulate about the initial condition of the universe can explain the widespread temporal asymmetries. Part of the reason is due to the structure of state space: there is an asymmetry of macrostate volumes (or dimensions) that emerges as a result of simple dynamics; it is part of the answer to the problem of time's arrow. Moreover, the Past Hypothesis explanation is perfectly compatible with statistical independence.

We have good reasons to think that superdeterministic theories, in contrast, will postulate something much more complicated than the Past Hypothesis as an initial condition. If such a superdeterministic theory is devised, we should also interpret its initial condition as a fundamental law of nature. (At the very least, it should be given a fundamental axiomatic status in the theory since it is not derived from other laws of the theory.) The widespread violations of Bell-type inequalities cry out for explanations. In such a superdeterministic theory, the initial condition is supposed to do the work of explaining why arbitrarily far away events are correlated with each other. We see no reason at all why such a theory (and especially its constraint on the state space) will be simple enough. At least we do not have any evidence that it will be simpler than the competing non-superdeterministic and non-local theories that are already on the market, such as Bohmian mechanics and GRW theory (see the survey chapters by Tumulka and Lewis in this volume).

Hence, there are significant differences between the superdeterministic theory that constrains its initial states to explain Bell-type correlations and a regular quantum theory that constrains its initial states (by the Past Hypothesis) to explain temporal asymmetries. However, these are differences in degrees and not of kind. If a superdeterministic theory aims to recover all quantum predictions, then it would be observationally equivalent to Bohmian mechanics and more or less equivalent to some versions of GRW theory. But we have good reasons to think that Bohmian mechanics and GRW theory are far simpler than the superdeterministic theory. Hence, the superdeterministic theory should receive much lower prior probability than either Bohmian mechanics or GRW theory.

Nevertheless, that does not mean we should assign zero credence to superdeterminism. Instead, I think we should follow Bell (1977) and be open-minded in a qualified way:

> Of course it might be that these reasonable ideas about physical randomizers are just wrong – for the purpose at hand. A theory may appear in which such conspiracies inevitably occur, and *these conspiracies may then seem more digestible than the non-localities of other theories.* When that theory is announced I will not refuse to listen, either on methodological or other grounds. (my emphasis)

If one constructs an empirically adequate superdeterministic theory that is simpler than a non-local theory such as Bohmian mechanics or GRW theory, we should be open to assign much higher credence in it. At the moment, no such theory is available.

12.5 Conclusion

In this short survey chapter, I introduced Bell's theorem by discussing a simple example. I focused on two strategies that attempt to avoid the conclusion of non-locality: (1) changing the axioms of classical probability theory and (2) embracing superdeterminism and allowing systematic violations of statistical independence. Both have to do in some way with the philosophy of probability. Neither seems promising. Nevertheless, understanding these ideas can help us come to a deeper understanding of Bell's theorem, its significance, and the relevance (or irrelevance) of the nature of probability.

12.6 Note Added

The chapter was completed in July 2020. Since then, an admirably clear article written by G.S. Ciepielewski, E. Okon, and D. Sudarsky (2020) has been posted on arXiv. Here, I comment on some of its features that are relevant to the point I made in Section 12.4. They present a superdeterministic model that exactly reproduces the quantum predictions with a set of strikingly simple dynamical laws and initial condition laws. The theory simulates the whole universe locally, by adding a copy of the universe (an internal space) at each point in physical space, and by stipulating a "pre-established harmony" that at t_0 the copies of the universe look exactly the same at different points in physical space. How things move in physical space derive from how things move in the internal spaces. In the internal spaces, things move according to Bohmian dynamics. Since each internal space occupies only a point in physical space, the fundamental dynamics is local from the perspective of physical space. As the authors acknowledge, their model shares features with Leibniz's *Monadology*. Hence, I think it is appropriate to call it *Leibnizian quantum mechanics* (LQM).

(More precisely, the idea of LQM is to (1) take a Bohmian universe of N particles moving in physical space represented by $\mathbb{R}^3$, (2) at each point x in physical space add an internal "configuration space" of $\mathbb{R}^{3N}$, (3) replace the universal wavefunction in Bohmian mechanics with a continuous infinity of wavefunctions, each defined separately in an internal space and each obeying the Schrödinger equation, (4) remove the N particles in physical space from the fundamental ontology, (5) in each internal space add a point representing the actual configuration in the internal space, whose history depends on the wavefunction defined in the internal space via a guidance equation, (6) specify a mass density in physical space from the configurations in the internal spaces, and (7) define a simple set of initial condition laws for the "pre-established harmony": the initial configurations are the same in all internal spaces and the initial wavefunctions are the same in all internal spaces, which can be expressed by two simple differential equations.)

The authors themselves acknowledge that their model is not a serious competitor to realist non-local quantum theories such as standard Bohmian mechanics. Nevertheless, we may wonder what the principled grounds are for its rejection. First, suppose we understand each internal space of LQM as representing the configuration space of some internal three-dimensional space at a point in physical space. Then, "inside" each internal space, there are N particles moving according to the usual Bohmian dynamics. If Alice lives in a world described by LQM, there are infinitely many exact copies of her, of which all except one are made out of particles moving in internal spaces. Hence, if Alice's self-locating credence is not too biased toward the exceptional one (that she is made out of particles moving in physical space), she would reason that most likely she lives in an internal space, whose dynamics is non-local. Second, it may be unclear what the fundamental physical space of LQM is. One could stipulate that it is just the physical space. However, if one is sympathetic to Albert (1996)'s point that it is something to be inferred and not stipulated, one runs into a difficulty. In the standard Bohmian case, one can adopt Chen (2017)'s criterion and infer that the fundamental physical space is the three-dimensional physical space. But in LQM, Chen's criterion suggests it is the internal three-dimensional spaces that should be regarded as physically fundamental, since it is the smallest space that allows a natural definition of the "multi-field." Hence, on some conception, inside the physically fundamental spaces, the dynamics is still non-local.

Finally and most importantly, even though LQM does not require overly complex initial condition laws or dynamical laws, its ontological additions make it much more complex than standard versions of Bohmian mechanics and GRW theory. LQM requires the addition of infinitely many more "universes" in addition to the physical space (and to whatever other standard internal spaces we need to postulate). The strictly additional universes, though "small," are exact copies of the physical space including all of its minute details. Unlike the "emergent" universes in the Everettian many-worlds interpretation, the infinitely many universes in LQM have a fundamental status. However, there is no such need to enlarge the ontology on standard Bohmian mechanics or GRW. The only advantage of a superdeterministic theory is its locality, which could be an advantage for seeking a fully relativistic theory. Unfortunately, in the case of LQM, it is local but it still requires a preferred rest frame, which disqualifies it from being fully relativistic. Locality without relativity does not compensate for the increase in complexity. The complexity of LQM lies not in its "nomology" (laws) but in its ontology. Hence, LQM is not an empirically adequate theory that is overall simpler and more attractive than a non-local theory such as Bohmian mechanics or GRW theory. The final point of Section 12.4 still stands.

Nevertheless, LQM is a rare example of an explicit superdeterministic model of the universe that reproduces the exact predictions of quantum mechanics. It adds a great deal of clarity for understanding the relative costs and benefits of maintaining locality and rejecting statistical independence. One may try to avoid complicating the laws by complicating the ontology instead; but either way one has to complicate some part of the theory. That could well be a generic feature of any superdeterministic theory that attempts to avoid the charge of non-locality.

12.7 Further Readings

For discussions of the issue of "realism" in Bell's proof, see T. Norsen, "Against 'realism'", *Foundations of Physics*, 37(3): 311–340, 2007; T. Maudlin, "What Bell did", *Journal of Physics A: Mathematical and Theoretical*, 47(42): 424010, 2014; and R. Tumulka, "The assumptions of Bell's proof", in M. Bell and S. Gao (eds.), *Quantum Nonlocality and Reality: 50 Years of Bell's Theorem* (Cambridge University Press, 2016). For discussions of non-locality, superluminal signaling, and relativistic invariance, T. Maudlin's

Quantum Non-locality and Relativity: Metaphysical Intimations of Modern Physics (Wiley, 2011) is a landmark monograph on the topic; for collapse models that demonstrate the compatibility of Lorentz invariance and non-locality, see R. Tumulka, "A relativistic version of the Ghirardi-Rimini-Weber model", *Journal of Statistical Physics*, 125(4): 821–840, 2006; and D. Bedingham et al., "Matter density and relativistic models of wavefunction collapse", *Journal of Statistical Physics*, 154(1–2): 623–631, 2014. For discussions of locality and non-locality in the many-worlds interpretation of quantum mechanics, see D. Wallace, *The Emergent Multiverse: Quantum Theory according to the Everett Interpretation* (Oxford University Press, 2012); and V. Allori et al., "Many worlds and Schrödinger's first quantum theory", *British Journal for the Philosophy of Science*, 62(1): 1–27, 2010. For discussions of parameter independence and outcome independence, see J.P. Jarrett, "On the physical significance of the locality conditions in the Bell arguments", *Noûs*, 18(4) 569–589, 1984, A. Shimony, "Search for a worldview which can accommodate our knowledge of microphysics", in J.T. Cushing and E. McMullin (eds.), *Philosophical Consequences of Quantum Theory* (University of Notre Dame Press, 1989), 62–76; R. Healey, "Chasing quantum causes, how wild is the goose?" *Philosophical Topics*, 20(1): 181–204, 1992; and T. Maudlin's *Quantum Non-locality and Relativity: Metaphysical Intimations of Modern Physics*, Ch.4. For discussions of causation and causal explanations, see J.S. Bell, "Bertlmann's socks and the nature of reality," *Le Journal de Physique Colloques*, 42(C2): C2–C41, 1981; M.L. Redhead, "Nonfactorizability, stochastic causality, and passion-at-a-distance", in J.T. Cushing and E. McMullin (eds.), *Philosophical Consequences of Quantum Theory* (University of Notre Dame Press, 1989), 145–153; R. Healey, "Chasing quantum causes, how wild is the goose?", and T. Maudlin's *Quantum Non-locality and Relativity: Metaphysical Intimations of Modern Physics*, Ch.5. For a survey of experimental tests and certain loophole-free tests of Bell's inequalities, see §4–§5 of W. Myrvold and A. Shimony, "Bell's theorem", *Stanford Encyclopedia of Philosophy* (2019). For a discussion of related matters from a relativity-centered perspective, see Chapter 8 by W. Myrvold in this volume.

Acknowledgment

I am grateful for helpful discussions with Craig Callender, Sheldon Goldstein, Mario Hubert, and Isaac Wilhelm, as well as written comments from Alan Hájek, Sabine Hossenfelder, Kelvin McQueen, Peter Morgan, Travis Norsen, Elias Okon, Timothy Palmer, and Alastair Wilson.

References

Albert, D.Z. (1992). *Quantum Mechanics and Experience*. Cambridge: Harvard University Press.

Albert, D.Z. (1996). Elementary quantum metaphysics. In J.T. Cushing, A. Fine and S. Goldstein (eds.), *Bohmian Mechanics and Quantum Theory: An Appraisal*. Dordrecht: Kluwer Academic Publishers, pp. 277–284.

Albert, D.Z. (2000). *Time and Chance*. Cambridge: Harvard University Press.

Allori, V., Goldstein, S., Tumulka, R. and Zanghì, N. (2010). Many worlds and Schrödinger's first quantum theory. *British Journal for the Philosophy of Science*, 62(1): 1–27.

Ashtekar, A. and Gupt, B. (2016). Initial conditions for cosmological perturbations. *Classical and Quantum Gravity*, 34(3): 035004.

Aspect, A., Dalibard, J. and Roger, G. (1982a). Experimental test of Bell's inequalities using time-varying analyzers. *Physical Review Letters*, 49(25): 1804.

Aspect, A., Grangier, P. and Roger, G. (1982b). Experimental realization of Einstein-Podolsky-Rosen-Bohm gedankenexperiment: A new violation of Bell's inequalities. *Physical Review Letters*, 49(2): 91.

Bedingham, D., Dürr, D., Ghirardi, G., Goldstein, S., Tumulka, R. and Zanghì, N. (2014). Matter density and relativistic models of wave function collapse. *Journal of Statistical Physics*, 154(1-2): 623–631.

Bell, J. (1971). Introduction to the hidden-variable question. *Foundations of Quantum Mechanics*, Proceedings of the International School of Physics 'Enrico Fermi', course IL, New York, Academic, pp. 171–181.

Bell, J.S. (1964). On the Einstein-Rosen-Podolsky paradox. *Physics*, 1(3): 195–200.

Bell, J.S. (1977). Free variables and local causality. *Epistemological Letters*, 17.3(15): 79–84.

Bell, J.S. (1981). Bertlmann's socks and the nature of reality. *Le Journal de Physique Colloques*, 42(C2): C2–C41.

Carroll, S.M. and Chen, J. (2004). *Spontaneous inflation and the origin of the arrow of time*. Available at: arXiv preprint hep-th/0410270.

Chen, E.K. (2017). Our fundamental physical space: An essay on the metaphysics of the wave function. *Journal of Philosophy*, 114: 7.

Chen, E.K. (2018). Quantum mechanics in a time-asymmetric universe: On the nature of the initial quantum state. *The British Journal for the Philosophy of Science*. Available at: https://doi.org/10.1093/bjps/axy068.

Ciepielewski, G.S., Okon, E. and Sudarsky, D. (2020). On superdeterministic rejections of settings independence. *The British Journal for the Philosophy of Science (forthcoming), arXiv:2008.00631*.

Clauser, J.F., Horne, M.A., Shimony, A. and Holt, R.A. (1969). Proposed experiment to test local hidden-variable theories. *Physical Review Letters*, 23(15): 880.

Einstein, A., Podolsky, B. and Rosen, N. (1935). Can quantum-mechanical description of physical reality be considered complete? *Physical Review*, 47(10): 777.

Feintzeig, B. (2015). Hidden variables and incompatible observables in quantum mechanics. *The British Journal for the Philosophy of Science*, 66(4): 905–927.

Fine, A. (1982a). Hidden variables, joint probability, and the Bell inequalities. *Physical Review Letters*, 48(5): 291.

Fine, A. (1982b). Joint distributions, quantum correlations, and commuting observables. *Journal of Mathematical Physics*, 23(7): 1306–1310.

Friederich, S. and Evans, P.W. (2019). Retrocausality in quantum mechanics. In E.N. Zalta (ed.), *The Stanford Encyclopedia of Philosophy*. Metaphysics Research Lab, Stanford University, summer 2019 edition. Available at: https://plato.stanford.edu/cgi-bin/encyclopedia/archinfo.cgi?entry=qm-retrocausality

Goldstein, S., Norsen, T., Tausk, D.V. and Zanghì, N. (2011). Bell's theorem. *Scholarpedia*, 6(10): 8378. revision #91049.

Hájek, A. (2019). Interpretations of probability. In E.N. Zalta (ed.), *The Stanford Encyclopedia of Philosophy*. Metaphysics Research Lab, Stanford University, fall 2019 edition. https://plato.stanford.edu/entries/probability-interpret/

Healey, R. (1992). Chasing quantum causes: How wild is the goose? *Philosophical Topics*, 20(1): 181–204.

Henderson, L. (2020). The problem of induction. In E.N. Zalta (ed.), *The Stanford Encyclopedia of Philosophy*. Metaphysics Research Lab, Stanford University, spring 2020 edition. https://plato.stanford.edu/entries/induction-problem/

Hossenfelder, S. and Palmer, T. (2020). Rethinking superdeterminism. *Frontiers in Physics*, 8: 139.

Jarrett, J.P. (1984). On the physical significance of the locality conditions in the Bell arguments. *Noûs*, 18(4): 569–589.

Kronz, F.M. (1990). Hidden locality, conspiracy and superluminal signals. *Philosophy of Science*, 57(3): 420–444.

Lewis, P.J. (2006). Conspiracy theories of quantum mechanics. *The British Journal for the Philosophy of Science*, 57(2): 359–381.

Malament, D.B. (2006). Notes on Bell's theorem. Available at: http://www.lps.uci.edu/malament/probdeterm/PDnotesBell.pdf.

Maudlin, T. (2011). *Quantum Non-Locality and Relativity: Metaphysical Intimations of Modern Physics*. West Sussex, UK: John Wiley & Sons.

Maudlin, T. (2014). What Bell did. *Journal of Physics A: Mathematical and Theoretical*, 47(42): 424010.

Maudlin, T. (2019). *Bell's Other Assumption(s)* (accessed on youtube).

Myrvold, Wayne, G.M. and Shimony, A. (2019). Bell's theorem. In E.N. Zalta (ed.), *The Stanford Encyclopedia of Philosophy*. Metaphysics Research Lab, Stanford University, spring 2019 edition. https://plato.stanford.edu/entries/bell-theorem/

Norsen, T. (2007). Against 'realism'. *Foundations of Physics*, 37(3): 311–340.

Penrose, R. (1979). Singularities and time-asymmetry. In S. Hawking and W. Israel (eds.), *General Relativity*. Cambridge: Cambridge University Press, pp. 581–638.

Pitowsky, I. (1989). *Quantum Probability-Quantum Logic*. Berlin: Springer.

Redhead, M.L. (1989). Nonfactorizability, stochastic causality, and passion-at-a-distance. In J.T. Cushing and E. McMullin (eds.), *Philosophical Consequences of Quantum Theory*. Notre Dame, IN: University of Notre Dame Press, pp. 145–153.

Shimony, A. (1989). Search for a worldview which can accommodate our knowledge of microphysics. In J.T. Cushing and E. McMullin (eds.), *Philosophical Consequences of Quantum Theory*. Notre Dame, IN: University of Notre Dame Press, pp. 62–76.

Shimony, A., Horne, M.A. and Clauser, J.F. (1976). Comment on 'The theory of local beables'. *Epistemological Letters*, 17.1(13): 1–8.

'T Hooft, G. (2014). *The cellular automaton interpretation of quantum mechanics*. Available at: arXiv preprint arXiv:1405.1548.

Tumulka, R. (2006). A relativistic version of the Ghirardi–Rimini–Weber model. *Journal of Statistical Physics*, 125(4): 821–840.

Tumulka, R. (2016). The assumptions of Bell's proof. In M. Bell and S. Gao (eds.), *Quantum Nonlocality and Reality: 50 Years of Bell's Theorem*. Cambridge, UK: Cambridge University Press.

Wallace, D. (2012). *The Emergent Multiverse: Quantum Theory According to the Everett Interpretation*. Oxford: Oxford University Press.

Wilce, A. (2017). Quantum logic and probability theory. In E.N. Zalta (ed.), *The Stanford Encyclopedia of Philosophy*. Metaphysics Research Lab, Stanford University, spring 2017 edition. https://plato.stanford.edu/entries/qt-quantlog/

13

QUANTUM DECOHERENCE

Elise M. Crull

13.1   Introduction: Three Quantum Puzzles

Saturn's moon Hyperion is one of the solar system's strangest bodies. Described in Bokulich (2008, p. 23) as "approximately three times the size of the state of Massachusetts and... roughly the shape of a potato," its irregular shape when coupled with inhomogeneous gravitational forces from Saturn and surrounding moons *should* leave Hyperion in a spatial orientation that is a coherent superposition over 57 degrees (Zurek and Paz, 1997). But this is not what is observed. Hyperion appears to occupy definite orientations – not a superposition of positions – as it chaotically tumbles around Saturn.

Or consider optical isomers like the sugar and ammonia molecules. Optical isomers have identical atomic composition and structure (therefore the same quantum state description) but dissimilar chirality (therefore different optical properties). In the ground state their position eigenstates are "left-handed" or "right-handed"; if these eigenstates are possible states, then coherent superpositions of left- and right-handedness are also possible states. However, while the ammonia molecule is typically observed in a superposition of chiral states, the sugar molecule is observed in a handed eigenstate. Why does nature act according to our expectations regarding the ammonia molecule's spatial position but contrary to our expectations in the case of sugar?

Lastly, consider Bohr's planetary atomic model of 1913. In it, electrons occupy definite energy states and perform instantaneous quantum jumps to transition to higher or lower energy orbitals. This model successfully explained certain atomic spectra despite its incorrect assumption that electrons always occupy an energy eigenstate instead of superpositions of such states.

These puzzles drawn from the macro-, meso-, and microscopic domains, respectively, all find an explanation in decoherence. Below these explanations are provided, but first decoherence and relevant concepts are defined (Section 13.2). Then (Section 13.3) the most widely adopted formalism for studying decoherence – that of density matrices – is sketched, followed by an overview of the four canonical models (Section 13.4). The latter half of the chapter examines the explanatory wealth of decoherence when brought to bear on foundational issues such as the question of why certain measurement bases seem to be preferred by nature (Section 13.5) – including the pointer basis of measuring apparatuses (Section 13.6) – the old familiar measurement problem (Section 13.7), and the emergence of classicality (Section 13.8). (For an introduction to the formalism of quantum mechanics including the concept of a basis, see chapter 2 of Albert (1992).) Because decoherence is prevalent in uncontrolled environments and extraordinarily effective in most circumstances, understanding its theoretical and experimental implications proves to be a crucial component in any conception of physical bodies and their interactions, philosophical or otherwise.

13.2 Definition and Basic Concepts

Quantum decoherence (hereafter simply decoherence) is a physical process resulting from a system's entanglement with an environment. At any given time, systems are interacting quantum mechanically (as opposed to thermally or mechanically) with external or independent degrees of freedom comprising an environment. The consequence of this interaction is generally entanglement with respect to certain degrees of freedom (hereafter DoFs), and this in turn enables the environment to decohere phases among components of the superposition in the basis or bases of entanglement. In analogy to classical wave mechanics, a quantum system is said to be in a coherent superposition when the phase relations between individual waves composing the superposed wave are *constant*. (Constant phase relations engender stable constructive and destructive interference patterns.) Decoherence can then be understood as the destruction (due to external interactions) of constancy among phase relations ergo the destruction of interference patterns.

Entanglement is often confused or used interchangeably with the term coupling. This confusion leads to misconceptions about the scope and nature of entanglement as a purely quantum phenomenon, as opposed to the broader class of phenomena related to coupling. Although coupling is, like entanglement, a result of interaction between two or more systems or between different DoFs within a single system, it is unlike entanglement in that the former is not a necessary consequence of interactions whereas the latter typically is. For example, the molecules of gas in a system at thermal equilibrium are not (on average) thermally coupled. These molecules are, however, interacting in a way that will generally lead to entanglement. They can be described independently at the level of classical interactions but are inseparable according to the quantum mechanical description. The confusion between coupling of systems and environments in the classical sense and entanglement as a (or according to Schrödinger, *the*) quantum mechanical property is in part why it took physicists so long to recognize decoherence was a quantum process, distinct from thermal dissipation.

Decoherence is not an interpretation of quantum mechanics nor is it new physics. It is important not to confuse the physical process itself with interpretations of quantum mechanics wherein decoherence plays a prominent role, e.g., the decoherent histories approach.[1] What is needed for decoherence to occur (within the usual Hilbert space formalism) is a set of three basic postulates, and their consequences when examined apart from several key idealizations. One may begin by assuming that (i) pure and mixed states can be represented by density operators in a Hilbert space, (ii) observables are defined as self-adjoint operators acting on that space, and (iii) the Schrödinger equation correctly describes unitary evolution of all closed systems. (Quantum mechanics must be assumed universally valid for the explanations from decoherence to have maximal significance. Since, however, the fundamentality of quantum theory is regarded by most to be the only viable working option, additional argumentation in support of this assumption shall not be given here.)

Add to these postulates the ubiquity of quantum interactions leading to widespread entanglement and hence the fictitiousness of any truly closed system (apart from the universe as a whole, perhaps) and one has the necessary ingredients for decoherence. In what follows, "system" can be read so weakly as to refer to one or more independent DoFs. For example, decoherence can occur within a single electron when treated as two separate systems: let the electron's spin be the system of interest, while the electron's translational DoFs form the environment. Correspondingly, what is considered the environment can be comprised by any suitable DoFs external to the system of interest.

Phase relations and superpositions play an important role in decoherence processes, and though these concepts were conceived in analogy with classical wave mechanics, they are importantly different in quantum systems. In the latter, phase relations mathematically express the degree to which amplitudes of individual waves constituting a superposition can be combined. Bear in mind that this analogy with classical wave mechanics is just that – an analogy. In classical mechanics, a wave packet is analyzable in terms of a superposition of the aggregate individual waves (e.g., electromagnetic field strength is just the sum of wave amplitudes at a spacetime point). In quantum mechanics, although mathematically the superposed state is still described as a sum of the individual component states, a quantum system in such a state may exhibit distinct physical properties.

This is precisely what happens in the puzzle of the sugar and ammonia molecules. In Joos and Zeh (1985), a seminal early work on decoherence, the authors investigated the paradox of optical isomers by entangling a parity eigenstate of each molecular system to a single unpolarized photon. They found that in the case of the sugar molecule, the photon-molecule composite system became strongly entangled to environmental photons. This entanglement led to the decoherence, or prodigious damping, of phase amplitudes of the sugar molecule's polarity, leaving only definite handed states stable under environmental influence. The more stable a state is under evolution, the more measurable it is. Hence, the failure to measure sugar in any state *other than* "left-handed" or "right-handed." Joos and Zeh calculated rudimentary values for the rate of decoherence of sugar's chiral states and found that it was on a timescale many orders of magnitude faster than the measurement process itself. In other words, before the measurement event on the sugar molecule could be completed (the detection of post-interaction photons indicating the molecule's polarity), decoherence had already practically irreversibly destabilized states in superpositions of polarity, effectively rendering the left- and right-handed eigenstates the only measurable states.

Decoherence also satisfactorily explains the behavior of the ammonia molecule. In this case, the molecule's spatial DoFs did *not* become entangled with the photon environment, and so no decoherence occurred in the ammonia molecule's basis of polarization. Superpositions of left- and right-handed states of ammonia remain stable under this type of environmental monitoring, and superposed states indeed are what one typically measures. As with sugar molecules, since a chiral state of ammonia has different optical properties than a superposition of chiral states, one can test whether the molecule is in a superposition by studying the specific effect it has on the polarity of incident light.[2] This illustrates the above point that superpositions of quantum systems can give rise to different observable states of affairs.

Of course, it is well known from the double-slit experiment that superpositions are possible states not just for mesoscopic systems like molecules but indeed for individual quanta: "self-interference" explains why spatial interference patterns emerge when particles are sent one at a time through a double-slit apparatus. (That is, as long as the distance between the slits is on the order of the de Broglie wavelength of the particles passing through.) If the particle's phase relations are constant in the position basis – in which case the wave packet as a whole does not disperse – the system will exhibit constructive or destructive interference effects whose strength depends on the character of the phase relations. Nonzero phases whose ratio is constant are defined as coherent superpositions. When system-environment entanglement allows for system monitoring by the environment, the system's phases are delocalized in the basis (or bases) being monitored, an effect which expresses itself in the suppression of interference phenomena. This is what one observes when a detector (that is, a device that can become entangled to and therefore record information about a particle's trajectory) is placed behind one of the slits: interference patterns are destroyed and classical statistical distributions emerge. Absent such a detector, interference patterns are observed, because superposition states of the particle remain coherent long enough to reach the detector and be registered as such.

While quantum theory does not provide a straightforward physical interpretation of superpositions, what is clear from the occurrence of interference phenomena is that a particle within this apparatus cannot be described in terms of a classical statistical ensemble – i.e., it cannot be described as following a classical trajectory – unless one supplements the three basic postulates given above with a collapse mechanism[3] or guiding wave[4] or another suitable physical event.

It is often mistakenly asserted that in order for decoherence processes to correctly explain certain phenomena like self-interference in a double-slit apparatus or the puzzle of the optical isomers, one must assume collapse of the wavefunction.[5] This is decidedly false. It is also frequently claimed that decoherence requires the Born rule (ergo the eigenstate-eigenvalue link). Because of these mistakes – conflation of physical collapse (which decoherence does *not* involve) with effective collapse (which it does – more details anon) and reliance on the Born rule for assigning definite values to measurement outcomes (which, as shall be argued below, is also unnecessary for decoherence) – many have bypassed decoherence theory and experiments without due appreciation for this process's explanatory wealth. Because these misunderstandings run deep even in the professional literature, the remainder of this section is dedicated to a precisification of why decoherence does not need any physical postulate beyond the three listed above.

Regarding the question of collapse, Bacciagaluppi (2012) lists two physical assumptions that collapse theories like GRW require but decoherence does not. It will be instructive to repeat them here. First, collapse theories require true, physical collapse of the wavefunction; decoherence only entails effective collapse, in that this process renders practically impossible the obtainment of measurement results other than eigenstates of a relatively stable basis. The experimental observation of coherence *revivals* within models of previously decohered systems lends credence to the claim that no physically irreversible mechanism is operating therein.[6]

The second assumption collapse theories require but decoherence does not is the existence of an as-yet undiscovered collapse mechanism that introduces nonlinearity apart from the system-environment Hamiltonian. Decoherence introduces no additional nonlinearity but "simply takes real, inevitable, interaction with external degrees of freedom to bear in the Hamiltonian" (Bacciagaluppi, 2012). To this add a third point, conceded by collapse theorists themselves. Because collapse theories do not explain the privileging of certain bases, decoherence is called upon to explain this piece of the dynamics *within* these interpretations. Indeed, all viable interpretations of quantum mechanics invoke decoherence as a crucial part of their explanatory package on the understanding that this process involves nothing beyond the standard formalism and is not by itself an interpretation. Decoherence is not "empirically complete" in the sense that while it explains why certain bases are more stable than others in a given environment (and hence the states therein more likely measurable, and measurable qua eigenstates of the stable basis), it does not explain why one *particular* apparently definite outcome is measured instead of another.

Hence, no physical collapse is needed so long as one stops short (as decoherence explanations do) of attempting to answer the question of why one obtains a specific outcome (as e.g. collapse theories do) instead of simply asking why one obtains an eigenvalue from a preferred basis. *Effective* wavefunction collapse is all that is required to explain why certain bases are preferred; this point is discussed in more detail in Section 13.5.

A similar situation applies to the Born rule and the eigenstate-eigenvalue link: as long as one stops short of declaring the results of measurements to be actually definite – which is a significant ontological step beyond declaring them only apparently definite – the Born rule can be understood as merely a guide to expectations and not a metaphysical claim requiring further justification.

13.3   The Formalisms of Decoherence

The Hamiltonian of a quantum system is unitary only as long as the system is closed; entanglement with an environment therefore destroys local unitarity while preserving it at the level of the newly entangled composite system. Because of this introduced non-unitarity, the maximum information obtainable about one of the entangled subsystems (without introducing collapse or some such physical mechanism) is the complete statistical information about its evolution in a certain basis, and these statistics are obtained using reduced density matrices. The natural utility of reduced density matrices applied to decoherence processes makes this formalism the favored one, and will now be briefly introduced.

If on the one hand a quantum system is said to occupy a pure state, this indicates the state is fully knowable and can be represented by a single vector (or superposition of vectors) in the system's Hilbert space. If on the other hand only statistical information is known about the state, it is referred to as a mixture, and can only be represented as a statistical distribution of pure states. However, this statistical distribution does not always describe a classical, or proper, ensemble wherein the system occupies a single state from among possible pure states but it is unknown which one. A mixed state density matrix may instead represent an *improper* ensemble, for which such an ignorance interpretation is incorrect. Entangled states in particular must be considered improper mixtures, as entanglement entails the existence of interference terms, or phase relations, among pure states comprising the ensemble.

It is important to stress that obtaining a diagonalized (interference-free) mixed state does *not* uniquely determine whether one is dealing with a proper or improper mixture. Although both proper and improper mixed states can be written in a variety of ways depending on the basis of measurement, as just described there are importantly different physical properties associated with each. For example, one might choose to write the density matrix for an improper mixture in the most convenient basis – the state's eigenbasis – in which case the density matrix will be diagonal and therefore formally identical to a proper mixture. However, diagonality in this basis is a particular feature of that basis. Because the system is an improper mixture, off-diagonal interference terms must exist in some other basis or bases, and this means that, barring any branching or collapsing, these terms *remain* part of the system's full description.

Give this formal underdetermination of mixed states, when a system's preparation is not fully controllable or not entirely known, one cannot say with any certainty that its density matrix represents a proper classical ensemble with the system occupying a definite pure state. In fact, due to the ubiquity of entanglement, it is likely that a given system (particularly "in the wild") is entangled with some other system, and so must be represented as an improper mixture.[7]

Before describing the aspect of this formalism most useful for decoherence – reduced density matrices – there is another important tool that must be introduced called the trace operation. The trace of a matrix is the sum of its diagonal elements; under the normalization condition, the trace of a density matrix must be equal to one. Reduced density matrices are obtained via *partial* traces over the density matrix of the composite system in order to extract statistical information about a single subsystem. Thus, a partial trace over the density matrix of an entangled system-plus environment with respect to the environment yields the reduced density matrix of the system alone (having "traced out" environmental DoFs). Again one must be careful when interpreting these mathematical results: whatever the form of the system's reduced density matrix – in particular, whether it is diagonalized or not – the physical state to which it corresponds cannot be pure (or a superposition of pure states), as it is in fact a subsystem of an entangled whole. In other words, a mixed state density matrix underdetermines whether the ensemble of states comprising it is proper or improper, and therefore whether the system definitely occupies a single state or not.

The density matrix formalism is the preferred approach in most of the literature for determining system-environment interaction, and thus has received the most detailed explanation here. However, it is important to acknowledge various other approaches adopted for exploring decoherence. For instance, some have studied decoherence from a perspective called coarse-graining, or closed-system approaches.[8] This approach still uses the density matrix formalism but advocates a different treatment of the "cut" between system and measuring apparatus. A few words about the term coarse-graining, which often arises in connection with the closed-system approach, are in order. The switch from using system Hamiltonians to density matrices might in a superficial way be considered coarse-graining as it is a move from fully determinate equations to statistical ensembles. The language of "coarse-graining" here is somewhat misleading, though, as there is no actual loss of information moving from Hamiltonian representations to that of density matrices. Additionally, the use of the density matrix formalism does not imply *necessary* coarse-graining, as there exists alternate formalism in which no coarse-graining occurs. For example, no coarse-graining exists in a powerful technique using Feynman path integrals: the restricted path integral (RPI) method.

The RPI method has become increasingly favored by experimentalists of late.[9] While RPIs provide an undeniably powerful tool, this technique essentially relies on mathematical maneuvers whose metaphysical implications are obscure. In particular, the RPI method (as with Feynmen path-integral approaches generally) involves treating the time evolution of a system as an integrand composed of all the possible trajectories (referred to as "channels" in decoherence research) of the system of interest. Obviously, issues arise as to the exact meaning of concepts like trajectories when applied to quantum systems, as one is necessarily dealing with nonlocal entities and quantized parameters. Even maximally coherent wave packets tracing out quasi-Newtonian trajectories in a system's phase space cannot be localized beyond the limit of the uncertainty relations, and so their evolution in accordance with approximately classical trajectories must be understood with significant qualification.

13.4   The Four Canonical Models

A vast array of physically interesting interactions have been successfully described using only four models, the so-called canonical models of decoherence. These models pair systems with both discrete and continuous DoFs with environments with discrete and continuous DoFs, giving the following permutations (listed as *system type-environment type*): the oscillator-oscillator model (continuous DoFs for both system and environment), the spin-oscillator model, also called the spin-boson model (discrete system DoFs, continuous environment DoFs), the spin-spin model (discrete DoFs for both), and the oscillator-spin model (continuous system DoFs, discrete environment DoFs). Below are tables representing common characteristics for these, where H_S designates the component of the total Hamiltonian (H_{tot}) due to the system's self-dynamics, H_E designates the environment's self-dynamics, and H_{int} is the interaction Hamiltonian for the given system and environment:

	Continuous DoFs	*Discrete DoFs*
System	• Mapped as harmonic oscillator in single-well potential • $H_S \approx H_{int}$, thus decoheres fastest in phase space • E.g., quantum Brownian particles • Mapped as continuum of N delocalized harmonic oscillators (bosonic field modes) • $H_E \ll H_S$, H_{int} gives weak-coupling limit	• Mapped as spin-1/2 particle in double-well potential • Possible self-entanglement requires complicated H_S • E.g., fermions and qubits • Mapped as continuum of spin-1/2 particles ("spin bath") • Highly localized energy modes achieved at extremely low T
Environment	• Dissipation and decoherence of system in this environment is practically irreversible; leads to increased delocalization of field modes • Assume coupling strength of field modes scales as $1/\sqrt{N}$ to ensure well-defined thermodynamic limit for $N \to \infty$ • E.g., gases	• $H_{tot} \approx H_S$ due to low T environment • All systems − continuous energy spectra (osc. systems) and discrete energy states (spin-1/2 systems) − are strongly decohered in spin bath • E.g., superconducting and quantum computing devices

All four canonical models of decoherence require an important assumption: the initial independence of the system of interest from the relevant environment. Prima facie this seems problematic: if entanglement and decoherence are as effective and ubiquitous as stated, how can it be appropriate to assume for a model's initial conditions that the central system is uncorrelated to the environment? Wasn't decoherence first appreciated as a consequence of *dropping* the idealization of a truly isolated system?

Though the initial assumption of unentangled systems and environments is a nontrivial one, it has been substantiated in a wide variety of models. Its earliest justifications can be found in Anglin et al. (1997) and Bose et al. (1999). In the former work, the authors convincingly argue that assuming a previously noninteracting system and environment is viable, especially in the most general case of uncontrolled interactions. Anglin et al. found that it is nearly always appropriate to characterize the environment as occupying an approximate pure state prior to interaction with the system due to the fact that uncontrolled environments (and sufficiently large controlled environments) have a prodigious number of DoFs which have been interacting among themselves prior to interaction with the system of interest. Thus, the environment is itself already typically decohered in the bases of interest and can be treated as approximately localized therein.

In the latter work, Bose and coauthors investigate decoherence in a Schrödinger cat-type situation where the "cat" is modeled by a macroscopic mirror sitting in an electromagnetic field cavity. They explicitly drop the assumption of an initially unentangled system and environment, and find that the final state of the composite system is similar to that of models which assume *no* initial correlation (Bose et al. 1999).

In sum, the devil remains in the details: the viability of this assumption depends importantly on the nature of a particular system-environment interaction, and on the internal dynamics of both. However, it seems safe to say that in uncontrolled environments such as encountered outside the quantum physics lab, this assumption is justified.

A related question might now arise: if entanglement leads to decoherence, and the effect of decoherence is to suppress quantum correlations beyond observability, how is it that correlations are nevertheless measured? For example, how does decoherence explain Bell-type experiments verifying quantum correlations between EPR pairs? To answer this, one must first remember that decoherence is *basis specific*. Second, recall that decoherence occurs at different rates in different bases depending on the nature of the environmental interaction. Thus, while it is indeed true that maximally entangled states like EPR pairs (engineered in carefully controlled environments as they must be) maintain coherence with respect to certain pre-selected bases long enough to be measured by distant detectors, this is not necessarily true for all bases of the pair, and certainly not true for all time. Indeed, were coherence in all bases easily maintained or preserved, quantum computing – which banks on the possibility of maintaining qubit coherence within the computer environment for certain periods of time and over some distances – would not present the enormous engineering difficulties it does. The possibility of a quantum computer depends upon the ability to engineer environmentally incorruptible qubits, and this is no trivial feat. A qubit is a quantum system with (effectively) two states that is capable not only of expressing binary language with its two eigenstates, but, due to its quantum nature, of encoding additional information via superpositions of its two states *and* via entanglement with other qubits. Such an enriched coding language of course dramatically increases computation power, but at cost: in order to keep the information coherent, qubits must be shielded from decoherence in an environment that is especially decoherence-friendly. Recall the table in Section 13.4: quantum computation research heavily relies upon spin-spin decoherence models, for in these models the system (the qubit) is mapped as a spin-1/2 particle in a double-well potential (as a result of which qubits are associated with complicated Hamiltonians that allow for the possibility of self-entanglement), and in addition, this system is surrounded by *other* qubits, so this environment is mapped by a spin bath (as a result of which, the entire quantum computer must be kept at extremely cold temperatures in order to effectively shield qubits from environmental or self-decoherence). Thus, quantum computers will require a number of extraordinary engineering feats to overcome the qubit's strong tendency to decohere in this arrangement, a process which scrambles information by (effectively) destroying bit integrity. Unsurprisingly, a particularly active area of investigation regarding the feasibility of quantum computers concerns the engineering of so-called "decoherence-free subspaces" wherein qubits avoid corruption. The literature on quantum computing is already prodigious, but for early discussions focusing on decoherence, see Deutsch (1985), Shor (1995), and Unruh (1995).

13.5   The Question of the Preferred Basis

The basis-specific nature of decoherence will allow for a satisfying explanation to one of quantum mechanics' oldest mysteries: that of the pointer basis. Why do measuring devices acting on quantum systems "point to" one specific, definite value when the wavefunction describing that system's state is probabilistic? Before answering this, however, it will be instructive to consider the more general underlying question: why do certain bases seem to be preferred over others (e.g., position for macroscopic systems, energy for microscopic systems) when no such preference is evident in the mathematical description?

Perhaps given the solution to the puzzle of optical isomers one can already guess at the answer here. Nature's apparent "preference" for certain bases of measurement is not due to as-yet undiscovered selection rules, but rather due to the relative stability of particular bases over others due to decoherence dynamics. The rate at which a system's phase relations become decohered in a particular environment depends on the strength and character of the system-environment entanglement. System DoFs that commute most effectively with environmental DoFs will become most quickly entangled with one another, and therefore most robustly decohered in the associated basis. The dynamical robustness of a given basis under environmental monitoring is another way of describing the effectiveness with which the system's interference terms in that basis are suppressed beyond measurability, leading to the extreme improbability of observing superpositions as opposed to eigenstates in that basis.

As a first approximation of how decoherence provides the dynamics undergirding preferred bases, recall the puzzle of Bohr's atomic model above. There it was asked how a model that did not take superpositions of energy states of the electron into account nevertheless explained empirical data like atomic spectra. It is now understood that by the time an electron's energy can be measured, it has become prodigiously decohered in the energy basis due to interaction and subsequent entanglement with the atomic nucleus. Thus, the electron's superposed energy states became prodigiously damped by environmental influence while its energy eigenstates remained largely unaffected. In other words, the energy basis of the electron remains most stable under environmental influence and so becomes the preferred basis.

As a higher order approximation, consider a quantum Brownian particle weakly interacting with a gas (a common instance of the oscillator-oscillator model). Here, the position of the quantum particle decoheres most rapidly, followed by momentum, and momentum in turn decoheres far more rapidly than the rate of thermal dissipation. Thus, one observes after a very short time the quantum Brownian particle tracing a quasi-Newtonian path in phase space (as stated in the above table). Generally speaking, although the canonical coordinate of the system might represent e.g. an electromagnetic variable instead of position, the system will nevertheless be most stable with respect to its position states, because that is the variable whose operator commutes with individual environmental modes of the gas, which are modeled in terms of *their* position coordinates. Thus, in the quantum Brownian motion model, the environment is said to continuously monitor the position of the system; when one takes a partial trace over position coordinates of just the environment, one obtains the system's reduced density matrix.

An explanation of the remaining quantum puzzle can now be given. Since Hyperion, an object uncontroversially considered "macroscopic," is nevertheless entangled with some environment (the gravitational field, for one), one expects decoherence to be in play. Indeed, in a paper titled "Why We

Don't Need Quantum Planetary Dynamics," Zurek and Paz (1997) investigate the dynamics of Hyperion *without* taking decoherence into account, and conclude that the moon should occupy "a very nonclassical superposition, behaving in a flagrantly quantum manner" (pp. 370–371). But as stated above, this is not what is observed. Unsurprisingly, when Zurek and Paz did include decoherence dynamics into their calculations, they found results in agreement with observation. Here, decoherence occurs most rapidly in position, probably due to immersion in an inhomogenous gravitational field whose $1/r^2$ dependence commutes with Hyperion's Hamiltonian.[10] This suppresses interference terms in the position basis so effectively that the moon follows an approximately Newtonian trajectory despite its underlying nonlinear quantum dynamics.

In this way decoherence not only provides the dynamics explaining why a certain basis is "preferred" (because it is most stable under interaction with a typical environment), but various models generate parameter-specific decoherence rates indicating the *degree* of a basis's stability through time as compared to other bases.

13.6   The Question of Pointer Positions

Zurek begins his seminal paper (Zurek, 1981) as follows:

> What does, in the real-world apparatuses, determine this apparently unique *pointer basis...* which records the corresponding relative states... of the system? Interaction with the environment is the key feature that distinguishes the here-proposed model of the apparatus from the manifestly quantum systems. We argue that the apparatus cannot be observed in a superposition of the pointer-basis states because its state vector is being continuously collapsed [sic]. It is the "monitoring" of the apparatus by the environment which results in the apparent reduction of the wave packet. Correlations between states of the pointer basis and corresponding relative states of the system are nevertheless preserved in the final mixed-state density matrix... (p. 1516)

Zurek demonstrates these claims by considering a reversible Stern-Gerlach apparatus (the incident beam of spin-1/2 particles is split into up- and down-streams by passing through an inhomogeneous magnetic field; the streams are then recombined by passing through a second inhomogenous magnetic field which is an inversion of the first) where a bistable atom measures the up-stream between the first and second magnet pairs. Zurek first calculates the composite wavefunction resulting from interaction of the spin–up stream (the system) with the atom (the apparatus). The final wavefunction is a pure state, indicating that there could not have been a real collapse of the wavefunction (ibid., p. 1518).

Zurek then calculates the composite wavefunction of the spin-up stream and the bistable atom in a different, arbitrary basis. Of this new, equally viable wavefunction Zurek writes "the illusion of a collapse may arise" (op. cit.), as this wavefunction is mixed – that is, it encodes maximal system-apparatus correlation, yet measurements will yield *apparently* definite outcomes. Analyzing these results Zurek writes (ibid., p. 1519):

> [T]he pointer basis of the apparatus *a* is chosen by the form of the apparatus-environment interaction: It is this basis which contains a reliable record of the state of the system. *S*. This in turn determines uniquely those relative states of the system which are correlated with the apparatus. Moreover, apparatus-environment correlations do not allow one to observe the *aS* combination in a superposition. Instead, it becomes a mixture diagonal in the basis constructed from the pointer-basis eigenstates... and the corresponding relative state of the system.

A caution that's been given above is worth emphasizing at this juncture: while decoherence explains the *apparent* definiteness of pointer positions, it is a separate claim – and one not substantiated by decoherence qua physical process alone – to insist that pointer positions are *truly* definite. The next two sections aim to clarify misunderstandings of decoherence that frequently arise in connection to this point. The first concerns what decoherence has to say about the measurement problem (broadly construed); the second concerns what is meant by classicality, and by the claim that decoherence explains its emergence.

13.7 Measurement Problem

The fact that decoherence does not solve the (entire) measurement problem has long been understood. As Joos wrote (Joos, 2000, p. 14): "Does decoherence solve the measurement problem? Clearly not. What decoherence tells us, is that certain objects appear classical when they are observed." More specifically, if in addressing the measurement problem one is careful to distinguish between a question about general outcomes ("why was the measurement result a definite state?") and a question about specific outcomes ("why was the measurement result *this* definite state?"), it becomes clear that while the former problem can be explained using decoherence, the latter cannot. For if decoherence could explain why a particular measurement yields the specific result it does, this would indicate decoherence contains the means for predicting the outcomes of quantum measurements. In which case we could pack up and go home...but here we remain.

Yet it is too crude to say decoherence does nothing to help alleviate certain longstanding questions often bound up with "the measurement problem." (The scare quotes are meant to signify the ambiguity with which this phrase is frequently deployed. Precision regarding what the perceived problem *is* is paramount for understanding how, and to what extent, decoherence has something to say about it.) For example, it has just been described how decoherence explains the problem of the preferred basis, and this is so not only in classical regimes but in a wide array of instances captured by the canonical models. It has also been suggested above that decoherence explains why superpositions are not observed in macroscopic systems like Hyperion, despite their intrinsic dynamics.

Consider the "measurement problem" in terms of both specific and general outcomes as applied to Schrödinger's cat. The general question is: why do we always observe the cat to be either alive or dead, and not a superposition thereof (a possible state according to the cat's wavefunction)? The specific question is: why did I just now observe an alive cat instead of a dead cat? As for the first, decoherence explains why the cat always appears to be *either* alive *or* dead: it's because the "alive-dead" basis is most stable under environmental decoherence. Or to put it another way, the interference between the alive state and the dead state is suppressed so effectively by entanglement with the environment that in practice one only *can* measure either the dead or alive eigenstates. But as for the second question, the quantum formalism (and consequently decoherence) alone provides no answer. And this is where various interpretations enter the story.

13.8 Emergence of Classicality

The issue of the emergence of an apparently classical world from a fundamentally quantum one seems to have found its resolution in decoherence, though to explain exactly how this happens depends not only on the particular system-environment under investigation, but on what one means by classicality. Here are but a few definitions: as Newtonian or quasi-Newtonian motion (characterized by Ehrenfest's theorem; Ehrenfest (1927)), as classical probability distributions or statistical ensembles (characterized by the Liouville regime; more below) in which probability distributions are mapped, as the limit $n \to \infty$, as the limit $\hbar \to 0$, and as mass $\to 0$.

These definitions are, individually, inadequate for universally characterizing the mythological "quantum-to-classical" border. That Ehrenfest's theorem provides neither a necessary nor sufficient condition for defining the classical regime is the central thesis of Ballentine et al. (1994). One is tempted in the case of the Liouville regime to interpret probabilities incorrectly: although in certain cases using Liouville's theorem one can recover a probability distribution identical to a classical statistical distribution, formal similarity does not always entail ontological similarity. Indeed, the quantum Liouville equation is a density matrix which, as discussed above, captures fully the statistics of a system's state-space but cannot be given a unique physical interpretation. Defining classicality as the limit where quantum number n approaches infinity will succeed in some but not all cases. This is proven in Messiah (1965) and Liboff (1984), both of whom draw upon the fact that the uncertainty relations set an insurmountable limit upon the assumption of continuous classical energy spectra. Neither will the limiting case of decreasing Planck's constant suffice, as this quantum-classical borderline is in many cases a singularity.[11] The naivety of relying on mass for one's definition of classicality is evident through examples of massive systems that nevertheless behave quantum mechanically. One such example (given by Joos in Giulini et al. 1996, p. 135) is the Weber bar, which is a tool for measuring gravitational waves. The bar itself must be quite massive (on the order of tons) for the detection of the waves, and yet the sensitivity of the device is such that it must detect displacements on the order of 10^{-21} meters, or approximately 50 billionths the radius of a ground-state hydrogen atom. As such, the Weber bar must be treated as a quantum oscillator in order to appropriately characterize its behavior, despite its size. Other examples of macroscopic quantum phenomena are the Josephson effect in superconducting (see for example Yu et al. (2002)) and, trivially, the commonplace laser pointer: the coherent superposition of numerous emitted monochromatic light quanta is the reason such beams retain focus at great distances.

Because none of the usual definitions of classicality are able to explain its emergence from quantum mechanics in full generality, discussions of the quantum-to-classical transition must never stray far from detailed information about specific system dynamics and their interactions with specific environments. Since decoherence models are designed to examine precisely these dynamics in wide-ranging situations, many consider this the most promising approach to the question of emergent classicality.

Notes

1 On the latter see Gell-Mann and Hartle (1996), Griffiths (2003), and Halliwell (1995).
2 For more detail, see sections 1–5 of Fortin et al. (2016).
3 See Lewis (this volume) for discussion of collapse theories.
4 See Tumulka (this volume) for discussion of theories of this kind.
5 For a longer discussion of this point as well as expanded discussion on other points of common confusion regarding decoherence, see Crull (2017).
6 See Narozhny et al. (1981) for theoretical groundwork and Kokorowski et al. (2001) as an entry point into fascinating work verifying decoherence models through experiments in atomic interferometry.
7 For discussion of the ubiquity of entanglement and decoherence, see Giulini et al. (1996) (especially Joos' contribution), Joos and Zeh (1985), and Zeh (1970).
8 The main works advocating and illustrating this approach to decoherence are a suite of papers by Castagnino, Lombardi, and Fortin, including Castagnino et al. (2007),Castagnino et al. (2010), and Lombardi et al. (2012).
9 For an excellent introduction to the theory and experimental practice of the path-integral approach to decoherence, see Mensky (2000). In Chapter 5, Mensky provides a nice comparison between the use of master equations used in tandem with the density matrix approach, and the path-integral approach. He argues that the two approaches are equivalent for nonselective measurements, which are precisely the sort of measurements occurring in nature.
10 Although Hyperion's decoherence in the position basis might best be explained through a combination of interactions (e.g., including the effects of scattering dust and planetary debris), the gravitational field is the most significant – and constant – factor.
11 See Batterman (1995), Batterman (2002), Berry (1994), Berry (2001), and Bokulich (2008) for fuller discussions of the failure of this approach to classicality.

References

Albert, D. (1992). *Quantum Mechanics and Experience*. Cambridge, MA: Harvard University Press.

Anglin, J., Paz, J. and Zurek, W. (1997, June). Deconstructing decoherence. *Physical Review A*, 55(6): 4041–4053.

Bacciagaluppi, G. (2012). The role of decoherence in quantum mechanics. In E.N. Zalta (ed.), *The Stanford Encyclopedia of Philosophy* (Winter 2012 edition). URL = http://plato.standford.edu/archives/win2012/entries/qm-decoherence/.

Ballentine, L., Yang, Y. and Zibin, J. (1994). Inadequacy of Ehrenfest's theorem to characterize the classical regime. *Physical Review A*, 50: 2854–2859.

Batterman, R. (1995). Theories between theories: Asymptotic limiting intertheoretic relations. *Synthese*, 103: 171–201.

Batterman, R. (2002). *The Devil in the Details: Asymptotic Reasoning in Explanation, Reduction and Emergence*. Oxford: Oxford University Press.

Berry, M. (1994). Asymptotics, singularities and the reduction of theories. In D. Prawitz, B. Skyrms and D. Westerstahl (eds.), *Logic, Methodology and Philosophy of Science IX*. Amsterdam: Elsevier, pp. 597–607.

Berry, M. (2001). Chaos and the semiclassical limit of quantum mechanics (is the moon there when somebody looks?). In R. Russell, P. Clayton, K. Wegter-McNelly and J. Polkinghorne (eds.), *Quantum Mechanics: Scientific Perspectives on Divine Action*. Vatican Observatory: CTNS Publications, pp. 41–54.

Bokulich, A. (2008). *Reexamining the Quantum-Classical Relation: Beyond Reductionism and Pluralism*. Cambridge: Cambridge University Press.

Bose, S., Jacobs, K. and Knight, P. (1999). Scheme to probe the decoherence of a macroscopic object. *Physical Review A*, 59(5): 3204–3210.

Castagnino, M., Fortin, S. and Lombardi, O. (2010). Is the decoherence of a system the result of its interaction with the environment? *Modern Physics Letters A*, 25(17): 1431–1439.

Castagnino, M., Laura, R. and Lombardi, O. (2007). A general conceptual framework for decoherence in closed and open systems. *Philosophy of Science*, 74: 968–980.

Crull, E. (2017). Yes, more decoherence: A reply to critics. *Foundations of Physics*, 47: 1428–1463.

Deutsch, D. (1985). Quantum theory, the Church-Turing principle and the universal quantum computer. *Proceedings of the Royal Society of London A*, 400: 96–117.

Ehrenfest, P. (1927). Bemerkung über die angenäherte Gültigkeit der klassischen Mechanik innerhalb der Quantenmechanik. *Zeitschrift für Physik*, 45(7–8): 455–457.

Fortin, S., Lombardi, O. and González, J.C.M. (2016, October). Isomerism and decoherence. *Foundations of Chemistry*, 18(3): 225–240.

Gell-Mann, M. and Hartle, J.B. (1996). Equivalent sets of histories and multiple quasiclassical realms. Available at: arXiv:gr-qc/9404013v3.

Giulini, D., Joos, E., Kiefer, C., Kupsch, J., Stamatescu, I.-O. and Zeh, H. (1996). *Decoherence and the Appearance of a Classical World in Quantum Theory* (1st edition). Berlin: Springer-Verlag.

Griffiths, R.B. (2003). *Consistent Quantum Theory*. Cambridge: Cambridge University Press.

Halliwell, J. (1995). A review of the decoherent histories approach to quantum mechanics. *Annals of the New York Academy of Sciences*, 755: 726–740.

Joos, E. (2000). Introduction. In P. Blanchard, D. Giulini, E. Joos, C. Kiefer and I.-O. Stamatescu (eds.), *Decoherence: Theoretical, Experimental, and Conceptual Problems*, Proceedings of the Bielefeld Workshop, Nov. 1998, Berlin: Springer, pp. 1–17.

Joos, E. and Zeh, H. (1985). The emergence of classical properties through interaction with the environment. *Zeitschrift für Physik*, B59: 223–243.

Kokorowski, D.A., Cronin, A.D., Robers, T.D. and Pritchard, D.E. (2001, March). From single- to multiple-photon decoherence in an atom interferometer. *Physical Review Letters*, 86(11): 2191–2195.

Liboff, R. (1984). The correspondence principle revisited. *Physics Today*, 37: 50–55.

Lombardi, O., Fortin, S. and Castagnino, M. (2012). The problem of identifying the system and the environment in the phenomenon of decoherence. In W.H. de Regt (ed.), *EPSA Philosophy of Science: Amsterdam 2009*. Dordrecht: Springer, pp. 161–174.

Mensky, M.B. (2000). *Quantum Measurements and Decoherence: Models and Phenomenology*. Dordrecht, Boston, MA & London: Kluwer Academic Publishers.

Messiah, A. (1965). *Quantum Mechanics*, Volume I and II. Amsterdam: North-Holland Publishing.

Narozhny, N., Sanchez-Mondragon, J. and Eberly, J. (1981). Coherence versus incoherence: Collapse and revival in a simple quantum model. *Physical Review A*, 23: 236–247.

Shor, P. (1995). Scheme for reducing decoherence in quantum memory. *Physical Review A*, 52: 2493–2496.

Unruh, W.G. (1995). Maintaining coherence in quantum computers. *Physical Review A*, 51: 992–997.

Yu, Y., Han, S., Chu, X., Chu, S.-I. and Wang, Z. (2002). Coherent temporal oscillations of macroscopic quantum states in a Josephson Junction. *Science*, 296: 889–892.

Zeh, H. (1970). On the interpretation of measurement in quantum theory. *Foundations of Physics*, 1: 69–76.

Zurek, W. (1981). Pointer basis of quantum apparatus: Into what mixture does the wave packet collapse? *Physical Review D*, 24: 1516–1525.

Zurek, W. and Paz, J. (1997). Why we don't need quantum planetary dynamics: Decoherence and the correspondence principle for chaotic systems. In B. Hu and D. Feng (eds.), *Quantum Classical Correspondence: Proceedings of the 4th Drexel Symposium on Quantum Nonintegrability, 1994*, Cambridge, MA. Drexel University: International Press, pp. 367–379.

Further Reading from the Editors

For a lively introduction to decoherence models from a pioneer in the field, see W. Zurek, "Decoherence and the Transition from Quantum to Classical – Revisited", *Physics Today*, June 2003. David Wallace's structuralist application of decoherence theory to Everettian quantum theory is discussed by Saunders in chapter 4.3; Wallace's mature presentation is D. Wallace, "Decoherence and Ontology (Or: How I Learned to Stop Worrying and Love FAPP)", in S. Saunders, J. Barrett, A. Kent and D. Wallace (eds.), *Many Worlds? Everett, Quantum Theory, & Reality* (Oxford University Press, 2010): 53–72. G. Bacciagaluppi, "The Role of Decoherence in Quantum Mechanics", *Stanford Encyclopedia of Philosophy* (2020) is a compendious resource on all philosophical aspects of decoherence. For a detailed book-length treatment, see M. Schlosshauer, *Decoherence and the Quantum-To-Classical Transition* (Springer-Verlag, 2007).

14

THE EVERETT INTERPRETATION: STRUCTURE

Simon W. Saunders

What does quantum mechanics tell us, if taken realistically, as a fundamental theory, which applies to everything? After a century of debate, it seems we are no nearer to an answer to this question. But in an important respect we know better what is in contention: it is the *Everett interpretation* of quantum mechanics, which is also known as the *many-worlds interpretation*. First advocated by Hugh Everett III in 1957, it is the only realist interpretation of quantum mechanics still standing. On every other approach the answer is that quantum mechanics tells us nothing when applied to everything because it makes *no sense* taken in this way; the theory must be changed. In the words of John Bell: "Either the wavefunction, as given by the Schrödinger equation, is not everything, or it is not right,"[1] seeking to reduce the options to either modifying the Schrödinger equation, or supplementing it with hidden variables. But "everything" for Bell meant "everything in the known universe"; dismissed was the alternative that the wavefunction (as given by the Schrödinger equation) describes *more* than the known universe – that it describes a quantum mechanical multiverse, a superposition of worlds, of which ours is only one.

The idea is fantastical, but quantum mechanics is a theory like no other, a revolution still in the making after all these years. The worlds are derived from the unitary formalism; they are not put in by hand. Measurement interactions, we know, lead to macroscopic superpositions – if the unitary dynamics operates untrammeled – where each state in the superposition contains a record of a sequence of events just as predicted by the measurement postulates, as Everett gave a simple model to show. The apparent conflict between quantum theory and locality, as codified in Bell's theorem, is removed: the interpretation extends without modification to relativistic quantum theory. It really is *interpretation* of the equations: it neither modifies nor supplements them, save that the Schrödinger equation is taken to be universal. It offers a radical and novel understanding of how determinism can be reconciled with indeterminism, solving puzzles in the philosophy of probability that have bedeviled the subject for decades. It provides a basis for quantum cosmology, free of a measurement problem.

These claims, of course, are all controversial. In a certain sense, if these arguments all stand up to scrutiny, the case for the Everett interpretation becomes *overwhelming*. No wonder they are strongly contested: there is simply too much at stake.

Much recent literature has been on the probability interpretation,[2] but there is another aspect to the development of the Everett interpretation in the last quarter-century that is just as important: in terms of decoherence theory. This frees the interpretation from dependence on the notion of measurement, and enhances the argument about records. It tells us how the worlds are composed. It allows the separation of the probability interpretation from the question of the structure of the state.

213

It provides an Everettian "tapestry of events," made out entirely in terms of categorical properties and relations, that now include phase relations and amplitudes, on which the probability interpretation is to be based.

This chapter is on this structural interpretation of the wavefunction, rather than the probability interpretation, which is the subject of Chapter 15. In particular, it is on the structure of the wavefunction as made out in terms of the quantum histories formalism. But it is also on Everett's writings, and especially his "automaton" argument, as published in 1957. I shall start with this, for it makes clearer the role of decoherence theory in going beyond Everett's writings (although as we shall see, there are hints of it in his "long" dissertation, eventually published in 1973 as "The theory of the universal wave function"[3]). Everett's much more widely read PhD thesis was one fifth the length, published under the title "'Relative State' formulation of quantum mechanics" (Everett, 1957). It was redacted from the 'long' dissertation at the insistence of his supervisor John Wheeler, who had long advocated Bohr's philosophy. It contained the argument from records, and thanks to the "Note added in proof" it conveyed the overall idea, but much was omitted. Everett was never to write on quantum mechanics again.

14.1   Everett's Automaton Argument

The strange gap between the determinism of the Schrödinger equation and the indeterminism evident at the observational level is bridged by the so-called *measurement postulates*. The first is given by the rule:

> **Born rule:** on measurement of $\hat{q} = \sum_k \lambda_k P_k$ on a system S prepared in the state $| \phi \rangle$, the outcome λ_k is obtained with probability $\frac{\langle \phi | P_k \phi \rangle}{\langle \phi | \phi \rangle}$.

(The P_ks are projection operators onto eigenstates of $\hat{q}$ with eigenvalues λ_k – this extends naturally to operators with continuous spectra. I will not always respect the distinctions between vectors, wavefunctions, and rays, and use the term "state" for all three. By the "amplitude" of a state, I mean its norm $\sqrt{\langle \phi | \phi \rangle}$.)

The Born rule has the flavor of a "correspondence rule" in logical empiricist philosophy of science, or perhaps an "operational definition": a rule that links theoretical concepts to observable ones (or theoretical terms to observation sentences). It provides the minimal interpretation necessary to submit quantum mechanics to test – provided, of course, recipes could be given for building appropriate measurement and state-preparation devices. The latter required purely classical concepts, according to one influential interpretation (Bohr's). That posed an obstacle to taking quantum mechanics as a fundamental theory: how then could it be based purely on classical concepts? (Bohr thought of quantum mechanics as a "generalization" of classical mechanics, rather than a theory distinct from it;[4] here Wheeler departed from Bohr, as he sought to quantize gravity, governing the macroscopic.)

The Born rule is the first and most important of the measurement postulates, but it says nothing about what happens to the state on measurement. It is usually supplemented by the:

> **Projection postulate**: for a repeatable experiment on S prepared in the state $|\phi\rangle$, if the outcome on measurement is λ_k, then S is left (up to normalisation) in the state $|\phi_k\rangle = P_k|\phi\rangle$.

The projection postulate is sometimes stated without the restriction to repeatability, but then it is very often false. Very often, in practice, the state of the system is changed in uncontrollable ways on measurement, or the system is even completely obliterated (so there is no state). But given that the macroscopic outcome is indeterministic, it is reasonable to suppose that the state of the measured system where it does still exist has changed indeterministically, and if it makes sense to assign a quantum state to the measurement apparatus as well, it too must be subject to indeterministic change. How is all this indeterminism at the level of the state consistent with the Schrödinger equation?

This is the infamous *measurement problem* of quantum mechanics, in what is probably its simplest guise. There are the two realist solutions already noted: add additional variables, or modify the Schrödinger equation. Everett offered a third and remarkable alternative: the *superposition* of all the indeterministic changes evolves deterministically, in accordance with the Schrödinger equation.

Everett's strategy to establish this conclusion was to show that a superposition of *records* of indeterminism could be obtained in this way, satisfying the Schrödinger equation – and specifically, records as could be encoded in a mechanical model. "The observer" was to be treated as a mechanical system interacting with the measured system, with the unitary equations applied to them both together. The answer to the question "What is observed?" is to be read off from this dynamical model, in terms of what can be laid down in records or in memory. This solved the problem, posed by Bohr, of how a system could be treated as both physically closed and yet under experimental control. Bohr had argued that physical closure renders observation impossible (unlike in classical physics, where the interactions needed for observation to be possible could be made arbitrarily small).[5] Everett's answer was to model the observer as within the closed system.

A great deal hangs on this argument. Is it acceptable to read off what is observed, from a direct dynamical model of observation, modeling the observer alongside the system under observation? Is Everett's approach in this respect a heightened realism, an extreme form of physicalism? No: as scientific method, it has impeccable credentials. It was important in the early modern period in establishing the Copernican system (a comparison Everett himself made), namely, in the analysis of what would be perceived, according to the theory, were the Earth in motion about the sun (showing there would be no great wind, no deviations in falling bodies). It was essential in the discovery of symmetries (think of Galileo's ship, and Faraday's cage). It was essential to Isaac Newton's method in the *Principia*. It played an essential role in Einstein's analysis of length contraction and time dilation in special relativity. Even Bohr extolled the general principle: "it is the theory that tells us what is observable."[6] The method is part and parcel of all the great discoveries in fundamental physics – save one. So is the practice of studying those theories at maximum strength, with the fundamental equations of each theory to have maximal scope – save one. Quantum mechanics is the exception. Von Neumann had made the first important step, of modeling the observer in quantum mechanics, in his *Mathematical Foundations of Quantum Mechanics*, published (in German) in 1932, but he was not yet ready to study the consequences of taking the Schrödinger equation to hold unrestrictedly. That honor fell to Everett.

Consider, for the sake of definiteness, a measurement of the z-component of spin of an atom of silver, as in the Stern-Gerlach experiment. The interaction Hamiltonian couples the magnetic moment of the atom in passaging a magnetic field with a certain symmetry, in such a way as to correlate the momentum of the atom of silver with the state of its component of spin, in that direction, either "+" or "−". The result is that the atom drifts in one of two opposite directions. A subsequent measurement of position is then directly correlated with the state of this component of spin. Depending on how the position measurement is performed (it may be made with wide latitude), the measurement may be repeatable. Suppose that it is. Then schematically, if the apparatus is to function as intended, when the initial state of the atom is $|\phi_+\rangle$ and the apparatus is in its "ready" state $|0\rangle$, it should be driven by the unitary dynamics governing the measurement, denote U_m, to display the outcome "+"; and when the initial state of the atom is $|\phi_-\rangle$ and the apparatus is in the ready state $|0\rangle$, it should be driven by U_m to display the outcome "−"; and in either case, for repeatability, the spin state of the silver atom should be unchanged. That is, U_m should satisfy the protocols:

$$|\phi_+\rangle \otimes |0\rangle \underset{U_m}{\rightarrow} |\phi_+\rangle \otimes |+\rangle \qquad (14.1a)$$

$$|\phi_-\rangle \otimes |0\rangle \underset{U_m}{\rightarrow} |\phi_-\rangle \otimes |-\rangle. \qquad (14.1b)$$

Simon W. Saunders

But it then follows, for any dynamics like this, that for an initial state

$$|\psi\rangle = c_+ |\phi_+\rangle + c_- |\phi_-\rangle \qquad (14.2)$$

where $c_\pm$ are complex numbers:

$$|\psi\rangle \otimes |0\rangle \underset{U_m}{\to} c_+|\phi_+\rangle \otimes |+\rangle + c_- |\phi_-\rangle \otimes |-\rangle. \qquad (14.3)$$

What could the RHS mean? Von Neumann's answer was that something very like "correspondence rules" were needed (although he did not use that terminology); for, he argued, mathematical expressions on their own, governing values of quantities, *never* amounted to a statement about what is observed. Needed was a further link to "experience": to a statement about what would be perceived or observed, or experienced; and this notwithstanding that mental events always had correlates in the physical (his famous "thesis of psychophysical parallelism", one of the core tenets of realism). According to von Neumann, correspondence rules were needed as much for Eq. (14.1a) and (14.1b), delivering probability one for seeing spin-up, given Eq. (14.1a), and probability one for seeing spin-down, given Eq. (14.1b), as they were needed given the superposition Eq. (14.3), when the probabilities are different from one. The rules were the measurement postulates (combined into what von Neumann called "Process 1"). Only in this way, said von Neumann, was the dynamical model including the observer "non-vacuous."

Everett, who studied von Neumann's book assiduously (it first appeared in English translation in 1955), took the thesis of psychophysical parallelism rather more to heart than did its author. If you want to know about the mental, he reasoned, sufficient to make sense of measurements, then model perception and memory explicitly. Suppose the apparatus to be a simple mechanical device, able to interact with the measured system in the sense of perception, but also able to store records of such perceptions in memory. Let the "ready" state of the apparatus with no records in memory be $| 0; \ldots\rangle$; in place of Eq. (14.1a) and Eq. (14.1b), require in addition that the positive spin outcome be recorded in memory as "+," and the negative as "−," and that the device resets to "0" That is, we suppose that we can build a mechanical device so that under the Schrödinger equation it satisfies the new protocols:

$$| \phi_+\rangle \otimes | 0; \ldots\rangle \underset{U_m}{\to} | \phi_+\rangle \otimes |+; \ldots\rangle \underset{U_0}{\to} | \phi_+\rangle | 0; +, \ldots\rangle \qquad (14.4a)$$

$$| \phi_-\rangle \otimes | 0; \ldots\rangle \underset{U_m}{\to} | \phi_-\rangle \otimes |-; \ldots\rangle \underset{U_0}{\to} | \phi_-\rangle | 0; -, \ldots\rangle \qquad (14.4b)$$

where U_m is the measurement process as before, and U_0 further records the outcome and resets the apparatus. "Experience" is now to be read off from what is registered and what is laid down in memory. What happens now, under these new protocols, when the initial state is the superposition $|\psi\rangle$ of (14.2)? The answer, from linearity, is:

$$|\psi\rangle \otimes |0; \ldots\rangle \underset{U_m}{\to} c_+ |\phi_+\rangle \otimes |+; \ldots\rangle + c_- |\phi_-\rangle \otimes | -; \ldots\rangle$$

$$\underset{U_0}{\to} c_+ |\phi_+\rangle \otimes | 0; +\ldots\rangle + c_- |\phi_-\rangle \otimes | 0; -\ldots\rangle.$$

The final state is a superposition of a record of positive z-component of spin, with a record of negative spin. On repeating the experiment (on the same microscopic system), we obtain:

$$c_1 |\phi_+\rangle \otimes | 0; |\ldots\rangle + c_- |\psi_-\rangle \otimes | 0; -\ldots\rangle$$

$$\underset{U_m}{\to} c_+ |\phi_+\rangle \otimes | +; +\ldots\rangle + c_- |\phi_-\rangle \otimes | -; -\ldots\rangle$$

$$\underset{U_0}{\to} c_+ |\phi_+\rangle \otimes | 0; ++\ldots\rangle + c_- |\phi_-\rangle \otimes | 0; --\ldots\rangle. \qquad (14.5)$$

The final state is a superposition of the record of a "+" outcome followed by a second "+" outcome, with the record of a "−" outcome followed by a second "−" outcome – where each of the latter is just what would have been obtained, on von Neumann's terms, by employing his Process 1 (essentially the projection postulate, extended so as to apply to the state of the measurement device).

What does it mean to have records of measurements *in a superposition* – for there to be two states, each describing a record of measurement, in a superposition? In itself the concept is hardly unfamiliar, in that every student of quantum mechanics is used to the idea of superpositions of contradictory properties at the microscopic level. It isn't even particularly mysterious when divorced from the measurement problem: there are superpositions of light signals and radio programs and TV channels in the electromagnetic field as well – even, or especially, when considered purely classically. This is not a problem. (But it can be made to *seem* mysterious – by insisting not that there is a superposition of radio programs, but that there is a radio program in a superposition – a point we shall come back to.)

Nor are we unused to multiple realities somehow in relation to each other. The world as I write these words is as real as can be, but for you it is some time ago – and your world, as you read these words, is as real as can be for you too, although it is far in the future for me. We have learned, although it is still a matter of philosophical controversy, how to understand these worlds as both existing, as being worlds at different times; Everett invites us to understand two different outcomes as both existing, as being worlds that are orthogonal at the same time.[7]

But if we may understand the idea of distinct processes taking place in a superposition, each as if the projection postulate had been applied, both processes happen with certainty. Where, in all this, is probability?

In the case of Eq. (14.5), the Born rule adds to the projection postulate (according to which just one of Eq. (14.4a) and Eq. (14.4b) is realized) the fact that the *probabilities* of the outcomes are $p = \frac{|c_+|^2}{|c_+|^2 + |c_-|^2}$ and $1 - p$, respectively. In the Everett interpretation, it follows from the Schrödinger equation that the *amplitudes* or the superposed outcomes are (up to a common multiple) $\sqrt{p}$, $\sqrt{(1-p)}$, respectively. In neither case do we as yet have a connection with any observable quantity. For this, what is needed, of course, are multiple experiments: a large number N of silver atoms all prepared in the same state Eq. (14.2), and independently measured in accordance with Eq. (14.4) (whether at the same time or at different times). The result will be a superposition of 2^N states, each a record of a unique sequence of measurement outcomes, each the same as that which would have been obtained by veridical observation, had that sequence resulted by chance, using the measurement postulates.

Everett called them *branches*. It is not too hard to see that the connection between amplitude and Born rule probability is retained for multiple experiments. The amplitude of each branch, at the end of N experiments, as determined by the unitary evolution alone (together with the initial state), equals the square root of the Born rule probability for that sequence of outcomes (just multiply together the probabilities for the results taken sequentially). Now consider the superposition of all those branches with the same relative frequency for the "+" outcome; not quite so obviously, the amplitude of this superposition is highly sensitive to the discrepancy, if any, between that relative frequency and the Born rule quantity for the "+" outcome, the quantity p. Let the discrepancy be ε; then the amplitude falls off exponentially as $\exp -N\varepsilon^2/\kappa$, where $\kappa = 4p(1-p)$ and N is, as before, the number of trials.[8] It is the first of a number of quantum Bernoulli theorem, the quantum analogues of the laws of large numbers: the amplitudes of branches with the "wrong" relative frequencies fall off exponentially quickly in the number of trials, in comparison with the amplitudes of Born rule compliant branches.

The squared amplitudes, in these essential respects, behave just like probabilities. Why the square? Everett had an answer to that question too. Let μ be a probability measure over branches. Then it should satisfy additivity:

$$|\psi\rangle = \sum |\phi\rangle \Rightarrow \mu\left[|\psi\rangle\right] = \sum \mu\left[|\phi\rangle\right]. \tag{14.6}$$

But then, if it is a function of the branch amplitude, it must be the square. (Let $\mu(\sqrt{x}) = f(x)$. Then from Eq. (14.6), $f(\Sigma|c_k|^2) = \Sigma f(|c_k|^2)$, so f is linear in x, $f = kx$ for some constant k. So $\mu(x) = f(x^2) = kx^2$. Here, k is fixed by normalization; Everett also showed that the phase is irrelevant.)

If now we may interpret the squared amplitudes of branches produced by measurements as the physical correlates of the concept of physical probability, we will have explained the Born rule. May we? That question we are postponing to Chapter 15. For the rest of this chapter, we consider rather how Everett's ideas relate to decoherence theory, and what difference that theory makes. But this needs some more history.

14.2   Realism about Measurements

The arguments thus summarized were all in Everett's "'Relative state' formulation of quantum mechanics." What of the relative state? Given an entanglement of the form (14.3), there is no choice of basis, respecting the tensor-product structure between the silver atom and the apparatus, in which the state is a product state; there is no way of attributing a unique pure state to the silver atom, or to the measurement apparatus. But from the total state and a pure state of the apparatus, a unique pure state of the silver atom can be defined. This is its relative state.

On a more deflationary way of putting it: given an entanglement, there are many correlations between states of subsystems. "Relative state" is useful terminology, and reminds us that the structure of the quantum state is relational. There is however a connection with the extended projection postulate. Let the relative state of $|\varphi\rangle$ in an entanglement $|\Psi\rangle$ be $|\psi\rangle$; then up to normalization, $|\varphi\rangle \otimes |\psi\rangle = P_{|\varphi\rangle} \otimes I|\Psi\rangle$. The collapse of the wavefunction, in terms of the extended projection postulate, is relativization of state, relativized to a state of the apparatus. (As Everett put it: "the discontinuous 'jump' into an eigenstate is only a relative proposition, dependent upon the mode of decomposition of the total wavefunction into the superposition, and relative to a particularly chosen apparatus-coordinate value" (Everett, 1957, p. 457).) It naturally generalizes to the relative state of a range of values of dynamical variables (and not just eigenvalues): replace $P_{|\varphi\rangle}$ by a projection P_Δ onto values of variables in some set Δ (a coarse-graining of the parameter space).

But this came later; well into the 1980s, the focus was on simple bipartite systems and the exact definition of a unique basis, in terms of which to decompose the total state, among them that given by the biorthogonal decomposition (or "Schmidt decomposition"). This was a harbinger of other ways of reading Everett's "Relative states" paper. For any entangled state $|\Psi\rangle$ of two subsystems, there exist orthonormal bases $\{|\varphi_k\rangle\}$ and $\{|\eta_k\rangle\}$ such that $|\Psi\rangle = \sum_k c_k |\varphi_k\rangle \otimes |\eta_k\rangle$. If $|c_k| \neq |c_j|$ for $k \neq j$, the bases are unique. (Equivalently, diagonalize the reduced density matrices of the two subsystems.) Dieter Zeh's early work on decoherence theory made use of biorthogonal decomposition (see e.g. Zeh, 1973), what Everett had called the "canonical representation") (Everett, 1973, p. 47). This was also the key to the "modal interpretation" (Dieks and Vermaas, 1998) (for which the failure of uniqueness eventually proved terminal). Zeh's most important contribution to decoherence theory, in collaboration with Ehrich Joos in 1985, was based on rather different ideas.

There were other distractions. Everett spoke of memories and records: what were they records of? Was the approach committed to a model of consciousness,[9] and were there only "appearances" of outcomes, rather than the outcomes themselves? Everett also spoke of a "memory trajectory of an observer" as being a "branching tree," suggesting there is only one observer in a superposition, – a question that Everett in the long dissertation called a "language difficulty." Perhaps the theory isn't committed to a genuine multiplicity after all?[10] In the two-slit experiment, is there one particle at both slits, or are there two particles? Everett also claimed the branches were non-interacting; is that true? And most concerning of all: the branches, the basis states entering into the superposition, were defined by the measurement interaction (in effect the protocols (14.1), (14.3)); but the approach is supposed to be realist, with measurements playing no special role. How was this basis to be defined without them?[11]

Improving on Everett's ideas in any of these ways seemed to require new assumptions, new postulates, new physics; but any step of that kind compromised the chief selling point of his approach – that it is quantum mechanics and nothing else. The problem of basis, the "preferred basis problem," was the most serious. If experiments play no special role (and there is no Born rule to specify the basis), what basis is to be used to define the branches? When does branching occur?

Bryce DeWitt, the first to take Everett's ideas seriously (the terminology "many worlds" is due to him), wrote on them in a number of articles (anthologized in DeWitt and Graham (1973)), but mostly avoided the question of basis. I have found only one comment that directly addressed this question:

> The student should perhaps be reminded again at this point that reality is not described by the state vector alone, but by the state vector plus a set of dynamical operator variables satisfying definite dynamical equations. Decompositions of the form [(14.4)] are not to be regarded as meaningful if they are merely abstract mathematical exercises in Hilbert space. Indeed such mathematical decompositions can be performed in an infinity of ways. Only those decompositions are meaningful which reflect the behavior of a concrete dynamical system.[12]

But the only mentioned systems were experiments, the only dynamical equations were for kinds of measurement interactions. "The many-worlds theory," according to DeWitt, rested on two postulates, both of which he attributed to Everett: the "postulate of mathematical content," concerning operator algebras and Hilbert-space theory, and the "postulate of complexity," namely, that "the world is decomposable into systems and apparata."[13] Here talk of "complexity" was a fig leaf: since when did a fundamental theory *postulate* the existence of measurement devices? Everett had made no such suggestion. What of a world without any people, without any devices?

We find in Everett's long dissertation a rather different set of ideas. He spent time on the definition of the entropy function in classical and quantum statistical mechanics, and on the concept of thermodynamic equilibrium. He pointed out that the unitary equations for many-particle systems did not imply that particles diffuse formlessly: electrons and protons in a box are not uniformly diffused, rather, they are diffused as atoms of hydrogen, and so on for more structured molecules and composites. He reminded us that the diffusion of the center of mass ('centroid') wavefunction for a large numbers of particles in a bound state is extremely slow, for even the smallest visible specks of matter. Everett's proposal was to use, as basis states, wavefunctions for centers of mass, as functions on three-dimensional space, well-localized in position and momentum.

What had always blocked this line of reasoning as an account of how macroscopic physics emerges from the unitary equations of quantum mechanics is that states in general do *not* have this form (and in particular: states following a quantum measurement do not have this form). But on Everett's approach the development of macroscopic superpositions is a feature, rather than a difficulty:

> The general state of a system of macroscopic objects does not, however, ascribe any nearly definite positions and momenta to the individual bodies. Nevertheless, any general state can at any instant be analyzed into a superposition of states each of which does represent the bodies with fairly well defined positions and momenta.

Everett concluded with a reference to von Neumann's construction of projections onto approximately well-localized regions of the one-particle phase space, what von Neumann had called "elementary building blocks of the macroscopic description of the world."[14]

So can the preferred basis be simply *stipulated*, chosen so that each basis state describes the macroscopic in recognizably classical terms? It is sometimes said that if there is to be a preferred basis, it must be *postulated* – written into the axioms of quantum theory – for the Everett interpretation to be well-defined. But neither stipulation nor postulation is needed: the state can be expanded in any basis, and using another, we do not obtain a new and different reality. It is the same reality, only divided up in a new way. Divide it then in the way that makes perspicuous its structure, perhaps one among several.

Are there no constraints? For example, is it true that the branches will be non-interacting? Everett had said that this followed from linearity of the Schrödinger equation alone – so, presumably, whatever basis is used – but here he was less sure-footed. It is true that from linearity, each state evolves as if the other is not there; given unitarity as well, if two such states are orthogonal, they remain orthogonal as they evolve in time. But that does not mean that on *subsequent* branching, the states thus produced will not interfere with each other. Take, for example, the two-slit experiment. On passage through the slits, the photon is in a superposition of two orthogonal states, each originating from one of the slits; they remain orthogonal under any unitary transformation, including their evolution to the screen. But there is interference at the screen: represent each as a superposition of spatially localized states, at the screen, and the latter are *not* all orthogonal to one another.

However, what the two-slit experiment also demonstrates is that overlap in configuration space is necessary for interference. This suggests that states describing large numbers of molecules all well-localized in space with well-defined velocities, even at the microscopic level, can no longer interfere. If the difficulty in getting states like this to overlap in configuration space needed spelling out, Everett could have cited David Bohm, who had written on it both in his then recent book *Quantum Theory*, and in the two-part paper on hidden variables that quickly followed.[15]

It is sometimes said that decoherence theory is needed to show that Everett's branch states do not interfere with each other, and it surely offers quantitative control; but the greater importance of decoherence theory lies elsewhere.

What of the comparison of the state to a tree-like structure? This was forced by the protocol (14.4), where the branching is defined by a measuring interaction for given initial state, and it fits with the analysis of chance: chance events are branching events. The recombination of branches has no such interpretation. But the same reasoning would seem to apply to states of macroscopic bodies differently localized in phase space: we do not expect a superposition of such states to unitarily evolve into a single localized state, for that would seem to require that they be finely synchronized (we shall come back to this shortly).

How *do* states well-localized in position and momentum unitarily evolve? When the mass is not too small or the time-scales are too large, the answer is: classically. To continue Everett's argument, it is not just that states of large numbers of particles well-localized in phase space do not interfere, and not only that at each instant in time they are recognizably interpretable in classical terms; it is that they *behave* classically:

> Each of these states then propagates approximately according to classical laws, so that the general state can be viewed as a superposition of quasi-classical states propagating according to nearly classical trajectories. In other words, if the masses are large or the time short, there will be strong correlations between the initial (approximate) positions and momenta and those at a later time, with the dependence being given approximately by classical mechanics.

A macroscopic body is then a propagating quasiclassical state, approximately obeying a classical Hamiltonian equation – it is dynamically defined. Notice now how the "language difficulty" is handled. A quasiclassical state propagating as such is a thing, so when a superposition develops, do not say "a thing is in a superposition," say rather "there is a superposition of things." (It is not a beam of light in a superposition of two directions, it is two beams in a superposition.) There is the sideways view of the state at each time, as a temporal sequence of superpositions, and there is the vertical view, of a superposition of temporal sequences, of "memory trajectories" or "propagating quasiclassical states" – each in accordance with classical equations.

Applied to the automaton, made up of mechanical parts (the servo-mechanism), Everett's model of the observer, it completed his argument:

> Since large scale objects obeying classical laws have a place in our theory of pure wave mechanics, we have justified the introduction of models for observers consisting of classically describable, automatically functioning machinery, and the treatment of observation is non-vacuous.[16]

The dynamical circle is closed, linking micro to macro, where the latter obeys classical equations, as follows from the Schrödinger equation, ensuring that the protocol (14.3) is satisfied. The measurement protocols are not only stipulated, also they have to be physically instantiated.

But although the circle is closed, it still depends on the concept of measurement; without it, what is the superposition, and what are the branches? DeWitt had a reason for his "postulate of complexity"; how to dispense with measurement interactions? But now there is an obvious candidate for an answer, prefigured by Everett: these propagating quasiclassical states have only *nearly* classical trajectories, and where that approximation is not satisfied, there is branching.

Everett did not take this further step; nor did he offer an argument for his claim of approximate classicality. But he could surely have provided one, drawing, if he wished, on Bohm's book, which included a chapter on the WKB approximation. (Everett cited this book and von Neumann's as his primary influences.) Here is a sketch based on Ehrenfest's theorem in the simple case of a single massive particle with position operator $\hat{x}$. The theorem states that in the state $|\psi\rangle$ and for the potential function $V(\hat{x},t)$ (using the notation $\langle \psi | \hat{x} | \psi \rangle = \langle \hat{x} \rangle_\psi$):

$$m\frac{d^2 \langle \hat{x} \rangle_\psi}{dt^2} = -\langle \nabla V(\hat{x}, t) \rangle_\psi .$$

It does not tell us that a particle behaves classically; failing an interpretation of the expectation value $\langle \hat{x} \rangle_\psi$ for a single system, in an arbitrary state, it tells us nothing at all. But in the special case where $|\psi\rangle$ is well-localized in position and velocity, and the gradient of the potential is approximately constant over such regions and slowly varying in time, we can replace the RHS by the gradient of the potential as a function of $\langle \hat{x} \rangle_\psi$, to obtain:

$$m\frac{d^2 \langle \hat{x} \rangle_\psi}{dt^2} = -\nabla V(\langle \hat{x} \rangle_\psi , t)$$

and whatever exactly $\langle \hat{x} \rangle_\psi$ means, it obeys the same equation as does the position of a classical particle; so long as the approximations hold, $\langle \hat{x} \rangle_\psi$ behaves like the classical particle position behaves. (This argument extends to systems of many weakly interacting bodies, each localized in position and momentum.[17]) And where it fails to hold, then there is branching of such trajectories, and in the branching, the quantum jump, and the appearance of randomness.

The nearly satisfied equations and the propagating quasiclassical states go together; neither is defined without the other. It is these patterns in the wavefunction that define the preferred basis in the Everett interpretation. They are dynamical patterns. They have to be derived from the unitary formalism, they cannot be "stipulated." In this respect the rules that defined the branches in the case of the measurement of spin – the measurement protocols – are a special case. They are given substance by the design and fabrication of a real physical system that instantiates them, according to the Schrödinger equation.

Worlds exist insofar as rule-based branches exist. *That* such branches evolve in accordance with those rules is basis-independent; *use* of a basis that assigns orthogonal states to each branch at each time is a matter of convenience. There is a parallel with the choice of coordinates in general relativity: one choice may be better adapted to the matter distribution, and describe it in a more perspicuous form, than another.

Might there be essentially just the *one* set of rules of this kind, one pattern, namely of states well-localized in phase space, satisfying classical Hamiltonian equations? With these as the building blocks, we can imagine recovering all of macroscopic physics without having to revisit its quantum origins. But that would already have seemed fanciful in Everett's time. Take for example, the property of rigidity; no choice of a potential function for a system of molecules can account for it in classical dynamical terms.

With more urgency, then, what are these rules and states? To what approximations do they hold, for what initial states and quantum Hamiltonians, when does branching occur, and of what are branches made?

14.3   Quantum Histories and Quasiclassical Domains

The list of classical or quasiclassical equations that have been derived from quantum mechanics is lengthy. Examples include Boltzmann's equation, the Focker-Plank equation, the Langevin equation, the Navier-Stokes equation, the Lindblad equation, the Zeh-Joos equation (quantum Brownian motion), the Caldira-Leggett model, and for sufficiently well-behaved potentials and short enough times, classical Hamiltonian equations. Of course, they were derived in a number of different ways; nevertheless, they are subject to theoretical checks, using the unitary formalism assisted as needed by the measurement postulates or their generalizations.

These can all, loosely, be called the business of decoherence theory, mostly developed independent of Everett's ideas (for example, in quantum optics in the 1960s, and in open quantum systems theory in the 1970s). But decoherence theory, on inspection, is itself a mass of different models, for different kinds of dynamical variables, in different coupling regimes and environments.[18] Some of them, spin-foam models for example, have little to do with ordinary matter. And those that do, even if embedded in or derived from non-relativistic quantum mechanics, do not in themselves speak for the Everett interpretation. The crucial question is whether the values of these quasiclassical variables, as they vary in time, obeying these phenomenological equations, can be derived from the unitary formalism in the way Everett had suggested: as superpositions of quasiclassical states where each of the latter propagates along definite trajectories, with branching a function of quantum dissipation and noise.

A proper answer to this question would require an investigation in each case taken separately, but they share much in common. They are mostly equations for many-particle systems. They are all derived by coarse-graining: the integrating out of some degrees of freedom, the definition of new "effective" degrees of freedom, as coarse-grained values of the old, including coarse-graining in time and the separation of "slow" from "fast" variables. The most versatile and widely used technique is probably the path-integral method, but there is a simple Hilbert-space framework as well: the quantum histories formalism. At low energies the two are translatable into each other.[19] From the quantum histories formalism the connection to Everett is direct.

Let us begin with a single time t_k. Let α_k (for fixed k) range over coarse-grained cells of some parameter space $\mathfrak{M}$ at time t_k (phase space, for example). Let P_{α_k} be the associated projectors, so that for $\alpha_k \neq \alpha'_k$, P_{α_k} and $P_{\alpha'_k}$ are orthogonal, with $\sum_{\alpha_k} P_{\alpha_k} = I$. When $\mathfrak{M}$ is a parameter space for commuting variables, the Cartesian product of their spectra, the associated families of projections are the basic tools of spectral theory. When $\mathfrak{M}$ is phase space, a more complicated construction is needed, as given by von Neumann's "elementary building blocks".

It is obvious how to coarse-grain: pass to sums of projection operators, projecting onto unions of subspaces. Let $\{\beta_k\}$ be a coarse-graining of $\{\alpha_k\}$ (so each cell β_k in the parameter space is the union of some of the α_ks); then the projection operator corresponding to β_k is:

$$P_{\beta_k} = \sum_{\alpha_k;\alpha_k \subset \beta_k} P_{\alpha_k}.$$

Sums of commuting projectors correspond to the union of the α_ks, and products to intersections.

All of this is entirely familiar. What is distinctive of the quantum histories formalism is to go over to the Heisenberg picture, and put what has so far been a purely notional time parameter to work. For each time t_k, define

$$P_{\alpha_k}(t_k) \stackrel{def}{:=} U(-t_k) P_{a_k} U(t_k)$$

where $U(t_k) := e^{-iHt_k}$ and H is the quantum Hamiltonian divided by Planck's constant. Consider now a sequence of times $t_N > t_{N-1} > \ldots > t_1$, and let $\alpha := \langle \alpha_N, \alpha_{N-1}, \ldots, \alpha_1 \rangle$ represent a corresponding sequence of parameters. Define the associated "chain" or "class" operators $C_a := P_{\alpha_N}(t_N)\ldots P_{\alpha_1}(t_1)$, in general the product of non-commuting operators. Let the initial state $t_k = 0$ be $| \psi \rangle$; then $C_a | \psi \rangle$ is obtained by: unitarily propagate $| \psi \rangle$ to t_1, project out the state $P_{\alpha_1} U(t_1) | \psi \rangle$, unitarily propagate to t_2, project out the state $P_{\alpha_2} U(t_2 - t_1) P_{\alpha_1} U(t_1) | \psi \rangle$, and so on – obtaining precisely the same state, at each time, as were a sequence of measurements performed, up to that time, yielding the outcomes $\alpha_1, \alpha_2, \ldots$, and applying the extended projection postulate (whether sequentially or just at the end). The Schrödinger-picture end-state obtained in this way has amplitude equal to the square root of the product of all the Born rule probabilities for that sequence of outcomes. Unitarily evolve it back to t_0 and we obtain the Heisenberg-picture state $C_a | \psi \rangle$, representing that history.

Each history a, as given by the sequence of parameters $\langle \alpha_N, \alpha_{N-1}, \ldots, \alpha_1 \rangle$, may seem to have nothing to do with any unitary equations (they may appear quite random, for example). But the *superposition* of all these propagating sequences, up to any time, is exactly the same as the unitary evolution of the total state to that time. That is:

$$\sum_{\langle \alpha_N, \alpha_{N-1}, \ldots, \alpha_1 \rangle} P_{\alpha_N} U(t_N - t_{N-1}) P_{\alpha_{N-1}} U(t_{N-1} - t_{N-2}) \ldots P_{\alpha_1}(t_1) | \psi \rangle$$

$$= \sum_a U(t_N) C_a | \psi \rangle = U(t_N) \sum_a C_a | \psi \rangle = U(t_N) | \psi \rangle$$

where the last equality follows from the identity $\sum_a C_a = I$. The Schrödinger-picture states $U(t_N) C_a | \psi \rangle$ are the states of Everett's branches at $t = t_N$.

Coarse-graining of projectors, defined by their sums, automatically extends to coarse-grainings of chain operators, defined by their sums. If β is a coarse-graining of a, denote $a \subset \beta$, then

$$C_\beta = \sum_{\alpha;\alpha \subset \beta} C_\alpha.$$

The operator $C_\alpha^\dagger C_\alpha$ is self-adjoint and positive: as such it defines a positive operator valued measure (POV measure), of the sort that now widely supplements the measurement postulates. (In the case of two-step histories, they were first known as "effects" (Davies, 1984).)

The Heisenberg picture is the natural one for determining the structure of the orbit of the quantum state, under time evolution, in four-dimensional terms. Relativistic quantum field theory, in both Lagrangian and axiomatic formulations, almost always uses the Heisenberg picture. We may think of the quantum state as fixed once and for all, and branch vectors $C_\alpha | \psi \rangle$ as components of this state, representing the corresponding histories. The perspective is four-dimensionalism, in the metaphysicians' sense, a "quantum block universe."[20]

What are all these histories, exactly? No use has as yet been made of the identification of the α's as coarse-grained values of position and momenta; the resolution of the identity could have been anything. But if it could be anything, the states $C_\alpha | \psi \rangle$ may not even be orthogonal, and there would be no guarantee that the interpretation of squared amplitude as probability is consistent with coarse-graining. In general, it is *not* true that

$$\left|\,C_\beta\,|\psi\rangle\right|^2 = \sum_{\alpha;\alpha\subset\beta} |\,C_\alpha\,|\psi\rangle|^2$$

(the so-called "sum rule").[21] But the two, very nearly, go together; when the $C_\alpha|\,\psi\rangle$s are orthogonal, the sum rule is satisfied (and when the sum rule is satisfied, the real part of $\langle\psi|\,C_{\alpha'}^\dagger C_\alpha|\,\psi\rangle$ for $\alpha\neq\alpha'$ vanishes). The "consistent histories interpretation" was developed in the hope that the sum rule would determine a preferred basis (the P_{α_k}'s for each k) for given state and Hamiltonian all on its own, and thus a well-defined probability measure over a space of histories, only one of which is realized, yielding a 'one-world' interpretation of quantum mechanics. Just how weak a condition consistency is was shown by Dowker and Kent (1996), who effectively put paid to that ambition.

The sum rule is recognizably Everett's additivity requirement, substituting the $C_\alpha|\,\psi\rangle$s for Everett's branches. (In the case of Everett's model of an automaton, if at each time a record is preserved of the sequence of outcomes prior to that time, the sum rule is automatically satisfied. Conversely, where two histories interfere, there can be no record of events in the two histories that differ.) For an example of coarse-graining, state, and Hamiltonian *violating* the sum rule, consider the two-slit experiment, and for projections at each time, coarse-grainings in space: projections onto the aperture Δ_1 at t_1, at each of the slits Δ_+ and Δ_- at t_2, and at a fixed region of the screen Δ_3 at t_3. The histories $\alpha_\pm := \langle\Delta_1, \Delta_\pm, \Delta_3\rangle$ sum as they must to $\beta := \langle\Delta_1, \Delta_+ \cup \Delta_-, \Delta_3\rangle$, but not the associated probabilities – indeed $C_{\alpha_+}|\,\psi\rangle$ and $C_{\alpha_-}|\,\psi\rangle$ are not orthogonal, and interfere at the screen.

Unlike orthogonality of Schrödinger-picture states at an instant of time, the orthogonality of Heisenberg-picture states representing histories involves the initial state and the Hamiltonian. A *decoherent history space* over a sequence of times $t_N > .. > t_k > \ldots > t_1$ is a quadruple $\langle|\,\psi\rangle, H, \mathfrak{M}, \{\alpha_k\}\rangle$ for which the states $C_\alpha|\,\psi\rangle$ are orthogonal. Such a space has the natural finite measure $\mu[\alpha] := \mu[\,C_\alpha|\psi\rangle] = |\,C_\alpha|\psi\rangle|^2$. Helping ourselves temporarily to the notion of probability, $\mu[\alpha]$ is the probability of history α. Let $\{\gamma\}, \{\delta\}$ be coarse-grainings of $\{\alpha\}$ for state $|\,\psi\rangle$, and let their composition, denote $\gamma * \delta$, be the sequence $\langle\gamma_n \cap \delta_n, \ldots, \gamma_1 \cap \delta_1\rangle$. If $\mu[\delta] \neq 0$, we may then define the conditional probability of γ relative to δ as:

$$\mu[\gamma/\delta] = \frac{\mu[\gamma * \delta]}{\mu[\delta]}. \tag{14.7}$$

These conditional probabilities include retrodictive probabilities of some past event, conditional on some future event, or conditional on some future sequence of events; or of a past, present, or future sequence of events, conditional on some event or sequence of events. (The two-state vector formalism is the special case of three-step histories in which the first and third time projector is one-dimensional, determining the probabilities of the second ('the present').) But equally, the formalism can be divested of this probability interpretation: these are ratios among squared amplitudes of state vectors, correlations among states, representing sequences of events, awaiting further interpretation.

Now for Everett's picture of a tree-like structure to the state. Given a decohering history space $\langle|\,\psi\rangle, H, \mathfrak{M}, \{\alpha_k\}\rangle$, it is always possible to define a new decohering history space $\langle|\,\psi\rangle, H, \mathfrak{M}, \{\epsilon_k\}\rangle$, where $\{\epsilon_k\}$ is a fine-graining of $\{\alpha_k\}$, with "branching structure" – in which histories only diverge to the future and never recombine.[22] (Branching structure, like decoherence, was ensured for Everett's automaton states, as defined by Eq. (14.3), because they encoded records of the past.) Formally, for any $t_k > t_j$, for any ϵ_k, ϵ_j,

$$\mu[\epsilon_j/\epsilon_k] = \frac{\left|P_{\epsilon_k}(t_k)\, P_{\epsilon_j}(t_j)\,|\psi\rangle\right|^2}{\left|P_{\epsilon_k}(t_k)\,|\psi\rangle\right|^2} \approx 0 \text{ or } 1. \tag{14.8}$$

The past of an event ϵ_k is therefore approximately unique. There is only one way, from a configuration at time t_k, of tracing a preceding sequence of configurations (think of a branching tree); Everett's concept of branching thus generalizes. But to what is this temporal asymmetry to be traced?

Evidently not to the unitary evolution, which is time-reversal invariant. We earlier saw reasons to expect branching in the case of states initially well-localized in position and momentum – that is, in the structure of the initial state. The point applies more generally, and it is the same as the explanation of the arrow of time in classical statistical mechanics: it is to be sought in the structure of the initial state.[23] Thus, if $|\psi\rangle$ at $t = 0$ for a given Hamiltonian H and coarse-graining $\{\alpha_k\}$ on $\mathfrak{M}$ defines a decoherent history space, hence with branching structure, $C_\alpha |\psi\rangle$ at $t = 0$ does not.

This point is worth perusing. What happens if we take a state like $C_\alpha |\psi\rangle$ as the initial state? Formally, as a Heisenberg-picture state, it is defined at $t = 0$, like $|\psi\rangle$; if we pass to the Schrödinger picture, and unitarily evolve it forward in time, it is the state $U(t) C_\alpha |\psi\rangle$, which for $t = t_N$ is the state of the Everett branch for the history α at t_N. It seems we have everything that we could wish for, a single history theory with a purely unitary time evolution. But no: while there is a single branch vector at $t = t_N$, there are innumerable others at earlier times – indeed a superposition of non-orthogonal branches, all with amplitudes and phases delicately adjusted as they unitarily evolve so that they all interfere with each other at t_N and all save one wink out of existence. Moving forward in time, after t_N, it is Everettian business as usual, and orthogonal branching, and no fine-tuning (unless, of course, the history space was fine-tuned to begin with).[24]

Put now to one side the probability interpretation, and view norms and ratios in norms as mere correlations, mere relations among amplitudes and relative states. View Eqs. (14.7) and (14.8) as an extension of Everett relativization at a single time to different times, and of correlations among states representing sequences of events. Understood in this way, the decoherent histories formalism provides a general language, a kind of four-dimensionalism, for describing the universal state, in which orthogonality of histories is as natural a criterion for a basis of states as is orthogonality of states in the case of a single time. As we have seen, it interestingly involves a direction in time, as determined by the initial state; what else does it involve?

The concept of *quasiclassical domain*, introduced by Murray Gell-Mann and Jim Hartle in the late 1980s,[25] is of a consistent history space for which the coarse-grained variables, the αs, approximately satisfy some *closed* system of equations, as they vary along each history; it is a history space made up only of certain kinds of sequences. Or more precisely (since any quantum history space contains all possible histories definable as sequences of values of the coarse-grained variables), those histories that do not conform to the equation have negligible norm in comparison to those that do. (We now see that the Born rule, in a way, falls in this category too.) It is then an open question as to whether and what kinds of quasiclassical domains may be found, with what kinds of equations, coarse-grained variables, initial states, and Hamiltonians.

For a metaphysical way of putting it, a quasiclassical domain is defined when the universal state can be written as a superposition of histories almost all of which are lawlike, that almost all of which obey a definite rule or equation. They are histories of propagating quasiclassical states, obeying definite rules – *just as envisioned by Everett*. But now Von Neumann's "elementary building blocks" are only one example.

I gave the punch line in advance: all the important, effective, non-relativistic equations for bulk matter, gasses, and fluids have now been obtained in this way.[26] Of course several of them predate the quantum histories formalism, and implicitly or explicitly rely on the measurement postulates; *but they can be cast* into the quantum histories formalism, and we know how to interpret the measurement postulates in the Everett interpretation. Generically, these quasiclassical equations are only approximately satisfied: departures involve branching, and in some cases, as in classically chaotic systems, pervasive branching. The equations themselves may involve dissipation and noise. No wonder, then, that branching and worlds cannot be defined axiomatically; they are not defined at the microscopic level at all. They are *emergent* structures, to be extracted from the unitary equations for large numbers of particles, using methods similar to those that apply across the board in the physical sciences.[27]

The implications are far-reaching. For the first time, the Everett interpretation (and arguably quantum mechanics) is freed from any dependence on classical physics (despite the name "quasiclassical," the equations that define a quasiclassical domain could in principle be entirely foreign to classical physics, the variables completely alien). It is no longer dependent on the concept of measurement; branching, and with it the preferred basis, is emergent structure, dynamically defined. The arrow of time in thermal and decoherence terms is aligned. Everett's automaton argument is much stronger: the automaton itself need no longer be a mechanical system, but could be made of anything. More importantly, what can be recorded in its memory, corresponding to the sequence of its relative states, is not just the statistics of quantum experiments, but the observable law-like behavior of everything else that is going on in the laboratory – the entire phenomenology of materials, fluids, and gases, all in excellent agreement with experiment. In these respects, the Everett interpretation is much better than either pilot-wave theories or dynamical collapse theories. They solve the measurement problem, but rarely even try to obtain quasiclassical phenomenology ("the classical limit") in their terms. (Of course, pilot-wave theory can always help itself to results obtained under the unitary formalism alone, since it too preserves the Schrödinger equation as universal – but thereby further illustrating the epiphenomenal character of the hidden variables. See also Rosaler (2015).)

14.4 Everett's "Note Added in Proof"

What to believe, in the face of all this evidence? Quantum mechanics may yet give way to a better theory, with an entirely different set of ideas. Doubts on the side of the probability interpretation may yet undermine the approach: see the companion Chapter 15. Experimental discoveries could as always change everything – of gravitationally induced state reduction, for example. Everett's place in history remains uncertain. But what if the superposition principle, and low-energy quantum mechanics, is here to stay, with no hint of any further, "hidden" variables?

Here is Everett's last word on the matter, in his "Note added in proof," added without Wheeler's permission, the one place where we know he spoke in his own voice. It is fitting to reprint it in full:

> *Note added in proof* – In reply to a preprint of this article some correspondents have raised the question of the "transition from possible to actual," arguing that in "reality" there is – as our experience testifies – no such splitting of observer states, so that only one branch can ever actually exist. Since this point may occur to other readers the following is offered in explanation.
>
> The whole issue of the transition from "possible" to "actual" is taken care of in the theory in a very simple way – there is no such transition, nor is such a transition necessary for the theory to be in accord with our experience. From the viewpoint of the theory all elements of a superposition (all "branches") are "actual," none any more "real" than the rest. It is unnecessary to suppose that all but one are somehow destroyed, since all the separate elements of a superposition individually obey the wave equation with complete indifference to the presence or absence ("actuality" or not) of any other elements. This total lack of effect of one branch on another also implies that no observer will ever be aware of any "splitting" process.
>
> Arguments that the world picture presented by this theory is contradicted by experience, because we are unaware of any branching process, are like the criticism of the Copernican theory that the mobility of the earth as a real physical fact is incompatible with the common-sense interpretation of nature because we feel no such motion. In both cases the argument fails when it is shown that the theory itself predicts that our

experience will be what it in fact is. (In the Copernican case the addition of Newtonian physics was required to be able to show that the earth's inhabitants would be unaware of any motion of the earth.)

It was Galileo, not Newton, who rebutted that criticism of the Copernican theory, on the basis of an incomplete and, at points, faulty conception of the physics. Everett's argument to show that we cannot be aware of branching ("splitting") was likewise incomplete: it does not rest on linearity alone. Everett, like Galileo, did not have all the physics needed to show that the appearances would be as they seem. But there is another comparison that is even more informative: between Everett's idea that all that there is are relative states and correlations, and the idea that all that there is are relative distances and relative velocities. The comparison is with Descartes. Both elevated a principle (the superposition principle; the principle of inertia) to universal status; both, in their different ways, had their writings suppressed. Both were transitional figures: neither was able to show, on dynamical grounds, what were the superposed worlds, what were the inertial motions. Both died young, their work unfinished. Each argued for his world-view in the same way: by a demonstration that to a mechanical being inhabiting such a universe, the world would seem exactly the same as it seems to us in the known universe – in Descartes' case, in *Le Monde*.

Notes

1 Bell (1987, p. 201).
2 See Saunders (this volume, Chapter 15). The *locus classicus* is Wallace (2012). For the major lines of debate, see also Saunders et al. (2010), with particular emphasis on the decision-theory strategy introduced by Deutsch (1999).
3 In DeWitt and Graham (1973). For the story of Everett's relationship with Wheeler, see Byrne (2010).
4 See Saunders (2005) for an extended discussion of this idea.
5 This was the main argument for the "complementarity" (mutual exclusivity) of causal and spatiotemporal descriptions on its first appearance (Bohr, 1928).
6 As Everett pointedly reminded us (Everett, 1957, p. 455).
7 See Saunders (1995, 1998) for more on the parallels between Everett's branching structure and four-dimensionalism in relativity theory.
8 See Saunders (2010, pp. 188–189) and Wallace (2012, p. 140) for further discussion.
9 As suggested by Albert and Loewer (1988), Lockwood (1989), Barrett (1999), and Zeh (2000).
10 A question repeatedly raised by Jeffrey Barrett; see e.g. Barrett (2011).
11 Additional concerns were raised about the interpretation of probability (in particular, the "branch counting rule"); see Saunders, this volume, Chapter 15.
12 DeWitt (1971, p. 210). See also Ballentine (1973, p. 233).
13 DeWitt (1970, p. 168).
14 von Neumann (1955, pp. 406–409).
15 Bohm (1951, ch. 6, 16, sec. 25), (1952, p. 178, fn. 18). Everett cited both these publications, albeit in other connections.
16 Everett (1973, pp. 89–90). "Non-vacuous" was patently a jibe at von Neumann.
17 This went unremarked in Hartle (2010).
18 See Crull (this volume, Chapter 13).
19 Path-integral methods of coarse-graining were introduced by Feynman and Vernon (1963), and in much of their work Gell-Mann and Hartle defined decoherence in terms of the decoherence functional and path integrals.
20 Saunders (1995). For background in metaphysics, see e.g. Sider (2001).
21 Due originally to Griffiths (1984).
22 Griffiths (1993); see also Wallace (2012, pp. 93–95).
23 See Shahvisi (this volume, Chapter 30) and Frigg and Werndl (this volume, Chapters 27 and 28).
24 For more on the Everett interpretation and the arrow of time, see Wallace (2012, Ch. 9).
25 Gell-Mann and Hartle (1990, 1993).
26 See, for example, Joos et al. (2003 Ch. 5).
27 The parallel was first drawn by Wallace (2003); it is developed at length in Wallace (2012, Ch. 1–3).

References

Albert, D. and Loewer, B. (1988). Interpreting the many-worlds interpretation. *Synthese*, 77: 195–213.

Ballentine, L.E. (1973). Can the statistical postulate of quantum theory be derived?—A critique of the many-universes interpretation. *Foundations of Physics*, 3: 229.

Barrett, J. (1999). *The Quantum Mechanics of Mind and Worlds*. Oxford: Oxford University Press.

Barrett, J. (2011). On the faithful interpretation of pure wave mechanics. *British Journal for the Philosophy of Science*, 62(4): 693–709.

Bell, J. (1987). *Speakable and Unspeakable in Quantum Mechanics*. Cambridge: Cambridge University Press.

Bohm, D. (1951). *Quantum Theory*. Englewood Cliffs, New Jersey: Prentice-Hall.

Bohm, D. (1952). A suggested interpretation of the quantum theory in terms of "hidden" variables. 1. *Physical Review*, 85: 166–179.

Bohr, N. (1928). The quantum postulate and the recent development of atomic theory. *Nature*, 121: 580–591. Reprinted in N. Bohr, *Atomic Theory and the Description of Nature*. Cambridge: Cambridge University Press, 1934.

Byrne, P. (2010). Everett and Wheeler, the untold story. In S. Saunders, J. Barrett, A. Kent and D. Wallace (eds.), *Many Worlds? Everett, Quantum Theory, and Reality*. Oxford: Oxford University Press, pp. 521–541.

Davies, B. (1984). *Open Quantum Systems Theory*. Oxford: Oxford University Press.

Deutsch, D. (1999). Quantum theory of probability and decisions. *Proceedings of the Royal Society of London*, A455: 3129–3137. Available at: arXiv.org/abs/quant-ph/9906015.

DeWitt, B. (1971). The many-universes interpretation of quantum mechanics. In *Proceedings of the International School of Physics 'Enrico Fermi', Course IL: Foundations of Quantum Mechanics*. Academic Press. Reprinted in DeWitt and Graham (1973).

DeWitt, B. and Graham, N. (1973). *The Many-Worlds Interpretation of Quantum Mechanics*. Princeton, NJ: Princeton University Press.

Dieks, D. and Vermaas, P., editors. (1998). *The Modal Interpretation of Quantum Mechanics*. Amsterdam: Kluwer.

Dowker, F. and Kent, A. (1996). .On the consistent histories approach to quantum mechanics. *Journal of Statistical Physics*, 82: 1575–1646.

Everett III, H. (1957). "Relative state" formulation of quantum mechanics. *Reviews of Modern Physics*, 29: 454–462. Reprinted in DeWitt and Graham (1973, pp. 141–150).

Everett III, H. (1973). Theory of the universal wave function. In B. DeWitt and N. Graham (eds.), *The Many-Worlds Interpretation of Quantum Mechanics*. Princeton, NJ: Princeton University Press, pp. 3–140.

Feynman, R. and Vernon, F.L. (1963). The theory of a general quantum system interacting with a linear dissipative system. *Annals of Physics*, 24: 118–173.

Gell-Mann, M. and Hartle, J.B. (1989). Quantum mechanics in the light of quantum cosmology. In W.H. Zurek (ed.), *Complexity, Entropy, and the Physics of Information*. Reading: Addison-Wesley.

Gell-Mann, M. and Hartle, J.B. (1993). Classical equations for quantum systems. *Physical Review D*, 47: 3345–3382. Available at: arXiv.org/abs/gr-qc/9210010.

Griffiths, R. (1984). Consistent histories and the interpretation of quantum mechanics. *Journal of Statistical Physics*, 36: 219–272.

Griffiths, R. (1993). Consistent interpretation of quantum mechanics using quantum trajectories. *Physical Review Letters*, 70: 2201–2204.

Halliwell, J. (2010). Macroscopic superpositions, decoherent histories, and the emergence of hydrodynamic behaviour. In S. Saunders, J. Barrett, A. Kent and D. Wallace (eds.), *Many Worlds? Everett, Quantum Theory, and Reality*. Oxford: Oxford University Press, pp. 99–120.

Hartle, J. (2012). Quasiclassical realms. In S. Saunders, J. Barrett, A. Kent and D. Wallace (eds.), *Many Worlds? Everett, Quantum Theory, and Reality*. Oxford: Oxford University Press, pp. 73–98.

Joos, E., and H.D. Zeh (1985). The emergence of classical properties through interaction with the environment. *Zeitschrift für Physik B: Condensed Matter*, 59: 223–243.

Joos, E., H.D. Zeh, C. Kiefer, D. Giulini, J. Kupsch, and I. Stamatescu (2003). *Decoherence and the Appearance of a Classical World in Quantum Theory*, 2nd Ed. Berlin: Springer-Verlag.

Lockwood, M. (1989). *Mind, Brain, and The Quantum*. Oxford: Basil Blackwell.

Rosaler, J. (2015). Is de Broglie-Bohm theory specially equipped to recover classical behavior? *Philosophy of Science*, 82: 1175–1187.

Saunders, S. (1995). Time, quantum mechanics, and decoherence. *Synthese*, 102: 235–266.

Saunders, S. (1998). Time, quantum mechanics, and probability. *Synthese*, 114: 405–444. Available at: arXiv.org/abs/quant-ph/0112081.

Saunders, S. (2005). Complementarity and scientific rationality. *Foundations of Physics*, 35: 417–447.

Saunders, S. (2010). Chance in the Everett interpretation. In S. Saunders, J. Barrett, A. Kent and D. Wallace (eds.), *Many Worlds? Everett, Quantum Theory, and Reality*. Oxford: Oxford University Press, pp.181–205.

Saunders, S., Barrett, J., Kent, A. and Wallace, D. (2010). *Many Worlds? Everett, Quantum Theory, and Reality*. Oxford: Oxford University Press.

Sider, T. (2001). *Four-Dimensionalism: An Ontology of Persistence and Time*. Oxford: Oxford University Press.

Von Neumann, J. (1932). *Mathematische Grundlagen Der Quantenmechanik*, translated by R.T. Beyer as *Mathematical Foundations of Quantum Mechanics*. Princeton, NJ: Princeton University Press (1955).

Wallace, D. (2003). Everett and structure. *Studies in the History and Philosophy of Physics*, 34: 87–105. Available at: arXiv.org/abs/quant-ph/0107144.

Wallace, D. (2012). *The Emergent Multiverse: Quantum theory according to the Everett Interpretation*. Oxford: Oxford University Press.

DeWitt, B., and Graham, N. (1973). *The Many-Worlds Interpretation of Quantum Mechanics*. Princeton, NJ: Princeton University Press.

Zeh, D. (1973). Toward a quantum theory of observation. *Foundations of Physics*, 3: 109–116. Revised version available at: arXiv.quant-ph/030615v1.

Zeh, D. (2000). The problem of conscious observation in quantum mechanical description. *Foundations of Physics Letters*, 13: 221–233. Available at: arXiv.quant-ph/9908084v3.

Further Reading from the Editors

For an accessible introduction to a contemporary Everettian vision, see S. Carroll, *Something Deeply Hidden* (E.P. Dutton, 2019). For a high-profile criticism of decoherence-based Everettianism, see T. Maudlin, "Can the World be Only Wavefunction?" in S. Saunders, J. Barrett, A. Kent and D. Wallace (eds.), *Many Worlds? Everett, Quantum Theory, & Reality* (Oxford and New York: Oxford University Press, 2010):121–143. A different sort of argument against the ontology of Everettian QM, based on probabilistic considerations, is given by D. Baker, "Measurement Outcomes and Probability in Everettian Quantum Mechanics," *Studies in History and Philosophy of Science Part B – Studies in History and Philosophy of Modern Physics*, 38 (2007):153–169. Everettians come in many flavors: discussion of a variety of approaches can be found in L. Vaidman, "Many-Worlds Interpretation of Quantum Mechanics," *Stanford Encyclopedia of Philosophy* (2014). An application to the metaphysics of modality is given by A. Wilson, *The Nature of Contingency: Quantum Physics as Modal Realism* (Oxford: Oxford University Press, 2020).

15

THE EVERETT INTERPRETATION: PROBABILITY[1]

Simon W. Saunders

The Everett interpretation of quantum mechanics is, inter alia, an interpretation of objective probability: an account of what probability really is. In this respect, it is unlike other realist interpretations of quantum theory or indeed any proposed modification to quantum mechanics (like pilot-wave theory or dynamical collapse theories); in none of these is probability itself the locus of inquiry. As for *why* the Everett interpretation is so engaged with the question of probability, it is in its nature: its starting point is the unitary, deterministic equations of quantum mechanics, and it introduces no hidden variables whose values are unknown.

Does it explain what objective probability is, or does it explain it away? If there are chances out there in the world, they are the *branch weights*. All who take the Everett interpretation seriously are agreed on this much: there is macroscopic branching structure to the wavefunction, and there are the squared amplitudes of those branches, the branch weights. The branches are worlds – provisionally, worlds at some time. The approach offers a picture of a branching tree with us at some branch, place, and time within. But whether these weights should properly be called "chances" or "physical probabilities" is another matter. For some, even among Everett's defenders, it is a disappearance theory of chance; there *are* no physical chances; probability only lives on as implicit in the preferences of rational agents, or as a "caring measure" over branches, or in degrees of confidence when it comes to the confirmation of theories or laws; probability has no place in the physics itself.[2] The interpretation was published by Hugh Everett III in 1957 under the name "'Relative state' formulation of quantum mechanics"; he named a much longer manuscript "Wave Mechanics Without Probability."[3]

Dissent on this point among Everett's defenders is significant. If the basic category of probability is to be abolished, Everett's approach can hardly claim to be an *interpretation* of quantum mechanics: for is not the theory couched in terms of the language of probability? Critics may well conclude that their work has been done for them, but for three reasons they should think differently.

First, because many of the criticisms that apply to probability in the Everett interpretation apply to every other half-way serious theory of physical probability. The difficulties may only be more vivid, more obvious, in the Everett interpretation; that may be to its credit.[4]

Second, because in philosophy of probability over the last three decades (and a mark of the influence of David Lewis's writings) a great deal of attention has been devoted to "reductive" theories of chance: theories that start with a non-chancy, "base" level of properties and relations ("Humean" properties and relations), on which probabilities, if any, are to supervene. Everett called his work "wave mechanics without probability" for good reason: it provides a non-chancy base level of categorical properties and relations. It fits this mold. Moreover, so much philosophical time and energy has only gone to show just how difficult the chance concept is in comparison to other more successful reductive projects (e.g., of causation, modality, persons, and even mind). It was with chance, famously, that Lewis feared his "Humean supervenience" project would fail. Naïve frequentism, the view that chances at t are relative frequencies up to time t, fails for well-known reasons;[5] the proposal that they may depend on future relative frequencies too, one of Lewis's main innovations, falls prey to "undermining" (see Section 15.5). But the Everett interpretation of quantum mechanics provides a different supervenience basis (branching structure) and new primitive relations (relations in phase and amplitude), and the differences appear to be decisive: armed with these, the usual criteria for reductive theories of chance can be met to perfection.

Third, among those criteria for a successful reductive theory there is one that has rarely been met even in isolation: the decision-theory link, the link between probability and rational belief. As distilled by Lewis and others, it is to explain, or justify, the Principal Principle – roughly speaking, the principle that if you know the chance of E at t is x, then your credence in E at t should be x. The theory of probability as it was developed by Blaise Pascal and Pierre-Simon Laplace had always included this link, for they based it on a principle of indifference, in turn based on symmetries. This worked well for games of chance, but less so for subsequent applications of probability theory to physics,[6] and in its development in terms of measure theory by Emile Borel at the end of the 19th century the link with the principle of indifference was lost. It has played little role in the philosophy of probability since. Yet nothing else – and certainly not observed relative frequencies – seems apt to provide a link between probability, as something out there in the world, and rationality.

In this context the observation made by David Deutsch in 1999 that given certain axioms of decision theory a principle of indifference must operate in quantum mechanics in the special case of experimental outcomes of equal amplitudes is of great importance. His further demonstration that by the use of various ancillary devices, the principle in effect forces the Born rule, the standard probability rule of ordinary quantum mechanics, is game-changing. The argument was substantially strengthened by David Wallace, making do with considerably weaker axioms of decision theory, supplemented instead by explicit appeal to the decoherence-based Everett interpretation. Wallace's book *The Emergent Multiverse* published in 2012 is a landmark in the foundations of both probability theory and quantum mechanics. Nothing comparable for the decision theory link has been achieved in any one-world theory of chance, however fanciful, let alone one based on any extant science.

It is not any old science. Quantum mechanics, by a long shot, is the most accurate, prolific, and unifying physical theory that has ever been seen – while yet, somehow, remaining the least understood. If the Everett interpretation is correct, it explains that fact as well. If in reality there is macroscopic branching satisfying the Schrödinger equation, and no hidden variables, no wonder quantum mechanics is so difficult to understand for those (the great majority) intent on a one-world interpretation. None of the usual paradoxes of quantum mechanics pose any problem to the Everett interpretation, and nor is it non-local in Bell's sense.[7]

15.1   The Connection with Uncertainty

Many of the conceptual questions that arise with probability on the Everett interpretation of quantum mechanics arise equally in one-world theories, but one stands out:[8] unlike in classical statistical mechanics and hidden-variable quantum theories, and unlike in a stochastic dynamical theory, there seems to be no room in the theory for the usual connection between chance and uncertainty. Knowing the quantum state and given a unitary deterministic process (the Schrödinger equation), in the absence of any additional and "hidden" variables, when a quantum experiment is performed the result is a superposition of all the outcomes, with varying phases and amplitudes. If that is all that there is, it seems we know everything there is to know – but then it seems there can be no place for *uncertainty*; and with no uncertainty, there is no probability either. It is difficult even to know what credences could possibly mean: degrees of belief in what, exactly?

I offer three answers, increasingly deflationary, all consistent with one another. The first is that the model of measurement on which the argument rests is only a fragment of a realistic analysis. The branching structure of macroscopic bodies involves much more – inter alia, it requires the quantum histories formalism and the theory of quasiclassical domains. In the latter formalism, there are superpositions of worlds understood as serial quasiclassical histories that extend to future times. The perspective is structural, a "quantum block universe." But from this perspective there is an obvious candidate for ignorance: we do not know which history is our own. Uncertainty is self-locating uncertainty, and degrees of belief are beliefs about one's location among all these histories.

A similar conclusion can be arrived at from a number of different philosophical directions, beginning from any of persons, persistence, language use, or identity. Take personhood: what are persons in terms of the Everettian reductive base? In a one-world setting a popular answer is that they are continuants, four-dimensional histories, spatially localized at each time (so spacetime worms or world tubes). The same can be taken over to branching structure (work backward, from arbitrary future times). Then given the usual attribution of speech acts to persons, ignorance in the face of branching is inevitable. If a *person* asks, at time *t*, prior to branching, what happens next *to her*, she cannot possibly know.[9]

This may seem like a cheat. Normally expressions that use indexicals like "here" or "now" or personal pronouns like "she" are tagged to places, times, persons, in a way that is causally informative. Prior to branching, how do persons that only differ with respect to future contingents presently pick themselves out? But they don't have to: the perspective is non-local in time, as seems appropriate to human agency.

A second argument for uncertainty is a stripped-down version of the first. According to this, uncertainty is still bound up with self-location, but it is localized to the chance process itself. This will need some stage setting. Consider a simple schematic measurement process, say the measurement of spin as in Chapter 14. As before let the apparatus or observer be initially in a "ready" state $|0\rangle$, and suppose that when presented with a spin system with positive spin in the state $|\phi_+\rangle$ it passes to a state in which it registers spin plus, denote $|+\rangle$, and presented with a state $|\phi_-\rangle$ it passes to a state in which it registers spin minus, denote $|-\rangle$. That is, the unitary evolution U_m (the subscript "m" is for measurement) should satisfy the following:

$$|\phi_+\rangle \otimes |0\rangle \xrightarrow[U_m]{} |\phi'_+\rangle \otimes |+\rangle \tag{15.1a}$$

$$|\phi_-\rangle \otimes |0\rangle \xrightarrow[U_m]{} |\phi'_-\rangle \otimes |-\rangle. \tag{15.1b}$$

(There is no need to assume the experiment is repeatable; hence the primes on the spin states following the measurement, which could be anything.) For an initial state

$$|\psi\rangle = c_+|\phi_+\rangle + c_-|\phi_-\rangle \tag{15.2}$$

where c_+ and c_- are non-zero complex numbers; it then follows from linearity that

$$|\psi\rangle \otimes |0\rangle \xrightarrow[U_m]{} c_+|\phi'_+\rangle \otimes |+\rangle + c_-|\phi'_-\rangle \otimes |-\rangle. \tag{15.3}$$

There are the two states at the end, in a superposition, two macroscopic branches; but there is only the one before the measurement. If we suppose, contrary to the preceding account, that persons are localized in time (as, say, in stage theory), and are fully described by a quantum state, like the initial state $|0\rangle$, then prior to the measurement it seems there can be no self-locating uncertainty.[10] But look again. Eq. (15.3) can equally be written as

$$c_+ \left|\phi_+\right\rangle \otimes \left|0\right\rangle + c_- \left|\phi_-\right\rangle \otimes \left|0\right\rangle \underset{U_m}{\rightarrow} c_+ \left|\phi'_+\right\rangle \otimes \left|+\right\rangle + c_- \left|\phi'_-\right\rangle \otimes \left|-\right\rangle . \qquad (15.4)$$

There is an observer in the state $\left|0\right\rangle$ correlated with the state $\left|\phi_+\right\rangle$, that has not yet interacted with it, who will unitarily evolve to record the value "+", and there is an observer correlated with $\left|\phi_-\right\rangle$, similarly without interaction, who will go on to record the value "-". Why is Eq. (15.3) preferable to Eq. (15.4)?

An added principle seems to be needed to rule out Eq. (15.4), perhaps a version of the principle of identity of indiscernibles, locally enforceable. For those so inclined, still an account of uncertainty can be given, shifting the period of uncertainty from immediately prior to the measurement, to a time immediately following. The observer, we may suppose, simply *closes her eyes*.[11] Given which, post-measurement there are undoubtedly two observers present, with different properties (denote $\left|0_+\right\rangle$, $\left|0_-\right\rangle$), neither as yet aware of the outcome. In place of Eq. (15.4), we have:

$$\left|\psi\right\rangle \otimes \left|0\right\rangle \underset{U_m}{\rightarrow} c_+ \left|\phi_+\right\rangle \otimes \left|0_+\right\rangle + c_- \left|\phi_-\right\rangle \otimes \left|0_-\right\rangle \qquad (15.5a)$$

$$\underset{U_O}{\rightarrow} c_+ \left|\phi_+\right\rangle \otimes \left|+\right\rangle + c_- \left|\phi_-\right\rangle \otimes \left|-\right\rangle \qquad (15.5b)$$

where U_m is as before the measurement process, yielding a superposition of macroscopically distinct outcomes, and U_O represents the further act of observation by the observer of what that outcome is. After Eq. (15.5a) but before Eq. (15.5b), there are unequivocally distinct observers, for the two states $\left|0_\pm\right\rangle$ ex hypothesi differ physically, with each ignorant of which state she is in. Eq. (15.4) gives momentary pre-measurement uncertainty, and Eq. (15.5a) gives momentary post-measurement pre-observation uncertainty. Arguably, anticipating uncertainty to come is a form of pre-measurement uncertainty.[12]

A third and still weaker notion of uncertainty appeals only to behavior.[13] Call "predictive behavior" behavior or action that (i) predicts E and (ii) commits to E, which selectively anticipates E to the exclusion of all other possibilities. Much of our behavior is predictive, in social interactions as in the natural world, as night follows day. But for other events, among them ruinous events, predictive behavior is often impossible – not because we cannot prepare for them but because we do not know when or whether they will occur. We cannot commit to more than one at a given time, by definition of "commitment." We can still commit to just one, so at least in that eventuality be fully prepared; in the same way, a stopped watch is right at least twice a day. Better, we do not commit to one eventuality at all; we prepare for all or several of them by spreading risk – by way of taking out insurance.

Consider now branching, in the special case where the branching structure is fully known, as in performing a simple quantum experiment. Let some of the ruinous events be among the outcomes of the experiment (say it is almost as bad as with Schrödinger's cat). Still predictive behavior is impossible, not because events E, E', E'' ... are unpredictable, as happening at unforeseen times, but because they are all happening at the same foreseen time. Any action initiated prior to measurement will take the same form in every branch, and if it is especially fitted to one, it is unfitted to the others. We can still commit to just one, so that at least in that branch we are fully prepared – in all the others that will not be so.

Resourcing events that occur unpredictably, at different times, is no different from resourcing events that occur predictably, in different branches. Predictive behavior in both cases is impossible. A limited transfer of resources, from the times and branches in which ruinous events do not intrude, to the times and branches in which they do, is the rational strategy. Uncertainty in the face of branching, on this approach, does not require lack of propositional knowledge, or lack of self-locating knowledge, but it is rather the lack of an ability. Knowing everything there is to know, we still cannot act predictively. It is because the same ability is lacking when predictions cannot be made, our usual predicament, that we think probability must involve ignorance. Degrees of belief are important because they are salient to action; so too, and more directly, are degrees of commitment.

15.2   The Connection with Statistics

A second standard objection to the identification of probability with branch weight is that the connection with statistics is wrong. To take the measurement of spin with protocol (15.1), after N repetitions, with the same initial state (15.2) prepared for each trial, the result is a superposition of 2^N branches, one for each possible sequence of outcomes, of varying weight depending on the sequence. Everett's insight was that for large N, the collective weight – the squared amplitude – of the superposition of all those branches with anomalous statistics falls off exponentially in N in comparison to the weight of all those with the correct statistics.[14] This is an example of a quantum law of large numbers (a "quantum Bernoulli theorem"). It is essential for the interpretation of branch weights as probabilities, but too much has been read into it; it does not, for example, imply that anomalous histories are not there – only that they have low collective weight. For those who understood the theorem as an attempt to *derive* probabilities from statistics, the attempt fails.[15] That was not, however, Everett's intent, which was rather to show parity with the way probability enters in classical statistical mechanics.

But is there parity? It is true that on any theory of probability there is a non-zero probability of anomalous statistics; but in a one-world setting, events of small enough probability may never happen. However, they *will* happen, eventually, in a one-world stochastic theory for sufficiently many trials, and likewise in a one-world deterministic theory for a world of finite volume, given sufficient time. In a spatially infinite world, they happen all the time, somewhere, on either theory. Given the size of the visible universe, and in the sure knowledge that it extends far beyond our event horizon, it is hard to set much store on the argument that the universe *might* in fact be small enough not to contain anomalous statistics as the basis of a key difference with probability in Everett's approach.

Denizens of anomalous branches, or of anomalous stretches of history in one-world theories, will be misled by the observed statistics of measurements. They will conclude that quantum mechanics (or at least the Born rule) is false. But they will simply be unlucky. We already have to live with this: the earth is large enough, and people are numerous enough, to play out the same argument. There are those among us who have been struck by lightning many times: how can they not believe, in their heart of hearts, that someone is out to get them? – and someone protecting them too. They are epistemically unlucky. We are used to this.

Questions of epistemology go over to questions of measurability. How can branch weights be measured? In the Everett interpretation, this is a purely *dynamical* question. It is the question of how amplitudes as they figure in a state of the form (15.2) can be reliably correlated with macroscopic indicators (a "probability meter"). The answer is that only the *ratio* of the amplitudes can reliably be measured, where "reliable" means: in branches with collective amplitude close to unity, as the number of trials becomes large. We are back to the law of large numbers, derived as before not from the axioms of probability theory, but from the unitary dynamics of quantum theory. In this way, the law of large numbers, and indeed the Kolmogorov axioms themselves, are derived as approximate, "high-level" laws. We are used to abstract theories of geometry, as opposed to physical geometry; there is likewise abstract probability theory, as opposed to physical.

In one-world chance theories, how chances are measured will depend on what those theories say those chances are, but no one expects to do any better than our actual practice. They will say: the best we can do is to measure relative frequencies, which will *probably* give the right probabilities, and we must rest content with that. However, if chances are squared amplitudes, the Everett interpretation doesn't just say this, it explains *why* they can only be measured in this way, by means of relative frequencies.[16]

There are those who criticize the Everett interpretation because of these limitative results; who insist that there must be a causal mechanism that will reliably bring about true beliefs about the amplitudes, where "reliable" does not involve probability; that it must be possible, if the unitary evolution is all that there is and the theory is to be empirically adequate, for ratios in branch weights to be unitarily, deterministically, driven in to the memory of some measurement device – and if this cannot be ensured then the Everett interpretation must be rejected.[17] But these are demands that no chance theory, if it is to comply with the axioms of probability, can satisfy.

15.3   Decoherence Theory

Two other objections will occupy us in this section and the next, both specific to the Everett interpretation. They both concern the reductive project itself, of giving an account of what probabilities are in terms of something that at first sight is not chancy. But first some stage-setting.

Everett offered up the picture of a branching structure to the wavefunction in which branching was defined by measurement interactions; he had no account of it otherwise. But as a realist interpretation of quantum theory, branching cannot arise *only* with quantum measurements, as if, fantastically, only a single world existed before quantum mechanics was discovered, and before any quantum experiments were performed. How, in the absence of measurement interactions, does branching arise, and with respect to what basis?

This was called the "preferred basis problem."[18] The solution lay in decoherence theory – roughly speaking, the theory of how the components of superpositions are subject to "effective" equations, yielding approximately classical behavior for the components, as a consequence of the unitary dynamics by which superpositions propagate as wholes. Where the rules break down (because only approximately satisfied), or where basis states propagate in the way of equations with dissipation and noise, each basis state evolves into further superpositions of basis states. Branching structure made out in this way is *emergent*: it involves approximations and the identification of salient dynamical variables, in much the same way that emergence is made out across the special sciences. But branching just is chancing; hence so too is chance. Physical probability is something emergent, along with classicality itself.[19] See, for more background, the companion Chapter 14.

All this structure, thus revealed, is needed to show that branch weights play the chance roles – to obtain the branch weights to begin with. But now if decoherence theory is used to define this structure, it had better not itself depend on any probability assumptions. It surely does talk of probabilities, however. Take, for example, the argument from Ehrenfest's theorem in the case of an initial state $|\psi\rangle$, well-localized in position and momentum. There we define a quantity $\langle \hat{x} \rangle_\psi$ that for sufficiently well-behaved potentials can be shown to satisfy classical equations. But the quantity $\langle \hat{x} \rangle_\psi$ has a probabilistic interpretation: it is the expectation value of the position operator $\hat{x}$ in the state $|\psi\rangle$, and it is measured by repeated experiments, invoking the Born rule.

For another example, take the concept of a quasiclassical domain, as first defined by Murray Gell-Mann and James Hartle in 1990. It is a history space "with probabilities peaked on quasiclassical histories." According to Adrian Kent, a prominent critic of the Everett interpretation, this shows that "the ontology is *defined* by applying the Born rule" (Kent 2010, pp. 337–338). Jonathan Halliwell, who has applied decoherent histories theory to numerous topics in quantum foundations, likewise speaks in probabilistic terms. For a quasiclassical domain defined by hydrodynamical variables:

> The final picture we have is as follows. We can imagine an initial state for the system which contains superpositions of macroscopically very distinct states. Decoherence of histories indicates that these states may be treated separately and we thus obtain a set of trajectories which may be regarded as exclusive alternatives each occurring with some probability. Those probabilities are peaked about the average values of the local densities. We have argued that each local density eigenstate may then tend to local equilibrium, and a set of hydrodynamic equations for the average values of the local densities then follows. We thus obtain a statistical ensemble of trajectories, each of which obeys

hydrodynamic equations. These equations could be very different from one trajectory to the next, having, for example, significantly different values of temperature. In the most general case they could even be in different phases, for example one a gas, one a liquid. (Halliwell, 2010, p. 111)

The criticism, after all this stage-setting, is this. Probability talk is ubiquitous in the literature on decoherence theory. In order to have meaning, probabilities have to be assumed from the outset. There is no reductive, Humean base-level of description, in terms of which probabilities can later be identified. When it comes to probability in Everettian quantum mechanics, the project of Humean supervenience cannot even get off the ground.

The point at issue, however, is not whether models of decoherence theory as *usually* derived and discussed are interpreted in terms of probability; grant that they are. Nor is it a surprise that these probabilities are interpreted in one-world terms (as in Halliwell's writings): from its inception the decoherent histories theory was supposed to provide a one-world interpretation of quantum theory, without any need of Everett's extreme ideas. The substantial objection can only be that these models *cannot be divested* of their probability interpretation, not even in the Everett interpretation.

Is this true? To take the case of Gell-Mann and Hartle's definition of a quasiclassical domain, here is a replacement formulation: it is a space of histories for which the amplitudes are strongly peaked on histories obeying a closed system of equations. "Strongly peaked amplitude" does not, prior to defining the branching structure of the state, have to be interpreted as "highly probable." Halliwell's summary can similarly be reworded, noting that the "average values of local densities" are picked out not by averaging the densities, but as the values of the local densities on those trajectories of comparatively large amplitude. In the case of Ehrenfest's theorem, it is *possible* to interpret $\langle\hat{x}\rangle_\psi$ operationally in terms of multiple measurements (assuming similar systems can be prepared in the same state $|\psi\rangle$); but it can *also* be interpreted realistically (so long as $|\psi\rangle$ is well-localized in position and momentum) – as the approximate support of the one-particle wavefunction $\langle x\,|\psi\rangle$ (or more accurately the wavefunction for the center of mass coordinate) moving about in three-dimensional space, as in unadorned dynamical collapse theories (amendments that introduce "primitive ontology" as something different from the quantum state are another matter).

To give another example, take the requirement of consistency among histories as defined in the previous chapter (the requirement that the so-called "sum rules" be satisfied). This is supposedly an a priori constraint on any probability theory on a history space. It forces the vanishing of the real part of the inner product of states of distinct histories, but that in turn can be directly related to the idea of interference between histories, which is hardly built into the concept of probability. And in fact the stronger condition, that real *and* imaginary parts vanish, is both more natural and far more widely used. As such it is the requirement that the structure to the quantum state, as defined in terms of quantum histories, is to be made out in terms of orthogonal vectors: in terms of a basis. Orthogonality is as useful to get at four-dimensional structure to the state as it is for its three-dimensional structure. Probability does not (yet) have to come into it.[20]

Is it that the derivations of equations in decoherence theory, absent their probability interpretation, simply cannot be understood? There is no doubt that a probability interpretation is an aid to understanding: the question is whether it is a ladder that can be kicked away. We have seen important episodes like this before. The investigation of stress, sheer, and strain in a mechanical medium and in mechanical terms was at the heart of Fresnel's work in optics, and, subsequently, Maxwell's discovery of his field equations. In order for terms in differential equations to be *understood*, they were mechanically analyzed. It took decades for electromagnetic theorists to free themselves from this mechanical crutch, first Maxwell, by the time of writing of the *Treatise*, then Hertz, and finally Lorentz. The result was electrodynamics divested of a mechanical interpretation of the fields, and an ether divested of any mechanical properties – to the point that it was little more than a preferred frame of reference.

The objection may concern *justification* rather than understanding, particularly if decoherence theory is used to derive the Born rule. According to Wojciech Zurek, an early pioneer of decoherence theory, concepts like "partial trace" and "reduced density matrix" cannot be used for that purpose "because their physical significance depends on Born's rule."[21] For a more recent critique:

> In order to neglect small values in favour of larger values, we have to establish that the magnitude of the corresponding variable is related to the entry's effect on the measurement to be performed. Since experimental testing and the entries in the density matrix are related in terms of the probabilities for measuring certain outcomes, in order to establish the negligibility of small entries in the density matrix we must introduce the Born rule.

There is another way to analyze the effect on the measurement to be performed, however: model that measurement device explicitly in the formalism. So long as we may interpret the quantum state that results, for example in terms of the pointer position, it can be established whether it depends on those small entries. The suggestion is perhaps that some tiny modification of an equation, or an approximate form of a solution, may have enormous experimental impact, so that failing its prior interpretation nothing can be neglected: only an exact solution will do. But given that for complex systems, or systems with large numbers of degrees of freedom, exact solutions are never available, the better conclusion to draw is that equations whose empirical meaning is sensitive to the tiniest approximation have isolated the wrong degrees of freedom, have made an incorrect series expansion, have chosen an unphysical topology.

As W.V. Quine said, a physical theory is tested as a whole; it is the exception when different parts of it can be isolated as independently testable. We may of course arrive at approximations in which small quantities should not have been neglected; we may make some or all of these mistakes; but there is no sure method for building up a theory from elements that are independently testable, so as to avoid them.

15.4   Branch Counting

Given the branching structure to the universal wavefunction, it is clear what is the intended interpretation of probability (ratios in branch weights). Are we sure there is no rival alternative? There is one that has been taken seriously even by those sympathetic to Everett's ideas: the *branch-counting rule*.[22] It is the rule that on any branching event, all outcomes, all histories that ensue are equiprobable. If from the repeated measurements a large slew of histories result, the number with a given relative frequency (divided by the total number) determines the probability of that relative frequency.

The result, for the Everett interpretation, is mayhem. To take again the measurement of spin with initial state (15.2), after N measurements the vast majority of states have relative frequency of "+" one half, and likewise "−" one half, *entirely independent of* $|c_+|^2$ and $|c_-|^2$. When their ratio differs significantly from unity, only a *tiny minority* of branches after N trials have the Born-rule compliant relative frequency.

The branch-counting rule makes nonsense of quantum mechanics but it appears to be suggested by the picture of branching. It spells trouble for a reductive approach to probability if the base provides two quite different candidates for the chance role. We know which one is wrong, on empirical grounds, but what makes the one *probability*, and not the other? Probability, it suddenly seems, may have to be taken as a primitive after all.

In point of fact (as shown by Wallace), this probability rule – call it "naive" branch counting – conflicts rather straightforwardly with axioms of probability. Thus, consider Eq. (15.5a), and suppose, in place of Eq. (15.5b), a second measurement is made at time t_1, but only in the branch with "+", of something else entirely, say position, producing two further branches, both with the outcome "+"

at t_2. In the branch with "−", no further measurement is made, so there is only one such branch at t_2. So what is the probability of "+" at t_2? At t_1 it is one-half, but at t_2 it is two-thirds. The latter cannot be obtained by updating in time, for it is in conflict with the sum rule:

$$\Pr(+; t_2) = \Pr(+; t_2/+; t_1)\Pr(+; t_1) + \Pr(+; t_2/-; t_1)\Pr(-; t_1) \qquad (15.6)$$

which follows from the probability calculus for histories when the probabilities of the plus and minus outcomes at t_1 sum to unity. Using naïve branch counting, Eq. (15.6) yields one-half, not two-thirds.

It follows that the naïve branch counting rule is not, in fact, a coherent probability rule at all. If this were the only alternative to the Born rule, there would be no problem of underdetermination as alleged. But there is another branch-counting rule that is just as intuitive: it is that on each branching event, the probabilities of each branch are the same, with probabilities for branches at subsequent times not equiprobable, but depending on how each branch came about − all its splittings − and conforming to sum rules of the form (15.6).[23]

This new branch-counting rule is just as hopelessly at odds with the Born rule; moreover, it is manifestly in conflict with locality (permitting super-luminary signaling) (McQueen and Vaidman, 2019). But that is only grist to the skeptic's mill: the skeptic is arguing that the branching structure to the state on measurement, with number equal to the number of possible outcomes, the central concept of the Everett interpretation, suggests an altogether inappropriate concept of probability, at odds with relativity as well as the Born rule; so much the worse for the Everett interpretation.

But is this branching number so defined central to the Everett interpretation? It was prior to decoherence theory, when appeal to experiments, with a definite number of possible outcomes, was the only way that branching was defined. But decoherence theory changed all that. Decoherence theory just is the theory of branching structure, but as it occurs naturally, independent of whether any experiments are performed. Moreover from this perspective, when measurements are made, branching number has nothing to do with the number of *readings* that can be made. Consider again the measurement of spin, and specifically, consider just one of the two protocols, say Eq. (15.1a). Is there just one way this experiment can come about − as a physical process? Consider all the quasiclassical processes going on − the thermal fluctuations, variations in pressure, Brownian motions, cascades of phonons, scattering of light − all of them involving branching, and all that just on opening the laboratory door. There are clearly *countless* different ways that the apparatus can obey Eq. (15.1a), evolving from macrostates in which it is "ready" to those in which it reads "+" even when the initial spin state that is measured is always $|\phi_+\rangle$. Here, "countless" means undefined: the number of branches, specified by what goes on in each branch during the process of measurement, is undefined.

This could be work in progress. It may be, for example, that there is a finest-grained history space that is a decohering history space, indeed, a quasiclassical domain − and we just don't happen to know how to approximate it. We use a convenient definition, and relative to this the branch number is fixed; the number is somewhat arbitrary to be sure, depending on the coarse-graining, but it might be thought of as our best guess on what the "correct" fine-graining is. If it can be shown that the probabilities thus defined (as ratios in finite numbers) are insensitive to this coarse-graining, including coarse-graining in time, the branch-counting rule would be definable after all − to the possible discredit of the Everett interpretation.[24]

The true nature of the difficulty only now comes into focus. For of all of these branches, thus defined, waiting to be counted: is it *all* of them, or only the ones with non-zero amplitudes? Neither choice makes any sense. If it is all of them, then the numbers are completely independent of the state; in what sense is this an interpretation of the structure of the state? But if only non zero amplitude branches are counted, the numbers will be discontinuous functions of the amplitudes. The tiniest variation in amplitude may make for arbitrarily large variations in branch number.[25] Yet continuity in the amplitudes (in the Hilbert-space norm, the norm topology) is essential to decoherence theory and the unitary dynamics.

The two objections to Everett's approach – that decoherence theory must be founded on a probability rule, and that one such rule is the branch-counting rule – can now be nicely brought together. Given the branch-counting rule as just stated, indeed no approximation can be made, no low-amplitude history neglected – which shows that the approximations used in decoherence theory must first be justified by the probability rule.

It is a Pyrrhic victory, twice over. Branch counting is hardly a rival to the branch-weight rule as an *interpretation* of branching structure, if it brings that branching structure crashing down. And of course the approximations used in a theoretical model must be justified by the probability rule, if they are to be consistent with it, and if that rule is posited in advance: for those used in decoherence are thus rendered meaningless.

The supervenience project goes precisely the other way. Branch numbers can always be defined using branch amplitudes, arbitrary to be sure, but in such a way that *ratios* in branch numbers are well-defined, free of any convention, in agreement with the Born rule.[26]

15.5   Undermining

The philosophical literature on reductive "Humean" theories of chance has led to a certain consensus. Such a theory should satisfy the following:[27]

(i)   The Principal Principle – one should set one's credence in E at t equal to the chance of E at t, no matter what else one knows, provided one has no magical (no "inadmissible") information from the future.

(ii)   Quantitative constraints – the chance of an event after it occurs is always 1 or 0; an event that has value 0 or 1 at one time retains that value for all subsequent times.

(iii) The chance-frequency link – the relative frequency of Es in an ensemble of systems all prepared in the same state approaches the probability of E in that state as the size of the ensemble increases, but the *possibility* of divergence in any finite ensemble remains, no matter how large the ensemble.

It has led to a consensus, in particular, that in a conventional Humean setting (meaning, inter alia, a one-world setting) no such theory has been found. Indeed, many conclude no such theory *can* be found, that there can be no "perfect" theory of chance.[28] The difficulty is that the link between chance and the reductive base must be slack – to countenance (iii) – so, for example, there must be possible worlds in which the statistics are the same and the chances different, or the chances the same and the statistics different; but no, the link has to be tight, the chances cannot drift far from actual events. If the two were distinct existences, there would be a world where E occurs at time t, but where the chance of E at some later time is not one, contrary to (ii).

Add to these the Basic Chance Principle (BCP) (Bigelow et al., 1993), the principle, roughly, that the chance for an event E at some future time may be the same, as determined by events up to some earlier time, whether or not E actually happens. Indexing to worlds and to times it is the principle:

(iv)   BCP – if the chance of E at world w at time t is $P_{tw}(E) > 0$, then there exists a world w' which (a) matches w up to time t, (b) contains E, and (c) satisfies $P_{tw}(E) = P_{tw'}(E)$.

If there were no such world w', then the chance of E at t could not be the same independent of whether or not E happens – for if the BCP fails, there is no world containing E for which the chance at t is the same. Yet it seems the BCP must fail – either that, or there are no patterns of events E, E', that yield distinct chance theories in any worlds w, w' in which E and E' occur; distinct, in particular, in that at some t, $P_{tw}(E) \neq P_{tw'}(E)$, and $P_{tw}(E') \neq P_{tw'}(E')$.

This is an instance of "undermining." A similar argument shows that (i), the Principal Principle, must fail; in effect, it shows that knowing the chance at t for such patterns *is* to have magical information about the future – is itself inadmissible – because what that chance is depends on what happens at future times.[29] But if there are no patterns of this kind, the very idea of a Humean reductive theory of chance is in trouble.

Back to Everett. A quasiclassical history, divested of amplitude and phase, just is one local pattern of events after another; each is a Humean tapestry of events, a Lewisian world w. The Everett interpretation thus provides a home for Lewis's metaphysics, but with these essential differences: the worlds are emergent structures, so no best-system analysis based on them can hope to give the fundamental laws (at most they may give the emergent laws); the worlds bear new and irreducible relations to each other, defined in terms of amplitude and phase; and the worlds have branching structure,[30] as defined by those amplitudes and phases.

On the most straightforward identifications, then, Lewisian worlds correspond to quasiclassical histories,[31] but the reductive base is the collection of all these histories in a superposition – a branching structure – or, in Lewisian terms, a collection of worlds. The chance theory for this collection, arranged in (derived from) this branching structure, is that chance events are branching events, and chances are ratios in branch weights. By that theory, for any world w at time t, the chances are determined only by its history up to time t, for that alone (with the unitary dynamics) determines the amplitudes of branching events thereafter. They are the same (as functions P_{tw} from an event space to chances) for all worlds w that match up to time t. There is just the one set of chances, at a branching event, regardless of subsequent branching events.

This theory of chance meets the principles (ii)–(iv) to perfection. The indexing to times in (ii) is automatically taken care of: chances are relations between branch weights, indexed to times, and they are retrodictively 0s and 1s because of branching structure (no recombination of branches – see Eq.(14.7)). (iii) is satisfied, reading "possibly" to mean there exist anomalous histories, albeit of vanishingly small amplitude. (iv) is satisfied: for *every* world w' that matches w up to time t, the chances at t of E are the same, for chances at t are determined by the prior history at t. There is no undermining: the conclusion of the undermining argument was that either the BCP is wrong or (roughly) present chances in world w are not determined by future events in w. The latter is now a feature of the chance theory, not a bug.

How is it a feature, and isn't there a cost? Yes: what was before an instability in what the chances really are (if dependent on future chance events, as well as past), now shows up as a mismatch of relative frequencies to branch weights. It has turned into the problem of anomalous statistics, already considered. There are worlds whose inhabitants will be misled by the observed statistics, the epistemically unlucky ones. In the overwhelming majority of worlds,[32] their inhabitants will be led to the right theory. The Born rule remains the best-systems analysis of all the branching worlds, but it does not supervene on the statistics in each world, rather it supervenes on – is a simple function of – the branch weights of all the worlds. Could the denizens of the unlucky worlds survey the entire branching structure, their favored best systems analysis would be the Born rule.

There remains (i), Lewis's famous Principal Principle. But here too there has been progress.

15.6   The Principal Principle

Failing an account of what chances are, as indexed to times, it was obscure why credence as indexed to times should conform to them.[33] It was already obscure why they should if chances are relative frequencies of events up to that time, although there at least the obscurity was dignified as a philosophical problem (the problem of induction). In this context, the recently proved Born-rule theorem is little short of a philosophical sensation. It is the derivation of the Principle Principal in the special case where the chances are identified as branch weights and chance processes are identified as branching processes. Alternatively, assuming the Princpal Princple, it shows that branch weights should be identified as probabilities. Either way, it shows why credences should conform to branch weights.

The result turns on the principle of indifference already announced, but it also depends on rational choice theory, and, specifically, on the operational techniques first introduced by Frank Ramsey and Bruno de Finetti in the early part of the last century, whereby credences are operationalized in terms of betting behavior. The Dutch book argument is a case in point: an agent was sure to lose money whatever the outcome of a bet, if their betting quotients did not conform to the axioms of probability theory. In Leonard Savage's (1954) *The Foundations of Statistics* (published by Princeton just as Everett began his studies), this took the form of a representation theorem: if the preferences of an agent among bets (of sufficient number and variety) conform to certain rules in decision theory, then there is an essentially unique credence function and a utility function, such that those preferences are the same as by ranking by expected utility. That one's credence and utility function should dictate a rank ordering is obvious, it is the converse, perhaps, that is surprising. Still, that credence function, other than qualifying as a bone fide probability distribution, could be anything, as likewise one's utilities.

Fairly obviously, it will be impossible to tie down the credence function further without knowing more about how the games are actually played, and what the agent knows about how the games are actually played – in short, without a physical theory governing those games, and agents who base their choices among games on that theory. But as before it is essential, if it is to serve its purpose, that that theory be divested of any probability interpretation: wave mechanics without probability to the rescue again. Moreover, it would be better if that theory is divested, even, of any talk of uncertainty, since that notion too is contested.[34] Given all of which, the remarkable result proved by Deutsch and Wallace is that wave mechanics, thus disinterpreted, is enough to tie down that credence function uniquely, to the point that it must conform to the branch weights. Agents, if rational, basing their choices on unitary quantum mechanics, and fully cognizant of the branching produced by each quantum game (as determined by the physics of the apparatus) and knowing the stake and rewards on each outcome, will order their preferences among games as if maximizing their expected utilities, for some utility function, using the Born rule. Difference in utilities will make a difference to their preferences, but their credence functions will always be the same.

The core of the proof is a symmetry argument.[35] Consider, for the last time, the measurement of spin. Suppose an agent is to bet on (apportion resources to) the spin-plus outcome of the experiment, where the measurement process satisfies Eq. (15.1), that is, according to the protocol:

$$P1: \ |\phi_+\rangle \otimes |0\rangle \underset{U_1}{\to} |+\rangle$$

$$|\phi_-\rangle \otimes |0\rangle \underset{U_1}{\to} |-\rangle$$

(where we assume that the agent doesn't care what happens to the state of the measured system $|\phi_\pm\rangle$ after the measurement, so it is just omitted). Let her credence in a spin-up outcome conditional on this protocol be $Cr(+/P1)$. Suppose now the protocol is changed to $P2$, according to which the same experiment is run, save that after the measurement the label of the plus outcome is replaced by "$-$," and vice versa. So the new protocol is:

$$P2: \ |\phi_+\rangle \otimes |0\rangle \underset{U_1}{\to} |+\rangle \underset{U_2}{\to} |-\rangle$$

$$|\phi_-\rangle \otimes |0\rangle \underset{U_1}{\to} |-\rangle \underset{U_2}{\to} |+\rangle \, .$$

Then $Cr(+/P1)$ should be equal to $Cr(-/P2)$, since they only differ in the (irrelevant) way the final outcome is labeled. But it follows that when the initial state is $|\psi\rangle$ as given by Eq. (15.2), the final state on $P1$ is (cf. Eq. (15.3)):

$$|\psi\rangle \otimes |0\rangle \underset{U_1}{\to} c_1 \, |+\rangle + c_2 \, |-\rangle$$

whereas on *P2* it is:

$$|\psi\rangle \otimes |0\rangle \underset{U_2 U_1}{\to} c_1 |-\rangle + c_2 |+\rangle$$

and in the particular case when $c_1 = c_2$ *the two states at the end of the measurement are exactly the same, whichever protocol is used.* Therefore, her credence in the "+" outcome, respectively "−" outcome, for this special initial state, should be the same whichever protocol is used; so

$$Cr(\pm/P1) = Cr(\pm/P2). \tag{15.7}$$

From our previous result in general we have:

$$Cr(\pm/P2) = Cr(\mp/P1).$$

So in the special case $c_1 = c_2$, it follows from Eq. (15.7):

$$Cr(\pm/P1) = Cr(\mp/P1)$$

and likewise for the protocol *P2*. *So her credences for the two outcomes, on either protocol, must be the same.*

Evidently, the argument at crucial moments appealed to normative principles. The two states at the end of the two experiments, for the two protocols, won't be *exactly* alike; but the agent shouldn't care about microscopic inessentials. In extending the argument to rational ratios of $|\alpha|$ and $|\beta|$, ancillary devices are needed, which produce additional branching; but the agent shouldn't care about that either, if the branches produced differ only in ways the agent doesn't care about. The agent should only care about what is physically realized in each state (what can be given a dollar value in each state), and so on.

Many of these normative judgments are based on pragmatic constraints, on the basis of that old adage, "ought" implies "can"; so contraposing, you ought not to care about branching per se, because you can't (branching as determined by decoherence theory is ubiquitous). Others are more purely normative. The end result is a demonstration of why an agent should care about branch weights, and why her credence should conform to them – and not, for example, to the number of four-leaf clovers on each branch, or the number of calibrations on the pointer dial.

Yet despite all these successes, the Born-rule theorem, for those convinced that the notion of probability in the Everett interpretation is otherwise unintelligible, has been found wanting. For them the theorem must carry the entire burden of probabilistic reasoning – whereby the explanation for the statistics of quantum experiments, normally provided by the Born rule, has to be provided instead by experimentalists' betting strategies. But how can someone's betting strategy explain why the radium atom has a half-life of less than a second? How can the Born rule, a rule of fundamental physics, be true by virtue of human behavior? Doesn't it follow that there is no such thing as probability in a universe without people? Moreover, are there not alternatives to using the branch weights, no matter if fanciful or practically impossible to implement, that may not yet be irrational?[36] But all this is to take the Born-rule theorem in isolation from the larger reductive project.

The theorem demonstrates that the particular role ordinarily but mysteriously played by physical probabilities, whatever they are, in our rational lives, is played in a wholly perspicuous and unmysterious way by ratios in branch weights and branching. It is by establishing that these quantities play all the other chance roles as well, that qualifies them as probabilities.

Notes

1 This is a companion piece to "The Everett Interpretation: Structure," Chapter 14 in this volume, to which I refer for more technical background.
2 This is the position of Deutsch (1999, 2016) and Greaves (2004) (under the rubric "the fission program"). It has been defended by Brown (2013) and Brown and Porath (2020).
3 As sent to Bohr and others in 1956. It eventually appeared as "The theory of the universal wave function" in DeWitt and Graham (1973).

4 For more in this vein see Papineau (2010).

5 See e.g. van Fraassen (1980), Ch.6, Wallace (2012 §4.5).

6 Although the concept of ergodicity played a similar role in Botlzmann's approach to probability. See also footnote 26.

7 The claim is sweeping but appears to be accurate. See Tipler (2014), Brown and Timpson (2016).

8 A related worry is that quantum mechanics under the Everett interpretation may not be testable. For a decision-theoretic approach in keeping with the "fission" program, see Greaves and Myrvold (2010) and, for a unified treatment including the Born-rule theorem, Wallace (2012, Ch.6). See also Deutsch (2016). Due to space constraint, I do not address these questions here; on the basis of the arguments that I shall advance, it follows branch weights should be identified as physical probabilities, whereupon quantum mechanics in the Everett interpretation should be tested just as any other probability theory.

9 See Saunders and Wallace (2008a, 2008b) and, for a contrary view, Tappenden (2008). For a systematic account, see Wallace (2012, Ch. 7) and for a comprehensive metaphysics, Wilson (2020).

10 As insisted e.g. by Kent (2010, pp. 346–347).

11 So-called "Vaidman uncertainty," first introduced in Vaidman (1998).

12 Tappenden (2011). See also Sebens and Carroll (2016), McQueen and Vaidman (2019) for recent derivations of the Born rule framed in terms of Vaidman ignorance, together with a locality assumption.

13 For the case of linguistic behavior, see Saunders (1998). The argument that follows is more in the spirit of Ismael (2017).

14 See Saunders, this volume, Chapter 14.

15 See e.g. Smolin (2019, p. 151); Wallace (2012 p.127) also calls Everett's argument "circular".

16 See Saunders (2010) for more in this vein.

17 This appears to be the position of Adlam (2014); see also Rae (2009).

18 Ballentine (1973) (although he did not use the terminology "preferred basis") . For more on decoherence theory and the preferred basis problem see Saunders, this volume, Chapter 14, and Crull, this volume, Chapter 13.

19 See Wallace (2003) and Saunders (2005). Branching as emergent structure is an important theme in Wallace (2012, Ch. 1–2).

20 This suggests consistency is a very weak condition, and so it is. See Section 14.4 of Chapter 14 for a review of the theory of quasiclassical histories divested of any probability interpretation.

21 Zurek (2005, p. 1); the quotation that follows is from Dawid and Thébault (2015, p. 8).

22 As first discussed by Graham (1973). It was at center stage in David Lewis's only foray into quantum mechanics (Lewis 2004).

23 See Lewis (2009). It is ruled out by Wallace's axioms (by "branching indifference" and "state superve- nience"), but also, critically, by certain "richness" axioms (in particular "problem continuity" and "solution continuity"; see 2012, pp. 170–171, 178, pp. 91–92), so ultimately, for reasons similar to the below.

24 However, such a stability condition seems likely to force the branch weight rule instead: see Saunders (2005, 2021).

25 While the number of branches of non-zero weight, for a sequence of states $|\phi_k\rangle$ that converges to $|\phi\rangle$ in the norm topology, may not converge to anything, the *weighted sum* converges smoothly. Thus let the region Δ_+, corresponding to the + outcome, the weighted sum in the state $|\phi_k\rangle$ is not $\sum_{\delta \subset \Delta_+}$, where the sum is over δ satisfying $\langle \phi_k | P_\delta | \phi_k \rangle \neq 0$, but $\sum_{\delta \subset \Delta_+} \langle \phi_k | P_\delta | \phi_k \rangle = \langle \phi_k | P_{\Delta_+} | \phi_k \rangle$, which converges to $\langle \phi | P_{\Delta_+} | \phi \rangle$ as $|\phi_k\rangle \to |\phi\rangle$.

26 Call it the "equi-amplitude rule" (Saunders 2021). It is similar to Boltzmann's procedure, in his combinatoric approach to probability, to obtain a discrete count of states; ratios in those numbers are independent of the choice of unit, when that unit is taken to zero, in agreement with Liouville measure.

27 The three criteria that follow are taken almost verbatim from Ismael (2009, pp. 421–422); the paragraph after is my attempt to paraphrase her subsequent argument.

28 The terminology is due to Schaffer (2003) (to be precise, a "perfect" theory in his sense is a reductive theory that satisfies the PP and the BCP; see below).

29 As Lewis eventually came to see (Lewis 1994), chances at *t* in a one-world best systems theory are themselves inadmissible at *t*. This points to a reconciliation of sorts between undermining and the PP (Lewis's "big, bad bug"; see Lewis (1986)). See also Suárez, this volume, Chapter 46.

30 For the connection with overlap and divergence, see Saunders (2010) and Wilson (2011, 2020). Here, I use "branching" as in the physics literature, neutral as to overlap vs divergence.

31 The entire branching structure – the collection of worlds closed under relations of amplitude and phase – also has a claim to be considered as a single Lewisian world. Or it might better be thought of as an "inner sphere" of worlds, in roughly Lewis's sense. See Wilson (2020).

32 In the sense of note 26.

33 For recent attempts to render the connection more transparent, see Hoefer (2007), Schwarz (2014), Myrvold (2021).

34 Nor should they be compromised on later *interpreting* the axioms in terms of ignorance and uncertainty. (Of course it will do no harm if they are thus strengthened; see Wilson (2013) for an argument of this kind.)

35 Here I closely follow Wallace (2012, pp. 151–152).

36 For criticisms of this form, see Albert (2010), Kent (2010), and Price (2010), and for discussion, see Saunders et al. (2010, pp. 391–406). For more recent criticisms, see Jansson (2016), Mandolesi (2018, 2019).

References

Adlam, E. (2014). The problem of confirmation in the Everett interpretation. *Studies in History and Philosophy of Modern Physics*, 47: 21–32.

Albert, D. (2010). Probability in the Everett picture. In S. Saunders, J. Barrett, A. Kent and D. Wallace (eds.), *Many Worlds? Everett, Quantum Theory, and Reality*. Oxford: Oxford University Press, pp. 355–368.

Ballentine, L.E. (1973). Can the statistical postulate of quantum theory be derived?—A critique of the many-universes interpretation. *Foundations of Physics*, 3: 229.

Bigelow, J., Collins, J. and Pargetter, R. (1993). 'The big bad bug: What are the Humean's chances? *The British Journal for the Philosophy of Science*, 44: 443–462.

Brown, H. (2013). Curious and sublime: The connection between uncertainty and probability in physics. *Proceedings of the Royal Society*, A369: 1–15.

Brown, H. and Porath, G. (2020). Everettian probabilities, the Deutsch-Wallace theorem and the Principal Principle. In M. Hemmo and O. Shenker (eds.), *Quantum, Probability, Logic: The Work and Influence of Itamar Pitowsky*. Berlin: Springer-Verlag.

Brown, H. and Timpson, C. (2016). Bell on Bell's theorem: The changing face of nonlocality. In M. Bell (ed.), *Quantum Non-Locality and Reality: 50 Years of Bell's Theorem*. Cambridge: Cambridge University Press.

Dawid, R. and Thébault, K. (2015). Many worlds: Decoherent or incoherent? *Synthese*, 192: 1559–1580.

Deutsch, D. (1999). Quantum theory of probability and decisions. *Proceedings of the Royal Society of London*, A455: 3129–3137. Available at: arXiv.org/abs/quant-ph/9906015.

Deutsch, D. (2016). The logic of experimental tests, particularly of Everettian quantum theory. *Studies in History and Philosophy of Modern Physics*, 55, 24–33.

DeWitt, B. and Graham, N. (1973). *The Many-Worlds Interpretation of Quantum Mechanics*. Princeton, NJ: Princeton University Press.

Everett III, H. (1957). "Relative state'" formulation of quantum mechanics. *Reviews of Modern Physics*, 29: 454–462. Reprinted in DeWitt and Graham (1973, pp. 141–150).

Everett III, H. (1973). Theory of the universal wave function. In B. DeWitt and N. Graham (eds.), *The Many-Worlds Interpretation of Quantum Mechanics*. Princeton, NJ: Princeton University Press, pp. 3–140.

Gell-Mann, M. and Hartle, J.B. (1990). Quantum mechanics in the light of quantum cosmology. In W.H. Zurek (ed.), *Complexity, Entropy, and the Physics of Information*. Reading: Addison-Wesley.

Graham, N. (1973). The measurement of relative frequency. In B. DeWitt and N. Graham (eds.), *The Many-Worlds Interpretation of Quantum Mechanics*. Princeton, NJ: Princeton University Press, pp. 229–253.

Greaves, H. (2004). Understanding Deutsch's probability in a deterministic multiverse. *Studies in History and Philosophy of Modern Physics*, 35: 423–456. Available at: philsci-archive.pitt.edu/archive/00001742/.

Greaves, H. and Myrvold, W. (2010). Everett and evidence. In S. Saunders, J. Barrett, A. Kent and D. Wallace (eds.), *Many Worlds? Everett, Quantum Theory, and Reality*. Oxford: Oxford University Press, pp. 264–306.

Halliwell, J. (2010). Macroscopic superpositions, decoherent histories, and the emergence of hydrodynamic behaviour. In S. Saunders, J. Barrett, A. Kent and D. Wallace (eds.), *Many Worlds? Everett, Quantum Theory, and Reality*. Oxford: Oxford University Press, pp. 91–120.

Hoefer, C. (2007). The third way on objective probability: A skeptic's guide to objective chance. *Mind*, 116: 549–596.

Ismael, J. (2009). A modest proposal about chance. *Journal of Philosophy*, 108: 416–422.

Ismael, J. (2017). An empiricist's guide to objective modality. In M. Slater and Z. Yudel (eds.), *Metaphysics and Philosophy of Science: New Essays*. New York: Oxford University Press

Jansson, L. (2016). Everettian quantum mechanics and physical probability: against the principle of "state supervenience". *Studies in History and Philosophy of Modern Physics* 53: 45–53.

Kent, A. (2010). One world versus many: The inadequacy of Everettian accounts of evolution, probability, and scientific confirmation. In S. Saunders, J. Barrett, A. Kent and D. Wallace (eds.), *Many Worlds? Everett, Quantum Theory, and Reality*. Oxford: Oxford University Press.

Lewis, D. (1986). *Philosophical Papers*, Vol. 2. Oxford: Oxford University Press.

Lewis, D. (1994). Humean supervenience debugged *Mind*, 103: 473–490.

Lewis, D. (2004). How many lives has Schrödinger's cat? *Australasian Journal of Philosophy*, 82: 3–22.

Lewis, P. (2009). Probability, self-location, and quantum branching. *Philosophy of Science*, 76: 1009–1019.

McQueen, K. and Vaidman, L. (2019). In defence of the self-location uncertainty account of probability in the many-worlds interpretation. *Studies in the History and Philosophy of Physics*, 66: 14–23.

Mandolesi, A. (2018). Analysis of Wallace's proof of the Born rule in Everettian quantum mechanics: formal aspects. *Foundations of Physics* 48: 751–82.

Mandolesi, A. (2019). Analysis of Wallace's proof of the Born rule in Everettian quantum mechanics II: concepts and axioms. *Foundations of Physics* 49: 24–52.

Myrvold, W. (2021). *Beyond Chance and Credence*: a theory of hybrid probabilities. Oxford: Oxford University Press.

Papineau, D. (2010). A fair deal for Everettians. In S. Saunders, J. Barrett, A. Kent and D. Wallace (eds.), *Many Worlds? Everett, Quantum Theory, and Reality*. Oxford: Oxford University Press.

Price, H. (2010). Decisions, decisions, decisions: Can Savage salvage Everettian probability? In S. Saunders, J. Barrett, A. Kent and D. Wallace (eds.), *Many Worlds? Everett, Quantum Theory, and Reality*. Oxford: Oxford University Press, pp. 369–90.

Rae, A. (2009). Everett and the Born rule. *Studies in History and Philosophy of Modern Physics*, 40: 243–250.

Saunders, S. (1998). Time, quantum mechanics, and probability. *Synthese*, 114: 405–444. Available at: arXiv.org/abs/quant-ph/0112081.

Saunders, S. (2005). What is probability? In A. Elitzur, S. Dolev and N. Kolenda (eds.), *Quo Vadis Quantum Mechanics*. Berlin: Springer, pp. 209–38. Available at: arXiv.org/abs/quant- ph/0412194.

Saunders, S. (2010). Chance in the Everett interpretation. In S. Saunders, J. Barrett, A. Kent and D. Wallace (eds.), *Many Worlds? Everett, Quantum Theory, and Reality*. Oxford: Oxford University Press, pp. 181–205. Available at: http://philsci-archive.pitt.edu/id/eprint/12441.

Saunders, S. (2020). The concept 'indistinguishable'. *Studies in History and Philosophy of Modern Physics*, 71: 37–55.

Saunders, S. (2021). Branch counting in the Everett interpretation. Forthcoming.

Saunders, S., Barrett, J., Kent, A. and Wallace, D., editors. (2010). *Many Worlds? Everett, Quantum Theory, and Reality*. Oxford: Oxford University Press.

Saunders, S. and Wallace, D. (2008a). Branching and uncertainty. *British Journal for the Philosophy of Science*, 59: 293–305. Available at: philsci-archive.pitt.edu/archive/00003811/.

Saunders, S. and Wallace, D. (2008b). Saunders and Wallace reply. *British Journal for the Philosophy of Science*, 59: 315–317.

Savage, L. (1954). *The Foundations of Statistics*. New York: John Wiley and Sons.

Schaffer, J. (2003). Principled chances. *British Journal for the Philosophy of Science*, 54: 27–41.

Schaffer, J. (2007). Deterministic chance. *British Journal for the Philosophy of Science*, 58: 113–140.

Schwarz, W. (2014). Proving the principal principle. In A. Wilson (ed.), *Chance and Temporal Asymmetry*. Oxford: Oxford University Press, pp. 81–99.

Sebens, C. and Carroll, S. (2016). Self-locating Uncertainty and the origin of probability in Everettian quantum mechanics. *British Journal for the Philosophy of Science*, 69: 25–74.

Smolin, L. (2019). *Einstein's Unfinished Revolution: The Search for What Lies beyond the Quantum*. London: Random House.

Tappenden, P. (2008). Comment on Saunders and Wallace. *British Journal for the Philosophy of Science*, 59: 306–314.

Tappenden, P. (2011). Evidence and uncertainty in Everett's multiverse. *British Journal for the Philosophy of Science*, 62: 99–123.

Tipler, F. (2014). Quantum nonlocality does not exist. *Proceedings of the National Academy of Sciences*, 111(31): 11281–11286.

Vaidman, L. (1998). On schizophrenic experiences of the neutron or why we should believe in the many-worlds interpretation of quantum theory. *International Studies in the Philosophy of Science*, 12: 245–261.

Vaidman, L. (2012). Probability in the Many-Worlds interpretation of quantum mechanics. In Y. Ben-Menahem and M. Hemmo (eds.), *Probability in Physics, The Frontiers Collection*. Berlin: Springer-Verlag.

Van Fraassen, B. (1980). *The Scientific Image*. Oxford: Clarendon Press.

Wallace, D. (2003). Everett and structure. *Studies in the History and Philosophy of Physics*, 34: 87–105. Available at: arXiv.org/abs/quant-ph/0107144.

Wallace, D. (2012). *The Emergent Multiverse: Quantum Theory according to the Everett Interpretation*. Oxford: Oxford University Press.

Wilson, A. (2011). Macroscopic ontology in Everettian quantum mechanics. *Philosophical Quarterly*, 61: 363–382.

Wilson, A. (2013). Objective probability in Everettian quantum mechanics. *British Journal for Philosophy of Science*, 64: 709–737.

Wilson, A. (2020). *The Nature of Contingency: Quantum Mechanics as Modal Realism*. Oxford: Oxford University Press.

Zurek, W.H. (2005). Probabilities from entanglement, Born's rule $p_k = |\psi_k|^2$ from envariance. *Physical Review*, A71: 052105–052129.

Further Reading from the Editors

For an accessible introduction to the probability problem in Everettian QM, see H. Greaves, "Probability in the Everett Interpretation," *Philosophy Compass* 2(1) (2007): 109–128. Recently the derivation of the Born rule given by C. Sebens and S. Carroll, "Self-locating Uncertainty and the origin of probability in Everettian quantum mechanics," *The British Journal for the Philosophy of Science* 69(1) (2018): 25–74 has been attracting plenty of attention; a prominent response is "Epistemic Separability and Everettian Branches: A Critique of Sebens and Carroll," R. Dawid and S. Friederich, *The British Journal for the Philosophy of Science* (forthcoming). For an alternative Everettian approach to probabilities, see A. Aguirre and M. Tegmark, "Born in an Infinite Universe: A Cosmological Interpretation of Quantum Mechanics," *Physical Review D* 84 (2011): 105002.

16

COLLAPSE THEORIES

Peter J. Lewis

The collapse postulate in quantum mechanics is problematic. In standard presentations of the theory, the state of a system prior to a measurement is a sum of terms, with one term representing each possible outcome of the measurement. According to the collapse postulate, a measurement precipitates a discontinuous jump to one of these terms; the others disappear. The reason this is problematic is that there are good reasons to think that measurement *per se* cannot initiate a new kind of physical process. This is the *measurement problem*, discussed in Section 16.1.

The problem here lies not with *collapse*, but with the appeal to measurement. That is, a theory that could underwrite the collapse process just described without ineliminable reference to measurement would constitute a *solution* to the measurement problem. This is the strategy pursued by *dynamical* (or *spontaneous*) collapse theories, which differ from standard presentations in that they replace the measurement-based collapse postulate with a dynamical mechanism formulated in terms of universal physical laws. Various dynamical collapse theories of quantum mechanics have been proposed; they are discussed in Section 16.2.

But dynamical collapse theories face a number of challenges. First, they make different empirical predictions from standard quantum mechanics, and hence are potentially empirically refutable. Of course, testability is a virtue, but since we have no empirical reason to think that systems ever undergo collapse, dynamical collapse theories are inherently empirically risky. Second, there are difficulties reconciling the dynamical collapse mechanism with special relativity. Third, the post-collapse state is not the same as the post-measurement state of standard quantum mechanics, raising the possibility that dynamical collapse theories do not solve the measurement problem after all. These challenges are described in Sections 16.3, 16.4, and 16.5, respectively.

Assuming that these challenges can be met, dynamical collapse theories can lay claim to being serious contenders for the correct description of the quantum world. And the description they provide has a number of interesting consequences. First, it makes indeterminism an irreducible fact about the physical world. It has been argued that this has important consequences for the foundations of statistical mechanics, for free will, and for consciousness. These claims are discussed in Section 16.6.

Second, many dynamical collapse theories imply that the quantum wavefunction is fundamental, and that particles are just a temporary "bunching up" of the fundamental wave-like entity. This understanding of the ontology of the quantum world gives rise to a new kind of vagueness, since a wave can be fuzzy around the edges in a way that a particle cannot. Furthermore, the quantum wavefunction is defined over a high-dimensional space, not over the three-dimensional space of experience, suggesting to some that three-dimensionality is an illusion. Dynamical collapse theories share these implications with other "wavefunction only" theories, such as Everettian approaches, and share some of them with wavefunction realist versions of Bohmian approaches. These implications of dynamical collapse theories are considered in Section 16.7.

Peter J. Lewis

16.1   The Measurement Problem

Quantum mechanics represents the state of a system in various ways, but the representation that is most perspicuous for understanding collapse theories uses a wavefunction. For a single particle, the wavefunction is a complex-valued function of three spatial dimensions and time: $\psi(x, y, z, t)$. The wavefunction changes over time according to a linear differential equation, the Schrödinger equation.

To begin with, let us stipulate that a particle is located in a particular spatial region if and only if all the corresponding wavefunction amplitude is contained in that region. (We will have reason to relax this stipulation later.) So, for example, the wavefunction shown schematically (in one dimension) in Figure 16.1a represents a particle in region A, and the one in Figure 16.1b represents a particle in a distinct region B.

The trouble with quantum mechanics starts from the *superposition principle*, which says that the sum of any two quantum states is also a quantum state. That is, if we take the function in Figure 16.1a and add it to the function in Figure 16.1b, we obtain another possible state of the particle, shown in Figure 16.1c. (We also need to rescale the function, due to the connection between the area under the curve and probability, to be explained shortly.)

States like Figure 16.1c are crucial to quantum mechanical explanations. But according to our earlier stipulation, the wavefunction in Figure 16.1c is not a state in which the particle is in region A, and it is not a state in which the particle is in region B. Nevertheless, when the position of a particle in such a state is measured, it is always found in one location or the other. Why should a state in which the particle is not in A and not in B generate a measurement result in which the particle is found in one of these locations?

The founders of quantum mechanics were aware of this problem. Born (1926) noted that the wavefunction can be used to generate the *probabilities* of the two outcomes: the square of the area under the wavefunction in region A is the probability that the particle will be found in region A, and similarly for region B. This is the *Born rule*. But the Born rule just quantifies the problem: if the particle is not in region A, why is there a 50% chance of finding it there?

The standard response, codified by von Neumann (1932), was to postulate that there are *two* dynamical processes by which the wavefunction changes over time. Between measurements, the wavefunction evolves continuously according to the Schrödinger equation, but during a measurement, the wavefunction jumps discontinuously into a state corresponding to a determinate location, with probabilities given by the Born rule. The latter is the *collapse postulate*.

Applied to a particle in the state shown in Figure 16.1c, the collapse postulate says that if the position of the particle is measured, there is a 50% chance that it will collapse to the state shown in Figure 16.1a, and a 50% chance that it will collapse to the state shown in Figure 16.1b. Hence even though the pre-measurement state is not one in which the particle is in region A, and not one in which it is in region B, the collapse postulate together with the post-collapse state can explain our measurement results, as well as the sense in which these results reveal the actual (post-measurement) location of the particle.

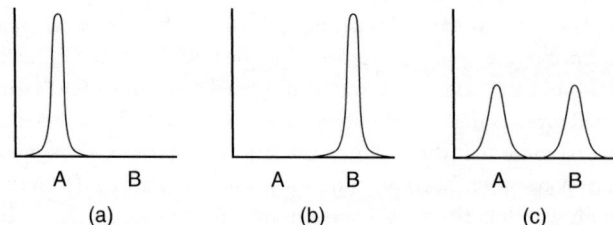

(a)   (b)   (c)

Figure 16.1 Three quantum states

248

As it stands, though, the collapse postulate is untenable. The continuous, linear Schrödinger evolution and the discontinuous, non-linear collapse process are incompatible: neither can be reduced to the other. So to be consistent, quantum mechanics must postulate a sharp division between those physical processes that count as measurement processes and those that do not. This seems like a tall order. What's more, since measuring devices are constructed out of particles that are not themselves being measured, the measuring device should operate according to the continuous Schrödinger evolution, and hence cannot instantiate the discontinuous collapse process. This is the much discussed *measurement problem*.

It is worth noting, though, that the problem lies not with collapse *per se*, but with tying the collapse process to measurement. If a collapse process could be described that yields the same outcome without the appeal to measurement, it would not be subject to the same critique. This is the approach to the interpretation of quantum mechanics pursued by dynamical collapse theories.

16.2 Dynamical theories of collapse

The basic approach was first described by Pearle (1976): instead of one dynamical law describing measurements and another applying between measurements, a single dynamical law applies at all times, deviating just slightly from the Schrödinger equation. The first fully developed theory along these lines was proposed by Ghirardi et al. (1986), and has become known as the GRW theory.

According to the GRW theory, the wavefunction for a single particle obeys a dynamical law that mostly coincides with the Schrödinger equation, except that there is a small chance per unit time that the wavefunction undergoes a spontaneous collapse process (or *hit*) in which it becomes highly localized around a point. More precisely, a hit multiplies the wavefunction by a narrow three-dimensional Gaussian (bell curve) in the coordinates of the particle concerned, where the width of the Gaussian is 10^{-5} cm.

The collapse process is indeterministic in two senses. First, whether a given particle undergoes a collapse is a random matter: there is a chance of 10^{-16} that any given particle will undergo a collapse in any given second. Second, if a particle undergoes a collapse, the location of the hit is random, with probabilities chosen so as to recover the Born rule: the chance that the Gaussian is centered in a particular region is given by the integral over that region of the pre-collapse wavefunction multiplied by the Gaussian.

For a single particle in the superposition state of Figure 16.1c, the effect of a hit is shown schematically (in one dimension) in Figure 16.2. In this state, there is a 50% chance of the hit being centered in region A, and a 50% chance of it being centered in region B. If it is centered in region A, then the effect is as shown: almost all the wave amplitude is now in region A, although a tiny amount (not shown to scale!) remains in region B, due to the fact that the "tails" of the Gaussian extend to infinity. It looks like this ought to be close enough for the particle to count as being in region A after the hit (although this question is addressed further in Section 16.5).

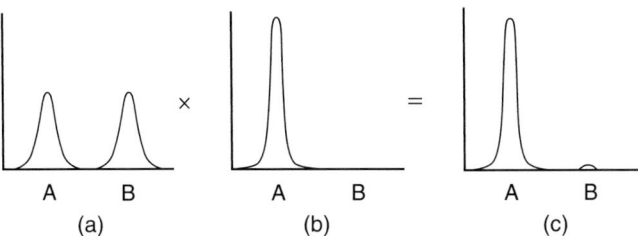

Figure 16.2 Collapse for a single particle

Peter J. Lewis

For a single particle, a collapse is extremely rare – about one every 100 million years. This is just as well, because experimental physics shows that individual particles (and small collections of them) always obey the Schrödinger equation very closely. But for a macroscopic object, the collapse rate can be appreciable: for an object containing of the order of 10^{23} particles, there will be around ten million collapse events per second. And if the positions of the particles are strongly correlated with each other, as is the case for a solid object, the collapse of a single particle is sufficient to localize all the particles.

This is shown schematically for two particles in Figure 16.3. The wavefunction for N particles is defined over a *configuration space* – a 3N-dimensional space of *configurations* of particles. For two particles, then, the wavefunction is a function of six spatial coordinates, three for each particle, plus time: $\psi(x_1, y_1, z_1, x_2, y_2, z_2, t)$. Figure 16.3 represents two of those dimensions: the horizontal axis represents one of the spatial dimensions of particle 1, and the vertical axis represents one of the spatial dimensions of particle 2. The shaded areas represent regions where the wavefunction amplitude is high.

For two particles whose positions are strongly correlated, a typical superposition state is as shown in Figure 16.3a: all the wave amplitude is concentrated in areas representing the particles as occupying the same spatial region. If particle 1 undergoes a hit, the wavefunction is multiplied by a function that is a Gaussian in the coordinates of particle 1 and a constant in the coordinates of particle 2, so it is large only in the "stripe" shown in Figure 16.3b. The result is that the post-hit wavefunction is large only in region A for *both* particles. The same goes for a hit on particle 2. That is, if either particle undergoes a collapse, both particles acquire locations.

Of course, a collapse for two particles is still exceedingly rare. But for a macroscopic solid object, a hit on one of the particles making up the object will occur on average every tenth of a microsecond. Furthermore, because of the forces binding the particles in a solid object together, the wavefunction will only be large in regions of configuration space in which the particles are close together. Due to these strong correlations between the positions of the particles making up the object, a collapse for one particle is sufficient to localize the whole object.

This is the heart of the dynamical collapse solution to the measurement problem. Suppose we start with a single particle in the superposition state of Figure 16.1c and we measure its position. Measuring its position requires us to correlate its position with something we can see, such as a pointer on a dial. But in doing that, we create a macroscopic object in a superposition of two distinct locations, pointing at "A" on the dial and pointing at "B." The GRW collapse process very rapidly reduces this superposition to one location or the other, and since the particle is correlated with the pointer, the particle too acquires a determinate location. Hence after the measurement, the particle is either in region A or in region B, and the pointer is pointing to the corresponding measurement result.

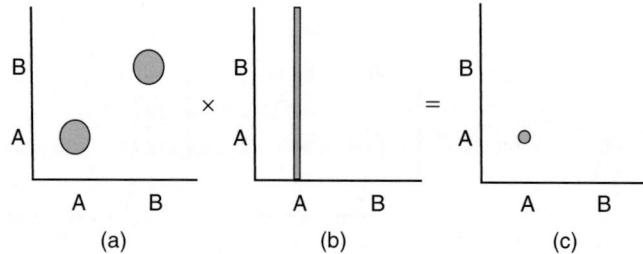

Figure 16.3 GRW collapse for two correlated particles

Note that there is no essential appeal to measurement in this account. When we correlate a superposition state of a microscopic system with the position of a macroscopic solid object, the GRW collapse process reduces the superposition state to one of its components, and it does so according to a single, precisely specified dynamical law.

To accommodate symmetrization requirements for identical particles, more recent dynamical collapse theories incorporate the collapse process as a non-linear correction to the Schrödinger equation, rather than as a distinct process, producing a collapse that is continuous rather than discrete (Ghirardi et al., 1990). Diosi (1989) and Penrose (1996) further suggest that the collapse mechanism may be connected to the role of gravity. Penrose notes that the existence of a macroscopic object in a superposition of two distinct locations entails, via gravity, a superposition of two distinct spacetime structures, and speculates that the latter superposition may be inherently unstable. At present, though, gravitational explanations of the collapse process have no empirical support. Indeed, there is no evidence for the existence of a collapse process at all – unless one takes the existence of definite experimental outcomes as evidence for collapse. One might take the lack of empirical support as an objection to the entire dynamical collapse project, so let us briefly consider the experimental situation.

16.3   Tests

Dynamical collapse theories make different empirical predictions from no-collapse interpretations of quantum mechanics, such as de Broglie-Bohm theory (Tumulka, this volume) and Everettian quantum mechanics (Saunders, this volume). So can't we just perform the relevant experiments to see which is right? Unfortunately, this is not at all straightforward.

There is no problem in principle. Dynamical collapse theories predict that a superposition of distinct locations for a macroscopic object is inherently unstable, and will rapidly evolve to some determinate location. No-collapse theories predict that such macroscopic superpositions are stable. And there are experiments that could in principle distinguish between an object in a superposition of locations and an object in one of those locations.

But in practice, these experiments are impossible to perform. Note first that a simple measurement of the position of the object won't suffice to distinguish the superposition from the determinate location. That is because both collapse and no-collapse interpretations have mechanisms to ensure that a measurement on a superposition state yields a determinate outcome, just as a measurement on a determinate location state does.

So we need something more subtle. What is required is an interference experiment, in which the two terms in the superposition are made to interact with each other, indicating that they are both present. Interference experiments are very sensitive to environmental effects. If a single outside particle becomes correlated with one term in the superposition but not the other, then the interference experiment fails. So to exhibit interference effects, a system has to be kept completely isolated from the environment. This is possible for microscopic systems, but is practically impossible for macroscopic systems.

Nevertheless, there has been considerable progress in demonstrating interference with larger and larger systems. Early attempts to detect quantum collapse centered on the behavior of superconducting quantum interference devices (SQUIDs). Using such devices, experimentalists are able to create and detect a superposition of a clockwise and a counter-clockwise electric current in a ring of approximately 1 cm diameter. Such devices do not reveal any collapse effects. However, the superposition in a SQUID, though macroscopic, is of distinct currents, not of distinct locations. The wavefunction for the electrons involved is distributed over the whole ring whether the current is clockwise or counter-clockwise. So a GRW collapse to a precise location is not a collapse to one *current* over the other, and GRW collapses would not be expected to have a measurable effect on the current (Rae, 1990).

Another important experimental development involves demonstrating interference effects for increasingly larger molecules. Interference can now be demonstrated for relatively large organic molecules, such as $C_{48}H_{26}F_{24}N_8O_8$ (Juffmann et al., 2012). This shows that such molecules can exist in superpositions of distinct location states, at least for the short amount of time it takes to traverse the apparatus. But even such a large molecule involves fewer than 5,000 fundamental particles, inducing a GRW collapse rate of $5,000 \times 10^{-16}$ per second, or about one every hundred thousand years – still far too rare to be detectable.

Further methods for detecting quantum collapses have been attempted (Leggett, 2002; Bassi et al., 2013). So far we have no empirical evidence to suggest that quantum collapses occur, but neither are the particular models of the GRW theory and its continuous variants ruled out. Therefore, empirical tests are inconclusive. However, there is an indirect empirical argument against dynamical collapse theories, namely that they conflict with another well-confirmed theory: special relativity.

16.4 Relativity

The original GRW theory conflicts with special relativity in two distinct ways. First, when a particle undergoes a hit, the probability distribution for the center of the collapse is based on the wavefunction distribution *at that time*. But according to special relativity, there is no absolute standard of simultaneity, so the state of a spread-out entity *at a time* is ill-defined. Second, the hit instantaneously multiplies the wavefunction over the whole of space by a Gaussian, and again this process is ill-defined according to special relativity. In particular, for correlated particles a hit on one particle has an instantaneous effect on the state of another particle, no matter how far apart, and instantaneous action at a distance is prima facie incompatible with special relativity.

However, progress has been made in constructing a dynamical collapse theory that is consistent with special relativity (see Myrvold, this volume, for further details). Tumulka (2006), following some remarks by Bell (1987, p. 205), suggests that the ontology of GRW-type theories should be understood as point-like: the spatiotemporal point at the center of a hit event is a "flash" of reality, and the spread-out wavefunction can be interpreted instrumentally as governing the probability distribution for these flashes. So for a single particle system, what exists is a set of point-like events, about one per hundred million years. Since the wavefunction is treated instrumentally, there is no real collapse of the wavefunction that could conflict with relativity. Furthermore, given a flash event at a particular space-time point, the probability distribution for the location of the next flash event is defined over surfaces that are relativistically invariant – that is, over surfaces such that $x^2 + y^2 + z^2 - c^2t^2$ is a constant, rather than simultaneity surfaces for which t is a constant.

Alternatively, Ghirardi et al. (1995) suggest that the appropriate ontology for GRW-type theories is a mass density distribution defined over three-dimensional space. In relativistic versions, the mass density in a small region of space is determined by the state in the past light cone of that region (Bedingham et al., 2014). Hence, a hit centered on a space-time point produces high mass density at that point, and a surrounding region of near-zero mass density that spreads outward at the speed of light as the collapse event enters the past light cone of surrounding points (Myrvold, 2016).

Hence for a single particle, versions of the GRW theory can be made consistent with special relativity. But for two or more particles, there is still the worry that for correlated particles, a hit on one particle can instantly affect the other, no matter how distant. Tumulka (2006, p. 350) suggests that this can be accommodated within special relativity by allowing that the direction of the influence is indeterminate. Consider two particles whose perfectly correlated z-coordinates are measured at space-like separated x-coordinates. In some frames of reference, the measurement of particle 1 occurs first, and causes particle 2 to acquire a determinate z-coordinate; in other frames of reference, the measurement of particle 2 occurs first, and causes particle 1 to acquire a determinate z-coordinate. Neither causal story is to be preferred, so no absolute standard of simultaneity is required.

Even so, both causal stories require faster-than-light causation. However, this is not in direct conflict with special relativity; rather, it just requires that the probabilities ascribed to the measurement outcomes depend on the frame of reference. In the frame in which particle 1 is measured first, one particular measurement outcome for particle 2 has probability 1 prior to measurement, whereas in the frame in which particle 2 is measured first, no measurement outcome for particle 2 has probability 1 prior to measurement (Myrvold, 2002, p. 461). Indeed, Myrvold (2016) prefers to say that there are non-local probabilistic correlations here, but no superluminal causation, on the grounds that the direction of causal relations cannot be indeterminate.

16.5   Tails

Consistency with special relativity is, therefore, a problem for dynamical collapse theories, but perhaps not an insuperable one. However, solving this problem is moot if, as some have claimed, the dynamical collapse approach does not even minimally solve the measurement problem.

The source of this concern is that, as mentioned above, for a particle in the superposition state 2(a), a collapse does not ensure that *all* the post-collapse wavefunction amplitude is in region A (or in region B), just that *most* of it is. The same goes for a macroscopic object. If a human being ends up in a superposition of occupying two distinct regions, for example as a result of correlating her location with the position of a particle in a superposition state, then collapses will very rapidly put her state *close* to one in which all her wavefunction amplitude is located in one of those regions. However, there remains a small but non-zero amplitude in the other location.

One form of the concern goes as follows. In the above example, the small term in the post-collapse state has exactly the same internal structure as the big term: it is the structure of a human being. Granted a rather plausible functionalism, it is the structure of a term, not its amplitude, which determines what it represents (Wallace, 2003). But then there is a human being in both locations, and dynamical collapse theories are ineffective at bringing about determinate measurement outcomes (Cordero, 1999).

Since the source of this problem is the non-zero "tails" of the Gaussian collapse function stretching to infinity in every direction, the most obvious solution is to eliminate the tails (Wallace, 2014). That is, if the collapse function were strictly *zero* at distances greater than 10^{-5} cm from the collapse center, then after a collapse there would be a wave term with the structure of a human being at only one location.

But there is another form of the concern about tails that is resistant to this solution. Between collapses, the state of a particle evolves according to the Schrödinger equation, and this means that even if collapse makes the wavefunction strictly zero outside a given region at a time, it has tails extending to infinity an instant later. So if what it is for a particle to be located in a region is that *all* its wave amplitude is located in that region (as we assumed earlier), then after a collapse the particle is still not located in any finite region of space, and dynamical collapse theories do not solve the measurement problem after all (Albert and Loewer, 1990).

One might take this as an additional motivation to adopt a "flashy" version of GRW, so that the ontology resides only at the precise instants of collapse. Albert and Loewer (1996) instead suggest solving this problem by relaxing the link between wavefunction amplitude and particle location: instead of demanding that *all* the wave amplitude be contained in a region for the particle to count as being located there, we only require that *almost* all of it is so located. This works, but at the cost of introducing a new kind of vagueness: there is presumably no fact of the matter about precisely how much of the amplitude needs to be in the region for the particle to count as located there. Whether this vagueness is problematic is considered in Section 16.7.

16.6   Chance

Quantum mechanics is often taken to be an indeterministic theory. But this is not a straightforward consequence of the theory: the de Broglie-Bohm and Everettian versions are deterministic at the fundamental level. Nevertheless, dynamical collapse theories really are irreducibly indeterministic: they incorporate objective chances into fundamental physical law (see Emery, this volume, and Suarez, this volume, for more on chance and determinism).

What consequences does this have? Albert (2000) suggests that the role of chance in dynamical collapse theories can solve a problem in the foundations of statistical mechanics. It is well known that for a given macrostate, the microstates exhibiting normal thermodynamic behavior vastly outnumber those exhibiting abnormal behavior, but it is far from clear how this asymmetry can be used to *explain* thermodynamic behavior (see Shahvisi, this volume, for further discussion). Albert notes that if there were some mechanism by which the states of systems were randomly and asymmetrically perturbed at the microscopic level, then normal thermodynamic behavior would be straightforwardly explicable via this asymmetry. Dynamical collapse theories entail this random, asymmetric perturbation, but other interpretations of quantum mechanics, notably de Broglie-Bohm and Everett, do not. Hence, dynamical collapse theories might gain indirect support from the foundations of statistical mechanics (Shenker, this volume, discusses some further ways in which quantum mechanics may bear on the foundations of statistical mechanics).

More controversially, some see the indeterminism of dynamical collapse theories as opening the door for a reconciliation of free will with physics. The trouble with such suggestions is that the collapse process is random, and randomness looks no more hospitable to free will than determinism. Still, Kane (1996) suggests that genuine indeterminacy, even if it is random, is essential to free will. The idea is that an agent must be ultimately responsible for their character if they are to be truly free, and a collapse in the brain at a suitable juncture, even if it is random, might be enough to secure ultimate responsibility.

Along similar lines, some have seen a connection between collapse and consciousness. Prior to the advent of dynamical collapse theories, Wigner (1961) famously speculated that consciousness might be required to explain the collapse postulate. More recently, Hameroff and Penrose (1996) suggest that dynamical collapses in the brain might explain consciousness, insofar as consciousness requires non-computability, and quantum collapse can introduce non-computability. It is worth noting, though, that these proposals concerning free will and consciousness add various highly contested philosophical claims to the already controversial status of collapse theories.

16.7   Ontology

If a dynamical collapse theory is true, what does this tell us about the furniture of the world? The clearest consequence is that particles are not fundamental. The fundamental law of a dynamical collapse theory governs the evolution of a wavefunction, and the wavefunction alone underlies all our empirical observations. After a collapse, the wavefunction becomes highly localized in three of its dimensions, and while that localization persists we can speak of a particle occupying a determinate position in three-dimensional space. But this "particle" is just a manner of speaking about the wavefunction.

However, there are several reasons to doubt that the ontology of dynamical collapse theories consists of the wavefunction and nothing but the wavefunction. First, as noted in Section 16.4, some attempts to reconcile dynamical collapse theories with special relativity replace the wavefunction with discrete, point-like events or with a mass density distribution over three-dimensional space. However, special relativity arguably does not rule out the view that the wavefunction is fundamental, provided one is willing to ascribe a distinct wavefunction to every spacelike hypersurface (Myrvold, 2002).

Second, the wavefunction for a system of N particles is defined over a 3N-dimensional configuration space. Hence, it seems that if the wavefunction is fundamental, then the appearance of the world as three-dimensional is somehow illusory (Albert, 1996). To avoid this conclusion, several commentators have again postulated that the fundamental ontology of dynamical collapse theories includes a mass density distribution over three-dimensional space whose value in a region can be derived from the wavefunction (Allori, 2013). The issue of dimensionality can also be used to motivate a flash ontology, since flashes are defined at points of three-dimensional space. However, it is also possible to argue that the wavefunction indirectly describes objects in three-dimensional space, bypassing these concerns (Lewis, 2013).

Third, some have argued that the wavefunction needs to be supplemented to counteract the effects of the novel quantum vagueness introduced by dynamical collapse theories (Section 16.5). This vagueness allows a particle to count as occupying a region when almost all its wave amplitude is in that region. It follows from this and the high-dimensional nature of the wavefunction that for a set of particles, each can individually count as occupying a given region, even though the set taken as a whole does not count as occupying the region. The same goes for macroscopic objects (Lewis, 1997). If objects are constituted by a mass density distribution rather than by a wave amplitude distribution, then this problem does not arise (Bassi and Ghirardi, 1999). However, it is not clear that there is really a problem here that needs solution: the strange properties of compound objects can be regarded as an inevitable consequence of the mismatch between our classical concepts and the quantum world (Lewis, 2003).

The ontological consequences of dynamical collapse theories remain an area of active debate.

Acknowledgment

I thank Wayne Myrvold and Alastair Wilson for very helpful comments on an earlier draft.

References

Albert, D.Z. (1996). Elementary quantum metaphysics. In J. Cushing, A. Fine and S. Goldstein (eds.), *Bohmian Mechanics and Quantum Theory: An Appraisal*. Dordrecht: Kluwer, pp. 277–284.

Albert, D.Z. (2000). *Time and Chance*. Cambridge, MA: Harvard University Press.

Albert, D.Z. and Loewer, B. (1990). Wanted dead or alive: Two attempts to solve Schrödinger's paradox. In A. Fine, M. Forbes and L. Wessels (eds.), *PSA 1990*, Vol. 1. Chicago: University of Chicago Press, pp. 277–285.

Albert, D.Z. and Loewer, B. (1996). Tails of Schrödinger's cat. In R. Clifton (ed.), *Perspectives on Quantum Reality*. Dordrecht: Kluwer, pp. 81–92.

Allori, V. (2013). Primitive ontology and the structure of fundamental physical theories. In A. Ney and D.Z. Albert (eds.), *The Wave Function*. Oxford: Oxford University Press, pp. 58–75.

Bassi, A. and Ghirardi, G.C. (1999). More about dynamical reduction and the enumeration principle. *British Journal for the Philosophy of Science*, 50: 719–734.

Bassi, A., Lochan, K., Satin, S., Singh, T.P. and Ulbricht, H. (2013). Models of wave-function collapse, underlying theories, and experimental tests. *Reviews of Modern Physics*, 85: 471–527.

Bedingham, D., Dürr, D., Ghirardi, G.C., Goldstein, S., Tumulka, R. and Zanghì, N. (2014). Matter density and relativistic models of wave function collapse. *Journal of Statistical Physics*, 154: 623–631.

Bell, J.S. (1987). *Speakable and Unspeakable in Quantum Mechanics*. Cambridge: Cambridge University Press.

Born, M. (1926). Quantenmechanik der Stoßvorgänge. *Zeitschrift für Physik*, 38: 803–827.

Cordero, A. (1999). Are GRW tails as bad as they say? *Philosophy of Science*, 66: S59–S71.

Diosi, L. (1989). Models for universal reduction of macroscopic quantum fluctuations. *Physical Review A*, 40: 1165–1174.

Emery, N. (this volume).

Ghirardi, G.C., Grassi, R. and Benatti, F. (1995). Describing the macroscopic world: Closing the circle within the dynamical reduction program. *Foundations of Physics*, 25: 5–38.

Ghirardi, G.C., Pearle, P. and Rimini, A. (1990). Markov processes in Hilbert space and continuous spontaneous localization of systems of identical particles. *Physical Review A*, 42: 78–89.

Ghirardi, G.C., Rimini, A. and Weber, T. (1986). Unified dynamics for microscopic and macroscopic systems. *Physical Review D*, 34: 470–491.

Hameroff, S.R. and Penrose, R. (1996). Conscious events as orchestrated space-time selections. *Journal of Consciousness Studies*, 3: 36–53.

Juffmann, T., Milic, A., Müllneritsch, M., Asenbaum, P., Tsukernik, A., Tüxen, J., Mayor, M., Cheshnovsky, O. and Arndt, M. (2012). Real-time single-molecule imaging of quantum interference. *Nature Nanotechnology*, 7: 297–300.

Kane, R. (1996). *The Significance of Free Will*. Oxford: Oxford University Press.

Leggett, A.J. (2002). Testing the limits of quantum mechanics: Motivation, state of play, prospects. *Journal of Physics: Condensed Matter*, 14: R415–R451.

Lewis, P.J. (1997). Quantum mechanics, orthogonality, and counting. *British Journal for the Philosophy of Science*, 48: 313–328.

Lewis, P.J. (2003). Four strategies for dealing with the counting anomaly in spontaneous collapse theories of quantum mechanics. *International Studies in the Philosophy of Science*, 17: 137–142.

Lewis, P.J. (2013). Dimension and illusion. In A. Ney and D.Z. Albert (eds.), *The Wave Function*. Oxford: Oxford University Press, pp. 110–125.

Myrvold, W.C. (2002). On peaceful coexistence: Is the collapse postulate incompatible with relativity? *Studies in History and Philosophy of Modern Physics*, 33: 435–466.

Myrvold, W.C. (2016). Lessons of Bell's theorem: Nonlocality, yes; action at a distance, not necessarily. In S. Gao and M. Bell (eds.), *Quantum Nonlocality and Reality – 50 Years of Bell's Theorem*. Cambridge. Cambridge University Press, pp. 238–260.

Myrvold, W.C. (this volume).

Ney, A. and Albert, D.Z., editors. (2013). *The Wave Function*. Oxford: Oxford University Press.

Pearle, P. (1976). Reduction of the state vector by a nonlinear Schrödinger equation. *Physical Review D*, 13: 857–868.

Penrose, R. (1996). On gravity's role in quantum state reduction. *General Relativity and Gravitation*, 28: 581–600.

Rae, A.I.M. (1990). Can GRW theory be tested by experiments on SQUIDS? *Journal of Physics A*, 23: L57–L60.

Tumulka, R. (2006). Collapse and relativity. In A. Bassi, D. Dürr, T. Weber and N. Zanghì (eds.), *Quantum Mechanics: Are There Quantum Jumps?* and *On the Present Status of Quantum Mechanics. AIP Conference Proceedings*, Vol. 844, pp. 340–352.

Tumulka, R. (this volume), "de Broglie-Bohm Theory".

Wallace, D. (2003). Everett and structure. *Studies in History and Philosophy of Modern Physics*, 34: 87–105.

Wallace, D. (2014). *Life and death in the tails of the GRW wave function*. Available at: arXiv:1407.4746.

Wigner, E.P. (1961). Remarks on the mind-body question. In I.J. Good (ed.), *The Scientist Speculates*. London: Heineman, pp. 284–302.

von Neumann, J. (1932). *Mathematische Grundlagen der Quantenmechanik*. Berlin: Springer.

Further Reading from the Editors

Readers may wish to start with the comprehensive G. Ghirardi and A. Bassi, "Collapse Theories," *Stanford Encyclopedia of Philosophy* (2020). A pair of review articles by Philip Pearle offer his perspective on the main problems that face collapse theories: P. Pearle, "How Stands Collapse I," *Journal of Physics. A, Mathematical and Theoretical* 40(12), (2007):3189–3204 and P. Pearle, "How Stands Collapse II," *Quantum Reality, Relativistic Causality, and Closing the Epistemic Circle* (Springer, 2009). For Penrose's version of the gravitational collapse hypothesis, see R. Penrose, *The Emperor's New Mind* (Oxford University Press, 1989). For a comparison between contemporary collapse theories and contemporary Bohmian mechanics, see V. Allori, S. Goldstein, R. Tumulka, and N. Zanghi, "On the Common Structure of Bohmian Mechanics and the Ghirardi-Rimini-Weber Theory," *The British Journal for the Philosophy of Science* 59(3), (2008):353–389.

17

BOHMIAN MECHANICS

Roderich Tumulka

17.1 Fundamental Laws of Bohmian Mechanics

Bohmian mechanics, also known as pilot-wave theory or de Broglie–Bohm theory, is a formulation of quantum mechanics whose fundamental axioms are not about what observers will see if they perform an experiment but about what happens in reality. It is therefore called a "quantum theory without observers," alongside with collapse theories (Bell (1987a); Lewis (this volume)) and many-worlds theories (Saunders (this volume); Allori et al. (2011)) and in contrast to orthodox quantum mechanics. It follows from these axioms that in a universe governed by Bohmian mechanics, observers will see outcomes with exactly the probabilities specified by the usual rules of quantum mechanics for empirical predictions. Specifically, Bohmian mechanics asserts that electrons and other elementary particles have a definite position at every time and move according to an equation of motion that is one of the fundamental laws of the theory and involves a wave function that evolves according to the usual Schrödinger equation. Bohmian mechanics is named after David Bohm (1917–1992), who was, although not the first to consider this theory, the first to realize (in 1952) that it actually makes correct predictions.

While Bohmian mechanics has been considered as a tool for visualization (Philippidis et al., 1979), for the efficient numerical simulation of the Schrödinger equation (Chattaraj, 2010), and other applications (Oriols and Mompart, 2012), the main interest in it arises from the fact that Bohmian mechanics provides a possible way how our world might be and might work.

In its non-relativistic form, the theory asserts the following: N material points ("particles") move in three-dimensional Euclidean space (denoted for simplicity as $\mathbb{R}^3$) in a way governed by a field-like entity that is mathematically given by a wavefunction ψ (as familiar from standard quantum mechanics). More precisely, the position $\mathbf{Q}_k(t)$ of particle number k at time t obeys Bohm's equation of motion

$$\frac{d\mathbf{Q}_k(t)}{dt} = \frac{\hbar}{m_k} \operatorname{Im} \frac{\psi^* \nabla_k \psi}{\psi^* \psi} (Q(t), t), \tag{17.1}$$

where $Q(t) = (\mathbf{Q}_1(t), \ldots, \mathbf{Q}_N(t)) \in \mathbb{R}^{3N}$ denotes the configuration of the particle system at time t, m_k is the mass of particle k, Im is the imaginary part of a complex number, $\psi : \mathbb{R}^{3N+1} \to \mathbb{C}^d$ is the wavefunction, and $\psi^* \phi$ denotes the scalar product in spin space $\mathbb{C}^d$,

$$\psi^* \phi = \sum_{s=1}^{d} \psi_s^* \phi_s. \tag{17.2}$$

For spinless particles ($d = 1$), ψ^* is the usual complex conjugate of ψ, and the factor ψ^* cancels out of (17.1); in that case, we can express ψ in terms of its modulus R and phase S as $\psi = R\,e^{iS/\hbar}$, and obtain $m_k^{-1} \nabla_k S$ for the right-hand side of (17.1). Yet another way of writing the right-hand side of

(17.1) is $\boldsymbol{j}_k/\rho$, where $\boldsymbol{j}_k = (\hbar/m_k)\mathrm{Im}(\psi^*\nabla_k\psi)$ is the quantity known in quantum mechanics as the probability current and $\rho = \psi^*\psi$ the probability density.

The wavefunction $\psi(q, t) = \psi(\boldsymbol{q}_1, \ldots, \boldsymbol{q}_N, t)$ evolves with time t according to the usual Schrödinger equation

$$i\hbar\frac{\partial\psi(q, t)}{\partial t} = -\sum_{k=1}^{N}\frac{\hbar^2}{2m_k}\nabla_k^2\psi(q, t) + V(q)\psi(q, t), \tag{17.3}$$

where $V : \mathbb{R}^{3N} \to \mathbb{R}$ is the potential function; for example, the Coulomb potential

$$V(\boldsymbol{q}_1, \ldots, \boldsymbol{q}_N) = \frac{1}{2}\sum_{j\neq k}\frac{e_j e_k}{|\boldsymbol{q}_j - \boldsymbol{q}_k|} \tag{17.4}$$

with e_j being the electric charge of particle j. The state of the system at time t is described by the pair $(Q(t), \psi(t))$, and Equations (17.1) and (17.3) together determine the state at any other time; thus, Bohmian mechanics is a deterministic theory.

By a mathematical fact known as the "equivariance theorem" (Bohm, 1952; Dürr et al., 1992), if the initial configuration $Q(0)$ is random with probability density $|\psi(q, t = 0)|^2$, then at every time t,

$$Q(t) \text{ is } |\psi(q, t)|^2 \text{ distributed.} \tag{17.5}$$

It is another basic rule of the theory that the configuration is indeed so distributed – this is the equivalent of the Born rule in Bohmian mechanics. As discussed in Section 17.9 below, this rule follows if it is assumed that the initial configuration of the universe is typical relative to the $|\Psi|^2$ distribution, where Ψ is the initial wavefunction of the universe.

We note that Bohmian mechanics is time-reversal invariant: if $t \mapsto (Q(t), \psi(t))$ is a solution of (17.1) and (17.3), then so is $t \mapsto (Q(-t), \psi^*(-t))$, and also Born's rule still holds for the time reverse. Likewise, Bohmian mechanics is invariant under Galilean boosts, spatial rotations, and space-time translations (see, e.g., Dürr et al. (1992)). I also note that equations of motion analogous to (17.1) have been devised for a wide class of Hamiltonians (Struyve and Valentini, 2009), not just non-relativistic Schrödinger operators as in (17.3). In particular, the equation of motion for the 1-particle Dirac equation (Bohm, 1953; Bohm and Hiley, 1993) asserts that the possible world-lines (i.e., particle paths in space-time) are the integral curves of the probability current 4-vector field $j^\mu = \overline{\psi}\gamma^\mu\psi$.

17.2 Example: The Double-Slit Experiment

In Bohmian mechanics, there is wave–particle duality in a literal sense: there is a wave, and there is a particle. In the double-slit experiment, the wave passes through both slits, whereas the particle passes through only one slit in each run. In Figure 17.1, a sample of 80 alternative trajectories is shown.

We may place a detecting screen at the right end of the figure. Since the trajectories in the sample are roughly $|\psi|^2$ distributed, their arrival points are also $|\psi|^2$ distributed by equivariance. One can see in Figure 17.1 that there are some locations in which more trajectories arrive – in fact, those are the locations where $|\psi|^2$ is large; thus, an interference pattern with bright and dark fringes is visible in the arrival points on the screen, in agreement with the empirical findings. This is how Bohmian mechanics explains the double-slit experiment, without any paradox remaining (Bell, 1980).

For the outcome of the experiment, it is crucial that the motion of the Bohmian particle is non-Newtonian, and that its trajectory can be bent also in the absence of external force fields. It is bent because the equation of motion (17.1) dictates it. Note that the motion of the particle, if it went through the upper slit, will depend also on the part of the wavefunction that went through the lower

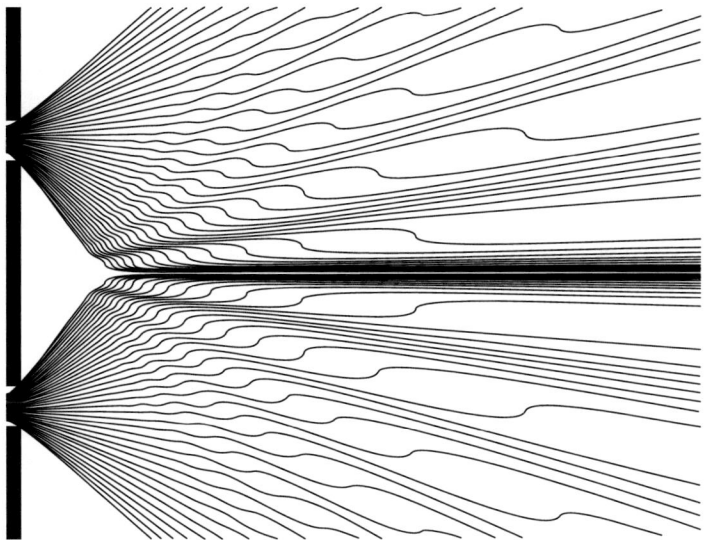

Figure 17.1 Several possible trajectories for a Bohmian particle in a double-slit setup, coming from the left. (Reprinted from Dürr and Teufel (2009), based on a figure in Philippidis et al. (1979))

slit. In particular, the motion of the particle depends on whether both slits are open, and would have been different if one slit had been closed. With only one slit open, the distribution of the arrival points on the screen would have created a different interference pattern. If both slits are open, then the particle, although it passes only through one slit, "knows" this because the wave passed through both slits. Note that the Bohmian explanation of the double-slit experiment involves no faster-than-light effects and no retrocausation. This is still true in Wheeler's (1978) "delayed choice" variant of the experiment, which was explained using Bohmian mechanics in Bell (1980).

17.3 Empirical Predictions

An analysis of Bohmian mechanics shows that its empirically testable predictions agree exactly with those of standard quantum mechanics, whenever the latter are unambiguous (Bohm, 1952; Bell, 1980; Dürr et al., 1992). Thus, Bohmian mechanics is a counter-example to the claim put forward by Niels Bohr (1935) (and often repeated since) that in quantum mechanics a single coherent picture of reality be impossible. In particular, it turns out that the statistics of outcomes of experiments are related to the operators known as "observables" in the same way as in standard quantum mechanics. The fact that these operators in general do not commute has often in orthodox quantum mechanics (OQM) been taken as a sign that a "realistic" picture be impossible, whereas Bohmian mechanics shows how it is, in fact, possible; see Section 17.7 for more detail.

The predictive rules of standard quantum mechanics include the collapse of the wavefunction; in fact, an effective collapse of the wavefunction of a system s comes out of the equations of Bohmian mechanics, although the wavefunction of the universe never collapses. Here is how. In a (hypothetical) universe governed by Bohmian mechanics, also observers and measurement apparatus are made of Bohmian particles, and are jointly governed by Equations (17.1) and (17.3). During the experiment, the initial wavefunction $\psi_0(q_s, q_a) = \varphi(q_s)\chi(q_a)$ of the system s and apparatus a together evolves according to (17.3) to $\psi_t(q_s, q_a)$. Suppose that whenever φ is an eigenfunction φ_α of the self-adjoint operator A with eigenvalue α, the final wavefunction is of the form

$$\psi_t(q_s, q_a) = \varphi_\alpha(q_s)\chi_\alpha(q_a), \tag{17.6}$$

where t is the duration of the experiment, and χ_α is a wavefunction of an apparatus displaying the outcome α; this would be the case in an "ideal quantum measurement" of A. Then it follows by the linearity of (17.3) for an arbitrary superposition $\varphi = \sum_\alpha c_\alpha \varphi_\alpha$ of eigenfunctions that

$$\psi_t(q_s, q_a) = \sum_\alpha c_\alpha \varphi_\alpha(q_s)\, \chi_\alpha(q_a). \tag{17.7}$$

Since the actual configuration $Q(t) = (Q_s, Q_a)$ is $|\psi_t|^2$ distributed, and since χ_α is concentrated on those configurations in which the apparatus displays the outcome α (say, by the position of a pointer), Q_a has probability $|c_\alpha|^2$ to be such a configuration; this probability value for obtaining the outcome α agrees with the rules of standard quantum mechanics. Moreover, since the various χ_α have macroscopically disjoint supports in configuration space, the packets $\psi_\alpha(q_s, q_a) = \varphi_\alpha(q_s)\,\chi_\alpha(q_a)$ will remain non-overlapping for the next 10^{100} years or more. But as long as $\psi = \sum_\alpha c_\alpha \psi_\alpha$ consists of non-overlapping packets, the motion of $Q(t)$, which is governed according to (17.1) by ψ and its derivatives at $Q(t)$, will depend only on one packet, the one containing $Q(t)$. Thus, it makes no difference for the future motion of all particles (at least for the next 10^{100} years) if we drop all other packets $\psi_{\alpha'}$ and replace ψ by ψ_α with α the actual outcome. The $|\psi_\alpha|^2$ distribution represents the conditional probability distribution of $Q(t)$, given that $Q(t)$ lies in the support of ψ_α (i.e., given that the outcome was α). That is why the wavefunction can be collapsed ($\psi \mapsto \psi_\alpha$), yielding an eigenfunction φ_α for the system s. I called it an "effective" collapse because one can practically replace $\psi \mapsto \psi_\alpha$ although the true wavefunction is still the uncollapsed ψ.

More generally (beyond ideal quantum measurements), in (spinless) Bohmian mechanics every system s can be attributed a wavefunction of its own, the *conditional wavefunction*

$$\psi^{(s)}(q_s) = \psi(q_s, Q_a) \tag{17.8}$$

obtained by inserting the actual configuration of the environment a (and normalizing if desired). (In the presence of spin, one uses a conditional density matrix, see Dürr et al. (2005).) In the situation discussed before, $\psi^{(s)} \propto \varphi$ before the measurement and $\psi^{(s)} \propto \varphi_\alpha$ afterward. The conditional wavefunction evolves according to the Schrödinger equation (17.3) if there is no interaction between s and a and s is suitably disentangled from a (viz., $\psi(q_s, q_a) = \varphi(q_s)\,\chi(q_a) + \psi^\perp(q_s, q_a)$ with the q_a-support of $\psi^\perp$ macroscopically disjoint from Q_a), but not in general. In a measurement-like situation, the conditional wavefunction undergoes a genuine collapse.

In OQM, one cannot form the conditional wavefunction as there is no Q_a. What is worse, since OQM insists that there are no variables (such as Q) in addition to the wavefunction, and since the post-measurement wavefunction (17.7) is a superposition of terms corresponding to all possible outcomes, there is nothing in the state according to OQM at time t that would represent the actual outcome − a problem known as the quantum measurement problem. The only ways out are (i) to introduce further variables (as in Bohmian mechanics), (ii) to deny the linearity of the time evolution of the wavefunction (as in collapse theories such as GRW (Bell, 1987a; Lewis, in this volume)), or (iii) to deny that there is a single outcome (as in many-worlds theories (Saunders, in this volume; Allori et al., 2011)).

17.4 Limitations to Knowledge and Control

In classical mechanics, one usually pretends that one can measure the state of a system (position and momentum of all particles) to arbitrary accuracy without disturbing it, and that one can in principle prepare a system in any state. In Bohmian mechanics, in contrast, there are sharp limitations to knowledge and control: inhabitants of a Bohmian universe cannot know the position of a particle more precisely than allowed by the $|\psi|^2$ distribution (Dürr et al., 1992), where ψ is its conditional wavefunction, and they cannot prepare a particle with a specific position $q \in \mathbb{R}^3$ and a wavefunction ψ other than a Dirac delta function (i.e., a wavefunction concentrated on a single point). Furthermore, they cannot measure the position at time t without disturbing the particle: its future trajectory will be significantly different from what it would have been without a measurement. A limitation to knowledge that applies to all versions of quantum mechanics is that wavefunctions cannot be measured (e.g., Cowan and Tumulka (2016)); i.e., if Alice prepares a particle with a wavefunction ψ of her choice then Bob cannot find out, from experiments on the particle, what ψ is. He could determine ψ, however, if Alice prepared many particles, each with the same wavefunction ψ. Further such limitations are described in Sections 17.6 and 17.7. Some researchers feel that theories that entail limitations to knowledge are poor; since, as just pointed out, every version of quantum mechanics must admit such limitations, such a sentiment seems inadequate.

Variables besides the wavefunction, such as the position variables $Q_k(t)$ in Bohmian mechanics, are traditionally called "hidden variables." This terminology has stuck but one should be aware that it is rather misleading: $Q_k(t)$ can in principle be measured at any time t to any desired accuracy (so it is hardly "hidden"), whereas ψ_t cannot – so it is really the wavefunction, which Bohmian mechanics has in common with OQM, which is a "hidden variable" (Bell, 1987a).

17.5   Differences from Orthodox Quantum Mechanics

OQM is the traditional understanding of quantum mechanics which goes back to the "Copenhagen interpretation" of Niels Bohr. It insists that quantum particles do not have trajectories, and more generally that there are no further variables besides ψ (for microscopic systems), so that ψ is the complete description of the state (e.g., Bohr (1935), von Neumann (1955, Chap. III.2), Landau and Lifshitz (1977, pp. 5–6)), whereas in Bohmian mechanics the state (i.e., the real factual situation) is represented by the pair (Q, ψ). However, it is assumed in OQM that macroscopic ("classical") quantities do always have sharp values and cannot be in a superposition, so there are also some kind of "hidden variables" in OQM, except that it remains vague which quantities exactly are supposed to be "classical," and which equations govern them. In Bohmian mechanics, a categorization of systems as "classical" and "quantum" may be convenient but is fundamentally unnecessary, as the entire universe is governed by the laws (17.1), (17.3), (17.5), so there is no problem about regarding the entire universe as quantum.

A deeper difference between OQM and Bohmian mechanics concerns a philosophical attitude that can be called "positivism" and whose central idea is that a statement is unscientific or even meaningless if it cannot be tested experimentally, that an object is not real if it cannot be observed, and that a variable is not well defined if it cannot be measured. Positivism suggests that science should limit itself to operational statements such as "if one performs experiment X then one obtains outcome Y with probability Z." That is why the usual axioms of quantum mechanics are about what observers will see. I regard this form of positivism as exaggerated and implausible: for example, events behind the horizon of a black hole may be unobservable to all outside observers but would very much seem to be as real as outside events. In fact, I regard positivism as *refuted* by the limitations to knowledge such as the impossibility of measuring wavefunctions. Nevertheless, in the physics literature one often reads arguments with a positivistic flavor, and OQM is particularly positivistically influenced. In contrast, Bohmian mechanics takes a "realist" attitude, according to which we should make the best hypothesis we can about what actually happens in reality. The reason we make these hypotheses is that we would like to find out how the world works (Maudlin, 2016).

A related philosophical difference between OQM and Bohmian mechanics is that OQM finds it acceptable to say contradictory things about what happens in reality as long as all operational predictions remain unaffected by the contradictions. This attitude, which makes OQM incoherent from a realist perspective, is often regarded by advocates of OQM as an aspect of complementarity; it is often accompanied by the attitude that everything we say about what happens in reality is anyway only metaphor and should not be taken seriously. In contrast, in Bohmian mechanics there are no contradictions and there is no need for metaphors, as we can easily and clearly say what actually happens in a Bohmian universe. That is because Bohmian mechanics provides a single coherent picture of reality.

A further difference concerns the status of observables. In OQM it is common to talk about observables as if they were quantities, as if they had values. It should not be surprising that paradoxes arise from such loose talk. In Bohmian mechanics, also this problem is absent as we are dealing with variables that actually have values; more about this in Section 17.7.

17.6   Non-Locality

We now turn to Bell's theorem (Bell, 1964, 1987b; Goldstein et al., 2011; Maudlin, 2011). It is sometimes claimed (e.g., Wigner (1983, p. 53), Hawking (1999)) that Bell's theorem excludes hidden-variable theories (such as Bohmian mechanics) because it implies that any hidden-variable theory has to be non-local (i.e., involve faster-than-light influences), and that would be in conflict with relativity. This picture is not quite right. Bell's theorem actually shows that the observed probabilities in certain experiments (Einstein-Podolsky-Rosen-Bell experiments) are incompatible with locality, so our world must be non-local, and every theory in agreement with experiment must be non-local. Since these observed probabilities are predicted by the predictive rules of quantum theory, one can also conclude that every theory in agreement with these rules must be non-local. Bohmian mechanics agrees with these rules, and it is non-local. Thus, Bell's theorem does not at all exclude Bohmian mechanics but, on the contrary, proves its non-local character inevitable. And non-locality need not conflict with relativity, as illustrated particularly by the relativistic Ghirardi–Rimini–Weber (GRW) theory (Tumulka, 2006; Maudlin, 2011).

The widespread misperception that Bell's theorem concerns only hidden-variable theories and has nothing to say about other approaches presumably originates from the fact that John Bell's original 1964 paper (Bell, 1964) cited the famous 1935 paper of Einstein, Podolsky, and Rosen (EPR, Einstein et al. (1935)) for showing (as it did) that locality implies the existence of hidden variables, so that Bell focused on excluding the remaining possibility of local hidden-variable theories in order to exclude all local theories. EPR's argument was and is often not sufficiently appreciated, and its relevance to Bell's argument often missed (see Goldstein et al. (2011); Maudlin (2014) for elaboration).

The non-locality shows up in Bohmian mechanics in the dependence of the velocity of particle 1 on the position of particle 2 according to (17.1); since this dependence is instantaneous, no matter how big the distance between the two particles, it is a faster-than-light influence. As a consequence, for devising a version of Bohmian mechanics in a relativistic space-time (Dürr et al., 1999), we need a temporal order between spacelike-separated space-time points, or a notion of simultaneity-at-a-distance. Such a notion can be mathematically expressed by a foliation (slicing) $\mathscr{F}$ of space-time into disjoint spacelike hypersurfaces (sometimes called "the time foliation"). In such versions of Bohmian mechanics that have been studied, it turns out that the probabilities of experimental outcomes do not depend on $\mathscr{F}$, so inhabitants of such a universe cannot find out empirically which foliation $\mathscr{F}$ is. Therefore in this theory, there is another limitation to knowledge: $\mathscr{F}$ cannot be determined experimentally. While positivistically inclined researchers find this objectionable, I find positivism objectionable. In contrast to the ether, which could simply be eliminated from classical electrodynamics, $\mathscr{F}$ cannot be eliminated from Bohmian mechanics without a breakdown of the equations. This situation suggests that $\mathscr{F}$ plays a legitimate role in the theory. The theory could still be regarded as relativistic if $\mathscr{F}$ is determined by a covariant law; proposals of such laws are outlined in Tumulka (2007); Dürr et al. (2014).

17.7   Observables

An analysis of the fundamental laws of Bohmian mechanics shows that there is no apparatus with which the inhabitants could measure the velocity of a particle without any prior knowledge of its wavefunction (Dürr et al., 2004b); this is another limitation to knowledge. In contrast, if the wavefunction is known, or is known to belong to a suitable class of wavefunctions (e.g., to be in the classical regime), then the velocity can be measured; in particular, there is no obstacle to measuring the velocity of a macroscopic object in the classical regime. Likewise, if many particles are given, and we know that each one has the same wavefunction, then the velocity of (almost) each one can be measured (as we then can determine the wavefunction). A specific procedure (Wiseman, 2007; Dürr et al., 2009) of this kind, so-called weak measurements of the velocities, has been carried out experimentally (Kocsis et al., 2011), resulting in a picture similar to Figure 17.1.

Furthermore, if we ask about the asymptotic velocity u as $t \to \infty$ (that the particle will ultimately have if from now on no external forces act on it) instead of the instantaneous velocity v at time t, then the situation is different: u, which is determined by Q and ψ, can be measured without knowledge of ψ, and the product mu (mass times asymptotic velocity) has the distribution known in standard quantum mechanics as the momentum distribution, i.e., (leaving out factors of $\hbar$) $|\hat{\psi}(k)|^2$ with $\hat{\psi}(k)$ the Fourier transform of $\psi(q)$. The Heisenberg uncertainty relation then means the following in Bohmian mechanics: while position and momentum (understood as mu) of a particle do have actual values, inhabitants cannot know both values with inaccuracies whose product is smaller than $\hbar/2$, even if they know the particle's wavefunction. (This is another limitation to knowledge.)

In contrast to position and momentum (mu), energy, angular momentum, and spin do not even have actual values (except for special wavefunctions, viz., eigenfunctions of energy or angular momentum or spin). Rather, the experiments that are commonly called "quantum measurements" of energy, etc., create random outcomes instead of revealing pre-existing quantities (as in the ordinary meaning of "measuring"). Thus, ironically, Bohmian mechanics is a "no-hidden-variables" theory for most observables.

(The outcomes are, in fact, determined by the initial wavefunction $\chi(q_a)$ of the apparatus, its initial configuration Q_a, the coupling Hamiltonian, and the object's initial wavefunction $\varphi(q_s)$ and configuration Q_s together; but they are in general not functions of φ and Q_s alone. See also Emery (in this volume) for a discussion of deterministic randomness.)

For example, a Stern–Gerlach experiment, the usual experiment for a quantum measurement of a component of spin, say σ_z, uses a magnetic field to direct the up-component of ψ into a different spatial region than the down-component, and then uses a detector to find out in which region the particle is. When the particle is found in the region containing the up-component, one says that the outcome is "spin up." This example shows several things: First, there is no need to introduce an actual value for spin components, as the Stern–Gerlach experiment is explained with positions and wavefunctions alone. Second, z-spin then is not a well-defined quantity in Bohmian mechanics, except when the wavefunction is an eigenstate of the Pauli matrix σ_z; the value of z-spin is only created during the experiment. Third, there can be different experiments, all of which are "quantum measurements" of the same observable σ_z, but which lead to different outcomes when applied to particles in the same Bohmian initial state (Q, ψ): indeed, inverting the field and interchanging the regions for "up" and "down" would still lead to the same statistics of "up" and "down" for a given ψ but not to the same outcome for fixed Q and ψ. (Concretely, the inverted field would deflect the z-down component of ψ upward, and in certain simple cases with $\psi = |x\text{-up}\rangle$, the Bohmian particle would be deflected upward whenever the initial position is in the upper half of the packet, resulting in outcome "down" with the inverted field but outcome "up" with the original field; see, e.g., Norsen (2014, Sec. V) for details.) And fourth, it further follows that it is, in fact, not possible to introduce an actual value S_z of σ_z such that every Stern–Gerlach experiment that qualifies as a quantum measurement of σ_z yields S_z as the outcome. See Norsen (2014) for elaboration about spin in Bohmian mechanics.

There are various "no-hidden-variables" theorems (Bell, 1966; Hemmick and Shakur, 2012). Some of them, such as the Kochen–Specker theorem, do not apply to Bohmian mechanics because Bohmian mechanics does not associate actual values with most quantum observables. However, for other theorems, such as the Bell inequality theorem, this does not matter much because we may choose a specific experiment (including, if necessary, the wavefunction and configuration of the apparatus) for each of the relevant observables, and Bohmian mechanics then does determine the outcome as a function of the particle's Q and ψ. Specifically, the Bell inequality theorem then shows that for an EPR pair $(Q = (\mathbf{Q}_a, \mathbf{Q}_b), \psi = 2^{-1/2}(|\uparrow\downarrow\rangle - |\downarrow\uparrow\rangle)\,\phi_a(\mathbf{q}_a)\,\phi_b(\mathbf{q}_b)$, where the supports of ϕ_a and ϕ_b are separated) on which Alice and Bob each carry out a Stern–Gerlach experiment about a spin component of their choice, say Alice first and Bob later, the function that yields Bob's outcome as a function of Q and ψ depends also on Alice's choice. That is, the theorem proves that Bohmian mechanics is non-local, which was clear already before Bell (Bohm, 1952, p. 186).

Let us return to a fact mentioned before: two different experiments, which have different outcomes when acting on the same Bohmian state (Q, ψ), may still have the same *statistics* of outcomes when fed with a random Q that is $|\psi|^2$ distributed. The statistics can, in fact, be expressed in terms of operators, more precisely of a POVM (positive-operator-valued measure) (Tumulka, 2009); this means in our case that with every possible outcome z of the experiment there is associated a positive self-adjoint operator $F(z)$ such that when the experiment is applied to a system with wavefunction ψ (with $\|\psi\| = 1$), the probability distribution of the random outcome Z is given by

$$\mathbb{P}_\psi(Z = z) = \langle\psi|F(z)|\psi\rangle \tag{17.9}$$

for every z (Dürr et al., 2004b). That is how operators come up as "observables" in Bohmian mechanics: they encode the probability distribution of Z as a function of ψ. The operators $F(z)$ have the property that they add up to the identity operator I, $\sum_z F(z) = I$. In the special case that the z are real numbers and the $F(z)$ are projections, the POVM $F(\cdot)$ can equivalently be represented by the self-adjoint operator

$$A = \sum_z z\,F(z) \tag{17.10}$$

of which it is the spectral decomposition (i.e., the z are the eigenvalues of A, and $F(z)$ is the projection to the eigenspace). That is why self-adjoint operators represent observables in many relevant cases.

Therefore, two different experiments can have equal statistics for every ψ, and this happens when they have the same POVM $F(\cdot)$. Since in orthodox terminology one says in this case that the two experiments "measure the same quantum observable," it can be difficult to appreciate that they can have different outcomes (in particular because in OQM one cannot ask the question whether they have different outcomes). For example, opponents of Bohmian mechanics have described an experiment whose POVM is the usual position observable but whose outcome is not the position of the particle, have claimed that position measurements disagree with the Bohmian position, and have called the Bohmian trajectories "surrealistic" (Englert et al., 1992; Dürr et al., 1993). In fact, this example only illustrates how misleading talk about "measurements" of "observables" can be: while ordinary particle detectors do find the particles at their Bohmian positions, this particular experiment does not measure position although it is a "quantum measurement" of the "position observable," meaning merely that the distribution of its outcomes agrees with $|\psi(q)|^2$.

(Here is another, simpler example of an experiment with this property. Consider two normalized one-dimensional wave packets ψ_1, ψ_2 centered around the locations x_1, x_2 respectively, with $x_1 < x_2$

and ψ_1 moving to the right and ψ_2 to the left, such that they exchange places in one time unit. On the two-dimensional subspace S of the superpositions $\psi = c_1\psi_1 + c_2\psi_2$ consider, as the analog of the position observable, the operator $A = x_1|\psi_1\rangle\langle\psi_1| + x_2|\psi_2\rangle\langle\psi_2|$, which is a coarse-grained position observable. Consider two experiments on a system with normalized initial wavefunction $\psi \in S$: (i) measure the (coarse-grained) position at time 0; (ii) let the system evolve to time 1, then measure the (coarse-grained) position and exchange $x_1 \leftrightarrow x_2$ in the result. Both experiments yield x_1 with probability $|c_1|^2$ and x_2 with probability $|c_2|^2$, so they are quantum measurements of the same observable (A at time 0). However, since Bohmian trajectories cannot cross, some of them start at time 0 near x_1 and end up at time 1 near x_1, and in that case the result of experiment (ii) does not agree with the position of the Bohmian particle at time 0. Note that there is nothing mysterious about the fact that the two experiments have the same distribution although they produce different outcomes.)

17.8   History

That quantum particles may have trajectories "guided" by a wave was proposed as early as 1923 by Albert Einstein (Wigner, 1980, p. 463) and John Slater (1975, p. 9), (Mehra and Rechenberg, 2001, p. 544) but not published. The equation of motion (17.1) was considered, without spin, by Louis de Broglie in 1926 (de Broglie, 1928) but later abandoned because he incorrectly believed that the theory made empirically wrong predictions. It was rediscovered independently by Nathan Rosen in 1945 (Rosen, 1945) and David Bohm in 1952 (Bohm, 1952). Bohm was the first to realize that the theory actually makes empirically correct predictions. The correct extension to particles with spin was given by John Bell in 1966 (Bell, 1966); Bell also contributed to the exploration and clarification of the theory (Bell, 1987b).

Bohm (1952) formulated the theory in an awkward way by introducing a second-order equation (the one obtained from (17.1) by taking a time derivative on both sides and expressing $\partial\psi/\partial t$ via the Schrödinger equation) as the equation of motion and demanding that the initial positions $\mathbf{Q}_j(0)$ and the initial velocities $V_j(0) = d\mathbf{Q}_j/dt$ be related according to a "constraint condition" identical to (17.1). Since it then follows mathematically that (17.1) is satisfied at all times, Bohm's prescription is actually equivalent to saying that $Q(t)$ is a solution of (17.1) at all times. Bohm also suggested introducing additional hidden variables for spin (Bohm and Hiley, 1993), which is an unnecessary move, as discussed above.

17.9   Typicality and the Origin of Randomness

Bohm realized, by means analogous to the considerations in Section 17.3, that inhabitants of a universe governed by Bohmian mechanics will observe statistics of outcomes of experiments in agreement with the predictive rules of standard quantum mechanics (for short, "the quantum statistics"), including the joint distribution of several experiments. A deeper analysis was provided by Dürr et al. (1992) with a result that parallels the law of large numbers in probability theory as follows. Recall that the outcomes of all experiments are determined by the initial wavefunction $\Psi = \Psi(0)$ and the initial configuration $Q = Q(0)$ of the universe (even if the necessary calculation cannot be carried out in practice). The result of Dürr et al. shows that, for given Ψ, most Q are such that inhabitants throughout the world history will observe the quantum statistics. Here, "most" means the following: we say that "most Q have property X" if

$$\int_S dq\, |\Psi(q)|^2 \text{ is close to } 1 \tag{17.11}$$

with S being the set of all Q with property X. That is why it suffices, as a fundamental law of Bohmian mechanics, to demand that Q be *typical* with respect to $|\Psi|^2$. This is close to saying that Q

was chosen randomly with distribution $|\Psi|^2$ but a little different. For example, the sequence of digits of π (314159265...) "looks random" in the long run although it cannot be said to *be random*. Put differently, for the purpose of the statistical looks of the sequence, π is a typical number in the interval $[0, 4]$. Likewise, for Q it is only relevant that it "looks as if chosen randomly with $|\Psi|^2$ distribution" and not whether it *is random* in a fundamental sense; and the "looks" refers to the macroscopic history of $Q(t)$. Thus, from the perspective of the inhabitants it is indistinguishable from being random, so the typicality of Q is the origin of randomness in a Bohmian universe.

It has been pointed out (Valentini and Westman, 2005) that the motion of the Bohmian configuration is often chaotic (even mixing, to be precise), with the consequence that any initial distribution $\rho \neq |\psi|^2$ on configuration space tends to come closer and closer to $|\psi_t|^2$ over time. It has further been suggested (Bohm, 1952; Valentini and Westman, 2005) that this phenomenon may explain the $|\psi|^2$ distribution. However, the real question concerns not so much the configuration space of a system as that of the universe; and there it does not seem as if a deeper insight were gained if somebody could show (which has not been shown) that also (say) a $|\Psi|^4$-typical or a $|\Psi|^0$-typical $Q(0)$ would lead to $|\psi|^2$-statistics for the outcomes of most experiments. While the $|\Psi|^2$ distribution is special since it is equivariant, the $|\Psi|^4$ and $|\Psi|^0$ distributions are just some among many conceivable distributions; in fact, if we found empirically (which we have not) that it was necessary to assume that $Q(0)$ was $|\Psi|^4$-distributed, then it would be a big puzzle needing explanation why it was $|\Psi|^4$ of all distributions instead of the natural, equivariant $|\Psi|^2$.

17.10   Identical Particles

The symmetrization postulate of quantum mechanics applies equally in Bohmian mechanics. It asserts that when particle i and particle j belong to the same particle species, the wavefunction is either symmetric or anti-symmetric against the permutation of their coordinates and spin indices,

$$\psi_{...s_i...s_j...}(...\boldsymbol{q}_i...\boldsymbol{q}_j...) = \gamma \, \psi_{...s_j...s_i...}(...\boldsymbol{q}_j...\boldsymbol{q}_i...) \tag{17.12}$$

with $\gamma = -1$ if the species is fermionic and $\gamma = +1$ if bosonic. It has been suggested in the literature (e.g., (Landau and Lifshitz, 1977, p. 227)) that the reason behind the symmetrization postulate is that there are no trajectories in quantum mechanics and therefore no fact about which particle at time t_1 is which particle at time t_2; however, Bohmian mechanics shows that the symmetrization postulate is compatible with the existence of trajectories.

However, if we take the particles seriously, as we do in Bohmian mechanics, then, for a system of N identical particles, the particles should not be numbered, as there is no fact in the physical world about which particle is particle number 1. Therefore, two configurations that differ only by a permutation, such as $(\boldsymbol{q}_3, \boldsymbol{q}_1, \boldsymbol{q}_2)$ and $(\boldsymbol{q}_1, \boldsymbol{q}_2, \boldsymbol{q}_3)$, should be regarded as two mathematical representations of the same physical configuration. Thus, the space of physical configurations is the space "$^N\mathbb{R}^3$" of *unordered* configurations (Leinaas and Myrheim, 1977), i.e., of permutation classes of N-tuples (leaving out for simplicity the N-tuples containing repetitions such as $\boldsymbol{q}_1 = \boldsymbol{q}_2$) or of N-element subsets of physical 3-space, as opposed to the space $\mathbb{R}^{3N}$ of *ordered* configurations. Now Bohm's law of motion (17.1) is such that for a wavefunction that is either symmetric or anti-symmetric, two initial configurations that differ only by a permutation evolve to later configurations at any time t that differ only by a permutation (in fact, the same permutation). Therefore, it defines a curve in $^N\mathbb{R}^3$, as it should. This would not be so for general wavefunctions (neither bosonic nor fermionic), and this situation can be regarded as a reason behind the symmetrization postulate (Bacciagaluppi, 2003; Dürr et al., 2006). Since $^N\mathbb{R}^3$ is a topologically non-trivial space, this approach to explaining the symmetrization postulate, which goes back to Jon M. Leinaas and Jan Myrheim (1977), is known as the "topological" approach.

17.11   The Classical Limit

In OQM, some systems have to be "classical" in order to even make sense of the theory. In Bohmian mechanics, in contrast, no problem would arise even if no system were in a classical regime. Nevertheless, such a regime actually governs many systems, particularly macroscopic ones. In OQM, the classical limit is problematical because of the unsolved problem of how definite properties (e.g., definite positions) of macroscopic objects arise, given that these objects consist of quantum particles (electrons and quarks) that can be in arbitrary superpositions. No such problem affects Bohmian mechanics: there, particles have trajectories, and the study of the classical limit concerns merely the clear mathematical question under which conditions these trajectories (or the center-of-mass trajectories of macroscopic bodies) will obey Newton's equation of motion

$$m_k \frac{d^2 \mathbf{Q}_k(t)}{dt^2} = -\nabla_k V(Q(t)). \tag{17.13}$$

See Allori et al. (2002) for an overview of results about this question. In brief, (17.13) tends to hold when two mechanisms prevail: decoherence (Crull, in this volume) and the tendency of wavefunctions to become locally plane waves.

17.12   Quantum Field Theory

Leaving aside the circumstance that in most quantum field theories the Hamiltonians are mathematically ill-defined (and that it is unknown how to define them rigorously (Wallace, in this volume)), Bohmian mechanics can be extended to quantum field theory in several ways. It has not been settled which Bohm-type model is the most convincing one. The proposed extensions differ in whether they use a *field ontology* or a *particle ontology*. For a field ontology (see Struyve (2010) for an overview), one assumes that instead of a particle configuration, the variable besides ψ is a field configuration (a function on 3-space). A particle ontology, in contrast, means to stick with a particle configuration. Also combinations are conceivable: Bohm (1952) proposed to use a particle ontology for fermions and a field ontology for the bosonic degrees of freedom. It seems not possible to set up a field ontology for fermions.

With a particle ontology, it seems natural to introduce the possibility of creation and annihilation of particles. Such models have been set up (Bell, 1986; Dürr et al., 2004a): for a given Hamiltonian H and configuration operators (a projection-valued measure $P(\cdot)$ on configuration space generalizing the position operators), the model is not unique, but uniquely selected by considerations of naturalness and simplicity. These models are no longer deterministic but instead stochastic, as the creation and annihilation events correspond to jump between sectors of configuration space corresponding to different particle numbers, and every jump toward a higher particle number has to be stochastic, or else $Q(t)$ could not be continuously distributed in the higher sector, so that $|\psi|^2$ could not be equivariant. Such considerations lead to a particular rate (probability per time) for the jump from q' to anywhere in the volume element dq, given by (Dürr et al., 2004a)

$$\sigma^\psi(q' \to dq) = \frac{2}{\hbar} \frac{\mathrm{Im}^+ \langle \psi | P(dq) H P(dq') | \psi \rangle}{\langle \psi | P(dq') | \psi \rangle} \tag{17.14}$$

with $x^+ = \max\{x, 0\}$.

It may be noteworthy that although Bohmian mechanics is deterministic, neither Bohm nor Bell nor the authors of Dürr et al. (2004a) found or find stochastic theories unacceptable. Rather, the simplest and most convincing theory of non-relativistic quantum mechanics (i.e., Bohmian mechanics) happens to be deterministic, and the simplest and most convincing theory of particle creation happens to be stochastic. Speaking of stochastic theories, I note that a variant of Bohmian mechanics for non-relativistic quantum mechanics with stochastic trajectories (a diffusion process) was proposed by Edward Nelson under the name "stochastic mechanics" (Nelson, 1985; Goldstein, 1987). While Nelson's initial hopes that this theory might somehow get rid of the wavefunction, and that it might provide a superior explanation of randomness, have not materialized, it does provide a coherent and working theory when understood in the same way as Bohmian mechanics. It has received less attention because it is more complicated than Bohmian mechanics (as it involves stochastic processes) while not being more convincing.

Another open question that comes up when setting up Bohmian versions of quantum field theories is whether the Dirac sea should be taken literally. (The Dirac sea is, roughly speaking, the picture that electrons cannot occur in a state of negative energy because all negative energy states are occupied in the wavefunction of the universe, and since electrons are fermions, a state cannot be occupied twice. Any unoccupied state of negative energy then appears like a positron of positive energy.) Some authors (e.g., Colin and Struyve (2007) and Deckert et al. (2019)) have proposed that electrons are real whereas positrons are literally just empty spots in a sea of electrons of negative energy. Others (Dürr et al., 2004a) have suggested that the situation is fundamentally symmetric between electrons and positrons, so that positrons are real particles in their own right, and that we are not surrounded by a sea of electrons that we usually do not notice; in such models, the Dirac sea has the status of a metaphor or mathematical analogy, but not of a reality.

17.13   Philosophical Questions

Bohmian mechanics is the subject of ongoing philosophical debate. One question concerns the ontological status of the wavefunction in Bohmian mechanics: can a function on configuration space be physically real? Or is the wavefunction to be regarded as nomological, i.e., as something like a law (roughly analogous to the Hamiltonian function in classical mechanics), instead of as a thing (Goldstein and Zanghì, 2013)? Another question concerns the possible choices in quantum field theory, such as field ontology versus particle ontology: which variant of the theory is the most convincing one? In some examples, different variants of Bohmian mechanics are empirically equivalent, so that no experiment can decide between them; it may nevertheless be possible, even natural, to prefer one over the other on theoretical or philosophical grounds (see Goldstein et al. (2005) for examples).

Another circle of philosophical questions concerns the following. Bohmian mechanics contains certain elements in its ontology, the particles with positions $Q_k(t)$, that immediately represent the distribution of matter in space. Such variables, and such elements of the ontology, are often called the *primitive ontology* (Dürr et al., 1992; Allori et al., 2014; Maudlin, 2016). I feel that for a fundamental physical theory to be acceptable, it must have a primitive ontology. This leads to questions such as: what exactly is required of a theory's ontology to make it acceptable as a fundamental physical theory? What kinds of things can constitute a primitive ontology? For example, Everett's many-worlds interpretation proposes that there is only wavefunction in the world; this does not seem to me to make enough sense as a fundamental physical theory (see also Maudlin (2010)), but this problem can be taken care of by introducing a suitable primitive ontology (Allori et al., 2011), which leads to a modified theory with many-worlds character that makes clear sense.

Acknowledgment

I thank Travis Norsen and the editors for comments on a draft of this chapter.

References

Allori, V., Dürr, D., Goldstein, S. and Zanghì, N. (2002). Seven steps towards the classical world. *Journal of Optics B*, 4: 482–488. http://arxiv.org/abs/quant-ph/0112005

Allori, V., Goldstein, S., Tumulka, R. and Zanghì, N. (2011). Many-worlds and Schrödinger's first quantum theory. *British Journal for the Philosophy of Science*, 62(1): 1–27. http://arxiv.org/abs/0903.2211

Allori, V., Goldstein, S., Tumulka, R. and Zanghì, N. (2014). Predictions and primitive ontology in quantum foundations: A study of examples. *British Journal for the Philosophy of Science*, 65: 323–352. http://arxiv.org/abs/1206.0019

Bacciagaluppi, G. (2003). *Derivation of the symmetry postulates for identical particles from pilot-wave theories.* Preprint https://arxiv.org/abs/quant-ph/0302099

Bacciagaluppi, G. and Valentini, A. (2009). *Quantum Theory at the Crossroads: Reconsidering the 1927 Solvay Conference.* Cambridge: Cambridge University Press. http://arxiv.org/abs/quant-ph/0609184

Bell, J.S. (1964). On the Einstein-Podolsky-Rosen paradox. *Physics*, 1: 195–200. Reprinted as chapter 2 of Bell (1987b).

Bell, J.S. (1966). On the problem of hidden variables in quantum mechanics. *Reviews of Modern Physics*, 38: 447–452. Reprinted as chapter 1 of Bell (1987b).

Bell, J.S. (1980). De Broglie–Bohm, delayed-choice double-slit experiment, and density matrix. *International Journal of Quantum Chemistry*, 14: 155–159. Reprinted as chapter 14 of Bell (1987b).

Bell, J.S. (1986). Beables for quantum field theory. *Physics Reports*, 137: 49–54. Reprinted as chapter 19 of Bell (1987b). Also reprinted on p. 227 in Peat, F.D. and Hiley, B.J., editors. (1987). *Quantum Implications: Essays in Honour of David Bohm.* London: Routledge.

Bell, J.S. (1986). Six possible worlds of quantum mechanics. In *Proceedings of the Nobel Symposium 65: Possible Worlds in Arts and Sciences* (Stockholm, August 11–15). Reprinted as chapter 20 of Bell (1987b).

Bell, J.S. (1987a). Are there quantum jumps? In C. W. Kilmister (ed.), *Schrödinger: Centenary Celebration of a Polymath.* Cambridge: Cambridge University Press, pp. 41–52. Reprinted as chapter 22 of Bell (1987b).

Bell, J.S. (1987b). *Speakable and Unspeakable in Quantum Mechanics.* Cambridge: Cambridge University Press.

Bohm, D. (1952). A suggested interpretation of the quantum theory in terms of "hidden" variables, I and II. *Physical Review*, 85: 166–193.

Bohm, D. (1953). Comments on an article of Takabayasi concerning the formulation of quantum mechanics with classical pictures. *Progress in Theoretical Physics*, 9: 273–287.

Bohm, D. and Hiley, B.J. (1993). *The Undivided Universe: An Ontological Interpretation of Quantum Theory.* London and New York: Routledge.

Bohr, N. (1935). Can quantum-mechanical description of physical reality be considered complete? *Physical Review*, 48: 696–702

Bricmont, J. (2016). *Making Sense of Quantum Mechanics.* Berlin: Springer

Chattaraj, P., editor. (2010). *Quantum Trajectories.* Boca Raton, FL: Taylor & Francis.

Colin, S. and Struyve, W. (2007). A Dirac sea pilot-wave model for quantum field theory. *Journal of Physics A: Mathematical and Theoretical*, 40: 7309–7342. http://arxiv.org/abs/quant-ph/0701085

Cowan, C.W. and Tumulka, R. (2016). Epistemology of wave function collapse in quantum physics. *British Journal for the Philosophy of Science*, 67(2): 405–434. http://arxiv.org/abs/1307.0827

Crull, E. *Quantum Decoherence.* In this volume.

de Broglie, L. (1928). La nouvelle dynamique des quanta. In Solvay Congress (1927), *Electrons et Photons: Rapports et Discussions du Cinquième Conseil de Physique tenu à Bruxelles du 24 au 29 Octobre 1927 sous les Auspices de l'Institut International de Physique Solvay.* Paris: Gauthier-Villars. English translation: The new dynamics of quanta, in Bacciagaluppi and Valentini (2009).

Deckert, D.-A., Esfeld, M. and Oldofredi, A. (2019). A persistent particle ontology for QFT in terms of the Dirac sea. *British Journal for the Philosophy of Science*, 70(3): 747–770. http://arxiv.org/abs/1608.06141

Dürr, D., Fusseder, W., Goldstein, S. and Zanghì, N. (1993). Comment on "surrealistic Bohm trajectories." *Zeitschrift für Naturforschung A*, 48: 1261–1262.

Dürr, D., Goldstein, S., Münch-Berndl, K. and Zanghì, N. (1999). Hypersurface Bohm–Dirac models. *Physical Review A*, 60: 2729–2736. Reprinted in Dürr et al. (2013). http://arXiv.org/abs/quant-ph/9801070

Dürr, D., Goldstein, S., Norsen, T., Struyve, W. and Zanghì, N. (2014). Can Bohmian mechanics be made relativistic? *Proceedings of The Royal Society A*, 470(2162): 20130699. http://arxiv.org/abs/1307.1714

Dürr, D., Goldstein, S., Taylor, J., Tumulka, R. and Zanghì, N. (2006). Topological factors derived from Bohmian mechanics. *Annales Henri Poincaré*, 7: 791–807. http://arxiv.org/abs/quant-ph/0601076

Dürr, D., Goldstein, S., Tumulka, R. and Zanghì, N. (2004a). Bohmian mechanics and quantum field theory. *Physical Review Letters*, 93: 090402. Reprinted in Dürr et al. (2013). http://arxivorg/abs/quant-ph/0303156

Dürr, D., Goldstein, S., Tumulka, R. and Zanghì, N. (2005). On the role of density matrices in Bohmian mechanics. *Foundations of Physics*, 35: 449–467. http://arxiv.org/abs/quant-ph/0311127

Dürr, D., Goldstein, S. and Zanghì, N. (1992). Quantum equilibrium and the origin of absolute uncertainty. *Journal of Statistical Physics*, 67: 843–907. Reprinted in Dürr et al. (2013). http://arxiv.org/abs/quant-ph/0308039

Dürr, D., Goldstein, S. and Zanghì, N. (2004b). Quantum equilibrium and the role of operators as observables in quantum theory. *Journal of Statistical Physics*, 116: 959–1055. Reprinted in Dürr et al. (2013). http://arxiv.org/abs/quant-ph/0308038

Dürr, D., Goldstein, S. and Zanghì, N. (2009). On the weak measurement of velocity in Bohmian mechanics. *Journal of Statistical Physics*, 134: 1023–1032. Reprinted in Dürr et al. (2013). http://arxiv.org/abs/0808.3324

Dürr, D., Goldstein, S. and Zanghì, N. (2013). *Quantum Physics Without Quantum Philosophy*. Heidelberg: Springer-Verlag.

Dürr, D. and Teufel, S. (2009) *Bohmian Mechanics*. Heidelberg: Springer-Verlag.

Einstein, A., Podolsky, B. and Rosen, N. (1935). Can quantum-mechanical description of physical reality be considered complete? *Physical Review*, 47: 777–780.

Emery, N. *Deterministic Chance*. In this volume.

Englert, B.-G., Scully, M.O., Süssmann, G. and Walther, H. (1992). Surrealistic Bohm trajectories. *Zeitschrift für Naturforschung A*, 47: 1175–1186.

Goldstein, S. (1987). Stochastic mechanics and quantum theory. *Journal of Statistical Physics*, 47: 645–667.

Goldstein, S. (2001). Bohmian mechanics. In E. N. Zalta (ed.), *Stanford Encyclopedia of Philosophy*, published online by Stanford University. http://plato.stanford.edu/entries/qm-bohm/

Goldstein, S., Norsen, T., Tausk, D.V. and Zanghì, N. (2011). Bell's theorem. *Scholarpedia*, 6(10): 8378. http://www.scholarpedia.org/article/Bell%27s_theorem

Goldstein, S., Taylor, J., Tumulka, R. and Zanghì, N. (2005). Are all particles real? *Studies in History and Philosophy of Modern Physics*, 36: 103–112. http://arxiv.org/abs/quant-ph/0404134

Goldstein, S. and Zanghì, N. (2013). Reality and the role of the wavefunction in quantum theory. In D. Albert and A. Ney (eds.), *The Wave Function: Essays in the Metaphysics of Quantum Mechanics*. Oxford University Press, pp. 91–109. Reprinted in Dürr et al. (2013). http://arxiv.org/abs/1101.4575

Hawking, S.W. (1999). *Does god play dice?* Available at: http://www.hawking.org.uk/does-god-play-dice.html

Hemmick, D.L. and Shakur, A.M. (2012). *Bell's Theorem and Quantum Realism: Reassessment in Light of the Schrödinger Paradox*. New York: Springer.

Kocsis, S., Braverman, B., Ravets, S., Stevens, M.J., Mirin, R.P., Shalm, L.K. and Steinberg, A. (2011). Observing the average trajectories of single photons in a two-slit interferometer. *Science*, 332(6034): 1170–1173.

Landau, L.D. and Lifshitz, E.M. (1977). *Quantum Mechanics: Non-Relativistic Theory*, Vol. 3 of *Course of Theoretical Physics*. Translated from the Russian by J.B. Sykes and J.S. Bell. Third edition. Oxford: Pergamon.

Leinaas, J.M. and Myrheim, J. (1977). On the theory of identical particles. *Il Nuovo Cimento*, 37B: 1–23.

Lewis, P. *Collapse Theories*. In this volume.

Maudlin, T. (2010). Can the world be only wave function? In S.Saunders, J. Barrett, A. Kent, and D. Wallace (eds.), *Many Worlds? Everett, Quantum Theory, and Reality*. Oxford: Oxford University Press, pp. 121–143.

Maudlin, T. (2011). *Quantum Non-Locality and Relativity: Metaphysical Intimations of Modern Physics* (3rd edition). Oxford: Blackwell.

Maudlin, T. (2014). What Bell did. *Journal of Physics A: Mathematical and Theoretical*, 47: 424010.

Maudlin, T. (2016). Local Beables and the foundations of physics. In M. Bell and S. Gao (eds.), *Quantum Nonlocality and Reality*. Cambridge: Cambridge University Press, pp. 317–330.

Mehra, J. and Rechenberg, H. (2001). *The Historical Development of Quantum Theory*, Vol. 1, Part 2. New York: Springer.

Nelson, E. (1985). *Quantum Fluctuations*. Princeton, NJ: Princeton University Press.

Norsen, T. (2014). The pilot-wave perspective on spin. *American Journal of Physics*, 82: 337–348. https://arxiv.org/abs/1305.1280

Oriols, X. and Mompart, J., editors. (2012). *Applied Bohmian Mechanics: From Nanoscale Systems to Cosmology*. Boca Raton, FL: Taylor & Francis.

Philippidis, C., Dewdney, C. and Hiley, B.J. (1979). Quantum interference and the quantum potential. *Il Nuovo Cimento Serie 11*, 52B(1): 15–28.

Rosen, N. (1945). On waves and particles. *Journal of the Elisha Mitchell Scientific Society*, 61: 67–73.

Saunders, S. *The Everett Interpretation*. In this volume.

Slater, J.C. (1975). *Solid-State and Molecular Theory: A Scientific Biography*. New York: Wiley.

Struyve, W. (2010). Pilot-wave theory and quantum fields. *Reports on Progress in Physics*, 73: 106001. http://arxiv.org/abs/0707.3685

Struyve, W. and Valentini, A. (2009). De Broglie-Bohm guidance equations for arbitrary Hamiltonians. *Journal of Physics A: Mathematical and Theoretical*, 42: 035301. http://arxiv.org/abs/0808.0290

Tumulka, R. (2006). A relativistic version of the Ghirardi–Rimini–Weber model. *Journal of Statistical Physics*, 125: 821–840. http://arxiv.org/abs/quant-ph/0406094

Tumulka, R. (2007). The 'unromantic pictures' of quantum theory. *Journal of Physics A: Mathematical and Theoretical*, 40: 3245–3273. http://arxiv.org/abs/quant-ph/0607124

Tumulka, R. (2009). POVM. In F. Weinert, D. Greenberger, and K. Hentschel (eds.), *Compendium of Quantum Physics*. Heidelberg: Springer-Verlag, pp. 480–484.

Valentini, A. and Westman, H. (2005). Dynamical origin of quantum probabilities. *Proceedings of the Royal Society London A*, 461: 253–272. http://arxiv.org/abs/quant-ph/0403034

von Neumann, J. (1955). *Mathematical Foundations of Quantum Mechanics*. Princeton, NJ: Princeton University Press.

Wallace, D. *The Quantum Theory of Fields*. In this volume.

Wheeler, J.A. (1978). The "past" and the "delayed-choice" double-slit experiment. In A.R. Marlow (ed.), *Mathematical Foundations of Quantum Mechanics*. Academic Press, pp. 9–48.

Wigner, E.P. (1980). Thirty years of knowing Einstein. In H. Woolf (ed.), *Some Strangeness in the Proportion: A Centennial Symposium to Celebrate the Achievements of Albert Einstein*. Reading, MA: Addison-Wesley, pp. 461–468.

Wigner, E.P. (1983). Review of quantum mechanical measurement problem. In P. Meystre and M. O. Scully (eds.). *Quantum Optics, Experimental Gravity, and Measurement Theory*. New York: Plenum, pp. 43–63.

Wiseman, H. (2007). Grounding Bohmian mechanics in weak values and Bayesianism. *New Journal of Physics*, 9: 165. http://arxiv.org/abs/0706.2522

Further Reading

A classic article is Bell's (1986) overview of different interpretations of quantum mechanics. An introductory article about Bohmian mechanics, particularly for philosophers, can be found in Goldstein (2001). Book-length introductions are provided in Dürr and Teufel (2009) and Bricmont (2016). Physical and mathematical applications of Bohmian mechanics are discussed in the collections (Chattaraj, 2010; Oriols and Mompart, 2012).

PART V

Quantum Field Theory

If we use precision of numerical prediction as our measure, quantum field theory (QFT) is the most successful scientific theory of all time. Perhaps surprisingly, this success has occurred despite wide-ranging disagreements concerning QFT's conceptual foundations. As with quantum theory in general, our mastery of QFT techniques and of their application to model individual systems continues to outstrip our understanding of the underlying foundational questions. Despite the many open conceptual questions, QFT has already been philosophically hugely rich. It has transformed our understanding of emergence and reduction, it has reshaped debates over realism and locality, and it offers us a radical new picture of the underlying nature of matter – as transient excitations in quantum fields – which looks set to overturn central foundational assumptions of mainstream philosophical ontology.

The first recognizable implementation of QFT was quantum electrodynamics (QED), developed initially through Dirac's 1927 quantum theory of the electromagnetic field and his 1928 wave equation for the electron; immediately the approach was immensely fruitful, yielding convincing explanations of the fine structure of the hydrogen atom spectrum and of Compton scattering as well as the spectacular theoretical discovery of antimatter. From the beginning, though, QFT was plagued by problematic divergences of key quantities; direct calculations of the higher-order perturbations for the energy of an electron's interaction with its own field generically give infinite results. Subsequent development of QFT has led to powerful new techniques for eliminating infinities in various theoretical contexts, and attention to QFT from a foundational perspective has often centered around the legitimacy of these *renormalization* methods. (Dirac, the pioneer of QFT, continued to reject renormalization as bad physics until his death in 1984.)

David Wallace's opening chapter presents a self-contained introduction to contemporary QFT, presenting it from first principles as a framework theory – a general system within which particular physical theories can be formulated – that generalizes classical continuum mechanics to the quantum setting. He emphasizes in particular the fact that all applications of QFT, through their reliance on renormalization, are rendered *effective field theories*, explicitly non-fundamental representations of a system that are in principle compatible with a number of distinct underlying fundamental realizers. This feature permits QFT to be applied very broadly, not only to high-energy physics but also to solid-state physics. However, it also makes it a very delicate matter to draw conclusions about fundamental reality from the way in which QFT predicts and explains, enormously successful though it is.

Sophisticated renormalization techniques were key to the first complete and adequate quantum theory of the electromagnetic interaction, the version of QED developed in the 1940s by Feynman, Schwinger, and Tomonaga, but renormalization did not obtain widespread acceptance by the theoretical community until the development of *renormalization group* methods in the second half of the 20th century, and it wasn't fully accepted until the discovery of *asymptotic freedom* in the 1970s; in asymptotically free theories high-energy interactions become weak, permitting a consistent perturbative treatment. The chapter by Porter Williams focuses on renormalization group methods including the path integral formalism. Apart from the pure physical interest of theories using the renormalization group, these techniques have historically been of special interest within the philosophy of science since renormalization group explanations operate at an extremely high level of abstraction; Williams' chapter and Batterman's chapter in Part IX discuss the philosophical consequences of the immense predictive and explanatory success of renormalization group methods.

One consequence of the historical suspicion of perturbative methods has been the existence of a parallel *algebraic* approach to QFT, which seeks to understand the theory from the top down by studying general properties of classes of consistent quantum theories. While perturbative field theory has been the source of the large majority of applications of the theory, algebraic approaches to QFT have remained high profile, motivated by the thought that if QFT is to be a fundamental theory it must be able to be given a consistent non-perturbative formulation. Laura Ruetsche's chapter is oriented within the algebraic tradition, and focuses on the key disputed concept of locality. Ruetsche argues that the apparent non-locality of orthodox quantum mechanics is cast in a new light within QFT, and that the existence of well-defined and consistent algebraic quantum field theories that are Lorentz covariant amounts to a constructive proof that quantum theory does not violate any well-motivated relativistic locality constraint.

One of the most striking consequences of QFT from the point of view of metaphysics is its apparently radical reimagining of the role of a particle. From Dirac's 1931 theory of the "negative energy sea", creation and destruction of particles has been a characteristic feature of QFT. Instead of being miniature hard spheres, or mass-endowed spacetime points, particles from the perspective of modern QFT are states of excitation of a quantum field. Doreen Fraser's chapter dissects the transformed role of the concept of a particle within QFT, and explores the many ramifications of adopting a physical worldview in which particles are no longer regarded as fundamental.

18

THE QUANTUM THEORY OF FIELDS

David Wallace

18.1 Introduction

Quantum theory, like Hamiltonian or Lagrangian classical mechanics,[1] is not a concrete theory like general relativity or electromagnetism, but a *framework theory* in which a great many concrete theories, from qubits and harmonic oscillators to proposed quantum theories of gravity, may be formulated. Quantum field theory, too, is a framework theory: a sub-framework of quantum mechanics, suitable to express the physics of spatially extended bodies, and of systems which can be approximated as extended bodies. Fairly obviously, this includes the solids and liquids that are studied in condensed matter physics, as well as quantized versions of the classical electromagnetic and (more controversially) gravitational fields. Less obviously, it also includes pretty much any theory of relativistic matter: the physics of the 1930s fairly clearly established that a quantum theory of relativistic *particles* pretty much has to be reexpressed as a quantum theory of relativistic *fields* once interactions are included. Therefore, quantum field theory includes within its framework the Standard Model of particle physics, the various low-energy limits of the Standard Model that describe different aspects of particle physics, gravitational physics below the Planck scale, and almost everything we know about many-body quantum physics. No more need be said, I hope, to support its significance for naturalistic philosophy.

In this chapter, I aim to give a self-contained introduction to quantum field theory (QFT), presupposing (for the most part) only some prior exposure to classical and quantum mechanics (parts of Sections 18.9–18.10 also assume a little familiarity with general relativity and particle physics). Since the normal form of that "introduction" in a physicist's education is two semesters of graduate-level courses, *this* introduction is inevitably pretty incomplete. I give virtually no calculations and require my reader to take almost everything on trust; my focus is on the conceptual structure of QFT and its philosophically interesting features. This chapter will not, realistically, equip its readers to carry out research in philosophy of QFT, but I hope it will help make sense of references to QFT in the philosophical and physics literature, serve to complement more technical introductions, and give some sense of just how important and interesting this field of physics has been in the last 50 years.

My account is in logical rather than historical order, and the physics described is all standard; I make no attempt to reference primary sources, though I give some suggested introductions to the literature at the end. I use units in which $\hbar$ and c are both set to 1; note that this means that mass and energy have the same dimensions, and that length has the dimensions of inverse mass, so that we can talk interchangeably of something happening at large energies/masses and at short length scales.

18.2   Warm-Up: Classical Continuum Mechanics

Classical mechanics represents the instantaneous state of a fluid by (inter alia) a pair of functions on space: a *mass density* $\rho(x)$ and a *velocity field* $\mathbf{v}(x)$. Their interpretations are straightforward and standard: the integral of $\rho(x)$ over some region R gives the total mass of the part of the fluid in R; the integral of $\rho(x)\mathbf{v}(x)$ over R gives the total momentum of that same part. ρ and $\mathbf{v}$ jointly satisfy well-known equations, which can be derived from first principles on phenomenological grounds and which characterize the fluid's dynamics in terms of a small number of parameters, such as its viscosity and compressibility. The equations are notoriously difficult to solve, but when they are solved (analytically, numerically, or under certain local approximations) they do an excellent job of describing the physics of fluids in terms of their continuously varying density and velocity fields.

Remarkably so: because real fluids don't have continuously varying densities and velocity fields. Whatever the "fundamental" story about the constituents of matter, it's beyond serious doubt that ordinary fluids like water or treacle are composed of discretely many atoms, which in turn have considerable substructure, and so – insofar as the language of classical physics is applicable at all – the "real" density of the fluid is varying wildly on length scales of $\sim 10^{-10}$ m and even more wildly on shorter length scales.

The issue is usually addressed early in a kinetic-theory or fluid-dynamics course. There, one is typically[2] told that these functions are defined by averaging over some region large enough to contain very many atoms, but small enough that the macroscopic parameters do not differ substantially across the width of that region. For a fluid that phenomenologically has fairly constant density and momentum over length scales of $\sim 10^{-4}$ m, for instance, that region needs to have a width L satisfying 10^{-4} m $\gg L \gg 10^{-10}$ m. But the merely kinematic task of defining these quantities does not capture the remarkable feature of fluid dynamics (and of emergence in physics in general: that there *exist* closed-form dynamical equations for these quantities, so that we can actually reason from current bulk features of the fluid to future bulk features without any need for additional information about the microphysics.

That is not to say that the microphysics is irrelevant to the fluid's behavior. When the assumptions under which the density is defined fail – that is, when there is *no* length scale both short relative to the scales on which the fluid varies, and long compared to atomic length scales – the fluid-dynamics description breaks down entirely and the underlying physics must be considered directly. Droplet formation has this feature, for instance: droplets break off from a stream of water when the width of the tube of water connecting the proto-droplet to the rest of the fluid becomes only a few atoms wide. So do shock waves: the width of the shock front depends on microphysics and not just on the phenomenological parameters. But outside these special cases, the relevance of the microphysics is purely that the viscosity and other parameters that characterize the fluid are determined – and can in some cases be calculated – from that microphysics.

So consider what we could learn about a given fluid if all we knew was fluid dynamics. We would have determined the coefficients in the equations directly by experiment, not by deduction from the microphysics, for we do not know what it is. We can predict from features *internal* to fluid dynamics that it will break down at some short length scales: droplet formation and shock waves lead to unphysical singularities if the continuum physics is exact. And we might (depending on how the details of the thought experiment are filled in) have reason external to the theory to expect such a breakdown. Furthermore, on the assumption that there is an underlying theory and that it reduces to fluid dynamics in some limit, we can reverse-engineer a few facts about that underlying theory – but the latter will be grossly underdetermined by the macro-level facts. Unless we have empirical access to droplet or shock-wave phenomena, or other means to probe the underlying theory directly, we will simply have to remain ignorant of that underlying theory. And if – with many real-life metaphysicians – we wish to set aside the emergent description of the world at large scales that fluid dynamics gives us and look for information about *fundamental metaphysics* contained within our physics, we will look in vain.[3]

This is pretty much the empirical situation that modern quantum field theory leaves us in.

18.3   Formal Quantum Theory of the Continuum

Suppose we set out to construct a *quantum* theory appropriate to describe a continuous entity – whether a field, or a solid object, or a fluid. Perhaps we know that the entity is not continuous at shorter scales; perhaps we even possess the physics applicable at those scales; perhaps we only suspect a breakdown of continuity; perhaps we believe there is no breakdown; but in any case we carry out our construction without using detailed information about the short-distance physics. (For simplicity, let's assume the continuum is spatially finite for now.)

As a concrete model, suppose that the continuum has one large-scale degree of freedom per spatial point: suppose, for instance, that it is a scalar field theory where the degree of freedom is field strength, or (less realistically) a solid body in one dimension where the degree of freedom is displacement from equilibrium in some fixed direction. (The most realistic solid-state equivalent would include *three* degrees of freedom per spatial point, since the solid can be displaced in any direction; I stick to one degree of freedom for expository simplicity.) In the rest of this chapter, I will use this model extensively for quantitative examples, but the qualitative features apply to pretty much all quantum field theories.

Observable quantities will be (or at least: will include) averages of this quantity over some region, so to any spatial region X we would expect to be able to assign an operator $\widehat{\varphi}_X$ representing that average. Furthermore, if X and Y are disjoint, we should expect $\widehat{\varphi}_X$ and $\widehat{\varphi}_Y$ to commute (recall that in quantum mechanics, operators commute iff the physical quantities they represent can simultaneously have definite values – something we would expect of physical quantities assigned to distinct subsystems).

If we partition space into, say, a grid of equal-size volumes X_i, each of length L, centered on points x_i, we can define simultaneous eigenstates of all the $\widehat{X}_i$; any such state $|\chi\rangle$ will define a real function $[\chi]_L$, given by

$$[\chi]_L(x_i) = \langle\chi|\widehat{X}_i|\chi\rangle \tag{18.1}$$

and interpolating between the x_i in some arbitrary smooth way. There will be many $|\chi\rangle$ corresponding to a given function $[\chi]_L$, corresponding to the many possibilities of short-distance physics on scales below L; we can think of each such $|\chi\rangle$ as representing a state of the continuum whose structure on scales large compared to L is given by $[\chi]_L$. Wave packets around such states are appropriate candidates to represent quasi-classical states of the continuum on such scales.

Formally speaking, we can consider the continuum limit of this theory, in which:

- To each point x of space is assigned an operator $\widehat{\varphi}(x)$, with any two such operators commuting;
- To each such $\widehat{\varphi}(x)$ can be assigned a conjugate momentum operator $\widehat{\pi}(x)$, with $[\widehat{\varphi}(x), \widehat{\pi}(y)] = i\delta(x-y)$;
- A simultaneous eigenstate $|\chi\rangle$ of all of the $\widehat{\varphi}(x)$ is represented by a function χ, with

$$\chi(x) = \langle\chi|\widehat{\varphi}(x)|\chi\rangle. \tag{18.2}$$

In this limit the degeneracy vanishes and each such χ picks out a *unique* state $|\chi\rangle$; continuing to proceed formally, we can represent an arbitrary state $|\Psi\rangle$ by

$$|\Psi\rangle = \int \mathcal{D}\chi\,\Psi[\chi]\,|\chi\rangle, \tag{18.3}$$

where $\int \mathcal{D}\chi$ is the *path integral* over all functions χ and $\Psi[\chi] \equiv \langle \chi | \Psi \rangle$ is a complex functional assigning a complex number to each real function.

- With the state space thus represented, we can write down a dynamics by means of a Hamiltonian like

$$\widehat{H} = \int \mathrm{d}x^3 \left(\frac{1}{2}\widehat{\pi}(x)^2 + (\nabla \widehat{\varphi}(x))^2 + \frac{1}{2}m^2 \widehat{\varphi}(x)^2 + V(\widehat{\varphi}(x)) \right) \qquad (18.4)$$

or (in practice usually more useful) via a Feynman path integral

$$\langle \chi_1 | \widehat{U}(t - t') | \chi_2 \rangle = \int_{\chi(t',x)=\chi_1(x)}^{\chi(t,x)=\chi_2(x)} \mathcal{D}\chi \, \exp(-iS[\chi]) \qquad (18.5)$$

where

$$S[\chi] = \int_{t'}^{t} \mathrm{d}t \int \mathrm{d}x^3 \left(\frac{1}{2}\dot{\chi}(x)^2 - \frac{1}{2}(\nabla \chi(x))^2 - \frac{1}{2}m^2 \chi(x)^2 - V(\chi(x)) \right). \qquad (18.6)$$

In either (18.4) or (18.6), the first three terms represent a linear (i.e., free) continuum theory, and the V term encodes self-interaction.

18.4 Particles

The formal limit discussed above is mathematically pathological, and understanding and resolving that pathology is key to understanding modern quantum field theory. But since the necessary discussion will be somewhat abstract, let's start off by proceeding formally and extracting some of the physical content of the continuum theory.

To begin with, let's consider free field theories, where $V = 0$ and so the Hamiltonian is quadratic. The paradigm of a quadratic Hamiltonian is the simple harmonic oscillator

$$\widehat{H} = \frac{1}{2} \left(\widehat{P}^2 + \omega^2 \widehat{Q}^2 \right), \qquad (18.7)$$

which can be solved exactly by introducing "annihilation" and "creation" operators $\widehat{a}, \widehat{a}^{\dagger}$ defined by

$$\widehat{a} = (2\omega)^{-1/2}\widehat{Q} + i(\omega/2)^{1/2}\widehat{P}; \quad \widehat{a}^{\dagger} = (2\omega)^{-1/2}\widehat{Q} - i(\omega/2)^{1/2}\widehat{P} \qquad (18.8)$$

and satisfying $[\widehat{a}, \widehat{a}^{\dagger}] = 1$. The ground state $|\Omega\rangle$ of the theory satisfies $\widehat{a}|\Omega\rangle = 0$ and has energy $\omega/2$, and the eigenstates of the theory are given by successive actions of $\widehat{a}^{\dagger}$ on the ground state:

$$|n\rangle \propto \left(\widehat{a}^{\dagger} \right)^n |\Omega\rangle ; \quad \widehat{H}|n\rangle = (n + 1/2)\omega |n\rangle . \qquad (18.9)$$

The Hamiltonian can be rewritten as

$$\widehat{H} = \omega \left(\widehat{a}^{\dagger}\widehat{a} + 1/2 \right) \qquad (18.10)$$

and the concrete mathematical form of the ground state, as a wavefunction in "position" space, is

$$\langle x|\Omega\rangle \propto \exp(-\omega x^2/2). \qquad (18.11)$$

(I put "position" in quotes because in this abstracted form of the simple harmonic oscillator there is no reason to require x to be interpreted as position in physical space.)

A general quadratic Hamiltonian with form

$$\widehat{H} = \frac{1}{2}\sum_{m,n} J_{nm}\widehat{P}_n\widehat{P}_m + \frac{1}{2}\sum_{n,m} K_{nm}\widehat{Q}_n\widehat{Q}_m \qquad (18.12)$$

can then be understood as a collection of *coupled* harmonic oscillators, and can always be diagonalized, by a linear change of variables, into a sum of *uncoupled* harmonic oscillators — *modes* — of different frequencies $\omega(k)$, with creation and annihilation operators $\widehat{a}^\dagger(k)$ and $\widehat{a}(k)$:

$$\widehat{H} = \sum_k \omega(k) \left(\widehat{a}^\dagger(k)\widehat{a}(k) + 1/2 \right) \equiv \sum_k \omega(k)\widehat{a}^\dagger(k)\widehat{a}(k) + \text{constant}. \tag{18.13}$$

These creation and annihilation operators satisfy

$$\left[\widehat{a}(k), \widehat{a}^\dagger(k) \right] = 1; \quad [a(k), a(l)] = \left[\widehat{a}(k), \widehat{a}^\dagger(l) \right] = 0 \text{ for } k \neq l. \tag{18.14}$$

We can formally treat a free field theory as the continuum limit of this oscillator sum. Indeed, the coordinate transform can be concretely calculated; the modes are labeled by spatial vectors k and given by

$$\widehat{a}^\dagger(k) = \mathcal{N} \int dx^3 \left((2\omega(k))^{-1/2} e^{-ik \cdot x} \widehat{\varphi}(x) - i(\omega(k)/2)^{1/2} e^{+ik \cdot x} \widehat{\pi}(x) \right) \tag{18.15}$$

where $\omega(k) = \sqrt{m^2 + k^2}$, and where the normalization constant $\mathcal{N}$ and the allowed values of k depend on the spatial extent of the system. The Hamiltonian of this system is formally infinite because of the constant term in (18.13) — the first of the infinities that occur because of the mathematical pathologies of the theory and which we will shortly have to tame — but formally speaking we can just remove the constant term by subtracting an infinite correction from the Hamiltonian, without affecting the physics. (If this makes you uncomfortable, good! — but bear with me a little longer.)

We expect the ground state of a set of coupled harmonic oscillators to be highly entangled with respect to the original variables, which in the case of our continuum theory is to say that the degrees of freedom at distinct points of space ought to be entangled. And so it turns out to be, as can most easily be seen by calculating $\langle \Omega | \widehat{\varphi}(x)\widehat{\varphi}(y) | \Omega \rangle$; we find that

$$\langle \Omega | \widehat{\varphi}(x)\widehat{\varphi}(y) | \Omega \rangle = \frac{K}{|x - y|} \exp(-|x - y|m). \tag{18.16}$$

(A similar expression holds for $\widehat{\pi}(x)$.) So if two points are separated by a distance $\ll 1/m$, Bell-type measurements of the fields will pick up significant Bell-inequality violation; on scales $\gg 1/m$, this will be negligible.[4] (The distance $1/m$ — or $\hbar/mc$, in more familiar units — is called the *Compton wavelength*; it is typically much larger than the scale on which we expect the continuum approximation to fail.) Unsurprisingly, this entanglement persists for interacting theories, and is a general feature of quantum field theories. It is a clue that even in particle physics, $|\Omega\rangle$ cannot be thought of simply as "empty space": it has a considerable amount of quantum structure. (A further clue comes from the formally infinite energy density of the vacuum.) Starting with the ground state, arbitrary states of the field can be created by superposing states of form

$$|n_k\rangle = \Pi_k (\widehat{a}^\dagger(k))^{n_k} |\Omega\rangle; \tag{18.17}$$

that is, states like this form a basis for the full Hilbert space of the theory.

To see the full physical significance of the harmonic-oscillator analysis of the continuum theory, let's define a subspace $\mathcal{H}_{1P}$ as spanned by the states

$$|k\rangle = \widehat{a}^\dagger(k) |\Omega\rangle; \tag{18.18}$$

that is, states in $\mathcal{H}_{1P}$ are arbitrary superpositions of singly excited modes. Since each of these states is an eigenstate of energy, $\mathcal{H}_{1P}$ is conserved under the Schrödinger equation, so that the physics of singly excited states is a self-contained dynamics in its own right.

Given expression (18.15) for $\widehat{a}(k)$, any such state can be written (non-uniquely) as

$$|\psi\rangle = \int \mathrm{d}x^3 (f(x)\widehat{\varphi}(x)\,|\Omega\rangle + g(x)\widehat{\pi}(x)\,|\Omega\rangle) \qquad (18.19)$$

for complex functions f, g, and in fact it is easy to show that the converse is also true. But relations like (18.16) tell us that the states $\widehat{\varphi}(x)\,|\Omega\rangle$ and $\widehat{\pi}(x)\,|\Omega\rangle$ are localized excitations of the continuum that are negligible at distances from x much larger than the Compton wavelength. So we can think of the states of $\mathcal{H}_{1P}$ as superpositions of singly localized excitations.

But that is exactly the concept of "particle" in quantum *mechanics*: quantum particles are not in general localized, but they can be expressed as superpositions of states that are localized (wave packets, say, or − formally − position eigenstates). This suggests that $\mathcal{H}_{1P}$ can be understood as a space of one-particle states – the *one-particle subspace* (hence, "1P") of the quantum field theory. This naturally suggests identifying multiply excited states like (18.17) as multi-particle states, and indeed reinterpreting any state of the theory as a superposition of multi-particle states. (Because the operators $\widehat{a}^\dagger(k)$ and $\widehat{a}^\dagger(l)$ commute, it is easy to show that these particles obey Bose statistics.)

As long as we continue to consider *non-interacting* field theories, this reinterpretation of a continuum theory as a multi-particle theory is exact, and indeed serves as an alternative construction of a quantum field theory: start with a one-particle quantum theory and construct from it the direct sum of the symmetrized N-fold tensor product for each N:

$$\mathcal{F}(\mathcal{H}_{1P}) = \mathcal{H}_0 \oplus \mathcal{H}_{1P} + \mathcal{S}\mathcal{H}_{1P} \otimes \mathcal{H}_{1P} \oplus \cdots \qquad (18.20)$$

(where $\mathcal{H}_0$ is a one-dimensional Hilbert space and $\mathcal{S}$ is the symmetrization operator, imposing Bose statistics), and define the field operators by inverting (18.15). The space thus constructed is called the (symmetric) *Fock space* of $\mathcal{H}_{1P}$ and the construction process itself is called *second quantization*; see, e. g., Saunders (1992) or Wald (1994) for contemporary presentations of it.

Given these equivalent ways of thinking of the theory, one sometimes hears talk of *field-particle duality*, of the idea that "field" and "particle" are equally valid ways of interpreting the underlying physics. But this talk of duality only really applies in the (ultimately physically uninteresting) case of theories without interactions.

If a *small* interaction term is introduced to the free Hamiltonian, we expect that the particle analysis of the theory remains approximately valid. The interaction term can then be naturally interpreted as introducing transitions between excited modes of the harmonic oscillators, which under the particle interpretation can be understood as scattering effects between particles, and its effects can be studied by means of perturbation theory. But this analysis will only ever be approximate: as the interaction strength increases, the particle description of the theory becomes less and less valid, and eventually will need to be abandoned altogether as a useful description of the theory. For this reason, we would (in my view) do better here to speak of "emergence" of particles from the continuum theory, rather than of duality. From this perspective, "particles" are certain excitations of the ground state of the continuum which, to a varyingly good degree of accuracy, approximately instantiate the physics of an interacting particle theory.[5]

It's worth stressing that this picture of particles plays out pretty much identically whether the underlying continuum quantum theory is the quantum theory of a solid or liquid, or a quantum field. In the former, particles are often referred to as *quasi-particles* (such as the phonon, the quantum of vibration) but as far as modern field theory is concerned, all particles are quasi-particles.

To discuss the field-particle relation any further, though, we need to address the fact that any talk of "small" or "large" interactions is completely undefined mathematically, as long as the pathologies of the continuum theory are unaddressed.

18.5   Effective Field Theories

These pathologies are trackable to the continuum theory's infinitely many degrees of freedom per spacetime volume. For a start, a Hilbert space like this is *non-separable*: it has no countable basis, and hence it naturally factors into uncountably many superselection sectors, which differ only by the (ex hypothesi unphysical) features of the theory on arbitrarily short length scales. More seriously, any attempt to calculate physical quantities by the normal methods of theoretical physics gives an infinite answer. This occurs even in the case of a non-interacting system: we have seen that the expected value of the Hamiltonian in (18.4) is formally infinite, and so is the path integral in (18.5), though in both cases we can formally treat this as an unobservable (infinite) correction to the energy or action. This rather unsatisfactory situation becomes intolerable as soon as interactions are introduced (in other words: as soon as the dynamics stops being trivial), at which point the formal machinery of quantum physics delivers infinities for pretty much any question we choose to ask.

Although there is an honorable tradition in mathematical physics (see the Further Reading) of trying to formulate a fully mathematically rigorous theory of the continuum that avoids these pathologies, while remaining well defined on all length scales, the current consensus in mainstream physics[6] is that the problems should be resolved by taking seriously the idea that we were in the first place only looking for a theory describing the continuum down to some length scale, and that the pathologies are caused by going to the continuum limit in the first place rather than remaining finite. To see why this might be, let's consider in more detail the path integral (18.3) that formally defines the inner product. This integral is supposed to be over *all* functions χ, or at the least over all square-integrable functions, and (restricting, for simplicity, to one dimension) an arbitrary such function can be written as

$$\chi(x) = \sum_{n=-\infty}^{n=\infty} \alpha_n \exp(-2\pi n i / R), \tag{18.21}$$

where R is the spatial extent of the system and $\alpha_{-n}^* = \alpha_n$. Therefore, the integral can be decomposed as

$$\int \mathcal{D}\chi = \prod_{n=0}^{\infty} \int d\alpha_n = \int d\alpha_0 \int d\alpha_1 \cdots. \tag{18.22}$$

And now it's fairly unsurprising that carrying out infinitely many such integrals gives an infinite answer. But if, say, $R = 10\,\mathrm{m}$, then the integrals over α_n for $n \gg 10^{11}$ integrate over functions varying rapidly on scales $\ll 10^{-10}\,\mathrm{m}$. Therefore, if we were only trying to construct a continuum theory describing features of the continuum on scales longer than that (if, for instance, we were studying a "continuum" which is actually made of discrete atoms of size $\sim 10^{-10}\,\mathrm{m}$, as in condensed matter physics) then it becomes reasonable to consider cutting off the integral, by discarding the integration over functions varying on those scales (that is, by discarding the integrals over α_n for $n > 10^{-11}$). Doing so removes the infinities from the theory.

If you're not suspicious of this process, you should be. It is one thing to set out to construct a theory applicable only above certain length scales; it is quite another to suppose that it can be done simply by discarding any influence of shorter length scale physics. After all, I could just as easily have chosen the cutoff length at $10^{-9}\,\mathrm{m}$, or imposed it by some other means than Fourier modes. The actual physics at the cutoff length scale is presumably extremely complicated, and there doesn't seem any reason not to expect those complexities to affect the form of the larger-scale physics. We might try to avoid this by supposing that the cutoff occurs at precisely the *physical* length scale at which the theory fails (at the atomic length scale in condensed matter physics, say) so that those degrees of freedom are unphysical anyway – but we still face the problem that our crude imposition of the cutoff is presumably far removed from the actual way in which a microphysical description fails.

It is one of the most remarkable features of quantum field theory – and one of the key discoveries of theoretical physics in the postwar period – that, on the contrary, the details of physics below the cutoff have *almost no empirical consequences* for large-scale physics. The details are mostly too technical for an article at this level, but the general idea can be understood as follows.

Consider again the dynamics (18.5). Expanding V as a power series in its argument,

$$V(x) = V_0 + \frac{1}{4!}\lambda_4 x^4 + \frac{1}{6!}\lambda_6 x^6 + \ldots \tag{18.23}$$

(where we assume V is symmetric under $x \to -x$, and ignore the x^2 term since it is already included in the Hamiltonian), we can see that the theory is specified by m^2 (corresponding, where a particle interpretation is valid, to the squared particle mass), V_0, the infinite number of coefficients $\lambda_4, \lambda_6, \ldots$, and, tacitly, by the method used to cut off the theory at short length scales, which we can schematically write as Λ. (We can think of Λ as denoting the length scale at which the cutoff is imposed as well as the details of the method by which it is imposed; by abuse of notation, I will also use Λ to denote the length scale alone.) Therefore, we can consider an infinite family of theories parametrized by these variables. I will write α to denote, collectively, all the variables except Λ; (α, Λ) then denotes a particular theory in this family. It will be helpful to think of α as a set of coordinates in a *theory space* $\mathcal{A}$; a theory is specified by a point in $\mathcal{A}$ together with a choice of cutoff.

In general, any two such theories will be physically distinguishable by some in-principle-measurable features. (For instance, any two theories with different cutoff procedures can be distinguished by probing the physics around the cutoff scale.) But recall that we are only really interested in using these theories to describe physics at scales large compared to the cutoff; close to the cutoff scale, our arbitrary assumptions about the nature of the cutoff make the theory untrustworthy. So trustworthy physical predictions arise only from the large-length scale features of these theories. With this in mind, define two theories T_1, T_2 as *IR-equivalent* (IR for "infra-red," physics jargon for long distance), $T_1 \sim_{IR} T_2$, if they make the same prediction values for the evolution of dynamical variables on length scales large compared to the cutoff, at least for quantum states which do not themselves vary sharply on length scales close to the cutoff. There are various ways of making this precise; for our purposes, a heuristic understanding will suffice.

(It is important to appreciate that IR-equivalence is a much finer notion than empirical equivalence. The "long-distance" features of a theory can be defined on scales of tens of nanometers (in condensed matter physics) or many orders of magnitude smaller (in particle physics); they do not correspond simply to directly observable features of the continuum.)

Now consider two cutoffs Λ, Λ' (with, for definiteness, $\Lambda' > \Lambda$). For any given $\mathcal{T}(\alpha, \Lambda)$, it turns out that changing the cutoff from Λ to Λ' can be compensated for by changing the other variables from α to some α', without in any way affecting the physical predictions of the theory at length scales large compared to Λ and Λ'. That is, there is some transformation $\mathcal{R}(\Lambda \to \Lambda')$ acting on theory space $\mathcal{A}$ such that

$$(\alpha, \Lambda) \sim_{IR} \left(\mathcal{R}(\Lambda \to \Lambda') \cdot \alpha, \Lambda' \right). \tag{18.24}$$

(Again, this can be proved with various degrees of generality and rigor; again, here I just assert it without proof.) The transformation $U(\Lambda \to \Lambda')$ is known as the *renormalization group*; it has a central role in modern quantum field theory and is discussed in more detail by Williams (this volume).

So there is in a sense redundancy in our family of theories: two apparently different theories may correspond to the same large-scale phenomena and so be interchangeable, as long as we regard the theories as in any case trustworthy only as regards large-scale phenomena. In fact the redundancy is considerably more dramatic than this: we can (in this particular example, and in fact in general) identify a "relevant" subset of the coordinates of $\mathcal{A}$ such that only the relevant coordinates have any significant effect on the physics. And this "relevant" subset is in general finite-dimensional. In the case of the three-dimensional scalar theory we have been considering, for instance, it is three-dimensional (the relevant coordinates can *loosely* be thought of as the zero of the potential V_0, the free-Hamiltonian parameter m^2, and the first coefficient λ_4 in the expansion of V, though as we will see in the next section, it's a bit subtler than that suggests.)

Therefore, for given cutoff Λ, we can specify a theory – at least as far as the long-distance content of the theory is concerned – just by giving the finitely many values of the relevant coordinates. And Λ itself can be chosen according to convenience or whim, for if we wish to replace it with Λ' we need only make a compensating change in those relevant coordinates.

This approach to "continuum" quantum physics (which, to repeat, is absolutely standard in modern physics, both in particle physics and in condensed matter physics) is known as *effective field theory*. The term also applies to individual theories: an effective field theory is a quantum field theory understood as applying only on length scales larger than some short-distance cutoff, and identified with an equivalence class of IR-equivalent theories defined by various specific cutoff schemes. All empirically relevant quantum field theories in physics, at present, are effective field theories. To quote the author of one popular textbook,

> I emphasize that Λ should be thought of as physical, parametrizing our threshold of ignorance, and not as a mathematical construct. Indeed, physically sensible quantum field theories should all come with an implicit Λ. If anyone tries to sell you a field theory claiming that it holds up to arbitrarily high energies, you should check to see if he sold used cars for a living. (Zee, 2003, p. 162)

Equally, though, the precise (or even approximate) value of Λ is irrelevant to the large-scale physics; all that matters is that it is much smaller than the length scales on which we deploy the field theory to model the world.

18.6 Renormalization

To get a better sense of how effective field theory works, let's look (schematically) at how we might calculate some physical quantity – say, the "four-point function" of the theory G, which gives the expectation value of quadruples of field operators (possibly at different times) with respect to the system's ground state. (G corresponds, loosely speaking, to the scattering amplitude between pairs of particles.) A standard approach to doing so in ordinary quantum mechanics is to decompose the Hamiltonian into a sum of a "free" and "interaction" term:

$$\widehat{H} = \widehat{H}_0 + \widehat{H}_{int}. \tag{18.25}$$

The idea is that we can solve the physics of $\widehat{H}_0$ exactly, and that $\widehat{H}_{int}$ can be treated as a small correction to $\widehat{H}_0$, whose effects can be estimated by the methods of perturbation theory.

If we try this for the Hamiltonian (18.4), it would be natural to take

$$\widehat{H}_0 = \int dx^3 \frac{1}{2} \left(\widehat{\pi}(x)^2 + \nabla\widehat{\varphi}(x)^2 + m^2\widehat{\varphi}(x)^2 \right) \tag{18.26}$$

and

$$\widehat{H}_{int} = \int dx^3 \widehat{V}(x). \tag{18.27}$$

After introducing a cutoff, both are well defined and finite; the former can be solved exactly to give a well-behaved theory of a non-self-interacting approximately continuous quantum system, along the lines of Section 18.3 (but with the formal moves of that section legitimated by the cutoff).

Suppose for now that V is purely quartic: that is, $V(x) = \lambda_4 x^4 / 4!$. Then what perturbation theory is *supposed* to deliver for us is an expression for G in powers of λ_4: something like

$$G = G_0 + G_1 \lambda_4 + G_2 (\lambda_4)^2 + \cdots \qquad (18.28)$$

If we calculate this power series to first order in λ_4 (known in physics as "tree-order," a reference to the Feynman-diagram notation used in practice to work out the power series), we obtain sensible, well-behaved answers (and answers independent of the cutoff scale Λ). But when we come to calculate the λ_4^2 term (the "one-loop correction," in physics terminology), we obtain a quite large result, a sum of terms the largest of which is proportional to the logarithm of the inverse cutoff length, $\log(1/\Lambda)$. (If we had tried to calculate this quantity formally in the continuum theory, without any cutoff, the answer would have been not just *large* but *infinite*). Terms this large invalidate the perturbative expansion and call into question the validity even of the "sensible" tree-order result. Evaluating subsequent terms in the power series likewise gives very large results.

It turns out, however, that these large results can largely be removed by absorbing them into the definitions of the parameters λ_4 and m^2. What this amounts to (roughly speaking, because I have omitted for expository simplicity the fact that the field strength also gets renormalized) is that we can rewrite the Hamiltonian as

$$\widehat{H} = \widehat{H}_0 + \widehat{H}_{int} = (\widehat{H}_0 + \widehat{\Delta}) + (\widehat{H}_{int} - \widehat{\Delta}) \equiv \widehat{H}'_0 + \widehat{H}'_{int} \qquad (18.29)$$

for some operator $\widehat{\Delta}$, such that

1. $\widehat{H}'_0$ has the same functional form as $\widehat{H}_0$, but with a new value $(m^2)^{ren}$ ("ren" for "renormalized") for the quadratic parameter, related to m^2 by an expression like

$$(m^2)^{ren} = m^2 + \alpha/\Lambda^2 + \text{(smaller terms)} \qquad (18.30)$$

   for some dimensionless quantity α.

2. $\widehat{H}'_{int}$ is small enough to treat as a perturbation of $\widehat{H}'_0$, at least for states in a certain energy range (which range is determined by the choice of $\widehat{\Delta}$).

3. That perturbative expansion is a power series expansion not in λ_4, but in a new parameter λ_4^{ren}, related to λ_4 by an expression like

$$\lambda_4^{ren} = \lambda_4 + \beta \log(1/m\Lambda) + \text{(smaller terms)} \qquad (18.31)$$

   for some dimensionless quantity β. The leading-order term in that expansion is the tree-order term from before (but using λ_4^{ren} and $(m^2)^{ren}$, not λ_4 and m^2).

This might seem a block to the applicability of the theory: to make calculations we need to know $(m^2)^{ren}$ and λ_4^{ren}, and we can only calculate them from m^2 and λ_4 via detailed knowledge of the cutoff mechanism and scale. And indeed this would be a block if we were presented with the theory by giving the original parameters m^2 and λ_4 (the so-called "bare parameters") as a gift from God. But in fact, we determine the parameters through experiment – and what the experiment gives us is the renormalized parameters, not the bare parameters. The latter are related to the measured parameters through a cutoff-dependent expression, but we don't in any case need them for calculations.

Furthermore, if we now include the higher order terms $\lambda_6 x^6$ (etc.) in $V(x)$, the result is that these terms:

1   Further renormalize m^2 and λ_4, adding corrections that are functions of Λ;
2   Make tree-order contributions to the calculation that are proportional to powers of L/Λ, where L is the length scale on which G is evaluated (and thus are extremely small if $L \gg \Lambda$);
3   Make loop-order contributions suppressed by successively larger powers of L/Λ;
4   Have no effect on the dynamics in the long-distance limit *except* to renormalize m^2 and λ_4.

We can now identify the renormalized parameters as two of the three parameters that (I claimed in the previous section) suffice to determine the theory up to IR-equivalence. The third parameter can be identified as the energy density of the ground state, related to the zero V_0 of the function V by an expression like

$$V_0^{ren} = V_0 + \gamma(1/\Lambda)^4 \tag{18.32}$$

but irrelevant to the physics except in the presence of gravity.

18.7   Scale Dependence, and Particles Again

I should stress one crucial feature of this renormalization process: it is *scale-dependent*. There is a certain amount of freedom in how to divide out the contribution of higher order terms in the perturbative expansion between (i) renormalizing the bare parameters, and (ii) contributing corrections to the tree-order calculations. In practice this is usually done by picking some scale at which the tree-order calculation (expressed in terms of the renormalized parameters) is *exact*. Calculations made at lengths close to this scale are well approximated by the methods of perturbation theory, but these methods become successively less effective at too-large or too-small lengths. The theory can still be understood as being *specified* by the parameters even at lengths far from the chosen scale – but the meaning of those parameters at that scale will be far from transparent.

To illustrate this, consider two examples from particle physics: quantum electrodynamics (QED) and quantum chromodynamics (QCD). In popular accounts, the former is the theory of electrons and photons, the latter the theory of quarks and gluons, but we will do better to think of the former as a theory of an electron *field* interacting with the photon field (i.e., quantized electromagnetic field) and the latter as a theory of a quark and gluon field interacting. (Even this is an imperfect way to think, due to gauge freedom: the electron and photon fields, or the quark and gluon fields, jointly represent the underlying physics, and the split between them is to some degree gauge dependent.)[7] In the absence of an interaction term, though, the usual particle account can be recovered: the electron is the particle associated with the electron field, etc.

In both theories (and simplifying slightly), one of the relevant parameters (relevant in the sense of our previous discussion, that is) is the strength of the interaction between the fields (electron-photon or quark-gluon); another, in the free-field limit, may be interpreted as the mass of the "matter" particle, i.e., the electron or quark. Schematically, let's call these parameters λ and m, respectively.

In both cases, the values of λ and m are scale-dependent. For QED, λ decreases at larger length scales. For length scales very large compared to the cutoff length, it is small enough that the interactions may be treated as a small perturbation of the free-field theory. In this regime, the particle interpretation of the theory remains approximately valid, and we can meaningfully interpret the theory as a theory of electrons scattering off one another with their interactions mediated by photons. At shorter and shorter length scales (corresponding to higher and higher energies), the electron-photon interaction becomes stronger (the mass also changes, though the exact form of that change is not relevant here).

Let's pause to reconsider the emergent status of electrons (originally discussed in Section 18.3) in this context. Recall that the particles of a free-field theory are created from the ground state of the free-field Hamiltonian by applying creation operators. But the definitions of the creation operators, and of the free-field ground state, depend on the parameters of the theory: on m directly, and on λ indirectly via its role in the renormalization process. Therefore, the one-particle Hilbert space constructed to analyze QED at high energies is a *different Hilbert space* from the one constructed at low energies. This ought to drive home the point that electrons cannot be thought of as fundamental building blocks of nature; they are simply a useful, but scale-relative, emergent feature of the underlying theory. But recall that this theory, too, should not be thought of as fundamental, given the tacit presence of the cutoff. In fact, QED actually imposes a minimum value for the cutoff length: the interaction strength λ continues to increase at smaller and smaller scales and eventually becomes infinite (the so-called "Landau pole"), indicating that QED ceases to be well defined for cutoffs smaller than this length. To be sure, the Landau pole occurs at a length far shorter than the point at which external reasons (like gravity) would lead QED to fail as a physical description of the world, but it is strong reason to think that there is no genuine continuum version of QED.

In QCD, conversely, the interaction strength λ *decreases* at shorter length scales (i. e. , higher energies). This means that in high-energy physics, the field can be viewed as describing a collection of weakly-interacting quarks. At low energies this description breaks down entirely. We can still describe the field in terms of particles, but now they are different particles: protons, neutrons, and various mesons and other hadrons. These particles can be thought of *very* loosely as bound states of the quarks; a somewhat more accurate statement would be that they are excitations associated not with the quark field but with various symmetrized products of that field.

18.8   Symmetry and Universality

Let's return again to the scalar field. In classical field theory, saying "this is a theory of an interacting scalar field" is nothing like enough to pick out a unique theory. Fundamentally different Lagrangians could be written down for a scalar field: some with φ^4 interactions, some with φ^6 interactions, some with combinations of both, and so forth. This is not merely a matter of specifying the coupling constants, but the actual functional form of the Lagrangian. Of course, any term written down will have to satisfy the symmetries of the scalar field (by definition; otherwise it would not be a scalar field), but specifying the symmetry is only the beginning of specifying the dynamics.

But we have seen that the situation is very different in quantum field theory. First, the entire class of *renormalizable* interactions is finite and small: indeed, it basically contains the (renormalized) λ_4 parameter. So as far as interactions on scales large compared to the cutoff are concerned, there is only really one way to write down a nontrivial dynamics. Second, and more profoundly, even if we exclude from the Lagrangian some (renormalizable or non-renormalizable) interaction terms by setting the associated parameter λ_n to zero, the parameter will move away from zero again if we shift the cutoff. To say that, for instance, the φ^6 term in the QFT Lagrangian is absent is to say something *cutoff-dependent*, and hence something that is not really of physical significance given the effective-field-theory approach to understanding quantum field theories.

This is not confined to scalar field theories, of course. In full generality, writing down the symmetries of a quantum field theory pretty much fixes the form of its dynamics, up to a very small number of parameters. This phenomenon – often called *universality* – goes some way to explaining the central role of symmetry in contemporary theoretical physics: once the symmetries of a quantum field theory are known, its dynamics are pretty much specified. (See Batterman, this volume, and Williams, this volume, for more on this subject and its connection to the renormalization group.)

18.9   Other Features of Quantum Field Theory

Pretty much everything I have discussed so far applies to *any* quantum field theory. In this section, I will mention some conceptually important features of QFT that apply in more specific theories. My focus will be on examples from particle physics; of necessity, the discussions will be brief, and I concentrate on examples which rely on specifically quantum-mechanical features of QFT (as opposed to, say, gauge theory, which plays a central role in the Standard Model but is to a large degree classical).

18.9.1   *Lorentz Covariance and the Classification of Particles*

In particle physics (but not in condensed matter physics), the imposition of Lorentz covariance places strong constraints on the form of a quantum field theory. Examples include the following:

Wigner's classification of particles: If we ask, independent of their origin as excitations of a field, what quantum states deserve the name "particles," we can argue – following Wigner (1939) – that they should correspond to irreducible representations of the Poincaré group. (A heuristic rationale: the one-particle subspace must transform under the Poincaré group; if it transforms reducibly, we can decompose it into a superposition of components that each transforms irreducibly. I don't know of a really careful conceptual analysis, though I make some suggestions in Wallace (2009).) Moreover, if we ask what linear field theories can be written down, we obtain the same result. (If we write down a field that transforms reducibly under the Poincaré group, the irreducible components get renormalized differently, becoming in effect different fields.) Either way, it seems at least highly plausible that the particle-like excitations of a field theory can be group-theoretically classified.

The result (skipping over some representations that seem unphysical) is as follows:

- Particles are classified completely by their mass (which can be zero or positive) and by their spin, which can have any positive integer or integer-plus-half value. This classification coincides, so far, with the nonrelativistic version of the same approach, which classifies particles via irreducible representations of the Galilei group Bargmann (1954). There are longstanding if somewhat contested arguments that particles of spin > 2 cannot consistently be associated with an interacting quantum field, but that lies beyond the group-theoretic analysis; see Bekaert et al. (2012) and references therein.
- Particles of nonzero mass and spin s have a $(2s + 1)$-dimensional internal space, again as in nonrelativistic mechanics.
- Particles of mass zero and spin > 0 have a two-dimensional internal space (so that photons, for instance, are spin-one particles but have only two orthogonal spin states for a given momentum). There are subtle connections between gauge symmetry and this reduction of a massless particle's internal degrees of freedom; again, they go beyond the group-theoretic analysis.

Antimatter: If a quantum field transforms under a representation of its internal symmetry group that has a natural complex structure (such as the standard representations of $U(1)$ or $SU(N)$), its one-particle Hilbert space separates naturally into matter and antimatter components. A consequence is that to all charged particles (and some uncharged particles, like the neutrino) is associated an antiparticle of the same mass but opposite charge and other quantum numbers. This is a purely relativistic effect with no nonrelativistic analog; see Wallace (2009) or (for a discussion from an algebraic-quantum-field-theory viewpoint) Baker and Halvorson (2009).

Discrete symmetries and the CPT theorem: A relativistic quantum field theory might in principle have three discrete symmetries in addition to any continuous internal symmetries and the continuous Poincaré symmetries: Parity (reflection in space), Time reversal (reflection in time), and Charge conjugation (exchange of matter for antimatter). The *CPT theorem* (also known as the TCP, CTP, PCT, and PTC theorem – I don't recall ever seeing TPC) establishes that any quantum field theory has a symmetry which can be identified as the product of all three transformations: that is, the transformed field at x, t is a function of the untransformed field at $-x$, $-t$ and the symmetry exchanges matter and antimatter. There is no requirement that the individual transformations are symmetries or even that they are well-defined transformations on the theory's Hilbert space.

The Spin-Statistics theorem: See below.

18.9.2   The Fermion/Boson Distinction

The scalar field theory used above as an example is specified (formally) by a function from points of space to operators such that pairs of spatially separated operators commute; insofar as it can be treated as weakly interacting, it can be interpreted as a theory of bosonic particles. But it is also possible to construct a quantum field (either in solid-state physics or in particle physics) where spatially separated operators *anti*-commute. The resultant theory, in the weakly interacting regime, can be analyzed in terms of fermionic particles. Such field theories are called *fermionic*, by contrast with the *bosonic* fields we have focused on so far.

The celebrated *spin-statistics theorem* establishes that a quantum field is bosonic if its associated particles have integer spin, and fermionic if it has (integer-plus-half) spin. In the Standard Model of particle physics, in particular, the fermionic fields are the quarks and leptons (electrons, neutrinos, and their heavier variants), which have spin 1/2; the bosonic fields are the force carriers (gluons, photons, the W and Z bosons) with spin 1, and the Higgs field, with spin zero (and, depending how the Standard Model is defined, possibly also the graviton, with spin 2). The theorem holds for (relativistic) spacetimes of dimension three or higher; conversely in two spacetime dimensions, there are examples where the same quantum field can be analyzed in terms of fermions in one regime and bosons in another Coleman (1985, ch. 6).

Formally speaking, fermionic fields can be treated very much like bosonic fields (other than a large number of minus signs that have to be kept track of in calculations). Conceptually, they seem dissimilar in important respects: for instance, the quantum state of a fermionic field cannot in any straightforward sense be understood as a wavefunctional on a space of classical field configurations. To the best of my knowledge there has been rather little discussion of fermionic fields in the philosophy literature on QFT.

18.9.3   Infra-red Divergences and the Large-Volume Limit

In my presentation of QFT so far I have assumed a spatially finite system. That seems reasonable enough in most applications of QFT, whether in solid-state physics (where the physical system to which QFT is applied is manifestly finite) and in particle physics (where the physical processes of interest are normally confined to a region of finite – and usually pretty small – extent). In each case, the physically reasonable implication is that the size R of the finite region will have little or no effect on the physics, provided it is much larger than the scales which we are studying.

However, in some physical applications – notably those involving zero-mass particles – this is not the case: loop-order calculations of physical quantities include terms that depend on R, and that diverge as $R \to \infty$. These *infra-red* divergences, analogously with the ultra-violet (short-range) divergences we have already considered, can be handled in one of two ways: either by developing, rigorously, a quantum theory of genuinely infinite systems, or by keeping R fixed but large (or otherwise regularizing the divergences, e.g., by adding a small mass term) and absorbing R-dependent terms into a renormalization of the physical parameters. The latter is the route taken in mainstream physics, and delivers useful physical insight into the origin of the infrared divergences: they occur because zero-mass particles can be created with arbitrarily low energy, and so a physical particle is surrounded by a cloud of very many – in the $R \to \infty$, infinitely many – such particles.

The infrared and ultraviolet divergences are *dis*analogous in one important way, though: while the short-distance cutoff is normally taken to describe a *physical* cutoff above which the theory cannot be trusted, the long-distance cutoff just reflects the finite extent of the part of physical reality we are trying to model, together with a physical assumption that the details of the boundary conditions on that region aren't physically significant for shorter-distance physics. Not unrelatedly, while (in my biased opinion) not much of conceptual value has been gained by trying to define continuum QFT on arbitrarily short length scales, mathematically rigorous considerations of spatially infinite QFTs have been conceptually very informative (see Ruetsche (2011) and references therein).

18.9.4 Symmetries and Symmetry Breaking

The duality between field and particle (in the noninteracting limit) suggests that a symmetry of a quantum field theory will be represented as a symmetry of the particle(s) associated with that theory, and so it often turns out. For instance, the quark field has the group $SU(3)$ as a dynamical symmetry, and transforms as the three-dimensional complex representation of that symmetry; correspondingly, quarks as particles have a three-complex-dimensional space of internal degrees of freedom. (This is usually described in popular accounts as there being three sorts of quarks – red, green, and blue – but in fact "red," "green," and "blue" are just arbitrary bases in the quark's internal space, and $(1/\sqrt{2})$(red + blue) or $(1/\sqrt{2})$(blue $-i$ green) are just as valid as quark states.) Similarly, there is an approximate symmetry called "isospin" in the effective field theory of the hadrons (the low-energy states of QCD) that shows up in the approximately equal masses of the proton and neutron.

However, this straightforward relation between field and particle symmetries only works if the theory's ground state $|\Omega\rangle$ is invariant under the symmetry group. If this is not the case – which entails that "the" ground state is degenerate – then we can construct a different set of particles by acting with creation operators on each ground state. A symmetry transformation under which the ground state is not invariant will transform between different (though indiscernible) sets of particles, rather than transforming a set of particle states among themselves. As a consequence, the symmetry is not visible in the dynamics of the particles, and so is said to be "spontaneously broken" ("hidden" might be a better term).

A spontaneously broken symmetry shows up in the phenomenology via constraints between the measured coupling constants. In the case of a global symmetry (one that acts the same way at every point of spacetime), it also shows up, via the presence of a "Goldstone boson," a massless particle which always occurs in the particle spectrum of such a system. (In the long-wavelength limit it corresponds to the symmetry that maps from one ground state to another.) For instance, the spatial translation symmetry is spontaneously broken in condensed matter physics by the lattice structure of the ground state of a solid body; the associated Goldstone boson is the phonon, the quantum of vibration. The pion (a two-quark low-energy excitation of the quark field) can be understood as a Goldstone boson of a spontaneously broken approximate symmetry of the hadrons; as such, its mass is low but not zero.

The physics is somewhat subtler when the spontaneously broken symmetry is local. In that situation, the *Higgs mechanism* causes the gauge fields associated with the local symmetry (which are normally massless) to acquire mass. The mechanism is sometimes heuristically described as the vector boson "eating" the Goldstone boson; there is some controversy in foundations of physics as to what a better description would be (Earman (2004); Struyve (2011); Friederich (2013)).

There are conceptually interesting mathematical subtleties involved with spontaneous symmetry breaking (global or local). In finite systems, it is known that the ground state is not genuinely degenerate because of the possibility of tunneling between symmetry-related states. The infinite-volume limit is required for true degeneracy, and in that limit some of the other assumptions of the theory break down. For discussion, see Ruetsche (1998) and references therein.

18.9.5   Non-renormalizable Interactions in Physics

My account of renormalization theory might give the impression that *non*-renormalizable interactions have no part to play in physics: either we are at energy levels very low compared to the cutoff (in which case their influence is swamped by the renormalizable interactions) or we are relatively close to the cutoff (in which case the theory is not reliable in any case). This is not quite correct. In particular, suppose we have a field theory with *no* renormalizable interactions. Then we would predict that:

1   The interactions will be dominated by that nonrenormalizable term which drops off least rapidly at greater length scales;
2   The interaction strength will be suppressed by some power of L/Λ, where L is the characteristic length scale of the interaction and Λ is the cutoff;
3   In particular, the interactions will be very weak at large lengths.

I give two important examples of this in practice. The first is the so-called *four-fermion theory*, where the only field is an uncharged spin-half fermionic field (taken to represent neutrinos, say). In the Standard Model, neutrino interactions are mediated by the W and Z bosons, but in situations where the energy levels of interactions are much lower than the W and Z masses, we can regard those masses as a cutoff on an effective field theory where the neutrinos interact directly. Any direct interaction between fermion fields is nonrenormalizable; the lowest order such term is a two-particle scattering term that allows pairs of neutrinos to scatter off one another. We would predict that this interaction is very small at interaction scales large compared to the Compton wavelength of the W and Z. And indeed this is what we find: the force that mediates neutrino interactions is called the weak interaction in particle-physics phenomenology.

The second example occurs in quantum gravity. There is no renormalizable interaction between the metric field and matter (or between the metric field and itself) but the lowest-order nonrenormalizable interaction term is the Einstein-Hilbert action term of general relativity (together with a cosmological-constant term, of which more below). The extreme weakness of the gravitational field is then explained by the fact that gravitational phenomena are studied on scales extremely large compared to the Planck length at which we expect full quantum gravity to impose a cutoff on field theory.

18.9.6   Quantum Field Theory on Curved Spacetime

Most of the theoretical development (and almost all the experimental data) in QFT assumes flat spacetime, either Newtonian (for condensed matter physics) or Minkowski (for particle physics). But quantum field theory can be formulated in at least some nonflat spacetimes, and doing so is the basis of important work in cosmology and in the physics of black holes, in particular in one of the most celebrated and surprising discoveries of the past 40 years: Hawking's discovery of black hole radiation (see Wallace 2017a and references therein for further details).

18.10   Outstanding Questions of Particle Physics

I have tried to indicate in this chapter just how successful and powerful quantum field theory is. But – optimistically from the point of view of exciting new physics – there remain some deep puzzles in the theory, in particular in its applications in particle physics. Here I identify three such puzzles; there are many others, but these are perhaps the most visible in contemporary physics. The first two are discussed in more detail by Barnes, this volume, and Williams, this volume.

18.10.1   Fine-Tuning of the Higgs Mass

My discussion of the scalar field glossed over one subtlety. I noted that the mass of the field is renormalized, so that the empirically accessible mass is related to the "bare" mass by a cutoff-dependent term. In general in quantum field theory, renormalizations like this are logarithmic in the cutoff, as in expression (18.31): the mass rescaling in a spin–half field, for instance, has abstract form

$$m^{ren} = m_0(1 + A\log(1/\Lambda)), \tag{18.33}$$

for some dimensionless A not usually too far from unity. Because of the slow scaling of the logarithm function, this means that if the bare mass is much less than the cutoff energy, so will the renormalized mass. But in the case of the scalar particle, the mass rescales according to equation (18.30), with additive corrections to the bare mass proportional to $1/\Lambda^2$. This means that we would expect the renormalized mass of a scalar particle to be of the same order as the cutoff energy, whatever the bare mass might be.

However, the Higgs boson – which, in the simplest versions of the Standard Model, is the particle associated with a scalar field – has a mass far below whatever the Standard Model's cutoff energy is. This is not a contradiction in the theory – a sufficiently careful choice of the bare mass can yield whatever value we like for the renormalized mass – but it seems to involve rather unattractive fine-tuning of the theory's parameters.

18.10.2   Fine-Tuning of the Cosmological Constant

We saw in Section 18.3 that the formal energy density of the vacuum of a free field theory is infinite. Adding a cutoff tames the infinity but still leaves a very large finite term, of order $(1/\Lambda)^4$. Interactions add further contributions, also of order $(1/\Lambda)^4$. Therefore, the expression for the total energy density ρ_{vac}, schematically, is

$$\rho_{vac} = V(0) + \text{free-field contribution} + \text{interaction renormalization}, \tag{18.34}$$

where $V(0)$, the classical vacuum energy density, is the value of the Lagrangian at zero field.

In nongravitational physics, none of this matters: the energy density has no effect on the physics and can be set to whatever value we find convenient (usually zero) by an appropriate choice of $V(0)$ without empirical consequence. But the stress-energy tensor of the field has the form

$$T_{\mu\nu} = T^{ren}_{\mu\nu} + g_{\mu\nu}\rho_{vac}, \tag{18.35}$$

where the "renormalized" stress-energy tensor $T^{ren}_{\mu\nu}$ is defined by the requirement that it vanishes for the ground state. And the Einstein field equation is

$$G_{\mu\nu} + g_{\mu\nu}C = 8\pi\,GT_{\mu\nu}, \tag{18.36}$$

where C is the cosmological constant. So it looks as if the vacuum expectation value should make an enormous contribution to the cosmological constant, albeit it's not fully clear how to interpret the field equation without a quantum theory of gravity (the simplest approach would be to take the right-hand side of the equation to be the *expectation value* of the stress-energy tensor; see Wald (1994) for further discussion of this hybrid theory). That contribution is more than 100 orders of magnitude larger than the observed value of the cosmological constant. Again, this is not a contradiction, as we can tune the classical vacuum energy density (or, equivalently, the bare cosmological constant) to whatever value we like; as with the Higgs mass, the problem is the extreme fine-tuning of the parameters that seems to be required. (There is an odd tendency, which I observe mostly in conversation, for philosophers of physics to draw a sharp distinction between a bare constant on the left-hand side of the Einstein field equation – where it is taken to pertain to spacetime – and the negative of the same constant on the right-hand side of the equation – where it is taken to pertain to matter. But, as I once heard Sean Carroll remark in response to one such comment, it is permissible to move terms from one side of an equation to the other!)

18.10.3   Quantum Gravity

A common claim about quantum gravity is that it is a puzzle that arises from the incompatibility of quantum mechanics with general relativity, and so would arise (in principle) whenever quantum effects and gravity apply simultaneously. This is not the perspective of most quantum field theorists: to them, the metric field is at least perturbatively perfectly well-behaved – albeit non-renormalizable – and can be handled in the effective-field-theory framework. (We have already seen that the extreme weakness of the gravitational field can be understood in effective-field-theory terms as a consequence of the non-renormalizability of the Einstein-Hilbert action.) Indeed, exactly this formalism is applied in the quantum-fluctuation calculations that underpin our theoretical models of the cosmic microwave background radiation, and so "quantum gravity," in the sense of a quantum-field-theoretic understanding of general relativity, has already passed at least a crude experimental test. (For the formalism, see Weinberg (2008, ch.10); for the conceptual discussion, see Wallace (2016)).

What most quantum field theorists mean by "quantum gravity" is the breakdown of effective-field-theory general relativity – and, it is usually assumed, the rest of the Standard Model of particle physics – around the Planck length, at $\sim 10^{-34}$ m. A quantum theory of gravity, in this sense, would be a genuinely finite theory from which particle physics, and general relativity, would emerge as effective field theories in appropriate long-distance regimes. (It is generally assumed that this theory would also tame the formal infinities that occur due to singularities in classical general relativity.)

Unfortunately, the great insensitivity of an effective field theory to the physical details of its high-energy cutoff, and the sheer energy scale of that cutoff, makes it very hard to gain evidence about the details of that theory: it "has imprinted few traces on physics below the Planck energy" (Bousso, 2002, p. 2). There is not the least hope that particle accelerators will ever probe the Planck scale; so far, only early universe cosmology seems to hold out any hope of giving us observational access to quantum gravity. One reason why string theory, loop quantum gravity, and other would-be quantum gravity programs have paid so much attention to the thermodynamics of black holes is that black hole radiation can be understood (via different choices of foliation) *either* as an effective-field theory result occurring at tolerably low energies, and so reasonably well understood, *or* as a fully quantum-gravitational effect; as such, black holes are a highly non-trivial consistency check on a putative quantum theory of gravity, above and beyond that given by ordinary effective-field-theory methods.[8]

18.11   Philosophical Morals

Quantum field theory, as the language in which a huge part of modern physics is written, is a natural setting for all manner of detailed questions in philosophy and foundations of physics, from the search for relativistic versions of dynamical-collapse and hidden-variable theories[9] to the correct understanding of the gauge principle. But in this last section I want to draw a more general moral. Contemporary philosophy of physics is for the most part focused on the so-called *fundamental* physics: that is, on those parts of physics which describe the world in full detail and at every scale, not simply in some emergent, approximate way in some regimes.[10] We currently have no fully worked-through fundamental physical theory. Indeed, we never have: there has never been a time when physicists had plausible ground to believe that they possessed any such theory. (Perhaps the closest point was at the turn of the 20th century, after the development of electromagnetism and thermodynamics, and before the twin revolutions of relativity and quantum theory.) So in practice philosophy of physics proceeds by taking a theory like classical or quantum particle mechanics, or classical general relativity, and studying it under the fiction that it is fundamental. It is tempting to imagine studying QFT on that basis too.

But quantum field theory is not that kind of theory. For all that our best and deepest physics is cast in its framework, it is *by its own nature* non-fundamental. An effective field theory – and, recall, all empirically successful quantum field theories are effective field theories – is defined through the methods of cutoff and renormalization, and does not even purport to fully describe the world. It is a remarkable irony that the Standard Model at one and the same time is the nearest we have ever come to a Theory of Everything, and is uninterpretable even in fiction as an exact description of the world. It is further irony that the theory itself tells us that it is compatible with an indefinitely large range of ways in which the deeper level physics might be specified. The fact that such a theory is (most physicists assume) quantum-mechanical allows us to say *something* about it, but quantum mechanics, like classical mechanics, is a framework theory and all manner of different theories fit within it.

Quantum field theory is a reminder to philosophers that physics, like other sciences, is hardly ever in the business of formulating theories that purport to describe the world on all scales. Those who wish to learn ontology from our best science, in the era of effective field theories, have two choices: recognize that deep and interesting metaphysical questions come up at all length scales in physics and are not confined to the "fundamental," or remain silent and wait, and hope, for a truly fundamental theory in the physics that is to come.

18.12   Further Reading

There are many textbooks on quantum field theory. Probably the best conceptually focused book-length account is Duncan (2012); other books that I have found helpful include Zee (2003), Banks (2008) (insightful but very terse), Peskin and Schroeder (1995) (the standard graduate-level textbook), and Weinberg (1995a) (not recommended as a first introduction). But tastes vary; get hold of several and see what suits your learning style. Coleman (1985) is not a textbook, exactly, but is highly insightful on a number of conceptual issues in QFT. All of these textbooks focus primarily on particle-physics applications of QFT.

For discussions of QFT more focused on solid-state physics, see Abrikosov et al. (1963) (old but classic) or Altland and Simons (2010) (much more up to date). For a more detailed account of the Standard Model, see Cheng and Li (1984) or Donoghue et al. (2014).

The methods of the renormalization group extend beyond QFT as understood as a quantum theory of the continuum and also have deep significance in classical statistical mechanics; for a very clear presentation of the renormalization group in this context, see Binney et al. (1992). (One word of caution here: do not confuse the *formal* analogy between classical statistical mechanics and QFT, with the *physical* similarities between QFT as applied to condensed matter systems and to particle physics.)

My favorite reference on quantum field theory in *curved* spacetime is Jacobson (2005); Wald (1994) is also excellent. Wald also provides an introduction to the broader issues of black hole thermodynamics, albeit now a little out of date; for more recent reviews from various perspectives, see Harlow (2016), Hartman (2015), and Grumiller et al. (2015). For philosophical considerations (going rather beyond QFT in each case), see Belot et al. (1999) and Wallace (2017a; 2017b).

For a philosophical defense (as opposed to simply an exposition, as here) of "mainstream" effective-field-theory QFT against the more rigorous, but empirically less successful, approach of trying to define continuum quantum mechanics exactly (nowadays usually called "algebraic quantum field theory," or "AQFT"), see Wallace (2006; 2011); for a response from the AQFT perspective, see Fraser (2009; 2011) (see also Baker (2016) for observations on the debate). Ruetsche (2011) provides a general introduction to AQFT methods and a route into the broader literature on philosophy of AQFT; other (more advanced) introductions are Haag (1996) and Halvorson (2007).

The question of particles in QFT has been extensively discussed in the philosophy literature (albeit mostly disjoint from the "emergent" attitude to particles I advocated in Section 18.3; see Fraser, this volume, and references therein).

Acknowledgments

Thanks to Doreen Fraser and Alastair Wilson for detailed and helpful comments on earlier drafts of this chapter.

Notes

1 See North, (this volume) for discussion of these formulations of classical mechanics.
2 See, e. g., Kenyon (1960, p. 3) or Kambe (2007, pp. 2–3).
3 See Wallace (2018a) for further discussion of this point.
4 See Chen (this volume) for discussion of the Bell inequality.
5 See Wallace (2012, ch. 2) and Rosaler (2015) for further discussion of this notion of instantiation.
6 For presentations, see e. g. Weinberg (1995b), Zee (2003), or Banks (2008).
7 See Wallace (2014) for further discussion.
8 See Wallace (2017a; 2017b) and references therein.
9 See Lewis (this volume) and Tumulka (this volume) for further discussion.
10 For more on this point, see French (this volume) and Wallace (2018b).

References

Abrikosov, A.A., Gorkov, L.P. and Dzyalohinski, I.E. (1963). *Methods of Quantum Field Theory in Statistical Physics*. New York: Dover. Revised English edition; translated and edited by R. A. Silverman.

Altland, A. and Simons, B.D. (2010). *Condensed Matter Field Theory* (2nd edition). Cambridge: Cambridge University Press.

Baker, D. (2016). The philosophy of quantum field theory. *Oxford Handbooks Online*. DOI: 10.1093/oxfordhb/9780199935314.013.33.

Baker, D. and Halvorson, H. (2009). Antimatter. *British Journal for the Philosophy of Science*, 60: 1–29.

Banks, T. (2008). *Modern Quantum Field Theory: A Concise Introduction*. Cambridge: Cambridge University Press.

Bargmann, V. (1954). On unitary ray representations of continuous groups. *Annals of Mathematics*, 59: 1–46.

Bekaert, X., Boulanger, N. and Sundell, P.A. (2012). How higher-spin gravity surpasses the spin-two barrier. *Reviews of Modern Physics*, 84: 987.

Belot, G., Earman, J. and Ruetsche, L. (1999). The Hawking information loss paradox: The anatomy of a controversy. *British Journal for the Philosophy of Science*, 50: 189–229.

Binney, J.J., Dowrick, N.J., Fisher, A.J. and Newman, M.E.J. (1992). *The Theory of Critical Phenomena: An Introduction to the Renormalisation Group*. Oxford: Oxford University Press.

Bousso, R. (2002). The holographic principle. *Reviews of Modern Physics*, 74: 825.

Cheng, T.-P. and L.-F. Li (1984). *Gauge Theory of Elementary Particle Physics*. Oxford: Oxford University Press.

Coleman, S. (1985). *Aspects of Symmetry: selected Erice Lectures*. Cambridge: Cambridge University Press.

Donoghue, J.F., Golowich, E. and Holstein, B.R. (2014). *Dynamics of the Standard Model* (2nd edition). Cambridge: Cambridge University Press.

Duncan, T. (2012). *The Conceptual Framework of Quantum Field Theory*. Oxford: Oxford University Press.

Earman, J. (2004). Laws, symmetry, and symmetry breaking: Invariance, conservation principles, and objectivity. *Philosophy of Science*, 71: 1227–1241.

Fraser, D. (2009). Quantum field theory: Underdetermination, inconsistency, and idealization. *Philosophy of Science*, 76: 536–567. Forthcoming in *Philosophy of Science*. Available at: philsci-archive.pitt.edu.

Fraser, D. (2011). How to take particle physics seriously: A further defence of axiomatic quantum field theory. *Studies in the History and Philosophy of Modern Physics*, 42: 126–135.

Friederich, S. (2013). Gauge symmetry breaking in gauge theories — in search of clarification. *European Journal for Philosophy of Science*, 3: 157–182.

Grumiller, D., McNees, R. and Salzer, J. (2015). Black holes and thermodynamics: The first half century. In X. Calmet (ed.), *Quantum Aspects of Black Holes: Fundamental Theories of Physics 178*. Switzerland: Springer, pp. 27–70.

Haag, R. (1996). *Local Quantum Theory: Fields, Particles, Algebras*. Berlin: Springer-Verlag.

Halvorson, H. (2007). Algebraic quantum field theory. In J. Butterfield and J. Earman (eds.), *Handbook of the Philosophy of Science: Philosophy of Physics, Part A*. Boston, MA: Elsevier, 731–864.

Harlow, D. (2016). Jerusalem lectures on black holes and quantum information. *Reviews of Modern Physics*, 88: 015002.

Hartman, T. (2015). *Lectures on quantum gravity and black holes*. Available at: http://www.hartmanhep.net/topics2015/gravity-lectures.pdf.

Jacobson, T. (2005). Introduction to quantum fields in curved spacetime and the Hawking effect. In A. Gomberoff and D. Marolf (eds.), *Lectures on Quantum Gravity*. New York: Springer, pp. 39–89. Available online at https://arxiv.org/abs/gr-qc/0308048.

Kambe, T. (2007). *Elementary Fluid Mechanics*. Singapore: World Scientific.

Kenyon, R.A. (1960). *Principles of Fluid Mechanics*. New York: The Ronald Press Company.

Peskin, M.E. and D.V. Schroeder (1995). *An introduction to Quantum Field Theory*. Reading, MA: Addison-Wesley.

Rosaler, J. (2015). "Formal" versus "empirical" approaches to quantum-classical reduction. *Topoi*, 34: 325–328.

Ruetsche, L. (1998). How close is 'close enough'? In D. Dieks and P.E. Vermaas (eds.), *The Modal Interpretation of Quantum Mechanics*. Dordrecht: Kluwer Academic Publishers, pp. 223–240.

Ruetsche, L. (2011). *Interpreting Quantum Theories*. Oxford: Oxford University Press.

Saunders, S. (1992). Locality, complex numbers and relativistic quantum theory. *Philosophy of Science Association*, 1: 365–380.

Struyve (2011). Gauge-invariant accounts of the Higgs mechanism. *Studies in the History and Philosophy of Modern Physics*, 42: 226–236.

Wald, R.M. (1994). *Quantum Field Theory in Curved Spacetime and Black Hole Thermodynamics*. Chicago: University of Chicago Press.

Wallace, D. (2006). In defence of naiveté: The conceptual status of Lagrangian quantum field theory. *Synthese*, 151: 33–80.

Wallace, D. (2009). QFT, antimatter, and symmetry. *Studies in the History and Philosophy of Modern Physics*, 40: 209–222.

Wallace, D. (2011). Taking particle physics seriously: a critique of the algebraic approach to quantum field theory. *Studies in the History and Philosophy of Modern Physics*, 42: 116–125.

Wallace, D. (2012). *The Emergent Multiverse: Quantum Theory according to the Everett Interpretation*. Oxford: Oxford University Press.

Wallace, D. (2014). *Deflating the Aharonov-Bohm effect*. Forthcoming; available online at https://arxiv.org/abs/1407.5073.

Wallace, D. (2016). *Interpreting the quantum mechanics of cosmology*. Forthcoming in "Guide to the Philosophy of Cosmology", Ijjas and Loewer (ed.).

Wallace, D. (2017a). *The case for black hole thermodynamics, part I: phenomenological Thermodynamics*. Available at: https://arxiv.org/abs/1710.02724.

Wallace, D. (2017b). *The case for black hole thermodynamics, part II: statistical Mechanics*. Available at: https://arxiv.org/abs/1710.02725.

Wallace, D. (2018a). Lessons from realistic physics for the metaphysics of quantum theory. *Synthse* 197: 4303–4318.

Wallace, D. (2018b). On the plurality of quantum theories: Quantum theory as a framework, and its implications for the quantum measurement problem. In S. French and J. Saatsi (eds.), *Scientific Metaphysics and the Quantum*. Oxford: Oxford University Press, forthcoming.

Weinberg, S. (1995a). *The Quantum Theory of Fields, Volume I: Foundations*. Cambridge: Cambridge University Press.

Weinberg, S. (1995b). *The Quantum Theory of Fields, Volume II: Modern Applications*. Cambridge: Cambridge University Press.

Weinberg, S. (2008). *Cosmology*. Oxford: Oxford University Press.

Wigner, E. (1939). On unitary representations of the inhomogenous Lorentz group. *Annals of Mathematics*, 40: 149.

Zee, A. (2003). *Quantum Field Theory in a Nutshell*. Princeton, NJ: Princeton University Press.

19

RENORMALIZATION GROUP METHODS

Porter Williams

It is a truism – in physics if not in philosophy – that in order to study physical behavior at a particular scale, one is best served by using degrees of freedom defined at, or suitably near, that scale. If one wants to study the classical behavior of a simple fluid on relatively long length scales ($\gg 10^{-10}$ m), for example, the appropriate degrees of freedom to use to describe the fluid are two continuous fields, the mass density $\rho(x)$ and the velocity field $\mathbf{v}(x)$. This despite the fact that the fluid itself is constituted by a large collection of discrete, atomic constituents, which in turn have their own subatomic structure. Remarkably, one can do a tremendous amount of empirically successful fluid dynamics ignoring almost entirely the more "fundamental," discrete description of the fluid and modeling its long-distance behavior using "effective" continuum degrees of freedom.

Even in purportedly fundamental physics the truism holds. For example, consider quantum chromodynamics, the theory that describes the physics of quarks and gluons interacting via the strong force. Although hadrons, such as protons and neutrons and pions, interact via the strong force and are, in a rather complicated sense, "composed" of quarks and gluons, a considerable amount of low-energy nuclear physics proceeds perfectly well by modeling nuclear processes in ways that *ignore* those "fundamental" quark and gluon degrees of freedom entirely, modeling physical processes in terms of "effective" hadronic degrees of freedom. If one wants to model the low-energy scattering of pions off of a fixed nucleon target, for example, one is better off eschewing any attempt to describe the scattering in terms of the "fundamental" quark and gluon degrees of freedom in favor of an "effective" description in terms of pions and nucleons. However, if one begins to scatter the pions off the nucleon target at energies sufficiently high that the pions begin to scatter inelastically (roughly, to probe the internal structure of the nucleon), then the theory involving only hadronic degrees of freedom becomes empirically and explanatorily inadequate. At those higher scattering energies, a set of degrees of freedom that include the quark and gluon fields becomes appropriate for modeling physics at that new energy scale.[1]

Theories that take into account only the degrees of freedom that are necessary for characterizing physical processes at some particular scale, but which (i) break down when pushed to scales beyond their limited domain of applicability and (ii) incorporate this inevitable breakdown into their mathematical framework, are called *effective theories*. Wallace (this volume, Chapter 18) has given a clarifying description of how this approach to physical theorizing plays out in quantum field theory (QFT), and the reader is strongly encouraged to read Wallace's chapter in conjunction with this one. The basic requirement that must be satisfied for the modeling strategy that underlies the effective theory to be successful might be called the "autonomy of scales": physical processes at long distances must exhibit minimal dependence on the shorter distance physics that the effective theory's description leaves out.[2]

In particle physics and condensed matter physics, the main set of tools for determining how physical processes at longer scales depend on short-distance physics – and thus for determining which degrees of freedom are optimal for characterizing physical processes at a particular scale – are renormalization group (RG) methods. As Steven Weinberg puts it,

> I think that this in the end is what the renormalization group is all about. It's a way of satisfying the Third Law of Progress in Theoretical Physics, which is that *you may use any degrees of freedom you like to describe a physical system, but if you use the wrong ones, you'll be sorry*. (Weinberg, 1983, p. 16)

Despite this seemingly sensible motivation, the process of renormalization in QFT has historically been treated as an ill-founded but necessary evil: a collection of distasteful tricks of dubious mathematical and physical standing, employed by practicing particle physicists in order to extract empirical predictions from the theory. This was the predominant attitude in the particle physics community from the 1930s through much of the 1960s, and this attitude contributed significantly to a widespread distrust in the entire framework of quantum field theory that took hold shortly after World War II and held sway until the early 1970s.

(One finds this description scattered throughout particle physicists' descriptions of the period; already in 1965, Kenneth Wilson lamented that "The Hamiltonian formulation of quantum mechanics has been essentially abandoned in investigations of the interactions of π mesons, nucleons, and strange particles" in favor of approaches using dispersion relations (Wilson, 1965, p. 445), while Steven Weinberg refers to the "the general distrust of quantum field theory that set in soon after the brilliant successes of quantum electrodynamics in the late 1940s" (Weinberg, 1983, p. 7), and David Gross begins his description of particle physics in the 1960s by recalling that "Field theory was in disgrace; S-matrix theory was in full bloom" (Gross, 2004, p.16). Cushing (1990) is a book-length treatment of the history of S-matrix theory, the main competitor framework to QFT in the 1960s.)

A major shift in the particle physics community's attitude toward QFT coincided with the invention and refinement of RG methods by Murray Gell-Mann and Francis Low, Leo Kadanoff, Michael Fisher, Kenneth Wilson, and others from the mid-1950s through the early 1970s. The RG was widely viewed as having put the earlier renormalization methods on secure physical foundations. It also provided a new set of mathematical tools that proved to be indispensable both to practicing particle physicists, who were engaged in extracting empirical predictions from particular quantum field-theoretic models, and to mathematical physicists, whose aim was to put the theoretical framework itself on secure mathematical footing. The development of the RG ultimately ushered in a transformation of the way physicists understand the conceptual foundations of quantum field theory itself: once thought to be a candidate fundamental theory, QFT is now widely understood to be just another effective theory.

(There is, however, an active research program aimed at determining whether a QFT describing gravitation may be well defined at arbitrarily high energies; see Niedermaier and Reuter (2006) for a review. If the asymptotic safety program is successful, that opens the door to the possibility that QFT may be a fundamental framework after all.)

This chapter proceeds as follows. I begin with a brief introduction to path integral methods, since the path integral formulation of quantum field theory is a natural home for examining the generality of RG methods. I then introduce in some detail the machinery of early renormalization theory before turning to RG methods; my aim is to illustrate the sense in which RG methods changed the way that we think about renormalization in QFT. I then turn to a major conceptual shift in the way that we think about QFT, brought on by the development of RG methods. The development of the RG has led naturally to a perspective in which individual quantum field theories live in a vast "space" of quantum field theories, and that one of the tasks of the theoretical physicist is to provide a map of theory space. I will conclude by considering the benefits that thinking in terms of theory space has for the task of defining and classifying QFTs.

Finally, a note on conventions: I use so-called "natural units" in which $\hbar = c = 1$. This means that energy and mass have the same dimensions, and length and time have dimensions of mass^{-1}. This makes possible two points that will prove useful going forward: we can characterize all physical quantities as having a positive or negative dimension of mass, and we can identify high energy scales with short distance scales. I will also use a "mostly minus" signature for the Minkowski metric, so that $\eta_{\mu\nu}x^{\mu}x^{\nu} = (x^0)^2 - (x_i x^i)$ and e.g. $(\partial_{\mu}\phi)^2 = (\partial_t\phi)^2 - (\nabla\phi)^2$.

19.1   Path Integrals

The path integral representation of a QFT provides a particularly natural home for RG methods. Although path integrals in QFT are notoriously plagued by mathematical difficulties, they are perfectly well-defined mathematical objects in the *nonrelativistic* quantum mechanics of finitely many particles. Since the physical ideas are largely the same in both cases, I will introduce path integrals in the latter, mathematically well-behaved setting before extending the discussion to QFT.

I will borrow a lovely example from Richard Feynman to motivate the notion of the path integral (Feynman and Hibbs , 1965/2010). Consider a double-slit experiment, in which one has (i) a source emitting individual particles, (ii) a detection screen, and (iii) a wall, between the source and the detector, with two small holes labeled 1 and 2. The aim is to calculate the probability that the particle will be detected at some particular point on the screen. The particle must go through either hole 1 or hole 2 in order to reach the screen, and so the total amplitude for the particle being detected at any particular point x on the screen is given by the sum of two amplitudes: one for the particle to pass through hole 1 and arrive at x and another for the particle to pass through hole 2 and arrive at x:

$$\psi_{12}(x) = \frac{1}{\sqrt{2}}(\psi_1(x) + \psi_2(x))$$

with the probability that the particle is detected at x given by $|\psi_{12}(x)|^2$.

Now suppose that between our source and our original wall we put two more walls, which we label A and B, and drill several holes in each of them. Now there are a number of paths that the particle can take from the source to the detection screen: it could go from the source to A_1, then to B_3, and then through hole 2; it could go from the source to A_5, then to B_2, and then through hole 1; and so on. Each of these paths has its own amplitude, and the total amplitude for the particle being detected at point x on the detection screen is the sum of all of them.

Now imagine inserting an increasing number of walls between the source and the detection screen, and drilling holes in each of the walls until there is nothing left. Eventually, there will be infinitely many walls, each with infinitely many holes, and thus infinitely many possible paths that will take the particle from the source to point x on the detection screen. In order to determine the total amplitude for the particle being detected at x, we will have to sum over the amplitudes for *all* of them.

This idea of summing the amplitudes for each of the possible ways that a physical system in one state could transition into another state is what underlies the path integral formalism. Though we motivated the idea of the path integral as a method for determining the probability for detecting a particle emitted from a source at some particular point x on a detection screen, the path integral formalism is more general: it gives an efficient means for calculating the probability that a physical system in a state $|\alpha\rangle$ at time $t = t_0$ will transition into a state $|\beta\rangle$ at some later time $t = T$.

More formally, suppose we would like to calculate the amplitude that a system in state $|\alpha\rangle$ will transition into $|\beta\rangle$ between $t = 0$ and $t = T$, where the dynamics governing the time evolution is contained in the Hamiltonian $\hat{H} = \frac{\hat{p}^2}{2m} + V(\hat{x})$:

$$\langle \beta, T | e^{-i\hat{H}T} | \alpha, 0 \rangle$$

We want to sum over all of the "paths" that the system could take from $|\alpha\rangle$ at $t = 0$ to $|\beta\rangle$ at $t = T$. We can do this in three steps:

1   Chop up the time interval $[0, T]$ into N equal time steps dt, with $\frac{T}{N} = dt$. This allows us to rewrite the time-evolution operator $e^{-i\hat{H}t}$ in the form $e^{-i\hat{H}t} = \underbrace{e^{-i\hat{H}dt} \times e^{-i\hat{H}dt} \times \cdots \times e^{-i\hat{H}dt}}_{N \text{ times}} =$

$$\prod^{N} e^{-i\hat{H}dt}$$

2   After each time step, insert a resolution of the identity operator $\mathbb{I} = \sum^{n} |\gamma_n\rangle\langle\gamma_n|$ in some appropriate basis. For example, in our motivating example above one would insert resolutions of the identity in the position basis $\mathbb{I} = \int dx_i \, |x_i\rangle\langle x_i|$. In the latter case, each insertion of the identity plays the role of inserting a new "wall" in which infinitely many holes have been drilled and through which the particle must pass after every dt. This allows us to write $\langle \beta, T | e^{-i\hat{H}T} | \alpha, 0 \rangle$ in the form

$$\prod^{N}_{i=0}(\int dx_i) \, \langle \beta | e^{-i\hat{H}dt} | x_N \rangle \times \langle x_N | e^{-i\hat{H}dt} | x_{N-1} \rangle \times \cdots \times \langle x_2 | e^{-i\hat{H}dt} | x_1 \rangle \times \langle x_1 | e^{-i\hat{H}dt} | \alpha \rangle$$

3   Finally, let the time steps dt become arbitrarily small by allowing $N \to \infty$. What one finds (though it should not be obvious from what we've done here; see e.g. (Zee, 2010, §I.2)) is that

$$\langle \beta, T | e^{-i\hat{H}T} | \alpha, 0 \rangle = \int Dx(t) e^{i \int_0^T dt \int d^3x [\frac{1}{2} m\dot{x}^2 - V(x)]},$$

where the "measure" is given by $Dx(t) = \lim_{N \to \infty} \prod^{N}_{i=0} dx_i$. The quantity $\mathcal{L}[x(t), \dot{x}(t)] = \frac{1}{2}m\dot{x}^2 - V(x)$ is called the *Lagrangian density*, and the integral of the Lagrangian density $\mathcal{S} = \int_0^T dt \int d^3x \mathcal{L}$ is called the *action*.

So far we have been dealing with a nonrelativistic quantum system, but all of this carries over directly to quantum field theory. (Modulo serious issues with its mathematical sensibility; see Glimm and Jaffe (1987). Most notably, the "measure" $Dx(t)$ typically does not exist in realistic models of QFT, leaving the path integral ill-defined.) One modification of presentation is that in the field-theoretic case, the Lagrangian density (more precisely, the action) takes the center stage: one typically specifies a quantum field theory by writing down a classical action and then "quantizing" it through one or another method of quantization. If one has in hand the action for a physical system then one can obtain the classical equations of motion for that system as the solution to the Euler-Lagrange equation; for this reason one often says that the action contains all the dynamical information about the physical system.

To connect this with QFT, consider a theory of a real scalar field with mass m and a self-interaction given by a ϕ^4 term:

$$\mathcal{L} = \frac{1}{2}\partial_\mu\phi(x)\partial^\mu\phi(x) - \frac{1}{2}m^2 - \frac{\lambda}{4!}\phi(x)^4$$

Using the equation derived above, we can use the action to calculate the amplitude that a quantum field will transition from one state $|\phi_0(x)\rangle$ to another state $|\phi_1(x)\rangle$ over a particular temporal interval:

$$\langle \phi_1(x) | e^{-i\hat{H}t} | \phi_0(x) \rangle = \int D\phi e^{i \int_0^T d^4x \mathcal{L}},$$

where $\int_0^T d^4x = \int_0^T dt \int d^3x$ is an integration over all of space and the temporal interval $[0, T]$, and the field $\phi(x)$ over whose histories we are integrating is constrained to take on the values $\phi(x) = \phi_0(x)$ at $t = 0$ and $\phi(x) = \phi_1(x)$ at $t = T$. The essential idea developed in the nonrelativistic case holds equally well here, although at a somewhat more abstract level: instead of possible "paths" of a quantum particle between two possible states of that particle, we are now integrating over all possible histories of states of a quantum field that would carry it between the states $|\phi_0(x)\rangle$ and $|\phi_1(x)\rangle$.

From a pragmatic perspective, the point of developing this machinery is to enable us to compute the amplitudes for transitions between different field states. Typically, one is interested in computing the amplitude that a specified state $|\phi_{in}\rangle$ of the field at $t = -\infty$ will transition into some other specified state $|\phi_{out}\rangle$ at $t = +\infty$, given a particular dynamics specifying how the excitations of the fields – roughly, particles – will interact. This information is contained in mathematical objects called correlation functions (equivalently, n-point functions). These correlation functions are of physical interest because they can be turned into amplitudes for the outcomes of scattering processes (S-matrix elements) via the Lehmann-Symanzik-Zimmerman (LSZ) reduction formula. The fact that the main situation of interest involves transitions between asymptotic states of the field at over the interval $[-\infty, +\infty]$ means that the path integral is typically an integration over all of space *and* time. We will therefore write the path integral as

$$\langle \phi_1(x) | e^{-i\hat{H}t} | \phi_0(x) \rangle = \int D\phi e^{i \int d^4x \mathcal{L}},$$

where the temporal interval over which we integrate is now $[-\infty, +\infty]$.

The path integral above describes the amplitude for a quantum field to transition between two generic states $|\phi_1(x)\rangle$ and $|\phi_0(x)\rangle$. Typically, in quantum field theory one is interested in $|\phi_1(x)\rangle = |\phi_0(x)\rangle = |\Omega\rangle$; speaking loosely of particles (the notion of a "particle" in QFT is subtle; see Fraser (this volume)), this gives the amplitude that the field begins in the vacuum state, out of which the field is "excited" in some spacetime region to create a particle that propagates for some period of time, and then the particle annihilates at some other spacetime region to return the field to its vacuum state. Thus, one is usually interested in correlation functions of the form $\langle \Omega | \phi(x_1)\phi(x_2) \cdots \phi(x_n) | \Omega \rangle$ which capture the correlation between the values of the field ϕ at distinct spacetime points. (These are called *vacuum expectation values*, or VEVs. They are just correlation functions that describe correlations of field values in one particular state of the field.) The simplest correlation function in quantum field theory is the *two-point function* $\langle \Omega | \phi(x_1)\phi(x_2) | \Omega \rangle$ which is called the *propagator*. Again speaking loosely, this gives the amplitude for an excitation of ϕ at x_1 to propagate to x_2 before annihilating and returning the field to the vacuum state $|\Omega\rangle$.

With these facts about correlation functions in hand, we are finally in a position to write down the path integral representation of a generic n-point function:

$$\langle \Omega | \phi(x_1)\phi(x_2) \cdots \phi(x_n) | \Omega \rangle = \mathcal{N} \int D\phi \, \phi(x_1)\phi(x_2) \cdots \phi(x_n) e^{i \int d^4x \mathcal{L}}$$

where $\mathcal{N}$ is a normalization factor. The obvious thing to do at this point is to try to *compute* one of these correlation functions. This will lead us naturally to the need for renormalization and, eventually, for RG methods.

19.2   Renormalization

One might naively think that computing correlation functions should be straightforward: just do the appropriate path integral! It turns out that things are not so simple, and that in physically interesting models – such as models describing quantum fields in four-dimensional spacetimes and in which the field excitations are allowed to interact – one is almost always forced to resort to various approximation methods. (Many of the difficulties one encounters can be traced back to the fact mentioned in note 4, namely that the "measure" in the path integral is typically ill-defined for realistic models of QFT.)

The most common of these is perturbation theory. When doing perturbation theory, one splits the path integral up into two parts: an integral over the "free" part of the dynamics $\mathcal{L}_{\text{free}} = \frac{1}{2}(\partial\phi)^2$ and an integral over the "interaction" part $\mathcal{L}_{\text{int}} = -\frac{1}{2}m^2\phi^2 - \frac{\lambda}{4!}\phi^4$. The free part of the dynamics is trivial, and one can do the path integral directly. The interacting part of the theory is where the difficulties arise. We cannot do the path integral over $\mathcal{L}_{\text{int}}$ directly, but one profitable approximation method is to treat $\mathcal{L}_{\text{int}}$ as a small perturbation of $\mathcal{L}_{\text{free}}$; this amounts to assuming that the *coupling* λ which determines how strongly the field excitations interact is sufficiently small that we can treat the full dynamics of the theory $\mathcal{L} = \mathcal{L}_{\text{free}} + \mathcal{L}_{\text{int}}$ as "close" to the theory devoid of interactions described by $\mathcal{L}_{\text{free}}$, and expand the path integral over $\mathcal{L}_{\text{int}}$ as a series approximation.

Slightly more explicitly, when doing perturbation theory one attempts to calculate (for example) the four-point function $G^{(4)}$ (speaking loosely, the amplitude for two excitations of the scalar field in some initial state to interact and scatter into another pair of excitations in some other final state) as follows:

$$\langle \Omega | \phi(x_1)\phi(x_2)\phi(x_3)\phi(x_4) | \Omega \rangle$$

$$= \mathcal{N} \int D\phi \, \phi(x_1)\phi(x_2)\phi(x_3)\phi(x_4) e^{i \int d^4x \mathcal{L}} \approx \sum^n \lambda^n \mathcal{I}_n,$$

where the $\mathcal{I}_n$ are integrals over the momenta of the particles, and since λ was assumed to be small at the outset, one can reasonably expect that successive terms in the series will become smaller. (In standard textbook presentations, one would use Feynman diagrams to keep track of the integrals $\mathcal{I}_n$ at each order of perturbation theory.) Consider the structure of the integrals $\mathcal{I}_n$ appearing in our perturbative calculation of $G^{(4)}$. It is well known that many of the integrals appearing in the series expansion $\sum^n \lambda^n \mathcal{I}_n$ are divergent, and that these divergences arise from including field excitations of arbitrarily high momenta in our calculations. We will begin with a schematic discussion of these divergences and their elimination, and then go through a concrete example.

The process of eliminating the divergences can be split into two steps: *regularization* and *renormalization*. One first adopts a method of *regularizing* the divergent integrals. The central idea of a regularization method is the following. Given a divergent integral $\mathcal{I}$ in the perturbative expansion of the path integral, one makes the integral a function of new parameter θ – the regulator – so that the integral now becomes $\mathcal{I}(\theta)$. In order to count as a successful regulator, there are two constraints that θ must satisfy: (i) finite values of the regulator θ must render $\mathcal{I}(\theta)$ finite, and (ii) if one were to remove the regulator by taking $\theta \to \infty$, the result must be the original divergent integral $\mathcal{I}$. A third, weaker constraint is that it is desirable to choose a regularization method that leaves the mathematical structure of the theory as unchanged as possible; we will see below that *some* modification of that structure is inevitable, and which aspects of that structure one is willing to modify will often depend on the physical processes one is trying to understand.

For example, one historically important regularization method – the so-called Pauli-Villars method – involves introducing new fictitious "ghost" particles of mass θ into the Lagrangian density. Pauli-Villars regularization does render divergent integrals finite while allowing one to recover the original divergent integral if the mass of the fictitious particles $\theta \to \infty$, and thus satisfies the two constraints. However, the method also requires the fictitious particles to violate the spin-statistics theorem, introduces a violation of the assumed unitarity of the time-evolution operator in quantum theory, and typically breaks any gauge symmetries the theory might possess. From the perspective of the third desirable constraint, it is thus a less than ideal regularization method. Another very useful regularization method involves replacing Minkowski spacetime with a Euclidean hypercubic lattice with lattice spacing $\frac{1}{\theta}$. This satisfies the two constraints by restricting integrals $\mathcal{I}(\theta)$ to include only momenta k such that $-\frac{\pi}{\theta} \leq k \leq \frac{\pi}{\theta}$, and sending the lattice spacing to zero then returns the originally divergent integral. It also preserves any gauge invariance the theory may have, but violates Poincaré invariance. One typically has to make such choices about which parts of the mathematical structure of the theory can be altered when choosing a regularization method. Ideally, the original mathematical structure of the theory will be restored when the regulator is eliminated, but in many cases this is a subtle question.

(The Nielsen–Ninomiya theorem, for example, captures a sense in which a lattice regulation introduces mathematical artifacts that cannot be eliminated by naively letting the lattice spacing $\frac{1}{\theta} \to 0$.)

The second step is renormalization. The renormalization procedure identifies the part of $\mathcal{I}(\theta)$ which causes the divergence as $\theta \to \infty$, then shifts the couplings in the action in such a way as to eliminate the divergent pieces of $\mathcal{I}(\theta)$. The modified action is called the *renormalized* action, while the original action is called *bare* and the shifts of the couplings are called *counterterms*. Schematically, the couplings are shifted from $\lambda_0 \to \lambda_R = \lambda_0 - \delta\lambda$, where $\delta\lambda$ is a counterterm chosen to cancel a divergent contribution the integrals $\mathcal{I}(\theta)$ appearing in the perturbative calculation of $G^{(4)}$. If one is lucky, then shifting finitely many couplings will eliminate *all* of the divergences that might appear in the perturbative calculation of *all* of the n-point functions of the theory (not only for the calculation of $G^{(4)}$!); in that case one says the theory is *renormalizable*.

Once one has shifted the couplings in the action to their renormalized values, one can then safely remove the θ-dependence of the integrals $\mathcal{I}_n$ by letting $\theta \to \infty$ *without* reintroducing the original divergence. After renormalization, all of the resulting integrals appearing in any perturbative calculation of any n-point function will be finite and independent of the regulator (assuming that the theory is renormalizable, of course). While the values of the masses and couplings in the bare Lagrangian are arbitrarily specifiable parameters, the values of the couplings in the renormalized Lagrangian are the *physical* values of those couplings and must be extracted from the experiment. Once those couplings have been measured at one scale for one physical process (e.g., $2 \to 2$ particle scattering at the scale $\mu \sim 100$ GeV), they can be used to calculate amplitudes for any other physical process at any other scale.

This two-step procedure of regularization and renormalization afforded a classification scheme for quantum field theories: if one can eliminate all of the divergences appearing in $\sum^n \lambda^n \mathcal{I}_n$ by making *finitely* many shifts of the bare couplings in the action, then the theory is *perturbatively renormalizable*. The theory is *perturbatively nonrenormalizable* if each higher order term in $\sum^n \lambda^n \mathcal{I}_n$ contains a divergence with a novel structure that, to be eliminated, would require a corresponding novel interaction to be added to the bare Lagrangian. Nonrenormalizable theories require an action containing infinitely many interactions to cancel all of the divergences in perturbative calculations of quantities of empirical interest, such as scattering amplitudes. It was long thought that nonrenormalizable QFTs were physically nonsensical, and the requirement that any physically meaningful QFT be perturbatively renormalizable was a guiding principle in particle physics through the 1970s.

In order to make this concrete, consider the simple ϕ^4 theory and imagine computing a particular scattering amplitude: the four-point function $G^{(4)}$ representing (again, loosely) the amplitude for two excitations of a scalar field to scatter off one another. The integrals $\mathcal{I}_n$ appearing in our series expansion will be integrals over allowed momenta of field excitations, and will generically take the form

$$\mathcal{I}_n \propto \lambda^n \int^{\infty} \frac{d^4k}{(2\pi)^4} \frac{i}{f(k^2)}$$

Consider only the first two terms of the series expansion of $G^{(4)} = \lambda \mathcal{I}_1 + \lambda^2 \mathcal{I}_2 + \mathcal{O}(\lambda^3)$. The first term in this series (the so-called "tree-level" term) is finite, and the integral is equal to $-i\lambda$. The second term – the so-called "one-loop" level – is in fact a sum of three integrals, each of which has the form

$$\mathcal{I}_2 \propto (-i\lambda)^2 \int^\infty \frac{d^4k}{(2\pi)4} \frac{i}{k^2 - m^2} \frac{i}{(C-k)^2 - m^2},$$

where the constant C represents one of three Mandelstam variables s, t, and u which are (roughly) a function of the square of the energy at which the particles are scattered. For sufficiently large values of the momentum k, the mass and Mandelstam variables can be neglected. In that high-momentum regime, the integrals $\mathcal{I}_2$ behave like $\int^\infty \frac{d^4k}{k^4}$, which diverges logarithmically (like $\lim_{k \to \infty} \ln(k)$) when one allows for the possibility that the field excitations might possess arbitrarily high momenta at some intermediate stage of the scattering process. Since the divergence comes from the high-momentum behavior of $G^{(4)}$, these are called *ultraviolet divergences*. It is also true that divergences arise from the arbitrarily *low*-momentum behavior of the integral. These are called infrared divergences; while they raise their own set of conceptual questions, I will neglect them in this chapter.

In order to render this integral mathematically sensible, one must first regularize it. The most physically transparent regularization simply ignores contributions of high-momenta field excitations to our scattering amplitude by sharply cutting off the integral at some large, but finite, value Λ. This makes $\mathcal{I}_2$ a function of Λ; explicitly, we obtain that $\mathcal{I}_2 = 2i \ln\left(\frac{\Lambda^2}{C^2}\right)$. Note that, as required above, this regularization reproduces the logarithmically divergent integral if we let $\Lambda \to \infty$. After regularization, the amplitude $G^{(4)}$ to one-loop reads:

$$G^{(4)} = -i\lambda + -i\lambda^2 \left[\ln\left(\frac{\Lambda^2}{s}\right) + \ln\left(\frac{\Lambda^2}{t}\right) + \ln\left(\frac{\Lambda^2}{u}\right) \right] + \mathcal{O}(\lambda^3),$$

where I have replaced the stand-in variable C with the Mandelstam variables for which it was a placeholder. What is important to note about this amplitude is that it now depends explicitly on the arbitrarily chosen value Λ for the cutoff. This should be alarming; what if one had chosen Λ^2? What about some $\Lambda' \ll \Lambda$? The n-point functions like $G^{(4)}$ contain all of the empirical content of QFT, and that content should not be a function of arbitrary choices of theorists. Ultimately, it will be RG methods that provide satisfying answers to these questions.

One natural way of eliminating the Λ-dependence of $G^{(4)}$ is to go out and extract from the experiment the value of the coupling λ by scattering particles at some particular energy scale corresponding to particular values of the Mandelstam variables s, t, and u. In fact, for simplicity one can set $s = t = u = \mu^2$ and imagine extracting the value of λ from a scattering experiment at the scale μ. Call this value of the coupling $\lambda_R(\mu)$ (R for "renormalized").

It turns out that one can replace the Λ-dependent coupling that we have been using – the bare coupling λ – with the measured, renormalized coupling $\lambda_R(\mu)$ in the calculation of the scattering amplitude $G^{(4)}$. What one finds is that

$$G^{(4)} = -i\lambda_R + i\lambda_R^2 \left[\ln\left(\frac{\mu^2}{s}\right) + \ln\left(\frac{\mu^2}{t}\right) + \ln\left(\frac{\mu^2}{u}\right) \right] + \mathcal{O}(\lambda_R^3)$$

The dependence on Λ has dropped out! One can safely take $\Lambda \to \infty$ without reintroducing divergences into our calculation, completing the second stage of the regularization and renormalization procedure sketched above.

However, the scattering amplitude now depends on the ratio between the energy scale at which the value of λ_R was measured and the energy scale at which we are scattering our particles, captured (roughly) by the Mandelstam variables s, t, and u. What's so special about μ? What would have happened if we chose to measure the value of the coupling at some other scale μ' and carried out our calculation with $\lambda_R(\mu')$ instead? A skeptic might think that while the introduction of $\lambda_R(\mu)$ eliminated the dependence of $G^{(4)}$ on any arbitrarily cutoff Λ, it replaced that with an equally arbitrary dependence on the scale μ at which we chose to measure the value of $\lambda_R(\mu)$. Attempting to answer this skeptic leads naturally to the renormalization group.

19.3   The Renormalization Group

Suppose that, unsettled by the skeptic, one wanted to see how the ϕ^4 theory with its coupling defined by its measured value at the scale μ – henceforth "the theory defined at μ" – relates to the theory defined at μ'. First consider the expression for the four-point correlation function in terms of $\lambda_R(\mu)$:

$$G^{(4)} = -i\lambda_R(\mu) + i\lambda_R^2(\mu)\left[\ln\left(\frac{\mu^2}{s}\right) + \ln\left(\frac{\mu^2}{t}\right) + \ln\left(\frac{\mu^2}{u}\right)\right] + \mathcal{O}(\lambda_R^3)$$

Suppose that one chose instead to use $\lambda_R(\mu')$, the value of the coupling measured at μ'. This amounts to making the simple replacement

$$G^{(4)} = -i\lambda_R(\mu') + i\lambda_R^2(\mu')\left[\ln\left(\frac{\mu'^2}{s}\right) + \ln\left(\frac{\mu'^2}{t}\right) + \ln\left(\frac{\mu'^2}{u}\right)\right] + \mathcal{O}(\lambda_R^3)$$

In order to see how these two quantities relate, subtract the latter from the former to express $\lambda_R(\mu')$ in terms of $\lambda_R(\mu)$:

$$\lambda_R(\mu') = \lambda_R(\mu) + 3\lambda_R(\mu)\ln\left(\frac{\mu'^2}{\mu^2}\right) + \mathcal{O}(\lambda_R^3)$$

For $\mu' = \mu - \mathrm{d}\mu$, this can be expressed as a differential equation for the way the coupling changes with scale:

$$\mu\frac{\mathrm{d}}{\mathrm{d}\mu}\lambda_R(\mu) = 6\lambda_R(\mu)^2 + \mathcal{O}(\lambda_R^3)$$

With a few lines of algebra, we have derived the *renormalization group equation* for the coupling $\lambda_R(\mu)$. The right-hand side of this question, often written $\beta(\lambda)$, is called the *beta function* for the coupling λ_R, and captures how the value of the coupling changes as a function of the energy scale at which it is measured. One point that becomes immediately apparent is that the notion of a "coupling constant" in QFT is a misnomer: the physical couplings that appear in the renormalized Lagrangian are functions of an energy scale. This may seem a startling result, but it is empirically quite well-confirmed. The fine-structure constant, for instance, is rather famously said to have the value $\alpha = \frac{1}{137}$. However, the renormalization group teaches us that it has the value *only at the relatively low energies at which we were historically able to measure it*. More recent experiments have confirmed the scale-dependent variation of α, with the value of α at the energy scale $\mu \sim 80$ GeV measured to be $\alpha(\mu) \approx \frac{1}{128}$ (Patrignani et al., 2016 and 2017 update).

Though the path to its derivation was relatively simple, the implications of this scale dependence for our understanding of QFT are vast. I will first develop these implications in a somewhat restricted context, then pair them explicitly with the path integral formulation of QFT that I said above was a natural home of renormalization group methods.

The first point to note is that *all* of the couplings in a QFT will vary with scale according to a renormalization group equation. For example, suppose that one added a (nonrenormalizable) interaction $\lambda_6\phi^6$ to the action of our scalar field theory; the coupling λ_6 would have its own beta function that would capture how that coupling changed as a function of μ. Similarly, there is also a renormalization group equation determining how the (effective) mass m of the scalar field varies with scale. In general, the beta function for any individual coupling g_i is a function of *all* of the couplings in the theory:

$$\mu \frac{\mathrm{d}}{\mathrm{d}\mu} g_i(\mu) = \beta_i(g_1, \cdots, g_n)$$

Of particular interest is the behavior of a theory's couplings at very high energies. It provides some guidance about whether the theory could have a well-defined continuum limit: an interesting mathematical question, albeit one of limited physical import.[3] The renormalization group equation tells us that there are three possible behaviors that a theory can have in that asymptotic regime:

1   The renormalization group flow hits a point $g^* \equiv (g_1^*, \cdots, g_n^*)$ at which the beta functions β_i for *all* of the couplings are zero. This is called a *fixed point*, and the QFT becomes *scale-invariant* at this point.

2   The beta function for at least one of the couplings in the theory is positive, entailing that the coupling *increases* with higher energies. If the coupling does not hit a fixed point, it eventually becomes infinite at some very large, but finite, energy scale μ_{Landau} and the theory is ill-defined; one says that its RG equations encounter a *Landau pole*. QFTs with this asymptotic behavior are called *trivial*, since they can only be well defined if the values of the couplings are zero. Conversely, as $\mu \to 0$ the couplings go to zero, and these theories are *infrared free*.

3   The beta functions for all of the couplings in the theory are negative, entailing that the couplings *decrease* with higher energies. Eventually, all of the couplings go to zero as $\mu \to \infty$, and the QFT hits a particular fixed point: the point $g^* = 0$. In this case, the QFT is said to be *asymptotically free*. Conversely, at low energies the couplings become large, eventually diverging at some long, but finite, energy scale; this is responsible for the phenomenon of *confinement* in quantum chromodynamics. (Strictly speaking, asymptotic freedom is just the special case of asymptotic safety where the fixed point $g^* = 0$. Its significance in modern particle physics makes it a special case worth highlighting.)

Analysis of the high-energy behavior of a theory's RG flow allows for a more general notion of renormalizability than was described above. Recall that a theory was perturbatively renormalizable if all divergences arising in the perturbative calculations of its *n*-point functions could be eliminated by shifting the values of finitely many of its couplings. What made these QFTs noteworthy was that they seemingly remained consistent up to arbitrarily high energies, in the sense that one could consistently let the cutoff $\Lambda \to \infty$ at the end of calculations. What the RG analysis of the possible high-energy behaviors of a QFT offers is a sense of *nonperturbative renormalizability*: a theory is said to be nonperturbatively renormalizable if its RG flow hits a fixed point when one examines the structure of the theory as $\mu \to \infty$. Interestingly, this is a sense of renormalizability that can apply to QFTs that are *not* perturbatively renormalizable; the Gross-Neveu model in 3 spacetime dimensions, for example, is perturbatively nonrenormalizable but asymptotes to a fixed point at high energies and has a perfectly sensible structure at arbitrarily high energies (Braun et al., 2011).

As I will discuss in more detail below, RG methods also naturally lead to a picture of a *space of QFTs* $\mathcal{A}$ that is coordinatized by the values of the couplings $g_1, \cdots, g_n$. The RG equations generate a "dynamics" on this theory space, determining how a theory – a point in $\mathcal{A}$ defined by a particular set of couplings defined at a particular energy scale – will "flow" through theory space as one changes the energy scale at which one is studying the structure of the QFT. (This notion of a "space of theories" is, in general, not mathematically well defined. See Douglas (2013) for an interesting discussion of theory space and the current state of our understanding of it.) The fixed point(s) of the theory play a particularly important role in making sense of this notion of a space of QFTs. In order to appreciate the power of this picture, let us return to the path integral framework introduced earlier and discuss a particular approach to the RG, the so-called "Wilsonian RG."[4]

Recall that in the perturbative calculation of $G^{(4)}$ above, we encountered integrals in the series expansion that were divergent. Those integrals were regularized by sharply cutting off our range of integration at some high but finite momentum Λ. At the level of the path integral, in which one integrates over entire histories of possible configurations of the field, this regularization procedure amounts to treating histories of the field $\phi(x)$ which may differ on length scales $L \leq \frac{1}{\Lambda}$ as equivalent. In the action, this amounts to replacing $\phi(x) \longrightarrow \phi_\Lambda(x)$ and replacing the "bare" couplings g with the scale-dependent, renormalized coupling(s) $g_R(\Lambda)$, emphasizing the fact that one is now dealing with a theory defined only up to the scale Λ. This automatically regularizes the divergent integrals that appear in perturbative calculations by excluding the high momenta that gave rise to the divergences. The resulting theory is defined only up to the scale of the cutoff Λ, but inevitably becomes inapplicable at that scale; such a QFT is called an *effective field theory* and the action that defines such a theory is called the *Wilsonian effective action* S_W. As (Wallace, this volume) describes, that QFTs are effective field theories is now the dominant attitude in the physics community toward the foundations of QFT. Although it should be clear from the preceding paragraph, perhaps it is worth emphasizing that in an effective field theory there is no problem of divergences whatsoever, and the renormalization of the couplings has nothing to do with the elimination of any infinities.

In the context of perturbative renormalization, the requirement that a theory's action does not include any nonrenormalizable interaction terms made sense: one must include infinitely many renormalized parameters to cancel all divergences in such a theory and since values of renormalized parameters must be taken from the experiment, these theories were useless for making predictions: one would need to perform infinitely many measurements before calculating anything at all. Steven Weinberg, writing just before the understanding of QFTs as effective field theories became widespread, exemplified this attitude:

> Throughout this history I have put great emphasis on the condition of renormalizability, the requirement that it should be possible to eliminate all infinities in a quantum field theory by a redefinition of a small number of physical parameters...it has always seemed to me that the requirement of renormalizability has just the kind of restrictiveness that we need in a fundamental physical theory. (Weinberg, 1977, p. 33)

In the effective field theory approach, there is no longer any justification for including only renormalizable terms in the action. As a result, a Wilsonian action will generally include *infinitely* many terms. For example, the action for the ϕ^4 theory we have been considering becomes

$$
\begin{aligned}
S_W &= \int d^4x \frac{1}{2}(\partial\phi_\Lambda)^2 + m^2\phi_\Lambda^2 + \lambda_4\phi_\Lambda^4 + \lambda_6\phi_\Lambda^6 + \cdots \\
&= \frac{1}{2}(\partial\phi_\Lambda)^2 + \sum_{n\geq 0}\lambda_n\phi_\Lambda^{2+n} + \sum_{n\geq 0}\lambda_n'(\partial\phi_\Lambda)^2\phi_\Lambda^n + \cdots = \mathcal{L}_\Lambda(\phi,\partial\phi,\ldots)
\end{aligned}
$$

One might think that this is sufficiently unwieldy to be useless. However, as we will see below, one of the virtues of RG methods is that they demonstrate that only a finite set of interactions – precisely the renormalizable ones! – contribute meaningfully to the low-energy physics to which we have experimental access.

It is part and parcel of the effective field theory approach that a theory is appropriate only for describing physical processes at energy scales $E \ll \Lambda$ and one doesn't trust the theory's description of physical processes at energy scales near the cutoff. However, an effective field theory defined at the cutoff Λ may still contain many high-energy degrees of freedom that are inappropriate for describing physics at some particular $E \ll \Lambda$. Suppose, for example, one wants to model the elastic scattering of protons at $E \sim 1$ GeV, but only has available to them quantum chromodynamics defined at (say) $\Lambda \sim 10^{15}$ GeV, which will contain quark and gluon fields as its fundamental degrees of freedom. This violates the truism with which I began this chapter. In order to achieve a more perspicuous description of the low-energy nuclear physics of interest, one ought to remove some of these high-energy degrees of freedom from the path integral by lowering the scale of the cutoff.

Essentially, one achieves this by doing the path integral a little bit at a time. Wilson's approach involves splitting the path integral of the "full-theory" into two parts: a "high-energy" component and a "low-energy" component. This is achieved by a splitting of the field $\phi(x)$ into $\phi(x) = \phi_H(x) + \phi_L(x)$. The high-energy component describes excitations of the field whose allowed energies E are contained in a momentum "shell" of width $d\Lambda$, i.e., $(\Lambda - d\Lambda) < E < \Lambda$. In order to lower the cutoff from $\Lambda \rightarrow \Lambda' = (\Lambda - d\Lambda)$, one simply computes the high-energy part of the path integral and leaves the low-energy part alone. (In general, this computation will be performed using the perturbative approximation sketched above.) The resulting QFT contains only the low-energy components of the field and is defined only up to the new, lower cutoff scale Λ'; one sometimes describes this process of integrating out the high-energy component of the action as "putting on blurry glasses" that prevent us from seeing excitations of the field on length scales shorter than $\frac{1}{\Lambda'}$.

It is integral to the tenability of the effective field theory approach that the empirical consequences of the high-energy part of the path integral can be *entirely* incorporated into the new, low-energy theory via a modification of its couplings. The specific manner in which the couplings must be modified as one changes the scale of the cutoff is, perhaps unsurprisingly, given by the beta functions of the couplings. This procedure of "integrating out" high-energy degrees of freedom can be iterated in the obvious way, with each change in the scale of the cutoff requiring a corresponding change in the value of the couplings of the theory, with the precise form of the change determined by the RG equations. The Wilsonian picture makes clear the physical import of the RG equations: the change of the couplings as a function of the energy scale should be understood as reflecting the way that the *physical effects* of field excitations at energies $\Lambda > \Lambda'$ are incorporated into a new Wilsonian action defined at a lower energy scale Λ'.

We can now join this description of the Wilsonian RG to our above discussion of the RG equations as defining a "flow" through a space of QFTs $\mathcal{A}$. In principle, one would like to begin by first pinpointing *all* of the fixed points of *all* QFTs. This is obviously impossible in practice – not only do we not know all the QFTs that could be consistently written down, but the mathematical techniques available to us are insufficient for identifying all of the fixed points even of the QFTs we *do* know how to consistently write down. Nevertheless, one can get the general idea by considering only free QFTs, which are the simplest possible fixed points. These will form an infinite set of isolated points in $\mathcal{A}$. This leads naturally to an interesting picture, somewhat common in the physics community, of what it means to *define* a quantum field theory (see, for example, Banks (2008, chapter 9)). One begins with a scale-invariant QFT defined by an action at a fixed point and "perturbs" it by adding an infinite sum of local interactions to its action, which will shift the theory away from the fixed point. (Though it is beyond the scope of this review, this will also break the scale invariance of the theory. The study of behavior arising from broken scale invariance is an important part of contemporary QFT.) Following the reasoning sketched above, one then cuts the theory off at some energy scale Λ. This generates a Wilsonian action $\mathcal{S}_W$, defined at a particular scale. One can then use the RG to analyze how this theory "moves" through the space $\mathcal{A}$ as one iteratively lowers the cutoff and changes the values of the couplings.

This definition of a QFT as a perturbation of a fixed-point action also offers a natural classification of the possible "perturbing" interactions appearing in the action into three categories: *relevant*, *marginal*, and *irrelevant*. The interactions are classified according to the RG behavior of their associated couplings in the neighborhood of the fixed point:

1   An interaction is *relevant* if its coupling *increases* at low energies, flowing *away* from the fixed point as one integrates high-energy degrees of freedom out of the path integral.

2   An interaction is *irrelevant* if its coupling *decreases* at low energies, flowing *back into* the fixed point.

3   An interaction is *marginal* if its coupling does not change with scale, and the interaction contributes equally to physical processes at all length scales.[5]

There are two important things to say at this point. The first answers the worry that although the Wilsonian action did not generate perturbative expansions that contained divergent integrals, this virtue was outweighed by the fact that it contained infinitely many interactions and so was too unwieldy to be useful. It also provides an *explanation* of why the criterion of renormalizability was successful as a constraint on theorizing through the 1970s.

In any given Wilsonian action $\mathcal{S}_W$, at most a finite number of interactions will be renormalizable, with the vast majority of the interactions in $\mathcal{S}_W$ rendering the theory nonrenormalizable; for example, the only two renormalizable interactions one can write down in the ϕ^4 theory in four spacetime dimensions are $m^2\phi^2$ and $\lambda\phi^4$ and in quantum electrodynamics, there is only one possible renormalizable interaction between the photon and electron fields. If the right way of understanding a QFT requires including infinitely many terms in the action, how could these theories have been so successful?

The answer goes as follows (see Duncan (2012, chapter 17.4) for details). Beginning with any arbitrary Wilsonian action, one can begin iteratively lowering the cutoff and studying the way that the couplings change. RG methods reveal that the couplings associated with all of the *nonrenormalizable* interactions will make contributions to low-energy physics in only two ways: (i) through a rescaling of the couplings associated with *renormalizable* interactions and (ii) through contributions to scattering amplitudes which suppressed by inverse powers of the high-energy cutoff Λ, rendering them negligible for describing low-energy physics. In terms of an RG flow in theory space $\mathcal{A}$: one finds that a generic Wilsonian action, which lives at the point on some infinite-dimensional manifold in $\mathcal{A}$ picked out by the values of the couplings associated with its infinitely many interactions, flows at low energies to a finite-dimensional manifold defined by the values of only its *renormalizable* interactions. The result is that at the low energies for which QFT has proved empirically adequate, only a finite number of interactions make non-negligible contributions to the computation of the low-energy scattering amplitudes that we have been able to test experimentally: precisely the renormalizable interactions! Thus, what seemed like an inexplicable bit of good luck to physicists in the middle of the 20th century is rendered straightforwardly explicable by RG methods.

The second point to say about the notion of theory space $\mathcal{A}$ is that it affords us a means for partitioning the space of QFTs into classes of theories defined by their behavior under RG transformations. As Batterman (this volume) explains, many different QFTs may flow to the *same* fixed point, and we say that these QFTs form a particular *universality class*; in principle, RG methods offer the possibility of partitioning the entirety of $\mathcal{A}$ into distinct universality classes. Though we are very far from the attainment of anything like such a classification of all QFTs, the fact that RG methods make such a classification possible, even in principle, illustrates the far-reaching import of RG methods for our understanding of QFT. While renormalization may have begun as a mathematically and physically dubious procedure for eliminating infinities from perturbative calculations, RG methods are now at the very heart of virtually every aspect of the modern understanding of QFT. Since I can hardly emphasize this centrality better myself, I will conclude with an update to Eddington's famous words about the second law of thermodynamics proposed by Vincent Rivasseau:

> The flow of the renormalization group holds, I think, the supreme position among the laws of Nature. If someone points out to you that your pet theory of the universe is in disagreement with Maxwell's equations - then so much the worse for Maxwell's equations. If it is found to be contradicted by observation - well, these experimentalists

do bungle things sometimes. But if your theory is found to be against the flow of the renormalization group I can give you no hope; there is nothing for it but to collapse in deepest humiliation. (Rivasseau, 2012, p. 28)

19.4   Further Reading

RG methods will form a significant portion of any modern introduction to QFT. My recommendations for further QFT reading essentially mirror those of Wallace (this volume), with the addition of Schwartz (2014) which contains a helpful discussion of several topics related to renormalization and RG methods. There are also several book-length treatments of RG methods in both QFT and classical statistical physics; I can recommend Collins (1984), Rivasseau (1991), Goldenfeld (1992), Cardy (1996), and Hollowood (2013). A longer survey of renormalization explicitly aimed at philosophers is Butterfield and Bouatta (2015).

RG methods have also been put to considerable philosophical use, most notably in debates concerning emergence and reduction. A considerable literature has developed around the discussions of renormalization and intertheoretic reduction in Batterman (2000) and Batterman (2002); see e.g. Butterfield (2014) or Reutlinger (2014) and references therein. Recently, it has also been suggested by Fraser (2016) and Williams (2017) that RG methods make possible a promising application of the *divide et impera* approach to scientific realism in QFT; see Ruetsche (2018) for a skeptical response.

Notes

1   See Jansson (this volume) for further discussion of the relationship between scale and explanation.
2   For a philosophical discussion of the issues raised by multi-scale modeling, see Batterman (2013).
3   See Li (2015) for a philosophical discussion of the uses of RG analyses in the mathematically rigorous setting of constructive QFT.
4   The *locus classicus* for this formulation is Wilson and Kogut (1974); although it is beyond the scope of this chapter, there are interesting relationships between several different formulations of the RG. See Schwartz (2014, chapter 23) for a textbook discussion.
5   Often a more sophisticated analysis will show a coupling to be marginal at the lowest order of perturbation theory, but when higher order corrections are included the coupling is shown to scale like a relevant or irrelevant interaction. In this case, the interaction is called *marginally relevant* or *marginally irrelevant*.

References

Banks, T. (2008). *Modern Quantum Field Theory*. Cambridge: Cambridge University Press.
Batterman, R. (2000). Multiple realizability and universality. *The British Journal for the Philosophy of Science*, 51(1): 115–145.
Batterman, R. (2002). *The Devil in the Details: Asymptotic Reasoning in Explanation, Reduction, and Emergence*. Oxford: Oxford University Press.
Batterman, R. (2013). The tyranny of scales. In R. Batterman (ed.), *Oxford Handbook of Philosophy of Physics*. Oxford: Oxford University Press, pp. 255–286.
Braun, J., Gies, H. and Scherer, D. (2011). Asymptotic safety: A simple example. *Physical Review D*, 83(8): 085012-1–085012-15.
Butterfield, J. (2014). Reduction, emergence, and renormalization. *The Journal of Philosophy*, 111(1): 5–49.
Butterfield, J. and Bouatta, N. (2015). Renormalization for philosophers. In T. Bigaj and C. Wüthrich (eds.), *Metaphysics in Contemporary Physics*, Vol. 104 of *Poznan Studies in the Philosophy of the Sciences and the Humanities*. Leiden, The Netherlands: Bill, pp. 437–485.
Cardy, J. (1996). *Scaling and Renormalization in Statistical Physics*. Cambridge: Cambridge University Press.
Collins, J. (1984). *Renormalization: An Introduction to Renormalization, the Renormalization Group and the Operator-Product Expansion*. Cambridge, UK: Cambridge University Press.
Cushing, J.T. (1990). *Theory Construction and Selection in Modern Physics: The S Matrix*. Cambridge: Cambridge University Press.
Douglas, M.R. (2013). Spaces of quantum field theories. *Journal of Physics: Conference Series*, 462: 012011-1–012011-15.

Duncan, A. (2012). *The Conceptual Framework of Quantum Field Theory*. Oxford: Oxford University Press.

Feynman, R. and Hibbs, A. (1965/2010). *Quantum Mechanics and Path Integrals*. New York: Dover. Edition emended by D. Styer.

Fraser, J. (2016). *What Is Quantum Field Theory? Idealisation, Explanation, and Realism in High Energy Physics*. PhD thesis, University of Leeds.

Glimm, J. and Jaffe, A. (1987). *Quantum Physics: A Functional Integral Point of View*. New York: Springer.

Goldenfeld, N. (1992). *Lectures on Phase Transitions and the Renormalization Group*. Reading: Addison-Wesley.

Gross, D.J. (2004). Asymptotic freedom and QCD–a historical perspective. *Nuclear Physics B-Proceedings Supplements*, 135: 193–211.

Hollowood, T.J. (2013). *Renormalization Group and Fixed Points in Quantum Field Theory*. London: Springer.

Li, B. (2015). Coarse-graining as a route to microscopic physics: The renormalization group in quantum field theory. *Philosophy of Science*, 82(5): 1211–1223.

Niedermaier, M. and Reuter, M. (2006). The asymptotic safety scenario in quantum gravity. *Living Reviews in Relativity*, 9(5): 1–173.

Patrignani, C. and et al. (2016 and 2017 update). Review of particle physics. *Chin. Phys. C*, 40(10): 100001.

Reutlinger, A. (2014). Why is there universal macrobehavior? Renormalization group explanation as noncausal explanation. *Philosophy of Science*, 81(5): 1157–1170.

Rivasseau, V. (1991). *From Constructive to Perturbative Renormalization*. Princeton, NJ: Princeton University Press.

Rivasseau, V. (2012). Quantum gravity and renormalization: The tensor track. *AIP Conference Proceedings 8*, 1444: 18–29.

Ruetsche, L. (2018). Renormalization group realism: The ascent of pessimism. *Philosophy of Science*, 85(5): 1176–1189. Chicago, IL: University of Chicago Press.

Schwartz, M. (2014). *Quantum Field Theory and the Standard Model*. Cambridge: Cambridge University Press.

Weinberg, S. (1977). The search for unity: Notes for a history of quantum field theory. *Daedalus*, 106(4): 17–35.

Weinberg, S. (1983). Why the renormalization group is a good thing. In A. Guth, K. Huang, and R.L. Jaffe (eds.), *Asymptotic Realms of Physics: Essays in Honor of Francis Low*. Cambridge, MA: The MIT Press, pp. 1–19.

Williams, P. (2017). Scientific realism made effective. *The British Journal for the Philosophy of Science*, 70(1): 209–237.

Wilson, K.G. (1965). Model hamiltonians for local quantum field theory. *Physical Review*, 140(2B): 445–457.

Wilson, K.G. and Kogut, J. (1974). The renormalization group and the ϵ-expansion. *Physics Reports*, 12(2): 75–199.

Zee, A. (2010). *Quantum Field Theory in a Nutshell*. Princeton, NJ: Princeton University Press.

20

LOCALITY IN (AXIOMATIC) QUANTUM FIELD THEORY: A MINORITY REPORT

Laura Ruetsche

20.1   Locality in Ordinary QM

My views about locality in ordinary, non-relativistic quantum mechanics (QM) aren't orthodox. They are relevant to my official topic: locality in quantum field theory (QFT). According to orthodoxy:

> CATECHISM.[1] Violation of the Bell Inequalities teaches us that nature is non-local.

I disagree. CATECHISM is misleading, inapt, and prone to rhetorical abuse.

Misleading: The Bell Inequalities are empirical commitments of a certain class of *hidden variable theories*. For present purposes, a hidden variable theory is any theory that is more profligate in its recognition of determinate quantum observables than is orthodox QM. The unhidden state of a quantum system is its quantum state ψ; orthodox QM recognizes as determinate only those observables whose values that state ascribes Born Rule probability 1. Thus sets of observables with no eigenstates in common can't be simultaneously determinate, and individual observables aren't determinate on systems whose states are non-trivial superpositions of their eigenstates. A hidden variable theory, generically, supplements the quantum state ψ with a hidden state λ; λ fixes values (or, if it's stochastic, a probability distribution over values) for observables not reckoned determinate by quantum orthodoxy. To derive the Bell Inequalities is to prove that a hidden variable theory, if it obeys certain constraints, makes empirical predictions that contradict those made by orthodox QM. There is nothing in the proof that requires that the hidden variables be located in the intersection of the past lightcones of two relatively spacelike separated regions – although commentators suggestively draw diagrams locating them there. Indeed, nothing in the proof requires relativistic spacetime structure or the association of observables with spacetime regions.[2] So while assumptions labeled "locality" may be, in the company of other assumptions, sufficient for deriving the Bell Inequalities, they are not necessary. Even suppressing the question of whether assumptions labeled "locality" are aptly labeled, CATECHISM misrepresents the logic of the situation.

Inapt: A metaphysical expectation that *it just can't be that if you touch the universe here, it wiggles over there*, is not a promising candidate for strict translation into the framework of mathematical physics. The special theory of relativity (STR – See Part II of this volume for detailed discussion), however, motivates a precise locality criterion. I don't mean the folklore – and it is just that (see Geroch, 2011) – prohibiting superluminal field or signal or causal propagation. STR isn't a theory about signals, causes, or the detailed dynamics of fields. It is a theory about theories, one requiring those set in Minkowski spacetime to exhibit Lorentz symmetries. *This* locality requirement can be precisely formulated. It is, however, inapplicable to ordinary non-relativistic QM or the hidden variable theories that would depose it. Not set in Minkowski spacetime, they simply aren't spacetime theories of the relevant sort. This – the literature has clearly shown – can't stop commentators from articulating locality requirements for these theories.[3] But we should recognize that this is playing a somewhat loose and improvisational game – and moderate our attitudes toward its outcomes accordingly.

Prone to Rhetorical Abuse: There are quantum theories to which the precisely formulated and physically motivated locality requirement of Lorentz symmetry applies. As the next section will elaborate, not only do we have examples of Lorentz-Invariant QFTs, the usual axioms framing axiomatic approaches to QFT *require* QFTs to respect Lorentz symmetry. This amounts to a constructive proof that QM and STR can peacefully coexist – worth noting because there's a reputable approach to quantum phenomena, Bohmian Mechanics, that threatens to disregard Lorentz symmetry (for a detailed discussion of Bohmian Mechanics, see Tumulka (this volume)). Bohmian Mechanics appears to posit a fundamental notion of distant simultaneity. If this posit is ineliminable from its dynamics, *Bohmian Mechanics is nonlocal in the precise sense of failing to be Lorentz-Invariant*. The sort of abuse to which CATECHISM is prone involves CATECHISM in an apology for this:

> Violations of the Bell Inequalities show that the *world* is non-local. It can be no criticism of a theory that it displays this feature of the world in an obvious way. (Maudlin, 2011, p. 111)

The apology equivocates. The loose sense of non-locality CATECHISM presents as established by the violation of the Bell Inequalities *differs markedly* from the precise and well-motivated Lorentz symmetry requirement that Bohmian Mechanics challenges.

20.2 Locality in QFT

Because the algebraic formulation of relativistic QFT has the resources to express precisely a variety of locality requirements, this section opens with an informal overview of the algebraic approach to quantum theories.[4] Typically the student makes her acquaintance with QM starting with the idea that quantum states are normed vectors in a Hilbert space $\mathcal{H}$; quantum observables are introduced as symmetric operators on this state space; they turn out to coincide with the self-adjoint elements of $\mathcal{B}(\mathcal{H})$, the collection of bounded operators on $\mathcal{H}$. The algebraic approach reverses this order of exposition. On the algebraic approach, one starts with a structured set of observables – for instance, the observables satisfying the canonical commutation or anti commutation relations constitutive of the quantum theory in question. One can use these canonical observables to generate an *algebra*. An algebra is just a collection of elements on which are defined ways to (i) add elements to one another, (ii) multiply elements by scalars, and (iii) multiply elements by one another; a collection of elements which is moreover closed under these operations. Put another way, an algebra is just a linear vector space ((i) and (ii)) equipped with a product operation (iii). Therefore to generate an algebra from canonical observables, one forms polynomials of those observables and (for the sake of closure) adds limit points. Having thus obtained an algebra of quantum observables, the algebraic approach defines states in its terms: states are well-behaved assignments of expectation values to elements of observable algebras.

The foregoing sketch suppresses many details: the structural properties algebraic operations mentioned must exhibit; the specification of additional structures requisite of a quantum observable algebras, including an adjoint operation and a norm that helps furnish a criterion of closure. Different ways of articulating these details result in different sorts of algebras. Some options are concrete, abstract, C*, von Neumann. For ease of exposition, what follows will assume that a quantum observable algebra is a concrete von Neumann algebras $\mathfrak{R}$ – that is, a collection of bounded operators on a concrete Hilbert space that is closed in the weak operator topology of that Hilbert space. (A sequence of operators A_n converges to an operator A in this topology if and only if for each pair of vectors $|v\rangle$, $|w\rangle$ in the Hilbert space, $|\langle v|(A_n - A)|w\rangle|$ goes to 0 as $n \to \infty$.) Our old friend $\mathfrak{B}(\mathcal{H})$ is a concrete von Neumann algebra.

Containing a projection operator for each finite-dimensional subspace of $\mathcal{H}$, $\mathfrak{B}(\mathcal{H})$ abounds in finite-dimensional projection operators, including projection operators that are *minimal* in the sense that only the 0 subspace of $\mathcal{H}$ is a proper subspace of their range. von Neumann algebras containing minimal projection operators are Type I von Neumann algebras. Not all von Neumann algebras assume this familiar form. If $\mathcal{H}$ is infinite-dimensional, $\mathfrak{B}(\mathcal{H})$ can have a weakly closed subalgebra $\mathfrak{R}$ whose only non-trivial projection operators have ranges that are infinite-dimensional subspaces of $\mathcal{H}$. Such an $\mathfrak{R}$ is also von Neumann algebra; its type is III. Type III von Neumann algebras are frequently encountered in QFT.

To take the algebraic approach to QFT, one sets up a systematic correspondence between local bounded regions O of a spacetime M, and algebras $\mathfrak{R}(O)$ of observables pertaining (in some sense) to those regions, and subjects the $O \to \mathfrak{R}(O)$ correspondence to physically motivated constraints, such as the axiom

> ISOTONY: if one spacetime region is contained in another, the algebra associated with the first is a subalgebra of the algebra associated with the second.

The closure over local bounded regions of this net of local algebras defines an algebra $\mathfrak{R}(M)$ for the entire spacetime M. In AQFT, the global state of the quantum field is a linear functional ϕ that takes observables in $\mathfrak{R}(M)$ to their expectation values. This global state ϕ is a state for all space and all time.

The literature contains a number of related but distinct ideas about how to use AQFT resources to express locality requirements. Before reviewing these, I must acknowledge an embarrassment. Thus far the AQFT framework has not been shown to accommodate the interacting QFTs making up the Standard Model for particle physics. To fit the sort of theories working physicists actually use, we may need to modify, radically reconceive, or even abandon the AQFT framework. This would necessitate a rethinking of locality principles defined in its terms. But since we have no crystal ball, let's proceed within the existing framework.

20.2.1 Locality of Observables and a Preliminary Conclusion

Here is a partial list of locality notions articulated in terms of AQFT's observable algebras.

1   The $O \to \mathfrak{R}(O)$ correspondence ensures that observables are *localized* in the sense that they are associated with local spacetime regions; the ISOTONY axiom guarantees this association respects relations of spacetime inclusion.

2   The LORENTZ COVARIANCE axiom requires a strongly continuous unitary representation of the identity-connected component of the Poincaré group to act on the Hilbert space of theory, and the VACUUM axiom requires the existence of a state invariant under this action. These axioms enforce (most of!) the Lorentz symmetry demanded by consistency with STR. (Discrete symmetries get left out, a complication I'll ignore here.)

3   There is also a zoo of conditions which are variously labeled locality/separability/causality requirements for local algebras. These include the following:

(a)   The MICROCAUSALITY axiom, which requires the commutativity of observables associated with relatively spacelike regions. This is conventionally glossed as making such observables jointly measurable, or such that the measurement of one fails to influence the measurement of the other. Such glosses muddy present waters, which concern only the structure of observable algebras, but not the interpretation of measurement. There is another way to motivate MICROCAUSALITY as an independence condition: space-like commutativity is necessary if it is to be possible for the global state to factorize into independent local states.

(b)   Even if factorization fails, *statistical independence* – the requirement that partial states on subsystem algebras can be extended to a common state on the total system algebra – which ensures that it makes global sense to attribute states to algebras associated with local regions. Statistical independence and MICROCAUSALITY are logically independent (Summers, 1990).

(c)   The split property, which a pair of algebras $(\mathcal{R}_1, \mathcal{R}_2)$ exhibits just in case there's a Type I von Neumann factor $\mathcal{M}$ such that $\mathcal{R}_1 \supset \mathcal{M} \supset \mathcal{R}_2$. Together with Microcausality, the split property implies that for spacelike regions $\mathcal{O}_1$ and $\mathcal{O}_2$, the algebra associated with the disjoint union of those regions is the tensor product of the algebras associated with the regions individually (see Summers, 2009, §4.5, for a discussion).

(d)   Many, many other conditions of complexity comparable to the split property.[5]

4   Categories (1)–(3) concern the local nature of observables and their distinctness/independence. One can also ask that the quantum field dynamics respect the relativistic spacetime structure. Field dynamics govern how quantum fields develop over time. Given initial data in the form of a quantum field defined on some timelike ($t = 0$) surface S, the dynamics determine the future evolution of that data. Relativistic spacetimes come equipped with causal structure; the causal future of S lies in its forward lightcone. Field dynamics respect this causal structure if they forbid initial data on a surface S to propagate outside of S's forward lightcone. Some commentators (see, for instance, Butterfield's useful 2007 review of relativistic causality in AQFT) have suggested that this follows from the SPECTRUM axiom, which requires the infinitesimal generators of the translation subgroup of the unitary representation of the Poincaré group to "lie in forward lightcone." But how fields propagate depends on details of their dynamics; these details can be such that SPECTRUM is consistent with superluminal propagation. For a broad range of relativistic quantum fields, a prohibition on superluminal propagation is arguably captured by the axiom[6]

> LOCAL PRIMITIVE CAUSALITY: if O lies in the domain of dependence of S – that is, if every causal curve through S intersects O – then $\mathfrak{R}(O)$ is a subalgebra of $\mathfrak{R}(S)$.

A state on $\mathfrak{R}(S)$ thus induces, by restriction to the relevant subalgebra, a state on $\mathfrak{R}(O)$. LOCAL PRIMITIVE CAUSALITY is independent of other axioms discussed so far (Haag and Schroer, 1962).

5   Closely related to, but distinct from, various notions of locality is the idea of non-holism (see Healey (this volume) for detailed discussion of various notions of holism in physics). One way to express this notion is by the condition of strong additivity: the algebra $\mathfrak{R}(\cup_j \mathcal{O}_j)$ of observables associated with the union of any family of open bounded spacetime regions $\mathcal{O}_j$ coincides with the algebra $\vee_j \mathfrak{R}(\mathcal{O}_j)$ generated by the subalgebras $\mathfrak{R}(\mathcal{O}_j)$ associated with each of the individual regions – a precise way of framing in algebraic terms the idea that the whole is not greater than the sum of its parts.

Two notes about this list. First, the axioms enforcing Lorentz symmetry are independent of other axioms (such as MICROCAUSALITY and LOCAL PRIMITIVE CAUSALITY) implementing other species of locality demand. This underscores the claim that STR secures very few of the intuitive locality desiderata whirling around in the literature. Second, relativistic quantum fields can satisfy *all* of the above conditions, and the models that are most relevant to applications (for instance, the free Klein-Gordon field and the Dirac field) do satisfy them. Thus a preliminary conclusion: *Nothing about the observables of AQFT gives any aid or comfort to the conventional wisdom that the quantum world is non-local.*

20.2.2   Holism of States in QFT

A variety of non-locality I'll call *holism* can arise from states that give correlations between observables associated with different (even very distant relatively spacelike) regions of spacetime. But this can happen in classical relativistic contexts as well. What is distinctive about QM is that states can be *quantum* entangled: the state cannot be expressed as a mixture of (normal) product states; thus, the correlations between subsystems cannot be interpreted as due to our ignorance of the actual uncorrelated states of the subsystems. The familiar spin singlet state exemplifies such entanglement.

Quantum entanglement deserves the name since it requires that the subalgebras of observables are non-commutative. Therefore, if the classical vs. quantum distinction is drawn in terms of commutative vs. non-commutative observable algebras, quantum entanglement is a new form of non-locality/holism that cannot arise in the classical setting. Moreover, this state-specific holism arises in QFT with a vengeance. Quantum entanglement is more dramatically endemic in QFT than it is in ordinary non-relativistic QM. For instance, in physically reasonable models of AQFT, the set of vector states that violate the Bell Inequalities for *any* pair of space like separated regions are generic, in the sense that they form an open dense set in the set of states (Halvorson and Clifton, 2000).

Quantum entanglement and the consequent holism (or "non-locality," if that is what one chooses to call it) do not, however, compromise the locality of quantum observables explicated by the requirements catalogued in Section 20.2.1. Quantum entanglement occurs in models of QFT where all of those requirements are satisfied. So the questions of entanglement holism and algebra non-locality should not be run together. As far as I can see, attempts to link them make recourse to a significant *additional* assumption: that collapse-like processes punctuate the quantum field dynamics.

20.2.3   Non-locality from Measurement Collapse

By "naive measurement collapse," I mean, for example, the *deus ex machina* posited by textbook solutions to the measurement problem, which simply assert that Schrödinger evolution breaks down during measurement, to be replaced by something we can understand as delivering systems to determinate outcome states. Dynamical collapse schemes of the sort discussed by Lewis (this volume) are not naive but physical theories. Naive measurement collapse is a miracle. If one injects miracles – even local ones – into classical relativistic theories, then non-local effects will emerge. Of course, one doesn't need quantum entanglement to get the glimmering of some form of action at a distance associated with measurement collapse. A single particle will do: finding the particle *here* changes the state of the particle *there*. However, in the case of quantum entangled systems, the choice of which observable to measure *here* does seem, *under the assumption of collapse*, to influence the nature of the outcome *there*. This sort of remote steering of events strikes many as spooky.

But to extract morals about spookiness from results obtained within the AQFT framework requires substantial, and often tendentiously collapse-laden, commentary on those results. Examples illustrating the gap between theorem and moral include the following:

- The Reeh-Schlieder theorem and its conventional commentary (see Redhead, 1995). Where $|\Omega\rangle$ is the vector implementing the Minkowski vacuum state and $\mathfrak{R}(O)$ the algebra associated with a region with non-empty spacelike complement, the theorem states that $|\Omega\rangle$ is *cyclic* with respect to $\mathfrak{R}(O)$: that is, acting on $|\Omega\rangle$ with elements of $\mathfrak{R}(O)$ yields a set of states dense in the space of possible states of the quantum field on Minkowski spacetime.[7] The commentary: "You can prepare an arbitrary global state by performing only interventions local to O." The commentary is spooky. It is also mediated by substantial assumptions about how to use the quantum formalism to model experimental interventions – for instance, the assumption that for each global state $|\psi\rangle$ and each $A \in \mathfrak{R}(O)$, there's something you can do in O to change $|\psi\rangle$ to $A|\psi\rangle$. But why would this be? Undergraduates being introduced to QM often assume such a change of state corresponds to an A measurement – but we try to disabuse them of this picture. A more sophisticated measurement model posits change of state due to *measurement collapse*. Even this more sophisticated posit falls short of a satisfactory justification of the commentary. One dissatisfaction is technical: because not every $A \in \mathfrak{R}(O)$ implements a measurement collapse, the posit fails to warrant consideration of the results of acting on $|\Omega\rangle$ with *arbitrary* elements of $\mathfrak{R}(O)$. Another dissatisfaction is conceptual. The posit relies critically on measurement collapses. In non-relativistic QM, naive measurement collapse is a scandal. In AQFT, I'll try to suggest in Section 20.2.4, it's inconsistent with the axioms.

• The Reeh-Schlieder theorem and MICROCAUSALITY imply that typical global states are eigenstates of *no* non-trivial local observables. The commentary: although algebras are as local as one can hope, there are no local *events/entities* in QFT (see Redhead, 1995, pp. 127–128). (This is significant because without a notion of local events, questions about the locality of causes and effects are difficult – impossible? – to pose precisely!) This commentary is mediated by another tendentious interpretive move: the assumption that what it takes to have a local event/entity is for the field to occupy an eigenstate of a local observable.

This pair of examples reinforces the pedestrian point that one typically needs to make substantive assumptions to invest formal results with interpretive significance. This makes it essential to identify and assess these assumptions. For future reference, note that the mediating assumptions in the two cases discussed above are mutually reinforcing: it's because we assume eigenstates correspond to determinate states of affairs that we entertain measurement collapse to eigenstates. And although the eigenstate-eigenvalue link has come in for a fair amount of abuse lately, I know of no concrete alternative proposals for how to make sense of quantum theory as a fundamental theory of determinate events. In modern Everett approaches, the determinateness isn't fundamental (for more on Everett approaches, see Saunders (this volume)). In Bohmian approaches, quantum theory isn't fundamental.

20.2.4   Signaling and Operations

Familiar no-signaling theorems seek to establish the impossibility of influencing the measurement statistics on one subsystem by means of choosing which observable to *measure* on another relatively spacelike subsystem. But is it possible that besides measurements there are other "physical operations" that will succeed in producing distant influences? This question has attracted a great deal of recent attention, with many results presented as results about locality/independence of AQFT algebras.[8] After describing some of this work, I explain why I think more needs to be done to articulate its significance for interpretive questions.

20.2.4.1   Lüders Rule

Given a quantum state ϕ on a von Neumann algebra $\mathfrak{R}$ and an observable A, the Lüders recipe for quantum conditional probabilities defines a state ϕ_{a_n} which encodes the probabilities ϕ assigns conditional on a measurement of a discrete observable A yielding the outcome a_n (selective measurement). Where E_n is the spectral projection corresponding to A eigenvalue a_n, $\phi_{a_n}(B) := \frac{\phi(E_n B E_n)}{\phi(E_n)}$ for all $B \in \mathfrak{R}$. ϕ_{a_n} is the unique normalized state on $\mathfrak{R}$ with features we'd expect of the conditional probabilities in question (see Bub (1977)). A natural generalization of the Lüders rule defines a state ϕ_A encoding probabilities conditional on A having been measured (non-selective measurement). Where $\{E_i\}$ is the complete set of A's spectral projections, $\phi_A(B) := \frac{\phi(\sum_i E_i B E_i)}{\phi(\sum_i E_i)} = \sum_i \phi(E_i B E_i)$. ψ_A is a mixture, weighted by Born Rule probabilities ϕ assigns possible A outcomes, of states obtained by Lüders-conditionalizing ϕ on those outcomes.

We can distinguish two accounts one might give of *why* ϕ_A is the appropriate way to represent system probabilities after an A measurement. The first account forges an intimate connection with the postulate of measurement collapse. According to that postulate, during a measurement yielding the result a_n, the measured system undergoes collapse to the A eigenspace associated with that outcome. The (selective) Lüders-conditionalized state ϕ_{a_n} coincides with the post-collapse state E_{a_n} corresponding to the outcome. The (non-selective) Lüders-conditionalized state ϕ_A is an unmysterious statistical mixture of collapse endpoints E_{a_i}. Thus the textbook account of quantum measurement collapse explains why the Lüders rule is an appropriate recipe for quantum conditional probabilities. Indeed, it's not unusual to see "projection postulate" used to refer both to the hypothesis that collapse occurs and to the hypothesis that the Lüders-conditionalized state describes the situation after measurement.

Those queasy about measurement collapse can avail themselves of a distinct account of why it is appropriate to follow Lüders rule. For even they can reasonably ask: what probabilities does the state ϕ assign conditional on obtaining a measurement outcome a_n? The Lüders rule gives the unique reasonable answer. Note that the spectral projections E_{a_i} are crucial ingredients of this reasonable answer. They're the mathematical representatives of the conditions – determinate measurement outcomes – for which the conditional probabilities *are* conditional probabilities. Note also that, rather than positing a physical change to the system's state ϕ, this account justifies a strategy for extracting conditional probabilities implicit in ϕ.

For both foes and fans of collapse, the pre-measurement state ϕ determines (via the Born rule) a probability distribution over possible outcomes of a measurement. Outcomes are represented by A's spectral projections E_{a_i}. These spectral projections determine a Boolean algebra of propositions attributing determinate values to quantum observables such that those propositions can be assigned truth values that behave classically. Thus the projection operators E_{a_i} pull double duty in quantum measurement theory: they represent determinate outcomes, and (with the pre-measurement state) they generate a probability distribution over those outcomes. They tell us not only what's happening, but also with what probability.

20.2.4.2   Schlieder's No Signaling Theorem; Its Apparent Limits

Schlieder's (1968) "no-signaling" theorem signals that, given MICROCAUSALITY, a non-selective measurement on one subsystem of a pair of spatially separated systems does not change the state of the other subsystem. What Schlieder shows is that when O_1 and O_2 are space-like separated and ϕ is a joint state over both regions, for any $A \in \mathcal{R}(O_1)$, the Lüders conditionalized state ϕ_A agrees with ϕ on $\mathcal{R}(O_2)$. The result suggests that the commutativity of AQFT observables associated with relatively spacelike regions precludes signaling. Therefore, the locality of observables appears to protect against entanglement + collapse blossoming into a potential to signal between space-like separated regions.

The "+ collapse" clause in the foregoing statement is significant. Angst about signaling arises for those who take ϕ_A to be the appropriate post-measurement state because they suppose a physical change of state to occur during measurement. Foes of collapse who regard ϕ_A as the appropriate summary of conditional probabilities implicit in ϕ entertain no physical changes that could send signals. Imagine a foe of collapse who was also a system 2 observer. What would warrant this observer in taking ϕ_A to codify measurement probabilities is *learning* that an A measurement had been carried out on system 1 – where that information would be communicated by conventional signals. So not only does the foe of collapse countenance no physical change, conventional signals are required to license $\phi \rightarrow \phi_A$ updating. The $\phi \rightarrow \phi_A$ transition prompts concerns about spooky quantum signaling only if it's understood to be a physical process.

Some regard Schlieder's result to be of limited relevance. The proof uses the generalized Lüders rule to model non-selective measurements. Although the Lüders recipe leads to well-defined conditional probabilities when the observable measured has a discrete spectrum, there may be technical impediments to using the recipe for algebras that (like the Type III algebras AQFT typically associates with local regions of spacetime) lack minimal projection operators.[9] One reaction to this is to suggest that to treat signaling in AQFT, we need to consider transitions more general than those wrought by Lüders conditionalization (Rédei and Valente, 2010, §1,2).

What sort of transitions? The AQFT literature on operational independence proposes *operations*, a generalization of familiar quantum measurement projections. An operation on an algebra $\mathfrak{R}$ is a *completely positive map* $T\colon \mathfrak{R} \to \mathfrak{R}$; it naturally induces a map $\phi \to T\phi$ on the space of states ϕ on $\mathfrak{R}$. (Technically, the induced map on the set of states is the dual of the operation on the algebra. For the sake of readability, I elide the distinction here. All operations considered here are normal; that is, continuous with respect to the natural topology of the von Neumann algebra. Normal operations map countably additive states to countably additive states.) A representation theorem due to Kraus (1983) tells us which elements of a Type I von Neumann algebra implement operations. $T\colon \mathfrak{B}(\mathcal{H}) \to \mathfrak{B}(\mathcal{H})$ is a normal operation iff there exist bounded operators $K_i \in \mathfrak{B}(\mathcal{H})$ s.t. $T(X) = \sum K_i^* X K_i$ for all $X \in \mathfrak{B}(\mathcal{H})$ and $\sum K_i^* K_i = 1$. Thus the generalized Lüders rule corresponds to an operation, with the spectral projections E_i of the measured observable playing the role of the bounded operators K_i. But in general, the implementers of Kraus operations needn't be complete sets of orthogonal projection operators but include sets of positive operators K_i that share with complete sets of orthogonal projection operators the feature that together with a state, they generate probabilities that sum to 1.

Schlieder's no-signaling theorem can be easily adapted to show that an operation of Kraus form on one subsystem leaves measurement statistics on a distant subsystem unchanged. Unfortunately, the representation theorem does not extend to the Type III case. There we lack a perfectly general account of the precise form operations take. There we might worry that non-Kraus operations could be used to signal.

20.2.4.3   Operations Research

Much of the AQFT signaling literature unfolds in the scope of the assumption that "Operations are mathematical representations of ... physical processes which take place as a result of physical interactions with the quantum system" (Summers, 2009, p. 135; cf. Redei, 2010, pp. 1045–1046). Where O is the spacetime region occupied by our laboratory this month and ϕ is the global state on the net of local algebras, the assumption (somewhat dramatized) is:

> OPERATION. Doing things in our lab this month corresponds to acting on ϕ with operations in $\mathfrak{R}(O)$. Thus if T is an operation in $\mathfrak{R}(O)$, there's something I can (in principle) do in our lab this month that will change ϕ to $T\phi$.

Having made this assumption, the literature seeks to aggravate or assuage worries about spooky action at a distance by articulating criteria for whether such changes are spooky, and producing results about whether various combinations of ϕs and Ts meet those criteria. Most results are framed as results about operational independence/separatedness/separability of *algebras*. They are also presented as assuaging spookiness worries: "local algebras in AQFT pertaining to space like separated regions typically do satisfy the above independence notions" (Rédei, 2010, p. 1050).

Rather than reviewing these results in detail, I want to step back to indicate some gaps between the results and interpretive conclusions about locality in AQFT.

What can it mean to change ϕ? The assumption OPERATION has received little by way of examination or direct motivation. Halvorson and Müger may be expressing a sensitivity to the lacuna when they introduce OPERATION as "a standard (perhaps somewhat justified) assumption" (2007, p. 757). Others refer readers to Kraus (1983) "for a detailed description and interpretation of the notion of operation" (Rédei, 2010, p. 1046). For me, it is hard to distill a defense of OPERATION from Kraus's thorough articulation of the mathematical theory of operations.

One reason OPERATION deserves both examination and direct motivation is that it appears to violate axioms of the algebraic framework. ϕ, a global state on the algebra $\Re(M)$ associated with the entire spacetime M, is time-independent. How then are we to understand claims that ϕ *changes*, for instance by undergoing a ϕ to $T\phi$ transition when I do something in my lab this month? A natural idea is to assume that M is foliated by *time slices*, space-like surfaces with a good claim to be understood as instants of time, and, where S and S' are time slices to the past and future of my lab this month, understand the change of state as a difference between the state on the algebra $\Re(S)$ associated with the earlier surface and the state on the algebra $\Re(S')$ associated with the later surface. (For technical reasons, the algebras in question will be associated with thin sandwiches containing the time slices in question.) The problem with this idea is that it violates LOCAL PRIMITIVE CAUSALITY. In cases where a natural foliation of a spacetime into time slices is available, those time slices will be Cauchy surfaces, regions whose domain of dependence coincides with the entire spacetime. So we can suppose S is a Cauchy surface. It follows from LOCAL PRIMITIVE CAUSALITY that the algebra $\Re(S)$ associated with the past time slice contains the algebra $\Re(S')$ associated with the future time slice. Thus a state on $\Re(S)$ induces a state on $\Re(S')$, and nothing that happens in O can change this. But such changes of state appear to be exactly what OPERATION demands.

OPERATION is a substantive assumption. It requires defense. Defenses invoke collapse-like processes reveal results about operational dependence to concern not simply the algebras of AQFT, but the consequences of percolating dynamical miracles through those algebras.

Why not selective operations? Contributors to the literature don't take OPERATION seriously enough. AQFT no-signaling results restrict attention to "non-selective operations," which preserve the identity, and fail if selective operations are countenanced. In standard Lüders measurements, the transition $\phi \rightarrow \phi_A$ to a statistical mixture of determinate outcome states is a non-selective operation; the transition $\phi \rightarrow \phi_{a_n}$ to a specific determinate outcome is a selective operation. In the context of concerns about quantum spookiness, restricting attention to non-selective operations is like sticking one's head in the sand to elude predators. For there's a strong case to be made (Maudlin (2011) makes it with characteristic force) that, even if they can't be used to signal, selective measurements in the presence of entanglement entail "outcome-outcome" dependence and constitute causation in the sense of counterfactual dependence. If we're going to worry about notions of locality STR doesn't impose, why stop with superluminal signals and not consider superluminal causation as well?

It is particularly mysterious why one would adopt OPERATION yet restrict attention to non-selective operations. Consider, once again, selective and non-selective measurements. If we entertain transitions to statistical mixtures of outcome states (non-selective measurements), how can we make sense of those except as statistical mixtures of transitions to specific outcome states (selective measurements)? The only justification for excluding selective operations I've found in the literature is a passage from Clifton and Halvorson where they suggest (to my mind, not very persuasively) that selective operations are "conceptual" rather than "physical." Here is the passage:

> A selective operation involves performing a physical operation on an ensemble followed by a *purely conceptual* operation in which one makes a selection of subensemble on the basis of the outcome of the physical operation ... [in] Non-selective operations, by contrast, ... the final state is not obtained by selection but purely as a result of the physical interaction between object system and the device that effects the operation. (2001, p. 10; see Rédei, 2010, pp. 1055–1056, for a typical endorsement)

Conditional on what? The appeal to operations generalizes the standard model of quantum measurement. A reason the generalization might be *technically* expedient is that local algebras are bereft of the minimal projection operators deployed in standard measurement models; this limits the relevance to AQFT of results presupposing those models. But what are the *conceptual* grounds for the generalization? For an operation T_L that enacts a Lüders conditionalization, the set E_i of projection operators implementing the operation affords an account of why $T_L\phi$ is the appropriate state to attribute a system whose pre-measurement state was ϕ. For fans of collapse, $T_L\phi$ corresponds to a statistical mixture of post-collapse states E_i; for foes of collapse, $T_L\phi$ is the appropriate statistical mixture of probabilities ϕ assigns conditional on outcomes E_i. Either way, we understand what event structure $T_L\phi$'s probabilities are probabilities for: the event structure given by the Boolean algebra generated by the E_i, where each E_i defines a truth valuation on that algebra, and thus a way the world can be.

Contrast a more general operation T, e.g., one of Kraus form implemented by sets of positive (but not projection) operators K_i. Why think $T\phi$ is the appropriate state to attribute a system whose pre-operation state was ϕ? Because the K_i don't correspond to endpoints of collapse, $T\phi$ resists description as a statistical mixture of endpoints of collapse. Because the K_i aren't naturally thought of as conditions a system might occupy, $T\phi$ resists description as a statistical mixture of states obtained by conditioning ϕ on possible conditions the system might be in. (If we were told $T\phi$ reflects conditional probabilities, we'd have a right to ask: conditional on what?) Collectively, the K_i together with a quantum state generate probabilities summing to 1. However, individually they defy ready interpretation in terms of determinate states of affairs. When T is an operation more general than a Lüders conditionalization, it's difficult to see what's happening when the system is described by $T\phi$, and so difficult to see why we ought to describe the system that way.

So it is fair to ask whether and why the generalization from measurement projections to non-projective operations is acceptable. Beyond noting the expedience of the generalization for AQFT (where appropriate projections aren't always to be had) and referencing a mathematical literature developing (e.g.) representation theorems for general operations, the AQFT signaling literature has done little to squarely face the issue. Exceptions include work by Busch and collaborators, which link operations to an unsharpness "rooted in a genuine quantum indeterminacy inherent in the measuring device" (Busch et al., 1996, p. 129).

While I don't think the above concerns pose an insurmountable challenge to research programs making the key assumption OPERATION, I do think they're natural and central things to worry about, and as such deserve more attention.

20.3   Conclusion

The CATECHISM that the quantum world is non-local notwithstanding, the observable algebras of AQFT satisfy a wide variety of locality demands. AQFT states, however, exhibit a distinctively quantum non-locality wrought by entanglement and made possible by the non-commutativity of quantum observables. I urge we use the term "holism" for this feature of entangled quantum states, so that we might both track and underscore the contrast with the unimpeachable locality of AQFT algebras. Scenarios subjecting entangled states to measurement collapses and similar dynamical interruptions threaten to ramify entanglement into the sort of spookiness exemplified by the conventional commentary on the Reeh-Schlieder theorem, that one can steer the universe into any global state one chooses simply by performing appropriate local measurements. Unsurprisingly, such commentaries and ramifications rely on significant assumptions additional to the axioms of the AQFT framework. These include assumptions arguably in violation of those axioms! Care must be taken when drawing morals about locality in AQFT, especially morals about locality and independence of observables. Articulating the assumptions underlying those morals and addressing any conceptual loose ends encountered is one way to make further progress on provocative issues in the foundations of quantum theories.

Acknowledgments

I am grateful to Alastair Wilson for extremely helpful editorial attention to this contribution. Moreover, I am indebted to John Earman for voluminous conversation about and guidance through locality and nearby topics, as well as for stern feedback on earlier drafts.

Notes

1 The two most widely adopted introductions to the philosophy of QM – Albert (1992) and Maudlin (2011) – insist on this point.
2 See, e.g., Fine (1992), which shows that the inequalities are equivalent to the assumption (not plausibly branded a "locality" assumption) that for every pair and triple of quantum observables, a well-defined joint probability distribution exists.
3 For some examples, see the essay collection (Cushing and McMullin, 1989).
4 For a more extensive introduction, aimed at philosophers, see Ruetsche (2011). Halvorson and Müger (2007) offer a more sophisticated foundations-oriented exposition. Two standard mathematical references for operator algebra theory are Bratteli and Robinson (1987, 1997) and Kadison and Ringrose (1983, 1986). Limitations of space prevent a thorough survey of this burgeoning research field, and I'll omit many technical details and niceties from the topics I do manage to cover.
5 Summers (1990) gives a magisterial review of work done to that date. For surveys of more recent vintage, see Summers (2009) and Halvorson and Müger (2007, §3).
6 Earman and Valente (2014, §3, 5) develop details of the capture.
7 And likewise for other global states of bounded energy; see Clifton and Halvorson (2001) for details.
8 Rédei (2010) gives a compact review with an interpretive moral.
9 Space prohibits an adequate treatment of this nuanced issue, one strand of which is Arveson 1967's demonstration that no suitable conditional expectation exists for such algebras.

References

Albert, D. (1994). *Quantum Mechanics and Experience*. Cambridge, MA: Harvard.
Arveson, W.B. (1967). Analyticity in operator algebras. *American Journal of Mathematics*, 89: 578–642.
Bratteli, O. and Robinson, D.W. (1987). *Operator Algebras and Quantum Statistical Mechanics I* (2nd edition). Berlin: Springer-Verlag.
Bratteli, O. and Robinson, D.W. (1997). *Operator Algebras and Quantum Statistical Mechanics II* (2nd edition). Berlin: Springer-Verlag.
Bub, J. (1977). von Neumann's projection postulate as a probability conditionalization rule. *Journal of Philosophical Logic*, 6: 381–390.
Busch, P., Lahti, P.J. and Mittelstaedt, P. (1996). *The Quantum Theory of Measurement*. Berlin: Springer.
Butterfield, J. (2007). Reconsidering relativistic causality. *International Studies in the Philosophy of Science*, 21: 295–328.
Clifton, R. and Halvorson, H. (2001). Entanglement and open systems in algebraic quantum field theory. *Studies in History and Philosophy of Modern Physics*, 32: 1–31.
Cushing, J.T. and McMullin, E., editors. (1989). *Philosophical Consequences of Quantum Theory*. South Bend, IN: University of Notre Dame Press.
Earman, J. and Valente, G. (2014). Relativistic causality in algebraic quantum field theory. *International Studies in the Philosophy of Science*, 28: 1–48.
Fine, A. (1982). Hidden variables, joint probability, and the Bell inequalities. *Physical Review Letters*, 48: 291.
Geroch, R. (2011). Faster than light. *Advances in Lorentzian Geometry, AMS/IP Studies in Advanced Mathematics*, 49: 59–69.
Haag, R. and Schroer, B. (1962). Postulates of quantum field theory. *Journal of Mathematical Physics*, 3: 248–256.
Halvorson, H. and Müger, M. (2007). Algebraic quantum field theory. In J. Butterfield and J. Earman (eds.), *Handbook of Philosophy of Science: Philosophy of Physics*, Vol. 2. Amsterdam: Elsevier, pp. 731–922.
Kadison, R.V. and Ringrose, J.R. (1983). *Fundamentals of the Theory of Operator Algebras*, Vol. 1. New York: Academic Press.
Kadison, R.V. and Ringrose, J.R. (1986). *Fundamentals of the Theory of Operator Algebras*, Vol. 2. New York: Academic Press.
Kraus, K. (1983). *States, Effects and Operations*. Berlin: Springer.

Maudlin, T. (2011). *Quantum Non-Locality and Relativity: Metaphysical Intimations of Modern Physics*. Hoboken, NJ: Wiley.

Rédei, M. (2010). Einstein's dissatisfaction with nonrelativistic quantum mechanics and relativistic quantum field theory. *Philosophy of Science*, 77: 1042–1057.

Rédei, M. and Valente, G. (2010). How local are local operations in local quantum field theory? *Studies in History and Philosophy of Modern Physics*, 41: 346–353.

Redhead, M. (1995). More ado about nothing. *Foundations of Physics*, 25: 123–137.

Ruetsche, L. (2011). *Interpreting Quantum Theories: The Art of the Possible*. Oxford: Oxford.

Schlieder, S. (1968). Einige Bemerkungen zur Zustandsänderung von relativistischen quantenmechanischen Systemen durch Messungen und zur Lokalitätsforderung. *Communications in Mathematical Physics*, 7(4): 305–331.

Summers, S.J. (1990). On the independence of local algebras in quantum field theory. *Reviews in Mathematical Physics*, 2: 201–247.

Summers, S.J. (2009). Subsystems and independence in relativistic microscopic physics. *Studies in History and Philosophy of Modern Physics*, 40: 133–141.

Further Reading from the Editors

Readers unfamiliar with locality considerations in non-relativistic quantum theory may wish to start with E. Keming Chen, Chapter 12 in this volume, and R. Healey, Chapter 47 in this volume. Once warmed up, a next step is N. Swanson (2017). A philosopher's guide to the foundations of quantum field theory. *Philosophy Compass*, 12(5): e12414. For more technical philosophical treatments, see N. Huggett, (2000). Philosophical foundations of quantum field theory. *The British Journal for the Philosophy of Science*, 51: 617–637 and the extremely comprehensive H. Halvorson and M. Müger. (2006). Algebraic quantum field theory. In J. Butterfield and J. Earman (eds.), *Philosophy of physics* (Amsterdam: Elsevier), 731–922. For the canonical treatment of the relevant physics, see R. Haag, *Local Quantum Physics* (Berlin: Springer Verlag, 1992).

21

PARTICLES IN QUANTUM FIELD THEORY

Doreen Fraser

The consensus view among philosophers of physics is that relativistic quantum field theory (QFT)[1] does not describe particles. That is, according to QFT, particles are not fundamental entities. This assessment is apparently at odds with the fact that QFT is the theoretical framework for particle physics. The Standard Model presents a taxonomy of particle species (e.g., electron, photon, quarks, neutrinos). At the Large Hadron Collider (LHC), protons with high energies are smashed together. The recent detection of the Higgs boson at the LHC was heralded as a major discovery because the existence of the Higgs boson was finally experimentally confirmed. Clearly, particles play a central role in both theory and experiment. How does the negative conclusion that QFT does not sustain a particle interpretation square with the positive role that the particle notion plays in particle physics?

The first part of this chapter lays out multiple lines of negative argument that all conclude that QFT cannot be given a particle interpretation. These arguments probe the properties of the "particles" in standard formulations of QFT and the limited applicability of "particle" representations. (Approaches that posit localized particles and then modify QFT to accommodate this posit – e.g., variants of Bohmian relativistic quantum theory – fall outside the scope of this review.) The second part of the chapter surveys proposals for non-fundamental roles that the particle concept plays in particle physics. The conclusion suggests directions for future philosophical research. The pressing question of what sorts of fundamental entities QFT does describe will be set aside. This is an outstanding question that is a topic of current research. The natural candidate is fields. However, the fields which feature in the QFT formalism are mathematical expressions; work needs to be done to determine the ontological properties of the entities that are represented by the mathematical field formalism. Several suggestions about how to carry out this project have been made. This is a challenging task, a point that has been underscored by arguments that the most straightforward field interpretation of QFT fails for reasons that are closely related to the reasons for which the quanta interpretation fails (Baker, 2009).

21.1 The Quanta Interpretation of Fock Space

QFT is formulated as a field theory: the mathematical formalism contains fields (i.e., mathematical expressions that associate operators with regions of spacetime). The question of whether QFT describes particles amounts to the question of whether the field formalism can be given an interpretation in terms of particles. For massive free bosonic systems in Minkowski spacetime and inertial observers, there is a natural way of giving the field formalism a particle interpretation. A system in this category can be represented using a Fock space representation of the equal-time canonical

323

commutation relations (ETCCRs) that, for a fixed value of mass m, is unique (up to unitary equivalence). This representation is constructed by effecting a positive-negative Fourier decomposition of a classical free field and then promoting the coefficients to Heisenberg picture operators.[2] A Fock representation for a free bosonic real scalar field with $m > 0$ on Minkowski spacetime possesses the following formal properties:

1. *Field Operators*: There exist well-defined annihilation and creation operators $a(\mathbf{k}, t)$, $a^\dagger(\mathbf{k}, t)$ where $\mathbf{k}^2 = k_0^2 - m^2$. $a(\mathbf{k}, t)$, $a^\dagger(\mathbf{k}, t)$ obey the ETCCR's

$$\left[a(\mathbf{k}, t), a(\mathbf{k}', t)\right] = 0, \left[a^\dagger(\mathbf{k}, t), a^\dagger(\mathbf{k}', t)\right] = 0, \left[a(\mathbf{k}, t), a^\dagger(\mathbf{k}', t)\right] = \delta^3(\mathbf{k} - \mathbf{k}') \quad (21.1)$$

At any given time t the quantum field $\phi(\mathbf{x}, t)$ can be defined as follows (where $\omega_{\mathbf{k}}^2 = k_0^2 = \mathbf{k}^2 + m^2$):

$$\phi(\mathbf{x}, t) = \int \frac{d^3 k}{(2\pi)^{\frac{3}{2}}\sqrt{2\omega_{\mathbf{k}}}} \left[a^\dagger(\mathbf{k}, t)e^{ik\cdot x} + a(\mathbf{k}, t)e^{-ik\cdot x}\right] \quad (21.2)$$

The conjugate momentum field $\pi(\mathbf{x}, t)$ is defined using $\pi(\mathbf{x}, t) = \frac{\partial \phi(\mathbf{x}, t)}{\partial t}$ and $\frac{\partial a(\mathbf{k}, t)}{\partial t} = \frac{\partial a^\dagger(\mathbf{k}, t)}{\partial t} = 0$:

$$\pi(\mathbf{x}, t) = \int \frac{d^3 k}{(2\pi)^{\frac{3}{2}}\sqrt{2\omega_{\mathbf{k}}}} \left[i\omega_{\mathbf{k}} a^\dagger(\mathbf{k}, t)e^{ik\cdot x} - i\omega_{\mathbf{k}} a(\mathbf{k}, t)e^{-ik\cdot x}\right] \quad (21.3)$$

Inverting and solving for $a(\mathbf{k}, t)$, $a^\dagger(\mathbf{k}, t)$ gives

$$a(\mathbf{k}, t) = \int \frac{d^3 x}{(2\pi)^{\frac{3}{2}}\sqrt{2\omega_{\mathbf{k}}}} e^{ik\cdot x} [\omega_{\mathbf{k}}\phi(\mathbf{x}, t) + i\pi(\mathbf{x}, t)] \quad (21.4a)$$

$$a^\dagger(\mathbf{k}, t) = \int \frac{d^3 x}{(2\pi)^{\frac{3}{2}}\sqrt{2\omega_{\mathbf{k}}}} e^{ik\cdot x} [\omega_{\mathbf{k}}\phi(\mathbf{x}, t) - i\pi(\mathbf{x}, t)] \quad (21.4b)$$

2. *No-particle State*: There exists a unique (up to phase factor) "no-particle state" $|0\rangle$ such that

$$a(\mathbf{k}, t)|0\rangle = 0 \quad \text{for all } \mathbf{k}$$

3. *Number Operators*: Number operators $N(\mathbf{k})$ can be defined for any t:

$$N(\mathbf{k}) = a^\dagger(\mathbf{k}, t)a(\mathbf{k}, t) \quad (21.5)$$

$$N(\mathbf{k})[a^\dagger(\mathbf{k}, t)^n|0\rangle] = n[a^\dagger(\mathbf{k}, t)^n|0\rangle]$$

where $n = \{0, 1, 2, \ldots\}$. When normalized, the n-particle state becomes $\frac{a^\dagger(\mathbf{k}, t)^n|0\rangle}{\sqrt{n!}}$.

In addition, for any t, the total number operator $N = \int d^3 k N(\mathbf{k}) = \int d^3 k a^\dagger(\mathbf{k}, t)a(\mathbf{k}, t)$ is well-defined. (That is, when the number operators are properly defined using a test function space $\mathcal{T}$ $N = \sum_{j=1}^{\infty} a^\dagger(f_j)a(f_j)$ converges in the sense of strong convergence on the domain of N where $\{f_j\}$ is an orthonormal basis of $\mathcal{T}$ and N exists only if N exists and is the same for every choice of orthonormal basis $\{f_j\}$ (Dell'Antonio et al., 1966, pp. 225–226).)

4. *Fock Space*: The one-particle Hilbert space $\mathcal{H}$ has as a basis the set of vectors generated from $|0\rangle$ by single applications of $a^\dagger(\mathbf{k}, t)$ (for any $\mathbf{k}$ satisfying $\mathbf{k}^2 = k_0^2 - m^2$). The Fock space $\mathcal{F}$ for $\phi(\mathbf{x}, t)$ is obtained by taking the direct sum of the n-fold symmetric tensor

product of $\mathcal{H}$: $\mathcal{F}(\mathcal{H}) = \oplus_{n=0}^{\infty}(\otimes^n\mathcal{H})$ (Wald, 1994, p. 192). $|0\rangle$ is cyclic with respect to the $a^\dagger(\mathbf{k}, t)$'s.

These formal properties naturally lend themselves to a particle interpretation. The total number operator N can be physically interpreted as an operator that counts the number of particles because eigenstates of N can be physically interpreted as possessing a determinate number of particles. $|0\rangle$ is interpreted as a state in which there are no particles because it is the unique ground state (i.e., state with the lowest energy) and the unique state that is invariant under the unitary operators that give a representation of the Poincaré group (and thus looks the same to all inertial observers) (Streater and Wightman, 2000, p. 21). $a^\dagger(\mathbf{k}, t)|0\rangle$ is a state containing a single particle of momentum $\mathbf{k}$ and mass m because it possesses the correct energy for a single non-interacting particle, $\sqrt{\mathbf{k}^2 + m^2}$. Similarly, each of the eigenvectors $a^\dagger(\mathbf{k}, t)^n|0\rangle$ is identifiable as an n-particle state because each possesses the correct relativistic energy. The particles can be aggregated: a system which possesses n particles can be combined with a system which possesses m particles to yield a system which possesses $n + m$ particles.

Quantum particles are particlelike in these respects: they can be counted, they can be aggregated, and they have the expected relativistic energies. However, there are also respects in which these quantum particles seem decidedly un-particlelike. For this reason, the term "quanta" (the singular form is "quantum") is typically used to refer to these quantum entities and the term "particle" is typically reserved for entities that satisfy our ordinary conception of particles (Teller, 1995, p. 29). The most obvious difference is that some physical states are superpositions of eigenstates of particle number. A further respect in which quanta differ from particles is that quanta are not capable of bearing labels. Consider a system of two quanta. It does not make sense to name one of these quanta "Fred" and the other "Sally." Put another way, it is not possible to make reference to *this* quantum in a way that distinguishes it from *that* quantum. Quanta cannot be considered individuals in this sense. Nevertheless, we can still count them and aggregate them. On this basis, Teller (1995, p. 30) argues that quanta are sufficiently particlelike to count as a species of particles. However, subsequent arguments undermine this basis for giving QFT a particle interpretation. Teller (1995, pp. 85–91) recognizes that in the Fock space for a massive free system for a Klein–Gordon field the position eigenstates are not localizable in a finite region of space, but holds out hope that a "salvage operation" can be undertaken for exact localization. The subsequent work described in Section 21.2.1 clarified the situation.

21.2   Negative Arguments: QFT Does Not Describe Particles

21.2.1   *Quanta Cannot Be Localized in Any Finite Region of Space*

A more serious respect in which quanta are not particlelike is that they cannot be localized in any finite region of space. Malament (1996) proves a theorem intended to formalize the "dogma" that "there cannot be a relativistic, quantum mechanical theory of (localizable) particles" (1). In other words, the unification of special relativity and quantum theory leads to a field theory. Malament establishes that, given a set of reasonable conditions on a relativistic quantum theory, the probability of finding a particle in any bounded region of space is zero. Halvorson and Clifton (2002) strengthen Malament's result by proving it under weaker assumptions. They also prove a theorem stating that, again given reasonable relativistic assumptions, number operators such as those in a Fock space representation cannot be defined on finite regions of space. Strictly speaking, then, quanta cannot be conceived of as occupying any finite region of space.

While some have taken these results to be sufficient grounds for concluding that quanta are not particlelike entities, others have argued that these results are not fatal to a particle interpretation (Fleming, 2004). I will not pursue this issue here because it turns out that there are other sources of trouble for the quanta interpretation of QFT.

21.2.2 Accelerating Observers Disagree on the Number of Quanta

An important way in which QFT differs from non-relativistic quantum mechanics is that in QFT systems generically possess an infinite number of degrees of freedom. In non-relativistic quantum mechanics (with a finite number of degrees of freedom), the Stone–von Neumann theorem typically guarantees that all irreducible Hilbert space representations of the canonical communication relations that satisfy natural assumptions are unitarily equivalent to the standard Schrödinger representation (Ruetsche, 2011, Chapter 2). (That is, there is a unitary map $U: \mathcal{H}^1 \rightarrow \mathcal{H}^2$ from the set of operators $\{O_i^1\}$ on the first Hilbert space to the set of operators $\{O_i^2\}$ on the second Hilbert space such that $U^{-1} O_i^2 U = O_i^1$ for all i. This map preserves the expectation values in all states.[3]) In QFT, the Stone–von Neumann theorem does not hold because the assumption of a finite number of degrees of freedom is violated. As a result, in QFT there are continuously many representations available as mathematically possible representations. The availability of these representations is a valuable expressive resource. Of course, each of these mathematically possible representations may or may not have physical significance.

One situation in which unitarily inequivalent representations seem to be physically relevant is for accelerating observers, a phenomenon known as the Unruh effect.[4] Imagine two observers in flat spacetime, Jack and Jill, who are both observing the same free system.[5] Jack is moving inertially (i.e., has zero acceleration) and Jill is accelerating uniformly. Jack employs the Fock space representation described in Section 21.1, with the Minkowski vacuum and number operators. For Jill, it is most natural to represent the system using Rindler spacetime coordinates. She performs a positive-negative frequency decomposition with respect to her Rindler spacetime coordinates (more precisely, she performs the positive-negative frequency decomposition on the right Rindler wedge, which is itself a spacetime (Ruetsche, 2011, Sec. 9.6)) and then obtains a Fock space representation by following the same procedure outlined in Section 21.1. Jill's "Rindler" Fock space representation is unitarily inequivalent to Jack's "Minkowski" Fock space representation. Jill's Rindler total number operator is not defined on Jack's Minkowski vacuum state; conventionally, applying Jill's Rindler total number operator to Jack's Minkowski vacuum state returns the value infinity. Jill's Rindler total number operator is not defined on any of Jack's Minkowski n-particle states either. Similarly, Jack's Minkowski total number operators are not defined on Jill's Rindler vacuum state or Rindler n-particle states.

What does this mean for the possibility of a particle interpretation of the system? Different conclusions have been defended. Clifton and Halvorson (2001) argue that Jack's and Jill's descriptions of the particle content of the system are *complementary* in the sense that "neither the Minkowski nor Rindler perspective yields the uniquely 'correct' story about the particle content of the field, and that both are necessary to provide a complete picture" (459). In experimental terms, the idea is that Jill's and Jack's particle detectors couple to different particle observables associated with the field. While Clifton and Halvorson's assessment that Jill's and Jack's particle notions are complementary could be used to defend a fundamental particle ontology by salvaging objective quanta descriptions from the apparently conflicting subject-dependent descriptions, Clifton and Halvorson are unequivocal in their rejection of the fundamental particle interpretation: "quantum field theory is 'fundamentally' a theory of a field, not particles" (459). Arageorgis et al. (2003) argue that Jack's and Jill's particle concepts are *incommensurable* in the sense that "Jack's and Jill's particle assessments aren't different descriptions of the same set of facts (however complicated), but descriptions of disjoint sets of facts" (Ruetsche, 2011, p. 219). On the surface, this appears to spell disaster for a fundamental particle ontology. However, this is not the end of the story. Arageorgis, Earman, and Ruetsche further argue (following Wald (1994)) that Jill's Rindler Fock space representation is not physically acceptable because it fails to satisfy a physical condition pertaining to stress-energy: that Jill's vacuum state on the right Rindler wedge be extendable to a state on the full Minkowski spacetime that satisfies the Hadamard condition, which guarantees that a stress-energy observable is well-defined (Ruetsche (2011, Sec. 10.3). Jack's Minkowski Fock space representation is the only representation that is physically acceptable.

On this reading of the situation, a fundamental particle ontology is not threatened. However, Arageorgis, Earman, and Ruetsche agree with Clifton and Halvorson that – for reasons unrelated to the Unruh effect – "the particle notion should be demoted in QFT from fundamental to derivative status" (165). Thus, on either reading, the Unruh effect does not undermine a fundamental particle ontology; however, this is cold comfort because other arguments are fatal to particles.

21.2.3   No Unique Fock Space Representation in General in Curved Spacetimes

General relativity introduces curved spacetime to treat gravity. The procedure for constructing a unique (up to unitary equivalence) Fock space representation for a free system of mass m that is outlined in Section 21.1 is particular to Minkowski spacetime. This procedure can be generalized to stationary spacetimes (i.e., spacetimes possessing a time translation symmetry). (See Ruetsche (2011, Sec. 10.2) and Wald (1994, Chapter 4).) However, the procedure cannot be generalized to non-stationary spacetimes. The obstacle is that the requirement that the vacuum state in the Fock representation be invariant under time translations plays a crucial role in picking out a unique vacuum state. In non-stationary spacetimes, there is no time translation symmetry to play this role. The absence of time translation symmetry also poses difficulties for ascribing physically appropriate energies to one-particle states. As a result, infinitely many unitarily inequivalent analog of a Fock space representation can be constructed for a free system of mass m in a non-stationary spacetime and there is no physical reason to privilege one (unitary equivalence class) of these analog Fock representations over the others. Thus, in generic curved spacetimes, the availability of a particle description is not guaranteed. Wald (1994) argues that, as a result, the particle concept should not be regarded as fundamental. Again, the Hadamard condition can be invoked to reduce the number of instances in which there is no unique physically privileged particle concept, but cases of non-uniqueness will remain. In particular, the Hadamard condition only secures uniqueness for a spacetime with a compact Cauchy surface; for an "open universe," the non-uniqueness remains (Wald, 1994, pp. 96–97).)

21.2.4   Interacting Systems Cannot Be Given a Quanta Interpretation

Another difficulty for the quanta interpretation of QFT is that the analysis in terms of quanta offered in the preceding paragraphs is restricted to free systems. Free systems are systems in which there are no interactions. As a consequence of Haag's theorem, an interacting system cannot be given a quanta interpretation by representing it using the Fock representation for the corresponding free field (Haag, 1955; Fraser, 2008). Fraser (2008) argues that the Fock representation cannot be generalized to interacting systems in a way that preserves the quanta interpretation. One approach would be to quantize a classical interacting field by carrying out the same mathematical construction that, for a free field, generates a Fock representation. This approach fails because the Fourier decomposition of a classical interacting field is not relativistically covariant, and therefore it is not a candidate for representing physical fields in relativistic QFT. A second strategy would be to pick out a unique (up to unitary equivalence) Hilbert space representation of the canonical commutation relations by stipulating that it shares the formal properties of the Fock representation laid out in conditions 1–4 in Section 21.1. This strategy fails because the Hilbert space representation that is picked out in this way cannot be physically interpreted in terms of quanta in the same manner as a Fock space representation for a free system. In particular, for all non-trivial interactions, the no-particle state does not coincide with the vacuum state and, typically, the argument that one-particle states have the energy expectation values that special relativity assigns to single-particle states is undercut. It is worth noting that this is, in a way, a more severe blow to the particle interpretation of QFT than the result that the quanta counted by number operators in Fock representations for free systems are not localizable. The unavailability of either the Fock representation or a physical analog of the Fock representation means that there is no support for even non-localizable quanta in interacting systems!

(Of course, a particle ontology could be restored by finding a way to give a particle interpretation of QFT that does not rely on Fock space, but this is not a plausible option until there is a concrete proposal for how this interpretation would proceed.)

21.2.5 The Role of Relativity

Multiple lines of argument corroborate the negative conclusion that QFT cannot be given an interpretation in terms of particles that are aggregable and localizable. There is a common element to all of the arguments surveyed in this section: relativity theory is a key determining factor in the availability of a particle interpretation of QFT. On the one hand, the special theory of relativity makes a quanta interpretation of Fock space possible by introducing the mass–energy relation. Without the mass–energy relation, there would be no grounds for interpreting the eigenstates of total number operator N as representing definite numbers of particles rather than merely more examples of discrete energy level states that are the hallmark of non-relativistic quantum mechanics. On the hand, what special relativity gives, special relativity also takes away. For free systems, relativistic assumptions are required to obtain the result that quanta are not localizable in a finite region of space. For interacting systems, Haag's theorem relies on relativistic premises;[6] the Fourier decomposition is not covariant under Poincaré transformations; and, in the formal analog of the Fock space representation, the no-particle state is not invariant under Poincaré transformations and special relativity typically does not supply the correct assignment of energies to one-particle states. (For a free system, special relativity and the linear field equation conspire to produce a quanta interpretation; for an interacting system, the combination of special relativity and the non-linear field equation is not so fortuitous.) Bain (2011) argues that it is specifically the absolute temporal structure of non-relativistic spacetime that is required to support quanta that are countable and localizable. General relativity undermines quanta from a different direction, by (in general) dispensing with the stationary spacetime structure that in the context of Minkowski spacetime picks out a unique Fock space representation.

21.3 Particles as Non-Fundamental Entities: A Survey of Options

There are many reasons to think that the arguments for the negative conclusion that QFT does not support an ontology that includes particles as fundamental entities are not the whole story. Particles do seem to play an important role in the phenomenology associated with QFT. Cloud chamber photographs are taken to show trails of particles such as electrons and positrons traveling on approximately localized trajectories. Experimental tests of QFT often take the form of scattering experiments, which are taken to involve colliding particles, such as the collisions of protons at the Large Hadron Collider (LHC). Particle decay has been the subject of other experimental tests. If particles are not fundamental entities, then how are we to understand these experiments?

The theoretical side of QFT also supplies a strong rationale for following up on the negative conclusions with a positive investigation of the non-fundamental roles played by the particle concept. The content of the Standard Model of particle physics is typically cashed out in terms of families of particles of different types (leptons, quarks, bosons) and their properties (mass, charge, etc.). In the history of particle physics, the development of models has been motivated by puzzles concerning particles and solutions to puzzles have involved proposing new particles. For example, the infamous introduction of spontaneous symmetry breaking and the massive Higgs boson into the electroweak theory was prompted by the (apparent) puzzle that spontaneously broken symmetries are necessarily accompanied by massless Goldstone bosons, which had not been observed. More generally, within particle physics and outside of it, the atomic hypothesis is widely regarded as being one of the best confirmed theoretical posits in all of science. Writing about thermal systems, Norton opines "a thesis of ontological reduction asserts that thermal systems just are systems of many molecules, spins, radiation modes, and so on. ... While the ontological thesis is quite ambitious, the evidence in its favour is so massive that, now, no one who doubts it is or should be taken seriously" (2014, p. 206). This assessment is not universally shared. For example, Healey argues that the "decompositional strategy" of decomposing matter into atomic and sub-atomic parts that is an element of the atomic hypothesis has "probably run its course" in particle physics (2013, p. 56). But, for proponents, it does raise a question: How are we to reconcile the empirically well-supported components of the atomic hypothesis with the conclusion that particles are not among the fundamental entities? For example, Perrin's famous argument for the truth of the atomic hypothesis involved agreeing measurements of Avogadro's number, which is taken to represent the number of particles in a mole of a substance.

In this section, proposals for the non-fundamental role played by particles will be surveyed. Presumably, the nature of the fundamental constituents of the ontology of quantum field theory is relevant for fleshing out most of these proposals, but the important issue of how to determine the fundamental ontology will be set aside. The notions of approximation and idealization will play a prominent role in the taxonomy. To fix our terminology, we will adopt the definitions introduced in Norton (2012). An *approximation* is "an inexact description of a target system" and "it is propositional" (209). An *idealization* is "a real or fictitious system, distinct from the target system, some of whose properties provide an inexact description of some aspects of the target system" (209). The simple example of a body falling in a weakly resisting medium treated using classical mechanics illustrates the difference (210). The equation for velocity as a function of time that neglects velocity $v(t) = gt$ is used as an approximation when it is applied as an inexact description of the falling body. An idealization is a body falling in a vacuum, which is (exactly) described by the same equation $v(t) = gt$.

21.3.1 Operationalism: Particles Are What Particle Detectors Detect

In an article entitled "Particles do not exist," Paul Davies defends an operational interpretation of quanta:

> There are quantum states and there are particle detectors. Quantum field theory enables us to predict probabilistically how a particular detector will respond to that state. That is all. That is all there can ever be in physics, because physics is about the observations and measurements that we can make in the world. We can't talk meaningfully about whether such-and-such a state contains particles except in the context of a specified particle detector measurement. (69)

Davies' slogan is "particles are what particle detectors detect" (75). On its own, this brand of operationalism about particles addresses the first motivation for a non-fundamental particle notion – namely, to account for the phenomenology of particle detection – but it does not address the other motivations. The other motivations require a connection between the experimental context of particle detectors and the theory. How does QFT enable us to predict the response of particle detectors? Moreover, in the context of a fundamental ontology for QFT which does not include particles, it seems mysterious why particle detectors would reliably behave as if they detect particles without some further explanation.

21.3.2 Particles Are an Approximation

Like Davies, Halvorson and Clifton (2002) regard talk about particles as mere talk which has no referents in the actual world. They maintain that talk about particles is a "useful fiction" (20). Halvorson and Clifton add to Davies' operationalism a proposal for grounding particle detector phenomenology in the theory of QFT. They are particularly concerned with addressing the nonlocalizability of *n*-particle states of free systems. The idea is that *what particle detectors actually detect approximately fits the description of particles*. *n*-particle states in Fock space are an inexact description of the states actually measured by particle detectors. Halvorson and Clifton use the algebraic QFT framework and take fields to be part of the fundamental ontology. What particle detectors actually measure is some local observable (i.e., an observable that is measurable within spacetime region O) which is represented by an element C of the local algebra of observables $\mathcal{R}(O)$. This local observable is not a particle observable because (by the Reeh-Schlieder theorem) it does not have an expectation value of zero in the vacuum state. However, this local observable gives the appearance of being a particle observable because FAPP ("for all practical purposes") it is observationally indistinguishable from an observable that is "almost localized" within O and does have zero vacuum expectation value. That is, if we allow ourselves some error bound δ on our measurement of C, then there exists some almost localized observable C' that has expectation values that are within the error bound δ of the expectation values of C.

Colosi and Rovelli (2008) offer a different account of how particles are an approximation. Their primary motivation is (pace Wald) to formulate a particle notion that is well-defined in curved spacetime, and thus could furnish a starting point for a theory of quantum gravity. Like Halvorson and Clifton, they take the states that particle detectors actually measure to be localized. Unlike Halvorson and Clifton, Colosi and Rovelli identify "local particle observables" associated with particle detectors. Rodríguez-Vázquez et al. (2014) further develop this proposal. Local particle number observables are defined for each particle detector at each time. Malament's theorem is evaded by rejecting the very first step in the construction of the Fock space representation: the local *n*-particle states at a specified time are *not* composed of exclusively the positive frequency modes identified in the positive-negative frequency decomposition; instead both positive and negative frequency modes are used (Rodríguez-Vázquez et al., 2014, pp. 119–120). The fact that time translations are not needed to uniquely decompose the classical field into positive and negative frequency modes means that local *n*-particle states at a specified time for a given particle detector can be defined in non-stationary spacetimes. The local *n*-particle operators are approximations of the global *n*-particle operators on the Fock space in the sense that the local *n*-particle operators converge to the global *n*-particle operators in the weak limit (Colosi and Rovelli, 2008). (That is, expectation values of the local *n*-particle operators for some region R converge to the expectation values of the global *n*-particle operators as $R \to \infty$.) The particulate properties of global Fock space are "an artifact of the simplification taken by approximating a truly observed local particle state with easier-to-deal-with Fock particles" (2). (Rodríguez-Vázquez et al. (2014) modify Colosi and Rovelli's construction in order to satisfy the spatial boundary conditions. A consequence of respecting the boundary conditions is that the local modes can only be chosen to be spatially localized at a given time and not for all times.)

Colosi and Rovelli pose the question of whether their construction supports an interpretation of QFT according to which particles are fundamental entities. Their own answer to this question is "partially a yes and partially a no," but they do conclude their paper with the statement that "[t]he world is far more subtle than a bunch of particles that interact" (15). The "partially a yes" refers to the interpretive option of regarding local *n*-particle observables as "complementary," but not in the robust sense of Section 21.2.2 of providing complementary descriptions of a common underlying ontology with a single set of objects; the observables are complementary in the minimal sense that "there is no reason to select an observable as 'more real' than the others" (15). Colosi and Rovelli provide compelling reasons that the local *n*-particle operators and states do not support a fundamental particle notion. In essence, the trade-off for dropping the requirement of time translation invariance that went into the construction of global Fock space is that there is no unique basis of local *n*-particle states. In contrast to the global *n*-particle states and operators defined on the Fock space, a set of local *n*-particle states and operators is defined with respect to a detector that makes measurements within some finite region. The Hamiltonians associated with different detectors do not commute. Colosi and Rovelli point out the consequence that "[w]hether a particle exists or not depends on what I decide to measure," which is reminiscent of Davies' operationalist slogan that "particles are what particle detectors detect" (15). However, in the introduction to the paper, Colosi and Rovelli hold up Davies' position as a target for their arguments. This underscores that the main goal of Colosi and Rovelli's paper is not to shed light on the ontology of particles, but to argue that Davies (and Wald) have overlooked mathematically possible representations for curved spacetime that retain some features of Fock space in modified form. For Colosi and Rovelli, whether the mathematical framework of local particle observables latches onto the fundamental ontology is not the most important consideration for determining whether to use this mathematical framework in quantum gravity.

The approaches of both Halvorson and Clifton and Colosi and Rovelli are best categorized as regarding particle concepts as approximate in Norton's sense. Both approaches establish that the global, non-localized Fock *n*-particle descriptions are inexact but approximately accurate descriptions of the states identified by each approach as exactly representing the target system. For Halvorson and Clifton, the local observables are measured by particle detectors. Though they do allude to particles being fictions, which sounds like idealization in Norton's sense, their analysis is aimed at establishing the usefulness of talk of particles rather than carving out a role for particles as fictional entities. Colosi and Rovelli take local *n*-particle states associated with detectors to be exact states, and their primary aim is not to defend particles as fictional entities, but to relate these states to global *n*-particle Fock states by approximation.

21.3.3   Particles Are an Idealization Introduced by Scattering Theory

The development of the notion of particles as an approximation that was traced in the previous sub-section was largely a response to no-go theorems for free systems surveyed in Section 21.2. The no-go results for interacting systems inspire a different approach to regarding particles as non-fundamental entities. Bain (2000) argues that scattering theory supports a variety of particle interpretation at asymptotically early times prior to an interaction and asymptotically late times after an interaction. Essentially, at these asymptotic times, the interacting system tends to a free system and the Hilbert space representation is suitably related to a Fock space representation for a free system in the infinite limit. (LSZ and Haag-Ruelle scattering theories establish a suitable relationship to the Fock space representation for a free system in the asymptotic limit without violating Haag's theorem. See Fraser (2006, Sec. 3.1.2) for details.) Bain proposes that "a 'particle' be considered a system that minimally possesses an asymptotic state (i.e., a system that is free for all practical purposes at asymptotic times)" (394). Naturally, this "particle" can also be regarded as localized for all practical purposes, along the same lines sketched in the preceding section (Bain, 2000, pp. 395–396).

This notion of particle is an idealization in Norton's sense. The free system described by the Fock space representation is a fictitious system. In the real world, there are always interactions. In the infinite limits of early and late times, the properties of the fictitious free system provide an inexact description of the real interacting system. In particular, the particulate properties of the free system provide an inexact description of the interacting system. (Additional layer of approximation: the properties of the free system are only particulate – i.e., localized – for all practical purposes.) The fictitious system must be posited because, as explained in Section 21.2.4, interacting systems do not admit a quanta interpretation. In Norton's terms, this particle notion is an idealization – a mere approximation is insufficient. Within this approach, particles have the status of fictitious entities.

21.3.4   Realism: Particles Are Emergent Entities

Another way of fleshing out the non-fundamental status of particles is to regard them as emergent. Wallace (2001) defends this interpretation. The animating idea of his interpretation is that bosons in relativistic QFT have the same ontological status as quasi-particles (e.g., phonons) in non-relativistic condensed matter physics. Phonons are vibrational modes of an atomic crystal. For strongly interacting systems, there are phenomena which are most effectively represented using phonons. For example, heat transport is modeled using localized phonons. Representations involving phonons are useful, but in condensed matter physics it is clear that phonons are not part of the fundamental ontology. The atoms in the crystal lattice are part of the fundamental ontology; the phonons represent collective excitations of the crystal lattice. Wallace argues that phonons are nevertheless real:

> Are quasi-particles real? They can be created and annihilated; they can be scattered off one another; they can be detected (by, for instance, scattering them off "real" particles like neutrons); sometimes we can even measure their time of flight; they play a crucial part in solid-state explanations. We have no more evidence than this that "real" particles exist, and so it seems absurd to deny that quasi-particles exist–and yet, they consist only of a certain pattern within the constituents of the solid-state system in question. (2010, p. 59)

Following Dennett, Wallace regards patterns as constituting real, emergent entities (or properties) when they prove useful in a theory, especially with respect to explanatory power and predictive reliability (2010, p. 58).

Wallace reverses the argument in the block quote to support the conclusion that particles in QFT should be classified as emergent. Using quasi-particles as a template, he argues that in order to grant particles emergent status it is only necessary to provide pragmatic definitions; a notion of particle that is approximately localized to an extent that it can be regarded as localized in practice is sufficient. Wallace's formal proposal for introducing particle states that are approximately localized is similar to Halvorson and Clifton's,[7] but the interpretation is different. Halvorson and Clifton regard particle states as merely approximations, while Wallace classifies particles as real, emergent entities. Wallace also points out that Haag-Ruelle scattering theory uses a similar formal framework for analyzing the particle content at asymptotically early or late times (2001, pp. 8, 38). However, the physical interpretation again differs from Bain's. Wallace considers particles not as merely idealizations, but as real, emergent entities.

An important source of motivation for regarding particles in particle physics as having the same status as quasi-particles is the effective field theory perspective on QFT. As Wallace stresses in his contribution to this volume, a compelling reason to regard particles as emergent is that particle representations are scale-relative. For example, in QED the interaction between electrons and photons increases in strength as the energy scale gets higher. Different particle representations with different particulate properties (e.g., mass) are required for different scales. A related consequence of the application of renormalization group methods in particle physics is that the fact that the Standard Model has been empirically well-confirmed by experiments at relatively low-energy scales is compatible with the truth of a whole class of models (with different Lagrangians) at much higher energy scales.[8] The Standard Model is effectively valid at relatively low-energy scales in the same manner that quasi-particle models in solid-state physics are effectively valid at relatively large distance scales. Fraser (2017) and Williams (2017) have recently argued that the effective field theory perspective supports a brand of selective scientific realism according to which, if there are features of a model (including particles) valid at a specified lower energy scale that are insensitive to assumptions about the unknown physics of higher energy scales (for example, how a high energy-momentum cutoff is implemented or which of the possible Lagrangians that satisfies basic constraints – e.g., symmetries – is used), then it is appropriate to adopt a realist attitude toward these features.

21.3.5   *Anti-realism: Particles Are Intuitive Pictures That Are Heuristically Useful But Do Not Represent the World*

Wallace appeals to quasi-particles to make the methodological point that, when doing metaphysics, pragmatic characterizations that involve approximations and idealizations can be used to pick out emergent entities. However, as Wallace points out, there are strong formal similarities between the mathematical representations of particles in QFT and quasi-particles in solid-state physics. There is a long history of frameworks for representing particles being passed back and forth between relativistic QFT and non-relativistic quantum mechanics applied to many-body systems. Dirac's "hole theory" of the electron was possibly inspired by ionic crystal models of conductors constructed by Frenkel in the 1920s, and Dirac's idea was certainly picked up in solid-state physics in the 1930s (Kojevnikov, 1999). Renormalization techniques developed for QED in the 1940s and the associated concept of dressed electrons were exported from QED to solid-state physics in the 1950s (Blum and Joas, 2016). For example, Bohm and Pines' electron gas model for metals introduces an effective heavy electron and plasmons. In the late 1950s and 1960s, spontaneous symmetry breaking (SSB) and associated quasi-particles were developed in models of superconductivity and then exported to particle physics. These analogies inspire another approach to fleshing out a particle interpretation for particle physics: regard quasi-particles as ontological templates for particles in particle physics (i.e., particles have the same (or similar) physical properties as quasi-particles). (In contrast, Wallace argues that quasi-particles and particles have the same ontological status, but does not argue that the entities are physically similar.) For the purposes of assessing the viability of this approach, grant that quasi-particles are real entities. Motivation for this approach can be derived from the thrust of the "no miracles" argument, that success in science is explained by getting something right about the world. However, analysis of the types of analogies drawn between non-relativistic quantum mechanics of many-body systems and QFT systems reveals that this approach fails to yield a particle interpretation for QFT. The approach is untenable because the analogies are formal analogies, but not physical analogies; that is, the analogies map elements of the mathematical frameworks that play similar mathematical roles, but do not map elements with similar physical interpretations. Insofar as intuitions about quasi-particles played a role in the successful development of particle physics, this approach once again leads to a conception of particles as fictions, albeit useful ones.

As an illustration of how this approach to interpretation pans out, consider the role that analogies played in the development of the Higgs model. In this case, what has come to be known as SSB was originally a feature of the Ginzburg-Landau and Bardeen-Cooper-Schrieffer (BCS) models of superconductivity. Anderson, Higgs, and others noted that a model with a massive Higgs boson and massive W and Z bosons could be constructed by drawing on analogies to these models of superconductivity. The quasi-particle in the superconductivity models that is the analog of the Higgs boson is the plasmon. Fraser and Koberinski (2016) analyze the intricate pattern of analogical reasoning and conclude that formal analogies are accompanied by substantial physical disanalogies. The analogies map space in superconductor models to spacetime in the Higgs model. The causal structure of the superconductor models is not preserved by the mapping. The mathematical frameworks of the models are similar – and thus support the formal analogies – but the physical interpretations of the mathematical framework differ in these crucial respects. As a result, the physical properties associated with particles and quasi-particles in the superconductivity models (e.g., composition, effective mass, causal-dynamical process of spontaneous symmetry breaking) do not carry over to the analog "particles" in the Higgs model. A root cause of these physical disanalogies between the superconductor and Higgs models is that the former are non-relativistic (framed using non-relativistic quantum statistical mechanics) and the latter are relativistic (framed using relativistic QFT). Once again, special relativity gets in the way of a particle interpretation of QFT.

The Higgs model has unquestionably proven to be a success, even if at some point in the future it gets replaced by a successor model. The physical properties of particles and quasi-particles in condensed matter physics were a source of physical intuitions for the physicists who originally developed the Higgs model, and continue to be used in pedagogical presentations of SSB for particle physicists. The situation is similar to that in electromagnetism in the 19th century: Thomson, Maxwell, and others successfully used fluid and mechanical models to develop the theory of electromagnetism, even though we now recognize that these models do not accurately represent the world (even approximately). The intuitive physical pictures associated with the fluid and mechanical models were useful fictions. Similarly, the intuitive physical pictures associated with the particles and quasi-particles of superconductivity models (and condensed matter physics models more generally) are useful fictions for the purpose of developing models and theoretical frameworks in QFT.

21.4 Conclusion

Multiple lines of argument support the conclusion that QFT does not describe a world that contains particles as fundamental entities. These arguments rely on special relativistic or general relativistic premises. However, particles continue to play important roles in the experimental and theoretical practice of particle physics. Different proposals have been made for non-fundamental roles for particles, inspired by different no-go arguments. Davies advocates a minimal operationalist account. Halvorson and Clifton and Colosi and Rovelli defend accounts according to which quanta descriptions afforded by the Fock space for a free system are approximations of exact descriptions of the target system. Halvorson and Clifton are motivated by arguments that quanta are not localized, while Colosi and Rovelli are motivated by these same arguments as well as Wald's argument that non-stationary spacetimes typically undermine the uniqueness that the Fock space representation possesses in Minkowksi spacetime. To address the obstacles to a particle interpretation for interacting systems, Bain argues that particles are idealizations – fictional entities that are associated with interacting systems that asymptotically approach free systems. Wallace acknowledges both the localizability and interaction barriers to a fundamental particle interpretation, but argues that the resulting approximate, idealized notion of particle is a notion that is appropriate for a real, emergent entity. Finally, I have added the proposal that particles be considered fictions in another sense: particles and quasi-particles in condensed matter physics supply physical intuitions that are useful for formulating mathematical frameworks for QFT, but which do not represent the world (even approximately).

These approaches to according particles a non-fundamental status in QFT entail different ontological commitments. The apparent references to particles in experimental and theoretical particle physics do not require an ontology that includes particles as either fictional or real entities. One consideration that could distinguish among these alternatives is which (if any) plays a role in the successful development of new physics. For example, Wald, Colosi and Rovelli, and Rodríguez-Vázquez and collaborators formulate their approximations with an eye toward quantum gravity. Of course, another possibility is that it will be the entities that do have a fundamental status in QFT that turn out to be relevant to developing new physics. Furthermore, the domain of applicability of particle concepts considered here is limited to massive, bosonic QFTs. There is also philosophical and foundational work remaining to be done in investigating other cases. For example, massless bosons in QED (photons) present mathematical and conceptual challenges which may also turn out to be relevant to quantum gravity if massless gravitons are involved.[9]

Acknowledgments

Thank you to David Wallace, Hans Westman, Roger Wilkinson, and Alastair Wilson for helpful comments. This research was supported by a Standard Research Grant from the Social Sciences and Humanities Research Council of Canada.

Notes

1 In this entry, QFT is taken to be special relativistic quantum theory. Non-relativistic condensed matter physics is considered separately below.
2 For a detailed discussion, see Fraser (2008). Fock space is also discussed in Wallace (this volume).
3 See Ruetsche (2013) for a more sophisticated discussion of the relationship between unitary equivalence and physical equivalence.
4 The Unruh effect can also be characterized without reference to particles. See Earman (2011) for a thorough discussion.
5 The exposition in this paragraph is derived from chapter 9 of Ruetsche (2011), which can be consulted for further details.
6 Evidence for this claim is that Haag's theorem does not hold in Galilean QFT, for which there are models of non-trivial interactions in which vacuum polarization does not occur (Lévy-Leblond, 1967).
7 Wallace identifies a set of states that are approximately localized in a finite region of space O in the sense that expectation values of field operators defined on regions outside O in states approximately localized within O will differ negligibly from the vacuum expectation value. These states are in practice localized in O because outside of O very high energies would be required to distinguish them from the vacuum state.
8 See Williams (this volume) for a more detailed discussion of renormalization group methods.
9 For a brief introduction aimed at philosophers see Swanson (2017), for a mathematical physics presentation see Strocchi (2013), and for recent research relating to quantum gravity see http://perimeterinstitute.ca/conferences/infrared-problems-qed-and-quantum-gravity.

References

Arageorgis, A., Earman, J. and Ruetsche, L. (2003). Fulling non-uniqueness and the Unruh effect: A primer on some aspects of quantum field theory. *Philosophy of Science*, 70: 164–202.

Bain, J. (2000). Against particle/field duality: asymptotic particle states and interpolating fields in interacting QFT (or: Who's afraid of Haag's theorem?). *Erkenntnis*, 53(3): 375–406.

Bain, J. (2011). Quantum field theories in classical spacetimes and particles. *Studies in History and Philosophy of Modern Physics*, 42(2): 98–106.

Baker, D.J. (2009). Against field interpretations of quantum field theory. *British Journal for the Philosophy of Science*, 60: 585–609.

Blum, A.S. and Joas, C. (2016). From dressed electrons to quasi-particles: The emergence of emergent entities in quantum field theory. *Studies in the History and Philosophy of Modern Physics*, 53: 1–8.

Clifton, R. and Halvorson, H. (2001). Are Rindler quanta real? Inequivalent particle concepts in quantum field theory. *British Journal for the Philosophy of Science*, 52: 417–470.

Colosi, D. and Rovelli, C. (2008). What is a particle? *Classical and Quantum Gravity*, 26(2): 1–22.

Davies, P.C.W. (1984). Particles do not exist. In S.M. Christensen (ed.), *Quantum Theory of Gravity: Essays in Honor of the 60th Birthday of Brice DeWitt*. Bristol: Adam Hilger Ltd., pp. 66–77.

Dell'Antonio, G.F., Dophlicher, S. and Ruelle, D. (1966). A theorem on canonical commutation and anticommutation relations. *Communications in Mathematical Physics*, 2: 223–230.

Earman, J. (2011). The Unruh effect for philosophers. *Studies in History and Philosophy of Modern Physics*, 42(2): 81–97.

Fleming, G. (2004). *Observations on hyperplanes: II. Dynamical variables and localization observables*. Available at: http://philsci-archive.pitt.edu/2085/.

Fraser, D. (2008). The fate of 'particles' in quantum field theories with interactions. *Studies in the History and Philosophy of Modern Physics*, 39: 841–859.

Fraser, D. and Koberinski, A. (2016). The Higgs mechanism and superconductivity: A case study of formal analogies. *Studies in the History and Philosophy of Modern Physics*, 55: 72–91.

Fraser, D.L. (2006). *Philosophical implications of the treatment of interactions in quantum field theory*. Ph.D. thesis, University of Pittsburgh. Available at: http://etd.library.pitt.edu/ETD/available/etd-07042006-134120/.

Fraser, J.D. (2017). *Renormalization and the Formulation of Scientific Realism*. Available at: http://philsci-archive.pitt.edu/14155/.

Haag, R. (1955). On quantum field theories. *Danske Vid. Selsk. Mat.-Fys. Medd.*, 29(12): 37.

Halvorson, H. and Clifton, R.K. (2002). No place for particles in relativistic quantum theories? *Philosophy of Science*, 69: 1–28.

Healey, R. (2013). Physical composition. *Studies in History and Philosophy of Science Part B: Studies in History and Philosophy of Modern Physics*, 44(1): 48–62.

Kojevnikov, A. (1999). Freedom, collectivism, and quasiparticles: Social metaphors in quantum physics. *Historical Studies in the Physical and Biological Sciences*, 29(2): 295–331.

Lévy-Leblond, J.-M. (1967). Galilean quantum field theories and a ghostless Lee model. *Communications in Mathematical Physics*, 4: 157–176.

Malament, D. (1996). In defense of dogma: Why there cannot be a relativstic quantum mechanics of (localizable) particles. In R.K. Clifton (ed.), *Perspectives on Quantum Reality*. Boston, MA: Kluwer, pp. 1–10.

Norton, J.D. (2014). Infinite idealizations. In M.C. Galavotti, E. Nemeth, and F. Stadler (eds.), *European Philosophy of Science - Philosophy of Science in Europe and the Viennese Heritage*. Switzerland: Springer International Publishing, pp. 197–210.

Rodríguez-Vázquez, M., del Rey, M., Westman, H. and Leon, J. (2014). Local quanta, unitary inequivalence, and vacuum entanglement. *Annals of Physics*, 351: 112–137.

Ruetsche, L. (2011). *Interpreting Quantum Theories*. New York: Oxford University Press.

Ruetsche, L. (2013). Unitary equivalence and physical equivalence. In R. Batterman (ed.), *The Oxford Handbook of Philosophy of Physics*. New York: Oxford University Press, pp. 489–521.

Streater, R.F. and Wightman, A.S. (2000). *PCT, Spin and Statistics, and All That*. Princeton Landmarks in Physics. Princeton, NJ: Princeton University Press. Corrected third printing of the 1978 edition.

Strocchi, F. (2013). *An Introduction to Non-Perturbative Foundations of Quantum Field Theory*, Vol. 158 of *International Series of Monographs on Physics*. Oxford: Oxford University Press.

Swanson, N. (2017). A philosopher's guide to the foundations of quantum field theory. *Philosophy Compass*, 12(5): e12414.

Teller, P. (1995). *An Interpretive Introduction to Quantum Field Theory*. Princeton, NJ: Princeton University Press.

Wald, R.M. (1994). *Quantum field theory in curved spacetime and black hole thermodynamics*. Chicago Lectures in Physics. Chicago, IL: University of Chicago Press.

Wallace, D. (2001). *The emergence of particles from bosonic quantum field theory*. Available at: arXiv:0907.5294 [quant-ph].

Wallace, D. (2010). Decoherence and ontology: or, How I learned to stop worrying and love FAPP. In S. Saunders, J. Barrett, A. Kent, and D. Wallace (eds.), *Many Worlds?: Everett, Quantum Theory, & Reality*. Oxford: Oxford University Press, pp. 53–72.

Williams, P. (2019). Scientific realism made effective. /em The British Journal for the Philosophy of Science, 70(1): 209–237.

Further Reading from the Editors

David Wallace's chapter, Chapter 18 in this volume, presents a complementary discussion of particle emergence in QFT. For further general introductions, see D. Baker, *The Philosophy of Quantum Field Theory* (Oxford Handbooks Online) and P. Teller, *An Interpretive Introduction to Quantum Field Theory* (Princeton, NJ: Princeton University Press, 1995). A more advanced discussion of philosophical themes arising from quantum field theory, including critical analysis of particle interpretations, is L. Ruetsche, *Interpreting Quantum Theories* (New York: Oxford University Press, 2011). For a survey of approaches inspired by Bohmian mechanics that take particles as primitive, see W. Struyve, "Pilot-Wave Approaches to Quantum Field Theory." *Journal of Physics: Conference Series* 306 (July 8, 2011): 012047, available at arXiv: 1101.5819. A philosophical discussion of such an approach can be found in M. Esfeld and D. Deckert, *A Minimalist Ontology of the Natural World* (London: Routledge, 2018).

PART VI

Quantum Gravity

Introduction to Part VI

A reasonable introduction to this topic should start with the basics: what is meant by "a theory of quantum gravity"? Why might we need one? Neither question has straightforward answers. The label "quantum gravity" attaches to any theory which claims either to quantize gravity, or to provide a unified framework from which quantum and gravitational theories may be derived, but the precise scope and aim of theories and frameworks varies widely. So does the motivation. It is often said that we need a theory of quantum gravity because quantum mechanics and general relativity are "incompatible", but closer inspection reveals this to be, at best, imprecise. As Dean Rickles highlights in his chapter, this alleged incompatibility amounts to a "feeling of unease" rather than the inconsistency of the two theories.

Nonetheless, the fact remains that quantum field theory, our best theory of matter, and general relativity, our best theory of gravitation and spacetime, appear to describe very different realities. In one theory, quantum matter moves on a fixed background, and, in the other, continuous classical matter interacts with a dynamical spacetime. It's not clear how we can combine these into a unified world picture. But it's not only the fan of unification who might seek a theory of quantum gravity. Theoretical problems, particularly about the very early universe, fall within the domains of both quantum field theory and general relativity, and seem to demand an understanding of how to combine the two. At the same time, there is no urgent empirical reason to demand a new theory; both quantum mechanics and general relativity have passed every empirical test we have set them. Among those who agree on the need for quantum gravity, there are methodological and theoretical divisions. Some seek a grand unified theory, others merely a quantum theory of gravity. A field by the name of "phenomenological quantum gravity" focuses on looking for empirical results that might constrain our theorizing. Increasingly, different approaches borrow ideas from each other and blur the lines between theories. But the simplest way to divide the field of quantum gravity is to look at whether a framework "starts from" particle physics or general relativity. This part contains one chapter on string theory, which takes the former approach, and a second on approaches that stem in one sense or another from general relativity.

String theory, explored by Richard Dawid in the second chapter of this part, is the most prominent approach to quantum gravity. The first major developments happened in the 1960s, and its early theoretical success drew a great many theorists to the field. It recovers the particles of the standard model from the oscillations of one-dimensional "strings" and, notably, derives a boson of the gravitational interaction, the graviton, in the same way. In so doing, it provides an elegant unified account of gravity and the other forces. Of course, from the perspective of general relativity, this unification is somewhat puzzling. Gravity is not a force like the forces of the standard model, and string theory in the above form only quantizes perturbative gravity against a fixed spacetime background. Nonetheless, the fact that it quantizes and unifies at all has been seen as hugely promising, and contemporary string theory goes well beyond the original perturbative approach. As the dominant theory, it has been extensively criticized, both for its methodology, which relies on "non-empirical confirmation", and for taking resources away from other approaches. But, as Dawid points out, despite criticism from within physics, philosophical treatment of string theory is remarkably sparse. This chapter is richly populated with suggestions for further research projects.

Chris Wüthrich discusses a number of other approaches in his chapter in this section. Unlike string theory, these theories all "take general relativity seriously" – that is, they start from the assumption that a theory of gravity will not have a fixed spacetime background. The challenges are formidable; if we move away from a perturbative approach, quantizing gravity is considerably more difficult. One approach, causal set theory, bypasses this problem by attempting to start from a clear foundation in terms of causal structure, and hoping for quantization later. That quantization has not yet been achieved, but the theory's foundation in a particular view of spacetime makes it attractive to many. Loop quantum gravity takes on the challenge directly and quantizes variables derived from the metric of general relativity. The result is "spin network states", which do not live in a background spacetime, but aim to give rise to the dynamical spacetime of general relativity.

All of these approaches raise philosophical questions. Some are standard philosophy of science questions concerning methodology in a domain where experiments are sparse. But there are also a number of fascinating questions about space and time that arise. The theories above do not contain anything quite like the spacetime of general relativity, and in some cases do not appear to posit fundamental spatiotemporal structure at all. We are forced to think of general relativity as an effective, rather than fundamental, theory, and thus of the smooth, continuous curved spacetime it posits as an emergent feature of the world. (Note that "emergent" here may mean something quite weak.) On some conceptions of spacetime, this is puzzling. Spacetime is not usually thought of as an effective or emergent structure; it seems fundamental to our conception of the world and our ability to do experiments. Nick Huggett explores these puzzles and some possible solutions to them in his chapter.

The final problem discussed here is the thorny "problem of time" in quantum gravity. Quantizing gravity can have the rather extraordinary consequence of, in a sense, eliminating time from the theory. Time translations appear as gauge transformations of the quantized theory. But we usually think of gauge transformations as non-physical – mere redescriptions of the same physical situation. Karim Thébault looks at the problem, and possible solutions to it, in his chapter. The issues discussed are deeply intertwined with thoughts about symmetry that run through Chapters 3 and 10, and provide an excellent example of the way in which distinctively philosophical questions about symmetries and representation can be central to an important theoretical problem.

22

THE DEVELOPMENT OF QUANTUM GRAVITY: FROM FEELINGS TO PHENOMENA

Dean Rickles

Programs of quantization derive their motivation both from a general philosophical desire (just as [with] the unitary field theories) not to permit a compartmental approach to the theories of physics, and from the more specific "feeling" that quantized sources of a field require also a quantized field.

(Peter Bergmann, Royaumont, 1959)

22.1   The Quantum Gravity Paradox

The problem of quantum gravity is surprisingly old (over 100 years old, in fact), and its history rich and rewarding.[1] It is perhaps the most stubbornly persistent problem in modern physics. In a nutshell, the problem involves the incompatibility between what is usually viewed as our physics of large, massive systems (general relativity [GR]) and small, light systems (quantum mechanics [QM]). One has the "feeling," as Bergmann puts it, that there shouldn't be a division in nature according to which gravity stands separate (as a fundamentally *classical* interaction) from the rest of the world's phenomena, which are described by quantum theory. It seems intuitively sensible that given the universal nature of gravity (i.e., the fact that any and all sources of mass–energy gravitate), when we have quantized sources of gravity the gravitational field itself must be "caught up" in this quantization like a quantum mechanical infection – to deny this would appear to imply that we inhabit a world with a split brain. (This infection process is modeled after the famous Bohr-Rosenfeld analysis of 1933 ("On the Question of the Measurability of the Electromagnetic Field Strengths," published in the proceedings of the Danish Academy of Sciences), where it was shown to be inconsistent to have a classical electromagnetic field coupled to a quantum mechanical object such as a measurement apparatus. The "feeling" that Peter Bergmann refers to in the opening quote is based on broadly analogous reasoning – though there are important *dis-analogies* due to the unavailability of gravitationally neutral charges.[2])

And yet, sensible though this seems, and despite 100 years of struggle, we still have no consistent theory capable of consistently describing quantum gravity in such a way that a picture of our world emerges – though several proposals come awfully close. We only appear to be able to make sense of GR independently of quantum theory and quantum theory independently of GR. If we try to gain a unified picture, things break down, much like Penrose's impossible triangle (see Figure 22.1). Experimentally, the situation has matched this impossible triangle-like configuration: if we try to get objects massive enough to enable measurement of gravitational strengths, their quantum mechanical

Figure 22.1 Penrose's impossible triangle as a model for the problem of quantum gravity: locally (e.g., focusing on GR or QM in isolation), we can make sense of picture, but not so globally in which the situation becomes paradoxical.

aspects effectively disappear thanks to the inevitable environmental decoherence. But if we try to get objects small enough to retain quantum coherence, the gravitational aspects are too weak to be measured.[3] This has led some, most notably Roger Penrose (see, e.g., Penrose (1986)), to argue that gravity is inextricably bound up with the quantum measurement problem: gravity *causes* the collapse of wavefunctions that become too macroscopic since superpositions of spacetime geometries (that would be associated with the macroscopic superpositions, e.g., of position states) are unstable.

There is an inescapably conceptual aspect to the problem of quantum gravity having to do with the very distinct ways that space and time are treated within the two ingredient theoretical frameworks. This is really at the heart of the paradox. In the case of GR, the geometrical structure of spacetime (manifesting as a gravitational field) has to be solved for, like any other field. However, in QM the geometry is fixed at the outset, and never varied once set. In GR, then, space and time are dynamical variables, with the geometrical structure appearing as one of many other degrees of freedom. In QM, they are instead parameters against which the evolution of other quantities occurs. One needs the fixed causal structure thus provided in order to preserve causality, unitarity, and other seemingly essential features of quantum theory. Though there is ongoing controversy about how best to define and understand it, this cluster of features is known as background independence and includes the so-called "problem of time."[4]

There are a host of related apparent inconsistencies feeding into the quantum gravity problem, as Rovelli and Vidotto colorfully explain:

> A good student following a general relativity class in the morning and a quantum-field-theory class in the afternoon must think her teachers are chumps, or havent been talking to one another for decades. They teach two totally different worlds. In the morning, spacetime is curved and everything is smooth and deterministic. In the afternoon, the world is formed by discrete quanta jumping over a flat spacetime ... that the morning teacher has carefully explained not to be features of our world. (Rovelli and Vidotto, 2015, p. 5)

Yet there is only one world, and so solving the problem of quantum gravity then involves devising a framework in which such conflicts are resolved. As we shall see, there are a variety of ways that one might tackle this, treating one or the other of QM or GR as fundamental, or treating neither as fundamental, or showing how there can be peaceful coexistence despite the apparent conflict.

In this chapter, we briefly review, in broad brushstrokes, some of the history of this problem, and draw attention to philosophically interesting features *en route*. In particular, we draw attention to two features that render the problem unusually difficult as a scientific problem, beyond the obvious technical hurdles. We refer the reader to the relevant chapters on GR and quantum theories for information on quantum gravity's ingredient theories.

22.2   The Dimensions of Quantum Gravity

Research in quantum gravity can very simply be defined as any attempt to construct a theoretical scheme in which ideas from GR and quantum theory appear together in some way. For example, Ashtekar and Geroch, in their review of quantum gravity, characterize quantum gravity as "some physical theory which encompasses the principles of both quantum mechanics and general relativity" (Ashtekar and Geroch, 1974, p. 1213). This leaves a fair amount of elbowroom for the form such a theory might take, and is clearly too broad to pin down specific theoretical schemes: we need to be clear on what the essential principles are that ought to be brought together and the exact manner in which they are thus brought together. We can make things a little more specific by noting that a universal property of any such scheme is the existence of units with dimensions of (L)ength, (M)ass, and (T)ime formed from combinations of the characteristic constants of the ingredient theories: Newton's constant G, Planck's constant $\hbar$, and the speed of light c. We naturally expect a quantum theory of gravity to include all three constants, and the construction of these so-called "Planck units" enables us to see the fundamental limits of the theory, demarcating its domains of validity/applicability just as the constants do separately with respect to the ingredient theories. While the speed of light sets a limit on the speed of information transmission in relativistic contexts, so the Planck units set limits on the smallest possible lengths, areas, volumes, time intervals, and masses, in quantum gravitational contexts. Armed with this knowledge, we are able to say with certainty at what scales quantum gravitational effects would be expected to manifest themselves directly:[5]

$$L_P = \sqrt{\frac{\hbar G_N{}^3}{c}} = 1.616 \times 10^{-35}\,\text{m} \tag{22.1}$$

$$T_P = \sqrt{\frac{\hbar G^5}{c}} = 5.59 \times 10^{-44}\,\text{sec} \tag{22.2}$$

$$M_P = \sqrt{\frac{\hbar c}{G_N}} = 2.177 \times 10^{-5}\,\text{g} \tag{22.3}$$

At these scales, both QM and GR are expected to play a non-negligible role,[6] and (if we accept that GR is a theory of spacetime geometry) it is this scale that we expect quantum geometry to dominate – by the mid-1950s the notion of the Planck length was understood by those working on the so-called canonical approach as a measure of the fluctuations of spatial geometry, leading to topological features such as multiple-connectedness known as "spacetime foam"; for those working on spacetime covariant approaches, the Planck length marked a natural boundary to the wavelengths of quantum fields and held the promise of taming the ultraviolet divergences that result from considering fields at ever small distances (see Section 22.3 for more on these two approaches). It is hoped that features of spacetime geometry (and perhaps topology) at these minuscule scales (such as "graininess" of spacetime) will nonetheless develop into features discernible at presently (or soon-to-be) detectable scales (see Section 22.5).

One of the basic questions to be faced by any proposed solution to the problem of quantum gravity is what role is played by the Planck units. Whatever response one gives to this question, the small scales point to the kinds of issue that quantum gravity proposals might tackle: especially interiors of black holes (black hole singularities), the conditions close to the big bang (the big bang singularity), and the high-frequency behavior of quantum fields (light cone singularities). It is thought that the fundamental limits on length scales imposed in quantum gravity (i.e., the discrete quantum geometry) might serve to "smear out" singularities of each of these kinds – most of the major approaches achieve exactly this.

Dean Rickles

22.3 The Ways of Quantum Gravity

We can describe the various solutions to the problem in a fairly simple way. Let us return to the simplistic definition of quantum gravity given by Ashtekar and Geroch:

> [QG is] some physical theory which encompasses the principles of both quantum mechanics and general relativity.

We can simply write this in terms of the schema: GR + QM = QG (where "QM" usually amounts to some quantization method). A common way of reading this schema is in terms of the fundamental constants characterizing the theoretical frameworks of GR and QM, as we intimated above, so that GR + QM = QG can be read as $cG + \hbar = cG\hbar$. That is, one must demonstrate how both Planck's action constant and gravitational fields can exist in one and the same world. However, there are contexts (such as semi-classical gravity or quantum fields on curved spacetime) in which we have the presence of all three constants that would not *strictly speaking* qualify as quantum gravity proper (i.e., in which gravity itself is quantized) – they still qualify as responses to the basic problem, however. Rather, these approaches consider the effects of strong (classical) gravitational fields on quantum phenomena (such as the behavior of quantum fields near black holes).[7]

In the following very general partitioning of approaches, we split in terms of which of the two ingredient theories (GR and QM) is more fundamental. "Fundamental" here does not necessarily mean ontologically fundamental, [8] but instead just that the principles of one are more significant in terms of determining the final features of the approach to quantum gravity that results, so that one *modifies* features of the other for example.

22.3.1 QM as Fundamental

Perhaps the most common approach to resolving the problem of quantum gravity involves treating QM as the most basic element of the world (see the figure below). According to such approaches, quantum theory *contains* classical GR in some sense, or less radically modifies (or "corrects") classical GR.

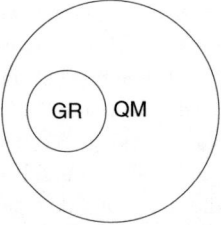

This can take a variety of forms, however. Many of the earliest attempts involved the direct quantization of GR as a classical field theory, so that the classical theory emerged in the appropriate limit. The main lines of research are of this type, depending on which of the various quantization methods one employs: covariant, canonical, or path integral. The most common approach along the lines of covariant quantization is based on the idea that gravitational interactions are mediated by a particle, the graviton. To get this picture, however, requires some violence to a certain view of GR as a "distinctive" theory in that it is a theory of spacetime as much as a theory of the gravitational interaction.

Covariant quantization approaches involve a four-dimensional formulation in which the symmetries of spacetime are preserved intact. Canonical approaches split this spacetime into space and time and have to demonstrate preservation of symmetries via more indirect means. The path-integral approach, due to Feynman, is really just a branch of covariant quantization in which one sums over

342

all four-dimensional metrics. There are, of course, interrelations between these formulations, but in the case of GR they are far from simple. Also of importance is the fact that the covariant (perturbative) approach takes place using the linear approximation to the full Einstein equations. The graviton concept itself only makes sense in this approximation.

Since the middle of the last century there has been a more or less philosophical clash of ideas with respect to whether the canonical or covariant approaches are the way to go. The canonical approach involves expressing GR in Hamiltonian form and then applying the standard prescription for Schrödinger quantization. This involves the singling out of a time (with a slicing into hypersurfaces representing instants) which seems to run counter to the spirit of GR. However, in a similar way, the covariant approach, while keeping space and time together as one, splits apart the metric into background terms and perturbations (in at least one approach.)[9] This is a different kind of violence to GR, attacking its so-called "background independence." However, if background independence is retained, as in the canonical approach, then a serious problem must be faced regarding the fact that since time translation is a gauge transformation in GR (being a diffeomorphism), the Hamiltonian vanishes, which in turn implies that the quantum theory is "timeless."

Superstring theory is the most popular approach of this kind (and still the most widely pursued approach to the problem of quantum gravity). Gravitation (and the results of GR) arises in string theory as a result of the presence of a massless spin-2 particle (the graviton) in the spectrum of a closed string (which all string theories contain). String theory itself is a sort of quantum field theory of strings, with no gravitation at a fundamental level. It is for this reason that some advocates of string theory often refer to string theory "predicting the existence of gravity." In a certain sense, this is perfectly true: it is a logical consequence of the theory. But, of course, in another sense it seems somewhat contrived since its presence constituted part of the reason for pursuing the theory in its current form after it had originally been used as a model for the strong interactions.[10]

22.3.2 GR as Fundamental

Though he initially believed that the quantum theory must modify GR, Einstein's later work on his "unified field theory" adopted the view that GR was the fundamental theory from which quantum phenomena (such as discreteness and the existence of particles) were to be derived. Here we find the converse of the problem above: getting a discrete theory/ontology from continuum theory/ontology.

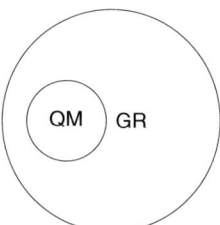

Such an apparently paradoxical problem was resolved to some extent by John Wheeler, through his enigmatic "charge without charge" ideas (in which the flow of charge in and out of a wormhole mimics the positive and negative poles of a discrete elementary charge).[11] Together with his students he got remarkably far with getting the structure of particles from a gravity treated via the path-integral approach (and classical gravity), including the existence of spin.[12]

More recent proposals along lines that prioritize GR come from the notion that gravitation might induce modifications of more basic aspects of QM. For example, Roger Penrose (2014) speaks of "gravitizing" QM so that the usual quantum mechanical principles (such as superposition) have exceptions in cases where significant masses in such superpositions are concerned (such superpositions have a "lifetime" due to instabilities in the spacetime geometry associated with superpositions of locations of masses). As we see in Section 22.5, there are recent almost-performable experiments that seek to examine such conditions.

22.3.3 Both as Fundamental

If both (classical) GR and quantum theory are equal partners in the world, then we must live in a hybrid world. In this case, there is no such thing as quantum gravity: it is not the kind of thing that needs to be quantized. Yet we still face the same puzzle (Bergmann's "feeling") that we started with: how can we have a world in which quantum mechanical systems interact with classical gravitational systems without the quantum mechanical properties seeping into the classical gravitational field? A world dappled in this way seems to be problematic.

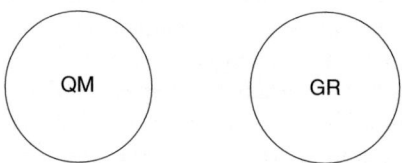

The problem has a mathematical aspect too since, if matter is treated quantum mechanically while gravity is treated classically, the two sides of the Einstein field equation appear to involve different structures: a c-number term and a q-number term. The earliest suggestion of this type, which offered a solution of the mathematical problem, was that of Christian Møller (1962), who argued that one could convert the right-hand (matter) side to a c-number by using the energy-momentum's expectation value instead. The problem with this is that it faces the measurement problem with renewed force: either there is a discontinuity in the gravitational field on measurement, or the spacetime geometry is put into a superposition of geometries in the style of the many-worlds interpretation. There are also arguments that suggest a potential violation of the uncertainty relations in such a world (see Mattingly (2009) for a review).

22.3.4 Neither as Fundamental

A new class of approaches having their roots in "effective field theory" views both GR (and so spacetime) and quantum field theory as valid within a certain range of (low) energy values, below the Planck energy (or, equivalently, for long wavelengths). These are known collectively as "emergent gravity" approaches. One way of conceptualizing such approaches is in terms of the search of the microscopic structure of spacetime. The idea is that the metric is a collective variable of some microscopic theory that does not possess the notion of a metric in its own fundamental ontology. This simply means that as one probes higher energies the metric does not appear, so rather than viewing classical GR as simply one of the classical limits of a quantum theory of gravity (e.g., understood via quantization), the theory and its features are instead emergent much as hydrodynamics (a continuum theory) is emergent from a particle theory. See Nick Huggett's entry in this volume for a discussion of related issues.

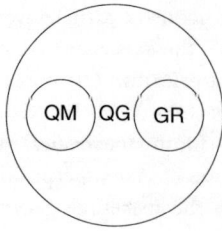

Although the fundamentality of GR is not asserted in this effective field theory approach, the low-energy part works perfectly well as a quantum field theory. What this shows us, rather than being a solution to the problem of quantum gravity (in the sense of an account of what happens when both gravitational and quantum effects are strong), is that at everyday distances there is no real conflict between quantum field theory and GR. This is important since it at least reveals that there is no problem with the ingredient theories at currently observable energies. However, it leaves us none the wiser about what happens at the more extreme energies that are expected to involve new physics with conceptual revisions.

22.4   Why Is the Problem of Quantum Gravity so Hard?

You do not win battles by debating exactly what is meant by the world 'battle'. You need to have good troops, good weapons, a good strategy, and then hit the enemy hard. The same applies to solving a difficult scientific problem.

(Francis Crick, *The Astonishing Hypothesis*)

Wolfgang Pauli once remarked to Bryce DeWitt, on hearing that the latter was working on the problem of quantum gravity: "That is a very important problem – but it will take someone really smart!" The efforts of very many smart people have not yet resulted in a resolution. Crick's notion of solving a difficult scientific problem as a battle might seem apt then: as scientific problems go, the problem of quantum gravity has to be one of the most difficult in the entire history of human thought. But *why* is it so hard? Part of the issue is that the shape of the problem ("the enemy") changes depending on time and context. We are, in fact, not even fighting the same battle that we were once fighting in the earliest days of quantum gravity: developments in the ingredient theories and background theories have revised the problem again and again. If the enemy keeps shifting, then it doesn't matter how good our troops, weapons, and strategies are: they simply might not be well-matched anymore. This isn't just an evolutionary arms race: the battlefield itself changes, as well as the enemy. As Peter Bergmann put it:

> Today's theoretical physics is largely built on two giant conceptual structures: quantum theory and GR. As the former governs primarily the atomic and subatomic worlds, whereas the latter's principal applications so far have been in astronomy and cosmology, *our failure to harmonize quanta and gravitation has not yet stifled progress in either front*. Nevertheless, the possibility that there might be some deep dissonance has caused physicists an esthetic unease, and it has caused a number of people to explore avenues that might lead to a quantum theory of gravitation, no matter how many decades away the observations. (Bergmann, 1992, p. 24, my emphasis)

That progress has not been stifled in either of the ingredient theories (GR and QT) is precisely a key part of the curious nature of the problem of quantum gravity – there has been *too much* progress which has the effect of modifying what is required in a solution. I expect that one can find similar instances in the history of physics, in which a unification is required between two theories while changes are going on in either or both,[13] but the timescales involved in QG (a century of struggle) and the pace of progress might make it qualitatively different in this sense.

So there is something peculiar about the problem, I think, making an already hard problem, a "really" hard problem. Referring to the Pauli quote: smartness might not be enough, if we are faced with a hydra that spouts more heads every time one is lopped off! Precisely because we do not have this direct experimental grasp, or specific phenomena to match up with, the problem itself is subject to fluctuations with respect to the ingredients of the problem: QM and GR (and background knowledge relevant to our understandings of these). This is precisely why we see analogies playing such a rich role in the history of quantum gravity: because we need to look for something concrete to latch onto. This much captures the non-stationarity of the problem.

There is a further problem (what I shall call the problem of *polysemicity*) in that the ingredient theories each allow for different formulations and interpretations which impacts on the possible solutions to the problem of quantum gravity. For example, if we agree with Steven Weinberg that the geometrical (gravity as genuine spacetime curvature) approach drove a "wedge" through physics, separating gravitation (GR) and elementary particle physics, then we face two very distinct paths: either treat gravity like a standard field theory, or treat gravity as something special. We face the same issue with respect to how we view quantum theory. As we have seen, the various quantization methods suggest very different solutions to the problem of quantum gravity. Other developments such as the discovery of spin, or the discovery of Yang-Mills theory radically altered the terms of the problem, offering up new (and very difficult) mathematical structures that now had to be incorporated in the theory, or offering up a host of new analogies respectively. We can further envisage that different interpretations (e.g., Bohmian versus Copenhagen) would suggest very different solutions and even outlaw certain approaches. (An additional, distinct aspect to the polysemicity problem concerns the *aims* of a quantum theory of gravity. The most obvious example which I won't discuss directly is the differing points of view of string theory and others on this point: does the QG problem involve unifying gravity with the other interactions or not? In this sense, string theory and, e.g., loop quantum gravity, though both claiming to offer *solutions* to the problem of quantum gravity, are really tackling different problems.)

22.5   To Phenomena and Experiment

Léon Rosenfeld was the first to attempt a direct quantization of the gravitational field (using both then-available methods: covariant and canonical) in the early 1930s. In his later life he was adamant that it was not proven that quantum gravity was even necessary since there were no "empirical clues" available (see Rosenfeld (1966)). All that existed were some (difficult) theoretical problems – such as the final stages of gravitational collapse and the physics near the big bang – and the feeling of unease described by Bergmann. Data have remained a problem throughout quantum gravity's life, and this has led to a proliferation of approaches, exacerbated by the shifts in the ingredient theories that we saw in the last section. Thus, a kind of (quite understandable) pragmatism guided the early neglect of quantum gravity research; the scales at which phenomena would be apparent were known then to be well out of reach of direct tests. In more detail: the characteristic "Planck length" is computed by dimensional analysis by combining the constants that would control the theory of quantum gravity into a unique length. As shown above, this is $l_p = \sqrt{\hbar G/c^3} = 1.6 \times 10^{-33}$ cm: a minuscule value, making gravity (effectively) a "collective phenomenon" requiring many interacting masses. That quantum gravitational effects will not be measurable on individual elementary particles is, therefore, quite clear.[14]

Dirac expressed this standpoint particularly clearly:

> There is no need to make the theory conform to GR, since GR is required only when one is dealing with gravitation, and gravitational forces are quite unimportant in atomic phenomena. (Dirac, 1966, p. 66)

There is a very simple way of exposing the problems that direct measurement of quantum gravitational effects might face. It is a basic fact that for any distance we wish to measure, the measurement probe must be at least as small as that distance. In the case of quantum gravitational measurements, the distances are absolutely tiny (and the energies huge), as we have seen. As we probe smaller distances, the probe's particles must be of ever higher momentum values (thanks to the uncertainty relations). But this higher momentum is a source of stress-energy and so is responsible for generating large spacetime curvature, so much so that the device will generate a black hole and be rendered unobservable – or at the very least, will distort spacetime to such a degree that measurement becomes a practical impossibility.[15]

However, a common confusion still persists which is no doubt a hangover from this early pragmatic stance and such arguments as these. This is the idea that in order to generate quantum gravity phenomenology (e.g., to detect the quantum properties of the gravitational field) one must be able to detect *individual* gravitons, and this therefore would require probing the Planck scale directly. For example, Freeman Dyson recently wrote:

> This talk is concerned with a different question, whether it is in principle possible to detect individual gravitons, or in other words, whether it is possible to detect the quantization of the gravitational field. (Dyson, 2012, p. 1)

But detecting individual gravitons and detecting the quantization of the gravitational field are two quite different things.[16] An analogy would be the conflation of the direct and indirect detection methods of gravitational waves: the decay rates of pulsars offer a means of detecting gravitational radiation beyond the kinds of direct detections one finds in the interferometer observations at LIGO and elsewhere. The pulsar data also count as evidence of gravitational radiation despite the indirectness. There are two recent avenues that rectify the parallel quantum gravity situation, by "going large" (using the heavens as an observatory) or by "going small."

Obviously, the major problem with the latter approach is precisely the seemingly killer combination Dyson is referring to: the extreme tininess of the Planck scale, the weakness of gravity, and the presence of decoherence, making it very difficult to generate quantum effects on objects that are nonetheless big enough to measure gravitational effects. However, technological rather than theoretical/conceptual advances look set to dislodge the century-old embargo on quantum gravitational data. What is required is simply enough control of quantum coherence (for masses in location superpositions) combined with enough sensitivity to measure the gravitational field for such (still very small) masses. Then it suffices to check whether the gravitational field is itself in a quantum superposition. Almost performable experiments have now been planned that make use of massive quantum oscillators capable of generating such location superpositions for objects as weighty as 1 ng together with micromechanical machines that can radically reduce the size of experimental equipment thus improving sensitivity to the required degree for gravitational field measurements at 1 ng (see, e.g., Smöle et al. (2016)).

The alternative ("going large") approach attempts to access the large energies by utilizing the large scale of the universe: that is, one can probe the minute Planck length by accessing the vastness of the universe. Such observations would be more along the lines of the measurements of binary pulsar decay. An example is the potential for Lorentz invariance violation as a result of photons traveling over "grainy" spacetime (which is treated as a lattice mathematically, and therefore possesses different symmetries to a continuum spacetime). Observationally, this would be measured as an energy-dependence on the time of arrival of the photons from very distant sources. Since some approaches to quantum gravity *predict* such features, such measurements have the power to confirm or rule out such approaches, thus bringing quantum gravity research in line with more standard scientific methodologies (featuring the selection and rejection of theories on the basis of empirical evidence). In fact, observations *were* made with no such effects discernible (a null result), thus reducing the space of possible theories (see Albert (2000): the MAGIC telescope collaboration). Though such data do not allow for a direct testing of specific approaches to quantum gravity (such as string theory), they do constrain along the lines of perfectly standard empirical results. They already demonstrate that quantum gravity research has progressed from Bergmann's vague feelings about how the world ought to be.

22.6 Conclusion

The problem of quantum gravity spent much of its earliest history at the mercy of wider changes with respect to the ingredient theories, general relativity, and quantum theory. Even once those theories settled down, quantum gravity remained firmly detached from experiments. This situation has only recently changed and promises to offer new phenomena to test proposed solutions to the problem which will enable us to make firmer statements about the more philosophical implications of these proposed solutions. However, we may still face the problem of polysemicity stemming from the very differing interpretations and formulations that the ingredient theories allow. However, this will at least ensure that philosophers of physics are kept busy in the coming years.

Notes

1 For a conceptually oriented sourcebook of the earliest phase of quantum gravity research, see Blum and Rickles (2017); for a more detailed early history, see Rickles (forthcoming). Shorter treatments can be found in Rickles (2011), Stachel (1999), and Rovelli (2002).
2 It is fair to say that much of the earliest work on quantum gravity was based on the attempt to treat gravity analogously to electromagnetism, and it was supposed by most that gravity would succumb to quantization with just a few modifications to the electromagnetic case.
3 Though as we see in the final section, this experimental roadblock is finally beginning to look more like a merely practical problem that might yield within a matter of years. There have also been proposals that attempt to draw experimental results applicable to quantum gravity, albeit in a more indirect way (known as quantum gravity phenomenology) by using astrophysical and cosmological features (such as patterns in "the cosmic microwave background radiation"), rather than terrestrial experiments.
4 See Pooley (2017) for a thorough recent review of background independence. See Thebault (2017) for a review of the problem of time.
5 Though as we see in Section 22.5, it is a mistake to think that quantum gravity is relevant *only* at these scales, and therefore, given their extreme magnitudes, out of bounds for the foreseeable future – note that the Planck mass, M_P, which might not look so extreme as the Planck time and length, must be localized within a region of space L_P^3, to generate directly observable quantum gravitational effects.
6 See Gorelik (1992) for more on the curious discovery of these units and their subsequent propagation into early quantum gravity research.
7 In fact, there has been at least some experimental work on the behavior of quantum systems with respect to classical gravity, e.g., in testing whether quantum particles obey the equivalence principle, as well as tests to determine quantum mechanical phase shifts induced by the Earth's gravitational field (this is the famous "COW" experiment, where "COW" stands for "Colella, Overhauser and Werner" – (Colella et al., 1975)).
8 Though it sometimes does, e.g., in the case of unified field theories in which classical GR is the *sole* source of ontology, with discrete quantum-like particles emerging from it.
9 The path-integral approach does not face these issues, and works by computing the amplitude to go from one 3-geometry to another 3-geometry by summing over all possible interpolating 4-geometries (the "histories," each weighted by a complex number amplitude). However, it faces its own issue in the form of a "measure problem" in which it isn't clear how to assign probabilities to outcomes in the path integral – in a nutshell, since the domain space is a space of all 4-manifolds (and, e.g., state which manifolds are equivalent and inequivalent) one needs to solve one of the most pressing problems in topology to resolve this measure problem.
10 For more on string theory, see Richard Dawid's entry in this volume. For a philosophically inclined history of the subject, see Rickles (2014).
11 Wheeler later dropped this interpretation of wormholes as elementary particles and viewed them instead as a consequence of a path-integral formulation of GR in which the fluctuations of the gravitational field induce a topologically multiply connected manifold. An earlier version of pulling quantum-like effects from topological trickery was Oskar Klein's five-dimensional theory of relativity, in which the fifth dimension was compactified down to a small circle, giving charge quantization through the circle's closure and periodicity (Klein, 1927).
12 Spin was long believed to be the sticking point for such unified field theory approaches since it seems to have no classical counterpart (the gravitational field appears to be purely bosonic: integer spin).
13 Certainly, the business of a scientific problem being buffeted around by external influences can be found in other areas (the early days of cosmology springs to mind, in which we have an initial phase of untestability) –

it is part and parcel of the growth of scientific knowledge to update as new information comes in. However, observations entered early on in cosmology, whereas quantum gravity has never settled enough to have a life of its own.

14 Bryce DeWitt devised rigorous arguments to show this to be the case so that the gravitational field itself does not make sense at such scales: the static field from such a particle (with a mass of the order 10^{-20} in dimensionless units) would not exceed the quantum fluctuations – the static field dominates for systems with masses greater than 3.07×10^{-6}. The gravitational field is from this viewpoint an "emergent" "statistical phenomenon of bulk matter" (DeWitt, 1962, p. 372).

15 Note that this same reasoning is often used to show that the notion of a spacetime point will no longer make physical sense in quantum gravity, since no physical significance can be attached to such a concept.

16 Cf. Sabine Hossenfelder's blog post "Quantum gravity phenomenology $\neq$ detecting gravitons" for a useful discussion on this point: http://backreaction.blogspot.com.au/2013/06/quantum-gravity-phenomenology-neq.html.

References

Albert, J. et al. (2008). Probing quantum gravity using photons from a flare of the active galactic nucleus Markarian 501 observed by the MAGIC telescope. *Physics Letters B*, 668(4): 253–257.

Bergmann, P. G. (1992). Quantization of the gravitational field, 1930–1988. In J. Eisenstaedt and A.J. Kox (eds.), *Studies in the History of General Relativity*. Basel: Birkhäuser, pp. 364–366.

Blum, A. and Rickles, D. (2017). *Quantum Gravity in the First Half of the Twentieth Century: A Sourcebook*. Max Planck Institute, Edition Open Access.

Colella, R., Overhauser, A.W. and Werner, S.A. (1975). Observation of gravitationally induced quantum interference. *Physical Review Letters*, 34: 1472–1474.

Crick, F. (1995). *The Astonishing Hypothesis: The Scientific Search for the Soul*. New York: Scribner.

DeWitt, B. (1962). Quantization of geometry. In L. Witten (ed.), *Gravitation: An Introduction to Current Research*. New York: John Wiley and Sons, pp. 266–381.

Dirac, P. (1966). *Principles of Quantum Mechanics*. Oxford: Clarendon Press.

Dyson, F. (2012). *Is a graviton detectable? Poincare prize lecture*. Available at: https://publications.ias.edu/sites/default/files/poincare2012.pdf.

Gorelik, G. (1992). First steps of quantum gravity and the Planck values. In J. Eisenstaedt and A. J. Kox. (eds.), *Studies in the history of general relativity* (Einstein Studies, Vol. 3). Boston, MA: Birkhäuser, pp. 364–379.

Klein, O. (1927). Zur Fünfdimensionalen Darstellung der Relativitätstheorie. *Zeitschrift für Physik*, 46: 188–208.

Mattingly, J. (2009). Mongrel gravity. *Erkenntnis*, 70(3): 379–395.

Møller, C. (1962). The energy-momentum complex in general relativity and related problems. In A. Lichnerowicz and M.A. Tonnelat (eds.), *Les theories relativistes de la gravitation*. Paris: Du Centre National de la Recherche Scientifique, pp. 15–29.

Penrose, R. (1986). Gravity and state-vector reduction. In R. Penrose and C. J. Isham (eds.), *Quantum Concepts in Space and Time*. Oxford: Clarendon Press, pp. 129–146.

Penrose, R. (2014). On the gravitization of quantum mechanics 1: Quantum state reduction. *Foundations of Physics*, 44(5): 557–575.

Planck, M. (1899). Über irreversible Strahlungsvorgänge. *Sitzungsberichte der Kniglich PreuS ischen Akademie der Wissenschaften zu Berlin*, 5: 440–480.

Pooley, O. (2017). Background independence, diffeomorphism invariance and the meaning of coordinates. In D. Lehmkuhl, G. Schiemann, E. Scholz (eds.), *Towards a Theory of Spacetime Theories*. Basel: Birkhäuser, pp. 105–143.

Rickles, D. (2011). Quantum gravity meets &HPS. In S. Mauskopf and T. Schmalz (eds.), *Integrating History and Philosophy of Science*. Boston Studies in the Philosophy of Science. Dordrecht: Springer, pp. 163–199.

Rickles, D. (2014). *A Brief History of String Theory: From Dual Models to M-Theory*. Berlin: Springer.

Rickles, D. (forthcoming). *Covered in Deep Mist: The Development of Quantum Gravity, 1916–1956*. Oxford: Oxford University Press.

Rosenfeld, L. (1966). Quantum theory and gravitation. In R.S. Cohen and J. Stachel (eds.), *Selected Papers of Léon Rosenfeld*. Dordrecht: D. Reidel Publishing Company, pp. 599–608.

Rovelli, C. (2002). Notes for a brief history of quantum gravity. In R.T. Jantzen et al. (eds.), *Proceedings of the Ninth Marcel Grossmann Meeting on General Relativity*. Singapore: World Scientific, pp. 742–768.

Rovelli, C. and Vidotto, F. (2015). *Covariant Loop Quantum Gravity*. Cambridge: Cambridge University Press.

Schmöle, J., Dragosits, M., Hepach, H. and Aspelmeyer, M. (2016). A micromechanical proof-of-principle experiment for measuring the gravitational force of milligram masses. *Classical and Quantum Gravity*, 33: 1–19.

Stachel, J. (1999). The early history of quantum gravity. In B.R. Lyer et al., (eds.), *Black Holes, Gravitational Radiation and the Universe*. Dordrecht: Springer, pp. 525–534.

Thebault, K. (2017). *The Problem of Time*. This volume.

Further Reading from the Editors

For the reader who would like to understand some issues without going into technical detail, Lee Smolin's. *Three Roads to Quantum Gravity* (New York: Basic Books, 2001) is a good popular summary of the search for quantum gravity. A number of relevant papers can be found in Callender, C. and Huggett, N., eds. *Physics Meets Philosophy at the Planck Scale* (Cambridge: Cambridge University Press, 2001). For some philosophers views on the quantization of gravity, see Callender, C. and Huggett, N. 'Why quantize gravity (or any other field for that matter),' (*Philosophy of Science*, 68(3): S382–S394, 2001), Curiel, E., 'Against the excesses of quantum gravity: A plea for modesty' (*Philosophy of Science*, 68(3): S424–S441, 2001) and Wüthrich, C., 'To quantize or not to quantize: Fact and folklore in quantum gravity,' (*Philosophy of Science*, 72: 777–788, 2005). *Beyond Spacetime*, edited by Huggett, Matsubara and Wuthrich (2020) is an excellent recent collection of papers.

23

STRING THEORY

Richard Dawid

23.1   Introduction

The basic idea of string theory goes back to the late 1960s. (For textbooks on string theory, see Polchinski (1999), Zwiebach (2007), and Becker et al. (2007). For a collection of texts on early string theory, see Cappelli et al. (2012). A philosophy-minded history of string theory is Rickles (2013c).) After its foundations had been laid from 1968 onward as a candidate theory for describing strong interaction (Veneziano, 1968), string theory was proposed as a universal theory of all interactions in 1974 (Scherk and Schwarz, 1974) and became popular after the formulation of the action of a quantized superstring (Green and Schwarz, 1984). Since then, string theory has played the role of the leading approach to a unified theory of all interactions.

String theory's history as a subject of philosophical investigation is much shorter. Apart from Weingard (1989), which introduced string theory to philosophers, not a single philosophical paper on string theory was written in the 20th century. After a second philosophical "suggestion" to look at the theory at the turn of the 21st century in Butterfield and Isham (2001), the theory became a subject of more extensive philosophical inquiry about a decade ago. Increasing activity in recent years may indicate the emergence of a fully fledged philosophical field of research.

23.2   The Role of String Theory

A consistent theory of quantum gravity is the main desideratum of contemporary fundamental physics. Observed phenomena in high-energy physics at this point can be very well accounted for by the standard model of particle physics. Observed gravitational phenomena are covered in a satisfactory way by general relativity. However, neither of those two theories on its own can provide a satisfactory description of the very early and dense states of the universe. What is needed is a consistent theoretical scheme that accounts for both types of phenomena, nuclear interactions and gravity.

There are two and a half ways to search for such a theory. The half option would be to retain the theories of general relativity and quantum field theory and just try to make them consistent with each other so that they can work in conjunction to describe regimes where both types of interaction are relevant.[1] No promising headway has been made in that direction and there are reasons to doubt that this is a viable way to go. The other two options start from one of the existing conceptual frameworks and build a theory from that starting point that can cover the other side as well. Canonical quantum gravity, loop quantum gravity, and related approaches start from gravity and aim at developing a quantized theory of gravity. String theory starts from the perspective of particle physics and aims to generalize it to include gravity.

In the eyes of string theorists, two general observations speak in favor of the latter approach. First, the principles of gauge field theory, on which the standard model of particle physics is based, put strong constraints on theory building. The requirement that the theory of quantum gravity should enforce that its low-energy effective theory is a gauge field theory therefore is a powerful guideline for theory building. Second, the approach seems to work in principle, which is by no means a trivial observation. An extension of the gauge principle to supersymmetry leads to a graviton, which suggests that quantum gravity is a natural extension of gauge field theory.

String theory started from a perturbative perspective, keeping the format of calculating Feynman diagrams but replacing the pointlike particles of quantum field theory by small one-dimensional objects, the strings. This modification is assumed to generate a finite theory, thereby offering a solution to the problem of the non-renormalizability of quantum field theories of gravity. Moreover, the fact that a quantized string necessarily contains a graviton as one of its oscillation modes renders the approach a natural framework for describing quantum gravity. The posit of extended elementary objects also seems consistent with the high-energy behavior one is led to expect from a theory of quantum gravity. The generation of black holes in particle collisions at the Planck scale enforces an upper limit to testing energy scales that fits well with the intrinsic fundamental length scale of the string. This vague initial argument was later fleshed out based on the minimal length scale imposed by string theory's duality structure (see below).

The seemingly innocent step toward extended fundamental objects carries huge conceptual implications. A quantized string theory that describes both fermions and bosons must be supersymmetric (that is, show a symmetry under transformations between fermionic and bosonic degrees of freedom). The quantization of this superstring only works in a consistent way in ten spatiotemporal dimensions. The six extra dimensions remain invisible at low energies. They are understood to run back into themselves like cylinder surfaces, thereby generating a topologically complex spacetime structure (a Calabi-Yau space).

Moving beyond a strictly perturbative perspective reveals that string theory contains higher dimensional objects of various dimensions called "branes" (Polchinski, 1995). Branes play a crucial role in one of the most remarkable features of string theory. In a number of cases, specific realizations of string theory that differ with respect to the kinds of symmetries they have, the radii of their compact dimensions, their spacetime topologies, the dimensionality of their higher dimensional objects, and other characteristics are dual to each other: they are empirically fully equivalent and the features of one formulation are related to features of its dual by a duality transformation. Duality relations connect all five types of superstring theory (Witten, 1995). Another duality relation connects string theory on AdS space (a space with constant negative curvature that corresponds to the solution of the Einstein equations for empty space with a negative cosmological constant) to a conformal string theory on the boundary (Maldacena, 1998), thereby revealing a deep connection between string theory and gauge field theory.

String theory in many respects is a very different kind of conceptual scheme than any other physical theory we know. In the following, I will give a brief survey of some of the most important novel characteristics of string theory that deserve serious philosophical attention. A number of them have by now been addressed extensively in a philosophical context. Others still lack the philosophical attention they deserve.

23.3 Specific Physical Characteristics of String Theory

23.3.1 *The Central Role of the Concept of Duality*

As already emphasized in the introduction, dualities play a pivotal role in string theory. They have also been at the center of the theory's philosophical analysis.

Dawid (2007, 2013a), Matsubara (2013), and Rickles (2011, 2017) identified string dualities as a serious problem for scientific realism. The extent to which dual theories differ from each other with respect to their fundamental ontologies is incompatible with ontological scientific realism (Dawid,

2007). For a number of reasons, dual formulations of string theory should be treated as different perspectives on one theory rather than as distinct theories: string theorists clearly treat them as one theory (Matsubara, 2013); only the entire web of dual formulations allows an adequate understanding of string physics (Dawid, 2013a); dual theories can be transformed into each other by duality transformations (Rickles, 2017). As pointed out by Castellani (2009), viewing duals as different perspectives on the same theory goes counter to the classical view that links theory individuation to a theory's ontological import.[2]

Dawid (2007, 2013a) and Matsubara (2103) argue that structural realism is in a better position than ontological realism to account for dualities but faces problems of its own. Dawid (2007) suggests that string theory might support a specific form of structural realism (consistent structure realism) that relies on the fact that string theory at a fundamental level is fully determined by consistency arguments. (It has no free parameters and no freedom of model choice.) Rickles (2017) also favors a structuralist approach. He emphasizes that duality transformations, in revealing physically irrelevant transformational degrees of freedom, are comparable to gauge transformations (see also de Haro et al., 2016). This analogy suggests that anything that can be changed under duality transformations is unphysical and no candidate for the theory's real content. Only characteristics invariant under duality transformations, such as the global symmetry group, should thus be acknowledged as real. Rickles argues that this leaves sufficient room for a structural realist take on string theory.

An interesting specific context for analyzing the anti-realist import of dualities is provided by S-duality. S-duality transforms a theory into a dual with inverse coupling constant. In string theory, S-duality arises as one of the pivotal duality relations that connect all five types of superstring theories. The fundamental strings of one theory show up as solitonic solutions (nonperturbative effects characterized by a topological charge) in its S-dual theory (Harvey and Strominger, 1995). The fundamental objects of one theory thus emerge as complex composite objects of its dual. String physicists Sen (1999) and Susskind (2013) have considered this characteristic of string theory to indicate that the theory does not allow for a reductionist understanding of the world. Objects don't stand in an unequivocal constituent-compound relations to each other. Sen's and Susskind's arguments support the incompatibility of string theory with ontological scientific realism.

McKenzie (2017) outlines a strategy for avoiding this conclusion in the (non-string theoretical) context of Montonen-Olive duality, a duality relation that holds in $N = 4$ supersymmetric conformal field theories: for a given coupling strength, one of the dual formulations can look more natural. This may justify the selection of a preferred fundamentality order in a world that is characterized by that specific coupling strength.

As emphasized by Susskind (2013), this defense of a preferred fundamentality ordering does not work with respect to string theory, however. The string coupling constant is no fundamental parameter that can be set to a certain value but corresponds to the value of a quantum field (the dilaton). Selecting the value of the string coupling therefore is a matter of the theory's dynamics. The value of the dilaton field can vary in spacetime, thereby creating a situation where one of the dual descriptions is simplest at one point in spacetime and the other description is preferable at another point.

Castellani (2017) raises a different issue. Understanding the solitonic objects in terms of composite states may seem plausible when viewing them from a perturbative starting point but is much less natural when analyzing them in terms of Noether charges and topological charges. The latter point of view would conceive of solitonic solutions as distinct elements of the particle spectrum without establishing a constituent-compound relation. Whether or not solitons are construed as compounds thus seems itself a matter of choice. In this light, reductionism may be considered to fail in the presence of any non-perturbative effects in quantum field theory irrespectively of the question as to whether or not duality relations arise.

Huggett (2017) addresses the issue of specifying the physical essence of dual theories by focusing on another duality relation. T-duality relates a theory with a given radius of a compact spatial dimension to a theory with inverse radius (measured in units of the string length). The duality transformation thereby transforms winding numbers of strings around that dimension into transversal momentum and vice versa. Based on the understanding (established above) that dual formulations don't amount to different theories but give the same physical description, Huggett distinguishes two possible ways to interpret this statement. According to the first interpretation, dual theories offer a translation manual. Moving from one description to its dual thus amounts to a transmutation of terms. This would allow attributing reality to a certain radius of a compact dimension since the duality transformation to a small radius would be compensated for by a transmutation of terms. The second interpretation holds that terms have the same meaning in both of the dual formulations. In that case, the real physical content of the theory has to be reduced to what remains invariant und duality transformations. The radius of a compact dimension would be indeterminate. Huggett rejects the first option because it offers no satisfactory way of viewing a different world with inverted radius based on the initial meaning of terms.

A different perspective on dualities is chosen by Dawid (2017). Dawid compares dualities to the traditional take on empirical equivalence in physics. Empirical equivalence, Dawid argues, was understood in the 20th century in terms of the flexibility of theory building due to the deployment of advanced mathematical and conceptual tools. Once those tools have been developed, making use of them allows for various conceptual representations of the same empirical data set. Dualities tell a very different story. In their case, new conceptual tools are deployed not to develop new theories but to demonstrate the empirical equivalence of existing theories that had previously been understood to substantially differ from each other. What is revealed by duality relations thus is a more constrained rather than a wider spectrum of theories. Dawid (2017) argues that this constitutes a significant shift with respect to the role of new conceptual tools in physics.

A technical analysis of gauge/gravity duality is carried out in de Haro, Teh, Butterfield (2016) and de Haro (2016).[3] Global symmetries are invariant under duality transformations but often have a different physical interpretation in the dual theory. Gauge transformations, to the contrary, do not survive duality transformations, which reflects the fact that they deal with unphysical degrees of freedom of a given formulation. The way in which global symmetries and gauge symmetries play out in gauge/gravity duality involves a number of subtle issues that are addressed in the papers.

23.3.2 The Emergence of Spacetime

The issue of duality leads up to another important philosophical discussion: the emergence of spacetime in a string-theoretical context. AdS/CFT duality relates a string theory on a specific background space to a conformal field theory at the boundary of this space. While there is strong support for the understanding that this relation has the status of an exact duality, it is an open question whether the duality can be generalized to a relation between any string theory and a gauge theory dual. Though some general arguments would suggest as much (see e.g. 't Hooft, 1993; Strominger, 2001), no manual for a general gauge/gravity duality has yet been found.

Some string physicists (see e.g. Seiberg, 2007; Horowitz and Polchinski, 2009) have expressed their understanding that gauge/gravity duality could be understood in terms of emergent spacetime. A number of philosophical papers have tried to square claims of emergence made by string theorists with the philosophical use of the concept of emergence. The core problem is the following. While emergence has a directedness, an exact duality relation is symmetric. The translation manual leads both ways without singling out one of the dual theories as more fundamental. There is no way to interpret one of the dual theories as a low-energy effective theory of the other.

Dieks et al. (2015) reject the applicability of the concept of emergence to exact duality relations for that reason. They suggest that, among the current discussions of dualities in the context of quantum gravity, only Verlinde's (2011) entropy approach to gravitation, which does not posit an exact duality but merely a duality relation at the level of effective descriptions, qualifies as a suggestion of emergent gravity. Rickles (2013b) expresses similar doubts about viewing gauge/gravity duality as a basis for the emergence of spacetime. He argues that indications of an emergent character of spacetime in the context of string theory are in line with more general reasoning in the context of quantum gravity and don't substantially rely on the issue of duality.[4] Teh (2017) widens the analysis by asking whether it is possible to identify directedness in duality relations by other means despite their formally symmetric character. He suggests that the most promising strategy to that end would focus on an explanatory advantage of one of the two duals. He comes to the conclusion that while specific explanations work better in one framework, there is no consistent pattern of favoring one of the duals over the other. This does not exclude, though, that a different formulation of the theory could be found one day, that has a clear overall explanatory advantage that justifies the understanding that other perspectives emerge from it.

Horowitz and Polchinski (2009) argue that, in the case of gauge/gravity duality, the symmetry between the duals may be less clear than in other cases. At the present point, one knows an exact description of the gauge theory while the exact formulation of the string theory is unknown. This fact would not amount to a fundamental difference between the two theories if it merely indicated a deficit in the present understanding of the string theory side. The asymmetry might be fundamental, however, if it were indeed impossible to reach a full formulation of string theory in any other way than by going to the gauge theoretical dual. In that case, string theory proper would be conceptually dependent on the gauge theory in a non-reciprocal way and therefore could justify the application of the concept of emergence.

23.3.3 *Black Hole Physics and Information Loss*

Closely related to the last point is the issue of information loss in black holes. It has been a longstanding problem for black hole physics to understand what happens once a black hole evaporates due to Hawking radiation. A conventional understanding of quantum physics would suggest that Hawking radiation cannot contain the information stored in objects that have fallen through the black hole horizon. However, thermodynamical principles would suggest that no information should have been lost once the black hole has vanished. String theory gives a clear answer based on AdS/CFT duality: since there is no information loss in the dual conformal field theory description, information must also be preserved in the gravitational system. While suggestions regarding the actual process of information conservation on the string theory side have been put forward (see e.g. Almheiri et al., 2013), no full understanding has been achieved so far. Black hole information is one of the most intensely investigated issues in string theory today. The substantial philosophical significance of those investigations has not yet been addressed in the philosophy of science.

23.3.4 *The Lack of Free Parameters and the String Theory Landscape*

At a fundamental level, string theory does not have any free parameters. Due to the web of dualities that connects all five types of string theory, this implies that, at a fundamental level, there is only one realization of a string theory that includes fermions and lives in more than two spacetime dimensions. This fact is one of the most remarkable features of the theory. Dawid (2007) has argued that this property of "theoretical uniqueness" is one specific reason for trust in string theory based on non-empirical theory assessment (see Section 23.4.2) and can provide the basis for a specifically string theoretical take on scientific realism (see Section 23.3.1).

For a while, string theory's uniqueness and lack of free parameters raised hopes that the theory might uniquely predict all parameter values of low-energy physics. An improved understanding of string vacua (Kachru et al., 2003) then established that one must expect a discrete but huge spectrum of vacua of the theory, represented by what is called the string theory landscape. This substantially reduced the expectations regarding string theory's predictive power and resulted in a more complex picture: the uniqueness at the fundamental level is compounded by a considerable amount of flexibility at the level of the theory's ground states. String theorists have been struggling with understanding the theory's predictive status under these circumstances (see e.g. Douglas, 2003). This situation raises a number of important philosophical questions. What is the significance of a theory's uniqueness at a fundamental level given the existence of a landscape? How should one understand the concept of empirical prediction under those circumstances? How should a continuous free parameter be compared to a large but discrete set of allowed parameter values? These and other questions still await analysis from a philosophy of science perspective.

23.3.5   *The Multiverse and Anthropic Reasoning*

One very important strand of philosophical analysis related to the string theory landscape is concerned with the multiverse and anthropic reasoning. While the multiverse is rooted in the cosmological concept of eternal inflation (Vilenkin, 1983), the string landscape is necessary for providing a physical basis for allowing different values of the cosmological constant and other parameters in each universe of the multiverse. String theory thus plays a crucial role in the setup that eventually leads to anthropic reasoning (Susskind, 2010). The philosophical issues that arise in this context are of fundamental importance but lie beyond the scope of this chapter – see Chapter 53 in this volume for further discussion.

23.4   The Meta-Level Issues of String Theory

String theory's philosophical relevance reaches beyond the theory's specific physical import. In a number of ways, string theory raises philosophical questions regarding the research process associated with the theory and, more generally, the role of theory in science.

23.4.1   *The Incompleteness of String Theory*

Nearly half a century of intense work on string theory has not resulted in anything close to a complete theory. As described above, the theory's formulation started from a perturbative perspective and then transcended that approach based on the discovery of duality relations. The most surprising result in this respect was the discovery of AdS/CFT correspondence, which indicated that a string theory could be empirically equivalent to a theory that was no string theory at all. Based on this insight, it is not clear anymore how central strings are to string theory. The insufficient grasp as to what string theory actually stands for raises the question whether one should call it a theory at all at the present stage. A variety of views on this issue have been expressed by string physicists. Historians of science – Camilleri and Ritson (2015a, 2015b) – have emphasized this wide spectrum as a core characteristic of the internal dispute on the theory's status. Roughly, the spectrum of positions can be viewed in terms of two conflicting considerations.

The first consideration emphasizes that string theorists so far have failed to grasp even the core principles of string physics in a conclusive way. This may be considered to suggest that what physicists have developed up to this point is less than an actual theory. David Gross (2015) has, in this vein, characterized string theory as a framework: a set of principles that determines the way theory building proceeds but does not fully specify empirical implications. Still, a framework in Gross' sense does constrain the spectrum of empirical data that can be modeled and therefore can be rejected or confirmed on an empirical basis.

The second consideration emphasizes the fact that the theory's development at a fundamental level seems driven entirely by consistency arguments without leaving room for conceptual choices on the way. This fact may be considered to indicate that even though the most fundamental principles of the theory have not yet been found, the conceptual posits that characterize the theory today are sufficient for uniquely determining the theory: string theory is whatever the consistent full set of implications of the posits that define string theory today amounts to. This perspective that treats string theory as a well-defined but ill-understood theory is suggested in Witten's (1996) spelling out of string theory's final theory claim or Polchinski's (2019) characterization of the theory's current status.

A more general philosophical analysis of what it takes to be a theory in fundamental physics at the time of string theory is a desideratum in the philosophy of science.

23.4.2   The Theory's Strong Status Despite the Lack of Empirical Confirmation and Completeness

The second main meta-level issue that has been raised with respect to string theory is the high degree of trust in the theory's viability many of its peers have developed during the last three decades. This degree of trust is remarkable, given the theory's highly incomplete state (see previous section) and the fact that it has not found any empirical confirmation up to this point. The physicist Lee Smolin (2006), the mathematician Peter Woit (2006), and philosophers Erik Curiel (2001),[5] Reiner Hedrich (2007a, 2007b), Roman Frigg, and Nancy Cartwright (2007) argue that this confidence is unfounded and indicates a problematic detachment of fundamental physics from empirical evidence. String theorists Joseph Polchinski (2007, 2019) and Mike Duff (2013) reject that criticism, pointing at inaccuracies of critical presentations of string theory, and argue for the reasonability of trust in string theory.

Dawid (2006, 2009, 2013a) puts forward the idea that the trust in string theory and a number of other physical concepts can be best understood by widening the concept of theory confirmation. Theory confirmation is hereby understood in a Bayesian sense as an increase of a theory's probability of being viable. The proposed wider concept of confirmation includes evidence that lies outside the theory's intended domain (i.e., it is not of the type that can be predicted by the theory in question) but nevertheless provides information about the outside world rather than merely about the theory's characteristics. A crucial role in the approach is played by the concept of scientific underdetermination (akin to what Lawrence Sklar (1975) and Kyle Stanford (2006) call transient underdetermination.) Observations about the research process may indicate that scientific underdetermination (which corresponds to the number of empirically distinguishable scientific theories that can be built based on the available empirical data) is severely limited. Strong limitations to scientific underdetermination can in turn increase the probability that the theory one has developed is viable. If that is the case, such observations about the research process amount to theory confirmation. Confirmation based on such observations is called non-empirical because it is not based on empirical tests of the confirmed theory. Dawid (2006, 2013a) proposes three main strategies of non-empirical confirmation. Dawid et al. (2015) and Dawid (2016) formalize this line of reasoning based on Bayesian confirmation theory.

While there is mostly agreement on the fact that physicists do use strategies of non-empirical confirmation, a number of aspects of the approach have been questioned and criticized. Smolin (2014) argues that non-empirical confirmation is too flexible for being reliable and further bolsters dominant research programs in an unwarranted way. Ellis and Silk (2014) argue that a widening of the concept of confirmation amounts to giving up core pillars of scientific checks and balances. Rovelli (2017) argues that the mismatch between the Bayesian notion of confirmation and the way the term confirmation is used in theoretical physics renders the message of non-empirical confirmation misleading from a physics perspective. Dawid (2019) responds to some of those criticisms. The role of non-empirical theory confirmation is also analyzed in Dardashti (2019) and Oriti (2019).

Rickles (2013a) argues for some degree of trust in string theory's viability without endorsing non-empirical confirmation. He suggests that the viability of string theory becomes more plausible due to the fertility of the mathematical concepts developed in its context. Matsubara and Johannsson

(2013) aim at striking a middle ground in their assessment of string theory. They suggest, following Frigg and Cartwright (2007), that string theory does not constitute a progressive research program in a Lakatossian sense at this point. Still, they argue that the string theorists' way of dealing with their theory looks reasonable from a Lakatossian perspective since, in the absence of a clearly progressive research program, focusing on the strategy that looks most promising is perfectly legitimate for the individual scientist.

23.4.3   The Theory's Universality and Its Final Theory Claim

The third important peculiarity of string theory pertains to its position within the overall fabric of physics. Theory building in physics may be viewed as a process of successful unification that leads from Newton's unified description of the movements of earthly and heavenly bodies to the conceptually unified relativistic description of all nuclear forces in the standard model of particle physics. The unification of nuclear forces and gravity, which is the aim of string theory, might be the final step in this series of unifications. Moreover, string theory conceptually does not allow for any form of amendment by additional theory. String theory therefore is a fully universal theory not only in virtue of the known set of phenomena that need to be accounted for but in virtue of its conceptual nature. Interestingly, string theory adds to this inherent universality claim an internal final theory claim that is based on a different line of reasoning. T-duality, one of the core string dualities, implies that the theory's own characteristic scale amounts to a minimal length scale. Any statements about a phenomenon at a smaller length scale can be understood as a statement about a phenomenon at a larger length scale by moving to the T-dual description. (See e.g. Witten, 1996.)

The universal character of string theory and its final theory claim raise at least three deeply philosophical questions.

First, the question arises whether a theory's finality can be meaningfully supported by arguments that rely on that very theory. Relying on a theory in an argument for that very theory's viability looks viciously circular. Dawid (2007, 2013b) acknowledges that this circularity devalidates a final theory argument when viewed in isolation. He argues, however, that a final theory claim can acquire some force in the context of a broader argument regarding the structure of limitations to scientific underdetermination. In such an argument, the lack of possible alternatives to a given theory is first established in a certain empirical context (for example at a certain energy scale) by conventional arguments of non-empirical theory assessment and then extended beyond that context based on a final theory argument.

Second, the question arises whether and in which way a theory's universal character or its prospects of being a final theory create a different overall understanding of the theory's scientific role and the mechanisms of scientific evolution. Dawid (2007, 2013a) relates string theory's final theory claim to its long-term incompleteness (see Section 23.4.1). The canonical understanding of the physical process assumes a finite – usually reasonably brief – time for the completion of a theory but projects an infinite series of theories superseeding each other due to an influx of empirical anomalies. String theory's final theory claim suggests that no further sequence of superseeding theories will be forthcoming. The incompleteness of the theory without indications of full redemption in the foreseeable future may be considered to suggest, however, that the time horizon for completing the (final) theory at hand has shifted toward infinity. In conjunction, those two shifts imply that while theory development should now be understood in terms of inner-theoretical conceptual evolution rather than theory succession, the prospects for an imminent completion of fundamental physics have barely improved.

Finally, the question arises as to whether and if so how string theory's incompleteness is compatible with a meaningful final theory claim. A number of string theorists have come to doubt this, which arguably has led to a reduced emphasis on string theory's final theory claims in recent years. A substantial analysis of this issue from a wider philosophy of science perspective is still missing.

23.4.4   *The Remarkable Relevance of String Theory as a Tool in Plasma Physics*

A further meta-level issue of significant philosophical interest is string theory's remarkable role as a tool for carrying out quantitative calculations of the strongly coupled quark gluon plasma. Those calculations are based on performing a gauge/gravity duality transformation on the strongly coupled gauge theoretical description of the system and then carry out calculations in the resulting gravitational theory, where they are much easier to do. As noted in Section 23.3.1, gauge gravity duality has only been established for a very specific class of gauge theories, conformal field theories. QCD, the theory describing the quark gluon plasma, is no conformal field theory, which means that no actual gravity dual has been identified. Quark gluon plasma calculations based on a duality transformation therefore can only be understood as a fairly rough approximation, which is in line with the fairly rough agreement of such calculations with actual data.

The method's success provides an interesting example of a theory that is considered to be a viable fundamental theory in one context and is utilized as a mere calculational tool in a different context. From a historical/sociological perspective, it is of interest how string theory's utility in that other context has widened the spectrum of physicists who know and deploy string theory in their work. Many of them do so without endorsing the theory as a fundamental theory of all interactions. In a more philosophical vein, one may raise the question whether the theory's success in plasma physics can be of any significance for understanding the theory's prospects as a fundamental theory. None of those issues has been addressed from a philosophy of science viewpoint so far.

23.5   Conclusion

String theory is a very different kind of conceptual scheme than any earlier physical theory. It is the first serious contender for a universal final theory. It is a theory for which previous expectations regarding the time horizon for completion are entirely inapplicable. It is a theory that generates a high degree of trust among its exponents for reasons that remain, at the present stage, entirely decoupled from empirical confirmation. Conceptually, the theory provides substantially new perspectives on the role of space and time, on the relation between a theory and its classical limits, on the uniqueness and the contingencies of a theory and its empirical implications, on the nature of black holes, and on many other core issues of physical theory.

All of the described shifts have a profound philosophical dimension. Some of them are specifically bound to string theory. The significance one attributes to those shifts thus may depend on one's assessment of the theory's prospects of being viable. But, at a general level, most of the shifts in perspective associated with string theory raise questions that transcend string theory proper by exemplifying the substantial changes that are forced upon scientists by fundamental physics today. Theory building in an environment of scarce empirical data, the overwhelming conceptual difficulties associated with developing a theory of quantum gravity, the conceptual issues that render the canonical view of theory succession questionable once one reaches the Planck scale, the emergence of space and time from more fundamental concepts, the seemingly holographic character of quantum gravity, the deep problems related to information in black hole physics: those are all issues faced by contemporary fundamental physics irrespectively of the fate of string theory.

String theory, in the eyes of many of its exponents, offers a number of reasons for assuming its viability as a theory or framework for addressing those issues. But even to those who doubt the theory's viability, it can serve as a case study for the ways in which core issues of contemporary physics transcend our traditional understanding of both what we should expect from a physical theory and how physics works as a discipline.

Notes

1 For a philosophical perspective on that strategy, see Wüthrich (2004).
2 See e.g. Coffey (2014) for a recent exposition of the latter view.
3 For a survey of gauge gravity duality, see also de Haro, Mayerson, Butterfield (2016).
4 For a philosophical view on the way the Einstein equations for background space in string theory can be extracted from consistency requirements on the propagation of individual quantum strings, see also Huggett and Vistarini (2015).
5 Curiel's paper does not focus on string theory but addresses quantum gravity in general.

References

Almheiri, A., Marolf, D., Polchinski, J. and Sully, J. (2013). Black holes: Complementarity or Firewalls? *Jhep*, 1302: 062.

Becker, K., Becker, M. and Schwarz, J. (2007). *String Theory and M-theory: a Modern Introduction*. Cambridge: Cambridge University Press.

Butterfield, J. and Isham, C. (2001). Spacetime and the philosophical challenge of quantum gravity. In C. Callender and N. Huggett (eds.), *Physics Meets Philosophy at the Planck Scale*. Cambridge: Cambridge University Press.

Camilleri, K. and Ritson, S. (2015a). The role of heuristic appraisal in Conflicting assessments of string theory. *Studies in History and Philosophy of Modern Physics*, 51: 44–56.

Camilleri, K. and Ritson, S. (2015b). Contested boundaries: The string theory debates and ideologies of science. *Perspectives on Science*, 23(2): 192–227.

Cappelli, A., Castellani, E., Colomo, F. and Di Vecchia, P. (2012). *The Birth of String Theory*. Cambridge University Press.

Castellani, E. (2009). Dualities and intertheoretic relations. In M. Suarez, M. Dorato and M. Redei (eds.), *EPSA Philosophical Issues in the Sciences*. Springer, pp. 9–19.

Castellani, E. (2017). Duality and particle democracy. *Studies in the History and Philosophy of Modern Physics*, 59: 100–108.

Curiel, E. (2001). Against the excesses of quantum gravity: A plea for modesty. *Philosophy of Science*, 68(proceedings): 424–441.

Dardashti, R. (2019). Physics without experiments? In R. Dardashti, R. Dawid and K. Thebault (eds.), *Why Trust a Theory?* Cambridge University Press, pp. 154–172.

Dawid, R. (2006). Underdetermination and theory succession from the perspective of string theory. *Philosophy of Science*, 73(3): 298–322.

Dawid, R. (2007). Scientific realism in the age of string theory. *Physics and Philosophy*, 11: 1–32.

Dawid, R. (2009). On the conflicting assessments of the current status of string theory. *Philosophy of Science*, 76(5): 984–996.

Dawid, R. (2013a). *String Theory and the Scientific Method*. Cambridge: Cambridge University Press.

Dawid, R. (2013b). Theory assessment and final theory claim in string theory. *Foundations of Physics*, 43(1): 81–100.

Dawid, R. (2016). Modelling non-empirical confirmation. In E. Ippoliti, T. Nickles and F. Sterpetti (eds.), *Models and Inferences in Science*. Springer, pp. 191–205.

Dawid, R. (2019). The significance of non-empirical confirmation in fundamental physics. In R. Dardashti, R. Dawid and K. Thebault (eds.), *Why Trust a Theory?* Cambridge: Cambridge University Press, pp. 99–119.

Dawid, R., Hartmann, S. and Sprenger, J. (2015). The no alternatives argument. *British Journal for the Philosophy of Science*, 66(1): 213–234.

de Haro, S. (2016). Dualities and emergent gravity: Gauge/gravity duality. *Studies in the History and Philosophy of Modern Physics*. Online first.

de Haro, S., Meyerson, D.R. and Butterfield, J.N. (2016). Conceptual Aspects of Gauge/Gravity Duality. *Foundations of Physics*, 46(11): 1381–1425.

de Haro, S., Teh, N. and Butterfield, J.N. (2016). Comparing dualities and gauge symmetries. *Studies in History and Philosophy of Science Part B: Studies in History and Philosophy of Modern Physics*, 59: pp. 68–80.

Dieks, D., van Dongen, J. and De Haro, S. (2015). Emergence in holographic scenarios for gravity. *SHPMP*, 52. 203–216.

Douglas, M. (2003). The statistics of string/M theory vacua. *JHEP* 0305: 046.

Duff, M. (2013). String and M-theory: Answering the critics. *Foundations of Physics*, 43: 182–200.

Ellis, G. and Silk, J. (2014). Defending the integrity of physics. *Nature*, 516: 321–323.

Frigg, R. and Cartwright, N. (2007). String theory under scrutiny. *Physics World* September 07, 15.

Green, M.B. and Schwarz, J.H. (1984). Anomaly cancellation in supersymmetric D=10 gauge theory and superstring theory. *Physics Letters*, B149: 117–122.

Gross, D. (2015). *What Is a Theory?, Talk at the Workshop "Why Trust a Theory?* Munich.

Gubser, S.S., Klebanov, I.R. and Polyakov, A.M. (1998). Gauge theory correlators from non-critical string theory. *Physics Letters*, B428: 105 (hep-th/9802109).

Harvey, J. and Strominger, A. (1995). The heterotic string is a soliton. *Nuclear Physics* B449: 535–552.

Hedrich, R. (2002). Anforderungen an eine physikalische Fundamentaltheorie. *Journal for General Philosophy of Science*, 33(1): 23.

Hedrich, R. (2007). Von der Physik zur Metaphysik, Ontos.

Horowitz, G.T. and Polchinski, J. (2009). Gauge/gravity duality. In D. Oriti (ed.), *Approaches to Quantum Gravity*. Cambridge: Cambridge University Press, pp. 169–186.

Huggett, N. (2017). Target space $\neq$ space. *Studies in the History and Philosophy of Modern Physics*, 59: 81–88.

Huggett, N. and Vistarini, T. (2015). Deriving general relativity from string theory. *Philosophy of Science*, 82(5), 1163–1174.

Johansson, L.-G. and Matsubara, K. (2011). String theory and general methodology: A mutual evaluation. *Studies in the History and Philosophy of Modern Physics* 42(3): 199–210.

Kachru, S., Kallosh, R., Linde, A.D. and Trivedi, S. P. (2003). De Sitter Vacua in string theory. *Physical Review*, D68: 046005.

Maldacena, J. (1998). The large N limits of superconformal field theories and supergravity. *Advances in Theoretical and Mathematical Physics*, 2: 231. Available at: arXiv:hep-th/9711200.

Matsubara, K. (2013). Realism, underdetermination and string dualities. *Synthese*, 190(3): 471–489.

McKenzie, K. (2016). Relativities of fundamentality. *Studies in the History and Philosophy of Modern Physics*. Online first.

Oriti, D. (2019). No alternative to proliferation. In R. Dardashti, R. Dawid and K. Thebault (eds.), *Why Trust a Theory?* Cambridge: Cambridge University Press, pp. 125–153.

Polchinski, J. (1995). Dirichlet-Branes and Ramond-Ramond charges. *Physical Review Letters*, 75: 4724. Available at: arXiv:hep-th/9510017.

Polchinski, J. (1999). *String Theory*, Vol. 2. Cambridge: Cambridge University Press.

Polchinski, J. (2007). All strung out? *American Scientist*, 95(1): 1.

Polchinski, J. (2019). String theory to the rescue. In R. Dardashti, R. Dawid and K. Thebault (eds.), *Why Trust a Theory?* Cambridge: Cambridge University Press, pp. 339–353.

Rickles, D. (2011). A Philosopher's look at string theory. *Studies in the History and Philosophy of Modern Physics*, 42(1): 54–67.

Rickles, D. (2013a). Mirror symmetry and other miracles in superstring theory. *Foundations of Physics*, 43(1): 54–80.

Rickles, D. (2013b). AdS/CFT duality and the emergence of spacetime. *Studies in the History and Philosophy of Modern Physics*, 44(3): 312–320.

Rickles, D. (2013c). *A Brief History of String Theory: From Dual Models to M-Theory*. Springer.

Rickles, D. (2016). Dualities: Same but different or different but same? *Studies in the History and Philosophy of Modern Physics*. Online first.

Rovelli, C. (2019). The dangers of non-empirical confirmation. In R. Dardashti, R. Dawid and K. Thebault (eds.), *Why Trust a Theory?* Cambridge: Cambridge University Press, pp. 120–124.

Scherk, J. and Schwarz, J.H. (1974). Dual models for nonhadrons. *Nuclear Physics*, B81: p118.

Sen, A. (1999). Duality symmetries in string theory. *Current Science*, 77(12): 1635.

Sklar, L. (1975). Methodological conservativism. *Philosophical Review*, 84: 384.

Smolin, L. (2006). *The Trouble with Physics*. Houghton Mifflin.

Smolin, L. (2014). Review of R. Dawid: "Why trust a theory?". *American Journal of Physics*, 82: 1105.

Stanford, K. (2006). *Exceeding Our Grasp - Science, History and the Problem of Unconceived Alternatives*. Oxford: Oxford University Press.

Strominger, A. (2001). The dS/CFT correspondence. *JHEP*, 0110: 034.

Susskind, L. (2010). The anthropic landscape of string theory. In B. Carr (ed.), *Universe or Multiverse?*. Cambridge: Cambridge University Press, pp. 247–253.

Susskind, L. (2013). String theory. *Foundations of Physics*, 43(1): 174–181.

Teh, N. (2013). Holography and emergence. *Studies in the History and Philosophy of Modern Physics*, 44(3): 300–311.

't Hooft, G. (1993). Dimensional reduction in quantum gravity. In A. Ali and D. Amati (eds.), *Salamfestschrift*. World Scientific, pp. 284–296.

Veneziano, G. (1968). Construction of a crossing-symmetric, Regge behaved amplitude for linearly rising trajectories. *Nuovo Cimento*, 57A: 190.

Verlinde, E. (2011). On the origin of gravity and the laws of Newton. *JHEP*, 1104: 029.

Vilenkin, A. (1983). The birth of inflationary universes. *Physical Review*, D27: 2848.

Vistarini, T. (2016). Holographic space and time: Emerging in what sense? *Studies in the History and Philosophy of Modern Physics*. Online first.

Weingard, R. (1989). A philosopher's look at string theory, reprinted in Callender, C. and Huggett, N. (2001). *Physics meets Philosophy at the Planck Scale*. Cambridge University Press.

Witten, E. (1995). String theory dynamics in various dimensions. *Nuclear Physics*, B443: 85–126. Available at: arXiv:hep-th/9503124.

Witten, E. (1996). Reflections on the fate of spacetime, reprinted in Callender, C. and Huggett, N., editors. (2001).*Physics meets Philosophy at the Planck Scale*. Cambridge University Press.

Witten, E. (1998). Anti- de Sitter space and holography. *Advances in Theoretical Mathematics Physics*, 2: 253 (hep-th/9802150).

Woit, P. (2006). *Not Even Wrong: The Failure of String Theory and the Continuing Challenge to Unify the Laws of Physics*. Jonathan Cape.

Wüthrich, C. (2004). *To Quantize or Not to Quantize: Fact and Folklore in Quantum Gravity*. Proceedings PSA 2004: Contributed Papers.

Zwiebach, B. (2004). *A First Course on String Theory*. Cambridge: Cambridge University Press.

Further Reading

A good popular introduction to string theory is Brian Greene's *The Elegant Universe* (W. W. Norton 1999). For an accessible technical introduction to string theory see Zwiebach, B., *A First Course on String Theory* (Cambridge: Cambridge University Press, 2004). For historical accounts of string theory see Cappelli, A., E. Castellani, F. Colomo, and P. Di Vecchia *The Birth of String Theory* (Cambridge University Press, 2012) and Rickles, D. *A Brief History of String Theory: From Dual Models to M-Theory* (Springer 2013). For a book length treatment of epistemic aspects of string theory, see Dawid, R. *String Theory and the Scientific Method* (Cambridge: Cambridge University Press, 2013). For thoughts about the spacetime of string theory, see Huggett, N. 'Target Space ≠Space' (*Studies in the History and Philosophy of Modern Physics* 59:81–88, 2017). Huggett, N. and Vistarini, T. 'Deriving General Relativity from String Theory' (*Philosophy of Science* 82(5): 1163–1174, 2015) looks at the prospects for recovering general relativity.

24

QUANTUM GRAVITY FROM GENERAL RELATIVITY

Christian Wüthrich

24.1 Introduction: The Need to Go Beyond General Relativity

General relativity (GR) is our best theory of gravity and one of the best confirmed theories in the history of science (Will, 2014). Why then do physicists believe that it needs to be replaced by a more fundamental theory of "quantum gravity"?[1] Standard arguments brought forth for such a need may be motivational, but can hardly be considered conclusive.

First, there is unification. But that the world conforms to this metaphysical ideal is certainly not an a priori truth, although there is some inductive evidence that unification is methodologically valuable in physics. Furthermore, a common interpretation of GR has it that GR shows that gravity is not a force, and hence not similar in kind to the three fundamental forces – weak and strong nuclear, and electromagnetic.[2] Given the distinction in kind, any need for unification becomes much less pressing. A second argument derives from Eppley and Hannah's thought experiment (Eppley and Hannah, 1977) and tries to establish not only that a quantum theory of gravity is needed, but that, additionally, it must be obtained by quantizing gravity. However, the argument contains loopholes – and the thought experiment seems physically impossible (Huggett and Callender, 2001; Wüthrich, 2005; Mattingly, 2006). Another argument, explicated e.g. by Doplicher et al. (1995) but originating earlier, claims that since no localization of a physical system by means of a detection by a photon can be made with an accuracy exceeding the Planck length, as this is the scale at which a photon, if compressed to a spacetime volume smaller than that scale, would collapse into a black hole and thus become unusable for detection services. However, this argument, even if successful, at best establishes that spacetime is operationally discrete in that there are limits beyond which we cannot detect its structure. It does manifestly not imply anything about the structure of spacetime itself (Wüthrich, 2005). As a fourth example, Peres and Terno (2001) have argued that a theory cannot accommodate classical and quantum interactions, on pain of a violation of central physical principles – specifically the correspondence principle. Their argument relies on an articulation of the problem in Koopmanian terms, which combine classical and quantum physics, and thus remains hostage to a particular formalism that is far from compulsory (Wüthrich, 2005).

Instead of proceeding from a presumption that these quantum considerations invade and in some sense invalidate GR, one could argue by identifying a fatal flaw in GR itself. The most popular target for this strategy is the fact that GR predicts the existence of singularities, which flag a breakdown of the mathematical representation as used in the theory. In light of the singularity theorems, which establish that under rather generic conditions, cosmological models of GR are singular, it looks as if we cannot dismiss singular spacetimes as mere unphysical artifacts; they are here to stay with GR. While it is reasonable to operate methodologically on the assumption that nature affords a regular mathematical description, there is of course no metaphysical guarantee that limitations to such a description are not objective features of the world. Perhaps the world just isn't amenable to scientific study in this way.

Similarly, one might construct an argument from an apparent insufficiency of thermodynamics: black holes potentially violate the second law, as we can conceivably drop an entropy package into a black hole such that this entropy then vanishes behind the hole's event horizon and is thus "lost" to the total entropy of the exterior universe. Consequently, we ought to generalize the second to account for the potential loss (Bekenstein, 1973). The simplest way to do that is to attribute an entropy to black holes, such that any such potential loss is at least made up by an increase of the hole's entropy. From statistical mechanics, we have come to expect that any physical system with an entropy has "microstates." Hence, black holes have microscopic states such that "entropy" becomes an applicable property. Thus, contrary to what classical GR asserts, black holes have additional properties beyond their mass, charge, and angular momentum. Therefore, the argument concludes, we need to go beyond GR to understand the micro-structure of black holes, i.e., their "quantum nature." However, this argument falls short on two accounts. First, it presupposes, perhaps incorrectly, the validity of the second law in cases where black holes are involved. Second, Bekenstein's attribution of entropy to black holes inappropriately depends on a fundamentally information-theoretic approach to physically – specifically when it infers to the existence of micro-structure from the existence of entropy (Wüthrich, 2018). Black holes may well be thermodynamic objects; but if this is so, it has to be for reasons other than those given by Bekenstein. If they are thermodynamic, they may require a quantum theory to describe the microstates that give rise to their entropy. As suggestive as this line of reasoning may be, it falls short from being conclusive.

However, even if all these arguments do not conclusively establish that GR needs to be supplanted by a more fundamental theory of quantum gravity, there is a single reason which does: it is a contingent, but extremely well-established fact that matter is quantum – not classical as GR assumes. For this reason alone, GR cannot be the last word on gravity and will eventually have to be replaced by a theory of gravity that incorporates the fact that matter is quantum. None of this has, prima facie, any implications for the structure of spacetime. However, as it turns out, the most promising approaches to a formulation of such a theory all have more or less radical consequences for spacetime.

Now with the need to go beyond GR established, there are various ways of reaching beyond GR. The next section (Section 24.2) briefly discusses a natural and conceptually simple extension of GR – semi-classical gravity – which basically plugs quantum fields into the right-hand side of the Einstein field equation. Section 24.3 then briefly introduces the first of the two approaches to full quantum gravity discussed here, causal set theory (CST). The remainder of this chapter is devoted to canonical quantum gravity and particularly to its main representative, loop quantum gravity, which will be introduced in Section 24.4. Section 24.5 is concerned with what I will call "the problem of spacetime" in the context of loop quantum gravity (see Thébault, this collection, for a review of the problem of time in canonical quantum gravity). There are of course other approaches to quantum gravity, most of which do not consciously start from GR and are thus not the subject of the present chapter (see e.g. Dawid, this collection).

24.2 Semi-classical Gravity

The most straightforward way to a theory of quantum gravity is by means of the *semi-classical approach to quantum gravity*. Semi-classical gravity treats gravity classically in the sense that it uses the framework of GR to describe, and assumes that the matter fields propagating in spacetime are quantum fields, described by an appropriate quantum field theory (QFT). The two are coupled to one another through the *semi-classical Einstein field equation*:

$$R_{ab} - \frac{1}{2}g_{ab}R = 8\pi \langle \psi | \hat{T}_{ab} | \psi \rangle, \tag{24.1}$$

where $\langle \psi | \hat{T}_{ab} | \psi \rangle$ is the expectation value of the stress-energy tensor of the quantum fields in a physically reasonable state $|\psi\rangle$. Already by the time Wald's classic *Quantum Field Theory in Curved Spacetime and Black Hole Thermodynamics* (Wald, 1994) appeared almost a quarter century ago, this approach could offer a fully satisfactory and mathematically rigorous theory for linear quantum fields in curved spacetime (p. 1, the construction follows in ch. 4).

However, semi-classical gravity faces severe difficulties and limitations. First, there is an ambiguity in the definition of $\langle \psi | \hat{T}_{ab} | \psi \rangle$, which could be resolved by a more fundamental theory, or be fixed by experiment (Wald, 1994, 98).[3] Although this speaks not against the *truth* of semi-classical gravity, it is a sign of its non-fundamentality. Second, $\langle \psi | \hat{T}_{ab} | \psi \rangle$ contains terms of fourth order in derivatives of the metric (and not just of second order), and the semi-classical Einstein equation will have new solutions, often with "runaway" character (Wald, 1994, 99). Third, standard energy conditions can be generically violated under physically reasonable conditions in semi-classical gravity (Curiel, 2016). However, these point-wise energy conditions can arguably be replaced by weaker counterparts serving a similar purpose, the so-called "quantum energy conditions," which merely prohibit that the energy density can be arbitrarily negative over long enough periods of time. As it turns out, these weaker conditions are generally not violated in semi-classical gravity (Fewster and Verch, 2015, §4.8.1). Finally, it is sometimes listed as a problem of the approach that it is generally impossible to compute $\langle \psi | \hat{T}_{ab} | \psi \rangle$. Although true, such more practical limitations befall all approaches to quantum gravity, and much of physics besides. For instance, the Navier-Stokes equations are of great conceptual utility, and arguably at least approximately true, even though their general solution is not known. In fact, the *classical* Einstein equation remains generally unsolved after a century.[4] Nevertheless, at least the first three kinds of technical difficulties may well reflect a deeper physical tension in the approach.

Recently, what has become known as the "firewall paradox" can be thought of as a new challenge to semi-classical gravity arising from black hole physics. If the argument by Almheiri et al. (2013) leading to the firewall paradox is accepted, then either (i) the dynamical evolution from matter falling into the black hole to outgoing Hawking radiation is not unitary, or (ii) a form of the equivalence principle is not true, or (iii) the usual semi-classical approach of QFT on (slightly) curved spacetime is not valid. The "paradox" is interesting because each of the options forces us to give up what appears to be an eminently reasonable and successful assumption behind well-confirmed physics. It is truly, as Raphael Bousso puts it, "a menu from hell" (Ouellette, 2012). While most physicists, including the authors of the original article, seem to favor discharging the equivalence principle into retirement – hence the moniker "firewall" as event horizons would then burn up infalling observers – it may thus be the case that semi-classical gravity in the sense introduced here is not valid.

No doubt semi-classical gravity deserves more philosophical scrutiny than it has so far received. But one may well ask – and not just in the light of the foregoing challenges – how a semi-classical mélange of physical principles could possibly justify that quantum physics and gravity are blended into a unified fundamental theory when the latter is generally expected to reject at least some of the dearly held principles on which the former is built. All this may indicate, as most physicists think, that semi-classical gravity is confined to nothing but a small, temporary, and incomplete extension of "old physics," and that therefore a bolder approach to quantum gravity is required, at least as a fundamental theory. It is the purpose of the remainder of this chapter to introduce two such attempts to articulate a quantum theory beyond GR with ambitions to be offer a more fundamental account. Common to these approaches is that they both take GR as their vantage point for quantum gravity.

Christian Wüthrich

24.3   Causal Set Theory

CST (Bombelli et al., 1987) takes the central insight of GR to be that, for causally sufficiently non-pathological relativistic spacetimes, the causal structure of a spacetime determines its full geometry up to a conformal factor (Malament, 1977).[5] In a popular slogan in the field, "spacetime = causality + size." This insight motivates taking the fundamental structures underlying relativistic spacetimes to be discrete causal sets, where a fundamental relation of causal connectibility underwrites the causal structure, and the discreteness fixes the conformal factor, i.e., it provides "size" information in the form of countable elementary, Planck-sized "chunks" of spacetime. A causal set is represented by an ordered pair of a set of elementary events and a binary relation $\prec$ of "causal precedence" which partially orders the set of elementary events. That the relation $\prec$ gives rise to a partial ordering of the basal events means that it is reflexive, antisymmetric, and transitive. Its antisymmetry rules out certain causal pathologies that famously afflict GR, such as the presence of closed timelike curves. The elementary events have no further physical properties beyond standing in relations of causal precedence with other events. Finally, the resulting structure is stipulated to be discrete. The CST research program has so far not delivered a *quantum* theory of gravity, and so remains a work in progress. The following discussion is thus confined to the classical theory.

The central question for CST is whether the postulated causal sets generically give rise to relativistic spacetimes, or an empirically indiscernibly close surrogate thereof. As stated in the previous paragraph, this is demonstrably not the case: almost all realistically large causal sets as defined above form so-called "KR orders" consisting of only three "generations" of elements with about half the elements in the middle generation and a quarter each in the first and last generations, giving us no realistic model of the history of the universe. Thus, additional conditions need to be imposed in order to restrict the set of admissible causal sets such that physically realistic models come to dominate. If this were accomplished, then CST could justifiedly claim to offer a viable, though still classical, theory of the structure underlying relativistic spacetimes. In this sense, additional, "dynamical" conditions are imposed. The most widely discussed approach is the classical sequential growth dynamics as introduced by Rideout and Sorkin (1999), which is based on the remarkable result that a small number of physically justifiable assumptions severely constrain the possible dynamics if the latter is understood as the totally ordered sequence of the "birthing" of elements accreting to a past-finite, future-infinite causal set. As there is nothing quantum about this proposed dynamical condition – it is a thoroughly classical prescription – classical sequential growth dynamics is offered as a stepping stone to a full quantum theory of gravity.

A central part of the proposal is that the total order of the sequence of "birthing" is not a physical aspect of the model, but rather an auxiliary construction to obtain the right kinds of causal sets, i.e., those that will generically give rise to past-finite relativistic spacetimes. It thus remains an undisputed option to interpret the resulting causal sets in the usual eternalist fashion honored by relativistic physics. However, leading causal set theorists Sorkin (2007) and Dowker (2014) have argued that this dynamics can be interpreted in A-theoretic terms, i.e., involving an ineliminable and substantive notion of passage, and thus vindicates a metaphysics of becoming compatible with relativistic physics. The metaphysics favored by Sorkin and Dowker most closely resembles that of a growing block. The price to be paid for an A-theoretic metaphysics, however, is that the relativistically kosher becoming cannot be global, i.e., spatially extended, on pain of violating the Lorentz symmetry that any relativistic theory must accommodate. Instead, we have what Sorkin dubbed an "asynchronous multiplicity" of localized becoming (2007, p. 158) as, metaphorically, in a tree which independently grows at the tips of its different branches such that it is meaningless to say that one tip objectively grew before the other. As Pooley (2013, 358n) has correctly noted, this view is close to Fine's (2005) "non-standard A-theory," which rejects that there are, fundamentally, absolute tensed facts, but instead insists that there are fundamental tensed facts which are relativized to inertial frames. Consequently, fundamental reality is fragmented.

Naturally, the arguments of Sorkin and Dowker have been taken up by philosophers, who have examined the claimed compatibility of a metaphysics involving a substantive, A-theoretic notion of becoming with relativistic physics (Butterfield, 2007; Earman, 2008; Wüthrich and Callender, 2017). It turns out that the usual dilemma foisted on an A-theoretic metaphysics by relativistic physics (Callender, 2000; Wüthrich, 2013) essentially survives into CST (Wüthrich and Callender, 2017): any notion of an objective, global becoming either answers to the A-theorist's explanatory request, or is compatible with Lorentz symmetry, but not both. The way in which the dilemma is resolved in the Sorkin-Dowker interpretation of causal set dynamics is in accepting that the objective sense of becoming or of a present is not global, but only local – "asynchronous." However, as it turns out, there remains the possibility of a bizarre metaphysics of a growing block with global becoming: whole swathes of the sum total of existence remain, sometimes for long periods on the cosmological clock, in an ontological indeterminate, liminal state between existence and non-existence (Wüthrich and Callender, 2017).

The main attraction of CST is that it offers, based on a central insight in GR, an ontologically clear picture – at least as long as one sticks to an eternalist, B-theoretic reading of its metaphysics and to a classical version of its dynamics. Unfortunately, this very achievement is at peril by the necessary transposition of the theory's main tenets into a quantum theory, which remains, as mentioned above, unaccomplished.

24.4   Loop Quantum Gravity

Another approach to developing a full quantum theory of gravity by starting out from GR subjects it to a so-called "canonical quantization" procedure.[6] This is generally a promising strategy, as the procedure has been applied to classical theories and successfully delivered effective quantum theories on other occasions. However, the application of canonical quantization is less straightforward in GR, as the fact that GR treats spacetime as a four-dimensional unit cannot be easily reconciled with the presupposition of canonical quantization that physical theories deal with three-dimensional systems which dynamically evolve over time. In order to compensate, as it were, for the required split of the four-dimensional structure of relativistic spacetime, constraints on the basic Hamiltonian variables arise. The resulting constraint equations are equivalent to the Einstein field equation; since the theory must presuppose a topology permitting a global time function, it is classically only equivalent to a part of GR and does therefore not fully capture it.

The first important choice for any program of quantization along these lines is to select a pair of canonical variables. A first promising set of variables based on the four-metric g_{ab} was proposed by Arnowitt et al. (1962): the three-metric q_{ij}, the "lapse" function $N = \sqrt{-g^{00}}$, and the "shift" function $N_i = g_{i0}$, where $a, b = 0, ..., 3$ are spacetime indices and $i, j = 1, 2, 3$ are merely spatial indices. The great advantage of these "ADM" variables is their intuitive geometric interpretation. The three-metric is the metric tensor induced by the four-metric on three-dimensional constant-t spacelike hypersurfaces, the lapse function represents the proper time elapsed between the t-hypersurface and the $(t + dt)$-hypersurface, and the shift function measures the displacement of the spatial coordinates between these two hypersurfaces in stationary coordinates. Unfortunately, this approach leads to a dauntingly hard form of the constraint equation, and progress has stalled along these lines.

A more promising canonical quantization re-expresses the geometry of classical relativistic spacetimes in terms of the connection A_a^i and its conjugate, a densitized triad E_i^a, rather than the metric-based ADM variables. These basic, so-called "Ashtekar variables" are then used to construct a "holonomy-flux algebra," which consists of holonomies of the basic connection, which use parallel transport around closed loops as a measure of the curvature of the connection, and its conjugate, fluxes constructed from the densitized triads. One then seeks a representation of this algebra in some appropriate Hilbert space and expresses the constraint equations in terms of the holonomies and the fluxes, which are then defined as operators on that Hilbert space. The elements of that Hilbert space which satisfy the constraint equations would then be those which represent the physically possible quantum states of the gravitational field. The "physical Hilbert space" would be the Hilbert space consisting of all and only those states which are physically possible in this sense.

Unfortunately, however, the so-called Hamiltonian constraint equation, thought to capture the dynamical content of the theory, has so far resisted its solution. The subspace consisting of those states, which satisfy all the other constraints, forms a Hilbert space as well and is known as the "kinematical Hilbert space" $\mathcal{H}_K$. As it turns out, there exists a useful orthogonal basis of $\mathcal{H}_K$ which permits a natural physical interpretation. The elements of this basis are the so-called "spin network states," which are the eigenstates of two important geometric operators, the "area" and "volume" operators. The spin network states in the basis are built up from a state, which can be interpreted to represent the quantum geometric vacuum, by iterated application of the holonomies as "creation" operators which raise the excitation level of the fluxes.

The spin network states are normally represented as graphs, which carry spin representations on their edges and vertices. The structure of the graph gives the network of adjacency relations between parts of the spin network and thus represents the "connectivity" of the basic structure. The spin representations are related to the eigenvalues of the geometric operators applied to parts of these spin networks. The representations sitting on the vertices are the eigenvalues of the volume operator, and those on the edges the eigenvalues of the area operator. These eigenvalues have a discrete spectrum with a non-zero, positive smallest value. Thus, the spin network basis lends itself to a natural physical interpretation in terms of geometric properties and suggests a fundamentally granular quantum structure that gives rise, in a yet to understand classical limit, to the smooth spacetimes of GR. The spin representation of the vertices then gives a measure of their volume, and the representations on the edges connecting two adjacent vertices give a measure of the area of their connecting "surface."

The spin network states do not, in general, solve the Hamiltonian constraint equation and are thus merely "kinematical." Consequently, they are routinely interpreted as "spatial" in that they are what underlies three-dimensional physical space as an aspect of four-dimensional spacetime. It thus appears as if at least the structure of manifest space might be straightforwardly explicable on this approach as a fundamentally discrete structure of granular parts, which combine by attaching to one another through their adjacency relations in a manner of Lego-like building blocks. Unfortunately, such view would be too simplistic and cannot be maintained. First, even at the kinematic level, we should not forget that generically, the state of the gravitational field is a *quantum superposition* of elements of the spin network basis of $\mathcal{H}_K$. Thus, even if the basis element affords an interpretation in straightforward terms, the generic state is one that superposes basis states of different, inconsistent geometric properties and so does not yield to a geometric interpretation. Second, in order to get the full ontological picture, the Hamiltonian constraint equation would have to be solved and the fully physical Hilbert space would have to be known.

As the canonical quantization procedure has thus encountered a formidable roadblock, physicists are seeking ways to circumvent the problem. They pursue two general strategies. The first simplifies the systems studied by the theory and thus reduces the numbers of degrees of freedom already at the classical level and then attempts a standard canonical quantization on the simplified theory. This succeeds, and it leads to what is interpreted to give us the cosmological sector of the theory – Loop Quantum Cosmology – as the symmetry-reduced classical system is isotropic and homogeneous.[7] As such, it should give us insight into the very early universe from a fundamental perspective, and hopefully solve at least some of the many challenges of present-day cosmology. Advocates claim that Loop Quantum Cosmology does indeed offer such insight: it is alleged to dissolve the initial singularity ("big bang") of the standard model of cosmology (Bojowald (2011, Ch. 7), but see also Wüthrich (2006, Ch. 6)), and to lead to an early inflationary period with a graceful exit without the additional postulation of an inflaton field or similar ad hoc stipulations (Bojowald, 2002). The general idea behind this research problem is now to successively introduce complexities such as anisotropies in a climb to a more realistic full theory. To what extent this endeavor will succeed is open.

The second avenue pursued by physicists is to abandon the canonical path half way through the procedure and replace the dynamics with a covariant approach, as advocated e.g. in Rovelli and Vidotto (2015). Instead of the Hamiltonian $\hat{H}$ of the canonical approach, the dynamics is expressed, hopefully equivalently, in terms of the corresponding transition amplitudes. In a quantum theory, the transition amplitude from an initial state to a final state in the form of a probability captures the dynamical content of a theory, as is computed as an integral over paths. In the absence of a spacetime background for a quantum system to propagate along "paths," the connection between an "initial" and a "final" spin network state is made through combinatorial "splittings," "persistings," and "joinings" of the granular structure, adorned with a probability for each. The resulting fundamental structure, which is what is assumed to ground classical spacetime, is a so-called "spinfoam."

Although many questions remain open, such as how to incorporate general matter fields into the picture, spinfoams have promising advantages. First, and in an echo to Loop Quantum Cosmology, one can also formulate a theory of quantum cosmology on this basis (Rovelli and Vidotto , 2015, §11.3). Second, the spinfoam approach permits a "derivation" of the Bekenstein-Hawking formula for the entropy of black holes (Rovelli and Vidotto , 2015, §10.4).

24.5   The Problem of Spacetime

Putting open problems and past accomplishments of LQG to the side, how should one conceive of the resulting fundamental structure, to the extent to which it is currently known and understood? In particular, how non-spatiotemporal is it, and how can classical smooth spacetime re-emerge from it? This section will address these philosophically central issues.[8]

As LQG is based on a quantization procedure which presupposes a foliation of spacetime into three-dimensional spaces totally ordered by a one-dimensional time, the destinies of space and time are not entirely parallel. Let us start with time. Like all other canonical approaches to quantum gravity, LQG suffers from a "problem of time."[9] This problem has two aspects. First, since the Hamiltonian operator $\hat{H}$, which generates the dynamics, also turns out to be a constraint (in the approach based on ADM variables as well as in LQG), we obtain as the basic dynamic equation for a physical state $|\Psi\rangle$

$$\hat{H}|\Psi\rangle = 0.$$

Since only the states $|\Psi\rangle$ which satisfy this constraint equation can be considered physically possible, it seems as if a physical state cannot change over time. All its truly physical properties, it seems, must be represented by operators which commute with the Hamiltonian and are thus constants of motion. Genuinely physical properties cannot change over time. This first aspect of the problem of time thus really is a *problem of change*. The second aspect of the problem of time at the quantum level is that in all approaches for which we have an explicit expression for $\hat{H}$, there is no time: it appears as if time, as a physical quantity, has simply fallen by the wayside. Although the problem of time requires much more careful scrutiny than it can be given here, it leaves us with the puzzling issue of how an apparently fully temporal world, buzzing and beaming with change, can emerge from what appears to be a fundamentally fully "static" or "frozen" structure.

Space also undergoes a change from GR to LQG, though nowhere near as complete as time seems to. As we noted in Section 24.4, the fundamental spin network states are certainly discrete and thus lack some of the structure of relativistic spacetime. However, there is more: the fact that smooth physical space is supposed to arise from a quantum superposition of the geometric spin network states such that the resulting state has no determinate geometric properties surely stands in need of explanation. Clearly, the quantum measurement problem rears its hydra head here once more, but the problem seems even deeper now that we are not dealing with a particle having no determinate position even though we always detect it somewhere, but instead with space(-time) itself not having any determinate geometric properties although we never fail to experience it in any other way. So how can it be that we have determinate and measurable geometric information of spacetime at our scales when the fundamental structure will not generally have corresponding geometric properties?

Based on Butterfield and Isham (1999) and as explained in more detail in Wüthrich (2017), an answer to this difficult question consists of two parts. First, one needs to identify the quantum states with approximately "classical," i.e. geometric, properties and articulate a physical mechanism that "drives" the system toward those states. These semi-classical states are thought to correspond to almost flat three-spaces with at most small quantum fluctuations. A promising way of identifying them among the states in $\mathcal{H}_K$ is the "weave state" approach using coherent states (Ashtekar et al., 1992). These weave states are (almost) eigenstates of the "volume" operator. This operator earns its name by virtue of the fact that its eigenvalues approximate the corresponding classical values for the three-volume of a region in spacetime as determined by the classical gravitational field. Moreover, these same states are (almost) eigenstates of the "area" operator, which corresponds to the classical property of the area of the two-surface of a spacetime region. The selection of weave states stands in need of justification, which is far from automatic (Wüthrich, 2017, §4.2). Such justification would be delivered, for instance, by a physical mechanism which systematically drives the kinematic states to the semi-classical weave states. Decoherence with an appropriate partition of the system's degrees of freedom into "salient" and "background" degrees of freedom.

The second step then consists in relating the weave states to the classical spacetimes. This will involve a limiting procedure, establishing the precise sense in which the quantum states approximate the relevant geometric properties of the emergent spacetime at sufficiently large scales. None of this yet solves the quantum measurement problem; but at least it gives us a template for how to understand the relation between the fundamental non-spatiotemporal structure and relativistic spacetimes.

This way of relating the fundamental with the emergent is thought to be broadly reductive, and hence the notion of "emergence" at play cannot be the non-reductive concept typically used in philosophy. Rather, it designates a relation expressing the novelty of the emergent vis-à-vis the fundamental that is nevertheless ontologically grounded in the latter in a way that is consistent with reduction. Emergence as defined by novelty and robustness, i.e., by non-fundamental behavior which is robust under irrelevant changes on the fundamental level, and which is logically independent of reduction, as articulated by Butterfield (2011a,b) and as developed by Crowther (2015, 2016) captures the relevant sense of emergence at stake.

In CST, there also exists a sketch of how one can recover aspects of spacetimes from the fundamental causal sets.[10] Here, just as in the case of LQG, it is not all aspects of relativistic spacetimes which are obtained in such recovery: for instance, the continuum is not recovered in either case, it is merely approximated, and qualitative aspects of "spatiality" and "temporality" perhaps remain lost. However, just as statistical mechanics does not precisely recover all of thermodynamics, it is not necessary to regain all aspects of classical spacetimes; it suffices to show how the fundamental structure, be it spin network or causal sets or whatever else, can play the relevant functional roles of spacetime, such as that of spatiotemporal localization or other empirically determinable geometric properties such as distances and durations. Lam and Wüthrich (2018) develop this functionalist strategy toward an understanding of the emergence of spacetime in the cases of CST and LQG and argue that the recommended functionalist attitude rejects the necessity to regain anything beyond the empirically salient functions of spacetime. In particular, any insistence on some allegedly irreducible spacetime "qualia" is considered deeply misguided; and attempts to constitute spacetime from elementary spatiotemporal building blocks of a primitive ontology is thought to be altogether unnecessary. In this sense, some qualitative features of spacetime may well be emergent in the stronger sense, i.e., in that they cannot be reduced to what is described by a quantum theory of gravity. But this is not a loss. The one and only assignment that must be completed by any candidate theory of fundamental physics is to show how the aspects of spacetime necessary to support the physics of our manifest world arise from the structures postulated by the theory. Make no mistake: this is a formidable task, which no program in quantum gravity can claim to have discharged to date.

Acknowledgments

I owe thanks to Karen Crowther, Nick Huggett, Niels Linnemann, and Tushar Menon for comments on earlier drafts. This work was partly performed under a collaborative agreement between the University of Illinois at Chicago and the University of Geneva and made possible by grant number 56314 from the John Templeton Foundation and its content are solely the responsibility of the author and do not represent the official views of the John Templeton Foundation.

Notes

1 Nota bene that this is not equivalent to *quantizing gravity*, since a theory that can deal with both quantum effects of matter and relativistic phenomena, i.e., a theory of quantum gravity, may be achieved by other means.

2 A less common interpretation of GR views gravity as the result of a massless spin-2 field and thus much more like a force.

3 Cf. Verch (2012) for a more recent, and more optimistic review. As it turns out, the undetermined renormalization parameters may serve to fix the vacuum energy, and thus solve the "dark energy" problem in cosmology (Dappiaggi et al. , 2008).

4 I am grateful to Rainer Verch for discussions on these points.

5 For a recent introductory survey, see Dowker (2013).

6 For a recent and accessible introduction, see Rovelli and Vidotto (2015), which also covers the covariant extensions of the theory. See also Chapter 22 of this volume.

7 For an introduction to Loop Quantum Cosmology, cf. Bojowald (2011).

8 For a recent review, cf. Matsubara (2017).

9 See also the contribution by Thébault to this volume and Huggett et al. (2013, §2), and references therein.

10 For the details of this, the reader is advised to consult Huggett and Wüthrich (forthcoming, Ch. 3) and Lam and Wüthrich (2018, §3).

References

Almheiri, A., Marolf, D., Polchinski, J. and Sully, J. (2013). Black holes: Complementarity or firewalls? *Journal of High Energy Physics*, 2013(2): 62.

Arnowitt, R., Deser, S. and Misner, C.W. (1962). The dynamics of general relativity. In L. Witten (ed.), *Gravitation: An Introduction to Current Research*. New York, London: John Wiley and Sons, pp. 227–265.

Ashtekar, A., Rovelli, C. and Smolin, L. (1992). Weaving a classical metric with quantum threads. *Physical Review Letters*, 69: 237–240.

Bekenstein, J.D. (1973). Black holes and entropy. *Physical Review D*, 7: 2333–2346.

Bojowald, M. (2002). Ination from quantum geometry. *Physical Review Letters*, 89: 261301.

Bojowald, M. (2011). *Quantum Cosmology: A Fundamental Description of the Universe*. New York: Springer.

Bombelli, L., Lee, J., Meyer, D. and Sorkin, R.D. (1987). Spacetime as a causal set. *Physical Review Letters*, 59: 521–524.

Butterfield, J. (2007). Stochastic Einstein locality revisited. *British Journal for the Philosophy of Science*, 58: 805–867.

Butterfield, J. (2011a). Emergence, reduction and supervenience: A varied landscape. *Foundations of Physics*, 41: 920–959.

Butterfield, J. (2011b). Less is different: Emergence and reduction reconciled. *Foundations of Physics*, 46: 1065–1135.

Butterfield, J. and Isham, C. (1999). On the emergence of time in quantum gravity. In J. Butterfield (ed.), *The Arguments of Time*. Oxford: Oxford University Press, pp. 111–168.

Callender, C. (2000). Shedding light on time. *Philosophy of Science*, 67: S587–S599.

Crowther, K. (2015). Decoupling emergence and reduction in physics. *European Journal of Philosophy of Science*, 5: 419–445.

Crowther, K. (2016). *Effective Spacetime: Understanding Emergence in Effective Field Theory and Quantum Gravity*. Cham: Springer.

Curiel, E. (2016). *On the cogency of quantum field theory on curved spacetime*. Unpublished Manuscript.

Dappiaggi, C., Fredenhagen, K. and Pinamonti, N. (2008). Stable cosmological models driven by a free quantum scalar field. *Physical Review D*, 77: 104015.

Doplicher, S., Fredenhagen, K. and Roberts, J.E. (1995). The quantum structure of spacetime at the Planck scale and quantum fields. *Communications in Mathematical Physics*, 172: 187–220.

Dowker, F. (2013). Introduction to causal sets and their phenomenology. *General Relativity and Gravitation*, 45: 1651–1667.

Dowker, F. (2014). The birth of spacetime atoms as the passage of time. *Annals of the New York Academy of Sciences*, 1326: 18–25.

Earman, J. (2008). Reassessing the prospects for a growing block model of the universe. *International Journal in the Philosophy of Science*, 22: 135–164.

Eppley, K. and Hannah, E. (1977). The necessity of quantizing the gravitational field. *Foundations of Physics*, 7: 51–68.

Fewster, C.J. and Verch, R. (2015). Algebraic quantum field theory in curved spacetimes. In R. Brunetti et al. (eds.), *Advances in Algebraic Quantum Field Theory*. Cham: Springer, pp. 125–189.

Fine, K. (2005). Tense and reality. In *Modality and Tense: Philosophical Papers*. Oxford: Clarendon Press, pp. 261–320.

Huggett, N. and Callender, C. (2001). Why quantize gravity (or any other field for that matter)? *Philosophy of Science*, 68: S382–S394.

Huggett, N., Vistarini, T. and Wüthrich, C. (2013). Time in quantum gravity. In A. Bardon and H. Dyke (eds.), *A Companion to the Philosophy of Time*. Chichester: Wiley-Blackwell, pp. 242–261.

Huggett, N. and Wüthrich, C. (forthcoming). *Out of Nowhere: The Emergence of Spacetime in Quantum Theories of Gravity*. Oxford: Oxford University Press.

Lam, V. and Wüthrich, C. (2018). Spacetime is as spacetime does. *Studies in the History and Philosophy of Modern Physics*, 68: 39–51.

Malament, D.B. (1977). The class of continuous timelike curves determines the topology of spacetime. *Journal of Mathematical Physics*, 18: 1399–1404.

Matsubara, K. (2017). Quantum gravity and the nature of space and time. *Philosophy Compass*, 12. Online.

Mattingly, J. (2006). Why Eppley and Hannahs thought experiment fails. *Physical Review D*, 73: 064025.

Ouellette, J. (2012). Alice and Bob meet the wall of fire. *Quanta Magazine*, December 2012. Available at: https://www.quantamagazine.org/20121221-alice-and-bob-meet-the-wall-of-fire/.

Peres, A. and Terno, D.R. (2001). Hybrid classical-quantum dynamics. *Physical Review A*, 63: 022101.

Pooley, O. (2013). Relativity, the open future, and the passage of time. *Proceedings of the Aristotelian Society*, 113: 321–363.

Rideout, D. and Sorkin, R.D. (1999). A classical sequential growth dynamics for causal sets. *Physical Review D*, 61: 024002.

Rovelli, C. and Vidotto, F. (2015). *Covariant Loop Quantum Gravity: An Elementary Introduction to Quantum Gravity and Spinfoam Theory*. Cambridge: Cambridge University Press.

Sorkin, R.D. (2007). Relativity theory does not imply that the future already exists: A counterexample. In V. Petkov (ed.), *Relativity and the Dimensionality of the World*. Dordrecht: Springer, pp. 153–161.

Verch, R. (2012). Local covariance, renormalization ambiguity, and local thermal equilibrium in cosmology. In F. Finster, O. Müller, M. Nardmann, J. Tolksdorf and E. Zeidler (eds.), *Quantum Field Theory and Gravity: Conceptual and Mathematical Advances in the Search for a Unified Framework*. Basel: Birkhäuser, pp. 229–256.

Wald, R.M. (1994). *Quantum Field Theory in Curved Spacetime and Black Hole Thermodynamics*. Chicago: University of Chicago Press.

Will, C.M. (2014). The confrontation between general relativity and experiments. *Living Reviews in Relativity*, 17(4): 1–117.

Wüthrich, C. (2005). To quantize or not to quantize: Fact and folklore in quantum gravity. *Philosophy of Science*, 72: 777–788.

Wüthrich, C. (2006). *Approaching the Planck Scale from a Generally Relativistic Points of View: A Philosophical Appraisal of Loop Quantum Gravity*. PhD thesis, University of Pittsburgh.

Wüthrich, C. (2013). The fate of presentism in modern physics. In R. Cuini, K. Miller and G. Torengo (eds.), *New Papers on the Present: Focus on Presentism*. Munich: Philosophia Verlag, pp. 91–131.

Wüthrich, C. (2017). Raiders of the lost spacetime. In D. Lehmkuhl, G. Schiemann and E. Scholz (eds.), *Towards a Theory of Spacetime Theories*. Basel: Birkhäuser, pp. 297–335.

Wüthrich, C. (2018). Are black holes about information? In R. Dawid, R. Dardashti and K.P.Y. Thébault (eds.), *Why Trust a Theory? Epistemology of Fundamental Physics*. Cambridge: Cambridge University Press, pp. 202–223.

Wüthrich, C. and Callender, C. (2017). What becomes of a causal set? *British Journal for the Philosophy of Science*, 68: 907–925.

Chapter 24 – Further Reading from the Editors

For an in introduction to causal set theory, see Fay Dowker's 'Introduction to causal sets and their phenomenology' (*General Relativity and Gravitation*, 45:1651–1667, 2013). Christian Wüthrich and Craig Callender address issues of time in causal set theory in 'What becomes of a causal set?' (*British Journal for the Philosophy of Science*, 68:907–925, 2017). For loop quantum gravity see Carlo Rovelli and Francesca Vidotto's *Covariant Loop Quantum Gravity: An Elementary Introduction to Quantum Gravity and Spinfoam Theory* (Cambridge University Press, Cambridge, 2015). Oriti, Daniele, ed. *Approaches to Quantum Gravity: Toward a New Understanding of Space, Time and Matter* (Cambridge University Press, 2009) contains a number of articles on loop quantum gravity alongside other approaches. A philosophical treatment can be found in Wüthrich's "Raiders of the lost spacetime" in Dennis Lehmkuhl, Gregor Schiemann, and Erhard Scholz, editors, *Towards a Theory of Spacetime Theories* (Birkhäuser, Basel, 2017 pp. 297–335).

25

SPACETIME "EMERGENCE"

Nick Huggett

Could spacetime be derived rather than fundamental? The question is pressing because attempts to quantize gravity have led to theories in which (arguably) there are either no, or only extremely thin, spacetime structures. Moreover, recent proposals for the interpretation of quantum mechanics have suggested that 3-dimensional space may be an "appearance" derived from the $3N$-dimensional space in which an N-particle wavefunction lives. In fact, I will largely assume a positive answer and investigate how it could be; in particular, I want to explicate the role of philosophy in producing a satisfactory explanation of spacetime, providing a roadmap for philosophical engagement with quantum gravity. First, I will explain why such a derivation can be described as "emergence."

25.1 Why Spacetime 'Emergence'?

Let's specify some terms. The general framework involves a pair of theories, one less fundamental, and one more fundamental from which it is putatively derived. For brevity, call the former "fundamental" (without implying that it is the "final" theory), and the latter "derived" (supposing that the putative derivation exists): ideal gas laws are derived, and the kinetic model fundamental, in these senses.

We are interested in cases in which the derived theory is spatiotemporal, but the fundamental theory, in some significant sense, is not. What cases are these? The derived theories are general relativity (GR) and quantum field theory (QFT); the former describes curved relativistic spacetime, the latter subatomic particles in flat relativistic spacetime (see Sections 25.3). Deriving the former means recovering solutions of the Einstein Field Equations , while deriving the latter means recovering particle scattering predictions, in suitable limits (so one expects corrections to these theories away from the limit). The spacetime of these theories can be referred to variously as "classical," "relativistic," "ordinary," or "empirical" (when one wants to emphasize that it is part of our best tested theories); or for brevity simply "spacetime."

The fundamental theories are proposed accounts of quantum spacetime (QS): for example, loop quantum gravity (see Chapter 24), string theory (see Chapter 24), causal set theory, non-commutative geometry, and group field theory (GFT).[1] Most are discussed elsewhere in this volume, so (except for GFT, below) I will not describe them further here (but see references in the final section). The goal is rather to explicate the general philosophical problem of deriving spacetime, which they have in common.

There are then two aspects to understanding derived spacetime. First, the sense in which the fundamental theory is non-spatiotemporal; what aspects of spacetime are missing? How should we understand the structures that are present? (If spacetime is present in the fundamental theory, the question of a derivation does not even arise.) These questions are not the focus of this essay; answers are sketched in Huggett and Wüthrich (2013). Second, the derivation; how is spacetime explained in terms of the fundamental structure? This question is our focus, and we will address it in two ways: how is such a thing even possible in principle, and how does it happen in a concrete case? (Other cases are addressed in Huggett and Wüthrich (forthcoming).)

Before we do, suppose that we have a derivation of spacetime in terms of a fundamental theory. Why should that situation be described as "emergence?" The term generally describes a situation in which a less-fundamental structure is *qualitatively* different from a fundamental one in some way. Different senses then arise from different specific accounts of qualitative difference. A particularly salient sense applies when the more fundamental theory cannot account for the less-fundamental theory at all, so they are autonomous: either there is no derivation (no systematic map from more to less fundamental) at all, or for some reason the map is a mere harmonic correspondence between distinct objects. Indeed, this sense – that the less fundamental does not supervene on the more fundamental – has been a characteristic sense of "emergence" in philosophy, at least since its revival in the 1990s. However, it is not the only sense with currency. For instance, Butterfield (2011) defends emergence as behavior that is robust, and novel with respect to the fundamental theory: novelty arises from the taking of limits, and his conception is logically distinct from others in the vicinity.

Or again, in the foundations of QS, Seiberg (2006) uses the term to mean that space and time "are not present in the fundamental formulation of the theory but appear as approximate macroscopic concepts." Such a sense is unlikely to agree with the philosophical one in specific cases, and it may or may not agree with Butterfield's, depending on how spacetime is recovered. Seiberg's core case is AdS/CFT duality (see Ch.23), in which whole dimensions of space are said to be derived; formally the derivation is of the kind Butterfield has in mind, but it is an equivalence, so there is no obvious sense in which there is a distinction between more and less fundamental (Teh, 2013). String theorists often have this specific model in mind when they apply the term, but others concerned with the foundations of QS tend to apply "emergence" more generally, following Seiberg's definition. The idea is that spacetime structures – whether they are Aristotelian, Newtonian, Galilean, Einsteinian, Euclidean, or (pseudo-)Riemannian – are so essential to all previous theories of physics that their absence is in itself a profound kind of qualitative difference.

As I said, what replaces them is discussed elsewhere (including this volume, and below), but it will be helpful to have something concrete in mind, to underline the chasm between a theory without spacetime and a theory with it. One might have in mind that a quantum spacetime is nothing but a kind of discrete spacetime, as a continuous energy spectrum might turn out to be discrete at fine discriminations. The conceptual gulf is then not so great. But, for instance, spacetime might be replaced by a set of objects with no essential spacetime meaning, and some structure (say that of a group – as in the case study below). A collection of structured objects is called a "space" in mathematics, but that does not mean it is ordinary space: in the relevant cases the objects are not identified with spacetime points, nor do they have the geometric or topological structures of space – all these things are derived. Space truly comes from "nowhere." It is not only hard, psychologically, to imagine such a thing at first, since existing theories assume space and time as their most basic postulates, it means a whole new understanding of nature, and spacetime's place within it is needed. No one has that understanding yet; the purpose of this article is to explicate the problems with developing it, while emphasizing their philosophical character.

Some different dimensions of emergence in this sense have been discussed. In the first place, Huggett and Wüthrich (2013) focused on the conceptual gap between spacetime and non-spatiotemporality. At one extreme, spacetime that is merely discrete is not conceptually far from ordinary, continuous spacetime, and emergence may not even apply. At the other, a theory whose basic elements are members of an algebra has a radically non-spatiotemporal ontology. However, the formal difficulty of deriving spacetime does not track such conceptual gaps. Or again, Oriti (2021) distinguishes "levels" of emergence in another way. At the lowest level are theories which do little more than allow quantum superpositions of classical spacetime states. More interesting are theories postulating non-spatiotemporal building blocks: if these are physical then we reach a low level of emergence. When many such atoms are present, there may be different "phases," analogous to gas or condensed liquid states; if only some of the phases are 'condensates' in which the blocks form spacetime, then we reach a higher level. The final level is reached if both spatiotemporal and non-spatiotemporal phases are physical, and if there is a literal transition between them – a process of "geometrogenesis," identified with the big bang in our phase of the Universe. (The big conceptual question of course being how to make sense of a *transition* from a non-temporal to a temporal state!) Our case study is an example of such a theory.

It is important to note that using "emergence" in any other than the central philosophical sense has been criticized (see Crowther in preparation), at least on the grounds of sewing confusion. In the remainder of this essay, I will attempt to finesse this issue by turning attention to the question of "explaining" spacetime. The qualitative gap between a fundamental theory without spacetime and a derived one with spacetime, will be seen to entail an explanatory gap: whether or not we describe the filling of this gap as "emergence," is not pertinent to the rather general account that I give.

In the remainder we will first (Section 25.2) turn to the general question of providing such a radical explanation, like that of spacetime in terms of the non-spatiotemporal, and study the issues in a historical analog. Then (Section 25.3) we will study how the lessons apply in a concrete case of QS. Finally, (Section 25.4) I will briefly mention some of the other issues that come up regarding the topic of spacetime emergence, and some of the relevant literature.

25.2   Explaining Spacetime

Let's suppose that we are faced with a (more) fundamental theory in which there are (to a substantial extent) no spatiotemporal quantities: perhaps one of the examples above, perhaps something as yet undiscovered. How, in general terms, could such a theory explain the appearance of (relativistic) spacetime? Well, the answer in similar cases is of course that the apparent, higher level quantities and "structures" (generally speaking) have to be *derived* from the fundamental, lower level ones. For instance, in the kinetic gas model, thermodynamical quantities such as temperature and pressure are (approximately) identified with the mean kinetic energy of and momentum transfer from atoms of gas; and then the thermodynamical law of proportionality between pressure and temperature (at fixed volume) is derived from the laws of elastic collision. From the fundamental theory in which temperature and pressure are not basic quantities are derived those thermodynamic quantities and their relations. Even though the gulf between theories with and without spacetime is a (far) larger one, the same basic model should apply to emergent spacetime: some suitable spatiotemporal quantities will have to be derived from non-spatiotemporal ones.

As Maudlin puts it,[2]

> one might [try to] derive a physical structure with the form of [spacetime quantities] from a basic ontology that does not postulate them. This would allow the theory to make contact with evidence still at the level of [spacetime quantities], but would also insist that, at a fundamental level, the local structure is not itself primitive.

However, he points out that "derivation" is ambiguous here, and that its weakest form will not really do.

> This approach turns critically on what such a derivation of something isomorphic to local structure would look like, where the derived structure deserves to be regarded as *physically salient* (rather than merely *mathematically definable*). Until we know how to identify physically serious derivative structure, it is not clear how to implement this strategy. (Maudlin, 2007, p. 3157: my emphasis).

The purpose of this section is to unpack the general notion of a "physically salient derivation," its meaning, nature, and grounds – and differentiate it from merely mathematical derivation. Such work is preliminary to understanding the special problem of physical salience for the emergence of spacetime, and how one should expect it to be resolved; that question is addressed in the next section, in a case study. We will start with an instructive historical example.

In his *Principles of Philosophy*, Descartes proposed that all physical processes were ultimately to be understood in terms of the arrangement, motions, and collisions – contact action – of particles of matter. This is the basic principle of the "mechanical philosophy," whose proponents contrasted it with Aristotelian or scholastic teleological science, in which even physical processes are explained by tendencies or "occult powers," driving systems to end states. For example, Aristotle explained (i) terrestrial gravity in terms of a tendency of certain elements toward their natural place at the center of the Universe, and (ii) the motions of the planets in terms of the natural tendency of heavenly matter (ether or quintessence) to rotate about that center. Descartes of course rejected such accounts: according to him (i) terrestrial gravity was the result of the greater centrifugal "force" on light matter than on dense matter, while (ii) the planets were carried by huge vortices of fine matter rotating around the Sun.

The theory of universal gravity developed by Newton (see Ch. 1) in his *Mathematical Principles of Natural Philosophy* does not offer such a mechanical account at all: instead a law is derived which ascribes a force between every pair of bodies in the Universe, proportional to their masses, and inversely proportional to the square of the distance between them. (i) Newton demonstrates that the aggregate gravitational effect on external bodies, of individual parts of matter arranged in a homogeneous sphere, is the same as if all the mass were concentrated at the center of mass. Terrestrial weight is then explained by the attraction that the Earth exerts on the matter in bodies. (ii) Regarding the planets, treated as point bodies because of their small sizes compared to interplanetary distances, Kepler's laws (up to corrections due to the mobility of the Sun, and the mutual gravitational interactions) can then be derived from the laws of mechanics and gravity. (This example will serve as our archetype of a derivation below.)

Famously, in his *Principles* Newton "feigns no hypothesis" – by which he means mechanical account – about the nature of gravity (though in his *Optiks* and elsewhere he critically considers such proposals). But it is not hard to see that a mechanical account of Newtonian gravity is hard to come by: its universality entails that bodies at opposite ends of the Universe exert equal forces on each other, so somehow collisions with the matter surrounding each body would have to be correlated. Thus the response of the mechanists to Newton was to accept the predictive accuracy of the inverse square law for the solar planets (given the precision and fertility of the law, they had little choice), but to deny its universality, as defying mechanical explanation. For them, accepting a non-mechanical attraction would be nothing less than reintroducing a banished occult power.

For example, as well as attempting a mechanical account in his *Tentamen*, Leibniz actively engaged with the Newtonians (and hence by proxy Newton) on just this issue. In the third letter of his *Correspondence* with Clarke he writes:

> If God would cause a body to move [round a] fixed center, without any [body] acting upon it . . . it cannot be explained by the nature of bodies. For, a free body does naturally recede from a curve in the tangent. And therefore . . . the attraction of bodies . . . is a miraculous thing, since it cannot be explained by the nature of bodies. (Alexander, 1998)

The point could not be clearer: according to the mechanical philosophy, physically only a collision with another body can explain a deviation from linear, inertial motion, so attraction at a distance would be unphysical, or miraculous in the contemporary idiom.[3]

In Maudlin's terms, Leibniz claims that the Newtonian theory of universal gravity allows the "mere mathematical derivation" of the observed system of motions of the planets, but that derivation is not "physically salient" because it violates the mechanical principle. Put yet another way, an instrumentally valid, predictively accurate, theory might fail to deliver physical explanations, because it fails to satisfy the principles (in a broad sense) of how more fundamental physical elements can combine to produce less fundamental structures: "principles of physical salience," we can say. In Leibniz's case the principle is explicit, but in general such principles are shown to exist by the possibility of taking this or that theory merely instrumentally (as opposed to blanket instrumentalism): accepting its claims about the observable, but denying the further claim that the observable is the literal "product" of unobservable elements described by the theory. Other examples include the principle that macroscopic structures should be co-located with their constituents, Hamilton's principle, (local) gauge principle, renormalizability principle, null energy condition; as well as other more "homely" principles about the application of theory to derive concrete observable consequences. If these are not (among the jointly) sufficient conditions for physical explanation then many familiar derivations are nothing but mathematics, and no reflection of an underlying physical story.

Of course, historically the failure of the Cartesians to produce a mechanical explanation (or contrary empirical evidence) led ultimately to the acceptance of Newtonian action-at-a-distance as physically explanatory, as part of physically salient derivations. (The difference from Aristotelian powers being that only a small number of fundamental forces are admitted; they are not proposed *ad hoc* whenever one wants to explain a new phenomenon.) As Newton himself suggested, $\mathbf{F} = m\mathbf{a}$ can be understood as a schema for the action of such fundamental forces. But, equally of course, subsequent developments swung the pendulum back, if not in favor of contact action between bodies, but in favor of local action: the work of Faraday and Maxwell revealed the electromagnetic field act according to local differential equations, and for its effects to propagate at the speed of light (in accordance with special relativity). Application of the local field principle to gravity led Einstein to general relativity, in which gravity does not act a distance, but instead propagates locally as a field. However, the development of quantum mechanics then swung the pendulum again, since it allows effects not easily understood in terms of the local propagation of fields in spacetime: entanglement, and the Bohm-Aharanov effect, for instance (see Ch. 42). While the "pendulum" does not strictly move in a plane, the oscillation between some kind of local action, and some kind of non-local action as permissible in physically salient derivations is clear enough – as is the centrality of principles concerning (non-)locality to the development of physics.[4]

I have told this story at some length, even though it does not directly concern the emergence of spacetime, because it offers a familiar template for an entirely unfamiliar situation. I draw an explicit parallel between the problems of physical explanation without the collisions of bodies, and without spacetime. As the former once seemed an *a priori* condition on physics, so can the latter today; indeed most, if not all, principles of physical salience presuppose classical spacetime in some way – consider the list given above. And if we understand how the former was replaced, we can understand how the latter may be too. We can, that is, understand how derived spacetime structures might come to "deserve to be regarded as physically salient." Specifically, considering the case of locality, we see the following.

First, the criteria for a formal derivation to be physically salient are theory dependent; we saw quite dramatically the changing status of locality principles. (While this idea that there are such "standards" of acceptable explanation can be found in Kuhn (1962), we need not draw the conclusions of scientific irrationality often attributed to him.)

Second, such principles are interwoven with our understanding of the theoretical content of the theory, the nature of its objects and structures; indeed, one could say that accepting principles of physical explanation is part of accepting an interpretation of a theory. So for instance, the principle of contact action is intimately connected with Descartes' account of space and matter, and his critique of Aristotelian powers, while similar stories hold of other principles. In short they are part of the conceptual, or philosophical background of a theory, dubbed the "relative *a priori*" by Friedman (2001), who offers a historically informed account of its development in various cases.[5]

Third, although the principles are thereby distanced from direct test, their ultimate epistemic warrant is empirical: the overall success of the theory in explaining and predicting phenomena. There is a vast literature on the question of the nature and epistemic authority of "empirical success"; of when and whether empirical evidence warrants belief in "theoretical" claims and explanations. So I will not elaborate on this contentious notion here, except to say that it is assumed here that there is a difference between taking an instrumentalist attitude to a theory and a more realist one.[6] However, it should not be the case that just *any* formal derivations from a successful theory are explanatory; not just anything can be a principle of physical salience. Philosophical analyses of theory change (like Kuhn's and Friedman's) address the question of how new principles are decided on, and so to what more general criteria they are answerable; again, I will largely defer to this literature. But for instance, the principles must be internally coherent, and systematize explanations within the theory, and they must also mesh with an accurate account of the relevant empirical data. (What they need not do is fit any psychologically comfortable, familiar picture.) So the philosophical aspect of developing a relative a priori is inherently *critical*, questioning whether existing or putative new principles in fact satisfy such broader criteria; it is not a matter of simply spinning a story around mathematics. Classic examples of this critical project include Newton's analysis of absolute and relative space, Poincaré's analysis of non-Euclidean physical geometry, and of course Einstein's analysis of space, time, and motion.

Fourthly, as Friedman again emphasizes, theories are not born as fully formed formalisms, simply awaiting principles of physical explanation, but rather formalism and interpretation are generally constructed simultaneously, each guiding the other in the search for a more fundamental account of the phenomena. Hence the philosophical project of articulating a relative a priori is carried out in tandem with the more formal project of providing a mathematical system; here Einstein (and his precursors') development and articulation of relativity is a paradigm.

Putting these four points together then, knowing how to "identify physically serious derivative structure" is one of the things discovered during the development of a new theory (if it is a radical departure from previous theory), and like the rest of the theory in order to account for phenomena, and *not* by reasoning of an absolute *a priori* kind. Therefore, in the search of a theory in which the existing principles of physical salience are inapplicable because there are no fundamental spacetime structures, we should expect to find new principles being proposed which will permit the physical explanation of empirical spacetime structures. Making explicit, analyzing, and critiquing such proposals is a philosophical activity of the first order. In the following section, we will see these points realized in a concrete case.

25.3 Case Study: The Emergence of Spacetime in Group Field Theory

Our case study is "Group Field Theory" (GFT).[7] A group generalizes the idea of geometric transformations, and specifically the pattern in which they combine: in the simplest case, clockwise rotations in the plane, a rotation of θ^o followed by a rotation of ϕ^o degrees equals a rotation of $(\theta + \phi)^o$, while a rotation of $(360 - \theta)^o$ will undo a rotation of θ^o. Abstracting away, a group is any collection of elements with an (associative) rule that maps any two into a third, and such that every element has an inverse, with which it maps to a special neutral element (the identity if the elements are in fact transformations). From the group point of view, the nature of the group elements is not the key thing, but simply which pairs are mapped onto which elements: the group *algebra*. For example, any group in which the elements map in the same way as planar rotations have the group algebra $SO(2)$.

In familiar cases, a physical field is a continuous distribution of some property – temperature, gravitational, or electromagnetic potential, say – over spacetime; we represent such a system by a function from points to mathematical quantities of some kind. GFT generalizes this conception, and considers a field that lives on the elements of a group; for instance, a map from the elements of $SO(2)$ to complex numbers. It is important in this picture not to think of the group elements as literal rotations in some plane of space, but rather as primitive points of some new "space," related just like the rotations; having this understanding is the point of our discussion of abstracting away from literal rotations. Indeed, the elements of $SO(2)$ form a circle, labeled by $0 \leq \theta < 360$, not a 2-dimensional plane at all. The question is how to derive ordinary space from such a group space.

The GFT that permits such a derivation utilizes the group $SO(1, 3)$ of Lorentz transformations – again abstracting, to view the group elements as primitive points, not literal transformations, but structured just like them. (NB: these transformations act on 4-dimensional spacetime, but they themselves form a 6-dimensional "space" since they include both translations and boosts in three spatial directions.) Because we want to recover a 4-dimensional spacetime there are four copies of $SO(1, 3)$, making a 24-dimensional "space". The field is a map from quadruples of group elements to the complex numbers: $\Phi(g_1, g_2, g_3, g_4)$ with $g_i \in SO(1, 3)$. Finally, the theory must be quantized (see Ch.18). Φ is replaced by a quantum operator $\hat{\Phi}^\dagger$, which "creates" a quantum of the field: if $|0\rangle$ represents the vacuum state, then $\hat{\Phi}^\dagger(g_1, g_2, g_3, g_4)|0\rangle$ represents a state in which a single quantum is present, at (g_1, g_2, g_3, g_4).[8]

So much for the basic structure of the theory, but how is spacetime derived? And is the derivation physically salient, or merely mathematical in the way we discussed previously. To frame that investigation, we first extract a simple model of physical explanation from our discussion of gravity.

Generally speaking, suppose that a less fundamental relation $\mathcal{L}$ says $f(A) = g(B)$. For example, Kepler's laws can be framed this way, in terms of the positions (over time) of the planets. Then, if (a) fundamental quantities X can be "aggregated" into $\alpha(X)$ and $\beta(X)$, such that (b) $f(\alpha(X)) = g(\beta(X))$ follows from fundamental laws, then $\mathcal{L}$ is *mathematically derived* (or defined). For instance, the positions of the parts of a planet can be "aggregated" to the center of mass, and then Kepler's laws follow from Newton's laws. ("Aggregating" is a deliberately broad concept: it could be summing, or averaging, or coarse-graining, or something else. The point is that in general, fundamental degrees of freedom are typically combined into fewer, effective degrees of freedom.)

Thus far, of course, the Cartesians were with Newton. His theory truly did allow an empirically accurate, mathematical derivation of the phenomena. But they denied that universal gravity physically explained them. So let us say that in addition, for physical explanation, (c) $\alpha(X)$ and $\beta(X)$ must be physically salient quantities in the sense explained above. Because it violated the mechanical principle, Leibniz claimed that gravitational attraction was not physically salient; although later physicists did accept it as such.

Our goal is to sketch the derivation of spacetime in GFT, and ask what new understanding of physical salience is required if we are to understand it not as purely formal, but as a physical explanation: satisfying not just (a) and (b), but also (c).

It is helpful to picture a quantum of the group field, $\hat{\Phi}^\dagger(g_1, g_2, g_3, g_4)|0\rangle$ as a tetrahedron, with the four faces "labeled" with the four "coordinates" of the point in group space: see Figure 25.1.

This representation immediately makes the quantum appear spatial, as a literal tetrahedron of space. Indeed, ultimately that interpretation will be used in the recovery of spacetime, but at this stage one should simply think of it as nothing more than an alternative formalism for expressing the state $\hat{\Phi}^\dagger(g_1, g_2, g_3, g_4)|0\rangle$. No physical interpretation is (yet) implied by this rewriting. Then, summarizing Gielen et al. (2013):

1   Given some additional physical assumptions about the field, only three of the coordinates are independent, determining the fourth. Given the group, in turn each of the three can be specified by a four-component quantity: g_i is specified by the four numbers a_i^μ ($i = 1, 2, 3$, $\mu = 0, 1, 2, 3$).

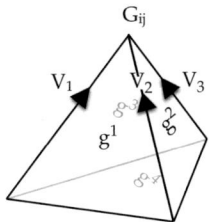

Figure 25.1 A pictorial representation of a GFT quantum, $\hat{\Phi}^{\dagger}(g_1, g_2, g_3, g_4)|0\rangle$.

2   If – and I stress the hypothetical nature of this statement – the corresponding tetrahedron were embedded in space, then (given appropriate symmetries) the a_i^{μ} determine vectors $\vec{V}_i$, defining the three edges of the tetrahedron, leading from one vertex, located at a spatial point p: see Figure 25.1. Linear combinations of them, $\vec{v} = \sum_i c^i \vec{V}_i$ can be interpreted as other spatial vectors.

3   Next, a symmetric 3×3 matrix, can be defined at p by $G_{ij}(p) \equiv \sum_{\mu} a_{i\mu} a_j^{\mu}$. This can be used to define the "dot product" of any two spatial vectors at p, $\vec{u} = \sum_i c^i \vec{V}_i$ and $\vec{v} = \sum_i d^i \vec{V}_i$, according to $\vec{u} \cdot \vec{v} \equiv \sum_{ij} G_{ij}(p) c^i d^j$. In other words, if the tetrahedron were thought of as living in space, then the GFT quantum would define a tiny piece of geometry for that space; a "metric," which determines the lengths of and angles between vectors in space (see Ch. 11).

4   Finally, suppose space were evenly filled with many such tetrahedra, such that the metric G_{ij} was homogeneous, the same everywhere. (Suppose that we measure distances with a ruler whose smallest gradations are millions of times bigger than the side of a tetrahedron, so that we cannot see that the metric is only given at discrete points.) Gielen et al. (2013) proved mathematically that then the field must be in a particular "coherent state." Such a state is a quantum superposition of every number of quanta: a superposition of one quantum, and two quanta, and ....[9]

The above (in its full detail) constitutes a mathematical definition of structure isomorphic to ordinary space: a collection of GFT quanta, represented as tetrahedra of space, and defining a metric as described, demonstrably yield a homogeneous space. In terms of our analytical framework, (a) defining the metric, positing space-filling tetrahedra, and taking the coherent state amount to aggregating the fundamental GFT degrees of freedom, while (b) the proof that the result is a well-defined homogeneous metric shows that the fundamental GFT laws entail that the desired less-fundamental relation holds. (In fact, things are much better even: spacetime geometry dynamics can also be derived, including Robertson-Walker metrics. But we will focus on space for simplicity.)

But of course, as far as physical explanation goes, there is a lacuna: why should *excitations* of a quantum field over a space whose points are group elements manifest as *chunks of physical space*? They may be isomorphic in some way (in some states) but that does not make them literally spatial, any more than (say) the numbers [0, 1] are space, even if they are identically structured. In particular, to constitute space the chunks must manifest themselves evenly across a macroscopic volume, rather than a microscopic region, or in disconnected islands, or observably far apart. Nothing in the basic theory dictates the distribution, but it (and more) is required if the defined GFT structure deserve to be regarded as physically salient, satisfying (c). The question arises because no spatial principles guide the explanation. GFT quanta fundamentally live "in" the group space, not in ordinary space at all; at (g_1, g_2, g_3, g_4) not at (x, y, z), so it seems a category error to even ask how they are co-located with spatial regions. Similarly, since there are two spaces, ordinary space and the group, notions of local or non-local action of quanta don't even apply – they refer to events in a single space. Just as we expected, when we seek to derive ordinary space in terms of a theory that does not posit it, existing principles of physical salience do not permit physical explanation. There is an explanatory gap.

As we saw in connection with universal gravity, such explanatory gaps are theory-relative, and new empirically successful theories come with principles of physical salience to fill them – when we accept the theory as something more than a successful instrument. Of course, we also saw that discovering such principles is a critical philosophical project, carried out simultaneously with the mathematical articulation of the theory. So we should not expect at this stage in the development of GFT to have a definitive statement of the principles; rather one should engage with the developing formal theory, to help critically articulate them.

For example, a first stab at the principle that would let us view the GFT derivation as a physical explanation is that "quanta occupy tetrahedra of space." But that can't be right for a number of reasons: the theory does not presuppose space, space is an appearance, so the tetrahedra constitute rather than occupy volumes; then we don't have space until there are many quanta, enough to compose a large volume; indeed, there is formal evidence that the quanta only make ordinary space in coherent states.[10] So even this short critical conceptual analysis leads us to a better proposal: "in a coherent state, quanta constitute evenly spread tetrahedral chunks of what appears empirically as space." Even that is only an illustration of the project that I have described; the ultimate correctness of this proposal depends on further development of the formal and conceptual aspects of GFT.

Finally, we should never forget that any such development ultimately stands or falls on the empirical success of the theory, and that at this stage it is far too early to claim definitive empirical success for any theory of quantum gravity. But as I argued, that is not a reason to refrain from philosophical analysis, for the formal and philosophical aspects need to be developed together, with hope for empirical success down the road. However, it does mean that one must view any philosophical conclusions as hypothetical: holding only on condition that the theory is vindicated. Philosophers might desire more, but that is not possible in scientific enquiry.

25.4 Further Exploration

This essay has focused on the general philosophical project (itself an aspect of a scientific project) of understanding the "emergence" of spacetime, how something apparently so basic to physics could have a physical explanation. In this final section I will give a partial list of works that either deal with specific cases, or develop other philosophical questions. In addition to papers mentioned, a store of video lectures and classes on the topic can be found at `www.beyondspacetime.net`.

25.4.1 Investigations of Specific Theories of QS

As mentioned the two core reviews of putative cases of spacetime emergence are Seiberg (2006) and Huggett and Wüthrich (2013); the former from a physics point of view, and the latter from a more philosophical one (the special issue of *Studies in History and Philosophy of Modern Physics* in which it appears contains several other papers of interest, some listed here). More specific works include:

1 *String Theory*: (See Ch. 23) gives an introduction. Teh (2013) discusses (negatively) whether the appearance of extra spatial dimensions in the bulk spacetime in AdS/CFT duality is a case of emergence. Huggett and Vistarini (2015) investigate the derivation of the field equations of GR in string theory, while Huggett (2017) argues that dualities imply that string theory does not include ordinary spacetime in its fundamental structures.

2 *Loop Quantum Gravity*: (See Ch. 24) gives an introduction, while Huggett and Wüthrich (forthcoming) will have a more detailed analysis. Rovelli (1998) is also recommended.

3 *Geometrogenesis*: a number of ideas concerning spacetime emergence, and especially that of "geometrogenesis" discussed above are developed in Oriti (2014).

4 *Causal Set Theory*: CST claims that spacetime grows one discrete point at a time; it is sometimes claimed that this picture entails a notion of temporal becoming, often thought missing from relativity. Huggett (2014) argues that such a view requires two times; that in which points are created, and that which they constitute. Wüthrich and Callender (2016) investigate this kind of picture critically, but suggesting a way in which it might be developed.

5 *Other Approaches*: Huggett, Lizzi, and Menon (2021) provides a nice introduction to non-commutative geometry, with some interesting remarks on the interpretation of the formalism. Bain (2013) investigates approaches to QG based on ideas from condensed matter physics. Shape dynamics has received little attention from philosophers, but Barbour (2012) contains a philosophically aware introduction.

This list is partial; in particular there are a number of other proposals for QS in the physics literature. The essay collections Callender and Huggett (2001), Huggett, Matsubara, and Wüthrich (2020), Oriti (2009), Rickles et al. (2006), and Wüthrich, Le Bihan, and Huggett (2021) contain numerous useful essays explaining different approaches, and in many cases discussing how spacetime is derived.

25.4.2   *Metaphysical Implications*

The implications of emergent spacetime for traditional metaphysical views are only starting to be explored – but given that so many are tied to a classical conception of spacetime, it should be expected that there is a great deal to learn. For example, Alastair Wilson has asked (in his talks) exactly how one should distinguish grounding from causation, especially in cases in which there is no spacetime: after all, traditional accounts of causation are deeply tied to spacetime. And again, Vistarini (2020) and Wüthrich (2020) note that Lewis' account of possible worlds assumes that they are spatiotemporal, and in different ways explore possible consequences.

Another important strand of enquiry that philosophers have taken up concerns the way in which ordinary spacetime might be grounded in the non-spatiotemporal. In particular, there are two recent proposals that go by the name "spacetime functionalism," though they differ in motivation and content.[11] On the one hand, Knox (2014) (although it doesn't explicitly use the term) has the interpretation of physical theory in mind, and proposes in very loose terms that spacetime is what plays the role of determining inertial trajectories: then, for example, she argues that for Newtonian gravity, Newton-Cartan spacetime plays that role, not Newtonian spacetime. On the other, Chalmers (2012, §7.5) has a neo-Carnapian constructive project: he leans to the view that spacetime concepts (such as shape or length) refer to whatever structures produce their typical spatiotemporal experiences. It's hard to bring the two functionalisms into close comparison, because Chalmers' is a view about ordinary concepts, while Knox's concerns a theoretical object. Nevertheless, Yates (2021) suggests that they differ according to whether "spacetime" will turn out to be anything like ordinary spacetime. According to Knox's account, even if fundamental physics is non-spatiotemporal, there may be some derived structure – presumably relativistic spacetime – that plays the appropriate role. However, Yates argues, if fundamental physics is ultimately what produces spatiotemporal experiences, then that will turn out to be the referent of "spacetime" for Chalmers, even if it is not spatiotemporal in any way familiar spacetimes are: much as water might not be H_2O on Twin Earth. These and other metaphysical matters are taken up in essays in Wüthrich, Le Bihan, and Huggett (2021).

Acknowledgments

This publication was made possible with the support of a grant from the John Templeton Foundation through the Beyond Spacetime project at the University of Illinois at Chicago. The opinions expressed are those of the author and do not necessarily reflect the views of the John Templeton Foundation.

Notes

1 Some but not all of these theories, or versions of them, go under the more specific heading of "quantum gravity" (QG); I use the more general term QS here. All such theories are seen as at least stepping stones to QG. (We will not explicitly discuss configuration space realism, though similar considerations apply. See Ney and Albert (2013).)
2 Admittedly in a different context: that of the derivation of 3-dimensional space from $3N$-dimensional configuration space. However, the same points apply to our situation. (He also uses "local beables" where I have inserted "spacetime quantities.")
3 Emphasizing this line of thought in Leibniz is somewhat misleading. More generally, he sought to reconcile Aristotelian metaphysics with mechanical physics.
4 Hesse (1961) tells the tale in far more detail.
5 I use "principles," but that term is potentially misleading, for it suggests an explicit, finite list of statements. Realistically, the rules that physicists adopt for permitted derivations are often implicit; they also amount to practical knowledge, so it is controversial whether they could even be codified in principle.
6 Arguably, parallel points about physical salience can be made for anti-realists. Constructive empiricists are committed to understanding a theory's claims literally, so the question of what physical explanations they propose still seems apt. And even a positivistic account of explanation must include rules about what patterns of derivation, and what idealizations and approximations, are permitted.
7 The following is based on Gielen et al. (2013); Oriti (2014). Note that GFT is a way of "second quantizing" loop quantum gravity, in which chunks of spacetime "foam" are created and annihilated.
8 It's worth emphasizing that the full mathematical apparatus of modern physics, including the calculus, applies in GFT; so one can proceed classically with an action, and in quantum mechanics with a path integral. In that regard, things are as usual.
9 They are also the most classical states of any quantum system, simultaneously minimizing position and momentum while respecting the Heisenberg uncertainty relations. Moreover, they are unitarily inequivalent to non-coherent states, strongly suggesting that there is a phase transition to such states: the geometrogenesis mentioned above (Oriti, 2014).
10 With perturbative quantum corrections.
11 Note that the functionalism here is not necessarily of the classic kind, in which entities sharing all and only the same *causal* powers are identified: e.g., Lewis (1972). Rather some more general variational co-dependency is meant; after all, in QS causality may not apply at all.

References

Alexander, H.G. (1998). *The Leibniz-Clarke Correspondence: Together with Extracts from Newton's Principia and Optics.* Manchester: Manchester University Press.
Bain, J. (2013). The emergence of spacetime in condensed matter approaches to quantum gravity. *Studies in History and Philosophy of Science Part B: Studies in History and Philosophy of Modern Physics*, 44(3): 338–345.
Barbour, J. (2012). Shape dynamics. An introduction. In Finster, F., Mller, O., Nardmann, M., Tolksdorf, J., and Zeidler, E. (Eds.), *Quantum Field Theory and Gravity.* Basel: Springer, pp. 257–297.
Butterfield, J. (2011). Emergence, reduction and supervenience: A varied landscape. *Foundations of Physics*, 41(6): 920–959.
Callender, C. and Huggett, N. (2001). *Physics Meets Philosophy at the Planck Scale.* Cambridge: Cambridge University Press.
Chalmers, D.J. (2012). *Constructing the World.* Oxford: Oxford University Press.
Friedman, M. (2001). *Dynamics of Reason.* Stanford: CSLI Publications.
Gielen, S., Oriti, D. and Sindoni, L. (2013). Cosmology from group field theory formalism for quantum gravity. *Physical Review Letters*, 111(3): 031301.
Hesse, M.B. (1961). *Forces and Fields.* Nashville: T. Nelson.
Huggett, N. (2014). Skeptical notes on a physics of passage. *Annals of the New York Academy of Sciences*, 1326(1): 9–17.
Huggett, N. (2017). Target space $\neq$ space. *Studies in History and Philosophy of Science Part B: Studies in History and Philosophy of Modern Physics*, 59: pp. 81–88
Huggett, N., Lizzi, F. and Menon, T., 2021. Missing the point in noncommutative geometry. *Synthese*, pp.1–34.
Huggett, N. and Vistarini, T. (2015). Deriving general relativity from string theory. *Philosophy of Science*, 82(5): 1163–1174.
Huggett, N. and Wüthrich, C. (2013). Emergent spacetime and empirical (in) coherence. *Studies in History and Philosophy of Science Part B: Studies in History and Philosophy of Modern Physics*, 44(3): 276–285.

Huggett, N. and Wüthrich, C. (forthcoming). *Out of Nowhere: The Emergence of Spacetime in Quantum Theories of Gravity.* Oxford: Oxford University Press.

Huggett, N., Matsubara, K., & Wüthrich, C. (Eds.). (2020). *Beyond Spacetime: The Foundations of Quantum Gravity.* Cambridge University Press.

Knox, E. (2014). Newtonian spacetime structure in light of the equivalence principle. *The British Journal for the Philosophy of Science*, 65(4): 863–880.

Kuhn, T.S. (1962). *The Structure of Scientific Revolutions.* Chicago: University of Chicago Press.

Lewis, D. (1972). Psychophysical and theoretical identifications. *Australasian Journal of Philosophy*, 50(3): 249–258.

Maudlin, T. (2007). Completeness, supervenience, and ontology. *Journal of Physics A: Mathematical and Theoretical*, 40:3151–3171.

Ney, A. and Albert, D.Z. (2013). *The Wave Function: Essays on the Metaphysics of Quantum Mechanics.* Oxford: Oxford University Press.

Oriti, D., editor. (2009). *Approaches to Quantum Gravity: Toward a New Understanding of Space, Time and Matter.* Cambridge: Cambridge University Press.

Oriti, D. (2014). Disappearance and emergence of space and time in quantum gravity. *Studies in History and Philosophy of Science Part B: Studies in History and Philosophy of Modern Physics*, 46: 186–199.

Oriti, D. (2018). Levels of spacetime emergence in quantum gravity. Available at: arXiv:1807.04875.

Rickles, D., French, S. and Saatsi, J.T. (2006). *The Structural Foundations of Quantum Gravity.* Oxford: Oxford University Press.

Rovelli, C. (1998). Loop quantum gravity. *Living Reviews in Relativity*, 1(1): 1.

Seiberg, N. (2006). *Emergent spacetime.* Available at: arXiv preprint hep-th/0601234.

Teh, N.J. (2013). Holography and emergence. *Studies in History and Philosophy of Science Part B: Studies in History and Philosophy of Modern Physics*, 44(3): 300–311.

Vistarini, T. (2020). Extending lewisian modal metaphysics from a specific quantum gravity perspective. pp. 304–337. doi:10.1017/9781108655705.017

Wüthrich, C. and Callender, C. (2016). What becomes of a causal set? *The British Journal for the Philosophy of Science*, 68(3): 907–925.

Wüthrich, C. (2020) When the actual world is not even possible. In Glick, D., Darby, G., & Marmodoro, A. (Eds.). (2020). *The foundation of reality: Fundamentality, space, and time.* New York, Oxford University Press, USA.

Yates, D. (2021). Thinking about spacetime. In Wüthrich, C., Le Bihan, B., and Huggett, N. (2021). *Philosophy Beyond Spacetime: Implications from Quantum Gravity.* Oxford: Oxford University Press.

Chapter 25 - Further Reading from the Editors

Nathan Seiberg's 'Emergent spacetime' in Henneaux, M., Sevrin, A. and Gross, D.J. eds. *Quantum Structure Of Space And Time, The-Proceedings Of The 23rd Solvay Conference On Physics* (World Scientific, 2007) gives a short and influential argument from a physics perspective. Modern philosophical debate starts with a pair of papers by Jeremy Butterfield and Chris Isham: 'On the emergence of time in quantum gravity,' in *The Arguments of Time*, J. Butterfield (ed.), (British Academy and Oxford University Press, 111–168, 1999) and 'Spacetime and the philosophical challenge of quantum gravity' in C. Callender and N. Huggett (eds.), *Physics Meets Philosophy at the Planck Scale* (Cambridge University Press, Cambridge, 2001). Huggett, N. and Wüthrich, C. 'Emergent spacetime and empirical (in) coherence'. (*Studies in History and Philosophy of Modern Physics*, 44(3):276–285, 2013) defends the empirical coherence of emergent spacetime.

26

THE PROBLEM OF TIME*

Karim P.Y. Thébault

26.1 Introduction

The "problem of time" is a cluster of interpretational and formal issues in the foundations of general relativity relating to both the representation of time in the classical canonical formalism and the quantization of the theory. The problem was first noticed by Bergmann and Dirac in the late 1950s, and is still a topic of intense debate in contemporary physics and philosophy of physics. The purpose of this short chapter is to provide an accessible introduction to the problem, and this will inevitably mean that many significant technical details will be obscured or over simplified. The most significant simplification that we will make is to focus exclusively on the *global* aspect of the problem of time. That is, we will, for the most part, restrict ourselves to the "disappearance of time" in theories invariant under global time reparametrizations. This restriction inevitably means that the important and philosophically rich subtleties relating to local time reparametrizations (refoliations) will not be considered in much detail. Furthermore, in the presentation below I have chosen to focus on a particular dialectic, drawn from the contrast between the views of Barbour and Rovelli, as a means to illustrate my own views. As such the treatment here is claimed to be neither comprehensive, nor entirely neutral. The reader in search of discussions of the problem of time seen from a wider viewpoint and described in its full technical splendor, is directed toward the research literature in physics and philosophy of physics (see further reading section).

26.2 Leibniz and the Problem of Time

Like many of the deepest conceptual problems in modern physics, important aspects of the problem of time in quantum gravity can be traced back to debates within early modern natural philosophy. Of particular importance is Leibniz's critique of the Newtonian absolute conception of time. In particular, Leibniz's assertion that "instants, consider'd without the things, are nothing at all; and that they consist only in the successive order of things" (Alexander 1998, pp. 26–27). The first half of this quote contains a negative critique of absolute time along the lines of what has been called "Aristotle's principle" – there cannot be changeless duration. The second half of the quote then puts forward the kernel of Leibniz's positive view – that time is an "order of successions." Although it is remarkably subtle, and arguably not entirely consistent, it will prove instructive to our discussion to consider Leibniz's positive metaphysics of time in a little detail. In particular, rather surprisingly, Leibniz's metaphysical vocabulary will be found to be well suited to distinguishing between modern approaches to the problem of time advocated by Julian Barbour and Carlo Rovelli. For the most part our discussion of Leibniz's metaphysics of time draws upon the magisterial scholarship of Richard Arthur (1985, 2014) other relevant sources will be indicated as appropriate.

---

* Forthcoming in the *Routledge Companion to the Philosophy of Physics* edited by Eleanor Knox and Alastair Wilson.

An instructive starting point distinction made in the contemporary literature (although not endorsed by Arthur) is between three "levels of reality" in Leibniz's mature metaphysics.[1] At the most basic level, what is real for Leibniz are simple substances which alone have true unity. These are the famously obscure monads. Next, we have the "phenomenal level" that is made up of *phenomena bene funda* – well-founded phenomena – that, due to pre-established harmony, are accurate reflections of the real and actual monadic states. Finally, we have the ideal level which, by contrast, is made up of *entia rationis* – abstract or fictional things – that include "phenomena" founded upon possible but non-actual monadic states. Crucially, although both the phenomenal and the ideal levels can include things which are infinite, all concepts that depend upon the continuum are only applicable to the ideal realm. Thus, if we were to define time as the real line, $\mathbb{R}$, then this concept of time could only be represented for Leibniz as an *entia rationis* and thus ideal. Furthermore, phenomenal things for Leibniz can only acquire their status as *phenomena bene funda* by their grounding upon the actual. They must always be understood as representations or perceptions of the monads of he actual world.

Leibniz's view of the ontological status of relations is subtle and significant for his view of time. Following Arthur (2014) (who is following Mugani (2012)) we can take Leibniz to believe that relations such as situation and succession supervene on intrinsic modifications of the monads: changes in their perceptions and appetition. Thus, with regard to time, real temporal relations of succession, both between the states of the same monad and the states of different monads, are founded upon the appetitive activity of individual monads. We can then consider notions of time relevant to each of the "three levels" defined above. On the fundamental level of the real monads or the actual world, all we have are changes of the monadic states which are coordinated via the principle of pre-established harmony. The phenomenal level consists of well-founded phenomena that arise as representations (perceptions) of the actual monads. At this level, we have both actual time ordering and also actual durations. Finally, at the ideal level of *entia rationis* we have possible time orderings and also possible durations. Rich and sophisticated though it may be, the Leibnizian metaphysics of time runs into two immediate problems. Each of which foreshadows an important aspect of the problem of time in quantum gravity. First, what determines the "order of succession" of phenomenal states needed to fix the actual time ordering? Second, what determines the duration measure needed to fix the "quantity of time" between actual phenomenal states?

In essence, the first question relates to the requirement for a monotonic parametrization of states. That is, an undirected labeling of temporal states by a parameter which is either always increasing or always decreasing. An earlier literature Rescher (1979) follows Russell (1900) in accusing Leibniz of vicious circularity that requires a non-relational concept of time at the basic monadic level. In contrast, Arthur (1985, 2014) convincingly argues that we can understand time at the monadic level purely in terms of a (non-circular) inter-monadic notion of temporal succession based upon compossibility. Temporal succession at the monadic level could then be taken to ground a total ordering of temporal states in the phenomenal realm.[2] This is sufficient to give us a monotonic parametrization of states and thus a non-directed model of time as an order of succession. Compossibility is not, however, sufficient to ground a directed ordering temporal states. Such a notion depends on further structure implicit in the monadic appetition. Since the question of directed time ordering is rather tangential to the problem of time we will set it aside and implicitly assume that "succession" has its undirected connotation. See Arthur (1985, 2014) for discussion in the context of Leibniz and Kiefer and Zeh (1995) for discussion in the context of the problem of time.

Let us now turn to the second question regarding duration. This issue seems particularly pressing for Leibniz. In fact, in the famous correspondence with Clarke, arguably the strongest critique Clarke gives of Leibnizian time is that "the order of things succeeding each other in time is not time itself, for they may succeed each other faster or slower in the same order of succession, but not in the same time" (Alexander 1998, p. 52). Leibniz's response in the correspondence is not entirely clear: he claims that the quantity of time could not become greater and yet the order of successions remain the same since "if the time is greater, there will be more successive and like states interposed" (Alexander 1998, pp. 89–90). On the one hand, although it does seem consistent for Leibniz to assert Aristotle's principle and flatly deny that the distinction Clarke is making corresponds to a difference. However, on the other, there is still an appreciable explanatory burden upon Leibniz to provide any means by which his phenomenal notion of duration can be quantified. A relational notion of time still requires a *determinate metric structure* in order for time to play its functional role in mechanics (De Risi 2007, p. 273). There is a strong hint toward a more satisfactory resolution in Leibniz's late writings. Arthur (1985, 2014), in particular, suggests that *Initia rerum mathemat carum metaphysica* (Loemker 1969) contains a line of response via the definition of temporal distances in terms of "maximally determined" or "simplest path" through interposed constituents (see also Vailati (1997, p. 136)). Even more tantalizingly, Leibniz wrote in 1680 that "the basis for measuring the duration of things is the agreement obtained by assuming different uniform motions (like those of different precise clocks)" (quoted in Arthur (2014, p. 206)). Furthermore, Rescher (1979, p. 66) suggests that based upon the principle of perfection we might expect that "nomic harmony" sufficient to establish a nature phenomenal measure of duration is a contingent feature of the actual world. These hints not withstanding, Leibnizian relationalism about time seems to have insufficient resources to give a relational basis for temporal distances.

Three points from our discussion of Leibniz's metaphysics of time will be of particular significance in what follows. First we have the idea that a relationalist about time, such as Leibniz, may still consistently assert that time orderings are fundamental. That is, a modern Leibnizian style relationalist about time will look to retain a monotonic parametrization of temporal states. Second, we have the Aristotle's principle that asserts that duration is inseparable from change, and thus that denies that changeless duration can exist. Finally, we have the metricity problem, notwithstanding Aristotle's principle, a relational notion of time still requires means to fix a determinate metric structure in order for time to play its functional role in mechanics.

26.3 Reparametrization Invariance

The formalism of Newton's system of mechanics is one of differential equations. In particular, Newtonian theory features equations between *rates of change of velocity* (i.e., acceleration) and forces (e.g., between gravitating bodies). When all goes well, these equations can be solved and the solutions are usually expressed as functions for the position of a body *over time*. Most iconically one can derive the elliptical orbits of the celestial bodes. Newtonian mechanics as written in terms of force laws and differential equations is extremely cumbersome in practice. One of the most important developments in 18th and 19th century mathematics was in *reformulating* Newtonian mechanics as a theory of *variational principles*. The essential idea is to represent the possible states of a mechanical system in an abstract high-dimensional space (a possibility space) and to represent possible histories as curves in this space. Physical possibilities are then picked out via restrictions on the curves. One of the most important variational formulations of mechanics is the 'Lagrangian formulation.' In this formulation the possibility space, labeled $T\mathcal{C}$, is made up of $6n$-dimensions where n is the number of particles in the system. If we have three particles then we would have 18 dimensions. To describe a particle, labeled with an index $i = 1, ..., n$, we specify the spatial position, q_i and velocity, $\dot{q}_i$. Since space is three dimensional, each of these quantities requires three numbers to be specified – it is *vectorial* – this is indicated by the bold typeface. All together we end up with $2 \times 3 \times n = 6n$. Paths, γ, through this $6n$ dimensional possibility space are mappings between the set of real numbers and $T\mathcal{C}$; i.e., we have that $\gamma : \mathbb{R} \to T\mathcal{C}$. We can pick out a privileged group of physical paths by use of an *action functional*,

$S(\gamma)$, which is defined via the integration of the *Lagrangian functional* along the path:

$$S(\boldsymbol{q}_i, t)] = \int_\gamma L(\boldsymbol{q}_i, t)dt \tag{26.1}$$

The physical paths are those that have an *extremal* action, $\delta S = 0$. This idea of an extremal action is a subtle one, and lies at the heart of all variational approaches to mechanics. Most significantly, variational principles of extremal action supply us with a nomological restriction of which curves in the possibility space are physically possible.[3]

The Lagrangian description of mechanics makes use of a temporal parameter in two senses: first, within the definition of the velocities, $\dot{\boldsymbol{q}}_i = \frac{dq_i}{dt}$; and second, within the time labeling of the curves in the possibility space – their *parametrization*. This representation of time conflicts with Leibniz's view since in this formalism time is something *more* than order of successions: the formalism allows us to represent distinct possibilities that have the same sequence of states of affairs but a different rate at which the states are passed through. This is precisely the possibility that Clarke asserts and Leibniz denies. We can think about this notion of duration is in terms of the existence of a *privileged temporal metric*. That is, an absolute temporal distance measure. Such a structure also implies an order of succession in terms of a monotonic parameterization of instantaneous states. Thus, the problem with Lagrangian mechanics from a Leibnizian relational viewpoint is one of excess temporal structure.

Fascinatingly for our purposes, not long after the Lagrangian formalism was developed (principally by Lagrange himself but also by Hamiltonian) a modification of the theory by Jacobi was made that allows more naturally for a Leibnizian viewpoint.[4] The first step is to expand our possibility space and treat time as an additional coordinate, $q_0 = t$, in a $6n + 2$ dimensional *extended possibility space*. Velocities in this space are then defined for all of the $\boldsymbol{q}_\mu$ by differentiation with respect to an arbitrary parameter τ so we have that $\boldsymbol{q}'_\mu = \frac{dq_\mu}{d\tau}$, $\mu = 0, ..., n$. This arbitrary parameter is also taken to vary monotonically along curves in extended configuration space: it is an arbitrary label for an (undirected) ordered succession. An important property of extended mechanics is that it is physically invariant under re-scalings of the parameter τ. Theories which display such a dynamic insensitivity to parameterization are said to be *reparametrization invariant*.

We can associate the time coordinate t (q_0) in extended mechanics with the value taken by a clock external to our mechanical system. In the case of an open system such an interpretation would seem appropriate; but what about if the system is a closed subsystem of the Universe? – or even the Universe as a whole? In this case there is clearly no physical basis for an external clock and as such we would look to eliminate q_0 as an independent variable. We can do this by the process of *Routhian reduction*.[5] Applying Routhian reduction to extended mechanics leads to a new *Jacobi formalism* that has a possibility space of the same dimensions as the Lagrangian formalism we started with, i.e., $6n$. The Jacobi formalism also features the same set of possible instantaneous states. The difference is that this process of expansion and reduction has lead to a further *constraint* that the total energy is zero. This constraint is directly related to the fact that our new *Jacobi formalism* has retained the reparametrization invariance of the extended formalism. In fact, it can be shown that for *any* theory that is reparametrization invariant, the *Hamiltonian function*, which represents the total energy, must be zero.[6] These two features – zero Hamiltonian and reparametrization invariance – are at the heart of the problem of time in quantum gravity. They are also, of course, directly related to the Leibnizian viewpoint on time. If time is *only* relational order of successions then we should demand that a theory of mechanics is reparametrization invariant and thus has a zero Hamiltonian.

26.4   The Global Problem of Time

A standard, and rather misleading, way of introducing the problem of time in quantum gravity is to make reference to a "deep conceptual conflict" between the treatment of time within the two great pillars of modern physics: quantum theory and general relativity. The problem, it is supposed, arises from forcing a background independent theory of spacetime onto the Procrustean bed of quantization with respect to a background time. For some vague, and rather mysterious, reason it is supposed that time simply *disappears* when we attempt to understand gravity within a quantum framework. Although such a view is difficult to countenance in any substantive sense, it does contain an important kernel of truth. Neither quantization nor gravity are fundamentally at the heart of the problem of time. Rather the problem arises generically from the manipulation of a particular class of classical theories, that includes general relativity, according to the standard formal steps that are preparatory for quantization. That is, the problem of time becomes apparent in the process of preparing *any* reparametrization invariant theory for quantization.

Here we have most in mind two different paths toward quantization: (i) Schrödinger's early and rather heuristic route via the Hamiltonian-Jacobi formalism; and (ii) the more rigorous canonical quantization techniques which were first developed by Dirac and von Neumann in the 1920s, and subsequently extended to the case of theories with constraints by Dirac in the 1950s (Dirac 1964). In each case one chooses a particular classical formulation of a theory and then applies a standard recipe to transform to the quantum domain. In each case, when we consider a reparametrization invariant theory the problem of time's disappearance becomes apparent *before* the transformation is applied.

The Hamiltonian formulation of mechanics makes use of a $6n$ dimensional possibility space: "phase space." Each point is made up of a pairing of a position and canonical momentum variable. Canonical momentum is a vectorial quantity possessed by each particle and given by the expression, $p_i = \frac{\partial L}{\partial \dot{q_i}}$ with $i = 1, ..., n$. Each point in our phase space can then be specified as $\Gamma \in x = (q_i, p_i)$ and curves, as before, can be taken to represent histories of physical systems.

It is possible within the Hamiltonian framework to provide the nomological restriction to physical curves that represent physically possible histories in terms of a variational principle. However, it is also possible to provide the relevant nomology in terms of the evolution of an *algebra of observables*. Any quantity that can be measured can be represented as a function that maps between points in phase space and the space of real numbers, $f : \Gamma \to \mathbb{R}$. These functions are called observables and together they form the algebra of observables, $\mathcal{O}(\Gamma)$. The most important observable is energy, and the function that represents total energy is the Hamiltonian function, H, that we met earlier. Now consider *any* other observable, f. If we want to know how f changes with time, in Hamiltonian mechanics all we have to do is calculate the *Poisson bracket* between H and f. This is simply given by:

$$\{f, H\} = \sum_{i=1}^{n} \frac{\partial f}{\partial q_i}\frac{\partial H}{\partial p_i} - \frac{\partial f}{\partial p_i}\frac{\partial H}{\partial q_i} = \dot{f} \qquad (26.2)$$

The observables form an algebra precisely because the Poisson bracket is a binary operation that takes any pairing of observable functions and returns a third. The Poisson bracket has a deep physical and mathematical significance and there is an important sense in which it is one of the key "heuristic structures" upon which quantum mechanics was constructed (Saunders 1993) – we will meet another in terms of the Hamilton-Jacobi principal function shortly.

Essentially the idea is that a classical algebra of observable functions, $\mathcal{O}(\Gamma)$, can be used as a platform upon which to construct a quantum algebra of observable operators, $\hat{A} \in \mathcal{A}(\mathcal{H})$. The former being defined upon phase space, Γ, (a smooth manifold), the latter being defined upon Hilbert space, $\mathcal{H}$ (a normed vector space equipped with an inner product). Just as points, x, in the phase space classical states, vectors in the Hilbert space, $|\psi\rangle$, represent quantum states. Whereas the Poisson bracket, $\{,\}$ plays the role of the binary operation in the classical observable algebra, the *commutator*, $[,]$, plays the role of the binary operation in the quantum observables algebra. The commutator takes two quantum observable operators and returns a third:

$$[\hat{A}_1, \hat{A}_2] = i\hbar\hat{A}_3 \qquad (26.3)$$

where $\hbar$ is Planck's constant dived by 2π. The crucial connection between the two algebras is encoded in the relation:

$$[\hat{A}_f, \hat{A}_g] = i\hbar \hat{A}_{\{f,g\}} \tag{26.4}$$

Formally speaking, quantization (the process of constructing a quantum from a classical theory) takes many forms. One of the best understood and most widely used is canonical quantization. This is the method of quantization that starts from the Poisson bracket and the Hamiltonian formalism and proceeds to the quantum regime by this identification between the two bracket structures (technically this is a Lie algebra morphism).

As mentioned above, canonical quantization in its original form is not applicable to theories where there are constraints on the phase space, and this of course includes reparametrization invariant theories, within which the Hamiltonian is itself a constraint. Dirac's methodology for the quantization of constrained Hamiltonian theories is rather too complicated to go into full detail in this short chapter.[7] However, by reference to the concept of an algebra of observables introduced above we can outline one key ingredient, and in doing so give a first, rather schematic, presentation of the problem of time. The idea is that for theories with constraints, all elements of the algebra of observables must have zero Poisson bracket with the constraints – they must *commute* with the constraints.[8] For reparametrization invariant theories, the idea is thus to consider a sub-algebra, $\mathcal{P} \subset \mathcal{O}$, made up of functions with *vanishing* Poisson bracket with the Hamiltonian $g \in \mathcal{P}$, where $g : \Gamma \to \mathbb{R}$ such that,

$$\{g, H\} = \dot{g} = 0. \tag{26.5}$$

Such functions correspond to observable quantities that do not change over time. The restriction to observables that commute with the Hamiltonian amounts to a restriction that *anything that is physically measurable cannot change*. We will call these functions *perennials* after the coinage of the Czech physicist Karel Kuchař – see Kuchař (1999) – they are also often referred to in the literature as Dirac observables. According to most standard accounts, the problem of time can be explained in terms the equivalence between the observables and perennials within the Hamiltonian formulations of reparametrization invariant theories. That is, when we recast theories such as the Jacobi theory – or in fact general relativity – into a Hamiltonian form, the argument goes there are "good formal reasons" to believe that the set of observables is equivalent to the set of perennials. In this sense, change is no longer part of our Hamiltonian theory! Clearly, whether or not we accept this argument depends very much upon the detailed analysis of these "good formal reasons." Such an analysis would require us to take a rather large detour into the technicalities of constrained Hamiltonian mechanics and we will not proceed in this direction here.[9] Rather we will consider the classical global problem of time in the context of the Hamilton-Jacobi formalism, such that the connection with quantum problem of time is immediately apparent.

The Hamiltonian system of mechanics that we introduced earlier has behind it a rich and beautiful geometrical structure. Much of this structure is in fact encoded in the Poisson bracket itself, and relates to ideas from symplectic geometry which we will not discuss here.[10] The most important geometrical idea is that of a *generating functional*. These are objects based upon which we can generate transformations of the *entire phase space* into different, but physically equivalent, canonical coordinates.[11] Such transformations are particularly useful when they are chosen such that the dynamics in the new coordinates takes a particularly simple form. More specifically, if we label the old coordinates (q_i, p_i) and the new coordinates (Q_i, P_i), then what we would like to find is a transformation such that $\dot{Q}_i = 0$. That is, the new position coordinates are constants of the motion. It can be proved that the generating functional that performs this task is given by a time dependent function of a mix of the old and new coordinates, $S_1(t, q_i, Q_i)$ that solves the *Hamilton-Jacobi equation*:

$$H\left(q_i, \frac{\partial S_1(t, q_i, Q_i)}{\partial q_i}\right) = \frac{\partial S_1(t, q_i, Q_i)}{\partial t} \tag{26.6}$$

The function S_1 is called the *principal functional*. The usual trick to solve the Hamiltonian equation is to use the *Ansatz* for the principal functional, $S_1(t, q, Q) = Et + W(q, Q)$. This reduces the problem to one of calculating the *characteristic functional*, $W(q, Q)$, that solves the equation:

$$H\left(\boldsymbol{q}_i, \frac{\partial W(\boldsymbol{q}_i, \boldsymbol{Q}_i)}{\partial \boldsymbol{q}_i}\right) = 0, \tag{26.7}$$

The last two equations give us a means to characterize the problem of time. According to one view, the hallmark of reparametrization invariant theories is that the Hamilton-Jacobi principal function, $S_1(t, \boldsymbol{q}_i, \boldsymbol{Q}_i)$, should be identified with the characteristic functional $W(\boldsymbol{q}_i, \boldsymbol{Q}_i)$.[12] That is we only have a "timeless" equation of the form:

$$H\left(\boldsymbol{q}_i, \frac{\partial S_1(\boldsymbol{q}_i, \boldsymbol{Q}_i)}{\partial \boldsymbol{q}_i}\right) = 0, \tag{26.8}$$

Given this equation as basic to the Hamilton-Jacobi formalism of reparametrization invariant theories there is then a precise sense in which even a Leibnizian notion of ordered succession is unavailable. In a theory of mechanics described by (26.6), the parameter t marks out a (one dimensional) ordered family of canonical transformations that trivialize the dynamics. In a theory of mechanics described by (26.8) there is only one such transformation, and thus all time itself is trivialised by our canonical transformation. Which of these two formalisms is a more adequate rendition of reparametrization invariant mechanics is thus a question of crucial importance and marks the divide between the two responses to the global problem of time as discussed in the next section.

Starting from the Hamilton-Jacobi formalism there is a reliable heuristic, dating back to Schrödinger, that takes us from the Hamilton-Jacobi principal functional to the wavefunction (Rund 1966, pp. 99–109).[13] Essentially, one considers families of hypersurfaces of constant value of the characteristic function as wavefronts propagating in configuration space with respect to the time parameter. The crucial step is then to interpret these wavefronts as surfaces of constant phase of a complex valued wavefunction on configuration space evolving with respect time. This means one takes the principal functional $S_1(t, q, Q)$ as the basis for a complex wavefunction $\Psi(t)$, defined on a Hilbert space labelled by the eigenvalues of a complete observables and the time parameter t. Applying this heuristic to Equation (26.6) leads directly to the Schrödinger equation:

$$\hat{H}|\Psi\rangle = i\hbar\frac{\partial|\Psi\rangle}{\partial t}. \tag{26.9}$$

On the other hand, starting from Equation (26.8), we are lead to an equation of the form:

$$\hat{H}|\psi\rangle = 0. \tag{26.10}$$

This is a simple form of the famous Wheeler-DeWitt equation.[14] The Wheeler-DeWitt equation provides us with a "frozen formalism" for quantum theory, and thus a problem of recovering time evolution. It describes a quantum system trapped in an energy eigenstate with the wavefunction a time independent function. How should we respond to this problem? Below we will discuss two options.[15] The first is to attempt to abstract an internal notion of time evolution based upon the Wheeler-DeWitt type formalism together with *classical* internal clocks. The second is to avoid passage to the frozen formalism in the first place. Each of these options will be discussed in the following section.

26.5 Finding Time Again

The staring point in our search for lost time is an iconic quote from the great 19th century German thinker Ernst Mach (1883):

It is utterly beyond our power to measure the changes of things by time. Quite the contrary, time is an abstraction, at which we arrive by means of the changes of things; made because we are not restricted to any one definite measure, all being interconnected.

Following the Mittelstaedt–Barbour (Mittelstaedt 1976; Barbour 1993) *interpretation* of Mach, we can take such quotes to motivate a view in which a consistent notion of time can be abstracted from the "changes of things" in a manner such that the inherently interconnected nature of every possible internal measure of time is accounted for. According to the Mittelstaedt–Barbour interpretation, we can understand this "second Mach's principle" as motivating a relational notion of time that is not merely ontologically parasitic on change, but also equitable, in that it can be derived uniquely from the motions of the entire system taken together. Thus, any isolated system – and, in fact, the universe as a whole – should have its own natural clock emergent from the dynamics. This form of relationalism involves a relative notion of duration as abstracted from change. For there to be a notion of time in this sense it is not enough to be merely a structure of temporal relations: our emergent time must also be unique and equitable. We cannot, therefore, merely identify an isolated subsystem as our relational clock, since to do so is not only non-unique but would also lead to an inequitable measure, insensitive to the dynamics of the clock system itself. Such sentiments are, to a large extent, consistent with the Leibnizian view of time discussed in the first section. Most obviously, we have Aristotle's principle of inseparability of duration from change. Furthermore, in assuming that there is a unique method for abstracting duration from change, we also assume that there is an absolute ordering within the change; otherwise the abstraction process would be underdetermined. This means that the second Mach's principle ultimately involves the assumption of temporal ordering structure equivalent to a monotonic time parametrization.

Recall from the first section, that a crucial problem that we identified was for the relationalist to fix a determinate metric structure such that time can play its functional role in mechanics. Fascinatingly, it is precisely in addressing this question that a relationalist response to the global problem of time can be formulated. In particular, in various formal and philosophical treatments spanning five decades Barbour (together with collaborators) has put forward a relationalist program for mechanics that is self-identified as in the spirit of both Leibniz and Mach.[16] We do not have space here to conduct a lengthly analysis of Barbour's views, rather we shall consider a particular formal step made in response to the classical global problem of time and discussed in Barbour and Foster (2008). First we re-consider the evolution equation for an observable function in a theory invariant under time reparameterizations but this time rewriting the total differential in terms of infinitesimal changes[17]:

$$\frac{\delta g}{\delta t} = \{g, H\} \tag{26.11}$$

Barbour and Foster insist that, contra Dirac, this equation should *not* be set to zero. Rather, we take the change in the observable to be real even though it may be arbitrarily parameterized. Furthermore, Barbour and Foster show that can express the change in the observable without reference to the parametrization at all by rewriting the equation as:

$$\delta g = \sqrt{\frac{\Sigma_i \delta \boldsymbol{q}_i . \delta \boldsymbol{q}_i}{2(E - V)}} \{g, H\} \tag{26.12}$$

with E the total energy and V the potential energy. This realizes exactly the idea that Leibniz seems to have had in mind: duration emerges as a harmonious aggregation of motions. Furthermore, in constructing a temporal metric from change we arrive at precisely the structure needed to complete Leibniz's relational project.

In contrast to the more moderate species of Leibniz-Barbour temporal relationalism, we can consider a more radical variant of relationalism which does not involve commitment to temporal ordering. In radical relationalism about time we assert that what it means for a physical degree of freedom to change is for it to vary with respect to a second physical degree of freedom; and there is no sense in which this variation can be described in absolute, non-relative terms. This radical relationalism about time is closely associated with the work of Rovelli[18] and to a large extent the mainstream view in the community of physicists working on problem of time in quantum gravity. The attractiveness of the view lies in its connection to a proposal for abstraction of internal notions of change within fundamentally timeless systems of equations (both classical and quantum). This proposal in its modern form is based upon the idea of "partial observables" and "complete observables" and can be illustrated explicitly using the Hamilton-Jacobi formalism discussed above.

Consider a globally reparametrization invariant description of n free particles. The Hamiltonian will take the simple form, $H = \Sigma_i \frac{p_i^2}{2m_i}$ for $i = 1, ..., n,$. Solving the Hamilton-Jacobi equation gives us an expression for the position variables, q_i, as functions of the constants of motion, Q_i and P_i, and time, t. This takes the form:

$$q_i(t) = Q_i + \frac{P_i}{m_i} t \tag{26.13}$$

These position variables do not commute with the Hamiltonian and so are clearly *not* perennials. This means that, on the standard view, they should not be considered observables. They do, however, prima facie, seem to have obvious physical significance since they represent the spatial degrees of freedom of our system of particles. To emphasize that such variables are physically significant but not fully observable, Rovelli calls them "partial observables." By definition a partial observable is 'a physical quantity with which we can associate a (measuring) procedure leading to a number' (Rovelli 2002, p. 2). The essence of the Rovelli internal clock prescription for dealing with the problem of time is to designate a subset of partial observables as internal clocks, and then use these clocks to construct "complete observables," that are both predicable and measurable, and which correspond to perennials.

We can illustrate the Rovelli prescription using our simple system as follows.[19] First, restrict to 1D so that each particle is represented simply by a single scalar position and momentum variable. Next, choose particle $i = 1$ as our clock and invert Equation (26.13) for t to get

$$t = \frac{m_1}{P_1}(q_1 - Q_1) \tag{26.14}$$

Re-insert this into (26.13) we get

$$q_a(q_1) = Q_a - \frac{P_a}{m_a}\frac{m_1}{P_1}(q_1 - Q_1), \tag{26.15}$$

for $a = 2, ..., n$. We then take $q_1 = \tau$, where $\tau \in \mathbb{R}$, to be the value of an internal clock, and define members of a family of "complete observables" in terms of $q_a(\tau)$ for some specified value of τ. Crucially, for any specification of τ we have, $q_a(\tau) : \Gamma \to \mathbb{R}$ and $\{H, q_a(\tau)\} = 0$, which means that the complete observables are perennials. Given this, one can proceed to construct a quantum theory based upon the classical algebra of complete observables. In the context of this quantum formalism the complete observables will be constructed as operators on a "physical" Hilbert space made up of states that solve the Wheeler-DeWitt equation.

There are number of conceptual and formal difficulties attached to the Rovelli proposal.[20] Perhaps the most philosophical interesting is whether the idea of things that are "measurable but not predictable" is coherent. Some authors[21] think not:

> ...a measurable quantity is always a complete observable, even pointers of a clock are observables and not partial observables. Now complete observables are defined with respect to non-measurable quantities...which we will simply call non-observables...
> (Thiemann 2007, p. 78)

The problem is that, if Thiemann is right and the partial observables are non-measurable, then we seem to loose our ability to use different values of the internal clocks to describe change. Rather, all we have are measurements of the complete observables which are (in a precise sense) temporally non-local. Furthermore, in denying the measurability of partial observables the internal time view arguably runs into the Leibnizian relationalist problem in explaining the determinate (local) metric structure of time at the functional level needed for mechanics. On the other hand, if Rovelli is right and the partial observables are measurable, although we do seem to have a good response to the metricity problem, we still need a more precise way of making sense of the ontological status of things that are "measurable but not predictable." These fascinating questions have received rather too little philosophical attention and are still, to a large extent, open. An important exception to this relative neglect are the various discussions of Rickles,[22] who in the context of advocating for his own structuralist position, makes the highly valuable observation that in essence the debate turns on the old metaphysical question of the relative ontological status of relations and relata – with Rovelli asserting (and Thiemann denying) the independent measurability of the relata.[23]

A further conceptual issue relates to the fact that the complete observables may be multivalued. That is, we are not-guaranteed that the partial observable chosen as the internal clock will be monotonically increasing. This is of course in direct conflict with the Leibniz-Babour form of relationalism discussed above. Given that we want to preserve temporal ordering structure, an alternative prescription for dealing with the problem of time is needed. Recent steps in this direction have been presented in a series of papers by Gryb and Thébault.[24] On the moderate temporal relationalist view defended by Gryb and Thébault, there is always assumed to exist a monotonically increasing time parametrization, but this parametrization is taken only to be defined up to smooth rescallings, and thus we do not have an absolute notion of duration.

The Gryb and Thébault view depends upon a particular interpretation of the Hamilton-Jacobi formalism discussed above. In particular, the view relies upon noting that the difference between Equations (26.6) and (26.7) above is entirely due to an extra time boundary term, namely: the transformation $S \rightarrow S + Et$. This does not affect the local equations of motion. At the classical level the two formalisms are observationally indistinguishable, the difference between them reducing to an interpretational choice regarding the energy being a constant of motion or constant of nature.[25] With this in mind, we are then free to *choose* which of the two Hamilton-Jacobi formalisms to base our quantum theory upon depending on the form of relationalism about time we which to adopt. Choosing the more moderate relationalism, and starting from (26.6), as discussed above, the resulting quantum formalism will retain a fundamental notion of time evolution, and we end up with a unitary evolution equation of the Schrödinger-type,

$$\hat{H} |\Psi\rangle = i\hbar \frac{\partial |\Psi\rangle}{\partial t}. \tag{26.16}$$

The classical algebra of observables through which the quantum formalism is defined are given by the partial observables. For our simple system these can be expressed in terms of equation (26.13). Clearly this equation traces out dynamical curves labelled by the arbitrary parameter t, which is of course itself not an observable. Rather, t is an independent parameter, and, as such, can be specified independently of quantities which are deemed measurable within the theory. The curves defined by (26.13) are reparametrization invariant even if the equation makes reference to the unphysical labelling parameter. Thus, the observables *are* invariant under the relevant global reparametrization symmetry. The "relational quantization" procedure developed by Gryb and Thébault can be motivated in the context of an analysis of globally time reparametrization theories via Faddeev-Popov path integral (Gryb and Thébault 2011), constraint quantization (Gryb and Thébault 2014) or Hamilton-Jacobi techniques (Gryb and Thébault 2016b). In each case, the resulting quantum formalism retains a fundamental notion of Schrödinger time evolution and in this sense the attractiveness of the proposal is obvious.

A severe limitation of relational quantization relates to the "local" problem of time that we have thus far been avoiding in our discussion. Theories such as the Jacobi theory are *globally* time reparametrization invariant and have a single Hamiltonian constraint that generates *global* time evolution. Refoliation invariant theories such as general relativity, by contrast, are *locally* time reparametrization invariant, and have an an infinite family of Hamiltonian constraints that generate *local* "many fingered" time evolution. Imagine a loaf of bread that we can irregularly cut up into a sequence of slices. The loaf is spacetime and the slices are instantaneous spatial surfaces. A *foliation* is then a parameterization of a spacetime by a time ordered sequence of spatial slices. Such a parametrization is local in the sense that it is defined for every point on every spatial slice. The symmetries of general relativity imply that all spacetimes that are related by *re*foliations are physically equivalent. In practice, application of the partial and complete observables programme to general relativity also suffers a number of limitations, such as those relating to integrality (Dittrich et al. 2015). However, in principle, there is no bar to applying the Rovelli prescription for constructing observables to theories invariant under local time reparametrization transformations. Thus, the partial and complete observables approach is a prospective solution to the local and global problem of time. On the other hand, relational quantization is geared specifically toward the solution of the global problem of time, and arguably in principle inapplicable to the local problem.

The viability of relational quantization as an approach to the the problem of time in any full theory of quantum gravity thus rests upon the adoption of a re-description of gravity in terms of a formalism that features a notion of preferred foliation. One attractive possibility along these lines is suggested by the *shape dynamics* formalism.[26] Within this formalism, the principle of local (spatial) scale invariance is introduced with the consequence of favoring a particular notion of simultaneity. This selects a unique global Hamiltonian and thus allows for relational quantization to be applied. Shape dynamics is based upon a re-codification of the physical degrees of freedom of general relativity via exploitation of a duality between the two relevant sets of symmetries. In the class of spacetimes where it is possible to move from one formalism to the other (those that are "CMC foliable") the physical degrees of freedom described by the two formalisms are provably equivalent, they are merely clothed in different descriptive redundancy.

Our two options for "finding time" thus both come with a mix of attractive and unattractive features. Arguably, relational quantization is more attractive than partial and complete observables on account of clearer ontological categories for the observables. And arguably partial and complete observables is more attractive than relational quantization on account of flexibility it gives in dealing with local time reparametrization invariance. In the end, the choice between the two is underdetermined by the choice between the relational ontologies of time, one with temporal ordering structure, one without. As is so often in science, future theoretical and empirical development is the only real prospect to decisively break such underdetermination.

Acknowledgments

I am particularly appreciative to Julian Barbour, Oliver Pooley, Dean Rickles, and Carlo Rovelli for discussion of various of the points above over the years. I also greatly profited from comments on an earlier draft from Eleanor Knox and discussion of Leibniz's view of time with Richard Arthur. Finally, and most significantly, in the above I have drawn heavily upon my work on the problem of time with Sean Gryb, to whom I owe an ever accumulating intellectual debt.

Notes

1 See for example Winterbourne (1982), Hartz and Cover (1988).
2 Arguably, compossibility seems to underdetermine the actual temporal ordering since it does not give us grounds to distinguish it from merely possible temporal orderings. See Cover (1997) and De Risi (2007).

3 See Smart and Thébault (2015) for discussion of extremal action principles and the metaphysics of laws of nature.

4 The classic formal treatments of the Jacobi formalism in the literature are Lanczos (1970, §5), Johns (2005, §11–12) and Rovelli (2004, §3.1).

5 A fuller discussion of Routhian reduction in general, and in this case in particular, is given in Lanczos (1970, §5) and Arnold et al. (1988, §3.s2).

6 More precisely, reparametrization invariance of the action by definition implies that the Lagrange density is homogeneous of order 1 in the velocities. This, via Euler's homogeneous function theorem, then implies that the Hamiltonian density must vanish; or, rather, that it be proportional to a constraint (Dirac 1964).

7 For the classic textbook discussions see Dirac (1964), Henneaux and Teitelboim (1992).

8 Being more precise, the observables must "weakly commute," meaning that the Poisson bracket must be zero only on the sub-manifold of phase space that the constraints define.

9 In essence, the question is whether of not we should understand Hamiltonian constraints as generating unphysical "gauge" transformations that do no change the physical state. While the majority opinion dating back to Dirac is that we should – see for example Rovelli (2004) – various authors have also put forward arguments that we should not (Kuchař 1991; Barbour 1994; Pons 2005; Pooley 2006; Barbour and Foster 2008; Gryb and Thébault 2014; Pitts 2014b).

10 See Thébault (2012b) for a discussion of the problem of time specifically in the context of symplectic mechanics.

11 Here and below we are following (Arnold 2013, §9). See also Lanczos (1970, §8).

12 See in particular Rovelli (2004, §3.2).

13 See Butterfield (2005) for philosophical discussion.

14 The full Wheeler-DeWitt equation for general relativity can be derived via an exactly analogous line of reasoning based upon the the Einstein-Hamilton-Jacobi equation (Peres 1962; DeWitt 1967).

15 See Isham (1992), Kuchař (1991), Anderson (2012) for discussion of further approaches.

16 A selection of Barbour's key works are: Barbour (1974), Barbour (1994), Barbour (2001a), Barbour (2001b), Barbour (2009), Barbour and Bertotti (1982), Barbour et al. (2014).

17 Again, we neglect the "weak equality" for simplicity of exposition.

18 See for example Rovelli (1990), Rovelli (1991), Rovelli (2002), Rovelli (2004), Rovelli (2007), Rovelli (2014).

19 See Dittrich (2006, 2007) for formal refinement of the procedure. Following the work of Dittrich the partial and complete observables proposal can be generalized to systems of multiple constraints via the idea of "partially invariant partial observables." This idea, combined with the notion of "weakly Abelian" constraints, allows for expression of complete observables of an arbitrary constrained system as an infinite power series.

20 The most important formal difficulties relate to invertibility and integrability. See Bojowald et al. (2011), Dittrich et al. (2015).

21 Rickles (2005, p. 26) contains a similar observation, made a little earlier.

22 See in particular Rickles (2007, pp. 161–171) and also related remarks in Rickles (2005, 2006a, 2006b, 2008, 2016).

23 For further discussion in the context of the physics literature see Tambornino et al. (2012), Rovelli (2014).

24 See in particular, Gryb and Thébault (2011, 2014, 2016b, 2016c).

25 See Gryb and Thébault (2016b) for extensive discussion of this point.

26 Shape dynamics was originally developed by Barbour and collaborators (Barbour 2003; Anderson et al. 2003; Anderson et al. 2005) and then brought into modern form in Gomes et al. (2011).

References

Alexander, H.G. (1998). *The Leibniz-Clarke Correspondence: Together with Extracts from Newton\'s Principia and Optics*. Manchester: Manchester University Press.

Anderson, E. (2012). Problem of time in quantum gravity. *Annalen der Physik*, 524(12): 757–786.

Anderson, E. (2017). *The Problem of Time: Quantum Mechanics versus General Relativity*, Vol. 190. Cambridge: Springer.

Anderson, E., Barbour, J., Foster, B.Z., Kelleher, B., and O'Murchadha, N. (2005). The physical gravitational degrees of freedom. *Classical and Quantum Gravity*, 22: 1795–1802.

Anderson, E., Barbour, J., Foster, B.Z. and O'Murchadha, N. (2003). Scale-invariant gravity: Geometrodynamics. *Classical and Quantum Gravity*, 20: 1571.

Arnold, V.I. (2013). *Mathematical Methods of Classical Mechanics*, Vol. 60. New York: Springer Science & Business Media.

Arnold, V.I., Kozlov, V.V. and Neishtadt, A.I. (2007). *Mathematical Aspects of Classical and Celestial Mechanics*, Vol. 3. New York: Springer Science & Business Media.

Arthur, R.T. (1985). Leibniz's theory of time. In Okruhlik, K. and Brown, J.R. (eds.), *The Natural Philosophy of Leibniz*, Vol. 29. Dordrecht: D. Reidel, pp. 263–313.

Arthur, R.T. (2014). *Leibniz*. Oxford: Polity Press.

Barbour, J. (1994). The timelessness of quantum gravity: I. The evidence from the classical theory. *Classical and Quantum Gravity*, 11(12): 2853–2873.

Barbour, J. (2001a). *The End of Time: The Next Revolution in Physics*. Oxford: Oxford University Press.

Barbour, J. (2003). Scale-invariant gravity: Particle dynamics. *Classical and Quantum Gravity*, 20: 1543–1570.

Barbour, J. (2009). *The nature of time*. Available at: arXiv preprint arXiv:0903.3489.

Barbour, J. and Foster, B.Z. (2008, August). *Constraints and gauge transformations: Dirac's theorem is not always valid*. ArXiv:0808.1223.

Barbour, J., Koslowski, T. and Mercati, F. (2014). Identification of a gravitational arrow of time. *Physical Review Letters*, 113(18): 181101.

Barbour, J.B. (1974). Relative-distance machian theories. *Nature*, 249(5455): 328.

Barbour, J.B. (1993). *Mach's principle: From Newton's bucket to quantum gravity. Proceedings, Conference, Tuebingen, Germany, July 26–30, 1993*. Birkhäuser. Boston, MA: Birkhaeuser (1995) 536 p. (Einstein studies. 6).

Barbour, J.B. (2001b). *The Discovery of Dynamics: A Study from a Machian Point of View of the Discovery and the Structure of Dynamical Theories*. Oxford: Oxford University Press.

Barbour, J.B. and Bertotti, B. (1982). Mach's principle and the structure of dynamical theories. *Proceedings of the Royal Society of London. A. Mathematical and Physical Sciences*, 382(1783): 295–306.

Belot, G. (2007, Jan). The representation of time and change in mechanics. In J. Butterfield and J. Earman (eds.), *Handbook of Philosophy of Physics*, Chapter 2. Amsterdam: Elsevier, pp. 133–227.

Belot, G. and Earman, J. (2001). Pre-socratic quantum gravity. In C. Callender and N. Hugget (eds.), *Physics Meets Philosophy at the Planck Scale*. Cambridge University Press. Cambridge pp 213–255.

Bojowald, M., Höhn, P.A. and Tsobanjan, A. (2011). An effective approach to the problem of time. *Classical and Quantum Gravity*, 28(3): 035006.

Butterfield, J. (2005). On hamilton-jacobi theory as a classical root of quantum theory. In Elitzur, A.C., Dolev, S., and Kolenda, N. (eds.), *Quo Vadis Quantum Mechanics?*. Berlin, Heidelberg: Springer, pp. 239–273.

Cover, J. (1997). Non-basic time and reductive strategies: Leibniz's theory of time. *Studies in History and Philosophy of Science Part A*, 28(2): 289–318.

De Risi, V. (2007). *Geometry and Monadology: Leibniz's Analysis Situs and Philosophy of Space*, Vol. 33. Springer Science & Business Media.

DeWitt, B.S. (1967). Quantum theory of gravity. 1. The canonical theory. *Physical Review*, 160: 1113–1148.

Dirac, P.A.M. (1964). *Lectures on Quantum Mechanics*. Yeshiva University, New York: Dover Publications.

Dittrich, B. (2006). Partial and complete observables for canonical general relativity. *Classical and Quantum Gravity*, 23: 6155.

Dittrich, B. (2007). Partial and complete observables for Hamiltonian constrained systems. *General Relativity and Gravitation*, 39: 1891.

Dittrich, B., Hoehn, P.A., Koslowski, T.A. and Nelson, M.I. (2015, 08). Chaos, Dirac observables and constraint quantization. arXiv:1508.01947

Earman, J. (2002). Thoroughly modern Mctaggart: Or, what Mctaggart would have said if he had read the general theory of relativity. *Philosopher's Imprint*, 2(3): 1–28.

Gambini, R., Porto, R.A., Pullin, J. and Torterolo, S. (2009). Conditional probabilities with Dirac observables and the problem of time in quantum gravity. *Physical Review D*, 79(4): 041501.

Gomes, H., Gryb, S., and Koslowski, T. (2011). Einstein gravity as a 3D conformally invariant theory. *Classical and Quantum Gravity*, 28: 045005.

Gryb, S. and Thébault, K.P.Y. (2011). The role of time in relational quantum theories. *Foundations of Physics*, 42(9):1–29.

Gryb, S. and Thébault, K.P.Y. (2014). Symmetry and evolution in quantum gravity. *Foundations of Physics*, 44(3): 305–348.

Gryb, S. and Thébault, K.P.Y. (2016c). Time remains. *British Journal for the Philosophy of Science*, 67(3): 663–705.

Gryb, S. and Thébault, K.P.Y (2016a). Regarding the 'hole argument' and the 'problem of time'. *Philosophy of Science*, 83(4): 563–584.

Gryb, S. and Thébault, K.P.Y. (2016b). Schrödinger evolution for the universe: Reparametrization. *Classical and Quantum Gravity*, 33(6): 065004.

Hartz, G.A. and Cover, J. (1988). Space and time in the Leibnizian metaphysic. *Noûs*, 22: 493–519.

Henneaux, M. and Teitelboim, C. (1992). *Quantization of gauge systems*. Princeton: Princeton University Press.

Isham, C. (1992). *Canonical quantum gravity and the problem of time*. Available at: Arxiv:gr-qc/9210011v1.

Johns, O. (2005). *Analytical Mechanics for Relativity and Quantum Mechanics*. Oxford: Oxford University Press.

Kiefer, C. and Zeh, H. (1995). Arrow of time in a recollapsing quantum universe. *Physical Review D*, 51(8): 4145.

Kuchař, K.V. (1991). The problem of time in canonical quantization of relativistic systems. In A. Ashtekar and J. Stachel (eds.), *Conceptual Problems of Quantum Gravity*. Boston: Boston University Press, p. 141.

Kuchař, K.V. (1999). The problem of time in quantum geometrodynamics. *The Arguments of Time*, Butterfield, J. (ed.). Oxford: Oxford University Press, 169–196.

Lanczos, C. (1970). *The Variational Principles of Mechanics*. New York: Dover Publications.

Loemker, L.E. (1969). *Gottfried Wilhelm Leibniz: Philosophical Papers and Letters*. Dordrecht: Kluwer.

Mach, E. (1883). *Die Mechanik in ihrer Entwicklung Historisch-Kritsch Dargestellt*. Leipzig: Barth. English translation: Mach, E 1960 *The Science of Mechanics*, Open Court, Chicago (translation of 1912 German edition).

Maudlin, T. (2002). Thoroughly muddled mctaggart: Or, how to abuse gauge freedom to create metaphysical monostrosities. *Philosopher's Imprint*, 2(4): 1–19.

Mittelstaedt, P. (1976). *Der Zeitbegriff in der Physik*. Germany: B.I.-Wissenschaftsverlag, Mannheim.

Mugani, M. (2012). *Oxford Studies in Early Modern Philosophy*, Vol. 6, Chapter Leibniz's theory of relations: A last word? Oxford University Press, pp. 171–208.

Peres, A. (1962). On Cauchy's problem in general relativity-ii. *Il Nuovo Cimento (1955–1965)*, 26(1): 53–62.

Pitts, J.B. (2014a). Change in Hamiltonian general relativity from the lack of a time-like killing vector field. *Studies in History and Philosophy of Modern Physics*, 47: 68–89.

Pitts, J.B. (2014b). A first class constraint generates not a gauge transformation, but a bad physical change: The case of electromagnetism. *Annals of Physics*, 351: 382–406.

Pons, J. (2005). On Dirac's incomplete analysis of gauge transformations. *Studies in History and Philosophy of Science Part B: . . .* , 36: 491.

Pons, J., Salisbury, D. and Sundermeyer, K.A. (2010). Observables in classical canonical gravity: Folklore demystified. *Journal of Physics A: Mathematical and General*, 222: 12018.

Pooley, O. (2006). A hole revolution, or are we back where we started? *Studies in History and Philosophy of Science Part B: Studies in History and Philosophy of Modern Physics*, 37(2): 372–380.

Rescher, N. (1979). *Leibniz an Introduction to His Philosophy*. Oxford: Blackwell Publishing.

Rickles, D. (2005). *Interpreting quantum gravity*. Available at: http://philsci-archive.pitt.edu/2407/.

Rickles, D. (2006a). Bringing the hole argument back in the loop: A response to Pooley. *Studies in History and Philosophy of Science Part B: Studies in History and Philosophy of Modern Physics*, 37(2): 381–387.

Rickles, D. (2006b). Time and structure in canonical gravity. In D. Rickles and S. French (eds.), *The Structural Foundations of Quantum Gravity*. Oxford: Oxford University Press, pp. 152–195.

Rickles, D. (2007). *Symmetry, Structure, and Spacetime*, Vol. 3 of *Philosophy and Foundations of Physics*. Amsterdam: Elsevier.

Rickles, D. (2008). Who's afraid of background independence? *Philosophy and Foundations of Physics*, 4: 133–152.

Rickles, D. (2016). Quantum gravity: A primer for philosophers. D. Rickles (Ed.) Routledge (Aldershot) In *The Ashgate Companion to Contemporary Philosophy of Physics*. Routledge, pp. 268–388.

Rovelli, C. (1990, October). Quantum mechanics without time: A model. *Physical Review D*, 42: 2638–2646.

Rovelli, C. (1991). Time in quantum gravity: An hypothesis. *Physical Review D*, 43: 442.

Rovelli, C. (2002). Partial observables. *Physical Review D*, 65: 124013.

Rovelli, C. (2004). *Quantum Gravity*. Cambridge: Cambridge University Press.

Rovelli, C. (2007). Comment on "Are the spectra of geometrical operators in loop quantum gravity really discrete?" by B. Dittrich and T. Thiemann.

Rovelli, C. (2014). Why gauge? *Foundations of Physics*, 44(1): 91–104.

Rund, H. (1966). *The Hamilton-Jacobi Theory in the Calculus of Variations: Its Role in Mathematics and Physics*. London: Van Nostrand.

Russell, B. (1900). *A Critical Exposition of the Philosophy of Leibniz, with an Appendix of Leading Passages*. Cambridge: Cambridge University Press.

Saunders, S. (1993). To what physics corresponds. In H. Kaminga and S. French (eds.), *Correspondence, Invariance, and Heuristics; Essays in Honour of Heinz Post*. Dordrecht: Kluwer, pp. 295–326.

Smart, B.T. and Thébault, K.P.Y. (2015). Dispositions and the principle of least action revisited. *Analysis*, 75(3): 386–395.

Tambornino, J. et al. (2012). Relational observables in gravity: A review. *SIGMA*, 8: 017.

Thébault, K.P.Y. (2012b). Symplectic reduction and the problem of time in nonrelativistic mechanics. newblock *The British Journal for the Philosophy of Science*, 63(4): 789–824.

Thébault, K.P.Y. (2012a). Three denials of time in the interpretation of canonical gravity. *Studies in History and Philosophy of Science Part B: Studies in History and Philosophy of Modern Physics*, 43(4): 277–294.

Thiemann, T. (2007). *Modern Canonical Quantum General Relativity*. Cambridge: Cambridge University Press.

Vailati, E. (1997). *Leibniz and Clarke: A Study of Their Correspondence*. Oxford: Oxford University Press.

Winterbourne, A. (1982). On the metaphysics of leibnizian space and time. *Gottfried Wilhelm Leibniz. Critical Assessments*, 3: 62–75.

Further Reading from the Editors

In order to understand the problem of time in quantum gravity, it is helpful to first understand related issues in classical theories. Belot, G. 'The representation of time and change in mechanics', in J. Butterfield and J. Earman (Eds.), *Handbook of Philosophy of Physics* (Elsevier, 2007) is a comprehensive introduction. Julian Barbour's 'The end of time: The next revolution in physics' (Oxford University Press, 2001) is a popular but subtle account of related issues. Anderson's *The Problem of Time: Quantum Mechanics versus General Relativity*, (Springer, 2017) is a comprehensive monograph devoted to the problem. Chris Isham's "Canonical quantum gravity and the question of time," in *Canonical Gravity: From Classical to Quantum* (Springer, Berlin, Heidelberg, pp. 150–169, 1994) is an extensive introduction to the quantum problem from a physicists perspective. A great deal of helpful contemporary philosophical work has come out of a collaboration between Sean Gryb and Karim Thébault. See, for example, their 'Time remains' (*The British Journal for the Philosophy of Science* 67, no. 3 (2016): 663–705).

PART VII

Statistical Mechanics and Thermodynamics

Introduction to Part VII

Statistical mechanics and thermodynamics are connected, obviously enough, in that they both provide tools for modeling thermal phenomena involving heat and pressure. But they are connected in a deeper way, in that they both describe physical phenomena at a high level of abstraction; their techniques enable us to eliminate the vast majority of a system's degrees of freedom from our model of that system while still predicting important aspects of its behavior with high confidence.

One perennial question in the philosophy of thermal physics is how statistical mechanics relates to thermodynamics: can the latter be reduced to the former, in any interesting sense? Puzzles about reduction and emergence in the philosophy of science have frequently taken the relation between statistical mechanics and thermodynamics as a case study, which by itself would be sufficient to motivate philosophical interest in these theories. But the study of the theories also feeds into numerous further philosophical questions: about the nature of objective probability, about abstraction in explanation, about the relation between information and physical reality, and about the arrow of time.

The modern science of (equilibrium) thermodynamics was discovered experimentally during the 19th century, and is the clearest example we have of a *phenomenological principle theory*: it consists of a set of (experimentally established) principles that constrain the possible transitions that a physical system can undergo. These principles dictate relations among high-level collective properties of the system, and neither they nor the additional parameters which are needed to characterize individual systems (such as the heat capacity of water vapor) were initially derived from theoretical considerations; they had to be experimentally determined. The first clear example was Boyle's law inversely relating volume and pressure for a quantity of gas at fixed temperature, which was discovered in 1662 using an apparatus of glass and mercury.

Statistical mechanics initially developed out of work on the kinetic theory of gases, which was being pursued alongside the emergence of thermodynamics. In the work especially of Maxwell, Boltzmann and Gibbs, it was shown that statistical mechanics allowed for partial explanations (albeit patchy and controversial ones) of thermodynamic phenomena. These striking explanatory achievements provided important evidence in favor of the atomic hypothesis, and they raised the community's hopes of a complete reduction of macroscopic phenomena to classical microphysics. Such a reduction of course still eludes us today, even if we restrict our attention to the reduction of thermodynamics; optimists argue that the outlines of its reduction to statistical mechanics are by now clear, while pessimists point to difficult – perhaps intractable – problems around such matters as entropy, ergodicity, reversibility, phase transitions, and the justification of the choice of measure over microstates. In addition to the chapters in this part, the chapters by Dizadji-Bahmani, Bangu, and Batterman in Part IX revisit the question of the viability of this reduction.

Contemporary statistical mechanics is dominated by two broad approaches, the relation between which remains disputed. The *Boltzmannian* approach has often been perceived as conceptually clearer, and it has been the setting for a resurgence of progress in the foundations of statistical physics in recent decades. It is Boltzmannian statistical mechanics that is employed by Albert and Loewer in their "imperialistic" Mentaculus proposal, which has been influential in metaphysics and the philosophy of science. (The Mentaculus promises a schematic explanation for all high-level physical regularities whatsoever in terms of a low-entropy constraint on the initial macrostate, a probability distribution over initial microstates, and the fundamental laws.) However, the *Gibbsian* approach (which theorizes directly about ensembles of systems, sometimes understood in probabilistic terms) tends to predominate in applications of statistical mechanics within physics, and it has always had supporters in foundations of physics who argue for its conceptual superiority. The initial two chapters by Frigg and Werndl introduce Boltzmannian and Gibbsian statistical mechanics respectively, through the lens of their different treatments of the key concept of *equilibrium*.

While statistical mechanics began as a statistical treatment of classical physics, in the 20th century it has been progressively adapted to fit a quantum-mechanical theoretical setting. Here the distinction between Boltzmannian and Gibbsian approaches comes again to the fore, with defenders of the Gibbsian approach maintaining that theirs generalizes more naturally to the quantum context. The chapter by Orly Shenker surveys the current state of quantum statistical mechanics and its relationship both to classical statistical mechanics and to non-statistical quantum theory.

The history of the universe is an approach to equilibrium writ large. At every scale that we can observe, systems that are closed (or near enough) display an increase of entropy; wherever a system decreases in entropy we can attribute it to an increase elsewhere. (Black holes may or may not present an exception to this rule.) The entropy asymmetry is one of the deepest facts that we have discovered about the cosmos, and how exactly to account for its source remains disputed. The chapter by Arianne Shahvisi sets out what we know about the entropy asymmetry, and shows how we can trace it back through cosmic history.

27

EQUILIBRIUM IN BOLTZMANNIAN STATISTICAL MECHANICS

Roman Frigg and Charlotte Werndl

27.1 Introduction

A gas that is confined to the left half of a container starts expanding as soon as the confining wall is removed and eventually spreads evenly over the entire available space. The state of being spread evenly is the *equilibrium state* and the process of expansion culminating in that state is the *approach to equilibrium*. It is one of the aims of statistical mechanics (SM) to give an exact characterization of equilibrium, and to explain why and how systems approach the state of equilibrium in terms of the dynamical laws that govern the individual molecules of which the gas is made up of. What is it about molecules and their motions that lead them to spread when the wall is removed? An important answer to these questions was suggested by Boltzmann (1909 [1877]), and variants of this answer are currently regarded by many as a promising option. As is customary in the current literature on SM, we refer to the approach that originates in Boltzmann's 1877 paper as Boltzmannian SM (BSM). In this chapter, we introduce different versions of BSM and discuss how they conceptualize equilibrium. Another approach that could be labeled "Boltzmannian" departs from the so-called Boltzmann equation. We set this approach aside and focus on what we call BSM.[1]

We begin by introducing the basic notions of BSM. A system in SM has the mathematical structure of a *measure-preserving deterministic dynamical system* (X, μ, T_t). In line with most of the literature on BSM we focus on deterministic systems; for a discussion of BSM from a stochastic point of view see Werndl and Frigg (2017a). X is the state space and it contains all possible *micro-states* of the system. In the case of an isolated gas in a box made up of n particles a micro-state is given by the positions and the momenta of every molecule in the gas. The measure μ specifies how large certain parts of X are. We assume that the measure is normalized: $\mu(X) = 1$. This is for reasons of mathematical convenience and no connection between this measure and probability is assumed at this point. The time evolution of the system is given by the *evolution function* $T_t : X \rightarrow X$ where t is time and the function satisfies the requirement $T_{t_1+t_2}(x) = T_{t_2}(T_{t_1}(x))$ for all micro-states $x \in X$ and all instants of time t_1 and t_2. Time can either be continuous ($t \in \mathbb{R}$) or discrete ($t \in \mathbb{Z}$).[2] Intuitively this means that it does not matter whether the time-evolution is carried out in one go or takes place in stages: if the process begins in micro-state x, then we end up in the same final state $y \in X$ irrespective of whether we evolve x forward in time by $t_1 + t_2$ at once or whether we first evolve it x forward by t_1 and then evolve the resulting state forward by t_2. The measure μ is assumed to be invariant under the dynamics, meaning that $\mu(T_t(A)) = \mu(A)$ for all measurable subsets A of X and all t.

The macro-condition of a system is specified by its *macro-state*. In the case of the gas this can be done, for instance, by specifying the volume, pressure and temperature of the gas. It is usually assumed that there are a finite number m of macro-states, and we denote these by M_i, $i = 1, ..., m$. The core posit of Boltzmannian SM is that macro-states supervene on micro-states, meaning that a change in the macro-state must be accompanied by a change in the micro-state (it is not possible, for instance, to change the pressure of a gas without also changing the state of motion of at least some of its molecules). Hence, to every given micro-state x there corresponds *exactly one* macro-state. This determination relation is not one-to-one; usually many different micro-states correspond to the same macro-state. One can now group together all micro-states that belong to the same macro-state M_i. The result of this grouping is the *macro-region* X_{M_i}. All X_{M_i} taken together form a partition of X, meaning that they do not overlap and jointly cover X.

The *Boltzmann entropy* of a macro-state M_i is defined as $S_B(M_i) := k \log[\mu(X_{M_i})]$, where k is the Boltzmann constant; the Boltzmann entropy of a *system* at time t, $S_B(t)$, is the entropy of the macro-state of the system at t: $S_B(t) := S_B(M_{x(t)})$, where $x(t)$ is the micro-state at t and $M_{x(t)}$ is the macro-state supervening on $x(t)$.

Among the macro-states of the system two are of particular importance, namely the equilibrium state and the macro-state at the beginning of the process, also referred to as the "past state." We introduce the special labels M_{eq} and M_p for these states (and choose a labeling of macro-states so that $M_p = M_1$ and $M_{eq} = M_m$). A crucial aspect of the standard presentation of BSM is that the equilibrium macro-region $X_{M_{eq}}$ is vastly larger than any other macro-region. In fact $X_{M_{eq}}$ is so large that it takes up most of X.[3] However, Lavis (2005, pp. 255–258, 2008, pp. 685–687) points out that in some situations the equilibrium macro-region is *larger than any other macro-state* without taking up most of X. We set this problem aside, but it is discussed in Werndl and Frigg (2015b). The size of the macro-region is generally seen as the crucial feature of equilibrium.

This raises two questions. First, why is equilibrium associated with the state that has the largest macro-region? The connection between equilibrium and large macro-regions is certainly not analytical: there is nothing in the *concept* of equilibrium tying it to the macro-state with the largest macro-region. Second, isolated systems, when left to themselves, end up in the equilibrium state. In our initial example the gas spreads until it fills the container evenly. Why do systems approach equilibrium? We discuss these problems in the next two sections.

27.2 Definitions of Equilibrium

Why is the equilibrium macro-state the state with the largest macro-region? A prominent answer, now known as the *combinatorial argument*, originates in Boltzmann (1909 [1877]).[4] The basic principle at work in Boltzmann's argument is that equilibrium is the macro-state that is compatible with the largest number of micro-states. Based on this notion Boltzmann constructs $X_{M_{eq}}$ as follows. Consider the same system of n identical particles as above, but now focus on the phase space χ of *one* of these particles (the space has six dimensions – the three position and three momentum coordinates of the particle). Only a finite portion χ_a of χ is accessible to the particle because the motion of the particle is constrained by the container walls and the total energy E of the gas. Now put a regular grid on χ_a such that the grid lines run parallel to the position and momentum axis, thereby dividing χ_a into a finite number of cells of equal size $\delta\omega$. This grid is also known as a coarse-graining of χ_a. Now label the grid cells ω_i for $i = 1, ..., l$. The so-called coarse-grained micro-state of a particle is given by specifying in which cell ω_j its micro-state lies. The micro-state of the entire gas is a specification of the micro-state of every particle in the system, which is therefore determined by n labeled points in χ_a. The coarse-grained micro-state of the gas, also known as an *arrangement*, is a specification of which state of the particle lies in which cell of the partition.

The crucial observation now is that a number of arrangements correspond to the same macro-state because the macro-properties of the system are determined solely by the number of particles in each cell, while it is irrelevant exactly which particle is in which cell. For instance, whether particle number 5 and particle number 7 are in cells ω_1 and ω_2 respectively, or vice versa, makes no difference to the macro-properties of the system as a whole because these do not depend on which particle is in which cell. Hence, all one needs in order to determine the macro-properties of the system is a specification of how many particles there are in each cell of the coarse-graining of χ_a. Such a specification is called a *distribution* and it can be written as a tuple $D = (n_1, \ldots, n_l)$, meaning that there are n_1 particles in cell ω_1, etc. The n_j are referred to as *occupation numbers*. This allows us to ask a crucial question: how many arrangements are compatible with a given distribution D? Some elementary combinatorial considerations show that

$$G(D) := \frac{n!}{n_1! \ldots n_l!} \qquad (27.1)$$

arrangements are compatible with a given distribution D (where '!' denotes factorials, i.e. $k! := k(k-1) \ldots 1$, for any natural number k and $0! := 1$). By construction $G(D)$ is a measure for how many micro-states are compatible with a macro-state: the larger the value of G, the more the arrangements are compatible with a given distribution D.

The number $G(D)$ matters to our original question about the size of $X_{M_{eq}}$. Since X is a Cartesian product of $6n$ copies of χ, each point in X corresponds to exactly one distribution D. Let us denote the region in X that corresponds to distribution D as X_D. One can then show that the size of that region is given by

$$\mu(X_D) = G(D)\,(\delta\omega)^n. \qquad (27.2)$$

Hence the size of the part of X that corresponds to D is directly proportional to $G(D)$: the larger $G(D)$, the larger X_D. So, by maximizing $G(D)$ — that is, by finding the distribution that is compatible with the largest number of (coarse-grained) micro-states — one at the same also finds the largest macro-region. Let D_{max} be that distribution. If one then adopts the above principle that equilibrium is the macro-state that is compatible with the largest number of micro-states, then, by definition, D_{max} is the equilibrium distribution and $X_{D_{max}}$ is the equilibrium macro-region: $X_{D_{max}} = X_{M_{eq}}$. Since D_{max} has the largest G of all distributions, it also corresponds to the largest macro-region, which provides the sought-after justification of the notion that equilibrium is the state with the largest macro-region.

It remains to find D_{max}. To solve this problem, Boltzmann makes two crucial sets of assumptions. The first concerns the energy of the particles. Boltzmann assumes that the energy of a particle only depends on which cell ω_j it is in, but *not* on the states of the other particles; that is, he neglects the contribution to the energy of the system that stems from interactions between the particles. He then also assumes that the energy of a particle only depends on which cell it is in but not on where it lies within the cell. Under these assumptions, the total energy of the system is given by $\sum_{j=1}^{l} n_j E_j$, where E_j is the energy of a particle in cell j. The second assumption is that there are many particles in each individual cell: $n_j \gg 1$ for all j. Under these assumptions one can prove that $G(D)$ reaches its maximum for

$$n_j = \alpha \exp(-\beta E_j), \qquad (27.3)$$

which is the (discrete) *Maxwell-Boltzmann distribution*, where α and β are constants depending on the nature of the system. So the largest macro-region in X corresponds to the Maxwell-Boltzmann and that is the equilibrium region.

This justificatory strategy faces both conceptual and technical challenges. The main conceptual problem is the absence of a connection with the thermodynamic notion of equilibrium. The following is a typical thermodynamics (TD) textbook definition of equilibrium: "A thermodynamic system is in equilibrium when none of its thermodynamic properties are changing with time [...]" (Reiss, 1996, p. 3). The problem is that there is no *conceptual* connection between this notion of equilibrium and the idea that the equilibrium macro-state is the macro-state that is compatible with the largest number of micro-states. This is a problem for anyone wishing to maintain at least some congruence between SM and TD.

Even if one was willing to set aside conceptual issues, there remains the technical problem that the assumptions made for the optimization are so strong that the domain of application of the theory is in effect limited to dilute gases. The combinatorial argument assumes that the energy of a particle depends only on the cell in which it is located. This assumption applies, strictly speaking, *only to systems with non-interacting particles, i.e., ideal gases* (Uffink, 2007; Frigg, 2008). Ideal gases are, perhaps, a good approximation for *dilute gases*, i.e., gases of low density, and so the argument may deliver the approximately correct results for such systems. However, the argument remains silent about systems with stronger inter-particle forces such as liquids and solids. This is a serious limitation, and no suggestions have been made so far as to how it could be overcome.

One might try to avoid at least the technical difficulties by reading the argument backward, as it were: postulate that the Maxwell-Boltzmann distribution is the equilibrium distribution (rather than deriving it from an optimization procedure), and then appeal to combinatorial considerations to establish that the corresponding macro-region is large. This move is motivated by the fact that Maxwell's (1860) original derivation of the Maxwell-Boltzmann distribution does not appeal to optimization procedures. Unfortunately, this is a blind alley. The Maxwell-Boltzmann distribution is in fact the equilibrium distribution only for a limited class of systems, namely for systems consisting of particles with negligible inter-particle forces. In general, systems with non-negligible interactions will have equilibrium distributions that are different from the Maxwell-Boltzmann distribution (Gupta, 2003). A closer look at Maxwell's derivation shows why: non-interaction enters via the postulate that the probability distributions in different spatial directions can be factorized, which is true only if there is no interaction between particles (see Uffink, 2007).

A different justification of the postulate that equilibrium corresponds to the largest macro-region appeals to the time-evolution of a system and argues that if a part of the state space is overwhelmingly large, then a state sooner or later has to move into that part and stay there for a long time. This, so the argument continues, is the defining feature of equilibrium, and therefore the largest macro-state is the equilibrium state. This reply is closely tied to the typicality approach that we discuss in the next section, where we will see that the basic assumptions of that approach are questionable.

As we have seen in the last section, the Boltzmann entropy is proportional to the logarithm of the measure of a macro-region and therefore the macro-state with the largest macro-region also has the highest Boltzmann entropy. We know from TD that, if left to itself, a system approaches equilibrium, and equilibrium is the maximum entropy state. Therefore the macro-state with the largest macro-region is the equilibrium state.[5]

This line of argument faces a number of difficulties. The first is that it attributes an entropy to systems out of equilibrium and compares entropy levels at different stages of a process. But thermodynamics does not attribute an entropy to systems out of equilibrium at all and therefore comparing the entropies of different macro-states (most of which will be non-equilibrium states) makes little sense from a thermodynamic point of view. But even if this problem could be resolved (through a suitable generalization of thermodynamics, for instance) there would remain a question why the fact that the thermodynamic entropy reaches a maximum in equilibrium would imply that this also holds for the Boltzmann entropy. To justify this inference, the assumption would need to be made that the thermodynamic entropy reduces to the Boltzmann entropy. However, it is far from clear whether that is the case. A connection between the TD entropy and the Boltzmann entropy has been established only for ideal gases where the so-called Sackur-Tatrode formula can be derived from BSM, which shows that both entropies have the same functional dependence on thermodynamic state variables. Yet for systems with interactions no such results are known (cf. Frigg and Werndl, 2011b). Furthermore, there are well-known differences between the TD and the Boltzmann entropy. For example, the TD entropy is extensive but the Boltzmann entropy is not (Ainsworth, 2012), and an extensive concept cannot reduce to a non-extensive concept (at least not without further qualifications).

A different route is taken in the long-run-fraction-of-time-account recently proposed by Werndl and Frigg (2015a, 2015b). This approach *defines* equilibrium in terms of how long a system spends in a certain state (rather than in terms of the size of its macro-region) and then proves a theorem establishing that the state is large in the requisite sense. Let $LF_M(x)$ be the long-run fraction of time that a system starting in micro-state x spends in macro-region X_M. If, for instance, $LF_M(x) = 0.43$, then the system starting in initial condition x spends 43% of the time in macro-state M in the long run. Equilibrium, then, is the macro-state in which the system spend most of its time for nearly all initial conditions. Formally, if a macro-state M satisfies the following condition, then it is the system's equilibrium macro-state: $LF_M(x) \geq \alpha$ for a real number $\alpha \in (0.5, 1]$ and for all $x \in Y$ where Y is a subset of X such that $\mu(Y) \geq 1 - \varepsilon$ for a very small real number $\varepsilon \geq 0$. An obvious question concerns the value of α. Often the assumption seems to be that α is close to one. This is a reasonable but not the only possible choice, and nothing hangs on the value of α. The introduction of ε accounts for the possibility that we should not expect *every* initial condition to approach equilibrium (Callender, 2001).

This definition is couched entirely in terms of time and remains completely silent about the size of X_M. This raises the question: is X_M large in the way BSM would have it? The answer to this question is given by the Dominance Theorem: If M is the equilibrium macro-state, then $\mu(X_M) \geq \alpha(1 - \varepsilon)$ (Werndl and Frigg, 2015b). The theorem is completely general in that no dynamical assumptions are made (in particular it is not assumed that the system is ergodic) and hence the theorem also applies to strongly interacting systems such as solids and liquids. It is important to note that the theorem makes the conditional claim that *if* there is an equilibrium, *then* X_M is large. As with all conditionals, the crucial and often vexing question is whether, and under what conditions, the antecedent holds.

27.3 The Approach to Equilibrium

Why do isolated systems, when left to themselves, approach equilibrium? To appreciate the thrust of this question we have to recall an important property of the phenomenology of the approach to equilibrium: irreversibility. We see gases spread, yet we never observe the reverse process of a uniformly distributed gas suddenly concentrating in the left half of the container. So it seems that the question ought to be: why do systems approach equilibrium *irreversibly*? The standard line is that this irreversible approach to equilibrium is a consequence of the second law of thermodynamics, which, roughly, states that entropy cannot decrease in isolated systems. However, as Brown and Uffink (2001) rightly point out, that systems prepared in a non–equilibrium state do approach equilibrium is not a consequence of the second law (or any other law of standard thermodynamics) and has to be added as an independent principle, which they call the "minus first law." This suggests that the task for non-equilibrium SM is to derive the minus first law of thermodynamics from BSM.

This is setting the bar too high. SM systems show Poincaré recurrence. That is, the system's micro-state will eventually return arbitrarily close to the original initial condition. But a system with Poincaré recurrence cannot possibly exhibit strict irreversible behavior (and that it may take a very long time until a system actually returns close to its initial condition does not make this point moot). Contributors to the discussion have acknowledged this fact (more or less explicitly) and there is an (at least tacit) agreement that what should be derived from SM is an *approximate* version of the minus first law. Different version of BSM differ in how they explicate the approximation.

Boltzmann (1909 [1877]) conceptualized the problem as one of showing that the system is over-whelmingly likely to be in equilibrium whenever one observes the system. To this end Boltz-mann defined macro-states in terms of distributions (he assumed that to every distribution D there corresponds a macro-state) and introduced the postulate that the probability of a macro-state is proportional to $G(D)$. Since, as we have seen above, $G(D)$ is largest for the equilibrium state, it follows immediately that equilibrium is the most likely state and he argued that "the system of bod-ies always evolves from a more unlikely to a more likely state" (Boltzmann, 1909 [1877]: 166; our translation). This view faces a number of problems (Frigg, 2010a). There is the question of how to justify the postulate that the probability of a macro-state is proportional to $G(D)$, and, more seriously, it is unclear where the tendency to move toward more probable states comes from. The probabil-ities of macro-states are unconditional probabilities, and as such they do not imply anything about the succession of macro-states, let alone that ones of low probability are followed by ones of higher probability.

A related approach (also originating in the work of Boltzmann) appeals to the notion of ergodicity. Intuitively, a system is ergodic if the fraction of time an arbitrary solution stays in a subset A of X equals the measure of A in the long run. More formally, let $LF_A(x)$ be the long-run fraction of time a system starting in initial condition x will spend in a set $A \subset X$ (that is, for time $t \to \infty$). A system is ergodic iff $LF_A(x) = \mu(A)$ for all subsets $A \subset X$ and for almost all initial conditions x (that is, except perhaps for some initial conditions that taken together form a set of measure zero). If, for instance, a certain set has measure $1/3$ and if the system is ergodic, then we know that in the long run it will spend $1/3$ of the time in that set.

Since ergodicity concerns any set $A \subset X$, it a fortiori also concerns macro-regions, and hence an ergodic system will spend a fraction of time in every macro-state that is proportional to the size of the corresponding macro-region. Since equilibrium is the state with the largest macro-region, an ergodic system will spend most of the time in equilibrium for almost all initial conditions (which implies, of course, that if the system is set in motion in a non-equilibrium micro-condition, it will soon enough move into the equilibrium macro-region). This is the justification of the minus first law of thermodynamics. It is an approximate justification in two ways. First, systems are said to spend most of the time in equilibrium but they are not said to never move out of equilibrium. In fact they can (and they will), and so ergodicity does not imply strict irreversibility. Second, ergodicity does not rule out that there are "bad" initial conditions, i.e., initial conditions that lie on trajectories that do not satisfy $LF_A(x) = \mu(A)$; ergodicity just requires that there are 'few' bad initial conditions in the sense that all bad conditions taken together form a set of measure zero.

The ergodic program faces two main challenges. The first, known as the *measure zero problem*, points out that the set of allowed "bad" states can actually rather large since measure zero sets can be rather big. As an example consider the set of the rational numbers. This set has measure zero in the real numbers, but there are great many rational numbers. This fact can be made visible by using alternative ways to assess the size of sets, such as Baire categories (Sklar, 1993, pp. 182–188). The second objection, the *irrelevancy challenge*, is that ergodicity is irrelevant because many real systems are not ergodic (Earman and Rédei, 1996). The force of this argument can be mitigated by introducing the notion of epsilon-ergodicity (Vranas, 1998). Intuitively, a dynamical system is epsilon-ergodic if it is ergodic on the vast majority of X, namely, on a set of measure $\geq 1 - \varepsilon$, where ε is a very small real number or zero. The class of systems that are epsilon-ergodic is larger than the class of systems that is ergodic and it is plausible that it comprises many realistic systems such as gases and some liquids (Frigg and Werndl, 2011a). However, there will be SM systems that are not epsilon-ergodic and it remains unclear how the ergodic program deals with them.

An alternative approach is developed by Albert (2000), who proposes to explain irreversible behav-ior in terms of transition probabilities: rather than assigning probabilities to macro-states, we should look at how likely a system is to transition into a right macro-state given that it is in a certain macro-state now. To this end Albert introduces the *statistical postulate*: given that the system is in macro-state M at time t, the probability of finding the micro-state of the system in a set $C \subset X_M$ at time t is $\mu(C)/\mu(X_M)$. One can separate the states in X_M into "good" and "bad" ones, with good ones being those lying on trajectories that move into macro-states of higher entropy (than M) once they leave X_M and bad ones being those lying on trajectories that move into lower entropy (than M) states. Thermodynamic behavior is then justified if one can show that for all macro-states (other than the equilibrium macro-state) it is the case that the system is overwhelmingly more likely to move toward a higher (Boltzmann) entropy macro-state than to a lower (Boltzmann) entropy macro-state. However, the statistical postulate also allows to calculate the probability that a system has moved into the current macro-state from a higher or a lower entropy macro-state. It turns out to be the case that whenever the system is overwhelmingly likely to evolve toward a macro-state of higher entropy in the future, it is also overwhelmingly likely to have evolved into the current macro-state from a past macro-state which also has higher entropy. So the formalism returns wrong transition probabilities.

Albert discusses this problem at length and suggests fixing it by first taking the system under investigation to be the entire universe and then adopting the so-called *Past Hypothesis*, the postulate that "the world came into being in whatever particular low-entropy highly condensed big-bang sort of macro-condition it is that the normal inferential procedures of cosmology will eventually present to us" (Albert, 2000, p. 96). The problem with the wrong transition probabilities is then solved by conditioning on the past state: given that the system is in macro-state M at time t, the probability of finding the system's micro-state in set $C \subset X_M$ at time t is $\mu(C \cap T_t(X_{M_p}))/\mu(X_M \cap T_t(X_{M_p}))$. Albert argues at length that if this rule is used to calculate probabilities, then a high-entropy future as well as a low-entropy past are overwhelmingly likely.

There are a number of concerns about this explanation. A crucial aspect of Albert's explanation is to apply SM to the universe as a whole. Earman argued that this project is doomed to failure because in the setting of standard cosmological models the past hypothesis is "not even false" (Earman, 2006, p. 400).[6] The Boltzmann entropy is a global quantity characterizing the macro-state of an entire system, in Albert's case the entire universe.

As Winsberg (2004) points out, just because the overall entropy of the universe increases, it need not be the case that the entropy in a small subsystem also increases and hence Albert's approach cannot explain the behavior of small systems such as gases in laboratories. Albert's argument relies on assigning probabilities to sets of microstates based on a certain algorithm, and there are questions about the justification of that algorithm (see Davey, 2008; Frigg, 2010a). Furthermore, Albert's justification that a high-entropy future and low-entropy past are overwhelmingly likely appeals to a dynamical assumption, which he calls the "scattering condition." And there is a question whether this condition holds true in SM systems.[7]

An alternative account explains the approach to equilibrium in terms of *typicality*. Consider an element e of a set Σ. Typicality is a relational property of e, which e posses with respect to Σ, a property P and a measure ν, often referred o as the "typicality measure." Intuitively, e is typical if most members of Σ have property P and e is one of them. More precisely, let $\Pi \subset \Sigma$ be the set of all elements that have property P. Then the element e is typical if $e \in \Pi$ and $\nu(\Pi)/\nu(\Sigma) \geq 1 - \varepsilon$, where ε is a finite but very small positive real number. The element of interest in SM is a micro-state x and the typicality measure is the Lebesgue measure μ. The typicality account comes in different versions that disagree about the choice of the set Σ and the selection of property P (for a discussion of these different accounts see Frigg (2009) and (2010b)).

The first account explains the approach to equilibrium in terms of equilibrium micro-states being typical. As we have seen above, the equilibrium macro-region is by far the largest macro-region. This implies that equilibrium micro-states are typical with respect to X because "reaching the equilibrium distribution in the course of the temporal evolution of a system is inevitable due to the fact that the overwhelming majority of microstates in the phase space have this distribution"

(Zanghì, 2005, p. 196; our translation). However, as Uffink (2007, pp. 979–980) points out, in general there is no reason to assume that points in an atypical set have to evolve into a typical set; typical states are not attractors of atypical states. The second account combines a typicality claim about equilibrium micro-states with a further typicality claim about the dynamics of the system. Goldstein champions such an account when he submits that for "a non-equilibrium phase point $[x]$ of energy E, the Hamiltonian dynamics governing the motion $[x(t)]$ would have to be ridiculously special to avoid reasonably quickly carrying $[x(t)]$ into $[X_{M_{eq}}]$ and keeping it there for an extremely long time — unless, of course, $[x]$ itself were ridiculously special" (Goldstein, 2001, pp. 43–44). Unfortunately, Goldstein offers no account of what it means for a dynamical law not to be "ridiculously special" and so the account remains underspecified (see Frigg and Werndl (2012) for a proposed completion of the account in terms of epsilon-ergodicity.) The third account, due to Lebowitz (1993a, 1993b), considers the internal structure of macro-regions (in much the same way as the transition probability approach we have seen above) and argues that micro-states lying on entropy-increasing trajectories are typical. This line of argument faces the same challenges as Albert's and it remains an unproven assertion that the internal structure of the macro-regions is as the account asserts (see Frigg, 2010b for a discussion of this issue).

The problem of the approach to equilibrium takes a different form in the long-run-fraction-of-time-account (Werndl and Frigg, 2015a, 2015b). In that account it is part and parcel of the notion of an equilibrium state that the system approaches that state and stays there most of the time; if a state does not have that feature, then it is not an equilibrium state. The crucial question then is: under what conditions does an equilibrium exist? Werndl and Frigg (forthcoming) point out that for an equilibrium to exist three factors need to co operate: the choice of macro-variables, the dynamics of the system, and the choice of the effective state space. They then prove a theorem providing general necessary and sufficient conditions for the existence of an equilibrium state. Intuitively, the theorem says that there is an equilibrium just in case the effective state space of the system is split up into invariant regions on which the motion is ergodic and the equilibrium macro-state takes up most of each region. This suggests a change in the way in which we think about equilibrium: rather than launching a search for one crucial factor (such as ergodicity or typicality), the focus should be on finding triplets of macro-variables, dynamical conditions, and effective state spaces that satisfy the conditions of the theorem. An example of such triplet is the dynamics of the Kac ring on the full state space with a homogeneity macro-variable (Werndl and Frigg, 2015b). With this theorem, the theoretical question when and under what conditions an equilibrium exists is solved. But the identification and classification of these triplets for concrete problems remains an open question.

27.4   Conclusion

We have reviewed a number of Boltzmannian definitions of equilibrium along with explanations of the approach to equilibrium. We want to conclude by drawing attention to two further issues. The first is the relation of BSM to Gibbsian SM. There are two competing theoretical approaches in SM, one associated with Boltzmann and the other with Gibbs, which offer different conceptualizations of equilibrium. This would not be a problem if the two formalisms were equivalent (for instance, in a similar way in which the Schrödinger and the Heisenberg picture in quantum mechanics are equivalent). Unfortunately they are not, and there is no obvious way to translate results from one framework into the other. But having two incompatible notions of equilibrium at work in SM is unsatisfactory and a sustained reflection on how the Boltzmannian and the Gibbsian approach relate to each other is necessary. Steps toward a better understanding of the relationship between these two approaches are made in Lavis (2005), Werndl and Frigg (2017b, 2020), and Frigg and Werndl (2019), but the problem is one that deserves more attention that it has received so far.

The second issue is the interpretation of probability. As we have seen, many explanations of the approach to equilibrium rely in one way or another on probabilities, and there is a question of how these should be interpreted. A number of suggestions have been made including Humean Best Systems (Loewer, 2001; Frigg and Hoefer, 2015), typicality measures (Werndl, 2013), frequencies (discussed but not endorsed in van Lith, 2001) and propensities (discussed but not endorsed in Clark, 2001). Another approach attempts to ground the probabilities of SM in quantum mechanics, and so they are interpreted in whatever way quantum probabilities are interpreted. See Frigg and Werndl (this volume b) for a discussion of these approaches. There is no consensus on this issue, and finding a coherent interpretation of SM probabilities remains a challenge.

Notes

1 For a discussion of this approach see Ardourel (2017), Brown et al. (2009), Uffink and Valente (2015) and Valente (2014).
2 For a simple and intuitive introduction to dynamical systems see Berkovitz et al. (2016). Detailed discussions of the framework of BSM can be found in Frigg (2008) and Uffink (2007).
3 See, for instance, Albert (2000, pp. 56–57), Bricmont (1995, p. 146); Goldstein (2001, pp. 43, 45); Goldstein and Lebowitz (2004, p. 57), Penrose (1989, p. 403) and Zanghì (2005, pp. 191, 196).
4 Classical presentations of the argument can be found in Ehrenfest and Ehrenfest (2002 [1912]) and Tolman (1979 [1938]: Ch. 4). For discussions see Frigg (2008) and Uffink (2007).
5 This strategy has been mentioned to us in conversation but is hard to track down in print. Albert's (2000) considerations concerning entropy seem to gesture in the direction of this third strategy.
6 For a further discussion of the past hypothesis see Wallace (2017), and for a discussion of its explanatory relevance see Callender (2004) and Price (2004).
7 For further discussions see Shahvisi (this volume) and Frigg (2008, Sec. 3.2.5).

References

Ainsworth, P.M. (2012). Entropy in statistical mechanics. *Philosophy of Science*, 79: 542–560.
Albert, D. (2000). *Time and Chance*. Cambridge, MA and London: Harvard University Press.
Ardourel, V. (2017). Irreversibility in the derivation of the Boltzmann equation. *Foundations of Physics*, 47: 471–489.
Berkovitz, J. et al. (2016). The ergodic hierarchy. In E.N. Zalta (ed.), *Stanford Encyclopedia of Philosophy* (Summer 2016 Edition). Available at: https://plato.stanford.edu/archives/sum2016/entries/ergodic-hierarchy/.
Boltzmann, L. (1909 [1877]). Über die Beziehung zwischen dem zweiten Hauptsatze der mechanischen Wärmetheorie und der Wahrscheinlichkeitsrechnung resp. den Sätzen über das Wärmegleichgewicht. In F. Hasenöhrl (ed.), *Wissenschaftliche Abhandlungen Leipzig*, Vol. 2. Leipzig: J. A. Barth, pp. 164–223.
Bricmont, J. (1996). Science of chaos or chaos in science? In P.R. Gross, N. Levitt and M.W. Lewis (eds.), *The Flight from Science and Reason* (Annals of the New York Academy of Sciences, Vol. 775). New York: The New York Academy of Sciences, pp. 131–175.
Brown, H., Myrvold, W. and Uffink, J. (2009). Boltzmann's H-theorem, its discontents, and the birth of statistical mechanics. *Studies in History and Philosophy of Modern Physics*, 40: 174–191.
Brown, H. and Uffink, J. (2001). The origin of time-asymmetry in thermodynamics: The minus first law. *Studies in History and Philosophy of Modern Physics*, 32: 525–538.
Callender, C. (2001). Taking thermodynamics too seriously. *Studies in the History and Philosophy of Modern Physics*, 32: 539–553.
Callender, C. (2004). There is no puzzle about the low-entropy past. In C. Hitchcock (ed.), *Contemporary Debates in Philosophy of Science*. Oxford, Malden, MA and Victoria: Blackwell, pp. 240–255.
Clark, P. (2001). Statistical mechanics and the propensity interpretation of probability. In J. Bricmont, D. Durr, M.C. Galavotti, G. Ghirardi, F. Petruccione and N. Zanghì (eds.), *Chance in Physics: Foundations and Perspectives* (Lecture Notes in Physics, Volume 574). Berlin, Heidelberg and New York: Springer, pp. 271–281.
Davey, K. (2008). The justification of probability measures in statistical mechanics. *Philosophy of Science*, 75: 28–44.
Earman, J. (2006). The past hypothesis: Not even false. *Studies in History and Philosophy of Modern Physics*, 37: 399–430.
Earman, J. and Rédei, M. (1996). Why ergodic theory does not explain the success of equilibrium statistical mechanics. *British Journal for the Philosophy of Science*, 47: 63–78.

Ehrenfest, P. and Ehrenfest, T. (2002 [1912]) *The Conceptual Foundations of the Statistical Approach in Mechanics.* Mineola/New York: Dover Publications.

Frigg, R. (2008). A field guide to recent work on the foundations of statistical mechanics. In D. Rickles (ed.), *The Ashgate Companion to Contemporary Philosophy of Physics.* London: Ashgate, pp. 99–196.

Frigg, R. (2009). Typicality and the approach to equilibrium in Boltzmannian statistical mechanics. *Philosophy of Science,* 76: 997–1008.

Frigg, R. (2010a). Probability in Boltzmannian statistical mechanics. In G. Ernst and A. Hüttemann (eds.), *Time, Chance and Reduction. Philosophical Aspects of Statistical Mechanics.* Cambridge: Cambridge University Press, pp. 92–118.

Frigg, R. (2010b). Why typicality does not explain the approach to equilibrium. In M. Suárez (ed.), *Probabilities, Causes and Propensities in Physics.* Dordrecht: Springer, pp. 77–93.

Frigg, R. and Hoefer, C. (2015). The best Humean system for statistical mechanics. *Erkenntnis,* 80: 551–574.

Frigg, R. and Werndl, C. (2011a). Explaining thermodynamic-like behaviour in terms of Epsilon-ergodicity. *Philosophy of Science,* 78: 628–652.

Frigg, R. and Werndl, C. (2011b). Entropy - A guide for the perplexed. In C. Beisbart and S. Hartmann (eds.), *Probability in Physics.* Oxford: Oxford University Press, pp. 115–142.

Frigg, R. and Werndl, C. (2012). Demystifying typicality. *Philosophy of Science,* 79: 917–929.

Frigg, R. and Werndl, C. (2019). Statistical Mechanics: A Tale of Two Theories. *The Monist,* 102: 424–438.

Goldstein, S. (2001). Boltzmann's approach to statistical mechanics. In J. Bricmont, D. Durr, M.C. Galavotti, G. Ghirardi, F. Petruccione and N. Zanghi (eds.), *Chance in Physics: Foundations and Perspectives* (Lecture Notes in Physics, Volume 574). Berlin, Heidelberg and New York: Springer, pp. 39–54.

Goldstein, S. and Lebowitz, J. (2004). On the (Boltzmann) entropy of nonequilibrium systems. *Physica D,* 193: 53–66.

Gupta, M.C. (2003). *Statistical Thermodynamics.* New Delhi: New Age.

Lavis, D. (2005). Boltzmann and Gibbs: An attempted reconciliation. *Studies in History and Philosophy of Modern Physics,* 36: 245–273.

Lavis, D. (2008). Boltzmann, Gibbs and the concept of equilibrium. *Philosophy of Science,* 75: 682–696.

Loewer, B. (2001). Determinism and chance. *Studies in History and Philosophy of Modern Physics,* 32: 609–629.

Lebowitz, J. (1993a). Boltzmann's entropy and time's arrow. *Physics Today,* 46: 32–38.

Lebowitz, J. (1993b). Macroscopic laws, microscopic dynamics, time's arrow and Boltzmann's entropy. *Physica A,* 194: 1–27.

Maxwell, J.C. (1860). Illustrations of the dynamical theory of gases. Part 1. On the motions and collisions of perfectly elastic spheres; Part 2. On the process of diffusion of two or more kinds of moving particles among one another. *Philosophical Magazine,* 19, 20: 19–32, 21–37. ser. 4, 19.

Penrose, R. (1989). *The Emperor's New Mind.* Oxford: Oxford University Press.

Price, H. (2004). On the origins of the arrow of time: Why there is still a puzzle about the low-entropy past. In C. Hitchcock (ed.), *Contemporary Debates in Philosophy of Science.* Oxford and Malden, MA: Blackwell, pp. 219–239.

Reiss, H. (1996). *Methods of Thermodynamics.* Mineaola, NY: Dover.

Sklar, L. (1993). *Physics and Chance. Philosophical Issues in the Foundations of Statistical Mechanics.* Cambridge: Cambridge University Press.

Tolman, R.C. (1979 [1938]). *The Principles of Statistical Mechanics.* Mineola, NY: Dover.

Uffink, J. (2007). Compendium of the foundations of classical statistical physics. In J. Butterfield and J. Earman (eds.), *Philosophy of Physics.* Amsterdam: North Holland, pp. 923–1047.

Uffink, J. and Valente, G. (2015). Lanford's theorem and the emergence of irreversibility. *Foundations of Physics,* 45: 404–438.

Valente, G. (2014). The approach towards equilibrium in Lanford's theorem. *European Journal for Philosophy of Science,* 4: 309–335.

van Lith, J. (2001). Ergodic theory, interpretations of probability and the foundations of statistical mechanics. *Studies in History and Philosophy of Modern Physics,* 37: 581–594.

Vranas, P.B.M. (1998). Epsilon-ergodicity and the success of equilibrium statistical mechanics. *Philosophy of Science,* 65: 688–708.

Wallace, D. (2017). The nature of the past hypothesis. In K. Chamcham, J. Silk and J.D. Barrow (eds.), *The Philosophy of Cosmology.* Cambridge: Cambridge University Press, pp. 486–499.

Werndl, C. (2013). Justifying typicality measures in Boltzmannian statistical mechanics. *Studies in History and Philosophy of Modern Physics,* 44: 470–479.

Werndl C. and Frigg, R. (2015a). Rethinking Boltzmannian equilibrium. *Philosophy of Science,* 82: 1224–1235.

Werndl C. and Frigg, R. (2015b). Reconceptualising equilibrium in boltzmannian statistical mechanics and characterising its existence. *Studies in History and Philosophy of Modern Physics* 49: 19–31.

Werndl C. and Frigg, R. (2017a). Boltzmannian equilibrium in stochastic systems. In M. Massimi and J.-W. Romeijn (eds.), *Proceedings of the EPSA15 Conference*. Berlin and New York: Springer, pp. 243–254.

Werndl C. and Frigg, R. (forthcoming). When does a Boltzmannian equilibrium exist? In D. Bedingham, O. Maroney and C. Timpson (eds.), *Quantum Foundations of Statistical Mechanics*. Oxford: Oxford University Press.

Werndl C. and Frigg, R. (2017b). Mind the gap: Gibbs versus Boltzmann. *Philosophy of Science*, 84(5): 1289–1302.

Werndl C. and Frigg, R. (2020) 'When do Gibbsian Phase Averages and Boltzmannian Equilibrium Values Agree?' Studies in History and Philosophy of Modern Physics 72: 46–69.

Winsberg, E. (2004). Can conditioning on the past hypothesis militate against the reversibility objections? *Philosophy of Science*, 71: 489–504.

Zanghì, N. (2005). I fondamenti concettuali dell'approccio statistico in fisica. In V. Allori, M. Dorato, F. Laudisa and N. Zanghì (eds.), *La Natura Delle Cose. Introduzione ai Fundamenti e alla Filosofia della Fisica*. Roma: Carocci, pp. 139-228.

Further Reading from the Editors

Many philosophers of physics get their first introduction to philosophy of thermal physics through D. Albert, *Time and Chance* (Cambridge, MA: Harvard University Press, 2000). For a rounded introduction, read this alongside L. Sklar, *Physics and Chance: Philosophical Issues in the Foundations of Statistical Mechanics* (Cambridge: Cambridge University Press, 1993). A scholarly treatment of Boltzmannian statistical mechanics is given by J. Uffink, "Boltzmann's Work in Statistical Physics", *Stanford Encyclopedia of Philosophy* (2014), and a vigorous defence of the Boltzmannian approach by S. Goldstein, "Boltzmann's Approach to Statistical Mechanics", in J. Bricmont, G. Ghirardi, D. Dürr, F. Petruccione, M.C. Galavotti, N. Zanghi (eds.) *Chance in Physics* (Heidelberg: Springer, 2001). A more conciliatory approach is taken by D. A. Lavis, "Boltzmann and Gibbs: An attempted reconciliation", *Studies in History and Philosophy of Science Part B: Studies in History and Philosophy of Modern Physics* 36(2), (2005): 245–273. R. Frigg, "A Field Guide to recent Work on the Foundations of Statistical Mechanics," in D. Rickles (ed.), *The Ashgate Companion to Contemporary Philosophy of Physics* (London: Ashgate, 2008): 99–196 provides a broad overview of recent work.

28

EQUILIBRIUM IN GIBBSIAN STATISTICAL MECHANICS

Roman Frigg and Charlotte Werndl

28.1   Introduction

Consider a macroscopic system such as a gas or a solid, and assume that this system is described by macroscopic variables such as temperature, pressure, volume and magnetization. Intuitively, a system has reached equilibrium when all change has come to a halt and the values of the macroscopic variables remain constant over time. How can equilibrium be characterized exactly, and why and how do systems that are not initially in an equilibrium state approach equilibrium? It is the aim of statistical mechanics (SM) to answer these questions in terms of the mechanical properties of the micro-constituents of these systems and the dynamical laws that govern their time evolution.

Gibbsian SM (GSM) offers answers to these questions by associating an *ensemble* with each physical system. Let X be the state space of a system of interest. In a mechanical n-particle system, for instance, the state space has $6n$ dimensions, three dimensions for the position of each particle and three dimensions for the corresponding momenta. An ensemble is specified by a probability density $\rho(x, t)$ over the state space of a system, where t is time (which can be either continuous or discrete) and $x \in X$. Physical observables are associated with real-valued functions $f : X \rightarrow \mathbb{R}$. These functions represent physical quantities, such as internal energy, magnetization, and polarization. The *ensemble average* (or sometimes "phase average") of an observable is:

$$\langle f \rangle = \int_X f(x)\rho(x, t)dx \qquad (28.1)$$

Gibbs refers to the fact that "the distribution [...] will remain unchanged" as the condition of 'statistical equilibrium' (1981 [1902], p. 6). In modern parlance this amounts to saying that for an ensemble to be in equilibrium the distribution has to be stationary, meaning that it does not depend on time: $\rho(x, t) = \rho(x)$. Hence, equilibrium in GSM is the property of an ensemble rather than of an individual system. However, experimental tests are carried out on an individual physical system and not on an ensemble, which raises the question of what one can infer from an ensemble about such a system. The Gibbsian answer is that ensemble averages correspond to empirical equilibrium values: when measuring a physical quantity represented by f in a system in equilibrium, then GSM tells us that the observed value will be $\langle f \rangle$, which does not depend on time because ρ is stationary.[1] This method is also known as *Gibbsian Ensemble Averaging* (or *Gibbsian Phase Averaging*).

Of the infinitely many stationary distributions three enjoy a special status: the microcanonical, canonical and grandcanonical distributions. The so-called *microcanonical distribution* describes the equilibrium of an ensemble consisting of systems in which the number of particles and the energy are both constants, and entire probability mass is evenly distributed on the hyper-surface of constant energy in X. The so-called *canonical distribution* is the equilibrium distribution of an ensemble of systems with a constant number of particles but varying energy. If both the particle number and the energy can vary, the relevant distribution is the so-called *grand canonical distribution*.[2]

GSM raises three questions about equilibrium. First, GSM introduces three special distributions and specifies which situations they describe. These distributions are immensely powerful and are used in many applications. But what makes them special? Why, say, choose the microcanonical distribution when describing an isolated system and not one of the infinitely many other stationary distributions over X? We address this issue in Section 28.2. Second, why does Gibbsian ensemble averaging work? GSM posits that what we observe are phase averages. In practice this is a successful method, but why do ensemble averages coincide with the values found in measurements performed on an actual physical system in equilibrium? There is no obvious connection between the two and if Gibbsian phase averaging is to be more than a black-box technique, we require an explanation of the connection between empirical values and ensemble averages. We address this question in Section 28.3. Third, a typical example of a non-equilibrium process is the following: a gas that is confined to the left half of a container starts expanding as soon as the confining wall is removed and eventually spreads evenly over the entire available space. The state of being spread evenly is the equilibrium state and the process of expansion culminating in that state is the approach to equilibrium. As formulated so far GSM is an equilibrium theory, which does not offer a description of the approach to equilibrium. As we shall see, there is in fact a serious obstacle to extending GSM to non-equilibrium processes within the theory itself. The question is how this obstacle can be overcome. We discuss various suggestions in Section 28.4.

28.2   Justifying Equilibrium Distributions

There is an arbitrary number of stationary distributions over X and a further criterion is needed to single out a particular distribution as the one that corresponds to the equilibrium of an ensemble in a certain situation. A first suggestion is to select equilibrium distributions on grounds of simplicity (Phillies, 2000, p. 112). But this criterion does not get off the ground as long as there is no hard and fast definition of simplicity, and even if a there was such a definition it would remain unclear why equilibrium should correspond to a simple distribution.

The *principle of indifference* says that as long as one has no reason to prefer one outcome over the other, one should not arbitrarily favor one outcome and instead assign equal probabilities to both. Tolman (1979 [1938], pp. 59–70) put this idea to use in SM when he introduced the principle that one should assign equal probabilities for a system to be in different parts of the state space when these parts have equal size (with respect to a relevant background measure on X). The microcanonical ensemble can then be seen as a straightforward result of the application of this principle. However, the principle of indifference suffers from a number of well-known problems (for a discussion see Salmon et al., 1992, pp. 74–77), and these problems also beset its application in GSM.

An important way to improve on the principle of indifference is an appeal to maximum entropy considerations. The Gibbs entropy of a distribution is

$$S_G[\rho] = -k \int_X \rho(x, t) \log[\rho(x, t)] dx, \qquad (28.2)$$

where k is the Boltzmann constant. The *maximum entropy principle* then says that the correct distribution for a certain situation is the one for which S_G is maximal with respect to the constraints imposed on the system in that situation (Kardar, 2007, p. 34). The last clause is essential because different constraints lead to different equilibrium distributions. If, for instance, one keeps both the energy and the particle number in the system fixed, one can show that S_G is maximal for the microcanonical distribution. The special status of certain distributions in GSM are then justified by pointing out that they are the ones that satisfy the maximum entropy principle.

This pushes the justificatory question one step back: why should we accept the maximum entropy principle? Jaynes (1983) argues that the probabilities in SM are subjective probabilities that represent our state of knowledge about the system (rather than any objective feature of the system itself). This allows him to appeal to Shannon's (1949) information theory to interpret entropy as a measure of the lack of information. A distribution with maximum entropy is then justified because it is maximally non-committal with respect to the available information. However, this justification of the maximum entropy principle remains controversial, both in GSM and in statistics more generally. Uffink (1995, 1996a) points out that the claim that a maximization procedure returns a unique distribution rests on unreasonably strong assumptions, and Howson and Urbach (2006, pp. 276–288) argue that a maximum entropy distribution is not non-committal.

A different route is taken by Malament and Zabell (1980), who aim to justify the microcanonical distribution by appeal to the dynamics of the system. A system is ergodic on X if and only if for all measurable functions on X and for almost all initial conditions $x \in X$ (that is, for all initial conditions except, perhaps, for ones that taken together form a set of measure zero) the infinite time average of the function is equal to the ensemble average.[3] Intuitively this means that if a set $A \in X$ has a measure of, say, 1/3, then almost all trajectories will spend 1/3 or their time in A in the long run. The significance of ergodicity in the current context lies in the fact that there is a uniqueness theorem for measures in ergodic systems. A measure λ on X is *absolutely continuous* with respect to the ergodic measure respect to a measure μ on X iff whenever μ assigns measure zero to a set then λ also assigns measure zero to that set. The uniqueness theorem says that if a system is ergodic with respect to the measure μ and λ is a measure that is (i) invariant under the dynamics of the system and (ii) absolutely continuous with respect to μ, then $\mu = \lambda$. Intuitively the theorem says that the ergodic measure is the only invariant measure in the class of all measures that are absolutely continuous with respect to the ergodic measure. Malament and Zabell postulate that any acceptable measure has to be absolutely continuous with respect to the ergodic measure. This is because if a subset of X is just a small displacement of another subset of X, then the probability of finding the micro-state of the system in one set should be close to that of finding it in the other, which, as they show, is equivalent to absolute continuity. This singles the microcanonical measure as the only correct measure.

This approach is premised on systems being ergodic, but many systems which are dealt with successfully by GSM fail to meet this condition. For instance, in a solid the molecules oscillate around fixed positions in a lattice and therefore can only access a small part of the state space (Uffink, 2007, p. 1017). Bricmont (2001) points out that neither the Kac Ring (Kac, 1959) nor a system of n uncoupled anharmonic oscillators of identical mass is ergodic and yet both exhibit equilibrium behavior. And, most notably, a system of non-interacting point particles is known not to be ergodic and many applications of SM are based on such systems (Uffink, 1996b, p. 381). This poses a problem because, as Earman and Rédei (1996, p. 71) point out, even if a system is just a "little bit" non-ergodic, then the uniqueness theorem fails completely, and with it Malament and Zabell's approach.

To circumvent this difficulty, Vranas (1998) suggested replacing ergodicity with what he calls ε-ergodicity. Intuitively, a system is ε-ergodic when it is ergodic on the vast majority of X, namely on a subset Y of X so that $\mu(Y) \geq 1 - \varepsilon$, where ε is very small real number or zero (and it is assumed that μ is normalized). Vranas points out that there is a middle ground between ergodicity holding and failing completely. He articulates a notion of two measures being ε close and proves a ε-version of the uniqueness theorem. This allows him to reformulate Malament and Zabell's approach for ε-ergodic systems. This is progress because the class of systems that are ε-ergodic is larger than the class of systems that is ergodic and it is plausible that it comprises many realistic systems such as gases and some liquids (Frigg and Werndl, 2011). However, there will be SM systems that are not ε-ergodic and it remains unclear how the current approach can deal with them. And, most notably, even if all these difficulties could be resolved, the approach only concerns the microcanonical distribution and it remains unclear how the canonical and the grandcanonical distributions can be justified.

Finally, Wallace (2001) argues that there is no reason to think that the classical definitions and derivations of equilibrium apply to quantum mechanical systems. Because of this, he argues, these definitions and derivations cannot be valid even in the classical regime. If they can be made to work for classical systems, then this must be an artifact of classical mechanics that has nothing to do with the actual world. Therefore, he argues, a new notion of quantum mechanical equilibrium is needed. Wallace does not articulate such a new notion, and it remains an open question how the new quantum mechanical equilibrium would look like. Finding a quantum mechanical justification of probabilities in SM is a worthwhile endeavor. Yet it remains unclear whether far-reaching conclusions about classical justification can be drawn. In particular, it is not clear why the fact that classical derivations of the equilibrium distributions do not work in quantum mechanics automatically implies that the justifications for the probability distributions in the classical regime also fail. Even if at the fundamental level the world is quantum, there can be different justifications for different levels of reality.

28.3   Why Does Gibbsian Ensemble Averaging Work?

Why does Gibbsian ensemble averaging work? That is, why do phase averages coincide with values measured in actual physical systems? Common wisdom justifies the use of phase averages by appeal to time averages.[4] GSM associates a physical quantity with a real-valued functions f on X. Performing a measurement takes time, and so a measurement device eventually registers a time average of f. Although making a measurement only takes a short time by human standards, the time is long compared to the time that a typical molecular process takes. For this reason, the measured value is approximately equal to the *infinite* time average of f. If one now assumes that the system is ergodic, time and ensemble averages are equal. This, so the argument concludes, shows that the ensemble averages one can calculate with GSM are the values obtained from measurements.

This argument faces a number of difficulties. First, from the fact that measurements take time it does not follow that the measurement outcome is a time average. Measurement devices could just as well record an instantaneous value at some point during the measurement. Second, even if it was the case that measurement outcomes reflected time averages, equating a *finite* time average with an *infinite* time average is problematic. Even if a measurement takes a long time compared to molecular processes, finite and infinite averages could be very different (Sklar, 1973, p. 211; Malament and Zabell, 1980, pp. 342–343). This gives cause for concern because ergodicity *only* holds for infinite time averages. But if infinite time averages are replaced by finite time averages one can no longer appeal to ergodicity to equate time and ensemble averages, and this pulls the rug from underneath the above argument. Third, as we have seen above, many systems which are dealt with successfully by GSM are not ergodic.

Khinchin (1960 [1949]) addresses the third problem by pointing out that full-fledged ergodicity is unnecessary. He sees the key to an understanding of GSM in a careful study of the class of functions that are relevant for SM. He submits that there are two restrictions. First, only systems with a large number of degrees of freedom are considered because SM studies systems that are made up of a large number of micro-constituents. Second, attention should be restricted to so-called *sum-functions*, i.e., functions that are a sum over one-particle functions. Restricting attention to Hamiltonian systems and assuming that the Hamiltonian is itself a sum function, Khinchin could prove that for every sum-function f there are constants k_1 and k_2 such that the measure of all states in X for which $|(f^*(x) - \langle f \rangle)/\langle f \rangle|$ is greater than $k_1 \, n^{-1/4}$ is smaller than $k_2 \, n^{-1/4}$, where $f^*(x)$ is the time average of f on a trajectory starting in x. This theorem is sometimes also referred to as "Khinchin's ergodic theorem."[5] Intuitively, it says that as n becomes larger, the measure of those regions of X where the time and the space average differ by more than a small amount tends toward zero.

This approach faces a number of challenges. First, like the original ergodic approach it associates measurement outcomes with infinite time averages and is therefore vulnerable to the same objections. Second, the motivation for focusing attention on sum-functions is that this is the relevant class of functions for SM. Batterman (1998, p. 191) points out that this is too narrow as there are functions of interest that do not have this form, in particular the Hamiltonian of systems in which particles interact with each other. Third, and most importantly, there is what Khinchin himself called the "methodological paradox" (Khinchin, 1960 [1949], pp. 41–43). The proof of Khinchin's theorem assumes that the Hamiltonian is a sum function, which is tantamount to assuming that the components of the system do not interact. However, non-interacting systems often do not reach equilibrium. If, for instance, the particles of a gas do not collide, the gas does not converge toward an equilibrium state. So Khinchin's theorem seems to describe the equilibrium of a class of systems that don't reach equilibrium! Khinchin suggested avoiding the problem by assuming that there are only short range interactions between the molecules. These are sufficient to bring the system into equilibrium and at the same time have no significant effect on averages.

This response seems *ad hoc* and a better solution is needed. This was the starting point for a research program now known as the *thermodynamic limit* which aims to establish "Khinchin-like" results for Hamiltonians *with* interaction terms. Results of this kind can be proven in the limit for $n \to \infty$, if also the volume V of the system tends toward infinity in such a way that the number-density n/V remains constant. Classic statements are Ruelle (1969, 2004); surveys and further references can be found in Compagner (1989), Luczak (2016) and Uffink (2007, pp. 1020–1028).

Malement and Zabell's approach (which we have introduced in Section 28.2) rejects the idea that we observe time averages and posits that if a system is in state $x \in X$ at time t and a measurement is performed at t, then the observed value is $f(x)$. We nevertheless observe phase averages if the function f exhibits small dispersion with respect to the microcanonical measure, meaning that the set of points in X where the value of f significantly differs from $\langle f \rangle$ is vanishingly small. If this is the case, then, at any given time, it is overwhelmingly likely that the micro-state of the system is one for which the value of f coincides with $\langle f \rangle$, which explains why Gibbsian ensemble averaging works. To justify that functions meet this criterion, Malement and Zabell refer to Khinchin's research program. However, as we have seen, there are questions about Khinchin's assumptions. Furthermore, it is not the case that for all observables f relevant in practice, f exhibits small dispersion with respect to the microcanonical measure (cf. Werndl and Frigg, 2017, 2020). Finally, as noted above, Malement and Zabell's approach addresses only the microcanonical ensemble and even if all internal issues could be resolved, there would still be a question about the justification of the use of phase averages in the context of the canonical and the grandcanonical distributions.

Frigg and Werndl (2019b, 2020) suggest that the solution to the problem lies in a reassessment of the status of GSM. The tacit assumption shared by all accounts discussed so far is that GSM is a true fundamental theory. Frigg and Werndl argue that this not so. The true fundamental theory is Boltzmannian SM, and GSM is an effective theory in that it offers a set of principles that make the calculation of outcomes possible in cases where the fundamental theory is too difficult to handle. Like all effective theories, GSM buys computational tractability at the cost of incompleteness, and for this reason it has a limited range of applicability. The crucial question therefore is under what circumstances Gibbsian SM yields correct results and when its procedures fail to deliver. Werndl and Frigg (2017, 2020) investigate the conditions under which Gibbsian phase averages coincide with Boltzmannian equilibrium values. They point out that these values need not coincide and identify three sufficient conditions under which they do agree. The first is that the relevant phase function has a small dispersion, i.e., assumes (almost) the same value almost everywhere; the second is that the relevant observable is the sum of one component observable defined on partition of the one-particle space where all values of the observable have the same probability; the third is that the state space is divided up in such a way that next to the largest macro-region (which corresponds to the Boltzmannian equilibrium) there are always two macro-states of equal size whose average equals the Boltzmannian equilibrium value.

28.4   Gibbsian Non-Equilibrium Theory

A gas that is confined to the left half of a container starts expanding when the confining wall is removed and eventually fills the available space evenly. This is a standard example of a system approaching equilibrium when initially prepared in a non-equilibrium state. In GSM the distribution $\rho(x, t)$ specifies the state of the Gibbsian ensemble and so one would expect that the approach to equilibrium is described in terms of changes in $\rho(x, t)$. The idea would be that there would be a distribution associated with the gas confined to the left half, which then evolves toward the stationary equilibrium distribution of maximum entropy for the entire available space.

Large parts of GSM are carried out against the background of Hamiltonian mechanics (HM), meaning that the time evolution of the system is assumed to be governed by Hamilton's equation of motion. Unfortunately, HM does not sit well with the expectations described in the last paragraph. First, the Gibbs entropy turns out to be a constant of motion (Zeh, 2001, pp. 48–49), meaning that it does not change over time. This precludes a characterization of the approach to equilibrium in terms of increasing entropy. Second, the Hamiltonian equations of motion preclude the evolution from a non-stationary to a stationary distribution. If, at some point in time, the distribution is non-stationary, then it will remain non-stationary for all times and, conversely, if it is stationary at some time, then it must have been stationary all along (van Lith, 2001a, pp. 591–592). Hence, either the system has always been in equilibrium, or else it will never be. This is serious problem for characterization of equilibrium in terms of stationary distributions because it implies that systems never approach equilibrium.

This has sparked a number of attempts to remedy the situation. A first suggestion is based on the method of *coarse-graining*. The method was introduced by Gibbs (1981 [1902], Ch. 12) and has since gained prominence in the literature on GSM (Penrose, 1970). The idea is to put a grid on the state space X consisting of cells of finite size (hence the expression "coarse-graining"), and then define the coarse-grained density $\bar{\rho}$ as the density that is uniform within each cell and takes the average of ρ in the cell as its value. One can then plug $\bar{\rho}$ into Equation 28.2, which yields the coarse-grained entropy $\bar{S}_G$. The coarse-grained density is not subject to the same restrictions as the original density and so it is, at least in principle, possible for $\bar{\rho}$ to evolve in such a way that it will converge to the maximum entropy distribution that does not change any more once it has been reached. This state is the *coarse-grained equilibrium* (Ridderbos, 2002, p. 69). The approach to the coarse-grained equilibrium is accompanied by an increase in the coarse-grained entropy, which, unlike the original Gibbs entropy, is not a constant of motion. The question then is under what circumstances $\bar{\rho}$ tends toward a coarse-grained equilibrium. The standard answer is that the dynamics of the system has to be *mixing*. Intuitively, a system is mixing if, in the long run, every subset A of X gets evenly scattered over X.[6] In sum, if one replaces ρ and S_G by $\bar{\rho}$ and $\bar{S}_G$, and assumes the dynamics is mixing, then the system reaches a coarse-grained equilibrium.

This program faces a number of challenges. First, mixing is a stringent requirement, which many systems fail to meet. In fact, mixing implies ergodicity and so if a system is not ergodic, it is not mixing either. We have seen above that ergodicity is not easy to come by. Second, the thrust of the argument in favor of coarse-graining is that one cannot empirically distinguish between ρ and $\bar{\rho}$ and so one is at liberty to use $\bar{\rho}$ rather than ρ if this resolves theoretical issues in the theory. Ridderbos and Redhead (1998) argue that this was too hasty a conclusion because there are in fact experimental contexts in which it is possible to empirically distinguish between ρ or $\bar{\rho}$. One such context is the so-called spin–echo experiment (Hahn, 1950), which studies how spins process in a magnetic field.[7] In the experiment a spin system is prepared in a non-equilibrium state, and this state turns out to be such that the system evolves into a coarse-grained equilibrium. If the system is then exposed to a radio pulse, the spins end up returning to their initial state, thereby moving out of the coarse-grained equilibrium. This, so the argument goes, shows that one can empirically distinguish between a real equilibrium and coarse-grained equilibrium, which invalidates the rationale for replacing ρ by $\bar{\rho}$. This criticism has not gone unchallenged. Ainsworth (2005) defends coarse-graining on the grounds that empirical equivalence is not necessary, and Lavis (2004) argues that the behavior of $\bar{\rho}$ and the coarse-grained entropy predicted by Ridderbos and Redhead is an artifact of the way in which they calculated these quantities.

Van Lith proposes to change the definition of equilibrium, but in a different way than the coarse-graining approach. She points out that the stationarity of ρ is actually too strong a requirement because all that is needed from a thermodynamic point of view is that the distribution be such that the phase average of every function in a physically relevant set of functions only fluctuates mildly around its average value (van Lith, 1999, p. 114). A distribution can meet this requirement even if it is not stationary. As in the coarse-graining approach the question arises under what circumstances averages behave in the required way, and the answer is also that mixing is a sufficient condition. But as already noted, mixing is a stringent condition. While mixing is not needed for van Lith's approach to equilibrium to take place (because her equilibrium condition is weaker), she does not propose an alternative condition that is sufficient for the approach to equilibrium. Furthermore, as van Lith herself points out (van Lith, 1999, p. 115), the proposal does not contain a recipe to get the Gibbs entropy moving and hence an approach to equilibrium is not accompanied by a corresponding entropy increase.

Interventionism takes the opposite route: it leaves the definitions of equilibrium and entropy intact and changes the dynamics of the system. So far we assumed that the systems under consideration are isolated. Blatt (1959) suggests that this is the source of the problem. Interventionism renounces this idea and submits that taking outside influences on the system into account offers a solution to the problem. The perturbations from outside the system in fact serve as a kind of "stirring mechanism" that drives the system into equilibrium. These outside influences are describable only in statistical terms and hence the system is no longer governed by the deterministic Hamiltonian equations. This removes the obstacle to the distribution converging to an unchanging maximum entropy distribution, and the entropy can increase. Blatt (1959) and Ridderbos and Redhead (1998) look at models of outside perturbations and report that under realistic assumptions systems indeed do approach equilibrium.

An obvious objection is that we are always free to consider a larger system consisting of our "original" system *and* its environment (for instance, we can consider a "gas plus container" system). Because Hamilton's equation are universal, they will govern that system, and so we are where we started. The interventionist can then reply that we can consider the environment of the environment (for instance, the laboratory in which the container is located) and consider outside influences, and the critic can make the same move as before. This leads to a regress that only ends when we consider the entire universe as one big system. Whether the interventionist really has to throw in the towel at this point depends on the status one assigns to the mechanical background theory, here HM. For those who see this HM as a truly universal theory this is the end of interventionism. But this understanding of mechanics is not uncontroversial. If, like Cartwright (1999), one thinks that we cannot legitimately claim that laws apply universally (and extends this skeptical argument to mechanics), then the argument against interventionism does not get off the ground.

Another approach that changes the dynamics of the system is the *stochastic dynamics* program, which replaces the (deterministic) Hamiltonian equations of motion by stochastic equations. Typically this is done by coarse-graining the phase space and then postulating a probabilistic law describing the transition from one cell of the partition to another cell. This removes the restrictions of HM and makes room for the distribution to spread and the entropy to increase. Classical statements of this approach are Mackey (1992), Penrose (1970), and Streater (1995). The main question facing this account is how the stochastic equations can be justified. Its probabilistic equations are often introduced as independent postulates on a case-by-case basis, and it is unclear whether they follow from a underlying fundamental law of motion.[8]

Brussels School (sometimes also "Brussels-Austin School") bases its analysis on the view that the notion of a precise state of a system (represented by a point x in X) is meaningless in a system with sensitive dependence on initial conditions. For this reason, deterministic laws specifying the time evolution of points should be replaced by an explicitly probabilistic description of the system in terms of open regions of the phase space. This program, if successful, can be seen as providing the sought after justification for the stochastic dynamics program mentioned in the last paragraph.[9]

Jaynes' approach, which we encountered in Section 28.2, also offers a response to the issues with non-equilibrium processes. Since ρ is a codification of a scientist's knowledge rather than a feature of the system itself, it is not bound by mechanical equations of motion. This can be exploited to change the distribution so that the entropy increases and the distribution spreads. This is done by iteratively updating the distribution every time an observation is made, ensuring that the distribution satisfies the maximum entropy principle after every observation (see Sklar (1993, pp. 255–257) for a description of the details of this process). However, Lavis and Milligan (1985) point out that this procedure does not necessarily lead to a *monotonic* entropy increase, as conformity with thermodynamics would require. Furthermore, if ρ is a reflection of our knowledge, then changes in ρ are reflections on changes in our knowledge. This, so a number of critics have argued, has the consequence that the approach to equilibrium depends on what we know. But, so the argument goes, this is absurd because the boiling of kettles or the spreading of gases is a consequence of how molecules behave and not of what we know about them (Redhead, 1995, pp. 27–28; Goldstein, 2001, p. 48).

28.5 Conclusion

We have introduced the Gibbsian notion of equilibrium and reviewed a number of issues that arise in its justification and use. To conclude we would like to draw attention to two further issues. The first issue is the interpretation of the probabilities in GSM. Gibbs himself seems to have thought about probabilities in a frequentist way, which is problematic for a number of reasons (Frigg, 2008, pp. 153–154). An alternative interpretation are time averages, but these do not fare better (von Plato, 1981; Dougherty, 1993). Bayesian probabilities face the objections mentioned in Section 28.2. So the interpretation of GSM probabilities remains a challenge. Second, the relation between GSM and Boltzmannian SM is poorly understood. Even though both frameworks are invoked in practical applications as well as in foundational discussions, it remains unclear how they, and in particular their notions of equilibrium, relate to each other. But having two incompatible notions of equilibrium at work in SM is unsatisfactory and a reflection on how the two approaches relate is necessary. Steps toward a better understanding of the relationship between the two frameworks are made in Lavis (2005) and Werndl and Frigg (2020), but the problem deserves further attention.

Notes

1 For a discussion of the Gibbsian formalism and its interpretations see Frigg and Werndl (2019a).
2 For formal definitions of these distributions see, for instance, Lavis and Bell (1999).
3 See Berkovitz et al. (2011) for a non-technical introduction to ergodic theory.
4 This view is discussed but not endorsed, for instance, in Bricmont (1996, pp. 145–146), Earman and Rédei (1996, pp. 67–69), Malament and Zabell (1980, p. 342), and van Lith (2001a, pp. 581–583).
5 For a discussion of the proof see Badino (2006), Batterman (1998, pp. 190–198) and van Lith (2001b, pp. 83–90).

6 For a non-technical introduction to mixing see Berkovitz et al. (2011).
7 For detailed discussions of the spin-echo experiment see Frigg (2008, pp. 159–164) and Sklar (1993, pp. 219–222).
8 For discussions of the stochastic dynamics program see Callender (1999, pp. 358–364), Sklar (1993, Chs. 6 and 7), and Uffink (2007, pp. 1038–1063).
9 For presentations and critical discussions of the ideas of the Brussels School see Batterman (1991), Bishop (2004), and Karakostas (1996).

References

Ainsworth, P. (2005). The spin-echo experiment and statistical mechanics. *Foundations of Physics Letters*, 18: 621–635.
Badino, M. (2006). The foundational role of ergodic theory. *Foundations of Science*, 11: 323–347.
Batterman, R.W. (1991). Randomness and probability in dynamical theories: On the proposals of the Prigogine School. *Philosophy of Science*, 58: 241–263.
Batterman, R.W. (1998). Why equilibrium statistical mechanics works: Universality and the renormalization group. *Philosophy of Science*, 65: 183–208.
Berkovitz, J. et al. (2016). The ergodic hierarchy. In E.N. Zalta (ed), *Stanford Encyclopedia of Philosophy* (Summer 2016 Edition). Available at: https://plato.stanford.edu/archives/sum2016/entries/ergodic-hierarchy/.
Bishop, R. (2004). Nonequilibrium statistical mechanics Brussels-Austin style. *Studies in History and Philosophy of Modern Physics*, 35: 1–30.
Blatt, J. (1959). An alternative approach to the ergodic problem. *Progress in Theoretical Physics*, 22: 745–755.
Bricmont, J. (1996). Science of chaos or chaos in science? In P.R. Gross, N. Levitt and M.W. Lewis (eds.), *The Flight from Science and Reason* (Annals of the New York Academy of Sciences, Vol. 775). New York: The New York Academy of Sciences, pp. 131–175.
Bricmont, J. (2001). Bayes, Boltzmann, and Bohm: Probabilities in physics. In J. Bricmont, G. Ghirardi, D. Dürr, F. Petruccione, M.C. Galavotti and N. Zanghì (eds.), *Foundations and Perspectives* (Lecture Notes in Physics, Vol. 574). Berlin and Heidelberg: Springer, pp. 4–21.
Callender, C. (1999). Reducing thermodynamics to statistical mechanics: The case of entropy. *Journal of Philosophy*, 96: 348–373.
Cartwright, N. (1999). *The Dappled World. A Study of the Boundaries of Science*. Cambridge: Cambridge University Press.
Compagner, A. (1989). Thermodynamics as the continuum limit of statistical mechanics. *American Journal of Physics*, 57: 106–117.
Dougherty, J. P. (1993). Explaining statistical mechanics. *Studies in History and Philosophy of Science*, 24: 843–866.
Earman, J. and Rédei, M. (1996). Why ergodic theory does not explain the success of equilibrium statistical mechanics. *British Journal for the Philosophy of Science*, 47: 63–78.
Frigg, R. (2008). A field guide to recent work on the foundations of statistical mechanics. In D. Rickles (ed.), *The Ashgate Companion to Contemporary Philosophy of Physics*. London: Ashgate, pp. 99–196.
Frigg, R. and Werndl, C. (2011). Explaining thermodynamic-like behaviour in terms of Epsilon-ergodicity. *Philosophy of Science*, 78: 628–652.
Frigg, R. and Werndl, C. (2019a). Can somebody please say what Gibbsian statistical mechanics says? *The British Journal for the Philosophy of Science*. Online first, doi: 10.1093/bjps/axy057.
Frigg, R. and Werndl, C. (2019b). Statistical mechanics: A tale of two theories. *The Monist*, 102: 424–438.
Frigg, R. and Werndl, C. (2020). Taming abundance: On the relation between Boltzmannian and Gibbsian statistical mechanics. In V. Allori (ed.), *Statistical Mechanics and Scientific Explanation: Determinism, Indeterminism and Laws of Nature*. Singapore: World Scientific, pp. 617–646.
Gibbs, J.W. (1981 [1902]). *Elementary Principles in Statistical Mechanics*. Woodbridge: Ox Bow Press.
Goldstein, S. (2001). Boltzmann's approach to statistical mechanics. In J. Bricmont, D. Durr, M.C. Galavotti, G. Ghirardi, F. Petruccione and N. Zanghi (eds.), *Chance in Physics: Foundations and Perspectives* (Lecture Notes in Physics, Volume 574). Berlin, Heidelberg and New York: Springer, pp. 39–54.
Hahn, E.L. (1950). Spin echoes. *Physics Review*, 80: 580–594.
Howson, C. and Urbach, P. (2006). *Scientific Reasoning. The Bayesian Approach*. Chicago and La Salle, IL: Open Court.
Jaynes, E.T. (1983). *Papers on Probability, Statistics, and Statistical Physics*, ed. by R.D. Rosenkrantz. Dordrecht: Reidel.
Kac, M. (1959). *Probability and Related Topics in Physical Science*. New York: Interscience Publishing.

Karakostas, V. (1996). On the Brussels Schools arrow of time in quantum theory. *Philosophy of Science*, 63: 374–400.

Kardar, M. (2007). *Statistical Physics of Particles*. Cambridge: Cambridge University Press.

Khinchin, A.I. (1960 [1949]). *Mathematical Foundations of Statistical Mechanics*. Mineola, NY: Dover Publications.

Lavis, D. (2004). The spin-echo system reconsidered. *Foundations of Physics*, 34: 669–688.

Lavis, D. (2005). Boltzmann and Gibbs: An attempted reconciliation. *Studies in History and Philosophy of Modern Physics*, 36: 245–273.

Lavis, D. and Bell, G. (1999). *Statistical Mechanics of Lattice Systems: Volume 1: Closed-Form and Exact Solutions*. Berlin: Springer.

Lavis, D. and Milligan, P. (1985). Essay review of Jaynes collected papers. *British Journal for the Philosophy of Science*, 36: 193–210.

Luczak J. (2016). On how to approach the approach to equilibrium. *Philosophy of Science*, 83: 393–411.

Mackey, M.C. (1992). *Times Arrow: The Origins of Thermodynamic Behaviour*. Berlin: Springer.

Malament, D. and Zabell, S.L. (1980). Why Gibbs phase averages work. *Philosophy of Science*, 47: 339–349.

Penrose, O. (1970). *Foundations of Statistical Mechanics*. Oxford: Oxford University Press.

Phillies, G.D.J. (2000). *Elementary Lectures in Statistical Mechanics*. Berlin: Springer.

Redhead, M.L.G. (1995). *From Physics to Metaphysics*. Cambridge: Cambridge University Press.

Ridderbos, K. (2002). The coarse-graining approach to statistical mechanics: How blissful is our ignorance? *Studies in History and Philosophy of Modern Physics*, 33: 6–77.

Ridderbos, T.M. and Redhead, M.L.G. (1998). The Spin-echo experiments and the second law of thermodynamics. *Foundations of Physics*, 28: 1237–1270.

Ruelle, D. (1969). *Statistical Mechanics: Rigorous Results*. New York: Benjamin.

Ruelle, D. (2004). *Thermodynamic Formalism: The Mathematical Structure of Equilibrium Statistical Mechanics*. Cambridge: Cambridge University Press.

Salmon, M. et al. (1992). *Introduction to the Philosophy of Science*. Indianapolis and Cambridge: Hackett.

Shannon, C.E. (1949). The mathematical theory of communication. In C.E. Shannon and W. Warren (eds.), *The Mathematical Theory of Communication*. Urbana, Chicago and London: University of Illinois Press, pp. 29-125.

Sklar, L. (1973). Statistical explanation and ergodic theory. *Philosophy of Science*, 40: 194–212.

Sklar, L. (1993). *Physics and Chance. Philosophical Issues in the Foundations of Statistical Mechanics*. Cambridge: Cambridge University Press.

Streater, R.F. (1995). *Stochastic Dynamics: A Stochastic Approach to Nonequilibrium Thermodynamics*. London: Imperial College Press.

Tolman, R.C. (1979 [1938]). *The Principles of Statistical Mechanics*. Mineola, NY: Dover.

Uffink, J. (1995). Can the maximum entropy principle be explained as a consistency requirement? *Studies in History and Philosophy of Modern Physics*, 26: 223–261.

Uffink, J. (1996a). The constraint rule of the maximum entropy principle. *Studies in History and Philosophy of Modern Physics*, 27: 47–79.

Uffink, J. (1996b). Nought but molecules in motion (Review essay of Lawrence Sklar: *Physics and Chance*). *Studies in History and Philosophy of Modern Physics*, 27: 373–387.

Uffink, J. (2007). Compendium of the foundations of classical statistical physics. In J. Butterfield and J. Earman (eds.), *Philosophy of Physics*. Amsterdam: North Holland, pp. 923–1047.

van Lith, J. (1999). Reconsidering the concept of equilibrium in classical statistical mechanics. *Philosophy of Science*, 66(Supplement): 107–118.

van Lith, J. (2001a). Ergodic theory, interpretations of probability and the foundations of statistical mechanics. *Studies in History and Philosophy of Modern Physics*, 37: 581–594.

van Lith, J. (2001b). *Stir in sillness: A study in the foundations of equilibrium statistical mechanics*. PhD Thesis, University of Utrecht. Available at: http://www.library.uu.nl/digiarchief/dip/diss/1957294/inhoud.htm.

von Plato, J. (1981). Reductive relations in interpretations of probability. *Synthese*, 48: 61–75.

Vranas, P.B.M. (1998). Epsilon-ergodicity and the success of equilibrium statistical mechanics. *Philosophy of Science*, 65: 688–708.

Wallace, D. (2001). Implications of quantum theory in the foundations of statistical mechanics. Manuscript, available in the PhilSci Archive at http://philsci-archive.pitt.edu/410/1/wallace.pdf.

Werndl, C. and Frigg, R. (2017). Mind the gap: Boltzmannian vs Gibbsian equilibrium. *Philosophy of Science*, 84(5): 1289–1302.

Werndl, C. and Frigg, R. (2020). When do Gibbsian phase averages and Boltzmannian equilibrium values agree? *Studies in History and Philosophy of Modern Physics*. Online first, https://doi.org/10.1016/j.shpsb.2020.05.003.

Zeh, H.D. (2001). *The Physical Basis of the Direction of Time*. Berlin and New York: Springer.

Further Reading from the Editors

An influential philosophical critique of the Gibbsian approach to statistical mechanics is C. Callender, "Reducing thermodynamics to statistical mechanics: The case of entropy", *Journal of Philosophy* 96 (1999): 348–373. Defences of the Gibbsian approach are given by D. Wallace, "The Necessity of Gibbsian Statistical Mechanics", (forthcoming), by D. Wallace, "The quantitative content of statistical mechanics", *Studies in History and Philosophy of Science Part B: Studies in History and Philosophy of Modern Physics* 52 (2015): 285–293, and by K. Robertson, "In search of the holy grail: how to reduce the second law of thermodynamics", *The British Journal for the Philosophy of Science* (forthcoming). See also K. Ridderbos "The coarse-graining approach to statistical mechanics: How blissful is our ignorance?", *Studies in the History and Philosophy of Modern Physics* 33 (2002): 65–77.

29

QUANTUM FOUNDATIONS OF STATISTICAL MECHANICS AND THERMODYNAMICS

Orly Shenker

29.1   Introduction

Statistical mechanics is often taken to be the paradigm of a successful inter-theoretic reduction, which explains the high-level phenomena (primarily those described by thermodynamics) by using the fundamental theories of physics together with some auxiliary hypotheses. In my view, the scope of statistical mechanics is wider since it is the type-identity physicalist account of all the special sciences. But in this chapter, I focus on the more traditional and less controversial domain of this theory, namely, that of explaining the thermodynamic phenomena (see Shenker, 2017c).

What are the fundamental theories that are taken to explain the thermodynamic phenomena? The lively research into the foundations of *classical* statistical mechanics suggests that using classical mechanics to explain the thermodynamic phenomena is fruitful (see overviews in Sklar, 1993; Uffink, 2007; Frigg, 2008; Hemmo and Shenker, 2012; Shenker, 2017a, 2017b; and references in all of them). Strictly speaking, in contemporary physics, classical mechanics is considered to be false. Since classical mechanics preserves certain explanatory and predictive aspects of the true fundamental theories, it can be successfully applied in certain cases (see Wallace, 2001; Ladyman and Ross, 2007). In other circumstances, classical mechanics has to be replaced by quantum mechanics (Emch, 2007 provides a comprehensive overview of quantum statistical mechanics). In this chapter I ask the following two questions:

I   How does *quantum* statistical mechanics differ from *classical* statistical mechanics? How are the well-known differences between the two fundamental theories reflected in the statistical mechanical account of high-level phenomena?

II   How does quantum *statistical* mechanics differ from quantum mechanics *simpliciter*?

To make our main points I need to only consider non-relativistic quantum mechanics (see Wallace, 2001, Section 1 on this point). Most of the ideas described and addressed in this chapter hold irrespective of the choice of a (so-called) interpretation of quantum mechanics, and so I will mention interpretations only when the differences between them are important to the matter discussed.

29.2   Quantum Mechanical Microstates

The main idea of classical statistical mechanics, for which a quantum mechanical counterpart is sought, is this. This idea should be understood in the most general terms and is shared by the

Boltzmannian and the Gibbsian frameworks, discussed in Section 29.3. According to the *ontology* of classical mechanics, every system is, at every moment, in a well-defined mechanical state (called *microstate* in statistical mechanics), and this state evolves according to the laws of mechanics. As used here, the term microstate has nothing to do with being small or being part of a whole; as an instructive example think of the microstate of the universe. Others sometimes use the term differently; for example, Goldstein et al. (2016) use "microscopic observable" as pertaining to a small subsystem.

When we prepare or observe a system we only have *epistemic* access to a *partial description* (or a *coarse-grained* description) of its microstate. This partial description pertains to some *aspects* of the system's microstate (*macrovariable* is a term often used to refer to such an aspect). Examples for such aspects are average kinetic energy of the particles, the energy distribution among the particles, or the time average of these quantities: all give only partial information concerning the system's actual microstates. The great achievement of statistical mechanics is the discovery that such partial knowledge, about an aspect of the prepared microstate of a system, is sufficient to make predictions concerning further aspects of future microstates, and that the thermodynamic properties correspond to those aspects and the thermodynamic regularities correspond to regularities governing those aspects. Of course, the evolution from one microstate to the next (and therefore from one aspect to the next) is governed by the dynamical laws pertaining to the particles, but the discovery is that we don't have to follow all the details of this microevolution to make useful predictions concerning macroscopic phenomena. Instead, the following ideas are used. In general, any given aspect of the *actual* microstate (in which a system is prepared, according to the ontology) is shared by many *counterfactual* microstates, which belong to the same equivalence set relative to that aspect (this set is usually called *macrostate*). Since the only thing we know about the actual microstate is that it belongs to some such macrostates, we remain ignorant about which of the microstates in that set is the actual one. Since the microstates in a given macrostate (that *share* a given aspect) *differ* in *other* aspects, their future evolutions may vary from each other quite significantly. And since we do not know which microstate in this set is the actual one, we cannot be sure how the actual microstate will evolve. To express this ignorance notions such as probability or typicality come into play.

(This main idea of statistical mechanics has sometimes been *mistakenly* understood as suggesting that we should be able to observe any aspect of interest and predict its evolution (e.g., the evolution of the aspect of the world's microstate that corresponds to the behavior of the stock market). Sadly, we can only observe *certain specific* aspects of the microstates in our environment, namely, those to which our sense organs and measuring devices are physically sensitive, that is, with which they become entangled in quantum mechanics due to the interaction Hamiltonian. The thermodynamic magnitudes are some of those aspects.)

Since the terminology in the literature is not uniform I shall use the term *macrovariable* to denote an *aspect* of a microstate, given by its *partial description*, and the term *macrostate* to denote a *set* of microstates that share the same macrovariable. (See discussions of the nature of macrovariables in, e.g., Lebowitz, 1993; Callender, 1999; Albert, 2000; Goldstein and Lebowitz, 2004; Earman, 2006; Frigg, 2008; Wallace, 2011.[1])

To find the quantum mechanical counterpart of this idea we first need to learn what can be the *quantum mechanical microstate*. An option that first comes to mind (but which is problematic as we shall see) is that the microstate of a system is its quantum state: a system is prepared in some definite quantum state, but since many quantum states are compatible with this preparation, we remain ignorant as to which of them is the actual one. (In hidden variables theories the microstate may be different.) The ignorance about the quantum microstate should be quantified, as in the classical case, by a probability distribution. In making predictions to these ignorance probabilities, one adds the Born rule concerning the probabilities for measurement outcomes in each of the possible quantum states. The two kinds of probabilities are in play in predicting measurement outcomes, and their combination is described by a *density matrix*. This case is called (following d'Espagnat, 1976) a *proper mixture*.

Two problems arise in this picture. The first is that typically thermodynamic systems are entangled with their environments, and therefore do not have separate definite quantum states that we could identify as their microstates. Consequently, although the reduced state of a subsystem, formally obtained by tracing out the environment, is represented by a density matrix, this density matrix *cannot* be understood as expressing ignorance. For this reason, this case is called an *improper mixture* (following d'Espagnat, 1976). One may argue that, since entanglement is a rule rather than an exception, pure states should not be understood as microstates in statistical mechanics (see Wallace, 2001; Linden et al., 2009). But discussing pure states is nevertheless useful for two reasons. First, the universe is arguably in an (unknown) pure state, from which the states of affairs with respect to subsystems are derived (see the considerations in Goldstein et al., 2016, Section 2). Second, in the prevalent case of decoherence interactions with the environment, the observations of a subsystem of interest may resemble, for practical purposes, those of a system in an unknown pure state. (In decoherence models the environment is described in probabilistic terms, and those probabilities arguably describe ignorance. I do not pursue this point here.)[2] Therefore in this chapter, I discuss both pure states and reduced states. I do not address POVMs (positive operator-valued measures), since it is hard to understand them realistically.[3]

The second problem in the above picture arises when we ask ourselves what it would mean to prepare a quantum state that has a certain macrovariable. Suppose that we prepare a collection of systems by measuring on each of them the observable M, of which the eigenvalues are M_0, M_1, etc., and collect only those with eigenvalue M_0 (with the corresponding eigenstate or eigensubspace). Unlike the classical case (and except in the special case described below) the result is that all the systems are prepared in exactly the same quantum state. (The states may differ in hidden variables, but I do not discuss those at the moment; see Peres (1993) on the notion of quantum mechanical preparation.) The ensuing evolutions of different systems prepared with the same quantum state will, to be sure, be probabilistic in the sense of the Born rule, but this is a result of quantum mechanics *simpliciter*, not of quantum *statistical* mechanics. (Wallace, 2013 also argues that there is no justification for the idea that quantum statistical mechanics involves putting probability distributions over quantum states just as classical statistical mechanics involves putting probability distributions over classical states.)

A way to prepare a system with a certain eigenvalue and nevertheless remain ignorant as to its actual quantum state (thus having quantum mechanical counterpart of the classical notion of macrostate) is to measure a degenerate observable in special circumstances as follows. In the general case, upon measurement of the observable A on some quantum state $|\psi>$, if the outcome is eigenvalue a_n with a degree of degeneracy g_n, then according to the projection postulate, the final state is a (normalized) superposition of all the g_n eigenvectors of A associated with a_n. But if the initial quantum state $|\psi>$ is itself one of the eigenvectors associated with a_n then the final state will remain unchanged and will not become a superposition of all the eigenvectors of a_n. This has the following consequence. Suppose that before A is measured, a non-degenerate observable B is measured non-selectively, and then a suitable Hamiltonian is applied on the resulting proper mixture, such that each of the possible eigenstates b_n of B evolves to one of the eigenstates of the degenerate eigenvalue a_n of A. In this case, the state of affairs after the measurement of A can be described in terms of a *proper mixture* of the eigenstates associated with a_n, in which each of the eigenstates associated with a_n can be treated as a microstate, the share eigenvalue a_n can be treated as a shared macrovariable, and the set of these eigenstates can therefore form a macrostate – much like in a classical preparation of a macrostate is the core idea of classical statistical mechanics.

Some interpretations of quantum mechanics leave room for ignorance in other ways. In the GRW theory (or I should say, a family of theories, see Ghirardi, 2016) the observer may be ignorant of the actual state of affairs brought about by unknown spontaneous localization of the wavefunction. In hidden variables theories the observer is ignorant of the hidden variable, even when the quantum state is known.

29.3   Quantum Mechanical Macrostates and Synchronic Functional Relations

The central idea of statistical mechanics (described above) is that in order to account for the thermodynamic regularities only partial information about the microstates is needed, concerning some *macrovariables* of them, that is, concerning the fact that they belong to certain *macrostates* in which all the microstates share the same macrovariables (but differ otherwise). A famous example of a *classical* macrovariable is the *Maxwell-Boltzmann energy distribution* (see Frigg, 2008): the very partial information about the microstate of an ideal gas, which says that the energy is distributed among the particles of the gas in the Maxwell-Boltzmann way, yields important explanations and predictions about the gas' behavior (e.g., that it will obey the *ideal gas law*; see Uffink, 2007). In certain conditions, however, (of high density or low temperature,) the classical approximation of Maxwell-Boltzmann energy distribution no longer captures the relevant aspects of the gas and therefore yields wrong predictions. In these circumstances, the quantum mechanical distinction between bosons and fermions becomes significant, and the so-called *Bose-Einstein* and *Fermi-Dirac statistics* must be applied to describe the distribution of particles between energy levels (see Emch, 2007, Sections 2.4–2.5).[4] It is natural to *prima facie* treat these statistics as analogous to the classical Maxwell–Boltzmann distribution, and to think of them as quantum mechanical macrovariables, that is, as aspects (given by partial descriptions) of the quantum state. However, as we saw in the previous section, the distinct non–classical features of quantum mechanics entail that the classical notion of macrovariables does not have a straightforward counterpart in quantum mechanics and that the idea that one can remain partly ignorant concerning the microstate even after a measurement has been carried out is applicable in the quantum domain only in special cases. For this reason, we can either focus on these special cases or opt for a completely different conceptual framework (e.g., that there is no quantum statistical mechanics above and beyond quantum mechanics simpliciter; see Section 29.6 and Wallace, 2013). I now turn to describe various approaches to quantum statistical mechanics that are offered in the literature.

To describe the different approaches to quantum statistical mechanics it is useful to distinguish between two kinds of regularities that are addressed by thermodynamics: One kind involves *synchronic functional relations*, especially those that hold in equilibrium, such as the ideal gas law, and the other involves *diachronic relations*, especially the approach to equilibrium and the second law of thermodynamics (for the distinction between the two latter ones see Brown and Uffink, 2001). In this section, I focus on synchronic relations and discuss the diachronic ones in the next section.

In *classical* statistical mechanics there are two main theories concerning synchronic functional relations, usually referred to as stemming from the works of Boltzmann and of Gibbs (see the chapters by Frigg and Werndl in this volume and references therein). In the Boltzmannian theory, the measured magnitudes are understood as macrovariables, that is, as aspects of the microstates that obtain during a measurement. In the Gibbsian theory, measured magnitudes are understood as weighted functions over the entire phase space. (For conceptual problems with the latter theory see Callender, 1999, Wallace, 2013; for an approach that combines both theories and thus accounts for both, see Lavis, 2005, 2008; Hemmo and Shenker, 2012, Chapter 11). In *quantum* statistical mechanics there are two corresponding views. The first view (called "individualist" by Goldstein and Tumulka, 2011) states that a system has a given thermodynamic property if it is *either* in a pure state with high amplitudes for eigenvalues corresponding to that thermodynamic property, *or* in a reduced state that entails a high probability for the corresponding thermodynamic observations (see Linden et al., 2009; Goldstein et al., 2016, and references in both). The second view (called "ensemblist" in Goldstein and Tumulka, 2011) states that a system has a given thermodynamic property if it is in an appropriate statistical state, given by a density matrix, which entails a certain expectation value. Perhaps Von Neumann's concept of thermal equilibrium is ensemblist (see Goldstein and Tumulka, 2011; Goldstein et al., 2010; Goldstein et al., 2016, Section 9.2), and Emch (2007) seems to follow this line as well. The conceptual problems in the classical Gibbsian approach carry over to the quantum mechanical domain, *mutatis mutandis* (see Wallace, 2001, 2013). From now on I focus mainly on individualist approaches.

An important example of a (synchronic) thermodynamic property, which needs to be explained in quantum statistical mechanics, is that of *being in thermal equilibrium*. In thermodynamics, thermal equilibrium has three main characteristics. First, the temperature is spatially uniformly distributed; I call this condition "*uniformity*" (in other kinds of equilibrium this condition is replaced by similar ones, e.g., spatially uniform pressure or spatially uniform chemical mixture). Second, the uniformity condition is stationary: the system reaches it and remains there indefinitely; I call this condition "*stability*." Third, when the uniformity and stability conditions are obtained, the *entropy* of the system is maximal (the notion of entropy is addressed below). Statistical mechanics adds a fourth condition: In the *individualist* theory of *classical* statistical mechanics it was argued (by Boltzmann, see Uffink, 2004) that the set of microstates of an ideal gas that satisfies the uniformity condition has the largest Lebesgue measure; I call this the "*big set*" condition. The Lebesgue measure of macrostates has since then been associated with entropy, thus connecting the third and fourth conditions, and attempts have been made to show that this set also satisfies the stability condition (I expand on these attempts in the next section). In the ensemblist approach, the "big set" condition is replaced by the condition of a dynamically invariant probability distribution.

The idea that the four conditions of uniformity, stability, entropy, and big set are interconnected has been generalized to the quantum mechanical realm. In the next section, I discuss the quantum mechanical counterparts of the stability condition, and in the present section, I will say a few words about the quantum mechanical counterparts of the big set and uniformity conditions, and (finally) of the notion of entropy.

An example of the big set condition in quantum statistical mechanics is given by Goldstein and Tumulka (2011). Consider a quantum state ψ, expressed as a superposition of energy eigenstates all of which are in a given energy shell. And suppose that there is a subspace H_{eq} whose dimensionality is almost that of the entire space H: $\dim(H_{eq})/\dim(H) \approx 1$. Then ψ satisfies the big set characterization of equilibrium if the projection P_{eq} of ψ onto H_{eq} is $<\psi|P_{eq}|\psi> \approx 1$. Of course, the only reason for focusing on the big set condition, to begin with, is the thought that it is connected to the other characteristics of thermodynamic thermal equilibrium. The quantum mechanical counterpart of the uniformity condition, in this context, could be that upon measurement of the appropriate observables on ψ one is extremely likely to end up with eigenvalues that correspond to a uniform temperature distribution (see the characterization of this desideratum in Goldstein et al. (2015, 2016). Linden et al. (2009) generalize the quantum mechanical uniformity condition.).

In order to examine the (alleged) connection between the big set condition and that of high entropy, we must first identify the quantum mechanical counterpart of the thermodynamic notion of entropy. In thermodynamics, entropy difference is usually understood as quantifying the change in the degree to which the energy of a system can be exploited to produce work. (This holds only if the second law of thermodynamics is universally true, that is, if Maxwellian Demons are impossible, a topic I discuss in Section 29.7.)[5] In mechanics, in general, the more *control* one has over the system's microstate, the more one can exploit its energy; and assuming that more information about a system's state contributes to its controllability, it is reasonable to say that the more *information* we have about a system's state, the more its energy is exploitable, and therefore it is natural to associate entropy with information (see more on control in this context in Wallace, 2014).[6] In classical (Boltzmannian) statistical mechanics, this idea is realized by associating entropy with the (logarithm of the) Lebesgue measure of a set of microstates. What would be an analog of the notion of entropy in quantum mechanics that can be understood along these lines? One possibility that immediately comes to mind is the (logarithm of the) dimension of the subspace corresponding to the degenerate eigenvalue of the thermodynamic observable (i.e., the degenerate eigenvalue that corresponds to a macrovariable, as explained in the previous section). The big set characterization of equilibrium in Goldstein and Tumulka (2011), Goldstein et al. (2016) and Linden et al. (2009) (described below) seems to go along these lines.

Another prevalent candidate for the quantum mechanical counterpart of thermodynamic entropy is the so-called *Von Neumann entropy*, Tr(ρlog ρ) (see, e.g., Peres, 1993, p. 270). Following Von Neumann (1932), all the arguments for this idea are grounded in thought experiments of the kind one finds in thermodynamics: in a cycle of operation, the entropy is balanced by adding the Von Neumann entropy at the right stage. In previous writings (Shenker, 1999; Hemmo and Shenker, 2006) I have shown that in systems with a finite number of particles the entropy balance is kept without this addition, and concluded that Von Neumann's entropy does not correspond to thermodynamic entropy since it does not satisfy the corresponding functional relations. An alternative line of argument, in support of the claim that Von Neumann's expression refers to thermodynamic entropy, could be that this quantity expresses the essential feature of thermodynamic entropy, namely, the degree of energy exploitability (Peres, 1993 seems to have this in mind; see, e.g., pp. 369–370. Entropy is a measure of the exploitability of energy only if the second law of thermodynamics is universally true.) This may work, as follows. The Von Neumann's entropy corresponds to the degree of uniformity of the probability distribution over the possible outcomes of the next measurement, and if higher uniformity corresponds to lower control over what will be the next quantum state of the system, and if we take a degree of control as an essential feature of thermodynamic entropy, then Von Neumann's entropy can be said to correspond to the thermodynamic entropy. It should be noted, however, that this quantity is more naturally seen as part of a Gibbsian ("ensemblist") understanding of thermodynamic properties than to a Boltzmannian ("individualist") notion.

29.4 Diachronic Functional Relations in Quantum Statistical Mechanics: "Finite-Time" Arguments Concerning the Approach to Equilibrium

The thermodynamic law of approach to equilibrium (which is part of, or assumed by, the second law of thermodynamics) says that systems, prepared in non-equilibrium states, will invariably evolve to equilibrium states fixed by the constraints on these systems, and remain in them indefinitely (see Brown and Uffink, 2001). In mechanics, however, due to the *velocity reversal invariance* of the fundamental microscopic dynamics, given any Hamiltonian, if there are entropy-increasing trajectory segments in the system's state space, then there are also entropy-decreasing ones, and so the thermodynamic law cannot be strictly true. The standard explanation for why we do not encounter anti-thermodynamic evolutions is that those are extremely atypical or unlikely. The meaning of probability in this context is a subject of ongoing debates in classical statistical mechanics, which carry over to quantum statistical mechanics, as we shall see. (For the debate on the meaning of *typicality* in this context and the difference between typicality and probability see Frigg, 2011; Frigg and Werndl, 2012; Goldstein, 2012; Pitowsky, 2012; Hemmo and Shenker, 2014).

In the classical domain, two kinds of accounts have been proposed for the law of approach to equilibrium, and both have quantum mechanical counterparts: I call them the *finite time* and *infinite time* accounts. The *finite time* account aims to prove that a system prepared in a microstate that belongs to some non-equilibrium macrostate is highly likely to evolve to equilibrium within a characteristic time interval. The *infinite time* account aims to prove that as time goes to infinity, *typical* systems spend *most* of their time in equilibrium, and therefore if a system is observed at some randomly sampled moment (as it were) then it will most likely be found in equilibrium, and if found out of equilibrium then it is most likely to be in a minimum entropy state from which its entropy will increase. I now expand on the finite time account and address the infinite time account in the next section.

Consider Figure 29.1. Suppose that a system is prepared in a microstate that is a member of the macrostate M_0, in which all microstates share the macrovariable M_0. And suppose further that the dynamics is such that the evolution of each microstate in M_0 takes the system to a microstate within region $B(t_1)$ (with the same Lebesgue measure, according to Liouville's theorem) so that if we measure the system at t_1, we shall find it in either macrostate M_0 or M_1 depending on its actual trajectory, with the probability for each such possibility given by the measure of the region of overlap of $B(t_1)$ with M_0 and M_1, respectively. Let us assume, as is usual, that increasingly larger macrostates (by Lebesgue measure) correspond to increasingly uniform temperature distributions so that the largest macrostate (M_2 in the figure) is the one in which the uniformity condition of equilibrium is satisfied (such assumptions are usually based on combinatorial arguments, following Boltzmann; see Uffink, 2007). In these mechanical terms the law of approach to equilibrium would say that the dynamics is such that, with time, the regions $B(t)$ have increasingly larger overlaps with increasingly large macrostates, where "large" is understood relative to some appropriate measure. Conceptually, the measure of overlap (that gives rise to probability) and the measure of the macrostates (associated with entropy) need not be the same; see Hemmo and Shenker (2012). For a description of what in statistical mechanics can be proven from mechanics by itself and what requires auxiliary hypotheses see Shenker (2017a, 2017b) and Hemmo and Shenker (2019). An example of a detailed finite time proof is Lanford's theorem (see Uffink and Valente, 2010, 2015).

A major difficulty in the finite times account (which has a counterpart in quantum statistical mechanics, discussed below) is that due to the *retrodiction invariance* of the underlying dynamics, proving that entropy is highly likely to increases toward the future entails that entropy is equally likely to increase toward the past: this is the so-called parity of reasoning (or minimum entropy) problem. (This holds if pairs of velocity-reversed microstates belong to the same macrostate. This condition is normally assumed implicitly.) Since this result makes the theory internally inconsistent (since entropy is at a minimum at every moment, but is not constant), it had to be solved by adding symmetry-breaking postulates to the theory, which allow only histories in which past entropy was low; in contemporary literature, this idea is often called the *past hypothesis* (see Feynman, 1965; Albert, 2000; Winsberg, 2004; Earman, 2006; Shahvisi, this volume).

What would be the quantum mechanical counterpart of the classical finite time account of the approach to equilibrium? Suppose that the quantum state ψ has high amplitude for some *nonequilibrium* eigenvalue, and the quantum state φ has high amplitude for the *equilibrium* eigenvalue (that satisfies the quantum mechanical counterparts of the four conditions above). A Schrödinger evolution from ψ at t_0 to φ at t_1 is deemed a quantum mechanical counterpart of an approach to equilibrium. However, due to the quantum-mechanical counterpart of the classical velocity reversal symmetry, to each such quantum mechanical thermodynamic evolution there exists a corresponding anti-thermodynamic evolution, from the complex conjugate $\varphi\star$ at t_0 to the complex conjugate $\psi\star$ at t_1. Why then don't we observe anti-thermodynamic evolutions? One intuition here employs

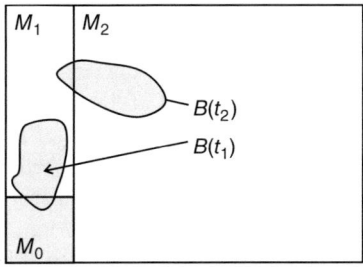

Figure 29.1 The interplay between dynamics and macrovariables

an analogy to the classical way of thinking: Every standard preparation gives rise to two kinds of quantum states: some states evolve thermodynamically and some evolve anti-thermodynamically; and for typical Schrödinger evolutions of thermodynamic systems, the former set is larger (by an appropriate measure). But this line of thinking requires that standard preparations give rise to two kinds of quantum states – whereas, as I said in Section 29.2, in general quantum mechanical preparations (by measuring some quantum observable M and selecting the systems that end up in some eigenvalue M_0) result in all the prepared systems being in exactly the same quantum state. One solution to this problem, described in Section 29.2, is to suppose that the thermodynamic magnitudes correspond to degenerate eigenvalues of quantum observables and that the preparations of thermodynamic systems are as outlined in Section 29.2. Given this assumption, the task, of proving a quantum mechanical counterpart of the law of approach to equilibrium, is to follow the dynamical evolution of this set of states and to find the probability (calculated with the suitable density matrix) of finding the system in eigenvalues that correspond to equilibrium.

Peres considers a different case in which there is ignorance concerning the *dynamics* (not the quantum state). He writes:

[A] quantum system prepared in a pure state remains pure when it evolves in a perfectly controlled environment. More generally, the entropy $S = -\text{Tr}(\rho\log\rho)$ remains invariant under the unitary evolution $\rho \to \rho' = U\rho U^\dagger$. On the other hand, if the environment is not perfectly controlled and is only statistically known, we must replace the evolution operator U with an ensemble of unitary matrices $U\alpha$, with respective probabilities $p\alpha$. The dynamical evolution then is [such] that the entropy never decreases. (Peres, 1993, p. 369)[7]

The parity of reasoning problem has its quantum mechanical counterpart: if – given an appropriately constructed initial quantum macrostate – evolution to quantum states with high amplitudes for equilibrium is highly likely, then it is equally likely that such was the quantum states in the past. A quantum mechanical symmetry-breaking *past hypothesis* is, then, needed. In this context, the temporal asymmetry of the projection postulate may become significant. Von Neumann (1932) thought that the time asymmetric nature of the projection postulate should be explained on the basis of the thermodynamic regularities (and so the former cannot explain the latter; *contra* the view described in Section 29.6). He wrote (ibid., p. 358): "It is desirable to utilize the thermodynamical method of analysis, because it alone makes it possible for us to understand correctly the difference between [Schrodinger's unitary transformation] and [the measurement transformation], into which reversibility questions obviously enter." (For more details on this argument by Von Neumann see Shenker, 1999, Hemmo and Shenker, 2006.)

29.5 Diachronic Functional Relations in Quantum Statistical Mechanics: "Infinite Time" Arguments

I now turn to the *infinite time* account of the thermodynamic law of approach to equilibrium. According to this approach, "for physical initial states ψ_0 of suitable macroscopic quantum systems, the system will spend most of its time in thermal equilibrium." (Goldstein and Tumulka, 2011). Given a quantum mechanical macrostate (in the above sense of the term), and assuming a Schrodinger evolution that is typical of thermodynamic systems,[8] most of the quantum states in this macrostate evolve in such a way that, as time goes to infinity, they spend most of their time in equilibrium (that is, in quantum states in which there is high amplitude for equilibrium eigenvalues). Presumably, this approach is made empirically significant by taking it to entail that if a system is observed at some random moment then it is highly likely to be found in equilibrium (in that sense).

A well-known problem in this account is that it entails that if one prepares a system in a non-equilibrium state, the approach to equilibrium could take *eons* and the account would still hold since it pertains to the *infinite time* limit (see Earman and Redei, 1996). Another problem is that (on this view) the probability of equilibrium is already high *immediately after* the system is found far from equilibrium, regardless of the system's dynamics and even if this requires superluminal speeds (see the debate in Allori, 2013; Allori, 2015; Hemmo and Shenker, 2015). These problems, known in the classical context, carry over to the quantum mechanical case. The infinite time approach is nevertheless prevalent, perhaps due to its advantages, especially the fact that it is not subject to the reversibility and parity of reasoning problems (since a system that spends most of the future in equilibrium also spent most of the past in equilibrium.)

Several recent papers in quantum statistical mechanics offer arguments in in support of this. Here I describe very briefly the general gist of two of these arguments and refer the reader to the literature for the details. One is by Goldstein et al. (2016). First, these writers show that *most* of the pure states in a given energy shell in Hilbert space satisfy a counterpart of the *uniformity condition* of equilibrium, that is, upon the measurement of the appropriate observables, there is a high probability of outcomes that correspond to thermodynamic equilibrium, in the sense of our uniformity condition above. Then they show that those quantum states that satisfy this uniformity condition also satisfy a counterpart of the *big set condition*: they are all close to a certain subspace H_{eq} which (for many systems) has the overwhelming majority of dimensions in the energy shell. Next, they need to prove the *stability condition*, which they understand as an *infinite time* theorem. To do so they decompose the Hilbert space into a sum of orthogonal subspaces (called "macrospaces," corresponding to classical sets of microstates that share macrovariables) using approximate commutativity, and identify one of them as H_{eq} , which has most of the dimensions. Since only a set of states with measure zero lies in subspaces of less than full dimension, most of the states in the energy shell will have their dominant part in H_{eq} and so most of those will also satisfy the (quantum counterpart of the) uniformity condition. They also prove a stronger theorem, namely, that small subsystems of pure state equilibrium systems, entangled with the rest of the system, are also in equilibrium in the sense that if we were to take a quantum measurement of a relevant observable in that subsystem, then the probability distribution over the measurement outcomes would agree with the thermal distribution.

The second infinite time argument I describe here is by Linden et al. (2009). These writers focus on the reduced state of a small system that is entangled with a heat bath and prove that the subsystem equilibrates for every one of its possible states and almost every possible state of the bath, and for (what they argue are) prevalent Hamiltonians. While the authors describe thermal equilibrium in terms that are close to our thermodynamic uniformity condition, this is not part of their main theorem; they consider equilibrium to be any state at which the system stays most of the time, which is the stability condition. Their proof connects the big set condition (concerning the relatively very small dimensionality of subspaces of small subsystems) with the stability condition: "Whenever the state of the whole system – and in particular of the bath – goes through many distinct states, any small subsystem reaches equilibrium," and the uniformity condition is supposed to be a special case of this general proof.

Both of the above-described results pertain to *most* states, namely, to a large majority of the states considered. Goldstein et al. (2016) write that "throughout this paper, 'most' means 'the overwhelming majority of' (or 'all except a small set') relative to the relevant uniform distribution" (Sections 1 and 4.1), and Linden et al. (2009) write: "In this situation, we have proved that for every state of the subsystem and *almost* every state of the bath, the subsystem equilibrates" (p. 8, our italics). In both cases, the idea is that if a condition is true of most cases, this suggests that the condition is also true of a concrete given system unless we have reason to expect otherwise. Indeed, as in the classical case, all the proofs of the approach to equilibrium in quantum statistical mechanics, both the finite time and infinite time ones, are valid only for *most* of the initial conditions. This also holds for our own proposals in Hemmo and Shenker (2001, 2003, 2005): These results hold only for systems in which the initial conditions lead to environmental decoherence, and according to prevalent decoherence models these results hold at best for most quantum states of the universe, given the right measure.

To establish such *most* proofs one needs to rely on a measure, and the idea is that the theoretical context suggests certain *natural* measures (see also Goldstein et al., 2010). What makes a measure natural for determining which set of states contains the *most*? The prevalent arguments in *classical* statistical mechanics for taking the Lebesgue measure to be natural for counting initial conditions rely on the fact that this measure has a special status in the diachronic regularities: for example, it is invariant under the dynamics according to Liouville's theorem. However, this dynamical fact is irrelevant for counting initial conditions (see Hemmo and Shenker, 2014, who also criticize based on similar grounds the preference of the quantum mechanical measure in Bohmian mechanics). But in quantum mechanics there seems to be another resource for choosing a measure: can the probabilities in quantum statistical mechanics be the result of the Born rule at the microscopic level? In the next section, I describe an attempt to achieve such a result.

29.6 Can Statistical Mechanical Probabilities Be Reduced to Quantum Mechanical Probabilities?

Albert (2000, Ch. 7) proposes an approach to quantum statistical mechanics in which the only kind of probabilistic statements are those derived from the micro-dynamics, namely, those of the Born rule. In this proposal, standard quantum mechanics is replaced with the GRW dynamics (Ghirardi et al., 1986; Bell, 1987). There are several versions of this dynamics (see Ghirardi, 2016; Lewis, this volume), but for our purpose, it is useful to describe it briefly as follows: Every quantum state φ_1 evolves for some time according to the Schrödinger equation to another quantum state φ_2, and then collapses spontaneously into a third state φ_3, which is a Gaussian superposition of positions centered around some point x. The probability that the spontaneous collapse will take place at any given moment is fixed by the temporal constant in the stochastic equation of motion, and the probability of the position x is fixed by the amplitude of x in φ_2.

Although the GRW spontaneous localizations are in position, when applied to macroscopic systems the result can also appear in the form of thermodynamic magnitudes, which can then be characterized in terms of their proximity to thermodynamic equilibrium (according to the four conditions of equilibrium mentioned above). In order to talk about the approach to equilibrium, I shall use the following terminology. Suppose that the eigenvalue a_{eq} of the observable A corresponds to an equilibrium state (e.g., it corresponds to the uniformity condition of thermal equilibrium), and suppose that given a certain Hamiltonian H the quantum state $\psi(t_1)$ evolves to $\psi(t_2)$. If, according to the Born rule, the probability for a GRW spontaneous collapse to a_{eq} is higher in $\psi(t_2)$ than in $\psi(t_1)$ then I shall say that $\psi(t_1)$ is a *thermodynamic* quantum state *relative to H* (Albert, 2000 uses the term "thermodynamically normal"). If, given the same dynamic evolution, the probability for a GRW spontaneous collapse to a_{eq} is lower in $\psi(t_2)$ than in $\psi(t_1)$, I shall say that $\psi(t_1)$ is an *anti-thermodynamic* quantum state *relative to H*. The proviso "relative to H" is necessary since I ascribe the property of being thermodynamic (or anti-thermodynamic) to the initial state $\psi(t_1)$, although this property is about the relation between $\psi(t_1)$ and the time evolved state $\psi(t_2)$.

Given these definitions, Albert makes the dynamical hypothesis, that the Hamiltonian H that governs thermodynamic systems has the following characterization: *Every* (not only most!) initial state $\psi(t_0)$ has high Born probability to collapse under the GRW dynamics to another quantum state $\psi(t_1)$ which is thermodynamic relative to H, regardless of whether or not $\psi(t_0)$ itself was thermodynamic relative to H. That is, with high probability the state $\psi(t_0)$ will collapse to a state $\psi(t_1)$ that will evolve deterministically, under the Hamiltonian H, to another quantum state $\psi(t_2)$, in which the amplitude for a GRW spontaneous collapse to a_{eq} will be higher than it is in $\psi(t_1)$.

Albert provides no proof for this hypothesis, and his only plausibility argument for it is based on the fact that observed systems are actually thermodynamic. (In comparison, Linden et al., 2009, for example, provide arguments for the plausibility of the particular sort of Hamiltonian they rely on.)

Thus Albert's theorem is not a *proof* from first principles that the world is thermodynamic, but only a *conjecture* or an *empirical generalization* that it is so. (See more on Albert's approach in Uffink, 2002; Sklar, 2015; Callender, 2016.)

29.7   Quantum Mechanical Maxwellian Demon

In 1867 J.C. Maxwell proposed a thought experiment, in which a tiny automaton (that came to be known as *Maxwell's Demon*) manipulates the individual molecules of a gas that is initially in uniform temperature, directing the relatively hotter (that is, faster) molecules to one side and the colder (i.e., slower) one to the other side, thus creating a temperature difference, which amounts to an entropy decrease. The result is a total decrease in the entropy of the universe, in violation of the second law of thermodynamics. Since, as things stand now, there is no general proof from first principles that the second law of thermodynamics is (probabilistically) universally true (in neither classical nor quantum statistical mechanics, neither finite nor infinite arguments, neither individualist nor ensemblist approaches), the question of whether a Maxwellian Demon is compatible with fundamental physics became an indirect route to asking about the universal validity of the second law. For this reason, numerous attempts have been made to "exorcise" the Demon, that is, show that it is *incompatible* with fundamental physics and hence the second law is true (see Earman and Norton, 1998, 1999; Leff and Rex, 2003). But all have failed, and in recent years the *compatibility* of Maxwell's Demon with classical mechanics has been proven (Albert, 2000; Hemmo and Shenker, 2010, 2011, 2012, 2013, 2016, 2020). Interestingly, some central attempts to exorcise the Demon that appear to rely on quantum mechanics (such as Zurek, 1984) have turned out to be grounded in classical statistical mechanical ideas (on Zurek's argument see Earman and Norton, 1999[9]). It has recently been shown that Maxwellian Demons are also compatible with quantum mechanics, either with or without collapse (Hemmo and Shenker, 2012, 2020). These results are significant in that they entail that attempts at a universal proof of the law of approach to equilibrium and the second law of thermodynamics are futile; there is hope only for proofs pertaining to special circumstances – namely, special Hamiltonians and specially prepared quantum states. This realization goes against deeply entrenched convictions, according to which the second law of thermodynamics expresses a universal truth. But if quantum mechanics is the fundamental theory, then this conviction will have to be relinquished.[10,11]

Acknowledgments

I am extremely grateful to Meir Hemmo for jointly working with me on many topics mentioned in this chapter, described in our joint publications. This research was supported by Israeli Science Foundation grant 1148/18.

Notes

1 In Hemmo and Shenker (2015) the terminology is a bit different, and 'macrostate 'is used to denote sets of microstates that share a macrovariable that is accessible for a given observer.

2 See Crull (this volume) for more details of decoherence.

3 See Wallace (2013, p. 14) and references there.

4 These cases raise interesting questions concerning individuation, see French and Redhead 1989, and for more recent studies Ladyman (2015).

5 See Fermi (1936); Hemmo and Shenker (2012, Ch. 1).

6 The spin echo experiments can be understood as a case where control over the microstate is available despite lack of information concerning it; see Hemmo and Shenker (2012, sec. 6.7).

7 I discuss the Von Neumann entropy $-\mathrm{Tr}(\rho \log \rho)$ in Section 29.6.

8 In its classical version this account is inspired and illustrated by Boltzmann's own arguments in which the dynamics is assumed to be ergodic, where ergodicity should be understood *á la* Birkhoff and Von Neumann, see Frigg (2008, sec. 3.2.4).

9 Zurek's argument relies on the Landauer-Bennett thesis that was disproved in Hemmo and Shenker (2013).

10 Scully et al. (2003) show that a quantum heat engine allows us to extract work from a single thermal reservoir, in violation of Carnot's principle.

11 This research was supported by a Lockheed-Martin Research Grant. The understanding of statistical mechanics presented here is the result of many years of collaboration with Meir Hemmo.

References

Albert, D. (2000). *Time and Chance*. Cambridge, MA: Harvard University Press.

Allori, V. (2013). Book review of *The road to Maxwell's Demon*. *International Studies in the Philosophy of Science*, 27(4): 453–456.

Allori, V. (2015). Response to authors. *International Studies in the Philosophy of Science*, 29(1): 94–98.

Bell, J.S. (1987). Are there quantum jumps? In J.S. Bell (ed.), *Speakableand Unspeakable in Quantum Mechanics*. Cambridge: Cambridge University Press, pp. 201–212.

Brown, H. and Uffink, J. (2001). The origins of time-asymmetry in thermodynamics: The minus first law. *Studies in History and Philosophy of Modern Physics*, 32(4): 525–538.

Callender, C. (1999). Reducing thermodynamics to statistical mechanics: The case of entropy. *Journal of Philosophy*, XCVI: 348–373.

Callender, C. (2016). Thermodynamic asymmetry in time. In E.N. Zalta (ed.), *The Stanford Encyclopedia of Philosophy* (Winter 2016 edition). Available at: https://plato.stanford.edu/archives/win2016/entries/time-thermo/.

d'Espagnat, B. (1976). *Conceptual Foundations of Quantum Mechanics*. Boston, MA: Addison-Wesley.

Earman, J. (2006). The past hypothesis: Not even false. *Studies in History and Philosophy of Modern Physics*, 37: 399–430.

Earman J. and Norton J. (1998). Exorcist XIV: The wrath of Maxwell's Demon. Part I. From Maxwell to Szilard. *Studies in History and Philosophy of Modern Physics*, 29: 435–471.

Earman J. and Norton J. (1999). Exorcist XIV: The wrath of Maxwell's Demon. Part II. From Szilard to Landauer and beyond. *Studies in History and Philosophy of Modern Physics*, 30(1): 1–40.

Earman J. and Redei M. (1996). Why ergodic theory does not explain the success of equilibrium statistical mechanics. *The British Journal for the Philosophy of Science*, 47: 63–78.

Emch, G. (2007). Quantum statistical physics. In J. Butterfield and J. Earman (eds.) *Philosophy of Physics*. Amsterdam: Elsevier, pp. 1075–1182.

Fermi, E. (1936). *Thermodynamics*. New York: Dover, 1956.

Feynman, R. (1965). *The Character of Physical Law*. Cambridge, MA: MIT Press.

Frigg, R. (2008). A field guide to recent work on the foundations of statistical mechanics. In D. Rickles (ed.), *The Ashgate Companion to Contemporary Philosophy of Physics*. London: Ashgate, pp. 99–196.

Frigg R. (2011). Why typicality does not explain the approach to equilibrium. In M. Suárez (ed.), *Probabilities, Causes and Propensities in Physics*. Wien, Synthèse Library 347, pp. 77–93.

Frigg, R. and Werndl, C. (2012). Demystifying typicality. *Philosophy of Science*, 79(5): 917–929.

Ghirardi, G. (2016). Collapse theories. In E.N. Zalta (ed.), *The Stanford Encyclopedia of Philosophy*. Available at: https://plato.stanford.edu/archives/spr2016/entries/qm-collapse/.

Ghirardi, G., Rimini, A. and Weber, T. (1986). Unified dynamics for microscopic and macroscopic systems. *Physical Review D*, 34: 470–479.

Goldstein S. (2012). Typicality and notions of probability in physics. In Y. Ben Menahem and M. Hemmo (eds.), *Probability in Physics*. Wien: Springer, pp. 59–72.

Goldstein, S., Huse, D.A., Lebowitz, J.L. and Tumulka, R. (2015). *Thermal equilibrium of a macroscopic quantum system in a pure state*. Unpublished manuscript. Available at: arXiv:1506.07494 [cond-mat.stat-mech].

Goldstein, S., Huse, D.A., Lebowitz, J.L. and Tumulka R. (2016). *Macroscopic and microscopic thermal equilibrium*. Unpublished manuscript. Available at: arXiv:1610.02312v1 [quant-ph].

Goldstein, S. and Lebowitz, J. (2004). On the (Boltzmann) entropy of nonequilibrium systems. *Physica D*, 193: 53–66.

Goldstein, S., Lebowitz, J.L., Mastrodonato, C., Tumulka, R., and Zanghi, N. (2010). Normal typicality and Von Neumann's quantum ergodic theorem. *Proceedings of the Royal Society London A*, 466(2123): 3203–3224.

Goldstein, S. and Tumulka R. (2011). On the approach to thermal equilibrium of macroscopicquantum systems. In P.L. Garrido, F. de los Santos, and J. Marro (eds.), *Non-Equilibrium Statistical Physics Today: Proceedings of the 11th Granada Seminar on Computational and Statistical Physics*, Granada, Spain AIP Conference Proceedings 1332, pp. 155–163.

Hemmo, M. and Shenker, O. (2001). Can we explain thermodynamics by quantum decoherence? *Studies in History and Philosophy of Modern Physics*, 32(4): 555–568.

Hemmo, M. and Shenker, O. (2003). Quantum decoherence and the approach to equilibrium. *Philosophy of Science*, 70(2): 330–358.

Hemmo, M. and Shenker, O. (2005). Quantum decoherence and the approach to equilibrium II. *Studies in History and Philosophy of Modern Physics*, 36: 626–648.

Hemmo, M. and Shenker, O. (2006). The Von Neumann entropy does not correspond to thermodynamics entropy. *Philosophy of Science*, 73(2): 153–174.

Hemmo, M. and Shenker, O. (2010). Maxwell's demon. *The Journal of Philosophy*, 107: 389–411.

Hemmo, M. and Shenker, O. (2011). Szilard's perpetuum mobile. *Philosophy of Science*, 78(2): 264–283.

Hemmo, M. and Shenker, O. (2012). *The Road to Maxwell's Demon*. Cambridge: Cambridge University Press.

Hemmo, M. and Shenker, O. (2013). Entropy and computation: The Landauer-Bennett thesis reexamined. *Entropy*, 15: 3387–3401.

Hemmo, M. and Shenker, O. (2014). Probability and typicality in deterministic physics. *Erkenntnis*, 80: 575–586.

Hemmo M. and Shenker O. (2015). Boltzmann's approach to probability: Letter to the editor. *International Studies in the Philosophy of Science*, 29(10): 91–92.

Hemmo, M. and Shenker, O. (2016). Maxwell's demon. *Oxford Online Handbook*. Oxford University Press. Available at: http://www.oxfordhandbooks.com/view/10.1093/oxfordhb/9780199935314.001.0001/oxfordhb-9780199935314-e-63?rskey=plUl7T&result=1.

Hemmo, M. and Shenker, O. (2019). Two Kinds of High Level Probability. *The Monist*, 102: 458–477.

Hemmo, M. and Shenker, O. (2020). Maxwell's Demon in Quantum Mechanics. *Entropy*, 22(3): 269.

Ladyman, J. (2015). Are there individuals in physics, and if so, what are they? In A. Guay and T. Pradeu (eds.), *Individuals across the Sciences*. Oxford: Oxford University Press, 193–206.

Ladyman, J. and Ross, D. (2007). *Everything Must Go*. Oxford: Oxford University Press.

Lavis, D. (2005). Boltzmann and Gibbs: An attempted reconciliation. *Studies in History and Philosophy of Modern Physics*, 36: 245–273.

Lavis, D. (2008). Boltzmann, Gibbs and the concept of equilibrium. *Philosophy of Science*, 75: 682–696.

Lebowitz, J. (1993). Boltzmann's entropy and time's arrow. *Physics Today*, 46(9): 32–38.

Leff, H.S. and Rex, A. (eds.) (2003). *Maxwell's Demon 2: Entropy, Classical and Quantum Information, Computing*. Bristol: Institute of Physics Publishing.

Linden, N., Popescu, S., Short, A.J. and Winter, A. (2009). Quantum mechanical evolution towards thermal equilibrium. *Physical Review E*, 79: 061103-1–061103-12.

Peres, A. (1993). *Quantum Theory: Concepts and Methods*. Dordrecht: Kluwer.

Pitowsky, I. (2012). Typicality and the role of the Lebesgue measure in statistical mechanics. In Y. Ben Menahem and M. Hemmo (eds.), *Probability in Physics*. Berlin: Springer, pp. 41–58.

Scully, M.O., Zubairy, M.S., Agarwal, G.S. and Walther, H. (2003). Extracting work from a single heat bath via vanishing quantum coherence. *Science*, 299: 862–864.

Shenker, O. (1999). Is -kTr(ρlnρ) the entropy in quantum mechanics? *British Journal for the Philosophy of Science*, 50: 33–48.

Shenker, O. (2017a). The foundations of statistical mechanics: Mechanics by itself. *Philosophy Compass* 12(12). doi: 10.1111/phc3.12465.

Shenker, O. (2017b). The foundations of statistical mechanics: The auxiliary hypotheses. *Philosophy Compass*. 12(12). 10.1111/phc3.12464.

Shenker, O. (2017c). Flat physicalism: Some implications. *Iyyun: The Jerusalem Philosophical Quarterly*, 66: 211–225.

Sklar, L. (1993). *Physics and Chance*. Cambridge: Cambridge University Press.

Sklar, L. (2015). Philosophy of statistical mechanics. In E.N. Zalta (ed.), *The Stanford Encyclopedia of Philosophy* (Fall 2015 Edition). Available at: https://plato.stanford.edu/archives/fall2015/entries/statphys-statmech/.

Uffink, J. (2002). Essay review of time and chance. *Studies in History and Philosophy of Modern Physics*, 33: 555–563.

Uffink, J. (2004). Boltzmann's work in statistical physics. In E.N. Zalta (ed.), *The Stanford Encyclopedia of Philosophy* (Winter 2008 Edition). Available at: http://plato.stanford.edu/archives/win2008/entries/statphys-Boltzmann/.

Uffink, J. (2007). Compendium to the foundations of classical statistical physics. In J. Butterfield and J. Earman (eds.), *Philosophy of Physics*. Amsterdam: Elsevier, pp. 923–1074.

Uffink, J. and Valente, G. (2010). Time's arrow and Lanford's theorem. *Seminaire Poincare*, XV: 141–173.

Uffink J. and Valente, G. (2015). Lanford's theorem and the emergence of irreversibility. *Foundations of Physics*, 45: 404–438.

Von Neumann, J. (1932). *Mathematical Foundations of Quantum Mechanics*. Trans. R.T. Beyer 1955. Princeton, NJ: Princeton University Press.

Wallace, D. (2001). *Implications of quantum theory in the foundations of statistical mechanics*. Unpublished manuscript.

Wallace, D. (2011). The logic of the past hypothesis. In B. Loewer, E. Winsberg and B. Weslake (eds.), *Time's Arrow and the Origin of the Universe: Reflections on Time and Chance: Essays in Honor of David Albert's Work* (forthcoming). Available at: philsci-archive.pitt.edu/8894/1/pastlogic_2011.pdf.

Wallace, D. (2013). Inferential vs. dynamical conceptions of physics. *quant-ph*. Available at: https://arxiv.org/abs/1306.4907.

Wallace, D. (2014). Thermodynamics as control theory. *Entropy*, 16: 699–725.

Winsberg, E. (2004). Can conditioning on the 'past hypothesis' militate against the reversibility objections? *Philosophy of Science*, 71: 489–504.

Zurek, W.H. (1984). Maxwell's demon, Szilard's engine and quantum measurement. In H.S. Leff and A. Rex (eds.), *Maxwell's Demon 2: Entropy, Classical and Quantum Information, Computing*. Bristol: Institute of Physics Publishing, pp. 179–189.

Further Reading from the Editors

The final chapter of D. Albert, *Time and Chance* (Cambridge, MA: Harvard University Press, 2003) identifies a potential explanation for thermodynamic irreversibility in terms of dynamical-collapse quantum mechanics. Other recent approaches to quantum foundations for statistical mechanics draw on unitary quantum mechanics: see in particular O. Maroney, "The Physical Basis of the Gibbs-von Neumann Entropy," *arXiv.org*, quant-ph/0701127 and D. Wallace, "Probability and Irreversibility in Modern Statistical Mechanics: Classical and Quantum," in D. Bedingham, O. Maroney and C. Timpson (eds.), *Quantum Foundations of Statistical Mechanics* (Oxford: Oxford University Press, forthcoming) along with other papers in the latter volume.

30

ENTROPY ASYMMETRY

Arianne Shahvisi

The second law of thermodynamics rules that entropy increases with time. All of the macroscopic processes with which we are familiar, including life itself, rely upon this entropy asymmetry. Yet the microscopic equations of motion are time-reversal invariant, so the irreversibility that is observed on the macroscopic scale stands in need of explanation. The conventional strategy is to employ the combinatorial arguments of Boltzmann, in addition to a postulate that the entropy of the early Universe was very low. The first part of this chapter will introduce entropy and its asymmetry, rehearse the Boltzmannian explanation for entropy asymmetry, and present some of the challenges it faces. The second part of the chapter will offer a brief cosmological history, in order to trace the origins of out-of-equilibrium thermodynamic systems, and the way these systems relate to the low-entropy initial state. Its aim is to characterize those cosmological features which are the progenitors of phenomena within modern thermodynamics and emphasize that there is also a cosmological explanation for the entropy processes we observe.

30.1   The Second Law of Thermodynamics

The second law of thermodynamics states that, for an isolated system subject to an irreversible thermodynamic process, there will be a net increase in entropy. More precisely, if the entropy at time t_1 is S_1 (where S_1 is below the maximum entropy for the system), then it is *vastly probable* that the entropy, S_2, at a later time, t_2, will be greater than S_1. So the entropy of a system is very likely to always increase with time until it reaches its *equilibrium* value, the state for which the entropy is maximal.[1]

A phase-space representation of a classical,[2] macroscopic gas (with particle number of about 10^{23}) is given by a single point $X = (q, p)$ within the $6N$-dimensional phase-space Γ, where q denotes the positions of the N particles, $q = (q_1, \ldots, q_{3N})$ and p denotes their momenta, $p = (p_1, \ldots, p_{3N})$. (For simplicity, we also assume that the particles are point particles, so that each has just six degrees of freedom: three for position and three for momentum.) The relationship between thermodynamic variables and micro-configurations of the system is many-to-one, and the phase-space can therefore be compartmentalized by grouping states that are macroscopically equivalent. This process of carving up the phase-space into cells of microstate points which realize the same macrostate is known as *coarse-graining*.[3]

For example, there will be a number of points in the phase-space which, though exhibiting different values for the positions and momenta of individual particles (different microstates), have the same net value for all of their macroscopic variables (identical macrostates). In other words, if you were able to change the position or momentum of a tiny number of particles in the system by a small amount, the overall pressure (say) would not change, even though the microstate of the system would be different. Of course, were our macroscopic measuring devices sufficiently sensitive, a change in pressure would be detected, but this point just highlights another conceptual difficulty: that of objectively defining the microscopic-microscopic boundary over what we know to be a continuum. If the device was sufficiently sensitive that it took account of exact microstates, then macroscopic properties become moot. One example of a case in which a microstate could change without any concomitant difference in the macro-variable would be to invert the momentum of two particles in opposite senses (respecting conservation laws) while keeping the magnitudes the same.[4]

Each macrostate can then be associated with a certain volume of phase-space that contains all the points which realize it. Here, we use the standard[5] (Lebesgue) measure, which states that the measure of the set of phase-space points which corresponds to a particular macrostate is given by the volume of phase-space that the macrostate occupies. The probability of a macrostate obtaining is then equal to the measure of its set of microstates, normalized by dividing by the measure of the entire phase-space. Macrostates which are realized by a greater measure of micro-configurations are therefore more probable, and the system is more likely to be found in these states.

Typically, there will be one particular coarse-grained cell of phase-space that is vastly larger than the others. This corresponds to the *equilibrium* macrostate and is the largest region because it is realized by the greatest total measure of configurations. Suppose we begin with a state corresponding to a phase-space point within one of the smaller macrostate regions and allow this to evolve forward with time. The system will very likely soon wander into the larger regions of the phase-space and reach the region with the greatest volume. Once there, it is extremely unlikely to leave this region since it constitutes almost all of the phase-space.[6]

The entropy of a macroscopic state is simply any appropriate measure of the volume of the region of phase-space corresponding to that macroscopic state. However, in order to ensure that the entropy of a composite system is the sum of the entropies of its constituent subsystems, such that $S(M_1 + M_2) = S(M_1) + S(M_2)$, i.e., to ensure that the entropy is *extensive*, it is expressed as the *logarithm* of the measure of phase-space points. (If the number of microstates of the first system is n_1 and the number of microstates of the second system is n_2 then the number of microstates of the combined system is $n_1 n_2$. So if the entropy was simply proportional to the number of microstates of a given system state, rather than its logarithm, the entropy of a combined system would be proportional to the product of their numbers of microstates and would not be extensive.)

There is a lively debate in the literature concerning which of the Gibbs or Boltzmann entropies is superior in terms of empirical utility and conceptual simplicity. Here, I will bypass these issues; the conception of entropy in this chapter is merely a placeholder. I favor the Boltzmann entropy because I find it has greater conceptual appeal.[7]

The Boltzmann entropy is given by:

$$S_B(M) = k_B \log(\Gamma_M)$$

I.e. the Boltzmann entropy, S_B, of a macrostate, M, is proportional (via Boltzmann's constant, k_B) to the logarithm of the region of phase-space corresponding to that macrostate. Returning to the reasoning above, this means that systems are overwhelmingly likely to increase their entropy toward a maximum (equilibrium) value. And this, according to Boltzmann, is how the second law is justified.

30.1.1   The Reversibility Objection and Its Solution

However, the explanation offered in the previous section is incomplete. It explains why, in general, the entropy of a system is overwhelmingly likely to increase. But the equations of motion which constrain the trajectories of classical particles within any system are *time-reversal invariant*. This means that, if the macrostate of a system at time t is $X(t)$, then the set of microstates (which correspond to the macrostate) of the system at an earlier time, $X(t - \delta t)$, is the time evolution by $2\delta t$ from the time-reverse of the set of microstates of the system at a later time $X(t + \delta t)$. Classically, the time-reverse of each of the microstates which make up $X(t)$ is generated simply by reversing the velocities of all the particles in the system. This means that the entropy of the macrostate and its time-reverse are equal since all that differs between them is the directions of the particle velocities in each of the microstates (i.e., they are still realized by the same *measure* of microstates, which is how we defined entropy). This implies that if the entropy of the macrostate $X(t)$ increases forward toward $X(t + \delta t)$, it also increases *backward* toward $X(t - \delta t)$.

So the above explanation for the operation of the second law, as it stands, states that entropy is overwhelmingly likely to increase in both the forward and backward temporal directions.[8] That is, just as the arguments of the previous section lead us to predict, with near-certainty, that all current systems will evolve toward higher-entropy states, so it also leads us to retrodict with near-certainty that all current systems evolved *from* higher-entropy states in the past. To make this explicit: if, say, we expect an ice cube to be more melted in the future, we should symmetrically infer that it evolved from a more melted state in the past.

One obvious way to address this "reversibility objection" and guarantee the correct form of the second law is to impose that the initial macrostate of the system corresponds to one of the smaller compartments of phase-space. This means that the initial entropy of the system was low, and is therefore very likely to increase toward its equilibrium value. Further, if the initial macrostate of the system was one of low entropy, and the entropy now is higher (as it is overwhelmingly likely to be), then it must be the case that the entropy of the system *decreases* as you look backward in time. In this way, the transition from a high-entropy state to a low-entropy state is forbidden by a fact of the matter about the entropy value of the initial state.

But the second law is observed *universally*. The fix just described works for the particular system in question, whose low initial entropy value is put in by hand, but we need a way of guaranteeing that the entropy of *all* systems was lower in the past.

The conventional strategy is to introduce a postulate which delivers the second law in its observed form, militates against the reversibility objection, and thereby grounds the existence of time-asymmetric thermodynamic behavior everywhere in the Universe. That postulate is as follows:

> The "Past Hypothesis": the postulate that the Universe began in a state of low entropy, that is, in a very small region of its phase-space.

The reversibility objection described in the previous section can therefore be assuaged by accepting that the macrostate of the early Universe was one of sufficiently low entropy to probabilistically entail, when combined with the combinatorial arguments made in the last section, the entropy-increasing processes that we observe.

Albert (2000, p. 96) coined the term "Past Hypothesis," but the idea can be traced to Boltzmann:

> That in nature the transition from a probable to an improbable state does not take place as often as the converse, can be explained by assuming a very improbable initial state of the entire universe surrounding us, in consequence of which an arbitrary system of interacting bodies will in general find itself initially in an improbable state. (Boltzmann, 1896–1898, p. 447)

We now appear to be in possession of an explanation which ensures that both our future-directed predictions and our past-directed retrodictions are aligned with what our intuitions tell us, informed by what we in fact observe, and with our records of the past.

(Not quite. It might be the case that those regions of the phase-space which contain microstates which lead to a low-entropy future are close to those regions of phase-space which contain microstates which came from a low-entropy past. Then the reasoning above would lead us to the false prediction that the entropy is very likely to decrease in the future. Albert counters this challenge by stating that these abnormal regions (i.e., those containing microstates which lead to a low-entropy future) are not only incredibly small but also "scattered, in unimaginably tiny clusters, more or less at random, all over the place" (Albert, 2000, p. 82). He offers no justification for this scattering condition, and I have been unable to find any.)

30.1.2 Challenges to the Conventional Explanation

The explanation just given – or something functionally equivalent – is now close to orthodoxy within the physics community. Nonetheless, it faces several major challenges:

1   **The Improbability of the Past State.** The Past Hypothesis refers to a state of low entropy which obtains at some early time: the "Past State." Since, according to the second law, entropy both (a) increases with time, and (b) is correlated with probability, we infer that the Past State was exceedingly improbable. In other words, the conventional explanation has the Universe starting in a tiny volume of its phase-space. This is problematic inasmuch as the explanation for the second law relies upon a state which is itself so overwhelmingly unlikely. Some of the sting can be drawn by noting that the Past State is independently confirmed by cosmology, nonetheless, a debate then ensues as to whether the Past State stands in (urgent) need of explanation (compare Callender (2004) and Price (2004)).

2   **Boltzmann Brains**.[9] Given this last point, another worry emerges. It begins with Boltzmann's suggestion that thermodynamic asymmetry might instead be explained by chance alone. Consider that the Universe is very old, and let us postulate that it did not start in a vastly improbable low-entropy state, but has instead been close to – or at – equilibrium for most of its existence (a much more likely state of affairs). At some point a fluctuation occurred by chance and took the macrostate of the Universe away from equilibrium, producing the low-entropy initial state which evolved into the world of our thermodynamic experience. According to this view, the Universe is now returning to equilibrium, and we are experiencing this as the operation of the second law. (And, of course, as living beings, we can necessarily only exist in such a low-entropy region, since life requires an entropy gradient. This is a version of the "Anthropic Principle.") This proposal has surprising consequences. A fluctuation sufficient to produce a low-entropy state which then evolves to produce our familiar world would need to depart quite dramatically from equilibrium, therefore the chance of this happening spontaneously would be very small indeed. However, a fluctuation which instead spontaneously reproduced our familiar world in its current state *with fake records* of the past, rather than the past having happened, would not require such an enormous fluctuation, since it would need only to get to the entropy state *now*, rather than the much lower (and therefore less probable) one in the early Universe.

We do not know which of these situations is correct (our evidence is the same in every case: the world that we observe *now* and the records we have) but according to the probabilities, it is overwhelmingly more likely that our records and memories are misleading: they came into existence by chance just a moment ago. Worse, it is not even necessary to have the entire Universe: one can settle for a much smaller fluctuation and instead have just a single brain – a "Boltzmann brain" – to replace my brain or yours and contain the mere impression of our familiar Universe along with all the requisite records and memories. And this is far more likely – considering probabilities alone – than the conventional explanation, with its vastly improbable Past Hypothesis. Fortunately, we do not consider probabilities alone! Our knowledge of the Universe is vast and growing: we can observe regions of the Universe distant in both space and time, and everything we have seen so far also has low entropy. In other words, we might expect that if our world was a mere fluctuation, there would be no more than is needed to bring about the illusion. Instead, it seems there is an enormous extravagance, in space and time, of this low-entropy region. The larger that region is, the lesser its probabilistic advantage over the conventional explanation. (If the region encompasses the entire Universe, and the fluctuation dates back to the time at which the Past State obtained, then the probabilities equalize.) This, of course, does nothing to counter the unsettling possibility that I am just a Boltzmann brain, but such a discussion belongs in dedicated work on skepticism and will be set aside here.[10]

3   **What about the Direction of Time Itself?** There is an important distinction to be made about what precisely the conventional explanation explains. A distinction must be made between the asymmetry *of* time, and the asymmetry *in* time (Price, 2013, pp. 193–194). The first refers to an asymmetrical property of *time itself*, while the second refers to the asymmetrical properties of *events that occur in time*. It is the second sense that is intended here: the task is to explain why the entropy gradients of all familiar processes are aligned with each other (in a direction that is conventionally labeled past-to-future). Even if the conventional explanation is successful in explaining this asymmetry (the fact that all the entropy gradients point one way rather than the other), it does nothing to combat any asymmetry that may be inherent within time itself. And there is perhaps good reason to doubt that any such innate asymmetry exists, for once the processes (which will exhibit thermodynamic asymmetry if they exhibit asymmetry at all) are subtracted, it seems impossible to isolate any vestigial background directionality. (One ostensible exception to this is the violation of charge conjugation parity (CP) symmetry) (Lees et al., 2012), but this is a non-thermodynamic violation of time-reversibility and is not relevant to the discussion here.) For that reason, I will set aside discussion of time itself here.

30.2   A Brief History of Heat: The Cosmological Origins of Non-equilibrium Systems

"What then is that precious something contained in our food which keeps us from death? That is easily answered. Every process, event, happening – call it what you will; in a word, everything that is going on in Nature means an increase of the entropy of the part of the world where it is going on. Thus a living organism continually increases its entropy ... and thus tends to approach the dangerous state of maximum entropy, which is death. It can only keep aloof from it, i.e., alive, by continually drawing from its environment negative entropy – which is something very positive as we shall immediately see" Schrödinger (1944, p. 6).

It is a triviality that all phenomena can be traced to the Big Bang, but cosmological constraints are rarely of sufficient relevance to be called upon to account for characteristics of the present-day Universe. In the case of thermodynamics, however, the operation of the second law in macroscopic processes in the current Universe is inextricably linked to early cosmology via the Past State. One can also trace entropy asymmetry in more tangible terms, in particular, in the processes of primordial nucleosynthesis and the formation of structure through gravitational collapse. In this section, I will describe the way in which supplies of low entropy came about, and how they enable familiar thermodynamic processes.

I will set $t_1 = 10^{-2}$ seconds after the Big Bang, when the Universe was a hot, dense plasma, and bracket whatever came before. This safely places what follows within the domain of uncontroversial, well-understood cosmology. When the Universe was 10^{-2} seconds old, its energy was of the order of ten MeV. Such conditions are easily simulated in particle accelerators, so our knowledge of the Universe at this age has its basis in experiment as well as theory and is not subject to the same uncertainty of earlier epochs at higher energies. My objective is to give a brief account of the thermodynamics of the Universe over the period that begins with t_1 and ends with the current Universe, 14 billion years later. However, I will start this account at the end of the story, rather than the beginning; instead of asking why the second law holds in general, I will instead ask why there is such a proliferation of non-equilibrium processes on *Earth*. This will bridge the cosmology to the thermodynamics of our direct experience.

Without the abundance of low-entropy photons emitted from the Sun, animals could not ingest low-entropy food sources and discard higher-entropy waste products. Therefore, life, or indeed any interesting set of events, would not be possible. All familiar thermodynamic processes entail an increase in entropy: chemicals reacting, animals respiring, brains cogitating. Had the Universe already reached a global maximum entropy-state, these increases in entropy would probably never occur; ours would be a Universe where little would happen. It is therefore the *potential* for entropy increase in the present-day Universe that underpins the existence of the second law processes which we observe.

This potential, in broad strokes, is a result of the fact that (a) 99.9% of the protons and neutrons in the Universe are *not* bound into iron nuclei, and are therefore not yet in their lowest-energy configuration, and (b) matter in the early Universe had near-perfect homogeneity and has therefore since been able to form bound structures capable of maintaining considerable temperature gradients. In what follows, I offer details of how these factors operate and the cosmological features that made them possible.

30.2.1 Low-Entropy Hot-Spots and How We Get the Benefit of Them

Out-of-equilibrium processes are prepared with, and subsist on, low-entropy sources of energy. The energy carried by an input of low-entropy national grid electrons keeps the freezer colder than the kitchen; we are able to turn Van der Waals forces between liquid molecules into a *lower* entropy hexagonal crystal lattice and therefore make ice from water, while the back of the freezer outputs high-entropy energy in the form of thermal radiation or direct conduction, ensuring the second law is not violated.

On Earth, this "low-entropy energy" is supplied by the Sun. There are negligible sources of energy on Earth which are not dependent on the Sun. Geothermal energy has its origins in the gravitational collapse which created the Earth, and nuclear fission takes advantage of the fact that primordial nucleosynthesis was prematurely aborted by the cooling of the Universe due to expansion. The Sun and Earth can be modeled as thermal black bodies, whose temperature difference, of over 5,000 degrees, is responsible for the creation and maintenance of non-equilibrium processes on Earth. When a photon emitted from a higher temperature source is absorbed by a body at a lower temperature, the energy of the incident photon is redistributed and emitted among several other photons, leading to an overall increase in entropy.

Therefore, if thermal radiation from the Sun is incident on an inanimate object on Earth, the object heats up – its atoms or molecules increase their translational or vibrational kinetic energies – and radiation of a higher entropy is emitted back into the atmosphere. If this was *all* that happened, as is the case on the surfaces of the other planets in our solar system, our varied biosphere could not exist. However, when thermal radiation is incident on the leaves of photosynthetic plants, thermal radiation is emitted back to the atmosphere as before, *but* the increase in entropy of the emitted radiation compensates for the *decrease* of entropy brought about by the production of organic molecules.

It is through the process of photosynthesis that energy from the Sun is fixed on Earth. The increase in entropy through additional thermal radiation is several times larger than the decrease in entropy due to the production of glucose, in keeping with the second law. In this way, plants fix carbon from the atmosphere into complex organic molecules which may be broken down to release energy through respiration, either within their own cells, or the cells of animals at higher trophic levels. Animals use this energy in the manufacture of organic molecules such as amino acids and fatty acids, which are necessary for the growth and repair of tissues. By the intake of this source of comparatively low-entropy energy (i.e., plants), and the release of higher-entropy energy, they too are able to support chemical reactions which entail localized decreases in entropy and permit the existence of complex, differentiated cells and tissues.

Photosynthesis, then, is a necessary and sufficient energetic starting point for the upkeep of the complex, out-of-equilibrium systems that constitute living organisms (see e.g., Catling et al., 2005;

Bounama et al., 2007). Its effects are far-reaching: after death, the carbon from plants and animals is processed into fossil fuels by anaerobic bacteria in the Earth's crust (followed by chemical alteration as a result of conditions of high temperature and pressure), which is then burned to provide energy for the preparation and maintenance of other out-of-equilibrium systems.

All of this is possible because the Sun is able, through nuclear fusion in its core, to maintain a temperature imbalance with respect to the Earth, which allows for the provision of low-entropy energy. Stars may be described as "low-entropy hot-spots."

> The sky is in a state of temperature imbalance: one small region of it, namely that occupied by the Sun, is at a very much higher temperature than the rest. This fact provides us with the required powerful low-entropy source. The Earth gets energy from that hot-spot in a low-entropy form (few photons), and it re-radiates to the cold regions in a high-entropy form (many photons). (Penrose, 1989, p. 415)

In order to understand how these hot-spots came into being, how it is possible for them to maintain their temperature gradients and the entropy changes associated with their formation, I will now sketch an outline of stellar birth and evolution.

30.2.2   Stars as Entropy Powerhouses

Were it not for the existence of life, stars would be the end of every thermodynamical story: hot, dense spheres of gas in hydrostatic equilibrium, against the backdrop of the cold, diffuse expanse of space. As it is, in the present epoch, stars – and, in particular, our Sun – are the only astrophysical objects that are relevant to our thermodynamics on Earth. It is therefore important to understand the way in which stars produce their vital source of low-entropy energy, and also the way in which broader cosmological constraints are imprinted in their thermodynamics, i.e., the way in which early-Universe events are causally efficacious in stars now.

Just under a billion years after the Big Bang, clouds (i.e., over-densities) of gas and dust within young galaxies began to accrue matter through gravitational attraction until hydrostatic equilibrium could no longer be supported. Eventually, masses were obtained at which the kinetic energy of the gas pressure was insufficient to counteract the gravitational potential. When this critical mass (the Jeans mass) was achieved, clouds began to collapse under their own gravity, radiating away the gravitational potential energy as thermal kinetic energy, in the form of high-entropy radiation due to collisions between the increasingly energetic particles.

This collapse process continued unimpeded for as long as this radiation could escape, and the rate at which the temperature of the contracting gas clouds rose was tempered by the outward flux of radiation. However, since contraction increases particle densities, eventually radiation could no longer escape, and developing proto-stars became opaque to their radiation. At this point, the temperature of the collapsing cloud rose rapidly, and the dust itself began to radiate in the infra-red region of the electromagnetic spectrum.

In the meantime, as the collapsing cloud contracted, a very high-density core developed (the inverse square law rules that the gravitational force is quadratically stronger as the center of mass is approached) in which the temperatures were substantially higher than in the less dense surrounding envelope. As the temperature of the core climbed, molecular hydrogen began to dissociate and ionize, absorbing some of the energy of the collapse. Eventually, the conditions of temperature and pressure were sufficient for hydrogen to undergo nuclear fusion in the core. Though requiring a phenomenally high activation energy (the Coulomb barrier) nuclear fusion is extremely exothermic, and results in the release of high-energy radiation, with an associated increase in entropy. One billion years after the Big Bang, stellar cores were the only environments in the Universe sufficiently hot to overcome this barrier.

Nuclear fusion occurs primarily by a series of reactions known as the proton-proton chain, in which the production of one atom of ^4He involves the release of eight energetic photons. The interactions between these photons and the electrons and protons in the core account for the kinetic energy which resists the gravitational collapse of the star. (Two energetic neutrinos are also released in the production of an atom of helium, but, since neutrinos interact incredibly weakly with matter, their energy cannot support the star against gravitational collapse.) Nuclear fusion reactions continue until the star's supply of core hydrogen has been depleted, at which point fusion of the helium product begins. This takes around 10 billion years for a solar-mass star: sufficient time for the evolution of photosynthetic plants. Eventually, a large star will burn its fusion products through to ^{56}Fe, which has the lowest mass per nucleon of any nuclide, and the third-highest binding energy per nucleon, and is therefore the most stable end-product of nucleosynthesis. Iron can therefore be taken as the *de facto* equilibrium state of baryonic matter with the corresponding highest entropy. (Both ^{58}Fe and ^{62}Ni have higher binding energies per nucleon than ^{56}Fe (Fewell, 1995), but they are less rarely produced by nucleosynthesis because the required additional neutrons are not readily available in stellar contexts.)

In this way, stars achieve hydrostatic "equilibrium" (not true equilibrium, but a reasonable approximation, even over millions of years), with the outward radiation pressure balancing the inward force of gravity. A solar-mass star may "burn" hydrogen in its core, and provide a continuous supply of radiation to maintain this hydrostatic equilibrium, for around 10 billion years.

Stars could not operate as entropy powerhouses in this way were it not for the *possibility* of gravitational collapse and nuclear fusion. Gravitational collapse could not have occurred if matter in the Universe was not in the first place dispersed and homogeneous. If, for example, the distribution of matter was already clumped, prior to the era of structure formation, collapse would not have occurred in the same way, and stellar nuclear fusion would never have been activated. As it is, the distribution of matter in the Universe, one hundred thousand years after the Big Bang, was extremely smooth. This means that, prior to gravitational collapse, matter was essentially in its highest gravitational potential energy configuration, and, given the entropy-increasing effect of gravitational collapse, its lowest entropy state. (The relationship between gravity and entropy is far from straightforward, but Wallace (2010) provides a helpful overview.)

It was also necessary that the interstellar medium was composed mainly of hydrogen, a relatively low-entropy state for matter. If all the baryonic matter in the Universe had already combined into iron, gravitational collapse would result in the formation of neutron stars and/or black holes. Although photons would also be emitted in these collapse processes, and entropy would also increase, the time-scales of emission, and the emission frequencies, would not be appropriate for the evolution and sustenance of Earth-like life. Primordial cosmology gives us very good reasons why the interstellar medium consisted almost entirely of hydrogen and helium, and these will be discussed in the next section.

30.2.3   *Nucleosynthesis: The Primordial Thermal Bath*

The Universe at $t_1 = 10^{-2}$ seconds consisted of a homogenous, slowly cooling thermal bath of fundamental particles in a state of local thermal equilibrium, mediated by particle interactions. It was in *equilibrium* because the particle interactions which maintain the balance between the creation and decay of these particles were sufficiently rapid (with respect to the expansion rate) that the particle species were at each point in equilibrium with each other. And the equilibrium state was *local* rather than *global* because the Universe was expanding. An expanding Universe cannot be in global equilibrium since its total phase-space has a degree of freedom for its size (which is, by definition, changing), and equilibrium requires that all thermodynamic parameters be time independent.

While heavy elements in the modern Universe have been generated through the fusion of less massive elements in stars and supernovae, light element isotopes of hydrogen, helium, and lithium had non-zero abundance even in infant stars. These elements have primordial origins; they are the result of thermonuclear reactions which took place in the first second of the Universe, in a process known as Big Bang nucleosynthesis.

A fraction of a second after the Big Bang, neutrons and protons interchanged by way of weak interactions, with the release of a neutrino. At around $t = 1$, neutrinos "froze-out" of equilibrium (the rate of the expansion of the Universe exceeded the rate of their interactions), fixing the ratio of neutrons to protons at around a fifth. However, neutrons are highly unstable, and they promptly began to combine into deuterium. The binding energy of deuterium is low, and the high temperatures meant that deuteron nuclei were disassociated by energetic photons upon formation. Eventually, the Universe cooled sufficiently that photons were no longer energetic enough to disrupt deuterium, and the "bottleneck" ended. Deuterium then acted as a catalyst in light element synthesis, and, almost all neutrons were bound into ^4He.

Twenty minutes in, the energies required to overcome the Coulomb repulsion between nuclei were no longer available, and nucleosynthesis terminated. At this point, the abundance (as a proportion of baryonic mass) of H was 75%, that of ^4He was 25%, and there were trace amounts of deuterium, lithium, and beryllium. No other elements existed; heavier elements would be manufactured in the cores of stars hundreds of millions of years later. That these abundances took their particular values when primordial nucleosynthesis terminated is critical to the way in which the evolution of the Universe later unfolded.

30.2.4   Conclusion

Iron is the most stable element; as long as there is non-ferric matter in the Universe, there cannot be true equilibrium, and a global maximum-entropy state has not been reached. However, as we have seen, the Universe expanded and cooled too quickly during its primordial nucleosynthesis stage for hydrogen to fuse all the way through to ^{56}Fe. In fact, just a quarter of the primordial hydrogen was converted to helium before energies fell below the fusion activation barrier. Subsequently, the primordial elements were preserved in high-energy, low-entropy states, a fact that is crucial to their role as fodder for nuclear fusion.

Now, owing to the potential energy lost in gravitational collapse, stars exist in which the conditions of temperature and pressure are sufficiently high that the enormous activation energies are easily overcome, and nuclear fusion reaction pathways take light elements to lower energy states through a series of thermonuclear reactions. In these nuclear fusion reactions, small amounts of mass are converted into enormous amounts of energy, which are released as radiation in the form of energetic photons. These photons have a significantly higher temperature than the black-body temperature of the surface of the Earth, therefore, when they are absorbed, their entropy is lower than the entropy of the radiation that is reemitted back into the atmosphere. Our non-equilibrium processes on Earth, described by statistical mechanics, subsist on this ration of low-entropy energy.

So it is that the low-entropy state that is postulated by the Past Hypothesis aligns with the historical narrative of entropy increase in our Universe, where a low-entropy initial state provided the "budget" required for the development of structure, and for the evolution of beings whose efforts toward explaining entropy asymmetry draw from the same account.

Notes

1 See Frigg and Werndl (this volume, a, b) for detailed discussions of alternative definitions of equilibrium in both Boltzmannian and Gibbsian statistical mechanics.
2 For readers interested in quantum statistical mechanics, see Shenker (this volume).
3 See Frigg and Werndl (this volume, a, p. 4) for a description of coarse-graining into macro-regions.
4 For more details on this and other difficulties with coarse-graining, see (Frigg, 2008, pp. 134–137).

5 The use of the Lebesgue measure is often justified on the basis that it is "natural", but there are associated problems, as outlined in, e.g., Davey (2008), and Sklar (2006).

6 For a more technical description of the approach to equilibrium, refer to Frigg and Werndl (this volume, pp. 5–9).

7 I refer the reader to Frigg (2008), Frigg and Werndl (2011), Jaynes (1964), and Lavis (2008) for the details and merits of both approaches.

8 The discrepancy between the time-asymmetric laws of classical mechanics and the time-asymmetric behaviour of macroscopic systems was first noted by Loschmidt in 1876.

9 Boltzmann (1895) credits the idea of chance fluctuations from equilibrium to his "old assistant, Dr Schuetz." The full concept of Boltzmann brains was first articulated in the work of Albrecht and Sorbo (2004).

10 Dorr and Arntzenius (2017) offer a more detailed explication of Boltzmann brains.

References

Albert, D.Z. (2000). *Time and Chance*. Cambridge: Harvard University Press.

Albrecht, A. and Sorbo, L. (2004). Can the universe afford inflation? *Physical Review D*, 70(6): 063528.

Boltzmann, L. (1895). On certain questions of the theory of gases. Nature, 51: 413.

Boltzmann, L. (1896–1898). *Vorlesungen Über Gastheorie*. Leipzig: J.A. Barth.

Bounama, C., Von Bloh, W. and Franck, S. (2007). How rare is complex life in the Milky Way? *Astrobiology*, 7(5): 745–756.

Callender, C. (2004). There is no puzzle about the low-entropy past. In C. Hitchcock (ed.), *Contemporary Debates in Philosophy of Science*. London: Blackwell, pp. 240–255.

Catling, D.C., Glein, C.R., Zahnle, K.J. and McKay, C.P. (2005). Why O_2 is required by complex life on habitable planets and the concept of planetary "oxygenation time". Astrobiology, 5(3): 415–438.

Davey, K. (2008). The justification of probability measures in statistical mechanics. *Philosophy of Science, 75*: 28–44.

Dorr, C. and Arntzenius, F. (2017). Self-locating priors and cosmological measures. In K. Chamcham, J. Silk, J.D. Barrow and S. Saunders (eds.), *The Philosophy of Cosmology*. Cambridge: Cambridge University Press, pp. 396–428.

Fewell, M.P. (1995). The atomic nuclide with the highest mean binding energy. *American Journal of Physics*, 63(7): 653–658.

Frigg, R. (2008). A field guide to recent work on the foundations of statistical mechanics. In D. Rickles (ed.), *The Ashgate Companion to Contemporary Philosophy of Physics*. London: Ashgate, pp. 99–196.

Frigg, R. and Werndl, C. (2011). Entropy – a guide for the perplexed. In C. Beisbart and S. Hartmann (eds.), *Probabilities in Physics*. Oxford: Oxford University Press, pp. 115–142.

Jaynes, E.T. (1964). Gibbs vs. Boltzmann entropies. American Journal of Physics, 33(5): 391–398.

Lavis, D. (2008). Boltzmann, Gibbs, and the concept of equilibrium. *Philosophy of Science*, 75: 682–696.

Lees, J.P., Poireau, V., Tisserand, V., Tico, J.G., Grauges, E., Palano, A. … and Kolomensky, Y.G. (2012). Observation of time-reversal violation in the B^0 meson system. *Physical Review Letters*, 109(21): 211801.

Loschmidt, J. (1876). *Über den Zustand des Wärmegleichgewichtes eines Systems von Körpern mit Rücksicht auf die Schwerkraft: I [-IV]*. Vienna: KK Hof-und Staatsdruckerei.

Penrose, R. (1989). *The Emperor's New Mind: Concerning Computers, Brains and the Laws of Physics*. Oxford: Oxford University Press.

Price, H. (2004). On the origins of the arrow of time: Why there is still a puzzle about the low-entropy past. In C. Hitchcock (ed.), *Contemporary Debates in Philosophy of Science*. London: Blackwell, pp. 219–239.

Price, H. (2013). Time's arrow and Eddington's challenge. In B. Duplantier (ed.), *Time: Poincaré Seminar 2010*, Vol. 63. Basel: Springer Science & Business Media, pp. 187–215.

Schrödinger, E. (1944). *What Is Life?* Cambridge: Cambridge University Press.

Sklar, L.. (2006). Why does the standard measure work in statistical mechanics? *Boston Studies in the Philosophy of Science*, 251: 307–320.

Wallace, D. (2010). Gravity, entropy, and cosmology: In search of clarity. *The British Journal for the Philosophy of Science*, 61(3): 513–540.

Further Reading from the Editors

H. Price, *Time's Arrow and Archimedes' Point* (New York, Oxford: Oxford University Press, 1996) is a classic extended treatment of temporal asymmetry in physics. H.R. Brown and J. Uffink, "The origins of time-asymmetry in thermodynamics: The minus first law," *Studies in History and Philosophy of Science Part B: Studies in History and Philosophy of Modern Physics* 32(4), (2001): 525–538 has become a standard reference point. A very thorough review of the foundations of statistical mechanics, including the entropy asymmetry, is J. Uffink, "Compendium of the foundations of classical statistical physics," *Handbook for Philosophy of Physics*, J. Butterfield and J. Earman (eds.) (Amsterdam: Elsevier, 2006). More recent proposals which locate the source of the entropy asymmetry in quantum theory are D. Wallace, "The logic of the past hypothesis," in *Time's Arrows and the Probability Structure of the World*, B. Loewer, E. Winsberg and B. Weslake (eds.) (Cambridge, MA: Harvard University Press, forthcoming) and K. Robertson, "Asymmetry, abstraction, and autonomy: Justifying coarse-graining in statistical mechanics," *The British Journal for the Philosophy of Science* 71(2), (2018): 547–579.

Section B

Themes

PART VIII

Explanation

Introduction to Part VIII

Explanation is a core topic in a standard philosophy of science class. This volume concerns the philosophy of *physics*. While the two are undoubtedly intertwined, why dedicate a section to explanation? As with many of the topics in the "Themes" half of this book, there is a growing understanding that the issues surrounding explanation in physics are often specific to physics. Indeed, they are sometimes specific to small sub-domains of physics. In the first half of the 20th century, standard physics examples were used to support universal theories of explanation. By contrast, the philosophy of explanation is now often marked by explanatory pluralism; with this pluralism comes an acknowledgment of the existence of field-specific explanation types.

Despite this pluralism, many philosophers of physics see explanation as deeply significant. Arguments for or against the inclusion of some object into the ontology of a theory often rest on inference to the best explanation; for example, arguments over substantivalism often rest on the explanatory value of space or spacetime. Explanation also looms large in discussions of emergence, reduction, and inter-theoretic relations.

The peculiarities of explanation in physics largely arise from its mathematized nature. A focus on physics led Hempel and Oppenheim to posit the deductive-nomological model of explanation in 1948. This model assumes that regularities can be explained by subsumption under laws – clearly more applicable in law-based physics than in sciences where universal generalizations are more difficult to find. But physics also hosts its own stock of explanation types which depend on particular mathematical models. One class of explanation involves the renormalization group, a particular formal process by which we can rescale the variables describing a system. Other forms of explanation look, for example, geometrical or topological; many explanations seem to appeal to the features of either spacetime or a more abstract space. These kinds of explanation raise questions about the relationship between physics and mathematics – some of them look like distinctively mathematical explanations of physical phenomena.

The mathematical character of certain explanations fuels debates in philosophy of mathematics, notably a modern form of the indispensability argument for platonism. But it also casts doubt on the possibility of applying the causal model of explanation in physics. Bertrand Russell famously claimed to be unable to find examples of causal relations in physics:

… the word 'cause' is so inextricably bound up with misleading associations as to make its complete extrusion from the philosophical vocabulary desirable … All philosophers, of every school, imagine that causation is one of the fundamental axioms or postulates of science, yet, oddly enough, in advanced sciences such as gravitational astronomy, the word 'cause' never appears. Dr James Ward … makes this a ground of complaint against physics … To me, it seems that … the reason why physics has ceased to look for causes is that, in fact, there are no such things. The law of causality, I believe, like much that passes muster among philosophers, is a relic of a bygone age, surviving, like the monarchy, only because it is erroneously supposed to do no harm. (Russell, B. 1913, "On the notion of Cause", *Proceedings of the Aristotelian Society*)

Yet the causal model of explanation can be seen as the more successful successor of the deductive-nomological model. Was Russell right that it (and the notion of cause altogether) simply don't apply in physics? Mathias Frisch tackles this question in his chapter, pointing out that our informal explanatory strategies present a challenge for the neo-Russellian. Sophisticated models of causal explanation may also give us some clues as to how to provide a somewhat unified account of explanation.

Juha Saatsi's chapter extends this discussion to distinctively non-causal explanation. He discusses paradigmatic cases of non-causal explanation such as geometrical explanations, symmetry-based explanations, and the explanations provided by accounts of the relationship between theories. Although most of these are clearly non-causal, accounts of explanation which highlight the importance of counterfactual dependence can capture most of them. These accounts share some of their structure with causal explanations.

Outside of philosophy of physics, the notion of mechanism has received a great deal of attention in the philosophical literature. Mechanistic accounts seem to do a good job of capturing, for example, neurological or biological phenomena. Mechanistic explanation is thus an increasingly popular topic focus point in the philosophy of science. Might it be applied in physics? Laura Felline argues in her chapter that it can be (although not universally), and that philosophy of physics has much to learn from the topic.

The final chapter of this section provides a link to the next. Lina Jansson discusses a particular aspect of explanations in physics (and elsewhere) associated with scale and inter-theoretic relations. It is clear that we often treat "higher level" (that is, generally, larger scale) descriptions as more explanatory than microscopic descriptions. Hilary Putnam famously pointed out that a microphysical explanation of why a round peg does not fit in a square hole is of much less value than a simple geometrical one. Getting the scale right appears to be an important part of explanatory practice, but one that is not accounted for on many traditional models of explanation. The explanatory value of higher level explanations has sometimes been touted as an argument against reductionism, for example by Robert Batterman. Jansson gives a nuanced description of the territory, one which allows for novel explanatory value at higher levels but need not draw any extreme anti-reductionist conclusions.

31

CAUSAL EXPLANATION IN PHYSICS

Mathias Frisch

Are there causal explanations in physics? Answers to this question range from the claim that there are no causal explanations in physics since the notion of cause plays no legitimate role in physics (and, perhaps, elsewhere) to the claim that all explanations in physics are causal in virtue of the fact that *all* explanations in general (or at least all scientific explanations) are causal. In addition to these two polar opposite positions, some philosophers have argued for pluralist views that allow for both causal explanations and non-causal explanations in physics.

Some of the arguments concerning the place of causal explanations in physics appeal to general conditions that any account of scientific explanation ought to satisfy. Thus, according to Carl Hempel's deductive nomological (*DN*) model of explanation, there are no causal explanations in physics simply because there are no genuinely causal explanations anywhere (Hempel and Oppenheim, 1948; Hempel, 1970). By contrast, on David Lewis's account of explanation any explanation of a specific event is causal (whether in physics or not) in virtue of providing us with information about the causal history of the event in question (Lewis, 1986). If all explanations are causal, explanations in physics are causal as well.

Other views appeal to putatively distinct features of theorizing in physics, either to argue that physics is particularly well suited for causal explanations or (perhaps more often) to argue that physics is especially inhospitable to causal notions and therefore to causal explanations. According to views of the latter kind, causal explanations may play a role in the so-called "special sciences" but such explanations sit ill with how physical theories represent the world.

In what follows I will discuss both what influential general theories of scientific explanation imply for the particular issue of causal explanation in physics and whether there are arguments distinct to physics concerning the status of causal explanations.

31.1 The *Deductive Nomological* Model

The theory of scientific explanation as a philosophical sub-discipline has its origins in Hempel's development of the *Deductive Nomological (DN)* model. According to the *DN* model, a scientific explanation is a deductively valid argument from true premises (constituting the explanans), which has the explanandum sentence as its conclusion (Hempel and Oppenheim, 1948; Hempel, 1970). This is the deductive part of the model. The nomological part states that the premises must contain at least one law essentially. The *DN* model was intended to provide an analysis of the concept of scientific explanation that answers to empiricist worries about causal and other modal notions and denies that there are properly causal explanations.

One of the many challenges the *DN* model faces is to provide an account of the concept of a *scientific law* that allows us to distinguish laws from accidental regularities in a manner acceptable to empiricists (Hempel, 1970).[1] Independently of whether and how this challenge can be met, physics with its generalizations of relatively broad scope appears to be particularly well suited for the *DN* model. Many explanations in physics involve derivations from equations that are taken to have broad or even universal validity, such as Newton's Laws, the Maxwell Equations, or the Schrödinger equation, and no matter how one ultimately tries to flesh out the concept of law, many of the basic equations of physics will clearly have to fall under it. It is less clear how well the *DN* model can be extended to other sciences and whether generalizations in the special sciences have enough of the requisite characteristic features to count as genuinely nomic constraints.

A number of widely discussed putative counterexamples to the *DN* model suggest that attempts to formulate a theory of explanation that avoids causal notions are ultimately unsuccessful. Perhaps the most prominent problem in this respect is the problem of explanatory asymmetries. There are many cases in which a derivation of an explanandum event *E* from laws *L* and initial conditions *I* strikes us as being explanatory, while the inverse derivation of *I* from *L* and *E* does not seem to be explanatory, even though both derivations satisfy the *DN* model. Consider Sylvain Bromberger's well-known example of the flagpole and its shadow (Bromberger, 1966; van Fraassen, 1980): While a derivation of the length of the shadow from the height of the flagpole and the Sun's angle in the sky together with the law of the rectilinear propagation of light seems to constitute an explanation of the length of the shadow, a derivation of the height of the flagpole from the length of the shadow does not seem to explain the flagpole's height. A plausible diagnosis of why the second derivation is not explanatory is that it purports to explain the height of the flagpole in terms of its effect, the shadow. And while it may be possible to *derive* the presence of a cause from the occurrence of its effects, the effects' occurrence does not *explain* the cause. Thus, the fact that there are explanatory asymmetries suggests that it might not be possible to develop a general non-causal theory of explanation.

The problem of explanatory asymmetries puts pressure on the claim that the *DN* model provides sufficient conditions for explanation. Michael Scriven and others have argued that the *DN* model also does not provide necessary conditions for explanations (Scriven, 1962). Scriven cites what appear to be paradigmatic cases of causal explanations, which do, however, not satisfy the *DN* model. For example, an adequate explanation of why the ink jar spilled might be that I bumped the jar with my elbow. This explanation is adequate, Scriven argues, even if we are not in a position to derive the ink jar's spilling from physical laws together with appropriate initial conditions. One might question the relevance of Scriven's case to the issue of scientific explanation, arguing that it is only an example of a common-sense explanation. But as we will see below, even within physics there are many instances of inferences proceeding from less than a full specification of initial and boundary conditions and many of these inferences appear to be paradigmatically causal and explanatory inferences.

31.2 Conserved Quantity Accounts of Causation

The lesson some philosophers have drawn from examples such as the ones discussed in the previous section is that the *DN* model ought to be abandoned in favor of a causal account of explanation. We ought to put "*cause*" back into "*because*," Wesley Salmon urged (Salmon, 1984, 2006). But what is it to causally explain a phenomenon? One answer to this question is given by the causal process account first proposed by Salmon (1984) and developed further in Phil Dowe's conserved quantity account (Dowe, 2000). Dowe distinguishes causal processes and causal interactions, which he defines as follows:

> CQ1. A causal process is a world line of an object that possesses a conserved quantity.
> CQ2. A causal interaction is an intersection of world lines that involves the exchange of a conserved quantity. (2000, 90)

Conserved quantities are those quantities, such as energy, momentum, mass, or charge, that are conserved according to our physical theories. Even more so than the *DN* account, the conserved quantity account appears to derive its inspiration primarily from physics, where conservation laws play a fundamental role. In fact, according to Noether's First Theorem, there is a conservation law associated with each continuous symmetry property of a system (see Brading and Castellani, 2003, especially the essay by Brading and Brown therein).

Even though process accounts of causation were designed partly with the problems of the *DN* model in mind, as an account of scientific *explanation* the conserved quantity account is arguably subject to many of the same counterexamples that plague the *DN* model (see Woodward, 2017). A central problem for the *DN* model is that there can be nomic connections between an explanandum event and other events that do not capture features explanatorily relevant to the occurrence of the explanandum. Similarly, a quantity conserved in a causal process or exchanged in causal interactions also need not be explanatorily relevant to the phenomenon. Consider a collision of two billiard balls, during which some very small amount of electric charge is exchanged, and let us assume that the balls' charge is conserved both before and after the collision. The motion of the billiard balls constitutes two causal processes joined by a causal interaction. Yet charge conservation does not explain the billiard balls' motion. The relevant conservation law is energy and momentum conservation. But what makes it the case that it is energy and momentum conservation and not charge conservation that causally explains the motion? One possibility is to appeal to counterfactual information at this point: energy-momentum conservation provides the correct explanation since the motion of the two balls varies with changes in the balls' initial momenta and energies but does not vary with changes to the balls' charges in the right way. But this requires that we supplement the pure conserved quantity account with counterfactual considerations.

The conserved quantity account does not, on its own, provide a distinction between cause and effect and hence, like Hempel's account, is faced with the problem of explanatory asymmetries. Dowe's solution is to supplement the account by appealing to Hans Reichenbach's *fork asymmetry* (Reichenbach, 1956). A *conjunctive fork*, as Reichenbach defines it, consists of three events A, B, and C, such that A and B are unconditionally correlated but conditionalizing on C renders A and B probabilistically independent. That is:

$$P(A\&B) > P(A)P(B) \tag{31.1}$$

$$P(A\&B|C) = P(A|C)P(B|C) \tag{31.2}$$

The event C, it is said, *screens off A from B*. Conjunctive forks can be temporally open or closed. If there is an event C occurring in the past that screens off A from B, but there is no screening-off event in the future of A and B, then this constitutes an open fork. If there is an event C in the past and in addition an event C' in the future of A and B that screen off A from B, we have a closed fork. Now, Reichenbach's fork asymmetry thesis consists in the claim that all open forks are open toward the future: there are no conjunctive forks for which only a future screening-off event exists. Conjunctive forks allow us to introduce a direction for causal processes, and hence, allow a distinction between cause and effect. Dowe defines the direction of causal processes as follows:

> The direction of a causal process is given by the direction of an open conjunctive fork part-constituted by that process; or, if there is no such conjunctive fork, by the direction of the majority of open forks contained in the net in which the process is found. (2000, 204)

Two aspects of this definition are worth noting. First, adopting a disjunctive criterion allows him to distinguish causes from effects by their location in a causal net, even when a process itself does not involve a causal fork. Second, in a departure from Reichenbach, Dowe wants to allow for the possibility of backward causation and hence allows forks to open toward the past. Dowe's motivation for this is that he wants to appeal to backward causal relations as providing a local causal explanation of the quantum mechanical correlations exhibited by entangled states, which cannot be given a common-cause explanation.

In order to address the problems of explanatory asymmetries and irrelevancies, the conserved quantity account has to take two additional kinds of explanatory information on board: probabilistic information and counterfactual information. This raises the question as to what ultimately is doing the explanatory work in the account: is it the appeal to a conserved quantity or rather the counterfactual and probabilistic dependencies? The significance of Reichenbach's conjunctive fork is that conjunctive forks instantiate what appears to be a fundamental explanatory relation: common-cause explanations. Correlations between two distant events *A* and *B*, which are not related as cause and effect, are explained by an event *C*, the common cause of *A* and *B*, which, Reichenbach postulates, screens off *A* from *B*. Thus, one might want to bypass the appeal to conserved quantities and construct an account of causation and a causal explanation directly in terms of counterfactual and probabilistic dependencies. I will take up this suggestion in Section 31.3.2. First, however, I want to discuss a powerful and influential challenge to any causal account of explanation in physics, which has the same empiricist roots as the *DN* model but was developed independently of the literature on scientific explanation.

31.3 Causation in Physics

31.3.1 *Mach's and Russell's Challenges to Causation*

One answer to the problem of explanatory asymmetries has been to argue for the need for a causal account of explanation with an underlying notion of causation rich enough to underwrite a distinction between causes and effects. But there is another argumentative strand in the literature that denies the applicability or at least the usefulness of causal notions in physics. This literature is not primarily focused on the notion of *explanation*, but focuses directly on the notion of *cause*. Arguments pointing to an alleged incompatibility of causal thinking with theorizing in physics can be traced back to the writings of Ernst Mach (Mach, 1900, 1905) and Bertrand Russell's famous article "On the Notion of Cause" (Russell, 1912). More recent defenders of such a view include Huw Price (Price, 1997; Price and Weslake, 2009), Hartry Field (Field, 2003), John Earman (Earman, 2011), and, to some extent, John Norton (J. D. Norton, 2003, 2007; J. Norton, 2009).[2] While Russell's attack was a broad attack on the notion of cause in general, recent neo-Russellians seem to follow Mach in focusing their attention on physics and argue that there is something distinct about physics that makes it especially inhospitable to causal notions and, hence, to causal explanations.

Three Machian or Russellian arguments for why causal notions cannot have a legitimate place in physics have been particularly influential:

(i) The notions of cause and effect are inherently vague in contradistinction to the mathematical precision of derivations in physics. This vagueness infects especially metaphysically "rich" notions of causal production, which also sit ill with a broadly empiricist outlook.

(ii) Causal explanations are legitimate in contexts in which we can isolate a small set of factors of interest as those responsible for a phenomenon or the occurrence of an event and implicit in the notion of cause is a distinction between causes and background conditions. This distinction cannot be drawn in physics, where the character of basic physical equations requires that we take a complete cross section of the backward light cone of an event as its cause.

457

(iii) The notion of cause is time–asymmetric, whereas the dynamical laws of the fundamental or the established theories of physics have the same character in both temporal directions and are bi–deterministic.

The Machian and Russellian challenges imply that there can be no causal explanations in physics. Thus, neo-Russellians either would have to adopt a different, non-causal account of explanation in physics or would have to argue that explanation is an inherently pragmatic, context-dependent notion that may serve an important purpose as an external add-on to scientific investigations but has no place in physics proper.

31.3.2 *Answering the Challenge: Structural Models and Interventionism*

The first argument presents a challenge to any broadly empiricist theory of scientific explanation. Yet it is a challenge that arguably can be – and in recent years has been – met by the Bayes net or structural model accounts of causation developed by Peter Spirtes and his co-authors (2000) and by Judea Pearl (2000, 2009). These accounts provide mathematically rigorous and precise representations of causal structures. On Pearl's account, a structural causal model (SCM) consists of:

(i) a directed acyclic graph (which can visually be represented in terms of a "blobs-and-arrows" diagram) over a set of variables $V=\{X, Y, \ldots\}$ consisting of both endogenous variables V_i and exogenous variables U_i;

(ii) structural equations $x_i = f_i(pa_i, u_i)$, which specify the value of each variable x_i in terms of the value of the variable's causal parents pa_i and random exogenous disturbances u_i;

(iii) and a probability distribution $P(u_i)$ over the values u_i of the exogenous variables U_i, which induces a probability distribution over all variables.

There are two aspects of the structural model account that are particularly relevant to the issue of explanation. First, it is part of the definition of SCMs that the exogenous variables are probabilistically independent. From this together with the assumption that a causal model is complete one can derive the causal Markov condition, which states that for every variable X in V, X is probabilistically independent of the variables in the set $(V - \text{Descendants}(X))$ conditional on the parents of X. The causal Markov condition is a generalized common-cause condition. SCMs, thus, underwrite common-cause explanations.

Second, SCMs make perspicuous the tight connection between the notions of cause and intervention or manipulation. A causal model provides us with information on how the values of variables change under external interventions into a system.[3] And causal discovery algorithms allow us to construct causal models from information about probability distributions over the values of variables characterizing the system and from information about the effects of interventions. James Woodward has shown how this formal framework can be developed into a philosophical account of causation and of explanation (Woodward, 2003). On Woodward's account, to explain a phenomenon is to exhibit systematic patterns of counterfactual dependency: explanations allow us to answer *what-if-things-had-been-different questions* or *w-questions* (Woodward, 2003, p. 191). Since counterfactuals are in the first instance to be interpreted in terms of possible interventions into a system, w–questions are primarily requests for causal information. But Woodward's account allows for non-causal explanations as well. If the counterfactual changes in question cannot be interpreted as possible interventions, then an answer to a w-question can provide a non causal explanation.

Like the *DN* model and the conserved quantity account, Woodward's account of explanation seems to be motivated by features of explanatory practices in physics. While the *DN* model focuses on mathematical derivability and the conserved quantity account zeroes in on one particular albeit important feature of physical theories, Woodward's counterfactual account emphasizes the fact that physical equations constitute relations among *mathematical functions*, which provide us with information on how changing the value of one variable affects the values of other variables. Physical theories allow us to answer w-questions by representing a phenomenon as embedded into a pattern of functional dependencies.

Frisch (2016) argues that many physical theories provide mathematical machinery that appears to be tailor-made for the structural model account. Any linear differential operator L associated with an inhomogeneous differential equation $Ly = f(x)$ with constant coefficients possesses a *fundamental solution* or *Green's function G*, which is a solution to the inhomogeneous differential equation $LG = \delta(x)$. The Green function is quite naturally interpreted in causal or interventionist terms. The Green's function "propagates a point-inhomogeneity" and thereby tells us what the contribution of introducing a disturbance or perturbation into a system at (x', t') is to the state of the system at some other point (x, t). Thus, Green's functions are a natural candidate in physics for the structural equations in SCMs.[4]

What about the two remaining Russellian challenges I distinguished above? An argument appealing to the considerations described in (ii) has been defended by Field (2003). Paradigmatically causal explanations, Field claims, point to a small number of factors through which one could (at least in principle) manipulate a phenomenon. Field and others take this observation to suggest that the distinction between causes and background condition is an essential component of our concept of cause. But, the argument continues, this distinction appears to be obliterated in physics. For a large class of equations in physics, it is the case that solving these equations requires as input initial data on a complete cross section of the backward light cone of the spatial region occupied by the system of interest – that is, complete data in a region from which influences could reach the system by traveling at most at the speed of light – nothing less will do.

But if the set of an event's causes becomes too large, we seem to be committed to claims that conflict with central intuitions concerning the assertability of causal claims. Field asks us to consider a scenario in which Sara puts out a fire with a water-hose while Sam sits next to Sara praying for the fire to go out. It seems obvious that Sara's spraying the fire with water but not Sam's praying is a cause of the fire going out. Yet Sam's praying is in the backward light cone of the fire's going out. Thus, if we interpreted an event's physical determination by cross sections of its backward light cone causally, then Sam's praying next to a fire would come out as a cause of the fire's extinction just as much as Sara's aiming a hose at the fire does. This, Field maintains, is absurd. Yet physics provides no additional means for distinguishing causally salient factors from other factors within an event's backward light cone.

In reply to this argument, Woodward (2007) has argued that it is important to keep distinct different representations at different levels of grain. In a putatively complete microphysical representation of the fire and its surroundings, the *precise microphysical realization* of Sam's prayer will come out as a cause of the precise microphysical realization of the fire's extinction. By contrast, a more *coarse-grained macrophysical representation* of the fire's extinction will be counterfactually independent of a broad class of changes to Sam's macrostate, including, for example, whether he prays or just sits and watches Sara's rushing to put out the fire. At both levels, our intuitive causal judgments are preserved (see also Frisch, 2016, ch. 3).

Taking a step back, Field's worry and the neo-Russellians' position more generally are motivated by a way of thinking about physics that is quite common in philosophy: physics, it is assumed, ultimately presents us with global dynamical models of fundamental laws. To many philosophers, it seems difficult to conceive how causal models and, in particular an interventionist conception of cause, can get a foothold within this conception.[5] Yet there is an alternative conception that arguably fits much of the day-to-day practice of physics considerably better than does the globalists' picture and that can readily accommodate causal reasoning. On this second conception, the laws of physics are understood as rules governing localized subsystems of the universe (Ismael, 2016). Viewed from this perspective, the debate concerning the place of causal notions in physics is intimately connected with debates concerning the aims of theorizing in physics.

Perhaps the most influential argument for the claim that causal notions cannot play a legitimate role in physics appeals to the asymmetry of the causal relation, which is often taken to coincide with a temporal asymmetry according to which effects do not precede their causes. Since, as it is claimed the basic laws of physics are time-reversal invariant and have the same character in both the past and future direction, time-asymmetric causal structures are an illegitimate and epistemically not justifiable add-on to our physical theories (Price and Weslake, 2009; Earman, 2011). Russell is often interpreted as making this claim when he says that "in the motion of mutually gravitating bodies, there is nothing that can be called a cause and nothing that can be called an effect; there is merely a formula" (Russell, 1912, p. 141).

The argument has an analog in discussions of the proper interpretation of Green's function formalism. At least in the case of systems governed by so-called *hyperbolic equations*, one can represent one and the very same system in terms of a "causal" Green's function and in terms of its temporal inverse, an "anti-causal" Green's function. The first representation suggests that disturbances in a system propagate into the future, while the second representation suggests causal propagation into the past. Interpreting both representations causally threatens to result in a contradiction. But nothing can legitimately distinguish between the two representations. Hence, neither ought to be interpreted causally.

Its popularity notwithstanding, the argument can be challenged. First, the argument applies only to deterministic theories and among these arguably only to time-symmetric theories.[6] Theories with probabilistic state-transition laws are inherently time-asymmetric, as shown by Satosi Watanabe (1965).[7] Thus, if quantum mechanics is understood as a fundamentally probabilistic theory, the argument's scope is limited to what by the argument defenders' own lights are the less fundamental theories of classical physics.

Second, in concluding that there is no place for time-asymmetric causal relations in a theory with time-reversal invariant laws, the argument presupposes that the content of physics is exhausted by its dynamical equations and ignores the role of initial or final conditions. Generally, there is an asymmetry between prevailing initial and final conditions: initial conditions are random while final conditions are not. And this asymmetry, some philosophers argue, is intimately related to the causal asymmetry (Arntzenius, 1992; Maudlin, 2007; Frisch, 2016, ch.5), since an initial randomness assumption allows us to engage in common-cause reasoning (see, e.g., Pearl, 2000). Thus an asymmetry between prevailing initial and final conditions allows us to introduce a causal asymmetry and allows us to distinguish between the causal and anti-causal Green's functions: Since a putatively causal model constructed from a representation of a system in terms of the anti-causal Green's function would violate the initial independence assumption – an anti-causal world would have to satisfy a final independence assumption instead of an initial independence assumption – this assumption allows us to pick out the causal Green's function as providing the causally correct representation.

Arguably causal reasoning, and in particular common-cause reasoning, is a central and ineliminable inference pattern in physics (and elsewhere) since it allows inferences based on local data rather than on full knowledge of the state of the world on a full initial or final value surface to which we often do not have access. As a particularly stark example consider the detection of gravitational waves in 2016. The extremely strong correlations between the signals detected in the two LIGO (Laser Interferometer Gravitational-wave Observatory) detectors in Washington and Louisiana are part of the evidence for the colliding black holes as the signals' common cause. Implicit in the inference from the detected signals to the collision event as their cause is the assumption that there was no "carefully calibrated" gravitational wave coming in from past infinity, converging on the location of the black holes and re-diverging, thereby mimicking a wave produced by the collapsing black holes. We rule out this alternative explanation of the signals detected by LIGO as utterly implausible, because a source free gravitational field that mimicked the field associated with the black hole event would have required absurdly strongly correlated initial conditions, in violation of the randomness assumption. By contrast, we do not find "absurdly strongly correlated" final conditions implausible: correlated final conditions are just what we would expect as joint effects of a common cause such as the collapsing of the black hole. If we wanted to derive the black hole event without appealing to causal assumptions from knowledge of the data on a complete final value surface, we would have to know the precise state of the universe in a sphere with a diameter of many light-years – something that is obviously impossible.[8]

Deterministic laws appear to undercut time-asymmetric common-cause inferences for another reason, however. Under determinism, if there is an event C in the past of two events A and B that screens A and B off from each other, then there will also be an event $C^\star$ that occurs after A and B and renders the two events conditionally independent (Arntzenius, 1992). This threatens our ability to apply Pearl-style SCMs to physics. In particular, if the existence of earlier screening-off events also implies the existence of later screening-off events, then we cannot rely on causal discovery algorithms to infer causal relations from probabilistic dependencies and independencies. A common reply to this worry is to point out that future screening-off events, unlike those in the past, will in general be highly non-natural and non-localized (see, e.g., Woodward, 2007). Demanding that appropriate physical variables represent localized and not highly gerrymandered events allows us to preserve the asymmetry induced by the initial randomness assumption.

The initial randomness assumption is the very same assumption that underwrites the temporal asymmetry of statistical mechanics. Does this mean that the causal asymmetry and the thermodynamic asymmetry have the same origin? Two viewpoints seem possible: one can take the initial randomness assumption as fundamental and as the common origin of both the causal and thermodynamic asymmetry. Alternatively, one can understand the initial randomness assumption to reflect a fundamental causal asymmetry: initial states are distributed randomly precisely because (and just in case when) these states do not have common causes in their pasts that result in correlations.

One open question for an account of causal explanations in physics is to what extent structural causal models can be applied to quantum systems. To be sure, common-cause explanations and a microscopic randomness assumption also play a prominent role in explaining many quantum phenomena. For example, we explain why pure absorptions of a photon by an atom are much rarer than pure emissions, by appealing to the fact that the former would require photons that are finely tuned to one of the atoms excitation energies. By contrast, the energy of an emitted photon is of course always "finely tuned" to the excitation energies of an atom as the emission's cause (Atkinson, 2006). But there are quantum phenomena that apparently cannot be represented in terms of a local causal model. In particular, quantum entanglement poses a special challenge for causal explanation. Outcomes of measurements on entangled states that are spatially separated from each other are correlated, but these correlations cannot be explained by a localized common-cause model satisfying the causal Markov condition, as Bell's theorem shows. One version of the theorem says that there are quantum phenomena for which there is no model satisfying local causality (Wiseman and Cavalcanti, 2015). Local causality is a screening-off condition that states that the outcome B at one wing of the experiment is independent of the outcome A and measurement settings a at the other wing, conditional on the state preparation c, the measurement setting at the first wing b, and any hidden variable λ:

$$P(B|A; a; b; c; \lambda) = P(B|b; c; \lambda)$$

One can respond to Bell's theorem by giving up the Markov condition and allow for quantum common causes that do not screen off their effects from each other. Another response is to give up the prohibition against superluminal causation. This could take the form of positing either a direct causal link between the two wings of the experiment or a partially retro-causal connection that "zigzags" down and up the light cone centered on the preparation event c (Price, 2012).

How to causally model entangled states remains an unsolved question, but recent years have seen an increasing number of attempts to extend the framework of SCMs to quantum mechanics (Wood and Spekkens, 2015).

31.4   Causal Imperialism

Neo–Russellians deny that causal explanations play a fundamental role in physics. We have just seen that it is possible to resist their arguments. If we allow that there are causal explanations in physics, should we conclude that all explanations are causal? David Lewis thought so. In Lewis (1986) he argues that to explain is to provide information about the causal history of an explanandum (see also Skow, 2014).

Despite their stark disagreement, neo–Russellians and "causal imperialists," as we might call them, share a commitment to what Woodward has called "the hidden structure strategy" (Woodward, 2017). Both views are committed to the existence of what Peter Railton has called an "ideal explanatory text" (Railton, 1981) that contains all the information relevant to a complete explanation of some phenomenon. While actual explanations may fall short of providing us with the complete information contained in the ideal explanatory text, they are explanatory in virtue of providing us with some information about the text.

For the neo–Russellian, the fundamental explanatory structures consist of microphysically complete dynamical models of the backward light cone of a given explanandum. While the neo–Russellian view is compatible with the claim that in some non-fundamental domains and for pragmatic reasons information about the ideal explanatory text may fruitfully be presented in causal terms, the view holds that ideal physical explanations are not causal. Causal imperialism turns this picture on its head and maintains that the underlying ideal explanatory structures are causal structures. Hence all explanations are causal in virtue of the fact that they provide information about this structure, even though the information provided in an actual explanation may not be presented in explicitly causal terms.

As Woodward has argued, a problem for the hidden structure strategy is to explain how hidden structures that are epistemically inaccessible to us can account for the explanatory import of the explanatory accounts we actually give (Woodward, 2017). For the neo–Russellian, the problem is that we seem to be able to provide successful causal explanations of phenomena even when the complete initial data that are part of the ideal explanatory text are in principle inaccessible to us.

The causal imperialist's version of the hidden structure strategy faces an analogous problem. There are apparently successful explanations of phenomena that do not identify causes of the phenomenon. How does pointing to a hidden causal structure make perspicuous the explanatory import of such an explanation and what accounts for the difference between such an explanation and one that does explicitly identify a phenomenon's causes? One may demand that an account of causal explanation be able to relate the explanatory role of such explanations to the function of causal information more generally. As we have seen, Pearl and Woodward's accounts of causation emphasize two features as the characteristic function of causal notions. First, knowledge of causal structures allows us to identify relationships amenable to manipulation and control; and second, common-cause reasoning enables us to draw inferences from one time to another even when we possess only incomplete knowledge of the state of a system at a time. Yet as Woodward also points out, not every explanation fulfills either of these functions.

Take an explanation of the heat capacity of metals and, in particular, of the fact that the heat capacity is much lower than predicted classically, which appeals to the Pauli exclusion principle. This explanation embeds its explanandum into patterns of functional dependencies in a manner that allows us to see how the heat capacity depends on particle statistics. In order to get the correct result, we need to model free electrons in the metal as satisfying the quantum mechanical Fermi-Dirac statistics and the exclusion principle. The explanation appeals to properties of the phase-space available to the electrons and allows us to answer how the heat capacity would change if the available phase space were different. Thus, the explanation allows us to answer w-questions but not by specifying counterfactuals that can be interpreted in terms of interventions or manipulations of the electron states.

Arguably, by classifying explanations such as this as causal, the causal imperialist obliterates what is an important distinction between different explanatory functions and epistemic goals. That the value of the heat capacity of metals follows from features of the available phase-space and is not something that, even in principle, is open to manipulation or control seems itself to be explanatorily relevant, just as it is crucial to an explanation of the length of the shadow that it can be manipulated by changing the flagpole's heights. This distinction is lost if we classify both these explanations as causal by virtue of the fact that they provide information about the causal history of a sample of metal or the flagpole.

31.5 Conclusion

Are there, then, genuinely causal explanations in physics? For many decades the majority response to this question among philosophers of physics appears to have been "no." By contrast, after the end of the hegemony of the *DN* model, there was considerable support for causal theories of explanation among philosophers of science and metaphysicians more generally. In recent years we may be witnessing a rapprochement of the opposing camps. On the one hand, structural causal models introduced a formally precise, arguably metaphysically "thin" yet non-reductive notion of cause into philosophy that may be acceptable even to empiricist-minded philosophers of physics. On the other hand, there has been a growing interest in varieties of non-causal explanations in general philosophy of science challenging monolithic causal accounts of explanation (Lange, 2016). These developments make room for pluralist positions that accord causal notions and causal explanation a legitimate and even crucial role in physics while at the same time allowing that there are genuinely non-causal explanations in physics and elsewhere. Explanatory pluralism raises several questions, however, which point in directions for future research. Allowing for different models of explanation reopens the problem of explanatory asymmetries: if it is not a necessary condition for explanations to identify causes of the explanandum, does the problem of explanatory asymmetries reemerge? Can the problem be solved if we take causal explanations to occupy a privileged position in a theory of explanation, as Woodward's counterfactual account appears to do? Do different explanatory strategies reflect different epistemic goals? Is it nevertheless possible to give a unifying account of different explanatory accounts or at the very least to identify features shared by different types of explanation? Here the close conceptual links between the notions of explanation and understanding (Regt and Dieks, 2005) may provide some clues that may help in identifying common features of different models of explanation.

Notes

1 On this question see also Salmon (2006), Earman (1986), Cartwright (1983) Van Fraassen (1989).

2 See also the essays collected in Price et al. (2007) and especially Woodward's essay (Woodward, 2007). For a critical discussion of neo-Russellian arguments see Frisch (2016). Norton (2009) is part of a critical exchange on the role of causal notions in the derivation of dispersion relations (Frisch, 2009a, 2009b). Russell himself, it is often forgotten, changed his mind about the role of causation. For example, in *The Analysis of Matter*, he says that "all science rests upon induction and causality" (Russell, 1992).

3 There are various different notions of interventions that have been proposed in the literature: arrow-breaking or hard interventions, either involving intervention variables (in Woodward's account) or not (as in Pearl's *do-calculus*, see Pearl (2000)) and non-arrow-breaking or soft interventions, as investigated in Eberhardt and Scheines (2007).

4 For a more critical view on the causal role of Green functions, see Smith (2013), which is a criticism of Frisch (2009a, 2009b). Frisch (2016) contains a reply to Smith.

5 Frisch (2016, ch. 4) argues that interventionist causal notions can be introduced even within a globalist conception of physics.

6 Dynamical equations with a damping term, which are common in linear response theory, possess a unique, causal Green's function. (See Frisch, 2016, ch. 6).

7 See also Callender (2000).

8 See also the discussion in Albert (2015), even though Albert does not identify the kind of inferences that an initial randomness assumption makes possible as causal inferences.

References

Albert, D. (2015). *After Physics.* Boston: Harvard University Press.

Arntzenius, F. (1992). The common cause principle. *PSA: Proceedings of the Biennial Meeting of the Philosophy of Science Association*, 1992: 227–237.

Atkinson, D. (2006). Does quantum electrodynamics have an arrow of time? *Studies in History and Philosophy of Science Part B: Studies in History and Philosophy of Modern Physics*, 37(3): 528–541. https://doi.org/10.1016/j.shpsb.2005.03.003.

Brading, K. and Castellani, E., editors. (2003). *Symmetries in Physics: Philosophical Reflections* (1st edition). Cambridge; New York: Cambridge University Press.

Bromberger, S. (1966). Why questions. In R.G. Colodny (ed.), *Mind and Cosmos: Essays in Contemporary Science and Philosophy.* Pittsburgh: Pittsburgh University Press, pp. 86–111.

Callender, C. (2000). Is time 'handed' in a quantum world? *Proceedings of the Aristotelian Society*, 100: 247–269.

Cartwright, N. (1983). *How the Laws of Physics Lie.* Oxford: Oxford University Press.

Dowe, P. (2000). *Physical Causation.* Cambridge; New York: Cambridge University Press.

Earman, J. (1986). *A Primer on Determinism.* Amsterdam: Springer Science & Business Media.

Earman, J. (2011). Sharpening the electromagnetic arrow(s) of time. In C. Callender (ed.), *The Oxford Handbook of Philosophy of Time.* Oxford: Oxford University Press, pp. 486–527. http://www.oxfordhandbooks.com. proxy-um.researchport.umd.edu/view/10.1093/oxfordhb/9780199298204.001.0001/oxfordhb-9780199298204-e-17.

Eberhardt, F. and Scheines, R. (2007). Interventions and causal inference. *Philosophy of Science*, 74(5): 981–995.

Field, H. (2003). Causation in a physical world. In M.J. Loux and D.W. Zimmerman (eds.), *The Oxford Handbook of Metaphysics.* Oxford: Oxford University Press, pp. 435–460.

Fraassen, B.C. van (1980). *The Scientific Image.* Oxford; New York: Oxford University Press.

Frisch, M. (2009a). Causality and dispersion: A reply to John Norton. *The British Journal for the Philosophy of Science*, 60(3): 487–495.

Frisch, M. (2009b). 'The most sacred tenet'? Causal reasoning in physics. *The British Journal for the Philosophy of Science*, 60(3): 459–474. https://doi.org/10.1093/bjps/axp029.

Frisch, M. (2016). *Causal Reasoning in Physics.* Cambridge: Cambridge University Press.

Hempel, C.G. (1970). *Aspects of Scientific Explanation: And Other Essays in the Philosophy of Science.* New York; London: Free Press.

Hempel, C.G. and Oppenheim, P. (1948). Studies in the logic of explanation. *Philosophy of Science*, 15(2): 135–175.

Ismael, J.T. (2016). *How Physics Makes Us Free* (1st edition). New York: Oxford University Press.

Lange, M. (2016). *Because Without Cause: Non-Causal Explanations in Science and Mathematics* (1st edition). New York: Oxford University Press.

Lewis, D. (1986). Causal explanation. In D. Lewis (ed.), *Philosophical Papers Vol. Ii.* Oxford: Oxford University Press, pp. 214–240.

Mach, E. (1900). *Die principien der wärmelehre. Historisch-kritisch entwickelt,.* Leipzig: J.A. Barth.

Mach, E. (1905). *Erkenntnis und Irrtum. Skizzen zur Psychologie der Forschung.* Leipzig: Barth.

Maudlin, T. (2007). *The Metaphysics within Physics.* Oxford; New York: Oxford University Press.

Norton, J. (2009). Is there an independent principle of causality in physics? *British Journal for the Philosophy of Science*, 60(3): 475–486.

Norton, J.D. (2003). Causation as folk science. *Philosophers' Imprint*, 3(4): 1–22.

Norton, J.D. (2007). Do the causal principles of modern physics contradict causal anti-fundamentalism? In P.K. Machamer and G. Wolters (eds.), *Thinking about Causes: From Greek Philosophy to Modern Physics.* Pittsburgh: Pittsburgh University Press, pp. 222–234.

Pearl, J. (2000). *Causality: Models, Reasoning, and Inference.* Cambridge; New York: Cambridge University Press.

Pearl, J. (2009). *Causality : Models, Reasoning, and Inference* (2nd edition). Cambridge; New York: Cambridge University Press.

Price, H. (1997). *Time's Arrow & Archimedes' Point : New Directions for the Physics of Time*. New York: Oxford University Press.

Price, H. (2012). Does time-symmetry imply retrocausality? How the quantum world says 'maybe'? *Studies in History and Philosophy of Science Part B*, 43(2): 75–83.

Price, H., Corry, R. and University of Sydney, and Centre for Time (2007). *Causation, Physics, and the Constitution of Reality Russell's Republic Revisited*. Oxford; New York: Clarendon Press; Oxford University Press.

Price, H. and Weslake, B. (2009). The time-asymmetry of causation. In H. Beebee, C. Hitchcock, and P. Menzies (eds.), *The Oxford Handbook of Causation*. Oxford: Oxford University Press, pp. 414–443.

Railton, P. (1981). Probability, explanation, and information. *Synthese*, 48(2): 233–256.

Regt, HW. De and Dieks, D. (2005). A contextual approach to scientific understanding. *Synthese*, 144(1): 137–170. https://doi.org/10.1007/s11229-005-5000-4.

Reichenbach, H. (1956). *The Direction of Time*. Berkeley: University of California Press.

Russell, B. (1912). On the notion of cause. *Proceedings of the Aristotelian Society*, 7: 1–26.

Russell, B. (1992). *The Analysis of Matter*. London: Routledge.

Salmon, W.C. (1984). *Scientific Explanation and the Causal Structure of the World* (1st US-1st printing edition). Princeton, NJ: Princeton University Press.

Salmon, W.C. (2006). *Four Decades of Scientific Explanation* (1st edition). Pittsburgh: University of Pittsburgh Press.

Scriven, M. (1962). Explanations, predictions, and laws. In H. Feigl and G. Maxwell (eds.), *Scientific Explanation, Space, and Time* (1st edition). Minneapolis: University of Minnesota Press, pp. 170–230.

Skow, B. (2014). Are there non-causal explanations (of particular events)?" *The British Journal for the Philosophy of Science*, 65(3): 445–467. https://doi.org/10.1093/bjps/axs047.

Smith, S.R. (2013). Causation in classical mechanics. In R. Batterman (ed.), *The Oxford Handbook of Philosophy of Physics*. Oxford: Oxford University Press, pp. 107–140.

Spirtes, P., Glymour, C.N. and Scheines, R. (2000). *Causation Prediction & Search 2e*. Boston: MIT Press.

Van Fraassen, B.C. (1989). *Laws and Symmetry*. Oxford; New York: Oxford University Press.

Watanabe, S. (1965). Conditional probability in physics. *Progress of Theoretical Physics Supplement*, 65: 135–160.

Wiseman, H.M. and Cavalcanti, E.G. (2015). Causarum investigatio and the two Bell's theorems of John Bell. March. Available at: http://arxiv.org/abs/1503.06413.

Wood, C.J. and Spekkens, R.W. (2015). The lesson of causal discovery algorithms for quantum correlations: Causal explanations of Bell-inequality violations require fine-tuning. *New Journal of Physics*, 17(3): 033002. https://doi.org/10.1088/1367-2630/17/3/033002.

Woodward, J. (2003). *Making Things Happen: A Theory of Causal Explanation*. Oxford: Oxford University Press.

Woodward, J. (2007). Causation with a human face. In H. Price and R. Corry (eds.), *Causation, Physics, and the Constitution of Reality: Russell's Republic Revisited*. Oxford: Oxford University Press.

Woodward, J. (2017). Scientific explanation. In E.N. Zalta (ed.), *The Stanford Encyclopedia of Philosophy* (Spring 2017). Metaphysics Research Lab, Stanford University. Available at: https://plato.stanford.edu/archives/spr2017/entries/scientific-explanation/.

Further Reading from the Editors

For an overview of the topic, see Frisch, Mathias "Causation in physics" *The Stanford Encyclopedia in Philosophy* (Fall 2020 Edition), Edward N. Zalta (ed.). The most well-developed account of causal explanation is James Woodward's. See his *Making Things Happen: A Theory of Causal Explanation* (Oxford University Press, 2003). Huw Price and Richard Cory's *Causation, Physics, and the Constitution of Reality: Russell's Republic Revisited* (Oxford: Oxford University Press, 2007) is an excellent collection with a great deal to say about physics. Understanding the debate about causal explanation in physics requires understanding causal anti-fundamentalism and skepticism in physics. John Norton has advanced this in a number of papers. See his "Causation as Folk Science" (*Philosophers' Imprint* 3(4): 1–22, 2003), "Do the causal principles of modern physics contradict causal anti-fundamentalism?" In *Thinking about Causes: From Greek Philosophy to Modern Physics*, P.K. Machamer and G. Wolters (eds.) (Pittsburgh University Press, 2007) and "Is there an independent principle of causality in physics?" (*British Journal for the Philosophy of Science* 60(3): 475–486, 2009).

32

NON-CAUSAL EXPLANATIONS IN PHYSICS

Juha Saatsi

32.1 Introduction

The nature of scientific explanation and understanding is of central interest to philosophers of science. How (or by virtue of what) does a given theory, model, or body of information explain? What is the relationship between what is being explained (the explanandum), and what is doing the explaining (the explanans)? These questions lie at the heart of general philosophy of science, where they connect in important ways to broader issues concerning, e.g., scientific realism and the nature of scientific inferences, evidence, and theory-choice (see Saatsi, 2017a for a recent review). Over the last decade or so these debates have become increasingly deeply informed by the history and philosophy of physics.

Since the demise of the Deductive-Nomological (DN) account of explanation in the 1970s and 1980s, the philosophy of explanation has been largely dominated by the idea that explanation is intimately associated with causation. Unificationist theories of explanation have been the only serious competitor to causal-mechanical accounts for much of the period spanning from Hempel and Oppenheim's epoch-making DN-account (1948) to the current state of the art (see Salmon, 1989 for a classic review). Some of the best developed and most influential contemporary accounts, e.g., by Woodward (2003) and Strevens (2008), have further reinforced the hegemony of causal theories of explanation (the latter putting a unificationist spin on the causal ideology).

Undoubtedly there are countless explanations in physics that nicely fit the intuitive causal ideology, according to which explanatory information is suitably regimented causal-mechanical information (see the chapters by Laura Felline and Matthias Frisch in this volume). But looking at the explanatory practices of physics more broadly, it is also easy to identify numerous examples of seemingly *non-causal* explanations. Detailed studies of such explanations have led to steadily growing interest in understanding exactly how these explanations work: how they explain, and how they differ from causal explanations. There are also substantial issues concerning the relationship between causal and non-causal explanations, and whether some phenomena can only be explained in non-causal terms. I will review some prominent answers to these questions, but first I will bring out the richness of our subject matter with a menagerie of explanations that have been regarded as non-causal.

32.2 A Menagerie of Non-causal (?) Explanations

I have divided the examples below under three headings, involving (i) geometry, (ii) symmetries, and (iii) inter-theoretic relations. These broad headings serve to regiment the discussion for presentational purpose first and foremost, and many of these explanations cut across this "classification" in interesting ways. For the sake of brevity, I will only sketch each case; fuller details can be found in the literature cited. Whether all of these explanations *really* are non-causal is itself a contested issue to be discussed in the next section.

32.2.1 *Explanations Involving Geometry*

Inertially moving particles in curved space can (say) decrease in distance over time due to the "shape of space". Since no forces are involved – inertial particles are moving straight along geodesics – one may want to say that nothing causes this change in the particles' relative distance. Rather, what explains it is the curvature of space. This exemplifies a simple geometrical explanation according to Nerlich (1979).

Geometrical explanations can also involve global features of space, such as its dimensionality, or orientability. Asymmetrical objects, like hands, can be enantiomorphic if they are embedded in an orientable space of appropriate dimensionality. No continuous movement in two-dimensional orientable space takes a left-handed rigid L-shape to match a corresponding right-handed one, for instance. Add a third dimension, or make your two-dimensional space non-orientable (e.g., Moebius strip), then such continuous transformations do exist. What determines – and thus arguably explains, in some sense – that two objects are enantiomorphic (or otherwise), are global features of space in which these objects are "embedded" (Nerlich, 1979).

In a somewhat similar manner, it has been suggested that in Newtonian mechanics an explanatory relationship holds between the dimensionality of space, on the one hand, and the stability of planetary orbits, on the other. Although there is disagreement here which way the explanation really runs – this can depend, e.g., on whether we are relationists or substantivalists about Newtonian space (see Callender, 2005) – the basic idea turns on the fact that natural generalizations of Newtonian gravitational potential to n-dimensional ($n > 3$) spaces implies the impossibility of stable periodical gravitational orbits (Ehrenfest, 1917; Buchel, 1969). If we regard the dimensionality of space as a specific global feature of the world that could have been different, while holding the rest of Newtonian dynamics fixed (by generalizing Newtonian gravitational potential in a natural way), then the stability of planetary orbits depends on – and thus is arguably explained by, in some sense – this global feature of the space in which the planets orbit (see e.g., Woodward, 2018).

The examples above involve features of physical space. Physical theories of course deal also with various kinds of more abstract spaces, the geometrical features of which can be explanatorily relevant as well. Kinematics (in contrast to dynamics) is sometimes characterized as "the study of the geometry of motion" – the study of the geometrical features of physical systems' possible configurations, relating e.g., displacement, velocity, acceleration, and time, without reference to forces or causes of motion. Some purely kinematic explanations are naturally contrasted with causal explanations.

For a paradigmatic toy example, consider the space of possible instantaneous configurations of a stick the origin of which is confined to a curved two-dimensional surface: think of a spherical surface, say, and its tangent vectors. Some of these states are related to one another by parallel transport, viz. moving the stick along the surface so that it is kept "straight" along the way. Now, why does the stick, having been parallel-transported over a closed loop on a curved surface, return to the starting point so that its direction differs by angle α from the initial configuration? The standard geometrical explanation turns on the way in which α is determined by – and thus arguably explained by, in some sense – the curvature of the surface and the area enclosed by the path. (For spherical surface, this is simply given by the equation $\alpha = r^{-2} \times A$, where A is the enclosed area and r is the radius.) These non-local geometrical features do not seem causally responsible for α, and none of the dynamical features or laws of nature (governing how the stick moves around in a specified manner) feature in the explanation either, making this a plausible example of non-causal explanation (Saatsi, 2018).

The ideology of this paradigmatic kinematic explanation arguably also applies to broad-ranging phenomena of "geometrical phase" of suitable *dynamical* systems, involving, e.g., the precession of

Foucault's pendulum and other similar exemplifications of the so-called Hannay's angle that shows up when a classical dynamical system travels a closed loop in a parameter space without returning to its original state. (The daily precession α' experienced by a Foucault pendulum, as a function of the latitude λ, is given by $\alpha' = -2\pi \sin \lambda$. For a geometrical explanation see, e.g., von Bergmann and von Bergmann, 2007.) It also generalizes to various other areas of physics where a system's parameter space has a rich enough structure to give rise to geometrical (an)holonomies upon which some explanandum phenomenon can depend. Interesting examples can be found in, e.g., classical optics, nuclear magnetic resonance, hydrodynamics, and quantum field theory (see Wilczek and Shapere, 1989). The much-studied phenomenon of the Berry phase in quantum mechanics is a particularly important and well-known example (see Batterman, 2003; Lyre, 2014; Saatsi, 2018).

The space of a physical system's possible configurations can also have global *topological* features that can be deemed to be explanatory in a non-causal way. For example, why do *all* double pendulums have at least four equilibrium configurations? Arguably a critical part of this explanation turns on the fact their configuration space is a torus – a fact that holds regardless of the specifics of pendulum design and dynamics, and even regardless of the specifics of the dynamical laws that connect force to acceleration (Lange, 2017). This is a very simple example of a topological explanation, well suited for philosophical discussion and analysis; more involved real-life explanations involving topological notions abound, ranging from the Aharonov-Bohm effect to topological phases of matter. (See Healey, 2007 for in-depth philosophical discussion of the Aharonov-Bohm effect. Little has been written, in general, about topological explanations in physics.)

32.2.2   Explanations Involving Symmetries

Among the numerous and varied examples of symmetry-based explanations in physics, one finds appeal to both discrete and continuous symmetries. I will give a paradigmatic example of each.

In many-particle quantum mechanics, a system's wavefunction must be either symmetric or anti-symmetric under particle permutations. This exemplifies a discrete symmetry. This fundamental symmetry underlies quantum statistics, and it plays an indispensable explanatory role in explanations of various physical phenomena. For instance, the Pauli Exclusion Principle (PEP) for fermions follows from this symmetry, dictating that no two electrons (say) can be in the same quantum state. Various explanations involving PEP – including, e.g., the stability of matter, various features of chemical bonding, and the halting of white dwarf collapse – can be viewed as non-causal explanations: these phenomena are arguably essentially explained, not by reference to any causal features of the world, but rather in terms of a fundamental global kinematic constraint on the possible physical states of a multi-particle quantum system (see Skow 2014; French and Saatsi, 2018).

For a specific example, consider the explanation of the solubility of salt, for instance. An important part of the explanation turns on the role of the so-called "Pauli repulsion" in determining the chemical bond-dissociation energy for the ionic bond between Na+ and Cl− in NaCl molecules. Despite its name, "Pauli repulsion" arguably really should not be thought of as involving a force of any kind, or even having a causal basis more broadly construed. Rather, it is grounded in PEP (and the more fundamental quantum symmetry that underlies PEP, ultimately deriving from the so-called permutation invariance of quantum mechanics). In a similar way, the so-called "degeneracy pressure", which halts the gravitational collapse of a white dwarf star, is arguably best understood in non-causal terms as a kinematic consequence of the same quantum symmetry (see French and Saatsi, 2018 for details.).

Moving on to the explanatory use of continuous symmetries in physics, consider the intimate connection that Noether's theorem brings out as holding between a system's conserved physical quantities, on the one hand, and its continuous symmetries, on the other. Assuming that we are operating in the framework of Euler-Lagrange equations of motion for fields or particles, the generality of

Noether's theorem is remarkable, applying to any symmetry of the system's Lagrangian, whether these involve external degrees of freedom (e.g., homogeneity and isotropy of space), or internal degrees of freedom (e.g., gauge symmetries).

Given the connection brought out by Noether's theorem, we know that if a particular physical system displays some conserved quantities, there is a corresponding symmetry and vice versa. Can one explain the other? Arguably so: a system's symmetries are often naturally viewed as determining – and hence, in some sense, explaining – those aspects of the system's behavior that we identify as the conserved quantities.

For a simple concrete example, consider a classical particle moving under a constant central force. (Its potential energy depends only on the radial coordinate.) Why is the particle's trajectory constrained to a plane? This regularity of the system's dynamics follows from the symmetries of the particle's kinetic and potential energy functions (which together give the Lagrangian, of course). Arguably these symmetries explain the system's constant of motion because Noether's theorem tells how the latter would have been different if, e.g., the potential energy function were different in some way. For instance, it tells us what the constants of motion would have been, had the potential energy function been time-dependent in a particular way, say.

In the physics literature, Noether's theorem is often also associated with the idea that conservation *laws* can be explained in terms of global symmetries. This connection is harder to make sense of without introducing substantial assumptions regarding the metaphysics of laws. Lange (2007) argues that Noether's theorem is altogether irrelevant for accounting for conservation laws and that symmetries can explain conservation laws only if they are modally stronger "meta-laws" that govern ordinary laws.

32.2.3 *Explanations Involving Inter-theoretic Relations*

Various explanatory practices in physics involve relating one theory to another in some systematic, understanding-inducing way. (We are interpreting 'theory' broadly here, to include mathematical models.) The idea of inter-theoretic reduction in Nagel's (1961) sense provides an example of this: if two theories are so related that the laws of the reduced theory can be logically deduced from the laws of the reducing theory, given some well-defined bridge-principles that relate the two theories' theoretical vocabularies, then we arguably have a DN-type explanation of the reduced theory by the reducing theory. Thus, we can understand in non-causal terms the applicability and effectiveness of the reduced theory as turning on logical connections between two theoretical structures and the synthetic identity claims stated by the bridge-principles.

Insofar as physics features reductions like this (or something close enough), and such reductions really do count as explanatory (see Dizadji-Bahmani et al., 2010), we have found yet another class of explanations that are plausibly non-causal. Admittedly purely deductive relations between theories in the spirit of Nagel's account are arguably very few and far between – as anyone familiar with the extensive debates on (anti-)reductionism in physics very well knows. (See, however, Dizadji-Bahmani et al., 2010; Schaffner, 2012.) Be that as it may, for our current discussion little hangs on this, since the kinds of inter-theoretic relations capable of supporting non-causal explanations can take many forms other than Nagelian relations of logical deduction. Similarly, whether or not we can uphold any of the various reductionist theses, it is undeniable that much of physics involves attempts to link in explanatory ways theories at different "levels." Physics unquestionably offers explanatory understanding aplenty regarding how, e.g., thermodynamic phenomena relates to (both classical and quantum) mechanics; how phenomena described in ray-optics relates to wave optics and electromagnetism; how classical mechanics relates to quantum mechanics; how chemical phenomena relates to quantum mechanics; how phenomena described in Newtonian gravitation relates to general relativity; and so on. The explanatory importance of numerous inter-theoretic relationships uncovered by physics is not eroded by the fact that reductionism faces important challenges in relation to various purported reductions of one theory to another. Furthermore, plausibly many of the explanations furnished by inter-theoretic relations in physics are non-causal, and not amenable to a causal-mechanical (or causal-role functionalist) treatment. I will now briefly sketch two examples.

In Newtonian Gravity (NG), inertial and gravitational mass are conceptually entirely distinct, yet they turn out to be identical. Why? The answer is given, of course, by examining the relevant phenomena in relation to General Relativity (GR). But it is not the case that GR shows how the two concepts of mass get identified: gravitational mass does not feature in GR *at all*, given that the concept of gravitational force makes no sense in GR. Rather, the explanation involves showing how NG relates to GR, so as to answer the question: "why if GR is true, [NG] is such an effective theory in some contexts?" (Weatherall, 2011, p. 429) Answering this question explains *away* the seeming coincidence of gravitational and inertial masses, thereby providing an answer to the why-question above. As Weatherall explains:

> The explanation consists in showing that if the world is to be understood as a model of GR, then any model of [NG] that successfully approximates the world, even in limited regimes, must satisfy the conditions that [the two masses are identical]. (p. 437)

Showing this requires bringing out in detail the relevant relationship between NG and GR, which is done in terms of showing mathematically how models of (geometrized) NG arise as a limiting case of GR in a rigorous sense of "flattening" the light-cone structure of GR.

My second example concerns the renormalization group explanation of critical phenomena that involve (second-order) phase transitions in macroscopic systems near their critical point. Thermo-dynamic properties near criticality obey power laws as a function of (reduced) temperature. Micro-physically very different systems can obey the same power laws, in which case they belong to the same "universality class," identified by the exponents of the relevant power laws. These exponents depend only on the system's spatial dimensionality and certain abstract features of the micro-interaction (e.g., the dimensionality of the "spin" parameter). Why do the macroscopic phenomena quantified by the power laws – obeyed by all systems that belong to the same universality class – only depend on these features?

The answer to this question involves mathematical relationships among extremely large classes of possible microphysical models of statistical physical systems. These relationships are brought out by employing "renormalization transformations", which form equivalence classes of those models (in the abstract mathematical space of possible models) that share the same long-distance physics. It can be shown that this space – which is assumed to include the models that correctly capture the actual microphysics of systems that physicists actually measure or simulate – has a rich mathematical structure that reveals the fact that while the power laws are strikingly independent of the specifics of the micro-physical interactions, they only depend on the abstract features mentioned above (Batterman, 2003; Morrison, 2015; Saatsi and Reutlinger, 2018). The explanation furnished by such renormalization group techniques is again arguably non-causal, turning on rich mathematical relationships in the space of possible models of physical systems of very many interacting parts. (However, see Sullivan, 2019 for a criticism of this verdict).

32.2.4   Other Examples?

Physics features numerous other examples of purportedly non-causal explanations, many of which do not fall neatly under the three broad headings above. For some further examples and "head-ings", see, e.g., Lange's (2009) and Pexton's (2014) discussion of *dimensional* explanations; Lange's (2017, Ch. 5) discussion of *"really statistical"* explanations; the discussion of some *quantum mechanical* explanations in Clifton (1998) and Healey (2017); and Batterman's (2002) discussion of *asymptotic* explanations.

32.3   How Do These Explanations Work?

How do the above examples of (allegedly) non-causal explanations work? Where does their "explanatory power" derive from? Philosophers have attempted to articulate the *explanatoriness* of explanations like these in various ways. While many are happy to do this in non-causal terms, others find ways to regard as causal at least some of the above explanations (e.g., Skow, 2016; Strevens, 2018). Whether or not a given explanation counts as non-causal will partly depend, of course, on how the foil, causal explanation, is construed. Even the most ardent current proponents of non-causal explanations regard some of the examples above as causal. For example, Lange (2017) views as causal the simple "geometrical" explanation involving inertially moving particles in curved space, while he regards as non-causal the explanation involving enantiomorphic figures in an orientable space (see the start of Section 32.2). According to Lange, the former explanation is causal, "since it works by specifying the causes (forces) that are acting on the dust particles (namely none) as well as the laws and conditions determining how things in the particles' locations would behave were they under the influence of no causes". (p. 403). The latter explanation Lange analyzes as a non-causal explanation "by constraint". (p. 126).

Given the disagreement about the status of many of the examples cited, one may ask what point there is to demarcate between causal and non-causal explanations? What exactly hangs on it? As I see it, a central aim of the philosophy of explanation is to faithfully capture the explanatory practice of science and to mark significant distinctions that scientists themselves draw between types of explanations. What do physicists regard as explanatory, and why? The contrast between causal vs. non-causal explanations can be a useful way to articulate such distinctions, the philosophical point of such a contrast being underwritten by the corresponding distinctions' significance to science. Scientists do, after all, care what *kinds* of explanations there are, or could be, and how these relate to one another (see Lange, 2017; Saatsi, 2018).

In the rest of this section, I will briefly look at some prominent analyses of non-causal explanations. I do not have space to get into the details of the particular accounts, but I will highlight two competing *trends* in the currently booming literature. One trend is exemplified by philosophers who emphasize the *independence* of the explanandum of non-causal explanations from specific causal features of the system in question: many non-causal explanations have been thought to work by bringing out (in one way or another) such independence. The second, competing trend is exemplified by philosophers who emphasize the (counterfactual) *dependence* of the explanandum on the explanans, in broad analogy with causal explanations. According to this line of thought, many non-causal explanations explain in a way analogous to causal explanations, but by tracking relations of non-causal dependence instead of causal dependence.

Let's begin with the first trend. There are many ways in which many non-causal explanations can be independent of, or abstract away from, causal features. According to Pincock (2007), some non-causal mathematical explanations – such as Euler's well-known graph-theoretic explanation of why one cannot traverse Koenigsberg's bridges in a particular way – work by "abstracting away from the microphysics." And not just *micro*physics: Euler's explanation is independent of all the physical details regarding the bridge's lengths, angles, and so on.

According to Batterman and Rice (2014), some scientific models are capable of explaining "universal patterns" among micro-physically heterogeneous systems by demonstrating that the micro-details distinguishing all these systems (as well as the explanatory model system) from one another, are *irrelevant* for the universal behavior being explained. These explanatory "minimal models" are extreme caricatures of the real system, but according to Batterman and Rice, they do not explain by virtue of capturing any "common features" between the different systems as explanatorily underlying the universal behavior. Rather, minimal model explanations arguably work by showing how micro-physically heterogeneous systems can partake in a "universal pattern" – belong to one and the same "universality class," such as those involved in renormalization group explanations – by virtue of some features of the phenomenon at stake that renders *irrelevant* the details that distinguish the model system and various real systems.

Lange (2017) offers a very different account of non-causal explanations that also in a way exemplifies the trend that emphasizes the independence of the explanandum from specific causal features. According to Lange, a broad range of non-causal explanations work "by describing how the explanandum arises from certain facts ('constraints') possessing some variety of necessity stronger than ordinary laws of nature possess" (p. 10). He presents varied examples of such "explanations by constraint", which also include some of the examples discussed above (Section 32.2), such as the topological double-pendulum explanation. Many other geometrical explanations would also count as an explanation by constraint. For an illustration, consider the example involving the parallel-transported stick along the spherical surface (cf. Section 32.2). The standard geometrical explanation of α is independent of contingent causal, dynamical, and nomological features of the world, thereby holding with mathematical or logical necessity. (Due to this modal strength of the "constraint", Lange would call this "distinctly mathematical" explanation. Other explanations "by constraint" involve grades of necessity strictly weaker than this, but nevertheless stronger than those of ordinary causal laws.) Lange argues that explanations by constraint explain by virtue of showing why the explanandum *had to be* the way it is, for reasons that are modally stronger than ordinary causal laws. The relative modal strength of explanatory constraints Lange in turn characterizes in terms of their independence from actual causal laws.

It is undeniably true that a distinctive feature of many non-causal explanations in physics is a kind of independence or abstraction away from causal features of the world. It does not automatically follow that this feature is what makes these explanations explanatory, however. In particular, philosophers exemplifying the second trend highlighted above have argued that various purported examples of non-causal explanations work by virtue of providing *dependence* information, despite their extreme level of abstraction from causal features of the world.

In this spirit, Jansson and Saatsi (2019) critically discuss the accounts of Pincock and Lange, for example, arguing that many of the key examples motivating their accounts also involve information about non-causal dependence relations and that this information furthermore is naturally construed as explanatory very much in the spirit of a popular counterfactual account of causal explanation (Woodward, 2003). The "minimal models" account of Batterman and Rice is critically analyzed by Lange (2015), who argues that their examples can be construed as involving information about essential common features shared by otherwise heterogeneous systems. Furthermore, one could argue that these features, although highly abstract, are explanatory by virtue of showing what the explanandum depends on. (This further claim is not part of Lange's brief, however.) This would be in tune with Saatsi and Reutlinger (2018), who emphasize the role of counterfactual dependence information specifically in connection with renormalization group explanations of critical phenomena. (See also Reutlinger, 2016; Woodward 2018.)

The proponents of counterfactual accounts to explanations have sought to generalize the ideology of causal explanation to non-causal cases quite generally. This is motivated by noting that in a wide range of examples given above there is some interesting, non-trivial modal information in play that seems to inform us of the dependence of the explanandum on some other factors. For example, arguably important symmetry-based explanations – including those mentioned above – provide answers to counterfactual *what-if-things-had-been-different* questions (French and Saatsi, 2018). Similarly, Saatsi (2018) argues that many geometrical explanations in physics – including those involving a geometrical phase – are naturally understood in counterfactual terms, and Saatsi (2017b) suggest likewise regarding explanations furnished by phase-spaces and attractors in dynamical systems theory. Bokulich (2008) appropriates the counterfactual approach to a model-based explanation of the semiclassical phenomenon of wavefunction scarring, and Reutlinger (2018) discusses various further cases in the same spirit.

Consider again, for example, the paradigmatic geometrical explanation involving a stick's parallel transport over a closed loop on a spherical surface. In Saatsi's (2018) analysis this explanation is best understood in counterfactual terms, supported by the equation that connects the shift in the stick's angle, α, on the one hand, and the area of closed loop and the curvature of the surface, on the other. What is it that renders this explanation non-causal (or mathematical, perhaps), if the ideology of explanatory counterfactual information is apt? According to Saatsi, the difference between this non-causal geometrical explanation and causal explanations boils down to the inapplicability of the (causal) notion of intervention in the non-causal case, due to some of the explanans variables being suitably non-local. Another key feature of this kind of non-causal explanation is that the explanatory connection between the explanandum and explanans variables is too intimate to count as a contingent causal connection (Woodward, 2018).

There is still a good deal of work to be done on the metaphysics and epistemology of non-causal explanations. (See Reutlinger, 2017 for a recent review of the philosophy of non-causal explanation, and Reutlinger and Saatsi (2018) for a collection of recent research.) The nature of purported non-causal dependence relations is murkier than their causal counterparts, for instance, and there is a prominent challenge faced by everyone in the debate: how to account for the directionality of explanation? This challenge is unsurprising in the light of the well-known challenges to the DN-account and the subsequent appeal to causation and unification as offering a response to these challenges. Adopting the ideology of counterfactual causal accounts of explanation, *sans* causation, clearly leaves open the question of what grounds the directionality of explanation. (For different responses, see e.g., Jansson, 2015; Jansson and Saatsi, 2019; Reutlinger, 2018.) The directionality challenge has been explicitly raised against the "minimal models" account (by Lange, 2015); counterfactual accounts (by Lange, 2019), and Lange's account of distinctly mathematical explanations (by Craver and Povich, 2017).

32.4   Acknowledgments

Many thanks to Alex Reutlinger and Eleanor Knox for their helpful comments on an earlier draft.

References

Batterman, R.W. (2003). Falling cats, parallel parking, and polarized light. *Studies in History and Philosophy of Modern Physics*, 34(4): 527–557.

Batterman, R.W. and Rice, C.C. (2014). Minimal model explanations. *Philosophy of Science*, 81(3): 349–376.

Bokulich, A. (2008). Can classical structures explain quantum phenomena? *The British Journal for the Philosophy of Science*, 59(2): 217–235.

Buchel, W. (1969). Why is space three-dimensional? *American Journal of Physics*, 37: 1222–1224.

Callender, C. (2005). Answers in search of a question: 'Proofs' of the tri-dimensionality of space. *Studies in History and Philosophy of Modern Physics*, 36(1): 113–136.

Clifton, R. (1998). Scientific explanation in quantum theory. Unpublished manuscript. Available at: philsci-archive.pitt.edu/archive/00000091.

Craver, C.F. and Povich, M. (2017). The directionality of distinctively mathematical explanations. *Studies in History and Philosophy of Science*, 63: 31–38.

Dizadji-Bahmani, F., Frigg, R. and Hartmann, S. (2010). Who's afraid of Nagelian reduction. *Erkenntnis*, 73(3): 393–412.

Ehrenfest, P. (1917). In what way does it become manifest in the fundamental laws of physics that space has three dimensions? *Proceedings of the Amsterdam Academy*, 20: 200–209.

French, S. and Saatsi, J. (2018). Symmetries and explanatory dependencies in physics. In A. Reutlinger and J. Saatsi (eds.), *Explanation Beyond Causation: Philosophical Perspectives on Non-Causal Explanations*. Oxford: Oxford University Press, pp. 185–205.

Healey, R. (2007). *Gauging What's Real: The Conceptual Foundations of Contemporary Gauge Theories*. New York: Oxford University Press.

Healey, R. (2017). *The Quantum Revolution in Philosophy*. New York: Oxford University Press.

Hempel, C.G. and Oppenheim, P. (1948). Studies in the logic of explanation. *Philosophy of Science*, 15: 135–175.

Jansson, L. (2015). Explanatory asymmetries: Laws of nature rehabilitated. *Journal of Philosophy*, 112(11): 577–599.

Jansson, L. and Saatsi, J. (2019). Explanatory abstractions. *British Journal for the Philosophy of Science*, 70(3): 817–844.

Lange, M. (2007). Laws and meta-laws of nature: conservation laws and symmetries. *Studies in History and Philosophy of Modern Physics*, 38(3): 457–481.

Lange, M. (2009). Dimensional explanations. *Noûs*, 43(4): 742–775.

Lange, M. (2015). On "minimal model explanations": A reply to Batterman and Rice. *Philosophy of Science*, 82(2): 292–305.

Lange, M. (2017). *Because without Cause: Non-causal Explanations in Science and Mathematics*. Oxford: Oxford University Press.

Lange, M. (2019). Asymmetry as a challenge to counterfactual accounts. *Synthese*, https://doi.org/10.1007/s11229-019-02317-3.

Lyre, H. (2014). Berry phase and quantum structure. *Studies in History and Philosophy of Modern Physics*, 48: 45–51.

Morrison, M. (2015). *Reconstructing Reality: Models, Mathematics, and Simulations*. Oxford: Oxford University Press.

Nagel, E. (1961). *The Structure of Science*. London: Routledge and Keagan Paul.

Nerlich, G. (1979). What can geometry explain. *The British Journal for the Philosophy of Science*, 30(1): 69–83.

Pexton, M. (2014). How dimensional analysis can explain. *Synthese*, 191(10): 2333–2351.

Pincock, C. (2007). A role for mathematics in the physical sciences. *Noûs*, 41(2): 253–275.

Reutlinger, A. (2016). Is there a monist theory of causal and non-causal explanations? *Philosophy of Science*, 83(5): 733–745.

Reutlinger, A. (2017). Explanation beyond causation? New directions in the philosophy of scientific explanation: Explanation beyond causation. *Philosophy Compass*, 12(2): e12395.

Reutlinger, A. (2018). Extending the counterfactual theory of explanation. In A. Reutlinger and J. Saatsi (eds.), *Explanation Beyond Causation: Philosophical Perspectives on Non-Causal Explanations*. Oxford: Oxford University Press, pp. 74–95.

Reutlinger, A. and J. Saatsi (2018), *Explanation Beyond Causation: Philosophical Perspectives on Non-Causal Explanations*. Oxford: Oxford University Press.

Saatsi, J. (2017a). Realism and the limits of explanatory reasoning. In J. Saatsi (ed.), *The Routledge Handbook of Scientific Realism*. London: Routledge, pp. 200–211.

Saatsi, J. (2017b). Dynamical systems theory and explanatory indispensability. *Philosophy of Science*, 84(December): 1–14.

Saatsi, J. (2018). On explanations from geometry of motion. *The British Journal for the Philosophy of Science*, 69(1): 253–273.

Saatsi, J. and Reutlinger, A. (2018). Taking reductionism to the limit: How to rebut the anti-reductionist argument from infinite limits. *Philosophy of Science*, 85: 455–482.

Salmon, W.C. (1989). Four decades of scientific explanation. In P. Kitcher and W.C. Salmon (eds.), *Minnesota Studies in the Philosophy of Science*, Vol. 13. Minneapolis: University of Minnesota Press, pp. 3–219.

Schaffner, K.F. (2012). Ernest Nagel and reduction. *The Journal of Philosophy*, 109(8/9): 534–565.

Skow, B. (2014). Are there non-causal explanations (of particular events). *The British Journal for the Philosophy of Science*, 65(3): 445–467.

Skow, B. (2016). *Reasons Why*. New York: Oxford University Press.

Strevens, M. (2008). *Depth: An Account of Scientific Explanation*. Cambridge, MA: Harvard University Press.

Strevens, M. (2018). The mathematical route to causal understanding. In A. Reutlinger and J. Saatsi (eds.), *Explanation Beyond Causation: Philosophical Perspectives on Non-Causal Explanations*. Oxford: Oxford University Press, pp. 96–116.

Sullivan, E. (2019). Universality caused: The case of renormalization group explanation. *European Journal for Philosophy of Science*, 9(3): 1–21.

von Bergmann, J. and von Bergmann, H. (2007). Foucault pendulum through basic geometry. *American Journal of Physics*, 75: 888–892.

Weatherall, J.O. (2011). On (some) explanations in physics. *Philosophy of Science*, 78(3): 421–447.

Wilczek, F. and Shapere, A. (1989). *Geometric Phases in Physics*. Singapore: World Scientific.

Woodward, J. (2003). *Making Things Happen: A Causal Theory of Explanation*. Oxford: Oxford University Press.

Woodward, J. (2018). Some varieties of non-causal explanation. In A. Reutlinger and J. Saatsi (eds.), *Explanation Beyond Causation: Philosophical Perspectives on Non-Causal Explanations*. Oxford: Oxford University Press, pp. 117–140.

Further Reading from the Editors

Bob Batterman's work has been hugely influential in uncovering examples of physics reasoning that don't fit traditional philosophical patterns. His "Falling cats, parallel parking, and polarized light," (*Studies in History and Philosophy of Modern Physics*, 34(4), 527–557, 2003) give s number of such examples. Marc Lange's *Because without Cause: Non-causal Explanations in Science and Mathematics* (Oxford: Oxford University Press, 2017) develops a unified and sophisticated account of non-causal explanation in physics and beyond. Reutlinger, A., and J. Saatsi, eds. *Explanation beyond Causation: Philosophical Perspectives on Non-causal Explanations* (Oxford: Oxford University Press, 2018) is a very helpful collection. While many authors have emphasized the divergence between causal and non-causal explanation, there is recent work articulating a more unified framework. Jansson, Lina, and Juha Saatsi's "Explanatory abstractions" (*The British Journal for the Philosophy of Science*, 70(3), 817–844, 2019) is a nice example of this.

33

MECHANISTIC EXPLANATION IN PHYSICS

Laura Felline

33.1   Introduction

The idea at the core of the New Mechanical account of explanation can be summarized in the claim that explaining means showing "how things work". This simple motto hints at three basic features of Mechanistic Explanation (ME): ME is an *explanation-how*, that implies the description of the *processes* underlying the phenomenon to be explained and of the *entities* that engage in such processes. These three elements trace a fundamental contrast with the view inherited from Hume and later from strict logical empiricism (see Creath, 2017), focused on epistemic and formal features of science and according to which issues concerning the kind of entities and processes that lie within a theory's domain are extraneous to science and belong instead to ontology or metaphysics. Philosophers belonging to the new mechanical philosophy believe that the received view of scientific explanation (Hempel, 2001), pivoting on the notion of law of nature,[1] overshadows this insight.

Since its origin in the 17th century, mechanical philosophy aimed to explain natural phenomena by reducing them to mechanisms. Traditional attempts to define the concept of mechanism involved the identification of a limited set of fundamental elements as, for instance, contact action, action at a distance, inertial motion (see e.g., Hesse, 1961), and, more recently, the transmission of a mark, or of a conserved quantity (see Frisch, this volume). The new mechanical philosophy rejects this austere characterization of mechanisms and ME, and aims at providing a novel, philosophically rigorous explication of the concept of mechanism and of its role in scientific explanation and practice.

ME has been adopted with profit in the philosophy of special sciences (for instance in biomedical sciences, e.g. in the explanation of chemical transmission at synapses ((Machamer et al., 2000), MDC henceforth); but also in social sciences, e.g., the three kinds of social mechanisms in Coleman's analysis of Max Weber's account of the role of the Protestant ethic in the growth of capitalism (Hedström and Swedberg, 1998)), where exceptionless regularities are rarely ever found. In physics, it is generally possible to formulate explanations in law-based form, with the result that the plurality of explanatory forms might be overlooked. This should not come as a surprise, given that physics was the main inspiration for logical empiricists and, in particular, Newtonian physics was a template for Hempel's formulation of the covering law model. However, this situation is unfortunate, since, we will argue, knowing how things work is often part of the explanation of physical phenomena. In this chapter, we provide an introduction to the basic features of ME, with a specific focus on its application to physics (Section 33.1). The main part of the chapter is devoted to the defense of two theses: on the one hand, some domains of physics are not compatible with mechanistic reasoning and explanation (Section 33.2); on the other hand, a comprehensive account of explanation in physics can't dispense with ME (Section 33.3).

33.2   ME: General Features

A ME of a phenomenon *P* requires the description, which may be more or less idealized, of the mechanisms underlying the occurrence of *P*. The novelty of the new mechanical philosophy with respect to other accounts of mechanism (e.g., Salmon's (1984), see also Frisch (this volume)) lies in its novel definition of mechanism as organized entities and activities. The literature displays many variations over this basic idea, here we report a minimal characterization of the concept[2]:

> "A mechanism for a phenomenon consists of entities (or parts) whose activities and interactions are organized so as to be responsible for the phenomenon." (Glennan, 2017, 24)

A familiar example might serve as an illustration. The pressure of a gas in equilibrium inside a piston at constant temperature increases when its volume decreases and *vice versa*. A ME of this phenomenon displays a kinetic model of the ideal gas, described as a mechanism composed of Newtonian atoms and molecules (entities), with constant, rapid, and random motion (activities) and perfectly elastic collisions (interactions). The gas' pressure corresponds to the force resulting from the totality of the particles' hits on the wall of the container. The smaller the volume occupied by the gas, the higher will be the number of particles per unit of volume, and that will result in more frequent hits to the wall of the container. It follows that a reduction of volume corresponds to an increase in pressure.

Mechanisms are organized in nested hierarchies, in the sense that a mechanism's components are themselves mechanisms and their behavior is therefore mechanistically explainable through the description of its components and their activities (Glennan, 2011; MDC, 5.1). However, a theory describing such a tiered structure of mechanistic composition, always bottoms-out with a stable set of mechanisms that are fundamental in that theory, in the sense that the behavior of the entities that compose such mechanisms provide the basic building blocks of MEs in said domains and cannot be explained within the theory itself. Bottoming-out is domain-specific: while the behavior of entities cannot be explained within a theory where they are fundamental, they might be explainable by the mechanisms of a lower-level theory. For instance, a gas in a piston is an elementary mechanism in classical thermodynamics; its behavior is not explainable within such a theory, but it can be explained by the lower-level mechanistic models displayed by the kinetic theory of gases.[3] As another example, in fluid dynamics, a fluid is a fundamental entity whose lack of resistance to deformation under the action of a force is unexplained. However, the lack of resistance to deformation is explained by the lower-level mechanistic decomposition: fluids are entities composed of particles whose bonding force is so weak, that it opposes a negligible resistance to any force cutting particles apart – therefore the lack of resistance to deformation.

Bottoming-out raises the ontological issue of whether and when the structure of the nested hierarchy ends. It is currently a matter of debate whether an ontologically fundamental mechanistic level can exist or whether there are mechanisms "all the way down." According to Glennan (2017, ch. 5.5), this is an empirical question, and the possibility of mechanisms "all the way down" cannot be ruled out *a priori*.

The formulation of a ME requires in general the isolation of the entities that are the components of the mechanism (structural decomposition) and of the activities in which entities engage (functional decomposition). In a satisfactory ME, these two elements must be adequately integrated (Bechtel and Richardson, 2010).

A mechanism is always a mechanism for a behavior. The same mechanical system may exhibit different behaviors, and which structural and functional decompositions are correct for an explanation depends on the behavior to be explained (see Craver, 2013).

Another novelty of the new ME with respect to its predecessors is the rejection of explanatory fundamentalism, i.e. the assumption that the best explanation is provided in terms of the most fundamental theories (Craver, 2007, p. 11, n. 13). Crucially, the new mechanical philosophy imposes no *a priori* constraint (other than basic standards of scientificity, e.g., coherence, empirical adequacy, predictive power or salience) to the features of such building blocks, which vary when the available fundamental mechanisms of a given domain are found insufficient to explain new discovered phenomena. For instance, consider the development of classical mechanical philosophy,[4] based on the assumption that matter is the sole physical entity and motion and transmission of impulse by contact are the fundamental activities (Kochiras, 2013). As new phenomena were discovered in optics, electricity, and magnetism, the explanatory power of this model turned out to be insufficient, so that corpuscles as centers of force capable of attracting and repelling at a distance were added to the stable set of fundamental entities and activities. Finally, Maxwell's theory established that electrical and magnetic phenomena could not be explained exclusively in terms of a theory of action (although at a distance) between bodies, but required a new kind of fundamental active physical entity: the electromagnetic field.[5]

Entities interact in virtue of their properties, and interactions are the carriers of change and production. There is a general consensus that there is no ME without change and production, which makes ME a productive account of causal explanation, as opposed to a relevance account (Hall, 2004; Glennan, 2011).

The third constitutive element of a mechanism, together with entities and activities, is organization: a set of entities and their activities does not constitute a mechanism if they are not arranged in the correct way. Spatiotemporal relations constitute the most common example of the organization of mechanisms part: in order for one idle wheel to transmit torque between two others, it is not sufficient that the wheels' teeth have the right size, but they must be spatially arranged in such a way that the teeth of the middle wheel fit together with the indentation of the others. Another example of an organizational feature is the kind of geometry (e.g., Euclidean or Minkowskian) instantiated by the spatiotemporal properties of a mechanism.

33.3   Limits of ME in Physics

As the familiar examples illustrated so far show, mechanisms can be found all over physics.[6] In this section, we are going to show that ME is inapplicable to some explanations in physics.

A significant limit for the mechanistic account is represented by the class of explanations that are independent of the micro-constitution or micro-dynamics of the systems displaying the behavior to be explained. Notable examples in this class are explanations based on conservation laws (for example the explanation of Archimedes' Principle, exploiting the conservation of energy (Lange, 2011)) and the universal behavior of systems near their critical points (the so-called renormalization group (RG) explanations (Batterman, 2002)[7]). However, what conclusion should be drawn from this limit is a matter of debate. Kuhlmann (2011), for instance, argues that RG explanations are actually MEs, therefore the fact that the new mechanistic account does not cover them shows that it is too restrictive. He therefore proposes a suitable modification of the account to include the class of what he calls "structural mechanisms".

Notice that, even granted that such explanations are not MEs, yet individual instances of the explanandum behavior taken under consideration are mechanistically explained in physics. For instance, the RG explanation of why, when near their critical points, fluids of different molecular constitution behave in a similar way might be non-mechanistic, but the explanation of the behavior of an individual fluid near its critical point is mechanistic. This fact traces a significant difference between said non ME and others that, as well as being independent of the micro-constitution or micro-dynamics of the systems displaying the behavior to be explained, are also understood as accounting for phenomena that are not even in principle mechanically explainable.

As an example, take the geometrical explanation of length contraction in Special Relativity (SR).[8] Three features separate this explanation from ME (Felline, 2015). The first two features are the abstraction and generality of the model: since in this explanation no microphysical or dynamical element is relevant, its interpretation as ME would define the model of a "universal" mechanism, implemented by every physical system irrespectively of its features – even of whether or not they are complex systems. Including such a kind of explanation, Felline argues, would imply a trivialization of the concept of mechanism, and ME. The third feature is a consequence of the first two. Janssen (2009) argues that the universality of SR's explanations justifies a kinematical interpretation of relativistic effects. Since (at least in its "orthodox" interpretation)[9] length contraction is not causally produced, therefore *a fortiori* it does not represent the behavior of a mechanism. On the other hand, length contraction and time dilation are the instantiations of a feature of the spatiotemporal organization of any complex system relativistically described. Non-MEs of these kinds are the so-called "structural explanations" and might include explanations in quantum theory, as for instance the explanations of non-locality (Dorato and Felline, 2011) and the uncertainty relations (Felline, 2015).

According to Kuhlmann and Glennan (2014), the vast majority of quantum phenomena, even when they depend on the underlying dynamical and constitutive details of the systems involved, is non-mechanistically explainable, for at least three reasons. First of all, entities in a mechanism interact in virtue of their properties, but quantum states do not associate determinate values to every property of a system. For instance, quantum states associated with a determinate spin value in the x-axis, are not associated with a determinate spin value in the y-axis, and vice versa. Following the standard interpretation of quantum theory (QT), such indeterminacy is ontological: quantum systems in a state with determinate spin value in the x-axis do not possess a determinate spin in the y-axis. According to Kuhlmann and Glennan, this situation is not compatible with ME: first of all, they argue, mechanisms are composed of objects with definite properties; secondly, components of a mechanism are interconnected via local interactions, therefore, spatiotemporal localization is an especially important property for mechanisms, so the indeterminateness of position is especially problematic for ME; third, entanglement[10] seems to undermine the possibility of decomposition into separate parts. However, Kuhlmann's and Glennan's conclusion can be contrasted on the basis of the arguably too restrictive characterization of the mechanism they provide. For instance, as the two philosophers anticipate, one might reject the assumption of localizability (Bechtel and Richardson, 2010). Also, the problem with entanglement seems to have its origin in the requirement that "parts have properties that are relatively stable over time and that at least theoretically these parts are subject to manipulation and isolation from the rest of the mechanism" ((Glennan, 2008, p. 378), quoted in (Kuhlmann and Glennan, 2014)) – but many mechanist accounts dispense from such a strict requirement of modularity (e.g., MDC; Bechtel, 2009; Illari and Williamson, 2012).

Finally, a notable case study was recently displayed in the debate over the physical status of the Higgs phenomenon, often called the Higgs "mechanism", through which particles gain mass in the Standard Model. In the late 1950s, the promising hypothesis was investigated that protons and neutrons gained mass through the spontaneous breaking of some symmetry. However, under such a hypothesis the theory seemed to predict the appearance of massless bosons (the Goldstone bosons), which evidence strongly suggests do not exist and that were therefore considered unphysical. During the 1960s, several independent researchers (Peter Higgs, after which the phenomenon was named, was between them) formulated a theoretical argument blocking the appearance of Goldstone bosons. The first step of such an argument consists in the introduction of additional fields, thanks to which the symmetry group of the Lagrangian becomes gauge invariant. Secondly, with a suitable choice of gauge one can obtain a unitary gauge where the Goldstone boson disappears and the resulting field possesses mass.[11] This merely formal account is often illustrated through a metaphor according to which the Gauge boson "has eaten the Goldstone boson and grown heavy" (Coleman, 1985, p. 123). However, many philosophers of science have been critical about this semi-popular explanation of the reality behind the phenomenon. Margaret Morrison (2003), for instance, argues that the Higgs mechanism, as Maxwell's aether, possesses great heuristic importance, but is not realistically interpretable. According to Earman " readers of Scientific American can be satisfied with these just-so stories. But philosophers of science should not be. For a genuine property like mass cannot be gained by eating descriptive fluff, which is just what gauge is." (Earman, 2004, p. 1239). Lyre (2012) argues that, despite its name, the Higgs mechanism does not provide a causal dynamical story necessary to ground a ME, while Stölzner (2016) relates the explanatory import of the formal account to its unifying power, rather than to its mechanical underpinning.

33.4   The Role of ME in Physics

In the previous section, we have shown how some explanations in physics are incompatible with the mechanistic account. It might be tempting to conclude that a law-based account of explanation is to be favored in this domain of science. After all, the argument would go, the familiar illustrative examples cited in Section 33.2 are translatable in the form of derivations from laws.

Contra such a conclusion, in this section we want to show that, in physics as well as in special sciences, a comprehensive account of scientific explanation and its role in scientific practice requires the appeal to the notion of ME.

One way of cashing out the contribution of a variety of explanation in science is the analysis of its role as a heuristic guide. MEs are guides to theory-testing, due to the ways in which they suggest manipulations, and to discovery, since "if one knows what kind of activity is needed to do something, then one seeks kinds of entities that can do it, and vice versa" (MDC, 17). On the contrary, an account that reduces explanation to the articulation of laws covering the explanandum does not display the same richness when applied to scientific practice (Bechtel and Abrahamsen, 2005). Under this perspective, attempts at explaining phenomena by showing how things work has traced the path of scientific change throughout the history of physics. As an example, Maxwell's mechanical model of electromagnetic waves has been of crucial heuristic importance for the discovery of the electromagnetic nature of light (Nersessian, 1984).

Another way of demonstrating the role of ME in physics is through the analysis of cases where ME reflects and clarifies the desiderata of a suitable explanation, in a far richer way than law-based theories of explanation. As an illustrating example of how the mechanistic account can be superior to law-based accounts of explanation, in the rest of this section we analyze a time-honored problem in quantum theory: the measurement problem.

We can summarize the measurement problem as the problem of accounting for the apparent determinateness of results in quantum measurement.[12] A widespread approach to this problem is the so-called "black-box" approach – that, rather than providing a genuine explanation, takes the notion of measurement as primitive and provides instructions for obtaining probabilities *via* the Born rule. The evolution of the quantum state in case of measurement is here described by the projection postulate, although it is typically unclear whether the latter should be interpreted as a physical process, or as a change in our epistemic state.[13]

Part of the attractiveness of the black-box approach consists in its avoidance of commitments over the kinds of entities and processes that constitute the world. However, without a description of measurement interactions, traditional black-box solutions fail to provide a clear-cut criterion for the application of the projection postulate. This, in turn, complicates the possibility of a solution to the many issues gravitating around the measurement problem. For instance, depending on whether or not the projection postulate is interpreted as a real physical process (the so-called "collapse of the wave-function"), the theory has different implications with respect to related problems of internal and external consistency (e.g., Wigner's friend scenarios (Wigner 1995 (1961)), or non-locality and the relationship with SR).

In this sense, the standard black-box solution does not provide a satisfactory explanation to the measurement problem – and yet, this limit seems not to be highlighted within the covering law

(Hempel, 2001, p. 281) or the unificationist models (Friedman, 1974, p. 18). Contrarily to law-based accounts of explanation, and as we are about to argue more in details, the mechanistic account shows why, far from being motivated uniquely by metaphysical (as opposed to scientific) interests, the so-called "interpretative" claims have a substantial role in a satisfactory explanation of the determinateness of our experience.

Recently, other accounts have adopted a refined black-box approach. In particular, antirealist interpretations of quantum theory (e.g., Fuchs et al., 2014) put forward an explanation away of the measurement problem where the determinateness of our experiences is seen as a fundamental but unproblematic brute fact. In such epistemic interpretations, the quantum state does not represent the outside world, but rather our mental state. For instance, according to QBism "quantum mechanics is a tool anyone can use to evaluate, on the basis of one's past experience, one's probabilistic expectations for one's subsequent experience" (Fuchs et al., 2014, p. 1).

If the quantum state only represents our epistemic state, then the question "why do we have determinate experiences?" does not arise in the first place: why should we expect, in a theory about mental states, anything else than determinate experiences?

Felline (2020) argues that this kind of antirealist approaches represents a Pyrrhic victory rather than a genuine solution to the measurement problem since they are achieved at the price of dragging quantum theory outside the domain of empirical sciences. One of the pernicious consequences of this move is that explanations in quantum theory do not submit to the same standards of scientificity (e.g., that explanatory claims imply successful empirical predictions) as the rest of empirical science.

In any case, this kind of antirealist solution to the measurement problem rejects a necessary precondition of the inquiry carried in this chapter, i.e., that physics, as, more in general, empirical science, can represent the outside world. As such, they are outside the scope of this chapter. In the following, we assume that a realist interpretation of the quantum theory, according to which the theory describes the outside world, is possible.

Contrary to black-box approaches, the majority of proposals deems necessary, for a comprehensive and coherent quantum theory, the explanation of the determinateness of measurement results. In turn, such an explanation requires the opening of the black box, i.e., the description of the interactions that characterize measurement processes and the clear interpretation of the projection postulate. In being guided by the "opening the black-box" prescription, the proposals go beyond a law-based view of scientific explanation and are driven instead by the same spirit that we consider at the core of ME and summarize with the expression "showing how things work". In this sense, the development of the debate over the measurement problem is driven by the search for a ME.

This being said, specific solutions come in a variety of proposals that require further analysis.

An important divide among the explanations provided by different interpretations of quantum theory concerns the role played by the wavefunction, an entity defined in $3N$-dimensional configuration space, where N is the number of particles considered.

In the most straightforward reading of quantum formalism, the wave-function is a concrete entity inhabiting configuration space.

The sorts of physical objects that wave-functions are, on this way of thinking, are (plainly) fields – which is to say that they are the sorts of objects whose states one specifies by specifying the values of some set of numbers at every point in the space where they live. (Albert, 1996, p. 278)

GRW theory (Ghirardi et al., 1986, see also Lewis (this volume)), Everettian interpretations (Saunders (this volume)), and Bohmian mechanics (see Tumulka (this volume)) are often interpreted in this way.

What kind of explanation is at stake in these theories? Can a description of the measurement process where the wavefunction plays a constitutive role, ground a ME? Yes, if we adopt – as we think we should – a mechanistic account that does not rule out *a priori* entities that do not live on a three-dimensional space. The above-cited theories describe different mechanisms underlying the occurrence of determinate measurement results. GRW implements stochastic dynamics, according to which the wavefunction randomly collapses into one of the terms of the superposition.

According to Everettian interpretations, the wavefunction does not collapse, but all branches of the superposition coexist after a measurement and all possible results are realized. Here, decoherence shows how classical behavior (i.e., the loss of interference between branches) is caused to emerge from dynamical processes described by the Schrödinger equation.

Finally, in Bohmian mechanics, the wavefunction is interpreted as acting on one "universal particle" like a physical field in configuration space (Albert, 1996, p. 278). In this mechanistic decomposition, the interaction between wavefunction and universal particle constitutes a new fundamental interaction.

It might be objected at this point that the acknowledgment of the wave-function as a mechanistic entity goes far beyond our most liberal pre-theoretical image of mechanism. Indeed, many accounts (e.g., Craver, 2007) describe mechanisms as systems in space-time. However, revolutionary extensions of the stable set of mechanistic elements are not only allowed but also envisaged by the new mechanistic account, as the above-cited example of the discovery of the electromagnetic field clearly illustrates. Before Maxwell, only material bodies were said to causally interact (by contact or at a distance) between each other. Maxwell's realist interpretation of the electromagnetic field forced the abandonment of this assumption and added the interaction between material bodies and physical space to the basic elements of a ME. In the same way, within wavefunction realist interpretations of QT the wavefunction is causally active, and should therefore be accepted as a basic element of a ME.

Besides being compatible with the explanations provided by wavefunction ontologies, and contrarily to law-based accounts of explanation, the mechanical account is highly informative about the role that explanation has in the foundations of quantum theory.

For instance, an explanatory deficiency of $3N$-dimensional wavefunction ontologies is that it is not clear how they cover the manifest image of everyday 3-dimensional objects with their dynamics, from a $3N$-dimensional object. If, in other words, the world is in $3N$-dimensions, how then do you get our three-dimensional world?

> the concern with these theories is that, because the wave-function lives on configuration space and not three-dimensional space, the explanatory scheme developed in classical theories in terms of a primitive ontology must be drastically revised. A new explanatory scheme is needed, and nobody has found one yet. (Allori, 2013, p. 13. Page numbering refers to the online version)

The new mechanistic philosophy sheds a clarifying light over the issue raised in the above quote: the relation that typically grounds the relationship of explanatory relevance between higher and lower mechanistic levels is composition; however, there is no straightforward compositional relation between the (higher level) three-dimensional entities of our experience and the (lower level) wavefunction in $3N$-dimensions.

In Bohmian mechanics, a way to overcome this problem is to introduce the notion of a *multi-field*, a configuration of which assigns properties to sets or fusions of n points, and to view n-particle wavefunctions as corresponding to multi-fields on ordinary three-space rather than to fields on the much larger configuration space of the system. (Belot, 2012, pp. 5–6)

In this interpretation, three-dimensional macroscopic objects are composed of three-dimensional microscopic objects, and this relationship makes it possible for the lower level mechanism to explain the behavior of the higher level entities. Another notable solution, this time in the context of Everettian decoherence-based explanations of determinateness, is put forward in (Wallace, 2003, 2010) – a strategy pivoting on a functionalist criterion for the identification of macro-objects, in concordance with the perspectivalist account of mechanistic description and explanation (e.g., Craver, 2013).

The direct rival of the wavefunction ontology is the so-called primitive ontology approach (Allori et al., 2008), based on the assumption that the wavefunction has a "nomological status", while the only real physical entities are the three-dimensional objects with a well-defined dynamics. Under this approach, spontaneous collapse interpretations can be articulated with an ontology of flashes (Bell, 1987) or density of stuff (Benatti et al., 1995). Many-worlds theories might be implemented with a similar three-dimensional density field ontology (Allori et al., 2011); however, the bulk of the work in this direction was developed in the framework of Bohmian mechanics.

In the version originally put forward in Dürr et al. (1992), the primitive ontology approach can't be covered by the mechanistic account. In fact, while in the latter only concrete objects of physical reality, their properties, and organization play an explanatory role, in this interpretation of Bohmian mechanics the wavefunction is not part of the basic ontology of the world, and yet it plays an irreplaceable explanatory role. Following this view, Goldstein and Zanghì (2013) describe the wavefunction as a "nomological entity" that "governs" the behavior of particles (going in this way beyond the ontologically neutral view of laws as exceptionless regularities). However, if we discard a naive literal reading of the "governing" talk, it is unclear how this metaphor should be interpreted. Notice that, since the explanatory role of the wavefunction as a nomological entity goes beyond the epistemic, formal role that laws as exceptionless regularities hold in the covering law (or the unificationist) model of explanation, this version of the primitive ontology approach does not naturally fit with law-based accounts of explanation, either. It is because of the unclear explanatory import of this version of the primitive ontology approach that, notwithstanding the apparent incompatibility of such an account with ME, its use as a counterexample to the ME account begs the question.

Moreover, other primitive ontologies exist that provide alternatives to the "governing" explanation, easily accountable as mechanistic, as for instance dispositionalist interpretations (e.g., Esfeld et al., 2013), which avoid any appeal to laws of nature, or Humean approaches (e.g., Callender, 2015), where laws of nature play a role that is openly conceded by the new mechanistic philosophy (Craver and Kaiser, 2013).

This short survey of solutions to the measurement problem shows that the mechanistic account is not only compatible with (some explanations provided by) contemporary approaches to quantum theory, but it is also highly informative with respect to the constraints and desiderata of a satisfactory solution to the measurement problem, and to the role that explanation has in the foundations of quantum theory.

To this conclusion, it might be countered that the measurement problem, together with the other issues belonging to the so-called "interpretation" of quantum theory, is a metaphysical, rather than a scientific, problem. However, such a categorical distinction between science and metaphysics is inadequate when applied to issues like the measurement problem, which, pivotal as it is for the achievement of a coherent quantum theory, can't be considered external to science. On the contrary, this survey illustrates how the new mechanistic account of explanation has the virtue of showing that and how the question of what kind of entities and processes underlie phenomena is indeed part and parcel of physics and science.

Acknowledgment

For suggestions and comments, I am very grateful to Valia Allori, Michael Cuffaro, Mauro Dorato, Meinard Kuhlmann, Matteo Morganti, Davide Romano, Emanuele Rossanese, and Alastair Wilson.

Notes

1 Unless otherwise specified, in this paper we indicate with the expression "law of nature" the non-metaphysically charged notion of exceptionless regularities that support counterfactuals. See Lange (this volume) for a variety of accounts of laws of nature.

2 Other definitions can be found e.g. in MDC, 3, Glennan (2002, S344), Bechtel and Abrahamsen (2005) and Illari and Williamson (2012).
3 See Part 7 of this volume for further discussion of these explanations.
4 See Part 1 for detailed discussion of classical mechanics.
5 For an overview of the development of the concept of field, see Mary Hesse's magisterial book "Forces and Fields" (2005).
6 See also (Kuhlmann, 2017) for other examples.
7 See Batterman (this volume) for more on universality.
8 See Maudlin, this volume.
9 Though see Brown and Read (this volume) for discussion of dissenting views.
10 See Healey (this volume).
11 For an historical account of the discovery see Baggott (2012). See Coleman (1985) for a technical introduction.
12 See Part 4 of this volume for more details.
13 See Maroney (this volume) for details on this dispute.

Bibliography

Albert, D.Z. (1996). Elementary quantum metaphysics. In J. Cushing, A. Fine and S. Goldstein (eds.), *Bohmian Mechanics and Quantum Theory: An Appraisal*, Vol. 184. Boston Studies in the Philosophy of Science. Berlin: Springer Science & Business Media, pp. 277–284.

Allori, V., Goldstein, S., Tumulka, R., and Zanghì, N. (2008). On the common structure of Bohmian mechanics and the Ghirardi–Rimini–Weber theory: Dedicated to Giancarlo Ghirardi on the occasion of his 70th birthday. *The British Journal for the Philosophy of Science*, 59(3): 353–389.

Allori, V., Goldstein, S., Tumulka, R., and Zanghì, N. (2011). Many worlds and Schrödinger's first quantum theory. *British Journal for the Philosophy of Science*, 62(1): 1–27.

Allori, V. (2013). Primitive ontology and the structure of fundamental physical theories. In A. Ney and D.Z. Albert (eds.), *The Wave Function: Essays on the Metaphysics of Quantum Mechanics*. Oxford: Oxford University Press, pp. 58–75. Available at: http://philsci-archive.pitt.edu/9342/1/AlloriWFOLast.pdf

Baggott, J. (2012). *Higgs: The Invention and Discovery of the 'God Particle'*. Oxford: Oxford University Press.

Batterman, R. (2002) "Asymptotics and the Role of Minimal Models," *The British Journal for the Philosophy of Science*, 53(1): 21–38.

Bechtel, W. (2009). Explanation: Mechanism, modularity, and situated cognition. In P. Robbins and M. Aydede (eds.), *The Cambridge Handbook of Situated Cognition*. Cambridge: Cambridge University Press, pp. 155–170.

Bechtel, W. and A. Abrahamsen (2005). Explanation: A mechanist alternative. *Studies in History and Philosophy of Science Part C: Studies in History and Philosophy of Biological and Biomedical Sciences*, 36(2): 421–441.

Bechtel, W. and R.C. Richardson (2010) [1993]. *Discovering Complexity: Decomposition and Localization as Strategies in Scientific Research* (2nd edition). Cambridge, MA: MIT Press/Bradford Books.

Bell, J.S. (1987). *Speakable and Unspeakable in Quantum Mechanics*. Cambridge: Cambridge University Press.

Belot, G. (2012). Quantum states for primitive ontologists. *European Journal for Philosophy of Science*, 2: 67–83.

Benatti, F., Ghirardi, G.C. and Grassi, R. (1995). Describing the macroscopic world: Closing the circle within the dynamical reduction program. *Foundations of Physics*, 25: 5–38.

Callender, C. (2015). One world, one beable. *Synthese*, 192(10): 3153–3177.

Coleman, S. (1985). *Aspects of Symmetry: Selected Erice Lectures*. Cambridge: Cambridge University Press.

Craver, C.F. (2007). *Explaining the Brain*. Oxford: Oxford University Press.

Craver, C.F. (2013). Functions and mechanisms: A perspectivalist view. In P. Huneman (ed.), *Functions: Selection and Mechanisms*. Dordrecht: Springer, pp. 133–158.

Craver, C.F. and Kaiser, M.I. (2013). Mechanisms and laws: Clarifying the debate. In R. Millstein, H-K. Chao, and S-T. Chen (eds.), *Mechanism and Causality in Biology and Economics*. Dordrecht: Springer, pp. 125–145.

Creath, R. (2017). Logical empiricism. In E.N. Zalta (ed.), *The Stanford Encyclopedia of Philosophy* (Fall 2017 Edition). Available at: https://plato.stanford.edu/archives/fall2017/entries/logical-empiricism/.

Dorato, M. and Felline, L, (2011). Scientific explanation and scientific structuralism. In A. Bokulich and P. Bokulich (eds.), *Scientific Structuralism*. Dordrecht: Springer, pp. 161–176.

Dürr, D., Goldstein, S. and Zanghì, N. (1992). Quantum equilibrium and the origin of absolute uncertainty. *Journal of Statistical Physics*, 67: 843–907.

Earman, J. (2004). Laws, symmetry, and symmetry breaking: Invariance, conservation principles, and objectivity. *Philosophy of Science*, 71: 1227–1241.

Esfeld, M., Hubert, M., Lazarovici, D. and Dürr, D. (2013). The ontology of Bohmian mechanics. *The British Journal for the Philosophy of Science*, 65(4): 773–796.

Felline, L. (2015). Mechanisms meet structural explanation. *Synthese*, First online: 30 May: 1–16.

Felline, L. (2016). It's a matter of principle: Scientific explanation in information-theoretic reconstructions of quantum theory. *Dialectica*, 70(4): 549–575.

Felline, L. (2020). Quantum theory is not only about information. *Studies in History and Philosophy of Modern Physics*, 72: 25–265.

Friedman, M. (1974). Explanation and scientific understanding. *The Journal of Philosophy*, 71(1): 5–19.

Fuchs, C.A., Mermin, N.D. and Schack, R. (2014). An introduction to QBism with an application to the locality of quantum mechanics. *American Journal of Physics*, 82(8): 749–754. https://arxiv.org/pdf/1311.5253.pdf.

Glennan, S.S. (2002). Rethinking mechanistic explanation. *Philosophy of Science*, 69(S): S342–S353.

Glennan, S.S. (2008). Mechanisms. In M. Curd and S. Psillos (eds.), *Routledge Companion to the Philosophy of Science*. New York: Routledge, pp. 376–384.

Glennan, S.S. (2011). Singular and general causal relations: A mechanist perspective. In P. McKay Illari, F. Russo, and J. Williamson (eds.), *Causality in the Sciences*. Oxford: Oxford University Press, pp. 89–817.

Glennan, S.S. (2017). *The New Mechanical Philosophy*. Oxford: Oxford University Press.

Goldstein, S. and Zanghi, N. (2013). Reality and the role of the wave function in quantum theory. In A. Ney and D.Z. Albert (eds.), *The Wave Function: Essays on the Metaphysics of Quantum Mechanics*. Oxford: Oxford University Press, pp. 91–109.

Hall, N. (2004). Two concepts of causation. In J.D. Collins, E.J. Hall and L.A. Paul (eds.), *Causation and Counterfactuals*. Boston, MA: MIT Press, pp. 225–276.

Hedström, P. and Swedberg, R. (1998). Social mechanisms: An introductory essay. In P. Hedström and R. Swedberg (eds.), *Social Mechanisms: An Analytical Approach to Social Theory*. Cambridge; New York: Cambridge University Press, pp. 1–31.

Hempel, C.G. (2001). *The Philosophy of Carl G. Hempel: Studies in Science, Explanation, and Rationality*. Oxford: Oxford University Press.

Hesse, M.B. (1961). *Forces and Fields: The Concept of Action at a Distance in the History of Physics*. Mineola, NY: Dover Publications.

Illari, P.M. and Williamson, J. (2012). What is a mechanism? Thinking about mechanisms across the sciences. *European Journal for Philosophy of Science*, 2(1): 119–135.

Janssen, M. (2009). Drawing the line between kinematics and dynamics in special relativity. *Studies in History and Philosophy of Science Part B: Studies in History and Philosophy of Modern Physics*, 40(1), 26–52.

Kochiras, H. (2013). The mechanical philosophy and Newton's mechanical force. *Philosophy of Science*, 80(4): 557–578.

Kuhlmann, M. (2011). Mechanisms in dynamically complex systems. In P.M. Illari, F. Russo and J. Williamson (eds.), *Causality in the Sciences*. Oxford: Oxford University Press, pp. 880–906.

Kuhlmann, M. (2017). Mechanisms in physics. In S. Glennan and P. Illari (eds.), *Routledge Handbook of the Philosophy of Mechanisms and Mechanistic Philosophy,* Abingdon: Routledge. pp. 283–295.

Kuhlmann, M. and Glennan, S. (2014). On the relation between quantum mechanical and neo-mechanistic ontologies and explanatory strategies. *European Journal for Philosophy of Science*, 4(3): 337–359.

Lange, M. (2011). Conservation laws in scientific explanations: Constraints or coincidences? *Philosophy of Science*, 78(3): 333–352.

Lyre, H. (2012). The just-so Higgs story: A response to Adrian Wüthrich. *Journal for General Philosophy of Science*, 43(2): 289–294.

Machamer, P., Darden, L. and Craver, C.F. (2000). Thinking about mechanisms. *Philosophy of Science*, 67(1): 1–25.

Morrison, M. (2003). Spontaneous symmetry breaking: Theoretical arguments and philosophical problems. In K. Brading and E. Castellani (eds.), *Symmetries in Physics: Philosophical Reflections*. Cambridge: Cambridge University Press, pp. 347–363.

Nersessian, N.J. (1984). Aether/or: The creation of scientific concepts. *Studies in History and Philosophy of Science Part A*, 15(3): 175–212.

Salmon, W.C. (1984). *Scientific Explanation and the Causal Structure of the World*. Princeton, NJ: Princeton University Press.

Stöltzner, M. (2017). The variety of explanations in the Higgs sector. *Synthese*, 194(2): 433–460.

Wallace, D. (2003). Everett and structure. *Studies in History and Philosophy of Modern Physics*, 34: 87–105.

Wallace, D. (2010). Decoherence and ontology. In S. Saunders, J. Barrett, A. Kent and David Wallace (eds.), *Many Worlds?: Everett, Quantum Theory, and Reality*. Oxford: Oxford University Press, pp. 53–72.

Wigner, E.P. (1995). Remarks on the mind-body question. In J. Mehra (ed.), *Philosophical Reflections and Synthe-ses*. Berlin, Heidelberg: Springer, pp. 247–260. First published in Wigner, E.P. (1961). *The Scientist Speculates*. Heinmann, London.

Further Reading from the Editors

In order to learn more about mechanisms and mechanistic explanation in general, try Craver, C.F. "When mechanistic models explain" (*Synthese*, 153(3), 355–376, 2006), Woodward, J. "Mechanisms revisited" (*Synthese*, 183(3), 409–427, 2011) and Illari, P.M., & Williamson, J. "What is a mechanism? Thinking about mechanisms across the sciences" (*European Journal for Philosophy of Science*, 2(1), 119–135, 2012). Glennan, Stuart, and Phyllis Illari, eds. *The Routledge Handbook of Mechanisms and Mechanical Philosophy* (Taylor & Francis 2017) is highly relevant, in particular, see Kuhlmann's "Mechanisms in Physics" in that volume. For more on the connec-tion between mechanistic explanation and structural explanation, see Felline, L. "Mechanisms meet structural explanation" (*Synthese*, 195(1), 99–114, 2018).

34

THE EXPLANATORY VALUE OF SELECTING THE APPROPRIATE SCALE(S)

Lina Jansson

34.1   The Problem

Let me start with a well-worn example of explanations given at different scales. The example is from Putnam (1975). Imagine that we have a rigid board with two holes. One hole is square with sides 1 inch in length. The other hole is circular with a diameter of 1 inch. A solid square peg with sides just less than 1 inch will fit through the square hole, but not through the circular one. Why?

The answer that Putnam (and many others) favors cites the rigidity of the board and the peg together with the geometrical features of the situation: the explanation in terms of macro-structures and properties. We could, in principle, have decided to present the blow-by-blow account of the molecular interactions of the peg and the board: for now, let us call this the explanation in terms of microstructures and properties. The intuition that Putnam pushes is that even if we did have such a derivation available to us it would not be explanatory (the weaker claim that it is merely a less good explanation will do as the starting point for our discussion).

At least at first glance, this seems like a case where we have gained something valuable by offering the explanation at the macroscale rather than the microscale. This observation is often captured by claiming that, for example, generality or abstraction is an explanatory virtue. However, often discussions do not go on to specify how this fits with (or even better, follows from) a broader account of explanation. Why is generality, abstraction, or selecting the appropriate scale(s) explanatorily valuable? The question that I will address in this paper is the following: what kind of explanatory value (if any) comes with selecting the appropriate scale(s)? This question will plunge us into the deep water of debates about emergence versus reduction.

However, even close to shore things are not as simple as they seem. So, let us start in the shallows with: what is scale?

34.2   Emergentism versus Reductionism and the Question of Scale

In the very simple case of the square peg and the round hole above, I made use of the most familiar notion of scale: the spatial one. As we move from the macro to the micro we expect to encounter smaller entities and behavior that takes place at shorter length scales. However, there are plenty of other scales. For example, we might decide to focus on behavior at higher or lower energies, at longer or shorter periods of time, etc. Given that a spatial scale is not the only scale available, is there anything special about the spatial scale?

Traditional reductionist would answer: yes. The spatial scale is special since it tends to coincide with a necessary condition of parthood for concrete objects. The molecules of the peg are (proper) parts of the peg. Proper parts of concrete entities are (at least in typical cases of parthood) expected to be spatially smaller than the entities that they compose.[1] The spatial scale turns out to be naturally favored by classical formulations of explanatory reductionism that make use of the notion of parthood to define what it is to have an explanatory reduction.

As part of their general definition of reduction Oppenheim and Putnam include the criterion that "any observational data explainable by . . . [the reduced theory] is explainable by . . . [the reducing theory]" (1958, p. 5). What makes something the type of reduction that they are interested in, a micro-reduction, is that the reducing theory "deals with the parts of the objects dealt with by. . . " (1958, p. 6) the reduced theory. With the assumption that proper parts of concrete entities are spatially smaller than the entities that they compose, calling these reductions *micro-reductions*, seems very appropriate.[2]

According to a traditional explanatory reductionist view, foundational physics can, *in principle*, explain everything that the explanations offered by the sciences outside of foundational physics, such as biology, psychology, chemistry, or for that matter non-fundamental physical theories such as thermodynamics, etc., can account for.[3] The "in principle" clause requires some spelling out. The idea is not that we actually have to hand replacement explanations from foundational physics nor that our current best candidates for what counts as foundational physics can in principle account for everything explained by all the other branches of science. The "in principle" codifies the expectation that there is a theory that would belong to foundational physics that could, if all the details were known, explain everything currently explained by different scientific theories in different scientific branches.

34.3 Emergentism versus Reductionism and the Question of Explanation

I have only offered a very brief and partial view of the debate between emergentism and reductionism here (see chapters in Part 9 of this handbook for a much more detailed discussion). However, it is enough to bring out the role played by the notions of explanation and scale.

> In trying to make emergence intelligible, it is useful to divide the ideas usually associated with the concept into two groups. One group of ideas are manifest in the statement that emergent properties are "novel" and "unpredictable" from the knowledge of their lower-level bases, and that they are not "explainable" or "mechanistically reducible" in terms of their underlying properties. The second group of ideas I have in mind comprises the specific emergentist doctrines concerning emergent properties, and, in particular, claims about the causal powers of the emergents. (Kim, 1999, p. 5)

There are several ways that we could approach the explanatory aspect of the emergentist versus reductionist debate. We could focus on case studies and whether we have good reason to take explanatory reductionism to be true in particular cases (for examples of this type of argument, see Chapters 36 and 37 in this handbook).[4] Central among the case studies in the philosophy of physics is the example of phase transitions; that is, phenomena such as water freezing or the magnetization of a ferromagnet.[5] These explanations have two interesting features. The first is that the phenomena that we are trying to understand – the behaviors of these systems near criticality – do not appear to be ones that can be understood at a *single* characteristic scale. Second, the explanations offered from the putatively more fundamental theories seem to involve potentially essential appeals to infinite idealizations. Since phase transitions do occur in finite systems, this can be taken to lend support against the claim that these phenomena have been reductively explained by the putatively more fundamental theories.[6]

In addition to arguments for the failure of specific reductive explanations, there are also arguments against the possibility of providing a reductive explanation for some questions of interest.

The challenge of multiple realizability to explanatory reductionism *properly understood,* concerns the ability of the *theory of the heterogeneous micro-realizers* to explain the *robustness* of the common behavior displayed by the systems at macro-scales. That is, the challenge is to explain the *autonomy* of upper-scale common behavior from lower-scale details. However, as we have seen, "disunified" explanations, while certainly telling us a lot about the behavior of individual systems, do not explain the autonomy in question. And, this is true even if we buy into the idea that someday we will have a completed physics – even if we dismiss explanatory difficulties as "merely mathematical" or as involving only "pragmatic" difficulties. (Batterman, 2018, p. 862)

I will return briefly to this question later. For now, however, I will set aside these much-discussed examples. There are two reasons for this. First, I take Knox (2016, 2017) to have convincingly argued that many of the interesting questions about the value of non-foundational explanations are not limited to these cases (even within physics). Her focus is rather on the potential for the selection of appropriate variables to make possible abstractions.[7]

Second, since the approach through case studies is already well-explored elsewhere, I would like to present a different route to the questions at stake here. I will seek to clarify what kind of explanatory value selecting the appropriate scale(s) *could* be on various broad accounts of explanation.

These two approaches are not, to my mind, competitors. A good account of explanation and of the explanatory value of selecting the appropriate scale(s) will be informed by case studies from science. Similarly, which account of explanation one favors will inform the understanding of the particular case studies.

34.4 Three Different Broad Accounts of Explanation

The answer to the question of what the explanatory value of selecting the appropriate scale(s) is will vary depending on the account of explanation that we consider. I will divide accounts of explanation into three different categories (while recognizing that the categories are not so sharp as to not admit of borderline cases that are difficult to place): ontic, pragmatic, and epistemic.[8]

On ontic accounts of explanation, explanations are (at least in the primary notion of explanation) in the world.

Proponents of this [ontic] conception can speak in either of two ways about the relationship between explanations and the world. First one can say that explanations exist in the world. The explanation of some fact is whatever produced it or brought it about ... Second, the advocate of the ontic interpretation can say that an explanation is something—consisting of sentences or propositions—that reports such facts. It seems to me that either way of putting the ontic conception is acceptable... (Salmon, 1989, p. 86)[9]

On these accounts, it makes sense to talk about one aspect of the world explaining some (other) aspect of the world. For example, the hob being turned on explains why the water in the pot is boiling. Typically (as we see in the quote from Salmon), ontic accounts also allow that there is an associated communicative notion of explanation where we display information about which aspect explains what (other) aspect of the world. However, the notion of communication is taken to be derivative on, and secondary to, the idea of explanations as holding in the world.

There are several candidates for what explanations are on an ontic account but some of the most prominent suggestions are that the explanation of some phenomenon should be identified with the cause(s) of that phenomenon (or some subset of the cause(s) of the phenomenon) or, alternatively, with the aspects of the world and the laws of nature that nomologically necessitate the phenomenon to be explained.[10] On this account, there is no great difficulty in discussing explanations as existing independently of us.[11]

Contrasting with the ontic account, we could put the human aspect of giving, seeking, and having explanations at the center of our account of explanation. This is what happens on the pragmatic and the epistemic views. Here, the hob being turned on might *cause* the water in the pot to boil and there might be a causal relation in the world linking these two events. However, on these views, explanations are not relations in the world.

On pragmatic accounts of explanation, the focus has shifted from taking an explanation to be a relation in the world to the notion of an explanation as an answer to some why-question in some context. Crucially, for pragmatic accounts, whether an explanatory relationship holds depends on the context and the interests of the explanation seeker.

> It is sometimes said that an Omniscient Being would have a complete explanation, whereas these contextual factors only bespeak our limitations due to which we can only grasp one part or aspect of the complete explanation at any given time. But this is a mistake. If the Omniscient Being has no specific interests (legal, medical, economic; or just an interest in optics or thermodynamics rather than chemistry) and does not abstract (so that he never thinks of Caesar's death *qua* multiple stabbing, or *qua* assassination), then no why-questions ever arise for him in any way at all – and he does not have any explanation in the sense that we have explanations. If he does have interests and does abstract from individual peculiarities in his thinking about the world, then his why-questions are as essentially context-dependent as ours. (van Fraassen, 1980, p. 130)

On these views, to be told that the hob is turned on might count as an explanation of the boiling of the water in the pot for someone in the right circumstances.

Given the context-dependence of the explanatory relation, it is tempting to take pragmatic accounts to be subjective accounts of explanation (in contrast to the objective ontic accounts). Pragmatic accounts of explanations are, however, compatible with a range of views about how dependent on subjective considerations explanations are.

If we take a pragmatic approach that focuses on individual psychological notions of understanding, then we naturally get an account where it is ultimately the interests and background of the individual that determine the success criteria for having an explanation.[12] On these types of pragmatic views what matters is not truth or accurate representation (of at least some aspects of the system of interest) as such. Although, we can make sense of it mattering that the explanation seeker *believes* that the explanation (or some aspects of the explanation) is true or accurately (enough) represents.

Of course, if we expect many features related to achieving understanding to be shared between most humans, we also expect to find a good deal of overlap in terms of the criteria by which explanations should be judged. If instead of focusing on the notion of individual understanding, we turn to the community level and the idea of intersubjectively shared criteria for explanatory understanding, then we get an account of explanation where the success criteria for explanation are determined by the group at which it is targeted. Here the success criteria for explanations are dependent on the explanatory context set by the collective standards and/or interests of the group and not merely on those of the explanation seeker.

Here, it is natural to take it to be the case that "… explanatory relevance is something that is not judged trans-historically (by something like a brute number of w-questions) but, rather, is a function of the current state of scientific knowledge" (Bokulich, 2012, p. 736).

However, pragmatic accounts can also be combined with a general commitment to accurate representation. On such views, we still have an account where contextual factors enter into the explanatory relationship itself. One such view is put forward by Potochnik (2010a, 2010b).

> Explanation is indelibly context-dependent. . . . But let me be clear about the extent of this context-dependence. Explanation is only context-dependent at the level of determining which of the many actual causal factors should be included in a particular explanation. If the process under investigation does not conform to some causal pattern, then that pattern cannot explain the outcome of the process. No amount of interest in a pattern can will it into existence. (Potochnik, 2010b, pp. 224–225)

Once we have included an ontic constraint (such as the accurate representation of part of the causal history) we have moved far from a pragmatic account based around a notion of understanding for an individual. We are now close to the final type of accounts of explanation that I want to consider. This group of accounts differs from pragmatic accounts in that they reject the through and through context-dependence of the explanatory relation. Explanatory relevance can be judged transhistorically on these accounts. However, they differ from the metaphysically focused ontic accounts in taking the role of the agent to be central to explanations (there are no explanations without agents and explanations are not, in the primary notion, relations in the world). On (what I will call) epistemic accounts of explanation context-dependence and interest relativity have a role to play in accounting for our explanatory practices, but they enter into the selection of the explanatory questions that we are interested in. Once we have selected the explanandum of interest (perhaps, including the relevant other options) there is no further role for interest relativity or context to play in determining whether a putative explanans E explains some explanandum; that is, whether an explanatory relation holds between a particular explanandum and a particular explanans is not context-dependent or dependent on the interests of particular subjects/communities. A prominent example of the kind of account that I have in mind here is that of Woodward (2003).[13]

> On my analysis, interest relativity enters into what we explain but not into the explanatory relationship itself. What we try to explain depends on our interests, but it does not follow that for a fixed explanandum M and fixed explanans E, whether E explains M is itself interest-dependent. Obviously, it is not puzzling and no threat to the "objectivity" of explanation that the explanans E may explain M but a different explanans E' may be required to account for M'. (Woodward, 2003, p. 230)

On Woodward's account, an explanation must provide accurate information about what would have happened (to the explanandum system) had things been different. While there is scope for interest relativity in selecting the relevant explanandum and the relevant other options that we are considering, there is no further interest relativity in whether or not some putative explanans counts as an explanation of the explanandum in question. Here, some explanans E either explains some explanandum M (once the relevant contrast class, etc. is set) or not, independently of any information about the interests of the agent (or community of agents) seeking an explanation.

I am calling this option epistemic since explanations here are, by definition, something that we (as epistemic agents) seek, receive, and give; they are not primarily simply in the world as in the metaphysically focused ontic accounts. However, there is no interest relativity or context-dependence in the explanatory relation itself, so the account is not pragmatic.

34.5 Three Different Accounts of the Value of Selecting the Appropriate Explanatory Scale(s)

With the three broad accounts of explanation on the table (ontic, pragmatic, and epistemic) we can return to the question of what value, if any, comes with selecting the appropriate scale(s) for some putative explanation.

On ontic accounts, explanations are (at least in the primary sense) in the world. Here the correct explanatory scale(s) is naturally thought of as being matched to the scale(s) at which we find the causal or nomological or other modal relations that are taken to be explanatory. If there is an explanatory *novel* value at different scales of explanation, on a purely ontic account (setting aside for now that typically a communicative component is recognized too), such value has to be accounted for by the existence of new ontic power (such as new causal or nomological relations) at different scales.

Here we can see the connection between the first and second group of emergentist ideas that Kim (1999) identifies. On ontic accounts of explanation, the first group of ideas demands the existence of the new ontic powers that the second group of ideas focuses on. Of course, this also raises the familiar worries about emergence in terms of causal power that, for example, Kim (1998) raises: how do we accommodate *novel* causal powers by the whole that are not already captured by the parts that the whole supervenes upon?

Within the context of the philosophy of physics, the idea that explanatory novelty demands the postulation of a new causal relation leads McGivern (2008, p. 70) to argue from the independent explanatory role of multi-scale properties for taking very seriously the idea that they are independently causal. This is also the motivation for his consideration of objections to Kim's arguments.

On pragmatic accounts, there are several options for where explanatory value could be found. However, the distinctive option available here but not on epistemic accounts (discussed below) is that the value comes with being sensitive to the interests of the explanation seeker. This is Sober's response to Putnam's case of the peg that I started this paper with.[14]

> Perhaps the micro-details do not interest *Putnam*, but they may interest *others*, and for perfectly legitimate reasons. Explanations come with different levels of detail. When someone tells you more than you want to hear, this does not mean that what is said fails to be an explanation. (Sober, 1999, p. 547)

Here, the explanatory novelty involved in selecting the right explanatory scale is simply one to do with matching our interest or disinterest in details at that scale. It does not demand postulating any new causal powers.

The epistemic account introduces a third option that is not available on (purely) ontic accounts. Here the explanatory value of selecting the right scale(s) of explanation can be captured in terms of making available explanatory information that is not available at explanations formulated at competing scales. The explanatory value associated with selecting the correct scale(s) of explanation can derive from our epistemic access to the worldly relations backing explanation. This is neither merely a matter of whether we are interested in the details nor does it demand the introduction of new ontic powers at different scales.

For example, on Woodward's account of explanation (standing in as our representative of epistemic accounts of explanation), the value of selecting the appropriate scale and the appropriate variables is ultimately to be accounted for in terms of the ability to answer what-if-things-had-been-different questions. The value here is neither distinctively metaphysical nor distinctively pragmatic. We can increase our *access* to answers about what-if-things-had-been-different questions both by being correct about what depends on what, by correcting mistaken assumptions about what depends on what, and by seeking presentations that we are well-equipped to grasp.[15]

To see this consider a simple example where we have a holonomic constraint on motion. Take a bead sliding down a frictionless static wire shaped as a helix (as in Figure 34.1). Let us say that we are interested in the motion of the bead between two points. If we approach this problem on a large enough scale to notice that the helix is of constant radius R and with a uniform pattern of unwinding then we could use a Lagrangian approach to quickly notice that the motion of the bead is a function of z alone (rather than, say, x, y, z).

If we instead approach this problem at a smaller scale, this information about the constraints is not always readily available. It is not, of course, that the system does not now satisfy the constraints

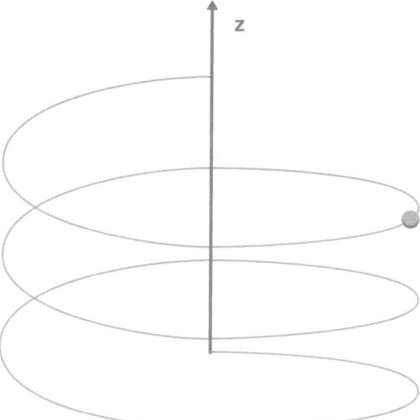

Figure 34.1 A bead sliding down a frictionless static wire shaped as a helix.

Figure 34.2 A Newtonian decomposition into gravitational and normal forces.

(it is, after all, the same system as earlier!), but these constraints are not apparent when we approach the problem at a very small scale through, for example, a Newtonian decomposition of forces and calculations of the force from gravity and the normal force (see Figure 34.2).

There are, plausibly, no new ontic powers at the larger scale here. Yet, it is perfectly objective that the motion does not depend on the x and y coordinates and that approaching the problem via a solution in terms of constraints more readily available at the larger than the smaller scale brings this out.

On Woodward's account of causal explanation (standing in as our representative of epistemic accounts of explanation), the value of selecting the right variables lies in avoiding *misrepresenting* the dependences involved by including variables that the explanandum does not in fact depend on.

> [T]he dependence ... should be such that (a) it explicitly or implicitly conveys accurate information about the conditions under which alternative states of the effect will be

realized *and* (b) it conveys *only* such information—that is, the cause is not characterized in such a way that alternative states of it *fail* to be associated with changes in the effect. (Woodward, 2010, p. 298)

Here, one of the challenging issues is to allow a principle such as the above, without ending up endorsing as explanatorily optimal explanations that list every possible explanans for the explanandum; typically, such an explanation would look unattractively disjunctive. For example, Franklin-Hall (2016) argues that Woodward's interventionist account does not have the resources to solve this problem without drawing on either pragmatic solutions or solutions in terms more suited to ontic accounts – such as metaphysical naturalness.[16]

Finally, let us return to the challenge from Batterman of explaining the autonomy of the upper-scale from the lower-scale. On an epistemic account, this autonomy does not have to be understood as a metaphysical autonomy of causal powers. The question here divides into three subquestions. First, do the properties involved in the lower-scale explanation provide the supervenience basis for the properties in the higher-scale explanation? As Potochnik (2010a, p. 62) has stressed, even if we have supervenience of properties it does not follow that we have supervenience in terms of the properties in *competitor explanations*. The higher-scale *explanation* could involve properties whose supervenience base is not found in the individual lower-scale *explanations* of the phenomenon of interest (although it might be found at the lower-scale). The explanations in terms of a Lagrangian approach with constraints or in terms of a direct Newtonian analysis through knowledge of the normal force would be an example such as this (from the normal force we could not uniquely recover the constraints). There is no particular challenge to global supervenience here, but the properties invoked in the Newtonian explanation do not provide a supervenience base for the properties in the Lagrangian competitor in the particular competitor explanations. Second, assuming that there is an, in principle, explanation of the autonomy of the higher-scale from foundational physics, we have the question of whether this derivation is as good an explanation as competitor ones. On an epistemic account, this will depend crucially on whether or not we ought to keep our general cognitive abilities fixed when evaluating the "in principle" derivation. Finally, we need to take a stance on questions such as whether the explanatory force stems only from application to nomologically possible scenarios. For example, Weslake (2010) has argued that if logically possible scenarios count, higher-scale explanations can apply to more situations and thus be better explanations than lower-scale ones.[17]

34.6 Conclusion

The main goal of this chapter has been to provide a broad overview of what the value of selecting the appropriate explanatory scale(s) could be on different accounts of explanation. The hope is to have clarified how the account of explanation favored influences the type of questions that we need to answer in order to address questions about explanatory reduction.

Notes

1 Rueger and McGivern (2010) argue that we should give up on this notion of parthood, in the context of understanding explanatory levels. We could, alternatively, give up on thinking of foundational physics as characterised by being a theory of entities that compose the entities of the other sciences. See for example, Ladyman and Ross (2007) for an alternative.
2 In Kim's (1998) definition we similarly find a focus on parthood.
3 In many versions primarily from philosophy of science the claim is more specific. The scientific laws, if any, are either part of foundational physics or can be accounted for by foundational physics. In versions primarily from within metaphysics, the focus is instead often on the reduction of macro-properties to micro-properties.
4 For example, Cartwright (1999), Batterman (2002, 2018), McGivern (2008), Morrison (2012) deny that we have good reason to take explanatory reductionism to be true (although their arguments differ).

5 These type of examples are extensively discussed by Batterman (2002, 2013). The discussion of how to understand these cases in relation to explanatory reductionism is very much ongoing. See, for example, Belot (2005), Butterfield (2011a, 2011b), Morrison (2012), Menon and Callender (2013), Reutlinger (2014).

6 See, for example, Menon and Callender (2013) for a discussion (but not endorsement) of this argument.

7 The focus on abstraction by being able to reduce the number of variables (and possibly equations) seems to me to be likely to have connections to the more metaphysical approach of Wilson (2010) and the claim that weak emergence can be understood by focusing on degrees of freedom.

8 There are other ways to classify different accounts of explanation. For some related classifications that have informed mine (but are not identical to it) see Jenkins (2008), Marcus (2014), and Bokulich (2016).

9 For a very similar more recent discussion see Strevens (2008, p. 6).

10 I am here deviating from Salmon's (1985) three categories.

11 Of course, subject to the caveat that it is not an explanation about us.

12 I take Faye's (2014) account to be of this kind.

13 Lange (2016) seems like another such view. I also take Strevens' discussion of idealisation, etc., to naturally belong here. I take my own view in Jansson (2015) to be of this kind as well. I also take the unificationist project of Kitcher (1989) and Friedman (1974) to involve relations of explanatory relevance that are not context-dependent.

14 For Sober (1999) himself this is somewhat more complicated since he does not advocate a pragmatic account of whether some explanans explains some explanandum but merely whether it explains it better or worse.

15 Ylikoski and Kuorikoski (2010) as well as Woodward (2016) emphasise all of these aspects.

16 I will leave it open here whether there is an account of naturalness that is neither metaphysical nor pragmatic. However, such appeals to naturalness are also part of, for example, Knox's (2016, pp. 55–57) treatment of the relation between statistical mechanics and thermodynamics.

17 Saatsi's chapter on non-causal explanations in this handbook contains examples of cases where it could be argued that non-nomological possibilities should be considered.

Bibliography

Batterman, R.W. (2002). *The Devil in the Details: Asymptotic Reasoning in Explanation, Reduction, and Emergence.* Oxford Studies in Philosophy of Science, Oxford: Oxford University Press.

Batterman, R.W. (2013). The tyranny of scales. In R.W. Batterman (ed.), *The Oxford Handbook of Philosophy of Physics.* Oxford: Oxford University Press, pp. 255–286.

Batterman, R.W. (2018). Autonomy of theories: An explanatory problem. *Noûs*, 52(4), 858–873.

Belot, G. (2005). Whose devil? Which details?. *Philosophy of Science*, 72(1), 128–153.

Bokulich, A. (2008). *Reexamining the Quantum-Classical Relation: Beyond Reductionism and Pluralism.* Cambridge: Cambridge University Press.

Bokulich, A. (2012). Distinguishing explanatory from nonexplanatory fictions. *Philosophy of Science*, 79(5), 725–737.

Bokulich, A. (2016). Fiction as a vehicle for truth: Moving beyond the ontic conception. *The Monist*, 99(3), 260–279.

Butterfield, J. (2011a). Emergence, reduction and supervenience: A varied landscape. *Foundations of Physics*, 41(6), 920–959.

Butterfield, J. (2011b). Less is different: Emergence and reduction reconciled. *Foundations of Physics*, 41(6), 1065–1135.

Cartwright, N. (1999). *The Dappled World: A Study of the Boundaries of Science.* Cambridge: Cambridge University Press.

Faye, J. (2014). *The Nature of Scientific Thinking: On Interpretation, Explanation, and Understanding.* London: Palgrave Macmillan.

Franklin-Hall, L.R. (2016). High-level explanation and the interventionist's 'variables problem'. *British Journal for the Philosophy of Science*, 67(2), 553–577.

Friedman, M. (1974). Explanation and scientific understanding. *Journal of Philosophy*, 71(1), 5–19.

Jansson, L. (2015). Explanatory asymmetries: Laws of nature rehabilitated. *Journal of Philosophy*, 112(11), 577–599.

Jenkins, C.S. (2008). Romeo, René, and the reasons why: What explanation is. *Proceedings of the Aristotelian Society*, 108(1):61–84.

Kim, J. (1998). *Mind in a Physical World.* MIT Press.

Kim, J. (1999). Making sense of emergence. *Philosophical Studies*, 95(1/2), 3–36.

Kitcher, P. (1989). Explanatory unification and the causal structure of the world. In P. Kitcher and W. Salmon (eds.), *Scientific Explanation.* University of Minnesota Press, pp. 410–505.

Knox, E. (2016). Abstraction and its limits: Finding space for novel explanation. *Nous*, 50(1), 41–60.

Knox, E., (2017), Novel explanation in the special sciences: Lessons from physics. The Proceedings of the Aristotelian Society, 117(2), 123–140.

Ladyman, J. and Ross, D., with D. Spurrett and J. Collier (2007). *Every Thing Must Go: Metaphysics Naturalized.* Oxford University Press.

Lange, M. (2016). *Because without Cause: Noncausal Explanations in Science and Mathematics.* Oxford University Press.

Marcus, R. (2014). How not to enhance the indispensability argument. *Philosophia Mathematica*, 22(3), 345–360.

McGivern, P. (2008). Reductive levels and multi-scale structure. *Synthese*, 165(1), 53–75.

Menon, T. and Callender, C. (2013). Turn and face the strange... Ch-Ch-changes: Philosophical questions raised by phase transitions. In R. W. Batterman (ed.), *The Oxford Handbook of Philosophy of Physics*. Oxford University Press.

Morrison, M. (2012). Emergent physics and micro-ontology. *Philosophy of Science*, 79(1).

Oppenheim, P. and Putnam, H. (1958). Unity of science as a working hypothesis. In H. Feigl, G. Maxwell and M. Scriven (eds.), *Minnesota Studies in the Philosophy of Science*, University of Minnesota Press.

Potochnik, A. (2010a). Levels of explanation reconceived. *Philosophy of Science*, 77(1), 59–72.

Potochnik, A. (2010b). Explanatory independence and epistemic interdependence: A case study of the optimality approach. *British Journal For the Philosophy of Science*, 61(1), 213–233.

Putnam, H. (1975). Philosophy and our mental life. In *Mind, Language, and Reality*. Cambridge University Press.

Rueger, A. and McGivern, P. (2010). Hierarchies and levels of reality. *Synthese*, 176(3), 379–397.

Reutlinger, A. (2017). Are causal facts really explanatorily emergent? Ladyman and Ross on higher-level causal facts and renormalization group explanation. *Synthese*, 194(7), 2291–2305.

Salmon, W.C. (1985). Scientific explanation: Three basic conceptions. PSA 1984, *Philosophy of Science*, 2, pp. 293–305.

Salmon, W.C. (1989/1990). *Four Decades of Scientific Explanation (foreword by Paul Humphreys)*. University of Pittsburgh Press.

Sober, E. (1999). The multiple realizability argument against reductionism. *Philosophy of Science*, 66(4), 542–564.

Strevens, M. (2008). *Depth: An Account of Scientific Explanation*. Boston: Harvard University Press.

van Fraassen, B. (1980). *The Scientific Image*. Oxford: Clarendon Press.

Weslake, B. (2010). Explanatory depth. *Philosophy of Science*, 77(2), 273–294.

Wilson, J. (2010). Non-reductive physicalism and degrees of freedom. *British Journal for the Philosophy of Science*, 61(2), 279–311.

Woodward, J. (2003). *Making Things Happen: A Theory of Causal Explanation*. Oxford University Press.

Woodward, J. (2010). Causation in biology: Stability, specificity and the choice of levels of explanation. *Biology and Philosophy*, 25(3), 287–318.

Woodward, J. (2016). The problem of variable choice. *Synthese*, 193(4), 1047–1072.

Ylikoski, P. and Kuorikoski, J. (2010). Dissecting explanatory power. *Philosophical Studies*, 148(2), 201–219.

Further Reading from the Editors

Bob Batterman's work has again been a great source of inspiration in this area. See his *The Devil in the Details: Asymptotic Reasoning in Explanation, Reduction, and Emergence* (Oxford Studies in Philosophy of Science, Oxford University Press, 2002), and his more recent "Autonomy of Theories: An Explanatory Problem," (*Noûs* 52(4), (2018): 858–873). For an account of the issues surrounding phase transitions see Menon, T., and Callender, C., "Turn and Face the Strange... Ch-Ch-Changes: Philosophical Questions Raised by Phase Transitions." In R. W. Batterman, ed., *The Oxford Handbook of Philosophy of Physics* (Oxford University Press, 2013). Relevant literature in the philosophy of science includes Bokulich, A. "Distinguishing Explanatory from Nonexplanatory Fictions" (*Philosophy of Science*, 79(5), 2012) and Strevens, M. *Depth: An Account of Scientific Explanation* (Harvard University Press, 2008).

PART IX

Intertheoretic Relations

Introduction to Part IX

While much of philosophy of physics explores individual physical theories in isolation, it is impossible to fully understand a theory without understanding that theory's limits of applicability and how it relates to theories which apply beyond those limits. This involves investigating the complex network of intertheoretic relations within contemporary physics, as well as the connections between contemporary theories and past theories, now superseded.

A familiar picture of science has it ordered in a hierarchy of levels, with chemistry and then biology building on the foundation provided by physics. The hierarchical picture may be applied within physics also: a simple implementation has high-energy physics as a foundation for nuclear physics, which is in turn a foundation for atomic physics, which is in turn a foundation for (variously) solid-state physics, plasma physics, hydrodynamics, statistical mechanics (and hence thermodynamics), and so on. The foundationalism that the hierarchical picture involves may be conceived in various ways – as theoretical, as ontological, as linguistic, or as conceptual – but however it may be conceived, the viability of a reduction of any higher level physics to any lower level physics is hotly disputed once we move beyond the simplest toy models.

Wherever reductive ambitions appear stymied for whatever reason, we typically encounter claims of emergence; this maxim holds as much in philosophy of physics as in philosophy of mind. The term "emergent" has many uses, but broadly speaking it may be characterized as combining the dependence of a higher level phenomenon on some lower level phenomena with a residual autonomy of the higher level phenomenon; this autonomy may then be further characterized in metaphysical or epistemological terms. Particularly, influential discussions of emergence in physics have included Anderson (1972), Batterman (2002), Wilson (2010), and Butterfield (2011).

The best developed framework for thinking about intertheoretic relations in physics remains the account of reduction developed by Nagel (1961). Nagelian reduction proceeds by constructing a theoretical model, using the lower level theory supplemented by suitable modeling assumptions (auxiliary assumptions) and definitions (bridge laws) of key higher level terms using expressions of the lower level theory. Dizadji-Bahmani's chapter in this part presents a thorough introduction to Nagelian reduction, including the author's own generalized account of reduction that merges Nagel's ideas with those of Schaffner. By weakening some of the requirements of the original Nagelian theory – for example, by abandoning the requirement that the construction of the higher level expressions be a matter of purely deductive reasoning – we obtain a model that can reasonably claim to capture the structure of those instances where we have obtained genuine understanding of a higher level phenomenon in terms of the underlying physics.

A major challenge to the possibility of reduction, whether on the Nagelian or some other model, comes from the phenomena associated with phase transitions. Batterman and others have over the last two decades argued that the involvement of an infinite limit in our best explanations of phase transitions blocks any derivation of these phenomena from statistical descriptions of the underlying microphysics. The example of phase transitions is explored in detail in Sorin Bangu's chapter, in which the author argues that discontinuous phase transitions provide one of the best candidates we have for an emergent phenomenon in physics, while also canvassing the various options for resisting the stronger claims in this vicinity.

Phase transitions (this time of the continuous variety) also feature as the lead example in Bob Batterman's discussion of *universality* – the striking phenomenon where particular forms of description or explanation cross boundaries of scale and constitution to apply in the same way to very different physical systems. Read in conjunction with Williams' chapter in Part V and Jansson's chapter in Part VIII, Bangu's and Batterman's chapters provide a comprehensive introduction to the intense contemporary debate over universality and the challenges it poses to reductionism.

The part concludes with Nina Emery's chapter on deterministic chance, which focuses on the relations between objective probabilities in physical theories at different levels. In general philosophy of probability, it is frequently assumed that a fundamental deterministic theory cannot support probabilistic phenomena at any higher level, or more generally that there cannot be non-trivial probabilities in higher level theories that are not encoded in probabilities at the lower level. These assumptions face significant challenges from some well-understood physical theories – Emery focuses on statistical mechanics and Bohmian mechanics – where a deterministic description at some lower level gives rise to an effectively probabilistic theory at some higher level; in each case, constraints arising from an objective physical limitation on the acquisition of evidence concerning the lower level play a crucial role in supporting the higher level probabilities.

Bibliography

Anderson, P. W. (1972). More is different. *Science*, 177(4047): 393–396.

Batterman, R. (2002). Asymptotics and the role of minimal models. *The British Journal for the Philosophy of Science*, 53(1): 21–38.

Butterfield, J. (2011). Less is different: Emergence and reduction reconciled. *Foundations of Physics*, 41(6): 1065–1113.

Nagel, E. (1961). *The Structure of Science: Problems in the Logic of Scientific Explanation*. New York: Harcourt, Brace & World.

Wilson, M. (2010). *Wandering Significance: An Essay on Conceptual Behaviour*. Oxford: Oxford University Press.

35

NAGELIAN REDUCTION IN PHYSICS

Foad Dizadji-Bahmani

Intertheoretic reduction is an important topic in analytic philosophy of science. There are a striking variety of reductive claims. Some claim that the very *modus operandi* of science is reductive, others that the history of science is replete with reductions, others still that the putative examples are not reductions after all, and yet others that intertheoretic reduction is not possible. Tied up with intertheoretic reduction are the notions of ontological reduction and reductionism, where, roughly speaking, the former is the reduction of objects or properties to others, and the latter is the claim that all of science reduces to physics. Yet, before one can consider whether or not reductions are ubiquitous, numerous, few, or impossible; whether science aims at reduction; whether all of science does reduce to physics; and so forth, *one must first settle what it is for one theory to reduce to another.*

Ernest Nagel gave the most well-known and widely discussed model of reduction. This chapter offers a critical discussion of Nagel's model, and developments thereof, in the context of physics. It is structured as follows. In Section 35.1, Nagel's model of reduction is set out in detail. In Section 35.2, I proffer a critical discussion of Nagelian reduction, and developments thereof.

35.1 Nagelian Reduction

Consider two theories, which I will refer to as the *to-be-reduced* and *reducing* theories. There are many different labels for these in the literature. Nagel used T_O and T_N respectively, where the indices "O" and "N" stand for "old" and "new." But this is just a device which reflects that, by Nagel's light, many intertheoretic reductions are between theories that have succeeded one another over time. Also it is worth noting that the philosophical use of the term "reduction" is different from how the term is used in physics. Roughly, in philosophy the less fundamental reduces to the more fundamental whereas in physics the more fundamental reduces to the less fundamental, e.g., in the context of "correspondence principles."

Nagel's fundamental idea is that reduction consists in deriving the laws of the to-be-reduced theory from the laws of the reducing theory. This idea is formally captured by postulating two criteria for a successful reduction: "Derivability" and "Connectability" (Nagel, 1961).

"Derivability" is the requirement that the laws of the to-be-reduced theory be derived from the laws of the reducing theory and some auxiliary assumptions. Schematically, we can think of the laws of the reducing theory and the auxiliary assumptions forming a set of premises from which the laws of the to-be-reduced theory are to be derived.

The laws of each theory are couched in terms of its own theoretical predicates. Clearly, for it to be *possible* to derive the laws of the to-be-reduced theory from the reducing theory and auxiliary assumptions, it must be the case that the theoretical predicates of the former be among the "set of premises." This is what the "Connectability" criterion requires.

Nagel introduces two kinds of intertheoretic reduction: "homogeneous" and "inhomogeneous." (Nagel sometimes refers to the latter as 'heterogeneous'. In the early 1949 paper, he also refers to this as a "qualitatively discontinuous" reduction (Nagel, 1949, p. 107).) A homogeneous reduction is when the set of theoretical predicates of the to-be-reduced theory are a subset of the set of the theoretical predicates of the reducing theory. In this sense, "homogeneous" reductions are ones where Connectability is straightforwardly satisfied. As an example of a homogeneous reduction, Nagel proposes that of Newtonian Mechanics to Special Relativity. Newton's laws are stated in terms of "mass," "force," "acceleration," "momentum" and so forth. Special relativity has these theoretical terms as a subset of its own set of theoretical terms. (Whether or not these *are* the same predicates is a controversial point. I am here just presenting Nagel's stance on the matter.)

Inhomogeneous reductions are those where the set of theoretical terms of the to-be-reduced theory, are *not* a subset of that of the reducing theory. To make the reduction possible – that is, to satisfy Connectability – the so-called "bridge-laws" are introduced. The *function* of bridge-laws is to connect the theoretical terms of the two theories. Formally, they indicate which terms of the reducing theory can be replaced by terms of the to-be-reduced theory in the derivation.

As an example of an inhomogeneous reduction, Nagel proposes that of thermodynamics (TD) to classical statistical mechanics (CSM). The laws of thermodynamics are couched in terms of thermodynamic predicates such as "temperature," "entropy," "heat capacity," and so forth. CSM does not have these theoretical terms. Or at least, they are not obviously the *same* predicates. So, for example, "entropy" does appear in CSM but, by Nagel's lights, it is not the same as thermodynamic entropy. To be able to derive the laws of TD from CSM, bridge-laws are needed. Nagel's exemplar is the bridge-law connecting temperature to mean kinetic energy.

In summary, the laws of the to-be-reduced theory are to be derived from the conjunction of the laws of the reducing theory, various auxiliary assumptions and, if needed, bridge-laws. If indeed they are, then the former theory is reduced to the latter.

For Nagel, reduction affords explanation. Indeed, the reduction of one theory to another *is* an explanation of the former by the latter, given the *deductive-nomological* model of explanation, which Nagel advocated. (See Frisch (this volume) for more details on the deductive-nomological model.) Nagel also considered whether reduction affords ontological simplification but was not committed to this. (Further discussion below.)

35.1.1   Toy Model

To facilitate a critical discussion of Nagel's model let's introduce a toy model. One of the most widely discussed (putative) reductions is that of thermodynamics to statistical mechanics. (See Frigg (2008).) One of the laws of thermodynamics is the Boyle-Charles law. The Boyle-Charles law states that pressure, p, of a gas is directly proportional to its temperature, T, at fixed volume V:

$$PV = kT \qquad (35.1)$$

This law is empirically well-confirmed for various gases over a range of pressures and temperatures. For the complete reduction of thermodynamics to statistical mechanics to succeed the Boyle-Charles law needs to be derived from statistical mechanics. Here is the canonical way that this is done *in physics*.

The main posit of CSM is that a gas in box of volume V consists of n number of particles (atoms or molecules) governed by Newtonian mechanics. The particles each have the same mass m, and a definite position x and velocity v. Further, there is a velocity distribution $f(v)$ specifying what portion of the particles move with v. Certain idealizations are then imposed: the particles have negligible volume (i.e., are represented as point particles), all collisions are perfectly elastic, and that $f(v)$ is isotropic. Defining pressure as force per unit area, it follows from the above that the pressure exerted by the gas on the box is directly proportional to the product of the mass, number, and average of the square of the velocity of the particles:

Given that kinetic energy E_{kin} is defined as $\frac{mv^2}{2}$ one obtains the following equation:

$$pV = \frac{2n}{3} < E_{kin} > \tag{35.2}$$

Associating temperature T with mean kinetic energy, and substituting this into the equation returns the Boyle-Charles law, as per equation 1.

How does this map onto the Nagelian model of reduction? The posits of the reducing theory – statistical mechanics – are supplemented by some auxiliary assumptions (point particles, elastic collisions, etc.). While there is some overlap in the conceptual frameworks of the two theories (for example, pressure is defined within classical statistical mechanics as well), this is only partial. Crucially temperature needs to be associated with a statistical mechanical property, mean kinetic energy. This then is a bridge-law, functioning to connect the properties of the two theories to allow the derivation to go through, required to meet the Connectability condition. (This is therefore an inhomogeneous reduction.) From these three components – the reducing theory, auxiliary assumptions, and bridge-law, the law of the to-be-reduced theory, the Boyle-Charles law, is derived, meeting the Derivability condition. If all the laws of thermodynamics were derived in a similar manner then this would constitute a Nagelian reduction of thermodynamics to statistical mechanics.

35.2   Critical Discussion of Nagelian Reduction

35.2.1   *Framework*

There several alleged problems with the framework in which Nagel's model is couched. One set of these might be dubbed the "syntactic view of theories" problem. Nagel was one of the leading logical empiricists and advocated what came to be called the "syntactic view," or the "received view," of theories. (The "received view" is more accurately a further specified position "within" the syntactic camp but I use the term interchangeably here.) Roughly speaking, in this view, a theory is taken to be an axiomatic system stated in a formalized language, which entails a set of theorems. One of the important distinctions of this view is the sharp distinction between observational and theoretical statements. (For classic statements and defenses of this view of theories see Nagel (1961) and Carnap (1967).)

The "received view" came to be widely rejected in the philosophy of science community. In part, this was due to the rejection of logical empiricism more widely. Critics of the syntactic view argued, among other things, that theories cannot be fully expressed in a formal language, that the requisite account of the relation between language and the World is not forthcoming, and that the distinction between observational and theoretical terms is not tenable. The so-called semantic view of theories came to replace the syntactic view as the dominant position in philosophy of science. (In this context, "semantic" is understood in the model-theoretic sense. See Winther (2016) and references therein.)

The problem for Nagelian reduction is then this: it is based on an untenable framework, and, as such, it is itself untenable. Statements to this effect can be found in Churchland (1985) and Bickle (1998). But there are, at least, three reasons which undercut this problem.

The first is that while it is the case the semantic "school" is dominant in contemporary philosophy of science, the debate between the syntactic and semantic "schools" is certainly not settled and there are controversies on both sides. For example, Halvorson argues that the semantic view of theories, if taken literally, leads to absurdities in that "it equates theories that are distinct, and it distinguishes theories that are equivalent" (Halvorson, 2012, p. 183). (See also Chakravartty (2001).) This suggests that it would be premature to simply discard the Nagelian model in light of this.

Second, there is a danger of a kind of "guilt by association." Even if there are reasons to reject the syntactic view of theories, it is not obvious this necessitates a rejection of the Nagelian model. Surely we would want to know what is supposedly wrong specifically with the model. In short, the inference from its being "based on" an untenable framework to itself being untenable is questionable. For example, were one to come to reject the syntactic view on the grounds that it the observational-theoretical distinction is untenable, this would not, I suggest, undermine the Nagelian model for nothing in the model rests on this distinction. (See Dizadji-Bahmani et al. (2010).)

Most importantly, the Nagelian model is "recovered" within the semantic view of theories anyway. The most thoroughly worked out program in the semantic "school" is the so-called "Munich School." In their definitive statement of the position, Balzer, Moulines, and Sneed take themselves to provide a version of Nagel's model within the structuralist mold. (Balzer et al. (1987). See also Rantala (1991).) Very roughly, the idea is to construe a theory as a class of models satisfying a set-theoretically constructed predicate. Some theory is then said to reduce to another just in case for every model in the class of the to-be-reduced theory, and one can construct an isomoprhic one from the class of models of the reducing theories. As noted by Endicott (1998) and Dizadji-Bahmani et al. (2010), the so-called "New Wave Reductionism" carries over the spirit of Nagelian reduction into a different framework (perhaps also placing more emphasis on process rather than logical structure).

35.2.2   The "Falsity" Problem

The typically older to-be-reduced theory is false, and the reducing theory is assumed to be true. But we cannot derive a false theory from a true one, so it seems that the Derivability requirement cannot be met. Call this the "Falsity" problem. This is one of the criticisms put forward by Feyerabend (1965), and is noted by Churchland (1985) and Bickle (1998).

While it *seems* as if one is deriving a false theory from a true one, this is not so. Rather, the false theory is deduced from the reducing theory in conjunction with some auxiliary assumptions, and these auxiliary assumptions are, strictly speaking, false.

35.2.3   Explanation

According to the Deductive-Nomological (DN) model, the explanation of the explanandum - the proposition to be explained - consists in the derivation of explanandum from a set of propositions at least one of which is a law of nature. The DN model allegedly faces problems several problems. While it is to provide necessary and sufficient conditions for explanation, there are alleged counterexamples to it, such as the "ink-pot" case, the "birth-control pill" case, and the "flagpole" case. (See Salmon (1985), Kitcher and Salmon (1989), and Frisch (this volume) for an overview.) The DN model of explanation is neither necessary nor sufficient criteria for explanation, and therefore it cannot be seen as *the* model of explanation. Of course, there are various other models of explanation which work better in different contexts. However, no model of explanation works in all cases either. (This suggests one adopt some kind of pluralism about explanation. See Jackson and Pettit (1992); Woodward (2010); van Bouwel and Weber (2008).) That the DN model of explanation does not cover all cases does not entail, however, that all instances which do fit the model are not explanatory. The Nagelian reductionist need only claim that there is explanatory import in actual cases of successful reduction and not that the DN model is the only correct model of explanation.

35.2.4   Derivability and Development of Nagelian Reduction

★★★One serious objection to Nagel's model is that Derivability is not going to satisfied in all but the most simple cases. The point is usually stated thus: for the vast majority of interesting cases it is not possible for technical reasons to deduce the exact laws of the to-be-reduced theory. If reduction consists in the deduction of the laws of the to-be-reduced theory then Nagelian reduction is *de facto* unrealizable. This is a point that was made very early on against Nagel's model by Feyerabend (1965) and has been repeated by many others since. (See Schaffner (1967) and Dizadji-Bahmani et al. (2010).)

By various people's lights "exact Derivability" is too strong a criterion for reduction and requires weakening. As it happens, Nagel himself (Nagel 1974) suggested some such weakening, although he did not provide much by way of detail. (See Richardson (2008).) Schaffner (1967) proposed a modification of Nagel's original model along these lines and his was more detailed. For Schaffner, if the exact law of the to-be-reduced theory cannot be derived, then what is needed is a derivation of a *strongly analogous* law. An important constraint is placed on it: the analog law must be empirically more adequate than the law of the to-be-reduced theory.

Dizadji-Bahmani et al. (2010) give a precise statement of Schaffner's modification of Nagel's model, which they dub the Generalized Nagel-Schaffner (GNS) model of reduction: reduction is the deductive subsumption of a corrected version of the to-be-reduced theory under the reducing theory in conjunction with auxiliary assumptions and bridge- laws.

The GNS model deals well with cases such as the derivation of the Second Law of Thermodynamics (SLT), which has been a central problem in physics and philosophy of physics. SLT states that the entropy of a thermodynamically isolated system cannot decrease. Forgoing a lot of the technical details, here is a sketch of how SLT is recovered in Boltzmannian statistical mechanics for, say, a gas. (See Frigg (2008) and Dizadji-Bahmani et al. (2010) for details. For recent developments see Werndl and Frigg (2015b), Werndl and Frigg (2015a), Shahvisi (this volume), and Frigg and Werndl (this volume a)) The central idea of Boltzmannian statistical mechanics is to partition the state space of the thermodynamic system of interest into macrostates (volumes of the state space corresponding to distinct macro-physical states). With respect to this, the Boltzmann entropy is defined as proportional to the log of the Lebesgue measure (which is the generalization of "ordinary" three-dimensional volume in higher dimensional state spaces) of that macrostate in which the system happens to be at that time. (See Frigg and Werndl (this volume a) for the technical details.) The aim is to then show that the dynamics of the system in conjunction with some other assumptions such as the Past Hypothesis,[1] entail that the entropy does not decrease. However, as Dizadji-Bahmani et al. (2010) point out:

> The thermodynamic entropy is static in equilibrium: once it has reached equilibrium, it does not change any more. The Boltzmann entropy, by contrast, fluctuates. This is generally deemed unproblematic because the fluctuations are very small and S_B stays close to the equilibrium value most of the time. (Dizadji-Bahmani et al., 2010, p. 397)

The GNS rational reconstruction of this case runs as follows. Taking the posit of Boltzmannian statistical mechanics in conjunction with some auxiliary assumptions, one shows that the target system's Boltzmann entropy fluctuates but stays very close to or at the maximum entropy value for most times. Invoking the bridge-law connecting the Boltzmann entropy to the thermodynamic entropy allows one then to derive an expression which is, in an intuitive sense, *strongly analogous* to SLT. Moreover, the law which is actually derived is empirically more adequate than SLT itself, as experiments show that, indeed, say, a gas in a box exhibits small short-lived fluctuations out of equilibrium.

35.2.5 *Bridge-Laws*

Bridge-laws have been deemed problematic for a variety of reasons. While their *function* is clear, it has been argued that recourse to them is unjustified or that their status is unclear. The most prominent arguments against bridge-laws are discussed in this section.

A well-known argument against Nagelian reduction pertaining to Nagel's recourse to bridge-laws is Feyerabend's "incommensurability" argument (Feyerabend, 1965). The crux of the argument is Feyerabend's claim that the meaning of theoretical terms is determined solely within the theory in which they are embedded. The consequence of this is that the "Connectability" criterion is impossible to satisfy according to Feyerabend and recourse to bridge-laws is nonsensical. However, Feyerabend's incommensurability thesis has been critiqued on various grounds. (See Preston (2016).) (It is worth noting that Feyerabend also argued against "Derivability." As Nagel correctly points out, one cannot maintain both these arguments simultaneously: that "Connectability" is met is a pre-requisite (conceptually prior) for a rejection of "Derivability." Two theories cannot be logically inconsistent if they are incommensurable.)

What *kind* of statements are bridge-laws? For example, how is "associated" to be interpreted in the temperature-mean kinetic energy bridge-law? The question of the status of bridge-laws has received the bulk of the philosophical attention. (See Sklar (1967); Churchland (1985); Kim (1992); Schaffner (1993);Schaffner (2006); Schaffner (2012); Esfeld and Sachse (2007); Klein (2009); Dizadji-Bahmani et al. (2010); van Riel (2011); Bangu (2011).)

Nagel considers three possible interpretations (statuses) of bridge-laws: meaning equivalence, conventional stipulations, or assertions about matters of fact (Nagel, 1961, pp. 354–355). The third interpretation breaks down further. A bridge-law could be: an identity (i.e., each predicate could refer to the same property), nomic correlation (i.e., the predicates could refer to nomically correlated properties), or, finally, a brute correlation (i.e., the predicates could refer to properties that are non-nomically correlated – what one might call a mere Humean regularity). Nagel himself was noncommittal between these different options.

The first two interpretations have been all but dismissed: bridge-laws are, surely, not semantic claims or matters of mere convention, it is claimed. For example, consider again the temperature – mean kinetic energy law. It is seemingly not the case that 'temperature' *means* "mean kinetic energy." Nor does it seem plausible to think of it as a *mere* convention: it seems that there is a fact of the matter about temperature being associate with mean kinetic energy, rather than, say, the square root of mean kinetic energy. The debate has centered around the third interpretation and in particular about whether bridge-laws can be interpreted as identities or some other kind of metaphysical relation.

The most widely discussed option in the matters of fact category has been identity. And the most prominent argument *for* identities has been from consideration of parsimony: that the properties or entities of the to-be-reduced theory are identical with some of the reducing theory then is the most parsimonious account. In this view, a reduction affords ontological simplification because it reduces the number of entities. The most prominent argument against identity is from consideration of multiple realizability. (See Section 35.2.8 below.)

For Schaffner bridge-laws are *reduction functions* expressing the co-extensionality of two properties but it remains an open question whether this is in virtue of a nomic connection, a mere contingent correlation (Humean regularity), or another metaphysical relation. Similarly, Dizadji-Bahmani et al. (2010) do not commit to a particular reading of bridge-laws other than it being a factual kind of relation between properties. There is widespread agreement that the status of bridge-laws is to be determined on a case by case basis, rather than commitment to one for all cases. That is, some bridge-laws may be identities, others nomic relations, and so forth.

One problem with taking bridge-laws to be matters of fact, in whichever guise, is that justifying them seems difficult. As Nagel (1961, p. 356) points out we cannot test bridge-laws independently, and thus there is an epistemological gap when it comes to bridge laws.

Churchland (1985, p. 11) argues that a "smooth" reduction (roughly, those involving neither counterfactual auxiliary assumptions nor boundary conditions) supports the bridge-law expressing an identity. (See also Bickle (1992).) Dizadji-Bahmani et al. argue that this epistemological gap is an instance of the Duhem problem.[2] But such underdetermination is rife: one is often unable to confirm hypotheses independently but rather only entire packages (consisting of theories and auxiliary assumptions). That the Duhem problem may come up in this context is not enough reason to reject Nagelian reduction (Dizadji-Bahmani et al., 2010, p. 408).

Another issue pertaining to bridge-laws is what determines their form. Why is it, for example, that temperature is "associated" with mean kinetic energy not mean kinetic energy squared? ("Associated" is here a neutral place-holding term for whatever the status of the bridge-law is.) This issue is hardly addressed (although see Dizadji-Bahmani et al. (2010)) and philosophers have tended to take as given the various bridge-laws that populate discussions about reduction and have really only concentrated on the question of their status, say an identity, nomic connection between two properties, etc.

What is needed is *warrant* for bridge-laws. This is determined by two things: *formal consistency* and *conceptual fit*. In order for a bridge-law to be *warranted* one must show that the properties it connects are formally consistent with one another and that their association fits conceptually, where this is determined with respect to the to-be-reduced and reducing theories respectively.

Reconsider the bridge-law connecting temperature with mean kinetic energy. This bridge-law is *warranted* because there is both *formal consistency* and *conceptual fit* between temperature and mean kinetic energy. Roughly speaking, there is *formal consistency* because the properties associated with one another, viz. temperature and mean kinetic energy, via the bridge-law share the relevant formal properties. Put colloquially, they "behave" in the same way. For example, they are both extensive properties, and both decrease as a function of pressure, and so forth. But *formal consistency* is not enough – one also needs to show that there is *conceptual fit*. In this case there is conceptual fit for temperature, as per thermodynamics itself, is directly proportional to internal energy of an isolated system.

35.2.6 *Emergence and Asymptotic Limits*

In very broad brush strokes, a property, entity, or phenomenon is emergent just in case it is novel with respect to, and not straightforwardly identifiable with, some more "fundamental" properties, entities, or phenomena. [Bedau, Mark, and Humphreys, Paul (2008). *Emergence: Contemporary Readings in Philosophy and Science*. Cambridge, MA: MIT Press.] The failure of an intertheoretic reduction is sometimes taken to indicate there being some kind of emergence. (Although see Butterfield (2011), wherein it is argued that emergence is not incompatible with reduction.)

There being emergence as a consequence of the failure of a particular reduction is not a problem for the Nagelian *model* of reduction (or indeed any model), *per se*, at least not without further argument. However, there is a seeming tension between emergence and reductionism – the thesis that all the sciences reduce to physics and that all of physics reduces to one final theory. (See Cat (2017).)

A much discussed example of a (putative) emergent phenomenon is phase transitions. It has been argued that the reduction of thermodynamics to statistical mechanics fails and that phase transitions are emergent. In what sense does the reduction with respect to phase transitions fail? In order to recover the first-order phase transitions of thermodynamics one needs to take the so-called "thermodynamic limit," letting the number of particles of the system to tend to infinity (or rather this is what is done, canonically). But doing so is philosophically problematic and/or technically flawed. (See Callender (2001), Liu (2001), Batterman (1995), Bangu (2009), Menon and Callender (2013), and Bangu (this volume).)

This, and similar examples of reductions involving "asymptotic limits," are discussed extensively by Batterman (1995; 2000) who, in short, argues that the requisite auxiliary assumption in each such derivation is unjustified. (See also Norton (2012).) It is obvious that if the auxiliary assumptions used in a derivation are not justified – that is, if they are not warranted – then this undermines the explanatory important of the reduction itself. It is interesting that the issue of warrant for auxiliary assumptions is largely neglected in the literature on reduction. I try to somewhat ameliorate that in the next section.

35.2.7 Auxiliary Assumptions

What counts as *warrant* for the auxiliary assumptions? Clearly this varies with the auxiliary assumptions: warrant differs for the auxiliary assumptions. However we can exemplify the notion of warrant by drawing on the derivation of the Boyle-Charles law above.

There are at least four kinds of auxiliary assumptions: Idealizations, Limits, Dynamical Assumptions, and Initial Conditions. (Notice that this list is not intended as mutually exclusive and exhaustive. The kinds of assumptions that a reduction involves may not fit neatly into the categories given here. However, I intend everything here to be suggestive enough to enable one to assimilate such assumptions into talk of warrant.)

Idealisations are those assumptions which idealize the "metaphysical picture" of the reducing theory. For example, treating gas molecules as point particles would be an idealization in this sense. Limits are mathematical simplifications involved in the derivation. For example, one may drop certain terms in an equation as being "negligible." Dynamical assumptions are those which make special probations about the dynamics of the system. For example, in the context of statistical mechanics, one approach to deriving the SLT is to assume ergodicity. (See Frigg (2008) for introduction and further references.) There are also assumptions about the initial conditions of the model from which one wants to derive the laws of the to-be-reduced theory. An example of this, again in the context of statistical mechanics, is the so-called *Past Hypothesis*, which, roughly speaking, is the assumption that the universe started in a low-entropy state. (See Frigg (2008) for introduction and further references.)

So what counts as warrant then for the different kinds of auxiliary assumptions? Let us start with warrant for the limiting assumptions. Were one to derive the same law from two sets of auxiliary assumptions, such that one derivation is more mathematically rigorous than the other, the former would constitute a better explanation. Batterman has written several influential papers on intertheoretic reductions (Batterman, 1995, 2000), involving asymptotic mathematical limits, i.e., non-converging limits. The thrust of Batterman's argument is that in these cases the putative reduction of the to-be-reduced theory is undermined. While Batterman's categorical attitude toward reduction – that either a theory reduces to another or it does not – is misplaced at least by my lights, the intuition that there is something amiss in using such asymptotic limits is right: asymptotic limits are an example of an auxiliary assumption for which there is little warrant because of the lack mathematical rigor.

As regards initial conditions, the more general – the less "special" – the initial conditions the better the explanation of the empirical success of the to-be-reduced theory based on it. Reconsider the derivation of the Boyle-Charles law but suppose that in order to have derived it was insufficient just to assume the auxiliary assumptions we did but that we also had to assume that the gas particles start in a restricted part of the container. This would undermine the explanatory import of the reduction, *ceteris paribus*, because it is in contradiction to the generality of the Boyle-Charles law. After all, the Boyle-Charles law is empirically adequate (to the degree that it is) for both gases that do and do not start in that special initial condition.

The *warrant* for Idealizations and Dynamical assumptions is more complicated to characterize. To do so it is helpful to again make use of possible world semantics. Call the possible world at which the reducing theory is true, PW_R. There is *warrant* for these auxiliary assumptions (i.e., idealizations and dynamical assumptions) just in case, and to the extent to which, making them less counterfactual with respect to PW_R, would entail a derivation of laws which are more empirically adequate at the actual world, PW_A, than the ones that are derived.[3] To see why this is right, contrast it with what *warrant* is *not*: the *warrant* for these auxiliary assumptions is not determined by how counterfactual they are with respect to PW_R. If this were the correct characterization, then making the auxiliary assumptions less counterfactual with respect to PW_R, would mean that we would have more *warrant* for them and by extension we should have a better explanation of the laws of the to-be-reduced theory. But, of course, we would not have a better explanation of the laws of the to-be-reduced theory – indeed we would not have an explanation of them at all – for we would fail to derive the laws! The auxiliary assumptions *need to be* counterfactual because the laws of the to-be-reduced theory are, recall, strictly speaking, false from the point of view of the reducing theory. The aim of the reduction was to explain the empirical success of a false theory, viz. the to-be-reduced theory, by deriving its laws, starting with the supposition that the reducing theory is true and in conjunction with the auxiliary assumptions and bridge-laws. So the counterfactualness of the auxiliary assumptions with respect to PW_R is not what grounds their *warrant*. So what is the *warrant* for these auxiliary assumptions? They are *warranted* just in case *making them less counterfactual with respect to PW_R we obtain empirically more adequate laws*. In so doing we would show that we have not derived the laws of the to-be-reduced theory surreptitiously. Conversely, if increasing the veracity of the auxiliary assumptions with respect to PW_R entailed empirically *less* adequate laws, then they are not *warranted*.

This is best illustrated by an example. Consider again the derivation of the Boyle-Charles Law. One idealization involved in the derivation is that the particles only have kinetic energy. Is this auxiliary assumption *warranted*? It is because were one to relax this assumption to include particles also having a weak interaction with one another – i.e., were one to make the assumption *less counterfactual* – one would derive a law relating the pressure and volume of a gas to its temperature which would be more empirically adequate. Indeed, what one would derive is Van der Waals equation, which is more empirically adequate than the Boyle-Charles law. Likewise the idealization that the gas consists of mono-atomic particles. Were one to model the gas as di-atomic, one would arrive an empirically better law for the relation between the pressure, volume and temperature for an actual di-atomic gas.

In summary: how good a reduction one has turns on the degree of warrant for the auxiliary assumptions. But this cannot be measured by how counterfactual the auxiliary assumptions are, for they typically *need* to be counterfactual for the law that one is deriving is typically false. What one needs is an assurance that the auxiliary assumptions are not just ad hoc, that they are not put in surreptitiously merely to yield the right result. The test for this is that in making them less counterfactual one derives empirically better laws.

35.2.8 *Multiple Realizability*

Another important topic that impacts on discussions of Nagelian reduction is that of multiple realizability. A property is multiply realized just in case it is realized by more than one "lower-level" property. For example, temperature is often cited as a multiply realized property: the temperature of a gas is realized by mean kinetic energy whereas the temperature of a solid is realized by mean kinetic and potential energy. Multiply realized properties would need to be accounted for by *disjunctive* bridge-laws in the Nagelian model.

However, a variety of arguments have been put forward to the effect that multiple realizability undermines reduction. A disjunctive bridge-law could not express an identity for obvious reasons. The discussion in Section 35.2.5 shows that bridge-laws *qua* identities are not a necessary condition for successful reduction, however. But blocking bridge-laws *qua* identities is not the only alleged problem with multiple realizability.

The most well-known and cited argument that involves multiple realizability comes from Fodor's "Special Sciences. Or: The Disunity of Science As A Working Hypothesis" (Fodor, 1974). This paper has been the touchstone for virtually all work about reduction and reductionism ever since. Fodor's arguments therein can be seen as a development of an argument presented by Putnam (1967). But while Putnam's was specifically concerned with multiple realizability in the context of philosophy of mind, Fodor was concerned with multiple realizability across the "special sciences," i.e., all the sciences excluding physics.

Fodor's is an argument against reductionism, the thesis that the "special sciences" reduce to physics. He is of the view that reductionism is ultimately an empirical thesis, yet given what we do know, Fodor argues, reductionism is likely to be false. It is important to be clear from the outset that Fodor's argument is not one which purports to show that a *particular* theory does not reduce to another. It is a general argument in that it threatens to undermine every putative instance of reduction where there is scope for multiple realization. In particular, it threatens to undermine every putative some-such reduction in physics.

It is useful to give a very rough sketch of Fodor's argument: if there is a multiply realized property in a "higher-level" theory then this is likely to block the putative reduction of it to a "lower-level" theory. It is very likely that there are such properties. Thus, it is very likely that theory will not reduce to the "lower-level" theory and, by extension, reductionism is likely to be false. (Throughout, Fodor's talk of "higher-level" and "lower-level" theories should be taken as the putative to-be-reduced and reducing theories. Likewise "higher-level" and "lower-level" properties, the properties of the to-be-reduced and reducing theories, respectively.) Why, by Fodor's lights, does a multiply realized property block a reduction? In case of multiply realizable properties the requisite bridge-laws would have to be disjunctive, but a disjunctive bridge-law could not even be a *law*, or so Fodor claims, for laws express relations between natural kind terms and no natural kind term is disjunctive. Similar concerns are voiced by Kim (1992): a multiply realizable property is "unfit to figure in laws, and is thereby disqualified as a useful scientific property" because of its "causal/ nomic heterogeneity" (p. 18). Moreover, Fodor argues that disjunctive bridge-laws undermine the explanatory import of a reduction.

Clearly a lot more needs to be said about all the notions involved in Fodor's argument but with this coarse characterization to hand, the various opposing positions can be cast. Contra Fodor, Lewis (1969), Sober (1999), Richardson (2008), Needham (2010), and Dizadji-Bahmani et al. (2010) have argued against the claim that multiple realizability undermines reduction.

Lewis argues that particular cases of multiple realizability are compatible with a contextually sensitive or "local" reduction and that therefore a "local" kind of reductionism is defensible despite multiple realizability. Richardson and Dizadji-Bahmani et al argue that Fodor's characterization of natural kind terms and laws of nature is problematic. Furthermore, bridge-laws are not laws of nature in the sense that, say, Schrodinger evolution of the wavefunction is, and indeed bridge-laws need not even express nomic connections between properties at all. (See Section 35.2.5, above.)

Sober engages with the notion of explanation, as it features in Fodor's argument. Sober's reading of Fodor's argument is that multiple realizability undermines explanation. The "special sciences," in virtue of having multiple realizable properties, have an explanatory power that cannot be captured by the "lower-level" science, and it is this that preserves their autonomy and undermines reductionism. It is not clear why having a disjunctive bridge-law *per se* undermines explanation:

> Are we really prepared to say that the truth and lawfulness of the higher-level generalization is inexplicable, just because [a] derivation is peppered with the word "or"? I confess that I feel my sense of incomprehension and mystery palpably subside when I contemplate [some such] derivation. Where am I going wrong? (Sober, 1999, p. 554)

Certainly nothing in the DN model rules out deductions involving disjunctive bridge-laws as bona fide explanations.

However, other authors have argued that Fodor's argument does pose a problem for the Nagelian model of reduction, and that it prompts a new model to be put forward. The thought is that reduction in the Nagelian sense is vulnerable to the multiple realizability argument but that just those instances which are undermined *are* genuine cases of reduction, intuitively speaking, so a new model of reduction is needed to do justice to intuition. Kim (2000) has argued for a new model, which he calls "functional reduction." Bickle (1996) has argued that the New Wave model of reduction avoids the problems stemming from multiple realizability.

A different kind of response to MR is to focus on the notion of "multiple realizability" itself. Indeed several authors argue that multiple realizability is a vague notion and that it can be explicated in a number of different ways. Polger (2008) argues that there are four different intuitions that underpin MR and Lyre (2009) argues, albeit metaphorically, that MR is itself multiply realized! Depending on the particular notion, reductionism is or is not undermined, it is claimed.

Related to the question of what MR really *is*, is the question of how much of it there is about. There are authors who think that (the various kinds of) MR is (are) ubiquitous in science. This is taken as indicative of the falsity of anti-reductionist arguments from MR for, so the counter arguments go, science is, as a matter of fact, full of reductions. Bickle (1996), Clapp (2001) and Endicott (2005) have arguments along these lines. A particular version of this argumentative strategy is that of Enc (1983), who argues that the best case of reduction, viz. thermodynamics to statistical mechanics, is one that involves a multiply realizable property, temperature. This, Enc claims, shows that MR cannot be a stumbling block to reduction. However, Lyre (2009) argues that the case of temperature is the best example (as opposed to mental properties, which are the standard example) of why multiple realizability *is* a problem for reduction!

Shapiro (2000) goes against this grain: he argues that there are good reasons to think that, far from being ubiquitous, the various putative instances of multiply realized properties are not in fact multiply realized after all. He does so by proffering a criterion for "genuine" multiple realizability, which turns on how lower-level properties realize their higher-order correlates. Given this criterion, Shapiro argues that there are no genuine cases of multiple realizability and that therefore, *de facto*, reductionism is not blocked.

A narrower, but similar argument, is made by Bechtel and Mundale (1999). The most cited putative instances of MR properties come from psychology/philosophy of mind. For example, pain is often cited as a property that is multiply realizable and it is this example that both Putnam and Fodor use to motivate their positions. However, Bechtel and Mundale (1999) argue that neuroscience, in fact, supports the claim that pain is not a multiply realizable property.

35.3   Conclusions

In this chapter, I have set out and critically discussed Nagel's model of intertheoretic reduction, and developments thereof, specifically the *GNS* model. I have also proffered some new work on bridge-laws and auxiliary assumptions.

Nagel's model, or rather, GNS, is, I think, a useful model in which to precisely frame questions about intertheoretic relations, and assess the prospects of reductionism.

Notes

1 See Albert (2000), Callender (2004).
2 See Duhem (1954) and Gilles (1993).
3 The counterfactualness of the auxiliary assumptions with respect to PW_R is itself something to be explicated in terms of possible world semantics. cf. Lewis (1969).

References

Albert, D.Z. (2000). *Time and Chance*. Cambridge, MA: Harvard University Press.
Balzer, W., Moulines, C. and Sneed, J. (1987). *An Architectonic for Science: The Structuralist Program*. Synthese Library. Dordrecht: D. Reidel Pub. Co.
Bangu, S. (2009). Understanding thermodynamic singularities: Phase transitions, data and phenomena. *Philosophy of Science*, 76: 488–505.

Bangu, S. (2011). On the role of bridge-laws in intertheoretic relations. *Philosophy of Science*, 78: 1108–1119.

Batterman, R.W. (1995). Theories between theories: Asymptotic limiting intertheoretic relations. *Synthese*, 103(2): 171–201.

Batterman, R.W. (2000). Multiple realizability and universality. *British Journal for the Philosophy of Science*, 51(1): 115–145.

Bechtel, W.P. and Mundale, J. (1999). Multiple realizability revisited: Linking cognitive and neural states. *Philosophy of Science*, 66(2): 175–207.

Bickle, J. (1992). Mental anomaly and the new mind-brain reductionism. *Philosophy of Science*, 59: 217–230.

Bickle, J. (1996). New wave psychophysical reductionism and the methodological caveats. *Philosophy and Phenomenological Research*, 56(1): 57–78.

Bickle, J. (1998). *Psychoneural Reductionism: The New Wave*. Cambridge: MA: MIT Press.

Butterfield, J. (2011). Less is different: Emergence and reduction reconciled. *Foundations of Physics*, 41(6): 1065–1135.

Callender, C. (2001). Taking thermodynamics too seriously. *Studies in History and Philosophy of Science Part B*, 32(4): 539–553.

Callender, C. (2004). There is no puzzle about the low-entropy past. In C. Hitchcock (ed.), *Contemporary Debates in the Philosophy of Science*. Oxford: Blackwell, pp. 240–256.

Carnap, R. (1967). *The Logical Structure of the World [and] Pseudoproblems in Philosophy*. London, Routledge K. Paul.

Cat, J. (2017). The unity of science. In E.N. Zalta (ed.), *The Stanford Encyclopedia of Philosophy* (Fall 2017 ed.). Stanford: The Metaphysics Research Lab, Stanford University.

Chakravartty, A. (2001). The semantic or model-theoretic view of theories and scientific realism. *Synthese*, 127(3): 325–345.

Churchland, P.M. (1985). Reduction, qualia and the direct introspection of brain states. *Journal of Philosophy*, 82(January): 8–28.

Clapp, L.J. (2001). Disjunctive properties: Multiple realizations. *Journal of Philosophy*, 98(3): 111–136.

Dizadji-Bahmani, F., Frigg, R. and Hartmann, S. (2010). Who is afraid of nagelian reduction? *Erkenntnis*, 73(3): 393–412.

Duhem, P. (1954). *The Aim and Structure of Physical Theory*, trans. from 2nd ed. by P. W. Wiener; originally published as La Thorie Physique: Son Objet et sa Structure. Princeton, NJ: Princeton University Press.

Enc, B. (1983). In defense of the identity theory. *Journal of Philosophy*, 80(May): 279–98.

Endicott, R.P. (1998). Collapse of the new wave. *Journal of Philosophy*, 95(2): 53–72.

Endicott, R.P. (2005). Multiple realizability. In *Encyclopedia of Philosophy* (2nd edition). Thomson Gale: Macmillan Reference. ISBN-13-9780028657882. ISBN-10-0028657888.

Esfeld, M. and Sachse, C. (2007). Theory reduction by means of functional sub-types. *International Studies in the Philosophy of Science*, 21(1): 1–17.

Feyerabend, P.K. (1965). On the "meaning" of scientific terms. *Journal of Philosophy*, 62(10): 266–274.

Fodor, J.A. (1974). Special sciences. or: The disunity of science as a working hypothesis. *Synthese*, 28: 97–115.

Frigg, R. (2008). A field guide to recent work on the foundations of statistical mechanics. In D. Rickles (ed.), *The Ashgate Companion to Contemporary Philosophy of Physics*. Aldershot: Ashgate Pub. Ltd.

Gilles, D. (1993). The Duhem thesis and the Quine thesis. In Philosophy of Science in the Twentieth Century. Oxford: Blackwell, pp. 98–116.

Halvorson, H. (2012). What scientific theories could not be. *Philosophy of Science*, 79: 183–206.

Jackson, F. and Pettit, P. (1992). In defense of explanatory ecumenism. *Economics and Philosophy*, 8(01): 1–21.

Kim, J. (1992). Multiple realization and the metaphysics of reduction. *Philosophy and Phenomenological Research*, 52(1): 1–26.

Kim, J. (2000). *Mind in a Physical World: An Essay on the Mind-Body Problem and Mental Causation*. Cambridge: MA: MIT Press.

Kitcher, P. and Salmon, W. (1989). *Scientific Explanation*. Minneapolis: University of Minnesota Press.

Klein, C. (2009). Reduction without reductionism: A defence of nagel on connectability. *Philosophical Quarterly*, 59: 39–53.

Lewis, D. (1969). Review of "art, mind, and religion". *Journal of Philosophy*, 66: 23–35.

Liu, C. (2001). Infinite systems in sm explanation: thermodynamic limit, renormalization (semi)-groups and ireversibility. *Philosophy of Science (Proceedings)*, 68: S325–S344.

Lyre, H. (2009). The "multirealization" of multiple realizability. In A. Hieke and H. Leitgeb (eds.), *Reduction - Abstraction - Analysis. Proceedings of the 31th International Ludwig Wittgenstein-Symposium in Kirchberg*. Heusenstamm Ontos Verlag, pp. 79–94.

Menon, T. and Callender, C. (2013). Going through a phase: Philosophical questions raised by phase transitions. In R. Batterman (ed.), *The Oxford Handbook of Philosophy of Physics*. Oxford: Oxford University Press, pp. 189–224.

Nagel, E. (1949). The meaning of reduction in the natural sciences. In R. Stauffer (ed.), *Science and Civilization*. Madison, WI: University of Wisconsin Press, pp. 98–123.

Nagel, E. (1961). *The Structure of Science: Problems in the Logic of Scientific Explanation*. Harcourt, Brace & World.

Needham, P. (2010). Nagel's analysis of reduction: Comments in defense as well as critique. *Studies in History and Philosophy of Science Part B*, 41(2): 163–170.

Norton, J. (2012). Approximation and idealization: Why the difference matters. *Philosophy of Science*, 79: 207–232.

Polger, T.W. (2008). Two confusions concerning multiple realization. *Philosophy of Science*, 75(5): 537–547.

Preston, J. (2016). Paul feyerabend. In E.N. Zalta (ed.), *The Stanford Encyclopedia of Philosophy* (Spring 2016 ed.). Stanford: The Metaphysics Research Lab, Stanford University. Available at: https://plato.stanford.edu/archives/fall2020/entries/feyerabend/

Putnam, H. (1967). Psychological predicates. In W.H. Capitan and D.D. Merrill (eds.), *Art, Mind and Religion*. Pittsburgh: Pittsburgh University Press, pp. 37–48.

Rantala, V. (1991). Review. *Synthese*, 86(2): 297–319.

Richardson, R.C. (2008). Autonomy and multiple realization. *Philosophy of Science*, 75(5): 526–536.

Salmon, W.C. (1985). Conflicting conceptions of scientific explanation. *Journal of Philosophy*, 82(11): 651–654.

Schaffner, K. (1993). Theory structure, reduction, and disciplinary integration in biology. *Biology and Philosophy*, 8(3): 319–347.

Schaffner, K. (2006). Reduction: The cheshire cat problem and a return to roots. *Synthese*, 151: 377–402.

Schaffner, K. (2012). Ernest Nagel and reduction. *Journal of Philosophy*, CIX: 534–565.

Schaffner, K.F. (1967). Approaches to reduction. *Philosophy of Science*, 34(2): 137–147.

Shapiro, L.A. (2000). Multiple realizations. *Journal of Philosophy*, 97(12): 635–654.

Sklar, L. (1967). Types of inter-theoretic reduction. *British Journal for Philosophy of Science*, 5: 464–482.

Sober, E. (1999). The multiple realizability argument against reductionism. *Philosophy of Science*, 66(4): 542–564.

van Bouwel, J. and Weber, E. (2008). A pragmatist defense of non-relativistic explanatory pluralism in history and social science. *History and Theory*, 47(2): 168–182.

van Riel, R. (2011). Nagelian reduction beyond the nagel-model. *Philosophy of Science*, 78(3): 353–375.

Werndl, C. and Frigg, R. (2015a). Reconceptionalising equilibrium in Boltzmannian statistical mechanics. *Studies in History and Philosophy of Modern Physics*, 49: 19–31.

Werndl, C. and Frigg, R. (2015b). Rethinking Boltzmannian equilibrium. *Philosophy of Science*, 82: 1224–1235.

Winther, R. (2016). The structure of scientific theories. In E.N. Zalta (ed.), *The Stanford Encyclopedia of Philosophy*. Stanford: The Metaphysics Research Lab, Stanford University. Available at: https://plato.stanford.edu/archives/spr2021/entries/structure-scientific-theories/

Woodward, J. (2010). Scientific explanation. In E.N. Zalta (ed.), *The Stanford Encyclopedia of Philosophy*. (Spring 2010 ed.). Stanford: The Metaphysics Research Lab, Stanford University. Available at: https://plato.stanford.edu/archives/win2019/entries/scientific-explanation/

Further Reading from the Editors

For further discussion of related issues in this volume see the chapters by Lina Jansson, David Wallace, and Porter Williams. For a general overview of intertheoretic relations in physics, see R. Batterman, "Intertheory relations in physics", *Stanford Encyclopedia of Philosophy* (2014). Two important recent papers on the notion of reduction in physics are J. Rosaler, "Local reduction in physics", *Studies in History and Philosophy of Science Part B: Studies in History and Philosophy of Modern Physics* 50 (2015): 54–69 and J. Rosaler, "Reduction as an a posteriori relation", *The British Journal for the Philosophy of Science* 70(1) (2019): 269–299. For connections to conceptions of reduction in metaphysics, see G. Rosen, "Metaphysical Dependence: Grounding and Reduction", in R. Hale and A. Hoffman (eds.), *Modality: Metaphysics, Logic, and Epistemology* (Oxford: Oxford University Press, 2010): 109–136 and D. Chalmers, *Constructing the World* (New York, Oxford: Oxford University Press).

36

PHASE TRANSITIONS

Sorin Bangu

36.1 Introduction

Water freezing is an immemorial human experience, but the study of such a *phase transition* phenomenon became systematic no earlier than a century and a half ago. This preoccupation takes center stage in the thermal physics community. In fact, according to David Ruelle, an illustrious member of this community, it is "the main physical problem which equilibrium statistical mechanics tries to clarify" (2004, p. 1). As he put it, the central question is straightforward: "When the temperature of water is lowered, why do its properties change first smoothly, then suddenly as the freezing point is reached?" Importantly, his assessment of the progress made so far is rather cautious: "While we have some general ideas about this, and many special results, a conceptual understanding is still missing." (2004, pp. 1–2).

Reserved appraisals like this play a significant role in motivating the philosophical interest in this issue. As the celebrated physicist Leo Kadanoff quipped, "the philosopher might wish to note that, strictly speaking, no phase transition can ever occur in a finite system" (2009, p. 778). Indeed, the problems generated by this claim constitute the backbone of the debate charted below. The puzzle becomes evident after we take into account modern physics' fundamental tenet that any bit of matter is constituted by a *finite* (albeit enormous, of the order of 10^{23}) number of atoms and molecules. The worrisome conclusion is now inescapable: such transitions between equilibrium phases don't even seem possible.

What follows is a (opinionated) survey of the views elicited by the conundrum above. We will encounter a broad range of themes and topics – emergence, reduction, approximation, idealization, representation, fiction, and experiment – a non-exhaustive list, as the debate is ongoing. (Interestingly, this debate began surprisingly recently with the pioneering works by Liu (1999, 2001), Batterman (2001), and Callender (2001).)[1]

36.2 Phase Transitions in Thermodynamics[2]

I begin by sketching the physics of phase transitions.[3] The presentation, although mainly technical, is not philosophically disengaged, since it also aims to sow the seeds for the interpretive disputes. The starting point in understanding the macroscopic, thermodynamical treatment of the phase transitions is the diagram in Figure 36.1.

The graph above describes the behavior of water as depending only on two thermodynamical parameters, pressure P and temperature T (the graph is a projection of the complete three-dimensional phase diagram, also including a volume axis V, onto the P-T plane). Points in this plane describe states of the system, and only P and T can be manipulated. Point O, the intersection of all three phase-boundary lines, is the "triple" point: the substance is in all three phases simultaneously.

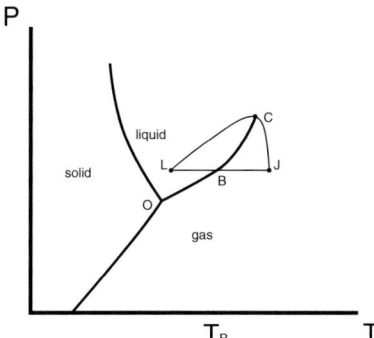

Figure 36.1 Phase diagram for water in terms of temperature and pressure

Point C is the "critical" point. It exists only at the liquid-gas separation; at it, and beyond it, the substance is no longer identifiable as a liquid or a gas, becoming a ("supercritical") fluid.

Let's manipulate the system as follows. Start at point L, within the liquid region, and by supplying thermal energy at constant pressure, quasi-statically increase the temperature. We go toward point B lying on OC. Before reaching B, the system is still in the liquid phase and increases its temperature; then it reaches a point (B), where its temperature remains constant for a while (T_B), although it keeps absorbing heat. We observe that gas bubbles are forming. At B, water reaches its "boiling" point (i.e., the boiling temperature T_B, given a certain pressure), and the gaseous phase becomes manifest, coexisting with the liquid phase. Then, as we keep heating it, the system's temperature begins to increase again; the system moves toward J, where the liquid water entirely turns into vapor. The density drops spectacularly (by a factor of 1,600). Such a LBJ transition is accompanied by the existence of latent heat (at B, the system had to keep absorbing energy from the heat bath to move out of B), and an equally spectacular abrupt increase in volume. Other parameters manifest discontinuities too, such as the entropy S. This kind of transition is called *discontinuous*, or *first-order*. (For a way to visualize the discontinuity, see Figure 36.3c, and the subsequent discussion.)

There is, however, another kind of transition from one equilibrium phase to another. Let us take the system along the path LJ again, but this time by going through the critical point. Start at L, gradually increase both the pressure and the temperature, then decrease them to reach J. Near C, we note that our sample of water becomes a hot, very compressed *fluid*, such that the liquid and gaseous phases are virtually indistinguishable. (They do not coexist anymore, as in the case of the discontinuous transition at B, since we now see "critical opalescence," a milky appearance of water not seen when both phases are present). No latent heat is involved, the density changes smoothly, and both the entropy and the volume are continuous; now a "signature" of the transition is given by other parameters becoming discontinuous, such as the heat capacity (others diverge). The system changes from one equilibrium phase to another by undergoing a *continuous* phase change. (It is important to keep in mind that the relevant physics is described here by combining two elements: a thermodynamical theoretical account involving some idealizations, and a sketch of some basic experimental manipulations. The relevance of these simplifications – e.g., that the substance reaches perfect equilibrium, that the measurements (of temperature) have unlimited precision, and so on – will be discussed again later. As we'll see, from the experimental point of view, B in the diagram is not really a "point.")

When compared to the discontinuous transition, the continuous transition generates a host of new (yet related) philosophical-conceptual puzzles. The continuous phase changes and, more generally, the phenomena taking place "at criticality" raise fundamental questions about the nature of an unexpected phenomenon called "universality." Experiments show that the numbers called "critical exponents", which characterize the systems' behavior near criticality are the same, hence "universal." (More precisely put, close to the critical point, thermodynamic functions have power-law forms, characterized by certain numbers – these are the "critical exponents.") This holds for a range of widely different substances and materials (including magnets, when transitioning, upon heating, between ferromagnetism and paramagnetism). Yet, it turns out, the theoretical derivation of these exponents fails even after supplementing statistical mechanics with some new theoretical tools, such as (Landau's) Mean Field Theory. Thus, a new approach was needed and it was partially borrowed from particle physics: the mathematics of the renormalization group theory (RGT) – in essence, a method developed by Kadanoff, Ken Wilson, and others, to analyze the flow generated by a scaling transformation on a space of Hamiltonians. While this is a fascinating philosophical topic, in what follows I will leave it aside (see Chapter 27 on Universality and Chapter 19 on Renormalization Group Methods). In fact, I will not discuss the continuous phase transitions partly because the conceptual difficulties encountered in their case already arise (although in a different guise) for discontinuous transitions (in essence, the appeal to infinities, as noted at the outset). Thus, from now on, my focus will be almost exclusively on the discontinuous transitions.

But what *is* a phase transition in thermodynamics? A key-concept employed to characterize it is that of "free energy", a (so-called) thermodynamic "potential". Typically two such potentials are used, the ("Helmholtz" and "Gibbs") free energies A and G defined, respectively, as

$$A = U - TS$$
$$G = H - TS$$

U is the internal energy of the system, H its enthalpy ($= U + PV$), T the absolute temperature, and S the entropy.

Introducing this concept allows a more general treatment of the phase change process. When two equilibrium phases are considered independently, the free energy function for each phase is well-defined and has continuous derivatives. Now, consider the manipulation of a system shown in Figure 36.2 below.

Start at D, within one (the initial) phase. Then increase the temperature while keeping the pressure constant. The key aspect to focus on here is what happens with the free energy function A when passing from one phase to another – at K. Along the segment DK, A describes one phase. When it reaches K, however, it does not merge *smoothly* into the dotted segment KE, but switches *suddenly* to the segment KN, where A characterizes the *other* phase. We can thus see that when the system passes through K, the tangent (the first derivative) of DE at K – that is, segment t_{DE} – becomes suddenly t_{FN}, the tangent of FN at K. The entire path DN is supposed to depict the free energy A, and K is a "kink", a sharp turn, on it. K is thus a mathematical point where the free energy function is not (infinitely) differentiable, hence *non-analytic* (i.e., a "singularity").

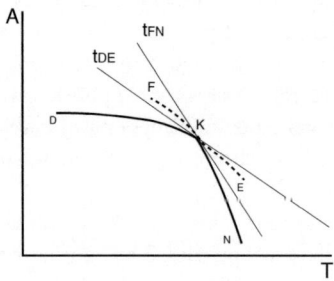

Figure 36.2 A singularity in the free energy

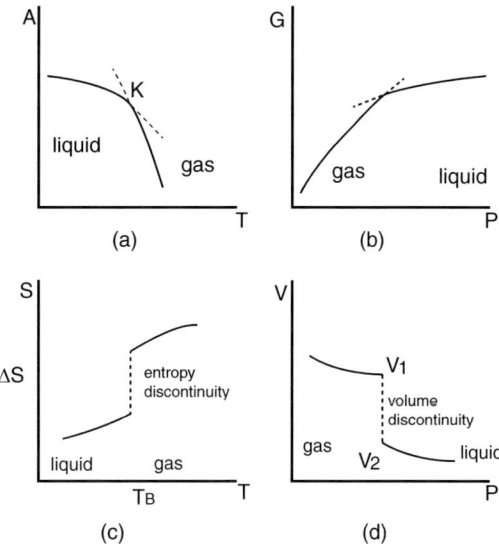

Figure 36.3 Singularities in Helmholtz and Gibbs free energies ((a) and (b)) correspond to discontinuities in entropy and volume, respectively ((c) and d)). The entropy discontinuity gives the amount of latent heat $T_B\Delta S$. Adapted from https://nptel.ac.in/courses/115/103/115103028/

As is clear from their definitions above, partial differentiation of A and G give other important parameters, such as the entropy and volume ($S = -(\partial A/\partial T)_P$, $V = (\partial G/\partial P)_T$) which, as noted, during a transition change abruptly; understood as first-derivatives of the free energy, they change discontinuously. Figure 36.3 below summarizes these observations:

Thus, such transitions are defined in terms of free energy as follows:

PHTD (definition of a phase transition in thermodynamics)	A first-order (discontinuous) phase transition occurs when the first-order derivatives of the free energy (e.g., volume, entropy) change discontinuously – or, equivalently, when free energy has a non-analytic point.

Note the current classification (originating in, but refining, Ehrenfest's): if the first-order derivative of the free energy is discontinuous, then we call the transition "first-order" (or, discontinuous). Continuous transitions are those in which the first-order derivatives are continuous, while the second-order (and higher) derivatives are discontinuous (and thus the parameters they represent are also discontinuous); or when certain parameters diverge to infinity.

36.3   Phase Transitions in Statistical Mechanics. The Coherence Question

Against this background, we should now look at the statistical mechanical treatment of a phase change. Since its inception, the general project of statistical mechanics was to provide a microscopic, average-based, account of the macroscopic behavior captured by thermodynamics – a project pursued under the assumption mentioned above, that matter in bulk, the object of study of thermodynamics, consists in an enormous, but finite, number of components. The idea enjoyed massive success, both conceptual-explanatory and experimental-predictive since the microscopic theory turned out to have resources to account for virtually all major results previously obtained in phenomenological thermodynamics.

One of these resources – perhaps the most important – is the concept of the (canonical) *partition function Z*, defined as

$$Z = \sum_i e^{-E_i/k_B T}$$

Its role in thermal physics is paramount since it ensures (via Boltzmann's constant k_B) the connection between the microscopic and the macroscopic. First, it is a micro-level notion; it assumes a microscopic and finite constitution of systems, being a sum over all the possible microstates of energy E_i. But it also provides expressions for free energies (A and G), which, in turn, by differentiation (as we saw) give expressions for all other macro quantities. The expression for the (Helmholtz) free energy is of particular interest in what follows:

$$A = -k_B T \; lnZ$$

Everything is now in place for us to pause and raise the following issue: what is the relation between the two levels of description, macro and micro, of a phase change? (Before we get to address this question, let's take note of a further technicality. As it was already implicit in the discussion of Figure 36.2 above, Z is usually defined for a homogeneous system, and at a fixed volume, while here we deal, at point B, with a situation where liquid *and* gas coexist. So, several additional assumptions are needed in order to ensure we use the appropriate partition function).

Now, to answer the compatibility, or *coherence question* above (as I'll call it hereafter), one has to examine how the thermodynamical definition PHTD fares when imported into the context of finite statistical mechanics – where the expression of the free energy is given above. As the mathematics shows, a non-analyticity of A is a non-analyticity of Z. (The same holds for G, since $G = A + PV$, and the product PV is expressible in terms of Z too; each of P and V is a function of lnZ). Yet Z is a finite sum of exponentials, each of which is an analytic function. Hence Z itself is analytic, so it *can't* have non-analytic points – i.e., corresponding to K in Figure 36.2. This amounts to saying that the same Z *can't* describe *both* phases. And this gives a more precise sense to Kadanoff's warning, that "no phase transition can ever occur in a finite [statistical mechanical] system."

But now the coherence question becomes a source of major conceptual-philosophical difficulties. The prima facie answer to it seems to be "no, the two descriptions are not compatible; when it comes to phase transitions, there is a deep incoherence at the heart of thermal physics." It looks like statistical mechanics is unable (in the strongest, mathematical, sense) to acknowledge the existence of phase transitions – so Ruelle's concerns at the outset are vindicated. The nature of this impossibility constitutes the central theme of reflection in the current philosophical literature.

At this stage in the argument, the standard narrative within thermal physics is that in trying to come to grips with the issue, physicists resorted to an idealization. They considered another statistical mechanical system, an idealized version of the finite system we find in nature. This idealized system is not realistic in that it contains an infinite number of particles N; it also occupies an infinite volume V, while its density N/V is finite. *This* system is obtained by taking a limit, the $N \rightarrow \infty$ "thermodynamic" limit. (Note, however, that some further conditions have to be satisfied in order for this limit to be well-defined; see Liu (1999, p. S96). For examples where the limit does not exist, see Goldenfeld (1992, pp. 25–28). Sklar (1993, pp. 78–81) gives an inventory of reasons why the limit is generally useful, and perhaps the most relevant here is that it makes the effect of fluctuations vanish). In these circumstances then, the mathematical theorem that blocks the existence of the singularity in Z does not apply, and it can be shown that a singularity exists (as we recall, the theorem states that a finite sum of exponentials, each of which is an analytic function, is analytic too). Historically, H. Kramers first proposed (in the late 1930s) the idea of such a limit and, contrary to the general belief at the time, hypothesized that the same partition function can describe *both* phases. (The issue was even put up to a vote at a congress in 1937, and the result is recalled to have been "inconclusive" – as Goldenfeld (1992, p. 80) tells the story.) The limit entered mainstream physics in the mid-1940s when L. Onsager (drawing on work by Kramers and Wannier, and others) showed for the first time that, in such a limit, the 2-dimensional Ising model undergoes a continuous phase transition: near criticality, the specific heat diverges logarithmically.[4] (Roughly, the Ising model consists of a lattice of spins, which can be in one of two states ("up" or "down"), and each spin is only allowed to interact with its nearest neighbors. Yang and Lee (1952) gave a now-classic proof that taking the limit yields discontinuous transitions too; also, it generalizes to other types of transitions.)

This hopefully sheds more light on why "the existence of a phase transition requires an infinite system" (Kadanoff, 2000, p. 238). Coherence is recovered but only in an asymptotic sense. Before we move on, recall that the issue of infinities also reappears in the treatment of continuous phase changes within the Renormalization Group Theory (RGT) in the derivation of the critical exponents. In a nutshell, this is so because in RGT rescaling works only for an infinite system; a finite system has a characteristic length scale, and once we get beyond it, invariance is lost. Moreover, the need for the limit in the RGT corresponds to the need for the limit in the statistical mechanical account just presented. For example, the "correlation length," a parameter characterizing the system near criticality, is given in terms of the second derivative of the free energy. This derivative, and hence the parameter itself, diverges near the critical point, and this can't happen in RGT unless the system is infinite.

So, as I hinted at the beginning, the attempts to tackle the infinity puzzle presuppose deep engagement with several central themes in the philosophy of physics and metaphysics. Below, I divided the discussion into two sections; although some overlapping is unavoidable, they approach the coherence question from somewhat different perspectives. The first series of issues centers on the notions of emergence, reduction, and approximation.

36.4 Emergence, Reduction, and Approximation

There are several, and quite sophisticated, ways to capture the otherwise elusive concept of "emergence". In the present context, however, D. Chalmers' characterization will hopefully suffice: "We can say that a high-level phenomenon is (...) emergent with respect to a low-level domain when the high-level phenomenon arises (in some sense) from the low-level domain, but truths concerning that phenomenon are not deducible even in principle from truths in the low-level domain." (2006, p. 244). (Chalmers actually defines "strong" emergence (in contrast to the epistemological notion of 'weak' emergence; for more on the distinction, see Wilson (2015)). Note also that Chalmers takes reduction to be essentially deduction, as does Butterfield (2011)).

If we operate the natural identifications, the "high-level phenomenon" would be the observed *qualitative* change of phase, captured (mathematically) by a singular point. The "low-level domain" would be the finite-component microstates from which the macro behavior "arises." Then one of "the truths concerning that phenomenon" – namely, that the free energy has a singularity – is "not deducible even in principle from truths in the low-level domain," i.e., the (mathematical) truths about the finite partition function.

Something like this line of thought has probably led some to agree with Prigogine's anti-reductionist stance, that "the existence of phase transitions shows that we have to be careful when we adopt a reductionist approach. Phase transitions correspond to emerging properties." (1997, p. 45). Lebowitz also calls them "paradigms of emergent behavior" (1999, p. S346). Similarly, some philosophers of science are sympathetic, in various degrees, to the view that phase transitions are, in some sense, emergent. Liu does not hesitate to characterize them as "truly emergent properties"

(1999, p. S92). Humphreys (1997), Primas (1998), Batterman (2002, 2005, 2013), Morrison (2012), Bangu (2015a), and (perhaps) Butterfield (2011) can also be read along these lines – but the reader should be warned that these authors develop their positions while taking into account complications and qualifications not mentioned here, hence their claims should be judged carefully in their proper contexts. Butterfield, for instance, calls for a more general revision of the orthodoxy, arguing that "although emergence is usually opposed to reduction, many examples exhibit both." (2011, p. 1066; one of the examples is that of phase transitions.)

Callender and Menon (2013) also discuss the emergence claim and in the end, it seems to me, dismiss it (yet see p. 205). In fact, earlier on, in his seminal 2001 paper, Callender advances an argument to this effect. The gist of it is the idea that even if strictly speaking statistical mechanics can't account for a phase transition in the sharp sense required by thermodynamics, it can "in some sense *approximate* a singularity" (Callender, 2001, p. 550; my emphasis). This amounts to trying to save coherence through approximation. The idea then is to show that a non-analytic function can be reached as a limit of analytic ones; the suggestion has been worked out in detail by Butterfield (2011). (Note here that Batterman (2005, p. 235) accepts, with reservations, the approximation approach, warning that the difficulties appear near the critical point: "We can make precise predictions about the extent of fluctuations in large but finite systems away from criticality despite (in fact, because of) our use of the infinite N idealization. And, we may be able to say in what sense our finite systems actually approximate a singularity at first-order phase transitions. One way of expressing this is to say that the idealization of infinite N is *controllable* in those contexts – away from critical points.")

The approximation strategy is undoubtedly attractive, and we'll return to (a version of) it later. Yet an important physicist, R. Baxter, after positively reviewing the main approximation methods physicists actually use, still voices the lingering suspicion that this strategy may only push the problem one step back: "Approximate methods [. . .] require considerable acts of faith in the assumptions made." (1989, p. 11) He also points out that "The approximation schemes [. . .] can give quite accurate values for the thermodynamic properties, *except near the critical point*." (1989, p. 10). Another difficulty here is that the appeal to approximation seems, in the end, essentially pragmatic. To invoke Primas' (1998, p. 90) analogy when commenting on this case, although a very-many-sized polygon can approximate a circle as closely as we like, and for all practical purposes, such a shape is still a polygon and not a circle! Finally, Norton (2012) can be read as a different argument in favor of the approximation approach. He distinguishes two "analytic activities" (p. 228) involved in this debate. On one hand, we can think of a phase transition as an approximation – an inexactly described property of a real, finite system. On the other, we may be tempted to "promote" this approximation "to an idealization" (p. 211), i.e. to posit a "limit system" (as Norton calls it[5]) and describe the transition as a property realized *exactly* by this system. Yet this second way of describing what we do (as analyzing the properties of this limit system), although part of the standard narrative, as we recall, is far from innocuous. By drawing on several clever examples, he warns that "the infinite systems often have properties very different from those of the finite system" (p. 215), and thus one (actually, Ruelle is named here) can only "hope (. . .) that the limit property and the limit system will agree." (p. 215). Hence, "since an infinite system can carry unexpected and even contradictory properties, the latter practice [the investigation of limit systems; my note] carries considerably more risk" (p. 228).

Returning to Callender's stance, his view is that the emergence issue appears only under the (problematic) assumption that the thermodynamical definition PHTD of a phase transition *must* be used within statistical mechanics: "the fact that thermodynamics treats phase transitions as singularities does not imply that statistical mechanics must too. To assume that would be to take thermodynamics too seriously. It will now come as no surprise that I believe the source of this 'emergence' is again the result of a too-literal translation from thermodynamics to statistical mechanics." (Callender, 2001, p. 550). In the next section, I examine the implications of this stance.

36.5 Inter-theoretic Relations, Representation, and Experiment

In my view, there are in fact two ways to construe Callender's profound insight, that we have the option to *deny* that statistical mechanics must import (and thus use) definition PHTD. One way (hereafter alternative (A1)) is to say that the general demand for compatibility between thermodynamics and statistical mechanics should not be unconditional. Coherence can't be a "must" since statistical mechanics supersedes and *corrects* classical thermodynamics – and phase transitions are a case in point. Thus, the coherence question receives a "no" answer, together with the qualification: "... and in asking this question one assumes, wrongly, that thermodynamics has to be entirely recovered within statistical mechanics."

Here, the chief reason for advancing this view is that PHTD is – from the perspective of *finite*-particle statistical mechanics – non-referential. From this theoretical perspective, phase transitions *can't* be singularities: if the definition is understood as referring to genuine physical discontinuities, then this definition fails to pick out anything in reality. Noteworthy, Callender and Menon (2013, pp. 216–217) seem to vacillate between (reluctantly) accepting and (vigorously) rejecting the idea of a real discontinuity. Norton, however, writes: "If the atomic theory of matter is true, then ordinary thermal systems of finitely many components cannot display discontinuous changes in their thermodynamic properties. The changes they manifest are merely so rapid as to be observationally indistinguishable from discontinuous behavior. Indeed, if we could establish that the phase transitions of real substances exhibit these discontinuities, we would have refuted the atomic theory of matter, which holds that ordinary thermal systems are composed of finitely many atoms, molecules, or components." (2012, p. 225) See also Shech (2013, p. 1175): "if systems are composed of finitely many particles, which is the case within the context of the atomistic theory of matter (...), then it makes no sense to talk of concrete discontinuities." However, I'm afraid that these remarks about the connections between the structure of matter and the discontinuities are not quite right, as we'll see below (more details are in Bangu (2019)).

To recap: on this view then, since there are no PHTD phase transitions to begin with, to object to the idea that it is *obligatory* for statistical mechanics to use PHTD amounts to saying that it must *not* use PHTD.

But, as I signaled previously (Bangu, 2015a, 2015b), one can't just leave the matter here. If (finite) statistical mechanics does not use PHTD, then *how* does it account for phase transitions? Does it have its own definition(s)? Callender and Menon seem willing to take seriously (!) the "several proposals for finite-particle accounts of phase transitions" (2013, p. 206) they find in the physics literature. More concretely, they review in some detail two such definitions (in their 2013, pp. 207–208), as well as an account in topological terms. However, in closing their respective discussions, and in contrast to the general optimistic tone of the presentation, they note several challenges still to be met by these approaches. They say, e.g., that "probably *none* of the definitions provide necessary and sufficient conditions for a phase transition" (2013, p. 210; my emphasis), and that "it is *unclear* what topological criteria will be necessary and sufficient to define phase transitions *if* any such criteria *can* be found." (2013, p. 217; my emphases). Concessions like these give, in the end, the feeling that their optimism is rather unwarranted. Moreover, the impression that they overstate the case strengthens upon a closer review of the relevant literature. For instance, two papers they cite in favor of the first alternative definition are Chomaz et al. (2001), and Touchette (2006) – but the latter is actually correcting the former. Moreover, for the topological approach, they cite Franzosi et al. (2000); yet in a later paper, two of these authors, Franzosi and Pettini (2004) admit serious limitations of their results. (Callender and Menon do not cite *this* paper. More details are in Bangu (2015a)). Indeed, there is work in this field; but this, in and of itself, doesn't prove Callender and Menon's main point. (For a discussion of more attempts to capture phase transitions in finite systems, see Ardourel (2017).)

The other alternative (A2) sketched here is motivated in part by the worries about the difficulties with alternative (A1). In light of the genuine possibility that these challenges *cannot* be met, one is entitled to reconsider the idea that the PHTD orthodox definition may actually be perfectly fine, where this means that there is a sense in which it is legitimate to talk about physical discontinuities, captured mathematically as singularities. (As Batterman points out: "My contention is that thermo-dynamics is correct to characterize phase transitions as real physical discontinuities and it is correct to represent them mathematically as singularities" (2005, p. 234).) Thus, while the thermodynam-ical definition PHTD is not taken *too* seriously indeed – since one fully realizes the difficulties of importing it into statistical mechanics – it is however taken seriously *enough* to be used in statistical mechanics. After all, and this is important, one who objects to the idea that the use of PHTD is *obligatory* in statistical mechanics does not thereby *also* object to the idea that this use is *permitted* – so one still leaves it open that statistical mechanics *can* use PHTD. As is clear, 'It's not obligatory for S to φ' does *not* rule out 'It is permitted for S to φ'.

Although this position is available in the logical space, an advocate of it faces a two-fold difficulty,[6] namely, to present an account that will be able (i) to explicate how the PHTD definition succeeds in being referential (*pace* (A1)), as well as (ii) to tame the concerns generated by the use of infinities.

The worries about (i) vanish once we use the terms carefully. A singularity or non-analytic kink in the thermodynamical free energy (Figure 36.3a, b) is a purely mathematical feature of the free energy functions, posited as an intersection of two continuous curves (Figure 36.2). It does correspond to a discontinuity, e.g., in the volume (Figure 36.3d), and, importantly, there is nothing un-physical about such a discontinuity (the atomic theory of matter is not threatened, the substance doesn't disappear and then reappear, etc.) To say that the volume (see Figure 36.3d) is "discontinuous" is just to say that for the time the system spends between V_2 and V_1 the amount of liquid decreases (quickly), and vapors take its place. Moreover, we can talk quantitatively, about how the ratio of liquid/gas diminishes along V_2V_1.[7] Thus, a genuinely physical process is represented in an eminently natural way, as "encapsulated" into a mathematical non-analytic point.[8]

Further difficulties may arise, however, from the observation that "we do not actually *measure* per-fect singularities" (Callender, 2001, p. 550; my emphasis). Thus, "the transition is neither 'smooth' nor 'singular'" (Liu, 1999, p. S103). Then, a point like K, if seen as the result of a measurement, doesn't actually represent anything: it is a "fiction" (Liu, 2001, p. S336) – hence if we recall the logic of (A1), statistical mechanics doesn't have to "take it seriously." Now, dispelling these concerns requires not a blank rejection of the fiction talk, but a clarification of *what kind* of fictions these singu-larities are. As everywhere in science, experimental measurements are necessarily of limited precision, the data sets coming out of the labs are never sharp, but in cluster form, curves are never smooth, etc. Essentially, the view that scientific theorizing and testing focus on the raw data measurements is rather naïve (and this obviously includes the theorizing in thermal physics too). Instead, *data* are gath-ered and then shaped into *phenomena* – and, importantly, I use these notions in the sense articulated by Bogen and Woodward in their now-classic 1988 paper, work in which they also clarify (in Sect. VIII) in what sense phenomena are *real*, a position to which I adhere here as well. The "shaping" takes place through statistical processing, "noise" elimination, etc. – in essence, by approximation procedures.

So, on the approach I'm sketching here, singularities *do* refer – to the real phenomena constructed from these measurements (see Bangu, 2019). Thus, it is actually misleading to call them "fictions" without qualification and dismiss them on this basis. Once we understand that they enter the story in this unproblematic way, the consequence of their introduction – the need for the infinite limit – should not create virtually any worries. As a directly observable occurrence, a phase change is just too elusive to be an object of scientific investigation. "What is going on" in the heated flask while liquid water is turning into vapors is *not yet* a phenomenon (in the specific sense above) to study – that is, until thermodynamics models it as a singularity. The infinite limit is then needed in statistical mechanics not as physically real, but as part of the modeling procedure shared by the two theories.

To conclude. A supporter of position (A1) interprets the advice not to take thermodynamics too seriously in a rather extreme way. She virtually dismisses it (in a "reductionist" fashion), putting all her hopes in the alternative definitions and approaches mentioned above. An advocate of position (A2) finds this unwise. She does not dismiss thermodynamics, and shares the working physicists' attitude, by recognizing that the two theoretical frameworks are interdependent. (As Zemansky and Dittman remarked: "we can never be sure that the assumptions [of the microscopic point of view] are justified until we have compared some deduction made on the basis of these assumptions with a similar deduction based on experimentally proven macroscopic point of view. (. . .) We look to the macroscopic point of view for guidance" (1997, pp. 5–6))

These two positions ((A1) and (A2)) also deal with the coherence question differently. When not declaring it ill posed, a proponent of (A1) answers it in the negative. On the other hand, an advocate of (A2) argues for a qualified "yes," suggesting that here lies what makes a phase transition not only uniquely intriguing philosophically (since it allows us to see that the two theoretical frameworks are in fact complementary, and peacefully coexisting), but also a somewhat unusual object of aesthetic appreciation. To give Kadanoff the last word, "this coupling of microscopic with macroscopic has an unexpected and quite breathtaking beauty" (2013, p. 183).

Acknowledgments

For comments on various drafts of this chapter, I would like to thank Alastair Wilson, Patricia Palacios, and Elay Shech. I'm also grateful to the physicist Lapo Casetti for enlightening conversations on these issues. I am solely responsible for the final version of the text.

Notes

1 Sklar (2000, p. 66) hints at some of the issues.
2 This section and part of the next draw on sect. 1 in Bangu (2019).
3 The presentation and the diagrams follow the standard textbook treatment (except for small modifications to stress certain conceptual points). See Zemansky and Dittman (1997).
4 See Onsager (1944).
5 See Shech (2015, p. 1066) for discussion.
6 Below I outline a view spelled out in Bangu (2009, 2015b).
7 Details in Reif (1965, p. 310) and Bangu (2019, p. 1927).
8 Shech (2014) discusses the representation issue.

References

Ardourel, V. (2017). The infinite limit as an eliminable approximation for phase transitions. *Studies in the History and Philosophy of Modern Physics*, 62: 71–84.

Bangu, S. (2009). Understanding thermodynamic singularities. Phase transitions, data and phenomena. *Philosophy of Science*, 76(4): 488–505.

Bangu, S. (2015a). Neither weak, nor strong? Emergence and functional reduction. In M. Morrison and B. Falkenburg (eds.), *Why More Is Different*. Verlag Berlin Heidelberg: Springer, pp. 153–168.

Bangu, S. (2015b). Why does water boil? Fictions in scientific explanation. In U. Mäki et al. (eds.), *Recent Developments in the Philosophy of Science*. Switzerland: Springer International Publishing, pp. 319–330.

Bangu, S. (2019). Discontinuities and singularities, data and phenomena: For referentialism. *Synthese*, 196: 1919–1937.

Batterman R. (2001). *The Devil in the Details*. New York: Oxford University Press.

Batterman, R. (2005). Critical phenomena and breaking drops: Infinite idealizations in physics. *Studies in History and Philosophy of Modern Physics*, 36: 225–244.

Batterman, R., editor. (2013). *Oxford Handbook for the Philosophy of Physics*. New York: Oxford University Press.

Baxter, R.J. (1989). *Exactly Solved Models in Statistical Mechanics*. London: Academic Press.

Bogen, J. and Woodward, J. (1988). Saving the phenomena. *Philosophical Review*, 97: 303–352.

Butterfield, J. (2011). Less is different: Emergence and reduction reconciled. *Foundations of Physics*, 41: 1065–1135.

Callender, C. (2001). Taking thermodynamics too seriously. *Studies in History and Philosophy of Modern Physics*, 32: 539–553.

Callender, C. and Menon, T. (2013). Turn and face the strange. Ch-ch-changes: Philosophical questions raised by phase transitions. In R. Batterman (ed.), *Oxford Handbook for the Philosophy of Physics*. Oxford: Oxford University Press, pp. 189–223.

Chalmers, D. (2006). Strong and weak emergence. In P. Clayton and P. Davies (eds.), *The Re-Emergence of Emergence*. New York: Oxford University Press, pp. 244–256.

Chomaz, P., Gulminelli, F. and Duflot, V. (2001). Topology of event distributions as a generalized definition of phase transitions in finite systems. *Physical Review E*, 64: 046114.

Franzosi, R. and Pettini, M. (2004). Theorem on the origin of phase transitions. *Physical Review Letters*, 92: 060601.

Franzosi, R., Pettini, M. and Spinelli, L. (2000). Topology and phase transitions: Paradigmatic evidence. *Physical Review Letters*, 84: 2774–2777.

Goldenfeld, N. (1992). *Lectures on Phase Transitions and the Renormalization Group*. Reading, MA: Perseus Books.

Humphreys, P. (1997). How properties emerge. *Philosophy of Science*, 64: 1–17.

Kadanoff, L. (2000). *Statistical Physics*. Singapore: World Scientific.

Kadanoff, L. (2009). More is the same; mean field theory and phase transitions. *Journal of Statistical Physics*, 137: 777–797.

Kadanoff, L. (2013). Theories of matter: Infinities and renormalization. In R. Batterman (ed.), *Oxford Handbook for the Philosophy of Physics*. Oxford: Oxford University Press, pp. 141–188.

Lebowitz, J.L. (1999). Statistical mechanics: A selective review of two central issues. *Reviews of Modern Physics*, 71: S346–S347.

Liu, C. (1999). Explaining the emergence of cooperative phenomena. *Philosophy of Science*, 66: S92–S106.

Liu, C. (2001). Infinite systems in SM explanations: Thermodynamic limit, renormalization (semi-) groups, and irreversibility. *Philosophy of Science*, 68: S325–S344.

Morrison, M. (2012). Emergent physics and micro-ontology. *Philosophy of Science*, 79: 141–166.

Norton, J. (2012). Approximation and idealization: Why the difference matters. *Philosophy of Science*, 79(2): 207–232.

Onsager, L. (1944). Crystal statistics. I. A two-dimensional model with an order-disorder transition. *Physical Review*, 65: 117–149.

Prigogine, I. (1997). *End of certainty*. New York: The Free Press.

Primas, H. (1998). Emergence in exact natural sciences. *Acta Polytechnica Scandinavica* Ma, 91: 83–98.

Reif, F. (1965). *Fundamentals of Statistical and Thermal Physics*. Boston, MA: McGraw-Hill.

Ruelle, D. (2004). *Thermodynamic Formalism*. 2nd ed. Cambridge University Press.

Shech, E. (2013). What is the paradox of phase transitions? *Philosophy of Science*, 80(5): 1170–1181.

Shech, E. (2014). Scientific misrepresentation and guides to ontology: The need for representational code and contents. *Synthese*, 192(11): 3463–3485.

Shech, E. (2015). Two approaches to fractional statistics in the quantum hall effect: Idealizations and the curious case of the Anyon. *Foundations of Physics*, 45(9): 1063–1110.

Sklar, L. (1993) *Physics and Chance*. Cambridge: Cambridge University Press.

Sklar, L. (2000). *Theory and Truth*. New York: Oxford University Press.

Touchete, H. (2006). Comment on "First-order phase transition: Equivalence between bimodalities and the Yang-Lee theorem." *Physica A: Statistical Mechanics and Its Applications*, 359: 375–379.

Wilson, J. 2015. Metaphysical emergence: Weak and strong. In T. Bigaj and C. Wuthrich (eds.), *Metaphysics in Contemporary Physics*. Amsterdam: Rodopi, pp. 251–306.

Yang, C.N. and Lee, T.D. (1952). Statistical theory of equations of state and phase transitions. I. Theory of condensation. *Physical Review*, 87: 404–409.

Zemansky, M. and Dittman, R. (1997). *Heat and Thermodynamics* (7th editon). New York: McGraw-Hill.

Further Reading from the Editors

A helpful overview of the philosophy of phase transitions is T. Menon and C. Callender, "Turn and Face the Strange... Ch-ch-changes: Philosophical Questions Raised by Phase Transitions," in R. Batterman (ed.), *The*

Oxford Handbook of Philosophy of Physics (Oxford: Oxford University Press, 2013). The classic presentation of the case for anti-reductionism from phase transitions is R. Batterman, *The Devil in the Details: Asymptotic Reasoning in Explanation, Reduction, and Emergence* (Oxford: Oxford University Press, 2001). A key response to this case has been J. Butterfield, "Less Is Different: Emergence and Reduction Reconciled," *Foundations of Physics* 41, (2011): 1065–1135. Further responses to arguments from phase transitions against reductionism can be found in J. Saatsi and A. Reutlinger, "Taking Reductionism to the Limit: How to Rebut the Antireductionist Argument from Infinite Limits," *Philosophy of Science* 85(3), (2018): 455–482 and in P. Palacios, "Phase Transitions: A Challenge for Reductionism?" *Philosophy of Science* 86(4), (2019): 612–640.

37

UNIVERSALITY

Robert W. Batterman

37.1   Introduction

In its broadest sense, "universality" is a technical term for something quite ordinary. It refers to the existence of patterns of behavior by physical systems that recur and repeat despite the fact that in some sense the situations in which these patterns recur and repeat are different. Rainbows, for example, always exhibit the same pattern of spacings and intensities of their bows despite the fact that the rain showers are different on each occasion. They are different because the shapes of the drops, and their sizes can vary quite widely due to differences in temperature, wind speed and direction, etc. There are different questions one might ask about such patterns. For instance, one might ask why the particular rainbow that I'm currently seeing exhibits the spacings and intensities of its bows that it does. Perhaps an answer to that question might need to refer to the particular sizes and shapes of the drops in the particular rain shower at this time. On the other hand, one might ask about how it is possible that despite the differences in the lower scale details about the sizes and shapes of drops in different rain showers, the spacings and intensities of the bows in the different rainbows are the same. This latter question concerns the explanation of the (universal) pattern of behavior. It is arguable that the answer to the former question (or even answers to the former question for all the different rainbows all taken together) cannot answer the second question. (See Batterman (2002).)

This paper examines what is, in the physics literature, the paradigm example of universality; namely, the so-called universality of critical phenomena—certain kinds of phase transitions that systems (e.g.,fluids and magnets) can undergo. In the next section I describe the phenomena. In Section 37.3 I lay out the mathematical and physical ingredients required for one to describe this universal pattern. These include the introduction of a variable called an "order parameter" that serves to represent a surprising change of symmetry as a system passes through a so-called critical point. Section 37.4 looks back to some early work on phase transitions to see how the concept of universality should be properly defined. One of the most important features, virtually ignored in recent philosophical discussions, is a kind of *stability* of macroscopic behavior under changes of microscopic details. *Any explanation of the possibility of universal behavior must account for this type of stability.* In Section 37.5 I briefly discuss how the renormalization group (RG) can explain the existence of universal behavior, in part by explaining the existence of this kind of stability.

37.2   Universality of Critical Phenomena: A Paradigm Case

One of the most striking examples of universal behavior concerns the pattern displayed by molecularly distinct fluids near their so-called critical points. It is worth spending a bit of time examining this *paradigmatic* example of universality.

It is commonplace that water can exist in three distinct phases: as liquid, as solid, and as vapor or gas. One can represent these different phases graphically using the thermodynamic variables, pressure (P), volume (V), and temperature (T). In Figure 37.1, the curves show how the pressure depends on the volume at different fixed values for the temperature. Consider the boiling region. This corresponds to the process that takes place when water boils in a kettle. Inside the kettle both vapor and liquid coexist. The special point (T_c, P_c) is called the critical point. It is special in the following way. Below the critical temperature ($T_c = 647\,\mathrm{K}$) and critical pressure ($P_c = 22.9\,\mathrm{MPa}$) one finds the region of liquid/vapor coexistence. Above that critical point, the kettle will no longer contain two distinct phases of water. There is an abrupt change in the makeup of the stuff in the kettle at that critical temperature and pressure. For carbon dioxide (as for many other fluids) the diagram looks exactly the same although there will be different values for the critical temperature and pressure. (For carbon dioxide, $T_c = 31.1\,\mathrm{C}(= 304.3\,\mathrm{K})$ and $P_c = 7.2\,\mathrm{MPa}$. Experimentally, it is a lot easier to realize the critical temperature and pressure of CO_2 than it is for water.) In saying that the diagram looks exactly the same for carbon dioxide as it does for water, the important thing is that the *shapes* of the dotted *lines* near (T_c, P_c) for both fluids are identical. So, while water and carbon dioxide are very different fluids as is evidenced by their very different critical temperatures and pressures, nevertheless near their critical points ("near criticality") they exhibit identical behavior. This is universal behavior realized by molecularly very different fluids.

A remarkable representation of the experimental fact of universality is provided by a figure in E. A. Guggenheim's 1945 paper entitled "The Principle of Corresponding States." When plotted in reduced coordinates $\left(\frac{\rho}{\rho_c}, \frac{T}{T_c}\right)$, the coexistence curves for eight different fluids near criticality all collapse onto the same curve. See Figure 37.2.

One can quantify the universal behavior by introducing a so-called "order parameter." For the transition between the boiling region with two coexistent phases and the region above, define the order parameter Ψ to be the difference between the densities of the liquid and the vapor in the kettle:

$$\Psi = |\rho_l - \rho_v|.$$

Then the relation

$$\Psi \propto \epsilon^\beta \tag{37.1}$$

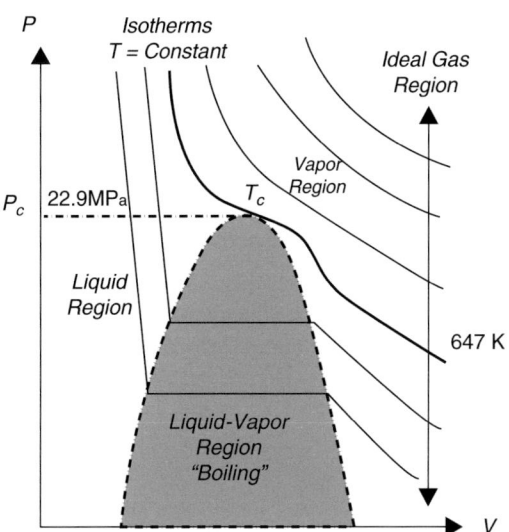

Figure 37.1 Cartoon PVT diagram for water (Kadanoff, 2013, p. 148)

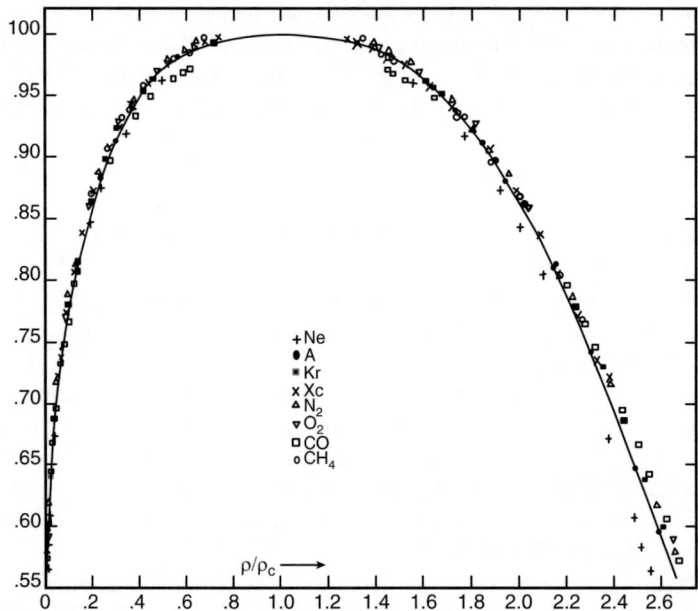

Figure 37.2 Universality of critical phenomena (Guggenheim, 1945)

describes the shape of the coexistence curve for a fluid. ($\epsilon = |(T_c - T)/T|$ and is a measure of how close a system is to its critical temperature in dimensionless units.) Universality is also expressed by the fact that β is the same for the different fluids. This is just what is represented in Figure 37.2.

Guggenheim's demonstration of the data collapse is indeed striking. But it is even more remarkable that the critical behavior of magnets exhibits the identical scaling relation. For a ferromagnet, the order parameter M is the net magnetization, and the relation

$$M \propto \epsilon^\beta \qquad (37.2)$$

holds as well with β identical to the value in equation (37.1). (The analogy between Ψ and M is very strong indeed: $M = |\rho_\uparrow(, \mathbf{r}) - \rho_\downarrow(, \mathbf{r})|$. It is the difference between the densities of up-spins and down-spins at locations $\mathbf{r}$.)

37.3   The Ingredients of Universality

37.3.1   *Order Parameters and Symmetries*

Exactly what kind of quantities are the order parameters Ψ and M? One can think of them as thermo-dynamic properties that allow us to characterize the qualitative behavior of systems near their critical points. That is, one can treat these as continuum properties on a par with pressure, temperature, and volume. As thermodynamic properties they describe the behavior to be expected as a system is cooled from a temperature above T_c to below T_c. In the case of Ψ we see that as that temperature is crossed, there is a spontaneous appearance of two new states of matter (liquid and vapor). In the case of M, the magnet exhibits zero net magnetization (it is in a so-called paramagnetic phase) and spontaneously gains a net magnetization as the critical temperature is crossed. In both cases, above T_c there exists a symmetry that is broken upon passing through T_c. For example, in a magnet in zero external magnetic field, there is rotational symmetry (no preferred direction) above T_c that is broken upon passing through T_c: All of a sudden there is a preferred direction of magnetization.

The concept of an order parameter was first introduced[1] by Landau in 1937 (Landau, 1965, pp. 193–216). The order parameter, as noted, captures the macro or continuum behavior and reflects the symmetry changes in a fluid as a parameter (temperature) is varied. But, from the point of view of statistical mechanics, one needs to think about Ψ and M in a different way. What, after all, is responsible for there being non-zero values of M below T_c? The answer has to depend on some kind of lower scale/microscopic features of the magnet—some fact about the arrangement of (magnetic) spins on a lattice. Michael Fisher puts this as follows:

> To assert that there exists an order parameter in essence says: "I may not understand the microscopic phenomena at all" (as was historically, the case for superfluid helium), "but I recognize that there is a microscopic level and I believe it should have certain general, overall properties as regards locality and symmetry: those then serve to govern the most characteristic behavior on scales greater than atomic." (Fisher, 1998, p. 654)

Once one sees the order parameter as coding for some feature of the *microstructure* of the magnet or fluid, one is in the domain of statistical physics. Now one needs to treat the order parameters as *averages* and one needs to consider the possibility of *fluctuations* in the values of the order parameters. Furthermore, thinking like this actually requires that one distinguish between the macroscopic scale (the scale of the continuum where the order parameter is simply a function of thermodynamic macroscopic properties like temperature and pressure), a mesoscale where fluctuations in aggregates of atomic scale properties may be important, and the atomic scale where what matters is the detailed natures of the atoms/molecules and spins. The order parameter lives in the intermediate region. Here is Fisher again:

> Significantly, in my view, Landau's introduction of the order parameter exposed a novel and unexpected *foliation* or level in our understanding of the physical world. Traditionally, one characterizes statistical mechanics as directly linking the *microscopic* world of nuclei and atoms (on length scales of 10^{-13} to 10^{-8} cm) to the *macroscopic* world of say, millimeters to meters. But the order parameter, as a dynamic, fluctuating object in many cases intervenes on an intermediate or *mesoscopic* level characterized by scales of tens or hundreds of angstroms up to microns (say, $10^{-6.5}$ to $10^{-3.5}$ cm). (Fisher, 1998, p. 654)

It is reasonable to ask why this works. Why is this foliation of the physical world successful and appropriate? The answer to this is important for understanding *both* how universal behavior is possible and for understanding how one can explain such universality. That is to say, the existence of mesoscale features of the world captured by the order parameter is a necessary condition for universality.

Although I cannot argue this point here, I believe that the spatial (and in other cases, temporal) foliation into small (short), meso, and large (long) scales is an important feature of universality in various sciences. This includes materials science, among others.[2]

37.3.2 *Length Scales*

Kadanoff describes a "very interesting and fundamental question" concerning the fact that the world

> shows an amazing variety of length scales: There is the Hubble radius of the universe, 10^{10} light years or so and the radius of our own solar system, 10^{11} meters roughly, and us—two meters perhaps, and an atom—10^{-10} meters in radius, and a proton 10^{-16} meters, and the characteristic length of quantum gravity—which involves another factor of about 10^{20}.
>
> How these vastly different lengths arise is a very interesting and fundamental question. . . . However, we can think about how one describes the region between these lengths. If

one is looking at scales between two fundamental lengths, there are no nearby characteristic lengths. Similarly in critical phenomena, whenever one looks [at] any event which
takes place between the scale of the lattice constant [the spacing between molecules or
spins] and the much larger scale of the coherence length, one is in a situation in which
there are no nearby characteristic lengths. (Kadanoff, 2000, p. 251)

This lack of characteristic length scales is crucial for both the existence of universal behavior and for its
explanation. In the above quote, Kadanoff refers to the "coherence length" sometimes also called the
"correlation length." For a system near criticality the correlation length becomes enormous and for
infinite systems, it diverges to infinity.[3] For a fluid system like the water in the kettle, the correlation
length is a measure of the average size of a region of vapor (say) in the kettle. Vapor molecules cluster
with other vapor molecules and liquid molecules cluster with other liquid molecules. (The analog
of this in a ferromagnet is that the spins on the lattice sites want to be next to spins pointing in the
same direction. So one has "droplets" of up-spins of a certain size and "droplets" of down-spins as
well.) As the water in the boiling region (or the spins in the magnet) heat up and approach critical
temperatures, the size of the different droplets get larger and larger. This means that, even though
the physical forces between the molecules (or spins) remain local, distant molecules (spins) become
correlated as a result of existing in the different droplets. In addition, one has large correlated droplets
of vapor inside droplets of liquid inside droplets of vapor . . . . See Figure 37.3.

In the phase diagram of Figure 37.1, the region near critical point inside the boiling region corresponds to this "fractal like" structure of droplets within droplets, etc. The reason for this is that near
the critical point, fluctuations are dominant and average values for the order parameters essentially
lose their meaning. Orthodox statistical mechanics is unable to describe the critical behavior because
there are fluctuations at all length scales from the macroscopic correlation length equal to the size of
the system (the kettle), and the microscopic distance of the range of the forces between molecules or
spins. These fluctuations "cannot probe all the details of the interatomic potential. Rather they only
see certain gross features of the potential: for example the amount of breaking of an exact symmetry
. . . or the distance from the critical point" (Kadanoff, 1971, p. 104).

Given all these details about what is happening at the micro-, meso-, and macro-scales in the
neighborhood of a critical point, we can now see how properly to define the universality of critical
phenomena.

Figure 37.3 Droplets inside droplets inside droplets . . . (Kadanoff, 1976, p. 12)

37.4   How Universality Is Defined

Kadanoff states the "hypothesis of universality" as follows:

> All phase transition problems can be divided into a small number of different classes depending upon the dimensionality of the system and the symmetries of the order state. Within each class, all phase transitions have identical behavior in the critical region, only the names of the variables are changed. (Kadanoff, 1971, p. 103)

Table 37.1 exhibits different values for the scaling exponent β for different phase transitions at criticality.[4] We can here see how the scaling exponent depends upon the dimensionality of the system. The ferromagnetic transition for a two dimensional film is in a different universality class than the three dimensional magnet. Notice also that despite remarkable differences in microstructural makeup, the liquid/vapor systems, superfluid helium, and the $d = 3$ ferromagnet are all in the same universality class; that is, the scaling exponent β appearing in equations (37.1) and (37.2) are the same (to within experimental error possible at the time).

A full description of a phase transition typically involves two field variables. Consider the ferromagnetic transition. One can take the external magnetic field h as a field that can drive a system from one coexisting phase (up-spins) to the other (down-spins). The second field is provided by the reduced temperature, ϵ, defined above. This field drives the system closer to or away from the critical point. These are the variables to which Kadanoff refers in his hypothesis of universality above. In terms of these field variables, the free energy for the system can be written as follows: $F = F(h, \epsilon)$. (The free energy is a measure of the internal energy of the system that is available to do work.) In differential form, this becomes

$$dF = <M> dh + <\mathcal{H}> d\epsilon, \tag{37.3}$$

where $<\mathcal{H}>$ is an energy. In this differential form we see that the fields h and ϵ are paired with two thermodynamically conjugate variables: $<M>$ and $<\mathcal{H}>$, respectively. For the liquid/vapor transition, the corresponding pairs of variables are the order parameter Ψ with its conjugate $(\mu - \mu_c)$ related to the chemical potential. When Kadanoff says that "only the names of the variables change" for systems in the same universality class these are the changes to which he refers:

$$(M, h) \leftrightarrow (\Psi, (\mu - \mu_c))$$

The ferromagnetic/paramagnetic phase transition and the liquid/vapor phase transition are in the same relatively small class.

37.4.1   Stability

Thus there is a relationship between different phase transitions problems that leaves invariant various features of those transitions. Here is Kadanoff again:

Table 37.1 Scaling exponents for different transitions

Phase Transition	Value of β
Mean Field Theory	1/2
$d = 2$ Ising Model Ferromagnet	1/8
$d = 3$ Liquid/Vapor CO_2, Xe	0.35
$d = 3$ Superfluid Helium ^{4}He	0.359
$d = 3$ Ising Model Ferromagnet	0.315

The theorist can discuss this relation in the following way: He imagines that yet another field is inserted into the free energy. Call that other field λ and the operator which is its thermodynamic conjugate, U. Here λ represents a parameter in the Hamiltonian. Continuous variation from $\lambda = 0$ to $\lambda = 1$ might represent the change in the Hamiltonian which takes us from the Ising model to the Heisenberg model, or from Ni to Fe or from a nearest-neighbor interaction to a next nearest-neighbor interaction. Therefore, the discussion of λ and its thermodynamic conjugate U is in effect the discussion of the relationship among different phase transition problems. (Kadanoff, 1971, p. 105)

Consider the ferromagnetic transition. Inserting this new parameter into the Hamiltonian means that the free energy is now a function, not only of the magnetic field h and the field ϵ, but also of the field λ:

$$F = F(\epsilon, h, \lambda).$$

In differential form we now have

$$dF = < M > dh+ < \mathcal{H} > d\epsilon+ < U > d\lambda. \tag{37.4}$$

As an example, consider the Hamiltonian for a nearest- neighbor Ising ferromagnet:

$$\mathcal{H}_\Omega = -J_{n.n.} \sum_{<ij>} \sigma_i\sigma_j - h \sum_{i\in\Omega} \sigma_i, \tag{37.5}$$

where Ω is a region of spins on a d-dimensional lattice, $< ij >$ signals nearest-neighbor pairs on the lattice, $J_{n.n.}$ characterizes the nearest-neighbor spin-spin coupling, and h is a (uniform) magnetic field.

Next, let

$$\lambda = \frac{J_{n.n.n.}}{J_{n.n.}},$$

be the ratio of the next nearest-neighbor spin coupling strength to that of nearest- neighbor coupling strength. Consider the variation of λ from $\lambda = 0$ to $\lambda = 1$.[5] This takes us from the nearest-neighbor to next-nearest-neighbor coupling. Let ϵ and h be defined as follows:

$$\epsilon = \frac{(T - T_c(\lambda))}{T_c(\lambda)} \quad \text{and} \quad h = \frac{\mu_\beta H_z}{KT_c}.$$

For $\lambda = 0$ write the order parameter as function, m_0 of h and ϵ:

$$< M >= m_0(h, \epsilon).$$

Similarly introducing the distance r between spins, one can write the spin-spin correlation function g_0 as follows:

$$< \sigma_z(0)\sigma_z(r) >= g_0(h, \epsilon, r).$$

Then Kadanoff (1971, p. 106) asserts that "[u]niversality implies for $\lambda \neq 0$":

$$< M >= am_0\left(\bar{h}, \bar{\epsilon}\right), \quad < \sigma_z(0)\sigma_z(r) >= \left(ab/d^3\right) g_0\left(\bar{h}, \bar{\epsilon}, \bar{r}\right),$$

where

$$\bar{h} = bh, \quad \bar{\epsilon} = c\epsilon, \quad \bar{r} = dr.$$

This means that *the functional forms of $< M >$ and $< \sigma_z(0)\sigma_z(r) >$ do not change as λ varies.*

Universality, then

> ...implies that the basic thermodynamic functions and correlation functions only depend on λ via a trivial change of variables. The functional form is the same as at λ = 0. However, the variables in these functions are changed in that h, ϵ, and M are each multiplied by parameters a, b, c, d which depend upon λ. (Kadanoff, 1971, p. 105)

What does this result mean? It means that the Hamiltonians of different systems—as different as[6] nickel and iron, or as different as[7] CO_2, Xe, and ^{4}He—can be perturbed into one another without changing the nature of the phase transition problem. (Not all such transitions will preserve the nature of the phase transition problem; in fact, a perturbation from an Ising Hamiltonian to a Heisenberg Hamiltonian, will take us from one problem to another.) This, in turn, means that the scaling behavior of the order parameter will remain unchanged as the transformation (by varying the value of λ) between Hamiltonians is effected. In other words, many of the details that genuinely distinguish a lattice of nickel from that of iron (different interatomic strengths, etc.) are irrelevant for the scaling behavior. This is best understood as a *stability result*. The class of systems represented by their Hamiltonians between which such λ-transformations can take place without effecting the scaling behavior of the order parameter (whether it is M, or Ψ, or whatever) is called a "universality class." It is defined as that set of systems between which such (perturbative) transformations hold.

This stability under perturbation is *the key property* of universality. To explain how universality is possible, then, it requires one to explain two features:

1   Why are the phase transitions *stable* under perturbation of the microscopic details of the systems (as encoded in their Hamiltonians)?
2   Why are the universality classes *dependent* upon the symmetry of the order parameter and the dimensionality of the systems? (Recall Kadanoff's "hypothesis of universality" quoted at the beginning of this section. It should be clear from the way this question is answered below that the kind of dependence on symmetry and dimension is not causal. Nor is the explanation of that dependence causal in nature.)

37.5   Explaining Universality

In this section I outline, very briefly, how the RG allows for an explanation of the two key features of universal behavior just mentioned. This explanation has been called into question by a number of commentators and I will cite these references below.

Recall the ingredients of universality. Near criticality, the correlation length is enormous and there is a droplets-within-droplets structure that exhibits self-similarity (i.e, it behaves like a fractal). As a result there are no characteristic scales between the atomic/lattice spacing and the continuum. Importantly, as noted, for an infinite system the correlation length actually diverges to infinity.

Kadanoff recognized that one could exploit the droplets-within-droplets fractal structure to change a Hamiltonian representing a system into a related *effective* Hamiltonian by a kind of coarse-graining procedure. This is now known as the Kadanoff block spin method.

The idea is to group or block spins or molecules and replace them with some kind of average. (It almost doesn't matter what averaging or coarse-graining scheme is used. This fact is just another signature of the *stability* of phase transitions under perturbation.) In Figure 37.4 there are nine spins per block and the "averaging" rule is to let the majority rule: If more spins in a block are up-spins (down-spins), replace those nine spins with a single "block spin" that points up (down). Now we have a lattice of block spins that looks pretty much like the original lattice but the spacings between the block spins is greater. Next one spatially rescales so as to put the new spins on the same lattice as the original. Finally one changes the block spins so that they exhibit coupling strengths as similar to the original coupling as possible. (This part of the procedure is to a certain extent an art. There is no

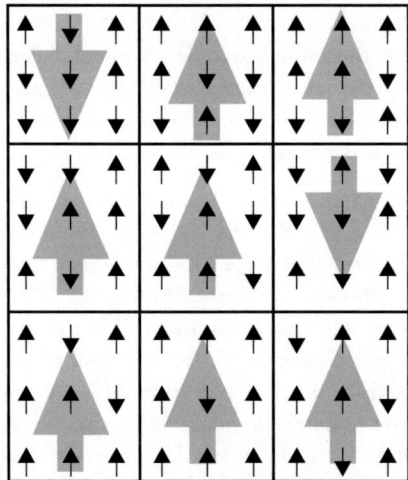

Figure 37.4 Blocking and averaging to yield a new (coarse-grained) effective system (Kadanoff, 2013, p. 172)

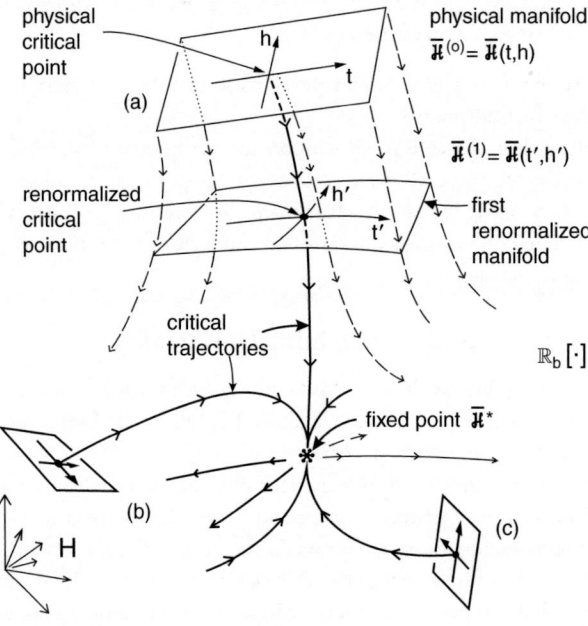

Figure 37.5 Fixed point and universality class (Fisher 1998, p. 673)

explicit recipe that one can follow.) We now have a new "renormalized" Hamiltonian corresponding to this *effective* system. Continued iteration of this procedure leads to a flow (the RG flow) on an abstract space of Hamiltonians. That is, it induces a dynamics that takes one Hamiltonian into another as the blocking procedure is repeated. See Figures 37.5 and 37.6.

One examines the dynamical flow on this abstract space and looks for potential fixed points. These are points which when acted upon by the transformation yield the same point. (If τ represents the transformation and p^* is a fixed point we will have $\tau(p^*) = p^*$.) A fixed point is a property of the transformation itself and all details of the systems that flow toward that fixed point have

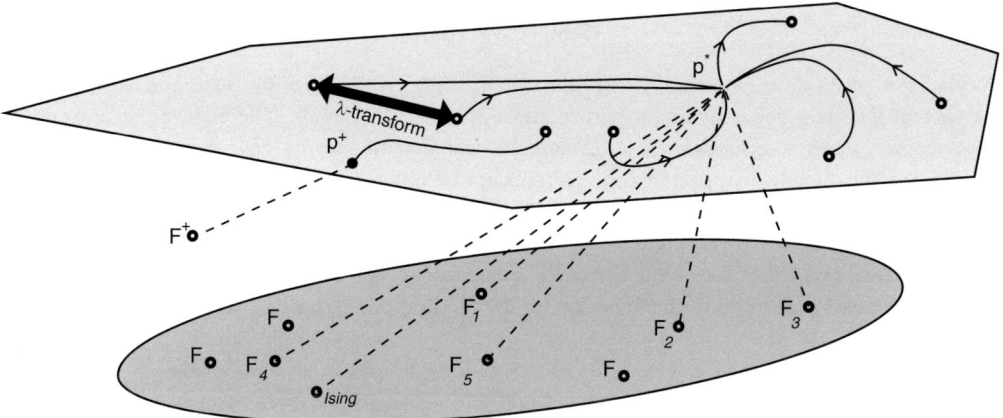

Figure 37.6 Fixed point, universality class, and λ-transformation

been eliminated. Those systems/models (points in the space) that flow to the *same* fixed point are in the same universality class—the universality class is delimited—and they will exhibit the same macro scaling behavior. (To put this another way: the universality class is the basin of attraction of the fixed point.) That macrobehavior, in particular, the determination of the scaling exponent β in equations (37.1) and (37.2) can be determined by an analysis of the transformation in the neighborhood of the fixed point.

Crucially, those systems that actually flow to a fixed point are at criticality. This means that they are infinite systems. As such, those systems are idealizations. But the infinite idealization is necessary if one is to locate the fixed points of the RG flow in the abstract space. This is because the correlation length must diverge to be able to infinitely iterate the RG transformation. Nevertheless, systems that are near criticality (real, large finite systems) will start off close to the critical systems and their behavior can be understood by examining the topography of the RG flow in the neighborhood of the fixed point. So, the RG along with the analysis of the topography in the neighborhood of the fixed point provides the explanation of near critical, real systems. It explains what is going on in the neighborhood of the critical point in the boiling region of Figure 37.1. (If it only explained the behavior of idealized infinite systems, it would not be such a big deal. Hardly worthy of a Nobel prize! Note also, contrary to some assertions in the literature, that the linearization analysis of the flow in the neighborhood of the fixed point is *not* part of the RG. It is a standard technique in dynamical systems theory for analyzing dynamics near attractor states.

So the fixed point delimits the class of systems that all exhibit similar behavior near criticality. It also, thereby, *justifies the existence* of the kind of perturbative stability Kadanoff describes in his discussion of what we've called the λ-transformations. See Figure 37.6. The justification of the existence of such transformations lies at the heart of the claim that minimal models like the Ising model need not accurately represent the actual features of systems in the universality class.[8] Finally, this analysis also demonstrates that the only important or relevant features (other than being near criticality) for this common behavior are the dimensionality of the system and the symmetry of the order parameter in the critical region.

37.6 Conclusion

As Kadanoff and others (specifically Griffiths (1970)) have emphasized, universal behavior reflects a stability of a certain behavior under perturbation of particular details. The concept of universality also depends upon a separation of scales and the fact that no nearby characteristic length (time) scale is present. Finally, it is part of the very concept of universality of critical phenomena that the (universality) class of systems depends upon the dimensionality of the system and the symmetry of the ordered state. Thus, any explanation for how the universality of critical phenomena can be possible *requires demonstrations that the systems are stable* under the appropriate (λ-)perturbation and that the only system features relevant to the behavior are the physical dimensionality and the symmetry of the ordered state.

The RG group can provide these two demonstrations. It does so by introducing a transformation on an abstract space of Hamiltonians corresponding to actual and possible systems and finding fixed points of that transformation. The Hamiltonians that flow to the same fixed point are exactly those between which the perturbative (λ) stability holds. They are also those critical systems that share dimensionality and the appropriate mesoscale symmetry.

Several philosophers have raised objections to this account of how the RG group can explain universality. For instance, Mark Lange (2015) and Alexander Reutlinger (2017) have argued that what universal behavior is explained by share common features among the members of the universality class. The account here aims to explain why such features (common spatial dimension and symmetry of the order state) are, in fact, the common features.[9]

A second class of objections by Mainwood (2006) and Franklin (2017) argue that the RG explanatory scheme I have just outlined is not the proper one. Rather, there is a field-theoretic understanding of the RG (rather than the one I described sometimes known as "real-space" RG) that is appropriate and explanatory.[10]

For further discussion of related issues in this volume see the chapters by Lina Jansson, David Wallace, and Porter Williams.

Acknowledgment

Thanks to Michael Miller and Porter Williams for helpful comments and discussions.

Notes

1 Michael Fisher (1998, p. 654) says it's fair to say Landau invented the order parameter.
2 See the discussion of intermediate asymptotics in Batterman (2002a). See also, Batterman (2018b).
3 I will have more to say about the role of infinite systems below.
4 Table 37.1 is from data presented in Kadanoff (1971, p. 102).
5 See also the paper by Robert Griffiths "Dependence of Critical Indices on a Parameter." Griffiths (1970)
6 See above quote from Kadanoff.
7 See Table 37.1 above.
8 See Batterman and Rice (2014).
9 See Batterman (2018a) for a response to this objection.
10 Again, see Batterman (2018a) for a response to this objection.

References

Batterman, R.W. (2002a). Asymptotics and the role of minimal models. *The British Journal for the Philosophy of Science*, 53: 21–38.
Batterman, R.W. (2002b). *The Devil in the Details: Asymptotic Reasoning in Explanation, Reduction, and Emergence.* Oxford Studies in Philosophy of Science. New York: Oxford University Press.
Batterman, R.W. (2018a). Universality and RG explanations. *Perspectives on Science*, 27(1): 26–47.

Batterman, R.W. (2018b). Multiscale modeling in inactive and active materials. Forthcoming in *Levels of Organization in the Biological Sciences*, Ed. Brooks, D., DiFrisco, J., and Wimsatt, W. (2021).

Batterman, R.W. and Rice, C. (2014). Minimal model explanations. *Philosophy of Science*, 81(3): 349–376.

Fisher, M.E. (1983). Scaling, universality and renormalization group theory. In F.J.W. Hahne (ed.), *Critical Phenomena*, Vol. 186 of *Lecture Notes in Physics*. Berlin. Summer School held at the University of Stellenbosch, South Africa; January 18–29, 1982, Berlin Springer-Verlag.

Fisher, M.E. (1998). Renormalization group theory: Its basis and formulation in statistical physics. *Reviews of Modern Physics*, 70(2): 653–681.

Franklin, A. (2018). On the renormalisation group explanation of universality. *Philosophy of Science*. DOI: 10.1086/696812.

Grifiths, R.B. (1970). Dependence of critical indices on a parameter. *Physical Review Letters*, 24(26): 1479–1482.

Guggenheim, E.A. (1945). The principle of corresponding states. *The Journal of Chemical Physics*, 13(7): 253–261.

Kadanoff, L.P. (1971). Critical behavior. Universality and scaling. In M.S. Green (ed.), *Proceedings of the International School of Physics "Enrico Fermi" Course LI*, vol. Course LI. New York: Italian Physical Society, Academic Press, pp. 100–117.

Kadanoff, L.P. (1976). Scaling, universality, and operator algebras. In C.Domb and M.S. Green (eds.), *Phase Transitions and Critical Phenomena*, vol. 5A. London: Academic Press.

Kadanoff, L.P. (2000). *Statistical Physics: Statics, Dynamics, and Renormalization*. Singapore: World Scientific.

Kadanoff, L.P. (2013). Theories of matter: Infinities and renormalization. In R.W. Batterman (ed.), *The Oxford Handbook of Philosophy of Physics*, chapter Four. New York: Oxford University Press, pp. 141–188.

Landau, L.D. (1965). On the theory of phase transistions. In D. Ter Haar (ed.), *Collected Papers of L. D. Landau*, chapter 29. New York: Gordon and Breach, Science Publishers, pp. 193–216.

Lange, M. (2015). On "minimal model explanations": A reply to batterman and rice. *Philosophy of Science*, 82: 292–305.

Mainwood, P. (2006). *Is More Different? Emergent Properties in Physics*. PhD thesis, Merton College, University of Oxford.

Reutlinger, A. (2017). Do renormalization group explanations conform to the commonality strategy? *Journal of General Philosophy of Science*, 48: 143–150.

Further Reading from the Editors

Alongside Batterman, the most prominent defender of the irreducibility of renormalization group explanations is Margaret Morrison: see especially M. Morrison, "Emergent physics and micro-ontology", *Philosophy of Science* 79(1) (2012): 141–166 and M. Morrison, "Complex systems and renormalization group explanations," *Philosophy of Science* 81(5) (2014): 1144–1156. J. D. Norton, "Approximation and idealization: Why the difference matters", *Philosophy of Science* 79(2) (2012): 207–232 draws a distinction between approximation and idealization and uses it to argue that infinite idealizations are dispensable in renormalization group explanations. A. Franklin, *Philosophy of Science* 86(5) (2019): 1295–1306 provides further arguments against the view that renormalization group explanations are irreducible. For discussion of related issues in this volume, see Chapter 34 by Lina Jansson, Chapter 18 by David Wallace, and Chapter 19 by Porter Williams.

38

CHANCE AND DETERMINISM

Nina Emery

On a very intuitive way of thinking, if it is already determined that some event will happen, then there is no non-trivial chance (no chance between 0 and 1) of it failing to happen, and if it is already determined that some event will not happen, then there is no non-trivial chance of it happening.[1] On this way of thinking, it does not make sense to claim both that it is already determined that Always Dreaming will win this year's Kentucky Derby *and* that the chance of Classic Empire winning instead is 1/2.

Nonetheless, it is becoming increasingly common for philosophers to claim that there are non-trivial chances in worlds where the fundamental dynamical laws are deterministic.[2] In such worlds, for any event e, at any time t, it is already (at t) either determined that e will happen or determined that e will not happen. But, according to these philosophers, there are at least some cases in such worlds where the chance (at t) of e happening is between 0 and 1. Call the chances that are supposed to exist in such worlds *deterministic chances* and the philosophers who think that they do in fact exist *compatibilists about chance and determinism*, or just *compatibilists*.

This entry surveys some arguments that motivate compatibilists (Section 38.1), with a focus on arguments that begin from the various roles that probabilities play in deterministic scientific theories – scientific theories with deterministic fundamental laws. I then discuss the extent to which deterministic chances, as established by such arguments, are compatible with existing metaphysical analyses of chance and with various pre-theoretic platitudes about the chance concept (Section 38.2).

Before we begin, a note about terminology. Most discussions of chance begin by claiming that chances are objective probabilities. But what is meant by "objective" in this definition is not always clear. In what follows, I will assume that chances are objective probabilities in the sense that they are wholly determined by the world as it is independently of us and our epistemic position within that world. Chances, in other words, do not depend on the sorts of beliefs and evidence that we have about the world, the types of creatures that we are, and the ways in which we reason. (Note that this understanding of "objective" may need to be modified insofar as one is interested in objective probabilities of various doxastic states or epistemic positions. I set such worries aside.)

It is widely accepted that whatever chances are, they are to be contrasted with *individual credences* (the degrees of belief of some particular agent). And indeed the above way of understanding "objective" captures this result. But – and this is more controversial – it also follows from the above way of understanding "objective" that *rational credences* (the degrees of belief some agent *should* have) also do not count as genuine chances. While the degrees of belief that an agent should have will of course depend on what the world is like, they will also depend, at least in part, on the evidence that agent has, the sort of creature she is, and the kind of reasoning she is capable of engaging in. It will depend, in other words, on the epistemic position she occupies. (A more difficult question is whether *evidential probabilities* – the degree to which the evidence available from a certain epistemic position supports some proposition – are genuine chances. It will depend on what one means by evidence and whether the evidence available from a certain epistemic position depends on the nature of the agent that occupies it.)

In what follows, I will assume that in order to establish compatibilism about chance and determinism one must do more than establish that there are rational credences or evidential probabilities in worlds where the fundamental laws are deterministic. Compatibilism requires that we establish that there are probabilities that are wholly objective, in the sense described above, in such worlds. This assumption will play an important role in the discussion of arguments for compatibilism in the next section.

38.1   The Case for Deterministic Chance

Why be a compatibilist? The type of argument I will focus on is this:

1   Non-trivial probabilities play role R in theory T (according to which the fundamental laws are deterministic).
2   In order to play role R, the probabilities in question must be objective.

It follows from these two premises that there are chances in (at least some) worlds where the fundamental laws are deterministic (namely those worlds in which theory T is true).

The two deterministic theories that are most frequently discussed by compatibilists are Boltzmannian statistical mechanics and Bohmian mechanics.[3] I focus on versions of the above schema involving these two theories in Sections 38.1.1 and 38.1.2 below. In Section 38.1.3, I will briefly gesture toward a somewhat different, but importantly similar sort of argument for compatibilism.

38.1.1   *Deterministic Chance in Boltzmannian Statistical Mechanics*

According to classical statistical mechanics, the fundamental laws are Newtonian and thus deterministic.[4] Given a complete specification of the initial state of a statistical mechanical system (i.e. given a complete specification of the position and momentum of each particle in the system at the initial time), those laws determine the behavior of that system at all other times.

Boltzmann argued for different approaches to statistical mechanics at different times in his career (see Frigg and Werndl, this volume, Chapter 27), and interpretations of each of the various approaches differ. As is relatively standard in the literature on deterministic chance, I will focus specifically on the Boltzmannian approach as spelled out in detail in Albert (2000). This version of the Boltzmannian approach to statistical mechanics[5] adds to these fundamental laws a probabilistic postulate. For R an arbitrary region of phase space and r an arbitrary subregion of R, the rule says:

> *The Boltzmannian statistical postulate.* The probability that a system starts off in r, given that it starts off in R, is just the measure (on the Lebesgue measure) of the points within R that are also within r.

Here the Lebesgue measure is the measure that is uniform over phase space with respect to position and momentum. (As discussed in Uffink, 2017, 4.1, a measure that is uniform over phase space with respect to energy generates a probabilistic rule that makes incorrect predictions.)

The Boltzmannian statistical postulate allows us to predict the microstate of a statistical mechanical system on the basis of its macrostate. Consider a gas enclosed in a box. There are ways of arranging the particles of the gas such that the gas is concentrated in one corner of the box. But – loosely speaking – there are far more ways of arranging the particles such that the gas is roughly evenly distributed throughout the box. (More carefully, if B is the region of phase space that corresponds to the initial macrostate of the box, the measure (on the Lebesgue measure) of the points within B that correspond to the gas being concentrated in the corner is very small compared to the measure of the points within B that correspond to the gas being roughly evenly distributed.) So the Boltzmannian statistical postulate tells us that it is very likely that the initial state of the gas is such that it is roughly evenly distributed throughout the box.

The Boltzmannian statistical postulate, combined with the fundamental laws, also allows us to predict the behavior of a statistical mechanical system over time, given only a specification of its macrostate. For R1 and R2 arbitrary regions of phase space, it follows from the Boltzmannian statistical postulate that the probability that a system that starts off in R1 will evolve into R2 is just the measure (on the standard Lebesgue measure) of the points in R1 that evolve into R2 according to the fundamental dynamical laws. Let H be the region of phase space that corresponds to the initial macrostate of a gas that starts off confined to one half of an empty box. There are points within H that lead to the gas remaining concentrated in that half of the box or even contracting to occupy a smaller volume. But the measure of such points is tiny compared to the measure of the points within H that lead to the gas expanding to occupy the available volume. It follows from the Boltzmannian statistical postulate that it is extremely likely that the gas will expand to occupy the available volume.

Of particular note is the fact that the Boltzmannian statistical postulate allows us to predict the data that was previously predicted by the second law of thermodynamics. It allows us to predict the fact that we rarely observe *anti-entropic behavior* – behavior in which a closed (or nearly-closed) system evolves from a higher entropy state into a lower entropy state. Within any non-gerrymandered region of phase space – including those regions that correspond to the sorts of largely isolated, macrophysical systems that we interact with on an everyday basis – the measure (on the Lebesgue measure) of the points within that region that will evolve into regions that correspond to an increase in the system's entropy is tiny. It follows from this fact, combined with the Boltzmannian statistical postulate, that it is extremely unlikely that we will observe anti-entropic behavior.

It is tempting to think that what has been said so far about the role that the Boltzmannian statistical postulate plays in generating predictions is sufficient to support an argument for compatibilism along the following lines:

3   Non-trivial probabilities determine what we ought to expect to happen in Boltzmannian statistical mechanics.
4   In order to determine what we ought to expect to happen, the probabilities in question must be objective.

But one should tread carefully here. Although premise 3 is uncontroversial, the plausibility of premise 4 depends crucially on what is meant by "objective." In particular, it is not clear that in order to determine what we ought to expect to happen, the probabilities need to be wholly objective in the sense described in the introduction. At least on the face of it, rational degrees of belief are good candidates for determining what we ought to expect to happen. And, as was argued in the introduction, rational degrees of belief are not wholly objective probabilities.

For this reason, it is important to recognize that most philosophers who argue for compatibilism on the basis of the role that probabilities play in Boltzmannian statistical mechanics emphasize that the Boltzmannian statistical postulate does not only play an important role in generating predictions It also plays a crucial explanatory role.[6]

Consider again the fact that we rarely observe anti-entropic behavior. Insofar as one considers just the fundamental dynamical laws of classical statistical mechanics, the absence of anti-entropic behavior is surprising. Those laws, after all, are *time-reversal invariant*—if it is nomologically possible for a system to evolve from state S1 to state S2, it is also possible for the system to evolve from S2 to S1. So if it is nomologically possible for a statistical mechanical system to evolve from a lower entropy state to a higher entropy state – as indeed it is; we observe such behavior all the time – then it is also possible for such a system to evolve from a higher entropy state into a lower one. So why don't we ever observe the latter sort of behavior? The Boltzmannian statistical postulate gives us a straightforward answer to this question: we don't ever observe such behavior because, although it is possible, it is extremely unlikely.

This further explanatory role gives rise to an argument for compatibilism based on the following premises.

5   Non-trivial probabilities explain relative frequencies in Boltzmannian statistical mechanics.
6   In order to explain these relative frequencies, the probabilities in question must be objective.

This argument, at first glance at least, is substantially stronger than an argument based on premises 3 and 4. First, notice that premise 6 is uncontroversial. It is one of the most basic assumptions that we make about the world that nothing about the evidence that we as inquirers have about the world, or about the types of creatures that we are and the ways in which we reason, plays a role in explaining the behavior of statistical mechanical systems. Insofar as there are probabilistic explanations of statistical mechanical phenomena, therefore, the probabilities involved must be objective.

As for premise 5, some philosophers who are resistant to the idea of deterministic chances point out that there are alternative explanations available for the explananda in question. Perhaps most obviously, one can explain the lack of anti-entropic behavior in any particular system simply by pointing to the exact microphysical state that system started in, combined with the fundamental laws.[7] But it is clearly part of standard scientific practice both historically and today to use probabilistic explanations of the sort described by premise 5.[8] If nothing else, that ought to make acceptance of premise 5 the default view. (It is perhaps also worth noting that using probabilities to explain relative frequencies in the way suggested by premise 5 is utterly prosaic. Think, for instance, of how natural it is to answer the question, "why don't we ever see a fair coin land heads 100 times in a row?" by pointing out that it is extremely unlikely for that sequence of flips to occur.)

The two arguments set out above are not the only ways of constructing an argument from Boltzmannian statistical mechanics that appeals to some version of premises 1 and 2. Other reasons for thinking that the probabilities generated by the Boltzmannian statistical postulate are genuinely objective may include the role that probabilities play: (i) in determining the truth (or assertibility) conditions of counterfactuals,[9] (ii) in underwriting various laws,[10] (iii) in the confirmation of a theory,[11] and so on. I leave it to the reader to construct and evaluate arguments for compatibilism based on these further roles for Boltzmannian probabilities.

38.1.2 *Deterministic Chance in Bohmian Mechanics*

According to Bohmian mechanics, the fundamental laws are Schrodinger's equation and the guidance equation. Taken together, these laws are deterministic. Given a complete specification of the initial state of a quantum system (i.e., given a complete specification of the initial position of each particle in the system and the initial wavefunction of the system), these laws allow us to predict the behavior of that system at all other times.

Standard approaches to Bohmian mechanics[12] add a probabilistic postulate to Schrodinger's equation and the guidance equation. For C, an arbitrary region of configuration space, and c an arbitrary subregion of C, the postulate says:

> *Bohmian statistical postulate.* The probability that a system starts off in c, given that it starts off in C, is just the measure (on the standard quantum measure) of the points within C that are also within c.

Where the standard quantum measure for a system that has initial wavefunction ψ is given by $|\psi|^2$.

The Bohmian statistical postulate allows us to predict the exact configuration of the particles in a quantum system based on less specific information about the configuration of those particles and the wavefunction of that system. Consider, for instance, a single particle that starts off located somewhere within region R and that has an initial wavefunction that is symmetric over the *x*-axis, which bisects R. It follows from the Bohmian statistical postulate that the probability that the particle starts off located above the *x*-axis is 1/2. If the initial wavefunction had instead been such that its amplitude was much higher above the *x*-axis than below the *x*-axis, then the probability that the particle started off located above the *x*-axis would have been very high.

The Bohmian statistical postulate also allows us to predict and explain the behavior of quantum mechanical systems based on a less than complete specification of the system's initial state. Let C1 and C2 be arbitrary regions of configuration space. It follows from the Bohmian statistical postulate that the probability that a system that starts in C1 at t1 will evolve into C2 at t2 is just the measure (on the standard quantum measure at t1) of the points in C1 that evolve into C2 according to the fundamental dynamical laws.

But of particular importance is the fact that the Bohmian statistical postulate allows us to predict the data that in the standard quantum mechanical formalism is predicted by Born's Rule.

> *Born's Rule.* The probability that a system with wave function ψ at t will be found in configuration c if we perform a measurement on it at t is given by $|\psi(c)|^2$.

This is because the fundamental dynamical laws of Bohmian mechanics are such that if the probability that a system will be found in some configuration c at some time t_0 is $\left|\psi_{t_0}(c)\right|^2$, then for all t, the probability that a system will be found in c is $|\psi_t(c)|^2$. It follows from this fact, combined with the Bohmian statistical postulate, that the relative frequencies of the outcomes of various experiments will match the probabilities given by Born's Rule.

As with the probabilities in Boltzmannian's rule, it is tempting to try to construct an argument for compatibilism just based on the role that the Bohmian statistical postulate play in generating these sorts of predictions. Such an argument would look like this:

7   According to Bohmian mechanics, non-trivial probabilities determine what we ought to expect to happen.
8   In order to determine what we ought to expect to happen, the probabilities in question must be objective.

But once again, one needs to tread carefully. Although premise 7 is uncontroversial, insofar as we adopt the understanding of "objective" outlined in the introduction, there is no reason to endorse premise 8. Rational credences can determine what we ought to expect, and on that understanding of "objective," rational credences are not genuinely objective probabilities.

For this reason, it is important that one also consider the role that the Bohmian statistical postulate plays in explaining the behavior of quantum mechanical systems. Insofar as one just considers the fundamental laws of Bohmian mechanics, the utility of Born's Rule is surprising. There are initial arrangements of particles that, when combined with the fundamental laws, lead to configurations to which Born's Rule assigns very low probability. But we rarely see such configurations. Why not? The Bohmian statistical postulate provides us with a straightforward answer to this question: we rarely see such configurations because, although they are possible, they are unlikely.

This further explanatory role for the probabilities in the Bohmian statistical postulate gives rise to the following argument for compatibilism:

9   According to Bohmian mechanics, non-trivial probabilities explain the behavior of quantum mechanical systems.

10   In order to explain the behavior of quantum mechanical systems, the probabilities in question must be objective.

Here premise 10 is uncontroversial. As for premise 9, it is not entirely clear that it is standard scientific practice to use probabilities to explain frequencies in Bohmian mechanics, if only because few practicing physicists endorse Bohmian mechanics and few physics texts discuss the theory in detail. Nonetheless, this premise presumably inherits some plausibility from the very same arguments that can be mustered to support premise 5 above. Insofar as this sort of probabilistic explanation is legitimate – and preferable to alternatives – in statistical mechanics, presumably the same is true in Bohmian mechanics.

As was discussed with respect to Boltzmannian probabilities, there are also other roles that the probabilities in the Bohmian statistical postulate are supposed to play and that are plausibly such that any probabilities that play that role must be wholly objective. These include the roles that Bohmian probabilities may play in determining the truth (or assertibility) conditions of counterfactuals, in confirming the theory, and so on. But once again, I leave it to the reader to investigate those further arguments in detail.

38.1.3   *Arguments from the Irrelevance of the Fundamental Laws*

A somewhat different type of argument for compatibilism that is worth discussing here starts from the observation that non-trivial probabilities play a certain role *whether or not* the underlying laws are deterministic, and then continues by claiming that in order to play that role the relevant probabilities must be objective. This sort of argument is sometimes put forward with respect to various roles that probabilities play in evolutionary theory (Sober, 2010), but it also comes up with respect to the roles that probabilities play in more prosaic contexts, like various kinds of gambling set-ups (Handfield and Wilson, 2014).

I mention these arguments here mainly to point out the ways in which the very same considerations that were discussed in Sections 38.1.1 and 38.1.2 will bear on these arguments. Along these lines, it is worth noting that the most obvious ways to construct such an argument will appeal to the roles that such probabilities play in prediction or explanation. So it seems plausible that the very same sorts of motivations and concerns that motivate the introduction of deterministic chance in Boltzmannian statistical mechanics and Bohmian mechanics will be relevant here.

In addition, notice that in the case of gambling set-ups, at least, the relevant deterministic chances can also be derived by placing a relative natural measure over the space of possible initial states of the system and interpreting that measure as a probability measure. For instance, for a fair coin, initial conditions that lead to heads and initial conditions that lead to tails are relatively evenly distributed throughout the state space and within any non-gerrymandered region of that space, roughly half of the possible initial conditions will be such that they lead to the coin landing heads.[13] It follows that insofar as you put a relatively natural measure over the space of possible initial conditions and interpret that measure as a probability measure, the probability of a fair coin landing heads will be 1/2. So it seems plausible that any metaphysical analysis of deterministic chance that is able to handle the chances in Boltzmannian statistical mechanics and Bohmian mechanics will also be able to handle chances insofar as they arise in various gambling set-ups.

38.2   Two Worries about Deterministic Chance

Now that we have a sense of the arguments that motivate compatibilism, we can address two common worries about deterministic chance. The first is a worry about whether we can understand deterministic chance in terms of any of the familiar metaphysical analyses of chance. The second is whether deterministic chance violates some sort of platitude about the chance concept.

38.2.1 *Metaphysical Analyses of Deterministic Chance*

Can we give an analysis of deterministic chance in terms that are familiar from the debate over the metaphysics of chance in general? If not, advocates of deterministic chance may be required to give either a novel or a disjunctive account of chance (according to which deterministic chances are distinctively different sorts of entities from other chances).[14] Perhaps an account of one of those two types is ultimately necessary, but at the very least it would be a significant cost. (Not everyone cares about these costs. For instance, Strevens's (2011) microconstant chance is an analysis of deterministic chance that makes such chances distinct from indeterministic chance.) Luckily for compatibilists, though, it looks as though deterministic chances as discussed in Section 38.1 can be accommodated by several of the leading contenders for metaphysical theories of chance.

Perhaps most obviously, deterministic chances can be easily accommodated insofar as one adopts some kind of frequency analyses of chance.[15] In particular, consider an *actual frequency analysis* of chance, according to which the chance of event of type E is just the actual relative frequency with which events of type E actually occur. In order to give this sort of analysis of deterministic chances as they arise in, for instance, Boltzmannian statistical mechanics one need merely establish that the actual relative frequency with which the events described in the Boltzmannian statistical postulate is given by the Lebesgue measure over the relevant region of phase space.

The actual frequency analysis of chance, of course, faces many objections.[16] One worry that might seem especially pressing in the present context is that such an analysis robs chances of their explanatory role. Insofar as chances just are relative frequencies, they cannot explain those relative frequencies; after all, nothing can explain itself. This seems especially worrisome since the explanatory role played by the probabilities in, e.g., the Boltzmannian statistical postulate, was a key reason for thinking those probabilities were genuine chances.

It is not clear, however, that this worry about the explanatory power of actual frequentist accounts of chance in general is much of a worry for actual frequentist accounts of deterministic chance. It depends, in particular, on what sorts of relative frequencies we are trying to explain. Insofar as one is a frequentist one cannot, for instance, use the Boltzmannian statistical postulate to explain the relative frequency with which systems that start off in some region of phase space R also start off in some subregion r. To do so would be to use one and the same fact to explain itself. But one can use the Boltzmannian statistical postulate to explain the relative frequency with which systems that start off in R1 at t1 evolve into R2 at t2. In that case, the explanandum and the explanans (which will appeal to the relative frequency within which systems that start off R1 start off in some particular subregion of R1) are distinct.

A different sort of worry that arises for an actual frequentist analysis of deterministic chance is that such an analysis seems unsuited to theories like Boltzmannian statistical mechanics or Bohmian mechanics insofar as those theories are supposed to describe the evolution of the Universe as a whole. In order for an actual frequentist to make sense of the Boltzmannian and Bohmian statistical postulate as applied to the Universe as a whole, they would need to make sense of a probability distribution over the initial state of the Universe in terms of the actual relative frequency with which the Universe starts off in various initial conditions. But on a natural way of thinking, there is just a single Universe and the Universe has a single initial state. So the actual relative frequency with which the Universe starts off in any particular initial condition is either 0 or 1.

In response to this worry, frequentists have two options. First, they can focus solely on relatively closed sub-systems of the Universe, like everyday statistical mechanical systems and gambling devices. Insofar as one restricts one's interest to such systems, one need not make sense of the actual relative frequency of various initial conditions of the Universe, only of the actual relative frequency of various initial conditions of these isolated systems. Second, they can insist that what is meant by "the Universe" in theories that describe the evolution of the Universe as a whole is not everything that there is (and was and will be). Perhaps, for instance, the Universe is just one of many closed systems, as in some contemporary multi-verse or "bubble Universe" views. Such a view would leave room for the relative frequency with which Universes start in a certain kind of initial condition to be between 0 and 1.

A second prominent metaphysical analysis of chance is the *best systems analysis,* according to which chances are those probability distributions that appear in the best way of systematizing the occurrent (non-modal, non-dispositional, non-casual) facts about the world.[17] What exactly makes one way of systematizing the world better than another varies, depending on which version of the best systems analysis you consider. But in general, the introduction of various probability distributions potentially provides a significant advantage in terms of informativeness with little cost in terms of complexity. Specifically adding the right probability distribution can provide a significant advantage in terms of what Lewis (1994) called fit – the extent to which the theory assigns high probability to events that do happen and low probability to events that do not.

Perhaps the best-known analysis along these lines is due to Albert and Loewer.[18] According to their view – which they call the *mentaculus* – probabilistic postulates like the Boltzmannian or Bohmian statistical postulates derive from a single probability distribution over the possible initial states of the Universe. In Boltzmannian statistical mechanics, for instance, the probability distribution is given by Lebesgue measure over the region of the initial state space of the Universe that corresponds to the Universe having very low entropy. They then argue that the best systemization of the occurrent facts about the world includes both the fundamental dynamical laws and this initial probability distribution. It follows, given a best systems analysis, that the probability distribution in question is a genuine chance distribution.[19]

The best systems analysis has no trouble making sense of a probability distribution over the initial state of the Universe. It does, however, face worries about the explanatory power of chances. On Albert and Loewer's account, for instance, the relative frequency of, for instance, anti-entropic behavior plays a role in making the probability distribution over the initial state of the Universe a part of the best systematization. Is it legitimate to also claim that the probability distribution over the initial state explains anti-entropic behavior? This is a difficult philosophical question which has seen significant recent attention in the context of the best systems analysis of laws.[20]

Finally, consider propensity analyses of chance, according to which the chance of a system in S1 at t1 evolving into S2 at t2 is given by the propensity (or tendency, or causal disposition) of systems in S1 evolving into S2 over the specified time interval.[21] Is it possible to give a propensity analysis of deterministic chance?

At first glance, it seems not. After all, propensities are diachronic and the chances that arise in the Boltzmannian and Bohmian statistical postulates are synchronic chances. But there is at least one clever way of understanding deterministic chances as diachronic, and thus leaving room for a propensity analysis – an approach found in Demarest (2016). Demarest agrees with Albert and Loewer about the structure of deterministic chance – all deterministic chances ultimately derive from a probability distribution over the possible initial state of the Universe. But she disagrees with Albert and Loewer's analysis of that initial probability distribution as a chance distribution deriving from the best system. Instead, Demarest thinks that the initial probability distribution is determined by a single chance event which brought the initial state into existence. This view is straightforwardly amenable to a propensity analysis of chance.

It should be clear, then, that there are options available for understanding deterministic chance that are in keeping with several of the leading metaphysical analyses of chance. These various analyses of deterministic chance may come with various costs, and substantive further philosophical work is required in order to establish their ultimate viability. But there seems little reason to think that advocates of deterministic chance will be required to give a novel or disjunctive account of chance.

38.2.2   *Deterministic Chance and the Chance Platitudes*

Some philosophers claim that deterministic chance violates important platitudes about the chance concept. I don't have space to go into all of the arguments of this form in detail, but here is one example that illustrates the sort of considerations at stake.

Consider the fact that there appears to be an important connection between chance and possibility along the following lines: if there is a non-trivial chance of something happening, it must be possible for that thing to happen and possible for it not to happen. If there is a 1/2 chance of Classic Empire winning the Kentucky Derby, for instance, it must be possible for Classic Empire to win and possible for him not to win.

Here is one way of capturing that platitude:

> *The chance-possibility platitude – incompatibilist's version.* If the chance, at world w, at time t, of proposition p is greater than zero, then there exists a world w' such that (i) w' matches w in laws, (ii) w' and w have the same microphysical history up until time t, and (iii) p is true at w'.[22]

It follows straightforwardly from this version of the chance-possibility platitude that there are no non-trivial chances in worlds where the fundamental laws are deterministic. If the chance of some proposition p is non-trivial, then the chance of p is greater than zero and the chance of $\sim$p is greater than zero. But if the laws of world w are deterministic then the microphysical history up until time t and the laws either determine that p is true or determine that p is not true. If they determine that p is true, then every world w' that matches w in the laws and in its microphysical history is a world in which p is true, and it follows that the chance of $\sim$p is not greater than zero. If they determine that p is not true, then every world w' that matches w in the law and in its microphysical history is a world in which $\sim$p is true, and it follows the chance of p is not greater than zero. Either way, the chance of p is not non-trivial.

The important thing to notice, however, is that the incompatibilist's version of the chance-possibility platitude is not the only version. The compatibilist who thinks that there are chances in Boltzmannian statistical mechanics, for instance, cannot adopt the incompatibilist's version of the chance-possibility platitude. But she can adopt the following version:

> *The chance-possibility platitude – BSM compatibilist's version.* If the chance, at world w, at time t, of proposition p is greater than zero, then there exists a world w' such that (i) w' matches w in laws, (ii) w' and w have the same macrophysical history up until time t, and (iii) p is true at w'.

Similarly, the compatibilist who thinks that there are chances in Bohmian mechanics cannot adopt the incompatibilist's version of the chance-possibility platitude. But she can adopt the following version:

> *The chance-possibility platitude – BM compatibilist's version.* If the chance, at world w, at time t, of proposition p is greater than zero, then there exists a world w' such that (i) w' matches w in laws, (ii) w' and w have the same wavefunction up until time t, and (iii) p is true at w'.

The challenge for anyone who wants to insist that deterministic chances are not genuine chances is to argue for the incompatibilist's version of the chance-possibility platitude over these alternative versions. At the very least, though, there is room for the compatibilist to claim that they have retained at least some aspect of the platitude in question.[23]

Similar points can be made regarding the supposed incompatibility of deterministic chance and the standard way of thinking about the connection between chance and credence (the *principal principle*)[24] and regarding the standard way of thinking about the connection between chance and laws.[25]

Here is a related worry that, while rarely made explicit in the literature, may be playing a significant role in motivating incompatibilism. It is very natural to think that the past is not chancy. That is to say, it is very natural to think that chance is always time-indexed, and for an arbitrary proposition p, if t is in the past, then the chance at t of p is either 0 or 1. But one consequence of giving up various pre-theoretic platitudes about chance, like the incompatibilist's version of the chance-possibility platitude, is that you leave open the possibility of non-trivial chances of events in the past. Consider, for instance, BSM compatibilist's version of the chance-possibility platitude. This version leaves open the possibility that there will be a non-trivial chance of microphysical events in the past.

Insofar as this is going to be an objection to deterministic chance more must be done to argue that compatibilists should in fact accept non-trivial chances of past events, not just that they might have to do so. But it is also worth pointing out that on a very standard view about the philosophy of time, and in particular on the sort of view that appears to fit best with contemporary physics, there are no objective or fundamental differences between times that are past, and those that are not. So anyone who builds their defense of incompatibilism on the claim that the past cannot be chancy (while the future may), has their work cut out for them not only in terms of spelling out further details regarding the nature and commitments of deterministic chance but also in defending their view against the standard approach to the philosophy of time.[26]

Notes

1 Lewis (1980, 1986); Popper (1992); Hájek (1996); Schaffer (2007).
2 Loewer (2001), Ismael (2009), Hoefer (2007), Glynn (2010), Handfield and Wilson (2014), Emery (2015), Frigg and Hoefer (2015).
3 For a more detailed treatment of Boltzmannian statistical mechanics see Frigg and Werndl (this volume, Chapter 27) and Shahvisi (this volume, Chapter 30), and for a more detailed treatment of Bohmian mechanics see Tumulka (this volume, Chapter 17).
4 *Pace* the concerns in Earman (1986) and Norton (2008). For further discussion see Suárez (this volume, Chapter 46).
5 For a discussion of probability in other versions of classical statistical mechanics see Sklar (1995).
6 See for instance Albert (2000), Goldstein (2001), Loewer (2001), Meacham (2005), North (2010), Emery (2015).
7 See Schaffer (2007) and Frigg (2008, p. 680) for explicit endorsements of this alternative strategy. A different way of providing an alternative explanation is to explain the behavior directly in terms of facts about the measure as in Maudlin's (2007a) typicality account.
8 This assertion is found in Albert (2000), Loewer (2001), Meacham (2005), North (2010), and elsewhere. See Strevens (2000) for an argument that the explanatory power of the probabilities found in, e.g. Boltzmann's rule, is required in order to explain the adoption of statistical mechanics over rival theories in the late 19th century.
9 See Albert (2000, 2012); Loewer (2007); Emery (2017).
10 See Loewer (2004) and Glynn (2010).
11 Ismael (2009) emphasizes the role that probabilities play in confirming deterministic theories.
12 See for instance Dürr et al. (1991), Albert (1992), and Tumulka (this volume, Chapter 17).
13 As discussed in Suárez (this volume, Chapter 46), this observation was originally due to Poincaré.
14 As an example of a metaphysical analysis of deterministic chance which is not obviously amenable to also handling indeterministic chance, and thus gives rise to a disjunctive account of chance as a whole, see Strevens (1999, 2013). Further discussion can also be found in Suárez (this volume, Chapter 46).
15 For more on frequency interpretations of chance see Suárez (this volume, Chapter 46).
16 See Hajek (1996) and Suárez (this volume, Chapter 46).
17 See Lewis (1994).
18 See Albert (2000, 2012); Loewer (2004, 2012).
19 Important criticisms of this approach include those found in Elga (2001), Frigg (2008), Frisch (2010), and Meacham (2010). Other versions of a BSA analysis of deterministic chance are found in Cohen and Callender (2009), Hoefer (2007) and Frigg and Hoefer (2015). This sort of view is also discussed in Maudlin (2007a, section 2).

20 See Maudlin (2007b), Loewer (2012), and Lange (2013).
21 For more on propensity analyses of chance see Suárez (this volume, Chapter 46).
22 This principle is called the realization principle in Schaffer (2007). It's a stronger version of the basic chance principle found in Bigelow et al. (1993).
23 The debate over which version of the chance-possibility platitude is the correct version is discussed in more detail in Emery (2015).
24 This sort of argument against deterministic chance clearly motivated Lewis (1986) and is explicit in Schaffer (2007) and Lyon (2011). Responses to it can be found in Meacham (2005), Hoefer (2007), and Handfield and Wilson (2014).
25 See Glynn (2010).
26 Thanks to Heather Demarest, Matt Duncan, and Al Wilson for helpful comments.

References

Albert, D.Z. (1992). *Quantum Mechanics and Experience*. Cambridge, MA: Harvard University Press.
Albert, D.Z. (2000). *Time and Chance*. Cambridge, MA: Harvard University Press.
Albert, D.Z. (2012). Physics and chance. In Y. Ben-Menahem and M. Hemmo (eds.), *Probability in Physics*. Berlin: Springer, pp. 17–40.
Bigelow, J., Collins, J. and Pargetter, R. (1993) The Big Bad Bug: What Are the Humean's Chances? *British Journal for the Philosophy of Science*, 44: 443–462.
Cohen, J. and Callender, C. (2009). A better best system account of lawhood. *Philosophical Studies*, 145(1): 1–34.
Demarest, H. (2016). The universe had one chance. *Philosophy of Science*, 83(2): 248–264.
Dürr, D., Goldstein, S., and Zanghi, N. (1992). Quantum Equilibrium and the Origin of Absolute Uncertainty. *Journal of Statistical Physics*, 67: 843–907.
Earman, J. (1986). *A Primer on Determinism*. Dordrecht: D. Reidel.
Elga, A. (2001). Statistical mechanics and the asymmetry of counterfactual dependence. *Philosophy of Science*, 68(S3): S313–S324.
Emery, N. (2015). Chance, possibility, and explanation. *The British Journal for the Philosophy of Science*, 66(1): 95–120.
Emery, N. (2017). The metaphysical consequences of counterfactual skepticism. *Philosophy and Phenomenological Research*, 94(2): 399–432.
Frigg, R. (2008). A field guide to recent work on the foundations of statistical mechanics. In D. Rickles (ed.), *The Ashgate Companion to Contemporary Philosophy of Physics*. London: Ashgate, pp. 991–996.
Frigg, R. and Hoefer, C. (2015). The best Humean system for statistical mechanics. *Erkenntnis*, 80(3): 551–574.
Frisch, M. (2010). Does a low-entropy constraint prevent us from influencing the past? In G. Ernst and A. Hutteman (eds.), *Time, Chance, and Reduction: Philosophical Aspects of Statistical Mechanics*. Cambridge: Cambridge University Press, pp. 13–33.
Glynn, L. (2010). Deterministic chance. *The British Journal for the Philosophy of Science*, 61(1): 51–80.
Goldstein, S. (2001). Boltzmann's Approach to Statistical Mechanics. In J. Bricmont, D. Dürr, M.C. Galavotti, G. Ghirardi, F. Petruccione and N. Zanghì (eds.), *Chance in Physics: Foundations and Perspectives, Lecture Notes in Physics*. Berlin: Springer-Verlag, pp. 39–54.
Hájek, A. (1996). 'Mises Redux'—redux: Fifteen arguments against finite frequentism. *Erkenntnis*, 45(2–3): 209–227.
Handfield, T. and Wilson, A. (2014). Chance and context. In A. Wilson (ed.), *Chance and Temporal Asymmetry*. Oxford: Oxford University Press, pp. 1–19.
Hoefer, C. (2007). The third way on objective probability: A Sceptic's guide to objective chance. *Mind*, 116(463): 549–596.
Ismael, J.T. (2009). Probability in deterministic physics. *The Journal of Philosophy*, 106(2): 89–108.
Lange, M. (2013). Grounding, scientific explanation, and Humean laws. *Philosophical Studies*, 164(1): 255–261.
Lewis, D. (1980). A subjectivist's guide to objective chance. In R.C. Jeffrey (ed.), *Studies in Inductive Logical and Probability*, Volume II. Berkeley: University of California Press, pp. 263–293.
Lewis, D. (1986). *Philosophical Papers*, Vol. 2. Oxford: Oxford University Press.
Lewis, D. (1994). Humean supervenience debugged. *Mind*, 103(412): 473–490.
Loewer, B. (2001). Determinism and chance. *Studies in History and Philosophy of Science Part B: Studies in History and Philosophy of Modern Physics*, 32(4): 609–620.
Loewer, B. (2004). David Lewis's Humean theory of objective chance. *Philosophy of Science*, 71(5): 1115–1125.
Loewer, B. (2007). Counterfactuals and the Second Law. In H. Price and R. Corry (eds.), *Causation, Physics, and the Constitution of Reality: Russell's Republic Revisited*. Oxford: Oxford University Press, pp. 293–326.

Loewer, B. (2012). Two accounts of laws and time. *Philosophical Studies*, 160(1): 115–137.

Lyon, A. (2011). Deterministic Probability: Neither Chance nor Credence. *Synthese*, 182(3): 413–432.

Maudlin, T. (2007a). What could be objective about probabilities? *Studies in History and Philosophy of Science Part B: Studies in History and Philosophy of Modern Physics*, 38(2): 275–291.

Maudlin, T. (2007b). *The Metaphysics within Physics*. Oxford: Oxford University Press.

Meacham, C.J.G. (2005). Three proposals regarding a theory of chance. *Philosophical Perspectives*, 19(1): 281–307.

Meacham, C.J.G. (2010). Contemporary Approaches to Statistical Mechanical Probabilities: A Critical Commentary - Part II: The Regularity Approach. *Philosophy Compass* 5 (12): 1127-1136.

North, J. (2010). An empirical approach to symmetry and probability. *Studies in History and Philosophy of Science Part B: Studies in History and Philosophy of Modern Physics*, 41(1): 27–40.

Norton, J.D. (2008). The dome: An unexpectedly simple failure of determinism. *Philosophy of Science*, 75(5): 786–798.

Popper, K.R. (1992). *Quantum Theory and the Schism in Physics*, Vol. 3. New York: Routledge.

Schaffer, J. (2007). Deterministic chance? *The British Journal for the Philosophy of Science*, 58(2): 113–140.

Sklar, L. (1995). *Physics and Chance: Philosophical Issues in the Foundations of Statistical Mechanics*. Cambridge: Cambridge University Press.

Sober, E. (2010). Evolutionary theory and the reality of macro probabilities. In E. Eells and J.H. Fetzer (eds.), *The Place of Probability in Science*. New York: Springer, pp. 133–161.

Strevens, M. (1999). Inferring probabilities from symmetries. *Noûs*, 32(2): 231–246.

Strevens, M. (2000). Do large probabilities explain better? *Philosophy of Science*, 67(3): 366–390.

Strevens, M. (2013). *Tychomancy*. Cambridge, MA: Harvard University Press.

Strevens, M. (2011). Probability out of determinism. In C. Beisbart and S. Hartmann (eds.), *Probabilities in Physics*. Oxford: Oxford University Press, pp. 339–364.

Uffink, J. (2017). Boltzmann's work in statistical physics. In E.N. Zalta (ed.), *The Stanford Encyclopedia of Philosophy*. Available at: https://plato.stanford.edu/archives/spr2017/entries/statphys-Boltzmann/.

Further Reading from the Editors

This chapter can be usefully read in conjunction with the chapters in this volume by M. Suárez (Chapter 46) and E. Keming Chen (Chapter 12). A useful overview of the debate over chance compatibilism is S. Bradley, "Are objective chances compatible with determinism?" *Philosophy Compass* 12(8), (2017): e12430. A systematic defence of a compatibilist conception of chance is C. Hoefer, *Chance in the World: A Humean Guide to Objective Chance* (Oxford: Oxford University Press, 2019). Physics-specific discussions of especial interest are J. Ismael, "Probability in Deterministic Physics," *Journal of Philosophy* 106(2), (2009) and W. Myrvold, "Deterministic Laws and Epistemic Chances" in *Probability in Physics*, Y. Ben-Menahem and M. Hemmo (eds.) (Berlin: Springer-Verlag, 2012): 73–85.

PART X

Symmetries

Introduction to Part X

It is difficult to overstate the importance of symmetry considerations in physics. Many of the chapters in the "Theories" half of this book were riddled with questions about symmetry. Leibniz's challenges to Newton's absolute space come about because of symmetries of the dynamics. Special relativity is, effectively, a theory about symmetry: to be special relativistic is to transform appropriately under the Lorentz transformations. Questions about the symmetry group of general relativity underpin some of the deepest problems of the theory. Moreover, this centrality of symmetry goes well beyond spacetime theories. Gauge symmetries are important in quantum mechanics, and to classical electromagnetism. But they become more crucial still when one considers the quantization involved in quantum field theory and the standard model, where sectors of the theory are defined by their symmetry groups. No topic in this book has quite so many ramifications for the others.

Most importantly to the physicist, symmetry often functions as a guide to theory building. If a theory is to be relativistically acceptable, it must be covariant under the transformations of the Poincaré group. But symmetry can be more than just a constraint on theory building. As Emmy Noether proved, symmetries lead to theories with associated conservation laws – global symmetries lead to conservation of quantities like momentum and energy. Local symmetries lead to the conserved quantities associated with the fundamental forces. Many theories of quantum gravity follow this blueprint, for example postulating supersymmetry in order to accommodate gravity. Sometimes, as with the creation of special relativity, observed empirical symmetries guide theory choice. In other cases, the considerations driving the choice of symmetry group are theoretical.

Alongside this importance of symmetry for theory creation stands its importance for theory interpretation. It is commonplace to put a great deal of emphasis on the invariant quantities of a theory: those that remain unchanged under the symmetries of the theory. This might be a matter of theory application; solving relativistic problems becomes much easier for a first year undergraduate once they understand the importance of relativity's invariant quantities. But it also seems to be a matter of interpretation; invariant quantities are the objective quantities which we might take to be part of the theory's ontology. Non-invariant quantities are "surplus structure", artifacts of the form of the theory or a particular representation.

Needless to say, the above claim needs a great deal of unpacking, and is hardly uncontroversial. Shamik Dasgupta's chapter looks at symmetry and superfluous structure from a metaphysical perspective. He asks what prevents us from taking a quantity like Newtonian absolute velocity seriously. What exactly is the symmetry that leads us to reject Newtonian absolute space? Can a "method of symmetry" really give us a methodology for metaphysics? How do the symmetries of our dynamics relate to empirical symmetries, and how should the latter constrain the metaphysics of our physics? As Dasgupta notes, much of the import of empirical symmetries is epistemic; we can either find a theory to match those empirical symmetries or accept that our theories are underdetermined. The import of symmetries is as much epistemic as it is metaphysical.

Given the importance of the topic, this chapter contains two introductory overviews. Jenann Ismael tackles symmetry from a somewhat more historical and physics-based perspective. She presents a picture on which symmetries provide clues to structure, but on which considerable subtlety is needed to apply them. After all, as both she and Dasgupta note, we may have access to symmetries of the dynamics and empirical symmetries, but we have no direct access to *metaphysical* symmetries. If Newtonian absolute space were really part of the furniture of the world, then Galilean boosts would not be a symmetry of this structure.

The remainder of this part dives into more theory-specific, and inevitably more technical discussion. Adam Caulton's entry on permutation symmetries focuses on the symmetry properties of groups of quantum particles. The wavefunctions of entangled quantum particles are either symmetric or anti-symmetric under particle exchange – particle permutations are a symmetry of the theory. What should we make of this? Should we interpret the permutation as a mere relabeling of the same situation, or can we make sense of distinct but indistinguishable situations in which particles are permuted? What does the result tell us about the metaphysics of quantum particles? Caulton gives a formal framework for thinking about these issues and connecting them to key metaphysical questions.

One class of symmetry is particularly crucial for theory building: gauge symmetries. The so-called "gauge theories" have local symmetries – the elements of the symmetry group are functions of space-time. A particular class of such theories, called Yang-Mills theories, contribute to the standard model of particle physics when quantized. These theories are particularly puzzling. On the one hand, quantities that are not invariant under the local gauge transformations seem to be surplus structure – their values make no difference to empirical predictions. On the other, the gauge symmetries themselves are essential to the nature of the theory. As Nic Teh puts it in his chapter, the descriptive fluff seems strongly non-idle. Teh demonstrates that the simple symmetry principles we applied in Newtonian mechanics won't work in these cases – we cannot simply eliminate the gauge-variant quantities. There are subtle lessons here for the way in which this kind of structure might represent reality.

Bryan Roberts discusses time reversal symmetry in his chapter. It's often claimed that most physics is time reversal invariant – the arrow of time does not appear to arise in such fundamental physics. But the issue of what we mean by time reversal is extremely subtle, and there are competing stories. Do we reverse the instantaneous states, or instead think about reordering them? Roberts discusses what we actually mean by the time reversal operation and defends an orthodox understanding of time reversal.

The last chapter in this part, written by Elena Castellani and Radin Dardashti, is about symmetry breaking, rather than symmetry itself. Clearly, the world around us has much *less* symmetry than the laws of nature. Crystals, for example, arise from translation invariant physics and yet have a discrete structure that isn't invariant under arbitrary translations. Much physics depends on accounts of symmetry breaking. Perhaps the most obviously important involves the way in which our current, highly structured universe arose from a much more symmetric early state. Castellani and Dardashti discuss various things that symmetry breaking might mean and distinguish the philosophical consequences of each.

39

SYMMETRY AND SUPERFLUOUS STRUCTURE: A METAPHYSICAL OVERVIEW

Shamik Dasgupta

39.1   The Method of Symmetry

Symmetry plays a number of central roles in modern physics. As the physicist Paul Anderson famously remarked, "it is only slightly overstating the case to say that physics is the study of symmetry" (1972, p. 394). Here I discuss just one role of symmetry: its use as a guide to superfluous structure, with a particular eye on its application to metaphysics.

What is symmetry? Generally speaking, a symmetry is an operation that leaves its object unchanged in a certain respect. Rotate a square piece of paper 90 degrees in its own plane and it will have the same extension through space; in this sense, the rotation is a symmetry of its extension. But we will focus on the symmetries of *physical theories*, not extensions. Roughly speaking, these are operations on entire physical systems that leave some aspects of the theory unchanged. Which aspects? That depends. Different symmetries preserve different aspects, but the important classes are those that leave the dynamical laws of the theory unchanged. These are known as *dynamical symmetries*.

For example, consider a theory that describes point particles with mass moving through Newtonian spacetime. Suppose the dynamical laws describing their motions are Newton's laws of motion and the inverse-square gravitational force law. In Newtonian spacetime, there is, in addition to a particle's velocity relative to another particle, its velocity *through space* – a quantity that reflects how far it moved (in a direction) over an interval of time independently of its motion relative to other particles.[1] Call this its *absolute* velocity. Now, imagine an operation that takes any possible world consisting of particles in Newtonian spacetime and delivers another world just like it except that everything is uniformly boosted by some velocity V, say 1 mph in a certain direction. At any given time in the boosted world, a particle's absolute velocity is equal to its absolute velocity in the original world at that time plus V. This operation – known as a uniform boost – leaves Newton's laws unchanged: if the original world satisfies those laws, the boosted world does too. (To verify this, note that those laws relate the quantities of mass, distance at a time, and acceleration; and all these quantities are preserved by a uniform boost.) Hence uniform boosts are dynamical symmetries of this theory.

In truth, this is too quick: as we will see, not *every* operation that leaves the laws unchanged counts a "dynamical symmetry." What other conditions must the operation satisfy? There is little consensus on how the term is to be defined, but we need not settle this here, for a uniform boost is a paradigm example of a dynamical symmetry of this theory, in the sense that every definition of "dynamical symmetry" counts it as such. Other paradigm examples include uniform translations through space (moving everything over in the same direction by the same distance) and uniform translations through time. Our discussion of dynamical symmetries can proceed for now on the basis of these paradigm cases.[2]

I said that symmetry is used as a guide to a superfluous structure. How? Call those features that are unchanged by the dynamical symmetries of a theory the *invariant* features of the theory. Other features are *variant* features of the theory. A uniform boost changes a particle's absolute velocity; hence absolute velocity is a variant quantity of the Newtonian theory above. By contrast, the invariant quantities include acceleration, relative velocity, and spatial distance at a time: all these quantities are unchanged by the dynamical symmetries of the theory. Then the core idea is that *the variant features of a theory are superfluous to it.*

What does this mean? For the purposes of mathematical physics, perhaps just this: that if some part of the mathematical formalism of a theory encodes a variant feature, it can be regarded as physically insignificant and ignored. But for the purposes of metaphysics, the idea is that variance is a sign of unreality – a reason to think that the feature is unreal. Suppose we believed the little theory above and then realized that absolute velocity is one of its variant quantities. The idea is that this would be a reason to think that there is no such thing as absolute velocity after all. To draw this conclusion, one would have to give up the idea that particles move around in Newtonian spacetime and look for an alternative theory of the structure of spacetime on which there is no quantity of absolute velocity. In this way, symmetry is used as a guide to the nature of spacetime.[3]

Metaphysicians should find this intriguing. For metaphysics is in deep need of method, and what we appear to have here is a "method of symmetry": take your best physics, find its dynamical symmetries and hence its variant features, and conclude that those features aren't real.[4] If this method can teach us about the nature of spacetime, we might hope it could generalize to other putative aspects of reality too. If a metaphysician proposes a theory on which reality contains something X, and X can be shown to vary under a dynamical symmetry, the method of symmetry would have us conclude that there is no such thing as X after all.

But how exactly does the method work? And to what other kinds of cases can it be applied? This chapter will survey various approaches to the method of symmetry.

To this end, I focus on the case study of absolute velocity in classical physics. Why trouble ourselves with this tired and fictional example? Wouldn't we be better off drawing metaphysical lessons from a more cutting-edge physics? Yes, ultimately. But if our aim is to understand the *method*, this sanitized example is useful insofar as it abstracts from distracting complications. This is no worse (and, perhaps, no better) than asking whether a white sneaker confirms the hypothesis that all ravens are black. In both cases there may be little interest in the example *per se*; the point is rather to illuminate the logic behind the method.

39.2 Justifications

Let's start by asking what justifies the method of symmetry. Suppose we discover that something is a variant feature of our best physics. Why is this a reason to think that it's not real? What is it about variant quantities that distinguishes them from invariant quantities in this regard? The literature contains at least three answers: that variant quantities are *not objective*, that they are *physically redundant*, and that they are *undetectable*. I'll discuss each in turn.

The idea that invariant quantities are objective has long been associated with symmetry. As Weyl put it, "objectivity means invariance" (1952, chapter 5). This idea was revived by Nozick, who proposed that "an objective fact is invariant under various transformations" (2001, p. 76). So, perhaps the reason to dispense with variant quantities is that they aren't objective.

What does it mean to say that something is "objective"? If it just means that it's real, then it's trivial that quantities that aren't objective aren't real. But then the question is why we should think that variant quantities aren't objective in this sense. That's just the question we started with; no progress has been made.

Another notion of objectivity is perspective independence. To say that a quantity is objective in this sense is to say that it has the same value from all perspectives so that all observers would agree in their measurements of it. Daston and Gallison (2007) argue that this concept played a central role in the thinking of Weyl and other early 20th century physicists. But important as it may be, this conception of objectivity cannot be what justifies our method of symmetry. The reason is that being objective in this sense doesn't have much to do with being invariant. True, velocity isn't objective in this sense since observers moving at a constant velocity relative to one another will disagree in their measurements of velocity: from their own perspectives, each will claim to be at rest while the other moves. But by the same token, acceleration isn't objective in this sense either: if two observers are accelerating relative to one another, each will claim to be at rest from their own perspective while the other is accelerating.

It's unclear, then, whether the concept of objectivity helps to justify the method of symmetry. What we need is a definition of "objectivity" that would apply to invariant but not variant quantities, and which doesn't just *mean* being real. I leave it as an open question whether such a conception can be made out.

Turn now to the idea that variant quantities are *physically redundant*. Here again, we must ask: redundant in what sense? One suggestion is that variant quantities are explanatorily redundant in the sense that they make no difference to how a physical system evolves in time.[5] The idea is tempting because *in*variant quantities do make a difference: two systems that differ initially in inter-particle distances would, according to our simple Newtonian theory, evolve differently over time because the gravitational forces would be different. By contrast, the idea is, absolute velocity isn't like that. Consider two possible physical systems related by a uniform velocity boost. At time t_0 they differ in facts about absolute velocity, yet at all subsequent times, they agree on all invariant quantities including relative positions, relative velocities, accelerations, and masses; hence, the initial difference in absolute velocity led to no difference later on. But as Sklar (1974, p. 180) pointed out, this is a mistake: at subsequent times the two systems differ regarding the absolute velocities of each particle! Hence absolute velocity does make a difference to how the system evolves over time after all. One might now say that subsequent differences in absolute velocity aren't *real* differences, but of course, that is the very conclusion we are trying to establish.

Another suggestion is that variant quantities are redundant in the sense that their specific values make no difference to whether the dynamical laws hold.[6] It is true that variant quantities are redundant in this sense: the variant quantities are, by definition, those whose values can be altered while preserving the truth of the dynamical laws. What is less clear is why this is a reason to think that such quantities aren't real. Why should we dispense with a quantity, just because its specific values don't make a difference to whether the dynamical laws hold?

To be sure, anyone justifying the method of symmetry will at some point reach bedrock and appeal to some basic epistemic principle such as induction, modus ponens, or what have you. It would then be unfair to ask them to justify induction! But the principle "dispense with quantities that don't make a difference to whether the dynamical laws hold" isn't epistemically basic in this sense. It may be a sound epistemic principle, but our question is why.

One might answer that if a quantity is redundant in this sense, one can formulate an alternative physics that makes no reference to the quantity. The latter, one might then say, is preferable because it uses "less structure." Compare Newtonian with Galilean spacetime: the idea would be that the latter has less structure than the former and should be preferred on that basis.[7] This is some progress insofar as parsimony is, arguably, a basic epistemic principle. The main challenge is to explain the notion of "less structure"; in particular, one must show that "structure," *so understood*, is what the epistemic principle of parsimony asks us to minimize. As Barrett (2015) and others have argued, it is not clear that there is a notion that does the job.[8]

That leaves the idea that the reason to dispense with variant quantities is that they are *undetectable*.[9] The chief virtue of this approach is that the epistemic principle it rests on is relatively basic and uncontroversial: namely, that in theory choice there should be a presumption against positing quantities that are undetectable. Of course, it may be impossible to dispense with an undetectable quantity without sacrificing other theoretical virtues, in which case one's all-things-considered best theory may retain it. But the point remains that a quantity's being undetectable is *a reason* to prefer a theory that does without it, and that is all the method of symmetry promised us. In what follows, then, I will focus on this approach.

39.3   From Symmetry to Undetectability

The main challenge facing this approach is to show that variant quantities are, in fact, undetectable. This is by no means trivial, but let me sketch one approach.

Consider absolute velocity. Why think it's undetectable? Galileo famously pointed out that physical systems related by a boost would look the same. On the inside, an airplane on the tarmac looks and feels exactly like a plane flying smoothly. Of course, things would look different if you looked outside. But if *everything* were uniformly boosted, including the plane and its surroundings, things would look exactly the same even if you looked out the window. To mark this point, let's say that entire possible worlds related by a boost are *observationally equivalent*.

Still, this doesn't imply that absolute velocity is undetectable. If we can't tell the difference between boosted systems with the naked eye, maybe that's just because our eyes aren't sensitive enough. Perhaps we could build a fancy measuring device that is sensitive to whether it's moving uniformly or at rest. But we can rule this out if uniform boosts are dynamical symmetries of the actual laws. The key is that a measuring device is a physical object and is therefore governed by those laws. So, suppose we built a device in the hope that it would display "Rest" on a computer screen if its absolute velocity is zero, and "Moving" otherwise. And suppose we turn it on and it displays "Rest." Can we infer that it's at rest? No! For consider a uniformly boosted world, in which the absolute velocity of the device is different. If uniform boosts are dynamical symmetries, the same laws are true in the boosted world. Hence the boosted world represents how the device would behave, given the laws governing it, if its absolute velocity differed. And yet the boosted system – by construction – is one in which the device displays "Rest" when we turn it on. After all, a uniform boost preserves all the relative positions of particles, and hence if the pixels on a screen display "Rest" then they'll display the same sign in the boosted system too. Hence the device would say "Rest" *whatever* its absolute velocity was – hardly a "measuring device" to be proud of![10]

This isn't to say that we can't measure velocity *relative* to other bodies. A car's speedometer is sensitive to its speed relative to the road in the following sense: given the dynamical laws governing it, what it displays on its screen is a function of its velocity relative to the road. What the argument purports to show is that, thanks to the symmetries of the laws, it's impossible to build a measuring device that's sensitive to *mere* differences in absolute velocity.

This "symmetry argument" is no ordinary skeptical argument. The ordinary skeptic argues that we can't tell whether we are brains in vats or embodied in a world that is more-or-less as it appears. One response is that so long as our minds are reliably connected to the world, our belief that we are embodied may count as knowledge. Another response is that the hypothesis that we are embodied is a better explanation of our perceptions – simpler, more elegant, and so on – and should therefore be favored on abductive grounds. Yet another response is that in ordinary contexts, the standards for knowledge are sufficiently low that we turn out to know we are embodied after all. But none of these are appropriate responses to the symmetry argument. For one thing, the symmetry argument demonstrates that measuring devices, and by extension, our minds, are *not* reliably connected to facts about absolute velocity. And the competing hypotheses – that our velocity is 1 mph in a certain direction, that it is 2 mph in that direction, and so on – are all equally simple and elegant, so nothing can be said abductively in favor of one hypothesis over the others. For the same reason, lowering the standards wouldn't favor any one of these hypotheses over the others. The symmetry argument is therefore much stronger than an ordinary skeptical argument. No surprise that it is taken seriously by physicists such as Feynman (1963, p. 15), who recognize that it is not just a philosopher's puzzle.

The symmetry argument as stated concerns absolute velocity, but does it generalize to show that *any* variant quantity is undetectable? The argument rested on three facts. First, that the same dynamical laws obtain in the boosted world – without this it wouldn't follow that the boosted system represents how our "measuring" device would behave, given the laws, if its absolute velocity differed. But this was guaranteed by the fact that uniform boosts are dynamical symmetries. So this will generalize for any variant quantity: by definition, there will be a symmetry operation that alters its values yet preserves the dynamical laws.

Second, the symmetry argument required that boosted physical systems are observationally equivalent. Now, we cannot expect this to generalize: we cannot expect that two worlds related by *any* dynamical symmetry of the laws will be observationally equivalent. At least, not given what I've said so far: I said that a dynamical symmetry is an operation that preserves the laws, and operations can do that while changing how the world looks: just consider an operation that maps each physical system to the "null" system containing nothing whatsoever, in which the laws are true vacuously. Still, I also said that *merely* preserving the truth of the laws doesn't suffice for being a dynamical symmetry. More conditions must be added. To generalize the symmetry argument, then, one must show that these extra conditions entail that systems related by a dynamical symmetry are observationally equivalent.

Whether this is true then depends on what definition of dynamical symmetry you work with. But insofar as our aim is to develop the method of symmetry, the natural move is to reverse engineer and *stipulate* that an operation counts as a dynamical symmetry only if worlds related by the operation are observationally equivalent. This could involve building the condition of observational equivalence directly into the definition of dynamical symmetry. Or it could involve providing some other condition – for example, that the operation preserves topological structure – which is then shown to imply this condition of observational equivalence. I will not decide between these two approaches but see Dasgupta (2016) for some considerations in favor of the former.

Third, the symmetry argument required that the hypotheses that distinguish the boosted worlds – that the device is moving at 1 mph to the north, that it is moving at 2 mph to the north, and so on – are equally simple, elegant, common-sensical, and so on; more generally, that they score equally well on every theoretical virtue. This is what distinguished the symmetry argument from an ordinary skeptical argument. To mark this, let us call the competing hypotheses "abductively equivalent," and by extension let us call the boosted worlds abductively equivalent too. Again, we cannot expect this to generalize: we cannot expect that two worlds related by *any* dynamical symmetry *as I've defined the term so far* will be abductively equivalent. But again insofar as our aim is to develop the method of symmetry, the natural approach is to reverse engineer and stipulate that an operation counts as a dynamical symmetry only if worlds related by it are abductively equivalent. As before, I leave open whether the definition contains this condition of abductive equivalence explicitly, or whether it contains some other condition from which abductive equivalence follows.

Of course, if you started off with a fixed idea of what "dynamical symmetry" meant, this reverse engineering may strike you as perverse. Given *your* definition, perhaps the argument doesn't generalize. Fine; there is no need to fight over terms. The fact remains that operations on worlds that satisfy our three conditions – they map worlds to worlds that (i) agree on the laws, (ii) are observationally equivalent, and (iii) are abductively equivalent – have the following property: any quantity altered by such a transformation can be shown, by the symmetry argument above, to be undetectable. Whatever we call these operations, metaphysicians in search of method should find them intriguing!

There may of course be other ways of arguing that variant quantities are undetectable. But for now let us run with the argument we have, which uses the engineered notion of symmetry defined by the three conditions above. I will call these dynamical symmetries, though you are free to substitute with another term if you wish.

39.4   Epistemic Possibility

I said that a dynamical symmetry is an operation on *possible worlds*. But possibility comes in many senses: metaphysical, epistemic, conceptual, logical, and others besides. What is the relevant sense of possibility?

This question is easy to overlook if one studies symmetries mathematically. There, one typically defines a mathematical space of *models* or *states* and then defines a symmetry to be a function on that space.[11] Insofar as the space was mathematically well-defined, there is no *mathematical* question of what entities the symmetry operates on. Still, our philosophical question remains. For a model is just a bit of math, and philosophical conclusions concerning *detection* or *reality* do not follow from math without interpretation. So the question remains as to what the models or states represent. The typical answer is that they represent *possible worlds* or *possible physical states*; our question is what the relevant sense of possibility is.

Metaphysicians may naturally reach for the notion of metaphysical possibility. A uniform boost would then be thought of as a function that maps a metaphysically possible world to one that differs only in facts about absolute velocity.

But a better candidate, I suggest, is the notion of epistemic possibility. On this approach, a dynamical symmetry is an operation on epistemically possible worlds in roughly David Chalmers' sense of a (maximally specific) way the world could be for all we know *apriori*.[12] I know that London once hosted the Olympic Games, but I don't know that *apriori*; hence there is an epistemically possible world in which London never hosted the games. Similarly, I know that water is H_2O, but I don't know that *apriori*; hence there is an epistemically possible world in which water is not H_2O.

To see how this works, consider again the case of Newtonian spacetime. Even if Newtonian spacetime isn't real – even if we *know* it isn't real – we don't know this *apriori*. So there are epistemically possible worlds in which spacetime is Newtonian. Since we can't know *apriori* what trajectories bodies make through Newtonian spacetime, there are many such worlds that differ in the trajectories of bodies. Indeed, there are many such worlds that agree on all relative motions and differ only in a uniform boost – after all, we cannot know a body's absolute velocity *apriori*. We can then understand a uniform boost to be a function on worlds like these: a function that takes an epistemically possible world in which there is a well-defined quantity of absolute velocity and maps it to an epistemically possible world in which all the velocities are uniformly transformed.

Epistemic possibilities have two advantages over metaphysical possibilities. First, our method of symmetry uses dynamical symmetries to establish *epistemic* results about what we can or cannot detect, and the framework of epistemic possibility is well-suited for this purpose. Indeed, in this framework, the argument in Section 39.3 that a variant quantity like absolute velocity is undetectable can be seen, more generally, as an argument that it is *unknowable*. Thanks to being variant, there are epistemically possible worlds that differ in its value; hence we cannot know its value *apriori*. What the argument in Section 39.3 then establishes is that we cannot know its value by observation either. The argument was that whatever output a "measuring" device produces, there are multiple epistemic possibilities in which the output is the same but the value of the quantity differs; hence we cannot know its value on the basis of the observed output either. This last step assumes that the multiple possibilities imply ignorance, an assumption that is trivial if the possibilities are epistemic but not if they are metaphysical.

Second, there may not be enough metaphysical possibilities to make sense of dynamical symmetries. Suppose absolute velocity is not real. And suppose that this is a necessary truth, the idea being that certain structural features of spacetime are necessary. Then there won't be metaphysical possibilities that differ only in a uniform boost, and so there won't be a function on metaphysical possibilities for a uniform boost to *be*. Yet we still want to say that boosts are symmetries; indeed, that may be our reason for thinking that absolute velocity isn't real in the first place! Epistemic possibilities avoid the problem: even if there are no metaphysical possibilities that differ in a uniform boost, there are epistemic possibilities that do.

39.5   Empirical Symmetries

The "method of symmetry," as we now have it, is this. We take our best physics. We look at its laws and find its dynamical symmetries and hence the variant quantities. We then argue that those quantities are unknowable. And we take that as a reason to think that the quantity isn't real.

This last step should now be clearer. There is no claim that being unknowable *logically implies* that the quantity isn't real: after all, there are epistemically possible worlds in which it is real. For the same reason, this is not a verificationist argument that talk of the quantity is meaningless. The idea is just that positing undetectable quantities is a bad-making feature of a theory, so that theories that dispense with the quantity are preferable in that respect. Of course, whether the latter theory is all-things-considered preferable then depends on assessing its other virtues and vices, and much argument might be had on that matter. Still, we have on our hands a principled method that yields *reasons* for various metaphysical theses. In metaphysics, that counts for a lot!

But if this is our method, it arguably generalizes to a broader class of operations. To see this, suppose that absolute velocity is *not* a variant quantity. To take a fictional example, suppose that color is a primitive property of bodies, and suppose there's a dynamical law to the effect that things take on a greenish hue when their absolute velocity is zero but not otherwise. Uniform boosts would not then be dynamical symmetries because they don't preserve this law: they map worlds in which the law holds to worlds in which objects take on a greenish hue when they have some non-zero velocity instead. Absolute velocity, then, would be an *invariant* quantity. Does this mean that we could measure whether something is at absolute rest by looking to see whether it is greenish? Unfortunately not. For if we see something greenish, we can infer that its absolute velocity is zero only on the basis of a *theory*, namely that green correlates with zero velocity. Contrast this with the theory that green correlates with some non-zero velocity, say 5 mph in a certain direction. If this theory were true, green would indicate this non-zero velocity instead. But how could we ever tell which theory is true? All we would ever *see* is that the bodies with a greenish hue are all at rest relative to one another, and moving relative to other things. True, this means that if there are absolute velocities, one of them would correlate with green. The trouble is, there'd be no way of telling which one it is.

Or suppose there was a velocity-dependent force law. Suppose it implies, of a particular system, that if its center of mass is at absolute rest then it'll evolve one way, while if the velocity is non-zero it'll evolve differently. And suppose the difference between the two evolutions is directly observable: a different pattern of ink on paper is produced in each case, say. In this case, again, uniform boosts wouldn't be dynamical symmetries, so absolute velocity would be an invariant quantity. Can we then measure absolute velocity by observing how the system evolves? No again, and for the same reason. We can only infer that a system exhibiting the first pattern is at absolute rest on the basis of the *theory* that that pattern correlates with absolute rest. Contrast this with a theory on which that pattern correlates with some non-zero velocity instead. Both theories are consistent with what we'd *see*, namely that all systems that display that first pattern are at rest relative to one another, and moving relative to systems that don't. But there'd be no way of telling what the absolute velocity of those former systems is.

The point is that *measuring* a quantity involves inferring its value on the basis of (i) a measurement outcome, and (ii) the physical laws that generated that outcome.[13] In Section 39.3 we assumed that we knew the laws; the problem was that the same outcome would be produced regardless of what the value of the quantity was. Here the problem is that we can't know what the laws are in the first place.

With these fictional laws, then, absolute velocity would be unknowable even though uniform boosts wouldn't be *dynamical* symmetries. Still, boosts would be symmetries in a broader sense. For one thing, there's an intuitive sense in which boosts preserve the "form" of these laws. In the case of color, all boosted worlds have a law of the form 'green correlates with V'; they just differ on the particular velocity V. Indeed, theorists of symmetry often focus on transformations that preserve the form in something like this sense.[14] But what exactly is form? For our purposes, the important fact is that we can never know which of the various color laws obtains. This requires that each law is abductively equivalent in the earlier sense. Consider the range of color laws: that green correlates with absolute rest, that it correlates with velocity 1 mph in a certain direction, that it correlates with velocity 2 mph in that direction, and so on. The idea behind the argument above is that each law is equally simple, explanatory, and so on, so there is no reason to believe one over the others. (You might object that we should favor the law that green correlates with absolute rest because it's simpler than the rest. But then you'd also have to say that the hypothesis that the center of mass of the universe is at absolute rest is simpler than its boosted alternatives, in which case boosted worlds aren't abductively equivalent and the argument that absolute velocity is unknowable in Section 39.3 doesn't go through. But surely we can't know our absolute velocity on this basis! What this shows is that the sense in which the rest hypothesis is simpler is epistemically insignificant.)

Thus, we can define the *empirical symmetries* of a theory just like dynamical symmetries with the one exception that the transformed world isn't required to have the *same* laws as the original world, just *abductively equivalent* laws. Since any law is abductively equivalent to itself, it follows that any dynamical symmetry is also an empirical symmetry. But the reverse is not the case: uniform boosts are empirical symmetries of the color law, but not dynamical symmetries. So the notion of empirical symmetry is more general, and the idea would be that any quantity that varies under the *empirical* symmetries of the laws is unknowable.

This then gives us a more expansive method of symmetry. We take our best physics, look at its laws and find its *empirical* symmetries, and argue that quantities altered by *those* symmetries are unknowable. As before, we then take this as a reason to think that the quantity isn't real. This method is more expansive insofar as it'll lead us to dispense with more quantities. If this leaves you feeling queasy, be assured that there are limits. Reichenbach's (1958) transformation that maps a non-Euclidean world to a Euclidean world with universal forces, for example, isn't an empirical symmetry. The worlds are observationally equivalent, but not abductively equivalent because they have different abductive virtues (the former is simpler, in one sense of the term, while the latter is closer to pre-theoretic belief about geometry).

Note that this move to empirical symmetries hangs on our decision, in Section 39.2, to think of the method of symmetry as eliminating variant quantities because they're *undetectable*. If we had eliminated them because they're *physically redundant*, this move to empirical symmetries would not make sense. For in the color law above, absolute velocity is *not* redundant in any of the senses surveyed there: it *does* make a difference to whether a body will take on the greenish tinge; its values *do* make a difference to whether the law holds; and so on. I said in Section 39.2 that this justification of the method of symmetry faces problems anyway; my point here is that if one justified the method this way, one would be restricted to using dynamical symmetries. The move to empirical symmetries makes sense only if thinks of the method of symmetry as eliminating *undetectable* quantities, for as we've seen these include those that are variant under the empirical symmetries, not just the dynamical symmetries.

This method of empirical symmetry can lead to substantial revisions in physics. Suppose we started with a theory on which green correlates with absolute rest. Suppose we then use the method of empirical symmetry and dispense with absolute velocity. We must then revise our theory. We can no longer say that green correlates with any particular velocity. Rather, we can only say that things that take on a greenish hue are all at rest relative to one another, and are moving relative to things that don't. This is a very different theory. For one thing, in the original theory, a body's state of motion *determines* whether it takes on a greenish hue, but in the revised theory this is not so. If we take a possible system in which two things are at rest relative to one another, our revised theory doesn't determine whether they'll have the greenish hue; it just determines that either both will or neither will. Thus, the method of empirical symmetry led from a theory that was deterministic to one that is not.

This raises the question of whether the new theory would be all-things-considered better than the old theory. True, the old theory had the vice of containing unknowable structure, but the new theory is indeterministic. Is this a vice? If so, is it worth it? More generally, if the method of empirical symmetry leads us to new physical theories with surprising features, will the new theory be all-things-considered better?

This can only be settled on a case by case basis, based on a close examination of the rival theories. But we can say this: we cannot object to the new theory *on the basis of observation*. In the case of color, we cannot complain that we *observed* bodies behaving in the deterministic manner described by the old theory. For, the method of empirical symmetry shows that we observed no such thing. It shows that we never saw that the greenish hue correlates with *absolute rest* rather than any other velocity, we just saw that things with the hue are at rest relative to one another. If we previously *thought* we had observed that green correlates with absolute rest, that was based on a mistaken characterization of the observations – perhaps we had a deeply ingrained belief that some reference body was in a state of absolute rest so that when we saw something at rest relative to it we mistakenly chalked that up to seeing it at absolute rest. In any event, the point is that an objection to the indeterminism cannot be based on observation; it must be based on considerations of another kind.[15]

More generally, the method of empirical symmetry assures us that the new theory is empirically adequate if the old theory was, for the new theory dispenses only with quantities that are undetectable. Thus, any surprising feature of the new theory cannot be said to be inconsistent with observation.

39.6   Metaphysical Applications

I have outlined two methods of symmetry, one based on dynamical symmetry and the other on empirical symmetry. Which method should we use? This depends on our purposes. Philosophers of physics sometimes ask what the appropriate spacetime structure is *for a particular set of laws*. For example, what is the natural spacetime structure for classical Newtonian mechanics? In this case, dynamical symmetries are more appropriate. We want to look at those worlds *in which the laws obtain* and see what structure we can dispense with. Insofar as the method is to dispense with undetectable structure, we can think of us as asking: conditional on these laws, what would be undetectable? Dynamical symmetries are perfectly suited for this purpose.

But we may alternatively ask what spacetime structure is indicated *by a certain class of observations*. Given the kind of observable phenomena there'd be in a Newtonian world, for example, we may ask what we should believe about the structure of spacetime if we observed those phenomena. In this case, the empirical symmetries are more appropriate. We want to look at the worlds that agree on those observational phenomena and ask what would be undetectable, and in doing so it makes sense to consider worlds with different laws if we couldn't know which of those laws obtained.

To the extent that metaphysicians engage with physics at all, the second question is the more pertinent. They want to know what's real. To this end, the method of symmetry has them ask which features are undetectable. But it would be odd to hear that as asking what would be undetectable conditional on a particular set of laws if we could never know that those laws obtain! Thus, metaphysicians should be particularly interested in the method of empirical symmetries.

What metaphysical questions might this method illuminate? Questions about the structure of spacetime are most familiar. But if we use empirical symmetries, not dynamical symmetries, we may be led to surprising conclusions. Consider classical Newtonian mechanics. And consider a time-dependent rotation: an operation that takes a physical system and puts it into constant rotation over time around some axis. This is not a dynamical symmetry. Indeed, that is the point of Newton's bucket argument: if the water in a bucket is flat, then according to Newton's laws it would go concave if spinning; hence a world that differs only in its state of rotation but not in the concavity of water in buckets would violate Newton's laws. This leads many commentators to the view that the natural spacetime structure for Newtonian mechanics contains a notion of absolute rotation.[16] Still, one might argue that time-dependent rotations are *empirical* symmetries of the theory. As Sklar (1974, p. 230) pointed out, we don't directly *see* that the concave water is rotating; what we see is that it's rotating *relative to water that remains flat*. More generally, we don't directly see which particular state of rotation correlates with inertial effects; we just see that bodies undergoing different inertial effects are rotating relative to one another. Thus there are many possible theories consistent with the data: that no inertial effects correlate with *zero* absolute rotation, that it correlates with a *non-zero* state of rotation R, and so on. *If* these theories are abductively equivalent, then time-dependent rotations are *empirical* symmetries and absolute rotation is unknowable after all. That is a big "if" (to put it mildly!) and I won't try to settle it here; the point is that for someone using the method of empirical symmetry, this is the pertinent question.

The upshot is that if one asks what the natural spacetime structure for Newtonian mechanics *as it is standardly written*, the answer might be that it must contain a notion of absolute rotation. But if one asks what spacetime theory would be best confirmed if one lived in a world with Newtonian observational phenomena, the answer might be the opposite.

I've focused on symmetries that act on spatiotemporal quantities such as boosts and rotations. But the same goes for symmetries that act on other quantities or fields. These include symmetries that act on mass and charge (see Dasgupta, 2013); permutation symmetries in quantum mechanics and statistical mechanics (see entries by Ladyman and Caulton, this volume), and gauge symmetries (see the entry by Teh, this volume). In all these cases we must distinguish the dynamical from the empirical symmetries. Suppose one of these operations is *not* a dynamical symmetry of some theory as standardly written. What that means is that *conditional on the theory* we can detect the quantity altered by the operation. But that leaves open the possibility that the operation is an *empirical* symmetry, in which case we can't detect the quantity because we can't know whether the theory is true.[17]

Notes

1 For a description of Newtonian spacetime see Samaroo (this volume).
2 For representative examples of definitions in the philosophical literature, see Earman (1989, pp. 45–46) and Belot (2013).
3 See Ismael, this volume, for a description of "Galilean" spacetime in which there is no quantity of absolute velocity.
4 Statements of this "method of symmetry" abound in the literature; for a small sample see Earman (1989, p. 46), Ismael and van Fraassen (2003), Roberts (2008), North (2009), Healey (2009), Baker (2010), Belot (2013), and Dasgupta (2016).
5 See for example Baker (2010).
6 Earman (1989, p. 46) might be read as advancing this idea.

7 See North (2009) and Belot (2013) for explicit statements of this idea. It may also be what Earman (1989, p. 46) had in mind.

8 See also Dasgupta and Turner (2017) for some doubts as to whether "minimizing structure" is the norm driving the move from Newtonian to Galilean spacetime.

9 This idea has been defended explicitly by Ismael and van Fraassen (2003), Roberts (2008), Dasgupta (2016), and Ismael (this volume).

10 This style of argument is sometimes gestured at by physicists such as Feynman (see his 1963, p. 15). But it was spelt out explicitly by Roberts (2008); my presentation here is based on Roberts'.

11 The use of mathematical spaces of models or states is widespread, but for some explicit examples see Earman (1989, p. 45), Brading and Castellani (2007, p. 1342), and Baker (2010, p. 1158).

12 See Chalmers (2012) and references therein for more on this notion of epistemic possibility.

13 See Albert (2015), who emphasizes that one also needs information about the initial state of the measuring device. But I bracket this here for simplicity.

14 See, for example, Brading and Castellani (2007, p. 1342).

15 I discuss this issue of indeterminism in Dasgupta (2021).

16 See for example Saunders (2013), Knox (2014), and Wallace (2020).

17 I am deeply grateful to Thomas Barrett, Daniel Berntson, and Alastair Wilson for their extremely helpful and timely comments on an earlier draft of this paper.

References

Albert, D.Z. (2015). The difference between the past and the future. In D.Z. Albert (ed.), *After Physics*. Cambridge, MA: Harvard University Press, 31–70.

Anderson, P. W. (1972). More is different. *Science*, 177(4047): 393–396.

Baker, D. (2010). Symmetry and the metaphysics of physics. *Philosophy Compass*, 5: 1157–1166.

Barrett, T. (2015). Spacetime structure. *Studies in History and Philosophy of Modern Physics*, 51: 37–43.

Belot, G. (2013). Symmetry and equivalence. In R. Batterman (ed.), *The Handbook of Philosophy of Physics*. Oxford: Oxford University Press.

Brading, K. and Castellani, E. (2007). Symmetries and invariances in classical physics. In J. Butterfield and J. Earman (eds.), *Philosophy of Physics*, Part B. Amsterdam: Elsevier, pp. 1331–1367.

Chalmers, D. (2012). *Constructing the World*. Oxford: Oxford University Press.

Dasgupta, S. (2013). "Absolutism vs Comparativism about Quantity" *Oxford Studies in Metaphysics*, 8 (2013): 105–148.

Dasgupta, S. (2016). Symmetry as an epistemic notion (twice over). *The British Journal for the Philosophy of Science*, 67: 837–878.

Dasgupta, S. (2021). How to be a relationalist. *Oxford Studies in Metaphysics: Volume 12*. Oxford: Oxford University Press, pp. 113–163.

Dasgupta, S. and Turner, J. (2017). Postscript. In E. Barnes (ed.), *Current Controversies in Metaphysics*. New York: Routledge, pp. 35–42.

Earman, J. (1989). *World Enough and Space-Time: Absolute versus Relational Theories of Space and Time*. Cambridge, MA: MIT Press.

Feynman, R. (1963). *Lectures on Physics: Volume 1*. Reading, MA: Addison-Wesley.

Healey, R. (2009). Perfect symmetries. *The British Journal for the Philosophy of Science*, 60: 697–720.

Ismael, J. (this volume). *Symmetry and Superfluous Structure: Lessons from History and Tempered Enthusiasm*.

Ismael, J. and B. van Fraassen (2003). Symmetry as a guide to superfluous theoretical structure. In K. Brading and E. Castellani (eds.), *Symmetries in Physics: Philosophical Reflections*. Cambridge: Cambridge University Press, pp. 371–392.

Knox, E. (2014). Newtonian spacetime structure in light of the equivalence principle. *The British Journal for the Philosophy of Science*, 65(4): 863–880.

North, J. (2009). The 'structure' of physics: A case study. *The Journal of Philosophy*, 106: 57–88.

Nozick, R. (2001). *Invariances: The Structure of the Objective World*. Cambridge, MA: Harvard University Press.

Reichenbach, H. (1958). *The Philosophy of Space and Time*. Mineola: Dover Publications.

Roberts, J. (2008). A puzzle about laws, symmetries, and measurability. *The British Journal for the Philosophy of Science*, 59: 143–168.

Saunders, S. (2013). Rethinking Newton's *Principia*. *Philosophy of Science*, 80: 22–48.

Sklar, L. (1974). *Space, Time, and Spacetime*. Berkeley: University of California Press.

Wallace, D. (2020). Fundamental and emergent geometry in Newtonian physics. *The British Journal for the Philosophy of Science*, 71(1): 1–32.

Weyl, H. (1952). *Symmetry*. Princeton, NJ: Princeton University Press.

Further Reading from the Editors

H. Weyl, *Symmetry* (Princeton, NJ: Princeton University Press, 1952) gives the view of a great physicist and mathematician. J. Earman, *World Enough and Space-Time: Absolute versus Relational Theories of Space and Time* (Cambridge, MA: MIT Press, 1989) provides an overview of spacetime structure and symmetries. More general thoughts about the connections between symmetry and metaphysics can be found in D. Baker, 'Symmetry and the Metaphysics of Physics' (*Philosophy Compass* 5: 1157–1166, 2010). A longer overview is G. Belot, 'Symmetry and Equivalence' in R. Batterman (ed.), *The Handbook of Philosophy of Physics* (Oxford: Oxford University Press, 2013). For a development of the views expressed in this chapter, see S. Dasgupta, 'Symmetry as an Epistemic Notion (Twice Over)' (*The British Journal for the Philosophy of Science* 67: 837–878, 2016).

40

SYMMETRY AND SUPERFLUOUS STRUCTURE: LESSONS FROM HISTORY AND TEMPERED ENTHUSIASM

Jenann Ismael

You know you've achieved perfection in design, not when you have nothing more to add, but when you have nothing more to take away. (De Saint-Exupery)

In this chapter, I will discuss a particularly powerful method for identifying and excising superfluous structure in a formalism. Superfluous structure is theoretical structure which plays no role in supporting dynamical behavior.

40.1 Mathematical Methods in Physics

In its most abstract form, an ontology is an account of fundamental degrees of freedom in nature. The metaphysician asks, "What are the independently varying components of nature, their internal degrees of freedom, and the configurations they can assume?" The rationalist metaphysician supposes that we have some form of rational insight into the nature of reality. The naturalistic metaphysician relies on observation and experiment. Her task is to infer ontology from data. Given an ontology and a set of laws, one can generate a range of possible behavior. (There are two ways of doing this. One can apply a combinatorial principle to the basic elements to obtain a set of possible configurations and then apply the laws to generate histories, or one can apply a combinatorial principle to generate all possible sequences of states and then use the laws to narrow those down to physically possible histories.) The naturalistic metaphysician faces an inverse problem: how does she infer backward *from* a range of observed behavior *to* underlying ontology?

Pre-scientific metaphysics was led by imagination. For the likes of Thales, Aristotle, Heraclitus, and Abelard, imagination was the primary tool in trying to develop insight into the nature of the world. But imagination has certain inherent limits. As complexity increases, visualizability decreases. In this setting, mathematics begins to emerge as an essential tool. For one thing, the evidence becomes difficult to survey. Mathematics is needed to provide compact ways of representing the vast bodies of data we now bring to bear on theorizing. For another thing, mathematics gives us new ways of recognizing and disclosing regularities in the data. From Galileo's discovery of the law of the pendulum (1602) to Kepler's discovery of the laws of planetary motion (1609, 1619), the most important discoveries turn out to be less a matter of finding new phenomena than of revealing hidden patterns in the data. And searching is not a matter of putting on boots and going out into the wild, but rather of sitting in the parlor in your slippers and looking through the data on paper for hidden patterns. Brahe was a scientist in the first sense. He had his eyes trained on the stars. Kepler was a scientist in this second sense; he had his eyes trained on the numbers. He saw patterns everywhere. There was the famous scheme with the geometric solids that was supposed to reveal God's geometrical plan for the universe, and his harmonic theory, which linked geometry, cosmology and astrology through a series of musical intervals.[1] He did discover the laws of planetary motion, but he also proposed that the ellipticities of the planet orbits were determined by tunes they hummed as they circled.

Once one has one's eyes on the right quantities, regularities in the phenomena emerge with clarity, but identifying those quantities takes special insight and can be a long and arduous process. Consider Galileo's discovery of the law of the pendulum. When he was 20 years old, Galileo noticed a lamp swinging overhead while he was in a cathedral. Curious to find out how long it took the lamp to swing back and forth, he used his pulse to time large and small swings and noticed something that no one else had apparently realized: viz. that the period of each swing was the same. People had seen swinging lamps before, but they did not have their eyes on the right quantities and (therefore) had not seen the regularity.

What led Galileo to see the regularity? Not just his personal genius. Galileo was educated against the background of Aristotelian science, but there was a complex process that led his attention to the right quantities. Kuhn (1996) remarks on this process:

"Seeing constrained fall, the Aristotelian would measure… the weight of the stone, the vertical height to which it had been raised, and the time required for it to achieve rest. Together with the resistance of the medium, these were the conceptual categories deployed by Aristotelian science when dealing with a falling body…. Galileo saw the swinging stone quite differently. Archimedes' work on floating bodies made the medium non-essential; the impetus theory rendered the motion symmetrical and enduring; and Neo-Platonism directed Galileo's attention to the motion's circular form. He therefore measured only weight, radius, angular displacement, and time per swing … given [these], pendulum-like regularities were very nearly accessible to inspection."[2]

The (sometimes extended) process of zeroing in the invariant relationships can be hidden from view because once the invariant relationships have been identified they get embodied in the classifications we employ. (Invariant relationships are, in general, relative to a set of transformations. The general concept of invariance is a way of capturing what remains fixed when some specified set of other things are allowed to vary.) To call a system a pendulum, for example, is already to assimilate it to a category of things that are characterized by these invariant relationships, but arriving at the *category* of a pendulum is a part of the process of discovery. 'Pendulum' not a category that *pre*-Galilean philosophers employed.

Our brains are attuned to recognizing patterns. We can see that the wall tiling in Figure 40.1 has a repeated pattern. We can see that the top half of the scene in Figure 40.2 is the same as the bottom half, reflected through an axis. We can see the rotationally symmetric pattern of the leaves in Figure 40.3.

These are quite simple patterns, but we can also see more complex ones, those in Figure 40.4.

In saying that we can *see* these patterns, I mean that the apprehension of the pattern is a perceptual phenomenon: it is a part of the processing of sensory information that requires no conscious effort or inference on our part. Recognizing first-order patterns (repetition, symmetry, regularity) in the visual field is something that the biological brain was made to do, and it does it very well. The regularity that Galileo discerned, and the sorts of regularities that get embodied in scientific laws quite generally, are not, however, first-order patterns in an array of visually presented properties. They are typically higher-order invariant relationships between measurable quantities. They are not the kind of thing one *sees* in the way one sees the regularity in the pattern of tiles or the symmetry in the photo of the reflected landscape.

Figure 40.1 Wall tiling with repeated pattern.

Figure 40.2 A reflected scene in water.

Figure 40.3 Circular patterns of cactus leaves.

Figure 40.4 Nested patterns of tiles and architectural details in a mosque.

Mathematical representation – whether in symbolic, geometric, or graphical form – discloses regularities hidden in the phenomena. It is unavoidable in the search for dynamical regularities since in that case the items whose relations are being examined are not presented simultaneously in perception. We need to write down a time-line, draw a trajectory, or otherwise represent a system's history in order to get a look at the relationships between its parts. We use diagrams, graphs, and equations strategically in presentations to highlight relationships that we want to call attention to. In the mathematical investigation that precedes discovery, different notations are used in an exploratory way. The physicist introduces notations that foreground some relationships and suppress others. When she hits on a good representation, patterns emerge and regularities disclose themselves to inspection. One way of describing what she is doing is that she is using mathematical representation to transform a higher-order pattern recognition problem into a first-order problem. She plays around with different ways of representing the phenomena until she finds one in which previously hidden regularities are *rendered* visible. Insodoing, she is using symbolic tools to transform a problem into one that the biological brain can handle. The idea that our ability to use symbolic tools in this way may be the cognitive innovation that underwrites a cascade of characteristically human abilities was originally developed by Karmiloff-Smith (Karmiloff-Smith, A. (1979). *A Functional Approach to Child Language*. Cambridge University Press, and (1986). "From Meta-Process to Conscious Access. *Cognition*, 23, 95–147"). The idea that symbolic tools allow us to represent knowledge in different formats allowing new kinds of cognitive operation and access has been further developed by a number of people.[3]

The process is not just useful for capturing patterns in motion over time. The recognition that two apparently very different kinds of motion are the same is also a matter of recognizing an abstract, higher-order similarity. So, for example, when Galileo compares the motion of a pendulum to that of objects moving on an inclined plane in his notebooks, or Newton compares the motion of a cannonball to that of planets orbiting the sun in the Principia both accompany the explanation with a diagram.

Metaphysics in the old days was a largely solitary project that made use of sparse, mostly qualitative data. Natural philosophers relied on first-order regularities in the behavior of visibly similar things (fire rises, massive objects fall), and a single powerful imagination produces a vision of the world that reproduces those regularities. Metaphysics in the days of mathematical physics, by contrast, uses large bodies of precise, quantitative data. Mathematical methods are employed to identify higher-order patterns hidden in the data. It is really with Newton that this method receives purest expression. It is obvious in retrospect, once you have your eyes on the right quantities where the regularities lie. But it took Newton's physical insight to see beyond the quantities that the impetus theory directed attention to, and fix on force, mass and acceleration as exhibiting the kinds of invariant relationships that could be captured in laws. Whitehead said of Newton that what was so characteristic of him as a natural philosopher was that he had "an eye attuned to the mathematicizable substructure of reality." The same could be said for theoretical physicists quite generally.

40.2 Symmetries as Clues to Superfluous Structure

The first task for the physicist is to capture the regularities. But once we have a precise quantitative description of the regularities in dynamical behavior in hand, another role emerges for mathematics. This one is less obvious and less often talked about in philosophical reflections on science. It is much more recent: it is really only in the 20th century that it has come into its own as an important technique in physical theorizing. It has less to do with the discovery of laws, and more directly to do with discerning the underlying ontology. This is where symmetries play an important role. What symmetries of the right kind do is help identify and excise structure (either structure in the formalism or structure that we've attributed to the object) that is playing no dynamical role. It turns out that here as well, mathematical methods prove much more powerful than the imagination by itself. Indeed, the imagination is playing catch up. We use these methods to arrive at a new formal structure (which we now think captures the intrinsic structure of the physical reality we are trying to represent), and we are left to try to accommodate our imagination. It is really with these methods that physics comes into its own not just as a way of discovering laws, but as a very distinctive way of doing *metaphysics*.

The first domain of application of these methods was spacetime physics. Here the focus is entirely on motion. We have a formalism in hand that captures the laws of motion and the goal is (i) to excise physically insignificant structure in the underlying space, and (ii) to do it cleanly, leaving in place structure that is needed to support the dynamics. Let us begin by looking at dynamical symmetries i.e., symmetries that preserve a theory's laws. (This contrasts with the symmetries of individual solutions. There are further distinctions to be made between continuous and discrete symmetries. Continuous symmetries correspond to continuous changes in the geometry of the system, described by continuous or smooth functions. Discrete symmetries correspond to non-continuous changes in a system, e.g., reflections or rotations through a fixed degree, permutations, or interchanges.) Let us focus on dynamical symmetries that map solutions onto empirically indistinguishable counterparts. Now, we can notice that if we regard those solutions as representing physically distinct situations, we are recognizing structure in the situations depicted by those solutions that is not just unobservable, but indiscernible in a much stronger sense. It generates no measurable effect, which is to say that it plays no role in the law-governed, observable dynamics of bodies, measuring instruments, etc. provided that the equations are formulated in a generally covariant manner. (That requirement forces us to make explicit the features of a system that are crucial to its dynamical behavior.)[4] If, on the other hand, we identify the physical situations represented by those solutions, effectively treating the structure that distinguishes them as a merely notational difference, we eliminate the dynamically irrelevant structure, and we are left with an account of the underlying structure that doesn't recognize physical differences between systems that don't produce differences in the observable, law-governed behavior of the system.

It is now fairly customary to reconstruct the history of spacetime physics as one in which the theoretical progression is driven by a progressive elimination of excess structure, using symmetries as a guide.[5] We start with a dynamics formulated in a space that is quite rich in structure, and then we use the dynamical symmetries to help us identify and whittle away structure that is playing no dynamical role.[6] Newton formulated his dynamics against the background of an absolute space and time with a very rich geometrical structure. Space was attributed the structure of E3. Time was attributed the structure of the real line, and the individual parts of space were thought to persist through time. It was observed that there are symmetries of the laws – Galilean boosts – that map solutions onto empirically indistinguishable solutions that were nevertheless regarded as physically distinct. That meant that Newtonian mechanics with absolute space and time recognizes physical structure that plays no dynamical role. We want to eliminate all and only structure not invariant under Galilean transformations. We find a way to do that by eliminating identity of points over time, retaining the structure to distinguish absolute acceleration but eliminating absolute velocity (retaining a distinction between inertial and non-inertial motion, but elimination a distinction between absolute rest and absolute velocity). There we have the excision of structure that is not playing a dynamical role. The result is Galilean spacetime.

Figure 40.5 shows Newtonian spacetime, with absolute space and time.

Figure 40.6 shows Galilean spacetime. As in Newtonian spacetime, time is absolute and the spatial geometry on every spatial hypersurface is E3. The difference is that there is no identity of points over time.

When phenomena proved resistant to Newtonian treatment, Newtonian physics was replaced by Special Relativity (STR). Since we have new laws, we have to repeat the process. If we want to know the intrinsic structure of the space that supports STR, we take the laws of STR and look at their symmetries. STR laws are invariant under Poincare transformations and Poincare transformations

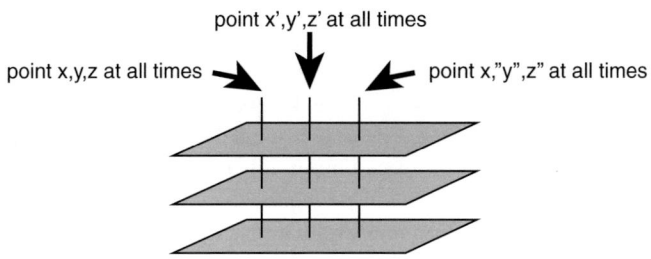

Figure 40.5 Newtonian spacetime.

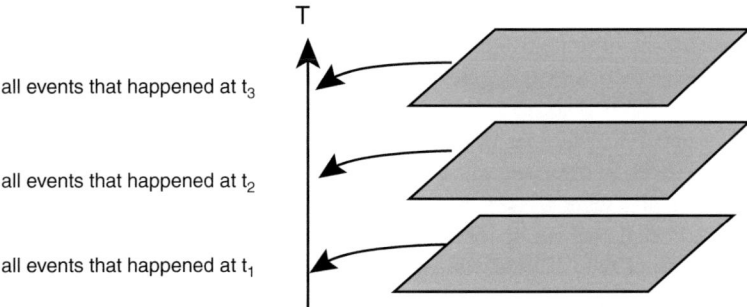

Figure 40.6 Galilean spacetime.

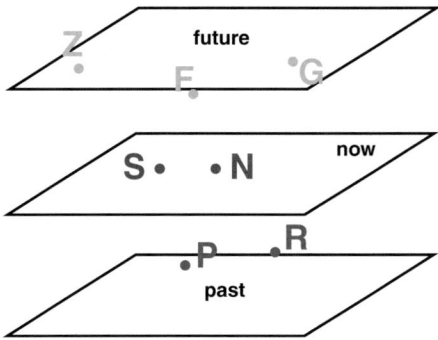

Figure 40.7 Planes of absolute simultaneity in Galilean spacetime.

map solutions onto empirically indiscernible counterparts, so we eliminate geometric structure not invariant under those transformations. The result is Minkowski spacetime.

Figure 40.7 shows Galilean spacetime. In Galilean spacetime, we don't have identity of points over time, but we have planes of absolute simultaneity.

Those have been eliminated in Minkowski spacetime. In its place, we have only the light cone structure.

The schema for arriving at a spacetime that excises structure not invariant under symmetries follows this pattern. Start with the space of solutions to the laws. Consider the symmetries of the laws that map solutions onto empirically indiscernible solutions. Those symmetries define an equivalence class of solutions that we now regard as physically equivalent. Structure not invariant under the symmetries (i.e., structure that distinguishes solutions in the same equivalence class) is interpreted as physically *in*significant. Structure that is invariant (i.e., structure shared by solutions in the same equivalence class) is regarded as physically significant. Of course, this is a rather indirect method of representation, and it is better when we can replace the equivalence classes with an intrinsic representation of that geometrical structure but it serves well as a way of capturing the logic of the reasoning. The state-space obtained in this way is known as the reduced state-space. I described this deliberately in the most general way. The only notions I used were (i) a very general notion of symmetry (the mappings of the solution space onto itself), and (ii) empirical indistinguishability. We could build more structure into the state-space and require that the transformations that are candidates for interpretation as equivalence relations preserve that structure. So, for example, we might require that transformations that are candidates for interpretation as equivalence relations preserve differences that produce measurable effects when coupled with neighboring theories, or we might have other reasons for thinking that differences between models related by a given transformation are physically real. Whatever additional requirements we impose, however, have to be motivated in physical terms.

What is characteristic of the move to a reduced state-space is a simplification of the ontology in the sense that there is a reduction in degrees of freedom that are recognized in the state of the systems being represented. Gordon Belot carries out the construction and shows how degrees of freedom get eliminated in detail, for the case of classical mechanics.[7] The state-space Newton used for his mechanics of N particles is 6N-dimensional. In constructing the reduced space of the Newtonian mechanics by identifying states related by Galilean symmetries, we reduce the theory's degrees of freedom by 10 (so the reduced state-space has 6N-10 degrees of freedom). Quantities like the position and the linear motion of the universe's center of mass have been eliminated. Acceleration, but not position or velocity, remains absolute.

The inference from "S and S$\star$ are related by a spacetime symmetry" to "S and S$\star$ are physically equivalent" has been called "Leibniz equivalence," because it is a modern way of motivating Leibnizian conclusions about the identity of spacetimes that exhibit the right kind of indiscernibility.[8] We can apply Leibniz equivalence to general relativity (GR). Unlike the cases we just considered, the dynamics of GR isn't defined against the fixed background of a single spacetime. Instead, many distinct spacetimes (with their own symmetries) are solutions of the same theory. Since we are interested in symmetries of the *laws* that map solutions onto *empirically indistinguishable* solutions, in GR we look to the diffeomorphisms. Applying Leibniz equivalence, a physical spacetime corresponds to a diffeomorphism equivalence class of mathematical spacetimes. The structure transformed by diffeomorphisms is then regarded as physically insignificant.

It turns out that this has one nice consequence. It permits the interpreter of GR to avoid the Hole Argument and the chanceless, unobservable indeterminism that the argument would entail if diffeomorphism-related solutions were regarded as representing physically distinct spacetimes. (If one does assume, contrary to the present approach, that there are distinct physical possibilities related by diffeomorphisms, a type of unobservable indeterminism arises in GR. This is because a diffeomorphism can sometimes leave a particular surface of simultaneity unchanged while shuffling around what happens at which point in the future of that surface. Since the shuffling changes nothing invariant under GR's symmetries, the indeterminism disappears on the present approach, where only such invariants are regarded as physically real.)[9] But the appeal to Leibniz equivalence also leads to new questions. So, for example, if we excise a structure that is not diffeomorphism invariant, then it would seem that the only quantities that remain are constant along gauge orbits. It turns out (as Earman (2002) pointed out) that these quantities don't evolve, so it seems we are left with the result that there is no temporal change. This initiates discussion of what that could mean. Maudlin has argued that this is an absurd result. Healey has argued that there's a mistake in Maudlin's reasoning. So this is the beginning rather than the end of the discussion. But that's the method. You apply it and see what you get. You use the symmetries as *clues* to the presence of excess structure (either as excess structure attributed to the world, or redundancy in the formalism), but they are *only* clues, part of the naturalistic metaphysician's toolkit. One has to work through the results of applying the method to see whether it generates a physically sensible conclusion, and it all has to be done judgment and care.

40.3   Remarks

A few remarks are in order here.

First, this sort of reasoning has become enshrined as part of the standard way for presenting the theoretical progression in spacetime physics. It goes along with the very big theoretical shift from thinking of spacetime as the fixed background against which all of physics takes place to thinking of it as part of the subject matter of physics. It canonizes the accepted justification for rejecting Newtonian spacetime in favor of Galilean, motivating Minkowski spacetime in STR, and guiding the interpretation of GR. But it is not an actual, on-the-ground historical description of the messy set of facts that drive the opinion of the individual scientist or the community as a whole. There was some of this kind of reasoning on the ground, but it was tangled up with less fruitful lines of reasoning, and it emerged with full clarity only in retrospect. (Anyone steeped in the actual history will notice that I didn't mention coordinate systems. These played a central, and mostly unfortunate role in the actual history.)[10] In representing the history of physics for "textbook" purposes, we tend to bring into relief arguments that lead us to what we now think is the right conclusion, forgetting the false starts and confusions, portraying the theoretical progression as a rational exchange of one theory for another. But that also shows that this kind of history has normative component that actual history does not. The importance of this reasoning has emerged through backward-looking filters. We now know which avenues of development turned out to be fruitful. The authority of this method of identifying and excising structure derives from its historical success in leading to those avenues.

Second, translating this reasoning into a technical setting is non-trivial. I used a very general notion of symmetry that is much wider than those that are used even in a classical setting when we talk of the symmetry group of a theory. Because of that, I needed to invoke a notion of empirical indistinguishability to narrow down the class of symmetries that are candidates for defining equivalence classes. Without that restriction, the directive to identify solutions related by symmetries of the laws would mean no theory would have more than one solution, so we would lose all of those physically real distinctions between solutions that represent *manifestly different* situations.[11] The problem, however, is that empirical indistinguishability is a notion that we impose from the outside. It is not rigorously defined in a general way. It may be that we don't need to insist on a hard line between observable and unobservable quantities so long as our theories have a certain kind of causal integration (Albert, 2015). If we have a device D whose pointer observables are uncontroversially observable, then any physical quantity that can be made, by any physical process, no matter how indirect, to influence the pointer observable on D will then count as producing observable differences. The procedure will fail only for a theory that postulates equivocally observable quantities that can't be linked to any unequivocally observable quantity by any procedure.

There *are* more specific, rigorously definable notions of symmetry on offer in classical theories (e.g., the generalized, or Hamiltonian symmetries, or variational symmetries).[12] The reason for not employing those in this capacity (i.e., as identifying those symmetries that alert us to the presence of excess structure) is (i) we don't want to tie the relevant notion of symmetry too tightly to any particular formalism, and (ii) they won't, in any case, do the work that we want them to do here.

Third, there are few who would deny that (i) if you have a formalism that is empirically adequate, and you can identify structure that is playing no physically meaningful role. But generating testable predictions either with respect to the phenomena that are the primary domain of that theory *or* in coupling with other theories to generate that fall outside that domain, it should be excised and (ii) that symmetries of the laws that map solutions onto empirically indistinguishable counterparts can be a good guide at identifying structure that plays no physically meaningful role. But it can be very difficult to implement this reasoning in the complex setting of a real theory. A clean separation of physically significant from physically insignificant structure is often very hard to achieve. It usually requires the development of new formalisms. And even when we can separate the structures, the interpretation of the results is difficult. The re-education of the imagination that was needed in the move from Newtonian to Galilean, and from Galilean to Minkowski spacetime, provide good examples. It is something that is even more amply illustrated in the case of modern gauge theories, so we turn now to those.

40.4   Gauge Theories

When we turn to non-spacetime theories such as electrostatics, electromagnetism, or quantum mechanics, we look to internal symmetries, i.e., transformations that alter the values of physical quantities that are supposed to represent the *internal degrees of freedom* pertaining to a system (or the matter in some volume of spacetime) rather than changing anything spatially or temporally external to it. We distinguish global internal symmetries from local symmetries. Global internal symmetries (such as the electrostatic potential, or phase transformations in quantum mechanics) act identically on the values of all points throughout spacetime. Local symmetries diverge from point to point. The best-known example of a theory with local gauge symmetry is classical electrodynamics. The vector potential is a four-dimensional vector; the time component is the familiar electric potential (V) while the space components collectively form the magnetic potential (A). Since physical predictions are fixed by the field tensor which is left unchanged if we add the gradient of a scalar, if we follow the kind of reasoning we applied above the most natural thing would be to regard structure not invariant under this transformation as physically unreal. The discovery and experimental confirmation of the Aharonov-Bohm effect, however, seemed to many to present an insuperable obstacle to that line of reasoning because it showed that changes in the vector potential lead to a measurable difference in the behavior of quantum particles passing by solenoids despite the zero-field in their vicinity (Vaidman, 2012). One *could* say that there is action at a distance from the field inside the solenoid, but if we want to avoid a new form of non-locality, we have to work a little harder to separate physically significant from physically insignificant structure. So begins the search for a new ontology for classical electromagnetism.

40.5   Where This Reasoning Leads

One of the most interesting aspects of this kind of reasoning from a philosophical point of view is that after getting rid of the surplus, we are left with quantities that aren't localized in point-sized regions of space. On Healey's view, the fundamental quantities are the "holonomies," defined on closed loops in spacetime (Healey, 2007).[13] (Holonomies are images of oriented closed curves that can be constructed from line integrals of the gauge potential along closed curves.) On another view – discussed by Maudlin[14] – they are gauge-invariant values of the connection, which are (in Maudlin's parlance) "hyper-local," so that there is no determinate matter of fact about whether distant spacetime points agree as to their value. Either way, the clues we get from symmetry in this setting push us in new ontological directions. They push against common sense and they push against programs in analytic metaphysics. One of the defining principles of David Lewis' metaphysical program, for example – which occupies a very large share of research in contemporary analytic metaphysics – is that the fundamental building blocks of the universe are point-sized events in spacetime.

More specifically, when we excise structure that is not invariant under local and global gauge transformations, we are left with fundamental quantities that aren't point-sized. Something similar happens in quantum mechanics with respect to phase transformations. In this case, when we "mod out" the structure not invariant under phase transformations, we are left with a failure of Lewis's Humean Supervenience program, because the state of a complex system of spatially separated parts does not supervene on the states of the regions in which the parts are located.[15] All of these point to the failure of the idea that the world is composed of spatially localized degrees of freedom, which jointly determine the state of the world as a whole.

When we have this kind of conflict – i.e., when the methods described here which use purely mathematical clues to isolate the basic degrees of freedom in nature push against the programs in analytic metaphysics – it forces us to choose allegiances. Do we let our ideas about ontology be led entirely by physics, or is there some other form of reasoning embodied in the methods of analytic philosophy that has authority? A first step in the direction of providing an answer to that question is getting a clear idea of what is driving ontology in physics. I think that if there is a method that can be discerned in the bewildering detail and specificity of discussions in physics about fundamental ontology, this is a big part of it. We find laws that capture the invariant relationships, and then we use symmetries to identify and excise structure that is not playing a dynamical role.

There is an important footnote to this debate. David Wallace has recently challenged the role that the Aharonov-Bohm effect has played in this discussion (Wallace MS). He argues that if we keep track of the complex scalar field to which the magnetic vector potential is coupled, we find gauge-invariant features of the scalar field and potential together that explain the effect and can be given a local characterization. If he is right, the interpretation of gauge –invariance as permuting physically insignificant structure be restored without non-locality, and the more radical conclusions can be avoided. As can be seen from the ongoing nature of these debates, these issues are unsettled, and we can hope to gain some clarity in years to come.

40.6   Difficulties of Application

Some general remarks about all of this worth emphasizing:

First, it is not a simple and straightforward judgment that some bit of structure is not playing an empirically meaningful role. There is no way around the case-by-case careful judgment about what is empirically significant and what is not. *All* of the phenomena have to be considered, including phenomena that are generated when a theory is coupled to a theory outside its primary domain of application. The Aharonov-Bohm effect provides a good example of this. The effect emerges only when an electromagnetic field is coupled to a complex vector field, and so it doesn't emerge until classical electromagnetism is combined with a theory that describes the behavior of quantum particles.

Second, as I've emphasized, even when we've become convinced that there is some structure that is not physically meaningful, excising that structure leaving in place all the physically meaningful structure is far from easy. The sanitized history of spacetime theories made it look as though we identify empirically superfluous structure and go in to do a clean little surgery to remove the offending formation, leaving everything else in place. In retrospect, it looks clear as day. But while we are working in a formalism in which the offending structure is bound up with living tissue, the discussion is complex and extremely hard to sort out. Hence the difficulty of the ongoing discussion of gauge theories.

Third, applying the method is *never* free of interpretive assumptions. There is no *purely* formal way of recognizing a redundant structure. As I mentioned above, if we simply identify all solutions related by symmetries of the laws, we end up with a single solution to every set of laws. Before we have a structure to which we can apply these sorts of symmetry probing considerations without that kind of triviality, we need a set of differences that can function as sufficient conditions for distinctness. That has to come from outside the formalism we are working with. I have suggested empirical distinguishability as a generic criterion because it captures the reasoning that we are trying to apply in the murkier, technically involved setting of modern theory. The reasoning is that ultimately we are only warranted in postulating unobservable structure that plays a role in supporting observable dynamical behavior, and so we should be getting rid of structure that doesn't play such a role. The use of symmetries to identify structure that plays no dynamical role, however, need not be applied in this general setting. It can also be powerful when applied in a setting in which we are willing to take certain structures for granted as representing real physical differences for purposes at hand, and we are interested in seeing whether some other sets of structures play a physically meaningful role (the role of a difference-maker) in producing differences at that level. The only way in which this reasoning can be applied in a *theory-neutral* way is if there is some basic set of differences that are regarded *pre-theoretically* as sufficient conditions for distinctness.

Finally, it is a little misleading to say that structures related by a symmetry of the laws never have any physical interpretation at all, as we will see in the next section.

40.7   Justification

There are difficult, open questions about why the method works. There is no a priori *logical* principle that guarantees that this method for limning the structure of our theories won't lead us to eliminate a structure that represents something real. We can certainly imagine worlds in which that would be so. Consider, for example, Newton's world with absolute time and space or the non-localized gauge potential properties view considered by Healey. These are perfectly intelligible empirical possibilities.

Given any rule that says "whenever S is a symmetry of the laws and S maps a onto empirically indistinguishable counterpart b, then a=b", we can immediately construct a counterexample in the form of a universe in which S permutes physically real elements but in which S is a contingent symmetry. Indeed combinatorial principles will entail that for any account of what the basic elements are, there will be some metaphysically possible distinct situations a and b that are related by S. If S is reflection through a plane P, and the laws are symmetric with respect to S, there will be bilaterally symmetric solutions to the laws. So the inference from "S is a symmetry of the laws and S maps a onto empirically indistinguishable counterpart b," to "a=b," would fail.

There are, nevertheless, two broad routes to rationalizing the method that suggests themselves as promising.

(i) The first is the one that I adverted to above: there is a weaker form of non-logical rationality, which we might call "evidential rationality," that shifts the burden of proof. We start from the principle that we aren't warranted in distinguishing elements related by certain kinds of symmetry that we know only indirectly, as elements in a structure interpolated behind experience, unless differences between those elements produce empirically distinguishable effects. Then we point out that someone who recognizes structure that maps solutions of the laws onto empirically indistinguishable solutions can be faulted for recognizing structure that they can have no evidence for. That is a strong intuition, but making it more than an intuition would require showing that that given a body of data that respects the symmetry is much more likely to have been produced by a situation in which a=b than a world in which a≠b, i.e., the probability of evidence obeying the symmetry being generated by an a=b world is higher than a a≠b world. The problem is that these kinds of probabilistic claims always depend on a measure over possible worlds and can be reversed with different choices of measure.

(ii) The second focuses on issues about reference and argues that if we have a theory in which we purport to refer to distinct elements that are not distinguishable by any observation, or their relations to anything observable (e.g., two points in absolute space at different times, or internal properties that don't make a discernibly different impact on measuring devices), we have not really managed to do so. We have not actually managed, that is to say, to get our referential hooks into two separate elements of reality. What we have, in that case, is actually two names (coined by rigidified definite descriptions, obtained from the Ramsey sentence of our theory) for the same thing (a coarse-grained object constituted by the pair or more generally, an equivalence class of weakly indiscernible objects). This way of rationalizing the method would have the advantage of reducing every case of recognizing too much structure in the world to a case of unacknowledged redundancy in the formalism. But it is a tough row to hoe because it will have to rest on questions about how it is that terms in a theoretical formalism come to represent (or refer to) some particular bit of the natural world. That is an issue that is very much contested and very hard to see one's way through.

The question of why the method works is bound up with the theoretical preference for simpler theories. There is a temptation to try to justify the method by appeal to Ockham's razor. That temptation should be resisted.[16] Instead of looking to Ockham's razor for justification of the method, better to look at the logic behind the reasoning directly and see if it captures the valid heart of Ockham's razor. It might be that the preference for simpler theories is a generalized version of the method discussed here. The idea here would be that whenever you are choosing between competing, empirically adequate formalisms, one of which is simpler (where simplicity is measured in the most obvious way, by counting up the number of basic degrees of freedom postulated), it is a good methodological bet to suppose that the more complex theory has some redundancy in the formalism.

This might not be the right way to put it to the extent that it suggests we are identifying structures that have no physical interpretation at all, i.e., part of the mathematical machinery that is empty of any physical significance. But that is not quite right. Rovelli makes the point elegantly in "Why Gauge?" (http://arxiv.org/abs/1308.5599). In the paradigm cases, the structures in question are not physically insignificant, but rather implicitly relativized to a reference frame that can be embodied in another physical object (like an observer or a measuring device). Frame-dependent quantities like "being simultaneous with," or "being at rest" are real and measurable (indeed, in some cases, directly observable), but they are not absolute. So, the problem was not that we were regarding as real structures that are "merely mathematical," but that we were regarding as absolute structures that are implicitly relational. The relations are invariant and when we couple to another object we can "measure" the non-invariant quantities directly (though we can't translate them into a description of the invariant quantities when we don't know our own situation, so to speak). So when we see differences in the values of those quantities, we are seeing differences in the perspective from which the object is viewed rather than differences intrinsic to the object. In a Minkowski spacetime, for example, observers traveling non-inertially relative to one another disagree on simultaneity in the way that observers in different locations disagree on what is nearby. Just so, Rovelli argues, observers who see changes in the values of gauge-dependent quantities are seeing changes in their relations to the gauge-invariant quantities that characterize the intrinsic properties of a system. None of this means that there aren't some symmetries that permute physically empty features of notation. It is only to suggest that gauge invariance may not be best understood that way.

The real authority of the method, in practice, rests on its historical success at leading to fruitful lines of development. It is one of the most distinctive methods of modern physics and the question of why it works is part of the more general puzzle of why physics works.

40.8   Conclusion

I have given the very broad-brush historical background and articulated the kind of reasoning behind the use of symmetries to zero in on the appropriate ontology for a physical theory, mostly by way of providing some context. The discussion now passes to the highly technical and highly specific issues about how to separate the physically significant from the physically insignificant structure in particular theories carried on by those in the trenches, so to speak. The issues are unavoidably complex, unavoidably empirical, and they demand looking at theories in full detail, one by one.

Acknowledgment

I would like to thank Jim Weatherall, Jeff Barrett, Richard Healey, Gordon Belot, Chris Smeenk, Giovanni Valente, Hans Halvorson, and other participants in the workshop on the Foundations of

Gauge Theories. This paper exists because of that invitation, and I learned a great deal from the discussion there. Special thanks to Alastair Wilson for insightful comments on an earlier draft.

Notes

1 See Field and Bowden (1994), Harman (1990), Barker (2002).
2 Kuhn (1996, p. 124).
3 For a sampling, see Karmiloff-Smith (1992), Clark (1993), Clark and Karmiloff-Smith (1993), Clark (1998), Dennett (1994).
4 See Ismael (1997), Norton (1993).
5 Baker (2010), Brading and Castellani (2003), Maudlin (2012), Friedman (1983).
6 See Dasgupta (this volume).
7 Belot (2001a, 2001b).
8 Norton (2015), Carroll (1997). See also the chapter by Pooley (this volume).
9 Norton (2011), ibid.
10 Michael Friedman (1983) contains an historical overview.
11 Van Fraassen and Ismael (2003) make a similar suggestion.
12 Castrillon Lopez and Marsden (2008), Cantwell (2002).
13 Healey (2007); also Leeds (1999), Maudlin (1998).
14 Maudlin (2007, pp. 78–103).
15 See Healey (this volume) for further discussion.
16 There are as many different versions of Ockham's razor as there are notions of simplicity; Sober (2015) is an excellent discussion of different versions of the principle and its various applications.

References

Albert, D. (2015). The technique of significables. In *After Physics*. Cambridge: Harvard University Press.
Baker, D.J. (2010). Symmetry and the metaphysics of physics. *Philosophy Compass*, 5(12): 1157–1166.
Barker, P. (2002). Constructing copernicus. *Perspectives on Science*, 10(2): 208–227.
Belot, G. (2001a). Symmetry and equivalence. In R. Batterman (ed.), *Oxford Handbook of Philosophy of Physics*. New York: Oxford University Press, pp. 318–339.
Belot, G. (2001b). The principle of sufficient reason. *Journal of Philosophy*, XCVIII: 55–74.
Brading, K.A. and Castellani, E. (2003). *Symmetries in Physics: Philosophical Reflections*. Cambridge: Cambridge University Press.
Cantwell, B. (2002). *Introduction to Symmetry Analysis*. Cambridge: Cambridge University Press.
Carroll, S. (1997). *Lecture notes on general relativity*. Available at: http://arxiv.org/abs/gr-qc/9712019.
Castrillon, L. and Marsden, J.E. (2008). Covariant and dynamical reduction for principal bundle field theories. *Annals of Global Analysis and Geometry*, 34: 263–285.
Clark, A. (1993). *Associative Engines: Connectionism, Concepts and Representational Change*. Cambridge, MA: MIT Press.
Clark, A. (1998). Magic words: How language augments human computation. In P. Carruthers and J. Boucher (eds.), *Language and Thought: Interdisciplinary Themes*. Cambridge: Cambridge University Press, pp. 162–183.
Clark, A. and Karmiloff-Smith, A. (1993). The Cognizer's innards: A psychological & philosophical perspective on the development of thought. *Mind & Language*, 8(4): 487–519.
Dennett, D. (1994). Labeling and learning (Commentary on Clark & Karmiloff-Smith 1994). *Mind and Language*, 8: 540–548.
Earman. (2002). Toroughly Modern Mctaggart: Or, what Mctaggart would have said if he had read the general theory of relativity. Philosopher's Imprint, 2: 1–28.
Field, V. and Bowden, M.E. (1994). Kepler's geometrical cosmology. *Annals of Science*, 51(1): 95–97.
van Fraassen, B. and Ismael, J. (2003). Symmetry as a guide to superfluous theoretical structure. In E. Castellani and K. Brading (eds.), *Symmetries in Physics: Philosophical Reflections*. Cambridge: Cambridge University Press, pp. 371–392.
Friedman, M. (1983). *Foundations of Space-Time Theories: Relativistic Physics and Philosophy of Science*. Princeton, NJ: Princeton University Press.
Harman, P.M. (1990). Kepler's geometrical cosmology. *International Studies in Philosophy*, 22(3): 103.
Healey, R. (2007). *Gauging What's Real*. Cambridge: Cambridge University Press.
Ismael, J. (1997). Curie's principle. *Synthèse*, 110(2): 167–190.

Karmiloff-Smith, A. (1992). *Beyond Modularity: A Developmental Perspective on Cognitive Science*. Cambridge, MA: MIT Press/Bradford Books.

Kuhn, T. (1996). *The Structure of Scientific Revolutions* (3rd edition). Chicago, IL: University of Chicago Press.

Leeds, S. (1999). Gauges: Aharonov, Bohm, Yang, Healey. *Philosophy of Science*, 66: 606–627.

Maudlin, T. (1998). Healey on the Aharonov-Bohm effect. *Philosophy of Science*, 65: 361–368.

Maudlin, T. (2007). *The Metaphysics Within Physics*. Oxford: Oxford University Press.

Maudlin, T. (2012). *Philosophy of Physics: Space and Time*. Princeton, NJ: Princeton University Press.

Norton, J. (1993). General covariance and the foundations of general relativity: Eight decades of dispute. *Reports of Progress in Physics*, 56: 791–861.

Norton, J.D. (2015). The hole argument. In E.N. Zalta (ed.), *The Stanford Encyclopedia of Philosophy* (Fall 2015 Edition). Available at: https://plato.stanford.edu/archives/fall2015/entries/spacetime-holearg/.

Rovelli, C. (2014). Why gauge? *Foundations of Physics*, 44(1): 91–104.

Sober, E. (2015). *Ockham's Razors: A User's Manual*. Cambridge: Cambridge University Press.

Vaidman, L. (2012). On the role of potentials in the Aharonov-Bohm effect. *Physical Review A*, 86: 040101.

Wallace, D. (MS). *Deflating the Aharonov-Bohm effect*. Available at: https://arxiv.org/abs/1407.5073.

Further Reading from the Editors

Michael Friedman's *Foundations of Space-Time Theories: Relativistic Physics and Philosophy of Science* (Princeton University Press, 1983) is a classic book on spacetime (and hence its symmetries) at a relatively high technical level. K. Brading and E. Castellani (eds). *Symmetries in Physics: Philosophical Reflections* (Cambridge University Press, 2003) is a classic collection of excellent papers. B. Van Fraassen and J. Ismael, "Symmetry as a Guide to Superfluous Theoretical Structure" in that volume is particularly relevant to the issues here. K. Brading and H. Brown, 'Symmetries and Noether's theorems', also in that volume, is also essential reading for those unfamiliar with the importance of symmetry principles.

41

PERMUTATIONS

Adam Caulton

41.1 Introduction: What Are Permutations?

Given any set X, consider the set of bijections $\pi : X \to X$. This set can be endowed with a natural group structure: the identity e maps each object in X to itself and the binary group operation $\circ$ is just functional composition, i.e., $(\pi_2 \circ \pi_1)(a) := \pi_2(\pi_1(a))$ for all $a \in X$. (In future, we will represent composition of group elements by simple concatenation; e.g., $\pi_2\pi_1$ instead of $\pi_2 \circ \pi_1$.) The resulting group is called the *symmetric group on X*, denoted S_X or $\mathrm{Sym}(X)$. The order (i.e., cardinality) of S_X is $|X|!$, where $|X|$ is the cardinality of X.

In the case where $X = \{1, 2, \ldots, N\}$ for some positive natural number N, the corresponding group is often called *the symmetric group on N symbols* and is denoted S_N. S_N is the typical focus when considering permutations of *any* collection of N objects, since we usually care only about the group structure, for which S_N is taken as the archetype.

Any subgroup of a symmetric group is called a *permutation group*. In a certain sense, the study of permutation groups encompasses *all* groups. This is due to Cayley's Theorem in that any group G is isomorphic to some subgroup of $\mathrm{Sym}(G)$. So permutations are ubiquitous! However, in this article I will restrict focus to the finite symmetric groups S_N, especially as they are realized or represented as groups of permutations on physical systems, especially particles. I will not have the space to go into detail about the theory of the symmetric and permutation groups. For this consult Sagan (1991) and Dixon and Mortimer (1996).

I just said that the kinds of permutations we will consider are those which act on particles, but in fact that characterization already assumes too much. For, it is a central interpretative dispute whether or not permutations, as they act on a theory's state space, should be interpreted as really permutations of physical particles at all, as opposed to a mere reshuffling of arbitrary labels for those particles. This distinction may bring to mind the distinction between active and passive coordinate transformations: roughly, active transformations represent genuine physical changes to the system, while passive transformations are mere relabelings, and so preserve the physical state. The dispute is over whether permutations, at least as they are usually used in physical theory, have an active interpretation at all. If they do not, then any permutation preserves the physical state, and so nothing—i.e., nothing except for our means of *representing* our target physical system—can really have been permuted.

I aim to articulate this dispute in some generality, and then to illustrate it for a number of physical theories. The theories I will consider are: classical Hamiltonian mechanics, classical statistical mechanics, and quantum mechanics. A recurring theme will be to what degree each version of the dispute may be settled. We will see that in no case is there a direct and unequivocal verdict to be gleaned from empirical evidence; but that there are, in some cases, more subtle considerations which may bear on the matter.

41.2   Permutations in Logic and Model Theory

Models, in roughly the sense of relational structures, offer a very general and systematic means of representing possibilities. It will therefore be helpful to briefly review the treatment of permutations in this general context, before we look at specific physical theories. (For comprehensive treatments of model theory, see Hodges, 1993; Button and Walsh, 2018.)

For the sake of simplicity, I will concentrate on first-order structures of the form $\mathfrak{A} = \langle A, \mathcal{R} \rangle$ and call them *models*. The *domain* of $\mathfrak{A}$, A, is just any old set. The *structure* of $\mathfrak{A}$, $\mathcal{R} = \langle R_1, \ldots, R_m \rangle$, is a sequence of relations on A; so, if R_i is an n-ary relation, then $R_i \subseteq A^n$. In the context of the semantics for a first-order language, each R_i is the extension assigned by $\mathfrak{A}$ to an n-ary predicate symbol. Following Quine and again for the sake of simplicity, I restrict attention to models whose structures comprise only relations; distinguished elements (extensions assigned to constant terms) and functions may be replaced without loss by relations in the usual way (see Quine, 1986, pp. 25–26).

41.2.1   *Permutations and Permutability*

Given any model $\mathfrak{A} = \langle A, \mathcal{R} \rangle$ with domain A and relations $\mathcal{R} = \langle R_1, \ldots, R_m \rangle$, any permutation $\pi : A \to A$ induces a *lift* π^* on models with the same domain A (Button and Walsh, 2018, §2.1 call this the *Push-Through* construction):

For each n-ary $R \in \mathcal{R}$ and all $a_1, \ldots, a_n \in A$: $\langle a_1, \ldots, a_n \rangle \in \pi^*R$ iff $\langle \pi(a_1), \ldots, \pi(a_n) \rangle \in R$. Then we may define $\pi^*\mathcal{R} := \langle \pi^*(R_1), \ldots, \pi^*(R_m) \rangle$ and $\pi^*\mathfrak{A} := \langle A, \pi^*\mathcal{R} \rangle$. π^* is called the *lift* of π, since while π acts on the objects *in* A, π^* acts on models, all of which have A as their domain. By construction, π constitutes an *isomorphism* between $\mathfrak{A}$ and $\pi^*\mathfrak{A}$.

Any permutation $\pi : A \to A$ is a *symmetry*, a.k.a. an *automorphism*, of $\mathfrak{A}$ iff $\pi^*\mathfrak{A} = \mathfrak{A}$, i.e., it *fixes*, or leaves invariant, the structure of $\mathfrak{A}$. The symmetries of $\mathfrak{A}$ form a subgroup $\text{Aut}(\mathfrak{A})$ of the group $\text{Sym}(A)$ of all permutations on $\mathfrak{A}$'s domain. Any model $\mathfrak{A}$ is *rigid* iff its only symmetry is the identity on A, i.e., $\text{Aut}(\mathfrak{A}) = \{e\}$. I will call a model *symmetric* iff it is not rigid, i.e., has a non-trivial symmetry, and *totally symmetric* iff every permutation on its domain is a symmetry, i.e., $\text{Aut}(\mathfrak{A}) = \text{Sym}(A)$. For example, the complex field $\mathbb{C}$ is symmetric, since $\pi(z) = z^*$ (each complex number is mapped to its complex conjugate) is a non-trivial symmetry, but not totally symmetric, since it has plenty of permutations that are not symmetries, such as $\pi(z) = -z$.

Using the symmetries of $\mathfrak{A}$ we may define the relation $\sim_{\mathfrak{A}}$ of *permutability in* $\mathfrak{A}$ on $\mathfrak{A}$'s domain, as follows: for any $a, b \in A$, $a \sim_{\mathfrak{A}} b$ iff there is some permutation $\pi : A \to A$ such that π is a symmetry of $\mathfrak{A}$ and $\pi(a) = b$. For example, the imaginary numbers i and $-i$ are permutable in $\mathbb{C}$. Permutability is the semantic counterpart of the syntactic notion of *absolute indiscernibility*: roughly, indiscernibility by monadic formulae that do not contain individual constants or equality. I will not go further into the issue of indiscernibility; interested readers should consult Quine (1976), Saunders (2003a, 2003b, 2013), Ketland (2011), Caulton and Butterfield (2012), Ladyman et al. (2012), and Muller (2015).

Permutability in $\mathfrak{A}$ is an equivalence relation, and so we can define permutability equivalence classes $[a]_{\mathfrak{A}} := \{b \in A : a \sim_{\mathfrak{A}} b\}$. Each equivalence class is the *orbit*, under the symmetries of $\mathfrak{A}$, of any one of its elements. I shall call any object $a \in A$ an *individual in* $\mathfrak{A}$ iff it is *fixed* by every symmetry of $\mathfrak{A}$, i.e., it is permutable only with itself (so $[a]_{\mathfrak{A}} = \{a\}$). For example, each rational number is an individual in the field $\mathbb{C}$, since any symmetry of $\mathbb{C}$ leaves the rational numbers untouched.

The notion of individuality just introduced is closely connected to the model-theoretic notion of definability. An element $a \in A$ is *definable* in $\mathfrak{A}$ if a uniquely satisfies some (first-order) monadic formula in the language that $\mathfrak{A}$ interprets. If a is definable in $\mathfrak{A}$, then a is an individual in $\mathfrak{A}$. In all finite models and some infinite models, definability and individuality coincide; but typically the expressive resources of the language will fall short of making every individual definable.

Given any model $\mathfrak{A} = \langle A, \mathcal{R} \rangle$ with $\mathcal{R} = \langle R_1, \ldots R_m \rangle$, we can construct another model whose objects are $\mathfrak{A}$'s permutability classes and whose structure closely resembles $\mathfrak{A}$'s. The idea is to obtain a model *all* of whose objects are individuals (i.e., not permutable with anything but themselves)—so that the new model we obtain is rigid—but which is in all other respects as much like $\mathfrak{A}$ as possible. This model is a kind of a *quotient model*: it is the quotient of $\mathfrak{A}$ *under $\mathfrak{A}$'s symmetries*. It is often denoted $\mathfrak{A}/\mathrm{Aut}(\mathfrak{A})$. It is defined as $\mathfrak{A}/\mathrm{Aut}(\mathfrak{A}) := \langle A/\sim_{\mathfrak{A}}, \tilde{\mathcal{R}} \rangle$, where $\tilde{\mathcal{R}} = \langle \tilde{R}_1, \ldots \tilde{R}_m \rangle$ and:

- $A/\sim_{\mathfrak{A}} := \{[a]_{\mathfrak{A}} : a \in A\}$;
- for all $R_i \in \mathcal{R}$ and all $a_1, \ldots, a_n \in A$: if $\langle a_1, \ldots, a_n \rangle \in R_i$, then $\langle [a_1]_{\mathfrak{A}}, \ldots, [a_n]_{\mathfrak{A}} \rangle \in \tilde{R}_i$;
- no other *n*-tuples belong to $\tilde{R}_i$ except those introduced in the step above.

If the original model $\mathfrak{A}$ is rigid, i.e., its only symmetry is the identity, $\mathrm{Aut}(\mathfrak{A}) = \{e\}$, then the quotient model $\mathfrak{A}/\mathrm{Aut}(\mathfrak{A})$ is isomorphic to $\mathfrak{A}$. If $\mathfrak{A}$ is symmetric, then passing to its quotient can do considerable violence to its original structure. The procedure would be catastrophic as a general policy in pure mathematics, where there is an abundance of symmetric structures. For example, if we quotient the field $\mathbb{C}$ under its symmetries, then (among many other disasters) addition fails to be a function: e.g., $[i]_{\mathbb{C}}$ sums with itself to *both* $[2i]_{\mathbb{C}}$ and $[0]_{\mathbb{C}}$. However, quotienting the state space of a physical theory in this way may be a much more sensible policy; later we will see what difference it makes.

41.2.2 *Lifted Permutations*

Recall that, given any model $\mathfrak{A} = \langle A, \mathcal{R} \rangle$, any permutation $\pi : A \to A$ on its domain induces a lift π^* which acts on models. I will call any such π^* a *lifted permutation* because, not only is it the lift of some permutation, it is itself a permutation in its own right, but one on models. It turns out that it is very helpful to consider these lifted permutations, chiefly because the central "objects" we find represented in the formalism of a physical theory are not objects in the normal sense at all; rather, they are states, which are rather like the models we have been considering, at least in that they are commonly deployed to represent the possibilities for some target physical system.

I will pursue a simple example. So consider all models of the form $\langle D, \{P\} \rangle$, whose domain is the three-element set $D := \{a, b, c\}$ and whose only structure is the monadic property P, which is some subset of D. Since each of the three objects a, b, c may or may not have the property P, there are $2^3 = 8$ such models. I will denote each model by $\mathfrak{A}_P$, so that e.g., the model such that $P = \{a, b\}$ is denoted by $\mathfrak{A}_{\{a,b\}}$, etc.

The eight models $\mathfrak{A}_{\varnothing}, \mathfrak{A}_{\{a\}}, \mathfrak{A}_{\{b\}}, \ldots, \mathfrak{A}_{\{a,b,c\}}$ can be put into a set, which I will call W (for "worlds"). We can now define a model $\mathfrak{W}$ whose domain is W (so, a model of models!). $\mathfrak{W}$ is a toy state space. The structure of $\mathfrak{W}$ will encode the permutation-invariant information about the eight models $\mathfrak{A}_{\varnothing}, \mathfrak{A}_{\{a\}}, \mathfrak{A}_{\{b\}}, \ldots, \mathfrak{A}_{\{a,b,c\}}$. Since each $\mathfrak{A}_P$ has as structure only one monadic property P, the permutation-invariant information about each $\mathfrak{A}_P$ is just *how many* of the objects from the common domain $\{a, b, c\}$ have the property P. So let's define $\mathfrak{W} = \langle W, \{\#\} \rangle$, where $\#$ is a function $\# : W \to \{0, 1, 2, 3\}$ that returns, for each $\mathfrak{A}_P$, the number $\#(\mathfrak{A}_P)$ of objects in D which lie in P. So, for example, $\#(\mathfrak{A}_{\varnothing}) = 0$, where nothing has the property P, and $\#(\mathfrak{A}_{\{a,b\}}) = \#(\mathfrak{A}_{\{a,c\}}) = \#(\mathfrak{A}_{\{b,c\}}) = 2$, where two things have the property P. Essentially, $\#$ yields an *occupation number*, i.e., the number of objects with the property P, according to the "world" in question.

Each "world" $\mathfrak{A}_P$ has the same domain $D = \{a, b, c\}$, containing three objects. There are $3! = 6$ permutations on this domain. Each such permutation π induces a lift $\pi^* : W \to W$ on the set of "worlds." For example, the permutation $\pi = (ab)$, which swaps a and b, induces a lift π^* such that, e.g., $\pi^*(\mathfrak{A}_{\{c\}}) = \mathfrak{A}_{\{c\}}$, $\pi^*(\mathfrak{A}_{\{a,b\}}) = \mathfrak{A}_{\{a,b\}}$, and $\pi^*(\mathfrak{A}_{\{a,b\}}) = \mathfrak{A}_{\{b,c\}}$. Clearly, each permutation π on D corresponds to a unique lifted permutation π^* on W, so there are six lifted permutations. In the language of group theory, the permutations π have a natural *group action* on $\mathfrak{W}$ which is *faithful*: all this means is that the lifted permutations have exactly the same group structure as the original permutations.

W contains eight "worlds," so it admits $8! = 40\,320$ permutations in total. Yet only six of them are lifted permutations, and it is only these six that have any interest for us. Since the model $\mathfrak{W}$ encodes only the permutation-invariant information about each "world" $\mathfrak{A}_P$, every lifted permutation is a symmetry of $\mathfrak{W}$. Since $\mathfrak{W}$ is a symmetric model, some of its elements, the "worlds" $\mathfrak{A}_P$, are permutable. It is easy to see which these are. $\mathfrak{A}_\varnothing$ is fixed by every lifted permutation, so it is not permutable (except with itself); i.e. it is an individual in $\mathfrak{W}$. Ditto for $\mathfrak{A}_{\{a,b,c\}}$. Meanwhile, $\mathfrak{A}_{\{a\}}$, $\mathfrak{A}_{\{b\}}$ and $\mathfrak{A}_{\{c\}}$ comprise a permutability class, as do $\mathfrak{A}_{\{a,b\}}$, $\mathfrak{A}_{\{b,c\}}$ and $\mathfrak{A}_{\{a,c\}}$. The eight "worlds" are therefore partitioned by the lifted permutations into four permutability classes. Each permutability class corresponds to one of the four possible occupation numbers (0, 1, 2 or 3) yielded by #. The quotient model $\mathfrak{W}/\mathrm{Aut}(\mathfrak{W})$ therefore has a domain with four objects, and the quotiented function $\tilde{\#}$ simply maps each one to the appropriate occupation number.

41.3   Related Metaphysical and Interpretative Disputes

Now that we have the idea of a model whose domain comprises states (or "worlds"), and of lifted permutations, we can generalize to states which do not necessarily take the form of models. All we really need is the idea that, of all the permutations definable on the set of states, only a very restricted subclass of them will be *lifted* permutations, i.e., understandable as having been induced by permutations on objects of some common domain shared by the states. Generally, the link between permutations on the objects and lifted permutations on the states is provided, as above, by a natural group action, or—as in the case of quantum mechanics—by a natural group representation. All we need is that, for any two permutations π_1, π_2 on the original set of objects, the lifted permutations obey $(\pi_2\pi_1)^* = \pi_2^*\pi_1^*$.

With this in mind, consider the general idea of a model $\mathfrak{W}$, describing a "space" of states (it may well be some kind of space, but it need not be), and for which every lifted permutation $\pi^* : W \to W$ is a symmetry of $\mathfrak{W}$, because the structure of $\mathfrak{W}$ articulates only what is permutation-invariant in the states. We can likewise generalize the idea of the quotient model $\mathfrak{W}/\mathrm{Aut}(\mathfrak{W})$, in which permutable states in W have been identified, and so is itself rigid. These general ideas will be useful in articulating some important rival interpretative positions, to which I now turn.

41.3.1   Haecceitism vs. Anti-Haecceitism

Our first pair of interpretative positions concern how the mathematical object $\mathfrak{W}$ is used to represent the possible states or histories of some target, physical system. To avoid confusion, I will continue to call the elements of $\mathfrak{W}$'s domain, which are purely mathematical, *states*, and I will call the physical possibilities that they aim to represent *possibilities*. We may now define:

(*Haecceitism*)   Any two distinct states permutable in $\mathfrak{W}$ represent distinct possibilities.

(*Anti-haecceitism*)   Any two distinct states permutable in $\mathfrak{W}$ represent the same possibility.

These names originate from medieval philosophy (the Latin term *haecceitas* roughly translates as *thisness*), but the positions were first introduced into the recent philosophical literature by Kaplan (1975). The positions defined above are closer to Lewis's (1986, p. 221) understanding of the terms.

Anti-haecceitism is a statement of a kind of permutation invariance: the possibility being represented is invariant under any symmetry of $\mathfrak{W}$, which is (typically but perhaps not always) a lifted permutation. It follows that, according to anti-haecceitism, the quotient model $\mathfrak{W}/\mathrm{Aut}(\mathfrak{W})$ provides a more perspicuous representation of the possibilities than $\mathfrak{W}$.

Haecceitism expresses a willingness to deny this permutation invariance. $\mathfrak{W}$ offers, on this view, a more perspicuous representation of the possibilities than its quotient $\mathfrak{W}/\mathrm{Aut}(\mathfrak{W})$. But the haecceitist should hope to do better than $\mathfrak{W}$, since $\mathfrak{W}$ attributes only permutation-invariant structure to the states. The haecceitist takes the possible facts to surpass merely what is permutation-invariant, so we might expect them to enrich $\mathfrak{W}$'s structure (but not its domain of states) with the means suitable to express facts about which object is which. (In classical and quantum mechanics, as we shall see, this can be conveyed by the "full" algebra of quantities, not restricted by permutation invariance.)

Are haecceitism and anti-haecceitism contraries? Not quite. They agree on the totally symmetric states (the states fixed by any lifted permutation), since each of those states is the sole occupant of its permutability class, and so $[w]_{\mathfrak{W}} = \{w\}$ is as good as w as a representative of any possibility. If *every* state in $\mathfrak{W}$'s domain is totally symmetric, then *no* two distinct states are permutable in $\mathfrak{W}$, $\mathfrak{W}$ and $\mathfrak{W}/\mathrm{Aut}(\mathfrak{W})$ are therefore isomorphic, and both haecceitism and anti-haecceitism are trivially satisfied. Moreover, the haecceitist and anti-haecceitist cannot in this case even disagree about whether $\mathfrak{W}$'s structure suffices for individuating any given state: if no two distinct states are permutable, then every one of $\mathfrak{W}$'s states is an individual in $\mathfrak{W}$, and so has a unique qualitative character. This point may seem recherché, but the situation of totally symmetric states arises in the quantum mechanics of bosons and fermions (see Section 41.6.2). In this case, it is hard to see how the dispute between haecceitism and anti-haecceitism could be adjudicated, a point emphasized by French (1989).

41.3.2   *Transcendental vs. Qualitative Individuality*

However, a more fine-grained dispute can be articulated, which survives even when all states are totally symmetric. This dispute concerns the question: What is really being permuted? Of course, the states of $\mathfrak{W}$ are being permuted, but these are *lifted* permutations, which are the lifts of permutations defined on the *objects* in the states' common domain, which I will call D. But the objects in D are just mathematical representatives—effectively, labels—so what do they represent? I see two salient proposals here, which are mutually exclusive but not jointly exhaustive: *transcendental individuality* (TI) and *qualitative individuality* (QI).

(*TI*)   Each label in D denotes some object of the target system, and that label denotes the same object in all states.

(*QI*)   The labels in D denote no object in particular of the target system.

The term "transcendental individuality" was coined by Post (1963) and also appears in Redhead and Teller (1991, 1992); I mean it in roughly their sense. The guiding idea is that facts about object identity from possibility to possibility transcend what is invariant under permutations. This is expressed, using $\mathfrak{W}$, by taking advantage of the fact that the states have a common domain, and stipulating that the same object in that domain always stands for the same object in the target system. TI entails haecceitism.

The term "qualitative individuality" is less commonly used in the literature. The guiding idea is that the objects in the states' common domain are nothing but placeholders. They serve as representative "hooks" on which to hang properties and relations, but any hooks will do. (Hence the "hooks" represent nothing "in particular.") Specifically, we should afford no significance to the fact that the same "hook" appears from state to state. To return to our example in Section 41.2.2, the objects a, b, c in the domain of each model $\mathfrak{A}_P$ are, under QI, mere "hooks" on which the property P is hung (or not hung). If permuting the "hooks" yields a distinct state in $\mathfrak{W}$, this is a distinction without a difference. So QI entails anti-haecceitism.

The dispute between TI and QI is more fine-grained than the one between haecceitism and anti-haecceitism, at least as they are defined here, since TI and QI still have a substantive disagreement in the case where all the states in $\mathfrak{W}$ are totally symmetric, i.e., invariant under any lifted permutation. (This harks back to the apparently recherché observation made in the previous section.) The TI/QI dispute has a counterpart in the modal metaphysics literature, where the associated positions usually go by the names "transworld identity" and "world-bound individuals." (The literature on this is enormous, but Lewis (1986, §§4.1–4.3) is a good place to start.)

41.4 Permutations in Classical Hamiltonian Mechanics

41.4.1 The Realization of Permutations on Joint Configuration Spaces

We begin with $\mathcal{Q}$, the configuration space for a generic elementary system. (Note: the system being represented need not *be* elementary; it's just being treated as such.) For a particle in d-dimensional Euclidean space, we may take $\mathcal{Q} = \mathbb{R}^d$. In any case, I'll assume that $\mathcal{Q}$ is a smooth manifold.

Representing an assembly of such systems, say N of them, standardly proceeds as follows. First we form the joint configuration space $\mathcal{Q}^N$, often written

$$\mathcal{Q}^N := \underbrace{\mathcal{Q} \times \ldots \times \mathcal{Q}}_{N} \tag{41.1}$$

to indicate that the joint configuration space's points are elements in the N-fold Cartesian product of $\mathcal{Q}$. There are standard ways to ensure that $\mathcal{Q}^N$ is endowed with the appropriate spatial structure; see e.g., Willard (1970, §3.8).

There are broadly two routes from here. The first is to define the joint phase space, and then consider lifted permutations of system labels. The second is to consider lifted permutations of system labels on $\mathcal{Q}^N$ *first*, and then proceed to the joint phase space. I will take the first route. The joint phase space is the cotangent bundle $T^*(\mathcal{Q}^N)$, equipped with a suitable symplectic form Ω, which is required to build the Hamiltonian equations of motion. The result is equivalent, as a phase space, to the N-fold Cartesian product of Γ (whose symplectic form is determined by that on Γ), so I will denote this joint phase space by Γ^N.

States in the joint phase space Γ^N may be denoted by $(\xi_1, \ldots, \xi_N)$, where each $\xi_i \in \Gamma$ is a single-system state. We can now define a natural *realization*, a.k.a. *group action*, of the group S_N of permutations on N symbols, as follows. For each permutation $\pi \in S_N$, define the lifted permutation $\pi^* : \Gamma^N \to \Gamma^N$ such that

$$\pi^*(\xi_1, \ldots, \xi_N) := (\xi_{\pi(1)}, \ldots, \xi_{\pi(N)}) \,. \tag{41.2}$$

The lift $^* : \pi \mapsto \pi^*$ is a realization of S_N precisely because it preserves the group structure of S_N, i.e., $(\pi_2\pi_1)^* = \pi_2^*\pi_1^*$. The realization is *faithful*, i.e., distinct permutations in S_N are sent by the lift to distinct maps on Γ^N. Moreover, any lifted permutations is a symmetry of Γ^N, since any lifted permutation preserves its manifold structure and symplectic form Ω (in other words, lifted permutations on Γ^N are canonical transformations).

41.4.2 Classical Permutation Invariance

We can add more structure to Γ^N by introducing an algebra $\mathcal{A}$ of quantities. It is standard to posit the Poisson algebra of all smooth real-valued functions $f : \Gamma^N \to \mathbb{R}$, equipped, via the symplectic form Ω, with an associated Poisson bracket $\{\cdot, \cdot\} : \mathcal{A} \otimes \mathcal{A} \to \mathcal{A}$. The resulting model $\langle \Gamma^N, \Omega, \mathcal{A} \rangle$ is the standard arena for classical N-particle mechanics.

This model is rigid, since any two distinct states differ on the values of some quantities. For example, if the states $(\xi_1, \xi_2), (\xi_2, \xi_1) \in \Gamma^2$ are distinct, then we must have $\xi_1 \neq \xi_2$; but in that case the two states will disagree on, e.g., the values they assign either to $\mathbf{q}_1$ (the position of system 1) or to $\mathbf{p}_1$ (the momentum of system 1), since ξ_1 and ξ_2, if distinct single-system states, must disagree on either position or momentum. However, we might be skeptical of the empirical accessibility, or even the physical meaning, of (e.g.,) the difference between $\mathbf{q}_1$ and $\mathbf{q}_2$. After all, what we observe is that there is (or isn't) a particle at some particular location, not that particle 1, as opposed to particle 17, is at some particular location. We may therefore wish to capture only the permutation-invariant features of each state; that is, the features of the state which are irrelevant to which particle is which.

To this end, we restrict the algebra $\mathcal{A}$ to the permutation-invariant quantities, which are the smooth functions $f : \Gamma^N \to \mathbb{R}$ such that

$$f(\pi^*(\xi_1, \ldots, \xi_N)) = f(\xi_1, \ldots, \xi_N) \,. \tag{41.3}$$

Obviously, quantities such as the position of system 5 or the momentum of system 17 will not be among the permutation-invariant quantities. But we may uniquely characterize an unlabeled cluster of N points in the single-system phase space Γ with only permutation-invariant quantities: this is as specific as we can be without getting into which particle is which. Call the restricted permutation-invariant algebra $\mathcal{A}_{PI}$.

While $\mathfrak{P} = \langle \Gamma^N, \Omega, \mathcal{A} \rangle$ is a rigid model, $\mathfrak{P}_{PI} := \langle \Gamma^N, \Omega, \mathcal{A}_{PI} \rangle$ has symmetries. These are precisely the lifts π^* of the permutations $\pi \in S_N$. So we find ourselves with a concrete instance of the "space" of states $\mathfrak{W}$ discussed in generality in Sections 41.2.2 and 41.3. In an anti-haecceitistic spirit, we may pass to the quotient model $\mathfrak{P}_{PI}/S_N = \langle \Gamma^N/S_N, \tilde{\Omega}, \tilde{\mathcal{A}}_{PI} \rangle$. Here, the symplectic form $\tilde{\Omega}$ and algebra $\tilde{\mathcal{A}}_{PI}$ are defined in the obvious way from Ω and $\mathcal{A}_{PI}$. (This construction mirrors the definition of the structure of $\mathfrak{A}/\mathrm{Aut}(\mathfrak{A})$, outlined at the end of Section 41.2.1.) The *reduced phase space* Γ^N/S_N is also straightforwardly defined (see Willard, 1970, §3.9 for details).

However, Γ^N has points on which the group action of S_N is not faithful; these are the states $(\xi_1, \ldots, \xi_N)$ such that $\xi_i = \xi_j$ for some $i \neq j$. These points form a boundary in Γ^N/S_N, and so Γ^N/S_N is not a manifold (it is what is called an *orbifold*). This generates a host of technical issues, not least of which is the fact that tangent spaces on the boundary have the "wrong" dimension, and so smooth vector fields on the bulk—such as Hamiltonian flows—are ill-defined at the boundary.

It is standard practice to avoid these issues by removing *collision configurations* from the joint configuration space $\mathcal{Q}^N$ *before* defining the joint phase space and its quotient under S_N. The collision configurations comprise the set $\Delta := \{(x_1, \ldots, x_N) \in \mathcal{Q}^N : x_i = x_j \text{ for some } i \neq j\}$, which has vanishing Lebesgue measure. So let $\Gamma_N := T^*(\mathcal{Q}^N \setminus \Delta)$ be our new joint phase space. The group of permutations S_N acts *freely* on this space (every state $\Xi \in \Gamma_N$ is such that $\pi_1^* \Xi \neq \pi_2^* \Xi$ for any two distinct permutations $\pi_1, \pi_2 \in S_N$), and the resulting quotient phase space Γ_N/S_N is a manifold, isomorphic as a phase space to $T^*((\mathcal{Q}^N \setminus \Delta)/S_N)$.

Note that the justification for removing the collision configurations was purely technical: it was to ensure a quotient phase space with nice properties. The physical justification—if there is one— is murkier: it is certainly suspect that classical systems should become impenetrable on grounds of a technicality. Certainly, if we expect the dynamics to make the particles impenetrable (presumably by means of some strong short-range repulsive force), then the excision is justified—but what about alternative dynamics? Besides, it must be emphasized that the excision of collision points is far from innocent: it typically leaves the resulting joint configuration space, and its quotient under S_N, topologically non-trivial. As will be mentioned later in Section 41.6.3, this fact is relevant when considering how to quantize the classical theory.

41.4.3 Individuation in Permutation-Invariant Classical Mechanics

We appear to have two very different classical N-particle theories. The first is associated with the model $\mathfrak{P} = \langle \Gamma_N, \Omega, \mathcal{A} \rangle$, and is suited to a proponent of transcendental individuality, and therefore haecceitism. The second theory is associated with the model $\mathfrak{P}_{PI}/S_N = \langle \Gamma_N/S_N, \tilde{\Omega}, \tilde{\mathcal{A}}_{PI} \rangle$, and is suited to the proponent of qualitative individuality, and therefore anti-haecceitism. There are distinct states in $\mathfrak{P}$ that become permutable if we restrict to the sub-algebra of permutation-invariant quantities; so haecceitism and anti-haecceitism seem to be genuine rivals here.

However, if the Hamiltonian of the system is among the permutation-invariant quantities, then the two theories are in fact equivalent, up to arbitrary stipulations. That is, one may recover a unique anti-haecceitistic trajectory from any haecceitistic one; and a unique haecceitistic trajectory, up to an arbitrary stipulation, from any anti-haecceitistic one. (This is argued by Leinaas and Myrheim, 1977, p. 5.) The determination of a unique anti-haecceitistic trajectory from a haecceitistic one is straightforward. In the opposite direction, we manage to recover $N!$ haecceitistic trajectories from any given anti-haecceitistic one by demanding that the haecceitistic trajectories, like the anti-haecceitistic one, are continuous in the associated configuration space. Crucial to this result is the fact that collision points in the joint configuration space have been removed. Otherwise, the demand for continuity fails to provide unique trans-temporal identifications between the haecceitistic states whenever any two or more particles occupy the same location.

Which one of the $N!$ trajectories should the haecceitist then choose as the correct one? No answer to this question is required. After all, not even an advocate of transcendental individuation will deny that it is a matter of mere convention which system label denotes which particle. The $N!$ haecceitistic trajectories correspond perfectly to the $N!$ ways to associate N system labels with the N particles.

Could the theories, and their associated metaphysical positions, be distinguished by the explanations they can provide for why it is that the Hamiltonian is permutation-invariant in the first place? The prospects are dim. Qualitative individuation explains permutation invariance by making it inevitable: since on this view there *is* no question which particle is which, any physical quantity (which the Hamiltonian surely is) must take the same value when system labels are permuted. But there is a convincing explanation under transcendental individuation too. Even though permutation invariance is not *inevitable* in this view—there is certainly no contradiction in supposing it to fail—there is every reason to expect it holds of the Hamiltonian. After all, by hypothesis the particles possess the same state-independent properties, such as their particular mass; so while there is here a physical difference between permuted states, it is hard to see how this is a difference that could make a difference to the total energy of the state.

41.5   Permutations in Classical Statistical Mechanics

In statistical mechanics, two new considerations come into play which seem promising with regard to adjudicating between our metaphysical positions. The first is that this theory's states are distributions over the states considered above, which we will now call *microstates*. Since qualitative and transcendental individuation differ over the nature of trans-state identifications, it may be thought that differences emerge once we have to consider many microstates at once, as when we define a distribution over them. The second consideration is that statistical mechanical predictions often involve *counting* microstates. Yet this is another point on which TI and QI differ: where TI sees many permutable microstates, QI sees only one.

So let us turn to two phenomena which at first glance appear to favor one metaphysical position over its rival; they are considered in more detail by Huggett (1999a). Surprisingly, the two cases seem *prima facie* to point in opposite directions: the first, appealing to equilibrium distributions, appears to favor haecceitism; the other, appealing to Gibbs' paradox, appears to favor anti-haecceitism.

The equilibrium distribution in question is the Maxwell-Boltzmann velocity distribution, derived through a combinatorial argument (see Frigg, 2008, §2.2), which shows it to correspond to the macrostate with largest Lebesgue volume in the joint phase space. The details are as follows. The single-system phase space Γ is divided into c occupied cells, each with Lebesgue volume ω and a characteristic mean energy E_i. Macrostates are then characterized by occupation numbers $(n_1, \ldots, n_c)$, where n_i is the population of particles whose states lie in the ith cell. In the haecceitistic theory, each macrostate defines a region of Γ_N consisting of the number

$$\frac{N!}{\prod_{i=1}^{c} n_i!} \tag{41.4}$$

of (typically disconnected) cells, each with Lebesgue volume ω^N. This volume is maximal, subject to the constraints of conservation of particle number ($\sum_i n_i = N$) and of energy ($\sum_i n_i E_i = E$), for the Maxwell-Boltzmann distribution

$$n_i = Ne^{-\beta(E_i - \mu)} , \tag{41.5}$$

where β and μ are constants, determined by the two constraints.

Seemingly crucial to the derivation is the fact that the number (41.4) of cells in the joint phase space is maximal for the Maxwell-Boltzmann distribution (41.5); but in the anti-haecceitistic theory, there is only *one* cell in the joint phase space corresponding to those occupation numbers. In the anti-haecceitistic theory there is always only one cell corresponding to any specification of occupation numbers!

However, we derive (41.5) on the anti-haecceitistic theory too. The trick is that, while the haecceitistic derivation relies on a variable number of cells of constant volume, the anti-haecceitistic derivation relies on a constant number of cells (namely, one) of variable volume. This variation in volume is due to quotienting. The cell in Γ_N / S_N corresponding to the occupation numbers $(n_1, \ldots, n_c)$ has Lebesgue volume

$$\frac{\omega^N}{\prod_{i=1}^c n_i!} . \tag{41.6}$$

This volume is maximal when the n_i obey the Maxwell-Boltzmann distribution (41.5).

As for the Gibbs paradox, there is some dispute about what exactly the paradox is (see e.g. Uffink, 2006, §5.2), but I will take the problem to be the search for a statistical mechanical account of the extensivity of the entropy of an isolated gas. For simplicity I will address an ideal gas. A naïve calculation using the microcanonical ensemble goes *via* the Lebesgue volume, in the joint phase space Γ^N, of the energy $E = \frac{3}{2} NkT$ hypersurface, which is the product of the spatial volume V^N, where each of the N particles in the gas is confined to volume V, and the surface "area" of the $(3N - 1)$-dimensional momentum hypersphere with radius $\sqrt{2mE} = \sqrt{3NmkT}$:

$$W_{naïve} = V^N \frac{2\pi^{\frac{3}{2}N}}{\Gamma(\frac{3}{2}N)} (3NmkT)^{\frac{3N-1}{2}} . \tag{41.7}$$

With the usual prescription for the entropy (either Boltzmann's or Gibbs,' assuming the microcanonical ensemble), one then obtains

$$S_{naïve}(N, V, T) = k \log W_{naïve} \approx Nk \log \left(\frac{VT^{\frac{3}{2}}}{\Phi} \right), \tag{41.8}$$

where Φ is some constant with dimensions $[volume][temperature]^{\frac{3}{2}}$. Extensivity demands that

$$S(\alpha N, \alpha V, T) = \alpha S(N, V, T) \tag{41.9}$$

for any $\alpha \in \mathbb{R}_+$, but this fails for $S_{naïve}$. For example, if our gas is separated into two chambers of equal volume $\frac{1}{2} V$ and density $\frac{N}{V}$ by a partition, and we assume that the total entropy is just the sum of the entropies of the gases either side of the partition, then gently removing the partition produces a rise of entropy by $Nk \log 2$. But this is a reversible thermodynamical process: since the gas is in equilibrium after the partition is removed, i.e., it has constant temperature and density throughout the volume V, one can separate the gases again, performing negligible work, just by gently reintroducing the partition.

Gibbs' (1902, Ch. 15) own solution to this problem was to work not with the "specific phases"—essentially, our haecceitistic microstates—but the "generic phases"—essentially, our anti-haecceitistic microstates. Gibbs in effect passes to the quotient phase space Γ_N/S_N. The consequence is a division by $N!$ of the naïve phase space volume $W_{naïve}$, and one obtains the Sackur-Tetrode equation

$$S(N, V, T) \approx Nk \log \left(\frac{VT^{\frac{3}{2}}}{N\Phi} \right), \tag{41.10}$$

which yields an extensive entropy. This seems to suggest that reconciliation with thermodynamics requires an anti-haecceitistic statistical mechanical theory. However, a haecceitistic derivation of (41.10) can also be given. Following a suggestion of Ehrenfest and Trkal (1921), developed by van Kampen (1984), the haecceitist demands that we take into account *all* molecules in the universe of the same kind as those in the chamber. Specifically, we must multiply the naïve phase space volume by the number of ways that N of them can appear in the chamber. Supposing the total number of such molecules in the universe is $M >> N$, this number is

$$\left(\begin{array}{c} M \\ N \end{array} \right) \approx \frac{M^N}{N!}, \tag{41.11}$$

and leads to an entropy differing from (41.10) by an irrelevant additive constant, also a multiple of N; so the entropy is again extensive.

So it would appear that we can save the phenomena regarding equilibrium distributions and the mixing of gases under both haecceitism and anti-haecceitism. Huggett (1999a) counsels metaphysical skepticism in response. Saunders (2013) points out, in the case of Gibbs' paradox, that the haecceitistic and anti-haecceitistic solutions differ over whether the gas is properly treated as an open or closed system. Regarding this point, note that the haecceitistic solution crucially involves an appeal to the total number of molecules of the same kind in the *entire* universe, and the assumption that those in the chamber comprise a tiny fraction of this total—all this despite the fact that the molecules inside the box cannot mix with those outside.

41.6   Permutations in Quantum Mechanics

There are broadly three routes to a permutation-invariant quantum theory for a constant number N systems. They are as follows:

1   *Implement permutation invariance on some corresponding permutation-non-invariant quantum theory for N systems.* This route leads to the theory of group representations; specifically, the irreducible representations of S_N. I will outline this route in Section 41.6.1.

2   *Quantize some corresponding permutation-invariant classical theory for N systems.* This route has permutation invariance built in at the outset, and leads to considerations of non-equivalent quantizations brought about by topological features of the underlying classical configuration space. I will briefly mention this route in Section 41.6.3.

3   *Restrict to the N-system subspace in some appropriate corresponding quantum field theory.* This route also has permutation invariance built in at the outset. Indeed, I know of no way of understanding the states of the quantum field (in the particle picture) in anything but an anti-haecceitistic way.

I will say the least about route 3, or about quantum field theory more generally. (For details about particle permutation symmetry in quantum field theory, see Greenberg et al. (1964), Stolt and Taylor (1970), Doplicher et al. (1974) and Ohnuki and Kamefuchi (1982); a philosophical presentation is given by Baker et al. (2015), with further references.) One would be forgiven for considering this route to be the most enlightening when it comes to the origins of permutation invariance, since (in contrast with route 1) our world is more accurately described by quantum field theory than many-particle quantum mechanics, and (in contrast with route 2) we ought to be cautious about gleaning insights into a quantum theory by appeal to some classical theory of which it happens to be the quantization. Nevertheless, I will press on with a brief outline of route 1, followed by even briefer comments on route 2.

41.6.1 The Representation of Permutations on Joint Hilbert Spaces

We begin with the quantum theory for a single system. Standardly, this is some separable Hilbert space $\mathcal{H}$ and an associated algebra of quantities, all of which are linear operators on $\mathcal{H}$. I will not go into the details here about where $\mathcal{H}$ and $\mathfrak{a}$ "come from," but simply take them as given.

Representing an assembly of such systems, say N of them, standardly proceeds as follows. First we form the joint Hilbert space and joint algebra

$$\mathcal{H}^N := \underbrace{\mathcal{H} \otimes \ldots \otimes \mathcal{H}}_{N} \, ; \qquad \mathcal{A} := \underbrace{\mathfrak{a} \otimes \ldots \otimes \mathfrak{a}}_{N} \, . \tag{41.12}$$

The model $\langle \mathcal{H}^N, \mathcal{A} \rangle$ forms the arena for the haecceitistic theory of N equivalent particles.

Lifted permutations may be defined on $\mathcal{H}^N$, and here we get into the group representation theory of S_N. There is an obvious map $U : S_N \to \mathcal{U}(\mathcal{H}^N)$ from the group S_N of permutations to the unitary operators $\mathfrak{U}(\mathcal{H}^N) \subset \mathcal{A}$ such that $U(\pi)$ implements the permutation π on joint states. We define U by its action on product states and then extend by linearity to all vectors in the joint Hilbert space. Given any permutation $\pi \in S_N$ we have, for any product state $|\psi_1\rangle \otimes \ldots \otimes |\psi_N\rangle \in \mathcal{H}^N$,

$$U(\pi)|\psi_1\rangle \otimes \ldots \otimes |\psi_N\rangle := |\psi_{\pi(1)}\rangle \otimes \ldots \otimes |\psi_{\pi(N)}\rangle \tag{41.13}$$

(compare with equation (41.2) in Section 41.4.1). U so-defined constitutes a representation of S_N in the technical sense, which is that for any $\pi_1, \pi_2 \in S_N$, $U(\pi_2\pi_1) = U(\pi_2)U(\pi_1)$, and so U is a group homomorphism. And furthermore, since each $U(\pi)$ is a unitary operator, U is a *unitary* representation of S_N.

Since two states differing only up to a global phase factor yield the same expectation values, one may reasonably wonder why we cannot instead demand only that $U(\pi_2\pi_1) = e^{i\omega(\pi_2, \pi_1)} U(\pi_2)U(\pi_1)$, so that we obtain a group homomorphism up to a phase factor. Such representations are called *projective* unitary as opposed to just unitary (or *linear* unitary). (Moreover, consistency puts demands on the form of ω: it must obey the *cocycle equation* $\omega(\pi_1, \pi_3) + \omega(\pi_2, \pi_1\pi_3) = \omega(\pi_2, \pi_1) + \omega(\pi_2\pi_1, \pi_3)$.) One answer, provided by Read (2003), is that merely projective representations of S_N violate locality: i.e.,the rigorous derivation of particle statistics in quantum field theory obeying a local dynamics (with a local Hamiltonian) are at odds with these merely projective representations. However, there is a related matter whether we should be thinking of permutations in terms of the group S_N at all. I will briefly return to this in Section 41.6.3.

For now assuming (41.13), let us delve briefly into the group representation theory of S_N; complete treatments may be found in e.g. Tung (1985, Chs. 3 & 5) and Sternberg (1994, Ch. 2). We find that the representation U decomposes into a direct sum of *irreducible* unitary representations D_λ, which I will call *irreps*. Irreps come in a variety of types, labeled by λ, according to how joint states transform under the lifted permutations $U(\pi)$, and typically there will be many copies of the same irrep in the decomposition of U. One type of irrep is D_+, for which $D_+(\pi) = 1$ for all $\pi \in S_N$, and which corresponds by definition to *bosonic* states. Another type of irrep is D_-, for which $D_-(\pi) = (-1)^{\deg \pi}$, where $\deg \pi$ is the number of pairwise swaps involved in the permutation π, and which corresponds by definition to *fermionic* states. If the number of particles is three or more, we find in addition to the bosonic and fermionic irreps a variety of multi-dimensional irreps, corresponding to what are known as *paraparticle* or *parastatistical* states.

Each copy of each irrep D_λ occurring in U acts on only a small subspace of the joint Hilbert space $\mathcal{H}^N$; these subspaces are called *irreducible invariant subspaces*, or i.i.s.s. (Messiah and Greenberg, 1964 call these i.i.s.s *generalized rays*.) Each i.i.s. is the smallest non-trivial subspace left invariant under action by the lifted permutations $U(\pi)$, which is what earns them and the associated irreps D_λ the designation "irreducible." Just as the representation U is the analog of a group action of S_N in the classical case, the i.i.s.s are the group representation analog of classical orbits. Bosonic and fermionic i.i.s.s are 1-dimensional, corresponding to the irreps $D_\pm$ yielding simple scalars: $D_\pm(\pi) = \pm 1$. Paraparticle i.i.s.s are multi-dimensional, and so their associated irreps D_λ are represented by matrices.

Each irrep is associated with characteristic large-particle-number behavior. Bosons are associated with Bose-Einstein statistics; fermions with Fermi-Dirac statistics. Both are distinguished from the classical Maxwell-Boltzmann statistics, seen in Section 41.5. Incidentally, the reason for this difference between classical and quantum statistics is often mistakenly linked to the dispute between haecceitism and anti-haecceitism. In fact, it is due to the ways states are "counted" in the two theories: the measure over states in the classical case is continuous (phase space volume) and in the quantum case it is discrete (dimension count). This is a point emphasized by Huggett (1999b) and Saunders (2006b, 2013).

41.6.2   Quantum Permutation Invariance

We may now define what it is for any quantity in the joint algebra to be permutation-invariant. A quantity $A \in \mathcal{A}$ is permutation-invariant iff: for any permutation $\pi \in S_N$ and any joint vector state $\Psi \in \mathcal{H}^N$,

$$\langle U(\pi)\Psi, AU(\pi)\Psi \rangle = \langle \Psi, A\Psi \rangle \tag{41.14}$$

(compare with equation (41.3) in Section 41.4.2). This condition is widely known as the *Indistinguishability Postulate*; as far as I know, the term originates with Messiah and Greenberg (1964). Since this condition holds for all vectors in $\mathcal{H}^N$, it may be rephrased as an operator identity. For any permutation $\pi \in S_N$:

$$U(\pi)^\dagger AU(\pi) = A , \qquad \text{i.e.} \qquad [A, U(\pi)] = 0 . \tag{41.15}$$

This condition provides a criterion for membership in the permutation-invariant algebra $\mathcal{A}_{PI} \subset \mathcal{A}$ of quantities.

Now appealing to a powerful theorem in group representation theory known as *Schur's Lemma*, we may deduce the following two facts. First, the eigenspaces of any permutation-invariant quantity $A \in \mathcal{A}_{PI}$ respects the decomposition of U into irreps, in the sense that A's eigenspaces are superspaces of i.i.s.s. It follows that A takes the same expectation value $\langle \Psi, A\Psi \rangle$ on any joint vector state Ψ lying in the same i.i.s. This is the quantum analog of the fact in classical mechanics that any permutation-invariant quantity takes the same value on every joint state in the same orbit—but this is trivial in the case of the bosonic or fermionic states, whose i.i.s.s are 1-dimensional.

Second, the permutation-invariant algebra $\mathcal{A}_{PI}$ acts *reducibly* on the haecceitistic joint Hilbert space $\mathcal{H}^N$, so that $\mathcal{H}^N$ naturally decomposes into a number of sectors $\mathcal{H}^N_\lambda$, each corresponding to an irrep type λ. In other words: any permutation-invariant quantity $A \in \mathcal{A}_{PI}$ becomes block-diagonalized by irrep type, so that transition amplitudes vanish, $\langle \Psi, A\Phi \rangle = 0$, for any joint states Ψ and Φ which belong to different sectors $\mathcal{H}^N_\lambda$. So, for example, any transition amplitude between bosonic and fermionic sectors vanishes. In the physicists' jargon, *symmetry type* (boson, fermion, etc.) is *superselected* by the restriction to permutation-invariant quantities. This second fact has no classical analog.

It may now be wondered: What is the quantum analog of passing to the quotient of $\langle \mathcal{H}^N, \mathcal{A}_{PI} \rangle$ under permutation symmetry; i.e., what is the arena for anti-haecceitistic quantum theory? We proceed in two stages:

(i)   First we restrict to the sector $\mathcal{H}_\lambda^N$ corresponding to some chosen irrep λ; i.e., we restrict attention to just one "symmetry type," e.g., the bosonic or fermionic states. This stage has no classical analog.

(ii)   Then, if required (i.e., if the irrep is multi-dimensional, as for paraparticles), we define a new joint Hilbert space $\mathcal{H}_\lambda^N / S_N$, any ray of which corresponds to an i.i.s. in $\mathcal{H}_\lambda^N$ (so that each generalized ray in $\mathcal{H}_\lambda^N$ is mapped to a ray $\mathcal{H}_\lambda^N / S_N$). Each permutation-invariant quantity $A \in \mathcal{A}_{PI}$ has a well-defined, unique counterpart $\tilde{A}_\lambda \in \tilde{\mathcal{A}}_\lambda$ which is a linear operator on $\mathcal{H}_\lambda^N / S_N$. (Essential here is the fact that A yields the same expectation value on any state in the same i.i.s. in $\mathcal{H}_\lambda^N$.) As would be expected, any lifted permutation is represented on $\mathcal{H}_\lambda^N / S_N$ as simple multiplication by ± 1, depending on the permutation and the irrep λ.

Note that stage (ii) is unnecessary for all of the elementary particles we believe to exist, which are all either bosons or fermions, since the bosonic and fermionic i.i.s.s are already 1-dimensional. The justification for stage (i) comes from the superselection of symmetry type by the imposition of permutation invariance on the algebra of quantities: $\mathcal{A}_{PI}$ acts irreducibly when restricted to each sector $\mathcal{H}_\lambda^N$. What is surprising is that we obtain not one but several anti-haecceitistic models $\langle \mathcal{H}_\lambda^N / S_N, \tilde{A}_\lambda \rangle$, each corresponding to its own symmetry type λ.

The restriction to either the bosonic sector $\mathcal{H}_+^N$ or the fermionic sector $\mathcal{H}_-^N$ is widely known as the *Symmetrization Postulate*. All evidence so far suggests that the postulate is true, though there is an interesting history here. Before the acceptance of quark color, it had been proposed, chiefly by O. W. Greenberg, that quarks might be paraparticles. (For more, see French, 1995.) We now believe quarks to be fermions, but this raises the general question why paraparticles, perfectly allowed under permutation invariance, are not found in nature. Dürr et al. (2007) make the case that de Broglie-Bohm theory may have the edge over its rivals here, since the requirement that corpuscle trajectories be determinate appears to rule out multi-dimensional irreps of S_N. Baker et al. (2015) point to a result in the framework of algebraic quantum field theory, that any local quantum field exhibiting paraparticle statistics is equivalent to one with bosonic or fermionic statistics plus a new internal degree of freedom with accompanying constraints; they argue on the basis of this that the issue of particle statistics may be a matter of mere convention, rendering the Symmetrization Postulate more a decision than a discovery.

The fact that, for bosons and fermions, the i.i.s.s under the lifted permutations are all 1-dimensional means that we have here a case where haecceitism and anti-haecceitism trivially agree: for these joint states are fixed (up to global phase) by any lifted permutation. It is commonly claimed of these states that they represent particles which are indiscernible by means of monadic properties, i.e., absolutely indiscernible. In fact this is the wide consensus in the quantum identity literature. (The consensus began with Margenau (1944) and continued with Post (1963), French and Redhead (1988), van Fraassen (1991, Ch. 11), Butterfield (1993), Saunders (2003a, 2003b, 2006a), French and Krause (2006, §4.2.1), Muller and Saunders (2008), Muller and Seevinck (2009), Ladyman and Bigaj (2010), Caulton (2012) and Huggett and Norton (2012).) However, this conclusion can be reached only on the assumption that the particles in question are denoted by the order of the factor Hilbert spaces in the tensor product $\mathcal{H}^N$, so that a permutation of the factor Hilbert spaces under each $U(\pi)$ is taken to correspond to a permutation of the particles. This is tantamount to a commitment to transcendental individuality, as defined in Section 41.3.2. A proponent of qualitative individuality would contend rather that the lifted permutations $U(\pi)$ correspond to no real permutation at all. Proposals along the lines of qualitative individuality may be found in Huggett and Imbo (2009), Dieks and Lubberdink (2011) and Earman (2015).

41.6.3   The Topological Approach to Quantum Statistics

Let us now briefly survey route 2: the quantization of the anti-haecceitistic classical theory. The crucial thing to know is that in standard quantum mechanics, in which we consider a configuration space that looks like $\mathbb{R}^n$, the quantum theory is essentially uniquely defined. That is, there is one and only one way (up to unitary equivalence) of representing the canonical commutation relations obeyed by position and momentum as Hermitian linear operators on a complex Hilbert space (this is the celebrated Stone-von Neumann theorem). However, uniqueness of quantization typically fails when considering other configuration spaces. Important for us is that uniqueness *does* fail when the configuration space is the anti-haecceitistic one $(\mathcal{Q}^N \setminus \Delta)/S_N$, introduced in Section 41.4.2. Due either to the removal of the collision configurations Δ, or to quotienting under S_N (or both), this configuration space is topologically non-trivial, and this opens the door to a number of inequivalent ways to define the associated quantum theory. (For all the gory details, consult Morandi (1992, Ch. 3) and Landsman (2016). The pioneers were Laidlaw and DeWitt (1971) and Leinaas and Myrheim (1977).)

Without going into detail, the gamut of available quantizations is intimately tied to a specific way in which the classical configuration space $\mathfrak{Q}$ fails to be topologically boring. Rival quantizations are in one-to-one correspondence with irreps of the group $\pi_1(\mathfrak{Q})$ of (homotopy equivalence classes of) closed loops on $\mathfrak{Q}$, which is non-trivial if $\mathfrak{Q}$ fails to be simply connected. In the case $\mathfrak{Q} = (\mathbb{R}^{Nd} \setminus \Delta)/S_N$, the anti-haecceitistic configuration space for N particles in $\mathbb{R}^d$, and $d \geqslant 3$ (three or more spatial dimensions), it turns out that $\pi_1(\mathfrak{Q}) = S_N$, and we are led, as above, to consider the irreps of S_N. We obtain the same outcome, namely a variety of quantum theories, each associated with one of the symmetry types: boson, fermion, or some kind of paraparticle.

If $d = 2$, however, we find that $\pi_1(\mathfrak{Q}) = B_N$, the *braid group* on N objects. The elements of the braid group are like permutations with memory: a pairwise swap implemented twice fails to be the identity. This gives B_N a much richer structure than S_N and leads to an expanded variety of quantizations. Among just the 1-dimensional irreps of B_N we find again bosons and fermions, but now also *anyons*, for which the joint state picks up an arbitrary phase factor $e^{i\theta}$ under pairwise swaps, not just the ± 1 for bosons and fermions. Anyonic statistics are now widely taken to account for the fractional quantum Hall effect, of which Wilczek (1990) offers an account and a collection of important papers. For more on the braid groups, consult Kassel and Turaev (2008). Hansson et al. (1992) consider quantizations for the case $d = 1$. Imbo et al. (1990) consider more exotic configuration spaces, which lead to yet more varieties of particle statistics, which they call *ambistatistics*.

41.6.4   What Can Be Settled?

Let me conclude with some remarks about the possibility of settling the dispute between haecceitism and anti-haecceitism, or between transcendental and qualitative individuality. First, it must be accepted that the permutation invariance of the Hamiltonian governing the system may be empirically settled: as we saw above, permutation invariance leads to the superselection of symmetry type, and different symmetry types give rise to different particle statistics (Bose-Einstein, Fermi-Dirac, etc.), which can be empirically tested. (Incidentally, Fermi-Dirac statistics are essential in accounting for the stability of bulk matter; see Dyson and Lenard (1967, 1968) and Lieb (1976, 1990).) But this fails to distinguish haecceitism from anti-haecceitism. First, permutation invariance, while mandated under anti-haecceitism, is still possible under haecceitism. Also, it is a mistake to believe that Maxwell-Boltzmann statistics arise under haecceitism, as is sometimes claimed. Rather, Maxwell-Boltzmann statistics arise for classical particles under both haecceitism and anti-haecceitism. Similarly, quantum statistics (Fermi-Dirac and Bose-Einstein) arise for quantum particles under both haecceitism and anti-haecceitism. Second, the only real-world cases of permutation-invariant collections of particles we know are bosonic or fermionic, for which the states are invariant under permutations. So the distinction between haecceitism and anti-haecceitism does not even get off the ground here.

Can we distinguish them according to the explanations they may provide for permutation invariance? Here we find ourselves in the same position as in the classical case, described at the end of Section 41.4.3. Under anti-haecceitism, and so QI, permutation invariance is compulsory. But under haecceitism, and so TI, permutation invariance is still to be expected: the particles are permutable by dint of their being intrinsically indistinguishable (i.e., having the same mass, spin, charge, etc.). This observation was made by French and Redhead (1988).

However, transcendental individuation makes sense only on route 1 (where permutation invariance is implemented on the haecceitistic joint Hilbert space), since it is only here that we have system labels in the first place. On either route 2 (quantizing the anti-haecceitistic configuration space) or route 3 (quantum field theory), the resources simply *do not exist* to differentiate between states that are permutable in the haecceitistic joint Hilbert space. QI must be the default option on these routes. Therefore, any advocate of TI in the quantum regime must provide convincing reasons for taking route 1, even though: (i) all routes lead, in the end, to equivalent theories, at least if the number of spatial dimensions exceeds 2; and (ii) route 3 would appear to be our best account of the origin of permutation invariance in quantum mechanics. It seems implausible, to say the least, that such convincing reasons could be found.

References

Baker, D., Halvorson, H. and Swanson, N. (2015). The conventionality of parastatistics. *British Journal for the Philosophy of Science*, 66: 929–976.

Brown, H., Sjöqvist, E. and Bacciagaluppi, G. (1999). Remarks on identical particles in de Broglie-Bohm theory. *Physics Letters A*, 251: 229–235.

Butterfield, J. (1993). Interpretation and identity in quantum theory. *Studies in History and Philosophy of Science*, 24: 443–476.

Button, T. and Walsh, S. (2018). *Philosophy and Model Theory*. Oxford: Oxford University Press.

Caulton, A. and Butterfield, J.N. (2012). On kinds of indiscernibility in logic and metaphysics. *British Journal for the Philosophy of Science*, 63: 27–84.

Dieks, D. and Lubberdink, A. (2011). How classical particles emerge from the quantum world. *Foundations of Physics*, 41: 1051–1064.

Dixon, J.D. and Mortimer, B. (1996). *Permutation Groups*. Berlin: Springer.

Doplicher, S., Haag, R. and Roberts, J.E. (1974). Local observables and particle statistics II. *Communications in Mathematical Physics*, 35: 49–85.

Dürr, D., Goldstein, S., Taylor, J., Tumulka, R. and Zanghì, N. (2007). Quantum mechanics in multiply-connected spaces. *Journal of Physics A*, 40: 2997–3031.

Dyson, F.J. and Lenard, A. (1967). Stability of matter I. *Journal of Mathematical Physics*, 8: 423–434.

Dyson, F.J. and Lenard, A. (1968). Stability of matter II. *Journal of Mathematical Physics*, 9: 1538–1545.

Earman, J. (2015). Some puzzles and unresolved issues about quantum Entanglement. *Erkenntnis*, 80: 303–337.

Ehrenfest, P. and Trkal, V. (1921). Deduction of the dissociation equilibrium from the theory of quanta and a calculation of the chemical constant based on this. *Proceedings of the Amsterdam Academy*, 23: 162–183.

van Fraassen, B.C. (1991). *Quantum Mechanics: An Empiricist View*. Oxford: Clarendon Press.

French, S. (1989). Identity and individuality in classical and quantum physics. *Australasian Journal of Philosophy*, 67: 432–446.

French, S. (1995). The esperable uberty of quantum chromodynamics. *Studies in History and Philosophy of Modern Physics*, 26: 87–105.

French, S. and Krause, D. (2006). *Identity in Physics*. Oxford: Oxford University Press.

French, S. and Redhead, M. (1988). Quantum physics and the identity of indiscernibles. *British Journal for the Philosophy of Science*, 39: 233–246.

Frigg, R. (2008). A field guide to recent work on the foundations of statistical mechanics. In D. Rickles (ed.), *The Ashgate Companion to Contemporary Philosophy of Physics*. London: Routledge, pp. 105–202.

Greenberg, O.W., Dell'Antonio, G.F. and Sudarshan, E.C.G. (1964). Parastatistics: Axiomatic formulations, connection with spin and TCP theorem for a general field theory. In F. Gursey (ed.), *Group Theoretical Concepts and Methods in Elementary Particle Physics*. London: Gordon and Breach, , pp. 403–407.

Hansson, T.H., Leinaas, J.M., and Myrheim, J. (1992). Dimensional reduction in anyon systems. *Nuclear Physics B*, 384(3): 559–580.

Hodges, W. (1993). *Model Theory*. Cambridge: Cambridge University Press.

Huggett, N. (1999a). Atomic metaphysics. *Journal of Philosophy*, 96: 5–24.

Huggett, N. (1999b). On the significance of the permutation symmetry. *British Journal for the Philosophy of Science*, 50: 325–347.

Huggett, N. and Imbo, T.D. (2009). Indistinguishability. In D. Greenberger, K. Hentschel and F. Weinert (eds.), *Compendium of Quantum Physics*. Berlin: Springer, pp. 311–317.

Huggett, N. and Norton, J. (2014). Weak discernibility for quanta, the right way. *British Journal for the Philosophy of Science*, 65: 39–58.

Imbo, T.D., Shah Imbo, C. and Sudarshan, E.C.G. (1990). Identical particles, exotic statistics and braid groups. *Physics Letters B*, 234: 103–107.

van Kampen, N. (1984). The Gibbs paradox. In W.E. Parry (ed.), *Essays in Theoretical Physics in Honour of Dirk ter Haar*. Oxford: Pergamon Press, pp. 303–312.

Kaplan, D. (1975). How to Russell a Frege-Church. *Journal of Philosophy*, 72: 716–729.

Kassel, C. and Turaev, V. (2008). *Braid Groups: Graduate Texts in Mathematics 247*. Berlin: Springer.

Ketland, J. (2011). Identity and indiscernibility. *Review of Symbolic Logic*, 4: 171–185.

Ladyman, J., and Bigaj, T. (2010). The principle of the identity of indiscernibles and quantum mechanics. *Philosophy of Science*, 77: 117–136.

Ladyman, J., Linnebo, O. and Pettigrew, R. (2012). Identity and discernibility in philosophy and logic. *Review of Symbolic Logic*, 5: 162–186.

Laidlaw, M.G.G. and DeWitt, C.M. (1971). Feynman functional integrals for systems of indistinguishable particles. *Physical Review D*, 3: 1375–1378.

Landsman, N.P. (2016). Quantization and superselection sectors III: Multiply connected spaces and indistinguishable particles. *Reviews of Mathematical Physics*, 28: 1650019.

Leinaas, J.M. and Myrheim, J. (1977). On the theory of identical particles. *Il Nuovo Cimento B*, 37: 1–23.

Lewis, D.K. (1986). *On the Plurality of Worlds*. Oxford: Blackwell.

Lieb, E.H. (1976). The stability of matter. *Reviews of Modern Physics*, 48: 553–569.

Lieb, E.H. (1990). The stability of matter: From atoms to stars. *Bulletin of the American Mathematical Society*, 22: 1–49.

Margenau, H. (1944). The exclusion principle and its philosophical importance. *Philosophy of Science*, 11: 187–208.

Messiah, A.M.L., and Greenberg, O.W. (1964). Symmetrization postulate and its experimental foundation. *Physical Review B*, 136: 248–267.

Morandi, G. (1992). *The Role of Topology in Classical and Quantum Physics*. Berlin: Springer.

Muller, F.A. (2015). The rise of relationals. *Mind*, 124: 201–237.

Muller, F.A. and Saunders, S. (2008). Discerning fermions. *British Journal for the Philosophy of Science*, 59: 499–548.

Muller, F.A. and Seevinck, M. (2009). Discerning elementary particles. *Philosophy of Science*, 76: 179–200.

Ohnuki, Y. and Kamefuchi, S. (1982). *Quantum Field Theory and Parastatistics*. Berlin: Springer.

Post, H. (1963). Individuality and physics. *The Listener*, 70: 534–537.

Quine, W.V.O. (1976). Grades of discriminability. *Journal of Philosophy*, 73: 113–116.

Quine, W.V.O. (1986). *Philosophy of Logic* (2nd edition). Cambridge, MA: Harvard University Press.

Read, N. (2003). Non-Abelian braid statistics versus projective permutation statistics. *Journal of Mathematical Physics*, 44: 558–563.

Redhead, M. and Teller, P. (1991). Particles, particle labels, and quanta: The toll of unacknowledged metaphysics. *Foundations of Physics*, 21: 43–62.

Redhead, M. and Teller, P. (1992). Particle labels and the theory of indistinguishable particles in quantum mechanics. *British Journal for the Philosophy of Science*, 43: 201–218.

Sagan, B.E. (1991). *The Symmetric Group: Representations, Combinatorial Algorithms, and Symmetric Functions* (2nd edition). Berlin: Springer.

Saunders, S. (2003a). Physics and Leibniz's principles. In K. Brading and E. Castellani (eds.), *Symmetries in Physics: Philosophical Reflections*. Cambridge: Cambridge University Press, pp. 289–307.

Saunders, S. (2003b). Indiscernibles, covariance and other symmetries: The case for non-reductive relationism. In A. Ashtkar, D. Howard, J. Renn, S. Sarkar and A. Shimony (eds.), *Revisiting the Foundations of Relativistic Physics: Festschrift in Honour of John Stachel*. Amsterdam: Kluwer, pp. 151–173.

Saunders, S. (2006a). Are quantum particles objects?. *Analysis*, 66: 52–63.

Saunders, S. (2006b). On the explanation for quantum statistics. *Studies in the History and Philosophy of Modern Physics*, 37: 192–211.

Saunders, S. (2013). Indistinguishability. In R. Batterman (ed.), *The Oxford Handbook in Philosophy of Physics*. Oxford: Oxford University Press, pp. 340–380.

Sternberg, S. (1994). *Group Theory and Physics*. Cambridge: Cambridge University Press.

Stolt, R.H. and Taylor, J.R. (1970). Correspondence between the first- and second-quantized theories of paraparticles. *Nuclear Physics B*, 19: 1–19.

Tung, W.-K. (1985). *Group Theory in Physics*. Singapore: World Scientific.

Uffink, J. (2006). Compendium of the foundations of classical statistical physics. In J. Butterfield and J. Earman (eds.), *Handbook of the Philosophy of Science: Philosophy of Physics*. Amsterdam: North Holland, pp. 923–1074.

Wilczek, F. (1990). *Fractional Statistics and Anyon Superconductivity*. Singapore: World Scientific.

Willard, S. (1970). *General Topology*. Reading, MA: Addison-Wesley.

Chapter 41 - Further Reading from the Editors

A relatively early paper on particles and permutation is Redhead, M. and Teller, P. 'Particle labels and the theory of indistinguishable particles in quantum mechanics' (*British Journal for the Philosophy of Science* 43, pp. 201–218, 1992) These issues were then taken up in papers by Simon Saunders: 'Physics and Leibniz's Principles', in K. Brading and E. Castellani, (eds.), *Symmetries in Physics: Philosophical Reflections* (Cambridge: Cambridge University Press, pp. 289–307, 2003) and 'Indiscernibles, covariance and other symmetries: the case for non-reductive relationism', in A. Ashtkar, D. Howard, J. Renn, S. Sarkar and A. Shimony (eds.), *Revisiting the Foundations of Relativistic Physics: Festschrift in Honour of John Stachel* (Amsterdam: Kluwer, pp. 151–173, 2003). Saunders' 'Indistinguishability', in R. Batterman (ed.), *The Oxford Handbook in Philosophy of Physics* (Oxford: Oxford University Press, pp. 340–380, 2013) provides an overview. The logical issues discussed have their origins in Quine, W. V. O. 'Grades of discriminability' (*Journal of Philosophy* 73, pp. 113–116, 1976). See also Huggett, N. and Norton, J. 'Weak discernibility for quanta, the right way' (*British Journal for the Philosophy of Science* 65, pp. 39–58, 2014) and Oliver Pooley's "Points, Particles and Structural Realism," in Rickles, French, and Saatsi, eds. *The Structural Foundations of Quantum Gravity* (Oxford University Press, 2006).

42

GAUGE THEORIES

Nicholas J. Teh

42.1 Introduction

The term "gauge" was introduced into the theoretical physics lexicon by means of Weyl's (1918) paper on gravitation and electricity, but it has since come to be used in myriad ways within the foundations of physics.[1] It will thus be helpful to begin by distinguishing between a broad and a narrow sense of the term.

In its broadest sense (cf. the characterization given by Redhead (2003)), a "gauge" refers to a mathematical structure that non-uniquely represents a physical scenario, because many distinct (but as we shall see: also in some sense "equivalent") gauges can represent the scenario equally well.[2] Such non-uniqueness is referred to as "gauge freedom," and if a symmetry relates the gauges then it is referred to as a "gauge symmetry." In this sense, the discussion of "gauge" in physics is of course a time-honored subject and does not turn on any of the novel features introduced by Weyl's theory and its descendants. To give a familiar example, many would say (along with Leibniz) that spatial translations of the entire universe do not correspond to physically distinct scenarios, and it thus counts as a "gauge symmetry" with respect to the "gauge quantity" of absolute position.

On the other hand, within the physics (especially the high-energy theory) community, "gauge" is typically used in a narrow sense that is meant to capture the novel features introduced by Weyl's theory. In this sense, a "gauge theory" is a type of field theory whose field configurations (or states) are acted on by a "local" gauge symmetry, i.e., a group whose elements are functions of spacetime. According to a standard interpretation, field configurations which are related by these gauge transformations are taken to represent the same physical scenario, at least locally. Thus, such fields are "gauge quantities" in the broad sense introduced above.

Within the class of theories picked out by the narrow sense of "gauge," there is an even more restricted subclass that stems from the work of Yang and Mills, and whose quantized version is the basis for the Standard Model: these are often referred to as *gauge theories of Yang-Mills type*. Yang-Mills type theories are distinguished by symmetries (and indeed a global invariant formulation) that take a particularly simple form, which in turn allows the dynamics of the theory to be straightforwardly specified.[3] As such, and because of their empirical utility, these theories constitute a focal case of gauge theory for many physicists. They will do so for the purposes of this survey, and henceforth I will simply refer to them as "gauge theories."

The literature on the philosophy of gauge theory is by now vast, and overlaps with various topics within the philosophy of science and metaphysics. To give a non-exhaustive list, there has been work on the empirical significance of gauge symmetry,[4] the relationship between gauge symmetry and determinism,[5] debates about the appropriate ontology for gauge theory,[6] the relationship between gauge theory and gravity,[7] the interpretation of spontaneous symmetry breaking,[8] the relationship between gauge symmetry and physical dualities,[9] and most recently on the special complications introduced by the presence of (fiducial and non-fiducial) spacetime boundaries in gauge theory.[10]

Needless to say, there is no room in a brief survey to consider these topics in any detail. Instead, I will adopt a different tack in order to give the reader a feel for the subject: I will first provide an exposition of an elementary but fundamental "problematic" that lies at the heart of much philosophical work on gauge theory. I will then argue that although the features of gauge theory that give rise to this "problem" are certainly of philosophical interest, the "problem" itself has been misdiagnosed.

Section 42.1 introduces just enough of the machinery (and standard interpretation) of gauge theory in order to formulate and discuss what I will call "the problem of descriptive fluff" for gauge theory. Sections 42.2 and 42.3 both argue that the problem of descriptive fluff rests on a false premise: the former by appealing to the full representational capacities of the theory, and the latter by showing that the relevant premise is untenable when classical gauge theory is interpreted in conjunction with quantum theory. Finally, Section 42.4 concludes with several morals.

42.2   The Problem of "Descriptive Fluff"

42.2.1   *Gauge Theory: A Relatively Informal Description*

According to one standard way of proceeding, a physical theory is specified by first describing the space of kinematically possible states according to the theory, and then proceeding to lay down a dynamical principle that restricts the kinematical possibilities to a subspace of dynamically possible states. I now give an overview of how this is done for gauge theories.

A gauge theory is a theory of fields. Thus, a "state" of the world according to such a theory is an assignment of (perhaps fairly abstract and complex types of) values to a region of spacetime, i.e., a field. Physicists typically specify these fields by beginning with the simplest "small patches" of spacetime, i.e., patches of spacetime that have a trivial topological structure like that of Euclidean space. It is furthermore assumed (although not always explicitly stated) that fields on larger regions of spacetime can be built up from fields on small patches by a suitable "gluing procedure"—a point that we will return to later. For now, the important thing to note is that the "space of kinematically possible states" for a field theory is always described with respect to a patch of spacetime: it is the space of fields (also called "field configurations") on that patch.

To see how such structure is specified in a particular model of gauge theory, let us first fix a spacetime manifold M: we will make the simplifying assumption that this is just a smooth manifold, since none of the points that we wish to discuss turn on the presence of further metric structure.

Let U be any small patch of M. Physicists typically lay out the kinematics of gauge theory by specifying a set of formal states on U called "gauge fields." In order to define these objects, we will need to introduce the "symmetry data" of the theory, viz. a semisimple Lie group G and its "infinitesimal" description or Lie algebra $\mathfrak{g}$. A *gauge field* is then defined to be a $\mathfrak{g}$-valued 1-form A, i.e.,

$$A : U \to \Omega^1(M) \otimes \mathfrak{g}.$$

In other words, it is a field that assigns a matrix of 1-forms to a small patch U. On this formal level, then, the space of kinematic states over U will be the "space" of gauge fields over U. (For the moment, we will be deliberately vague about what exactly such a "space" consists of, but we shall return to this point in some detail in the next section.) We are now also in a position to describe how a gauge symmetry transformation $g : U \to G$ acts on a gauge field A:

$$g : A \mapsto A' = gAg^{-1} - (dg)g^{-1}. \tag{42.1}$$

In the simplest case of Maxwell gauge theory (where $G = U(1)$), this transformation becomes the familiar $A \mapsto A' = A - d\chi$, where A is a 1-form field and χ is a smooth function.

We will use the term "gauge orbit" to denote an equivalence class $[A]$ of gauge fields with respect to the gauge symmetry action, i.e., $A \sim A'$ just in case A is related to A' by (42.1). There is, in both the physics and the philosophy literature, a standard interpretation of the gauge fields and the gauge symmetry:

> **(Standard)** On a small patch of spacetime, the gauge orbits are in a 1-1 correspondence with the true/physical states described by the theory; thus the gauge fields in an orbit are in a many-to-one correspondence with the physical state corresponding to that orbit.

Why is (Standard) widely held? To understand this point, we will need to consider the dynamics of a gauge theory. Such dynamics is typically specified in a Lagrangian form, and then converted (by means of a standard mapping called the Legendre transformation) to a Hamiltonian form in order to obtain a more perspicuous analysis. We now review the rough outline of this procedure.

The first part of the story, i.e., the transition from Lagrangian to Hamiltonian dynamics, is highly general, and will be familiar to the reader from elementary classical mechanics. Just like any other classical field theory, the dynamics of a gauge theory can be specified by means of a Lagrangian (density) $L(q, \dot{q})$, where q and $\dot{q}$ schematically represent the field configurations and time derivatives of fields, respectively. By means of the Euler-Lagrange equations, the theory's equations of motion can then be derived from L; requiring that the states satisfy these equations of motion picks out the subspace of dynamically possible states within the space of kinematically possible states.

The Legendre transformation is then used to map this Lagrangian dynamics to its Hamiltonian formulation: in particular, it maps $(q, \dot{q})$ to $(q, p := \partial L / \partial \dot{q})$ (i.e., the field configurations and their conjugate momenta), and it maps the Lagrangian L to a Hamiltonian $H(q, p)$. The target space for this mapping (i.e., the space of kinematically possible (q, p)) is called the *phase space P*, and it is equipped with a geometric structure called the Poisson bracket. Smooth and invertible maps of P that preserve the Poisson bracket structure are called *canonical transformations*. One of the virtues of the Hamiltonian formalism is that it allows us to formulate the theory's dynamics in a geometric way: given H, the relevant equations of motion can be formulated in terms of the Poisson bracket structure. Thus, a solution to the equations of motion is given by a family of canonical transformations that is parameterized by time: this geometric "flow" traces out curves in P as the time parameter varies.

The twist in the story arises when we consider gauge theories, for which the Legendre transformation map is not surjective. Instead, its image is a submanifold $C \subset P$ called the *constraint surface*, which is defined by the constraint equations $G_j(q, p) = 0$, $j = 1, \ldots, n$ for a set of generators (of canonical transformations) $\{G_j\}$. Thus, the dynamically possible states of the theory all lie within this constraint surface.[11]

The structure of the generators $\{G_j\}$ is very rich, and contains much information of physical interest. For the purpose of justifying (Standard), we will be interested in generators that have the property of being "first class," meaning that their Poisson bracket with any of the other generators is a linear combination of generators (and so in particular vanishes on the constraint surface). They also have a perspicuous geometric characterization, viz. as generators of "first-class" canonical transformations within the constraint surface, as shown in Figure 42.2 below.

The relevance of "first class" generators for the justification of (Standard) consists in two facts, viz. (i) given some dynamical solution corresponding to the initial data (q_0, p_0), a first class canonical transformation can always be used to generate another solution corresponding to (q_0, p_0); and (ii) the first class canonical transformations are the Hamiltonian representation of the gauge symmetry transformations (42.1). Note that because of (ii), it makes sense for us to refer to such canonical transformations as "gauge transformations" and to a submanifold of states related by gauge transformations as a "gauge orbit." By conjoining (i) and (ii), we arrive at the conclusion that, locally, treating gauge fields related by gauge transformations as *physically distinct* states leads to an indeterministic dynamics for a gauge theory. Whence the conventional justification of (Standard): such indeterminism seems

597

spurious (in contrast to e.g., the case of evolution past a Cauchy horizon in General Relativity), and it should thus be eliminated by adopting (Standard).

It is beyond the scope of this overview to further probe this justification for (Standard). We will thus assume that both the justification, and (Standard), are reasonable. Correspondingly, the physical observables of the theory are taken to be gauge-invariant functions, i.e., functions that are constant on the gauge orbits. For instance, in Maxwell gauge theory, one of the gauge-invariant observables is $F := dA$, which is clearly invariant under (42.1) and thus constant on gauge-orbits.[12]

Given (Standard), one secures a deterministic interpretation of gauge theory. However, this is not enough for some authors—in addition to understanding the theory as deterministic, they would furthermore like the phase space P of the theory to be modified so that the theory's formalism more perspicuously exhibits such determinism. Such authors counsel pursuing the formal strategy of replacing the constraint surface in P with the space of gauge orbits: the resulting space of states is called the reduced phase space $\hat{P}$. More generally, let us refer to the strategy of replacing gauge fields with gauge orbits (a strategy that can be implemented at the purely kinematic level) as **(Reduce)**. In the next section, we will consider a philosophical problematic to which (Reduce) is an apparent—but ultimately unsatisfactory—solution.

42.2.2 *The Problem of Descriptive Fluff*

Let us now assume the interpretive principle (Standard). The term "descriptive fluff" or "surplus structure" (see e.g., Redhead (2003); Earman (2003)) is commonly used in the literature to refer to fact that, in typical formulations of gauge theory, there is a many-to-one relationship between the formal representations of physically possible states (i.e., gauge fields), on the one hand, and the true physically possible states, on the other hand; thus, these formulations contain descriptive fluff, or *"fluff"* for short. Put in another way, "fluff" can be characterized as any of the theory's state structure over above the structure of gauge equivalence classes. 'Fluff' is sometimes taken to have a pejorative connotation, i.e., it signals that there is something amiss about the theory; however, for our purposes, it will be helpful to use the term in a strictly neutral way.

We will instead package the pejorative content into two further claims, which we will jointly refer to as "the problem of descriptive fluff." First,

> **(Idle)**: The presence of "fluff" in gauge theory shows that the theory possesses idle structures (gauge fields and gauge transformations), i.e., structures that have play no role in describing physical reality.

And second, the claim that such idle fluff is unacceptable in a philosophically or physically perspicuous formulation of a physical theory. For instance, Redhead (2003) says that it "leaves us with a mysterious, even mystical, Platonist-Pythagorean role for purely mathematical considerations in theoretical physics."

In what follows, I will focus on showing that (Idle) should be rejected, thus providing one way of dissolving the problem of descriptive fluff. But one might reasonably ask: why attempt to fend off the problem in this way when the strategy that we called (Reduce) in the last section appears to sidestep the problem altogether? In (Reduce), a gauge theory is reformulated so that it only quantifies over the gauge orbits and not any of the putatively idle structures (i.e., gauge fields and gauge transformations) so the problem clearly does not arise. The reason is that I shall be arguing that (Idle) should be rejected because fluffy structures *do* play a role in describing physical phenomena. And since fluff is characterized as structure over above the gauge orbits, any successful argument of this kind would also show that (Reduce) is not in general a viable strategy for sidestepping the problem.

In order to make progress, let us distinguish two ways in which fluff might play a role in describing physical phenomena, thus showing (Idle) to be false. The first—weak—role is one in which the fluff is practically indispensable for modeling a certain phenomenon (e.g., because the relevant idealizations involve gauge fields). The second—strong—role does not concern pragmatics. We will consider two examples of this in Section 42.3: the first is a case in which the fluff can be shown to play a representational role, and the second is a case in which the fluff functions as a condition of possibility for consistently formulating an intertheoretic relation.

We turn to a discussion of the weak role in the remainder of this section. There are (at least) three respects in which fluff might be said to be practically indispensable for modeling physical systems. First, from the perspective of Lagrangian field theory, gauge-dependent descriptions are an essential tool for understanding how gauge fields couple to other kinds of fields; indeed the gauge-invariant Lagrangian of the entire theory is built up by combining such non-gauge-invariant field descriptions. Second, the space of gauge orbits is typically highly singular, and analytically intractable (which is why the process of constructing a reduced phase space was referred to as a "formal" strategy in the previous section)—thus, for many applications, it is more convenient to use the non-reduced phase space P. Third, physicists are very familiar with the idea that a convenient choice of gauge often makes equations easier to solve. But there is also a deeper and more thorough-going version of this point that deserves our attention, and which arises in attempts to rigorously analyze the partial differential equations of gauge theory. It is famously difficult to prove the existence, uniqueness, and stability of solutions to such equations (for reasonable classes of initial data and boundary conditions), and many of the most powerful techniques for analyzing these equations are not gauge-invariant. Thus, in one strict sense of "model" (in which a model of a theory is a rigorous analytical solution to a theory's equations of motion) it is reasonable to say that, as far as we presently know, fluff is required in order to construct many models of gauge theory.

42.3 Fluff Is Strongly Non-Idle

42.3.1 *A Role in Global Kinematic Structure*

We now provide our first example of a scenario in which fluff is strongly non-idle. The example (and others like it) is based on the existence and interest of topologically non-trivial global field configurations in gauge theory, and it demonstrates that fluff is necessary in order to describe (what many physicists would standardly take to be) some of the representational capacities of gauge theory.[13]

Let us begin by recalling a theme that we mentioned briefly in Section 42.2.1: physicists typically assume that the following "gluing constraint" holds of the relationship between global field data on a region R of spacetime, and local field data on a set of small patches $\{U_i\}$ that cover R, i.e., (a) given any local field data (that matches on the overlaps of the patches), it should be possible to consistently glue this data into global field data; and (b) given any global field data on R, it should be possible to construct this data by gluing together some set of local field data.

We will now consider an elementary example of how (Reduce) results in a space of fields that fails to satisfy (b). In this example, spacetime will be a sphere, i.e., $M = S^2$, and we will only consider two local patches H_1 and H_2, which intersect in a belt around the equator, as depicted in Figure 42.1. We shall use the notation $H_{12} := H_1 \cap H_2$.

According to (Reduce), the space of fields over H_i is the set of gauge orbits over H_i; in other words, a field is a gauge equivalence class $[A]$. Thus, in order to show that (Reduce) violates (b), it suffices to exhibit a field configuration over S^2 that cannot possibly result from gluing together the field data over H_1 and H_2, respectively. Such a global field configuration (indeed a solution) exists and in fact predated the advent of Yang-Mills theory by almost 20 years—it is called the Dirac monopole (Dirac, 1931).

Roughly speaking, the monopole field is constructed by means of the following procedure: let the local field data on H_i be the gauge fields, along with the gauge transformations $g_i : H_i \to U(1)$, where $i = 1, 2$. On the overlap H_{12} the data of the monopole solution is constructed by requiring that the following gluing conditions are satisfied:

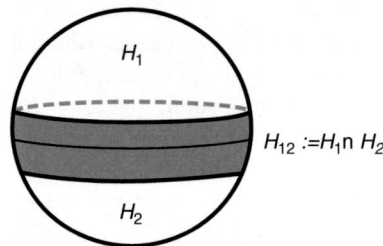

Figure 42.1 Two patches H_1 and H_2 intersecting in a belt H12 around the equator of a spherical spacetime.

$$g_2 = g_{12}\, g_1, \qquad A_1 = g_{12}^{-1}\, A_2\, g_{12} + g_{12}^{-1}\, dg_{12}, \qquad (42.2)$$

where $g_{12} : H_{12} \rightarrow U(1)$. In virtue of these gluing relationships, one obtains a global field 2-form field configuration F on S^2 that is locally exact, i.e. $F = dA_i$, but which is globally non-exact (in the sense that it cannot be expressed as the exterior derivative of a 1-form). Furthermore, F is a solution of an equation of motion in the sense that $dF = 0$.

Such non-trivial global field configurations show that (Reduce) violates (b). First, notice that the gluing relations (42.2) required to construct such global fields are not part of the local field data of (Reduce), because these relations are expressed in terms of the gauge transformations, which (Reduce) eliminates by taking gauge equivalence classes. Relatedly, a local field configuration of (Reduce), i.e., a gauge equivalence class, is topologically trivial (in the sense of being globally exact) and such data is in itself inadequate to build up topologically non-trivial field data such as that of the monopole solution: this point reinforces the necessity of retaining "gauge symmetry" data in the space of fields in order to preserve (b).

How then can the gauge symmetry data be incorporated into the space of fields? An idea that has been repeatedly emphasized in the recent literature on gauge theory (see Benini et al. (2015) and references therein) is that the space of states over a patch of spacetime should not be described as a set of states (or indeed equivalence classes of states), but rather as a groupoid (i.e., a category all of whose morphisms are isomorphisms) whose objects are gauge fields and whose morphisms are gauge transformations between the gauge fields: call this the *groupoid of gauge fields*. This groupoid is illustrated schematically in Figure 42.2.

Notice that such a conceptualization of a theory's space of states is not at odds with (Standard). Indeed, the standard interpretation of the groupoid of gauge fields incorporates (Standard) by laying it down that isomorphic objects represent the same physical state; however, by formalizing these as *distinct objects* of a category, it also leaves room for including the gauge transformation data as part of this "space," thus providing the expressive resources to glue this data into non-trivial global states such as the monopole. Indeed, various rigorous results show that on a minimal formal conception of what it means to "glue" local field data into global field data, this general strategy manages to capture all non-trivial global field data on non-contractible manifolds (Benini et al., 2015).

Let us summarize the point of this section. Given (i) what we called the "gluing constraint" on fields; and (ii) the use of topologically non-trivial global field configurations to represent physical scenarios, I argued that fluff (i.e., the structure of gauge fields and gauge transformations) is not idle since it is required to construct global field representations out of local field data. Notice further that this argument is consistent with holding (Standard). Of course, this argument can be challenged by rejecting either (i) or (ii). However, both of these options are unappealing. To reject (i) is to not only reject a commonplace intuition about "fields" among physicists, but also to impoverish the structure of the theory by omitting from it important information about the relationships between models on local regions of spacetime and models on global regions of spacetime. And to reject (ii) is to reject the important role that topological solitons have played, both in developing empirically adequate models within the context of condensed matter physics, and in spurring new theoretical developments (especially in the subject of "dualities") within quantum field theory (QFT) and string theory.

42.3.2 A Role in Local Dynamical Structure

In the previous section, we considered a reason for rejecting (Idle Fluff) that was based on the kinematics of gauge theory and its potential for representing physical scenarios on topologically non-trivial regions of spacetime, i.e., that involved global descriptions of the theory's states. In this section, we will instead focus on a reason for rejecting (Idle Fluff) that involves the dynamics of gauge theory, but for which it suffices to consider only local descriptions of the theory's states.

First, recall that the constrained Hamiltonian analysis of Section 42.2.1 invoked the dynamical data of a gauge theory, because the theory's Lagrangian was used to describe the constraint surface in the larger phase space P. Recall too that upon performing this analysis, we arrived at the following (infinitesimal) description of the theory's gauge symmetry: the gauge symmetry generators $G_a(q, p)$ satisfy the relation

$$\{G_a, G_b\} = C_{ab}^c \, G_c \tag{42.3}$$

where $\{\cdot, \cdot\}$ is the Poisson bracket and the coefficients C_{ab}^c are the structure constants associated with the Lie algebra. (Indeed, because the coefficients are constant, this algebra is "closed" on the entire phase space P.)

There exists a purely classical generalization of this gauge symmetry called the (Hamiltonian) BRST symmetry (named after Becchi, Rouet, Stora and Tyutin): it arises by, first, generalizing the structure constants to structure functions $C_{ab}^c(q, p)$; and second, introducing new anti-commuting pairs of phase space coordinates called "ghosts." With this generalization in place, one can define a new "BRST generator" Ω, whose algebraic properties capture in an invariant form the information about the structure functions, including higher-order relations between these structure functions. In particular, Ω contains the original structure of the gauge symmetry algebra (42.3) with which we began.[14]

As one might expect, the algebraic properties of Ω allow one to recover all the information that one would typically think of as encoded in the gauge symmetry, including the typical operation of defining gauge-invariant observables. More specifically, the cohomology group $H^0 := (\text{Ker } \Omega/\text{Im } \Omega)^0$ is isomorphic to the set of (equivalence classes of) gauge-invariant observables of the theory. This sort of result assures us that the representational capacities of (Reduce) are captured by the BRST symmetry.

However, we would like to demonstrate that the BRST symmetry contains "physical information" that goes beyond that of (Reduce), and thus constitutes a counter-example to (Idle Fluff). In order to do so, we will argue that (i) the algebraic properties of the BRST generator captures physically relevant information about gauge theory; and that (ii) these algebraic properties stem from the fluffy structure, i.e., the gauge fields and gauge transformations.

We begin by explaining point (ii). In Henneaux and Teitelboim (1988), it was demonstrated that the algebraic information (i.e., cohomology groups) encoded in the BRST generator could be understood in terms of the properties of the constraint surface in the original (non-reduced) phase space P. In particular, by defining an antisymmetric (exterior) derivative along the gauge orbits (cf. Figure 42.2) in the constraint surface, one can interpret the ghosts as 1-forms along these gauge orbits, thereby reproducing the algebraic properties of the BRST generator in terms of the (again, cohomological) properties of this special derivative. The moral of this point for us should be clear: the algebraic information encoded in BRST symmetry stems from the (local) geometry of states which are related by gauge transformations, and the new "ghost coordinates" can in a loose sense

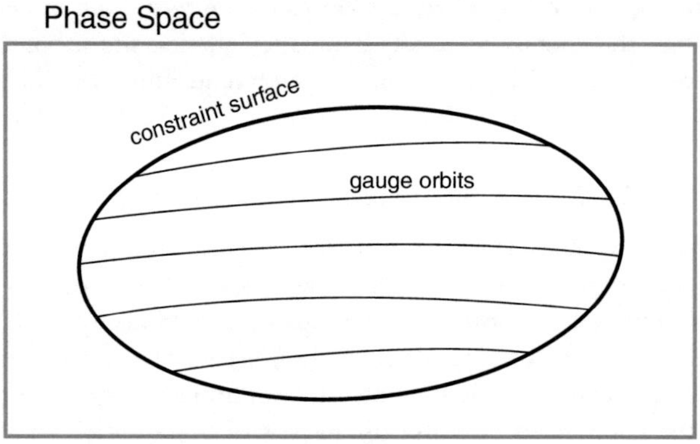

Phase Space

Figure 42.2 Gauge orbits within a constraint surface in phase space.

be interpreted as higher-order relations between such gauge transformations. On the other hand, pursuing (Reduce) would be to collapse the geometry of a gauge orbit into a point, thus omitting the information captured by the algebraic properties of the BRST generator.

Next, let us demonstrate point (i), viz. that the algebraic properties of the BRST generator contain physically relevant information about gauge theory. At this point in the argument, it will be helpful to remind the reader a moral urged by Belot (1998) in his discussion of the Aharanov-Bohm effect: there he argues that theories cannot be interpreted in isolation from each other, and that "...understanding intertheoretic relations is a crucial component of the articulation of the content of individual physical theories." Belot further explains that the Aharanov-Bohm effect illustrates this moral in the specific case where considering the intertheoretic relationships (quantization and the classical limit) between the theories of quantum mechanics and classical mechanics, on the one hand, forces us to relinquish a certain interpretation of electromagnetism, on the other hand.

A similar moral is relevant in explaining why BRST symmetry is physically relevant: here the point is not to consider the implications of the intertheoretic relationship of "quantization" for the interpretation another theory, but rather to consider the part of a theory's structure that determines whether the quantization can even be defined to begin with. To the extent that we understand the intertheoretic relationship of "quantization" that obtains between a QFT and the classical field theory from which the QFT is perturbatively defined, it is commonly thought that there are general consistency conditions that need to be met by the classical theory in order for this relation (and the quantum side of the relata) to be defined, on pain of the quantized theory being non-unitary or having a state space of indefinite norm. In the case of gauge theories, the structural obstruction to meeting these conditions is referred to as a "gauge anomaly," and the further algebraic (cohomological) information contained in the BRST generator determines whether or not a gauge theory has a "gauge anomaly." Thus, fluff is non-idle because it determines whether or not the worlds described by a classical gauge theory and quantum theory are radically distant in the space of physically possible worlds, and such modal information is typically regarded by physicists as being "physical."

Developments of BRST symmetry in the Lagrangian formalism have shown that this essentially fluffy structure contains additional physical information about various phenomena, including the question of whether or not a classical gauge theory is perturbatively renormalizable, and the question of how one can consistently add interactions to a free gauge theory.

42.4 Conclusion and Morals

Let us take stock. In this survey, I described what physicists and philosophers typically have in mind when they speak of the "redundancies," "descriptive fluff," or "surplus structure" of gauge theories. I then addressed the main reason for which various thinkers have found such fluff to be philosophically problematic, i.e., because of the claim that it plays no role in describing physical phenomena. As we have seen, this claim turns out to be false for several—quite different—reasons. And these reasons are also reasons to reject the naive strategy of omitting fluff by taking gauge equivalence classes.

Thus, the original problem of descriptive fluff rests on a false premise. However, this by no means shows that the philosophical literature has failed to point us to any interesting issues. Nothing could be further from the truth, since the ways in which fluff captures "physical content" are subtle and reveal novel ways in which the formal structure of a theory can represent facts about relationships between local and global models, the conditions of possibility for intertheoretic relationships, and the limits of certain forms of physical modality. These topics have hardly been explored by philosophers, and much remains to be done.

Finally, it is worth noting that the naive strategy of taking gauge equivalence classes is not the only way of trying to omit "fluff" from a gauge theory. In light of recent work on physical dualities (i.e., "equivalences" between two theories), it is well-known that (i) some gauge theories are "dual" to theories that have no gauge fields (and gauge symmetry), and thus do not have fluff; and (ii) some gauge theories are "dual" to different gauge theories, i.e., both theories have fluff, but they have different fluff.

Notes

1 Yang points out that initially the English term for *Eich Invarianz* was "calibration invariance" and only about a decade later the translation "gauge invariance" was introduced (Yang, 2005, p. 528).

2 In fact, Redhead defines "gauge" as a representation and takes "non-uniqueness" to be a further property of the representation; but this use does not square with contemporary practice.

3 To be more precise, I am using "Yang-Mills type" theory to include cases that have a different action and dimensionality from the theory that Yang and Mills proposed (e.g., Chern-Simons theory) and also cases that have different symmetry groups (the most important condition on the suitability of such a symmetry group being that its Lie algebra admits of a non-degenerate bilinear form that can be used to construct a Lagrangian density).

4 Brading and Brown (2004), Greaves and Wallace (2014), Teh (2015).

5 Earman (2002), Lyre (2009).

6 Healey (2007), Maudlin (2007), Myrvold (2011).

7 Wallace (2015), Teh (2016).

8 Earman (2003), Smeenk (2006), Friederich (2013).

9 Read (2016), de Haro et al. (2016).

10 Gomes (2019), Mathieu et al. (2020).

11 Expressing these ideas in terms of constrained Hamiltonian dynamics is of course only one way to tell the story; for a perspective on gauge that uses the more modern idiom of the covariant phase space, see Gomes (2019); Mathieu et al. (2020).

12 More generally, the gauge-invariant observables of a gauge theory are generated by the holonomies of the theory. Observables can also be described algebraically, in the more modern idiom of ordinary differential cohomology.

13 This discussion is a much abridged version of the argument of Nguyen et al. (2020), to which I refer the reader for further details. There are also examples that do *not* rely on topologically non-trivial configurations: see Gomes (2019) and Mathieu et al. (2020) for a discussion of such examples.

14 We note that it is possible to understand these BRST structures as emerging naturally from a "derived" enhancement of the space of solutions (the "derived critical locus") picked out by a variational problem. This strategy is developed and pursued in Mathieu et al. (2020).

References

Belot, G. (1998). Understanding electromagnetism. *The British Journal for the Philosophy of Science*, 49(4): 531.

Benini, M., Schenkel, A. and Szabo, R.J. (2015). Homotopy colimits and global observables in abelian gauge theory. *Letters in Mathematical Physics*, 105(9): 1193–1222.

Brading, K. and Brown, H.R. (2004). Are gauge symmetry transformations observable? *British Journal for the Philosophy of Science*, 55(4): 645–665.

de Haro, S., Teh, N. and Butterfield, J. (2016). On the relation between dualities and gauge symmetries. *Philosophy of Science*, 83(5):1059–1069.

Dirac, P.A.M. (1931). Quantised singularities in the electromagnetic field. *Proceedings of the Royal Society of London A: Mathematical, Physical and Engineering Sciences*, 133(821): 60–72.

Earman, J. (2002). Gauge matters. *Proceedings of the Philosophy of Science Association*, 2002(3): 209–220.

Earman, J. (2003). Rough guide to spontaneous symmetry breaking. In K.A. Brading and E. Castellani (eds.), *Symmetries in Physics: Philosophical Reflections*. Cambridge: Cambridge University Press, pp. 335–346.

Friederich, S. (2013). Gauge symmetry breaking in gauge theories—in search of clarification. *European Journal for Philosophy of Science*, 3(2): 157–182.

Gomes, H. (2019). Gauging the boundary in field-space. *Studies in History and Philosophy of Science Part B: Studies in History and Philosophy of Modern Physics*, 67: 89–110.

Greaves, H. and Wallace, D. (2014). Empirical consequences of symmetries. *British Journal for the Philosophy of Science*, 65(1): 59–89.

Healey, R. (2007). *Gauging What's Real: The Conceptual Foundations of Contemporary Gauge Theories*. Oxford: Oxford University Press.

Henneaux, M. and Teitelboim, C. (1988). BRST cohomology in classical mechanics. *Communications in Mathematical Physics*, 115(2): 213–230.

Lyre, H. (2009). Gauge symmetry. In D. Greenberger, K. Hentschel and F. Weinert (ed.), *Compendium of Quantum Physics*. Berlin/Heidelberg: Springer, pp. 248–255.

Mathieu, P., Schenkel, A., Teh, N.J. and Wells, L. (2020). Homological perspective on edge modes in linear Yang-Mills theory. *Letters in Mathematical Physics*, 110: 1559–1584.

Maudlin, T. (2007). *The Metaphysics within Physics*. Oxford: Oxford University Press.

Myrvold, W.C. (2011). Nonseparability, classical, and quantum. *The British Journal for the Philosophy of Science*, 62(2): 417.

Nguyen, J., Teh, N.J. and Wells, L. (2020). Why surplus structure is not superfluous. *The British Journal for the Philosophy of Science*, 71(2): 665–695.

Read, J. (2016). The interpretation of string-theoretic dualities. *Foundations of Physics*, 46(2): 209–235.

Redhead, M. (2003). The interpretation of gauge symmetry. In K.A. Brading and E. Castellani (eds.), *Symmetries in Physics: Philosophical Reflections*. Cambridge: Cambridge University Press, pp. 124–139.

Smeenk, C. (2006). The elusive Higgs mechanism. *Philosophy of Science*, 73(5): 487–499.

Teh, N.J. (2015). Galileo's gauge: Understanding the empirical significance of gauge symmetry. *Philosophy of Science*, 83(1): 93–118.

Teh, N.J. (2016). Gravity and gauge. *British Journal for the Philosophy of Science*, 67(2): 497–530.

Wallace, D. (2015). Fields as bodies: A unified presentation of spacetime and internal gauge symmetry.

Weyl, H. (1918). Gravitation and electricity. *Sitzungsberichte der Königlich Preussischen Akademie der Wissenschaften*, 26: 465–480.

Yang, C. (2005). *Selected Papers (1945–1980), with Commentary*. World Scientific series in 20th century physics. Singapore: World Scientific.

Further Reading from the Editors

For an understanding of electromagnetism as a guide to other gauge theories, it is worth starting with Gordon Belot's 'Understanding electromagnetism' (*The British Journal for the Philosophy of Science*, 49(4): 531, 1998). Issues discussed there are expanded to other gauge theories under a particular interpretation in Richard Healey's *Gauging What's Real: The Conceptual Foundations of Contemporary Gauge Theories* (Oxford University Press, 2007). A thorny and central issue concerns the empirical significance of gauge symmetries. See Brading, K. and Brown, H. R. 'Are gauge symmetry transformations observable?' (*British Journal for the Philosophy of Science*, 55(4): 645–665, 2004), Greaves, H. and Wallace, D. 'Empirical consequences of symmetries' (*British Journal for the Philosophy of Science*, 65(1): 59–89, 2014) and Teh, N. J. 'Galileo's gauge: Understanding the empirical significance of gauge symmetry' (*Philosophy of Science*, 83(1): 93–118, 2015).

43

TIME REVERSAL

Bryan W. Roberts

43.1 Introduction

Time reversal is a wonderfully strange concept. It sounds like science fiction at first blush, and yet plays a substantial role in the foundations of physics. For example, time reversal is often used to describe the "arrow of time," by allowing one to say how evolving to the future is different from evolving to the past. Most fundamental laws of physics are thought to be time reversal invariant; so, when Cronin and Fitch discovered evidence that time symmetry is violated, it provided crucial new insight into the burgeoning Standard Model of particle physics (Christenson et al., 1964; Roberts, 2015b). Time reversal is a cornerstone of many important concepts in physics, from the Wigner (1932) derivation of Kramers degeneracy (which plays an important role in low-temperature physics and superconduction), to the boson-fermion superselection rule (Wick et al., 1952), to the Feynman-Wheeler interpretation of antimatter (Arntzenius and Greaves, 2009). Through the Sakharov conditions, the understanding of time reversal is also thought to play a role in explaining the apparent large-scale asymmetry between matter and antimatter in the universe (Sakharov, 1967).

This chapter introduces one little corner of the rich literature on time reversal,[1] which deals with the question of what time reversal means. We'll begin with a presentation of the standard account of time reversal, with plenty of examples, followed by a popular non-standard account. I will then argue that, in spite of recent commentary to the contrary, the standard approach to the meaning of time reversal is the only one that is philosophically and physically viable. I conclude with a few open research problems about time reversal.

43.2 The Standard Account

Spatial rotation can be understood by studying rotated physical objects, and spatial translation by studying translated physical objects. How then are we to understand time reversal? It doesn't seem possible to physically "reverse time." One imagines an impassioned philosopher of physics pulling hard on their hair while exclaiming, *What would it even mean to reverse time?*

To guide intuitions, a common initial response is to imagine a film of a body in motion, like a billiard ball bouncing around a frictionless billiard table, and then to imagine that the film is reversed: this reversed film is then said to display the time-reversed motion. But as North (2008) points out, it is hard to see how the precise properties of that reversed motion follow from a simple "film" thought experiment.

How can one make the reversed description precise? How does one know it's correct? The standard account is the following.

43.2.1 Two Components

There are two components to the standard account of time reversal. The first is *order (in time) reversal*. Suppose we represent the changing states of a physical system by a curve $s : \mathbb{R} \to \mathcal{P}$ through some set of states $\mathcal{P}$, with initial state $s(0) \in \mathcal{P}$. One thing that time reversal ought to do is turn around the temporal order in which such states occur. The standard time reversal transformation does this by transforming the temporal parameter t as follows:

$$t \mapsto -t.$$

This flips time around a particular initial moment $t = 0$. That may seem a bit arbitrary at first. However, most of the time, it isn't philosophically or physically significant. One could instead write $t \mapsto -t + c$, and thereby flip time around the moment $t = c$. But this new transformation is related to the old one by a translation (forward or backward) by c in time, called a *time translation*. In the local physics of isolated systems, it turns out that time translation is always a symmetry, related to the conservation of energy. When that is the case, these two time reversal transformations can always be viewed as two different ways of representing the same transformation.

A more interesting question is: why is $t \mapsto f(t) = -t + c$ appropriate for time reversal, and not a transformation like $t \mapsto g(t) = e^{-t}$, which also reverses the order of events in time? This particular transformation $g(t)$ can be excluded by demanding that time reversal is an involution, meaning that $f(f(t)) = t$. The idea is just to take seriously what it means to be a "reversal": by applying it twice, you get back to where you started. This assumption can be found in Sachs (1987) and Roberts (2012), and was explored extensively by Peterson (2015). More generally, it turns out that if one demands that time reversal is an order-reversing involution that is *linear*, so as not to "stretch time out unevenly," then the only possible transformation is of the standard form $t \mapsto -t + c$, just as is standardly assumed.[2]

The second component of time reversal, on the standard account, is the *(instantaneous) time reversal operator*. When one views a film of a classical billiard ball in reverse, both the momentum appears to have reversed direction: a ball with momentum to the right becomes one with momentum to the left, and so on.[3] The operation that implements these instantaneous changes is called the *time reversal operator T*. Some common instantaneous properties and their transformation rules under the time reversal operator are indicated in Figure 43.1. A central foundational question is then: how does one know which properties are preserved and which are reversed? We will discuss this question over the course of this article; for now, let us simply try to summarize what the standard account says about time reversal.

Combining the two components, the standard account of time reversal is the following. Begin with a state space $\mathcal{P}$, and let $s(t)$ be a curve through $\mathcal{P}$ that represents the evolution of a physical system in time. The standard time reversal transformation takes $s(t)$ to a new curve $Ts(-t)$, which reverses the order of events, and also applies the time reversal operator $T : \mathcal{P} \to \mathcal{P}$ to the instantaneous state $s(t)$ at each time t. Thus, on the standard account, time reversal is a transformation of dynamical trajectories, such as those associated with solutions to a law of nature, which reverses the order of states, but also adjusts instantaneous properties like momentum and spin in an "appropriate" way.

Reversed		Preserved	
Momentum:	$p \mapsto -p$	Position:	$q \mapsto q$
Magnetic field:	$B \mapsto -B$	Electric field:	$E \mapsto E$
Spin:	$\sigma \mapsto -\sigma$	Kinetic energy:	$p^2/2m \mapsto p^2/2m$
Position wavefunction:	$\psi(x) \mapsto \psi(x)^*$	Transition probability:	$|\langle \psi, \phi \rangle|^2 \mapsto |\langle \psi, \phi \rangle|^2$

Figure 43.1 Some properties of the time reversal operator T.

To keep the language clear, I will systematically use the phrase *time reversal operator* to refer to the operator $T : \mathcal{P} \to \mathcal{P}$ on instantaneous states, and the phrase *time reversal transformation* to refer to the transformation of dynamical trajectories, which also includes the reversal of order in time.

43.2.2   Time Reversal Invariance

Although our central discussion is about what time reversal means, an essential component of the debate involves what it means for the laws describing a physical system to be temporally symmetric, or *time reversal invariant*.

Most physical theories can be identified with a state space $\mathcal{P}$, as well as a set of S of preferred possible trajectories, which are the "solutions" to some law. For example, in the Newtonian description of a point particle of mass m in the presence of a force F, the states are the possible positions of the particle in space, and the trajectories are the curves with acceleration a satisfying Newton's Second Law, $F = ma$.

Consider a set of curves through a state space $\mathcal{P}$, and let S be the subset of the curves through $\mathcal{P}$ that are solutions to some law. For a bijection φ on the set of curves through $\mathcal{P}$ to be a *symmetry* or an *invariance* of the law means: if some curve is a solution to the law, then the φ-transformed curve is a solution, too. Another way to put this is: given a law with a solution set S, φ is a symmetry of the law if and only if $\varphi(S) = S$. Time reversal invariance is just a special case of this: for a law to be *time reversal invariant* means that a curve is a solution to the law only if the time-reversed curve is a solution as well.

The reader should be warned that, in some treatments of this topic, a symmetry of a law is defined to be a passive transformation that "preserves the form" of that law. This is, unfortunately, a rather vague way to put it, and in practice it is also prone to error. When carried out correctly it usually amounts to the same thing. However, especially when one is a newcomer to the subtleties of time reversal, the reader is encouraged to stick to the more precise characterization above. Concrete examples can be found below.

43.2.3   Examples

In the context of a physical theory, the standard account of what time reversal means can be given in more precise terms, and even argued for. To really dig into this, it's important to work some examples. Some encouragement for those readers who do not follow the mathematical details: don't lose hope! One can skip the technicalities of these examples and still understand the philosophical argumentation of the sections to follow. The main message of these examples is that, on the standard account of time reversal, its exact meaning can only be understood once we identify how the state space of the theory represents the world.

43.2.3.1   Example: Newtonian Mechanics

In classical mechanics applied to point particles, the state space is typically $\mathcal{P} = \mathbb{R}^{3n}$, and a state is a vector $x \in \mathbb{R}^{3n}$ describing the positions of n particles in space at a moment. The dynamical trajectories $x(t)$ of a physical system are curves in that state space that satisfy Newton's equation, $F = m \cdot (d^2/dt^2)x(t)$, for a collection of masses $m \in \mathbb{R}^n$ and a total force $F \in \mathbb{R}^3$. The time reversal operator $T : \mathcal{P} \to \mathbb{R}$ is trivial, in that it is the identity operator, since time reversal is assumed not to transform instantaneous positions. As a result, the time reversal transformation just reverses the order of trajectories:

$$x(t) \mapsto x(-t).$$

Time reversal invariance in this context means that, given some force F, if $x(t)$ is a solution to Newton's equation, then so is its time-reverse $x(-t)$. One can easily check that, if the force vector

F is a function only of position in space, then Newtonian mechanics is time reversal invariant on the standard account. However, for more exotic forces, it is not generally guaranteed that classical mechanics is time reversal invariant, in spite of frequent commentary to the contrary (Roberts, 2013b).

43.2.3.2   Example: Hamiltonian Mechanics

The Hamiltonian approach to classical mechanics is a theory built using differential geometry.[4] Here, things are a little more interesting. The state space is a $2n$-dimensional manifold $\mathcal{P}$ together with a symplectic form Ω, which are sometimes together called a *phase space*. The dynamical trajectories $s(t)$ are the curves that satisfy Hamilton's equations,[5] for some smooth function $h : \mathcal{P} \rightarrow \mathbb{R}$ called the Hamiltonian. By Darboux's theorem, a state $\xi \in \mathcal{P}$ in a symplectic manifold admits a neighborhood in which there is a coordinate system for which $\xi = (q_1, \ldots, q_n, p_1, \ldots, p_n)$. I will write this as $\xi = (q, p)$ for short.

Often, though not always, the q components in Darboux coordinates are interpreted as position values, while the p components are interpreted as momentum. In that case, the time reversal operator is a function $T : \mathcal{P} \rightarrow \mathcal{P}$ that preserves position and reverses momentum, $T(q, p) = (q, -p)$. The time reversal transformation then has the following effect on dynamical trajectories:

$$q(t), p(t) \mapsto q(-t), -p(-t).$$

That is: the same positions are occupied in the time-reversed description, but in the reverse order, and with the instantaneous momentum of each state turned around.

However, it is equally possible to use the q's to represent momentum and the p's to represent position, or some other quantity entirely. So, in Hamiltonian mechanics, time reversal really cannot be given explicit meaning until one has interpreted how a state represents reality, and in particular what the q's and p's represent. Nevertheless, there is a general feature that all time reversal transformations in Hamiltonian mechanics are assumed to share (e.g. in Abraham and Marsden, 1978, Definition 4.3.12), which is that they are *antisymplectic*. This means that the push-forward T^* of T reverses the sign of the symplectic form, $T^*\Omega = -\Omega$. Reversing p in Darboux coordinates is one example of an antisymplectic transformation: writing the symplectic form in terms of the wedge-product as $\Omega = dq \wedge dp$, we find that $(q, p) \mapsto (q, -p)$ induces a transformation $\Omega \mapsto -\Omega$.

Viewing time reversal in Hamiltonian mechanics as some particular antisymplectic bijection $T : \mathcal{P} \rightarrow \mathcal{P}$, there is an easy way to interpret all the other antisymplectic transformations: they are a combination of time reversal plus some other (non–time-reversing) symmetry. To see this, let $A : \mathcal{P} \rightarrow \mathcal{P}$ be any other antisymplectic transformation. Then $S = A \circ T^{-1}$ is symplectic, and $A = S \circ T$. In practice, such a transformation $S \circ T$ might represent parity and time reversal (PT), or spatial translation and time reversal, or any other symplectic transformation followed by the reversal of time.

Time reversal invariance in this context means that if $\xi(t)$ is a solution to Hamilton's equation with Hamiltonian h, then so is the time-reversed curve $T\xi(-t)$, where $T : \mathcal{P} \rightarrow \mathcal{P}$ is the antisymplectic time reversal operator. This is easily checked to be equivalent to the statement that the time reversal operator leaves the Hamiltonian invariant, in that $h \circ T(\xi) = h(\xi)$ for all $\xi \in \mathcal{P}$.

43.2.3.3   Example: Quantum Mechanics

The quantum mechanical expression of time reversal is very similar to that of classical Hamiltonian mechanics. Let the space of (pure) states be a Hilbert space $\mathcal{P} = \mathcal{H}$. The dynamical trajectories $\psi(t)$ are the curves that satisfy the law of unitary evolution $\psi(t) = e^{-itH}\psi$ for some self-adjoint H and initial state $\psi \in \mathcal{H}$; in differential form this is just the Schrödinger equation $i(d/dt)\psi(t) = H\psi(t)$. As in Hamiltonian mechanics, the explicit definition of time reversal in quantum mechanics depends on what we take the states ψ to represent. Suppose we take them to represent "position wavefunctions" in the following sense. Let $\mathcal{H} = \mathcal{L}_2(\mathbb{R}^n)$ be the Hilbert space of square-integrable functions ψ : $\mathbb{R}^n \rightarrow \mathbb{C}$. Define the operators Q, P as in the Schrödinger representation, by $Q\psi(x) := x\psi(x)$ and $P\psi(x) = i(d/dx)\psi(x)$ (on the appropriate domains). If we interpret Q and P as the "position observable" and "momentum observable," respectively, then the time reversal operator is defined by $T\psi(x) = \psi(x)^*$, and the time reversal transformation is given by:

$$\psi(t) \mapsto \psi(-t)^*.$$

One can check that transformation has the effect of preserving position and reversing momentum: $TQT^{-1} = Q$ and $TPT^{-1} = -P$.

Of course, the formal operators Q and P do not have to be interpreted as position and momentum, just like in classical Hamiltonian mechanics. For example, in the momentum representation, one would interpret this operator Q as momentum. However, as in classical Hamiltonian mechanics, it is generally assumed that all time reversal operators share a property, which is that of being *antiunitary*. An antiunitary operator A is one that satisfies $A^*A = AA^* = I$, but which is antilinear: $A(a\psi + b\phi) = a^*\psi + b^*\phi$, for all vectors ψ, ϕ and for all $a, b \in \mathbb{C}$. This implies that $\langle A\psi, A\phi \rangle = \langle \psi, \phi \rangle^*$, for all $\psi, \phi \in \mathcal{H}$. Once we are convinced that time reversal preserves Q and reverses P, then it follows that T must be antiunitary.[6] And, in an irreducible representation of the canonical commutation relations in Weyl form, this T is in fact the unique antiunitary operator (up to a multiplicative constant) that preserves Q and reverses P (Roberts, 2017, Proposition 2). However, some identification of what the observables represent is needed in order to make this kind of argument.

Once we do fix some antiunitary operator T as time reversal, it turns out here too that the antiunitary operators are precisely those that can be written as unitary (non-time-reversing) operator combined with T. Namely, given an antiunitary A, we define $U = AT^{-1}$, when it follows that U is unitary and $A = UT$.

A complication introduced in quantum theory is the existence of internal degrees of freedom associated with observables like spin. For example, consider the standard representation of the Pauli matrices on a 2-dimensional Hilbert space $\mathcal{H}$:

$$I = \begin{pmatrix} 1 & \\ & 1 \end{pmatrix}, \quad \sigma_1 = \begin{pmatrix} & 1 \\ 1 & \end{pmatrix}, \quad \sigma_2 = \begin{pmatrix} & i \\ -i & \end{pmatrix}, \quad \sigma_3 = \begin{pmatrix} 1 & \\ & -1 \end{pmatrix}.$$

In this context, the time reversal operator is defined by,

$$T\psi := \sigma_y \psi^*,$$

where the vectors ψ are written in the eigenstate basis associated with the σ_3 operator. This operator T is easily verified to have the property of reversing the spin observables, in that $T\sigma_i T^{-1} = -\sigma_i$ for each $i = 1, 2, 3$. Conversely, in an irreducible representation of the canonical anticommutation relations, this T is in fact the unique antiunitary operator (up to a multiplicative constant) that reverses all the spin observables in this way (Roberts, 2017, Proposition 3).

In quantum mechanics, time reversal invariance means that if $\psi(t)$ describes a unitary solution to the Schrödinger equation for some Hamiltonian H, in that if $\psi(t) = e^{-itH}\psi$ for some $\psi \in \mathcal{H}$, then the time-reversed trajectory $T\psi(-t)$ is a unitary solution too, in that $T\psi(-t) = e^{-itH}\phi$ for some $\phi \in \mathcal{H}$. This turns out to be equivalent to the statement that the antiunitary time reversal operator T commutes with the Hamiltonian, $TH = HT$.

Why think of time reversal in this way? In one sense, the answer is easy: we want an operator that preserves position, and reverses momentum and spin. But why do we want that? There are in fact systematic ways to answer this. For example, if one demands a time reversal transformation that is non-trivial, in that it allows at least one non-zero Hamiltonian that is time reversal invariant, then this is already enough to guarantee that the time reversal operator is antiunitary (Roberts, 2017, Proposition 1). Symmetry arguments from the homogeneity and isotropy of space can then be used to determine the transformation rules for position, momentum and spin. The result of this analysis is a derivation of the standard time reversal transformation for quantum mechanics. I will not go into further details here; they can found in the paper cited above.

43.2.3.4   Example: Relativistic Field Theory

Most relativistic field theories can be described with a globally hyperbolic Lorentzian manifold[7] (M, g_{ab}), together with a collection of classical or quantum fields satisfying some field law. Classical electromagnetism is one such example: there is a vector field J^a representing the 4-current, and an antisymmetric rank-2 tensor field F_{ab} representing the Maxwell-Faraday field, which together satisfy Maxwell's equations.

Such theories often have a Hamiltonian formulation, in which case time reversal can be understood using the same techniques described above for classical and quantum mechanics. However, Malament (2004) has proposed another way to understand time reversal in this context, which is both natural and powerful. In short, suppose we identify the direction of time with a systematic choice of light cone lobes to represent the "future." Then we can treat time reversal as the reversal of these lobes, as in Figure 43.2. It is then often possible to identify the physical fields that plausibly exhibit some dependence on the choice of a future direction, and use this to work out how they transform when these lobes are reversed.

In more precise terms: a *temporal orientation* for (M, g_{ab}) is an equivalence class of smooth timelike vector fields with the property that every pair of its elements ξ^a, χ^b "points in the same direction," in that $g_{ab}\xi^a\chi^b > 0$. Every globally hyperbolic spacetime admits exactly two temporal orientations, corresponding to the two lobes of each light cone. In practice, it is convenient to describe an orientation by choosing a representative smooth vector field τ^a from the equivalence class, so long as we remember that any other choice would have served just as well. Then we can understand a time-reversing transformation to be one that reverses the temporal orientation, $\tau^a \mapsto -\tau^a$. The effect of this is that, if an event p "happens before" an event q in a spacetime structure (M, g_{ab}, τ^a), in that p is connected to q by a future-directed timelike curve, then 'q happens before p' in the time-reversed structure $(M, g_{ab}, -\tau^a)$

This lead North to propose that time reversal means: "*Only* flip the temporal orientation vector field" (North, 2008, p. 212). However, it's important to add that one must occasionally do other things as well, in order to distinguish time reversal from (say) parity-time reversal. To exclude the latter, we must also require that the time reversal transformation reverse total orientation, $\varepsilon_{abcd} \mapsto -\varepsilon_{abcd}$. When combined with time reversal, this ensures that no 3-volume element ε_{abc} associated with a spatial surface reverses orientation; that is, it ensures that only time (and not space) is reversed.[8]

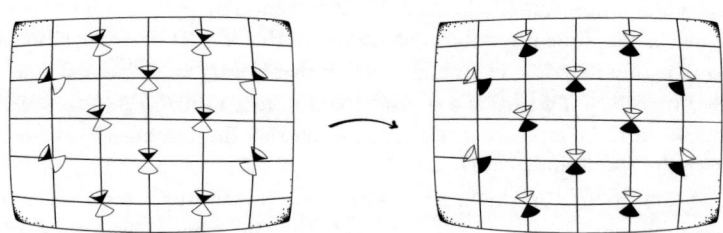

Figure 43.2 Time reversal viewed as reversal of temporal orientation: the black lobe represents the 'future direction' in each case.

The remaining step is to figure out how this definition of time reversal transforms a matter field. Suppose we associate each spacetime structure of the form $(M, g_{ab}, \tau^a, \varepsilon_{abcd})$ with a matter field, represented by the vector field ξ^a, which we define to be $\xi^a := k\tau^a$ for some $k \in \mathbb{R}$. Then the time reversal transformation $(\varepsilon_{abcd}, \tau^a \mapsto -\tau^a) \mapsto (-\varepsilon_{abcd}, -\tau^a)$ induces a transformation $\xi^a \mapsto -\xi^a$ of the matter field, in that if $(M, g_{ab}, \varepsilon_{abcd}, \tau^a)$ is associated with ξ^a, then the transformed structure $(M, g_{ab}, -\varepsilon_{abcd}, -\tau^a)$ is associated with $-\xi^a$. This transformation occurs because the matter field is "linked" to the direction of time by its very definition. Malament (2004) points out that the fields of electromagnetism share this property. The student (and everyone) will do best to study Malament's lucid article directly; I will not go into more details here. Instead I will just state the conclusion: on a natural understanding of the Maxwell-Faraday tensor F_{ab}, and the associated electric field E^a and magnetic field B^a in some reference frame, the effect of the time-reversal transformation defined above is to preserve E^a, while reversing F_{ab} and B^a:

$$F_{ab} \mapsto -F_{ab} \qquad B^a \mapsto -B^a \qquad E^a \mapsto E^a.$$

This characterization of time reversal notably requires a temporally and totally oriented spacetime.[9] Without such a global notion of future and past, there is, of course, no global define time reversal, either. Peterson (2015) has suggested this is a concern about the approach. However, it is possible to give a similar analysis of time reversal for non-temporally-oriented spacetimes, too. For example, given a non-temporally-orientable spacetime, one can define time reversal with respect to a point by considering a (possibly small) region around that point that is temporally orientable. Or, one can construct the universal covering space, which can be used to represent all the same local physical facts, and which is, in general, temporally orientable (Hawking and Ellis, 1973, §6.1). Thus, although "flipping the temporal orientation" is a guiding idea here, temporal orientability is really no barrier to defining time reversal.

43.3   The Albert-Callender "Pancake Account"

The above discussion presents the standard account of time reversal, and some of the arguments that have been given for it. In this section, I will present (and then critique) a creative non-standard account due to Albert and Callender, which has recently become popular among some philosophers.[10]

43.3.1   *Critique of the Standard Account*

Some of the properties of the instantaneous time reversal operators T discussed above (and in the table of Figure 43.1) are intuitively reasonable: the time reverse of a billiard ball appears to warrant reversing both velocity and momentum. Some are also automatic, such as the fact that the velocity dx/dt of a trajectory $x(t)$ reverses sign under the transformation $t \mapsto -t$.

However, at least to the newcomer, other properties of the standard account are less intuitive: the spin and magnetic field reverse, and position wavefunctions are conjugated. At least on a superficial level, these transformations do not arise from reversing the sign of the "little t" index describing order in time. So, why do these changes occur? We have seen some positive reasons in the examples above: they can be derived by thinking about how these objects depend on the temporal orientation of spacetime, or by demanding a time reversal transformation that is non-trivial. But let's also consider a more disruptive alternative: to reject the standard account as untenable.

David Albert responded to the textbook definition of time reversal in this way:

> [T]he books identify precisely that transformation as the transformation of "time-reversal." ... The thing is that this identification is *wrong*. Magnetic fields are *not* the sorts of things that any proper time-reversal transformation can possibly turn around.

> Magnetic fields are not — either logically or conceptually — the *rates of change* of anything. ... [Time reversal] can involve nothing whatsoever other than reversing the *velocities of the particles*. (Albert, 2000, p. 20)

Craig Callender independently came to the same conclusion,[11] writing that "David Albert... argues — rightly in my opinion — that the traditional definition of [time-reversal invariance], which I have just given, is in fact gibberish. It does not make sense to time-reverse a truly instantaneous state of a system" (Callender, 2000, p. 254).

Their proposal, instead, is that time reversal is nothing more than the reversal of the order of states in time. This is to adopt the first component of the standard account of time reversal, that the order or trajectories is reversed, but not the second, by rejecting the use of a time reversal operator T on instantaneous states. Equivalently, Albert and Callender adopt a time reversal operator T that is the identity operator, $T = I$, which only transforms instantaneous states trivially to themselves.

The Albert-Callender objection is natural given a certain perspective on the passage of time. According to it, all moments are stacked up like a stack of inert pancakes, so that time reversal consists solely in a reversal of the order of pancakes in the stack (Figure 43.3). As Malament (2004) puts it, this view presumes that physical phenomena described at each moment "just lie there" at any given instant, without any connection to the direction of time, unless they happen to be defined (like velocity) in a way that depends on the time parameter t. The ordinary description of a magnetic field is not like this, nor is any other quantity appearing in the left column of Figure 43.1. As a result, Albert and Callender conclude, the instantaneous description of these quantities should be left unchanged by time reversal.

43.3.2 Consequence for Time Symmetry

The "pancake objection," if accepted, leads to a radical revision of how most people currently understand time reversal. For example, all of the interesting applications described in the introduction to this chapter would evaporate, or at least no longer be disassociated with the "true" meaning of time reversal according to Callender and Albert. In particular, the widely-held belief that most isolated systems are time reversal invariant would no longer be true. Albert and Callender themselves emphasize this fact; for example, Albert writes:

> [Electromagnetism] is *not* invariant under time-reversal. Period. And neither (it turns out) is quantum mechanics, and neither is relativistic quantum field theory, and neither is general relativity, and neither is supergravity, and neither is supersymmetric quantum string theory, and neither (for that matter) are any of the candidates for a fundamental theory that anybody has taken seriously since Newton. (Albert, 2000, p. 14)

Indeed, even classical Hamiltonian mechanics fails to be invariant under time reversal on this perspective, in all but the most trivial cases. It's instructive to review this example in some mathematical detail: suppose we have a classical phase space $\mathcal{P} = \mathbb{R}^{2n}$ with a Hamiltonian function $h : \mathcal{P} \to \mathbb{R}$. Time reversal invariance means, as usual, that if a curve is a solution to Hamilton's equations with this Hamiltonian, then so is its time-reverse, now viewed as the mere order reversal $t \mapsto -t$ of states. So, in local coordinates (q, p), suppose a system is time reversal invariant in this non-standard sense:

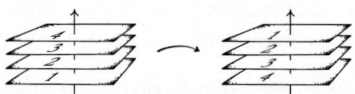

Figure 43.3 The Albert-Callender "pancake" approach to time reversal: order reversal of inert spatial slices with no temporal properties.

given a solution $(q(t), p(t))$ to Hamilton's equations, its "pancake time reverse" $(q(-t), p(-t))$ is a solution as well, for all times $t \in \mathbb{R}$:

$$\frac{d}{dt}q(t) = \frac{\partial}{\partial p}h(q, p) \qquad\qquad \frac{d}{dt}q(-t) = \frac{\partial}{\partial p}h(q, p)$$

$$\frac{d}{dt}p(t) = -\frac{\partial}{\partial q}h(q, p) \qquad\qquad \frac{d}{dt}p(-t) = -\frac{\partial}{\partial q}h(q, p)$$

Since these equations hold for all $t \in \mathbb{R}$, they continue to hold if we apply the substitution $t \mapsto -t$ to the two equations on the right. But, combining this with the equations on the left, we find that both $\partial h/\partial q$ and $\partial h/\partial p$ are equal to their negatives, and hence,

$$\frac{\partial}{\partial q}h(q, p) = \frac{\partial}{\partial p}h(q, p) = 0,$$

which implies that $h(q, p) = c$ is constant across all points $(q, p) \in \mathcal{P}$. This means that, on the "pancake" account of Albert and Callender, the only time reversal invariant system is the trivial one in which energy is constant across all states.[12] This is, in particular, a system in which nothing ever changes. So, a consequence of the pancake perspective on time reversal is that time reversal invariance holds of only the most trivial physical descriptions, for which no change over time is possible.

43.3.3 *Philosophical Implications*

In spite of these radical consequences for physics, many have recently entertained the Albert-Callender perspective and explored its philosophical consequences. For example, Farr and Reutlinger (2013) suggest it could be used in support of the Russellian "deflationary" account of causation, although they ultimately endorse the standard view of time reversal. Allori (2015) adopts the Albert-Callender perspective wholesale, and uses it to argue that electromagnetism contains an interpretive paradox. Similarly, Castellani and Ismael (2016) adopt it in their response to the claim that time reversal is a counterexample to Curie's Principle[13] (Roberts, 2013a).

In this last case, to save Curie's Principle, Castellani and Ismael argue that it is reasonable to reject the standard account of time reversal and adopt the Albert-Callender perspective instead:

> In physical terms, time reversal should leave the states intrinsically untouched and just change their order. If we cleave to that understanding of time reversal, none of the counterexamples Roberts offers constitutes a failure of [Curie's Principle]. (Castellani and Ismael, 2016, p. 1011).

I am not convinced. In the next sections, I will identify a few objections to the Albert-Callender "pancake" perspective on time reversal, and thus to the philosophical agendas that make use of it above.

43.4 Against the Pancake Account

This "pancake" account of time reversal is a creative perspective on the nature of time, and certainly worth exploring. However, I will argue in this section that it is not an equally legitimate competitor to the standard account: it suffers from at least five concerns, all of which are avoided by the standard account of time reversal.

43.4.1 *Concern 1: Questionable Motivation*

The main motivation for the Albert-Callender view is that the only instantaneous properties that depend on the direction of time are "rates of change," such as the linear change in position with respect to time (velocity), or rotational change in position with respect to time (angular velocity). What is the justification for this claim? In physics, as well as in ordinary language, there appear to

Figure 43.4 Brave or cowardly? This instantaneous property of the soldier appears to depend on the direction of
time.

be many instantaneous properties that do depend on temporal direction, and which are not rates of
change of anything in time.

For example, consider a soldier racing toward a deadly dragon (Figure 43.4). At some moment,
the soldier might be said to have the instantaneous property of "bravery." But what happens when
this description is transformed to its time reverse, in which the soldier is back-pedaling away from
the dragon? Now the soldier in the same moment might be justifiably described as "cowardly." In
other words, the instantaneous properties of 'bravery' and "cowardice" display a dependence on the
direction of time, in spite of not being necessarily being characterized by a rate of change.

The same goes for physics: some degrees of freedom, such as an electron's spin or a magnetic field
in some reference frame, are not necessarily the rate of change of anything, and yet may be linked
to the direction of time. To deny that this is the case requires a positive argument. The burden of
providing such an argument is a challenge for the supporter of the pancake account.

43.4.2   *Concern 2: Momentum in the Wrong Direction*

Further conceptual difficulties arise on the pancake account. For example, consider the description
of a state $\xi = (q, p)$ in classical Hamiltonian mechanics, with q interpreted as position and p as
momentum. By only reversing the order of states in time, and not the instantaneous states $\xi = (q, p)$,
the pancake account entails a reversal of velocity dq/dt, but not of momentum. In other words,
velocity and momentum after time reversal point in the opposite directions!

This problem is avoided in the standard account, where we reverse momentum too, as shown
in Figure 43.5. Recognizing this issue, North (2011, p. 317) suggests that Albert might avoid this
problem by taking momentum to be a "non-intrinsic, non-fundamental quantity" satisfying $p = mv$.

This is effectively to deny the Hamiltonian description of classical mechanics, where (q, p) can represent a great many things independently of the dynamics. And, it is simply not always the case that momentum always has the form $p = mv$ (see endnote 3 above).

43.4.3   Concern 3: No Non-Trivial Examples

In the 20th century, it was discovered that certain decay events (for example, a decay from a kaon into an antikaon) occur with a different transition probability than the reverse decay event (from an antikaon into a kaon). This is a rare and unusual phenomenon: almost all known decay events have a transition probability that is the same in both directions. On the standard account, having equal transition probabilities in both directions is equivalent to time reversal invariance; Roberts (2015b) called this "Kabir's principle." This provides a way to explain the strangeness of decay events in things like neutral kaons: although most interactions are time symmetric, neutral kaon decay is asymmetric in time.

On the pancake account, no comparable explanation of the difference between these interactions is available. As discussed in Section 43.3.2, the pancake account treats all non-trivial processes as equally asymmetric in time. There are no non-trivial examples of time symmetric systems. As a result, this transformation is of little physical interest if we wish to identify the distinction between temporal symmetries and asymmetries in realistic physical systems. Thus, a further burden on the pancake account is to explain why it is useful.

43.4.4   Concern 4: Radical Underdetermination

A further issue is that the pancake account of time reversal, at least as it has so-far been presented, is radically underdetermined. The claim that time reversal transforms a trajectory to one "with the temporal order inverted" (Callender, 2000, p. 253) is compatible with infinitely many transformations. One commonly identifies $t \mapsto -t$ as an order-reversing transformation. But $t \mapsto \sqrt[3]{a - t^3}$ is an order-reversing transformation as well, and indeed it is even an involution, meaning that applying it twice yields the identity transformation. Callender suggests that his time reversal transformation moves objects such that "if their old coordinates were t, their new ones are $-t$, assuming the axis of reflection is the coordinate origin" (Callender, 2000, p. 253). But what justifies this? An emphasis on mere order reversal is certainly not enough. Thus, a challenge for the pancake supporter is to determine which of the many order-reversing transformations really counts as time reversal.

In order to guarantee an order-reversing transformation of the form $t \mapsto -t$, one might think in more detail about which properties of matter and spacetime are linked to the direction of time, as we did when we argued for this property in Section 43.2.1. There we asserted things like "time reversal is an involution" and "time reversal does not stretch time out unevenly." But once one has started down this road, it is hard to see why other reasonable principles guiding the definition of time reversal

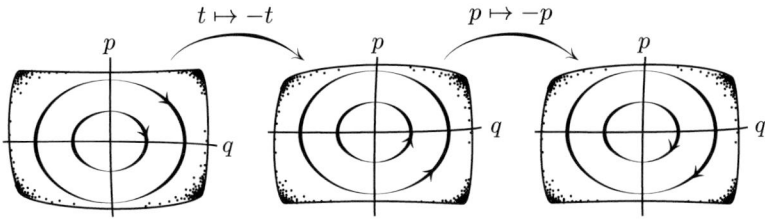

Figure 43.5 Time reversal of a harmonic oscillator's phase space: the pancake account only includes the first (order-reversing) transformation, leading to momentum and velocity in opposite directions; the standard account avoids this by including the second (instantaneous state) transformation as well.

should be excluded. Can we also now ask that time reversal be a transformation that allows for some non-trivial systems to be time symmetric? Can we ask that time reversal respect the dependence of a physical field on light cone structure? If the answer is yes, then the defenses of Malament (2004) and Roberts (2017) lead straight to the standard definition of time reversal. There appears to be little to prevent this, once reasonable steps are taken to avoid the underdetermination of the pancake account.

43.4.5   *Concern 5: Relativistic Fields*

It is particularly hard to make sense of the pancake account in the context of fields on a relativistic spacetime. For example, the toy matter field of Section 43.2.3.4, defined by $\xi^a = k\tau^a$, appears to be somehow forbidden on the pancake account: on that view, a vector field that is not a rate of change of anything must "just lie there" under time reversal. But the vector field ξ^a manifestly does not, at least on Malament's account of time reversal.

Arntzenius (2004) has pointed out a further difficulty, that saying electric and magnetic fields 'just lie there' at a given instant suggests they do not transform under Lorentz boosts, in spite of the well-known empirical fact that they do. In fact, the situation is even worse: on Malament's reconstruction, electric and magnetic fields can only be meaningfully defined with respect to a temporal orientation, and so there is no way to reverse one without inducing a transformation of the other. The pancake account of time reversal appears to not even be wrong in this context: it is incoherent.

In spite of this difficulty with the pancake account, some authors still demand "taking it seriously" in the context of electromagnetism (c.f. Allori, 2015, p. 4). Arntzenius and Greaves (2009, §3.4) have made a valiant attempt to do so, by proposing that we interpret at least Albert as holding the background assumption that "spacetime is in fact Newtonian, velocities are spatial 3-vectors, and so are the electric and magnetic fields," a view that they attribute to him on the basis of personal correspondence (Arntzenius and Greaves, 2009, p. 568).

However, it is not clear that even this is enough: in order to make the view coherent, we must define electric and magnetic fields in such a way that they (a) correctly describe electromagnetic phenomena, and (b) do not depend on the direction of time for their definitions. I do not see how this is possible on the pancake account. If the charge moves in one direction along a straight conducting wire, then it is associated with a magnetic field that curls around in a particular orientation. If the charge moves in the other direction, the magnetic field has the opposite orientation. So, to in order to correctly describe this magnetic field, even on a Newtonian spatial slice, we need to know the direction of motion of a charged particle. And the direction of motion, even on the Albert-Callender account, depends on the direction of time, however we formulate it mathematically.

43.5   Conclusion and Open Research Questions

This chapter has tried to give an overview of the standard account of time reversal, and to present a critique of a popular competing view. Much remains to be learned about the foundations of time reversal, even on this narrow topic of its meaning. We thus conclude by mentioning a few open research questions, of varying degrees of difficulty, which are suitable for researchers or for student research essays.

1   *Time vs Motion reversal.* Ballentine (1998, p. 377) wrote, following Wigner (1931, p. 325), that "the term 'time reversal' is misleading, and the operation... would be more accurately described as motion reversal." Evaluate this claim.

2   *Curie's Principle.* Evaluate the Castellani and Ismael (2016) response to the critique of Curie's Principle given by Roberts (2013a, 2015a) and Norton (2016).

3   *Time reversal in Open Systems.* Most derivations of the meaning of time reversal assume a context of local physics for isolated systems; see e.g. the discussion of Section §43.2.1. What can be said

about the physics of large-scale or open systems? On the latter, see Maroney (2010, §2.2) for a helpful start.

4   *Deriving the Lagrangian time reversal operator.* Formulate an assumption about what time reversal ought to mean in the context of Lagrangian mechanics on a tangent bundle TM (c.f. De León and Rodrigues, 1989, §7). Then derive the basic mathematical properties of the time reversal operator, in the spirit of Malament (2004) and Roberts (2017). Check whether its image under the Legendre transformation is antisymplectic on the cotangent bundle T^*M with its canonical symplectic form.

5   *Spacetime approach to quantum time reversal.* Consider any rigorous formulation of quantum field theory on curved spacetime (c.f. Wald, 1994; Haag, 1996; Araki, 1999). Derive the meaning of time reversal in this context by examining what transformation is induced when temporal orientation is reversed. *Suggestion:* As a starting point, consider the result of Varadarajan (2007, Lemma 9.9).

Acknowledgments

Prepared for *The Routledge Companion to the Philosophy of Physics*, Eleanor Knox and Alistair Wilson (eds.). With thanks to I. Ufuk Tasdan for helpful comments. This article was written with the support of a Philip Leverhulme Prize and of the Inter-University Centre, Dubrovnik.

Notes

1   Some other fascinating issues in the time reversal literature that I will not cover here include the implications and significance of time reversal symmetry (Earman, 2002; Price, 1996; Farr, 2020); time reversal symmetry violation in the weak sector (Sachs, 1987; Bigi and Sanda, 2009; Roberts, 2015a,b; Gołosz, 2017)) the Feynman "going backwards in time" interpretation of antimatter (Earman, 1967; Arntzenius and Greaves, 2009); and the significance of the CPT theorem (Greaves, 2010; Greaves and Thomas, 2014).

2   More precisely: let $f : \mathbb{R} \to \mathbb{R}$ satisfy (i) *involution:* $f(f(t)) = t$; (ii) *linearity:* $f(t) = at + c$; and (iii) *order reversal:* $t < t' \Leftrightarrow f(t) > f(t')$. Then $t = f(f(t)) = a^2 t + ac + c$, hence $(a^2 - 1)t + (a+1)c = 0$ for all $t \in \mathbb{R}$. So, $a = \pm 1$, and $c = 0$ whenever $a = 1$. But only $a = -1$ is order-reversing, and hence $f(t) = -t + c$ (see Roberts, 2017, §3.3).

3   Note that, although velocity dq/dt and momentum p are proportional for structureless Newtonian particles, $dq/dt = (1/m)p$, this is not always the case. For example, in electromagnetism, velocity may take the form $dq/dt = (1/m)(p + a(q))$ for some vector potential a. So, when one is being careful, the reversal of velocity and of momentum should be treated as separate concepts.

4   A classic reference for this perspective on classical mechanics is Arnold (1989), or the more recent Marsden and Ratiu (2010). I adopt their notation in this chapter. Philosophers of physics influenced by the Chicago School sometimes use Penrose's abstract index notation instead, for which see Geroch (1974).

5   In geometric expression, this says that the trajectories are the integral curves of the vector field X for which $\iota_X \Omega = d\Omega$. In (Darboux) coordinate form, Hamilton's equations are the system of $2n$ equations $(d/dt)q_i(t) = (\partial/\partial p)h(q, p)$ and $(d/dt)p_i(t) = -(\partial/\partial q)h(q, p)$, for each $i = 1, \ldots, n$.

6   Proof: given that $[Q, P] = i$, and that T preserves Q and reverses P, it follows that $TiT^{-1} = T[Q, P]T^{-1} = [TQT^{-1}, TPT^{-1}] = -[Q, P] = -i$, which implies that T cannot be unitary. Adopting the assumptions of Wigner's theorem, it follows that T is antiunitary.

7   For an introduction to the foundations of relativity theory, the *locus classicus* is Malament (2012). In this section, I adopt the notation and terminology from this book, including Penrose "abstract index" notation for tensors in this section, since it is so standard in this context — with apologies that it is a different tensor notion than the one used above.

8   The 3-volume element associated with a future-directed timelike vector ξ^a is given by $\varepsilon_{abc} := \xi^n \varepsilon_{anbc}$. Moreover, reversing temporal orientation induces a change in sign in the future-directed vector $\xi_a \mapsto -\xi_a$. We thus find that by reversing the sign of both ε_{abcd} as well as the temporal orientation, we induce a transformation $\varepsilon_{bcd} \mapsto -(-\varepsilon_{bcd}) = \varepsilon_{bcd}$, and so the 3-volume is preserved.

9   Not all spacetimes are temporally orientable, and it may be viewed as an open question as to whether our spacetime has one (Earman, 1974).

10   An early expression of a similar view can also be found in Horwich (1989).

11 This work drew on Callender's (1997) dissertation. See also Horwich (1989, §3) for a related view.

12 This argument is a "one liner" from the geometric perspective on Hamiltonian mechanics: let $(\mathcal{P}, \Omega)$ be a symplectic manifold, and suppose that X and $-X$ both describe the Hamiltonian vector field generated by $h : \mathcal{P} \to \mathbb{R}$. This means that, $dh = \iota_X \Omega = \iota_{(-X)} \Omega = -\iota_X \Omega = -dh$. The exterior derivative dh thus vanishes, which implies that h is a constant function.

13 Curie's Principle is the claim that "when certain causes produce certain effects, the elements of symmetry of the causes must be found in the produced effects" (Curie, 1894, p. 394); English translation in Brading and Castellani (2003, §17).

References

Abraham, R. and Marsden, J.E. (1978). *Foundations of Mechanics* (2nd edition). Reading, MA: Addison-Wesley.

Albert, D.Z. (2000). *Time and Chance*. Cambridge, MA: Harvard University Press.

Allori, V. (2015). Maxwell's paradox: The metaphysics of classical electrodynamics and its time-reversal invariance. α*nalytica*, 1: 1–19. Postprint: http://philsci-archive.pitt.edu/12390/.

Araki, H. (1999). *Mathematical Theory of Quantum Fields*. Oxford: Oxford University Press.

Arnold, V.I. (1989). *Mathematical methods of classical mechanics* (2nd edition). New York: Springer.

Arntzenius, F. (2004). Time reversal operations, representations of the lorentz group, and the direction of time. *Studies in History and Philosophy of Modern Physics*, 35(1): 31–43.

Arntzenius, F. and Greaves, H. (2009). Time reversal in classical electromagnetism. *The British Journal for the Philosophy of Science*. 60(3): 557–584. Preprint: http://philsci-archive.pitt.edu/4601/.

Ballentine, L.E. (1998). *Quantum Mechanics: A Modern Development*. Singapore: World Scientific.

Bigi, I.I. and Sanda, A.I. (2009). *CP Violation*. Cambridge: Cambridge University Press.

Brading, K. and Castellani, E. (2003). *Symmetries in Physics: Philosophical Reflections*. Cambridge: Cambridge University Press.

Callender, C. (1997). *Explaining Time's Arrow*. PhD thesis, Rutgers, The State University of New Jersey, New Brunswick.

Callender, C. (2000). Is time 'handed' in a quantum world?, *Proceedings of the Aristotelian Society*, Aristotelian Society, pp. 247–269. Preprint: http://philsci-archive.pitt.edu/612/.

Castellani, E. and Ismael, J. (2016). Which Curie's principle? *Philosophy of Science*, 83(2): 1002–1013. Preprint: http://philsci-archive.pitt.edu/11543/.

Christenson, J.H., Cronin, J.W., Fitch, V.L. and Turlay, R. (1964). Evidence for the 2π decay of the k_2^0 meson. *Physics Review Lettets*, 13(4): 138–140.

Curie, P. (1894). Sur la symétrie dans les phénomènes physique, symétrie d'un champ électrique et d'un champ magnétique. *Journal de Physique Théorique et Appliquée*, 3: 393–415.

De León, M. and Rodrigues, P.R. (1989). *Methods of Differential Geometry in Analytical Mechanics*, Vol. 158 of *North-Holland Mathematical Studies*. Amsterdam: Elsevier Science Publishers B.V.

Earman, J. (1967). On going backward in time. *Philosophy of Science*, 34(3): 211–222.

Earman, J. (1974). An attempt to add a little direction to "the problem of the direction of time". *Philosophy of Science*, 41(1): 15–47.

Earman, J. (2002). What time reversal is and why it matters. *International Studies in the Philosophy of Science*, 16(3): 245–264.

Farr, M. (2020). Causation and time reversal. *The British Journal for the Philosophy of Science*, 71(1): 177–204. http://philsci-archive.pitt.edu/12658/.

Farr, M. and Reutlinger, A. (2013). A relic of a bygone age? Causation, time symmetry and the directionality argument. *Erkenntnis*, 78(2): 215–235. Preprint: http://philsci-archive.pitt.edu/9561/.

Geroch, R. (1974). *Geometrical quantum mechanics*. Unpublished notes. Available at: https://uchicago.app.box.com/s/lsj4uf3mqbv8fvnaml3hfx7femoftjt7.

Gołosz, J. (2017). Weak interactions: Asymmetry of time or asymmetry in time? *Journal for General Philosophy of Science*, 48(1): 19–33.

Greaves, H. (2010). Towards a geometrical understanding of the CPT theorem. *The British Journal for the Philosophy of Science*, 61(1): 27–50. Preprint: http://philsci-archive.pitt.edu/4566/.

Greaves, H. and Thomas, T (2014). On the CPT theorem. *Studies in History and Philosophy of Science Part B: Studies in History and Philosophy of Modern Physics*, 45: 46–65.

Haag, R. (1996). *Local Quantum Physics: Fields, Particles, Algebras* (2nd edition). Berlin/Heidelberg: Springer.

Hawking, S.W. and Ellis, G.F.R. (1973). *The Large Scale Structure of Space-Time*. New York: Cambridge University Press.

Horwich, P. (1989). *Asymmetries in Time: Problems in the Philosophy of Science*. Cambridge, MA: The MIT Press.

Malament, D.B. (2004). On the time reversal invariance of classical electromagnetic theory. *Studies in History and Philosophy of Modern Physics*, 35: 295–315. Preprint: http://philsci-archive.pitt.edu/1475/.

Malament, D.B. (2012). *Topics in the Foundations of General Relativity and Newtonian Gravitation Theory*. Chicago: University of Chicago Press.

Maroney, O.J.E. (2010). Does a computer have an arrow of time? *Foundations of Physics*, 40(2): 205–238. Preprint: http://philsci-archive.pitt.edu/3533/.

Marsden, J. and Ratiu, T. (2010). *Introduction to Mechanics and Symmetry: A Basic Exposition of Classical Mechanical Systems* (2nd edition). Berlin/Heidelberg: Springer.

North, J. (2008). Two views on time reversal. *Philosophy of Science*, 75(2): 201–223. Preprint: http://philsci-archive.pitt.edu/4960/.

North, J. (2011). Time in thermodynamics. *The Oxford Handbook of Philosophy of Time*, Edited by Craig Callender. Oxford: Oxford University Press, pp. 312–350. Preprint: http://philsci-archive.pitt.edu/8947/.

Norton, J.D. (2016). Curie's truism. *Philosophy of Science*, 83(5): 1014–1026. Preprint: http://philsci-archive.pitt.edu/10926/.

Peterson, D. (2015). Prospects for a new account of time reversal. *Studies in History and Philosophy of Modern Physics*, 49: 42 – 56. Preprint: http://philsci-archive.pitt.edu/11302/.

Price, H. (1996). *Time's Arrow and Archimedes' Point: New Directions for the Physics of Time*. New York: Oxford University Press.

Roberts, B.W. (2012). *Time, symmetry and structure: A study in the foundations of quantum theory*. PhD thesis, University of Pittsburgh. Dissertation: http://d-scholarship.pitt.edu/12533/.

Roberts, B.W. (2013a). The simple failure of Curie's principle. *Philosophy of Science*, 80(4): 579–592. Preprint: http://philsci-archive.pitt.edu/9862/.

Roberts, B.W. (2013b). When we do (and do not) have a classical arrow of time. *Philosophy of Science*, 80(5): 1112–1124.

Roberts, B.W. (2015a). Curie's hazard: From electromagnetism to symmetry violation. *Erkenntnis*, 81(5): 1011–1029. Preprint: http://philsci-archive.pitt.edu/10971/.

Roberts, B.W. (2015b). Three merry roads to T-violation. *Studies in History and Philosophy of Modern Physics*, 52: 8–15. Preprint: http://philsci-archive.pitt.edu/9632/.

Roberts, B.W. (2017). Three myths about time reversal in quantum theory. *Philosophy of Science*, 84: 1–20. Preprint: http://philsci-archive.pitt.edu/12305/.

Sachs, R.G. (1987). *The Physics of Time Reversal*. Chicago: University of Chicago Press.

Sakharov, A.D. (1967). Violation of CP-invariance, C-asymmetry, and Baryon asymmetry of the Universe. In *The Intermissions... Collected Works on Research into the Essentials of Theoretical Physics in Russian Federal Nuclear Center, Arzamas-16*, World Scientific, pp. 84–87. Translation published in 1998, from *Zhurnal Eksperimental'noi i Teoreticheskoi Fiziki: Pis'ma v Redaktsiyu*, Vol. 5, NQ 1, pp. 32–35.

Varadarajan, V.S. (2007). *Geometry of Quantum Theory* (2nd edition). New York: Springer Science and Business Media, LLC.

Wald, R.M. (1994). *Quantum Field Theory in Curved Spacetime and Black Hole Thermodynamics*. Chicago, IL: University of Chicago Press.

Wick, G.C., Wightman, A.S. and Wigner, E.P. (1952). The intrinsic parity of elementary particles. *Physics Review*, 88: 101–105.

Wigner, E.P. (1931). *Gruppentheorie und ihre Anwendung auf die Quantenmechanik der Atomspektren*. Braunschweig: Vieweg. English translation (1959). *Group Theory and Its Application to the Quantum Mechanics of Atomic Spectra*. New York: Academic Press.

Wigner, E.P. (1932). Über die Operation der Zeitumkehr in der Quantenmechanik. *Nachrichten der Gesellschaft der Wissenschaften zu Göttingen Mathematisch-Physikalische Klasse*, 31, pp. 546–559.

Further Reading

This chapter has tried to give an overview of the standard account of time reversal, and to present and evaluate a critique of a popular competing view. Much remains to be learned about the foundations of time reversal, even on this narrow topic of its meaning. For further reading, the reader is encouraged to dive into the references in this chapter! But let me highlight a few works in particular: D. Albert, *Time and Chance* (HUP, 2000) and C. Callender, "Is Time 'Handed' in a Quantum World?", *Proceedings of the Aristotelian Society* (2000) launched the debate that this article focuses on and both deserve careful study. A related area of research on CPT symmetry has been opened up by H. Greaves, Towards a Geometrical Understanding of the CPT Theorem', *The British Journal for the Philosophy of Science* 61(1) (2010): 27-50 and N. Swanson, "Deciphering the algebraic CPT theorem", *Studies in History and Philosophy of Modern Physics* 68 (2019): 106–125.

44

SYMMETRY BREAKING

Elena Castellani and Radin Dardashti

44.1 Introduction

Symmetry breaking is ubiquitous in almost all areas of physics. It is a feature of everyday phenomena as well as in more specific contexts within physics when considering elementary particles described by quantum fields, quantum mechanical descriptions of condensed matter systems or general relativistic descriptions of the entire universe. In all of these, symmetry breaking plays an essential role. However, one should be careful in understanding "symmetry breaking" as this *one* thing, e.g. this one mechanism, you can find in all the various physical systems. The reason for this is that the notion of symmetry breaking is very broad, in the sense that many very different scenarios are covered under this name, and also very misleading, as there is often not much that is really being "broken."

Symmetry and symmetry breaking are, in a sense, the two faces of the same coin. In terms of the scientific notion of symmetry, i.e., invariance under a group of transformations, this can be made very precise. On the one hand, a symmetry of a given order can be seen as the result of a higher-order symmetry being broken to a lower-order symmetry, where the order of a symmetry is the order of the corresponding symmetry group (that is, the number of independent symmetry operations which are the elements of the group). This can be said of any symmetry apart from the "absolute" symmetry, including all possible symmetry transformations. Note that nothing with a definite structure could exist in a situation of absolute symmetry, since invariance under all possible transformation groups means total lack of differentiation. For the presence of some structure, a lower symmetry is needed: in this sense, symmetry breaking is essential for the existence of a structured "thing."

On the other hand, the breaking of a given symmetry generally does not imply that no symmetry is present; what happens is just that the final configuration is characterized by a lower symmetry than the initial configuration. In other words, the original symmetry group is broken to one of its subgroups. The relation between a group (the unbroken symmetry group) and its subgroups play thus an important role in the description of symmetry breaking.[1]

To find some orientation in this rather confusing state of affair it is useful to specify three aspects of symmetry breaking: (i) What is the *entity* that has the symmetry that is being broken?, (ii) What is the *symmetry* that is being broken? and (iii) What is the *mechanism* by which it is broken? Depending on the answers you give to each of these questions, various subtleties can arise, which have led to an intricate and interesting range of philosophical questions.

(i) System vs Law: In Nature, crystals provide a paradigmatic representation of this "symmetry/symmetry breaking" interplay. The many striking symmetries of their morphology and structure are the remains of the breaking of the symmetry of the initial medium from which they originate, that is, a hot gas of identical atoms. This medium has a very high symmetry, the equations describing it being invariant under all rigid motions as well as under all permutations of the atoms. As the gas cools down, the original symmetry breaks down and the physical system takes up a stable state with less symmetry: this is the final crystal, with its peculiar morphology and internal lattice structure.[2]

Crystals are physical objects. In general, when considering the meaning and functions of symmetry and symmetry breaking it is important and useful to distinguish between the systems, i.e., physical objects and phenomena, and the physical laws governing their behavior.

Historically, symmetry breaking in physics was first considered in relation to properties of objects and phenomena. This is not surprising, since the scientific study of symmetry and symmetry breaking originated with respect to the manifest symmetry properties of familiar spatial figures and physical objects (such as, first of all, crystals). Indeed, the symmetries and dissymmetries of crystals occasioned the first explicit analysis of the role of symmetry breaking in physics,[3] due to Pierre Curie in a series of papers devoted to the study of symmetry and symmetry breaking in physical phenomena toward the end of the 19th century.

Curie was motivated to reflect on the relationship between physical properties and symmetry properties of a physical system through his studies of such properties as the pyro- and piezo-electricity of crystals (which were directly related to their structure, and hence their symmetry properties). In particular, he investigated which physical phenomena are allowed to occur in a physical medium (for example, a crystalline medium) endowed with specified symmetry properties. By applying the techniques and concepts of the crystallographic theory of symmetry groups, he arrived at some definite conclusions. In his own words (Curie 1894):

(a)   When certain causes produce certain effects, the symmetry elements of the causes must be found in their effects.[4]

(b)   A phenomenon may exist in a medium having the same characteristic symmetry or the symmetry of a subgroup of its characteristic symmetry. In other words, certain elements of symmetry can coexist with certain phenomena, but they are not necessary. What is necessary, is that certain elements of symmetry do not exist. *Dissymmetry is what creates the phenomenon.*

Thus, intending the phenomenon as the "effect" and the medium as the "cause," the conclusion is that the symmetry of the medium cannot be higher than the symmetry of the phenomenon.[5] Given that the media in which phenomena occur generally start out in a highly uniform (and therefore symmetric) state, the occurrence of a phenomenon in a medium requires the original symmetry group of the medium to be lowered (broken) to the symmetry group of the phenomenon (or to a subgroup of the phenomenon's symmetry group).[6] In such sense, symmetry breaking is what "creates the phenomenon" as claimed by Curie. For this analysis, Curie is credited as the first one to have recognized the important heuristic, or more generally, methodological role of symmetry breaking in physics.

While Curie considered the concept of symmetry breaking with regard to objects and phenomena, in modern physics the focus has turned to the symmetries of the laws. This will be the focus in the rest of this Chapter. The physical system under consideration can be described by the Lagrangian or Hamiltonian and so we will often be speaking of the symmetry "of" the Lagrangian or the Hamiltonian. There are still many issues that may affect the possibility and the kind of symmetry breaking that can occur. For one, there are differences in symmetry breaking depending on whether it is a classical, a quantum mechanical or a quantum field theoretical description of the physical system. Another issue, although related, is whether the description has a finite or an infinite number of degrees of freedom. Finally, also of relevance is the dimension of the system under consideration, as there are certain theorems addressing the possibility or impossibility of symmetry breaking given certain dimensions.[7]

(ii) Kind of Symmetry: Once we have specified what exhibits the symmetry, the occurrence of symmetry breaking depends also on what the symmetry is that is supposed to be broken. Depending on whether the symmetry is continuous or discrete, a spacetime or internal symmetry or whether the symmetry is global or local will affect the way and kind of symmetry breaking that is possible.[8]

(iii) Breaking Mechanism: This leaves us with the third aspect of symmetry breaking, namely the mechanism by which it is broken. There are broadly speaking three kinds of symmetry breaking mechanisms: explicit symmetry breaking (Section 44.2), anomalous symmetry breaking (Section 44.3) and spontaneous symmetry breaking (Section 44.4). We will now discuss each of these in more detail and consider some of the subtleties involved.

44.2   Explicit Symmetry Breaking

A simple illustration of explicit symmetry breaking is given by starting with some Hamiltonian $\mathcal{H}_0$ which is invariant under a symmetry group G and adding to it an additional term $\mathcal{H}_{ESB}$, such that $\mathcal{H}_0 + \mathcal{H}_{ESB}$ is not invariant under G anymore. In such cases, the symmetry of $\mathcal{H}$ is *explicitly broken* by $\mathcal{H}_{ESB}$ whatever the cause of it may be.

A much discussed example for this kind of symmetry breaking is the Heisenberg ferromagnet given by the Hamiltonian

$$\mathcal{H} = -\frac{1}{2}\sum_{i \neq j} J\, S_i \cdot S_j, \tag{44.1}$$

where S_i is a three-dimensional spin operator on lattice site i and J is a positive constant which is only non-zero for neighboring sites. The Hamiltonian is invariant under the SO(3) rotation symmetry. Now by turning on an external magnetic field B the Hamiltonian becomes

$$\mathcal{H} = -\frac{1}{2}\sum_{i \neq j} J\, S_i \cdot S_j - B\sum_i S_i, \tag{44.2}$$

which is not invariant under the SO(3) rotation symmetry anymore. The external magnetic field has explicitly broken the symmetry of the original Hamiltonian by introducing a "preferred" direction, namely the direction of the magnetic field. The spin of the electrons will align accordingly.

Note that the breaking of the symmetry in the previous example is simply due to an external magnetic field. As such it is not a conceptually interesting case of symmetry breaking. However, there are also other sources of explicit symmetry breaking. In some circumstances one may have experimental or theoretical reasons to introduce a small term to the Lagrangian, which breaks some symmetry. For instance Lee and Yang (1956) predicted, on the grounds of theoretical development, that the parity symmetry could be violated in the weak interactions. This was subsequently experimentally confirmed by Wu et al. (1957). Now, one may argue that this breaking of the symmetry is not really a breaking of the symmetry at all, since it was already there and we just did not know. In some sense the breaking just represents the epistemic change of situation at a certain time: we just assumed the symmetry of the Lagrangian to be bigger than it actually was, and we were shown to be wrong about it.[9]

Historically, it was precisely this kind of change to first trigger the interest in the meaning of the symmetry breaking of laws in physics. Just before the discovery of the violation of parity, in the 1952 seminal book *Symmetry* by Weyl, the issue was still not considered. For Weyl, any form of symmetry breaking was in the phenomena and due to contingency: in his own words, "If nature were all lawfulness, then every phenomenon would share the full symmetry of the universal laws of nature [...]. The mere fact that this is not so proves that *contingency* is an essential feature of the world" (Weyl 1952, p. 26).

The discovery of the violation of parity, soon followed by the observation of other violations of the discrete space and time symmetries,[10] brought a change in the above "contingency view." The symmetry violation of a law, such as the parity violation, could now be intended in the sense that what was thought to be a non–observable turned out to be actually an observable, a view particularly defended by Lee himself. From the more general viewpoint of the issue of how to interpret physical symmetries, this is a corollary of an epistemic stance on symmetries, ascribing their significance to the presence of unobservable (or irrelevant features) in the physical description.[11]

Finally, in some circumstances the reason we did not know about the symmetry breaking term is that it was broken only by a small term and one may wonder why does such a small breaking term appear. This initiated the wide-ranging theoretical and philosophical discussion of what counts as a natural parameter.[12]

44.3   Anomalous Symmetry Breaking

Let us turn to another kind of symmetry breaking, which has so far not received much philosophical discussion, namely anomalies. Anomalies label instances, where the symmetry of the classical theory turn out not to remain the symmetry of the corresponding quantum theory. While "anomaly" may sound very serious, maybe even something that can give rise to scientific revolutions, the name should rather be understood as the consequence of the bafflement physicists found themselves in when they realized that quantum fluctuations can break classical symmetries.[13] A more suitable name may be "quantum mechanical symmetry breaking."

Whether this is something that needs to be cured or not depends most crucially on the kind of symmetry that is being broken in the transition. That the symmetry can break in the transition from classical to quantum becomes obvious if we take a path integral perspective on quantum theory.[14] Let us consider the symmetry transformation of some field ψ:

$$\psi \rightarrow \psi' = U\psi. \tag{44.3}$$

If this is a symmetry of the Lagrangian, then

$$\mathcal{L}(\psi) \underset{U}{\rightarrow} \mathcal{L}(\psi'). \tag{44.4}$$

However, for the quantum theory to be invariant under the symmetry transformation, we need $\int D\psi \, e^{i\int \mathcal{L}(\psi)d^4x}$ to be invariant. But, as we know, any coordinate transformation

$$D\psi \underset{U}{\rightarrow} \mathcal{J}D\psi' \tag{44.5}$$

introduces a Jacobian $\mathcal{J}$. Thus, once the Jacobian is non–unit, the symmetry does not translate to the quantum theory. The actual calculation of $\mathcal{J}$ is quite involved in quantum field theory and requires regularization as the Jacobian diverges.[15]

A nice simple example, which already illustrates the far-reaching consequences of anomalies is given by the Schwinger model.[16] You start with a classical massless charged particle ψ coupled to an electromagnetic field A_μ

$$\mathcal{L}_\mathcal{S} = \bar{\psi}(i\gamma_\mu D^\mu)\psi - \frac{1}{4}F_{\mu\nu}F^{\mu\nu}, \tag{44.6}$$

where $iD_\mu = i\partial_\mu - eA_\mu$. The Lagrangian has a chiral symmetry, i.e., you may rotate the left handed component of the spinor independently from the right-handed component and still keep the Lagrangian invariant. However, the chiral symmetry does not survive quantization as the Jacobian gives rise to a term, which effectively introduces an additional term to $\mathcal{L}_\mathcal{S}$. This additional term explicitly breaks the chiral symmetry, leading to a non–interacting theory with massive photons.

Broadly construed, there are now two ways anomalies have been interpreted, depending on whether a global or a gauge symmetry is broken by the quantum fluctuations. Global symmetries that do not survive the quantization lead to possible new effects that are now interpreted as predictions of the theory. This was the case with the first appearance of anomalies in particle physics, namely the problem of understanding the decay rate of the neutral pion ($\pi^0 \rightarrow \gamma\gamma$). The derived decay rate, which was based on the assumption that the chiral symmetry holds, was in disagreement with observations.[17] This deviation was later shown by Adler (1969) and Bell and Jackiw (1969) to be due to the breaking of chiral symmetry through one-loop calculations.

Unlike for global symmetries, the anomalies that arise for gauge symmetries, the so-called gauge anomalies, can be troublesome. The main reason for this is that gauge symmetries allow one to dispose of negative norm states, which otherwise would render the theory "inconsistent." Thus, unlike the appearance of anomalies for global symmetries, gauge anomalies need to be cured. This imposes strong *theoretical* constraints both on existing theories and on any theory to be developed. For instance, it happens that the particle content of the standard model is appropriately "tuned" to cancel any possible gauge anomaly. If for instance there would be more quarks than leptons, certain gauge anomalies would not cancel. Similarly, strong constraints on the charges of the various particles and their relations are set due to the need to cancel the anomaly.[18]

This gives rise to many interesting philosophical questions, that have not yet received any treatment, with obvious methodological implications for theory development. There is an intricate interplay between inconsistencies of the theory, their quantum origin, and an apparently fine-tuned particle content of the standard model universe. Anomaly cancellation leads to strong restrictions on possible representations of the gauge group for grand unified theories or, similarly, to the need for bosonic string theory to be 26-dimensional and superstring theory to be 10-dimensional. Much of this depends on how problematic the "inconsistency" is and remains an issue that warrants further discussion.[19] As said, this is another illustrative example of the methodological role of symmetry breaking, or more generally symmetry considerations in fundamental physics.

44.4 Spontaneous Symmetry Breaking

The philosophical discussion of symmetry breaking has mainly been focused on spontaneous symmetry breaking (SSB), which is a rich and subtle topic. SSB occurs when the law governing the behavior of a system has a symmetry which is not shared by its ground state or vacuum. Especially in the context of perturbative quantum field theory, where states are built up from the vacuum, the specification of its symmetry properties is crucial. Depending on whether the symmetry of the law is shared by the vacuum or not, one speaks of different modes or realizations of the symmetry.

Consider a unitary representation of the symmetry group of, say, the Lagrangian. Then, the distinction between the modes relies on a result by Fabri and Picasso (1966), which states that there are, roughly speaking, only two possibilities:[20]

- The symmetry of the Lagrangian leaves also the vacuum invariant, i.e.,
  $U|0\rangle = |0\rangle$ (Wigner-Weyl mode)
- The symmetry of the Lagrangian does not leave the vacuum invariant:[21]
  $U|0\rangle \neq |0\rangle$, and

  - The symmetry is global (Nambu-Goldstone mode),
  - The symmetry is local (Higgs mode).

Symmetries realized in the Wigner mode can only be broken explicitly or anomalously. We will now turn to the other two modes, which are instances of spontaneous symmetry breaking. A typical and simple illustration of SSB is made in terms of the following Lagrangian, describing a real scalar field ϕ with a quartic interaction:[22]

$$\mathcal{L} = \frac{1}{2}(\partial_\mu \phi)(\partial^\mu \phi) - \frac{1}{2}\mu^2\phi^2 - \frac{1}{4}\lambda\phi^4, \tag{44.7}$$

with $\lambda > 0$. The Lagrangian is invariant under the *discrete* symmetry transformation

$$\phi \rightarrow -\phi. \tag{44.8}$$

For $\mu^2 > 0$ we have a unique minimum with a vacuum expectation value $\langle 0|\phi|0\rangle = 0$, which is invariant under (44.8) (i.e., the symmetry is realized in the Wigner-Weyl mode). For $\mu^2 < 0$, however, the potential exhibits a degenerate vacuum leading to two minima at $\langle 0|\phi|0\rangle = \pm v$ with $v = \sqrt{\frac{-\mu^2}{\lambda}}$. The Lagrangian remains invariant under the discrete symmetry (44.8), which however is not shared by any specific vacuum. This is an example of spontaneous symmetry breaking, since one may think of the field "spontaneously choosing" one of the vacua, as it has no reason to prefer one over the other.

Now, calling it "spontaneous" may be misleading and some call it rather hidden or secret symmetry.[23] To see why this may be more accurate, let us imagine the field in one of the degenerate vacua, say $\langle 0|\phi|0\rangle = +v$, and consider the construction of the particle spectrum from this vacuum. For this purpose it is useful to redefine the scalar field shifted toward v, i.e.,

$$\psi(x) \equiv \phi(x) - v, \tag{44.9}$$

where we now have $\langle 0|\psi|0\rangle = 0$. Plugging (44.9) into (44.7) yields

$$\mathcal{L} = \frac{1}{2}(\partial_\mu \psi)(\partial^\mu \psi) - \lambda v^2\psi^2 - \lambda v\psi^3 - \frac{1}{4}\lambda v^2\psi^2. \tag{44.10}$$

Note that (44.10) is not invariant under the discrete symmetry of (44.8), and the field living in the chosen vacuum is not able to "recognize" the more fundamental symmetry of (44.7). It is in this sense that the symmetry (44.8) is hidden or a secret symmetry. Note also that the field ψ has a mass $m = 1/2\lambda v^2 = \sqrt{-2\mu^2}$, which therefore differs from the mass of the scalar field ϕ in the unbroken phase.

This simple example exhibits several features which are characteristic for spontaneous symmetry breaking. First, the existence of a degenerate non-zero vacuum expectation value. Second, any of such vacuum states is not invariant under the symmetry of the Lagrangian, but the symmetry transformation relates each vacuum state to each other. Third, on expanding around the chosen vacuum the original symmetry remains hidden.

Nevertheless there are additional features of SSB that do not occur in this simple example. These are special features which occur when you move from discrete to continuous symmetry (Nambu-Goldstone mode) and from global to local symmetry (Higgs mode). Let us consider the Lagrangian in equation (44.7) with complex scalar fields, i.e.,

$$\mathcal{L} = (\partial_\mu \phi)(\partial^\mu \phi^*) - \mu^2\phi\phi^* - \lambda(\phi\phi^*)^2, \tag{44.11}$$

where $\phi = \phi_1 + i\phi_2$ and $\mu^2 < 0$. This Lagrangian is now invariant under the global continuous transformation

$$\phi \rightarrow e^{i\theta}\phi. \tag{44.12}$$

The minima of the potential (see Figure 44.1) are now given by $\phi\phi^* = -\mu^2/2\lambda$, which corresponds to infinitely many possible vacuum states the field can "spontaneously" fall into. This infinite degenerate vacuum is a standard feature of SSB for continuous symmetries. Let us for convenience choose the specific vacuum state $\langle\phi_1\rangle = \sqrt{\frac{-\mu^2}{2\lambda}} = v/\sqrt{2}$ and $\langle\phi_2\rangle = 0$. This solution is now related to all other solutions via (44.12). If we now expand around this arbitrarily chosen solution

$$\phi(x) = \frac{1}{\sqrt{2}}\left(v + \psi(x) + i\eta(x)\right) \tag{44.13}$$

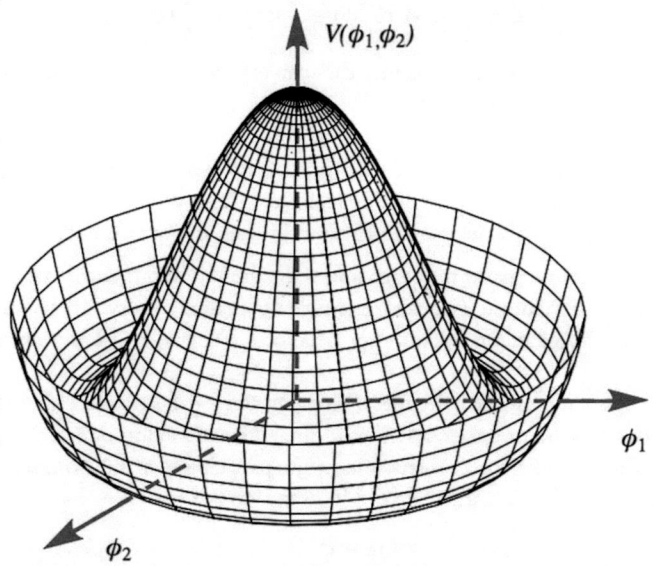

Figure 44.1 Plot of potential in equation (44.11). Figure under CC Attribution-Share Alike 3.0 Unported license.

the Lagrangian becomes

$$\mathcal{L} = \frac{1}{2}(\partial_\mu \psi)^2 + \frac{1}{2}(\partial_\mu \eta)^2 - \lambda v^2 \psi^2 + \text{cubic and quartic interaction terms.} \quad (44.14)$$

That is, you have an interacting theory of one massive scalar $\psi(x)$ (corresponding to modes along the radial direction in Figure 44.1) and one massless scalar η (along the angular direction in Figure 44.1). The existence of this massless scalar field, called Goldstone boson, is a generic feature of the spontaneous breaking of global continuous symmetries (according to the Goldstone theorem). The theorem states that there are as many Goldstone bosons as there are broken group generators.

Let us now turn from the Nambu-Goldstone realization to the Higgs realization, i.e., let us require the global symmetry transformation from (44.12) to be a local one.

$$\phi \rightarrow e^{ie\theta(x)} \phi. \quad (44.15)$$

The Lagrangian (44.11) is not invariant under the local transformation. For it to be invariant under local transformations one needs to follow the standard procedure to replace the derivative with the covariant derivative $D_\mu = \partial_\mu + ieA_\mu$, where A_μ is a gauge field, and add the standard kinetic term for the gauge field. If then at the same time $A_\mu \rightarrow A_\mu - \partial_\mu \theta(x)$, the Lagrangian will be invariant. Note that the local transformation does not allow a mass term for the gauge field. However, as we are still in the same potential as before, the argument follows analogously with the only difference that no massless particle appears after the expansion around some arbitrarily chosen vacuum. The massless degree of freedom associated with the Goldstone boson now appears as an additional degree of freedom of the gauge field (longitudinal polarization) making it massive. This result is known as the Higgs mechanism proposed by Peter Higgs and others and it provides the mechanism by which mass is generated in the standard model of particle physics.[24]

There is a well-known theorem of lattice gauge theory, which, however, explicitly forbids the possibility of a spontaneously broken local symmetry (Elitzur 1975). This has led to some discussion as to how the spontaneous breaking of a local symmetry does not violate this theorem. For instance, Smeenk (2006) and Friederich (2013) have addressed this issue in more detail relying also on an approach by Fröhlich et al. (1981), where the Higgs mechanism is accounted for in an entirely gauge-invariant approach, i.e., where it is shown that the origin of the Higgs mechanism does not rely on the breaking of a local symmetry.

The philosophical literature on SSB has to a large extent made use of the algebraic formulation of quantum field theory to address the various puzzles that arise for SSB. The reason for this is, as Earman (2003, p. 344) states, that "the algebraic formulation of Quantum Field Theory, though useless for calculations, helps to clarify foundational issues." However, introducing the formalism would go beyond the scope of this entry and Earman (2003) already provides a useful short introduction to the formalism relevant to understand it in the context of SSB.

44.5 Spontaneous Symmetry Breaking and Phase Transitions

So far there was no mention of phase transitions. However, they are intricately related to spontaneous symmetry breaking since, in many cases of phase transitions, the system undergoes a change in symmetry as well. Usually, the symmetry of the high energy phase is larger than the symmetry of the low energy phase. More precisely, the system has some *order parameter* (e.g., the magnetization), such that the expectation value of it at the ground state breaks the symmetry (this is the SSB component). The order parameter is also a function of some *control parameter* (e.g., the temperature), which allows it to transition to the broken or unbroken phase (this is the phase transition component). While a detailed analysis would go beyond the scope of this entry, we will illustrate the relation considering examples from above.

Let us first return to the Heisenberg ferromagnet. We saw how we could explicitly break the symmetry of the Hamiltonian with the help of an external magnetic field. The point being, that all electron spins will align along the direction of the magnetic field and thereby break the SO(3)-symmetry. However, this breaking also occurs, now spontaneously, by taking the infinite-volume limit, while letting the magnetization $B \to 0$. In this limit the ground state expectation value does not vanish, thereby not sharing the SO(3)-symmetry of (44.1). By increasing the temperature above the Curie temperature the system transitions to the unbroken phase, where the magnetization vanishes, again realizing the full symmetry of the system. In the broken phase, there is the interesting feature that the symmetry generators that are broken, if applied to the ground state, provide an infinitely degenerate set of ground states.[25] In the infinite-volume limit, the thus obtained degenerate ground states actually belong to different Hilbert spaces implying unitarily inequivalent representations of the commutation relations. However, this characteristic feature of SSB seems to sensitively depend on the infinite-volume limit. The need for this idealizing assumption, i.e., to take the infinite limit, has attracted significant discussion within the philosophical literature. Can a genuine physical effect depend on a necessary idealizing assumption (Earman 2004) or is it possible to de-idealize this assumption in certain ways (Butterfield 2011, Landsman 2013, Fraser 2016, Landsman 2017)?

While phase transitions find a natural habitat in the context of condensed matter systems, this is less obvious in the particle physics context. As we mentioned, the control parameter, i.e., temperature, allows one to go from e.g., the unbroken phase to the broken phase in a ferromagnet. But what plays the role of the control parameter in the Higgs mechanism discussed above? Is there a need to be one?

This has led to a discussion regarding the ontological status of the Higgs mechanism, which involves also difficulties in understanding gauge symmetries in general (see Nic Teh, this volume). As the Higgs mechanism was presented above, one might consider it as "a mere reshuffling of degrees of freedom" (Lyre 2008, p. 130). The degrees of freedom associated with the massless Goldstone boson provide the needed degrees of freedom to make the gauge boson massive. On the contrary, another view[26] considers the Higgs mechanism in strict analogy to the ferromagnet case, where the two phases are determined by $\mu^2 > 0$ corresponding to the unbroken phase and $\mu^2 < 0$ corresponding to the broken phase. The transition between these two phases then could have occurred physically during the cooling of the early universe. The formal analogy between spontaneous symmetry breaking in these different contexts has received further discussion recently in Fraser (2012) Fraser and Koberinski (2016).

44.6   Conclusion

In this chapter we have given an introduction to the ubiquitous concept of symmetry breaking as it is used in physics and to some of the philosophical discussions it generated. As we saw, symmetry breaking may occur in many different theories with different implications depending on the mechanism that underlies it. One obvious philosophical implication follows directly from its use in the context of laws rather than systems. In the law context a larger interpretational gap needs to be overcome to translate the formal implementation of symmetry breaking to what it corresponds to in the real world. Many of the philosophical issues that arise here are then directly related to the issues that arise for symmetries more generally. That is, the assessment of the impact of symmetry breaking in the context of e.g., global or gauge symmetries is strongly interlinked with the formal, methodological, epistemological and ontological analysis of these respective symmetry concepts.

However, there are certain issues more generally concerned with symmetry breaking. As we mentioned, symmetry breaking is a necessary ingredient for the existence of some phenomenon. We, nevertheless, wish to present theories in symmetric form. Spontaneous symmetry breaking is then an ingenious way to account for the lack of symmetry in the real world, while keeping the symmetry of the laws. One contentious way of looking at it, is to ask why we would want to impose a symmetry on the laws of nature that is not observed. Consequently, if symmetry is preferred over asymmetry, then occurrences of asymmetry are in need of explanation. But why is it that we prefer symmetry over asymmetry in the first place?

Notes

1 Stewart and Golubitsky (1992) is a clear illustration of how a general theory of symmetry breaking can be developed by addressing such questions as "which subgroups can occur?" and "when does a given subgroup occur?"

2 On this point, and more generally on the role of symmetry breaking in the formation of nature's patterns from the smallest scales to the largest, see for example Stewart and Golubitsky (1992), chapter 3. See also Shubnikov and Kopstik (1974).

3 The terminological use was the following: "dissymmetry" indicated that some of the possible symmetries compatible with the physical constraints were not present, while "asymmetry" was used to mean the absence of all the possible symmetries compatible with the situation considered. We will follow this usage, here.

4 This is the statement which has become known as "Curie's principle." On the current debate of Curie's Principle, see Castellani and Ismael (2016); Norton (2016) and Roberts (2016).

5 For example, the characteristic symmetry of a magnetic field is that of a cylinder rotating about its axis: this means that, for a magnetic field (the effect) to exist, the medium (the cause) must have a symmetry lower or equal to that of a rotating cylinder.

6 See for example Curie's description of such physical effects as the "Wiedemann effect" and the "Hall effect."

7 E.g., Coleman (1973) proves that spontaneous symmetry breaking does not occur in two-dimensional quantum field theories.

8 E.g., the aforementioned theorem by Coleman in endnote 7 holds for continuous but not for discrete symmetries.

9 The issue of whether symmetry breaking is something occurring, for instance temporally, in nature or just representing a change in our epistemic state at a certain time will also come up in the context of spontaneous symmetry breaking and phase transitions (see Sections 44.4 and 44.5).

10 Namely, the violation of the combination of charge conjugation and parity (CP symmetry) and, therefore, the violation of time inversion (T symmetry) in virtue of the CPT theorem. See Roberts, this volume.

11 (Lee 1981) explicitly claims that "the root of all symmetry principles lies in the assumption that it is impossible to observe certain basic quantities? (p. 178). See on this (and, more generally, on the relation between symmetry, equivalence and irrelevance) Castellani (2003). Dasgupta (2016) defends an epistemic interpretation of symmetry on a similar basis as Lee.

12 The original paper by 't Hooft (1980) introduced the idea of naturalness and its relation to symmetry breaking. Subsequently, the notion of naturalness was also considered in different contexts. See Williams (2015), Wells (2015) for more detailed philosophical discussions.

13 As Zee (2010, p. 271) puts it: "[the field theorists] were so shocked as to give this phenomenon the rather misleading name "anomaly," as if it were some kind of sickness of field theory."

14 This observation is due to Fujikawa (1980).

15 Early on, this led theorists to believe that the anomaly may only be due to the choice of regularization. However, one can show that it is actually independent of this choice. See Jatkar (2016) for calculations using different regularization schemes.

16 See Schwinger (1951) for the original paper, Peskin and Schroeder (1995, Sect.19.1) for a detailed discussion and Holstein (1993, p. 144) for an elementary discussion.

17 See Weinberg (1995, Sect. 22.1) for the historical background.

18 See Schwartz (2014, Sect. 30.4) for a detailed discussion of these constraints.

19 This becomes apparent when Guidry (2008, p. 281) speaks of it as a "theoretical prejudice": "The current theoretical prejudice is that gauge theories with incurable anomalies are incorrect because they cannot be perturbatively renormalized."

20 See Nair (2005, Ch.11) and Aitchison (1982, Sect. 6.1) for a nice discussion on this.

21 More accurately, $U|0\rangle$ does not exist in the Hilbert space.

22 This example can be found in many textbooks on quantum field theory; see e.g., Coleman (1988, Sect. 5.2) or Guidry (2008, Sect. 8.2), which we are following here.

23 See Aitchison (1982, p. 69) and Coleman (1988, Ch. 5).

24 See Higgs (1964), Guralnik et al. (1964) and Englert and Brout (1964). There is extensive historical discussion regarding the development of the Higgs mechanism. See for instance Guralnik (2009).

25 See Arodz et al. (2012, Ch. 1).

26 See Wüthrich (2012) for details.

References

Adler, S.L. (1969). Axial-vector vertex in spinor electrodynamics. *Physical Review*, 177(5): 2426.

Aitchison, I.J. (1982). *An informal introduction to gauge field theories*. Cambridge: Cambridge University Press.

Arodz, H., Dziarmaga, J. and Zurek, W.H. (2012). *Patterns of Symmetry Breaking*, Vol. 127. Berlin/Heidelberg: Springer Science & Business Media.

Bell, J.S. and Jackiw, R. (1969). A PCAC puzzle: $\pi \, \gamma\gamma$ in the σ-model. In M. Bell, K. Gottfried and M. Veltman (eds.), *Quantum Mechanics, High Energy Physics And Accelerators: Selected Papers of John S Bell (With Commentary)*. Singapore: World Scientific, pp. 367–381.

Butterfield, J. (2011). Less is different: Emergence and reduction reconciled. *Foundations of Physics*, 41(6): 1065–1135.

Castellani, E. (2003). Symmetry and equivalence. In K. Brading and E. Castellani (eds.), *Symmetries in Physics: Philosophical Reflections*. Cambridge: Cambridge University Press, pp. 425–436.

Castellani, E. and Ismael, J. (2016). Which Curie's principle? *Philosophy of Science*, 83(5): 1002–1013.

Coleman, S. (1973). There are no goldstone bosons in two dimensions. *Communications in Mathematical Physics*, 31(4): 259–264.

Coleman, S. (1988). *Aspects of Symmetry: Selected Erice Lectures*. Cambridge: Cambridge University Press.

Curie, P. (1894). Sur la symétrie dans les phénomènes physiques, symétrie d'un champ électrique et d'un champ magnétique. *Journal de physique théorique et appliquée*, 3(1): 393–415.

Dasgupta, S. (2016). Symmetry as an epistemic notion (twice over). *The British Journal for the Philosophy of Science*, 67(3): 837–878.

Earman, J. (2003). Rough guide to spontaneous symmetry breaking. In K. Brading and E. Castellani (eds.), *Symmetries in Physics: Philosophical Reflections*. Cambridge: Cambridge University Press, pp. 335–346.

Earman, J. (2004). Curie's principle and spontaneous symmetry breaking. *International Studies in the Philosophy of Science*, 18(2–3): 173–198.

Elitzur, S. (1975). Impossibility of spontaneously breaking local symmetries. *Physical Review D*, 12(12): 3978.

Englert, F. and Brout, R. (1964). Broken symmetry and the mass of gauge vector mesons. *Physical Review Letters*, 13(9): 321.

Fabri, E. and Picasso, L. (1966). Quantum field theory and approximate symmetries. *Physical Review Letters*, 16(10): 408.

Fraser, D. (2012). Spontaneous symmetry breaking: Quantum statistical mechanics versus quantum field theory. *Philosophy of Science*, 79(5): 905–916.

Fraser, D. and Koberinski, A. (2016). The higgs mechanism and superconductivity: A case study of formal analogies. *Studies in History and Philosophy of Science Part B: Studies in History and Philosophy of Modern Physics*, 55: 72–91.

Fraser, J.D. (2016). Spontaneous symmetry breaking in finite systems. *Philosophy of Science*, 83(4): 585–605.

Friederich, S. (2013). Gauge symmetry breaking in gauge theories—in search of clarification. *European Journal for Philosophy of Science*, 3(2): 157–182.

Fröhlich, J., Morchio, G. and Strocchi, F. (1981). Higgs phenomenon without symmetry breaking order parameter. *Nuclear Physics B*, 190(3): 553–582.

Fujikawa, K. (1980). Path integral for gauge theories with fermions. *Physical Review D*, 21(10): 2848.

Guidry, M. (2008). *Gauge Field Theories: An Introduction with Applications*. New York: John Wiley & Sons.

Guralnik, G.S. (2009). The history of the Guralnik, Hagen and Kibble development of the theory of spontaneous symmetry breaking and gauge particles. *International Journal of Modern Physics A*, 24(14): 2601–2627.

Guralnik, G.S., Hagen, C.R. and Kibble, T.W. (1964). Global conservation laws and massless particles. *Physical Review Letters*, 13(20): 585.

Higgs, P.W. (1964). Broken symmetries, massless particles and gauge fields. *Physics Letters*, 12(2): 132–133.

Holstein, B.R. (1993). Anomalies for pedestrians. *American Journal of Physics*, 61(2): 142–147.

Hooft, G. (1980). Naturalness, chiral symmetry, and spontaneous chiral symmetry breaking. In A. Jaffe, C. Itzykson, G. 't Hooft, H. Lehmann, I. M. Singer, P. K. Mitter, R. Stora (eds.), *Recent Developments in Gauge Theories*. New York and London: Plenum Press, 135–157.

Jatkar, D.P. (2016). Introduction to anomalies. In R. Rangarajan and M. Sivakumar (eds.), *Surveys in Theoretical High Energy Physics-2*. Berlin/Heidelberg: Springer, pp. 141–186.

Landsman, N.P. (2013). Spontaneous symmetry breaking in quantum systems: Emergence or reduction? *Studies in History and Philosophy of Science Part B: Studies in History and Philosophy of Modern Physics*, 44(4): 379–394.

Landsman, N.P. (2017). *Foundations of Quantum Theory: From Classical Concepts to Operator Algebras*. Berlin/Heidelberg: Springer Nature.

Lee, T.D. (1981). *Particle physics and introduction to field theory*. Amsterdam: OPA.

Lee, T.-D. and Yang, C.-N. (1956). Question of parity conservation in weak interactions. *Physical Review*, 104(1): 254.

Lyre, H. (2008). Does the Higgs mechanism exist? *International Studies in the Philosophy of Science*, 22(2), 119–133.

Nair, V.P. (2005). *Quantum Field Theory: A Modern Perspective*. Berlin/Heidelberg: Springer Science & Business Media.

Norton, J.D. (2016). Curie's truism. *Philosophy of Science*, 83(5): 1014–1026.

Peskin, M.E. and Schroeder, D.V. (1995). *An Introduction to Quantum Field Theory*. Boulder, CO: Westview Press.

Roberts, B.W. (2016). Curie's hazard: From electromagnetism to symmetry violation. *Erkenntnis*, 81(5): 1011–1029.

Schwartz, M.D. (2014). *Quantum Field Theory and the Standard Model*. Cambridge: Cambridge University Press.

Schwinger, J. (1951). On gauge invariance and vacuum polarization. *Physical Review*, 82(5): 664.

Shubnikov, A. and Kopstik, V. (1974). *Symmetry in Science and Art* (transl. from Russian by ED Archard; edited by D. Harker). New York: Plenum Press.

Smeenk, C. (2006). The elusive higgs mechanism. *Philosophy of Science*, 73(5): 487–499.

Stewart, I. and Golubitsky, M. (1992). *Fearful Symmetry: Is God a Geometer?* Oxford: Blackwell.

Weinberg, S. (1995). *The Quantum Theory of Fields*, Vol. 2. Cambridge: Cambridge University Press.

Wells, J.D. (2015). The utility of naturalness, and how its application to quantum electrodynamics envisages the standard model and Higgs boson. *Studies in History and Philosophy of Science Part B: Studies in History and Philosophy of Modern Physics*, 49: 102–108.

Weyl, H. (1952). *Symmetry*. Princeton: Princeton University Press.

Williams, P. (2015). Naturalness, the autonomy of scales, and the 125 gev Higgs. *Studies in History and Philosophy of Science Part B: Studies in History and Philosophy of Modern Physics*, 51: 82–96.

Wüthrich, A. (2012). Eating goldstone bosons in a phase transition: A critical review of lyre's analysis of the higgs mechanism. *Journal for General Philosophy of Science*, 43(2): 281–287.

Zee, A. (2010). *Quantum Field Theory in a Nutshell* (2nd edition). Princeton: Princeton University Press.

Further Reading from the Editors

The reader who would like to start with a classic paper could do worse than P. Higgs, 'Broken symmetries, massless particles and gauge fields' (*Physics Letters* 12(2), 132–133, 1964). Philosophical papers relating to the Higgs mechanism, include J. Earman, 'Curie's principle and spontaneous symmetry breaking' (*International Studies in the Philosophy of Science* 18(2–3), 173–19, 2004) and C. Smeenk, 'The elusive Higgs mechanism' (*Philosophy of Science* 73(5), 487–499, 2006). K. Brading, and E. Castellani, eds. *Symmetries in Physics: Philosophical Reflections* (Cambridge: Cambridge University Press, 2003) contains several papers on symmetry breaking and Brading, K., E. Castellani, and N. Teh, "Symmetry and Symmetry Breaking", *The Stanford Encyclopedia of Philosophy* (Winter 2017 Edition), Edward N. Zalta (ed.) is a helpful and updated resource reviewing the literature on symmetry breaking. Slightly more recent work includes J. Fraser, 'Spontaneous symmetry breaking in finite systems' (*Philosophy of Science* 83(4), 585–605, 2016).

PART XI

Metaphysics

Introduction to Part XI

Physics and metaphysics have always had a turbulent relationship. Even before Newton's eyebrow-raising denial, in the General Scholium to the *Principia* (1713), that he was committed to any metaphysical hypotheses, Locke had argued that the underlying nature of matter did not need to be understood in order for the study of natural phenomena to make progress; not long afterward Hume encouraged us to "commit to the flames" any metaphysical doctrines that do not contain "experimental reasoning concerning matters of fact and existence" (Hume 1748). In spite of such naysayers, during the early modern period a diverse range of metaphysical hypotheses were framed and offered as potential foundations for the dazzling success of classical mechanics, from a Cartesian plenum through Gassendi's atoms, Leibnizian monads, and Boscovich's forces to the Kantian synthetic a priori. As physics matured, the rise of positivism and its associated conventionalism in the 19th century (as well as the emerging challenges to the classical physics which Kant had hubristically declared a necessary precondition of any possible experience) generated renewed skepticism about metaphysics and its potential to contribute to physics.

Mach's influential program to eliminate unobservable elements from physical theory influenced a number of key players in the revolutions of physics at the start of the 20th century, with the result that an anti-metaphysical stance was baked in from the very beginning to the community's conceptions of general relativity and quantum theory. While metaphysically minimal approaches to these theories remain active research programs (Barbour's Machian program in the case of GR, and a variety of approaches to quantum theory including the Copenhagen interpretation, QBism, and Healey's pragmatism), these minimalist approaches have been joined in the intervening years by formulations in terms of a realist metaphysics. These realist formulations aim to provide the theories in question with a clear ontology and an explicit account of the dependencies between the elements of that ontology. Representative examples of contemporary metaphysically realist approaches to philosophy of physics may be found in the chapters by Jill North, Tim Maudlin, Simon Saunders, Roderich Tumulka, and Peter Lewis in preceding parts, and a broad form of scientific realism is now mainstream opinion in philosophy of physics.

Once we move beyond the core metaphysical question of scientific realism about our physical theories, we may ask more specific metaphysical questions about the central theoretical concepts of physics. A very natural place to start is with laws of physics, which are treated in the chapter by Lange. The terminology of "laws" recurs in numerous places in physics, from mechanics to optics to thermodynamics, but we also talk about the content of physical theory more generally as comprising the "laws of physics." The notion of laws and lawlikeness has also played a central role in philosophy of science, being implicated in the inductive method and in the nature of scientific explanation. Lange's chapter untangles the different roles that the concept of law plays in physics, and outlines and defends his own influential proposal to think of laws as truths about the physical world that are especially stable under counterfactual suppositions: laws tell us how things would be over a wide range of physically possible scenarios.

While some laws, including the laws of classical physics and of Bohmian mechanics, are deterministic (or near enough), physics also encompasses a wide range of theories that posit indeterministic laws. Nonetheless, these theories do not leave physical reality unconstrained; typically an indeterministic theory involves a range of possible histories for a system and an objective probability (or chance) distribution over them. Mauricio Suárez's chapter explores the nature of physical chance, providing an introduction to the history of probabilistic theories in physics, and complements Emery's chapter in arguing that both indeterministic and deterministic fundamental theories may give rise to genuine physical probabilities.

The characteristic feature of quantum theory, entanglement, gives rise to connections between spatially distant systems that have seemed to many authors to call out for an explanation that draws on novel metaphysics. An entangled system is represented in quantum theory by a state which cannot be factored into distinct states for distinct components of the system; authors such as Teller and Redhead have argued that this failure of factorizability reveals that quantum theory is committed to a form of metaphysical holism, in that the whole system is theoretically prior to its parts. In the limiting case, a completely adequate description of any part of reality may require the ascription of a quantum state to the entire universe. Richard Healey's chapter surveys the various holist claims that have been made in physics, with particular attention to quantum-mechanical holism and the motivation that Bell's theorem might be thought to provide for it.

Physical quantities by and large are *dimensional* quantities, and are measured through systems of units corresponding to their dimension(s). Acceleration has a dimension of ms^{-2}, for example, so a statement of the value of some acceleration must be relativized both to a unit for distance and to a unit of time. The chapter by Susan Sterrett presents a comprehensive introduction to the philosophical study of dimensionality and to the techniques of dimensional analysis which find uses throughout physics. The discussion encompasses the transition from the old SI unit scheme to the new SI scheme of units (generally regarded as conceptually cleaner) and the chapter closes by raising a key question – how is it that dimensional analysis can be so theoretically useful? – and answering it through the author's suggestion that an appropriate system of units encodes part of the physical content of the theories to which it is adapted.

The final chapter in this part is Steven French's introduction to the role of fundamentality in physics, and survey of the candidates for fundamental status in our best current theories. Drawing on his own recent work and that of McKenzie, French argues that developments in modern quantum field theory present a radical challenge to metaphysical orthodoxy, suggesting a replacement of the picture of a fundamental level of physical stuff with a picture on which abstract theoretical elements including symmetry groups can play a fundamental role in explaining physical phenomena. The chapter may fruitfully be read in conjunction with the chapters in other parts by Fraser, Batterman, Saatsi, Dasgupta, and Teh.

References

Hume, D. (1748). *An Enquiry Concerning Human Understanding*. London: A. Millar.
Newton, I. (1713). *Philosophiæ Naturalis Principia Mathematica* (2nd edition). Cambridge: Cornelius Crownfield. https://cudl.lib.cam.ac.uk/view/PR-ADV-B-00039-00002/1

45

LAWS

Marc Lange

45.1   Laws of Nature and Laws in Physics

On the standard view in metaphysics, there are three kinds of facts. First, there are the "broadly logical" necessities: facts that absolutely could not have been otherwise. These include the fact that triangles have three sides and the fact that either you are now sitting down or it is not the case that you are now sitting down. The rest of the facts are "contingent." They divide into two classes: the "natural necessities," which follow logically from the laws of nature alone, and the "accidents," which do not. Among the accidents are that all of the coins in my pocket today are silver-colored and that all solid gold cubes are smaller than a cubic meter. (For the sake of argument, let's suppose that these are truths.) The natural laws, according to our best current science, include that all gold is electrically conductive and that electric charge is conserved. According to the standard view in metaphysics that I am describing, both laws and accidents are contingent: Just as magnetic monopoles could possibly have existed and material bodies could possibly have been accelerated from rest beyond 3×10^8 meters per second (contrary to natural law), so a solid gold cube larger than a cubic meter could have existed (contrary to accidental fact).

Notice that the accidental regularity concerning gold cubes is just as general, universal, and exceptionless as the law that all solid cubes of uranium-235 are smaller than a cubic meter. (Large clumps of U^{235} undergo nuclear chain reactions, as in an atomic bomb.) Notice also that a law may currently be undiscovered (though I can't give you an example of one of those!) and that, after it has been discovered, it need not be officially called a "law" (as with the axioms of quantum mechanics, Bernoulli's principle, and Maxwell's equations). Some things that are still called laws (such as Newton's law of gravity and Bode's law) may not currently be regarded as genuine laws (or even as facts at all). ("Bode's law" is the following relation among the radii of the planets' orbits, measured in multiples of the radius of the Earth's orbit: the orbit of the nth planet from the Sun is equal to 0.4 plus the nth term in the sequence 0, 0.3, 0.6, 1.2, 2.4.... This relation happens to be very accurate up to Uranus ($n = 8$), if we count the asteroid belt as $n = 5$.)

Physics is concerned with some of the laws of nature – including, arguably, the most fundamental ones. (See French (this volume) for more on the notion of fundamentality.) There are many kinds of laws of physics (at least, according to various physical theories). Some (putative) laws of physics are force laws, such as the Lorentz force law and Coulomb's law of electrostatics. By contrast, dynamical laws (such as Newton's second law of motion) relate a body's motion to the forces on it (or local fields or potentials). Some laws of physics relate certain fields to others (e.g., Faraday's law of electromagnetic induction) or fields to charges. By contrast with laws concerning forces or other causes of motion, there are laws of kinematics (such as the relation between a body's instantaneous acceleration at each moment during some interval and the body's change in velocity across that interval). By contrast with the dynamical laws, there are laws of statics giving the conditions for static equilibrium.

There are gas laws, laws of fluid flow, and laws of heat flow. One physical law may specify that certain quantities are proportional to others, whereas another law may give the value of the constant of proportionality. Whether the "second law of thermodynamics" is genuinely a law of nature, or is instead an accidental regularity (reflecting certain accidental features of the universe's early history), remains controversial. (See Shahvisi (this volume) for more details of this controversy.) Some laws of physics concern the general structure of spacetime, and others concern the dispositions of particular kinds of inhabitants of spacetime (such as the law that gold does not form oxides). Some laws (such as the force laws) are arguably causal (see Cartwright, 1979), whereas others (such as variational principles, symmetry principles, and conservation laws) are not plausibly interpreted as specifying causal relations.

In philosophy, natural law ties into a host of perennial metaphysical and epistemological topics, including causation, chance, confirmation, counterfactuals, determinism, dispositions, emergence, explanation, models, natural kinds, necessity, properties, reduction, unification, and universals. The concept of a law of nature is often used by philosophers in their accounts of these other matters, helping to explain not only what these other things are but also how they are interconnected.

Some philosophers have even denied the "standard view" ("There are three kinds of facts. . . ") with which I began. "Scientific essentialists" (such as Ellis, 2002; Bird, 2007) regard laws as metaphysically necessary: it is part of electric charge's essence that it involves the causal power to exert and to feel forces in accordance with certain particular laws. Cartwright (1983) has argued that some processes are ungoverned by laws and that statements of the laws of nature are not even truths – at least when they are interpreted as describing exceptionless regularities, though perhaps they are true when interpreted as describing causal powers. Giere (1999) and van Fraassen (1989) contend that the philosophical tradition has been led astray in employing the concept of natural law to rationally reconstruct science.

In this chapter, I will confine myself to two questions (and even then, I can do little more than ask them). First, what difference does it make, in scientific reasoning, whether some truth is believed to be a law or an accident? Second, what is it about the world that makes some fact a law rather than an accident? Ideally, the answer to the second question should account for the answer to the first question. If these questions cannot be answered satisfactorily within the "standard view," then perhaps something more radical will be necessary. I will also consider whether some laws of physics (such as symmetry principles and conservation laws) may have a different status than others (such as force laws).

45.2   What Laws Do

How do laws differ from accidents in their scientific roles? To begin with, an accidental truth just happens to obtain. A gold cube larger than a cubic meter could have formed, but the requisite conditions happened never to arise. In contrast, it is no accident that a large cube of uranium-235 never formed, since the laws governing nuclear chain reactions prohibit it. In short, things *must* conform to the laws – the laws have a kind of *necessity* ("natural necessity", which is weaker than logical, conceptual, mathematical, and metaphysical necessity, according to the standard view) – whereas accidents are just giant coincidences.

For instance, had Bill Gates wanted to build a large gold cube, then there would have been a gold cube greater than a cubic meter. (In terms of the "possible worlds" that Lewis (1973) describes: in all of the closest possible worlds where Gates wants to build a large gold cube, there is a gold cube exceeding a cubic meter.) But even if Gates had wanted to build a large cube of uranium-235, all U-235 cubes would still have been less than a cubic meter. The laws govern not only what actually happens, but also what would have happened under various circumstances that did not actually happen. The laws underwrite various facts expressed by subjunctive (counterfactual) conditionals, i.e., statements of the form "Had *p* been the case, then *q* would have been the case" (where *p* is false). That's why scientists use the laws in figuring out what Earth would have been like, had it been farther from the Sun. In contrast, for any accident a, there exists some p that is "naturally possible" (i.e., consistent with all of the laws' logical consequences) such that a would not still have held, had p been the case. For example, had there been a gold cube exceeding a cubic meter (a natural possibility), then it would not have been the case that all gold cubes are smaller than a cubic meter.

Counterfactuals are notoriously context-sensitive. In Quine's famous example, the counterfactual "Had Caesar been in command in the Korean War, he would have used the atomic bomb" is correct in some contexts, whereas in others, "...he would have used catapults" is correct. What is preserved under a counterfactual supposition, and what is allowed to vary, depends upon our interests in entertaining the supposition. But facts about who was in command in the Korean War are accidents rather than laws. According to many philosophers (notably Goodman (1983)), even if context influences whether some accident is preserved under a given counterfactual supposition, a law's preservation is much less context-sensitive: in any context, the laws tell us what would have happened, under any natural possibility p. (Lewis (1973, 1986) is a notable dissenter, as we will see.)

Because of their necessity, laws have an explanatory power that accidents lack. For example, that all sodium salts, when ignited, burn with a yellow flame is explained by the fact that this regularity is a law and therefore must obtain. By contrast, that some regularity is not a law – and so does not have to obtain – fails to help to explain why it does obtain.

The distinction between laws and accidents makes itself felt not only metaphysically, but also epistemologically – namely, with respect to inductive reasoning. We believe that it would be a mere coincidence if all of the families on our block have two children. So we regard the fact that all of the block's families that we have checked so far have two children as failing to provide any evidence in favor of believing that the next one to be examined also has two children. To know that all of the block's families have two children, we would have to examine every single family on the block. In contrast, a candidate law is confirmed by evidence differently: that one sample of a given chemical substance melts at 383K (under standard conditions) confirms, for every unexamined sample of that substance, that its melting point is 383K (in standard conditions).

That the very same claims play all of these special scientific roles – in connection with necessity, counterfactuals, explanations, and inductive confirmations – would suggest that science draws an important distinction here, which philosophers characterize as the difference between laws and accidents. However, it is notoriously difficult to capture the laws' "special roles" precisely. Take counterfactuals. The mathematical function relating my car's maximum speed on a dry, flat road to its gas pedal's distance from the floor is not a law (since it reflects accidental features of the car's engine). (At least, it is not a law of physics. Could there be a highly specialized science of my car? In its current state?) Yet this function supports counterfactuals regarding the car's maximum speed had we depressed the pedal to one-half inch from the floor. Likewise, past instances exhibiting my car's pedal-speed function confirm the function's holding of certain unexamined cases (though not ones where the car's engine is altered). Moreover, my car's pedal-speed function (together with the road's condition and the pedal's current position) explains the car's current maximum speed.

So even if a fact's lawhood makes a difference to science, it is difficult to identify exactly the difference it makes. Furthermore, even if it is true that in any context, the laws tell us what would have happened under any natural possibility, this does not allow us to pick out the laws, since it uses the laws to pick out the relevant range of counterfactual suppositions. It is circular to specify the laws as exactly the truths that would still have held under any counterfactual supposition that is logically consistent with the laws themselves.

What if we allow a set of truths containing some accidents to pick out the relevant counterfactual suppositions: those that are logically consistent with every member of that set? Take, for instance, a logically closed set of truths that includes the accident that all gold cubes are smaller than a cubic meter but omits the accident that all of the coins in my pocket are silver-colored. Here's a counterfactual supposition that is consistent with every member of this set: had there been either a gold cube that is *larger* than a cubic meter or a coin in my pocket that is *not* silver-colored. What would the world then have been like? In many conversational contexts, we would deny that of the two accidents I have mentioned, the one in the set ("All gold cubes are smaller than a cubic meter") would still have held. (Perhaps it is the case for neither accident that it would still have held.) The same sort of argument could presumably be made regarding any logically closed set of truths that includes *some* accidents but not *all* of them. Given the opportunity to pick out the range of counterfactual suppositions convenient to itself, the set nevertheless is not invariant under all of those suppositions.

Here, then, is my rough suggestion for the laws' distinctive relation to counterfactuals. Take a set of truths that is "logically closed" (i.e., that includes every logical consequence of its members) and is neither the empty set nor the set of all truths. Call such a set *stable* exactly when in every context, every member g of the set would still have held had p been the case, for each of the counterfactual suppositions p that is logically consistent with every member of the set. My rough suggestion: g is a natural necessity exactly when g belongs to a stable set. (For a more careful discussion, see Lange, 2000, Lange, 2009.)

What makes the natural necessities special is their stability: *taken as a set*, they are invariant under as broad a range of counterfactual suppositions as they *could* logically possibly be. No set containing an accident can make that boast (except for the set of all truths, for which the boast is trivial). Because the set of laws (and their logical consequences) is non-trivially as invariant under counterfactual perturbations as it could be, there is a variety of *necessity* corresponding to it; necessity involves possessing *maximal* invariance under counterfactual perturbations. No necessity is possessed by an accident, even one (such as my car's gas pedal-maximum speed function) that would still have held under many counterfactual suppositions. The notion of "stability" distinguishes laws from accidents, accounts for the laws' necessity, and gives us a way out of the notorious circle that results from specifying the natural necessities as the truths that would still have held under those counterfactual suppositions consistent with the natural necessities.

The stable sets form a natural hierarchy; for any two stable sets, one must be a proper subset of the other. The proof is as follows:

1   Suppose (for *reductio*) that Γ and Σ are stable, t is a member of Γ but not of Σ, and s is a member of Σ but not of Γ.
2   Then ($\sim s$ or $\sim t$) is logically consistent with Γ.
3   Since Γ is stable, every member of Γ would still have been true, had ($\sim s$ or $\sim t$) obtained.
4   In particular, t would still have been true, had ($\sim s$ or $\sim t$) obtained.
5   So had ($\sim s$ or $\sim t$) obtained, then t & ($\sim s$ or $\sim t$) would have held, so $\sim s$ would have held.
6   Since ($\sim s$ or $\sim t$) is logically consistent with the stable Σ, no member of Σ would have been false had ($\sim s$ or $\sim t$) obtained.
7   In particular, s would not have been false, had ($\sim s$ or $\sim t$) obtained.
8   Contradiction from 5 and 7.

The hierarchy non-trivially of stable sets may contain more than just two members: the set of all laws and its proper subset containing exactly the broadly logical necessities. In particular, some proper subsets of the laws of physics may serve as constraints on the other laws by forming their own stable sets. For instance, perhaps the conservation laws, explained by the symmetry principles, constrain the kinds of force laws that there could be. Feynman might be interpreted as suggesting that the force laws and conservation laws stand in this relation:

When learning about the laws of physics you find that there are a large number of complicated and detailed laws, laws of gravitation, of electricity and magnetism, nuclear interactions, and so on, but across the variety of these detailed laws there sweep great general principles which all the laws seem to follow. Examples of these are the principles of conservation. . . . (Feynman, 1967, p. 59)

On this view, the fact that the laws of two distinct forces are alike in conserving energy is no coincidence. Rather, the conservation laws supply a common explanation. Moreover, since the conservation laws belong to a stable set that omits the force laws, the conservation laws have a stronger variety of necessity than the force laws they constrain; the conservation laws would still have held even if there had been additional types of forces. Our discovery that all force laws known so far follow a given putative conservation law constitutes some evidence in favor of the hypothesis that were there additional force laws, then they would too.

On this view, the conservation laws play all of the special scientific roles characteristic of laws – in connection with necessity, counterfactuals, explanations, and inductive confirmations – in a way that makes them transcend the force laws. (Perhaps the laws of statics, the laws of kinematics, and the laws giving spacetime's characteristic properties also belong to strata of physical law autonomous from the force laws.) The conservation laws, in turn, are explained by symmetry principles that function as meta-laws – that is, laws about the first-order laws of nature. The symmetries would still have held even if the first-order laws had been different. As Wigner says:

[F]or those [conservation laws] which derive from the geometrical principles of invariance it is clear that their validity transcends that of any special theory – gravitational, electromagnetic, etc. – which are only loosely connected. (Wigner, 1972, p. 13)

In other words, Wigner contends that those symmetries are not coincidences of the particular kinds of forces there happen to be, and so the associated conservation laws transcend the idiosyncrasies of the force laws.

45.3 What Laws Are

Lewis (1973, 1986) gives the most sophisticated "Humean" or "regularity" account of natural law. According to Lewis, facts about laws "supervene" on the spacetime geometry and the spatiotemporal mosaic of instantiations of the properties belonging to a certain elite class. (That is, two possible worlds cannot differ in their laws without differing in their spacetime geometry or their mosaics.) These elite properties are the properties meeting the following conditions:

- They are perfectly natural – that is, each property represents a respect of perfect similarity, as in (for instance) the property of possessing exactly 1 statcoulomb of positive electric charge – unlike, for instance, the disjunctive property of being an emerald or greater than three inches long. That is, they are among the "sparse" properties – non-gerrymandered ones, not mere shadows of predicates. (See Armstrong, 1978, Lewis, 1983.)
- They are categorical – that is, "Humean": they involve no modalities, propensities, chances, laws, counterfactuals, dispositions. . . (Scientific essentialists deny that there are any such properties; see Ellis, 2002.)
- They are qualitative in the sense that they do not involve the property, which (according to some philosophers, such as Adams, 1979) a given thing intrinsically possesses, of being the particular individual thing that it is. (Such a property is called a "haecceity.") For example, the property of being identical to Jones is not elite.
- They are possessed intrinsically by spacetime points or occupants thereof.

Also supervening on the Humean mosaic are facts about single-case objective chances, such as this atom's having a 50% chance of undergoing radioactive decay in the next 13.81 seconds. Consider the deductive systems of truths regarding instantiations of elite properties and claims regarding the objective chances at various times that certain elite properties will be instantiated at later times (where the system says A only if it also says that A never had any chance of not obtaining). These systems, Lewis says, compete along three dimensions:

1   informativeness (in excluding or in assigning chances to possible arrangements of elite-property instantiations),
2   simplicity (e.g., in the number of axioms and the order of polynomials therein, as expressed in terms of natural properties, spacetime relations, and chances), and
3   fit (which is greater insofar as the actual course of elite-property instantiations receives a higher probability).

These three criteria stand in some tension. Greater informativeness can be achieved by adding facts to the system, which often brings a loss of simplicity. (Often, but not always: see the discussion of "vacuous laws" just below, where I explain how Coulomb's law brings greater informativeness *and* greater simplicity as compared to just some of its instantiated cases.) Likewise, if property P is instantiated at time t_2, then by adding to a system the claim that c is the chance at t_1 of P being instantiated at t_2, we may add informativeness (though not as much as we would had we added that P is instantiated at t_2) and we may add fit (though not as much as we would had c been greater).

Perhaps some single system is by far the best on balance in meeting these three criteria. Perhaps which system "wins" the competition is relatively insensitive to any arbitrary features of our sense of simplicity or our rate of exchange among the three criteria. In that case, the laws of nature are the contingent generalizations belonging to the best system, and the facts about chances at a given moment are whatever the best system (and the history of elite-property instantiations until that moment) entails them to be.

Lewis's account has the virtue of using only Humean resources to distinguish between laws and accidents. It also nicely accommodates vacuous laws. Take Coulomb's law, which specifies the electrostatic force between any two point charges long at rest. Suppose we replace Coulomb's law in the best system by a generalization that agrees with Coulomb's law except in the case of a point body of exactly 1.234 statcoulombs at exactly 5 centimeters from a point body of exactly 6.789 statcoulombs. If there never exists such a pair of bodies, then the replacement generalization is true, just like Coulomb's law. However, it is not as simple as Coulomb's law, since it treats this hypothetical pair of bodies as a special case. So the best system contains Coulomb's law, replete with uninstantiated cases.

On the other hand, it might be wondered whether the laws really do supervene on the Humean mosaic. Couldn't two possible worlds involve the same Humean mosaic, but whereas in one world it is a law that all F's are G, this regularity is accidental in the other world? Perhaps there it is a law that all F's have 99.99% chance of being G. Or suppose there had been nothing in the entire history of the universe except a single electron moving uniformly forever. (Presumably, this impoverished world is naturally possible.) Lewis's account apparently dictates that the laws would then have included "At all times, there exists only one body." But intuitively, perhaps, the laws of nature would have been no different had there been only a single lonely electron (i.e., in the closest possible world where there is nothing but a single electron). Only the universe's initial conditions would have been different. In some possible lonely electron worlds (such as the closest one), the laws say that like electric charges repel, whereas in other possible lonely electron worlds, the laws say that like charges attract. The laws thus fail to supervene on the mosaic. (For more sophisticated arguments for nomic non-supervenience, see Tooley, 1977, pp. 669–672 and Carroll, 1994, pp. 60–68.)

On Lewis's behalf, it might be replied that were there nothing but a lonely electron, then a great many actual laws would be vacuous (such as Coulomb's law, not to mention "All emeralds are green"). They would then be true, trivially, but what would there be to make them laws? Furthermore, if laws fail to supervene on the Humean base, then how could we ever know – even if all observable facts were available to us – what the laws are?

Clearly, we have here a major philosophical dispute. Lewis regards the laws as arising "from below," out of the Humean mosaic; they are constituted by that mosaic. Non-Humean accounts, in contrast, take the laws as governing the universe and so as imposed on the Humean mosaic "from above"; the laws are facts over and above the facts they govern. Armstrong (1983), Dretske (1977), and Tooley (1977) have proposed broadly similar non-Humean accounts, according to which the laws are irreducible, contingent relations among universals. That emeraldhood (a universal) stands in a relation of "nomic necessitation" to greenness (another universal) metaphysically demands that all emeralds are green (a regularity).

Any account of natural law must account for the laws' special scientific roles in connection with induction, counterfactuals, and explanation. Armstrong, Dretske, and Tooley argue that if a law is merely a regularity (even one belonging to the best system), then for the law that all F's are G, together with Fa, to explain why it is the case that Ga would amount either to Fa&Ga explaining itself, or to spatiotemporally remote instances of the regularity explaining this one (and vice versa), or to the entire Humean mosaic (including irrelevant events) effectively figuring in the explanation. So a regularity view cannot account for the laws' explanatory power. Lewis replies that to explain a fact just is to place it within the simplest, most comprehensive system of the world, i.e., to locate it in relation to the "best system." In contrast, without some further, independent characterization of the relation of "nomic necessitation," Lewis (1986, p. xii) says, it is unclear why Ga should be explained by Fa and such a relation's holding between F-ness and G-ness. This relation is merely stipulated to be explanatorily potent.

Although Armstrong, Dretske, and Tooley deem a relation of nomic necessitation to be contingent, they also say that it would still have held had things been different in some naturally possible way; relations among universals are not vulnerable to being overturned by such counterfactual perturbations among the particulars they govern. Hence, the laws would still have been laws, had I missed my bus to work this morning. Lewis replies that once again, non-Humeans are merely stipulating their lawmaker to have whatever properties they believe it must have in order to account for scientific reasoning. Furthermore, Lewis believes that in a deterministic world (which Lewis believes to be metaphysically possible), the counterfactual supposition that I missed my bus to work this morning requires a "small miracle" (a single localized violation of the actual laws) in order for this departure from actuality to be accommodated in the least disruptive fashion: without modifying the past by including the causal antecedents that this supposed event must (naturally necessarily) have. Hence, the laws would have been different, had I missed my bus to work this morning; some actual laws would have been violated (by the "miracle"), so not all actual laws would still have been true, and since the laws must at least be facts, not all actual laws would still have been laws. On Lewis's view, then, the laws are not "held sacred" under counterfactual suppositions and this behavior on their part is best explained by a Humean view, according to which there is no great metaphysical gulf separating laws from accidents.

Armstrong contends that Lewis's account cannot explain why the law that all emeralds are green underwrites the fact that had there been another emerald, then all emeralds would still have been green. That the best system includes the fact that all emeralds are green gives us no basis, in supposing that there were another emerald, for believing that it would be green. We are arbitrarily extending the regularity to cover a new case. Lewis (1973, p. 74) replies that part of the logic of counterfactual reasoning is that the best system is especially influential in determining which possible world where there is another emerald is closest to the actual world. A scientific essentialist such as Bird (2007, p. 96), on the other hand, turns Armstrong's objection to Lewis against Armstrong himself. Whereas Armstrong believes that a certain relation's holding among universals forces a regularity onto the world, making that regularity (naturally) necessary, a scientific essentialist argues that a relation's holding contingently among universals has no necessity to impart to a regularity; a regularity isn't made necessary by virtue of following from a relationship among universals unless that relationship is itself necessary. (Lewis (1983, p. 366; cf. van Fraassen, 1989, p. 106) agrees with this critique of Armstrong: that Armstrong's posited relation is called "nomic necessitation" does not give it the power to confer necessity upon a regularity.) A brute, contingent relation of nomic necessitation is insufficient to sustain counterfactuals; there is no reason why a contingent relation among universals should still obtain, had there been another emerald. A metaphysically necessary relation is required.

Yet some counterfactual conditionals seem sustained partly by mere accidents, such as counterfactuals regarding the car's maximum speed had we depressed the pedal to one-half inch from the floor. Furthermore, the laws' metaphysical necessity would make the laws true in every possible world. How, then, do we account for the truth of counterfactuals with naturally impossible antecedents? For example, if it is an accident that all of the wires on the table are made of copper, then (in some conversational contexts, at least) it is true that had copper been electrically insulating, then all of the wires on the table would have been useless. Likewise, physicists tell us that the existence of living things is the result of exquisite coordination among the laws of nature: had the electromagnetic force been a little stronger relative to the strong nuclear force, then nuclei larger than carbon would not have been stable. If laws are metaphysical necessities, then it is difficult to understand what makes these counterlegals true.

A wide variety of other accounts of lawhood have been proposed. All must answer Euthyphro-style questions about lawhood and its relation to the laws' characteristic scientific roles: Are the laws necessary (or explanatory or invariant under counterfactual perturbations or inductively confirmable...) by virtue of being laws, or are they laws by virtue of being necessary (or explanatory or invariant under counterfactual perturbations or inductively confirmable...)? On Maudlin's (2007) view, lawhood is a metaphysical primitive, responsible for all of the laws' characteristic scientific roles. (But if we see lawhood as a metaphysical primitive, then it is more difficult for us to say – except by stipulation – why lawhood is associated with these characteristic scientific roles.) On Kment's (2014) view, laws support counterfactuals and possess a characteristic variety of necessity by virtue ultimately of their distinctive explanatory power; explanatory relations play a special role in determining which possible worlds are closest to the actual world. On Lange's (2009) view, facts about which subjunctive conditionals hold are the lawmakers; they are responsible for distinguishing laws from accidents and for making the laws necessary. Roberts (2008) regards lawhood as arising from the epistemic norms governing the role of certain measurement procedures in epistemic justification in science; in any given scientific context, the laws derive their lawhood from being the reliability conditions for the legitimate measurement procedures. Cohen and Callender (2009) have defended an amended version of Lewis's account that does not rely on a privileged class of natural properties and thereby allows laws in special sciences (including branches of physics dealing with macroscopic, non-fundamental properties) to count as laws for the same reason as laws of fundamental physics do. Undoubtedly, philosophers will continue to develop various accounts of what laws are that aim to explain what laws do.

References

Adams, R.M. (1979). Primitive thisness and primitive identity. *The Journal of Philosophy*, 76: 5–26.
Armstrong, D.M. (1978). *Universals and Scientific Realism*, Vol. 1. Cambridge: Cambridge University Press.
Armstrong, D.M. (1983). *What Is a Law of Nature?* Cambridge: Cambridge University Press.
Bird, A. (2007). *Nature's Metaphysics*. Oxford: Clarendon.

Carroll, J. (1994). *Laws of Nature*. Cambridge: Cambridge University Press.

Cartwright, N. (1979). Causal laws and effective strategies. *Nous*, 13: 419–437.

Cartwright, N. (1983). *How the Laws of Nature Lie*. Oxford: Clarendon.

Cohen, J. and Callender, C. (2009). A better best system account of lawhood. *Philosophical Studies*, 145: 1–34.

Dretske, F. (1977). Laws of nature. *Philosophy of Science*, 44: 248–268.

Ellis, B. (2002). *The Philosophy of Nature: A Guide to the New Essentialism*. Montreal and Kingston: McGill-Queen's University Press.

Feynman, R. (1967). *The Character of Physical Law*. Cambridge, MA: MIT Press.

Giere, R. (1999). *Science without Laws*. Chicago: University of Chicago Press.

Goodman, N. (1983). *Fact, Fiction and Forecast* (4th edition). Cambridge, MA: Harvard University Press.

Kment, B. (2014). *Modality and Explanatory Reasoning*. New York: Oxford University Press.

Lange, M. (2000). *Natural Laws in Scientific Practice*. New York: Oxford University Press.

Lange, M. (2009). *Laws and Lawmakers*. New York: Oxford University Press.

Lewis, D. (1973). *Counterfactuals*. Cambridge, MA: Harvard University Press.

Lewis, D. (1983). New work for a theory of universals. *Australasian Journal of Philosophy*, 61: 343–377.

Lewis, D. (1986). *Philosophical Papers, Volume II*. New York: Oxford University Press.

Maudlin, T. (2007). *The Metaphysics within Physics*. Oxford: Oxford University Press.

Roberts, J. (2008). *The Law-Governed Universe*. Oxford: Oxford University Press.

Tooley, M. (1977). The nature of law. *Canadian Journal of Philosophy*, 7: 667–698.

Van Fraassen, B.C. (1989). *Laws and Symmetry*. Oxford: Clarendon.

Wigner, E. (1972). Events, laws of nature, and invariance principles. In *Nobel Lectures: Physics 1963–70*. Amsterdam: Elsevier, pp. 6–19.

Further Reading from the Editors

For an accessible introduction to the topic of laws and related concepts in general metaphysics of science, see M. Schrenk, *Metaphysics of Science* (London: Routledge, 2016). A classic general critique of the concept of law of nature is B. Van Fraassen, *Laws and Symmetry* (Oxford: Clarendon Press, 1989). N. Cartwright, *How the Laws of Physics Lie* (Oxford: Clarendon Press, 1983) criticizes philosophical theories of laws in the specific context of fundamental physics. For a contemporary Humean approach to the laws of physics, see J. Ismael, "How to be Humean," in *A Companion to David Lewis*, B. Loewer and J. Schaffer (eds.). (Oxford: John Wiley and Sons, 2015). For a contemporary anti-Humean approach, see T. Maudlin, *The Metaphysics within Physics* (Oxford: Oxford University Press).

46

CHANCE

Mauricio Suárez

46.1 The History of Chance: Physics and Metaphysics

Probability, as we know and use it nowadays, was born in the 17th century, in the context of disputes within the Catholic Church regarding the nature of evidence. It was born as a dual, or Janus-faced concept (Hacking, 1975), endowed with both ontological and epistemic significance. Arnauld, Pascal, and Leibniz emphasized its epistemological salience, while Huygens, Bernoulli, and later, Laplace and Poincaré focused on the ontological implications. The hybrid nature continues to this day.

In this chapter, I am concerned with the application of the ontological dimension of probability to physical chance. It is therefore to Huygens that I turn in this section for some historical background. Yet, in addressing contemporary debates, it often helps to be reminded that probability remains stubbornly hybrid. Thus, the foundations of decision theory (e.g., in Pascal's wager) require some antecedent objective chances; and more generally the cogency of subjective probability requires objective probabilistic independence (Gillies, 2000). Similarly, single case chances in the sciences have often been supposed to be essentially subjective or to require some subjective or otherwise pragmatic rules of application or analyses (Howson and Urbach, 1989 (1993), p. 346; Lewis, 1986). Yet, such analyses often arguably presuppose the reality of objective chance. Not surprisingly, the essential duality of probability, as we shall see, becomes characteristic of debates on the nature of physical and quantum chance.

It is worth recalling that historically a certain sense of metaphysical chance predates – and in fact contributes to – the genesis of probability. And although our full contemporary notion of lawful chance does not arise until the end of the 19th century, the practice of employing statistical measures to represent objective or ontological chance is already well established in the 17th century. The connection between ratios in populations and a primitive sense of "probability" is already present in Fracastoro and other renaissance scholars (Hacking, 1975, Ch. 3). But objective chance first fully emerges in the work of Christian Huygens (1657/1714), who is perhaps the first to distinguish different statistics in a population. Huygens' defense of the distinction between the average mean age of a population and its life expectancy implicitly deploys estimates for the objective chance of any individual to live up to a certain age. The difference between the mean and the expectation is of course critically important for very skewed distributions, or those with a large standard deviation, but remains largely invisible in well-behaved (i.e., symmetrical and smooth) distributions over homogenous populations.

For a discrete random variable X, its expected value is calculated as a weighted average, with the weights representing probabilities, as follows: $E(X) = \sum_{i=1}^{n} p(x_i) A(x_i)$, where x_i is the ith value of the discrete random variable X, and p_i is its probability. In the case of a continuous random variable, we compute its value as: $E(x) = \int_{-\infty}^{+\infty} xf(x)\, dx$, where $f(x)$ is the probability density function for the random variable x.

The relevant philosophical question concerns the interpretation of $p(x)$, and $f(x)$. Huygens assumes that these functions describe objective chances since he models them after a lottery, i.e., the typical game of chance at the time (Hald, 2003, p. 108). The chances of a lottery game are arithmetic (assuming the equiprobability of drawing any one ticket rather than another). Hence the only thing that matters is the relative proportion of tickets with the same "value" in the overall pack. In the case of life expectancy, which we may also take to be the result of some underlying probability distribution over some discrete variable (age of death) defined over a population, it is the proportion of people in each subdivision of age. And this is thus implicitly taken to be just as objective as the arithmetic proportion of tickets of each kind in a pack. The question, however, is what precise objective property of people (the elements of the population) this probability picks out. From this point onwards, it becomes possible to distinguish "objective probability," as the formal concept, from "chance," as whatever objective property in the world the formal concept picks out.

Similar conceptions of objective chance underpin Laplace's later work (Laplace, 1814/1951). Laplace is sometimes celebrated as the champion and pioneer of a purely epistemic conception of probability, according to which the underlying dynamical laws of the universe are deterministic and probability represents only a certain degree of ignorance or lack of knowledge regarding initial conditions. But this is arguably a misrepresentation of Laplace's philosophy of probability, which combines both ontological and epistemic aspects. Laplace explicitly defines probability as the ratio of actual to total equipossible cases $\left(\text{the so-called classical definition of mathematical probability as a ratio: } \frac{\#positive\ cases}{\#equipossible\ cases}\right)$. The definition is fulfilled by any proportion of an attribute in an actual class, and Laplace was given to generalizing it to situations where the cases considered are not equipossible because they are not equiprobable. But even to state this requires an antecedent notion of objective equiprobability or chance – which Laplace is content to deploy at leisure.

46.2   Chance and the Interpretation of Probability

The most important philosophical question then concerns the interpretation of objective probability – and, most particularly, the question regarding the property of statistical populations that any statement of objective probability effectively picks out. Philosophers have grappled with this issue in different ways. Two main interpretations of objective probability that have emerged are the frequency and the propensity interpretations. The frequency interpretation was most explicitly championed by Von Mises (1928) and Reichenbach (1949). It is driven largely by empiricist concerns to keep the concept of chance firmly grounded in experience and equates chance with stable frequencies in repeatable sequences of experimental outcomes. The propensity interpretation, on the other hand, is often associated with Popper (1959) although it has marked antecedents in late 19th-century thought (Peirce, 1910). It is rather driven by an abductive understanding of chance attribution as an explanatory practice and equates chance with the tendencies in chancy objects to generate certain outcomes. (More precisely, in Popper's (1959) and Gillies' (2000) theories, with the dispositions of chance setups to yield stable frequencies of such outcomes in the long run). Of course, both ratios or proportions in populations, and dispositions and tendencies have a much longer philosophical history; their explicit association to probability and chance is, however, more of a fin de siècle development.

Hence the frequency interpretation assumes that a probability statement is meaningful if and only if it refers – implicitly if not explicitly – to a sequence or class of outcome events of an experimental set up of a certain kind. The statement of probability is then to be understood as the statement of the proportion of the outcome events in that sequence that possess a certain attribute. Hence, consider the attribute A in an appropriate finite sequence of observed outcome events $S = \{s_1, s_2, \ldots, s_n\}$, where we assume without loss of generality that n is even. A certain subset forming an appropriate subsequence is $S' = \{s_1^A, \ldots, s_m^A\}$, with $m \leq n$, containing all and only those elements in S that possess the attribute A. The probability of A in S, according to the frequency interpretation, is simply the ratio of positive cases in S' to all cases in S. Thus, if the rule that picks out the elements in the subsequence is, for example, one that selects each odd placed element in the original sequence, this is in effect $\frac{1}{2}$, since it is guaranteed to pick out half of the original members.

The above notion is simple, in line with Laplace's classical definition, and seems straightforward to apply. However, it gives rise to a very large number of decisive difficulties regarding: (i) the rule that picks out the subsequence, (ii) the 'appropriateness' of the sequences, (iii) the fact that the sequences are finite, and (iv) the role that frequencies, vis-à-vis probabilities, play in scientific practice. (For a summary of these and other objections see Hájek, 1997 and 2009). They all come to the fore when we consider a real-life ordinary case of physical chance – such as the chance of heads up in tossing a regular coin. If the tossing device is genuinely random, and the coin is fair, we expect this to be 12. Yet, there is no rule that picks out the subsequence S' of tosses with the relevant attribute ("heads up"); this is precisely part of what makes the generating device a random one. Hence there is no simple prescription for any rule that will do the required job. (In Von Mises' terms, 1928, p. 24, there is no place-selection rule).

Secondly, nothing can prevent an accidentally biased series of outcomes with the relevant attribute in any finite sequence. This is evident if we consider a short experimental run of ten coin tosses: the likelihood of obtaining precisely five heads is in practice less than one, however fair the coin. Yet, any other frequency may not be representative but accidental. The difficulty does not go away however long we let the experiment run for, for the sequence is finite – as it inevitably must be given the limited time span of any real experiment. This has severe implications for the probability of single events, which on this theory are strictly meaningless. Thus there is on this view no "probability of the battle of Waterloo," or "probability that an atom will decay this minute," etc.

As a possible solution, if the sequence is well-behaved, the frequency of the attribute may possess a limit, and we can take the limit to be the probability. So only certain sequences will do, namely, those that have a stable limiting frequency of the attribute in question ("collectives" in Von Mises's terminology, 1928, p. 11). But, and here comes the third set of issues, the probability is now identified with a frequency in an infinite sequence, or with a mathematical limiting property of the sequence. Both solutions are problematic for an empiricist conception of chance since they do not identify probability with any actual frequency in a sequence. The former identifies it with a hypothetical entity (an infinite sequence of experimental outcomes); the latter identifies it with an abstract mathematical property (a limit).

Finally, there are issues related to explanation (see e.g. Emery, 2015). Probabilities in physics and ordinary life are routinely employed to explain sequences of observable data. The probability for a coin to land heads explains the long run or limiting frequency; the probability of a given chemical element to decay (its half-life) explains the long-run frequency of decay in any sample of the given chemical material; and so on. Yet, on the frequency interpretation, probabilities are frequencies; and it is very hard to see how frequencies can explain other frequencies (except perhaps in the trivial and unenlightening sense of subsuming them as sub-sequences).

This last problem points toward the alternative objective interpretation of chance as propensity – a dispositional property of the experimental or chance set up that gives rise to well-behaved sequences or collectives. The view expresses an abandonment of any strict or reductive empiricism. In this view, probabilities are linked to the dispositional properties of chancy systems, or entire experimental setups, and these are not themselves necessarily observable or empirically accessible. (Note: the view is not however incompatible with a mild form of empiricism that recommends chances to be estimated from empirical data; and for evidence to be brought to bear for or against any given chance attribution). While the propensity interpretation of probability overcomes the previously described difficulties for frequencies (in some cases trivially since it does not identify probability with any frequency in any sequence), it nonetheless has problems of its own. The most notable one is "Humphreys' paradox," which concerns the interpretation of inverse probabilities. For any well-defined conditional probability $P(A/B)$ its inverse $P(B/A)$ is also well defined; yet a propensity is asymmetrical precisely because it is explanatory, and most explanations are asymmetric. Several scholars have argued, following Humphreys (1985), that probabilities cannot thereby be identified with propensities, but must be conceptually distinguished from them (see Suárez, 2014, for a review).

While these disputes about chance in the first instance concern its conceptual analysis – what Carnap (1950) refers to as "external questions" – they can also become rather substantial, requiring an assessment of both the coherence of each account, and its fit with both experimental data and, more generally, scientific practice. Not only have such philosophical disputes played an enormously important role in the history of probability, but they continue to play an enormously important role in contemporary debates regarding the nature of physical chance. Philosophers of physics often appeal to probability and its interpretation as part of their intended solution to many present-day conceptual puzzles. And, as it happens, it matters greatly what kind of underlying interpretation they hold. I here make a preliminary case for a type of propensity interpretation, but I mainly aim to show that chance may be fruitfully applied in different areas of physics regardless of underlying assumptions about determinism.

46.3   Chance in Deterministic Physics

Pierre-Simon Laplace first introduced the thesis of universal determinism, which he regarded a consequence of the dynamical laws of Newtonian mechanics. Newton's second law, in particular, defines a configuration of positions and forces at any given moment in time, and its formulation in a differential equation with respect to time allows us to calculate the dynamical evolution of a system for any arbitrary future time: $\vec{F} = m\frac{dx^2}{dt^2}$. Laplace also came up with what is nowadays known as "Laplace's demon": the thought that if universal determinism is true then for a fully omniscient intelligence, who could know the present and past state of the universe in its entire detail, "nothing would be uncertain, and the future just like the past would be laid out before her eyes" (1814, p. 4). If universal determinism is true, the past state of the universe is the total cause of its present state, and its present state is the total cause of any of its future states. Therefore, full knowledge of the state of the universe at any stage in its evolution guarantees full knowledge of its state at any other stage. In such a universe, endowed with universal deterministic dynamics, nothing would be left to chance. There would be no role for ontological probability because there would be no objective physical chance. Call this Laplace's thesis (though it is unclear that it is in fact due to Laplace): the only reason there are probabilities in classical physics is that our cognitive limits as human beings require them. Probability becomes a necessary tool for prediction for those less than omniscient intelligences like ours: It measures our lack of knowledge or ignorance of the actual conditions of the universe, thus allowing us to compute future states within the bounds of our ignorance.

Laplace's thesis has exerted a profound influence on the philosophy of probability, as well as scientific theorizing about chance. Many contemporary metaphysical accounts of chance (such as e.g. Lewis, 1986) are heavily in its debt. Yet, the thesis can be and has been contested. There are three main objections. Firstly, it is unclear that Newtonian dynamics in fact entails universal determinism. Secondly, even if it does, it is unclear that the rest of physics, never mind the rest of science, has dynamical laws akin to Newton's second law. Thirdly, it is unclear that universal determinism rules out ontological probability anyway. The third argument is obviously most relevant to our discussion, but the first two also have some interest.

Earman (1986) notoriously introduced the view that Newtonian mechanics is far from trivially deterministic (the view has antecedents in Born, 1969). His main examples were related to time-reversed unboundedly accelerated objects, also known as "space invaders" (see Hoefer, 2003, for a review). These objects are theoretically possible in classical mechanics, yet it is completely undetermined at what stage, if any, in the evolution of the world they come into being.

Norton (2003) introduced what is nowadays the best-known example of a Newtonian system with an indeterministic dynamics – the so-called "Norton's dome." This is an imaginary concrete object that obeys the laws of Newtonian mechanics – by definition. Yet, as can be purportedly demonstrated by performing a thought experiment on it, it is an openly indeterministic system, since it admits more than one possible state evolution (in fact an infinite number of possible future state evolutions) consistent with its present state. The dome is (Norton, 2008, p. 787) a radially symmetric surface with a shape defined by: $h = (2/3g)r^{3/2}$, where r is the radial distance coordinate in the surface of the dome, h is the vertical distance below the apex at $r = 0$, and g is the constant acceleration of a free mass of unit value in the vertical – i.e., downward – gravitational field surrounding the surface (Figure 46.1).

The thought experiment involves placing a point-like body of unit mass on the apex, and letting it evolve freely in time. Newtonian mechanics entails that the acceleration of this point-like unit mass is given by: $\frac{F}{m} = \frac{d^2r}{dt^2} = r^{1/2}$. This dynamical law has not one but two solutions, namely:

i) $r(t) = 0$, which entails that the point-like mass remains at rest at the apex for any future time; and

ii) $r(t) = \begin{cases} (1/144)(t-T)^2, \text{ for } t \geq T \\ 0, \text{ for } t \leq T \end{cases}$, which curiously entails that, after some arbitrary time T, the point-like mass starts to descend along any arbitrary radial direction down the dome's surface.

Norton (2008) makes the point that while one could lay out a probability distribution over the alternative radial directions down the dome (where each direction has the same probability) it does not seem possible to similarly lay out a probability distribution over time intervals $[0, T]$ such that the descent will begin within the given interval of time. Since $T \rightarrow \infty$, each such interval should receive probability zero, thus making it certain that no descent takes place, contrary to both common sense and the mathematical solution. Thus, it is not only indeterministic whether the point-like mass rests indefinitely, or descends; it also fails to be determinable at what time it will move if it does. Norton argues moreover that this precise moment cannot be determined even up to a certain probability (because time is modeled on the real number continuum, so the only consistent ascription of probability to any given interval within Newtonian mechanics is exactly zero). Yet, while it is true that Newtonian mechanics provides no prescription of probabilities for either Earman's "space invaders" or Norton's "motion down the dome," it is nonetheless always possible to impose a suitable measure. For example, a monotonically increasing measure that makes it increasingly more probable as time goes on, until a certain finite time greater than the start of motion time is reached,

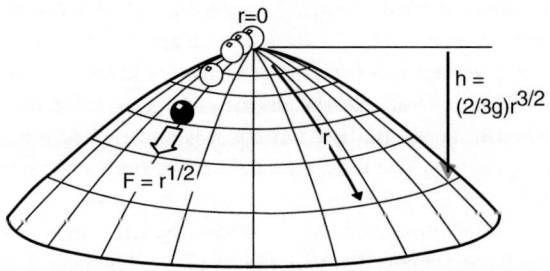

Figure 46.1 Norton's dome. (Redrawn from Norton, 2003)

then apportions whatever probability remains to the infinite amount of time left. Norton himself (2003, p. 10, footnote 8) proposes a different measure in agreement with exponential decay. This is sufficient to show that Laplace's thesis is false – objective chances are not incompatible with Newtonian dynamics. (Opponents of the compatibility of chance and determinism are likely to demur; in particular, they are likely to impose additional external constraints on the measures so as to rule out non-trivial chances for any motions on Norton's dome – yet, it remains relevant that those constraints are external, and that nothing in Newtonian mechanics per se seems to require them).

Secondly, there are of course "classical" theories other than Newtonian mechanics. Earman (1986, Ch. IV) argues that in fact the most hospitable environment for determinism is not Newtonian mechanics but the special theory of relativity. But, again, while the theory does not provide probabilities, e.g., for world-lines, nothing seems to preclude imposing them from outside the theory. As for classical statistical mechanics, the debate has centered upon whether it reduces thermodynamics and its arrow of time (see Chapter 29 in the present volume). The issue of reduction is tangential to our purposes, but the presumption that statistical mechanics is deterministic is of course not. There are some arguments to the effect that statistical mechanics is compatible with objective chance, and some classical phenomena – such as Brownian motion – seem to presuppose essential stochasticity in the motion of free particles. There is no space to pursue the matter further here, but many authors through the years have argued that statistical mechanics not only fails to be fully deterministic in Laplace's sense but in fact requires some probabilistic or stochastic assumptions to get its predictions off the ground (see Clark, 1987; or the fascinating discussion initiated by Albert, 2000).

Even if we were to suppose that both relativity and statistical mechanics are fully deterministic, Laplace's thesis does not follow unless Newtonian mechanics is so too. For all these theories assume that in the relevant limit (of small displacements in a flat Minkowski spacetime and microscopic particle-free motions), Newtonian mechanics does apply (to the slow motion of bodies in a flat spacetime relative to one another, and to the motion and interactions of free single-particle systems). These theories are therefore required to accept the possibility of deterministic chance in the limit. So Earman's and Norton's arguments cut to the bone of Laplace's thesis for all "classical" theories that accept Newtonian mechanics is the relevant limit. In such classical approximations, Laplace's theory cannot be true unless Newtonian mechanics precludes objective chance. (I am assuming that none of these classical approximations fundamentally replaces classical physics in its proper domain).

There is yet a third argument against Laplace's thesis. It is somewhat related to the previous two but works entirely within deterministic Newtonian physics. That is, suppose for the sake of argument that the universal dynamics of Newton's laws is indeed fully deterministic. It is then true that the present state of the universe determines every future state. And it is indeed true that the full and complete initial state of the universe suffices to fix completely every later state of the universe. It does not yet follow that there is no room for chance. Poincaré (1896) was perhaps the first to observe that a distribution function over the initial values of the dynamical variables of a deterministic system can give rise to probability distributions over the evolved values of related dynamical variables, provided some assumptions regarding the continuity and smoothness of both initial distribution and dynamics are met.

It stands to reason that if the initial distribution function characterizes or represents our lack of knowledge, the final chance distribution represents an epistemic probability. But as Poincaré himself noted, the initial distribution function typically characterizes not ignorance, but the actual frequencies of the initial variables. The dynamics then generate a final chance distribution that there is every reason to believe is objective (Poincaré, 1912). Not only that; Poincaré showed that – modulo the assumptions – the final chance distribution function is a characteristic of the system which is quite independent of the specific initial distribution over frequencies. Hence it is possible to assume any arbitrary initial function that fulfills the conditions in order to calculate the objective final probability distribution (what has come to be known as the "method of arbitrary functions"). Most games of chance satisfy Poincaré's continuity and smoothness assumptions. In a game of roulette – Poincaré's own example – the long-run probability of a red or black outcome is the same, irrespective of the frequency distribution over the direction and strength of the initial throw of the ball on the roulette – as long as the forces impinged in the initial throws satisfy the smoothness and continuity assumptions.

Strevens (2013) builds on Poincaré's theorem to argue that the causal mechanisms in the chance set up by themselves dynamically generate the resulting objective chance distribution. For instance, the dynamics of the shaking of a die in a cup is such that the resulting distribution of velocities and positions of the die as it leaves the cup satisfies all the dynamical conditions (microconstancy and microlinearity, in Strevens' terminology) to generate the familiar 1/6th chance for each side landing up – and this is so regardless of the precise initial conditions as the die is thrown into the cup. In other words: objective chance is a dynamical epi-phenomenon of complexity – quite independently of whether the underlying dynamics is deterministic or not.

46.4   Chance in Indeterministic Physics

Quantum mechanics (QM) is widely assumed to provide the paradigm examples of physical chance. It is supposed to furnish a radically distinct description that replaces classical mechanics at the fundamental level. Its inception 90 years ago certainly ushered in a golden era for physical indeterminism, and the amazing empirical successes of QM have often been assumed to sound the death knell for Laplace's thesis – by simply showing classical physics to be false. The uncertainty principle, as usually understood, prevents any quantum system from possessing values of conjugate observables simultaneously. Thus, no quantum particle may possess e.g. precise position and momentum simultaneously. More generally a system in a superposition state of eigenstates of a particular observable, may not be said to have any precise value of the observable in question – instead, QM predicts very precisely the probabilities for the different values of the observable. What value it ultimately has on measurement can only be left to "chance."

This kind of stochastic chance was introduced into QM by Max Born (1926) with his celebrated probability rule – according to which the normalized square modulus of the amplitude of the wavefunction provides the precise probabilities for the different values of the relevant observable. Its introduction was notoriously resisted, e.g. by Schrödinger and Einstein. The latter is famously supposed to have quipped something to the effect that: "God does not play dice" (Pais, 1982, Ch. 25). And not surprisingly, given the long shadow cast by Laplace's thesis, all these authors ipso facto rebelled against the indeterministic character of QM – and attempted to restore determinism instead. The most sophisticated such attempt has proven to be David Bohm's theory, nowadays known as Bohmian mechanics (see Chapter 15 in the present volume). It provides a Hamiltonian reformulation of QM in terms of 'hidden' variables. In Bohm's theory, quantum systems possess values of all their dynamical properties all the time, although these values are not knowable with precision. The uncertainty principle is thus understood as a statement not of ontological indeterminacy or chance, but epistemic limitation – it purports to show what limits there are on our knowledge of the evolution of a system at any time, given some initial uncertainty as to what the original values of its dynamical properties are. Laplace's shadow looms large here too: for an omniscient being, there would no uncertainty at any stage, since the Bohmian equations of motion are entirely in keeping with the deterministic character of classical Newtonian dynamics.

In other words, much discussion of stochastic chance in QM is predicated upon an understanding of classical dynamics that very much aligns it with Laplace's thesis. Both defenders and detractors of quantum chance share the view that a deterministic completion of QM in terms of hidden variables would compromise, if not simply eliminate, quantum chance. Yet, as noted in the previous section, classical determinism is not in fact incompatible with objective chance: Laplace's thesis may be false even if determinism is true. Not surprisingly, I shall argue, some of the discussions on the nature of quantum chance have similarly gone awry. Stochastic quantum chance is an explicit axiom in some interpretations of QM (such as collapse interpretations). But even those interpretations that do not make it explicit or axiomatic (such as Bohmian mechanics and the many-worlds interpretation), nonetheless allow quantum chance.

Collapse theories explicitly deny that the dynamical laws of QM are deterministic. Physical laws fix the evolution of the states of systems (where the state of a system provides a catalog of all its properties and their values at a given time). Now, according to collapse interpretations, quantum states are unlike classical states in that they are subject to two different kinds of evolution. The first kind of evolution is governed by the Schrödinger equation, which is a deterministic equation over the wavefunction: given the wavefunction at any time, Schrödinger evolution fixes uniquely the wavefunction at any later time. Yet, this is not a classical deterministic evolution because the wavefunction is not a literal or univocal description of the ontology of the quantum system (extant approaches include the "flash" and "mass density" ontologies – see e.g., Esfeld and Gisin, 2014 – and on neither of them does the wavefunction in fact represent a wave). Rather, as noted previously, Born's probability rule only lets us calculate probabilities for outcomes of measurements out of the wavefunction.

The standard rule for the interpretation of the wavefunction is the so-called eigenstate-eigenvalue ("e/e") link, according to which a quantum system may be said to possess a value of the property represented by a self-adjoint operator $\hat{O}$ if and only if the system is in an eigenstate of $\hat{O}$. For most states, this means that the system lacks a value for most of the relevant dynamical properties (all those represented by operators that do not commute with $\hat{O}$). Collapse interpretations then postulate the second kind of openly non-deterministic evolution in order to account for the fact that measurements of any dynamical property on a quantum system routinely obtain definite results. This is the "collapse" dynamical rule: a near-instantaneous evolution of the system that takes its state to the eigenstate of the relevant operator with a certain probability.

Collapse interpretations differ on how, when, and how often this type of indeterministic evolution takes place. The original collapse interpretation of Von Neumann (1932) invokes a principle of psycho-physical parallelism to suggest that collapse takes place whenever the measurement apparatus is apprehended (perceived) by a conscious observer. It is the interaction of the mind with matter that forces the indeterministic evolution. The Ghirardi-Rimini-Weber (GRW) interpretation asserts that collapses of the wavefunction occur spontaneously. The relaxation and free time parameters are sufficiently regular and sudden that no measurement interaction in the real world can ever detect a system in a state other than a 'collapsed' one (Ghirardi et al. 1986). In the Quantum State Diffusion approach collapses take place whenever a system interacts with its complex environment. Since, on this view, systems are typically open (Percival, 1999), environmental interaction is also typical, the many degrees of freedom of the environment dominate, and regular stochastic evolutions on the states of quantum systems are induced. Regardless of these differences, all collapse theories are committed to stochastic quantum chances (Frigg and Hoefer, 2007; Suárez, 2007).

Other interpretations of QM reject any collapse postulate or indeterministic evolution. They assert that the Schrödinger equation has no exceptions and Schrödinger evolution is the only kind. Most prominent among this is the Everett relative state formulation – sometimes known as the many-worlds interpretation. It too provides its own interpretation of the wavefunction and its connections with property values. Many worlds views assert the reality of a universal wavefunction – a giant superposition of tensor product states of the different interacting parts of the microscopic and macroscopic world alike. The appearances of definiteness are recovered in each branch of the universal wavefunction. Hence there is no indeterminism or collapse, and the quantum probabilities merely represent the weights that different appearances carry in the universal wavefunction. Still, questions must be raised about the meaning of these "weights." Putnam (2005) argues that many-worlds interpretations lack the resources to account for such weights as probabilities. Defenders of the many-worlds approach have tended to respond to such worries by appealing to decision-theoretic arguments (Deutsch, 1999; Wallace, 2010). But it is not at all clear that such appeals ultimately do away with quantum chance. For a start, it is implausible that such decision-theoretic arguments correspond to the subjective probabilities of any particular situated agent. More importantly, it is symptomatic that appeals to decision-theoretic reasoning often presuppose rather than eliminate objective chances. This is an objection that any attentive historian of probability will find famil-iar. Pascal founded modern decision theory with his wager (Hacking, 1972). But in order to show that theism was superior on decision-theoretic grounds he needed to make substantial assumptions regarding both the natural chance of God's existence and the objective utility derived from salvation. Contemporary defenses of decision-theoretic grounds for wavefunction realism often mirror Pascal's difficulties: objective quantum chances are presupposed rather than derived (Jansson, 2016). If so, far from avoiding stochastic quantum chances, many-worlds interpretations sneak them in through the back door.

The one version of QM that was constructed with the explicit aim of eliminating or rendering otiose any ontological quantum chance is Bohmian mechanics (Bohm, 1952; Bohm and Hiley, 1993). Yet, as I already noted, the argument from Bohmian mechanics against chance runs perilously close to the non-sequitur that assimilates the reality of chance to underlying indeterminism. Bohmian mechanics asserts that the only dynamical law is the Schrödinger equation – thus the wavefunction evolves deterministically. However, Bohmian mechanics also asserts that the quantum state is not the full state of a quantum object, which significantly includes hidden variables. These have their own deterministic dynamics. Poincaré's method of arbitrary functions then applies, so long as the initial values of the hidden variables are not uniquely distributed but meet the usual continuity assumptions. The frequencies of those values then suffice to generate objective probability distributions over the system's final values via the deterministic dynamics. In other words, it follows that any statistical distribution over the initial values of such hidden variables can generate objective chance distributions down the road (Suárez, 2015 argues further for an interpretation of these as manifesting underlying dispositions).

46.5 Conclusions

Objective chance appears to play a critical role in physics. Yet, Laplace's thesis states that in classical physics chance is rendered otiose to an omniscient being. The probability may only represent the cognitive shortcomings of an epistemically limited agent – his or her lack of knowledge. Despite its profound influence, Laplace's thesis does not hold in general. Classical physics does not require determinism; and determinism does not preclude chance. It follows that chance cannot be eliminated or done away with by simply re-formulating or modeling stochastic phenomena within classical physics. On the contrary, physical chance can be objective regardless of the dynamical character of physical laws. No wonder that the debate regarding the nature of chance – its metaphysics – shows no sign of abating. It certainly matters what physical chance is, for it impacts greatly our understanding of the underlying physics.

References

Albert, D. (2000). *Time and Chance*. Cambridge, MA: Harvard University Press.

Bohm, D. (1952). A suggested interpretation of the quantum theory in terms of hidden variables. *Physical Review*, 85: 166–193.

Bohm, D. and Hiley, B. (1993). *The Undivided Universe*. London: Routledge.

Born, M. (1926). Quantenmechanik der Stossvorgänge. *Zeitschrift für Physik*, 38: 803–827. Reprinted as "On the quantum mechanics of collisions", in Wheeler and Zurek (eds.), *Quantum Theory and Measurement*. Princeton, NJ: Princeton University Press, pp. 52–55.

Born, M. (1969). Is quantum mechanics in fact deterministic? In M. Born (ed.), *Physics in My Generation*. New York: Springer, pp. 78–83.

Carnap, R. (1950). *Logical Foundations of Probability*. Chicago: IL: Chicago University Press.

Clark, P. (1987). Determinism and probability in physics. *Proceedings of the Aristotelian Society*, 61: 185–210.

Deutsch, D. (1999). Quantum theory of probability and decisions. *Proceedings: Mathematical, Physical and Engineering Sciences*, 455: 257–283.

Earman, J. (1986). *A Primer on Determinism*. Dordrecht: Reidel.

Emery, N. (2015). Chance, possibility, and explanation. *British Journal for the Philosophy of Science*, 66(1): 95–120.

Esfeld, M. and Gisin, N. (2014). The GRW flash theory: A relativistic quantum ontology of matter in space-time. *Philosophy of Science*, 81: 248–264.

Frigg, R. and Hoefer, C. (2007). Probability in GRW theory. *Studies in History and Philosophy of Science*, 38(2): 371–389.

Ghirardi, G.C., Rimini, A. and Weber, T. (1986). Unified dynamics for microscopic and macroscopic systems. *Physical Review*, 34D: 470–491.

Gillies, D. (2000). *Philosophical Theories of Probability*. London: Routledge.

Hacking, I. (1972). The logic of Pascal's Wager. *American Philosophical Quarterly*, 9(2): 186–192.

Hacking, I. (1975). *The Emergence of Probability*. Cambridge: Cambridge University Press.

Hájek, A. (1997). 'Mises Redux' – Redux: Fifteen arguments against finite frequentism. *Erkenntnis*, 45: 209–227.

Hájek, A. (2009). Fifteen arguments against hypothetical frequentism. *Erkenntnis*, 70: 211–235.

Hald, A. (2003). *A History of Probability and Statistics*. New York: John Wiley and Sons.

Hoefer, C. (2003). Causal determinism. *Stanford Encyclopedia of Philosophy* (revised entry 2018). https://plato.stanford.edu/entries/determinism-causal/

Howson, C. and Urbach, P. (1989). *Scientific Reasoning: The Bayesian Approach*. La Salle, IL: Open Court. 2nd edition (1993).

Humphreys, P. (1985). Why propensities cannot be probabilities. *The Philosophical Review*, 94: 557–570.

Huygens, C. (1657). *De Ratiociniis in Ludo Aleae*, Ex officinia J. Elseviiri, translated into English and published as (1714) *The Value of All Chances in Games of Fortune; Cards, Dice, Wagers, Lotteries, &c. Mathematically Demonstrated*. London: S Keimer.

Jansson, L. (2016). Everettian quantum mechanics and physical probability: Against the principle of 'state supervenience'. *Studies in History and Philosophy of Modern Physics*, 53: 45–53.

Laplace, P.S. (1814). *Essai Philosophique sur le Probabilités*. Paris: Gauthier-Villars. Translated into English by F.W. Truscott and F.L. Emory, and published as Laplace, P.S. (1951), *A Philosophical Essay on Probabilities*. New York: Dover.

Lewis, D. (1986). A subjectivist's guide to objective chance. In D. Lewis (ed.), *Philosophical Papers*, Vol. II. Oxford: Oxford University Press, pp. 83–132.

Norton, J. (2003). Causation as folk science. *Philosophers' Imprint*, 3(4), November 2003, pp. 1–22.

Norton, J. (2008). An unexpectedly simple failure of determinism. *Philosophy of Science*, 75(5): 786–798.

Pais, A. (1982). *Subtle Is the Lord: The Science and the Life of Albert Einstein*. Oxford: Oxford University Press.

Peirce, C. (1910). On the doctrine of chances, with later reflections. In J. Buchler, (ed.), *Philosophical Writings of Peirce*. New York: Dover, pp. 157–173.

Percival, I. (1999). *Quantum State Diffusion*. Cambridge: Cambridge University Press.

Poincaré, H. (1896/1912). *Le Calcul des Probabilités*. Paris: Gauthier-Villars.

Popper, K. (1959). The propensity interpretation of probability. *British Journal for the Philosophy of Science*, 10: 25–42.

Putnam, H. (2005). "A philosopher looks at quantum mechanics (again). *British Journal for the Philosophy of Science*, 56(4): 615–634.

Reichenbach, H. (1949). *The Theory of Probability*. Los Angeles: University of California Press.

Strevens, M. (2013). *Tychomancy*. Cambridge, MA: Harvard University Press.

Suárez, M. (2007). Quantum propensities. *Studies in History and Philosophy of Modern Science*, 38: 418–438.

Suárez, M. (2014). A critique of empiricist propensity theories. *European Journal for the Philosophy of Science*, 4(2): 215–231.

Suárez, M. (2015). Bohmian dispositions. *Synthese*, 192(10): 3203.

Von Mises, R. (1928/1939). *Probability, Statistics, and Truth*. London-New York: Allen und Unwin. Reprinted (1957), New York: Dover.

Von Neumann, J. (1932/1955). *Mathematical Foundations of Quantum Mechanics*. Princeton, NJ: Princeton University Press.

Wallace, D. (2010). How to prove the Born rule. In S. Saunders, J. Barrett, A. Kent and Wallace, D. (eds.), *Many Worlds?: Everett, Quantum Theory, and Reality*. Oxford: Oxford University Press, pp. 227–263.

Mauricio Suárez

Further Reading from the Editors

This chapter should be read in conjunction with Chapter 38 by N. Emery in this volume on Chance and Determinism. For a beginner's guide to the philosophy of chance, see T. Handfield, *A Philosophical Guide to Chance: Physical Probability* (Cambridge University Press, 2012). For a survey of contemporary applications of objective probabilities in physics, see D. Wallace, "Probability in Physics: Stochastic, Statistical, Quantum," in A. Wilson (ed.), *Chance and Temporal Asymmetry* (Oxford: Oxford University Press, 2014). On the relationship between probability and symmetry, see M. Strevens, "Inferring Probabilities from Symmetries," *Noûs*, (2004). For discussions of the method of arbitrary functions in contemporary physics, see M. Strevens, *Tychomancy* (Cambridge MA: Harvard University Press, 2013) and M. Suárez, *The Philosophy of Probability and Statistical Modelling* (Cambridge: Cambridge University Press, 2020).

47

HOLISM

Richard Healey

47.1   Introduction

In slogan form, holism is the obscure thesis that the whole is more than the sum of its parts. Physics has been taken to exhibit holism of various kinds, associated with different ways of trying to make this thesis clear. As a first step, consider the claim that a whole has features that cannot be reduced to features of its parts. If we take the features of a physical system to include its behavior, and reduction to involve explanation, then we arrive at a thesis of explanatory or methodological holism. A methodological holist maintains that it is impossible to arrive at an adequate understanding of some physical system's behavior by analyzing that system into its constituent parts and appealing to the laws that apply to them.

Alternatively, reduction may be considered more a matter of metaphysics than epistemology. On a more metaphysical approach, holism is the view that the existence or features of some physical whole are not determined by the existence and features of its constituent parts. Some believe that quantum systems display this kind of holism. (Metaphysical holism may be compatible with the denial of methodological holism: some parts with their features may not determine what whole they compose and so what laws they obey.) After a section discussing methodological holism in physics, the rest of this entry considers whether our physical theories imply such metaphysical holism.

47.2   Methodological Holism

Methodologically, holism stands opposed to reductionism, somewhat as follows.

> *Methodological Holism*: An understanding of a certain kind of complex system is best sought at the level of principles governing the behavior of the whole system, and not at the level of the structure and behavior of its constituent parts.
> *Methodological Reductionism*: An understanding of a complex system is best sought at the level of the structure and behavior of its constituent parts.

Methodological reductionists favor an approach to (say) condensed matter physics which seeks to understand the behavior of a solid or liquid by applying quantum mechanics (say) to its constituent molecules, atoms, ions, or electrons. Methodological holists think this approach is misguided: As one condensed matter physicist put it "the most important advances in this area come about by the emergence of qualitatively new concepts at the intermediate or macroscopic levels — concepts which, one hopes, will be compatible with one's information about the microscopic constituents, but which are in no sense logically dependent on it" (Leggett, 1987, p. 113).

It is surprisingly difficult to find methodological reductionists among physicists. The elementary particle physicist Steven Weinberg, for example, is an avowed reductionist. He believes that by asking any sequence of deeper and deeper why-questions one will arrive ultimately at the same fundamental laws of physics. But this explanatory reductionism is metaphysical in so far as he takes explanation to be an ontic rather than a pragmatic category. On this view, it is not physicists but the fundamental laws themselves that explain why "higher level" scientific principles are the way they are. Weinberg (1992) explicitly distinguishes his view from methodological reductionism by saying that there is no reason to suppose that the convergence of scientific explanations must lead to a convergence of scientific methods.

47.3   Metaphysical Holism

The metaphysical holist believes that the nature of some wholes is not determined by that of their parts. One may distinguish three varieties of metaphysical holism: ontological, property and nomological holism.

> *Ontological Holism*: Some objects are not wholly composed of basic physical parts.
> *Property Holism*: Some objects have properties that are not determined by physical properties of their basic physical parts.
> *Nomological Holism*: Some objects obey laws that are not determined by fundamental physical laws governing the structure and behavior of their basic physical parts.

All three theses require an adequate clarification of the notion of a basic physical part. One way to do this would be to consider objects as basic, relative to a given class of objects subjected only to a certain kind of process, just in case every object in that class continues to be wholly composed of a fixed set of these (basic) objects. Thus, atoms would count as basic parts of hydrogen if it is burnt to form water, but not if it is converted into helium by a thermonuclear reaction. But this way excludes consideration of the metaphysician's time-slices and the physicist's point events (for example) as basic (spatio)temporal parts of an object. What counts as a part, and what parts are basic, are matters best settled in a particular context of enquiry.

Weinberg's (1992) reductionism is opposed to nomological holism in science. He claims, in particular, that thermodynamics has been explained in terms of particles and forces, which could hardly be the case if thermodynamic laws were autonomous. In fact thermodynamics presents a fascinating but complex test case for the theses both of property holism and of nomological holism.

One source of complexity is the variety of distinct concepts of both temperature and entropy that figure in both classical thermodynamics and statistical mechanics. Another is the large number of quite differently constituted systems to which thermodynamics can be applied, including not just gases and electromagnetic radiation but also magnets, chemical reactions, star clusters and black holes. Both sources of complexity require a careful examination of the extent to which thermodynamic properties are determined by the physical properties of the basic parts of thermodynamic systems.

A third difficulty stems from the problematic status of the probability assumptions that are required in addition to the basic mechanical laws in order to recover thermodynamic principles within statistical mechanics. An important example is the assumption that the micro-canonical ensemble is to be assigned the standard, invariant, probability distribution (see the chapters by Frigg and Werndl in this volume.) Since the basic laws of mechanics do not determine the principles of thermodynamics without some such assumptions (however weak), there may well be at least one interesting sense in which thermodynamics establishes nomological holism.

47.4   Property/Relational Holism

While some form of ontological holism has occasionally been considered (see Section 47.7), the variety of metaphysical holism most clearly at issue in quantum mechanics is property holism. But to see just what the issue is we need a more careful formulation of that thesis.

First the thesis should be contextualized to physical properties of composite physical objects. We are interested here in how far a physical object's *physical* properties are fixed by those of its parts, not in some more general determinationist physicalism. Next, to arrive at an interesting formulation of property holism we must accept that this thesis is not only concerned with monadic properties, and not concerned with all properties. The properties of a whole will typically depend upon relations among its proper parts as well as on properties of the individual parts. But if we are permitted to consider all properties and relations among the parts, then these trivially determine the properties of the whole they compose. For one relation among the parts is what we might call the complete composition relation — that relation among the parts which holds just in case they compose this very whole with all its properties.

Let us call a canonical set of properties and relations of the parts which may or may not determine the properties and relations of the whole the *supervenience basis*.[1] To avoid trivializing the theses we are trying to formulate, only certain properties and relations can be allowed in the supervenience basis. The intuition as to which these are is simple — the supervenience basis is to include just the qualitative intrinsic properties and relations of the parts, i.e., the properties and relations which these bear in and of themselves, without regard to any other objects, and irrespective of any further consequences of their bearing these properties for the properties of any wholes they might compose. Unfortunately, this simple intuition resists precise formulation. It is notoriously difficult to say precisely what is meant either by an intrinsic property or relation, or by a purely qualitative property or relation.[2] And the other notions appealed to in expressing the simple intuition are hardly less problematic. But, imprecise as it is, this statement serves already to exclude certain unwanted properties and relations, including the complete composition relation, from the supervenience basis.

Finally, we arrive at the following opposing theses:

Physical Property Determination: Every qualitative intrinsic physical property of and relation among some physical objects from any domain D subject only to type P processes supervenes on qualitative intrinsic physical properties and relations in the supervenience basis of their basic physical parts relative to D and P.

Physical Property Holism: There are some physical objects from a domain D subject only to type P processes, not all of whose qualitative intrinsic physical properties and relations supervene on qualitative intrinsic physical properties and relations in the supervenience basis of their basic physical parts (relative to D and P).

If we take the real state of some physical objects to be given by their qualitative intrinsic physical properties and relations, then physical property determination says (while physical property holism denies) that the real state of wholes is determined by the real state of their parts.

There is some residual unclarity in the notion of supervenience that figures in these theses. The idea is familiar enough — that there can be no relevant difference in objects in D without a relevant difference in their basic physical parts. I take it that the modality involved here is not logical but broadly physical or perhaps metaphysical. One might try to explicate the notion of supervenience here in terms of models of a true, descriptively complete, physical theory. At issue is whether such a physical theory has two (kinematically possible) models which agree on the qualitative intrinsic physical properties and relations of the basic parts of one or more objects in D but disagree on some qualitative intrinsic property or relation of these objects.

Teller (1989) has introduced the related idea of what he calls relational holism.

Relational Holism: There are non-supervening relations — that is, relations that do not supervene on the non-relational properties of the relata. (p. 214)

Within physics, this specializes to a close relative of physical property holism, namely:

Physical Relational Holism: There are physical relations between some physical objects that do not supervene on their qualitative intrinsic physical properties.

Physical property holism entails physical relational holism, but not vice versa. For suppose that F is some qualitative intrinsic physical property or relation of one or more elements of D that fails to supervene on qualitative intrinsic physical properties and relations in the supervenience basis of their basic physical parts. We may define a (non-intrinsic) physical relation R_F to hold of the basic physical parts of elements of D if and only if F holds of these elements. Clearly R_F does not supervene on the qualitative intrinsic physical properties of these parts. So physical property holism entails physical relational holism. But the converse entailment fails. For let R_G be a physical relation that holds between the basic parts of some elements in D when and only when those elements are in the relation S_G. R_G may fail to supervene on the qualitative intrinsic physical properties of these basic parts, even though all qualitative intrinsic physical properties and relations of elements of D (including S_G) supervene on the qualitative intrinsic physical properties and relations of their basic parts.

Physical relational holism seems at first sight too weak to capture any distinctive feature of quantum phenomena: even in classical physics the spatiotemporal relations between physical objects seem not to supervene on their qualitative intrinsic physical properties. But when he introduced relational holism Teller (1987) maintained a view of space-time as a quantity: In this view, spatiotemporal relations do in fact supervene on qualitative intrinsic physical properties of ordinary physical objects, since these include their spatiotemporal properties.

47.5 Holism in Classical Physics?

At least classically, spatial relations provide the only clear examples of qualitative intrinsic physical relations required in the supervenience basis for physical property determination/holism: other intrinsic physical relations seem to supervene on them. But if one thought a spatially localized object had a determinate value for a magnitude like mass only by virtue of its mass relations to other such objects elsewhere then one might decide to include those relations in the supervenience basis also (see Dasgupta (2013)).

The assumption that all physical processes are completely described by a local assignment of values to magnitudes forms part of the metaphysical background to classical physics. In Newtonian space-time, the kinematical behavior of a system of point particles under the action of finite forces is supervenient upon ascriptions of particular values of position and momentum to the particles along their trajectories. This supervenience on local magnitudes extends also to dynamics if the gravitational or other forces on the particles arise from fields defined at each space-time point.

The boiling of a kettle of water is an example of a more complex physical process. It consists in the increased kinetic energy of its constituent molecules permitting each to overcome the short range attractive forces which otherwise hold it in the liquid. It thus supervenes on the assignment, at each space-time point on the trajectory of each molecule, of physical magnitudes to that molecule (such as its kinetic energy), as well as to the fields that give rise to the attractive force acting on the molecule at that point. Such phase changes present problems when modeled within statistical mechanics (see Bangu's chapter in this volume and Butterfield (2011).

There is no hint of holism in examples like these. Instead, each seems to illustrate classical physics's conformity to physical property determination, insofar as physical properties and relations of the constituent particles or molecules determine the physical properties and relations of the physical system they compose. But these are not the only relevant physical systems here. There is also the field that gives rise to the forces to which these constituents are subjected.

A classical field is a physical system in which one or more physical magnitudes take values at points of space-time. Each actual value may be considered a property of that space-time point or of the part of the field that occupies it: in either case, the physical system that bears the property may be considered a basic part of a larger whole, of the field or the whole of space-time, respectively.

As an example of a process in Minkowski space-time (the space-time of Einstein's special theory of relativity,[3]) consider the propagation of an electromagnetic wave through empty space. This supervenes upon the values of local magnitudes — the components of the electromagnetic field tensor — at each point in space-time. Like the previous examples, this process apparently conforms to the principle of

(Spatiotemporal) Separability: Any physical process occupying space-time region R supervenes upon an assignment of qualitative intrinsic physical properties at space-time points in R.[4]

But one may question whether an assignment of values to basic physical magnitudes at space-time points amounts to or results from an assignment of qualitative intrinsic properties at those points. Take instantaneous velocity, for example: this is usually defined as the limit of average velocities over successively smaller temporal neighborhoods of that point. This provides a reason to deny that the instantaneous velocity of a particle at a point supervenes on qualitative intrinsic properties assigned at that point. Similar skeptical doubts can be raised about the intrinsic character of other "local" magnitudes such as the density of a fluid, the value of an electromagnetic field, or the metric and curvature of space-time (see Butterfield (2006)).

One response to such doubts is to admit to a minor consequent violation of spatiotemporal separability while introducing a weaker notion, namely

Weak Separability: Any physical process occupying space-time region R supervenes upon an assignment of qualitative intrinsic physical properties at points of R and/or in arbitrarily small neighborhoods of those points; along with a correspondingly strengthened notion of

Strong Nonseparability: Some physical process occupying a region R of space-time is not supervenient upon an assignment of qualitative intrinsic physical properties at points of R and/or in arbitrarily small neighborhoods of those points.

Strong nonseparability implies physical property holism in a physical system, some of whose basic physical parts are or occupy space-time points. But no such holism need be involved in a process that is nonseparable, but not strongly so, as long as the basic parts of the relevant system are themselves taken to be associated with neighborhoods rather than points.

Any physical process fully described by a local space-time theory will be at least weakly separable. For such a theory proceeds by assigning geometric objects (such as vectors or tensors) at each point in space-time to represent physical fields, and then requiring that these satisfy certain field equations. But processes described by theories of other forms will also be separable, including theories of collision which assign magnitudes to particles at each point on their trajectories. Of familiar classical theories, it is only theories involving direct action between spatially separated particles which involve nonseparability in their description of the dynamical histories of individual particles. But such processes are weakly separable within space-time regions that are large enough to include all sources of forces acting on these particles, so that the appearance of strong nonseparability may be attributed to a mistakenly narrow understanding of the space-time region these processes actually occupy.

The propagation of gravitational energy according to general relativity apparently involves strongly nonseparable processes, since gravitational energy cannot be localized (it does not contribute to the stress-energy tensor defined at each point of space-time as do other forms of energy — see Part III of this volume for further details). But even a non-locally-defined gravitational energy will still be supervenient upon the metric tensor defined at each point of the space-time, and so the process of its propagation will be weakly separable.

The definition of spatiotemporal separability becomes problematic in general relativity, since its application requires that one identify the same region R in possible space-times with different geometries. But while there is no generally applicable algorithm for making a uniquely appropriate identification, some identification may appear salient in a particular case. For example, one can meaningfully discuss whether or not the field is the same everywhere in the region outside the solenoid in the Aharonov-Bohm effect[5] with an increased current flowing, even though the size of the current will have a (tiny) influence on the geometry of that region. Note that the definition of nonseparability does not require that one identify the same point in space-times of distinct geometries.

While strictly outside the domain of classical physics, quantum phenomena such as the Aharonov-Bohm effect may be thought to manifest holism due to the failure of spatiotemporal separability even in *classical* electromagnetism. What lies behind this thought is that a satisfactory explanation of the effect apparently requires attribution of intrinsic electromagnetic properties to *loops* in space-time (see Healey (2016a), section 10). But electromagnetism is still weakly separable here, since one can take these loops to be arbitrarily small. So classical electromagnetism manifests no physical property holism, even in phenomena like the Aharonov-Bohm effect.

Separability would be a trivial notion if no qualitative intrinsic physical properties were ever assigned at space-time points or in their neighborhoods. But this would require a thorough-going relationism that took not just geometric but all local features to be irreducibly relational (cf. Esfeld (2004)).

47.6   Quantum Holism?

The main reason to believe quantum systems exhibit holism is quantum entanglement. In the first instance entanglement is a relation between not physical but mathematical objects representing the states of quantum systems. Different forms of quantum theory represent quantum states of various systems by different kinds of mathematical object. So the concept of quantum entanglement has been expressed by a family of definitions, each appropriate to a specific form and application of quantum theory (see Earman (2015)). The first definition (Schrödinger (1935)) was developed in the context of applications of ordinary non-relativistic quantum mechanics to pairs of distinguishable particles that have interacted, such as an electron and proton.

A hydrogen atom may be represented in ordinary non-relativistic quantum mechanics as a quantum system composed of two subsystems: an electron e and a nuclear proton p. When isolated, its quantum state may be represented by a vector Ψ in a vector space H constructed as a tensor product of spaces H_p and H_e used to represent states of e,p respectively. The states of e,p are then defined as entangled if and only if

$$\Psi \neq \Psi_p \otimes \Psi_e$$

for every pair of vectors Ψ_p, Ψ_e, in H_p and H_e respectively. This definition naturally generalizes to systems composed of n distinguishable particles. But alternative definitions seem preferable for a collection of indistinguishable particles — of electrons or of photons for example (see Ghirardi et al. (2002), Ladyman et al. (2013), Ladyman (this volume)).

It follows that the states of the electron and proton in an isolated hydrogen atom are entangled. But one may also represent the hydrogen atom as composed of a center-of-mass subsystem C and a relative subsystem R represented by vector states Ψ_C, Ψ_R in H_C, H_R respectively such that

$$\Psi = \Psi_C \otimes \Psi_R$$

If the state of the hydrogen atom is represented by Ψ then the states of quantum subsystems C,R are not entangled but the states of quantum subsystems p,e are entangled. This illustrates the important point that one cannot draw metaphysical conclusions from a mathematical condition of quantum entanglement without first deciding which quantum systems are physical parts composing some physical whole. It may seem natural to take the physical parts of a hydrogen atom to be an electron and a proton. But note that the state of an isolated hydrogen atom is usually represented by Ψ_R, and not by Ψ or Ψ_e.

Viewed as basic physical parts of a hydrogen atom represented by state Ψ, its electron and proton may be considered entangled component subsystems since Ψ cannot be expressed as a product of vectors representing the state of each. The electron and the proton may each be assigned a mixed state, but this pair of states does not uniquely determine the state Ψ: these quantum state assignments violate the following condition.

> *State Separability:* The state assigned to a compound physical system at any time is supervenient on the states then assigned to its component subsystems.

This may occasion no surprise if a system's quantum state merely specifies its chances of exhibiting various possible properties on measurement. But it may have metaphysical significance if its quantum state plays a role in specifying a system's categorical properties — its real state, so that the real state separability principle is threatened.[6]

> *Real State Separability Principle*: The real state of the pair AB consists precisely of the real state of A and the real state of B, which states have nothing to do with one another.[7]

His commitment to this last principle is one reason why Einstein denied that a physical system's real state is given by its quantum state (though it's not clear what he thought its real state consisted in). But according to (one variant of) the rival Copenhagen interpretation, the quantum state gives a physical system's real dynamical state by specifying that it contains just those qualitative intrinsic quantum dynamical properties to which it assigns probability 1. On this last interpretation, violation of state separability in quantum mechanics leads to physical property holism: it implies, for example, that a pair of fundamental particles may have the intrinsic property of being spinless even though this is not determined by the intrinsic properties and relations of its component particles.

If an entangled vector state of a pair of quantum systems violates state separability then there are measurements of dynamical variables (one on each subsystem) whose joint quantum probability distribution cannot be expressed as a product of probability distributions for separate measurements of each variable. Quantum theory predicts such a probability distribution for various types of spatially separated measurements of variables including spin and polarization components on a pair of entangled physical entities assigned such a state, and many of these distributions have been experimentally verified.[8]

Bell (1964, [2004]) considered a class of theories that introduce additional (the so-called) hidden variables λ and so permit a more complete description of such systems than that provided by their entangled quantum state. He reasoned that in order to reproduce all quantum predictions for the possible outcomes of spatially separated measurements on such an entangled pair, any such theory must yield a probability of 0 or 1 (conditional on the value of λ) in order to satisfy the locality condition that neither outcome depends on the choice of distant measurement. He then proved that the probabilistic predictions of any such local hidden variable theory must satisfy particular (Bell) inequalities violated by predictions of quantum theory for certain entangled state assignments.

In later work Bell (1990, [2004]) generalized this result to any theory of a certain type meeting a condition he called Local Causality which, he argued, quantum mechanics does not meet. He motivated this condition by appeal to an intuitive principle closely related to Einstein's (1948) principle of

> *Local Action*: If A and B are spatially distant things, then an external influence on A has no immediate effect on B.

Howard and Teller sought to defend Local Action against Bell's argument. Teller (1989) took violation of Bell's inequalities to be a manifestation of Relational Holism. Howard (1989) instead blamed their violation on the failure of the following separability condition:

> *Howard Separability*: The contents of any two regions of space-time separated by a non-vanishing spatiotemporal interval constitute separable physical systems, in the sense that (1) each possesses its own, distinct physical state, and (2) the joint state of the two systems is wholly determined by these separate states.

Henson (2013) and others have questioned this line of reasoning, including the conclusion that its appeal to holism or nonseparability helps one to understand how these correlations involving entangled systems come about without any action at a distance that violates relativity theory, Local Causality or Einstein's principle of Local Action.

While diverging from the Copenhagen prescription mentioned above, some modal interpretations[9] take real states of systems to be closely enough related to quantum states that entangled systems' violation of (quantum) state separability implies some kind of holism or nonseparability. Van Fraassen (1991, p. 294), for example, sees his modal interpretation as committed to "a strange holism" because it entails that a compound system may fail to have a property corresponding to a tensor product projection operator $P \otimes I$ even though its first component has a property corresponding to P. In fact, a clearer case of holism would arise in a modal interpretation that implied that the component lacked P while the compound had $P \otimes I$: *ceteris paribus*, that would provide an instance of physical property holism.

Healey (1989, 1994) offered a modal interpretation and used it to present a model account of Bell's puzzling correlations which portrays them as resulting from the operation of a process that violates both spatial and spatiotemporal separability. He argued that, on this interpretation, the nonseparability of the process is a consequence of physical property holism; and that the resulting account yields genuine understanding of how the correlations come about without any violation of relativity theory or Local Action. But subsequent work by Clifton and Dickson (1998) and Myrvold (2001) cast doubt on whether the account can be squared with relativity theory's requirement of Lorentz invariance. More recently Healey (2016b) has given a different account of how quantum theory may be used to explain violations of Bell inequalities consistent with Lorentz invariance and Local Action. This account involves no metaphysical holism or nonseparability.

In the context of unitary (Everettian) quantum theory, Wallace and Timpson (2010) advocate a form of space-time state realism according to which the properties of a region of space-time composed of two or more disjoint subregions fail to be determined by the properties of those subregions. Noting the implied physical property holism and failure of spatiotemporal separability, they defend this possibility against possible objections and argue for its positive advantages in the relativistic domain.

Esfeld (2001) takes holism, in the quantum domain and elsewhere, to involve more than just a failure of supervenience. He maintains that a compound system is holistic in that its subsystems themselves count as quantum systems only by virtue of their relations to other subsystems together with which they compose the whole.

47.7   Ontological Holism in Quantum Theory?

As applied to physics, ontological holism is the thesis that there are physical objects that are not wholly composed of basic physical parts. Views of Bohr, Bohm and others may be interpreted as endorsing some version of this thesis. In no case is it claimed that any physical object has nonphysical parts. The idea is rather that some physical entities that we take to be wholly composed of a particular set of basic physical parts are in fact not so composed: This may be because these do not exhaust their parts or because the entities actually *have* no independently existing proper parts.

According to Bohr's version of the Copenhagen interpretation,[10] one can meaningfully ascribe properties such as position or momentum to a quantum system only in the context of some well-defined experimental arrangement suitable for measuring the corresponding property. He used the expression "quantum phenomenon" to describe what happens in such an arrangement. In his view, then, although a quantum phenomenon is purely physical, it is not composed of distinct happenings involving independently characterizable physical objects — the quantum system on the one hand, and the classical apparatus on the other. And even if the quantum system may be taken to exist outside the context of a quantum phenomenon, little or nothing can then be meaningfully said about its properties. It would therefore be a mistake to consider a quantum object to be an independently existing constituent of the apparatus-object whole.

Bohm's reflections on quantum mechanics (Bohm, 1980; Bohm and Hiley, 1993) led him to adopt a more general holism. He believed that not just quantum object and apparatus, but any collection of quantum objects by themselves, constitute an indivisible whole. This may be made precise in the context of Bohm's (1952) interpretation of quantum mechanics (see Tumulka (this volume)) by noting that a complete specification of the state of the "undivided universe" requires not only a listing of all its constituent particles and their positions, but also of a field associated with the wave-function that guides their trajectories. If one assumes that the basic physical parts of the universe are just the particles it contains, then this establishes ontological holism in the context of Bohm's interpretation. But in an alternative view of the ontology of the Bohm theory, the wave function does not represent a physical field but merely specifies the dynamical law obeyed by the particles.

Some (Howard, 1989; Dickson, 1998) have connected the failure of a principle of separability to ontological holism in the context of violations of Bell inequalities. Howard considers his own Howard Separability principle to be a natural transposition of Einstein's Real State Separability to field theory, maintaining that Einstein uses this as a principle of individuation of physical systems, without which physical thought "in the sense familiar to us" would not be possible. Howard himself contemplates the possible failure of this principle for entangled quantum systems, with the consequence that these could no longer be taken to be wholly composed of what are typically regarded as their subsystems. Dickson (1998), on the other hand, uses these same concerns expressed by Einstein (1948) to argue that such holism is not "a tenable scientific doctrine, much less an explanatory one" (p. 156).

One may try to avoid the conclusion that experimental violations of Bell inequalities manifest a failure of Local Action by invoking ontological holism for events. The idea would be to deny that these experiments involve distinct, spatiotemporally separate, measurement events, and to maintain instead that what we usually describe as separate measurements involving an entangled system in fact constitute one indivisible, spatiotemporally disconnected, event with no spatiotemporal parts.[11] But such ontological holism conflicts with the criteria of individuation of events inherent in both quantum theory and experimental practice.

Notes

1 See, for example, Dean Rickles "Supervenience and Determination", *Internet Encyclopedia of Philosophy* http://www.iep.utm.edu/superven/ .
2 See, for example, Weatherson and Marshall (2017).
3 See Part II of this volume for more details.
4 For this and related notions of separability, see Healey (2016a).
5 See, for example, Feynman's *Lectures on Physics, Volume II* Section 15.5.
6 See Leifer (2014) for more on these alternative readings of the quantum state.
7 This principle was stated (in German) by Einstein in a letter to Schrödinger dated June 19th, 1935.
8 See, for example, Aspect et al. (1981, 1982), Henson et al. (2015), Giustina et al. (2015).
9 See Lombardi and Dieks (2017), especially section 6.

10 See the essays in Bohr (1934).
11 Butterfield (1992, pp. 41–42) attributes this suggestion to David Lewis in conversation: see also Lange (2002, pp. 286–297)

References

Aspect, A., Dalibard, J. and Roger, G. (1982). Experimental test of Bell's inequalities using time-varying analyzers. *Physical Review Letters*, 49: 1804–1807.

Aspect, A., Grangier, P. and Roger, G. (1981). Experimental realization of Einstein-Podolsky-Rosen-Bohm *Gedankenexperiment*: A new violation of Bell's inequalities. *Physical Review Letters*, 49: 91–94.

Bell, J.S. (1964). On the Einstein-Podolsky-Rosen Paradox. *Physics*, 1: 195–200: reprinted in Bell (2004), 14–21.

Bell, J.S. (1990). La nouvelle Cuisine. In A. Sarlemijn and P. Krose (eds.), *Between Science and Technology*. New York: Elsevier Science Publishing Company, pp. 97–115: reprinted in Bell (2004), 232–248.

Bell, J.S. (2004). *Speakable and Unspeakable in Quantum Mechanics* (2nd edition). Cambridge: Cambridge University Press.

Bohm, D. (1952). A suggested interpretation of the quantum theory in terms of 'hidden variables'. I. *Physical Review*, 85: 166–179.

Bohm, D. (1980). *Wholeness and the Implicate Order*. London: Routledge & Kegan Paul.

Bohm, D. and Hiley, B.J. (1993). *The Undivided Universe*. New York: Routledge.

Bohr, N. (1934). *Atomic Theory and the Description of Nature*. Cambridge: Cambridge University Press.

Butterfield, J. (1992). David Lewis meets John Bell. *Philosophy of Science*, 59: 26–43.

Butterfield, J. (2006). Against *Pointillisme* about mechanics. *British Journal for the Philosophy of Science*, 57: 655–689.

Butterfield, J. (2011). Less is different. *Foundations of Physics*, 41: 1065–1135.

Clifton, R. and Dickson, M. (1998). Lorentz invariance in modal interpretations. In D. Dieks and P. Vermaas (eds.), *The Modal Interpretation of Quantum Mechanics*, Dordrecht: Kluwer Academic, pp. 9–47.

Dasgupta, S. (2013). Absolutism vs comparativism about quantity. *Oxford Studies in Metaphysics*, 8: 105–148.

Dickson, M. (1998). *Quantum Chance and Non-Locality*. Cambridge: Cambridge University Press.

Dieks, D. and Lombardi, O. (2017). Modal interpretations of quantum mechanics. In E.N. Zalta (ed.), *The Stanford Encyclopedia of Philosophy* (Spring 2017 Edition). Available at: https://plato.stanford.edu/archives/spr2017/entries/qm-modal/.

Earman, J. (2015). Some puzzles and unresolved issues about quantum entanglement. *Erkenntnis*, 80: 303–337.

Einstein, A. (1948). Quantum mechanics and reality. *Dialectica*, 2: 320–324. (This translation from the original German by Howard 1989, 233–234.)

Esfeld, M. (2001). *Holism in Philosophy of Mind and Philosophy of Physics*. Dordrecht: Kluwer Academic Publishers.

Esfeld, M. (2004). Quantum entanglement and a metaphysics of relations. *Studies in History and Philosophy of Modern Physics*, 35: 601–617.

Ghirardi, G.C., Marinatto, L. and Weber, T. (2002). Entanglement and properties of composite systems. *Journal of Statistical Physics*, 108: 49–122.

Giustina, M. et al. (2015). Significant-Loophole-free test of Bell's theorem with entangled photons. *Physical Review Letters*, 115: 250401.

Healey, R.A. (1989). *The Philosophy of Quantum Mechanics: An Interactive Interpretation*. Cambridge: Cambridge University Press.

Healey, R.A. (1994). "Nonseparability and Causal Explanation", *Studies in History and Philosophy of Modern Physics* 25, 337–374.

Healey, R.A. (2016a) Holism and nonseparability in Physics. In E.N. Zalta (ed.), *The Stanford Encyclopedia of Philosophy* (Spring 2016 Edition). Available at: https://plato.stanford.edu/archives/spr2016/entries/physics-holism/.

Healey, R.A. (2016b) Locality, probability and causality. In M. Bell and S. Gao (eds.), *Quantum Nonlocality and Reality–50 Years of Bell's Theorem*. Cambridge: Cambridge University Press, pp. 172–194.

Henson, B. et al. (2015). Loophole-free Bell inequality violation using electron spins separated by 1.3 kilometers. *Nature*, 526: 682–686.

Henson, J. (2013). Nonseparability does not relieve the problem of Bell's theorem. *Foundations of Physics*, 43: 1008–1038.

Howard, D. (1989). Holism, separability and the metaphysical implications of the Bell experiments. In J. Cushing and E. McMullin (eds.), *Philosophical Consequences of Quantum Theory: Reflections on Bell's Theorem*. Notre Dame, Indiana: University of Notre Dame Press, pp. 224–253.

Ladyman, J., Linnebo, Ø. and Bigaj, T. (2013). Entanglement and non factorizability. *Studies in History and Philosophy of Modern Physics*, 44: 215–221.

Lange, M. (2002). *An Introduction to the Philosophy of Physics*. Oxford: Blackwell Publishing.

Leggett, A.J. (1987). *The Problems of Physics*. New York: Oxford University Press.

Leifer, M.S. (2014). Is the quantum state real? An extended review of ψ-ontology theorems. *Quanta*, 3(1): 67–155.

Myrvold, W. (2001). Modal interpretations and relativity. *Foundations of Physics*, 32: 1773–1784.

Schrödinger, E. (1935). Discussion of probability relations between separated systems. *Proceedings of the Cambridge Philosophical Society*, 31: 555–563.

Teller, P. (1987). Space–time as a physical quantity. In R. Kargon and P. Achinstein (eds.), *Kelvin's Baltimore Lectures and Modern Theoretical Physics*. Cambridge, MA: The MIT Press, pp. 425–447.

Teller, P. (1989). Relativity, relational holism, and the Bell inequalities. In J. Cushing and E. McMullin (eds.), *Philosophical Consequences of Quantum Theory: Reflections on Bell's Theorem*. Notre Dame, IN: University of Notre Dame Press, pp. 208–223.

Van Fraassen, B.C. (1991). *Quantum Mechanics: An Empiricist View*. Oxford: Clarendon Press.

Wallace, D. and Timpson, C. (2010). Quantum mechanics on spacetime I: Spacetime state realism. *British Journal for the Philosophy of Science*, 61: 697 727.

Weatherson, B. and Marshall, D. (2017). Intrinsic vs. extrinsic properties. In E.N. Zalta (ed.), *The Stanford Encyclopedia of Philosophy* (Fall 2017 Edition). Available at: https://plato.stanford.edu/archives/fall2017/entries/intrinsic-extrinsic/.

Weinberg, S. (1992). *Dreams of a Final Theory*. New York: Vintage Books.

Further Reading from the Editors

One of the first clear philosophical articulations of holistic implications of quantum theory is P. Teller, "Relational holism and quantum mechanics," *British Journal for the Philosophy of Science*, 37 (1986): 71–81. A contemporary debate is ongoing between the monist approach of J. Ismael and J. Schaffer, "Quantum holism: Nonseparability as common ground", *Synthèse* 197 (2020): 4131–4160 and the pluralist approach of C. Calosi and M. Morganti, "Interpreting quantum entanglement: Steps towards coherentist quantum mechanics", *The British Journal for the Philosophy of Science* (forthcoming). E. Adlam, "Spooky action at a temporal distance", *Entropy* 20(1), (2018): 41 considers the prospect of physical theories which are holistic with respect to cross-temporal patterns.

48

DIMENSIONS

Susan G. Sterrett

48.1 Introduction

This chapter concerns dimensions as the term is used in the physical sciences today. Quantities of the same kind have the same dimension. That two quantities have the same dimension, however, does not necessarily mean they are of the same kind. (An example of two different quantities that have the same dimension: heat capacity and entropy.) A more precise definition of dimension will be given within, which will reveal that the dimension of a quantity is not determined for a single quantity in isolation, but is determined relative to a system of quantities and the relations that hold between them.

The concept of dimension is considered to be more generally applicable than to just the physical sciences: some have speculated about the conditions under which the concept might (or might not) apply in social sciences such as sociology, anthropology, and economics (de Jong, 1967; McGuire, 1986; Wormser, 1986; Barenblatt, 2003; Barnett II, 2004; Folsom and Gonzalez, 2005; Grudzewski and Rosłanowska-Plichcińska, 2013). In physics, the concept of dimension is already institutionalized in that dimensions play a role in the foundation and development of some of the systems of units used in physics, particularly the SI (*Le Système International d'Unités*), also known as the International System of Units: dimensions are used in organizing the system of quantities that the units are to be associated with (BIPM, 2014, Section 1.3). In terms of logical priority, dimensions are logically prior to units. (Although the formulations of the concept of dimension we use do refer to units (Fourier, 1878; Barenblatt, 2003; Sterrett, 2009), all that the concept of dimension relies upon regarding units is the possibility of the existence of a system of units of a certain kind.)

Probably the most well-known use of dimensional analysis in physics, aside from checking that a formula does not violate dimensional homogeneity, is to reveal physical relationships. The principle that equations of physics must be dimensionally homogeneous can be used to deduce relationships between quantities. Dimensional analysis is somewhat like a logical or mathematical technique in that respect. The advantage of using dimensional analysis is that it requires less effort and information than the usual methods of mathematical derivation would require. This chapter indicates why that is so. Its aim is philosophical: to explain the role of dimension, so that it will become comprehensible that the basis for knowledge of such physical relationships as the use of dimensional analysis uncovers has already been incorporated in the notation of dimensions, quantities, and units employed in physics. Then it will be seen that methods using dimensional analysis, while elegant and impressively effective, are but a way of more fully developing the consequences of information that has already been built into the notation used in physics. We shall see that dimensional analysis is a deductive method of proving consequences that follow from the laws of physics (collectively) built into the system of units developed for quantitative sciences. It works as it does only when the system of units has the feature of being a "coherent system of units."

Another powerful application of dimensional analysis is to identify physically similar systems: a system's behavior described in terms of dimensioned quantities can usually be redescribed using a function whose arguments are dimensionless parameters. Informally, a system's behavior can be more succinctly and perspicaciously seen to follow from a set of dimensionless ratios, each of which expresses the interrelatedness of some of the quantities. The number of possible dimensionless parameters is unlimited and the set is nonunique; the set of dimensionless parameters is usually chosen such that the dimensionless parameters have some physical significance, such as a ratio of forces.

Dimensional analysis is employed in the method of physically similar systems to identify a suitable set of dimensionless parameters. A dimensionless parameter is a product of quantities (some may have negative exponents) such that in the corresponding dimensional formula, the exponent of each dimension is zero (e.g., Mach number, Reynolds number, Richardson number, and indefinitely many others) (Buckingham, 1914; Pankhurst, 1964; Sterrett, 2017a, 2017b). Mach number is one of the simplest examples, as it is the ratio of two velocities: the ratio of the velocity of something (e.g., fluid flow, a projectile) to the velocity of sound in that fluid, both with reference to the flow at the existent fluid conditions. The corresponding dimensional formula would be $[L] [L]^{-1}$, or $[L]^0$. For a physical system whose behavior is parameterized by Mach number, all the systems physically similar to it will exhibit that same behavior at the same numerical value of the Mach number. Similarity of systems is always relative to some kind of behavior (e.g., kinetic, dynamic, electrical, thermal, magnetohydrodynamic). Hence claims that two systems are physically similar systems always need to make clear what sort of behavior is claimed to be corresponding in both systems.

One useful application to take advantage of the characterization of a system's behavior in terms of dimensionless parameters is thus to build a system one can observe or experiment upon that bears this relation of "physically similar system" to a system that one wishes to study but that is experimentally inaccessible for some reason – and then to design an experiment or test such that the value of the dimensionless parameter(s) in the test and the actual system are the same. Thus one can conduct experiments on a system that is more convenient to experiment upon.

Of course in practice, one usually has to settle for approximations, but that does not affect the statement of the criterion of similarity; it just means something less than exact similarity has been achieved – just as when building a reproduction, exact identity is seldom if ever achieved. The criterion of similarity of system behavior is identity of dimensionless parameters – just as everyone ought to recognize the familiar point that similarity of triangles is a matter of identity of ratios. Similar triangles have the same angles, even if in practice when we draw triangles, we can only achieve approximating the angles. Stating the criterion in terms of parameters that are equal (invariant, if one figure is thought of as obtained via a transformation of the other) is a matter of stating similarity conditions correctly. Though two exactly similar systems might never exist in the world, it is still the case that similarity of systems is defined in terms of identity of dimensionless parameters. Period.

The method of physically similar systems can be used to establish the bases for the correlations that exist between analogue models and what they model in many areas of engineering and applied physics. It is not uncommon to build several different models of the same thing one is trying to model, in order to get good models of several different kinds of phenomenon associated with it. Being able to show that a certain physical system is equivalent to another with respect to a certain kind of behavior (dynamical, kinematical, buckling, and so on) can be useful in exploring theoretical questions as well. The potential of the method of physically similar systems is not nearly as well recognized for theoretical investigations, however.

Philosophers Robert Batterman (2002, 2010) and Marc Lange (2009) address the question of the explanatory role of dimensional analysis in science. Batterman's discussion follows along the general lines of Barenblatt (esp Barenblatt, 1996 on asymptotics), and Lange's of Bridgman. Each makes a novel programmatic suggestion about dimensional analysis in relation to explanation in philosophy of science. However, neither appeals to the features emphasized in this chapter, i.e., what is built into the notation and concept of dimensions and the method of dimensional analysis when a coherent system of units is used in physics. This chapter aims to show the overarching significance of these features.

Mario Bunge's aim in Bunge (1971) is closer to the aim of this paper. However, his approach is much more formal and set-theoretical. Also, his characterization of coherence of a system of units seems very different from that used in this paper, which makes it difficult to determine Bunge's view about the role of coherence to my points here.

48.2 Dimensions in Physics

Dimensional analysis does not have the institutional status of a discipline. As a result, when it is discussed and taught, it is usually done so in response to a need or opportunity arising in the context of a specific field or discipline. Consequently, references and publications that discuss dimensional analysis occur in a variety of disciplines (e.g., physics, philosophy, fluid dynamics, thermodynamics, mechanical engineering, and metrology), are often specialized treatments, and are authored by a diverse group of professionals and academics with differing backgrounds, goals, and training. Unfortunately, not all are correct concerning what they state about the theory of dimensions; some curating is required on the part of the reader who ventures into this literature.

In theoretical physics and philosophy of physics, the most well-known classic work is Percy Bridgman's *Dimensional Analysis* (Bridgman, 1922), which consists of five lectures given in 1920. Bridgman in turn acknowledges the work of Edgar Buckingham, whose most well-known paper on the topic, "On Physically Similar Systems: Illustrations of the Use of Dimensional Equations," which appeared in *Physical Review* in 1914, is also a classic of dimensional analysis (Buckingham, 1914). Since many readers will be familiar with, and rely upon, Bridgman (1922), it is mentioned here in order to point out how and where the views in this chapter diverge from it.

Bridgman's stated aim is to provide the fundamentals of dimensional analysis, his motivation being to "remove" some misconceptions about dimensional analysis that abound. In most uses of dimensions in physics, the existence and applicability of a certain set of dimensions are simply taken for granted. Occasionally, the nature and role of dimensions are even conflated with those of units. In contrast, there are some discussions of dimensions and dimensional analysis in physics in which the notion of dimension is explicitly examined and deliberately applied. According to his stated aim, Bridgman means to be doing the latter. Yet, Bridgman proceeds by assuming the existence of a system of measurement, including units. The topic of how dimensions are involved in the process of developing a system of units does not arise.

More recent books that are cited in the philosophy of physics literature include I. G. Barenblatt's *Scaling, Self-Similarity, and Intermediate Asymptotics* (1996), *Dimensional Analysis* (1987), and *Scaling* (2003), which, its author says "follows [Bridgman's *Dimensional Analysis*] in its general ideas." (Barenblatt, 1996, p. 14, 2003, p. 37) Barenblatt, too, recognizes the power and unrealized potential of dimensional analysis, writing that "using dimensional analysis, researchers have been able to obtain remarkably deep results that have sometimes changed entire branches of science" (Barenblatt, 1996, p. 1). He emphasizes that the power is in the use of dimensions rather than in advanced mathematics: "The mathematical techniques required to derive these results turn out to be simple and accessible to all." (Barenblatt, 1987, p. 1).

Barenblatt describes dimensional analysis as encapsulated in one simple idea, an idea about features of physical law, rather than about systems of units or about dimensions: "physical laws do not depend on arbitrarily chosen basic units of measurement" (Barenblatt, 2003, p. 9). From this, he says, follows the requirement on physical laws that they possess what he terms "generalized homogeneity or symmetry." Buckingham called it dimensional homogeneity, and credited Fourier with the idea; both Buckingham and Fourier pointed out that the requirement yields a system of linear equations (one for each dimension, expressing the constraint of homogeneity). The idea serves the same kind of purposes as grammatical rules about how to use passive and active verbs, or about subject and verb agreement do: the most obvious is to check that one's construction of a statement or equation is correct. De Clark (2016) remarks on this use of dimensional analysis: "When a result is found by inspection to violate this rule [dimensional homogeneity], this signals that a mistake must have been made in the course of the derivation. This is systematically used as a first check on results, and it seems fair to say that it constitutes the most widespread use of dimensional analysis in physics today" (p. 295). I mention, but have not emphasized, this use, as it sometimes encourages conflating the application of the principle of dimensional homogeneity with checking units. But there are other uses, too: just as, if one is deciphering a sentence that is encoded or is missing letters here and there, grammatical constraints help one infer results from partial information, so the principle of dimensional homogeneity is also helpful in deriving relationships expressed in physical equations using the principle of dimensional homogeneity.

Barenblatt's treatment of dimension follows Fourier's. However, like Bridgman, Barenblatt does not describe or address what is involved in the *development* of a system of units. Barenblatt is explicit about the existence, or possibility, of different classes of systems of units. He defines what it means for systems of units to be "of the same class" (Barenblatt, 2003). He states criteria for being *sufficient* – "sufficient for measuring the properties of the class of phenomena under consideration" – before a set of fundamental units can be called a system of units. But, as in Bridgman, there is very little room in the discussion for the role of dimension in the construction or development of a system of units to arise. Here I mean to do more than report on the current treatment of dimension in these classic works on dimensional analysis used in physics and philosophy of physics. I mean to go a bit beyond them by showing the fundamental role that the notion of dimension has in the construction of systems of units[1] – as well as the role it has in applying dimensional analysis (the requirement of dimensional homogeneity) – to obtain theoretical and practical results in physics.

There are other works regarded as classics in different academic communities. Here Birkhoff's (1960) *Hydrodynamics: A Study in Logic, Fact and Similitude* is worth mentioning in that Birkhoff regards inspectional analysis (proposed by Tatiana Ehrenfest-Afanassjewa (1916, 1926)) as more general than dimensional analysis. Palacios's *Dimensional Analysis* (1964) also developed some of Ehrenfest-Afanassjewa's ideas and is admired by some philosophers for its logical rigor. Mario Bunge (1971) has also provided a formal treatment. Other classics often cited in physics venues are Langhaar (1951), Pankhurst (1964), and Gibbings (2011). In the history of mathematics and metrology, John Roche's *The Mathematics of Measurement*, (1998) has become an important study.

48.2.1 Dimensions and Systems of Units

Critical evaluation of the nature and role of dimensions only seldom arises in the normal practice of science, but there are some kinds of developments in the history of science around which such discussions tend to cluster. They are: (i) when a new kind of quantity is to be included in the foundations of science (a key past example here would be temperature (Fourier's *Analytical Theory of Heat* (1878))), and (ii) when the foundations of a system of units are being reformulated (a past example here would be Giorgi's resolution of two conflicting systems of measurement (the electrostatic cgs and the magnetostatic cgs). The Giorgi system is considered the first forerunner of the current SI system. An especially important aspect of Giorgi's suggestion was to show that "the 'absolute' system of practical units could be combined with the three mechanical units meter, kilogram, and second to constitute a single coherent four-dimensional system of units" (Teichmann, 2001).

We have just seen another such point in history: the major revision of the International System of Units that took effect in May 2019. Though it is not the first, and may not be the last, such revision, it is of singular significance (BIPM, 2018). It is referred to as the Revised SI. It is of singular significance in that it achieves the centuries-old aim of a measurement system in which all the units are defined in terms of fundamental constants of nature.

The reason that the role of dimensions in constructing and developing a system of units will be discussed in this chapter, in spite of the absence of such discussions in the classic works mentioned above, is that in order to understand why dimensional analysis yields the information that it does, it is important to understand what is built into a system of units. The crucial feature built into a system of units that accounts for much of the power of the method of dimensional analysis is *coherence of a system of units.*[2]

One way to characterize coherence of a system of units is that "a system of units is coherent if the relations between the units used for the quantities are the same as the relations between the quantities in the fundamental equations of science." Here fundamental equations of science, including definitions, are the source of the relations between quantities (BIPM, 2014, section 1.1). Here is a very simple example: since velocity is proportional to length, and inversely proportional to time, with a constant of proportionality of unity, the requirement that the system be a coherent system of units requires that the units for velocity, length, and time must bear the same relations to each other that the quantities do. The same point about relations expressed by definitions holds for relations expressed by fundamental laws of nature, too. So, as de Clark illustrates using Newton's second law, which states that force equals mass times the second derivative of x with respect to time, $F = md^2x/d^2t$ "where F stands for the force on the object, m its mass, x its position and t for time – all understood as the physical quantities themselves. [...] The dimensions of a derivative are the same as that of a normal ratio; therefore, the dimensions of force are $[F] = MLT^{-2}$" (de Clark, 2016, p. 295). These are constraints we impose on the system of units in the course of developing it. Notice that there is an element of choice involved, i.e., applying the requirement of coherence of a system of units actually does some work in formulating a system of units. For, other than coherence of a system of units, there is no constraint that forces the choice of a constant of proportionality of 1 in this example, for instance.[3]

A different way to characterize the coherence of a system of units is in terms of base units and derived units. The coherence of a system of units on this characterization is stated as a requirement placed on the selection of base units: a requirement "to select basic units which would produce derived units by multiplication or division without introducing numerical coefficients; . . . " (Burdun, 1960, pp. 913–914). In this chapter, we will use the former characterization of coherence of a system of units, as it does not presume there must be base units in a system of units.

Once it can be presumed that the system of units in use in physics is a coherent system of units (in the sense that relations between units are the same as relations between quantities), dimensional analysis can be employed to spin out consequences that seem to have been generated out of practically nothing. This is because, as explained in earlier works (Sterrett, 2009), "if it is known that the system of units is coherent, it follows that the numerical equation has the same form as the fundamental relations [which are relations between quantities]." Thus, the requirement of dimensional homogeneity is about the logic of the interrelation of quantities used in physics, and it contains content gained empirically since the content of many of the fundamental laws of physics was obtained empirically.

A system of units constructed in this way thus has built into it many interdependencies between all the various units; equations expressing these relationships will thus have the same form as the interdependencies between quantities encoded in the fundamental laws or relations of physics. In a nutshell: the relations between units in a coherent system of units will reflect the relations between quantities (given by fundamental physical laws) and dimensional analysis (using the principle of dimensional homogeneity) will help spin out these relationships between quantities. Constructing a system of units is done as a whole; a system of units is much more than a collection of units constructed individually for each quantity. The units are interrelated, and they are interrelated in accordance with fundamental physical laws identified in the process of constructing that system of units (Sterrett, 2019).

48.2.2 *Dimensions and Quantity Equations*

We have seen that in a coherent system of units, the relations between units of the system reflect relations drawn from physical laws. This constraint can be applied even before identifying the value of any of the units (as will be explained below). Coherence of a system of units will be relative to the physical laws that one identifies as fundamental laws of physics, so the choice of physical laws on which to base a system of units is logically prior to the system of units. It is at this point in the logical reconstruction of a system of units that dimensions are involved.

It might seem puzzling at first to talk about laws or equations prior to having identified a system of units if one thinks that laws or equations must be expressed numerically. However, there is in fact a way to talk about kinds of quantities such as length, mass, time, charge, and temperature even before the units in which to measure them have been chosen: equations of physics can be regarded as expressing relations between quantities, or *quantity equations*. Alfred Lodge (1888, pp. 281–283) articulated this approach in the late nineteenth century. Lodge pointed out that, understood as quantity equations, the fundamental equations of a science "are independent of the mode of measurement of such quantities; much as one may say that two lengths are equal without inquiring whether they are going to be measured in feet or metres; and, indeed, even though one may be measured in feet and the other in metres." An article written to serve as a reference on the topic, "Units and Dimensions" (by Lodge's student Guggenheim), argues that "we are entitled to multiply together any two entities, provided our definition of multiplication is self-consistent and obeys the associative and distributive laws" (Guggenheim 1942). (Here it may be helpful to recall that it is possible to add, multiply, divide and take square roots using only an unmarked compass and straightedge (no ruler) in elementary school geometry.) The term "quantity equation" currently appears in the International Vocabulary of Metrology (BIPM 2012), which gives its meaning as "mathematical relation between quantities in a given system of quantities, independent of measurement units" (VIM 1.22 at http://jcgm.bipm.org/vim/en/1.22.html)."

The question of what counts as a quantity and how relations between quantities are determined is important to understanding what quantity equations are. The notion of dimension is used here. In this paper, we shall use brackets around an uppercase letter to indicate we are talking about dimension, rather than a specific quantity or a unit. Using brackets is a very common way to indicate dimension, although the SI (International System of Units) uses a different notation to indicate dimension instead (BIPM, 2014, Section 1.3 "Dimensions of Quantities," Sterrett, 2009, p. 812).

Fourier uses the notion of dimension to exact some discipline on his analysis, as he attempts to develop fundamental equations for a new science. He writes "the terms of one and the same equation could not be compared if they had not the same component of dimension. We have introduced this consideration into the theory of heat, in order to make our definitions more exact, and to serve to verify the analysis; it is derived from primary notions on quantities;" (Fourier, 1878, section 160, p. 128). The notion of dimension is introduced by considering how magnitudes change when the size of units is changed. For him, dimension is *with respect to a unit* for some kind of quantity; an example here is finding the dimension of the quantity "surface conducibility h" with respect to length (Fourier, 1878, section 161, p. 130). He reasons the exponent of dimension is -2. In our notation, this part of the formula would be written as $[L]^{-2}$ Fourier lays out the method for determining exponents of dimension in a way that is generalizable and thus provides a template, or canonical format.

Note that we might have two different quantities that yield the same exponents of dimension. This definitively illustrates the need for the notion of dimension in addition to the notion of quantity, and underscores the point that dimensions are not quantities. Lodge remarked on the fact that dimension does not determine the quantity, using the example of work and moment of force. Work and moment of force are two distinct quantities that have the same dimension. That two different quantities have the same dimension is not a reductio of the notion of dimension, however, pace some commentators, e.g., Emerson (2005). There is nothing objectionable about it; it is not inconsistent with the concept of dimension or of quantity.

Fourier uses the exponents of dimension in a statement expressing the constraint of dimensional homogeneity, i.e., that all the terms in the equation have the same dimension. He was presuming there would be a system of units with three fundamental units, one each for length, temperature, and duration. Each term in a physical law regarded as a quantity equation (which presumes no specific system of units) must have the same exponents of dimension, so there will be three equations (e.g., one for exponents with respect to length, one for exponents of dimension with respect to temperature, and one for exponents of dimension with respect to duration.) We would speak of the dimension for length, the dimension for temperature, and the dimension for duration.

The dimensional equation associated with a quantity equation is often not as informative about the physical situation as the quantity equation is. That is not a shortcoming, however, since its purpose is different. Dimensional equations might be thought of as showing grammatical features of physical equations, analogous to helping us meet grammatical constraints such as subject-verb agreement in an English sentence. The constraint, of course, is dimensional homogeneity, and it is a constraint on equations of physics that express relations between quantities. Consider a simple formula from mechanics: distance traveled equals acceleration times the square of the elapsed time, often written as $s = a\,t^2$. The dimension associated with distance would be simply [L]; the dimension associated with acceleration would be $[L][T]^{-2}$ (in Fourier's way of putting things, the exponent of the dimension of acceleration with respect to the unit for length is 1, and the exponent of the dimension of acceleration with respect to the unit for time is -2). The dimension for the square of elapsed time is, of course, $[T]^2$. The dimensional equation is thus $[L] = [L][T]^{-2}[T]^2$. If we were to write an equation for the exponents of dimension with respect to the unit of length we would get $1 = 1$, and for time, we would get $0 = -2 + 2$. Both of these are equalities, confirming that the formula is dimensionally homogenous.

Fourier's development of the notion of dimension was occasioned by his development of a theory of heat, which called for adding a new kind of quantity not included in mechanics. But it was not just an analytical theory of heat that called out for a new unit not already included in mechanics. The notions of quantity equations, coherence of a system of units, and dimensions were important in reaching a resolution in one of the most important advances toward the current SI system of units: the change from the CGS systems in absolute units, which used three base units, to a four-dimensional one, which went beyond the three-unit system used in mechanics and made room for a new unit specifically associated with electrodynamic phenomena.

48.2.3   Absolute Units

The word "absolute" in the term "absolute units" is meant to contrast with "relative" or "comparative." The historical context in which absolute units first arose was Humboldt's work aggregating magnetic measurements taken around the globe, which were relative measurements; he used the value of the instrument reading taken at the equator as a reference for other measurements taken with the same instrument. Gauss and Weber (Gauss, 1832/1995) devised a method of providing measurements of the earth's magnetism in dimensions used in mechanics: mass, length, and time (Main, 2007). The term absolute units is sometimes used even now to refer to a system of units that makes use of only units in these three dimensions. Such systems are sometimes referred to as 3D systems, and the base and derived units of CGS are referred to as "mechanical units."

Absolute units were also seen as having the benefit of relating a concept in a newer, comparatively unknown science about which there were still many questions about quantities and measurement, to the known science of mechanics about which there was much confidence: all the units of the new science could be conceived of in terms of units that were already familiar from mechanics.

Two additional systems of units, each of which used the three base units used in mechanics, arose: CGS-M for magnetic phenomena and CGS-E for electrical phenomena. All three – CGS, CGS-M, and CGS-E – were coherent systems of units, but they differed in the set of physical equations that served as the fundamental laws on which to base the coherence of their system. The dimensions of a quantity (e.g., charge) could differ depending on which of the systems one was working in, which was the source of much discussion. (Cornelius, 1964, 1965a, 1965b, 1965c) There has been a resurgence of interest in dimensional analysis among philosophers of science, and some rigorous critical-historical scholarship on the topic has appeared in just the last few years. One recent study notes that work on dimensional formulae by Maxwell is often left out of editions of his collected works, and describes some of the confusions that have been left behind as a result (Mitchell, 2017). In a scholarly critical-historical study of debates related to the dimensions of the magnetic pole in the nineteenth century, de Clark (2016) points out that the discussants sometimes did not even agree as to whether a purported problem concerning dimensions actually constituted an inconsistency or not (de Clark, 2016). As for related philosophical works, Nadine de Courtenay's (2015) "The double interpretation of the equations of physics" gives a critical-historical account from the standpoint of the kinds of equations used in physics, explaining the significance of the change from proportionalities to numerical equations. This is a very valuable piece of work for philosophers of science, as it explains how systems of units are constructed, and how the problem raised by the change made in the modern era to the use of numerical equations is now "hidden." These three recent papers provide socio-cultural and intellectual historical context that lend insight into the nature and significance of the theory of dimensions, and provide historical context that richly supplements other works on the topic, including Sterrett (2009, 2019, 2021) and the classic works on dimensional analysis mentioned earlier.

48.2.4   *Role of Dimension in the SI*

That the development of a coherent system of units involves identifying scientific equations on which the system is to be based, that there is sometimes a choice involved in the form of the equations chosen and that the choice can make a difference to the features of a system of units is evident from a recent official description of the SI (BIPM, 2014.) First, as to the need to identify scientific equations:

> In order to establish a system of units, such as the International System of Units, the SI, it is necessary first to establish a system of quantities, including a set of equations defining the relations between those quantities. This is necessary because the equations between the quantities determine the equations relating the units, ... (BIPM, 2014)

Thus, it is concluded that "in a logical development of this subject, the choice of quantities and the equations relating the quantities comes first, and the choice of units comes second." (BIPM, 2014, Section 1.1)

The relations between quantities are just the usual equations of science and engineering, the kind of relations found in textbooks and used on a daily basis in labs and engineering office. These have been identified in various standards and an effort made to consolidate them in an international standard (ISO 80000).

In the SI, the role of quantities, units, and dimensions is presented in official publications from the BIPM, as follows.

> physical quantities are organized in a system of dimensions. Each of the seven base quantities used in the SI is regarded as having its own dimension" [...] "All other quantities are derived quantities, which may be written in terms of the base quantities by the equations of physics. The dimensions of the derived quantities are written as products

of powers of the dimensions of the base quantities using the equations that relate the derived quantities to the base quantities. (BIPM, 2014, Section 1.3)

To a philosopher keen on identifying sources of knowledge in the practice of science, the preceding passage indicates even more than that the relations between the units are made to be the same as the relations between the quantities according to the equations of physics, science, and engineering with which we are familiar. It tells us that these relations are encoded in the notation for dimensions of the quantities used in physics. Thus, in using the notation for dimension in the SI and the principle of dimensional analysis to deduce equations, we are actually deducing some consequences of the equations of physics used in establishing the system of units as a coherent system of units.

Anyone who understands how mathematical derivations can likewise yield a great deal of interesting constructions and theorems from just a few postulates and definitions will understand what is meant here. Dimensional analysis is a genuine supplement to the methods of pure mathematics, though, in that it employs the notation of dimensions, and hence can take advantage of what is encoded, or built into, that notation, to obtain results from much less information. More generally, the fact that physical equations (equations of physics), must satisfy the requirement of dimensional homogeneity, provides an explanation of how mathematical equations as used in physics can be informative about the world. The answer does not lie in mathematics alone, but in an understanding of how physical equations, in conjunction with the use of a coherent system of units, will provide information about the world.

48.2.5 *Natural Units*

Not all science is carried out using the SI system, however. There are a number of alternatives loosely grouped under the rubric "natural unit systems." These systems of units employ non-SI units, and dimensional analysis is not always readily applicable when using equations expressed in terms of them (as explained in van Remortel, 2016).

One reason for favoring the use of such natural unit systems in certain subspecialties of physics is that the past SI systems of units are inconvenient for very large or very small scales. A more philosophical discomfort with the SI that led to preferring alternatives is that the determination of values for units in the SI relies on "precision measurements of standard prototypes [such as the standard kilogram or, in the past, the standard meter], objects, or systems that define a physical unit" (van Remortal, 2016), whereas, in natural unit systems, units are defined with reference to some fundamental physical constant. In some fields such as particle physics and general relativity, it is much more convenient to express results, theoretical as well as experimental, in reference to a fundamental physical constant of nature, such as the speed of light or mass of some subatomic particle (BIPM, 2014, Section 4.1).

48.2.6 *Role of Dimension in New SI*

The ideal that natural units aim at remains a desirable goal: a system in which units are defined in terms of fundamental constants of nature (Barrow, 1983). There have already been a number of changes in how specific units have been defined over the years. Here the meter is an iconic example, going from being defined with reference to the earth, then to a prototype kept in a vault, and ultimately to its present definition in terms of the velocity of light (BIPM, 2017). At the other end of the extreme is the kilogram, which, until the very recent change that took effect in May 2019, was still defined in terms of an artifact, which was a prototype kilogram kept in a vault. However, all the work required to enable a major overhaul of the way units are defined has been carried out and was implemented in the latest revision of the SI (May 2019). The ideal of defining all units in terms of fundamental constants of nature (per Box 48.1) has thus finally been achieved (Quinn, 2011).

Box 48.1 Definition of the International System of Units (from Bureau International des Poids et Mesures)

2.2 Definition of the SI

As for any quantity, the value of a fundamental constant can be expressed as the product of a number and a unit.

The definitions below specify the exact numerical value of each constant when its value is expressed in the corresponding SI unit. By fixing the exact numerical value, the unit becomes defined, since the product of the *numerical value* and the *unit* has to equal the *value* of the constant, which is postulated to be invariant.

Quotients of SI units may be expressed using either a solidus (/) or a negative exponent ()

For example,
$m/s = m\,s^{-1}$
$mol/mol = mol\,mol^{-1}$

The seven constants are chosen in such a way that any unit of the SI can be written either through a defining constant itself or through products or quotients of defining constants.

The International System of Units, the SI, is the system of units in which

- **the unperturbed ground state hyperfine transition frequency of the caesium 133 atom, $\Delta \nu_{Cs}$, is 9 192 631 770 Hz,**
- **the speed of light in vacuum, c, is 299 792 458 m/s,**
- **the Planck constant, h, is 6.626 070 15 × 10^{-34} J s,**
- **the elementary charge, e, is 1.602 176 634 × 10^{-19} C,**
- **the Boltzmann constant, k, is 1.380 649 × 10^{-23} J/K,**
- **the Avogadro constant, N_A, is 6.022 140 76 × 10^{23} mol^{-1},**
- **the luminous efficacy of monochromatic radiation of frequency 540 × 10^{12} Hz, K_{cd}, is 683 lm/W,**

where the hertz, joule, coulomb, lumen, and watt, with unit symbols Hz, J, C, lm, and W, respectively, are related to the units second, metre, kilogram, ampere, kelvin, mole, and candela, with unit symbols s, m, kg, A, K, mol, and cd, respectively, according to Hz = s^{-1}, J = kg m^2 s^{-2}, C = A s, lm = cd m^2 m^{-2} = cd sr, and W = kg m^2 s^{-3}.

The value of the fundamental constants are set in terms of the units to be defined, and the units are collectively defined in terms of the fundamental constants as a result. For example, as shown in Box 48.1, the velocity of light is no longer a measured quantity. (Rather, its value is set in units of second and meter, which has the effect of defining the meter in terms of it as a fundamental constant of nature. All seven units that have historically been designated as base units are collectively defined in terms of the seven fundamental constants shown in Box 48.1.) Since the value of the fundamental constants is set, rather than measured, the values of the fundamental constants have zero uncertainty.

The official wording of the revised SI system, including the definition of the units, took effect in May 2019 (Box 48.1). Here we are especially interested in how dimensions will be treated since the switch to defining all units in terms of fundamental constants. The Ninth SI Brochure reads as follows:

2.3.3 Dimensions of quantities

Physical quantities can be organized in a system of dimensions, where the system used is decided by convention. Each of the seven base quantities used in the SI is regarded as having its own dimension.

48.2.7   *Importance of Dimensions in Philosophy of Science*

Dimensions, units, and quantities are distinct notions. We have seen above how they are related in the design of coherent systems of units; the account involves the equations of physics. When the use of a coherent system of units can be presumed, dimensional analysis is a powerful logico-mathematical method for deriving equations and relations in physics, and for parameterizing equations in terms of dimensionless parameters, which allows identifying physically similar systems. The source of the information yielded by dimensional analysis is not yet well understood in the philosophy of physics. This chapter has attempted to reveal not only the role of dimensions in applications of dimensional analysis to obtain information by involving the principle of dimensional homogeneity but also the role of dimensions in encoding information about physical relationships in the language of dimensions, specifically via the feature of coherence of a system of units.

Philosophers of mathematics and philosophers of science have been concerned to address the question of the effectiveness of mathematics in science. Given the role of dimensions as explained in this chapter, an important part of the answer is that the notation of dimensions is a powerful means of including the content of physical theories into a system of units, and of providing the means of deducing valuable information about the consequences of them, collectively, afterward. Thus, no philosophical analysis of the question of the applicability of mathematics to science is complete without including dimensions and dimensional analysis in the picture.

Notes

1 De Courtenay (2015) is a rare and valuable paper in that it is devoted to a critical examination of the construction of a system of units; the treatment is broadly in accordance with the view in Roche (1998).
2 I have argued for this point regarding similarity and dimensional analysis in Sterrett (2009).
3 The element of choice involved is emphasized in Lodge (1888) and discussed in Sterrett (2009).

References

Barenblatt, G.I. (1987). *Dimensional Analysis*. Trans. from the Russian by Paul Makinen. London: Gordon and Breach Science Publishers.

Barenblatt, G.I. (1996). *Scaling, Self-Similarity, and Dimensional Analysis*. Cambridge Texts in Applied Mathematics. Cambridge: Cambridge University Press.

Barenblatt, G.I. (2003). *Scaling*. Cambridge: Cambridge University Press.

Barnett II, W. (2004). Dimensions and economics: Some problems. *Quarterly Journal of Austrian Economics*, 7(1): 95–104.

Barrow, J.D. (1983). Natural units before planck. *Quarterly Journal of the Royal Astronomical Society*, 24: 24–26.

Batterman, R.W. (2002). *The Devil in the Details: Asymptotic Reasoning in Explanation, Reduction, and Emergence*. Oxford: Oxford University Press.

Batterman, R.W. (2010). On the explanatory role of mathematics in empirical science. *British Journal for the Philosophy of Science*, 61(1): 1–25.

BIPM (Bureau International des Poids et Mesures). (2014). *SI Brochure, The International System of Units (SI), aka Le Système international d'unités (SI)* (8th edition, 2006; updated 2014). Available at: bipm.org.

BIPM (Bureau International des Poids et Mesures). (2016). *Draft SI Brochure, The International System of Units (SI), aka Le Système international d'unités (SI)* (9th edition, 2016). Available at: bipm.org.

BIPM (Bureau International des Poids et Mesures). (2017). *The BIPM and the Evolution of the Definition of the Metre*. Available at: bipm.org downloaded from http://www.bipm.org/en/measurement-units/history-si/evolution-metre.html on July 4, 2017.

BIPM (Bureau International des Poids et Mesures). (2019). *SI Brochure, The International System of Units (SI), aka Le Système international d'unités (SI)* (9th edition, 2019). Available at: bipm.org.

Birkhoff, G. (1960). *Hydrodynamics: A Study in Logic, Fact, and Similitude*. Princeton, NJ: Princeton University Press.

Bridgman, P. (1922). *Dimensional Analysis*. New Haven, CT: Yale University Press.

Buckingham, E. (1914). On physically similar systems; illustrations of the use of dimensional equations. *Physical Review*, 4(4): 345–376.

Bunge, M. (1971). A mathematical theory of the dimensions and units of physical quantities. In M. Bunge (ed.), *Problems in the Foundations of Physics*, Vol. 4. New York: Springer, pp. 1–16.

Burdun, G.D. (1960). International system of units. *Measurement Techniques*, 3(11): 913–919.

Cornelius, P., de Groot, W. and Vermeulen, R. (1964). Quantity equations and system variation in electricity. *Physica*, 30: 1446–1452.

Cornelius, P., de Groot, W. and Vermeulen, R. (1965a). Quantity equations, rationalization and change of number of fundamental quantities I. *Applied Scientific Research, Section B*, 12(1): 1–17.

Cornelius, P., de Groot, W. and Vermeulen, R. (1965b). Quantity equations, rationalization and change of number of fundamental quantities II. *Applied Scientific Research, Section B*, 12(1): 235–247.

Cornelius, P., de Groot, W. and Vermeulen, R. (1965c). Quantity equations, rationalization and change of number of fundamental quantities III. *Applied Scientific Research, Section B*, 12(1): 248–265.

de Clark, S.G. (2016). The dimensions of the magnetic pole: a controversy at the heart of early dimensional analysis. *Archive for the History of the Exact Sciences*, 70: 293–324.

de Courtenay, N. (2015). The double interpretation of the equations of physics. In O. Schlaudt and L. Huber (eds.), *Standardization in Measurement: Philosophical, Historical and Sociological Issues*. London: Pickering & Chatto Publishers. pp. 53–68.

de Jong, F. and Quade, W. (1967). *Dimensional Analysis for Economists. With a Mathematical Appendix on the Algebraic Structure of Dimensional Analysis*. Amsterdam: North-Holland Publishing.

Ehrenfest-Afanassjewa, T. (1916). On Mr RC Tolman's principle of similitude. *Physical Review*, 8(1): 1–7.

Ehrenfest-Afanassjewa, T.A. (1926). Dimensional analysis viewed from the standpoint of the theory of similitudes. *Philosophical Magazine*, Series 7, 1: 257–272.

Emerson, W.H. (2005). On the concept of *dimension*. *Metrologia*, 42: L21–L22.

Folsom, R.N. and Gonzalez, R.A. (2005). Dimensions and economics: Some answers. *The Quarterly Journal of Austrian Economics*, 8(4): 45–65.

Fourier, J.F. (1878). *The Analytical Theory of Heat*. Trans. with notes by Alexander Freeman. London: Cambridge University Press.

Gauss, C.F. (1832/1995). *The Intensity of the Earth's Magnetic Force Reduced to Absolute Measurement*. Trans. from the German by Susan P. Johnson (1995). The treatise "Intensitas vis magneticae terrestris ad mensuram absolutam revocata" was read by Gauss at the Goettingen Gesellschaft der Wissenschaften (Royal Scientific Society) on December 15, 1832, and printed in Volume 8 of the treatises of this society, pp. 3–44. The translation from the Latin to the German was provided by Herr Oberlehrer Dr. Kiel in Bonn. Edited by E. Dorn, Leipzig, Wilhelm Engelmann Verlag, 1894.

Gibbings, J.C. (2011). *Dimensional Analysis*. London: Springer Verlag.

Grudzewski, W.M. and Rosłanowska-Plichcińska, K. (2013). *Application of Dimensional Analysis to Economics*. Amsterdam: IOS Press.

Guggenheim, E.A. (1942). Units and dimensions. *The Philosophical Magazine* 33(222): 479–496.

Lange, M. (2009). Dimensional explanations. *Noûs*, 43: 742–775.

Langhaar, H.L. (1951). *Dimensional Analysis and Theory of Models*. New York: Wiley & Sons.

Lodge, A. (1888). The multiplication and division of concrete quantities. *Nature*, 38: 281–283.

Mitchell, D.J. (2017). Making sense of absolute measurement: James Clerk Maxwell, William Thomson, Fleeming Jenkin, and the invention of the dimensional formula. *Studies in History and Philosophy of Modern Physics*, 58: 63–79.

Palacios, J. (1964). *Dimensional Analysis*. Trans. from the Spanish by P. Lee and L. Roth. London: Macmillan.

Pankhurst, R.C. (1964). *Dimensional Analysis and Scale Factors*. London: Chapman & Hall.

Quinn, T. (2011). *From Artefacts to Atoms: The BIPM and the Search for Ultimate Measurement Standards*. New York: Oxford University Press.

Roche, J. (1998). *The Mathematics of Measurement*. New York: Springer.

Sterrett, S.G. (2009). Similarity and dimensional analysis. In A. Meijers (ed.), *Handbook of the Philosophy of Science*, Volume 9: Philosophy of Technology and the Engineering Sciences. Amsterdam: Elsevier, pp. 799–824.

Sterrett, S.G. (2017a). Physically similar systems: A history of the concept. In L. Magnani and T. Bertolotti (eds.), *Springer Handbook of Model-Based Science*. Leipzig: Springer International Publishing, pp. 377–412.

Sterrett, S.G. (2017b). Experimentation on analogue models. In L. Magnani and T. Bertolotti (eds.), *Springer Handbook of Model-Based Science*. Leipzig: Springer International Publishing, pp. 857–878.

Sterrett, S.G. (2019). Relations between units and relations between quantities. In N. de Courtenay, O. Darrigol and O. Schlaudt (eds.), *The Reform of the International System of Units (SI): Philosophical, Historical, and Sociological Issues*. London: Routledge, pp. 99–124.

Sterrett, S.G. (2021). Scale modelling. In D.P. Michelfelder and N. Doorn (eds.), *Handbook of the Philosophy of Engineering*. New York: Routledge, pp. 394–407.

Main, S.R.C. (2007). Gauss' determination of absolute intensity. In D. Gubbin and E. Herrera-Bervera (eds.), *Encyclopedia of Geomagnetism and Paleomagnetism*. Dordrecht: Springer, pp. 278–279.

McGuire, D.P., Pearson, J.J. and Dobbert, M.L. (1986). Dimensional analysis in cultural anthropology: Reply to Wormser. *American Anthropologist*, 88(3): 703–706.

Teichmann, H. (2001). *Celebrating the Centenary of SI (1901 - 2001): Giovanni Giorgi's Contribution and the Role of the IEC*. International Electrotechnical Commission, Geneva, Switzerland, p. 17.

van Remortel, N. (2016). The nature of natural units. *Nature Physics*, 12: 1082.

Wormser, A.J. (1986). Dimensional analysis: A critique and cautionary note. *American Anthropologist* New Series, 88(2): 448–452.

Further Reading

N. DeCourtenay, "The Double Interpretation of the Equations of Physics and the Quest for Common Meanings," in ed. O. Schlaudt and L. Huber, *Standardization in Measurement: Philosophical, Historical and Sociological Issues* (London: Routledge, 2015) is a rare and valuable study of the role of dimensions and the coherence of a system of units in the establishment of a framework for measurement that allows the use of numerical equations in physics. J. Roche, *The Mathematics of Measurement* (New York: Springer, 1998) is an unparalleled intellectual history on the topic. E. Buckingham's classic "On physically similar systems: illustrations of the use of dimensional equations," *Physical Review* 4(435) (1914) is foundational and bears repeated study. His work on dimensions preceded and informed the more well-known book by Percy Bridgman on the topic. S. de Clark, "The dimensions of the magnetic pole: a controversy at the heart of early dimensional analysis," *Archive for History of Exact Sciences* 70 (2016): 293–324 is a fascinating and enlightening historical study. J. D. Barrow, "Natural Units before Planck," *Quarterly Journal of the Royal Astronomical Society* 24: 24–26 (1983) explores the history of the idea of a privileged natural unit.

49

FUNDAMENTALITY

Steven French

49.1 Introduction

The idea that there is some fundamental "level" or "ground" where our description of the world bottoms out has acquired the status of 'the received view' in metaphysics (a classic statement of this view can be found in Oppenheim and Putnam (1958); for a more recent critical defense, see Cameron, 2008). Typically this view is cashed out in terms of some set of 'basic building blocks' populating this level, which sits at the bottom of a hierarchy ordered according to some set of compositional principles. These fundamental building blocks are thus taken to have some form of "ultimate" ontological priority with regard to everything else in the hierarchy. In this chapter I shall consider two kinds of threats to this view: the first comes from arguments against the idea of such a level in general, whereas the second concerns the nature of these occupants. As we'll see, both these threats become entwined in the context of modern physics but I'll conclude with a suggestion as to how this "received view" may be maintained in this context.

My discussion can be situated in the context of a vigorous debate over the relationship between metaphysics and science. In particular, it has been claimed that much of current metaphysics is too far removed from modern science and dependent on intuitions and "aprioristic" reasoning (see Ladyman et al., 2007, Ch. 1). Instead, it is argued, metaphysics should be more 'naturalistic' in the sense of drawing upon and responding to the results of science. In response, some metaphysicians have insisted that metaphysics has to do with what is *possible*, rather than what is actual and so should not be required to accommodate the impact of recent scientific developments (Lowe, 1998). In what follows the broad framework that I shall adopt with regard to this relationship will be that set out in (French and McKenzie, 2012, 2015): on the one hand, if metaphysics is to be understood as saying something about reality, then the implications of modern science and, in particular, physics need to be properly appreciated and this in turn will impact on certain "paradigmatic" metaphysical accounts, such as the received view, above; on the other, one does not have to accept that non-naturalistic metaphysics should be dismissed or even "discontinued" as some have pressed (Ladyman et al., op. cit.), since it can still serve as a kind of "toolbox" from which various devices and maneuvers can be extracted and put to use. In what follows I hope to illustrate both aspects of this framework.

49.2 The Idea of the Fundamental

Let us begin by considering the alternative to the 'received view': there is no such 'bottoming out'. There are two obvious ways to conceive of such an alternative: first, that any ontological priority is not ultimate; that is, nothing is fundamental. If one likes the picture of reality as organized into hierarchical levels (a contested picture for sure) one can understand this in terms of the hierarchy either not bottoming out or not topping out, or both – it just "keeps on going" "down" or "up" or in both directions, as it were. The first option generates what is typically called a "gunky" ontology; the second a "junky" one, while the third yields a "hunky" ontology (see Tahko 2018, pp. 5–6). The "gunky" view seems to have received more attention, perhaps because, in the context of the history of physics, it seems to be a "live" possibility (thus to jump ahead to the discussion of the second issue above, Saunders has suggested that reality could be structural "all the way down" (Saunders, 2003)). It is obviously more difficult to similarly naturalize the "junky" and "hunky" ontologies as "the Universe" seems a natural "topping out" point. One could perhaps advert to recent considerations of the "multiverse" in cosmology but leaving aside the issue of whether this truly meshes with a junky metaphysics, the suggestion currently remains extremely speculative.

The second way of conceiving reality as not involving anything fundamental is to deny the idea of ontological priority itself. Doing so across the sciences is obviously problematic: although most commentators acknowledge that establishing the reduction of biological theories and models, say, to chemical or physical ones is notoriously difficult, most will also agree that there is an ontological reduction, from proteins to molecules and thence to atoms and elementary particles. And this reduction will obviously follow the "chain" of priority. There is more to be said, of course, but here I will simply assume such a reduction in what follows.[1]

What sorts of arguments, then, might be deployed in defense of the denial of ontological priority? Schaffer, famously, has offered an inductive one, from the history of science:

> The history of science is a history of seeking ever-deeper structure. We have gone from "the elements" to "the atoms" to the subatomic electrons, protons and neutrons, to the zoo of "elementary particles," to thinking that the hadrons are built out of quarks and now we are promised that these entities are really strings, while some hypothesize that the quarks are built out of preons (in order to explain why quarks come in families). Should one not expect the future to be like the past? (Schaffer, 2003, p. 503)

Stated as such, the argument is really sweeping, taking us from the Greeks, through Dalton, to Thomson, Chadwick, Gell-Mann, and beyond. Perhaps it is too sweeping for many tastes, covering too many different kinds of putative fundamental entity, conceptualized in too many different kinds of ways, from "the elements" to strings. In that case, just consider the history of the last 50 years or so and the way in which order was brought to the so-called "particle zoo" of the 1960s via the quark model. Originally this posited just three kinds of quarks – "up," "down" and "strange" – which, together with their corresponding anti-particles, compose the multitude of hadrons. Within a year a fourth was introduced, with the flavor "charm" and with the subsequent addition of "top" and "bottom" quarks we now have six. Each comes in three "colors" and in addition there are six kinds of leptons, three of which – the electron, the muon, and tau – are charged and three – the corresponding neutrinos – are not. Thus we have 24 elementary fermions – 18 quarks and 6 leptons. Plus their anti-particles. Plus the various bosons – the photon, the W+, W− and the Z, the gluons and, of course, the Higgs boson. Even if we dismiss the bosons as "mere" force carriers, and ignore color, we're left with six quarks and six leptons, fueling speculation that there is a further "level," occupied by "preons," perhaps, as Schaffer mentions, although there is little evidence for them so far (Pati and Salam, 1974).

Whether or not such evidence eventually emerges from the Large Hadron Collider, say, drawing on the history of science like this is notoriously problematic. Even if we focus only on the recent history of physics, we might wonder whether we *should* expect the future to be like the past, at least when it comes to this search for "ever-deeper structure." To insist that we should strike some as resting on an implicit and speculative assumption that goes beyond the kind of naturalistic metaphysics that focuses on our current best theories (McKenzie, 2011, p. 246). However, even though one may be unconvinced by Schaffer's argument as originally stated, with its broad historical sweep, it may retain some force in the form of the more narrowly focused alternative. Of course one might still object that as presented even the historically narrower version of the argument rests on an assumption that is perhaps hard to justify as it stands, namely that of a principle of ontological economy that drives a reduction in the number of kinds of entities that should be regarded as fundamental. What justifies such a principle? One could insist that reality itself is economical with regard to the number of kinds of things that there are but for this to be naturalistically acceptable, it would have to be grounded in the physics itself, rather than simply assumed, or at least so it could be argued. Furthermore, given this, why should we assume that such a principle will always be "in play," as it were, and thus applied to future physical theories?

Given these concerns, are there non-historical arguments that can be mounted against the "received view?" McKenzie offers what she calls "internal" arguments against this view, formulated within the perspective of particular theories (ibid.). Thus, for example, consider the (in)famous "bootstrap model" of the strong interactions (see Chew, 1962) in which the idea that some particles are "elementary" and hence fundamental, is dropped, yielding what was referred to as a "nuclear democracy." Of course, as McKenzie explicitly acknowledges, this model was subsequently abandoned but nevertheless as she emphasizes it offers a useful case study, not least for demonstrating that fundamentality questions may be empirical in character. Nevertheless, the threat to the "received view" is blunted somewhat: the bootstrap model yielded reiterations of the same *kinds* of particles and hence still endorsed a form of fundamentality, albeit in terms of *properties*, rather than particles *per se* (ibid., p. 254).

In a similar vein, Callender has urged Schaffer to look, not to the history of science, but to even more recent physics for support, in particular to 'effective quantum field theory' (Callender 2001; see Chapters 18 and 19 for more details on effective field theories). I'll return to the impact of quantum field theory (QFT) on fundamentality claims in the next section but here we have a QFT applicable at a particular energy level approximating arbitrarily well another QFT at a higher energy level. The former is said to be an "effective" field theory for the latter, which is more fundamental. With an infinite number of energy levels, or scales, we obviously get an infinite "tower" of such theories, a prospect that Cao and Schweber (1993) argue may actually be true and which, again, obviously meshes with a "gunky" metaphysics, insofar as we begin, as it were, with the zero-energy levels and proceed to higher and higher levels without "bottoming out" (for criticisms see Huggett and Weingard, 1995; see also McKenzie 2017a for further nuanced considerations). Schaffer himself acknowledges (op. cit., pp. 504–505) that this example offers a "cautionary tale" to the fundamentalist and concludes that the empirical evidence is neutral between the two conceptions of the "hierarchy of nature": fundamentality and infinite descent (ibid., p. 505). Hence, he suggests, we should remain agnostic.

A further challenge to the "received view" might be drawn from the consideration of certain kinds of "dualities" found in QFT and string theory (see Dawid (this volume) for more details of string theory's dualities). Thus McKenzie considers the Montonen–Olive or electric-magnetic, duality according to which, in particular theoretical contexts, equivalent field formulations can be constructed in which electric and magnetic "charges" exchange roles (for details see Polchinski, 2017). According to one such formulation, electrons are elementary particles and magnetic monopoles are "emergent" composite topological solitons; according to the other, the latter are elementary and the former are the composite solitons. These equivalent formulations can be seen as "complementary perspectives" on the same theory and hence if we again read "elementary" as being "fundamental," which particles are fundamental depends on the perspective adopted. Of course, as McKenzie notes, we have as yet no evidence for magnetic monopoles (2017a), so this duality remains conjectural (as do indeed string theory and associated dualities in general), but as a physical possibility, it represents a further challenge to standard views of fundamentality, generating claims that which particle is taken to be fundamental should be seen as being only as a matter of computational convenience.

Of course, one must be careful not to be too hasty in making such assertions since it can be argued that the computational convenience is in fact facilitated by, rather than underpinning the relevant fundamentality claims (McKenzie 2017a, p. 8). Nevertheless, duality raises some interesting issues: with the electric and magnetic charge couplings as constants, each permitted pair of values defines a different model of the theory. Thus we get a whole spectrum of possible scenarios as these values vary – in those scenarios where the charge coupling is small we may regard electrons as fundamental and in those where the magnetic charge is small, it is the monopoles that are fundamental. And indeed, to any scenario of the former kind, there will correspond a scenario of the latter kind. But these, of course, are scenarios that lie at the ends of the spectrum, and in between, there are scenarios for which the couplings are comparable – in these scenarios we have to say that there are no fundamentality facts at all (ibid., p. 9). Thus, duality offers another physical possibility encompassing a non-fundamental ontology.

The upshot, then, is that the status of the "received view" of fundamentality must be evaluated against a background of relevant theoretical assumptions, both physical and metaphysical (see McKenzie 2017a).

49.3   Populating the Fundamental

Let us now turn to the second question, namely, what is it that we take to populate the fundamental level?

Much of the discussion around the "received view," particularly in the metaphysics literature (for an overview, see for example, Tahko and Lowe, 2016, esp. section 6.4; a useful critical survey can also be found in Schaffer, 2003), take those entities that are designated as "fundamental" to be particles, typically understood in a broadly classical sense; i.e., as little lumps of "stuff," banging into one another and which compose other, derivative entities in accord with some set of mereological principles. However, this is precisely the kind of "high school" metaphysics that the likes of Ladyman et al. (2007, Ch. 1, esp. section 1.6) dismiss as utterly inadequate in the context of modern science. And indeed, as we've already seen, Callender has also noted in this precise context that this crude "particle picture" just won't wash these days and that when physicists talk of "particles" it is shorthand for field quanta (op. cit.; and as is well-known, extracting a robust particle metaphysics from quantum field theory is beset with difficulties; see Wallace (this volume) and Fraser (this volume) for details of conceptions of particles as non-fundamental in QFT). Placing fields at the fundamental level might seem to raise concerns for the fundamentalist, given that they are characterized by an infinite number of degrees of freedom but as Callender remarks, that they offer a "horizontal" infinity does not imply a "vertical" one and hence there is no obstacle to a field-theoretic "bottom" layer in some hierarchy (Callender 2001). Of course, that still leaves the further issue of how that hierarchy composes (as McKenzie notes, we can still find a distinction between fundamental and derivative fields within the QFT framework; 2011, fn 9).

However *that* issue is resolved there is another that bears more directly on the fundamentalist thesis. Quantum fields evolve according to "unitary" dynamics. This means that the dynamics is linear (and hence states enter into superpositions) and "norm-preserving," in that the relevant probabilities sum to 1. It then follows that the fundamental laws will be those that continue to hold and retain that unitary nature even at the smallest spatial scales and hence the highest energy scales – indeed, even as the energy tends to infinity. Satisfying that requirement in general turns out to be hugely tricky, but it can be met in the case of a certain class of theories, namely those that are "asymptotically free" in the sense that the interaction couplings (which encode the strength of the interactions) upon which the relevant probabilities depend, tend to zero as the energy goes to infinity. It then turns out that asymptotically free theories cannot incorporate either too many kinds of fields or too few (see Coleman and Gross, 1973; Gross and Wilczek, 1973).

Thus in the case of quantum chromodynamics (the QFT applicable to the strong nuclear force), there can be no more than 16 kinds of fermionic field (the quarks) and 8 bosonic (see above). Generalizing this, McKenzie argues that if we take the fundamental *kinds* in the world to be given by the types of quantum fields, then a "Goldilocks Principle" holds, according to which there must be some number, greater than zero, of bosonic fields (strictly, non-Abelian gauge bosons), some upper limit on the number of fermionic fields and those two numbers must be related (2017b). As she goes on to note, this is a highly significant constraint on the fundamental kinds that there are. But then, if such a constraint can be understood as, at least, a partial explanation, then it cannot be the case that what kinds are instantiated in the world is a "brute" fact, in the sense of that which requires no further explanation. And if such a fact cannot be considered to be "brute," then, the argument goes, the associated kinds cannot be regarded as fundamental.

Now, of course, as McKenzie notes, there are various ways in which the advocate of the "received view" can try to avoid such a conclusion. She might, for example, draw on Humean metaphysics which insists that all there is to the world is a mosaic of modally unconstrained facts, with statements of laws, symmetry principles, and the like understood as mere descriptions of regularities in this mosaic. Thus she could insist that such principles as unitarity, while regarded as "constraints" by physicists themselves, have no such force metaphysically speaking and thus should not be understood as explanatory. That may seem a high price to pay for many, particularly given Humeanism's other problems in accommodating scientific practice (Hall, 2015).

Alternatively, she might argue that such partial explanations have no place in metaphysics but given that our identification of quantum fields as the fundamental entities already commits us to some form of naturalized metaphysics, it might seem odd to reject out of hand in our metaphysics that which has explanatory force in our best science. Finally, and relatedly, she might simply insist that the fundamental should not be characterized in terms of the explanatorily brute in this manner. Thus Barnes (2012) has argued that the distinction between the fundamental and the derivative should be "pulled apart" from that between the ontologically dependent and independent. (This allows her to propose an understanding of emergence in terms of that which is ontologically dependent yet also fundamental.) If ontological independence is identified with being explanatorily brute, then this might offer a framework in which one could accommodate entities that are deemed to be fundamental but are not the ultimate explanans. But then, that would generate a curious kind of "disconnect" with fundamentality's characterization in terms of that which has ontological priority, say, since typically if x is ontologically prior to y one would expect consideration of the behavior of x to feature in the explanation of the behavior of y.

Of course, there is more to say on all of the above but at the very least, a tension arises between the core precepts of QFT and what seems to be a plausible feature of fundamentality. One option would be to retain that feature and insist that the fundamental should be identified with the "ultimate" explanans; that is, the fundamental level is populated by whatever explains the occupants of the hierarchy above it. Now, as McKenzie makes clear, her Goldilocks Principle and the unitarity that underpins it only offers a *partial* explanation, so the question arises: can we move toward something more complete?

Perhaps we can. Consider the Standard Model, for example (strictly this is not asymptotically free but it may be asymptotically "safe" in the sense that the interaction couplings tend to finite values rather than zero). The role played by certain symmetry principles in generating this model is well-known and the nature of such principles, in particular, the way they may be seen to constrain the relevant laws, has been discussed quite extensively in the philosophy of physics literature (see Brading and Castellani, 2003). Here I want to focus on the point that certain so-called "fundamental" properties effectively drop out of such principles and hence it is the latter that should be regarded as fundamental, in the above explanatory sense.

Let us begin with "Permutation Symmetry" which effectively encodes the fact that it does not matter to the relevant measurement outcomes whether the particles of the system are permuted, or not.[2] As a constraint this divides up the state space into self-contained sectors, each corresponding to a certain fundamental kind of particle and yielding a particular form of quantum statistics, the two most well-known being fermions, which obey Fermi-Dirac statistics and bosons, obeying Bose-Einstein statistics (there are others, corresponding to so-called parastatistics which do not appear to be realized in nature). Thus, the most fundamental kinds that we considered above and into which the particle zoo can be divided, effectively drop out of the action of this particular symmetry. Secondly, the underlying framework of the Standard Model is, of course, QFT, which is a relativistic theory, so the second set of symmetries that needs to be considered are those of Minkowski space-time. These are the translations, rotations, and "boosts" that are captured mathematically by the Poincaré group and as Wigner famously showed (Wigner, 1939), this yields a classification of all "elementary" particles, in terms of their mass and spin. Thus these fundamental properties also effectively drop out of this particular symmetry.

Finally, the Standard Model itself is a gauge theory, represented mathematically by the group $SU(3) \times SU(2) \times U(1)$. This gauge-theoretic aspect refers to the way in which the Lagrangian of a system – which basically captures the dynamics of that system – remains invariant under a group of transformations, where the "gauge" denotes certain redundant degrees of freedom of that Lagrangian (see Teh (this volume) for more details of gauge theories). The generator of this group of transformations represents a field and when such a field is quantized, we get the so-called gauge bosons, also mentioned above. Thus, consider electrodynamics, for example: here the relevant gauge symmetry group associated with the property of charge is labeled $U(1)$ and the requirement of gauge invariance yields a particular gauge boson, namely the photon. Thus, the photon also "drops out" of the imposition of this further symmetry. This requirement can then be extended to the other forces in physics and so, for the weak nuclear force, we have the $SU(2)$ symmetry group associated with isospin, a property of protons and neutrons, and the strong nuclear force associated with $SU(3)$ which operates on the color property of quarks.

Clearly, as indicated, these symmetries play a crucial role as part of the framework of the Standard Model and one can take that role to be explanatory. So, consider again the bosonic and fermionic kinds, which "drop out" of the Permutation Symmetry. Again, that "dropping out" can be read in an explanatory fashion, and again we have (at least) a partial explanation in McKenzie's terms. Likewise, we can offer an explanation of the core properties of the purportedly "fundamental" particles and even of the existence of the force-carrying bosons such as the photon. Taking these together, the partial explanations add up to at least something approaching a complete explanation of the kinds of particles we observe, their properties, and the relevant gauge particles by which they interact (there are still features and properties left unexplained, such as neutrino oscillations and their associated mass, for example).

Of course, one might ask for further details as to the type of explanation that is involved here but fortunately, recent work can be called upon in response: Lange has argued that he is able to incorporate 'explanation via constraints', such as unitarity and symmetry principles, into his framework of non-causal explanations, according to which explanatory power accrues from a form of necessity that is stronger than that possessed by laws of nature (Lange, 2016). French and Saatsi have suggested that Woodward's counterfactual account can be extended to these sorts of situations, as the different possibilities encompassed by these symmetries allow us to entertain "what if things had been different?" scenarios (French and Saatsi 2018). Thus, in this respect at least, there seems to be no obstacle regarding symmetries as explaining things. Now, of course, there is always an issue as to where one takes the explanatory "buck" to stop but one can argue that such symmetries – regarded as "meta-laws" that constrain the relevant laws, as they typically are in physics – should be understood as not themselves being candidates for an explanation. And if the fundamental is identified with the "ultimate" explanans, in this sense of that which itself is not up for explanation, then we should take these symmetries as populating the fundamental level.

Now, further concerns obviously arise. In what sense can a symmetry principle be regarded as a fundamental element of reality? This breaks down into two: In what sense can a symmetry principle be regarded as a fundamental element of (physical) *reality*? In what sense can a symmetry principle be regarded as a *fundamental* element of (physical) reality?

Let us consider the first. Consider the equivalent question for laws: obviously if one is a Humean, then laws simply do not feature in one's metaphysical pantheon as noted above. But if one is not, if one feels that laws are instantiations of relations among universals, for example, or have some primitive status, or are part of the structure of the world, then it can be argued that they do have the requisite ontological status as elements of reality (see Lange (this volume) for more details of these different conceptions of lawhood). Likewise with symmetries, then, which Wigner, for example, regarded as meta-laws: ontologically they can serve as such elements. Indeed, a certain version of "Ontic Structural Realism" takes the structure of the world to consist of, or more bluntly, to *be* (in part) laws and symmetries, appropriately inter-related (French, 2014).

Nevertheless, one might still worry whether such principles can be considered *physical*. They are typically described mathematically via the formalism of group theory, as indicated above, and there is an easy slide from acknowledging that descriptive aspect to reifying the mathematics and understanding such principles as Platonic entities (see the debate between Cao and French and Ladyman in Cao, 2003; French and Ladyman, 2003). But such a slide should be resisted. One option is to identify the physical with the causal but causality is famously problematic, particularly in the context of modern physics (French and Ladyman, 2003; French, 2014 Ch.8; also Frisch (this volume)). An alternative is simply to insist that 'the physical' is that which can be related to empirical results, in some sense, where that relationship obviously needs careful spelling out (French and Ladyman, 2003; French, 2014, Ch. 8; McKenzie, 2014, p. 1100). Taking this line, symmetries can be regarded as physical since they yield determinate properties that can be measured directly or indirectly. Indeed, the further "part" of the structure of the world not mentioned in the brief characterization of Ontic Structural Realism above is the determinate properties (and associated measurement results) that effectively "pin down" the structure given by the laws and symmetries as the structure of the world.

Now let us consider the second question. This goes to the heart of how we should conceive of fundamentality. Again we recall that the "received view" holds that the fundamental level is occupied by the "basic building blocks" (whatever they are) with the "building" captured by some mereological framework. However Wilson has usefully compared that which should be considered as fundamental to the *axioms* of a theory (Wilson, 2014) and drawing on such suggestions, Tahko has argued that we should drop this mereological approach to fundamentality and instead base our account on the idea of "ontological minimality," in the sense that the fundamental level should simply be taken to consist of ontologically minimal elements, with no commitment to any mereological framework (and here he specifically mentions the possibility of including symmetries at the fundamental level; Tahko 2018). The further question then is whether symmetries can be *minimal* in Tahko's sense and hence fundamental.

McKenzie thinks not. Focusing on the priority aspect of fundamentality she first argues that neither of the more obvious ways of capturing that aspect – via relations of supervenience and dependence, respectively – can do the job. Supervenience, she states, is simply not fit for purpose, whereas dependence places both symmetries and particles on an equal footing. We recall that the given symmetries are described via group theory and the afore-mentioned particle properties, or the kinds "boson"/"fermion," are represented mathematically by the relevant group-theoretic representations.

The sense in which these properties and kinds "drop out" of the symmetry is precisely the sense in which a particular group's representations are derived from the group. But now consider the relation of supervenience, which holds that x supervenes on y iff *necessarily*, differences in x entail differences in y: for the particle properties to supervene on the symmetry, then, the former would need to be instantiated in every possible world in which we have the latter. However, not all of the possible representations and associated properties are instantiated in a given world, such as this one (there is an infinite number of "paraparticle" representations associated with Permutation Symmetry, for example, that are not manifested in this one!). Hence the particle properties and kinds do not supervene on the symmetries and the latter cannot be considered to be prior, in these terms, to the former (Wolff, 2012; McKenzie, 2014, p. 1097).

Now consider dependence: clearly, the representation is dependent on the group, since the former is given by and obtained from, mathematically, the latter. Since the relevant representation is then taken to capture the associated properties and the group represents the relevant symmetry, one can conclude that the properties must be dependent on the symmetry. But recall what was said above: for that symmetry, represented group-theoretically as just noted, to be regarded as *physical*, it must yield determinate, measurable properties, *via* the relevant representation. Hence the reference to the latter cannot be avoided if the given symmetry is to be taken to be more than just a feature of mathematics and a fundamental element of physical reality. In that sense, then, particle properties and symmetries (conceived of as physical) must be taken to be on a par ontologically (McKenzie, op. cit., p. 1101).

Perhaps there is some alternative device in the "toolbox" that will do the trick. It has been argued, for example, that the relation between symmetries and properties or kinds should not be understood as one of dependence but that of determinable-determinate (French, 2014, Ch. 10). Thus, for example, Permutation Symmetry can be taken to be the relevant determinable and the bosonic kind, as represented by the appropriate irreducible representation, one of that determinable's determinates, just as "scarlet" is a determinate of the determinable "red." Now there might be an immediate objection that the occupants of the fundamental level must be determinate, and so determinables cannot be fundamental, but it is hard to defend that line in a non-question begging manner and Wilson (2012) has argued that determinables are in fact perfectly acceptable as fundamental features of reality.

However, that doesn't resolve the above issue of priority. Consider: a world with only determinables in its fundamental level would be a modally indeterminate world. A world that just had Permutation Symmetry as fundamental, would be one with the possibility, but never the actuality of course, of the infinite number of different kinds of particles that are allowable. To remove that indeterminacy and obtain a specific possible world we need to incorporate the requisite determinates as well – bosonic and fermionic in the case of the actual world (French, 2014, p. 285). One can think of such determinates as "existential witnesses" in this sense of yielding a specific possible world (Wilson, 2012). But that means, of course, that both the relevant symmetries and certain representations – and hence certain particle kinds and their properties – must be included in the fundamental level (French, op. cit.).

49.4 Conclusion

Where does this leave us? Clearly, the "received view" with its basic building blocks lying at the seat of some physical hierarchy ordered according to standard mereological principles has come under threat. The nature of such elements, as given by modern physics, forces a revision that expands the fundamental level to include symmetry principles (and laws). Of course, one might still try to insist that once we get beyond that level, the standard picture returns but a simple reflection on the role of Permutation Symmetry in chemical bonding, for example, dashes that hope. Still, given that we should be committed to a naturalistic metaphysics, this is all exactly as it should be – we (metaphysicians as well as philosophers of science) should pay attention to the results of modern science, not least in the form of our "best" theories, when we consider such issues. However, that does not mean eschewing the results of metaphysical thought. Specific devices such as that of "determinable" as well as more general maneuvers such as shifting to a notion of ontological minimality can be regarded as tools to be appropriated in articulating an account of what it is for something to be "fundamental" that is both scientifically sensitive and metaphysically nuanced.

Notes

1 See Dizadji-Bahmani (this volume) for discussion of the concept of reduction.
2 See Caulton (this volume) for more details on permutation symmetry and identity; also French and Krause (2006).

References

Barnes, E. (2012). Emergence and fundamentality. *Mind*, 121: 873–901.
Brading, K. and Castellani, E., editors. (2003). *Symmetries in Physics: Philosophical Reflections*. Cambridge: Cambridge University Press.
Callender, C. (2001). Why be a fundamentalist?: Reply to Schaffer. Paper presented at the Pacific APA, San Francisco 2001. Available at: http://philsci archive.pitt.edu/archive/00000215/.
Cameron, R.P. (2008). Turtles all the way down: Regress, priority and fundamentality. *The Philosophical Quarterly*, 58: 1–14.
Cao, T. (2003). Structural realism and the interpretation of quantum field theory. *Synthese*, 136: 3–24.
Cao, T.Y. and Schweber, S. (1993). The conceptual foundations and philosophical aspects of renormalization theory. *Synthese*, 97: 32–108.
Chew, G. (1962). *S-Matrix Theory of Strong Interactions*. New York: W.A. Benjamin.
Coleman, S. and Gross, D.J. (1973). 'Price of Asymptotic Freedom'. *Phys. Rev. Lett*. 31: 851–854.
French, S. (2014). *The Structure of the World*. Oxford: Oxford University Press.
French, S. and Ladyman, J. (2003). Remodelling structural realism: Quantum physics and the metaphysics of structure: A reply to Cao, *Synthese*, 136: 31–56.
French, S. and McKenzie, K. (2012). 'Thinking outside the (tool)box: Towards a more productive engagement between metaphysics and philosophy of physics' with K. McKenzie. *The European Journal of Analytic Philosophy*, 8: 42–59.
French, S. and McKenzie, K. (2015). 'Rethinking outside the toolbox: Reflecting again on the relationship between philosophy of science and metaphysics', with K. McKenzie. In T. Bigaj and C. Wuthrich (eds.), *Metaphysics in Contemporary Physics*, Poznan Studies in the Philosophy of the Sciences and the Humanities, Rodopi, pp. 145–174.
French, S. and Saatsi, J. (2018), Symmetries and explanatory dependencies. In J. Saatsi and A. Reutlinger (eds.), *Explanation beyond Causation*. Oxford: Oxford University Press, pp. 185–205.
Gross, D.J. and Wilczek, F. (1973). Ultraviolet behaviour of non-Abelian gauge theories. *Physical Review Letters*, 30: 1343–1346.
Hall, N. (2015). Humean reductionism about laws of nature. In B. Loewer and J. Schaffer (eds.), *A Companion to David Lewis*. Oxford: Wiley, pp. 262–277.
Huggett, N. and Weingard, R. (1995). The renormalization group and effective field theories. *Synthese*, 102: 171–194.
Ladyman, J., Ross, D., et al. (2007). *Every Thing Must Go: Metaphysics Naturalized*. Oxford: Oxford University Press.
Lange, M. (2016). *Because without Cause: Non-Causal Explanations in Science and Mathematics*. Oxford: Oxford University Press.
Lowe, E.J. (1998). *The Possibility of Metaphysics*. Oxford: Oxford University Press.
McKenzie, K. (2011). Arguing against fundamentality. *Studies in History and Philosophy of Science Part B*, 42(4): 244–255.
McKenzie, K. (2014). On the fundamentality of symmetries. *Philosophy of Science*, 81: 1090–1102.
McKenzie, K. (2017a). Relativities of fundamentality. *Studies in History and Philosophy of Science Part B*, 59: 89–99.
McKenzie, K. (2017b). Against brute fundamentalism. *Dialectica*, 71: 231–261.
Oppenheim, P. and Putnam, H. (1958). Unity of science as a working hypothesis. In H. Feigl et al. (eds.), *Minnesota Studies in the Philosophy of Science*, Vol. 2, Minneapolis: Minnesota University Press; also in *The Philosophy of Science*, R. Boyd, P. Gasper, and J.D. Trout (eds.). London: MIT Press, pp. 405–427, 1991.

Pati, J.C., and Salam, A. (1974). Lepton number as the fourth "color". *Physical Review D*, 10: 275–289.

Polchinski, J. (2017). Dualities offields and Strings. *Studies in History and Philosophy of Modern Physics*, 59: 6–20.

Saunders, S. (2003). Structural realism, again, *Synthese*, 136: 127–133.

Schaffer, J. (2003). Is there a fundamental level? *Nous*, 37(3): 498–517.

Tahko, T.E. (2018). Fundamentality and ontological minimality. In R. Bliss and G. Priest (eds.), *Reality and Its Structure: Reality and Its Structure: Essays in Fundamentality*. Oxford: Oxford University Press, pp. 237–253.

Tahko, T.E. and Lowe, E.J. (2016). Ontological dependence. In E.N. Zalta (ed.), *The Stanford Encyclopedia of Philosophy* (Winter 2016 Edition) Available at: https://plato.stanford.edu/archives/win2016/entries/dependence-ontological/.

Wigner, E. (1939). On unitary representations of the inhomogeneous Lorentz group. *Annals of Mathematics*, 40: 149–204.

Wilson, J. (2012). Fundamental determinables. *Philosophers Imprint* 12. Available at: http://www.philosophersimprint.org/012004/.

Wilson, J. (2014). No work for a theory of grounding. *Inquiry*, 57(5–6): 535–579.

Wolff, J. (2012). Do objects depend on structures? *British Journal for the Philosophy of Science*, 63: 607–625.

Further Reading from the Editors

A. Ney "The Politics of Fundamentality," in A. Aguirre, B. Foster and Z. Merali (eds.), *What Is Fundamental?* (Springer: 2019) is a nuanced discussion of the role of the concept of fundamentality in the political case for physics funding. E. Adlam, "Fundamental?", in A. Aguirre, B. Foster and Z. Merali (eds.), *What Is Fundamental?*, (Springer: 2019) challenges standard assumptions about the necessity of a fundamental lowest physical level. For the state of the art in the general metaphysics of fundamentality, see T. Sider, *The Tools of Metaphysics and the Metaphysics of Science* (Oxford: Oxford University Press, 2020). For skepticism about both a priori and naturalistic metaphysics of the fundamental, see K. McKenzie, "A Curse on Both Houses: Naturalistic versus A Priori Metaphysics and the Problem of Progress," *Res Philosophica* 97(1), (2020): 1–29.

PART XII

Cosmology

Introduction to Part XII

It is one of the ironies of modern physics that over the past two centuries we have generally had a much better understanding of the physics of the very small than we have had of the very large. Our view of the lifecycle of the universe, and of its ultimate fate, has undergone a number of complete upheavals over recent decades and is still very much in flux, with radical ideas taken seriously at the cutting edge of the field. The cosmology part evades neat categorization into either a theory or a theme, since while the chapters in this part have a common target – physical reality on the largest scales – they explore such heterogeneous approaches to understanding their target that the unity between them is more thematic than material.

The further back in time we go, the less confident we are in our cosmological theories, and once we ask after the explanation for the beginning of time itself – when we ask why there is something rather than nothing – we seem to have transgressed the boundaries of any possible physical explanation. Nonetheless, a surprising variety of non-trivial explanations of the existence of anything at all have been offered both by philosophers and by physicists. Sean Carroll's chapter canvasses this range of answers, and comes to a broadly deflationary conclusion: there's no clear reason to think either that there can be an explanation of the existence of physical reality, or that there needs to be such an explanation.

The very idea of the beginning of time is conceptually problematic, and it is cast into further doubt by the way in which the role of time appears to be transformed within the quantum theories of spacetime explored in quantum gravity research. Does physical time even make sense from the point of view of the whole cosmos? Can we extract a measure of time suited to cosmological purposes? If we can, does such a measure of time bear any resemblance to our intuitive or philosophical notions of time? The chapter by Callender and McCoy addresses these questions; the authors conclude that contemporary cosmology leaves the status of time very much an open question.

The notorious problem of the fine-tuning of the universe for life highlights a very striking fact: numerous parameters in our best cosmological theories have values which are not explained from within the theory itself – their values must be obtained from measurement – and which, according to our best assessments, are such that slight variations in their values would produce a universe which is not hospitable to life, or indeed to the formation of any complex material structures. Such fine-tuning would be of interest even if it applied only to one or two parameter values, but it seems to be very widespread within cosmology; Rees (1999) identifies six dimensionless numbers – including the dimensionality of physical space, the ratio of strengths of the gravitational and electromagnetic forces, and the cosmological constant – each of which seems to be fine-tuned. Fine-tuning has variously been denied, declared real but not in need of any explanation, explained by appeal to a supernatural designer, explained by appeal to some exotic cosmological natural selection process, or explained within the context of a multiverse hypothesis. The chapter by Barnes presents a comprehensive survey of the current state of evidence for the fine-tuning hypothesis, and assesses the initial viability of some candidate explanations.

As well as metaphysical questions about the explanation of the values of cosmic parameters, of large-scale cosmic structure, of cosmic time, and of the existence of a cosmos in the first place, cosmology raises distinctive epistemological questions. Of special interest in recent years have been theories of dark matter and dark energy, which jointly provide an impressive account of the evolution of the mass distribution at large scales within our universe as part of the so-called "Λ-CDM cosmological model", involving a small positive cosmological constant and cold (slowly-moving) dark matter. However, neither dark matter nor dark energy has ever been directly detected and the legitimacy of theories invoking them is hotly disputed within the astrophysical community; Melissa Jacquart's chapter gives an up-to-date presentation of the debate.

Disagreements over dark matter and dark energy are diagnostic of deeper epistemological problems raised by doing physics on the very largest scale: our evidence is greatly impoverished, and our access to only one "run" of the universe makes probabilistic theories inherently difficult to test. The final chapter of the volume, by Sibylle Anderle, highlights these epistemological difficulties and offers a moderately optimistic assessment of the overall evidential status of astrophysics. The future seems bright, at least, since new sources of evidence including the results of gravitational wave astronomy look set to transform the evidential situation of the field over the coming decades.

Reference

Rees, M. (1999). *Just Six Numbers: The Deep Forces that Shape the Universe*. London: Weidenfeld & Nicolson.

50

WHY IS THERE SOMETHING, RATHER THAN NOTHING?

Sean M. Carroll

Science and philosophy are concerned with asking how things are, and why they are the way they are. It therefore seems natural to take the next step and ask why things *are* at all – why the universe exists, or why there is something rather than nothing (Holt, 2012; Leslie and Kuhn, 2013).

Ancient philosophers didn't focus too much on what Heidegger (1959) called the "fundamental question of metaphysics" and Grünbaum (2009) has dubbed the "Primordial Existential Question." It was Leibniz, in the 18th century, who first explicitly asked "Why is there something rather than nothing?" in the context of discussing his Principle of Sufficient Reason ("nothing is without a ground or reason why it is") (Leibniz, 1714). By way of an answer, Leibniz appealed to what has become a popular strategy: God is the reason why the universe exists, but God's existence is its own reason, since God exists necessarily. (There is a parallel with Aristotle's much earlier invocation of an unmoved mover, responsible for motion in the universe without itself being moved by anything else (Aristotle, 2008).)

Subsequent thinkers were less impressed by this move. Hume (1779) explicitly dismissed the idea of a necessary being, and both he (Hume, 1748) and Kant (Kant, 1781) doubted that the intellectual tools we have developed to understand the world of experience could sensibly be extended to an explanation for existence itself. In their inimitable styles, Bertrand Russell (Russell and Copleston, 1948) shrugged off the question with "I should say that the universe is just there, and that's all," while Ludwig Wittgenstein (1922) suggested there were some things about which we should remain silent: "It is not *how* things are in the world that is mystical, but *that* it exists." More recently, Parfit (2004) argued for a middle ground between this kind of "Brute Fact" view and the idea that the universe is necessary, by suggesting that one or more features of our universe may pick it out as somehow special, even if they don't imply necessity.

The naturalness of the impulse to ask why the universe exists does not imply that the question is coherent or answerable. Reality is unique – even if there are in some sense many existent "worlds," we can take reality to be the single collection of all such worlds. It might be the case that the property of "having a reason why" applies to facts within reality, but not to all of reality itself.

A major obstacle to addressing this question is the difficulty we have in putting aside strategies and assumptions that have served us well in less sweeping inquiries. Our experience of the world, which is confined to an extraordinarily tiny fraction of reality, doesn't leave us well equipped to think in appropriate ways about the question of its existence. On the contrary, it is very difficult to resist the temptation to treat the universe as just another thing, like an anteater or a smartphone, whose existence can be accounted for in relatively familiar ways. We should be constantly on guard not to insist on conventional answers for such a singular question.

Nevertheless, we can make some progress on the question of why reality exists, both by carefully considering what it might mean to obtain a convincing answer, and by looking at what modern physics and cosmology have taught us about the nature of the universe whose existence we are trying to explain. The most promising answer to date is that the existence of the universe is unlikely to be the kind of thing that has a reason why.

50.1 What Does "Why" Mean?

Since at least the time of Aristotle, philosophers have developed elaborate taxonomies of different kinds of causes or explanations or reasons why various things are true. For our limited purposes here, it should suffice to distinguish between how the universe came to be and what *mechanism* (if any) might have brought it into being – corresponding roughly to Aristotle's efficient cause – and the *reason* why (if any) it exists – corresponding to the final cause. Aristotle conceived of final causes teleologically, as ends or purposes. Here we're being a little broader, expanding the category to include anything that would qualify as a "reason why." (See Part VIII of this volume for a discussion of a variety of approaches to explanation in physics, and Skow (2016) for a recent philosophical investigation of reasons why.) Let us label these notions "mechanisms" and "reasons" for short. The categories are not exclusive. For some purposes, pointing to a mechanism might be enough to answer a question about why something happened; for others, a deeper kind of reason might be sought.

"What mechanism brought the universe into existence?" is conceivably a scientific question. Keeping in mind that "there is no such thing" is a perfectly plausible answer, attempts to identify a mechanism that brought the universe into being should, at the very least, be informed by our best current science, and ideas from contemporary physics have significantly affected what kind of answer we might reasonably expect.

"What reason explains why anything exists at all?" is another matter entirely. Aristotle treated final causes as a fundamental metaphysical category, an irreducible feature of the architecture of reality. Modern physics sees things differently. Rather than being a story of effects and their associated causes, the universe is described by patterns, called the laws of physics, that relate conditions at different times and places to each other, typically by differential equations. (See Lange [this volume] for more discussion of the nature of laws.) The difference between the two conceptions is that the former naturally associates things that happen with a deeper kind of reason why they do, while on the latter view every "why" question is definitively answered by "the dynamical laws of nature and the initial conditions of the universe." The idea that laws simply describe patterns, rather than actively governing what is allowed, is known in the philosophy literature as a "Humean account" of the laws of nature (Hall, 2015).

Scientists are still happy, of course, to talk about explanations and reasons why, in at least two contexts: when accounting for some particular state of affairs in the context of a higher-level (emergent) description, and when pointing to underlying principles as providing explanations for properties of the universe or its dynamics. Fundamental physics explains states of affairs by reference to the initial conditions of the universe, but in emergent theories involving coarse-graining of microscopic degrees of freedom, it can be natural to point to specific effects as arising from individual causes (Russell, 1913; Norton, 2003; Hitchcock, 2007; Carroll, 2016). When it comes to properties of the world rather than states of affairs, explanations often take the form of appeal to symmetries or other deeper principles: we say that conservation of charge is explained by the symmetry of gauge invariance in electromagnetism. The emergent nature of causality can be traced in part to the fact that the entropy of the universe was very low in the past, which gives us great leverage over associating past "causes" with present "effects," in a way that isn't available for future events

(Albert, 2000; Loewer, 2011). Fundamental physics often deals with small numbers of degrees of freedom, where entropic considerations aren't relevant and there is no arrow of time.[1]

What does this mean for the existence of the universe? If cause-and-effect language as applied to states of affairs can usefully emerge at higher levels but is absent in fundamental physics, looking for the "cause" of the universe would be a pointless endeavor. By construction, the universe is the most fundamental thing there is.[2] (I'm using "universe" here to refer to the entirety of physical reality. No judgment is implied about whether things other than physical reality can be usefully said to exist. This definition of universe would include all branches of the wavefunction in Everettian quantum mechanics, and all parts of a cosmological multiverse.) The best we can ask is whether we can imagine laws of nature that fully account for how the universe behaves, even at the earliest moments, or whether we are forced to look outside of reality itself in search of some kind of cause. While we don't currently know the once-and-for-all laws of nature, nothing that we do currently understand about physics implies any necessary obstacle to thinking of the universe as a fully law-abiding, self-contained system. In this case, there would be no such thing as the "cause" or "reason why" the universe exists, even if such notions are appropriate when talking about why a glass falls to the floor or why do fools fall in love. The latter examples are embedded within larger explanatory contexts, while reality is not.

In the second sense of explanation, accounting for properties of nature by appeal to deeper principles, we might hope to find purchase. That is, there might be something special about the way our universe is, which we could then point to as the reason why it exists. Perhaps it is the minimal imaginable universe, or the most symmetric, or the most elegant, or even the only possible universe (presumably subject to some reasonable conditions). If some such principle were to be found, we would still have to worry that the question was simply kicked up a level: *why* does the universe satisfy this particular criterion? Demonstrating that reality was the simplest or most beautiful example among a certain class of possible realities might gently scratch some explanatory itch, but we would be left wondering why such a principle should be given credit for bringing the universe into existence at all. Why shouldn't the universe be ugly or baroque? It would be different if our universe were the only possible one, but as we will see that possibility has its own problems.

This kind of worry generalizes into a concern about *explanatory regression*: given any purported reason why reality exists, why is that reason valid? One option, following Leibniz and others, is that we reach a level at which further explanation is not required, because something is necessarily true. At the other end of the spectrum, explanations might bottom out with a brute fact: something that simply is the case, without further reason, even though it didn't necessarily have to be that way. Arguably there is an in-between stance, where there is something that isn't strictly necessary, but nevertheless satisfies some principle (perhaps of simplicity or beauty) that qualifies as at least a partial explanation. We should be aware of all of these possibilities while examining how our universe might ultimately be explained.

Given these considerations, there is a list of options that might conceivably qualify as an answer to "Why is there something rather than nothing?":

- *Creation:* There is something apart from physical reality, which brings it into existence and/or sustains it. This hypothetical entity is often identified with God in the literature, but there is not necessarily any strong connection with a traditional theistic conception of the divine.
- *Metaverse:* Just as we can sometimes explain events within the universe by appeal to a causal web describing the universe as a whole, perhaps what we think of as reality is part of a larger context, a metaverse that could help explain the existence and properties of our universe. (We're imagining here something more profound than the traditional cosmological multiverse, which is just a universe in which conditions are very different in different regions of spacetime.)

- *Principle:* There is something special about reality, in that it satisfies some underlying principle, perhaps of simplicity or beauty.
- *Coherence:* Perhaps the concept of "nothingness" is incoherent, and the possibility of reality not existing was never actually a viable option.
- *Brute fact:* Reality itself simply exists, in the way that it does, without further explanation.

We can keep these alternatives in mind as we consider further background issues, before coming back to evaluating them at the end.

50.2   What Do "Something" and "Nothing" Mean?

One place where science has exerted an impact on the question is in our definitions of "something" and "nothing." In olden times, we might have described the universe as a collection of stuff (matter, energy, fields), distributed through space and evolving with time. We can then distinguish between two issues:

1   Why is there stuff? Why is there anything inside the universe, rather than just empty space?
2   Why is there space at all? Why is there anything we would recognize as "a universe?"

For the first question, the relevant notion of "nothing" is "empty space," while for the second it is the non-existence of reality altogether. Clearly it's the second question that most people have in mind when they ask why there is something rather than nothing, but answers to the first question (which are much easier to imagine obtaining) have often been passed off as relevant to the second.

Newtonian mechanics provides a precise mathematical formalization of this picture. (See Samaroo (this volume) for more details.) In the absence of external intervention, Newtonian absolute space is eternal, since the equations of motion can be extended infinitely far into the past or future. There is no natural context in which to talk about the creation of the universe, without explicitly invoking divine intervention or something equivalent. (Newton himself thought of God as creating the universe and sustaining its existence, even occasionally intervening when appropriate, but the need for anything outside the universe was rejected by Laplace and other subsequent Newtonian thinkers.)

With the advent of special relativity, space and time are combined into spacetime, and in general relativity spacetime becomes dynamical and responsive to the presence of matter and energy. The basic paradigm remains the same, with one important exception: spacetime itself can begin or end, in a Big Bang or Big Crunch singularity, and indeed the simplest models of our observed universe suggest that there was such a singularity in the past. General relativity therefore diverges from Newtonian absolute space and time in allowing for a universe with a beginning, a first moment in time. It is tempting to think of this as a transition from nothing to something; it is also tempting to think that a universe that begins calls out for an explicit cause more than an eternal universe would – otherwise why did it begin? We'll talk more about the wisdom of giving into these temptations in the next section.

The bigger shift came with the introduction of quantum mechanics. Here we face the problem that there is no consensus about what the ultimate ontology of quantum mechanics actually is; there are various competing "interpretations," which really amount to distinct physical theories. Common to them all is the idea of a wavefunction of a system, which provides us with the probability of obtaining specified measurement outcomes. In an Everettian or Many-Worlds approach, that wavefunction is all there is, and it splits into branches describing effectively separate worlds when subsystems become entangled and decohere. (See Wallace (2012) and Saunders (this volume).) In a hidden-variables approach such as the de Broglie-Bohm theory, we posit additional degrees of freedom whose evolution is simply guided by the wavefunction. (See Dürr et al. (2009) and Tumulka (this volume).) Here I will focus on the wavefunction-only ontology.

Non-relativistic quantum mechanics isn't that different from Newtonian physics, as far as something vs. nothing is concerned. There are a fixed number of particles, and the wavefunction describes the probability of observing particular values of their positions or momenta or other variables. Relativity requires that we move to quantum field theory, which is a particular version of quantum mechanics in which the classical variables that are quantized to give a wavefunction are a set of field values throughout space, rather than positions or momenta of individual particles (see Part V of this volume). The allowed states of the theory include a "vacuum," defined as the lowest-energy state, and excited states describing collections of particles. But the notion of the vacuum is subtle, as "empty space" isn't quite the same as "nothing there." Even in the emptiest lowest-energy state, there are still field degrees of freedom at every point in space, in a particular quantum configuration. These degrees of freedom are highly entangled with each other, and can be probed by measurement devices. For example, the Unruh effect describes the phenomenon by which an accelerated observer in the vacuum will detect a thermal bath of particles (Fulling and Matsas, 2014). Even more impressively, the Reeh-Schleider theorem establishes that any global quantum state of the system as a whole can be reached (to arbitrary precision) by starting with the vacuum and acting with some operator confined to a small region of space (Redhead, 1994). In other words, because field degrees of freedom in different regions of space are entangled in the vacuum, operating on the ones in any particular region can effectively produce any possible state of the theory.

There can also be multiple kinds of vacua in a single quantum field theory: a true vacuum that is the lowest-energy state, and false vacua that have no particles in them, but whose energy density is higher than in the true vacuum. (See also Castellani and Dardashti (this volume).) Due to the phenomenon of spontaneous symmetry breaking (Brading et al., 2017), the most symmetric vacuum (in which the expectation value of all the quantum fields vanishes) is generally not the true vacuum. Cosmological evolution plausibly involves a transition from a symmetric vacuum state, free of particles, to a collection of particles in a background given by a lower-energy vacuum. In some models, this evolution could dynamically favor matter over antimatter, helping to explain the current asymmetry in our observed universe. Such a scenario has given rise to the pithy saying that there is something rather than nothing because "nothing is unstable" (Wilczek, 1980; Krauss, 2013), if we allow ourselves the freedom to define "nothing" as "a symmetric false-vacuum state." This has nothing at all to do with the origin of the universe itself, and certainly nothing to do with why there is a quantum wavefunction in the first place.

In the context of creation of something from nothing, we must also face the issue of "quantum fluctuations." (For a discussion of the very different senses in which this term can be used, see Boddy et al. (2015).) It is often said that the quantum vacuum is filled with fluctuating virtual particles, and even that these particles sometimes pop into real existence, as in Hawking radiation from black holes (Page, 2004). This is a misleading description, arising from a tendency to speak as if wavefunctions represent statistical ensembles of classical particles, rather than true quantum states. A quantum state is simply a quantum state, and a true vacuum state will be stationary, with nothing "fluctuating" at all. Hawking particles can be emitted by black holes because a state with a black hole is not the vacuum state, and the wavefunction of a black hole state naturally evolves into one with particles radiating away as the black hole shrinks.

The situation diverges from our Newtonian intuition even more dramatically when we turn to quantum gravity, in which spacetime itself has a wavefunction (see Part VI of this volume). In that case there is no single "spacetime," there are only approximate notions of spacetime that apply in a classical limit. We do not yet have a full comprehensive theory of quantum gravity, but we can nevertheless try to make progress on the basis of general properties of gravity and quantum mechanics individually. One consequence of quantum gravity is that the distinction between "empty space" and "space filled with stuff" is blurred, practically to invisibility. An intriguing modern idea is that spacetime itself can be defined in terms of the entanglement between a set of abstract quantum degrees of freedom (Van Raamsdonk, 2010; Maldacena and Susskind, 2013; Cao et al., 2016).

The other relevant consequence of quantum gravity is for the beginning of the universe. Classically, there is good reason to believe that general relativity breaks down at a Big Bang singularity, which provides a boundary to spacetime in the past. (The Big Bang isn't a point in space, but should be thought of as a spacelike surface; one is tempted to say "a moment in time," except that spacetime is singular so "time" is not well-defined.) But in a world with quantum mechanics, the breakdown of a classical theory simply means that we shouldn't be taking the classical limit as an accurate description of the situation at that point.

A theorem by Borde et al. (2003) demonstrates that spacetimes with an average expansion rate greater than zero must be geodesically incomplete in the past. This is sometimes offered as an argument that the universe had a beginning, but that is incorrect. Trivially, the average expansion rate could be zero, as it would be in a bouncing cosmology. More importantly, the theorem only applies to classical spacetimes, so at most it could indicate where the classical approximation breaks down, not where the universe begins.

The best we can say is that our current incomplete understanding of quantum gravity is fully compatible with both the possibility that the universe has lasted forever, and that it had a first moment in time. We will look at both possibilities more carefully in the next section. To understand why there is something rather than nothing, we certainly have to understand why there is a physical world described by a quantum wavefunction at all, and we might possibly have to understand how such a universe could "come into existence" out of nothing.

50.3   The Possibility Question: Can the Universe Simply Be?

We can now turn to the question proper: why is there something rather than nothing? The first issue to be addressed is whether physical reality *requires* something external to itself to account for its existence: either something to sustain it, if the universe exists eternally, or something to bring it into existence, if the universe had a beginning. We can consider each scenario in turn.

For definiteness let's imagine that some form of quantum mechanics is the correct description of the physical world at its most fundamental level. That might not be true, but arguably the lessons we learn will generalize to other ontologies. A quantum state $|\Psi\rangle$ is a vector in a Hilbert space $\mathcal{H}$. (A Hilbert space is essentially just a vector space with an inner product defined between vectors.) We posit a Hamiltonian operator $\widehat{H}$ that defines the energy of a state, and then the dynamics of the theory are described by Schrödinger's equation

$$\widehat{H}|\Psi\rangle = i\hbar\frac{\partial}{\partial t}|\Psi\rangle. \tag{50.1}$$

This equation applies to the dynamics of any isolated quantum system, including relativistic quantum field theories and presumably quantum gravity; all one has to do is specify the right Hilbert space and Hamiltonian. (We assume the universe is isolated, or else we should be including whatever influences it as part of the universe.)

The Schrödinger equation has an immediate, profound consequence: almost all quantum states evolve eternally toward both the past and the future. Unlike classical models such as spacetime in general relativity, which can hit singularities beyond which evolution cannot be extended, quantum evolution is very simple. Any state can be written as a superposition of states of definite energy (eigenstates), in terms of which the Schrödinger equation implies that the magnitude of each coefficient remains constant, while the phase orbits at a fixed velocity (at least for time-independent Hamiltonians). In Hilbert space, the entire evolution of the universe simply describes eternal motion in a straight line within some high-dimensional space that is topologically a torus (Carroll, 2008).

If this setup describes the real world, there is no beginning nor end to time. This is not to say that there is no Big Bang in the usual sense; only that it is not a true physical singularity as it would be in classical general relativity, nor does it represent the first moment of the universe. As far as physics is concerned, such a universe would be completely self-contained, existing perpetually without any external cause. One can still question whether or not an uncaused eternal universe is intellectually satisfying, but there is no physical or cosmological obstacle to its existence.

This situation applies to "almost all" quantum states, because there is an exception: states with exactly zero energy. Then Schrödinger's equation collapses to

$$\widehat{H}|\Psi\rangle = 0, \tag{50.2}$$

which in the context of quantum gravity is known as the Wheeler-DeWitt equation (DeWitt, 1967). An equation of this form arises directly from a straightforward attempt to apply the usual rules of canonical quantization to general relativity. There is nothing special here about the quantization procedure; the Wheeler-DeWitt equation simply reflects the fact that general relativity is invariant under reparameterizations of the time coordinate, a feature which exists even in the classical theory (Rovelli, 2015). We should not imagine that we have any firm reason to expect that an equation of this form *must* be the foundational relation of quantum gravity; it is certainly plausible that the more general form (50.1) governs the evolution of the quantum state of the universe, and that the symmetries of general relativity are approximate and emergent in the classical limit.

If the Wheeler-DeWitt equation (50.2) is correct, it presents us with an immediate challenge, known as the "problem of time": there is no time parameter in the equation, so what is "time" supposed to mean? (See Thébault (this volume) for more details.) One might think that such an equation is ruled out by experiment, since we experience the passage of time in the real world. But time might be emergent, rather than fundamental. (Emergent time can still be "real," just as a fluid is still real even if it emerges out of the collective behavior of atoms. See Huggett (this volume) for more discussion.) In other words, we can imagine factorizing Hilbert space into a tensor product of the form

$$\mathcal{H} = \mathcal{H}_C \otimes \mathcal{H}_U, \tag{50.3}$$

where $\mathcal{H}_C$ describes some "clock" subsystem of the universe, and $\mathcal{H}_U$ describes everything else. Then it's possible to define variables such that the state restricted to $\mathcal{H}_U$ is entangled with the clock, so that the rest of the universe seems to "evolve" according to some emergent equations of motion (Page and Wootters, 1983; Banks, 1985; Isham and Butterfield, 1999). This procedure faces a somewhat under-appreciated problem, however, known as the "clock ambiguity": one can factorize Hilbert space in many different ways, obtaining different kinds of emergent time evolution (Albrecht and Iglesias, 2008). It is far from clear how this ambiguity should be resolved, and consequently unclear whether the Wheeler-DeWitt equation by itself can serve as the basis for a well-formed theory of quantum gravity.

Because time is emergent in such models, it might only stretch over a finite interval, so the universe might not be eternal (though in some models it still could be). The Wheeler-DeWitt equation has therefore been used as the basis for models in which the universe has an earliest moment of time (Vilenkin, 1983; Hartle and Hawking, 1983). Sometimes, such universes are said to "come into existence out of nothing." This is a misleading way of putting it, as it implies a temporal process that begins with nothing and ends with the universe. But if the universe doesn't exist, there is no time, and hence there are no processes. It is better, instead, to reserve temporal vocabulary for that portion of reality over which time actually exists. The question is not whether a universe could pop into existence out of nothingness, but whether a universe with a beginning can be entirely described by an appropriate set of laws of physics without the help of any external cause. The answer is that, by itself, the existence of an earliest moment to time is no obstacle to describing the physical universe in completely consistent, self-contained terms. There is therefore no requirement, at least as far as physics is concerned, that existence have an identifiable cause independent of physical reality, whether the universe stretches infinitely far back in time or only a finite interval.

This exposition has been somewhat technical, but we can construct a more intuitive explanation based on the concept of conservation of energy. Energy conservation can be thought of as the idea that the energy of a system at one moment is exactly the same as it was at a moment immediately before, and will be at a moment immediately after. If the energy is nonzero, therefore, it follows that time must extend in both directions – the energy must "go somewhere" in time. But the Wheeler-DeWitt equation describes a universe whose total energy is exactly zero, one where gravitational energy is precisely equal in magnitude and opposite in sign to the energy of matter and other sources. (This does not represent a delicate fine-tuning; it is automatically true in a universe where space is closed, for example a three-dimensional sphere.) Such universes are precisely the kind where time need not flow forever – effectively, a non-existent universe has the same energy (zero) as an existent one obeying the Wheeler-DeWitt equation – and which can therefore have a beginning. Whether or not we live in such a universe is still an open question.

These scientific considerations could be countered by an insistence that differential equations might *describe* what the universe does, but they don't explain the *reason why* it does those things. That is true as far as it goes (why these equations, rather than some other ones? why equations at all?), but it is sometimes extended to a demand that such an explanation *must* exist. Demands of this sort often refer to Leibniz's Principle of Sufficient Reason (PSR) or a modern version thereof: everything must have a reason or explanation, including the universe itself. To avoid an infinite regress, one can suggest that while the universe itself is *contingent* (it did not have to exist in its own right), the ultimate explanation for it can be found in a *necessary* being (Davidson, 2015). Necessary beings, so the idea goes, don't themselves require any further explanations or causes.

From a modern perspective, arguments of this sort are not very convincing, as the justification for the PSR is somewhat antiquated. (Leibniz's own justification relied heavily on his view of God, and there remains a correlation between acceptance of the PSR and theism, though it is not a necessary connection.) Once we think of the laws of nature as describing patterns rather than causal forces, and the notion of cause and effect as being appropriate to higher-level emergent descriptions of the world rather than the fundamental level, the PSR loses its luster. It is sometimes defended as a prerequisite for understanding and talking about the universe at all: if things happen without reasons, how can we possibly make any sense of the world? (Cosmologists do sometimes imagine universes in which typical living creatures are "Boltzmann Brains," random fluctuations out of the surrounding high-entropy chaos, and such models may indeed be rejected on the basis of cognitive instability (Carroll, 2017).) But the requirement that the world be orderly and intelligible is much weaker than the demand that everything has a cause or reason behind it; there is a sizable gap between the PSR as usually understood and "anything goes." In particular, somewhere in between is the idea of an orderly universe which follows impersonal, unbreakable patterns – precisely the kind of universe that is described by modern physics. Such a property is more than enough to allow for sensible investigation and discussion of how the world is, without implying the existence of anything outside the world; as we've seen, there is no shortage of ways the physical world could be both orderly and self-contained.

The idea of a necessary being is similarly unconvincing; there is a vast theological literature on the subject, and I will only offer the barest discussion here. One route to arguing in its favor would be a "cosmological argument," as briefly given above: everything requires a reason, and a necessary being would ultimately ground such reasons. If the PSR (or an equivalent assumption) isn't fundamental, this argument is deprived of its force, as the existence of the universe wouldn't require a reason in the conventional sense. An alternative is an "ontological argument," pioneered by Anselm of Canterbury and offered up in modified versions frequently since. The basic idea is to argue that we can conceive of a most perfect being, and to exist is more perfect than to not exist, therefore the most perfect being necessarily exists (Oppy, 2017). Of the several different objections one might raise, a strong one is to point out that we actually *can't* conceive of a most perfect being, as the notion of "perfection" is not rigorously defined. It has long been recognized that ontological arguments rarely convince skeptics of the need for a necessary being; as Alvin Plantinga admits about his own proof, it serves as a demonstration of the logical consistency of belief in God, not a requirement for doing so (Plantinga, 1974). The skeptics seem to be on firm ground; as Hume emphasized, there is no being whose non-existence would entail a logical contradiction, and we have no difficulty in conceiving of worlds in which no such being existed.

The idea of a universe created by a greater being, for some specific purpose or having some particular properties, seems somehow more *satisfying* than a universe that existed without a brute fact. (Our idea of satisfying explanations has, needless to say, been trained on our experience within a tiny fraction of reality, not on the existence of the whole of reality itself; but we work with what we have.) Moreover, the presence of regularities such as the laws of nature is itself something we might want to explain, even if it alone is sufficient to render the universe intelligible. We are therefore welcome to search for evidence for such an extra-universal entity, using the conventional methods of science and reason. But there is no logical or empirical reason why such an entity must exist; the universe can just be.

50.4   The Naturalness Question: Why This Particular Universe?

Even if the universe can simply be, we can ask why it exists in this way – what explains the specific properties of the laws of nature and the arrangement of stuff in the cosmos. This has been a longstanding goal of scientists; as Einstein (2000) famously put it, "What really interests me is whether God had any choice in the creation of the world." This phrasing points to an even farther-reaching ambition: to not only reveal the reason why the universe is this particular way, but potentially to discover that this way is unique, that there is literally no other way the universe could have been. Whether we classified such a prospective discovery as providing the reason why the universe exists, or removing the need for such a reason, it would certainly satisfy the goal of understanding its existence.

Taken at face value, this ambition seems hopeless. There are an infinite number of self-consistent quantum-mechanical systems that are different from our actual universe. (In terms of Schrödinger's equation, these different theories correspond to different sets of eigenvalues for the Hamiltonian operator.) And there are presumably an infinite number of ways the laws of physics could have been that aren't quantum-mechanical at all. (I am glossing over a distinction one might draw between "conceivable" and "possible." The latter could be a narrow category than the former, if we added extra requirements onto the condition of possibility, such as logical consistency or agreement with the laws of nature. Neither of these criteria is relevant here; any Hamiltonian leads to logically consistent laws of physics, and the laws of nature are precisely what are being decided upon.) A more sensible hypothesis might be that the universe and its laws of nature are the simplest that they could be, given that they also satisfy some other condition – something as specific as describing a quantum-mechanical four-dimensional spacetime with local laws of physics, or something as broad as the existence of intelligent observers.

It is interesting to speculate whether the laws of physics governing reality as we experience it are in some (to be specified) sense maximally elegant, at least given some basic requirements. The general trend of scientific discovery over the last few centuries has been to explain disparate complex phenomena in terms of comparatively simple and powerful frameworks. Newtonian mechanics unified a wide range of phenomena in classical physics; Maxwell's electromagnetism provided a single explanation for light, radiative heat, electricity, and magnetism; Darwinian evolution brought diverse species under the umbrella of a single history of life on Earth; Einstein's relativity and modern quantum field theory used the power of symmetries to provide a simple account of numerous features of the laws of physics; and today we know that the wide variety of phenomena in our everyday lives can be thought of as different manifestations of just a few elementary particles interacting through a handful of forces. Perhaps this trend can continue to an ultimate point where we find that all of the laws of physics applying to our universe can be encapsulated in a single succinct principle.

It's an attractive prospect, which may or may not be true. Many observed features of both fundamental physics and cosmology seem to be arbitrary, from the large-scale structure of stars and galaxies to the masses of elementary particles. Many physicists now suspect that the laws of physics in our observable universe are just one possibility among a very large "landscape" of physically realizable possibilities, known as "vacua" (since they are local minimum-energy states), each of which features different particles, forces, couplings, and even numbers of spatial dimensions (Susskind, 2003). There is the separate issue of whether such possibilities are actually realized in the form of a multiverse. This is a model-dependent question; within inflationary cosmology it doesn't seem difficult to imagine that all or most of the vacua are realized (Linde, 1986; Guth, 2000; Hawking and Hertog, 2006), though the issue isn't settled (Banks et al., 2004; Johnson and Larfors, 2008). There may also be a relationship between the cosmological multiverse and the many worlds of Everettian quantum theory (Bousso and Susskind, 2012; Nomura, 2012).) In string theory, estimates for the size of this landscape throw around numbers of the form 10^{500}. One might take the attitude that the underlying equations of string theory are somehow maximally elegant, even if the specific low-energy manifestation of them that we observe is not. But the lesson is that at present the idea that the ultimate laws are as simple as possible is a hope, not something suggested by the evidence. Moreover, the prospect still faces the challenge of explanatory regression, as one would left to explain why the underlying laws should be so simple.

Another strategy is to point to the existence of intelligent life in the universe. The "anthropic principle" (Barrow and Tipler, 1986; Hogan, 2000) is the idea that certain features of our observable environment are best explained by realizing that things had to be that way in order to allow for the existence of intelligent life. There are various versions of this kind of reasoning, the most respectable of which uses anthropics as a kind of environmental selection: given an ensemble of many different kinds of conditions (regions of space, branches of the quantum wavefunction, or truly separate universes), we are guaranteed to find ourselves in the subset of those conditions that allow for intelligent life. Could we explain the reason why we live in this universe, rather than some other kind of universe, on purely anthropic grounds? (See Barnes (this volume) for more discussion of these strategies.)

There is a substantial obstacle here, over and above the evident difficulty in understanding what conditions actually allow for the existence of life. First, given some parameter such as the mass of a particular particle or the average energy density of the universe, there will be a range of values that are anthropically acceptable. In the absence of any other considerations, we would predict that the parameter in question should look "typical" within that range. For example, the vacuum energy (or cosmological constant) is much smaller in magnitude than the naive value it might have, near the Planck scale of quantum gravity (Carroll, 2001). But if it were large and positive, the vacuum energy would cause such rapid acceleration of space that galaxies couldn't form, making it hard for life to exist; if it were large and negative, the universe would recollapse so rapidly that there would be no time for life to evolve. Such reasoning was used by Weinberg to predict, against the expectations of many theoretical physicists, that astronomers would eventually detect a nonzero value for the vacuum energy (Weinberg, 1987). A decade later, that's exactly what they did, with the observed value seeming roughly typical within the allowed range (Riess et al., 1998; Perlmutter et al., 1999).

For other parameters, however, this anthropic expectation predicts something very different from the real universe. An obvious example is the low entropy of the early universe

(Penrose, 1979; Carroll, 2014), which is many orders of magnitude smaller than what it would need to be in order for life to exist. More generally, the universe simply seems to have far more stuff in it than any reasonable anthropic criterion would imply; there are more than a trillion galaxies, with of order a hundred billion stars and planets in each of them, none of which is necessary for our existence here. If the universe were minimal subject to the existence of intelligent life, why wouldn't it take the form of a relatively small collection of atoms in otherwise empty space, enough to come together to form a small number of stars and planets, before eventually dispersing back into the void?

The fact that our universe doesn't look as minimal as it possibly could, even conditioning on the existence of life, is a strike against one potentially promising answer to the question of why there is something rather than nothing: that every possible world actually exists, and ours is simply the one in which we happen to find ourselves (Lewis, 1986; Tegmark, 2014). It's hard to know precisely what the set of all possible worlds looks like, and even harder to imagine putting a measure on it from which one could extract probabilistic anthropic predictions. Nevertheless, it seems reasonable that most intelligent observers would find themselves in worlds that were much less profligate with matter and energy than ours is, in a version of the Boltzmann Brain problem. The safest tentative conclusion to draw is that the properties of our particular universe cannot be solely attributed to the fact that intelligent observers exist within it, even if some particular properties may be.

50.5   The Reason Question: Why Does Anything Exist at All?

Having laid this groundwork, we can at last turn to the question of why anything exists at all. Let's consider the five options previously mentioned – creation, metaverse, principle, coherence, and brute fact – and briefly evaluate each of them.

• *Creation.*

The idea that our reality was brought into existence by some being outside of reality is perhaps the most intuitively appealing explanation for its existence. For one thing, even if the universe could exist as a brute fact, existence is arguably not what we would expect; as Swinburne (1996) has put it, "It is extraordinary that there should exist anything at all. Surely the most natural state of affairs is simply nothing."

"Natural" presumably isn't the right idea in this context; by definition, whatever reality is, it's natural. What is meant is probably something the the effect that non-existence is simpler or easier than existence, and for some reason is therefore to be expected. In part this expectation comes from our experience within the universe, where things typically need to be created and perhaps maintained. Consciously or not, we have in mind a metaphorical reality-chooser, who contemplates all the different ways the universe could have been (perhaps including non-existent) and makes a simple and elegant choice. Similarly, it is sometimes suggested that the regularities we label "laws of nature" are inexplicable in the absence of some entity that ensures those laws are obeyed, as if the reliability of such laws implies the existence of a legislative body or a law-enforcement agency. It is precisely this kind of intuitive and metaphorical reasoning that we should be suspicious of in this context. Ideas that become ingrained from our experience with the everyday world may not extend in any useful way to the very unique question of the existence of reality.

While a creator could explain the existence of our universe, we are left to explain the existence of a creator. In order to avoid explanatory regression, it is tempting to say that the creator explains its own existence, but then we can ask why the universe couldn't have done the same thing. (All else being equal, a self-explaining and necessary universe would be a simpler overall package than a self-explaining and necessary creator who then created the universe. But to most advocates of this general strategy, necessity seems like a more natural property to attribute to a supernatural creator than to the natural universe.) Thus we are left to identify the creator as a necessary being, in contrast with the contingent nature of our universe. But as we argued at the end of Section 50.3, the idea of a necessary being doesn't really hold together; there just isn't any such thing.

The conclusion is that invoking a creator does not provide us any escape from the need to posit something that simply exists because it does, without further reasons to which we can appeal. And if that is the case, there is no reason not to include all of reality in that category, without additionally imagining a creator at all. The existence of a creator of the universe should be judged on ordinary empirical grounds (does it provide a useful explanatory account of observed features of what we see?), not on *a priori* arguments for its necessity.

- *Metaverse.*

Cosmologists use the word "multiverse" to refer to something that is actually more prosaic than it sounds: a single connected spacetime, but with regions ("universes") where conditions are very different from each other. In other contexts, including philosophical writing as well as the popular media, the word is sometimes applied to the multiply-branched wavefunction of Everettian quantum mechanics. We have in mind something a bit more grandiose: a collection of truly distinct realities (noninteracting, not stemming from a common past, not necessarily with the same laws of physics), one of which is our own. We therefore label such a collection the "metaverse."

The cosmological multiverse can provide a context in which anthropic reasoning becomes appropriate, and may therefore help explain why our observed region of space has some of the properties that it does. Could a metaverse somehow explain why our universe exists at all? The hope here would be that the metaverse provides a context in which explanatory language becomes appropriate, just as cause-and-effect talk becomes useful in the emergent higher-level descriptions we apply to our everyday environment.

There is a problem, however: unlike the straightforward cosmological multiverse, which can arise naturally in theories of inflationary cosmology, different elements of the metaverse are not actually connected to each other by dynamical processes or influences. One reality is not created from another one, so it is hard to envision a sense in which the collection of realities explains the existence or properties of any individual member. The best we might do is to imagine a maximal metaverse, in which every reality exists, and then try to account for the specific one in which we live – but as noted above, this program is faced with significant obstacles. Moreover, the metaverse still faces a severe problem with explanatory regression, as we would be left trying to explain the existence of multiple realities rather than just one.

This is not to argue that such a metaverse could not (in some sense) exist. Even if such a thing is forever beyond our empirical reach, positing its existence could conceivably play an explanatory role in understanding the properties of our actual universe (though I personally am skeptical). But as the metaverse itself has no reason to be a necessarily existent thing, and because dynamical processes within it cannot causally account for the creation of our universe (as the different elements of the metaverse are noninteracting by hypothesis), it does not directly provide an answer to the question of why there is something rather than nothing.

- *Principle.*

Aside from an actual being or metaverse that could account for the existence of reality, we might imagine that the best explanation takes the form of a principle that picks out our universe among all the conceivable ones. Perhaps our universe is the simplest subject to certain conditions, or perhaps all possible realities actually exist. Such an answer would again face the explanatory regression problem, as noted, but the existence of such a principle could arguably serve to soften the blow of the universe not being unique or necessary. The biggest obstacle is that it's hard to see, given what we know about the actual universe, what such a principle could possibly be. Future scientific discoveries could reveal such an answer.

- *Coherence.*

One option we haven't considered in detail is the idea that "nothing exists" might not, despite the seeming naturalness of the formulation, actually be a coherent idea. We are once again challenged to think beyond our experience of objects within the physical world, where it is very sensible to imagine that an existent thing (say, the desk on which I am typing) could counterfactually have not existed. It is tempting, if only by analogy, to imagine the same thing about the universe as a whole; Parfit (2004) refers to the possibility that "nothing would have been."

But what does "been" really mean in such a construction? The absence of anything existing isn't quite the same as "that which exists" being identified with "nothing," as "existing" isn't something that "nothing" can sensibly do. Things can fail to exist within reality, but it does not immediately follow that reality itself could have not existed. (Could we assign a probability to that having been true?) So perhaps the universe exists simply because there was no coherent alternative.

Suggesting this possibility requires that we look beyond the naive metaphor of a reality-chooser, among whose potential options was to choose "nothing." Perhaps our language and modes of thought are tricking us, and existence is something that is metaphysically unavoidable. In that case some form of reality would be necessary, even if the specific form were left unexplained; we would still face the challenge of understanding our actual universe.

- *Brute fact.*

Every attempt to answer the question "Why is there something rather than nothing?" ultimately grounds in a brute fact, a feature of reality that has no further explanation. The universe is not unique, and there are no necessary beings; even if we decide that the concept of nothingness is incoherent, at least some properties of our particular universe are ultimately contingent. By the standards of modern science, it is extremely hard to see what could possibly qualify as a final and conclusive "reason why" the universe exists.

Perhaps the absence of such a reason shouldn't be surprising. As our knowledge of the universe improves, questions that once seemed urgent can become un-asked, as we realize that the context in which they were posed was not appropriate. In Kepler's time, the question of why there were precisely six planets (Mercury, Venus, Earth, Mars, Jupiter, and Saturn) was a natural one to ask, and he proposed a model where the five Platonic solids were inscribed between their orbits. Today we know that there are more than five planets, that the definition of a "planet" is controversial, and that a certain amount of randomness is involved in accounting for the actual distribution of bodies in the Solar System. In the 1930's, Eddington attempted to derive a numerical formula that would explain why the fine-structure constant α of electromagnetism should be exactly 1/136, as it was suspected to be at the time; when experiments improved the measured value to something closer to 1/137, he adjusted his formula in response (Eddington, 1974b). Today we know that α is not the reciprocal of any integer. It may still be true, of course, that there exists a subtle and elegant formula yet to be discovered that exactly reproduces the value of α, but in modern physics where electromagnetism is subsumed into a broader context of quantum field theory and electroweak unification, the search for such a formula is not a priority.

The universe could be the same way. Perhaps at bottom its existence and specific features include brute facts that are in some sense completely arbitrary; or perhaps there is a deeper principle that explains why it is precisely this universe, and the only brute fact is the validity of that principle. We are always welcome to look for deeper meanings and explanations. What we can't do is demand of the universe that there be something we humans would recognize as a satisfactory reason for its existence.

Acknowledgments

This research is funded in part by the Walter Burke Institute for Theoretical Physics at Caltech and by DOE grant DE-SC0011632.

Notes

1 See Shahvisi (this volume).
2 For more discussion of fundamentality, see French (this volume).

References

Albert, D.Z. (2000). *Time and Chance*. Cambridge, MA: Harvard University Press.

Albrecht, A. and Iglesias, A. (2008). The clock ambiguity: Implications and new developments. Fundamental Theories of Physics, 172(2012): 53–68. arXiv:0805.4452 [hep-th].

Aristotle (2008). *Physics*, VIII 4-6. R. Waterfield (trans.). Oxford, UK: Oxford University Press.

Banks, T. (1985). T C P, quantum gravity, the cosmological constant and all that... *Nuclear Physics B*, 249: 332.

Banks, T., Dine, M. and Gorbatov, E. (2004). Is there a string theory landscape? *JHEP*, 0408: 058 doi:10.1088/1126-6708/2004/08/058 [hep-th/0309170].

Barrow, J.D. and Tipler, F.J. (1986). *The Anthropic Cosmological Principle*. Oxford, UK: Clarendon.

Boddy, K.K., Carroll, S.M. and Pollack, J. (2015). Why Boltzmann brains don't fluctuate into existence from the De Sitter vacuum. In K. Chamcham, J. Silk, J.D. Barrow and S. Saunders (eds.), *The Philosophy of Cosmology*. Cambridge University Press; arXiv:1505.02780 [hep-th].

Borde, A., Guth, A.H. and Vilenkin, A. (2003). "Inflationary space-times are incomplete in past directions," *Physical Review Letters*, 90: 151301. [arXiv:gr-qc/0110012].

Bousso, R. and Susskind, L. (2012). The multiverse interpretation of quantum Mechanics. *Physics Review*, D85: 045007. arXiv:1105.3796 [hep-th].

Brading, K., Castellani, E. and Teh, N. (2017). Symmetry and symmetry breaking. In E.N. Zalta (ed.), *The Stanford Encyclopedia of Philosophy* (Winter 2017 Edition). Available at: https://plato.stanford.edu/archives/win2017/entries/symmetry-breaking/.

Cao, C., Carroll, S.M. and Michalakis, S. (2016). Space from Hilbert Space: Recovering Geometry from Bulk Entanglement. *Physical Review D*, 95: 024031; [arxiv:1606.08444].

Carroll, S.M. (2001). The cosmological constant. *Living Reviews in Relativity*, 4: 1. [astro-ph/0004075].

Carroll, S.M. (2008). What if time really exists? arXiv:0811.3772 [gr-qc].

Carroll, S.M. (2014). In what sense is the early universe fine-tuned? In B. Loewer, E. Winsberg and B. Weslake (eds.), *Time's Arrows and the Probability Structure of the World*. Cambridge, MA: Harvard University Press. [arXiv:1406.3057].

Carroll, S.M. (2016). *The Big Picture: On the Origins of Life, Meaning, and the Universe Itself*. New York: Dutton.

Carroll, S.M. (2017). Why Boltzmann brains are bad. to appear in *Current Controversies in the Philosophy of Science*, S. Dasgupta and B. Weslake, eds.; Milton Park, UK: Routledge (2020), pp. 7–20. [arxiv:1702.00850].

Davidson, M. (2015). God and other necessary beings. In E.N. Zalta (ed.), *The Stanford Encyclopedia of Philosophy* (Spring 2015 Edition). Available at: https://plato.stanford.edu/archives/spr2015/entries/god-necessary-being/.

DeWitt, B.S. (1967). Quantum theory of gravity. I. The canonical theory. *Physics Review*, 160: 1113–1148.

Dürr, D., Goldstein, S., Tumulka, R. and Zanghi, N. (2009). Bohmian mechanics. In D. Greenberger, K. Hentschel, and F. Weinert (eds.), *Compendium of Quantum Physics*. New York: Springer, pp. 47–55; arXiv:0903.2601 [quant-ph].

Eddington, A.S. (1946). *Fundamental Theory*. Cambridge, UK: Cambridge University Press.

Einstein, A., letter to E. Straus, quoted in A. Calaprice (2000). *The Expanded Quotable Einstein*. Princeton, NJ: Princeton University Press.

Fulling, S.A. and Matsas, G.E.A. (2014). Unruh effect. *Scholarpedia*, 9(10): 31789.

Grünbaum, A. (2009). Why is there a world at all, rather than just nothing? *Ontology Studies*, 9: 7–19.

Guth, A.H. (2000). Inflation and eternal inflation. *Physics Report*, 333: 555. doi:10.1016/S0370-1573(00)00037-5 [astro-ph/0002156].

Hall, N. (2015). Humean reductionism about laws of nature. In B. Loewer and J. Schaffer (eds.), *A Companion to David Lewis: Blackwell Companions to Philosophy*. Hoboken, NJ: Wiley-Blackwell, pp. 262–277.

Hartle, J.B. and Hawking, S.W. (1983). Wave function of the universe. *Physical Review D*, 28: 2960.

Hawking, S.W. and Hertog, T. (2006). Populating the landscape: A top down approach. *Physical Review D*, 73: 123527. doi:10.1103/PhysRevD.73.123527 [hep-th/0602091].

Heidegger, M. (1959). *Introduction to Metaphysics*. New Haven: Yale University Press.

Hitchcock, C. (2007). What Russell got right. In H. Price and R. Corry (eds.), *Causation, Physics, and the Constitution of Reality: Russell's Republic Revisited*. Oxford: Oxford University Press, pp. 45–65.

Hogan, C.J. (2000) Why the universe is just so. *Review of Modern Physics*, 72: 1149. [astro-ph/9909295].

Holt, J. (2012). *Why Does the World Exist? An Existential Detective Story*. New York: Liveright Publishing.

Hume, D. (1748). *An Enquiry Concerning Human Understanding*, Part III. Cambridge, MA: The Harvard Classics.

Hume, D. (1779). *Dialogues Concerning Natural Religion*, Part IX. New York: Penguin Classics.

Isham, C.J. and Butterfield, J. (1999). On the emergence of time in quantum gravity. In J. Butterfield (ed.), *The Arguments of Time*. Oxford: Oxford University Press.

Johnson, M.C. and Larfors, M. (2008). An obstacle to populating the string theory landscape. *Physical Review D*, 78: 123513. doi:10.1103/PhysRevD.78.123513 [arXiv:0809.2604 [hep-th]].

Kant, I. (1781). *Critique of Pure Reason*, A602/B630. Cambridge, UK: Cambridge University Press.

Krauss, L.M. (2013). *A Universe from Nothing: Why There Is Something Rather Than Nothing*. New York: Atria Books.

Leibniz, G.W. (1714). Principles of nature and grace founded on reason. G.H.R. Parkinson and M. Morris (trans.). In G.H.R. Parkinson (ed.), *Leibniz: Philosophical Writings*. Indianapolis: Hackett Publishing Company, pp. 195–204.

Leslie, J. and Kuhn, R.L., editors. (2013). *The Mystery of Existence: Why Is There Anything at All?* Hoboken: Wiley-Blackwell.

Lewis, D. (1986). *On the Plurality of Worlds*. Hoboken: Blackwell.

Linde, A.D. (1986). Eternally existing selfreproducing chaotic inflationary universe. *Physics Letters*, B175: 395–400.

Loewer, B. (2011). The emergence of time's arrows and special science laws from physics. *Interface focus*, rsfs20110072.

Maldacena, J. and Susskind, L. (2013). Cool horizons for entangled black holes. *Fortschritte der Physik*, 61: 781; [arXiv:1306.0533 [hep-th]].

Nomura, Y. (2012). The static quantum multiverse. *Physics Review*, D86: 083505. arXiv:1205.5550 [hep-th].

Norton, J.D. (2003). Causation as folk science. *Philosophers' Imprint*, 3: 1–22.

Oppy, G. (2017). Ontological arguments. In E.N. Zalta (ed.), *The Stanford Encyclopedia of Philosophy* (Summer 2017 Edition). Available at: https://plato.stanford.edu/ archives/sum2017/entries/ontological-arguments/.

Page, D.N. (2004). Hawking radiation and black hole thermodynamics. *New Journal of Physics*, 7: 203; [hep-th/0409024].

Page, D.N. and Wootters, W.K. (1983). Evolution without evolution: Dynamics described by stationary observables. *Physical Review D*, 27: 2885.

Parfit, D. (2004). Why anything? Why this? In T. Crane and K. Farkas (eds.), *Metaphysics: A Guide and Anthology*. Oxford: Oxford University Press.

Penrose, R. (1979). Singularities and time asymmetry. In S.W. Hawking and W. Israel (eds.), *General Relativity: An Einstein Centenary Survey*. Cambridge, UK: Cambridge University Press, pp. 581–638.

Perlmutter, S. et al. [Supernova Cosmology Project Collaboration] (1999). "Measurements of Omega and Lambda from 42 high redshift supernovae," *Astrophysical Journal*, 517: 565. [astro-ph/9812133].

Plantinga, A. (1974). *The Nature of Necessity*. Oxford: Oxford University Press.

Redhead, M. (1994). The vacuum in relativistic quantum field theory. *PSA: Proceedings of the Biennial Meeting of the Philosophy of Science Association*, 1994: 77–87.

Riess, A.G. et al. [Supernova Search Team] (1998). Observational evidence from supernovae for an accelerating universe and a cosmological constant. *Astronomical Journal*, 116: 1009. [astro-ph/9805201].

Rovelli, C. (2015). The strange equation of quantum gravity. *Classical and Quantum Gravity*, 32(12): 124005. [arXiv:1506.00927 [gr-qc]].

Russell, B. (1913). On the notion of cause. *Proceedings of the Aristotelian Society*, 13: 1–26.

Russell, B. and Copleston, F.C. (1948). God's existence: A debate. In J. Leslie and R.L. Kuhn (eds.), *The Mystery of Existence: Why Is There Anything at All?* New York: Wiley-Blackwell, pp. 53–56.

Skow, B. (2016). *Reasons Why*. Oxford: Oxford University Press.

Susskind, L. (2003). The anthropic landscape of string theory. In B. Carr (ed.), *Universe or Multiverse?* Cambridge, UK: Cambridge University Press, pp. 247–266. [hep-th/0302219].

Swinburne, R. (1996). *Is There a God?* Oxford: Oxford University Press.

Tegmark, M. (2014). *Our Mathematical Universe*. New York: Knopf.

Van Raamsdonk, M. (2010). Building up spacetime with quantum entanglement. *General Relativity and Gravitation*, 42: 2323 [International Journal of Modern Physics D, 19: 2429 (2010)]; [arXiv:1005.3035 [hep-th]].

Vilenkin, A. (1983). The birth of inflationary universes. *Physical Review D*, 27: 2848.

Wallace, D. (2012). *The Emergent Multiverse: Quantum Theory according to the Everett Interpretation*. Oxford: Oxford University Press.

Weinberg, S. (1987). Anthropic bound on the cosmological constant. *Physical Review Letters*, 59: 2607.

Wilczek, F. (1980). The cosmic asymmetry between matter and antimatter. *Scientific American*, 243(6): 82–90.

Wittgenstein, L. (1922). *Tractatus Logico-Philosophicus*, 6.44. C.K. Ogden (trans.). Mineola, NY: Dover Books.

Further Reading from the Editors

A classic philosophical perspective on explaining existence is D. Parfit, "Why anything? Why this?" In T. Crane and K. Farkas (eds.), *Metaphysics: A Guide and Anthology* (Oxford: Oxford University Press, 2004). The possibility of a substantive explanation of why there is anything at all is further explored by B. Skow (2010), "The dynamics of non-being", *Philosophers' Imprint* 10(1): 1–14. For applications of multiverse thinking to the question, see M. Tegmark, "The Mathematical Universe", *Foundations of Physics* 38 (20080: 101–150 and L. Susskind, "The Anthropic Landscape of String Theory", in *Universe or Multiverse*, ed. B Carr (Cambridge: Cambridge University Press, 2007): 247–266. For a recent public controversy on the prospects of physics for explaining existence, see D. Albert "Review of *A Universe from Nothing*, by Lawrence M. Krauss", *New York Times*, 25th May edition (2012).

51

TIME IN COSMOLOGY

Craig Callender and C. D. McCoy

Readers familiar with the workhorse of cosmology, the hot big bang model, may think that cosmology raises little of interest about time. As cosmological models are just relativistic spacetimes, time is understood as it is in relativity theory, and all cosmology adds is a few bells and whistles such as inflation and the big bang and no more. The aim of this chapter is to show that this opinion is not completely right—and may well be dead wrong. In our survey, we show how the hot big bang model invites deep questions about the nature of time, how inflationary cosmology has led to interesting new perspectives on time, and how cosmological speculation continues to entertain dramatically different models of time altogether. Together these issues indicate that the philosopher interested in the nature of time would do well to know a little about modern cosmology.

Different claims about time have long been at the heart of cosmology.[1] Ancient creation myths disagree over whether time is finite or infinite, linear or circular. This speculation led to Kant complaining in his famous antinomies that metaphysical reasoning about the nature of time leads to a "euthanasia of reason." But neither Kant's worry nor cosmology becoming a modern science succeeded in ending the speculation. Einstein's first model of the universe portrays a temporally infinite universe, where space is edgeless and its material contents unchanging. One of the more popular versions of the big bang model is temporally finite: it begins with a bang and ends with a crunch. Later, Bondi and Gold's rival to the hot big bang, the steady-state theory, is, like Einstein's universe, premised on the idea that the large scale features of the universe don't change in space or time (the so-called "perfect cosmological principle"). Rather more exotic ideas followed: Milne, for example, abandoned relativity theory and posited *two* times nonlinearly related to one another (Kragh, 1996, 61ff.). Today the speculations continue. Often motivated by quantum considerations, one finds plenty of alternative cosmologies to the big bang model's contemporary successor, the ΛCDM model. These alternatives range from cyclic time universes, to string-theoretic "branes" colliding in a high-dimensional space, to eternally inflating multiverses chock full of bubble universes (Kragh, 2011).

The reason for so much variety in cosmological possibility is simple: the universe is really big and what we can see is really small. We are, in other words, faced with severe observational limitations in the context of cosmology (Ellis, 2007) and hence a severe underdetermination of cosmological modeling (Butterfield, 2014). According to relativity theory we can at best only detect signals from events within or on our past light cone—this is the "observable universe" in the maximal sense. According to the standard big bang model the observable universe is further limited physically by the existence of a surface of last scattering. Before the time of this event (decoupling) light was tightly coupled to matter such that it could not travel in space and time to our telescopes here and now. This essentially means that we don't have observational access to the very early universe. Since the entire universe is most likely much, much larger than the observable universe, what we can verify observationally gives us a relatively weak handle on the nature of the cosmos at large. Thus, many genuinely scientific models can plausibly fit the same observations, since the universe may well be very different outside the observational scope of our telescopes.

Furthermore, our astronomical and cosmological observations themselves are heavily theory- and model-laden. Astronomy reveals just how much our judgments of distances, angles, and even objects depend on background assumptions about the nature of astrophysical objects, the geometry of the universe, etc.[2] And a related additional consideration is that confidence in our own theories runs low in various cosmological regimes. For example, the big bang model posits that the universe began with an initial singularity. But should we trust general relativity at the high temperatures and energies of the big bang itself? And does the addition of inflation to the big bang model modify this expectation? How might the big bang story be modified by quantum effects from an expected future theory of quantum gravity? All this uncertainty about how to model the unobservable universe opens up space for cosmological models that feature different perspectives on time, which means that speculation about time in cosmology flourishes as much as—if not even more than—ever, despite the subject now having more and better empirical data than it has enjoyed before.

After a brief treatment of time in the standard model, we hope to give the reader a sense of some of the live options available for time in cosmology which are of philosophical interest. Although time may be infinite, space is not; as a result, our discussion necessarily will be abbreviated and selective.

51.1   Time in the Standard Model

The current standard model of cosmology, as mentioned, is usually called the ΛCDM model. It extends the classic hot big bang model, which includes only normal matter and radiation, by including exotic cold dark matter (CDM), the more exotic dark energy (Λ, the usual symbol for Einstein's cosmological constant), and a stage of inflation in the early universe that seeds the initial perturbations which give rise to structure in the universe.[3] The theoretical core of the standard model is the general theory of relativity, which models the universe as a four-dimensional spacetime manifold M endowed with a pseudo-Riemannian metric g and matter-energy distribution T.

Observations suggest that the universe looks pretty much the same in every direction. Supposing that the same holds at every other location in the universe leads one to posit the cosmological principle: the assumption that the universe is spatially homogeneous and isotropic (Beisbart and Jung, 2006). Applying this principle to general relativity requires presuming that there is a congruence of timelike curves in the spacetime manifold M which foliates it into a one-parameter (λ) family of spacelike hypersurfaces Σ_λ of constant intrinsic curvature κ (which can be positively curved like a sphere, flat like a plane, or negatively curved like a hyperbolic surface). The matter-energy distribution T in such a spacetime takes the form of a perfect fluid with constant density ρ and pressure p on the hypersurfaces of constant curvature. The collection of relativistic spacetimes that meet these conditions are usually known as the Friedman-Robertson-Walker (FRW) or Friedman-Lemaître-Robertson-Walker (FLRW) spacetimes.

One usually thinks of the members of the family of hypersurfaces Σ_λ as representing space and λ as representing time. Relativity, of course, permits the choice of another congruence of timelike curves which also foliate M into "spaces." So the sense in which λ represents time is relative to the choice of congruence that gives rise to it as a parameter. Under the assumption of the cosmological principle and choosing a foliation such that the corresponding spaces have constant curvature reduces Einstein's field equations to the following two, usually called the Friedman equations:

$$\dot{H} + \frac{4}{3}\pi(\rho + 3p) + H^2 = 0; \tag{51.1}$$

$$\dot{\rho} = -3H(\rho + p), \tag{51.2}$$

where H, the Hubble parameter, represents the expansion rate of space, and overdots represent time derivatives with respect to the chosen standard of time. Thus the first equation concerns the acceleration of the expansion of space ($\dot{H}$); the second is a continuity equation for the "cosmic fluid," which consists of radiation, dust, dark matter, dark energy, etc.

There is a natural standard of time to use in a non-vacuum FRW spacetime: the proper time of anything not moving with respect to the cosmic fluid. Hypothetical objects that are at rest with respect to it are called fundamental observers. Galaxies, for example, are approximately fundamental observers, since they do not move much with respect to the background fluid. The proper time of the cosmic fluid, which we will denote by t, is generally known as *cosmic time*.

Nevertheless, although the decomposition from above naturally picks out a particular time function (t), the theory is still entirely relativistic. We chose to foliate spacetime into slices of constant spatial curvature, but we could just as well have chosen an infinity of other foliations. It is therefore only when we choose the natural foliation and use cosmic time that one can claim that the universe is 13.8 billion years old—this duration is determined by the proper time of a hypothetical fundamental observer from the beginning of time until now. By contrast, to photons the world just began. Relativity doesn't choose sides over who is right about time, fundamental observers, arbitrarily moving observers, or photons. All are.

51.2 Breaking Up Spacetime

One of the central lessons of contemporary physics is that space and time, traditionally thought to be conceptually independent, are replaced by a unified spacetime. It may come as a surprise to many readers to discover that some physicists and philosophers want to use cosmology to undo this revolution and break up spacetime back into space and time. By "break up" we mean a genuine divorce, at the level of basic theory, and not merely a provisional separation for pragmatic calculational convenience (which is done all the time in numerical relativity). The thought behind the machinations is that cosmology actually justifies resurrecting a Newtonian-like absolute time.

Indeed, one finds the idea arising just after relativity was born and espoused by some of cosmology's biggest stars. For instance, just three years after general relativity was discovered and only one year after experimentally confirming it, Sir Arthur Eddington in 1920 wrote that "absolute space and time are restored for phenomena on a cosmical scale," hinting even that this cosmic time is God's time (Eddington, 1920 [1987], p. 168). Sir James Jeans writes:

> It was natural to try in the first instance to retain the symmetry between space and time which had figured so prominently in [special relativity], but this was soon found to be impossible. If the theory of relativity was to be enlarged so as to cover the facts of astronomy, then the symmetry between space and time which had hitherto prevailed must be discarded. Thus time regained a real objective existence, although only on the astronomical scale, and with reference to astronomical phenomena. (Jeans, 1935, p. 21)

That figures as significant as Eddington and Jeans suggest ditching the core postulates of relativity may be somewhat shocking. Both seemed to want to restore an "intuitive" notion of time, although admittedly Eddington gave off mixed signals on that matter (Canales, 2015).

Perhaps more isn't made of this rejection because it's not entirely clear whether either had a full rejection of relativity in mind. Eddington says that relativity is "reduced to a local phenomenon." Did he really want to abandon the core conceptual insights of a theory that he helped confirm in local physics (i.e., the solar system) when applied to the universe? Similarly, Jeans's talk of time regaining real objective existence *but only on the astronomical scale* sounds somewhat obscure conceptually: what exactly is "real" and "objective" about time on the astronomical scale and, more importantly, what does such a time have to do with the intuitive time of our experiences?

Today those desirous of a sundering of spacetime are somewhat clearer. Their motivations are diverse. Some file "philosophical" grounds for divorce, and others make their case for the sake of new physics. The first group, like Eddington and Jeans, wish to free "intuitive" metaphysical models of time from the clutches of relativity (Lucas and Hodgson, 1990; Crisp, 2008; Swinburne, 2008). The idea that the present is special, that it is what exists or what becomes, is famously threatened by relativity.[4] Finding a scientifically respectable absolute time according to which the world comes into being would give these prospective divorcees the "proof" that they need. The second group is instead worried about quantum mechanics and, more specifically, its famous measurement problem. Some solutions to this problem pick out a preferred foliation of spacetime. Some hope that cosmology reveals this preferred foliation in its choice of a natural time function (Roser and Valentini, 2014).

What is this special time function? Hawking proved that as long as a solution of Einstein's field equations is "causally stable" (loosely put, it doesn't permit time travel or anything close to time travel) the spacetime described possesses a global function whose value increases along every future directed timelike or null worldline (Hawking, 1969). These time functions are a dime a dozen however: typically there are an infinite number definable in any causally stable spacetime. Enter cosmology. The hope is that cosmological considerations will narrow down these time functions to a smaller set, or better, to something unique.

The key idea is to look at the (approximate) symmetries of the cosmological spacetimes. Observations of the 2.7K cosmic microwave background (CMB) radiation and galaxy counts support the claim that at a large scale the universe is amazingly spatially isotropic. Deviations exist, but they are slight. The inference to global homogeneity is more theoretical. If the universe were spatially inhomogeneous, then for it to look so isotropic to us would mean we are very specially placed. Since to us modern Copernicans it seems plausible that our location is not special, we assume spatial homogeneity too, i.e., we naturally assume the cosmological principle mentioned above. In the context of general relativity, that gives us the constant (spatial) curvature spacetimes of FRW, described above, and FRW's special time function t, the cosmic time.

However, we should surely not insist on perfect isotropy and homogeneity, since our universe is definitely not perfectly isotropic and homogeneous. So a time function somewhat less tailored to FRW is desirable. One especially popular geometrical choice is York or *constant mean curvature* time (York, 1972). To understand York time, consider the collection of relativistic spacetimes (M, g). Some of these spacetimes can be foliated by everywhere spacelike three-dimensional hypersurfaces. The four- dimensional geometry obviously constrains the geometry on these spatial leaves; in fact, it induces on each hypersurface Σ a spatial three-dimensional metric h (which determines the intrinsic curvature of Σ) and an "extrinsic curvature" K (which, roughly speaking, describes how Σ is embedded in the entire spacetime M). Thinking of these in dynamical terms, e.g. in the initial value formulation of general relativity, you can regard h as a configurational, position-like variable and K as its conjugate momentum (which therefore says something about h's time development). Now, on a select set of these spacetimes, we can find some that permit a foliation where the mean extrinsic curvature associated with each spatial slice is constant. The mean, or average, extrinsic curvature is given by the trace of K, $\operatorname{Tr} K$. For these spacetimes each leaf of the foliation is a hypersurface whose mean (extrinsic) curvature is equal to a constant, i.e., $\operatorname{Tr} K = n$. York time is then defined as a time function $t: M \to I \subset \mathbf{R}$ such that its level sets $t^{-1}(I)$ are the hypersurfaces with mean curvature equal to n, for every $n \in I$ (Andersson et al., 2012). In short, each "tick" of the York clock reads a value of constant mean curvature. (The global spatial homogeneity of FRW spatial hypersurfaces implies that they have constant mean curvature, so York time coincides with cosmic time in FRW spacetimes.)

Constant mean curvature is an interesting topic in differential geometry and has many physical applications. York time too has a number of virtues, especially when it comes to numerical relativity and actual calculation. But in the present context what makes constant mean curvature and York time attractive is that under certain conditions it is provably unique (Marsden and Tipler, 1980; Rendall, 1996). For instance, restrict attention to globally hyperbolic spacetimes with a compact Cauchy hypersurface, a vanishing cosmological constant, and satisfying the so-called strong energy condition. Then, except for some outlier cases (Tr $K = 0$), the York time is provably unique. And as those three conditions are all geometrical in nature, one has a foliation that is in some sense geometrically unique (more below).

From a relativistic perspective, there is nothing surprising going on here. With York time one has just picked a special foliation tied to the curvature properties of certain spacetimes, and in some sufficiently special spacetimes this choice is natural. But one is no more forced to pick out a constant mean curvature foliation than to pick a different foliation, such as Misner's -log t, where t is the above FRW time (Misner, 1969). With York time in hand, however, those with the motivations described above now issue the divorce papers to spacetime. For tensers, no less than existence itself unfolds according to York time. And for those worried about quantum mechanics, the proposal is to identify the foliation quantum mechanics uses with York time. Physics or metaphysics marches to the beat of York. As for relativity, it can be abandoned. For the physics within the domain of special relativity, one sides with Lorentz over Einstein. For larger scales, one uses only the solutions of Einstein's equations that permit York time. The full possibilities allowed by general relativity are viewed as unnecessary extravagance.

Many commentators have questioned this identification of York time with metaphysical or quantum time. We cannot do justice to all the discussion here (see, for instance, Bourne, 2004; Callender, 2017). Perhaps the largest question that needs answering is simply: Why? Why think that metaphysical time or quantum time "cares" about constant mean curvature? Nothing in either metaphysics or (non-gravitational) quantum mechanics suggests any connection to curvature, let alone *constant mean curvature*. Why would metaphysical or quantum time care about maintaining an average? Here we will confine ourselves to just a couple of remarks.

First, the claims of the uniqueness of York time seem overblown—at least from the perspective of taking it seriously as a physically (as opposed to calculationally) preferred time. The uniqueness results occur only after quite a severe shrinking of the space of Einstein's solutions. The assumptions about a compact spatial slice, global hyperbolicity, the behavior of energy, and the cosmological constant may not hold. Probably they don't. But that is not all. Uniqueness is one thing, but existence is another. When does the York time exist? There are in fact many spacetimes that meet the above conditions but where York time does not exist. The "geometric invariance" of York time obtains only in a special class of spacetimes, one hardly demanded by observation.

Second, we suspect that many of the philosophical devotees of York time pick it simply because it seems to them to yield an attractive intuitive picture of cosmological evolution—that it is more or less like the familiar old absolute time of Newton or of their manifest *weltbild*. To chip away a little at this, we'll here note two relevant facts that we haven't seen in the philosophical literature. One, what makes York time great for numerical relativity but a bit weird for metaphysics is its avoidance of singularities. York time is definable in universes which, like ours (at least classically), contain black holes harboring singularities. The demand for constant mean curvature makes the foliation avoid these singularities, exponentially slowing York time to avoid them and lumping them all together at the "end of time." As a result, a singularity in a black hole born "today" (according to "intuitive" time) and one born a million years in your future (again, according to "intuitive" time) happen at the same moment of York time (Smarr and York, 1978) (see Figure 51.1). Two, during inflation the universe is said to enter into a de Sitter phase, that is, be describable by de Sitter spacetime. In the usual coordinates of de Sitter, all the $t =$ constant slices do indeed have constant curvature, which is nice; however, they all have the same value. The York clock seems to stop in the de Sitter phase!

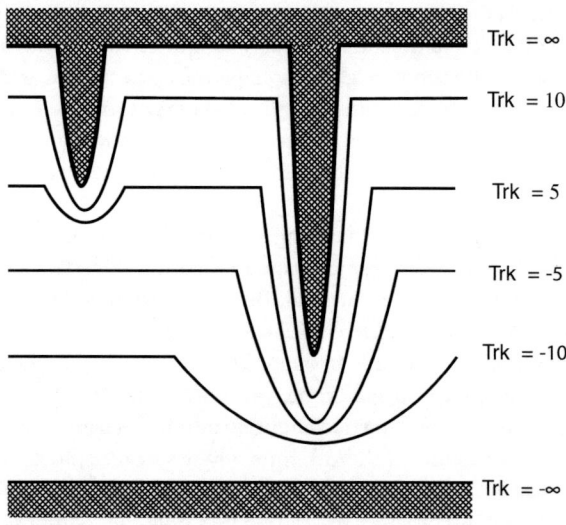

Figure 51.1 A depiction of York time in a spacetime with singularities. Adapted from similar figures in Qadir and Wheeler (1985)

None of these small observations constitute a knock–down objection to identifying York time with metaphysical or quantum time. But they should suggest to those hoping to rescue "intuitive time" that it's not really so much like our old friend absolute time as they might wish.

51.3 Cosmological Speculations

As we just saw, the use of distinguished foliations in cosmology invites deep questions about the nature of time and the significance of relativity. Putting those questions aside, the nature of time is pretty simple according to the standard big bang model: the universe began a finite amount of time to the past and will continue to expand indefinitely and ever faster to the future under the influence of dark energy; in addition, the entropy of its matter-energy distribution increases irresolutely toward this future; and, well, that's about it.

More specifically, due to the singularity theorems of Penrose, Hawking, and Geroch in the 1960s, the standard big bang model seems firmly committed to a past initial singularity (Smeenk, 2013). This basic consequence is intuitively suggested in typical expanding FRW universes since, assuming a positive cosmological constant and gravitating matter, the scale factor goes to zero and the matter density blows up as one goes back in time. The Hawking-Penrose singularity theorem makes the argument precise and therefore seems to provide an answer to the age old question about whether time is finite or infinite, the subject of Kant's famous first antinomy: time is finite toward the past but infinite toward the future (for flat and open cosmologies). Moreover, the geometrical modeling used in the big bang model allows one to skirt one of Kant's main worries, as the past is finite even though there is no "first" moment, just as there is no first positive real number.

That straightforward picture, however, is under threat from speculation hoping to address various outstanding theoretical problems. Modifications to the above picture—some quite dramatic—are frequently suggested based on cosmological applications of research programs in quantum gravity (e.g. superstring theory, loop quantum gravity) and from attempts to better understand inflation (e.g. eternal inflation). Indeed, despite its successes, inflation brings with it its own problems, and new developments of the general theory are regularly being proposed to escape these worries. These modifications essentially arise from what would be new physics. But modifications to the straightforward picture are often also connected to solutions of other perceived problems with the standard cosmological model. For instance, many physicists find the initial singularity predicted by the singularity theorems of general relativity to be a sign of the theory breaking down. They suspect that quantum effects will avoid these singularities. Similarly, many find the posit that the universe began in extremely low-entropy distasteful, or at least in need of explanation (Carroll, 2010). Can a cosmological model be devised that makes the past low-entropy a natural product of cosmological evolution as opposed to a special posit? Philosophers have debated whether these problems are really so dire.[5] Whether they are or not, eliminating one or both has led to the development of interesting new cosmological models that part ways in various respects with the standard picture of time. Here we briefly highlight some of the most prominent ideas that have gained currency.

51.3.1   *Multiverses and the Past*

One major development is the concept of a multiverse. This idea, that the universe given in the standard model is just one of a plurality of different universes, has become mainstream in cosmology and theoretical physics more generally. Interest in the multiverse derives from influential arguments that claim that the multiverse is a natural consequence of inflationary (Steinhardt, 1983; Vilenkin, 1983; Linde, 1986) and string-theoretic scenarios (Susskind, 2007). On the scenario known as eternal inflation, for example, a physical mechanism creates "bubble universes"—causally disconnected "pocket" universes—out of an inflating background spacetime. Insofar as one takes inflation to be an important part of the contemporary standard model of cosmology (the ΛCDM model), as cosmologists do, one has at least some reason to take the multiverse seriously. And clearly the idea of a multiverse opens up some intriguing new possibilities for the nature of time, possibilities that have only recently been explored.

The basic picture in eternal inflation is that the universe does not "enter into a de Sitter phase," as said above. Instead, the universe *is* essentially de Sitter space. The de Sitter universe is the relativistic spacetime analog of a sphere, just as Minkowski space is the relativistic spacetime analog of a plane. As such, it is the maximally symmetric spacetime with positive curvature (although it can be given a flat spatial slicing like the $k = 0$ FRW spacetime). In cosmology the accelerated expansion of de Sitter space is driven by a cosmological constant-like "inflaton" field in some appropriate potential. If spacetime were merely de Sitter, then nothing much would happen of course, apart from continuous exponential expansion. Quantum fluctuations of the inflaton field, however, can lead patches of spacetime to tunnel to another state and cease inflating, in such a way that a bubble universe may be born in a hot, dense big bang-like state. As bubble universes are continuously born, the space in between the universes continues to expand (although there is generally some probability that bubble universes may collide; these collisions may even leave a detectable footprint in, for example, the CMB). Thus one has an eternally inflating (toward the future) background spacetime which spawns a multiverse of big-bang-universe-like bubble universes, each of which has its own arrow of time, beginning, and (potentially, if it recollapses) ending.[6]

While many questions and concerns arise about the universe, let's focus on the traditional one about whether time is finite or infinite. Singularity theorems à la Penrose, Hawking, and Geroch do not obtain in eternally inflating spacetimes. These theorems make assumptions, typically about the behavior of matter-energy, that are not true in eternal inflation. So we don't have the same assurance that the world begins with a bang in an inflationary scenario as we do in a classical non-inflationary spacetime. Nonetheless, many cosmologists believe that the world still begins with a past singularity. They appeal to influential arguments by Vilenkin, Borde, and Guth that show that the geodesics of the inflationary multiverse scenario (eternal inflation) are past-incomplete, and hence that the inflationary multiverse had a beginning (Farhi and Guth, 1987; Borde and Vilenkin, 1996; Borde et al., 2003). These arguments have been called "singularity theorems" in analogy with the original classical singularity theorems. On the basis of these theorems, it would seem that the fundamental picture of time in the multiverse remains substantively the same as in the standard model, namely, finite to the past and infinite to the future (although of course there is the novelty associated with the birthing of pocket universes, which introduces times local to each universe).

But is that judgment correct? The eternal inflation "singularity theorems" are of quite a different character from those of Penrose, Hawking, and Geroch. They do not, for instance, make any assumptions about the behavior of energy, making them surprisingly general, nor do they apply to all past directed geodesics (past comoving geodesics are allowed to escape to past infinity). The main difference is that the geodesic incompleteness of eternally inflating spacetimes is not due to spacetime singularities but rather due to the spacetime being "incomplete," i.e., extendible. This observation and the physical plausibility of a steady-state-like eternal inflation have led Aguirre and Gratton, for example, to propose an extension which would make eternal inflation "two-sided" and geodesically complete. In their model eternal inflation proceeds in opposite temporal directions from a null boundary between the two regions, each causally-disconnected half of the entire spacetime birthing bubble universes to their respective futures (Aguirre and Gratton, 2002, 2003). Nonetheless, there is a clear sense in which time is infinite in both directions, since the arrows are connected at the boundary. Thus Aguirre and Gratton's clever maneuver would seem to partially restore Kant's old antinomy about the age of the universe, at least for eternally inflating spacetimes, since such spacetimes may seemingly be either finite or infinite toward the past.

51.3.2   *Bouncing Through the Big Bang*

Dissatisfaction with inflationary theory, a distaste for singularities, or an aversion to multiverses has led many to desire an alternative to the mainstream views in cosmology. Some propose that a big bang singularity can be avoided by a physical mechanism which causes a "bounce" before a singularity is reached. Thereby an old, collapsing (epoch of the) universe leads to a new (epoch of the) universe, one which looks just like the beginning of a classical big bang universe. Although the proposals flowing from the aforementioned motivations differ, they often have similar ideas so it is common to lump them together as the class of "bouncing" or "cyclical" cosmological models.[7] On some models this "bounce" happens only once; on other models it repeats and the universe's growing and shrinking resembles that of a linked set of sausages. In either case, what we thought of as the past-finite universe is only one epoch of a past-infinite universe of multiple epochs.

There are even more exotic possibilities though. Penrose proposes a model he calls the conformal cyclic cosmology: he glues together the "end" of what we thought was our universe, namely the stretched out accelerated infinite expansion, with the beginning of what we thought was our universe, the big bang. He is able to make this identification due to a conjecture that both the beginning and the end are conformally invariant. That is, despite appearing counterintuitive because of their size differences, the metrics describing the beginning and the end are conjectured to be the same except for scale—distances are different but angles are the same. In this theory, what one takes to be inflation is actually the *end* of a previous universe. This process repeats in a cycle and hence is an example of an infinitely recurring universe. Unlike in the "sausage link" model, where contracted big bangs are glued together, the conformal rescaling allows Penrose to glue together large expanded regions with small contracted regions (Figure 51.2).

It should be noted that bouncing and cyclical models do have a long history in cosmological thought, and indeed they have periodically appeared as scientific proposals in the 20th century.[8] Thus the recent spate of them is not entirely a novelty. What is novel is the ever-increasing breadth of these proposals and the ingenious application of theoretical resources. What makes these models more than just an exercise in pure reasoning is that they can be used to make predictions about (what in the standard model would be called) the early universe, in particular the primordial spectrum of

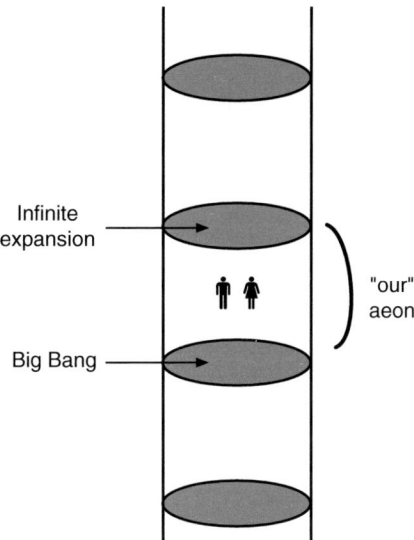

Figure 51.2 Aeons of time in a conformal cyclic universe

inhomogeneities that gave rise to the formation of stars and galaxies. Whether any of the predictions of these higher speculations will be confirmed (or are even confirmable) of course remains to be seen. But for philosophers of time, at least, it is worth noting that for now many of these remarkable proposals provide viable alternatives to the default view that time is finite to the past.

51.3.3   *Eliminating the Past Hypothesis*

To explain the thermodynamic arrow of time, most physicists and philosophers feel that one must, in some sense or other, assume that the early universe began in a state of extremely low-entropy. Making this assumption, typical initial states will then be overwhelmingly likely to evolve toward higher entropy states. Distinguished physicists such as Boltzmann and Feynman saw the need for such an initial posit and subsequent philosophers have generally agreed with it. Following Albert (2000), this assumption is often dubbed the "past hypothesis."[9]

As noted already, several physicists and philosophers feel that explaining the low-entropy state of the very early universe is one of the great mysteries still remaining in physics. The goal is to find a dynamical origin of the past hypothesis. If one can show that dynamical laws necessitate a low-entropy state in our local past, then we can get rid of the otherwise ad hoc past hypothesis. To this end, several physicists have in recent years proposed cosmological models that do not require the invocation of the past hypothesis (Aguirre and Gratton, 2003; Carroll and Chen, 2004; Barbour et al., 2014; Goldstein et al., 2016). Some of these approaches are closely connected to the topics of the previous subsections: the Carroll and Chen (2004) model, like the Aguirre and Gratton (2003) model,[10] is an eternally inflating multiverse with dual arrows of time, while the Barbour et al. (2014) model is a version of a "one bounce" cosmology driven by gravity.

Cosmological considerations of course have always been at the heart of discussions of time's arrow. Boltzmann's famous cosmological story was in fact not entirely unlike the bouncing universe scenarios. Thermal fluctuations to states of low-entropy are in effect past boundaries for two "universes," each with an oppositely directed arrow of time, pointed in the direction of increasing entropy (Figure 51.3). Later, approximately 30 years after the Hubble expansion of the universe was discovered, the cosmologist Thomas Gold suggested that expansion caused the thermodynamic arrow

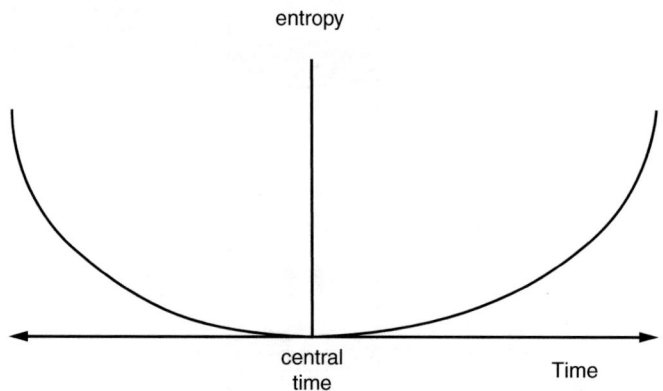

Figure 51.3 The arrow of time aligning with increasing entropy in "two-sided" universes

of time (Gold, 1962). Boltzmann's theory was roundly dismissed when he proposed it, as he wrote well before the advent of modern cosmology. And Gold's idea was criticized by many, as the link between expansion and thermodynamic entropy wasn't as tight as Gold hoped.

In the model of Goldstein et al. (2016), we see something like Boltzmann's original idea being floated again, possibly also connected to something like Gold's original idea. They examine a toy model, a large classical gas of non-interacting particles in Euclidean space. Measuring the size of 3d space occupied by the gas, they find that normal evolution takes the gas to a central time where the gas occupies a uniquely minimum size. They further show that the entropy associated with this measure of system size grows in both temporal directions away from this central time. Each solution comes with its own past hypothesis, if you like, at its central time. No special additional posit is necessary. Interestingly, this is compatible with a Boltzmannian picture of statistical mechanics, one where most microstates realizing low-entropy macrostates evolve into states realizing higher entropy macrostates. This picture is a lot like Boltzmann's original idea, except that the matter and size of the space it occupies grows without bound and hence "bounces" once, unlike in Boltzmann's model. Insofar as the model is connected to expansion, there is also an echo of Gold's idea here too. Gold's idea faltered because there was no necessary link between expansion and entropy increase. Expansion can be an isentropic process, so it is hard to see how it could cause the thermodynamic arrow. Whether the present idea suffers a similar fate, however, remains to be seen.

Notes

1 (Dainton 2010) is a good philosophical introduction to the philosophy of space and time. (Smeenk 2013) provides an accessible review on the topic of time in cosmology. (Harrison 2000) is a classic physics text on cosmology that is more oriented toward history and philosophy than usual. (Kragh 1996) and (Longair 2006) are useful historical references for 20th century cosmology.

2 See Anderl's contribution to this volume for more details on issues with astrophysical evidence.

3 For more on dark energy and dark matter, see Jacquart's contribution to this volume. For more on inflation, see Ijjas's contribution to this volume.

4 See Eagle's contribution to this volume for more on this threat.

5 See, e.g., (Earman 1995) for discussion of singularities and (Callender 2004) and (Price 2004) on the low-entropy posit.

6 See Ijjas's contribution to this volume for more on eternal inflation.

7 Prominent examples include those discussed in (Khoury et al. 2001); Steinhardt and (Turok 2007); (Ashtekar 2009); (Penrose 2010). A recent review of bouncing models is (Battefeld and Peter 2014).

8 A fine survey of this history can be found in (Kragh 2011, Ch. 8).

9 See Shahvisi's contribution to this volume for more on the past hypothesis and entropy increase.

10 See (Vilenkin 2013) for discussion of these models in relation to the arrow of time.

References

Aguirre, A. and Gratton, S. (2002). Steady-state eternal inflation. *Physical Review D*, 65: 083507.

Aguirre, A. and Gratton, S. (2003). Inflation without a beginning: A null boundary proposal. *Physical Review D*, 67: 083515.

Albert, D. (2000). *Time and Chance*. Cambridge, MA: Harvard University Press.

Andersson, L., Barbot, T., Béguin, F. and Zeghib, A. (2012). Cosmological time versus CMC time in spacetimes of constant curvature. *Asian Journal of Mathematics*, 16(1): 37–88. http://projecteuclid.org/euclid.ajm/13316 63451.

Ashtekar, A. (2009). Loop quantum cosmology: An overview. *General Relativity and Gravitation*, 41: 707–741.

Barbour, J., Koslowski, T. and Mercati, F. (2014) Identification of a gravitational arrow of time. *Physical Review Letters*, 113: 181101.

Battefeld, D. and Peter, P. (2015). A critical review of classical bouncing cosmologies. *Physics Reports*, 571: 1–66.

Beisbart, C. and Jung, T. (2006). Privileged, typical, or not even that?—Our place in the world according to the Copernican and the cosmological principles. *Journal for General Philosophy of Science*, 37: 225–256.

Borde, A., Guth, A. and Vilenkin, A. (2003). Inflationary spacetimes are incomplete in past direction. *Physical Review Letters*, 90: 151301.

Borde, A. and Vilenkin, A. (1996). Singularities in inflationary cosmology: A review. *International Journal of Modern Physics D*, 5: 813–824.

Bourne, C. (2004). Becoming inflated. *British Journal for the Philosophy of Science*, 55: 107–119.

Butterfield, J. (2014). On under-determination in cosmology. *Studies in History and Philosophy of Modern Physics*, 46: 57–69.

Callender, C. (2004). There is no puzzle about the low-entropy past. In C. Hitchcock (ed.), *Contemporary Debates in the Philosophy of Science*. Hoboken: Wiley-Blackwell, chapter 12, pp. 240–255.

Callender, C. (2017). *What Makes Time Special?* Oxford: Oxford University Press.

Canales, J. (2015). *The Physicist & the Philosopher*. Princeton, NJ: Princeton University Press.

Carroll, S. (2010). *From Eternity to Here*. New York: Plume.

Carroll, S. and Chen, J. (2004). *Spontaneous inflation and the origin of the arrow of time*. ArXiv Eprint. http://arxiv.org/abs/hep-th/0410270.

Crisp, T. (2008). Presentism, eternalism, and relativity physics. In W.L. Craig and Q. Smith (eds.), *Einstein, Relativity and Absolute Simultaneity*. New York: Routledge, pp. 262–278.

Dainton, B. (2010). *Time and Space* (2nd edition). Durham: Acumen Press.

Earman, J. (1995). *Bangs, Crunches, Whimpers, and Shrieks*. Oxford: Oxford University Press.

Eddington, A. (1920 [1987]). *Space, Time and Gravitation*. Cambridge: Cambridge University Press.

Ellis, G. (2007). Issues in the philosophy of cosmology. In J. Earman and J. Butterfield (eds.), *Philosophy of Physics*, Vol. B of *Handbook of the Philosophy of Science*. Amsterdam: Elsevier, pp. 1183–1285.

Farhi, E. and Guth, A. (1987). An obstacle to creating a universe in a laboratory. *Physics Letters B*, 183: 149–155.

Gold, T. (1962). The arrow of time. *American Journal of Physics*, 30: 403–410.

Goldstein, S. Tumulka, R. and Zanghì, N. (2016). Is the hypothesis about a low entropy initial state of the Universe necessary for explaining the arrow of time? *Physical Review D*, 94: 023520.

Harrison, E. (2000). *Cosmology: The Science of the Universe* (2nd edition). Cambridge: Cambridge University Press.

Hawking, S. (1969). The existence of cosmic time functions. *Proceedings of the Royal Society A: Mathematical, Physical and Engineering Science*, 308: 433–435.

Jeans, J. (1935). Man and the universe. In J. Jeans (ed.), *Scientific Progress*. London: Allen & Unwin, pp. 1–12.

Khoury, J., Ovrut, B., Steinhardt, P. and Turok, N. (2001). Ekpyrotic universe: Colliding branes and the origin of the hot big bang. *Physical Review D*, 64: 123522.

Kragh, H. (1996). *Cosmology and Controversy*. Princeton, NJ: Princeton University Press.

Kragh, H. (2011). *Higher Speculations*. Oxford: Oxford University Press.

Linde, A. (1986). Eternally existing self-reproducing chaotic inflationary universes. *Physics Letters B*, 175: 395–400.

Longair, M. (2006). *The Cosmic Century*. Cambridge: Cambridge University Press.

Lucas, J. and Hodgson, P. (1990). *Spacetime and Electromagnetism*. Oxford: Oxford University Press.

Marsden, J. and Tipler, F. (1980). Maximal hypersurfaces and foliations of constant mean curvature in general relativity. *Physics Reports*, 66: 109–139.

Misner, C. (1969). Absolute zero of time. *Physical Review*, 186: 1328–1333.

Penrose, R. (2010). *Cycles of Time*. London: Bodley Head.

Price, H. (2004). On the origins of the arrow of time: Why there is still a puzzle about the low-entropy past. In C. Hitchcock (ed.), *Contemporary Debates in the Philosophy of Science*. Hoboken, NJ: Wiley-Blackwell, pp. 219–239.

Qadir, A. and Wheeler, J.A. (1985). York's cosmic time versus proper time as relevant to changes in the dimensionless "constants", K-meson decay, and the unity of black holes and big crunch. In E. Gotsman and G. Tauber (eds.), *From SU(3) to Gravity*. Cambridge: Cambridge University Press, pp. 383–394.

Rendall, A. (1996). Constant mean curvature foliations in cosmological spacetimes. *Helvetica Physica Acta*, 69: 490–500.

Roser, P. and Valentini, A. (2014). Classical and quantum cosmology with York time. *Classical and Quantum Gravity*, 31: 245001.

Smarr, L. and York, J. (1978). Kinematic conditions on the construction of spacetime. *Physical Review D*, 17: 2529–2551.

Smeenk, C. (2013). Time in cosmology. In A. Bardon and H. Dyke (eds.), *A Companion to the Philosophy of Time*. Oxford: Wiley-Blackwell, pp. 201–219.

Steinhardt, P. (1983). Natural inflation. In G. Gibbons, S. Hawking and S. Siklos (eds.), *The Very Early Universe*. Cambridge: Cambridge University Press, pp. 251–266.

Steinhardt, P. and Turok, N. (2007). *Endless Universe*. New York: Doubleday.

Susskind, L. (2007). The anthropic landscape of string theory. In B. Carr (ed.), *Universe or Multiverse*. Cambridge: Cambridge University Press, pp. 247–266.

Swinburne, R. (2008). Cosmic simultaneity. In W.L. Craig and Q. Smith (eds.), *Einstein, Relativity and Absolute Simultaneity*. New York: Routledge, pp. 244–261.

Vilenkin, A. (1983). The birth of inflationary universes. *Physical Review D*, 27: 2848–2855.

Vilenkin, A. (2013). Arrow of time and the beginning of the universe. *Physihcal Review D*, 88: 043516.

York, J. (1972). Role of conformal three-geometry in the dynamics of gravitation. *Physical Review Letters*, 28: 1082–1085.

Further Reading from the Editors

In this volume, Chapter 25 by N. Huggett on time as emergent and Chapter 26 by K. Thébault on the problem of time are both relevant here. C. Smeenk, "Time in Cosmology", in A. Bardon and H. Dyke (eds.), *The Blackwell Companion to the Philosophy of Time* (Oxford: Blackwell, 2013): pp. 201–219 is an additional general survey of the present topic. J. Earman, *Bangs, Crunches, Whimpers and Shrieks* (New York: Oxford University Press, 1995) provides a review of relevant technical material. C. Callender, *What Makes Time Special?*. (Oxford: Oxford University Press, 2017) explores the question of what distinguishes time from other physical dimensions. S. Carroll, *From Eternity to Here* (New York: Plume, 2010) is a big-picture overview directed at a popular audience.

52

THE FINE-TUNING OF THE UNIVERSE FOR LIFE

Luke A. Barnes

52.1   Introduction: Fine-Tuning in Physics

When a physicist says that a theory is *fine-tuned*, they mean that it must make a suspiciously precise assumption in order to explain a certain observation. This is evidence that the theory is deficient or incomplete. As a simple example, consider a geocentric model of the Solar System. Naively, at any particular time, the Sun and planets could be anywhere in their orbits around the Earth. However, in our night sky, Mercury is never observed to be more than $28°$ from the Sun, and Venus is never seen more than $47°$ from the Sun.

Can a geocentric model explain this observation? Yes, but only by adding a postulate. In Ptolemy's geocentric model, Mercury and Venus travel on epicycles, and those epicycles are centered on a line joining the Earth to the Sun (Figure 52.1). This explains the data, so the model does not fail. However, in the context of the model, this assumption is unmotivated and suspiciously precise. Given only that the planets and Sun orbit the Earth, there is no reason to expect such an arrangement.

This fine-tuning of the geocentric model doesn't necessarily mean that it is wrong, but it should make us wary. We should search for a model in which the data is explained more naturally: Mercury and Venus are never seen too far from the Sun because the planets orbit the Sun, not the Earth.[1]

Similar arguments play an important role in modern cosmology and particle physics. A standard cosmology textbook case for cosmic inflation goes as follows (e.g., Peacock, 1998). In the standard model of cosmology, the geometry of the universe can be negatively curved, flat, or positively curved, depending on whether the density universe is less than, equal to, or greater than the *critical density*. In this model, two facts seem to be in tension with each other. Firstly, the matter in the universe causes the density of the universe to evolve away from critical. Secondly, observations tell us that the density of the universe is very close to critical.

What about in the past? If we extend the model back to nucleosynthesis, about 1 second after the beginning, then the density of the universe must be within one part in 10^{15} of the critical density in order to still be close to critical today. The further we push back, the closer the constraint: at the Planck time, it is one part in 10^{55}. As with Ptolemy's model, the standard model of cosmology can explain the data, but only with an unmotivated and suspiciously precise assumption. We must simply assume that the density of the universe was extremely close to critical in its earliest moments. This motivates inflationary models, in which a early period of accelerating expansion drives *toward* critical density.

A second example comes from particle physics (Dine, 2015). The observed mass of the Higgs particle can be written in terms of a "bare" value and quantum corrections. These quantities are

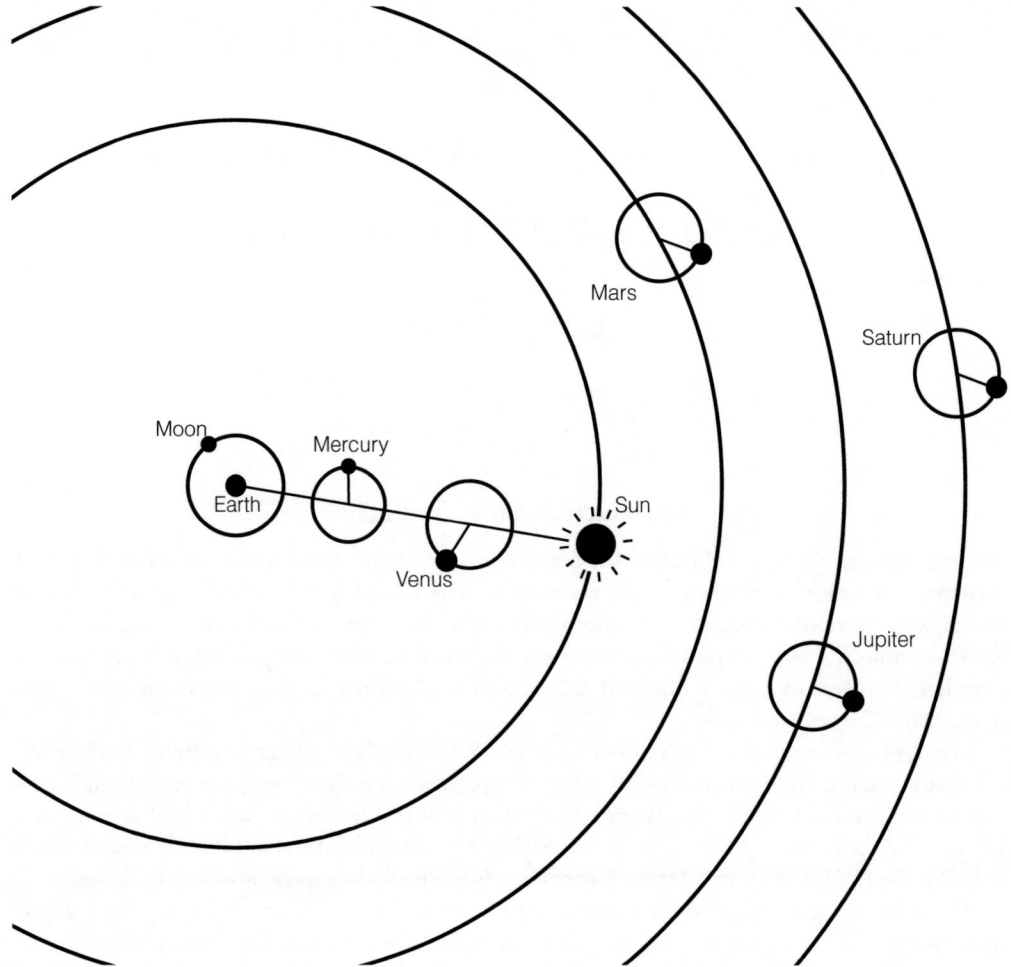

Figure 52.1 The Ptolemaic model of the Solar System. All the planets travel on epicycles (smaller circles) in orbit around the Earth, but in the case of Mercury and Venus it must be assumed that the centers of their epicycles are fixed to a line that connects in the Earth and the Sun. In this way, Mercury and Venus are never seen too far from the Sun. But in the context of the model, this assumption is unmotivated and suspiciously precise. This presents an opportunity for another model to explain this data more naturally: Mercury and Venus are never seen too far from the Sun because the planets orbit the Sun, not the Earth

independent in the model. However, the size of the quantum corrections diverges quadratically with the scale up to which the effective theory can be trusted. Dine says,

> if the cutoff is the Planck scale, this correction is enormous . . . about 34 four orders of magnitude larger than [the observed value], corresponding to a fine-tuning of the bare parameters against the radiative correction at the part in 10^{34} level.

Again, the model can explain the observed value, but only by making the unmotivated and suspiciously precise assumption that the bare value almost perfectly cancels out the quantum corrections (see also Donoghue, 2007). Particle physicists tend to call this situation "unnatural" rather than fine-tuned, but it's a similar idea. As Dine notes, "naturalness has for many years been a guiding principle in the search for physics beyond the standard model."

The assumptions underlying these arguments have been the subject of much theoretical attention, but the logic is quite widely accepted. The cosmological constant problem, the flatness problem, the big- and little- hierarchy problems of particle physics (see Jacquart in this volume) and the strong CP (charge-parity) problem (see Ijjas, Chapter 53) can be framed as fine-tuning problems.

One particular case of fine-tuning is particularly striking. The data in question are not the precise measurements of cosmology or particle physics, but a more general feature of our universe: it supports the existence of life. Before we look at this in more detail, it will be helpful to place fine-tuning in the context of Bayesian approaches to testing physical theories.

52.2 Bayesian Accounts of Fine-Tuning

The Bayesian approach to probability theory views probabilities as quantifying the *degree of plausibility* of some proposition, given other propositions. Bayesians have argued that the familiar probability axioms of Kolmogorov (1933) (or similar) also apply to degrees of plausibility. This can be shown via Dutch book arguments, representational theorems that trace back to Ramsey (1926), or (more common among physicists) the theorem of Cox (1946), which proposes that degrees of plausibility obey some intuitive desiderata (see also Jaynes, 2003; Caticha, 2009; Knuth and Skilling, 2012).

In the Bayesian approach, physical theories are tested as follows. Let,

- T = the proposed theory to be tested. As a concrete example, T may represent a set of symmetry principles, from which we can derive the mathematical form of a Lagrangian (or, equivalently, the dynamical equations), but not the values of its free parameters.
- U = our observations of this universe.
- B = everything else we know. For example, we treat the findings of mathematics and theoretical physics as given, so these are included in B. As I have defined it for our purposes here, the information in B does not give us any information about which possible world is actual. The theoretical physicist can explore models of the universe mathematically, without concern for whether they describe reality.

We then would like to know how plausible T is, in light of everything that we know UB. If the *posterior* probability $p(T|UB)$ — read "the probability of T given U and B" — were to descend to us on a cloud from the heavens, then our job would be done. Alternatively, we may need some help in calculating the posterior, and so we turn to Bayes Theorem,

$$p(T|UB) = \frac{p(U|TB)\,p(T|B)}{p(U|B)}. \tag{52.1}$$

If the theory in question has free parameters, which we generically denote α_T, then we must take into account our lack of knowledge of these parameters in evaluating the *likelihood* of the data given the theory $p(U|TB)$. We can think of this as dividing the theory into a large number of sub-theories, each with a different value of the free parameters. To calculate the likelihood, we need to average over these sub-theories — this is known as marginalizing over nuisance parameters. Sub-theories that can account for the data bring the average up, and sub-theories that can't bring the average down.

As a simplistic model, suppose a free parameter varies uniformly over a range R, but only a small range $\Delta\alpha_T$ is consistent with the data. Then the theory's likelihood is penalized by a factor $\Delta\alpha_T/R$. The smaller the range of free parameters that accounts for the data, relative to the range of the parameters dictated by the theory, the more the likelihood is penalized. Fine-tuning can be translated directly into improbability within a Bayesian approach (see also Aguirre, 2007; Fowlie, 2014; Barnes, 2017, and references therein).

52.3 The Fine-Tuning of the Universe for Life

Part of exploring any physical model is calculating the effect of varying its free parameters. As we have seen, this is necessary for calculating the likelihood of the data given the theory (via marginalizing), and so this can tell us whether the theory is fine-tuned or not. Beginning in the 1970s, physicists noted that seemingly small changes to the fundamental constants of nature and the initial conditions of the cosmos not only brought our models in conflict with precise measurements; they described universes in which no life form could exist. The complexity and stability required by any known or thus far conceived form of life could be rather easily erased.

This *fine-tuning of the universe for life* was first investigated by Carter (1974), Silk (1977), Carr and Rees (1979), Davies (1983), and Barrow and Tipler (1986), and has been reviewed recently by Hogan (2000), Barnes (2012), Schellekens (2013) and Lewis and Barnes (2016). We will consider a few examples.

52.3.1 *The Cosmological Constant*

The cosmological constant problem is described in the textbook of Burgess and Moore (2006) as "arguably the most severe theoretical problem in high-energy physics today, as measured by both the difference between observations and theoretical predictions, and by the lack of convincing theoretical ideas which address it." The problem is as follows. Quantum field theory describes particles as configurations of a field. There is a particular configuration of the field that corresponds to a state with zero particles; this is known as the vacuum state. Because the field is still there, we can ask: how much energy is contained in the vacuum?

The absolute energy of the field doesn't effect the interactions of the standard model of particle physics, which depend only on energy differences. But gravity, on Einstein's theory, responds to the absolute amount of energy. In a homogeneous and isotropic universe, vacuum energy has the same effect as Einstein's cosmological constant. When cosmologists speak of the cosmological constant, they usually mean the sum of the "bare" cosmological constant in Einstein's equation and all the forms of energy in the universe that behave in the same way. This is the quantity that is constrained by cosmological data. In Planck units ($\hbar = G = c = 1$) and expressed as a density, the observed cosmological constant has the value $\rho_\Lambda \approx 1.2 \times 10^{-123}$.

We can estimate the contribution to the energy in the vacuum from a given quantum field. Loosely speaking, even in the vacuum state, virtual particles will be created and annihilate, forming loops in a Feynman diagram. The vacuum energy depends on the energy scale up to which we trust the theory to describe this process. Even if we only consider well-understood fields (e.g., the electron field) up to energy scales that have been thoroughly investigated by experiment (say, ~ 100 GeV), the contribution to the vacuum energy is $\sim 10^{-68}$, or 55 orders of magnitude larger than the observed value. If we extend the range of our theory up to a popular energy scale where new physics is expected, the supersymmetry scale, then the contribution to the vacuum energy is $\sim 10^{-64}$. If we extend all the way to the Planck scale, where cannot trust our theories because they do not account for quantum gravity effects, the contribution to the vacuum energy is ~ 1, 123 orders of magnitude larger than the observed value.

This is a fine-tuning problem. Quantum field theory and general relativity *can* explain the small observed value of the cosmological constant, but only by supposing that the different (positive and negative) contributions to the vacuum energy from each quantum field happen to cancel each other to 123 decimal places. This requires an unmotivated but suspiciously precise coincidence between a number of independent factors.

As an example of fine-tuning *for life*, the cosmological constant problem is a near-perfect storm.

• It's actually several problems. Each quantum field – electron, quark, photon, neutrino, etc. – adds a very large (positive or negative) contribution to the vacuum energy of the universe.

- General Relativity won't help. Einstein's theory links energy and momentum to spacetime geometry. It does not dictate what energy and momentum exists in the universe. Universes that are no good for life are perfectly fine by the principles of General Relativity.
- Particle physics probably won't help. All particle physics processes, being described by quantum field theory, depend only on energy differences; only gravity responds to absolute energies. Thus, particle physics is largely blind to its effect on cosmology, and thereby life.
- It isn't just a problem at the Planck scale, so quantum gravity won't necessarily help. As noted above, we don't need to trust quantum field theory all the way up to the Planck energy in order to see the cosmological constant problem. It is entrenched firmly within well-understood, well-tested physics.
- Alternative forms of dark energy have very similar problems. They usually posit some other kind of field, and so the problem of the vacuum energy of the field remains, unchanged and unsolved. See Jacquart (this volume, Chapter 53) for more discussion of dark energy.
- We can't aim for zero. Before the accelerated expansion of the universe was discovered 1998, it was thought that some principle or symmetry would set the cosmological constant to zero. Even this was a speculative hope, and it has since evaporated.
- The quantum vacuum has observable consequences, and so cannot be dismissed as mere fiction. In particular, an electron in an atom feels the influence of the quantum vacuum (the *Lamb shift*). Our theory works beautifully for electrons and atoms. Why doesn't cosmic expansion feel the influence of the quantum vacuum?
- The cosmological constant has a very obvious and definitive effect on the necessary conditions for life. A positive cosmological constant causes the expansion of the universe to accelerate, freezing structure formation. Make the cosmological constant a few orders of magnitude larger and structure formation freezes before *anything* has formed. The universe will be a thin, uniform hydrogen and helium soup, a diffuse gas where the occasional particle collision is all that ever happens. A very simple way to make a universe lifeless is to make it devoid of any structure whatsoever. Alternatively, a negative cosmological constant causes the universe to recollapse. If the cosmological constant were -10^{-68}, then the universe would recollapse $\sim \sqrt{10^{68}}\, t_{\text{Planck}} \sim 10^{-10}$ seconds after the big bang.

52.3.2 *The Parameters of the Standard Model*

The standard model of particle physics has 25 free parameters which are constrained by experiment. Many of these play a crucial role in providing the complexity required by life.

The Higgs field "gives mass" to the fundamental particles of the standard model. We can write their masses in terms of the vacuum expectation value (vev) of the field (v) as $m_f = \Gamma_f v/\sqrt{2}$, where Γ_f is the particle's dimensionless Yukawa parameter. As with vacuum energy, quantum corrections to the bare Higgs vev are predicted to be of the same order as the scale up to which we trust the theory. The observed value of $v = 1.0 \times 10^{-17}$ is unnaturally small.

Similarly small changes to v significantly affect how particles interact and bind. Damour and Donoghue (2008) refines the approach of Agrawal et al. (1998) by considering nuclear binding, and conclude that unless $0.78 \times 10^{-17} \lesssim v/m_{\text{Planck}} \lesssim 3.3 \times 10^{-17}$ hydrogen is unstable to the reaction $p + e \to n + v_e$ (if v is too small) or else there is no nuclear binding at all (if v is too large).

Similarly, the strengths of the fundamental forces are subject to anthropic constraints. For example, unless $\alpha \lesssim (m_d - m_u)/141$ MeV, the electromagnetic contribution to the mass of the proton causes it to be heavier than the neutron, making the proton unstable (Hogan, 2000; Hall and Nomura, 2008). If the strong force were a few percent weaker, the deuteron would be unbound (Pochet et al., 1991). The first step in stellar burning would require a three-body reaction to form helium-3. This requires such extreme temperatures and densities that stable stars cannot form: anything big enough to burn is too big to be stable (Barnes and Lewis, 2017).[2] Weaken the strong force by a few more percent, or increase the strength of electromagnetism, and carbon and all larger elements are unstable (Barrow and Tipler, 1986). The parameters of the standard model must walk a tight-rope in order to form stable nuclei and support stable stars.

52.3.3 The Dimensionality of Spacetime

Spacetime is the arena in which physics takes place. At the length scales relevant to nuclei, atoms, stars, and the observable universe, spacetime is described by three dimensions of space and one of time. It is often straightforward to write down our familiar laws of nature in any number of dimensions. For example, in m time dimensions (t_i) and n space dimensions (x_j), the wave equation is,

$$\sum_{i=1}^{m} \frac{\partial^2 \rho}{\partial t_i^2} = c_s^2 \sum_{j=1}^{n} \frac{\partial^2 \rho}{\partial x_j^2}, \tag{52.2}$$

for the scalar wave variable ρ, and wave speed c_s.

Given that we can theoretically explore such universes, what would they be like? This question has been addressed by Ehrenfest (1917), Whitrow (1955), Barrow and Tipler (1986), and Tegmark (1997). It has been known for some time that Newtonian gravity only predicts stable planetary orbits in three space dimensions (Bertrand's theorem). With four space dimensions, for example, slightly non-circular orbits are spiraled, not elliptical — they would send the planet into the star or off into empty space. The same applies to atomic orbits described by the Schrodinger equation — there is no stable ground state.

We can also vary the number of time dimensions. In such a universe, an observer will have their own clock that measures time along their worldline; but what would they experience? Tegmark (1997) notes that linear partial differential equations, of which the wave equation is one example and by which many known laws can be approximated locally, have interesting properties when there is more than one time dimension. In our universe, we can approximately predict the behavior of a physical system into the future on the basis of knowledge of our immediate environment. (I don't necessarily mean *predict* in a mathematical sense. A bird "predicts" the path of a flying insect to catch it.) But if there were more (or less) than one time dimension, then the problem would be mathematically ill-posed, being infinitely sensitive to the initial conditions. The behavior of one's environment could not be predicted using only local, finite accuracy data, making storing and processing information impossible.

52.4 The Multiverse

Fine-tuning in physics serves as impetus to search for a better theory, one which can account for the facts in a more natural way, without unmotivated assumptions. But what could naturally explain a *life-permitting* universe?

Perhaps we won the cosmic lottery: a life-permitting universe exists, despite the seemingly overwhelming odds, because the universe as a whole consists of a vast, variegated ensemble of sub-universes — a *multiverse*.

A viable multiverse model needs a few ingredients. The first is a physical theory that goes beyond the standard models by promoting the constants of nature and initial conditions to dynamic variables. We have some hints about how to do this. The strengths of the fundamental forces of particle physics are a function of energy, and seem to converge at an energy far above our current experiments. This has led to the development of Grand Unified Theories (GUT), in which the strong nuclear force, weak nuclear force, and electromagnetism are manifestations of a single, unified force (see Raby,

2010). At low energy, the greater symmetry of the unified field is *spontaneously broken*: the strengths of the forces are not written in stone in the fundamental equations, but rather are a frozen accident.

There are other ways to promote the constants to variables. In string theory, there is a *landscape* of solutions to the fundamental equations, with the familiar "constants" of physics written into the various folds and holes of the extra, *compactified* spatial dimensions (Schellekens, 2013). They become free parameters of the *solution* to the equations, rather than appearing in the equations themselves.

The second ingredient of a multiverse theory is a cosmological mechanism to create domains of the universe with different values of the "constants." The leading contender today is cosmic inflation: in its earliest moments, the universe expanded at an accelerating rate, driving it toward critical density and laying down the seeds of cosmic structure.

The successful predictions of inflation require only that our observable universe inflated, but it has been argued that inflation will naturally produce a multiverse (see Linde, 2015). Most inflationary models posit a form of energy called an *inflaton field* that drives the expansion of the universe. The physics of a quantum field is codified in its *potential*: the dynamics is analogous to a ball rolling on a hill, and the shape of the hill tells us how the motion of the ball depends on the value of the field. For an inflaton field to cause accelerating expansion, it must be rolling slowly on a very flat section of the potential. Inflation ends when the field rolls off the flat section, usually into a valley. As the field oscillates around the bottom of the valley, reheating begins: the energy in the inflaton field is transferred into ordinary matter and radiation, beginning the hot big bang phase.

But the field is a quantum field, and so will not evolve deterministically (depending on your interpretation of quantum mechanics). Somewhat simplistically, consider an inflating region of the universe, in which the inflaton field dominates the energy of the universe and is slowly rolling. While in most of the region the field will roll into the valley and inflation will end, there is a finite probability that the field in some sub-region will evolve to a state further up the slope. This part of the universe will inflate for longer. Because this sub-region keeps growing in size, it will soon be larger than the original region, and so inflation will always continue somewhere. Given a sufficiently large initial inflating region, post-big-bang pockets form in an inflating background.

If the energy scale of reheating is above the symmetry breaking scale of the fields in the universe, then the symmetry will break differently in different sub-universes. This creates a population of sub-universes with different "constants" and big bang "initial" conditions.

The final ingredient is a selection effect (Wall and Jenkins, 2003; Bostrom, 2002; Neal, 2006). Consider the prediction of cosmic microwave background (CMB) anisotropies in the standard big bang model, from which cosmologists infer the values of various cosmic parameters. Like any thermodynamic system, there are fluctuations in the recombining plasma. So, in a sufficiently large universe, the probability of *someone* observing the CMB that we see approaches one regardless of the values of the cosmic parameters. If we tested physical models by calculating the probability that *some* observation in the universe matches our actual observations, then any values of the cosmic parameters would do in an infinite universe. We couldn't infer their values from observations. A multiverse would make this problem even worse.

To resolve this problem, remember that we don't just know that *some* observation has taken place, but that a *particular* observer has made an observation. Even if *some* observer sees a misleading CMB, the vast majority won't, justifying our inference. We apply this to the multiverse: that *some* region of the universe permits life is a good start but not sufficient. What will a *typical* observer see?

The anthropic prediction by Weinberg (1987) of the cosmological constant provides an excellent test case. Given a large enough variety of sub-universes with different values of the cosmological constant, somewhere will have a value that permits structure to form. In such an ensemble, asks Weinberg, what cosmological constant would a typical observer see? There is nothing in fundamental physics as we know it that singles out $\rho_\Lambda = 0$ as a privileged value, so we assume for the moment that for values of ρ_Λ much smaller than the "natural" Planck scale, the multiverse produces a roughly uniform distribution of values. Then (considering positive values of ρ_Λ for now) what is the largest value of ρ_Λ that permits the formation of structure? Weinberg's analytic calculation gives an upper limit of $\rho_{\Lambda,\max} \approx 550\rho_0 \approx 3 \times 10^{-121}$, where ρ_0 is the present cosmic mass density. Weinberg made this prediction before observation showed that $\rho_\Lambda \approx 1.2 \times 10^{-123}$.

A typical observer would expect to observe a vacuum energy roughly comparable with the anthropic upper bound. It can't be larger, of course — there are no observers in those sub-universes to make an observation. Weinberg's calculation gives the upper bound as being two orders of magnitude above the actual value, which is close enough to take the calculation seriously. In Barnes et al. (2018), we have repeated Weinberg's calculation using more sophisticated supercomputer models of galaxy formation.

If we had observed a value that was ten orders of magnitude smaller than the upper bound, then we would conclude that one of the assumptions in our model is probably wrong. We would look for a dynamical or symmetry-based explanation, rather than an anthropic one.

This kind of case for the multiverse has been criticized as speculative and untestable. But it should be remembered that such considerations are almost unavoidable in cosmology. Just as the astronomer must understand their telescope before they can understand what they see through it, when a cosmologist models the universe, they are inevitably modeling a system that contains themselves. We cannot pretend to stand outside the universe. Selection effects cannot be ignored. We are not Dr Frankenstein; we are the monster. We have woken up in a laboratory and are trying to understand how it made us.

We can test the multiverse using Bayesian probability theory. In this case, the "data" to be explained is the constants of nature. If the fine-tuning for life implies that almost all observers in the multiverse would observe similar constants to what we observe, then this could provide a major advantage for a multiverse hypothesis over theories in which the constants are free parameters (Aguirre, 2007; Barnes, 2017).

One way in which a multiverse theory can fail spectacularly is known as the *Boltzmann Brain problem*. Physical theories predict observations, and so a multiverse model should — in principle — be able to predict what *kind* of observer we would expect to be. One striking feature of our status as observers is that we formed through a long, consistently entropy-increasing process: gravitational collapse into galaxies and stars, stellar burning and supernovae, planet formation, and biological evolution. In some multiverses, including Boltzmann's original multiverse (Boltzmann, 1895), most observers form via a chance statistical fluctuation. Without a consistent thermodynamic arrow of time, they will not observe records of the processes that formed them (Hartle, 2004). They will observe as much free energy around them as is required for their existence as observers, and almost certainly no more.

To be clear, this is not the philosophical "brain-in-a-vat" problem: how can I know whether I'm a Boltzmann brain with false memories? This is a more straightforward "theoretical prediction meets observation" scenario: a cosmological theory predicts that a typical observer will be a Boltzmann brain, *and* will observe that they are a Boltzmann brain. And that prediction is wrong. Whether multiverse models can naturally avoid this problem is an open question; see, among many others, Page (2006); Linde (2007); Banks (2007); de Simone et al. (2010); Aguirre et al. (2011); Nomura (2011); Boddy and Carroll (2013); Albrecht (2015); Boddy et al. (2015).

Misgivings about the whole multiverse project are hardly surprising. Are the tests of multiverse theories enough to make it scientific? Unobservable sub-universes are very different to unobservable quarks: we can constrain the properties of quarks via experiment, but every other sub-universe in the multiverse could disappear tomorrow and we would never know. The meager tests of the multiverse "prove nothing," say Ellis and Silk (2014), "Fundamentally, the multiverse explanation relies on string theory, which is as yet unverified, and on speculative mechanisms for realizing different physics in different sister universes. It is not, in our opinion, robust, let alone testable."

A potentially tricky hurdle is the *measure problem*, about which there is an extensive literature. Many multiverse theories imply or assume that there are an infinite number of other sub-universes. Given a finite population, deriving probabilities is straightforward: what fraction of observers see a value of ρ_Λ as small as the one we observe? But in an infinite multiverse, we cannot simply count sub-universes.

In particular, once we have a useful definition of an observer, it seems that we should treat them all on equal footing. Think of this as permutation symmetry — having arbitrarily numbered all the observers (or observer moments), we should be able to shuffle the labels without changing the prediction of the model. But there is no assignment of probabilities to an infinite number of possibilities that respects this symmetry. This is often taken as incentive to assign different probabilities. But it could be argued, and with considerable force, that this means that an infinite multiverse theory cannot justify probabilities and so cannot make predictions. "In an infinite universe," says Olum (2012), "everything which can happen will happen an infinite number of times, so what does it mean to say that one thing is more likely than another?" These are open questions; see, among many others, Vilenkin (1995); Garriga et al. (2006); Aguirre et al. (2007); Vilenkin (2007a,b); Gibbons and Turok (2008); Page (2008); Bousso et al. (2009); de Simone et al. (2010); Freivogel (2011); Bousso and Susskind (2012); Garriga and Vilenkin (2013); Carroll (2017); Page (2017).

52.5 After Physics

In physics, fine-tuning problems afflict theories that seem to be successful, that is, that can account for the data. The problem is not a falsified prediction, as one might expect from a discounted or discarded theory. Recall the lesson of Ptolemy's model. Within the set of possible geocentric planetary systems, an uncomfortably large proportion look very different to our Solar System. A fine-tuning problem is raised by a large set of alternate possibilities. This suggests an interesting thought experiment.

Suppose there is an ultimate theory of physics. At a future International Meeting of Really Important Physicists, Alberta Einstein walks to the chalkboard, scribbles a few equations, and fundamental physics comes to an end. Like chess pieces who had discovered the laws of chess, no deeper rules exist.

By hypothesis, this theory would be consistent with all scientific data. But we may still glimpse a large set of alternate possibilities, and so a kind of fine-tuning problem remains. Even if it contains no free parameters, Alberta's chalkboard will show one particular mathematical equation or structure. We will be faced by a very old question: why *this* universe? Of all the ways the world could have been, why this way? Of all the mathematically consistent chalkboards of equations, why Alberta's?

Obviously, the answer is not yet-another chalkboard of equations. Neither is it more observations of this universe. This is not the kind of question that physics can answer, because we can't prove from any set of equations that they describe reality. Theories don't predict their own success. But if not physics, then what? What do we do when fundamental physics is over?

Perhaps we stop asking questions. Maybe reality doesn't have any ultimate reason for why it is the way it is. Explanations of the physical world reach the ultimate laws, and stop. This is the supposition of *naturalism*: the natural world is all there is. For a modern defense, see Carroll (2016).

Alternatively, Tegmark (1998) has defended the "the ultimate ensemble theory," that "physical existence is equivalent to mathematical existence." The actual world is not chosen from a set of mathematical possibilities; rather, all mathematical possibilities are equally real, and we are self-aware substructures (SASs) within a particular mathematical structure. A metaphysician might worry about the dissolution of the line between *abstract* and *concrete*. The physicist who tries to test Tegmark's idea via its prediction that "the mathematical structure describing our world is the most generic one that is consistent with our observations" faces a problem: we need a probability distribution over the set of mathematical structures, but a probability distribution is itself a mathematical structure. Tegmark says that probabilities are "merely subjective," but our subjective states of mind are mathematical substructures, too.

By contrast, axiarchism (Leslie, 1989) and theism (e.g., Swinburne, 2004; Collins, 2009) argue that beneath the mathematical structure of our universe is a *reason*: our universe is morally valuable, particularly its embodied, free, conscious agents. Just as Tegmark promotes possibilities to reality on mathematical grounds, axiarchism does so on moral grounds: the world exists because it is good. Theism proposes that God exists necessarily in some sense, and the physical world is the result of God's free choice to create a morally valuable world.

For each of these alternatives, the fine-tuning of the universe for life plays an important role. For Tegmark, the complexity required by any SAS explains why we see this universe/mathematical structure, rather than a simpler one. For axiarchism and theism, fine-tuning for life shows how these ideas could have explanatory power. Given the seemingly extraordinarily small proportion of possibilities that permit the existence of embodied moral agents, the axiarchist and theist can understand something of why Alberta's blackboard is the one has gone to all the bother of existing. Further examination of these alternatives takes us beyond the philosophy of physics.

Acknowledgments

This publication was made possible through the support of a grant from the John Templeton Foundation. The opinions expressed in this publication are those of the author and do not necessarily reflect the views of the John Templeton Foundation.

Notes

1 This isn't how it happened historically, but it does illustrate the principle.
2 The fine-tuning required for stable, life-powering stars has been clarified by recent work by Adams (2008); Barnes (2015); Adams (2016); Adams and Grohs (2016, 2017).

References

Adams, F.C. (2008). Stars in other universes: Stellar structure with different fundamental constants. *Journal of Cosmology and Astroparticle Physics*, 8: 010.
Adams, F.C. (2016). Constraints on alternate universes: Stars and habitable planets with different fundamental constants. *Journal of Cosmology and Astroparticle Physics*, 2: 042.
Adams, F.C. and Grohs, E. (2016). *On the habitability of universes without stable deuterium*. arXiv:1612.04741.
Adams, F.C. and Grohs, E. (2017). Stellar helium burning in other universes: A solution to the triple alpha fine-tuning problem. *Astroparticle Physics*, 87: 40.
Agrawal, V., Barr, S.M., Donoghue, J.F. and Seckel, D. (1998). Anthropic considerations in multiple-domain theories and the scale of electroweak symmetry breaking. *Physical Review Letters*, 80: 1822–1825.
Aguirre, A. (2007). Making predictions in a multiverse: Conundrums, dangers, coincidences. In B.J. Carr (ed.), *Universe or Multiverse?* Cambridge: Cambridge University Press, pp. 367–386.
Aguirre, A., Carroll, S.M. and Johnson, M.C. (2011). *Out of equilibrium: Understanding cosmological evolution to lower-entropy states*. arXiv preprint: 1108.0417.
Aguirre, A., Gratton, S. and Johnson, M.C. (2007). Hurdles for recent measures in eternal inflation. *Physical Review D*, 75: 123501.
Albrecht, A. (2015). Tuning, ergodicity, equilibrium, and cosmology. *Physical Review D*, 91: 111.
Banks, T.A. (2007). *Entropy and initial conditions in cosmology*. arXiv preprint: hep-th/0701146.
Barnes, L.A. (2012). The fine-tuning of the universe for intelligent life. *Publications of the Astronomical Society of Australia*, 29: 529564.
Barnes, L.A. (2015). Binding the diproton in stars: Anthropic limits on the strength of gravity. *Journal of Cosmology and Astroparticle Physics*, 12: 050.
Barnes, L.A. (2017). Testing the multiverse: Bayes, fine-tuning and typicality. In Chamcham et al. (eds.), *The Philosophy of Cosmology*, forthcoming with Cambridge University Press.

Barnes, L.A. and Lewis, G.F. (2017). Producing the deuteron in stars: Anthropic limits on fundamental constants. *Journal of Cosmology and Astroparticle Physics*, 07: 036.

Barrow, J.D. and Tipler, F.J. (1986). *The Anthropic Cosmological Principle*. Oxford: Clarendon Press.

Barnes, L.A. et al. (2018). Galaxy formation efficiency and the multiverse explanation of the cosmological constant with EAGLE simulations. *Monthly Notices of the Royal Astronomical Society*, 477 (3): 3727–3743.

Boltzmann, L. (1895). On certain questions of the theory of gases. *Nature*, 51: 413–415.

Boddy, K.K., and Carroll, S.M. (2013). *Can the Higgs Boson save us from the menace of the Boltzmann brains?* arXiv preprint: 1308.4686.

Boddy, K.K., Carroll, S.M. and Pollack, J. (2015). *Why Boltzmann brains dont fluctuate into existence from the De Sitter Vacuum*. arXiv preprint: 1505.02780.

Bostrom, N. (2002). *Anthropic Bias: Observation Selection Effects in Science and Philosophy*. New York: Routledge.

Bousso, R., Freivogel, B. and Yang, I.-S. (2009). Properties of the scale factor measure. *Physical Review D*, 79: 063513.

Bousso, R. and Susskind, L. (2012). Multiverse interpretation of quantum mechanics. *Physical Review D*, 85: 045007.

Burgess, C. and Moore, G. (2006). *The Standard Model: A Primer*. Cambridge: Cambridge University Press.

Carr, B.J. and Rees, M.J. (1979). The anthropic principle and the structure of the physical world. *Nature*, 278: 605–612.

Carroll, S.M. (2016). *The Big Picture: On the Origins of Life, Meaning, and the Universe Itself*. New York: Dutton.

Carroll, S.M. (2017). *Why Boltzmann brains are bad*. arXiv preprint: 1702.00850.

Carter, B. (1974). Large number coincidences and the anthropic principle in cosmology. In M.S. Longair (ed.), *Confrontation of Cosmological Theories with Observational Data*. Dordrecht: D. Reidel, pp. 291–298.

Caticha, A. (2009). Quantifying rational belief. *AIP Conference Proceedings*, 1193: 60–68.

Collins, R. (2009). The teleological argument: An exploration of the fine-tuning of the universe. In W.L. Craig and J.P. Moreland (eds.), *The Blackwell Companion to Natural Theology*. Oxford: Blackwell Publishing, pp. 202–281.

Cox, R.T. (1946). Probability, frequency and reasonable expectation. *American Journal of Physics*, 17: 1–13.

Damour, T. and Donoghue, J.F. (2008). Constraints on the variability of quark masses from nuclear binding. *Physical Review D*, 78: 014014.

Davies, P.C.W. (1983). The anthropic principle. *Progress in Particle and Nuclear Physics*, 10: 1–38.

de Simone, A., Guth, A.H., Linde, A., Noorbala, M., Salem, M.P. and Vilenkin, A. (2010). Boltzmann brains and the scale-factor cutoff measure of the multiverse. *Physical Review D*, 82: 063520.

Dine, M. (2015). *Naturalness under stress*. arXiv: 1501.01035.

Donoghue, J.F. (2007). The fine-tuning problems of particle physics and anthropic mechanisms. In B.J. Carr (ed.), *Universe or Multiverse?* Cambridge: Cambridge University Press, pp. 231–246.

Ehrenfest, P. (1917). Can atoms or planets exist in higher dimensions? *Proceedings of the Amsterdam Academy*, 20: 200.

Ellis, G.F.R. and Silk, J. (2014). Scientific method: Defend the integrity of physics. *Nature*, 516: 321–323.

Fowlie, A. (2014). CMSSM, naturalness and the "fine-tuning price" of the very large Hadron Collider. *Physical Review D*, 90: 015010.

Freivogel, B. (2011). Making predictions in the multiverse. *Classical and Quantum Gravity*, 28: 204007.

Garriga, J., Schwartz-Perlov, D., Vilenkin, A. and Winitzki, S. (2006). Probabilities in the inflationary multiverse. *Journal of Cosmology and Astroparticle Physics*, 1: 017.

Garriga, J. and Vilenkin, A. (2013). Watchers of the multiverse. *Journal of Cosmology and Astroparticle Physics*, 5: 037.

Gibbons, G.W. and Turok, N. (2008). Measure problem in cosmology. *Physical Review D*, 77: 063516.

Hall, L. and Nomura, Y. (2008). Evidence for the multiverse in the standard model and beyond. *Physical Review D*, 78: 035001.

Hartle, J.B. (2008). *The physics of now*. arXiv preprint: gr-qc/0403001.

Hogan, C.J. (2000). Why the universe is just so. *Reviews of Modern Physics*, 72: 1149–1161.

Jaynes, E.T. (2003). *Probability Theory: The Logic of Science*. Cambridge: Cambridge University Press.

Knuth, K.H. and Skilling, J. (2012). Foundations of inference. *Axioms*, 1(1): 38–73.

Kolmogorov, A. (1933). *Foundations of the Theory of Probability*. Berlin: Julius Springer.

Leslie, J. (1989). *Universes*. London: Routledge.

Lewis, G.F. and Barnes, L.A. (2016). *A Fortunate Universe: Life in a Finely Tuned Cosmos*. Cambridge: Cambridge University Press.

Linde, A. (2007). Sinks in the landscape, Boltzmann brains and the cosmological constant problem. *Journal of Cosmology and Astroparticle Physics*, 01: 22.

Linde, A. (2015). *A brief history of the multiverse*. arXiv: 1512.01203.

Neal, R.M. (2006). *Puzzles of anthropic reasoning resolved using full non-indexical conditioning*. arXiv: math/0608592.

Nomura, Y. (2011). Physical theories, eternal inflation, and the quantum universe. *Journal of High Energy Physics*, 11: 63.

Olum, K.D. (2012). Is there any coherent measure for eternal inflation? *Physical Review D*, 86: 1–6.

Page, D.N. (2006). *Susskinds challenge to the Hartle-Hawking no-boundary proposal and possible resolutions*. arXiv preprint: hep-th/0610199.

Page, D.N. (2008). Cosmological measures without volume weighting. *Journal of Cosmology and Astroparticle Physics*, 10: 025.

Page, D.N. (2017). *Bayes keeps Boltzmann brains at bay*. arXiv preprint: 1708.00449.

Peacock, J.A. (1998). *Cosmological Physics*. Cambridge: Cambridge University Press.

Pochet, T., Pearson, J.M., Beaudet, G. and Reeves, H. (1991). The binding of light nuclei, and the anthropic principle. *Astronomy and Astrophysics*, 243: 1–4.

Raby, S. (2010). Grand unified theories. In Review of Particle Physics, Nakamura, K. et al. *Physical Review DJ. Phys.* G37: 075021.

Ramsey, F.P. (1926). Truth and probability. In R.B. Braithwaite (ed.), *The Foundations of Mathematics and Other Logical Essays*. London: Kegan, Paul, Trench, Trubner & Co., pp. 156–198.

Schellekens, A.N. (2013). Life at the interface of particle physics and string theory. *Reviews of Modern Physics*, 85: 1491–1540.

Silk, J. (1977). Cosmogony and the magnitude of the dimensionless gravitational coupling constant. *Nature*, 265: 710–711.

Skilling, J. (2014). Foundations and algorithms. In M. Hobson et al (eds.), *Bayesian Methods in Cosmology*. Cambridge: Cambridge University Press, pp. 3–35.

Swinburne, R. (2004). *The Existence of God*. Oxford: Clarendon Press.

Tegmark, M. (1997). On the dimensionality of spacetime. *Classical and Quantum Gravity*, 14: L69.

Tegmark, M. (1998). Is "the theory of everything merely the ultimate ensemble theory? *Annals of Physics*, 270(1): 1–51.

Vilenkin, A. (1995). Making predictions in an eternally inflating universe. *Physical Review D*, 52: 3365.

Vilenkin, A. (2007a). Freak observers and the measure of the multiverse. *Journal of High Energy Physics*, 1: 092.

Vilenkin, A. (2007b). A measure of the multiverse. *Journal of Physics A*, 40: 6777.

Wall, J.V. and Jenkins, C.R. (2003). *Practical Statistics for Astronomers*. Cambridge: Cambridge University Press.

Weinberg, S. (1987). Anthropic bound on the cosmological constant. *Physical Review Letters*, 59: 2607–2610.

Whitrow, G.J. (1955). Why physical space has three dimensions. *The British Journal for the Philosophy of Science*, VI: 13.

Further Reading from the Editors

An early classic in the field is J. Barrow and F. Tipler, *The Anthropic Cosmological Principle* (Oxford: Clarendon Press, 1986). Clear presentations for a popular audience can be found in M. Rees, *Just Six Numbers: The Deep Forces That Shape the Universe* (New York: Basic Books, 2001), P. Davies, *The Goldilocks Enigma: Why Is the Universe Just Right for Life?* (London: Penguin Books, 2006), and G. Lewis and L. Barnes, *A Fortunate Universe: Life in a Finely Tuned Cosmos* (Cambridge: Cambridge University Press, 2016). More scientific details can be found in B. Carr (ed.), *Universe or Multiverse?* (Cambridge: Cambridge University Press, 2010) and D. Sloan et al. (ed.), *Fine-Tuning in the Physical Universe* (Cambridge: Cambridge University Press, 2020). J. Leslie, *Universes* (London: Routledge, 1989) is an influential philosophical defence of the cogency of fine-tuning reasoning; a contemporary defence is J. Hawthorne and Y. Isaacs, "Fine-tuning fine-tuning", in M. Benton, J. Hawthorne and D. Rabinowitz (eds.), *Knowledge, Belief, and God: New Insights in Religious Epistemology* (Oxford: Oxford University Press, 2018): 136–168. An important recent treatment of self-locating epistemology in cosmological contexts is F. Arntzenius and C. Dorr, "Self-locating priors and cosmological measures", in K. Chamcham, J. Barrow, S. Saunders and J. Silk (eds.), *The Philosophy of Cosmology* (Cambridge: Cambridge University Press, 2017): 396–428.

53

DARK MATTER AND DARK ENERGY

Melissa Jacquart

53.1   Introduction

The current best current cosmological model poses an intriguing puzzle. According to this model, a vast majority of the mass-energy content of our Universe is currently unobservable, except through indirect means. All of the observable luminous matter – including stars, galaxies, interstellar dust, and gas – makes up only a small fraction of the mass-energy in the Universe. While this sounds like an incredible claim, it is almost universally accepted by astrophysicists and cosmologists.

Contemporary understanding of the Universe is based fundamentally on astronomical observations and the current best theory of gravity. From observations, cosmologists have proposed a model for the history and structure of the Universe. This model is commonly referred to as the concordance, or "Standard Model" of cosmology, and includes elements such as Big Bang Theory, inflation, nucleosynthesis, and reionization.[1] This Standard Model is considered to be in agreement with all available observational data. It originates from our current best theory of gravity, general relativity (GR), as described by Einstein's Field Equations (see Part III of this volume for discussions of GR). By accounting for various initial assumptions and observational evidence, the Field Equations become the Friedmann–Lemaître–Robertson–Walker (FLRW) models.[2] More specifically, the Lambda Cold Dark Matter (ΛCDM) model is the current best parameterization of one of the FLRW models to describe the large-scale structure formation in the Universe. None of the parameter values in the model are fixed by the theory; instead, their values are determined based on our observational evidence. According to this model, the total mass-energy of the Universe contains only about 4% baryonic matter (ordinary visible matter such as stars and gas). The rest of the Universe is made up of 24% dark matter and 72% dark energy.

These new mysterious entities, dark matter (DM) and dark energy (DE), are accepted by most cosmologists as a well-established part of our Universe (Hinshaw et al., 2009). However, some see the inclusion of these strange elements to be an *ad hoc* addition to the model in order to ensure fit with the empirical data (López-Corredoira, 2014, 2017). Why should the mass directly observed in stars and gas account for such a small part of the Universe's content? From observations, cosmologists have inferred the existence of DM and DE, yet they know very little about their nature since they can only be observed indirectly.

The logical structure of these observational inferences is of great concern both to cosmologists and to philosophers. If the Standard Model of cosmology is correct, and there is this strange mass and energy, where is it, and what is its nature? How can we be confident DM and DE exist and were not simply invented to ensure observations are in agreement with GR? Should one take DM and DE as an indicator that GR is wrong? Given that belief in the existence of DM and DE stems from empirical observations, I begin with a review of the observational evidence. Much of the philosophical debate takes issue with this empirical evidence and what one is warranted to conclude regarding the nature of DM and DE. I then examine some of the other philosophical issues surrounding the discovery of DM and DE, including underdetermination of theory, theory change & theory choice, and the role of computer simulations in modern astrophysics and cosmology.

53.2 Observational Evidence

Claims about the existence of DM and DE are primarily motivated by observational evidence. In astrophysics and cosmology, there are two relevant notions of observational evidence: direct and indirect (see Anderl, this volume, Chapter 54 for further discussion of the nature of evidence in astrophysics more generally). There is direct observational evidence for many kinds of entities – stars and galaxies can be seen in optical wavelengths using telescopes, and X-ray telescopes can be used to directly observe interstellar gas via its emissions. The devices used to make these observations rely for their operation on physical theories (such as electromagnetic theory) that are remote from the theory being tested (that is, the theory of gravity). In the context of the search for DM and DE, what would count as direct evidence would be the detection of DM or DE particles (e.g., using something like a photodetector), or testing for them through a means independent from GR. However, thus far there has only been indirect observational evidence. Indirect observations involve predicting the interactions of the entity in question with other systems by using the relevant physical theory and directly observing the effects of those interactions. This inferential process relies on a more substantive assumption of the truth of the theory being tested, therefore confidence in the evidence depends on the degree of confidence in the underlying theory. Indirect evidence is often considered to be secondary, or less powerful when compared to direct evidence. Thus far, however, there has been no direct detection of DM and DE; their existence is inferred only from indirect observations. Nevertheless, most cosmologists consider there to be enough indirect evidence to establish that DM and DE exist.[3]

53.2.1 *Observational Evidence for Dark Matter*

DM is currently unobservable at any electromagnetic wavelength and has only been detected via its gravitational interactions (hence, "dark" matter). There are five observations that are considered to be the strongest indicators of the existence of DM.[4]

Beginning with observations of smaller-scale phenomena, one line of evidence comes from observations of single disc galaxies and their rotation curves. A galaxy rotation curve is a plot of the orbital speeds of stars or gas in a galaxy against their radial distance from that galaxy's center. The amount of matter in a given galaxy determines the curve for rotational speed as a function of the distance from the galactic center. In calculating a theoretically expected rotation curve (using GR), astronomers initially based their calculations on the visible mass – stars and gas – in a galaxy. They expected the velocity to decrease when moving from the center of mass of a galaxy to its outer edges, with the outer edges of the galaxies having a slower rotation, since not as much mass is present. However, in the late 1960s when astronomers (such as Vera Rubin) began to collect and plot actual observational data, they found it did not match their calculated expected curve. The galaxies in fact had a relatively constant, high-velocity rotation curve to their outer edges where few stars are visible. For galaxies to rotate in the way the observations indicated (and maintain consistency with GR), there must be significantly more mass in the outer edges of the galaxy than the mass they were able to see. To account for this discrepancy between observations and theory, it was postulated that there must be a halo of DM surrounding every disc galaxy.

This mass discrepancy is also present in larger structures in the Universe, such as clusters of hundreds of galaxies that are bound together gravitationally. While studying the Coma Cluster in 1933, Fritz Zwicky wanted to determine its gravitational mass using the virial theorem – an equation relating the average kinetic energy of a system to its total potential energy. He then compared the inferred gravitational mass to the mass of the luminous matter in the galaxies. Based on his calculations, there was not enough luminous matter present to hold the Coma Cluster together gravitationally. He concluded that there must be more mass there that he could not see holding the galaxy together (Zwicky, 1937). On even larger scales still, such as the large-scale structure formation surveys of the Sloan Digital Sky Survey, astronomers observe structural patterns that could not be held together gravitationally (at least according to GR) by the amount of visible matter alone. The observed structure formations also support DM being "cold" (i.e., slow with respect to the speed of light), rather than "hot" or "warm," as only CDM obtains structural properties in agreement with the observations.[5]

In addition to these three structural lines of evidence, a fourth, and perhaps most compelling, line of evidence for DM comes from 2006 observations of the Bullet Cluster (Clowe et al., 2006). The Bullet Cluster is actually two clusters of galaxies that have undergone a collision. In between galaxies in galaxy clusters, there is a vast amount of gas. During a collision, the gas particles in the clusters heat up from the collision and cause an increased brightness in X-ray emissions. From the observations of X-ray emissions, astronomers can determine where the gas is located, as well as how energetic it is. However, the visible matter in the galaxies in the clusters are not significantly affected by the collision, and they essentially pass through and form themselves in two separated regions (with the gas in between).

Astronomers can map the matter distribution in these kinds of collisions through the effects of gravitational lensing.[6] The matter in the Bullet Cluster distorts background galaxy images, and by measuring that distortion astronomers can measure the location of the cluster's mass. Given that the gas accounts for the vast majority of the baryonic (visible) matter, the lensing would be expected to follow the gas. However, the gravitational lensing effects are strongest in two regions near the visible galaxies. This is evidence that most of the matter in the cluster is near the galaxies. Since DM is considered to interact even less frequently than baryonic matter, during the collision the DM from one cluster would pass by the other objects in the cluster. By including DM in the calculations, astronomers obtain the mass in the right distribution to predict the gravitational lensing they see in the observations.[7] The dynamical interaction found in the Bullet Cluster is some of the clearest evidence that DM of some form exists (Gates, 2010).

The final line of evidence comes from the observations of the Cosmic Microwave Background (CMB) – the electromagnetic radiation left over from the epoch of recombination. The observation of the CMB is a landmark discovery in cosmology as it provides evidence in favor of many features including a Big Bang origin of the Universe and a very nearly flat geometry of the Universe. The observed patterns in the CMB also offer evidence for the existence of DM. Fluctuations in the CMB are typically explained as the result of two competing forces acting on the matter. The first is an attractive gravitational force, and the second is an outward pressure caused by photons. This competition results in variations, or oscillations, of dense regions in the CMB. These oscillations can be presented in the form of a power spectrum of the CMB. The peaks in the power spectrum, in particular, are sensitive to the matter density of the Universe, and are consistent with predictions that include DM, as well as those that do not. Data from the Cosmic Background Explorer (COBE), Wilkinson Microwave Anisotropy Probe (WMAP), and Planck satellite measured the CMB power spectrum through the first peak oscillation and suggested that the measured peaks match predications made with DM included in the model (Natarajan, 2016).

53.2.2   Observational Evidence for Dark Energy

Cosmologists have inferred from the empirical data that the Universe is pervaded by a relativistic energy density that carries negative pressure, driving the expansion of the Universe at an accelerated rate. This energy field is smoothly distributed (in that it is everywhere throughout the Universe) and persistent (in that the density remains approximately constant as the Universe expands). It is referred to as "energy" because energy fields exhibit a similar nature (i.e., smoothly distributed and persistent). It is "dark" in that it is not directly detectable; rather, its existence has been inferred through indirect observational means. DE appears in the Standard Model as the cosmological constant, Λ, representing the value of the energy density of the vacuum of space.[8]

Observations from CMB are taken to be a line of evidence for the existence of DE as well as being evidence for DM. Cosmologists use the CMB to measure the shape of the Universe (flat, no curvature; open, negative curvature; or closed, positive curvature), because the shape affects the magnitude of the slight variations seen in the CMB. By measuring these variations, they concluded that the Universe is very nearly flat. However, the exact shape of the Universe depends on the total mass-energy content; to have a flat Universe, the mass-energy density of the Universe must be equal to the critical density. Mass-energy content calculations based on the baryonic matter made up only a small portion of the mass-energy needed to have a flat Universe. Even when estimates of DM are included, in order for observations to match what is seen in the CMB, there is still about 72% of the required mass-energy unaccounted for. This is missing mass-energy is thought to be DE.

The second key line of evidence for DE comes from observations of Type Ia supernovae (SNIa). While cosmologists had determined the Universe is expanding (Hubble, 1929; Freeman et al., 2001), 1998 observations of these supernovae led to claims that the expansion is taking place at an accelerated rate (Riess et al., 1998). SNIa begins as white dwarf stars and accretes matter until they reach the Chandrasekhar limit (a mass 1.4 times the mass of the Sun) and explode. Since the Chandrasekhar limit is the same value everywhere in the Universe, the supernovae explode with roughly the same amount of energy and therefore have similar luminosities.[9] Distant SNIa were investigated by cosmologists because of their relationship between intrinsic luminosity and the length of time it takes for a supernova's brightness to decline after reaching peak luminosity. By measuring the brightness of SNIa, one can determine how far away the object is. From the relationship between the distance to an object and its redshift, one can determine how fast an object is receding. Observations showed that the distant supernovae were dimmer than expected, which meant that the supernovae were actually further away than what would be predicted if the Universe were expanding at a constant rate.[10] This led most cosmologists to conclude that the luminosity distance is dominated at low redshift by an accelerating component, and in order to account for the observed acceleration of the expansion rate of the Universe, cosmologists appealed to DE having the property of a strong negative pressure (acting repulsively).

Finally, the third line of evidence comes from baryon acoustic oscillations (BAO), very large-scale oscillations in the density of baryonic matter, whose magnitude helps measure the expansion history of the Universe. The overdensities in the distribution of matter in the Universe occur at regular intervals and therefore provide a means to measure distance. The BAO measurements allow for comparison between the observations of current acoustic waves to that of the acoustic waves at the time of recombination from the CMB. Drawing on the Doppler effect, BAO observations provide another way to measure the distance between objects.[11] These observations also point to the Universe expanding at an accelerated rate (Seo, 2003).

While these are considered the cornerstone observations in support of DE, recent work in cosmology attempts to find other means by which to support the existence of DE. The 2011 WiggleZ survey from the Australian Astronomical Observatory (Blake et al., 2011) attempts to measure galaxy redshifts, and analyze the galaxy distributions in order to learn more about the nature of DM, as well as support the hypothesis of the Universe's accelerated expansion independently of the SNIa data. Another approach is to look for late-time Integrated Sachs–Wolfe effect (ISW) in the CMB, as it would be a direct signal of DE in a flat Universe (Crittenden and Turok, 1996). Others still are attempting to test evidence of DE through observational Hubble constant data (Ma and Zhang, 2011).

53.3   Realism about Dark Matter and Dark Energy

Given the nature of the observational evidence described above, it does not seem unreasonable for astronomers to infer the existence of DM. The observations seem to indicate that there is something that behaves like baryonic matter by interacting gravitationally, yet is not directly observable. The mystery, then, is determining DM's basic physical properties. In principle, it is not necessary that DM be composed of some heretofore unknown kind of matter – it could be partially composed of standard baryonic matter, which for some reason cannot be observed. However, there are strong reasons to believe that DM is composed of a new fundamental particle since otherwise the laws governing the behavior of baryonic matter would seemingly have to be complex and disunified. Candidates include axions, sterile neutrinos, WIMPs, and self-interacting dark matter.[12] While the exact constitution and properties of DM are undetermined, most astronomers are nonetheless committed to its existence. This raises classical philosophical questions connected to scientific realism, and whether this positive epistemic attitude toward DM is justified.

Scientific realism is a commitment to the truth or approximate truth of scientific theories, and the entities posited by the theory. Scientific realism with respect to DM amounts to the belief that DM is *as real* as the stars and gas we can observe. However, no matter how much observational techniques improve, DM may turn out to never be directly observed. Realists do not tie their belief in the existence of a theoretical entity to its observability. Antirealists, on the other hand, are skeptical of the existence of DM, considering it to be an important part of the ΛCDM theoretical understanding of our observations, without any further metaphysical commitment. In general, antirealists do not hold realist commitments to entities they cannot directly observe. DM may pose a special problem for realists since they are committed to the existence of an unobservable entity that makes up a significant portion of the Universe's matter content (Shapere, 1993).

An intermediary position is *entity realism* (Cartwright, 1983, 1989; Hacking, 1983, 1989). On this view, we should be realists about entities about which we have significant causal knowledge, for example, those things that can be routinely manipulated in the laboratory. This has the benefit of allowing one to be a realist about entities such as subatomic particles, without committing to realism about DM, at least until more is known regarding its constitution and properties. Since we know little about DM's causal properties beyond its gravitational interaction, it is not yet the kind of entity that we could, hypothetically, reliably manipulate in the laboratory, and therefore falls outside the scope of entity realism.

The same issues regarding scientific realism apply to DE. However, the nature of DE is even more mysterious. While DE is posited as an explanation for the evidence that the Universe is expanding at an accelerated rate, we know little else about its composition and properties. DE is represented in the Standard Model by the cosmological constant, Λ. However, there are issues connected to determining the value of this constant, as well as specifying what physical feature, entity, or force the constant represents, which are addressed in the next section.

53.4   The Cosmological Constant Problem

Minimally, the empirical evidence seems to require that there be some element included in the FLRW equation to account for the accelerated expansion of the Universe. This element is a non-zero Λ, or cosmological constant term, which is taken to represent DE. There are two major research goals surrounding DE. One aims to refine the value of the DE constant and obtain the most precise equation to describe the data. The other attempts to answer the question of what the nature of DE is such that it *causes* the acceleration.

The main candidate for how to understand DE is as a true cosmological constant, or a vacuum energy – a fixed amount of energy associated to every region of space, which remains constant in time (i.e., does not dilute with expansion, and so is unlike other types of energy and matter). The vacuum energy is perfectly smooth and constant throughout the Universe and has a pressure and stress-energy density such that it has a negative value in the equation. In this view, the cosmological constant Λ is considered to represent this vacuum energy in the FLRW models.[13] Another candidate option for understanding DE is as "quintessence," a scalar field that fills the Universe, and changes very slowly as time passes. On the quintessence view, the Universe is filled with a new kind of dynamical energy fluid or field, leading to the accelerated expansion effects. In principle, quintessence does not directly rely on the FLRW model, however, it has been primarily studied within this context.[14] On both these models, DE has a uniform, extremely low density everywhere in space. Therefore, it may be possible to directly detect DE, since it would be present in local regions of the Universe.

The value of the cosmological constant was first measured in 1998 through the supernovae data described above, indicating Λ to be very small. Some physicists note the similarity between the Λ vacuum energy and the vacuum energy predicted by quantum field theory (QFT). However, this small value of Λ conflicts with the value of the vacuum energy predicted by QFT (see Chapter 18, Section 18.10.2, for more details). As a result, the disagreement between the cosmological constant vacuum energy density and the predictions suggested by QFT (zero-point energy) may be problematic. This is referred to as the Cosmological Constant Problem (CCP). In light of the empirical evidence for a positive cosmological constant, CCP concerns understanding why the cosmological constant is so small relative to the vacuum energy density calculated in QFT, but not exactly zero, as indicated by the accelerating expansion (Smeenk, 2014).[15]

Philosophers have attempted to contribute to solving the CCP by analyzing the different possible interpretations of Λ consistent with the evidence for it. Earman (2003) argues that much of the problem with Λ stems from the fact that most cosmologists are strongly committed to interpreting the empirical data within the context of the FLRW cosmological models, which requires either a positive cosmological constant ($\Lambda > 0$) or else something standing in for a positive Λ mimicking this behavior (such as quintessence). As such, there are two senses in which the cosmological constant can be a constant. While different interpretations of the Λ may be empirically indistinguishable, the theories that embody different Λ may not be. Another approach examines the relations between different fundamental physical theories (GR and QFT). Rugh and Zinkernagel (2002) consider the way in which a commitment to a physically real vacuum energy may influence the way in which the problem is defined. Others attempt to understand the assumptions at play in different proposals to address the CCP. Nobbenhuis (2006) distinguishes three different ways to understand the CCP: as a question about (1) why is the cosmological constant so small, (2) why is it not exactly equal to zero, and (3) why is its energy density today of the same order of magnitude as the matter energy density. These questions offer a schema in which to categorize proposals in hopes of gaining insight to advantages or drawbacks to different approaches to solve the problems. Bianchi and Rovelli (2010) on the other hand, find the arguments that the nature of DE is mysterious to be unconvincing, or ill-founded. They take the phenomena of an accelerated expanding Universe to be clearly predicted and well-described by GR. They also argue that identifying the cosmological constant with the QFT vacuum energy density is a mistake. More recently, Schneider (2020a, b) discusses how to best understand the CCP as a *problem* as well as what role the CCP serves within the physics community.

53.5 Underdetermination of Theory by Evidence

Underdetermination of theory by evidence is the problem that, for any body of evidence supporting a theory, there will be other theories that are logically compatible with the same body of evidence. As such, there may not be good empirical grounds for choosing one theory over another. In order to select one theory over another, other considerations need to be brought in − such as simplicity or explanatory power.[16] As a result of the limitations of empirical observations, the correct cosmological model of our Universe can be considered to be underdetermined by the evidence as well. Certain extra-theoretic considerations, such as consistency with our best theory of gravity, led to the inclusion of the previously unknown entities DM and DE in the FLRW models. In order to have a model consistent with empirical data, previously unknown entities like DM and DE were posited. However, some philosophers argue that the inclusion of DM and DE in the FLRW models is too *ad hoc* a theoretical posit, which indicates that one should favor a different theory of gravity. The question is, are cosmologists warranted in preferring GR over rival gravitation theories?

Vanderburgh (2003) argues that DM in fact highlights a weakness of GR. Thus far, DM has been indirectly detected by its gravitational effects. However, in order to claim its detection, one needs to assume a theory of gravity. Evaluating the empirical adequacy of a gravitational theory (on scales of a single galaxy or larger) the mass disruption in the dynamical system must be known. However, because of the astrophysical dynamical discrepancy (i.e., our observational data of single galaxy rotation curves not matching our original expected, non-DM, rotation curve), we do not know the actual mass distribution. In order for astronomers to infer that mass distribution from observations, they must already assume some gravitational law. (i.e., if assuming GR, then the only way to obtain the observed dynamics is that there must be other unobserved matter). However, that law of gravity cannot legitimately be assumed, as which gravitational law ought to be taken to apply is the very issue under consideration. This is referred to as the "dark matter double bind." On these larger scales there is not currently, and perhaps cannot be, an empirical basis on which to decide among rival gravitational theories. The evidential status of GR is thus, according to Vanderburgh, considerably weaker than is usually supposed. The only way to pick between competing theories of gravity is by appeal to methodological criteria of theory choice. Even when considering the other lines of evidence (rotation curves, velocity dispersions, X-ray temperatures, and gravitational lensing), Vanderburgh (2005) argues that, while the different methods give roughly agreeing results, they still measure the mass discrepancy by assuming GR applies to the systems.

Kosso (2013), on the other hand, argues that the Bullet Cluster will not fall prey to the dark matter double bind, since gravitational lensing is a direct consequence of Einstein's Equivalence Principle (EEP); a complete gravitational theory is not needed in order to derive the gravitational lensing effects. Thus, though it may still be indirect evidence, the gravitational lensing seen in the Bullet Cluster can offer an independent reason to believe that DM exists. Sus (2014), however, argues that on a careful analysis of the empirical evidence, the EEP alone cannot support the claim that gravitational lensing in the Bullet Cluster constitutes evidence for DM. Likewise, Vanderburgh (2014) argues that even in the case of the Bullet Cluster, there is still the need to assume GR (or some theory of gravity) in order to infer the precise mass distributions from the observations.

There are alternative gravitational theories that aim to account for the empirical evidence without the introduction of DM and DE. One of these views proposes modifying Newtonian dynamics so that the missing mass is not required to account for the evidence. However, the Modified Newtonian Dynamics (MOND) approach is viewed as contentious given many astrophysicists consider GR to be well established as the theory of gravity (Dodelson, 2011). They therefore regard the adoption of MOND as unjustified. Yet the advocates of MOND claim that their model is as good (if not better) as ΛCDM for describing observed galaxy dynamics (Milgrom, 1983; Famaey and McGaugh, 2013; McGaugh, 2014). By and large, the astrophysical community has favored maintaining GR and ΛCDM, rather than abandoning them for alternatives.

53.6 Theory Change and Theory Choice

Given the underdetermination problem and the existence of possible alternatives to ΛCDM, why do astrophysicists accept the model that posits DM and DE? One way of understanding this is to analyze the issue by examining the role theory choice plays in contemporary astrophysics. Regardless of underdetermination, in order for research to proceed a theory must be selected. A solution, then, might consist in focusing on how each theory is empirically supported, which may resolve the appearance of empirical equivalence. This can be achieved through understanding the broader theoretical framework within each theory is embedded, allowing analysis of how each receives indirect empirical support (Massimi and Peacock, 2014). Additionally, it may be useful to compare this case of DM and DE to other historical cases, such as the irregularity of Uranus' orbit or precession of Mercury's perihelion to determine what lessons can be learned about theory change and theory choice (Lahav and Massimi, 2014).

Alternatively, the conflict between ΛCDM and MOND might be best understood as two incommensurate paradigms, indicative of an approach to a Kuhnian scientific crisis, and a matter of acquiring enough anomalies to induce a paradigm shift (McGaugh, 2014). In the meantime, we might appeal to seemingly objective criteria of theory choice (accuracy, consistency, scope, simplicity, and fruitfulness) (Kuhn, 1977). However, these criteria are criticized as being imprecise and in conflict with each other. They are not sufficient to determine theory choice and depend on sociological considerations. Regardless, in order to continue to conduct research, the scientific community may see the theory as the best choice to make right now. This has led some philosophers and physicists (Ruphy, 2011; López-Corredoira, 2014) to offer a social hypothesis as the real justification for the prevalent use of the ΛCDM model in cosmology. It is not that the ΛCDM model is empirically well justified. Rather, cosmologists favor it due to the sheer amount of time and allocation of resources (both financial and intellectual) that have already been invested. More recently, Michela Massimi (2018) provides another alternative for making sense of the respective success claims of ΛCDM and MOND, proposing an analysis of the debate in terms of multiscale modeling. She suggests that the ΛCDM – MOND debate demonstrates the challenges of multiscale modeling in the context of cosmological scales (2018).[17]

Acceptance of ΛCDM can also depend on the characterization of empirical success. For instance, the Newtonian ideal of empirical success of a theory involves agreeing on measurements from diverse phenomena. Harper (2012) argues that this kind of reasoning is appealed to in support of DE. By tracing these dependences, one can assess the extent to which different measurements depend on independent assumptions. The success of a theory, then, is related to the degree to which a variety of independent lines of evidence constrain the parameters (such quantity and distribution of DM and DE) in the theory. In the case of DE, there is a convergence of accurate measurements of parameters by diverse phenomena. Given the variety in observations and assumptions behind those observations, there is surer footing regarding DE not as an *ad hoc* auxiliary hypothesis, but rather as an accepted background assumption.

In the background of these discussions is the issue of testability of scientific theories. While philosophers have raised numerous concerns regarding the Popperian falsification criterion, falsifiability is taken very seriously as a good criterion for scientific research by a majority of cosmologists. As such, there is concern regarding the testability of the ΛCDM model, and if tests aim at confirmation or falsification.[18] Rather than thinking of DM and DE as a response to falsifying observations of a gravitational theory, cosmologists interpret the observations as the discovery of DM and DE.

As such, Popperian "conventionalist stratagems" may be at play in favoring the ΛCDM model (Merritt, 2017). Alternatively, while testability may be preferred, it may not be required as it may not be feasible in practice given the lack of experimental access and large scales of cosmology. As such, there may be a need for a shift in the methodological and accepted epistemic standards (Kragh, 2014).

53.7   Models and Computer Simulations

Models and computer simulations serve as investigative tools to provide further insight into the nature of DM and DE. As computational power has increased, cosmologists have produced complex simulations of galaxy collisions, cluster interactions, and the structural history of the entire Universe. These models and simulations play a critical role in the justificatory reasoning process in astrophysics. Yet their use has also led to three standout problems for ΛCDM, particularly in the context of DM. First, the Millennium Run simulations quite closely match the observed large-scale structure of the Universe, which is taken to support the ΛCDM model (Boylan-Kolchin et al., 2009). However, there are cases where the simulations and observations disagree. MOND-based computer simulations have highlighted a discrepancy between the ΛCDM-predicted structure properties of galaxies, and actual astronomical observations. This discrepancy is between the observed DM density profiles of low-mass galaxies, and the density profiles predicted by ΛCDM-based cosmological N-body simulations (referred to as the Cusp/Core Problem) (Weinberg et al., 2015).

While some of the DM is accounted for in DM halos surrounding galaxies, this is only accounted for a portion of it. The second problem is referred to as the missing satellites problem. The computer simulations based on cold DM models predict large numbers of subhalos (~100–1,000 for a galaxy the size of our Milky Way). However, the Milky Way only has 23 known satellites. If the models are correct, there are a significant number of satellites yet to be observed. One possibility is that the galaxies are undetectable because they are composed entirely of DM (Weisberg, et al., 2018). Analyzing this in the context of modeling and simulation, one can assess the problem in terms of whether the models offer robust predictions. Some cosmologists cast the problem in these terms (Bullock, 2013), and drawing on the philosophical work of robustness analysis may be beneficial (Gueguen, 2019). Finally, ΛCDM simulations predict not only how many galaxies there should be, but also their masses. Even if some of the missing satellites are composed entirely of DM, the model still predicts satellites that are simply too massive to lack any visible matter, and "too big to fail" to form.

How should this conflict between observations and computer simulations be understood? By their very nature, models and computer simulations are necessarily incomplete representations – they contain idealizations and approximations, and simplify the features in the system being modeled. Given this, there is a need to understand what justifies their use to make claims about the nature of the real world (Jacquart, 2016). Given their critical role, there is a need to examine the use of computer simulations in the modern methodology of astrophysics, and evaluate how simulations contribute to the epistemic warrant of dark entities (Jacquart, 2020; Weisberg et al., 2018). There is also an interesting relationship between the philosophical issues for computer simulations, experiments, and theory. Ultimately, some issues discussed in this chapter could dissolve should the physics community directly detect DM or DE particles. Experiments such as ATLAS at the Large Hadron Collider may provide such a path. De Baerdemaeker (2020), for example, examines how particle physicists justify that experiments should be able to detect dark matter. She argues in favor of a "method-driven" logic for the justification of their effectiveness as dark matter searches.

Acknowledgments

I thank Lucas Dunlap for helpful comments on an earlier draft of this essay. This work was supported by the National Science Foundation award no. SES-1557138.

Notes

1 Cosmologists use "concordance model" to refer to the currently accepted cosmological model, the Standard Model of cosmology with the specified contributions of different types of matter that are in agreement with current best observations. This Standard Model of cosmology is distinct from the Standard Model of partial physics.

2 The cosmology literature often refers to FLRW models as a class of *exact solutions* (rather than class of models) to Einstein's field equations specified by the FLRW metric. That is, any specification of parameter values, such as curvature (Ω_k) and mass-energy density (Ω_m), which is consistent with the field equations, is referred to as a solution to the model. In this context, the ΛCDM model is a parameterization of the perturbed FLRW models. The reader should understand the FLRW solution as a set of models; any specification of parameters yields a particular model, such as the ΛCDM model. For a detailed discussion of the construction of the ΛCDM model from the FLRW models and Einstein's Field Equations, including its initial assumptions, observational evidence, and tests, see Ellis (2014), Hamilton (2014), Smeenk (2013).

3 See Spergel (2015), Matarrese et al. (2011), and Gates (2010).

4 See Trimble (1987) for complete history of the observational evidence and constraints on DM, as well as de Swart (2017).

5 See Bertone et al. (2005) for further discussion.

6 It is worth noting that gravitational lensing is a consequence of GR, and thus contributes to the indirect nature of the resulting evidence for DM.

7 This observation also is in agreement with DM being a weakly interacting massive particle (WIMP) rather than a massive astrophysical compact halo object (MACHO).

8 The cosmological constant, Λ was originally included by Einstein in his field equations to maintain a static universe (which he later considered a mistake). Contemporary cosmology has reintroduced the idea of the cosmological constant, now with a positive value, to account for the observed accelerated expansion of the universe.

9 These astronomical objects with a known absolute magnitude are called "standard candles."

10 In fact, these SNIa observations suggested that the expansion of the universe has been accelerating since around a redshift of $z \sim 0.5$.

11 Given their capacity to measure distances, BAO act as "standard rulers," see Bassett and Hlozek (2010).

12 See Spergel (2000); and Bertone et al. (2005) for review of these candidates, their evidence, and constraints.

13 When Einstein first introduced Λ, he didn't think of it as "energy." Rather, he thought of it as a modification of the way spacetime curvature interacted with energy. However, this turns out to be the same thing as vacuum energy. For a detailed history of the cosmological constant in modern physics, see Earman (2001).

14 See Uzan (2010) for extended discussion of DE candidates.

15 Prior to the 1998 supernovae observations, there was a different version of the CCP, now called the "old" CCP. Namely, why isn't there a cancellation mechanism that leads to $\Lambda = 0$? See Weinberg (1989) for detailed discussion of the old CCP.

16 Issues related to the underdetermination of general relativity include the underdetermination of global properties of spacetime geometry (Manchak, 2009, 2011) and the role of the cosmological principle in deriving general relativity (Ellis, 2007; Beisbart, 2009; Butterfield, 2014). See Wilson 1980 for a more general discussion.

17 See the special issue on "Dark Matter & Modified Gravity" in the Journal, *Studies in History and Philosophy of Science Part B* on for further philosophical discussion on these and other issues connected to modified gravity and dark matter debates.

18 See López-Corredoira (2017) for details on the tests for and problems of the ΛCDM Model.

References

Bassett, B. and Hlozek, R. (2010). Baryon acoustic oscillations. In P. Ruiz-Lapuente (ed.), *Dark Energy: Observational and Theoretical Approaches*, Cambridge: Cambridge University Press, pp. 246–278. Available at https://www.cambridge.org/us/academic/subjects/physics/cosmology-relativity-and-gravitation/dark-energy-observational-and-theoretical-approaches?format=HB

Beisbart, C. (2009). Can we justifiably assume the cosmological principle in order to break model underdetermination in cosmology? *Journal for General Philosophy of Science*, 40(2): 175–205.

Bertone, G., Hooper, D. and Silk, J. (2005). Particle dark matter: Evidence, candidates and constraints. *Physics Reports*, 405(5): 279–390.

Bianchi, E. and Rovelli, C. (2010). *Why all these prejudices against a constant?* Available at: arXiv preprint arXiv:1002.3966.

Blake, C. et al. (2011). The wigglez dark energy survey. *Monthly Notices of the Royal Astronomical Society*, 418(3): 1707–1724.

Boylan-Kolchin, M. et al. (2009). Resolving cosmic structure formation with the millennium-ii simulation. *Monthly Notices of the Royal Astronomical Society*, 398(3): 1150–1164.

Bullock, J. (2013). Notes on the missing satellites problem. In D. Martnez-Delgado and E. Mediavilla (eds.), *Local Group Cosmology*. Cambridge: Cambridge University Press, pp. 95–122.

Butterfield, J. (2014). On under-determination in cosmology. *Studies in History and Philosophy of Modern Physics*, 46: 57–69.

Cartwright, N. (1983). *How the Laws of Physics Lie*. Oxford: Oxford University Press.

Cartwright, N. (1989). *Nature's Capacities and Their Measurement*. Oxford: Clarendon Press.

Clowe, D. et al. (2006). A direct empirical proof of the existence of dark matter. *The Astrophysical Journal Letters*, 648(2): L109.

Crittenden, R.G. and Turok, N. (1996). Looking for a cosmological constant with the rees-sciama effect. *Physical Review Letters*, 76(4): 575.

De Baerdemaeker, S., 2021. Method-driven experiments and the search for dark matter. *Philosophy of Science*, 88(1), pp. 124–144.

de Swart, J.G., Bertone, G. and van Dongen, J. (2017). How dark matter came to matter. *Nature Astronomy*, 1(3): 1–9.

Dodelson, S. (2011). The real problem with mond. *International Journal of Modern Physics D*, 20(14): 2749–2753.

Earman, J. (2001). Lambda: The constant that refuses to die. *Archive for History of Exact Sciences*, 55: 189–220.

Earman, J. (2003). The cosmological constant, the fate of the universe, unimodular gravity, and all that. *Studies in History and Philosophy of Modern Physics*, 34(4), 559–577.

Ellis, G.F.R. (2007). Issues in the philosophy of cosmology. In J. Butterfield and J. Earman (eds.), *Philosophy of Physics*. Oxford: Elsevier, pp. 1183–1286.

Ellis, G.F.R. (2014). On the philosophy of cosmology. *Studies in History and Philosophy of Modern Physics*, 46: 5–23.

Famaey, B. and McGaugh, S. (2013). Challenges for λcdm and mond. *In Journal of Physics: Conference Series*, 437(1): 012001. IOP Publishing.

Freedman, W.L., Madore, B.F., Gibson, B.K., Ferrarese, L., Kelson, D.D., Sakai, S., Mould, J.R., Kennicutt Jr, R.C., Ford, H.C., Graham, J.A. and Huchra, J.P., 2001. Final results from the Hubble Space Telescope key project to measure the Hubble constant. *The Astrophysical Journal*, 553(1): 47.

Gates, E. (2010). *Einstein's Telescope: The Hunt for Dark Matter and Dark Energy in the Universe*. New York: WW Norton & Company.

Gueguen, M. (2019). On robustness in cosmological simulations. *Philosophy of Science*, 87(5): 1197–1208.

Hacking, I. (1983). *Representing and Intervening: Introductory Topics in the Philosophy of Natural Science*. Cambridge: Cambridge University Press.

Hacking, I. (1989). Extragalactic reality: The case of gravitational lensing. *Philosophy of Science*, 56(4): 555–581.

Hamilton, J.-C. (2014). What have we learned from observational cosmology? *Studies in History and Philosophy of Modern Physics*, 46: 70–85.

Harper, W. (2012). *Isaac Newton's Scientific Method: Turning Data into Evidence about Gravity and Cosmology*. New York: Oxford University Press.

Hinshaw, G. et al. (2009). Five-year wilkinson microwave anisotropy probe★ observations: Data processing, sky maps, and basic results. *The Astrophysical Journal Supplement Series*, 180(2): 225.

Hubble, E. (1929). A relation between distance and radial velocity among extra-galactic nebulae. *Proceedings of the National Academy of Sciences*, 15(3): 168–173.

Jacquart, M. (2016). *Similarity, Adequacy, and Purpose: Understanding the Success of Scientific Models*. Ph.D. thesis, The University of Western Ontario.

Jacquart, M. (2020). Observations, simulations, and reasoning in astrophysics. *Philosophy of Science*, 87(5), pp. 1209–1220.

Kosso, P. (2013). Evidence of dark matter, and the interpretive role of general relativity. *Studies in History and Philosophy of Modern Physics*, 44(2): 143–147.

Kragh, H. (2014). Testability and epistemic shifts in modern cosmology. *Studies in History and Philosophy of Modern Physics*, 46: 48–56.

Kuhn, T.S. (1977). Objectivity, value judgment, and theory choice. In A. Bird and J. Ladyman (eds.), *Arguing about Science*, New York: Routledge, pp. 74–86.

Lahav, O. and Massimi, M. (2014). Dark energy, paradigm shifts, and the role of evidence. *Astronomy & Geophysics*, 55(3): 3.12–3.15.

López-Corredoira, M. (2014). Non-standard models and the sociology of cosmology. *Studies in History and Philosophy of Modern Physics*, 46: 86–96.

López-Corredoira, M. (2017). Tests and problems of the standard model in cosmology. *Foundations of Physics*, 47(6): 711–768.

Ma, C. and Zhang, T.-J. (2011). Power of observational hubble parameter data. *The Astrophysical Journal*, 730(2): 74.

Manchak, J. (2009). Can we know the global structure of spacetime? *Studies in History and Philosophy of Modern Physics*, 40: 53–56.

Manchak, J. (2011). What is a physically reasonable spacetime? *Philosophy of Science*, 78: 410–420.

Massimi, M. (2018). *Three problems about multi-scale modelling in cosmology*. Available at: arXiv preprint arXiv:1804.07704.

Massimi, M. and Peacock, J. (2014). What are dark matter and dark energy? In M. Massimi (ed.), *Philosophy and the Sciences for Everyone*, Chapter 3. New York: Routledge, pp. 33–51.

Matarrese, S., et al. (2011). *Dark Matter and Dark Energy: A Challenge for Modern Cosmology*, Vol. 370. New York: Springer Science & Business Media.

McGaugh, S. (2014). A tale of two paradigms: The mutual incommensurability of λcdm and mond. *Canadian Journal of Physics*, 93(2): 250–259.

Merritt, D. (2017). Cosmology and convention. *Studies in History and Philosophy of Modern Physics*, 57: 41–52.

Milgrom, M. (1983). A modification of the Newtonian dynamics as a possible alternative to the hidden mass hypothesis. *The Astrophysical Journal*, 270: 365–370.

Natarajan, P. (2016). *Mapping the Heavens*. New Haven and London: Yale University Press.

Nobbenhuis, S. (2006). Categorizing different approaches to the cosmological constant problem. *Foundations of Physics*, 36(5): 613–680.

Riess, A. and et al. (1998). Observational evidence from supernovae for an accelerating universe and a cosmological constant. *The Astronomical Journal*, 116: 1009–1038.

Rugh, S.E. and Zinkernagel, H. (2002). The quantum vacuum and the cosmological constant problem. *Studies in History and Philosophy of Modern Physics*, 33(4): 663–705.

Ruphy, S. (2011). Limits to modeling: Balancing ambition and outcome in astrophysics and cosmology. *Simulation Gaming*, 42(2): 177–194.

Seo, H.-J. and Eisenstein, D.J. (2003). Probing dark energy with baryonic acoustic oscillations from future large galaxy redshift surveys. *The Astrophysical Journal*, 598(2): 720.

Shapere, D. (1993). Astronomy and antirealism. *Philosophy of Science*, 60(1): 134–150.

Smeenk, C. (2013). Philosophy of cosmology. In R. Batterman (ed.), *The Oxford Handbook for the Philosophy of Physics*. Oxford: Oxford University Press.

Smeenk, C. (2014). Cosmology. In M. Curd and S. Psillos (eds.), *The Routledge Companion to Philosophy of Science*. New York: Routledge, pp. 609–620.

Schneider, M.D. (2020a). Betting on future physics. *The British Journal for the Philosophy of Science*.

Schneider, M.D. (2020b). What's the problem with the cosmological constant? *Philosophy of Science*, 87(1): 1–20.

Spergel, D.N. (2015). The dark side of cosmology: Dark matter and dark energy. *Science*, 347(6226): 1100–1102.

Spergel, D.N. and Steinhardt, P.J. (2000). Observational evidence for self-interacting cold dark matter. *Physical Review Letters*, 84(17): 3760.

Sus, A. (2014). Dark matter, the equivalence principle and modified gravity. *Studies in History and Philosophy of Modern Physics*, 45: 66–71.

Trimble, V. (1987). Existence and nature of dark matter in the universe. *Annual Review of Astronomy and Astrophysics*, 25(1): 425–472.

Uzan, J.-P. (2010). Dark energy, gravitation and the Copernican principle. In P. Ruiz-Lapuente (ed.), *Dark Energy: Observational and Theoretical Approaches*. Cambridge: Cambridge University Press, pp. 3–50.

Vanderburgh, W.L. (2003). The dark matter double bind: Astrophysical aspects of the evidential warrant for general relativity. *Philosophy of Science*, 70(4): 812–832.

Vanderburgh, W.L. (2005). The methodological value of coincidences: Further remarks on dark matter and the astrophysical warrant for general relativity. *Philosophy of Science*, 72(5): 1324–1335.

Vanderburgh, W.L. (2014). On the interpretive role of theories of gravity and "ugly" solutions to the total evidence for dark matter. *Studies in History and Philosophy of Modern Physics*, 47: 62–67.

Weinberg, D.H. et al. (2015). Cold dark matter: Controversies on small scales. *Proceedings of the National Academy of Sciences*, 112(40): 12249–12255.

Weinberg, S. (1989). The cosmological constant problem. *Reviews of Modern Physics*, 61(1): 1.

Weisberg, M., Jacquart, M., Madore, B. and Seidel, M. (2018). The dark galaxy hypothesis. *Philosophy of Science*, 85(5): 1204–1215.

Wilson, M. (1980). The observational uniqueness of some theories. *The Journal of Philosophy*, 77(4): 208–233.

Zwicky, F. (1937). On the masses of nebulae and of clusters of nebulae. *The Astrophysical Journal*, 86: 217.

Further Reading from the Editors

Some overlapping issues are discussed in the review by C. Smeenk, "Philosophy of cosmology," in R. Batterman (ed.), *The Oxford Handbook of Philosophy of Physics* (Oxford University Press, 2013). A survey of numerous potential reasons for skepticism about the current standard model of cosmology is M. López-Corredoira, "Tests and problems of the standard model in cosmology," *Foundations of Physics* (2017): 1–58. An argument that dark matter searches are methodologically novel is given by S. Baerdemaeker, "Method-Driven Experiments and the Search for Dark Matter," *Philosophy of Science* (forthcoming). For an epistemological assessment of the prospects of the ΛCDM model, see S. De Baerdemaeker and N. Mills Boyd, "Jump ship, shift gears, or just keep on chugging: Assessing the responses to tensions between theory and evidence in contemporary cosmology," *Studies in History and Philosophy of Science Part B: Studies in History and Philosophy of Modern Physics* (forthcoming).

54

EVIDENCE IN ASTROPHYSICS

Sibylle Anderl

For millennia, astronomy consisted of passively and repeatedly observing stars and planets on their daily, monthly, or annual trajectories across the sky, punctuated by transient phenomena, such as novae, comets, eclipses, or occultations. Laws of the heavenly bodies were accordingly first based on direct visual perception, later aided by mechanical instruments such as quadrants and then augmented with simple telescopes in the early 17th century. Observational records were in the form of short tables, sketches, and drawings. Such observational data was mainly used as evidence for hypotheses on the mechanisms behind the changes of the visual night sky. The strong practical motivation consisted in applications to navigation and the making of calendars. Astronomy turned into astrophysics after the year 1814 when Joseph Fraunhofer first observed dark lines in the solar spectrum.[1] A few decades later, these were interpreted by Robert Bunsen and Gustav Kirchhoff as absorption features also seen in laboratory experiments on common terrestrial elements known to chemists and physicists. This observation paved the way for a fundamental insight: the physics that we know is operating on Earth is the same physics that is governing observable phenomena and processes in the distant Universe. Since then, the accessible empirical evidence for our physical understanding of the Universe and its phenomena has been ever growing. This growth has been intimately coupled with technological developments that continue to open up new sources of information for astrophysics beyond the visual spectrum (e.g. Harwit, 1981; Jaschek, 1989). For instance, infrared astronomy has its roots only in the 1830s, radio astronomy began in the 1930s, X-ray astronomy was initiated in the late 1940s, and we just witnessed in 2016 the beginning of another branch of astronomy with the first direct detection of gravitational waves (Abbott et al., 2016). Cosmology in particular has made its transition into an empirical discipline rather late after the general predominance of speculative theory at the beginning of the last century. Observational missions e.g., the Planck satellite run by the European Space Agency (ESA), which yielded a precise measurement of the cosmic microwave radiation, have offered a wealth of high-precision cosmological data scientists a few decades ago could only have dreamt of.

Today, astrophysics and cosmology have increasingly become "big-data sciences." Huge surveys create tremendous data volumes at much higher rates than ever before, so far only comparable to the recent situation in the big experiments of particle physics. For instance, the goal of the Gaia astrometric spacecraft, which was launched in 2013 by ESA, was to measure the distances, positions, and movements of more than 1 billion stars in our Milky Way. When the mission was planned it was already clear that the vast amount of data Gaia creates for this job will exceed the amount of data ground stations were capable of receiving. Accordingly, on-board triage of the data stream was necessary (see Gaia Collaboration, 2016) – an epistemologically interesting situation, as it prevents the possibility of a fundamental re-analysis of the raw data. Gaia is not particularly exceptional in terms of its data generation and processing. Many more astronomical projects and missions are on their way, which will create an unprecedented flow of astronomical data. The increasingly complex resulting datasets are themselves based on complex background and modeling assumptions, and may thereby be interrelated in more or less opaque ways. This development has raised new epistemological challenges for the understanding of scientific evidence. The same is true with respect to the development of more and more complex simulations, whose pivotal role today is hard to separate from the creation and interpretation of astronomical evidence. In view of these changes in research methodology, the question of what constitutes evidence in astrophysical research and how such evidence is used in order to understand the Universe we live in is more pressing than ever.

This being said, traditionally the concept of evidence in astrophysics has been based on observations: evidence for a certain hypothesis usually refers to observational data, processed to a level at which the data can be directly related to the hypothesis. Different philosophical accounts of the concept of evidence have pointed out the role of background knowledge for the bearing of evidence (for reviews see e.g. Kelly, 2016 or Achinstein, 2014)[2]: whether something is evidence for some hypothesis may depend on the prior probability distribution (Bayesianism), the paradigm (Thomas Kuhn), or the inductive method (confirmation theory according to Carnap.)[3] This background knowledge does not only consist of theoretical assumptions concerning the relation between evidence and the hypothesis but also on assumptions on the theory of data generation and data processing.

Concerning the role of empirical evidence in astrophysics (Section 54.1) it is an interesting point that this discipline is confronted with the need for characteristic background assumptions that arise from its predominantly observational status and the concomitant bias problems due to the restricted availability of data of a given quality for the objects of their studies. This point will be discussed in Section 54.1.2, focusing on the understanding of classes of cosmic events and phenomena. Before that, the role of empirical evidence for the understanding of singular cosmic events and phenomena will be introduced in Section 54.1.1, which is, in view of its lack of statistical complication, the simplest type of astrophysical problem that already illustrates a specific astrophysical methodology. This case is exemplified by considering the nature of the very massive object at the center of our Milky Way galaxy. This methodology may be increasingly influenced by the fact that astronomy is entering the "big-data-stage" – an aspect that is discussed in Section 54.1.3. Finally, Section 54.2 will discuss the evidential role of numerical modeling.

54.1 Empirical Evidence

54.1.1 *Understanding Singular Events and Phenomena*

The simplest form of astrophysical research problems concerns the understanding of historical one-time events: We receive information from a certain spatiotemporally defined cosmic location and try to understand what is going on there and what has led to the specific circumstances we witness: Why does the outflow of this very young protostar change direction two arcseconds north of the source? Why does this galaxy have such a disturbed morphology? Due to the finite speed of light, depending on the distance of the observed location from the Earth, the information we receive stems from some point of time in the nearer or more distant past.

This fact sometimes inspires the comparison of astrophysics with archaeology and indeed, the investigation of past cosmic processes shows methodological similarities to what is known as "historical sciences" (designating work from geology and biology, as well as paleontology and archaeology). Carol Cleland investigated the epistemology of the historical sciences in 2002. According to her, in these fields, the explanation of existing natural phenomena in terms of long-past causes has to come to terms with the fact that a given astrophysical hypothesis cannot be tested by controlled laboratory experiments (see also Anderl, 2016). Instead, puzzling traces of long-past events are observed and then hypotheses are searched for in order to explain them by finding a common cause for these "historical" traces.[4] Accordingly, such traces provide evidence for what has happened in the past in a way that is similar to having successful predictions provide evidence in an experimental context. Cleland states that the quality of a given hypothesis depends on the extent and diversity of the body of traces it can explain. In principle, this offers a way in which one can directly compare different hypotheses in terms of their relative probabilities.

The most direct way to decide between rival hypotheses is however provided by the so-called "smoking gun observations." Ben Jeffares (2008) defines these as follows: "Take multiple observations of evidence: $[o^a, o^b, o^c]$. Now take two hypotheses, H_1 and H_2. If H_1 accounts for $[o^a + o^b]$ but is incompatible with $[o^c]$, and H_2 accounts for all three observations $[o^a + o^b + o^c]$, then $[o^c]$ is the 'smoking gun' that discriminates between two hypotheses about a historical event." The effect of a smoking gun observation is to discredit one hypothesis while ideally supporting another. Indeed, a good deal of astrophysical observations is aiming at such smoking gun observations.

A recent example of this approach centers around the question concerning the nature of the apparently very massive object at the center of the Milky Way.[5] Measurements of stellar orbits close to the galactic center, variable X-ray emission, and strongly variable polarized near-infrared emission from the central radio source Sagittarius A* provide evidence for the interpretation that the massive object is a supermassive black hole. One possible alternative to the black hole scenario would be the existence of a fermion (e.g., neutrinos) ball at that location (Viollier et al., 1992; De Paolis et al., 2001). The fermion ball scenario could explain the low luminosity of Sagittarius A*, which would then be due to the existence of a spatially-extended, but very high-density mass at that location. From the observed stellar orbits, the total mass required to force the stars on their trajectories can however be estimated, as well as properties of the mass distribution (Ghez et al., 2008; Gillessen et al., 2009a,b). These dynamical constraints, together with the requirement that the fermion ball does not just collapse to a black hole under its own gravity, make this scenario less and less likely. In addition, it does not explain what happens with in-falling baryonic matter, which would drive the fermion ball to a black hole sooner or later. Furthermore, it does not explain the appearance of Sagittarius A* in X-ray and near-infrared emission, which hint at a very small spatial extent of the emitting source. The observed close stellar orbits, in particular, would classify as smoking gun observation in this example, as they are not compatible with the hypothesis of a fermion ball, while they agree with the hypothesis of a supermassive black hole with a currently low accretion rate and with a mass of about 4 million times that of the sun.

It is important to see that the relevant background theory in order to secure the interpretation of observable traces as evidence for certain past causes contains the laws from physics and chemistry. In the example, such background theories are general relativity as well as theories on potentially relevant quantum processes. Jeffares (2008) stressed that, in contrast to Cleland's line of arguments, even the historical sciences rely on assumed regularities – general claims about the relation between prior and later facts, provided by the experimental sciences, that go beyond hypotheses about particular times or places– which weakens Cleland's claim of their methodological otherness from the experimental sciences. The above example shows that he is right with his argument, given that the understanding of one-time cosmic events also rests on assumed regularities, such as the laws of physics and chemistry, which are (in most cases) already experimentally established. Cosmic regularities beyond physical or chemical laws are however often not crucial at this stage of studying a singular phenomenon. Accurately disentangling the roles of contingent environmental features and features that point at general properties of the cosmic phenomenon under study (such as the typical evolution of a black hole), is impossible based on the observations of a single example only.[6] This task requires a statistical analysis of samples of individual instances of phenomena of the same type that are large enough to allow for the establishment of statistically significant trends.[7]

54.1.2   *Understanding Classes of Events and Phenomena*

The goal of understanding the Universe entails the search for cosmic regularities. Astrophysicists try to understand phenomena and processes such as the life-cycle of interstellar matter, the typical lives of stars, or the co-evolution of both stars and gas in the form of galaxies. Ordering and interpreting these cosmic regularities are complicated by the lack of experimentation, as the relevant spatial and temporal scales and the physical conditions involved are usually impossible to simulate under controlled laboratory conditions. Such experiments would however be useful to disentangle the influence of contingent environmental conditions and generic features of types of cosmic phenomena. This lack is compensated for by the "cosmic laboratory" (e.g., Pasachoff, 1977): the fact that, at any given time, the Universe provides a huge variety of examples of a given phenomenon, caught at different evolutionary states and subject to different environmental conditions. Whatever cosmic configuration one could possibly think of, consistent with known underlying laws of nature – it should exist somewhere in the Universe due to its vast extent and diversity. The remaining challenge would be to find and identify it. A targeted search for specific configurations is however usually not even necessary in view of the available statistical methods. Using this cosmic laboratory to extract evidence for hypotheses on cosmic regularities requires the smart application of statistical techniques, similar to problems in the social sciences (see e.g., Dunning, 2012) where it is also difficult to control for confounding variables. The most straightforward strategy is to select a representative sample of a certain cosmic phenomenon, where hopefully the influence of contingent environmental factors will cancel out for the sample as a whole. Then one can look for correlations and explore whether these observed correlations can be understood as evidence for certain hypotheses on the type of phenomenon under study.

The crucial pieces of background information for such inferences concern the potential presence of selection bias. In astrophysics, such selection bias may stem from a number of structural constraints of astronomical observations. First of all, a reasonable comparison of observations requires the same angular resolution, as everything inside the telescope beam (the opening angle of the smallest resolvable angular unit) will be smoothed. Accordingly, the size of the physical region that is resolved in a particular astronomical observation depends primarily on three quantities: the wavelength of the radiation being observed, the aperture size (diameter) of the telescope being used, and not least the distance of the cosmic region being studied. For instance, only close objects can be observed at a very high spatial resolution for a telescope/detector combination having a given angular resolution. Accordingly, the question arises as to whether the knowledge that comes from detailed observations of our proximate cosmic environment is also representative of other regions of the more distant Universe. Empirical evidence derived from such sample studies rests on the validity of assumptions of representativity, which tends to be endangered by constraints concerning the comparability or availability of observations. Another issue concerns the fact, that the further we look, the deeper we see into the cosmic past. This is due to the finite speed of light combined with the vast distances involved. The previous point on angular resolution means that we see the distant past more blurred, less resolved, and this immediately hinders the comparison with observations from the local Universe. Furthermore, it is not immediately obvious how to relate phenomena from the distant past to those at present times in an evolutionary sequence. Evidential reasoning here also contains a theory on the evolution of the Universe as a whole as background knowledge, as this evolution sets environmental conditions at each point of time. Observing the same type of object over cosmic timescales therefore relies on wide-ranging background assumptions, be they instrumental, empirical, practical, or theoretical.

As mentioned before, hypotheses about types of cosmic phenomena can in turn be used in the explanation of the specific appearance of singular instances of this phenomenon. Evidence gained from the observation of classes of objects can therefore also act as evidence for or against more specific hypotheses to explain one-time events or phenomena and vice versa. This approach is not possible, however, when we try to understand the Universe as a whole, because we only have empirical access to one Universe and cannot make claims on any type-specific behavior based on the study of ensembles of similar universes. If there was no other way for deriving general cosmological laws or regularities, this would qualify cosmology as prototype of a historical rather than a law- or regularities seeking science. Chris Smeenk (2014), however, stresses that many laws used in cosmology are local dynamical laws, such as the field equations of general relativity, in the sense that they primarily apply to subregions of the Universe. Therefore, in their application to several subregions, they do show the "multiplicity" required to call them "laws." Problems arise, however, for distinctively cosmological laws that cannot be understood in a local way because they intrinsically describe global properties of spacetime – Smeenk's example is a law stating that the Universe is spatially bounded. He points out that in these cases the available evidence is not sufficient to determine the best model of global spacetime geometry due to two reasons: Firstly, because the maximum signal speed fundamentally limits the size of the accessible part of the Universe, secondly, because general relativity does not provide strong enough constraints on global spacetime geometry unless we accept assumptions like the "cosmological principle"[8] – the idea that spacetime is statistically homogenous and isotropic on large scales.

54.1.3 *Evidence and Big-Data Science*

After particle physics has set the stage for big-data science with CERN's Large Hadron Collider (LHC), astrophysics has been another field entering the regime of data production that raises data generation rates and storage as well as the processing and analyzing requirements to a whole new level.[9] The planned Square Kilometer Array (SKA), which will be an array of telescopes in South Africa and Australia constituting the world's largest radio telescope, will produce 600 petabytes (PB) of archived raw data per year (for comparison: the LHC created a total of about 50 PB of archived raw data in 2016 and is predicted to increase to 150 PB per year for its Run 3 starting in 2021). The data generation rate will amount to five times the data generation of the global internet in 2016. The technological challenge astrophysics and particle physics are now preparing for concerns of exascale (1 exabyte is 1 million petabytes) computing and data storage.

This practice of big-data science raises new methodological challenges concerning the use of this data as evidence to answer scientific questions. In particular, algorithms have to be developed in order to search the data for characteristic patterns, often in real time, to decide which data to keep, due to restricted storage. Data selection criteria are usually based on background theoretical assumptions, and high selectivity of these criteria may create some tension with the additional goal of an exploratory approach. For the case of the LHC, Koray Karaca (2017a) describes how data acquisition, based on selection criteria, is theory-laden, determined by considering the testable predictions of the theoretical models discussed. According to Karaca, such a "model of data acquisition that specifies and organizes the experimental procedures necessary to select the data according to a predetermined set of selection criteria" belongs to a different class of models than the ones described in the "hierarchy of models" account put forward by Patrick Suppes (1962) and Deborah Mayo (1996), namely, models of theory, models of experiment, or models of data. As a system of data selection according to a set of predetermined criteria, such a model cannot be understood, as he argues, in the context of classical procedures to statistically test some theoretical hypothesis. However, Karaca (2017b) also describes how an exploratory data selection procedure is successfully implemented in the Atlas experiment at the LHC despite the aforementioned theory-ladenness of selection criteria.

The key idea here is to construct the theoretical input in a way that it includes a wide variety of different theoretical predictions. C.D. McCoy and Michaela Massimi (2017) describe the important role of simplified models for this goal, as these extensions to the standard model only focus on a few additional decisive parameters for hypothetical "Beyond Standard Model" particles. The general underlying problem is the variety of theoretical options that exist as possible extensions to the standard model of particle physics: The uncertainty of where to look and what to look for makes it difficult to ensure not to miss anything in the choice of data acquisition selection criteria. Therefore, data selection needs to rest on assumptions that are general enough to make as few model-dependent restrictions as possible. Both model-independent and data-driven approaches are needed, according to McCoy and Massimi. This situation as analyzed for the case of particle physics is very similar to the case of cosmology when it comes to theories extending the Dark Energy-Dark Matter paradigm and explicating the nature of dark matter and dark energy.[10]

Classical confirmation theory is not able to capture this unanticipated, data-driven aspect of modern science: The traditional procedure of deducing empirical consequences from theoretical models and confronting these with the available empirical evidence does not match scientific practice anymore when it specifically comes to the extensions of the respective standard models in cosmology and particle physics. Big-data science demonstrates the need for theoretical guidance in deriving empirical evidence from the available data, as well as the need for general exploratory strategies in order to mitigate the problems caused by a theory-driven pre-selection of empirical evidence. At the same time, the option of being left with mere correlations emerging from certain patterns in the data, without hypotheses on underlying mechanisms needs to be addressed as well (e.g., Kitchin, 2014). This complex interplay of available theoretical hypotheses, which includes additional ingoing background knowledge and data selection/reduction strategies, that determines the extraction of evidence from vast amounts of data will be an exciting field for further epistemological analysis of astrophysical research in the coming years.

54.2   Computer Models as Evidential Resources

Astrophysics, like most other scientific disciplines, is currently grounded on numerical modeling. In most cases when empirical evidence is brought up in favor of or against some hypotheses, numerical models based on background knowledge from physics and chemistry together with reasonable ranges of parameter values, boundary-, and initial conditions are used to relate the hypothesis in question with the existing observational evidence (usually) in terms of a causal explanation: Why do we observe what we observe? Or more precisely: What is the best causal explanation of our obtaining the evidence that we have in fact obtained? The role of a particular simulation could then be understood as to provide an explanatory connection between evidence e and hypothesis h, as required in Achinstein's (2014) definition of scientific evidence. Alternatively, the evidential status of simulations can also be understood in an error-statistical view. Deborah Mayo (1996) develops an account of evidence that is based on the idea of a severe test for a hypothesis: e is good evidence if h passes a severe test with result e, and if the probability that the test results in e is much higher if h was true than if it was not. Wendy Parker (2008) has used Mayo's account to understand the practice of model evaluation, while Mayo herself applied her analysis to the question of how experimental inquiry can provide good evidence for scientific hypotheses. The core qualification for obtaining good evidence is the ability of scientists to deal with sources of experimental error, which could undermine the result. Parker now takes simulations as putative test procedures that may or may not yield results that fit the hypothesis h. This allows her to transfer Mayo's analysis to cases where computer simulations are used to provide evidence regarding the behavior of natural systems if the simulation would indeed be unlikely to deliver results that fit with h if h was false. The question of when results of a simulation provide good evidence leads Parker to an analysis of sources of error in computer simulations. Parker lists seven groups of such sources: Study design errors, substantive modeling errors, data processing errors, solution algorithm errors, numerical errors, programming errors, and hardware-related errors.

This analysis is further expanded upon in multiple discussions on the epistemology of simulations and experiments (see for example Winsberg, 2010, Franklin, 1986, and reference therein).

In her paper, Parker discusses whether results from simulations can deliver good evidence for scientific hypotheses about real-world target systems. Her examples are models of the Earth's atmosphere and its climate. Here the question is, whether a particular modeling result can be seen as evidence for a given prediction concerning future climate evolution. If this approach is used to understand the status of astrophysical models, it is however necessary to also account for the central role of existing empirical observations in this picture. A typical astrophysical research question would be whether a simulation can reproduce empirical observations based on a particular physical or chemical scenario. For instance, there are many cosmological simulations. A well-known example is the "Millennium Simulation" (Springel et al., 2005), which "rebuilds" the Universe based on a specific cosmological model. The model outputs are then required to statistically reproduce the structure of the observed Universe. If this is the case, this match can be seen as evidence for the empirical adequacy of the incorporated model. Parker critically discusses the possibility of empirically establishing the reliability of simulation results as an alternative strategy for devising lower-level severe tests in Mayo's sense for the localization of errors. In cosmological simulation as mentioned above, the probed agreement of model results with empirical observations however seems to have a different status than that equivalent of a mere error strategy, because it is the validity of the implemented physical theory (e.g., the specific cosmological model such as the Lambda Cold Dark Matter (ΛCDM) model or the Modified Newtonian dynamics (MOND) model) that is the main research question. Thus the possibility of using the wrong theory seems to be more than just one possible source of error among others.

Parker is however right in her conclusion that the practice of simulation model evaluation would profit from a more error-statistical perspective. The complexity of astrophysical simulations and the potential relevance of physical processes on many different spatial and temporal scales requires numerical tricks due to the limitations of our computational abilities, which can neither be motivated by theory nor be delivered by physical intuition (see e.g., Bailer-Jones, 2000; Ruphy, 2011). A thorough analysis of the epistemological conditions that need to hold for a complex simulation to yield evidence for an equally complex astrophysical hypothesis will be a fruitful task, not only for philosophers of science but for astrophysicists as well.

Notes

1 For a description of these historical events see e.g. Nath (2013) or Hearnshaw (2014).
2 In this article we will refrain from choosing one particular definition. In astrophysics the use of the term varies among sub-fields and communities: Some astrophysical articles are based on Bayesian definitions (for examples see, e.g., Brewer, 2008) while others may be closer to an error-statistical view (see, e.g., the discussion on the proper interpretation of BICEP2 2014 results, Ade et al. (2014) and papers referring to it). Therefore, in this article we prefer a methodological approach where different roles of evidence in astrophysics are highlighted.
3 For an overview on confirmation theory see e.g. Hájek and Joyce (2014).
4 Lewis (1986) used the collapse of a star to become a compact object (like a white dwarf or a neutron star) as example of a non-causal explanation in astrophysics. However, contrary to his description, today's textbooks would say the collapse is stopped by the pressure of degenerate matter – typical astrophysical explanations are causal explanations.
5 See for example, Eckart et al. (2017). See also the scientific background on the Nobel Prize in Physics 2020 (The Noble Committee for Physics, 2020).
6 If information on the type of the observed phenomenon is at hand, it can however help in the evaluation of rival hypotheses. If the compact mass at the center of our galaxy is seen as one instance of the general occurrence of compact masses at the center of most other spiral galaxies, the range of inferred masses in other systems yields additional constraints on the physical processes plausibly leading to their creation
7 In order to secure this, correlation theory is frequently used, see e.g. Rodgers and Nicewander (1988). The threshold for statistical significance is then often assumed to be 3 sigma.
8 See, e.g., Weinberg (1972), Chapter 14.1. For a philosophical discussion see, e.g., Beisbart (2009).

9 The novelty holds for several levels that are often distinguished with respect to big data: volume of data, velocity of data creation, diversity of data (structured, semi-structured, unstructured) and the cross-correlated nature of many datasets and their contents, see e.g. Garofalo et al. (2017).

10 See Jacquart (this volume), for more details on dark matter and dark energy.

References

Abbott, B.P., et al. (LIGO Scientific Collaboration and Virgo Collaboration) (2016). Observation of gravitational waves from a binary black hole merger. *Physical Review Letters*, 116(6): 061102.

Achinstein, P. (2014). Evidence. In M. Curd and S. Psillos (eds.), *The Routledge Companion to Philosophy of Science.* London, New York: Routledge, pp. 381–392.

Ade, P.A.R. et al. (BICEP2 Collaboration) (2014). BICEP2 I: Detection of B-mode polarization at degree angular scales. *Physical Review Letters*, 112(24): 241101.

Anderl, S. (2016). Astronomy and astrophysics. In P. Humphreys (ed.), *The Oxford Handbook of Philosophy of Science.* New York: Oxford University Press, pp. 652–670.

Bailer-Jones, D.M. (2000). Modelling extended extragalactic radio sources. *Studies in History and Philosophy of Modern Physics*, 31(1): 49–74.

Beisbart, C. (2009). Can we justifiably assume the cosmological principle in order to break model underdetermination in cosmology? *Journal for General Philosophy of Science*, 40: 175–205.

Brewer, B.J. (2008). *Applications of Bayesian Probability Theory in Astrophysics.* PhD thesis, arXiv:0809.0939v1.

Cleland, C. (2002). Methodological and epistemic differences between historical science and experimental science. *Philosophy of Science*, 69: 474–496.

De Paolis, F., et al. (2001). Astrophysical constraints on a possible neutrino ball at the Galactic Center. *Astronomy & Astrophysics*, 376: 853–860.

Dunning, T. (2012). *Natural Experiments in the Social Sciences.* Cambridge: Cambridge University Press.

Eckart, A., et al. (2017). The Milky Way's supermassive black hole: How good a case is it? *Foundation of Physics*, 47(5): 553–624.

Franklin, A. (1986). *The Neglect of Experiment.* Cambridge: Cambridge University Press.

Gaia Collaboration (2016). The Gaia mission. *Astronomy & Astrophysics*, 595, A1:1–36.

Garofalo, M., Botta, A. and Ventre, G. (2017). Astrophysics and big data: Challenges, methods, and tools. *Proceedings of the International Astronomical Union*, S325(12): 345–348.

Ghez, A., et al. (2008). Measuring distance and properties of the Milky Way's central supermassive black hole with stellar orbits. *The Astrophysical Journal*, 689(2): 1044–1062.

Gillessen, S., et al. (2009a). Monitoring stellar orbits around the massive black hole in the Galactic center. *The Astrophysical Journal*, 692(2): 1075–1109.

Gillessen, S., et al. (2009b). The orbit of the star S2 around SGR A* from Very Large Telescope and Keck data. *The Astrophysical Journal*, 707(2): L114–L117.

Hájek, A., and Joyce, J.M. (2014). Confirmation. In M. Curd and S. Psillos (eds.), *The Routledge Companion to Philosophy of Science.* London, New York: Routledge, pp. 146–159.

Harwit, M. (1981). *Cosmic Discovery.* New York: Basic Book, Inc.

Hearnshaw, J.B. (2014). *The Analysis of Starlight: Two Centuries of Astronomical Spectroscopy* (2nd edition) Cambridge: Cambridge University Press.

Jaschek, C. (1989). *Data in Astronomy.* Cambridge: Cambridge University Press.

Jeffares, B. (2008). Testing times: Regularities in the historical sciences. *Studies in History and Philosophy of Biological and Biomedical Science*, 39: 469–475.

Karaca, K. (2017a). Lessons from the Large Hadron Collider for model-based experimentation: The concept of a model of data acquisition and the scope of the hierarchy of models. *Synthese* https://doi.org/10.1007/s11229-017-1453-5.

Karaca, K. (2017b). A case study in experimental exploration: Exploratory data selection at the Large Hadron Collider. *Synthese*, 194(2): 333–354.

Kelly, T. Evidence. In E.N. Zalta (ed.), *The Stanford Encyclopedia of Philosophy* (Winter 2016 Edition). Available at: https://plato.stanford.edu/archives/win2016/entries/evidence/.

Kitchin, R., (2014). Big data, new epistemologies and paradigm shifts. *Big Data & Society*, 1(1): 1–12.

Lewis, D. (1986). Causal explanation. In D. Lewis (ed.), *Philosophical Papers Volume 2*, New York: Oxford University Press, pp. 214–240.

Mayo, D. (1996). *Error and the Growth of Experimental Knowledge.* Chicago: University of Chicago Press.

McCoy, C.D. and Massimi, M. (2017). Simplified models: A different perspective on models as mediators. *European Journal for the Philosophy of Science*, 1–25 https://doi.org/10.1007/s13194-017-0178-0.

Nath, B.B. (2013). *The Story of Helium and the Birth of Astrophysics*. New York: Springer.

Nobel Prize Committee for Physics (2020). Theoretical Foundations for Black Holes and the Supermassive Compact Object at the Galactic Centre. https://www.nobelprize.org/uploads/2020/10/advanced-physicsprize2020.pdf

Parker, W. (2008). Computer simulation through an error-statistical lens. *Synthese*, 163(3): 371–384.

Pasachoff, J.M. (1977). *Contemporary Astronomy*. New York: Saunders College Publishing.

Rodgers, J.L. and Nicewander, W.A. (1988). Thirteen ways to look at the correlation coefficient. *The American Statistician*, 42(1): 59–66.

Ruphy, S. (2011). Limits to modeling: Balancing ambition and outcome in astrophysics and cosmology. *Simulation & Gaming*, 42(2): 177–194.

Smeenk, C. (2014). Cosmology. In M. Curd and S. Psillos (eds.), *The Routledge Companion to Philosophy of Science*. London, New York: Routledge, pp. 609–620.

Springel, V., et al. (2005). Simulating the joint evolution of quasars, galaxies and their large-scale distribution. *Nature*, 435: 629–636.

Suppes, P. (1962). Models of data. In E. Nagel, P. Suppes, and A. Tarski (eds.), *Logic, Methodology, and Philosophy of Science: Proceedings of the 1960 International Congress*. Stanford University Press, pp. 252–261.

Viollier, R.D., et al. (1992). Halos of heavy neutrinos around baryonic stars. *Physics Letters B*, 297(1–2): 132–137.

Winsberg, E. (2010). *Science in the Age of Computer Simulation*. Chicago: The University of Chicago Press.

Further Reading from the Editors

A complementary overview of philosophical issues raised by astrophysics is S. Anderl, "Astronomy and astrophysics," in P. Humphreys (ed.), *The Oxford Handbook of Philosophy of Science* (New York: Oxford University Press, 2016): 652–670; see also C. Smeenk, "Cosmology," in M. Curd and S. Psillos (eds.), *The Routledge Companion to Philosophy of Science* (London, New York: Routledge, 2014): 609–620. A vigorous defense of testability in cosmology is George Ellis (2014). "On the philosophy of cosmology," *Studies in History and Philosophy of Science Part B: Studies in History and Philosophy of Modern Physics*, 46(1): 5–23. Another recent discussion of the notion of testability, with particular reference to multiverse theories, is H. Kragh (2014). "Testability and epistemic shifts in modern cosmology," *Studies in History and Philosophy of Modern Physics*, 46: 48–56. A case for epistemic caution concerning modeling in astrophysics is made by S. Ruphy (2011). "Limits to modeling: balancing ambition and outcome in astrophysics and cosmology," *Simulation & Gaming*, 42(2): 177–194.

INDEX

Note: **Bold** page numbers refer to tables; *italic* page numbers refer to figures and page numbers followed by "n" denote endnotes.

Abrikosov, A.A. 293

absolute space 6–7, 10, 33, 39–40, 46, 48–49, 52, 61–63, 76, 128, 130–131, 134, 549, 569, 694, 709

absolute time 46–49, 51, 61–63, 67, 386, 574, 709–712

acceleration 16, 467, 500, 553; absolute 40–41, 61, 569; constraint force 26; dimension of 672; Einstein equivalence principles 127–134; existence of 6–7, 9; gravitational force 25; gravity and 127–134; of mass 9; in Newton's second law of motion 38; particle's 22, 39; relation between force and 12; space-time structure 68–69; uniform 128

Achinstein, P. 749

action 299; bare couplings 302; of diffeomorphism 147; Einstein–Hilbert 78, 290, 292; Einstein–Maxwell system 78; extremal 389; fields of force 18; local 661–663; natural group 580, 581, 583, 584, 589; Newtonian 50; physical 16; predictive behavior 233; renormalized 302; theory's 306; Wilsonian 306–308

additional beables theories 99, 103–108; De Broglie–Bohm theory 103–106; hypersurfaces 104, *104*; modal interpretations 106–108

Adler, S.L. 624

Aguirre, A. 714, 715, 726, 727

Aharonov–Bohm effect 378, 468, 660

Ainsworth, P. 420

Albert, D.Z. 253, 254, 408, 409, 434, 435, 441, 543, 612, 613, 715; Albert's theorem 435

Alexander, H.G. 46

algebraic quantum field theory (AQFT) 293

Allori, V. 267, 613

Almheiri, A. 365

Altland, A. 293

Anderle, S. 689

Anderson, P.W. 497, 551

Anglin, J. 206

anthropic principle 442, 700

Arageorgis, A. 326, 327

Archimedes' principle 478

Ardourel, V. 519

Aristotle 692; principle 386, 388, 393

Armstrong, D.M. 641, 642

Arnowitt, R. 367

Arntzenius, F. 29, 616

Arnztenius, F. 107

Arthur, R.T. 387

Ashtekar, A. 194, 341, 342

Ashtekar variables 367

Aspect, A. 184

astrophysics: computer models as evidential resources 749–750; empirical evidence 745–749; evidence and big-data science 748–749; understanding classes of events and phenomena 747–748; understanding singular events and phenomena 745–746

asymmetry: causal 460–461; entropy (*see* entropy asymmetry); explanatory, problem of 455–457; temporal 194–195, 605, 615; thermodynamic 461

A-theory 96; arguments 86; moving spotlight 87; notion of really happening 88; presentism 86–87; puzzle for STR and 90–92

autonomy of scales 296

Bacciagaluppi, G. 203

background independence 340, 343

Bain, J. 328, 331, 334

Baker, D.J. 287

Ballentine, L. 209

Ballentine, L.E. 616

Bangu, S. 518, 519

Banks, T. 293

Banks, T.A. 726

Barbour, J. 7, 46, 49, 50, 51, 52, 53, 55, 56, 386, 397, 715

Barbour, J.B. 50

Bardeen–Cooper–Schrieffer (BCS) models 333

Barenblatt, G.I. 667

Barenblatt, I.G. 668, 669

bare parameters 284–285, 720

Bargmann, V. 287

Barnes, E. 683, 722

Barnes, L.A. 290, 722

Barrett, T. 554

Barrow, J.D. 722, 724

Basic Chance Principle (BCP) 239

Batterman, R. 308, 309, 453, 470, 518

Batterman, R.W. 471, 472, 497, 634, 667

Bayes Theorem 721

Bechtel, W.P. 509

Becker, K. 351

Bedingham, D. 114

Bekaert, X. 287

Bekenstein–Hawking formula 369

Bekenstein, J.D. 78

Bell inequality theorem 263

Bell, J.S. 70, 71, 75, 100, 108, 182, 184, 186, 195, 213, 259, 262, 264, 267, 311, 624, 661, 662; theorem 60, 117, 182, 184, 185–188, 189, 195, 262, 461, 634

Belot, G. 155, 293, 602

Benatti, F. 113

Berkovitz, J. 107

Berndl, K. 104, 107

Bernoulli 644; principle 635; theorem 217, 234

Bertotti, B. 7, 49

Bertrand's theorem 724

beta function 304

Bickle, J. 502, 509

big bang 715; cosmological constraints 443, 445, 707–708, 712, 723, 733; geometrogenesis 376; inflation 731; initial conditions 725; multiverse of 713; nucleosynthesis 446–447, 731; reionization 731; singularity 341, 368, 694, 696, 714

Binney, J.J. 293

Birkhoff, G. 669

bitopological space 167

black-box approach 415, 480–481

black holes 68, 261, 363, 446, 746; behavior of quantum fields 342; collapsing of 460–461, 746; entropy to 364, 369, 402; Hawking radiation 695; in particle collisions 352; physics and information loss 355, 359, 365; radiation, discovery of 290, 292; singularities 341, 711; thermodynamics 293, 364

Blatt, J. 420

Boddy, K.K. 726

Bohm, D. 182, 220, 265, 266, 267, 268, 663

Bohmian mechanics 650; classical limit 266–267; differences from orthodox quantum mechanics 261; double-slit experiment (example) *258, 258*–259; empirical predictions 259–260; fundamental laws of 257–258; history 265; identical particles 266; limitations to knowledge and control 260–261; non-locality 262; observables 262–264; philosophical questions 268; quantum field theory 267–268; typicality and origin of randomness 265–266

Bohmian potential 182

Bohr, N. 181, 214, 259

Bokulich, A. 472

Boltzmann Brain problem 701, 726

Boltzmann entropy 404, 406–407, 409, 440, 503

Boltzmann equation 403

Boltzmannian statistical mechanics 402; approach to equilibrium 407–410; definitions of equilibrium 404–407; deterministic chance in 537–539, 541, 542, 544

Boltzmann, L. 55, 403, 404, 407, 409, 715, 716

Borde, A. 696, 713

Borel, É. 231

Born, M. 248, 650

Born-rule theorem 240, 242

Bose–Einstein and Fermi–Dirac statistics 76, 428, 589, 591, 684

Bose, S. 76, 206

Boson 589–591, 685; Gauge 479, 627, 684; Goldstone 289, 329, 479, 626–627; Higgs 291, 323, 333, 680; W and Z 288, 290, 333

Bouatta, N. 309

Bousso, R. 727

Boyle, R. 17

branes 352, 707

Brassard, G. 116

"bridge-laws" 497, 500, 501, 503–505, 507, 508

Bridgman, P. 668, 669

Brighouse, C. 153, 155

Brown, H. R. 74, 78, 116, 134, 137, 172, 407

BRST symmetry (Becchi, Rouet, Stora and Tyutin) 601–602

Brussels School ("Brussels-Austin School") 421

B-theory: arguments 87; propositional temporalism 87

Buckingham, E. 668

Budden, T. 136

Bunge, M. 668, 669

Bunsen, R. 744

Burgess, C. 722

Butterfield, J.N. 153, 155, 309, 314, 351, 352, 370, 375, 497, 518, 579

Callender, C. 384, 385, 518, 519, 612, 613, 642, 681, 682, 689

Camilleri, K. 356

canonical coordinates 24, 26, 207, 391

canonical distribution 415
canonical models of decoherence 205
canonical quantization procedure 367, 368
canonical representation 218
canonical transformations 392, 583, 597
Cao, T.Y. 681
Cappelli, A. 351
Cardy, J. 309
Carnap, R. 501, 647
Carr, B.J. 722
Carroll, S.M. 50, 51, 54, 78, 194, 292, 689, 715, 726, 727
Cartan, E. 33, 41, 168
Carter, B. 722
Cartesian coordinates 24, 28, 46–47, 65 66
Cartwright, N. 357, 358, 420, 636
Castellani, E. 353, 550, 613, 616
Caulton, A. 549, 579
causal explanation: causal imperialism 462–463; causation in physics 457–462; conserved quantity accounts of causation 455–457; deductive nomological model 454–455; Mach's and Russell's challenges 457–458; structural models and interventionism 458–462
causality 8, 10–11, 16, 19; conditions 111–112; local primitive 314, 319, 662; relativistic 112, 114, 314
causal quantum theory 114–115
causal set theory (CST) 364, 366–367
cause, notion of 457–458, 463
Cayley's Theorem 578
ΛCDM model 690, 707, 708, 713, 731
Chalmers, D. 556
chance: in deterministic physics 647–650; history of 644–645; in indeterministic physics 650–651; and interpretation of probability 645–647; physics and metaphysics 644–645; *see also* deterministic chance
charge 456, 592, 616, 635, 639, 671, 673, 682, 684; conservation 456; of non-locality 182, 196
Chen, E.K. 60, 182
Cheng, T.P. 293
Chen, J. 194, 715
Chomaz, P. 519
Christoffel symbols 131–133
Churchland, P.M. 501, 502, 504
Clapp, L.J. 509
Clarke, S. 1, 7, 39, 148, 387, 388, 389
classical permutation invariance 583–584; individuation in permutation-invariant classical mechanics 584–585; realization of permutations on joint configuration spaces 583
classical mechanics: equivalent formulations 27–30; Hamiltonian mechanics 24–25; Lagrangian mechanics 23–24; Newtonian mechanics 22–23; plane pendulum (example) 25–27, *26*
Clauser, J.F. 188
Clifton, R. 107, 319, 325, 326, 327, 331, 332, 334, 470, 662

Clifton, R.K. 330
clock ambiguity 697
closed-system approach 204
coarse-graining approach 204, 439
Cohen, J. 642
coherence length 528
Coleman, S. 288, 293
collapse locality loophole 114–115
collapse of wave-function 480
collapse theories: causal quantum theory 114–115; chance 254; dynamical theories of collapse 249–251; GRW collapse *250*; measurement problem 248–249; in Minkowski spacetime 111–114; ontology 254–255; relativity 252–253; for single particle *249*; stochastic quantum state evolution 109–110; tails 253; tests 251–252
Collins, J. 309
Colosi, D. 330, 331, 334
combinatorial argument 404
Compagner, A. 418
complexity 55
Compton wavelength 279–280, 290
configuration space 23, 25, 29, 255, 481; anti-haecceitistic 591, 592; arbitrary region of 539–540; defined 250; internal 195, 196; joint 583–585; ; Newtonian 46, 49, 53; overlap in 220; relative 46
conformal structure 164
connection: affine 125, 132–133, 136, 162–164, 168, 170–172;
conservation: of charge 692; of energy 8, 456, 606, 698; laws 440, 456, 469, 478, 549, 636, 638–639; of momentum 8–9, 17, 18, 456
constitutive principle 13–15
constraint surface 597–598, 601, *602*
continuous spontaneous localization (CSL) theory 100, 108
coordinate-independent approaches: dynamical metric fields 80–81; non-dynamical metric fields 79–80
coordination, notion of 13
coordinative definitions 13, 16
Copenhagen interpretation 181, 261, 633, 661, 663
correlation functions 300, 304, 530–531
correlation length 517, 528, 531, 533
cosmic microwave background (CMB) 292, 349n3, 710, 713, 725, 733–735
cosmic time 690, 709–710, 747
Cosmological Constant Problem (CCP) 721–723, 735–736
cosmological principle 708, 710, 748
cosmology 689–690; bouncing through big bang 714–715; breaking up spacetime 709–712, *712*; ΛCDM model in 738–739; concordance or "Standard Model" 731; cosmological speculations 712–716; eliminating past hypothesis 715–716; inflationary 700, 702, 707; multiverses and past 713–714; "one bounce" 715;

standard model of 719, 731; time in standard model 708–709
Costa de Beauregard, O. 102
coupling constant, 304
Cox, R.T. 721
CPT theorem 287–288
critical exponents 513–514, 517
critical points 470, 478, 513–514, 517–518, 524–526, 528–529, 532–533
Crowther, K. 370
Crull, E. 182
Curiel, E. 357
Curie, P. 621
Curie's principle 613, 616, 630n4
curved space 4 71, 65, 467
Cusp/Core Problem 739

Damour, T. 723
Darboux's theorem 608
Dardashti, R. 357, 550
dark energy (DE): cosmological constant problem 735–736; models and computer simulations 739; observational evidence 734–735; realism about 735; theory change and theory choice 738–739; underdetermination of theory by evidence 737–738
dark matter (DM): cosmological constant problem 735–736; Lambda Cold Dark Matter (ΔC CDM) model 731; models and computer simulations 739; observational evidence 732–733; realism about 735; theory change and theory choice 738–739; underdetermination of theory by evidence 737–738
"dark matter double bind" 737
Dasgupta, S. 549, 555, 634
Davies, P. 329, 331, 722
Dawid, R. 337, 352, 353, 354, 355, 357, 358
de Broglie–Bohm theory 103, 117, 182, 183, 254, 257, 258, 259, 260, 261, 262, 263, 264, 265, 312, 316, 378; Bohmian quantum field theories 106; causal structure 103–106
de Broglie, L 265
Deckert, D.A. 267
De Clark, S.G. 669, 673
decoherence: business of decoherence theory 222; canonical models 205; free subspaces 207; quantum (*see* quantum decoherence); theory 235–237
de Courtenay, N. 673
deductive nomological (DN) model 454–455
degeneracy pressure 468
de Haro, S. 354
Demarest, H. 543
Demopoulos, W. 19
Dennett, D. 332
de Simone, A. 726, 727
determinism 151, 155, 636; and chance (*see* deterministic chance); defining, hole argument 155–157; universal 647

deterministic chance: arguments from irrelevance of fundamental laws 541; in Bohmian mechanics 539–541; in Boltzmannian statistical mechanics 537–539, 541, 542, 544; and chance platitudes 544–545; metaphysical analyses of 542–543
Deutsch, D. 207, 231, 241
Dewar, N. 156
DeWitt, B. 219, 221, 345
Dickson, M. 107, 662, 663
Dieks, D. 70, 106, 107, 354
diffeomorphisms: active and passive 146–147; hole 147
dilatational momentum 52
Dimensional Analysis 668–669
dimensional homogeneity, principle of 669–670, 676
dimensions: absolute units 672–673; importance in philosophy of science 675–676; natural units 674; and quantity equations 671–672; role in new SI 674–675; role in SI 673–674; and systems of units 669–670
Dine, M. 720
Diosi, L. 251
Dirac monopole 599
DiSalle, R. 10
discontinuous or first-order 513
Dixon, J.D. 578
Dizadji-Bahmani, F. 502, 503, 504, 508
Donoghue, J.F. 293, 723
Doplicher, S. 363
Dowe, P. 455, 457
Dowker, F. 224, 366, 367
dressed electrons 333
Dretske, F. 641
3D systems 672
Duff, M. 357
Duhem, P. 12
Duncan, T. 293
Durr, M.C. 260, 262, 265, 267, 483
dynamical approach to spacetime theories: coordinate-independent approaches 79–81; Dynamical Approach, Reprise 77; general relativity 77–79; geometrical approach 74–77; Maxwell's electrodynamics 71–73; symmetries and dynamical approach 73–74
dynamical symmetries 73, 551
Dyson, F. 347

Eagle, A. 60
early universe 194, 292, 337, 368, 439, 441–442, 444–445, 628, 700, 707–708, 714–715
Earman, J. 10, 39, 71, 74, 76, 107, 124, 145, 148, 150, 151, 153, 168, 289, 326, 327, 457, 479, 598, 627, 647, 649
Eddington, A.S. 1, 70, 709, 710
effective field theory 283, 286, 289–290, 292–293, 306–307, 332–333, 344–345, 681
effective theories 296, 418

Ehlers, J. 172
Ehrenfest-Afanassjewa, T.A. 669
Ehrenfest, P. 209, 236, 587, 724; theorem 209, 221, 235, 236
eigenstate-eigenvalue 651
Einstein, A. 9, 13, 14, 15, 33, 59, 60, 68, 70, 71, 72, 73, 76, 90, 123, 124, 125, 126, 127, 128, 129, 130, 131, 132, 133, 134, 135, 136, 137, 138, 145, 146, 168, 171, 186, 262, 265, 290, 291, 292, 343, 364, 365, 374, 378, 379, 650, 661, 663, 699, 707, 723, 727
Einstein–Cartan theory 168
Einstein equivalence principles (EEPs): gravitational fields as fundamental 130–131; heuristic role:midway to general principle of relativity 128–129; spacetime free of gravity 132–134; two construction sites instead of one 127; unifying gravity and inertia 131–132
Einstein's equations 134, 343, 352, 365, 711, 722
electromagnetism: classical 549, 573, 610, 660; Maxwell's 699; "premetric" version 172; in 19th century 334
Ellis, G.F.R. 357, 726
Emch, G. 429
emergence 354–355, 370, 517–518, 636; and asymptotic limits 505; of classicality 209–210; of particles 280; phase transitions 517–518; spacetime (*see* spacetime)
The Emergent Multiverse (Wallace) 231
Emerson, W.H. 672
Enc, B. 509
Endicott, R.P. 502, 509
energy: conservation of 8, 456, 606, 698; dark (*see* dark energy (DE)); free *514,* 514–517, 520, 529–530, 726; gravitational 126, 659, 698; high 273, 286, 298, 305–308, 333, 351–352, 445, 447, 497, 595, 627, 722; internal 414, 505, 514, 529; kinetic 23, 26, 49–51, 53, 56, 376, 426, 445, 500–501, 504–505, 507, 658, 733; low 99, 226, 275, 289, 296, 306–308, 332–333, 344, 352, 354, 355, 627, 700, 725; mass 113, 130, 328, 339, 731, 734; Planck 292, 344; potential 23, 26, 29, 48–49, 51, 393, 445–447, 469, 507, 733; stress 80, 145, 291, 326–327, 346, 365, 659, 736; vacuum 291, 700, 722–723, 726, 736
ensemble or mean density operator 110
entropy 429; asymmetry (*see* entropy asymmetry); to black holes 364, 369, 402; Boltzmann 404, 406–407, 409, 440, 503; high 409, 429, 441, 444–445, 698; low 194, 402, 409, 439, 441–447, 543, 700, 713, 715–716; Von Neumann 430
entropy asymmetry: cosmological origins of non-equilibrium systems 443–447; history of heat 443–447; second law of thermodynamics 439–440
EPR argument 186
equilibrium 402, 404; Boltzmannian statistical mechanics 407–410; definitions of 404–407,

420; distributions, justifying 415–417; Gibbsian notion of 421; and large macro-regions, connection between 404; thermal, Von Neumann's concept of 429; thermodynamic 219
equivalence principle(s) 123, 365; as bridge between theories 138–139; Einstein equivalence principles 127–134; gravity and acceleration 127–134; local validity of special relativity 134–138; manifold of principles 125–126; strong equivalence principles 134–138; weak equivalence principle 126–127
equivalent formulations: mathematical differences 27–28; metaphysical differences 29–30
equivariance of Born-rule distribution 103
equivariance theorem 258
Esfeld, M. 153, 662
eternal inflation 356, 713
Euler–Lagrange, or simply Lagrange, equations 23
European Space Agency (ESA) 744
Evans, P.W. 102
Everettian quantum mechanics 115–116, 181, 213, 236, 251, 693, 702; probability (*see* probabilities); quantum histories and quasiclassical domains 222–226; realism about measurements 218–222; structure 213–214
Everett III, H. 213, 214, 215, 216, 218, 222, 223, 224, 227, 230, 235, 238, 254, 316
evidence: and big-data science 748–749; concept of 745; empirical, astrophysics 745–749; observational, DE and DM 732–735; underdetermination of theory, DE and DM 737–738
existence: of absolute space and absolute time 61; of acceleration 6–7, 9; brute fact 703; coherence 703; creation 701–702; of DM and DE 731–735; of equilibrium state 410; of force 61; of gravitational field 131–132, 343; of life 445, 700–701, 721; metaverse 702; particular universe 699–701; of phase transitions 516–517; of physical reality 689; principle 702–703; reasons 692–694; "something" and "nothing" 694–696; of spacetime 152; of system of measurement 668; of system of units 666; of universal behavior 524, 528; universe 696–699; of universe 692, 693, 698
exotic smooth structures 167
explanation 457, 508, 637, 639, 641; Boltzmann Brains 442; causal (*see* causal explanation); challenges to conventional 442–443; improbability of Past State 442; Nagelian reduction in physics 502; non-causal (*see* non-causal explanations); question of, emergentism *vs.* reductionism and 488–489; renormalization group (RG) 470, 478; structural 479; three different broad accounts of 489–491

Fabri, E. 624
Farr, M. 613
Feintzeig, B. 190

fermion: and boson distinction 288, 352, 428, 582, 589–592, 684–686; existence of 746; four-fermion theory 290; kinds of 683

Feyerabend, P.K. 502

Feynman, R. 273, 352, 555, 638

fibre bundle 23, 44n19

field configurations 596

Field, H. 457, 459

Fine, A. 188, 190, 366

fine-tuning of universe for life: after physics 727–728; Bayesian accounts of fine-tuning 721; cosmological constant 722–723; dimensionality of spacetime 724; fine-tuning in physics 719–721; multiverse 724–727; parameters of standard model 723–724; Ptolemaic model of Solar System *720*

firewall paradox 365

Fisher, M.E. 297

FitzGerald, G.F. 70

flat space 65

Fletcher, S.C. 124, 154, 157

force 18; and acceleration, relation between 12; action fields of 18; constraint 26; definition of 9, 12; existence of 61; generalized 29; gravitational 25, 470; and motion, connection between 17

fork asymmetry thesis (Reichenbach) 456

Foster, B.Z. 393

The Foundations of Statistics (Savage) 7241

frame-dependency 576

Franklin, A. 534

Franklin-Hall, L.R. 494

Franzosi, R. 519

Fraser, D. 293, 333, 628, 634

Fraser, J. 309, 333

Fraunhofer, J. 744

free mobility, principle of 13

Freivogel, B. 727

French, S. 582, 634, 684

Friederich, S. 289, 627

Friedman equations 708

Friedman–Lemaître–Robertson–Walker (FLRW) spacetime 708

Friedman, M. 13, 14, 15, 379

Friedman–Robertson–Walker (FRW) spacetime 708

Friedrichs, K.O. 33, 41, 102

Frigg, R. 357, 403, 404, 407, 409, 410, 411, 418, 421

Frisch, M. 459

Frohlich, J. 76, 627

Fuchs, C.A. 116

Fundamentality 679–682

fundamental observers 709

fundamental physics 60, 145, 215, 242, 292, 296, 357–359, 370, 383, 435, 550, 624, 642, 692–693, 700, 725, 727

Galilean principle of relativity 10, 54, 126

Galilean spacetime 33, 35–38, *36, 37, 63, 569, 570*; Newton's first law 37–38; Newton's law of universal gravitation 38; Newton's second law 38; Newton's third law 38

Galileo Galilei 1, 16, 17, 18, 19, 33, 34, 36, 40, 54, 63, 64, 67, 69, 71, 75, 91, 126, 127, 134, 227, 287

Garriga, J. 727

gauge anomalies 602, 624

gauge bosons 479, 627, 683, 684

gauge fields 289, 351–352, 596–601, 603, 626

gauge freedom 595, 2857

gauge symmetry 287, 572, 595–597, 600–601, 603, 624, 684

gauge theories 550; fluff is strongly non-idle 599–602; problem of "descriptive fluff" 598–599; relatively informal description 596–598; role in global kinematic structure 599–601; role in local dynamical structure 601–602

gauge theories of Yang–Mills type 595

Gauss, C.F. 672

Gell-Mann, M. 225, 235, 236, 297

generalized coordinates 23

generalized forces 29

generalized positions 23–25, 29

general principle of relativity 17, 129

general relativity (GR) 123–124; causal set theory 366–367; loop quantum gravity 367–369; need to go beyond general relativity 363–364; problem cases for geometrical approach 78; problem of spacetime 369–371; semi-classical gravity 364–365; two miracles 78–79

geodesic 9, 49–50, 53, 75, 123, 130, 132–134, 137–138, 162, 166–168, 170–172, 467, 696, 713–714

geometrized Newtonian gravitation or Newton–Cartan theory 41

geometrogenesis 383

Geroch, R. 341, 342, 712, 713, 714

Ghins, M. 136

Ghirardi, G.C. 113, 181, 249, 252

Ghirardi–Rimini–Weber (GRW) theory 100, 108, 194–196, 249, 252, 262, 427

Gibbings, J.C. 669

Gibbons, G.W. 727

Gibbsian ensemble averaging or Gibbsian phase averaging 414

Gibbsian notion of equilibrium 421

Gibbsian statistical mechanics (GSM): ensemble averaging work 417–419; justifying equilibrium distributions 415–417; non-equilibrium theory 419–421

Gibbs, J.W. 419, 421, 585

Gibbs' paradox 585–587

Gielen, S. 380, 381

Giere, R. 636

Gillies, D. 645

Ginzburg–Landau model 333

Girvin, S.M. 76

Glennan, S.S. 477

Glimm, J. 299

gluing constraint 599–600
Goldenfeld, N. 309, 433
Goldilocks Principle 683
Goldstein, S. 185, 426, 429, 430, 483, 716
Goldstone boson 289, 329, 479, 626–627
Goldstone theorem 626
Gold, T. 715, 716
Graham, N. 219
grandcanonical distribution 414–415, 417–418
Grassi, R. 109, 113
Gratton, S. 714, 715
gravitational field 132; existence of 131–132, 343; as
    fundamental 130–131
gravitational force 25, 445, 470, 551, 733
Greaves, H. 616
Greenberg, O.W. 587
Gross, D. 356
group field theory (GFT) 379–382, *381*
Grumiller, D. 293
Grünbaum, A. 691
Gryb, S. 395
Gupt, B. 194
Guth, A. 713

Haag, R. 294, 327, 328
Haag's theorem 327, 328, 331
haecceity 639
Hall effect 76, 591
Halliwell, J. 235, 236
Halvorson, H. 287, 294, 319, 325, 326, 327, 330,
    331, 332, 334
Hameroff, S.R. 254
Hamiltonian 25, 608; classical mechanics
    (*see* classical Hamiltonian mechanics);
    constraint equation 368; mechanics *24*, 24–25;
    or canonical equations 25; principle 378;
    standard account 608
Hamilton, J.-C. 6
Hannay's angle 468
Harlow, D. 293
Hartle, J.B. 225, 235, 236
Hartman, T. 293
Hawking, S.W. 167, 262, 712, 713, 714
Hawley, K. 91, 94
Healey, R.A. 116, 470, 634, 662
heat, history of: cosmological origins of non-
    equilibrium systems 443–444; low-entropy
    hot-spots and how we get the benefit of them
    444–445; nucleosynthesis: the primordial
    thermal bath 446–447; stars as entropy
    powerhouses 445–446
Hedrich, R. 357
Hehl, F.W. 172
Heidegger, M. 691
Heisenberg, W. 29, 73, 102, 111, 181, 182, 223,
    263, 324, 410, 531
Heller, M. 166
Hemmo, M. 107, 431, 433

Hempel, C.G. 454
Henneaux, M. 601
Henson, B. 662
Hertz, H. 71, 236
"the hidden structure strategy" 462
hidden-variables theories 100, 182, 261, 661
Higgs mechanism 479, 626
Higgs, P.W. 626
Hilbert, D. 75, 78, 110, 201, 204, 219, 222, 279,
    280, 281, 286, 287, 288, 290, 312, 313, 324, 326,
    327, 367, 368
historical sciences 745
Hobbes, T. 17
Hoefer, C. 153
Hogan, C.J. 722
hole argument: argument against substantivalism
    145–146; conclusion of 147–150; defining
    determinism 155–157; diffeomorphisms
    146–147; relativity principle 145; responses to
    150–154
holism: in classical physics 658–660; metaphysical
    656; methodological 655–656; nomological 656;
    ontological holism in quantum theory 662–663;
    property/relational holism 657–658; quantum
    holism 660–662
Hollowood, T.J. 309
holonomy 367
homogeneity of function 52
horizontal stacking 54
Horowitz, G.T. 355
Hossenfelder, S. 192, 193
Howard, D. 662
Howson, C. 416
Huggett, N. 338, 344, 353, 374, 375, 382, 383,
    585, 587
Hume, D. 10, 11, 691
Humphreys, P. 518, 647; Humphreys' paradox 646
Huygens, C. 17, 18, 644, 645
Hydrodynamics: A Study in Logic, Fact and Similitude
    (Birkhoff) 669
hyperbolic equations 460
hyperplanes of simultaneity 90

"ideal explanatory text" 462
indifference, principle of 415
indistinguishability postulate 589
inertia 6, 11, 16; Galilean 134, 137; gravity and,
    unification of 125–126, 128–129, 131–132, 134,
    138; law of 62, 133; Newtonian 17–18, 133, 138
inertial frames 10, 48, 54, 70–72, 137–138, 145, 366
inertial mass 18, 50, 126–127, 130, 470
inertial motion 9–12, 15, 17, 18, 38, 72, 75–76,
    123, 128, 130, 227, 377, 476
inertial trajectories 10, 37, 172, 383
inflation: big bang 731; cosmic 719, 725; cosmology
    700, 702, 707; eternal 356, 712–714
information: black hole 355, 359; causal 458, 462;
    conservation 355; counterfactual 456, 457, 473;

dependence 472; loss of 204, 355, 359, 365, 416; permutation-invariant 580–581; physical 601, 602; probabilistic 457
infrared divergences 288–289, 303
Initial Projection Hypothesis 194
International System of Units 666, 669, 671, 675
interpretation of quantum mechanics 100, 425
interpretation of quantum theory 483
intertheoretic relations 497–498, 599, 602–603
interventionism 420, 458–462
intrinsic derivative 50
invariant: gauge 479, 573, 576, 598–599, 601, 627; irreducible subspaces 589; Lorentz 75, 108–109, 113, 312; permutation-invariant 580–581, 583–585, 587, 589–591; Poincaré 71, 73–74, 76–80, 137, 302; reparametrization 389–392, 394–396; scale-invariant 48, 52, 55, 305, 307; time-reversal 225, 258, 439–440, 460, 538, 550, 605, 607–609, 612–613
Isham, C. 351, 370
Ismael, J. 613, 616
isospin 289, 684

Jackiw, R. 624
Jacobi's principle 51
Jacobson, T. 78, 293
Jaffe, A. 299
J'anossy, L. 70
Janssen, M. 74, 75, 76, 77, 129, 131, 132, 137, 146, 479
Jansson, L. 453, 534
Janus point theories 52–56
Jaynes' approach 421
Jaynes, E.T. 416
Jeans, J. 709, 710
Jeffares, B. 746
Johansson, L.-G. 357
Joos, E. 202, 209, 218
José, J.V. 27
Josephson effect 210

Kabir's principle 615
Kadanoff block spin method 531
Kadanoff, L.P. 297
Kane, R. 254
Kant, I. 8, 11, 13, 14, 691, 707
Karaca, K. 748
Kastner, R. 102
Kent, A. 114, 115, 224
Ketland, J. 579
Khinchin, A.I. 417
Khinchin's ergodic theorem 417–418
Kiefer, C. 387
Kim, J. 492, 508, 509
kinematical relativistic effects of length contraction and time dilation 73
Kirchhoff, G. 744
Kleinian conception of geometry 79
Kment, B. 642

Knox, E. 41, 81, 137, 157, 172, 489
Koberinski, A. 333
Kochen–Specker theorem 263
Kolmogorov, A. 721
Koslowski, T. 53, 56
Kottler, F. 128, 131, 134
Kraus representation 109
Król, J. 166
KR orders 366
Kuchař, K.V. 391
Kuhlmann, M. 478
Kuhn, T.S. 378, 564

Ladyman, J. 579
Lagrange, J.L. 6
Lagrangian 23; density 102, 299, 301; formalism 389, 602; mechanics 22–29, 275; QFT 286
Lam, V. 153, 370
Lambda-CDM model 690, 707, 708, 713, 731
Landau pole 286, 305
Lanford's theorem 431
Lange, L. 10
Lange, M. 76, 469, 470, 471, 472, 534, 642, 667
Langhaar, H.L. 669
language difficulty 218, 220
Laplace, P.-S. 231, 644, 647
Laplace's demon 647
Large Hadron Collider (LHC) 328, 748
Larmor, J. 70, 73
Laudisa, F. 100
Lavis, D. 410, 420, 421
Law of Universal Gravitation, Newton's 38
laws: concept of 633, 635, 636; of nature 476, 635–636; in physics 635–636; what laws are 639–642; what laws do 636–639
laws of motion: methodological character of 10–16; as outcome of conceptual analysis 16–19
Lebowitz, J. 410
Lee, T.-D. 622
Legendre transformation 597
Lehmann–Symanzik–Zimmerman (LSZ) reduction formula 300
Leibniz–Clarke correspondence 1
Leibniz equivalence 148, 571
Leibniz, G.W. 1, 7, 17, 33, 34, 39, 40, 124, 378, 387, 388, 389, 393, 549, 571, 691
Leibnizian quantum mechanics (LQM) 195–196
Leibnizian spacetimes 33, 39–40
Leinaas, J.M. 266
Lewis, D. 87, 155, 231, 240, 454, 462, 508, 543, 636, 639, 641, 642
Lewis, G.F. 722
Lewis, P.J. 183, 193, 240, 633
Liboff, R. 210
light: backward 92, 457, 459, 462; behavior of 63–64, 66; bending of 68; Fodor's 508; Nagel's 499–500; role of 70; speed of 64, 66, 170, 252, 341, 378, 459, 674, 733, 745, 747

light cone 88; forward light cone 93, 314; future light cone 63–64, 66, 167; past light cone 63–64, 66, 93, 113–116, 252, 707

Li, L.F. 293

Lim, E. 78

Linde, A. 726

Linden, N. 430, 433

Liouville's theorem 210, 431, 434

Local Causality 661

locality in ordinary QM 311–312

locality in QFT: holism of states in QFT 315; locality of observables and preliminary conclusion 313–314; Lüders Rule 316–317; non-locality from measurement collapse 315–316; operations research 318–320; Schlieder's no signaling theorem; its apparent limits 317–318; signaling and operations 316–320

Loewer, B. 253

Lombardi, O. 106

loop quantum gravity 292, 338, 346, 351, 364, 367–369, 374, 382, 712

Lorentz coordinates 66–67, 67, 75, 101

Lorentz covariance: antimatter 287; discrete symmetries and CPT theorem 287–288; spin-statistics theorem 288; Wigner's classification of particles 287

Lorentz, H.A. 70, 73, 75, 123, 134, 312, 313

Low, F. 297

Luczak J. 418

Lyre, H. 480, 509

Mach, E. 1, 46, 123, 130, 171, 172, 392, 393, 457

Mach's principle 130, 171

Mackey, M.C. 421

macrospaces 433

macrostate 426; coarse-graining 439; different 439; identical 439; quantum mechanical 428–430, 432

Maidens, A. 153

Mainwood, P. 534

Malament, D.B. 10, 90, 102, 325, 416, 418, 610, 611, 612, 616, 617

Malament's theorem 330

many-worlds interpretation 213, 651

marginalizing over nuisance parameters 721

marginally relevant or marginally irrelevant 310n5

Massimi, M. 748, 749

Mathematical Foundations of Quantum Mechanics (Von Neumann) 215

Matsubara, K. 352, 353, 357

matter: baryonic 446, 731, 733–735, 746; condensed matter physics 275, 281–283, 287, 289–290, 293, 297, 332, 334, 383, 601, 620, 627, 655; dark (*see* dark matter (DM)); fields 71, 74, 76–80, 114, 137, 168, 170, 172, 364, 369; luminous 731, 733; visible 731, 733, 739

Mattingly, D. 78

Maudlin, T. 151, 152, 185, 192, 262, 633, 642

maximum entropy principle 415–416, 421

Maxwell–Boltzmann distribution 405

Maxwell–Huygens spacetimes 40–41, 44n22

Maxwell, J.C. 71, 72, 73, 131, 236, 308, 334, 378, 401, 455, 478, 479, 482, 673

Maxwell's Demon 435

Maxwell's equations 71–72, 131, 136, 308, 455, 610

Mayo, D. 748, 749

McCoy, C.D. 689, 748, 749

McGivern, P. 492

McKenzie, K. 353, 681, 683, 684, 685

measurement: collapse, non-locality from 315–316; concept of 221, 226; existence of system of 668; postulates 214; problem 209, 248–249; quantum measurements problem 260; realism about 218–222

"the mechanical philosophy" 16

mechanical units 672

mechanisms: collapse 202–204, 247, 251; concept of 476, 479; Higgs 289, 479–480, 626–627; physical 203, 370, 713–714; structural 478; universal 479

mechanistic explanation (ME): description 476–477; general features 477–478; limits of 478–480; role of 480–483

Melia, J. 155

Menon, T. 77, 81, 518, 519

Mermin, N.D. 116

Messiah, A. 210

metaphysical holism 656

metaphysical possibility, notion of 556

metaphysics 633–634; absolute rotation, notion of 560; deterministic chance 542–543; empirical symmetries 557–559; epistemic possibility 556–557; equivalent formulations 29–30; fundamental question of 691; holism 656; justifications 552–554; metaphysical applications 559–560; method of symmetry 551–552; physics and 644–645; spacetime emergence 383; from symmetry to undetectability 554–556

method of arbitrary functions 650

methodological holism 655–656

methodological paradox 418

methodological reductionism 655

microcanonical distribution 414

microstate in statistical mechanics 426

Milky Way 744, 745

"Millennium Simulation" 750

Miller, D.J. 70

Milligan, P. 421

Minkowski spacetime, collapse theories in: causality conditions *111,* 111–112; collapse theory ontology *112,* 112–113; stability of vacuum 113; using only standard degrees of freedom 114

missing satellites problem 739

modal interpretations 106–108; infinitely many degrees of freedom 107–108; relativistic causal structure 107

models: canonical models 205–207; "CDM cosmological model" 690; deductive nomological model 454–455; deductive nomological (DN) model 454–455; hole argument 149–150, 152, 154; Lambda Cold Dark Matter (ΛCDM) model 750; Modified Newtonian dynamics (MOND) model 750; Penrose's impossible triangle 339, *340*; spin-oscillator model 205; "Standard Model" of cosmology 731; structural models 458–462; toy model 500–501
Modified Newtonian dynamics (MOND) model 750
Moller, C. 344
moment of time, definition of 92
momentum: conservation of 8–9, 17, 18, 456; dilatational 52; phase space 25; in wrong direction, time reversal 614–615
Moore, G. 722
Morrison, M. 479, 518
Mortimer, B. 578
motion: and force, connection between 17; inertial 9–12, 15, 17, 18, 38, 72, 75–76, 123, 128, 130, 227, 377, 476; laws of 10–19, 38
Mugani, M. 387
Muller, F.A. 579
multiple realizability, notion of 509
multiverse: and anthropic reasoning 356; of big bang 713; cosmology 713–714; fine-tuning of universe for life 724–727
Mundale, J. 509
Munich School 502
Myrheim, J. 266
Myrvold, W.C. 74, 81, 107, 109, 113, 185, 253, 662

Nagel, E. 469, 497, 499, 501, 502, 504, 509
natural units 298, 674
Needham, P. 508
Nelson, E. 267
Nelson, W.M. 70
Nerlich, G. 467
Newton–Cartan spacetimes 40–41
Newton–Huygens spacetime 44n22
Newton, I. 1, 6, 8, 10, 13, 17, 18, 22, 23, 24, 33, 37, 38, 39, 40, 46, 48, 54, 55, 61, 63, 68, 71, 124, 127, 131, 132, 133, 134, 153, 227, 358, 377, 378, 388, 455, 551, 560, 568, 569, 574
Newtonian mechanics 22–23; forces and acceleration 6–7; formulations 8–9; foundations and philosophy 6; geometrical techniques 7; laws of motion 8, 10–19; substantivalism and relationism 7; theory and spatiotemporal framework 9–10
Newtonian spacetimes 2, 33, 39–40, *62, 569*
Newton's bucket experiment 40, 61, 63, 65, 68
Newton's laws of motion: first law 22–23, 37–38, 61–62, 64; second law 22, 25, 38, 41, 51–52, 607, 635, 647, 670; third law 23, 38
New Wave Reductionism 502

Nielsen–Ninomiya theorem 302
Noether, E. 1
Noether's First Theorem 456, 468–469
"no-hidden-variables" theorems 263
"nomic necessitation" 642
nomological holism 656
Nomura, Y. 726
non-causal explanations: examples 470; explanations 471–473; involving geometry 467–468; involving inter-theoretic relations 469–470; involving symmetries 468–469; menagerie of 466–470; universal patterns 471
non-locality: Bohmian mechanics 262; charge of 182, 196; from measurement collapse 315–316; minimal models 472–473
non-realist and pragmatist interpretations 116–117
non-relativistic quantum theory 181–183
nonselective operation 109
Norsen, T. 263
North, J. 6, 7, 605, 614, 633
Norton, J.D. 75, 77, 124, 145, 146, 148, 150, 152, 329, 331, 332, 457, 518, , 616, 648, 649
Norton's dome 648, *648*
nothingness, concept of 703
nuclear democracy 681

objectivity 491, 552–553
occupation numbers 405
Ohanian, H.C. 136
Okon, E. 114
Olum, K.D. 727
one-loop level 303
ontological holism in quantum theory 662–663
Oppenheim, P. 679
order parameter 524, 525
Oriti, D. 357, 375, 383
oscillator-oscillator model 205

Page, D.N. 727
Palacios, J. 669
Palmer, T. 192, 193
Pankhurst, R.C. 669
paramagnetic phase 526, 529
paraparticle or parastatistical states 588
parastatistics 684
Parfit, D. 691, 703
Parker, W. 749, 750
parthood, notion of 488
particle in quantum mechanics 280
particles in QFT: accelerating observers disagree on number of quanta 326–327; anti-realism 333–334; are approximation 330–331; are idealization introduced by scattering theory 331–332; interacting systems cannot be given a quanta interpretation 327–328; negative arguments 325–328; as non-fundamental entities 328–334; no unique fock space representation in general in curved spacetimes 327; operationalism

329; quanta cannot be localized in any finite region of space 325; quanta interpretation of fock space 323–325; realism 332–333; role of relativity 328

particles, notion of 331, 332, 334

particle zoo 680

Pascal, B. 231

Past Hypothesis 194–195, 409, 431, 432, 441, 442, 447, 503, 506; eliminating 715–716, *716*

Pauli exclusion principle (PEP) 462–463, 468

Pauli repulsion 468

Pauli–Villars method 301

Pauli, W. 70, 135, 136, 137

Paz, J. 208

Pearle, P. 109, 113, 114, 249

Pearl, J. 458

Penrose, R. 194, 251, 254, 340, 343, 421, 712, 713, 714

Peres, A. 363

perfect cosmological principle 707

perfection, notion of 699

permutations 578; absolute indiscernibility, notion of 579; in classical Hamiltonian mechanics 583–585; in classical statistical mechanics 585–587; haecceitism *vs.* anti-haecceitism 581–582; lifted permutation 580–581; in logic and model theory 579; and permutability 579–580; in quantum mechanics 587–592; related metaphysical and interpretative disputes 581–582; transcendental *vs.* qualitative individuality 582

Peskin, M.E. 293

Peterson, D. 611

Pettini, M. 519

Pexton, M. 470

Pfister, H. 50

phase space 25, 390, 597, 608; coarse-grained cell 439–440; constraint surface *602*; harmonic oscillator's *615*; joint phase space 583–586; momentum 25; non-reduced 599, 601; reduced 584, 599; six-dimensional 404, 439; velocity 23

phase transitions: emergence, reduction, and approximation 517–518; inter-theoretic relations, representation, and experiment 519–521; (canonical) partition function 516; in statistical mechanics, coherence question 515–517; in thermodynamics 512–515, *513, 514, 515*

philosophers of physics 3

physically salient derivation, notion of 377

Picasso, L. 624

Pincock, C. 471

Pitowsky, I. 188, 190

Pitts, J.B. 81

Planck length, notion of 341

Planck units 341

plane pendulum 25–27, *26*

Podolsky, B. 262

Poincaré, H. 1, 12, 70, 71, 73, 76, 77, 78, 79, 80, 123, 287, 313, 328, 644, 649–650, 652

Polchinski, J. 351, 355, 357

Polger, T.W. 509

Pooley, O. 74, 77, 79, 134, 147, 153, 366

Popper, K. 645

positivism 261

Post, H. 582

Potochnik, A. 490, 494

Prange, R.E. 76

preferred basis problem 235

pregeodesics 162

Price, H. 457

Primas, H. 518

primitive ontology approach 268, 483

Primordial Existential Question 691

principal functional 392

Principal Principle 239–240, 544

Principia, Newton's 1

principle of Local Action, Einstein's 662

Principle of Sufficient Reason (PSR) 691, 698

probabilities 189, 236, 238, 242; Born rule 217; branch counting 237–239; chance and interpretation of 645–647; connection with statistics 234–235; connection with uncertainty 231–233; decoherence theory 235–237; principal principle 240–242; quantum 188–190; undermining 239–240

problem of time 340, 697; finding time again 392–396; global problem of time 389–392; Leibniz and problem of time 386–388; quantum measurement 100; reparametrization invariance 388–389

propagator 300

property holism 657–658

proportionality, principle of 496n17

propositional temporalism 87, 96

proton-proton chain 445

Putnam, H. 453, 487, 507, 651, 679

pythagorean theorem 28

qualitative individuality (QI) 582

quantum algorithm 181

Quantum Bayesianism or QBism 116

quantum decoherence: definition and basic concepts 201–203; emergence of classicality 209–210; formalisms of decoherence 203–205; four canonical models 205–207; measurement problem 209; quantum puzzles 200; question of pointer positions 208–209; question of preferred basis 207–208

Quantum Dialogue: The Making of a Revolution (Beller) 182

quantum electrodynamics (QED) 273

quantum energy conditions 365

quantum field theory (QFT) 76, 273–274; algebraic 293; Bohmian mechanics 267–268; effective field theories 281–283; fermion/boson distinction 288; fine-tuning of cosmological constant 291–292; fine-tuning of Higgs mass

291; formal quantum theory of continuum 277–280; infrared divergences and large-volume limit 288–289; locality in (*see* locality in QFT); Lorentz covariance and classification of particles 287–288; non-renormalizable interactions in physics 290; outstanding questions of particle physics 290; particles in (*see* particles in QFT); philosophical morals 292–293; quantum field theory on curved spacetime 290; quantum gravity 292; renormalization 283–285; scale dependence, and particles again 285–286; symmetries and symmetry breaking 289–290; symmetry and universality 286; warm-up: classical continuum mechanics 276–277

Quantum Field Theory in Curved Spacetime and Black Hole Thermodynamics (Wald) 365

quantum gravity (QG) 349n3; causal set theory 366–367; definition of 342; dimensions of 341; loop quantum gravity 367–369; need to go beyond general relativity 363–364; paradox 339–340; Penrose's impossible triangle 339, *340*; phenomena and experiment 346–347; problem of 345–346; problem of spacetime 369–371; semi-classical gravity 364–365; ways of 342–345

quantum holism 660–662

quantum measurement problem 100, 260

quantum measurements 263

quantum mechanical preparation, notion of 427

quantum mechanics (QM): possibility of settling 591–592; quantum permutation invariance 589–590; relativistic constraints (*see* relativistic constraints, quantum mechanics); representation of permutations on joint Hilbert spaces 588–589; topological approach to quantum statistics 591

Quantum Mechanics: Historical Contingency and the Copenhagen Hegemony (Cushing) 182

Quantum Mechanics with Spontaneous Localization (QMSL) 108

quantum probabilities 188–190

quantum spacetime (QS) 374; causal set theory 383; geometrogenesis 383; loop quantum gravity 382; string theory 382

quantum state *248*, 695; anti-thermodynamic 432, 434; causal structure 103; collapse theories 108, 112; decoherence 193, 200, 433; diffusion approach 651; Everettian interpretations 115; evolution 109, 183, 432, 697; Hilbert space 696; holism 660–662; measurement problem 100; mechanical microstate 426–428, 431; modal interpretations 106; non-realist and pragmatist interpretations 116–117; particles 287; Schr¨odinger equation 231, 434, 696–697; superposition principle 248; thermodynamic 432, 434

Quantum Theory (Bohm) 220

quantum theory (QT): causal 114–115; Everettian, or many-worlds approaches 100, 700; in fields (*see* quantum field theory (QFT)); of gravity

346, 363, 366–367, 370; interpretation of 483; non-relativistic 181–183; ontological holism in 662–663; permutation-invariant 587; realist interpretations 230, 235; relativistic 60, 102–103, 213, 323, 325

quasiclassical domain 225, 235

quasi-particles 280

Quine, W.V.O. 14, 15, 237, 579

radiation: cosmic microwave background 710, 733, 744; discovery of black hole 290, 292; Hawking 290, 355, 365, 695; thermal 444

Ramsey, F.P. 241, 721

Raymond-Robichaud, P. 116

Read, J. 80, 136, 137, 157

realism: about dark energy 735; about dark matter 735; about measurements 218–222; particles in QFT 332–333

realization principle 547n22

"real-space" RG 534

real state separability principle 661

Redhead, M.L.G. 419, 420, 582, 634

reduction, Nagelian: auxiliary assumptions 506–507; Bridge-Laws 503–505; critical discussion of 501–; derivability and development of 502–503; emergence and asymptotic limits 505; emergentism *vs.* reductionism 487–489; explanation 502; "falsity" problem 502; framework 501–502; multiple realizability 507–509; toy model 500–501

Reeh–Schlieder theorem 315, 316, 320, 330, 695

Rees, M.J. 689, 722

Reichenbach, H. 13, 16, 456, 457, 558, 645

relational holism 657–658

relational least action, principle of 54

relationism: best matching 50–52; Janus-Point theories 52–54; Newtonian dynamics in shape space 47–49; from Newtonian to relational dynamics 49–50; phenomena explained by relational dynamics 54–56; relational arenas 46–47

relative state interpretation 100

relativistic constraints, quantum mechanics: additional beables theories 99, 103–108; collapse theories 99, 100, 108–115; Everettian interpretations 99, 100, 115–116; non-realist and pragmatist interpretations 116–117; quantum measurement problem 100; quantum theories in relativistic spacetimes 102–103

relativistic interval 65

relativistic spacetimes 101–102; alternative geometrical structure 168–169; alternative manifold structure 165–168; alternatives 171–173; quantum theories in 102–103; scope 160; standard determination of geometrical structure 170–171; standard determination of manifold structure 169–170; standard geometrical structure 162–164; standard manifold structure 161–162

relativity: absolute space and time 62, *62*;
A-theory 86–88, 96; Cartesian coordinates 66;
conciliatory approaches 92–94; Galilean or Neo-
Newtonian space-time 63; General Relativity
68–69; light cone structure 63–64, *64*; Lorentz
coordinates 66, *67*; Minkowski space-time
64–65, 67; Newton's bucket experiment
61–62, 65, 68; puzzle for STR and A-theory
90–92; special relativity from A to B 88–90;
supplementing responses 94–96
relativity principle 130, 145
renormalizability, notion of 305, 308
renormalization group (RG) 282; bare coupling
302–304; beta function 304; effective field
theory approach 306–307; effective theories
296; equation 304–305; explanations 470, 478;
infrared divergences 303; natural units 298;
path integrals 298–300; Pauli–Villars method
301–302; perturbation theory 301; perturbative
renormalization 306; quantum field theory
296–297; regularization and renormalization
301–302; ultraviolet divergences 303; Wilsonian
RG 307–308
Renormalization Group Theory (RGT) 517
renormalized action 302
Rescher, N. 387
Reutlinger, A. 309, 473, 534, 613
Rice, C.C. 471, 472
Richardson, R.C. 508
Rickles, D. 337, 351, 352, 353, 355, 357, 395
Ridderbos, T.M. 420
Rideout, D. 366
Riemann, B. 14
Rimini, A. 181
Ritson, S. 356
Rivasseau, V. 308, 309
Robb, A.A. 172
Roberts, B.W. 615, 616, 617
Roberts, J. 642
Roche, J. 669
Rosaler, J. 226
Rosenfeld, L. 346
Rosen, N. 262, 265
rotation 16, 40, 46, 50, 51, 258, 379–380, 560,
568, 684
Rovelli, C. 100, 330, 331, 334, 340, 357, 382, 386,
394, 395, 396
Ruelle, D. 418, 512
Ruetsche, L. 274, 290, 293, 309, 326, 327
Russell, B. 1, 12, 387, 452, 457, 691
Rynasiewicz, R. 153, 155

Saatsi, J. 472, 473, 634, 684
Sackur-Tatrode formula 406
Sagan, B.E. 578
Salart, D. 115
Saletan, E.J. 27
Salmon, N. 151

Salmon, W.C. 455
Samaroo, R. 6, 9, 15
Saunders, S. 34, 40, 91, 153, 280, 579, 587, 633, 680
Savage, L. 241
Savitt, S.F. 95
scale(s): accounts 491–494, *493*; autonomy of
296; broad accounts 489–491; emergentism *vs.*
reductionism 487–489; length 527–528; problem
487; and question of explanation 488–489;
question of scale 487–488
Scaling (Barenblatt) 668
Scaling, Self-Similarity, and Intermediate Asymptotics
(Barenblatt) 668
Schack, R. 116
Schaffer, J. 680, 681
Schaffner, K.F. 503, 504
Schellekens, A.N. 722
Schlieder, S. 317
Schrödinger, E. 29, 108, 201, 214, 215, 217,
223, 225, 231, 249, 315, 390, 410, 431, 443,
455, 539, 650
Schrödinger time evolution, notion of 395
Schroeder, D.V. 293
Schur's Lemma 589
Schwartz, M.D. 309
Schweber, S.S. 103, 681
Schwinger, J. 273
scientific law 455
scientific underdetermination 357
second law of thermodynamics 439–440;
challenges to conventional explanation 442–443;
reversibility objection and solution 440–442
second quantization 280
Seiberg, N. 375, 382
self-aware substructures (SASs) 727
Sen, A. 353
separability principle 659–661, 663
Shahvisi, A. 402
Shannon, C.E. 416
Shapiro, L.A. 509
Shenker, O. 431, 433
Shimony, A. 185, 192
Shor, P. 207
Silk, J. 357, 722, 726
Simons, B.D. 293
simultaneity 16, 38, 91; absolute 90–92, 94, 96,
127, 252–253, *570*; hyperplanes of 90; relation of
distant 101, 103, 107, 109, 117, 312; relativity of
69, 73; slices 63–64, 68–69
singularities: big bang 341, 368, 694, 696, 714;
black holes 341, 711; theorems 713
Sklar, L. 553, 560
Skow, B. 94, 692
Slater, J.C. 265
Smeenk, C. 627, 748
smoking gun observations 746
Smolin, L. 357
Sober, E. 508

Sorkin, R.D. 366, 367

space 34–35; absolute 10, 39; bitopological 167; concept of 10; configuration (*see* configuration space); curved 4 71, 65, 467; flat 65; (symmetric) Fock space 280; invaders 647; kinematical Hilbert space 368; macrospaces 433; of theories 305

space like interval 117, 184, 252, 281, 315, 317–319

spacetime 33; concept of 10; connection to geometry 78; dynamical approach to (*see* dynamical approach to spacetime theories); emergence 374–376; existence of 152; explaining 376–379; foam 341; Galilean (*see* Galilean spacetime); in group field theory (case study) 379–382; GR problem of 369–371; investigations of specific theories of QS 382–383; Leibnizian 33, 39–40; light cone structure *64*; Maxwell–Huygens spacetimes 40–41; metaphysical implications 383; Minkowski 111–114; Newton–Cartan 40–41; Newton–Huygens 44n22; Newtonian 33, 39–40; QG problem of 369–371; relativistic (*see* relativistic spacetimes); space *34, 34–35*; symmetry 73

special relativity 59–60; *see also* relativity

special sciences 454

special theory of relativity (STR) 88, 90–92

spin-boson model *see* spin-oscillator model

spin-echo experiment 420

spinfoam 369

spin network states 368

spin-oscillator model 205

spin-statistics theorem 288, 301

Spirtes, P. 458

Square Kilometer Array (SKA) 748

stability, notion of 638

Stachel, J. 132, 146, 153

standard account: Hamiltonian mechanics (example) 608; Newtonian mechanics (example) 607–608; Quantum mechanics (example) 608–610; relativistic field theory (example) 610–611; time reversal invariance 607; two components 606–607

Standard Model 76, 275, 288, 292, 328, 351, 358, 550, 605, 626, 722–723, 749

Stark, J. 127

statespace 22–25, 28–29

statistical mechanics (SM): diachronic functional relations 430–434, *431*; "finite-time" arguments 430–432; "infinite time" arguments 432–434; probabilities reduced to quantum mechanical probabilities 434–435; quantum mechanical macrostates and synchronic functional relations 428–430; quantum mechanical Maxwellian Demon 435; quantum mechanical microstates 425–428; and thermodynamics 401–402

Stein, H. 7, 33, 39

Stern–Gerlach experiment 103, 215, 263–264

Sterrett, S. 634

Stevens, S. 77, 81

stochastic dynamics program 420–421

stochastic Tomonaga–Schwinger picture 111

Stolt, R.H. 587

Stone-von Neumann theorem 326

Streater, R.F. 421

Strevens, M. 466, 542, 650

string theory: black hole physics and information loss 355; central role of concept of duality 352–354; emergence of spacetime 354–355; incompleteness of string theory 356–357; lack of empirical confirmation and completeness 357–358; lack of free parameters and string theory landscape 355–356; landscape 355; meta-level issues 356–359; multiverse and anthropic reasoning 356; physical characteristics 352–356; role of 351–352; theory's universality and final theory claim 358; as tool in plasma physics 359

strong energy condition 711

strong equivalence principles (SEPs): different versions of 135–137; Einstein's version of 134–135; role of 137–138

structural explanations 479

Struyve, W. 106, 289

Suárez, M. 634

substantivalism 7, 124, 146–148, 150–151, 452

subsystems: electron and nuclear proton 660; equilibrium systems 433; formation of 55; Newtonian gravity 52, 54; quantum 660; relative state 218; spacelike 316; statistical independence 314

Sudarsky, D. 114

sum-functions 417

sum rule 224, 236

superdeterminism 190–195

superposition principle 248

supervenience, notion of 657

Suppes, P. 748

Susskind, L. 353, 727

Swann, W.F.G. 70

Swinburne, R. 701

Symmetrization Postulate 590

symmetry 549–550; as clues to superfluous structure 568–571; concept of 621, 628; difficulties of application 574; dynamical 73, 551; gauge theories 572–573; justification 574–576; mathematical methods in physics 563–568, *565, 566, 567*; reasoning 573–574; remarks 571–572; spacetime 73; and universality 286

Symmetry (Weyl) 622

symmetry breaking: anomalous symmetry breaking 623–624; breaking mechanism 621; explicit symmetry breaking 622–623; kind of symmetry 621; and phase transitions 627–628; spontaneous symmetry breaking 624–628, *626*; system *vs.* law 620–621

syntactic view or received view of theories 501

system of units 666

systems: 3D 672; equilibrium 433; non-equilibrium 443–447; physical 8, 136, 390, 417, 467, 470, 498, 524, 551, 553–555, 578, 599, 615, 620, 659, 662–663; quantum 100, 182, 192, 201–202, 205, 207–208, 222, 349n7, 432, 461, 479, 650–651, 655, 660–663
systems of units 669–670

Taylor, J.R. 587
Tegmark, M. 724, 727
Teh, N. 355, 382, 634
Teitelboim, C. 601
Teller, P. 325, 582, 634, 658, 662
temperature 376, 379, 404, 414, 444–445, 447, 500–501, 504–505, 512–514, *513,* 525 529, 671–672
Terno, D.R. 363
Thébault, K.P.Y. 338, 395
theory confirmation 357
thermodynamics: asymmetry 461; black holes 293, 364; equilibrium 219; limit 418, 505; phase transitions 512–515; potential 514; quantum state 432, 434; second law of 439–440, 439–443; statistical 401–402
Thomson, J. 10, 333, 680
'THooft, G. 192, 194
time: concept of 387, 388, 393, 709; cosmic 690, 709–710, 747; in cosmology (*see* cosmology); problem of (*see* problem of time); in standard model 708–709; symmetry 612–613; translation 606
time like interval 117, 369, 386
time reversal: Albert–Callender "Pancake Account" 611–613; open research questions 616–617; operator 606, *606*; against pancake account 613–616; standard account 605–611; temporal orientation *610*
Timpson, C.G. 116, 662
Tipler, F.J. 722, 724
Tolman, R.C. 415
Tomonaga, S. 273
Tomonaga–Schwinger picture 102
Tooley, M. 641
topological approach 266
trace operation 204
trajectory: with equal interval lengths *67*; inertial 10, 37, 172, 383; or world-line 37
transcendental individuality (TI) 582
transition maps or changes of coordinates 161
translation 46, 50–51, 595; relative time 108; space-time 258, 327, 605; time 327, 330–331, 338, 343, 606
"tree-level" term 303
Trkal, V. 587
Tumulka, R. 182, 183, 252, 262, 312, 429, 430, 633
Turok, N. 727
twin paradox 67, *67,* 163

Uffink, J. 407, 409, 416, 418
ultraviolet divergences 289, 303, 341
uncertainty, notion of 233
universality 286, 513, 524, 534; of critical phenomena (paradigm case) 524–526, *525, 526*; defined 529–533; explaining 531–533, *532, 533*; ingredients of 526–528; length scales 527–528, *528*; order parameters and symmetries 526–527; scaling exponents 529, **529**; stability 529–531
universality class 531
universality of critical phenomena 524
Unruh effect 326–327
Unruh, W.G. 207
Urbach, P. 416

vacua 700
vacuum: energy 291, 700, 722–723, 726, 736, 743n13; quantum 695, 723; stable 109, 113–114; state 109, 113–114, 300, 315, 326–327, 330, 332, 380, 625, 695, 722; symmetric 695
vacuum expectation values (VEVs) 300
Vaidman uncertainty 245n11
van Fraassen, B.C. 636, 662
van Lith, J. 420
Varadarajan, V.S. 617
velocity: absolute 39–40, 48, 89, 551–559, 569; angular 613; asymptotic 263; field 276, 296; phase space 23–24, 37–38
velocity decomposition theorem 48
Verlinde, E. 354
vertical stacking 54
Vidotto, F. 340, 369
Vilenkin, A. 713, 727
Vistarini, T. 382
von Laue, M. 128, 132, 133
Von Mises, R. 645
Von Neumann entropy 430
von Neumann, J. 10, 216, 248, 312, 313, 318, 326, 390, 429, 430, 432, 651; concept of thermal equilibrium 429
Vranas, P.B.M. 416

Wald, R.M. 280, 291, 293, 326, 327, 334
Wallace, D. 79, 81, 181, 231, 241, 273, 287, 292, 293, 296, 309, 332, 333, 417, 446, 534, 573, 662
Wallis, J. 18
Watanabe, S. 460
wave function: collapse of 2, 203, 249, 253; conditional 260; "nomological entity" 483; particle's 236, 263, 374, 482; position 609, 611; universal 195, 237, 651
weak equivalence principle (WEP) 126–127
Weatherall, J.O. 7, 10, 81, 154, 157
Weber, M. 476, 672
Weber, T. 181, 210
Weinberg, S. 292, 293, 297, 346, 656, 725, 726
Weingard, R. 351

Werndl, C. 403, 404, 407, 410, 411, 418, 421
Weyl Curvature Hypothesis 194
Weyl geometry 171
Weyl, H. 33, 135, 171, 194, 595, 622
Weyl's theorem 171
Wheeler-DeWitt equation 697
Wheeler, J.A. 214
Whitrow, G.J. 724
Wigner, E. 254, 287, 639, 684
Wigner, E.P. 254, 287, 616, 639, 684
Will, C.M. 126
Williams, P. 273, 282, 290, 309, 333, 534
Wilsonian effective action 306
"Wilsonian RG" 305
Wilson, J. 497, 686
Wilson, K.G. 297, 307
Wilson, M. 21, 153
Witten, E. 357

Wittgenstein, L. 691
Woit, P. 357
Woodward, J. 458, 459, 466, 491, 520
Wren, C. 18
Wüthrich, C. 368, 370, 374, 375, 382, 383

Yang, C.-N. 622
Yang-Mills theories 550
Yu, Y. 210

Zabell, S.L. 416, 418
ZanghI, N. 483
Zee, A. 293
Zeeman, E.C. 172
Zeeman's theorem 173
Zeh, H.D. 202, 218, 387
Zurek, W.H. 208, 237
Zwiebach, B. 351